W9-BBS-226

PRENTICE HALL

Algebra 2 with Trigonometry

Bettye C. Hall
Mona Fabricant

Consulting Authors

Jan Fair ❖ Robert Kalin
Sadie C. Bragg ❖ Mary Kay Corbitt
Jerome D. Hayden

Prentice Hall dedicates this mathematics program to all mathematics educators and their students.

PRENTICE HALL

Needham, Massachusetts

PRENTICE HALL

Algebra 2 with Trigonometry

AUTHORS

Bettye C. Hall
Director of Mathematics
Houston Independent School District
Houston, Texas

Mona Fabricant
Professor of Mathematics
Queensborough Community College
Bayside, New York

CONSULTING AUTHORS

Jan Fair
Director, CAPP Project SUCCESS
A Math/Science California Academic Partnership Program
Lompoc/San Luis Obispo/Santa Maria, California

Robert Kalin
Professor Emeritus
Florida State University
Mathematics Education Program
Tallahassee, Florida

Jerome D. Hayden
Mathematics Department Chairman, K-12
McLean County Unit District 5
Normal, Illinois

Sadie C. Bragg
Professor of Mathematics and Associate Dean of Curriculum
Borough of Manhattan Community College
The City University of New York
New York, New York

Mary Kay Corbitt
Associate Professor of Mathematics
Valdosta State College
Valdosta, Georgia

CONSULTANTS

Beva Eastman
Associate Professor of Mathematics
William Paterson College
Wayne, New Jersey

Mary Dell Morrison
Mathematics Instructor (Retired)
Columbia High School
Maplewood, New Jersey

Harris S. Schultz
Professor of Mathematics
California State University, Fullerton
Fullerton, California

John M. Erickson
District Mathematics and Science Coordinator
Hopkins Public Schools
Hopkins, Minnesota

Jesse A. Rudnick
Professor of Mathematics Education
Temple University
Philadelphia, Pennsylvania

Walter Young
Formerly Professor of Mathematics
University of the District of Columbia
Washington, D.C.

Stephen Krulik
Professor of Mathematics Education
Temple University
Philadelphia, Pennsylvania

Rex Schweers, Jr.
Professor of Mathematics
University of Colorado
Greeley, Colorado

REVIEWERS

Calvin T. Long
Professor of Mathematics
Washington State University
Pullman, Washington

Francis Yu-Chaw Meng
School of Computer Science
Rochester Institute of Technology
Rochester, New York

Jerome T. Filipek
Mathematics Instructor
Dallas Independent School District
Dallas, Texas

Photo credits appear on page 916.

ISBN 0-13-026642-6

11 12 13 14 15 16 07 06 05 04 03 02

PRENTICE HALL

Cover Design: Martucci Studio
Chapter Opener Design: Function Thru Form

STAFF CREDITS
Editorial: Rosemary Calicchio, Enid Nagel, Debra Berger, Mary Ellen Cheasty, Michael Ferejohn, Tony Maksoud, John Nelson, Alan MacDonell, Ann Fattizzi
Design: Laura Jane Bird, Art Soares
Production: Amy Fleming, Lorraine Moffa, Suse Cioffi
Photo Research: Libby Forsyth, Emily Rose, Martha Conway
Publishing Technology: Andrew Grey Bommarito, Gwendollynn Waldron, Deborah Jones, Monduane Harris, Michael Colucci, Gregory Myers, Cleasta Wilburn
Marketing: Everett Draper, Julie Scarpa, Michelle Sergi
Pre-Press Production: Laura Sanderson, Natalia Bilash, Denise Herckenrath
Manufacturing: Rhett Conklin, Gertrude Szyferblatt
National Consultants: Susan Berk, Charlotte Mason

Contents

Functions and Graphs

CONNECTIONS: *Architecture • Engineering • Transportation • Education • Medicine • Carpentry • Construction • Chemistry • Earth Science • Cost Analysis • Biology • Cost Function • Geometry • Banking • Design*

Systems of Equations and Inequalities

CONNECTIONS: *Transportation • Electricity • Real Estate • Consumerism • Technology • Sports • Politics • Education • Market Research • Geometry • Finance*

5 Matrices and Determinants

CONNECTIONS: *Business • Consumerism • Geometry • Nutrition*

Polynomials

CONNECTIONS: *Geometry • Manufacturing • Business • Travel • Sports • Storage • Agriculture • Textiles • Interior Design • Construction*

Rational Expressions

CONNECTIONS: *Technology • Geometry • Transportation • Metallurgy • Navigation • Physics • Sports • Money • Carpentry • Masonry*

Irrational and Complex Numbers

CONNECTIONS: *Geometry • Number Theory • Physics • Electricity • Sports • Construction*

9 Quadratic Functions

CONNECTIONS: *Business • Physics • Construction • Manufacturing • Number Theory • Electricity • Optics*

10 Polynomial Functions

CONNECTIONS: *Geometry • Engineering • Technology • Number Theory • Physics*

11 Conic Sections

CONNECTIONS: *Drafting • Storage • Civil Engineering • Physics • Sports • Space Science • Forestry • Analytic Geometry • Astronomy • Astronautics • Geometry • Number Theory • Physical Fitness*

12 Exponential and Logarithmic Functions

CONNECTIONS: *Biology • Physics • Archaeology • Optics • Sets • Number Theory • Chemistry • Astronomy • Business • Electricity*

13 Sequences and Series

CONNECTIONS: *Business • Horticulture • Travel • Coin Collecting • Technology • Investment • Geometry • Consumerism • Sales • Art • Physics • Finance • Economics • Advertising • Botany • Combination Theory*

14 Probability and Statistics

CONNECTIONS: *Banking • Government • Sports • Manufacturing • Construction • Parades • Quality Control • Education • Marketing • Consumerism • Sales • Merchandizing • Agriculture*

15 Trigonometric Functions and Graphs

CONNECTIONS: *Physics • Construction • Boating • Geometry • Acoustics • Space Engineering • Forestry • Rescue • Navigation • Surveying*

16 Trigonometric Identities and Equations

CONNECTIONS: *Geometry • Physics • Coordinate Geometry • Navigation • Music • Sound • Design*

Developing Mathematical Power

Using Your Algebra Book

Algebra is considered the language of mathematics, and mathematics is a vital part of your life. Changes in society and in the use of technology require that you have a strong background in mathematics. The emphasis is not only on algebra skills, but on developing your *mathematical power*. What is mathematical power?

- The knowledge and understanding of mathematical ideas, concepts, and procedures
- The ability to solve problems within mathematics and in other subject areas
- The ability to use mathematical tools and techniques
- The ability to use mathematics in your world beyond the classroom
- The ability to use mathematics to communicate
- The ability to reason and think critically
- The understanding and appreciation of the nature and beauty of mathematics

Using your algebra book effectively will help you to develop your mathematical power. Let us introduce you to the parts of your algebra book that can help you to develop the aspects of mathematical power.

- Each chapter begins with an introduction that connects algebra to a familiar topic, and it also provides you with a project to complete independently or with classmates.
- The **Capsule Review** reviews mathematics skills and concepts you will need for the lesson.
- Completely worked-out examples contain instruction to help you to successfully complete your assignments.
- New concepts are highlighted for easy reference.
- The **Technology** logo and graphing calculator screens throughout the book identify when using technology is beneficial to and appropriate for the development of concepts. In addition, **Technology Lessons** provide you with alternative methods for exploring algebra. A graphing calculator is a valuable tool to use with this course. It has all the capabilities of a scientific calculator, plus you can plot and graph functions on the display screen, view formulas and solutions in their entirety, zoom in and trace graphs, and more.
- **Problem Solving Strategy** lessons, identified by blue borders, help you to develop techniques and strategies for problem solving in a variety of situations.
- Practice Exercises build a solid foundation of algebra skills . . .
  Thinking Critically—helps improve your reasoning skills.
  Applications—connects algebra to daily life and to other topics.
  Developing Mathematical Power—provides you with activities that may take you outside your classroom, exploring and using mathematics in today's world.
  Mixed Review—will help you to remember the skills and concepts already learned.
  Writing and **Reading in Algebra** activities help you to acquire communication skills.
  Integrating Algebra features provide a look at how algebra is used in daily life.
  Test Yourself provides you with a tool for self-evaluation. Answers to these and other selected exercises can be found in the back of the book.

Remember . . . mathematics is not a spectator sport! Actively participating, reading carefully, and completing your assignments will contribute to your success in algebra and in other mathematics courses you take in the future.

1 Real Numbers and Equations

A deal on wheels?

Developing Mathematical Power

When you lease a car, you generally use it for a number of years at a fixed monthly payment and then return it to the leasing agency. With some leases, you can buy the car at a prearranged price at the end of the lease's term.

If you buy a car, you might take a bank loan for part of the purchase price. When the loan is paid off, you own the car.

Project

Check newspaper ads and visit auto dealers to find the best price for your dream car. Research the conditions of leasing the car versus taking a bank loan to buy it. Then decide which would be the better deal for you.

The Set of Real Numbers

Objective: To identify the elements of the set of real numbers and represent them on the number line

Numbers are used constantly in everyday life. Whenever a price, a distance, a weight, a temperature, a speed, a pressure, or a time of day is discussed, numbers are employed. Indeed, numbers are used any time questions are asked about how many, how far, how fast, how often, or when. These questions can all be answered with real numbers, whether they are whole numbers, integers, fractions, or decimals.

Capsule Review

The following sets of numbers are *subsets* of the set of integers:

natural numbers: $\{1, 2, 3, 4, \ldots\}$ **whole numbers:** $\{0, 1, 2, 3, 4, \ldots\}$

The natural numbers are also called *positive integers.* Corresponding to each positive integer is a negative integer such that when the two are added the result is zero.

The set of integers consists of the negative integers, the positive integers, and zero.

integers: $\{\ldots -2, -1, 0, 1, 2, \ldots\}$

For example, -34 is a member of the set of integers. In set notation you can write

$$-34 \in \{\ldots, -2, -1, 0, 1, 2, \ldots\}, \text{ or } -34 \in \{\text{integers}\}.$$

The symbol $\in$ means that -34 "is an element of" the set of integers.

True or false?

1. $100 \in$ {natural numbers}

2. $0 \in$ {natural numbers}

3. $-57 \in$ {whole numbers}

4. $999 \in$ {integers}

5. $84 \in$ {whole numbers}

6. $-246 \in$ {integers}

7. The natural numbers are a subset of the integers.

8. The whole numbers are a subset of the natural numbers.

A **rational number** is a number that can be expressed as the quotient of two integers. Thus, $\frac{3}{5}$ is a rational number because it is of the form $\frac{n}{d}$, where n and d are integers and d is not zero. A rational number can also be expressed as either a terminating decimal or a nonterminating repeating decimal.

The rational number $\frac{3}{5}$ can be written as 0.6, a terminating decimal, by dividing the denominator into the numerator. The rational number $\frac{25}{99}$ can be written as 0.252525..., a nonterminating repeating decimal. A repeating decimal can be indicated by a bar, $0.\overline{25}$. Other examples of rational numbers are $-\frac{6}{1}$, $\frac{0}{5}$, and $\frac{11}{2}$.

In order to demonstrate that a number is rational, you must show that it can be expressed as the quotient of two integers.

EXAMPLE 1 **Show that the terminating decimal 0.625 is rational by writing it as the quotient of two integers.**

$$0.625 = \frac{625}{1000} = \frac{5}{8}$$

The next example shows how to express a nonterminating repeating decimal as the quotient of two integers.

EXAMPLE 2 **Show that the repeating decimal $0.\overline{63}$ can be written as the quotient of two integers.**

Let $N = 0.636363\ldots$

$100N = 63.636363\ldots$ *Multiply both sides by 100.*

$99N = 63$ *Subtract the first equation from the second.*

$$N = \frac{63}{99} = \frac{7}{11}$$

An **irrational number** is a number that neither repeats nor terminates. Numbers such as 0.212112111211112..., $\sqrt{3}$, and π are irrational numbers. The set of rational numbers and the set of irrational numbers have no elements in common and are said to be *mutually exclusive*.

The rational numbers and the irrational numbers are subsets of the **real numbers.** As shown in the diagram, the set of real numbers is the union of the set of rational numbers and the set of irrational numbers.

The Real Number System

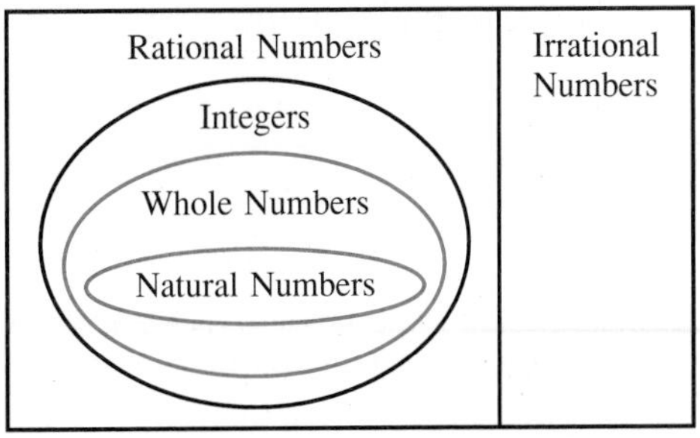

Rational Numbers
Integers
Whole Numbers
Natural Numbers
Irrational Numbers

EXAMPLE 3 **Use the diagram of the real number system on page 3 to name the subset(s) of the real numbers to which -2, 3, and $\sqrt{7}$ belong.**

-2: integers, rational numbers, real numbers
3: natural numbers, whole numbers, integers, rational numbers, real numbers
$\sqrt{7}$: irrational numbers, real numbers

For every real number there is a point on the number line called the **graph** of the number, and for every point on the number line there is a real number. That is, there is a **one-to-one correspondence** between the real numbers and the points on a number line. The real number is called the **coordinate** of the point. The irrational numbers $\sqrt{3}$ and π are labeled on the number line below, as are the rational numbers -3.5, -2, 0, and 3.

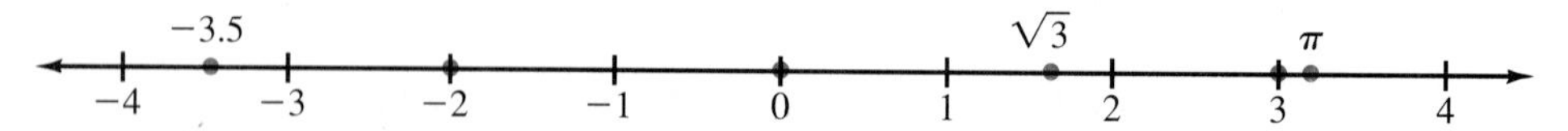

EXAMPLE 4 **Draw a number line and graph each of these real numbers: 0.5, -2.5, 3.1, $\sqrt{2}$, 1, -3, $-\sqrt{5}$. Label each point with its coordinate.**

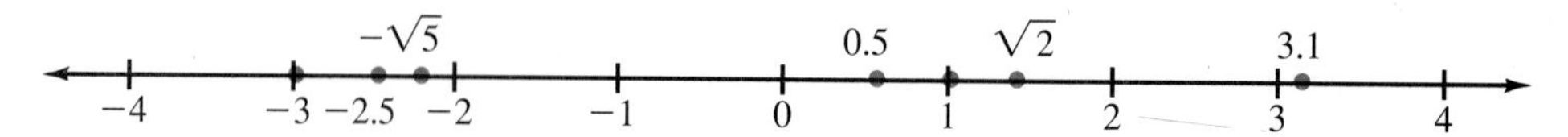

The location of the graphs of two real numbers relative to one another on the number line depends on their *order*.

For any two real numbers a and b,

> $a > b$ is read "a is greater than b."
> If $a > b$, the graph of a is to the right of the graph of b.
>
> $a < b$ is read "a is less than b."
> If $a < b$, the graph of a is to the left of the graph of b.
>
> $a = b$ is read "a is equal to b."
> If $a = b$, the graph of a is the same as the graph of b.

For example,

> Since $3 > 1$, the graph of 3 is to the right of the graph of 1.
> Since $-3 < -1$, the graph of -3 is to the left of the graph of -1.

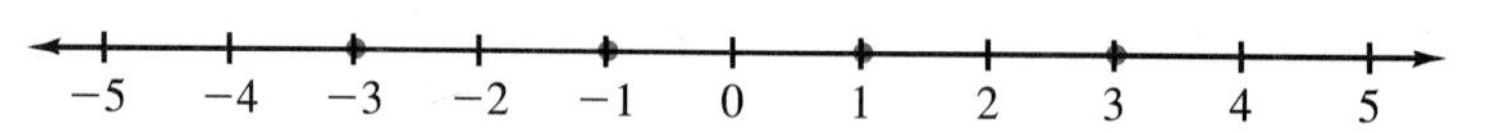

How many real numbers exist between two other reals? Consider $0.\overline{8}$ and 0.9. Is 0.89 between them? Is 0.889 or 0.898898889...? No matter how close together two real numbers seem to be, another real number can be found between them. For this reason, the set of real numbers is said to be *dense*.

EXAMPLE 5 **What relationship exists between each of the following pairs of numbers:** **a.** $9.56 \underline{\ ?\ } 9.07$ **b.** $-8.666\ldots \underline{\ ?\ } -5$ **c.** $\frac{1}{2} \underline{\ ?\ } 0.5$

a. $9.56 \underline{\ ?\ } 9.07$ — $9 = 9$, but $0.56 > 0.07$. Therefore, $9.56 > 9.07$.

b. $-8.666\ldots \underline{\ ?\ } -5$ — $-8 < -5$. Therefore, $-8.666\ldots < -5$.

c. $\frac{1}{2} \underline{\ ?\ } 0.5$ — $1 \div 2 = 0.5$. Therefore, $\frac{1}{2} = 0.5$.

CLASS EXERCISES

Use the diagram on page 3 to name the subset(s) of the real numbers to which each of the following belongs.

1. 7 **2.** -8 **3.** 3 **4.** $-5.777\ldots$ **5.** $\sqrt{5}$

Determine the correct relationship between each pair of numbers. Use the symbols >, <, or =.

6. $-3 \underline{\ ?\ } -8$ **7.** $8.2 \underline{\ ?\ } 8.03$ **8.** $39 \underline{\ ?\ } -2$

9. $0 \underline{\ ?\ } 5$ **10.** $0 \underline{\ ?\ } -4.80480080008\ldots$ **11.** $6.0 \underline{\ ?\ } 6.00$

12. Name the point which corresponds to each of these coordinates: -3.5, π, 7, -1, and 2.

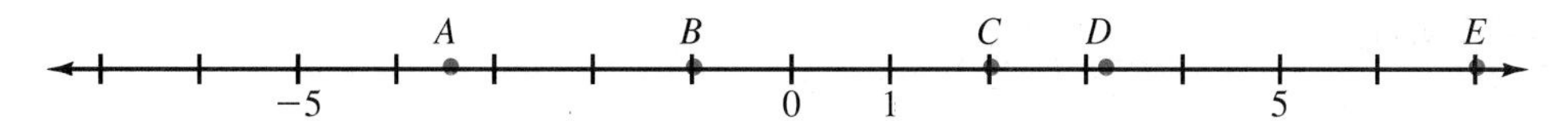

Show that each decimal can be written as a quotient of two integers.

13. 0.34 **14.** $0.\overline{7}$ **15.** $0.\overline{26}$

PRACTICE EXERCISES

Use the diagram on page 3 to name the subsets of the real numbers to which each of the following belongs.

1. 4 **2.** $\sqrt{2}$ **3.** π **4.** -6 **5.** 0

6. 103 **7.** $0.666\ldots$ **8.** -0.5 **9.** 2.5 **10.** $2\sqrt{6}$

Draw a number line and place each number on the line in its approximate location.

11. 0 **12.** $-\sqrt{24}$ **13.** -2 **14.** $2\frac{1}{2}$ **15.** $4\frac{2}{3}$ **16.** π

Give the coordinate that indicates the approximate location of each point.

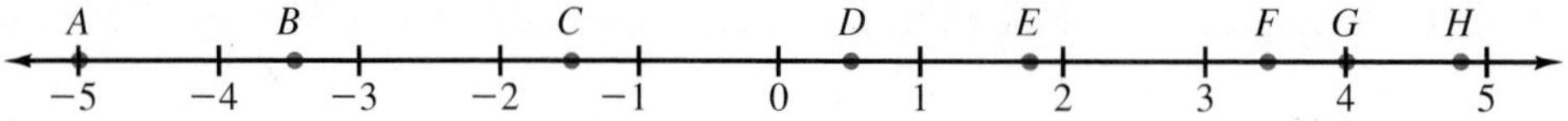

17. point A **18.** point B **19.** point C **20.** point D

21. point E **22.** point F **23.** point G **24.** point H

Write a comparison statement for each pair of numbers using <, =, or >.

25. $-7 \underline{\ ?\ } -9$ **26.** $3 \underline{\ ?\ } 3$ **27.** $14 \underline{\ ?\ } \sqrt{14}$

28. $\sqrt{6} \underline{\ ?\ } \sqrt{10}$ **29.** $0 \underline{\ ?\ } -0.333\ldots$ **30.** $0.8 \underline{\ ?\ } \frac{4}{5}$

Show that each decimal can be written as a quotient of two integers.

31. 0.32 **32.** 1.8 **33.** 0.22 **34.** 2.125

35. $0.\overline{6}$ **36.** $0.\overline{17}$ **37.** $0.\overline{45}$ **38.** $0.\overline{135}$

Convert each fraction to decimal form.

39. $\frac{3}{8}$ **40.** $\frac{7}{25}$ **41.** $\frac{6}{11}$ **42.** $\frac{2}{7}$

True or false?

43. $50.098 > 50.0098$ **44.** $-36 > -34$ **45.** $0.3 = 0.3000$

46. $0.008 = 0.8$ **47.** $-8 < -80$ **48.** $\frac{3}{4} = 0.75$

49. $-2 \in$ {whole numbers} **50.** $-6 \in$ {irrational numbers}

51. $\sqrt{3} \in$ {real numbers} **52.** $-7 \in$ {integers}

53. π is not a rational number. **54.** 0 is a rational number.

55. $4.343434\ldots \in$ {rational numbers}

56. $0.121221222\ldots \in$ {irrational numbers}

57. If a number is rational, then it is an integer.

58. If a number is an integer, then it is rational.

List the numbers in order from least to greatest.

59. $-17, -9.5, -20, -8$ **60.** $4, 2, 1, \sqrt{2}$ **61.** $-0.03, -0.003, -0.3, -3$

62. $1\frac{3}{10}, 1\frac{7}{25}, 1\frac{1}{2}, 1\frac{9}{50}$ **63.** $\frac{7}{8}, \frac{9}{16}, \frac{3}{4}, \frac{15}{24}$

Write each repeating decimal as a ratio of two integers.

64. $0.\overline{142857}$ **65.** $0.3\overline{17}$ **66.** $2.23\overline{6}$

If the statement is true for all real numbers a and b, mark it *true*. If the statement is *not* true, mark it *false* and give a numerical example to show that it is not true.

67. If $a > b$, then $a^2 > b^2$.

68. If $a^2 > b^2$, then $a > b$.

69. If $d < 0$, then $d^2 < 0$.

70. If $c > d$, then $c + a > d + a$.

71. If $c > d$, then $ac > ad$.

72. If $a = b$, then $a^2 = b^2$.

73. If $a^2 = b^2$, then $a = b$.

74. If $d^2 > 0$, then $d > 0$.

75. The product of two rational numbers is a rational number.

76. The product of two irrational numbers is irrational.

77. The sum of two whole numbers is a whole number.

78. The difference of two whole numbers is a whole number.

Applications

Use a real number to represent each situation.

79. Sports A gain of 9 yards

80. Meteorology 16 degrees below freezing on the Fahrenheit scale

81. Navigation 28 meters below sea level

82. Finance A debit of $2818

83. Sports A batting average when a player has made 7 hits in 20 times at bat (Batting averages are expressed as 3-place decimals.)

84. Finance A credit of $75.

HISTORICAL NOTE

In everyday usage, the word *infinite* means immense or beyond measure. A nineteenth century mathematician, Georg Cantor (1845–1918), developed a method of sizing sets with infinite numbers of elements. Using a one-to-one correspondence, Cantor proved that some infinite sets are larger than others. For example, the set of real numbers is larger than the set of integers and yet both sets have an infinite number of elements.

Investigate Georg Cantor and the branch of mathematics known as set theory.

1.2

Operations with Real Numbers

Objective: To perform operations on real numbers

On four successive plays a football team gained 7 yards, lost 10 yards, lost 8 yards, and gained 14 yards. What was the team's net gain? In order to answer this question, the sum of 7, -10, -8, and 14 must be found.

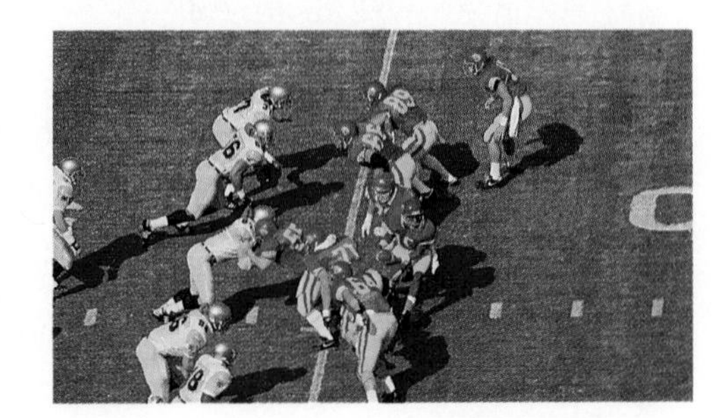

Capsule Review

EXAMPLE **Add:** **a.** $5\frac{2}{3} + 9\frac{5}{6}$ **b.** $0.25 + 0.09 + 7.2$

a. $5\frac{2}{3} + 9\frac{5}{6} = 5\frac{4}{6} + 9\frac{5}{6}$

$= 14\frac{9}{6}$

$= 15\frac{3}{6} = 15\frac{1}{2}$

b. $0.25 + 0.09 + 7.2$

$$\begin{array}{r} 0.25 \\ 0.09 \\ +7.2 \\ \hline 7.54 \end{array}$$

Add.

1. $4.1 + 12.25$ **2.** $28 + 3.7$ **3.** $0.005 + 0.05$

4. $6\frac{1}{2} + 3\frac{1}{8}$ **5.** $14\frac{4}{5} + 2\frac{2}{5}$ **6.** $13\frac{1}{3} + 1\frac{2}{3}$

7. $7.875 + 6.125$ **8.** $10.01 + 1.001$ **9.** $100.01 + 2.89$

The *absolute value* of a real number is its distance from zero on the number line. The absolute value of any real number a is designated by $|a|$.

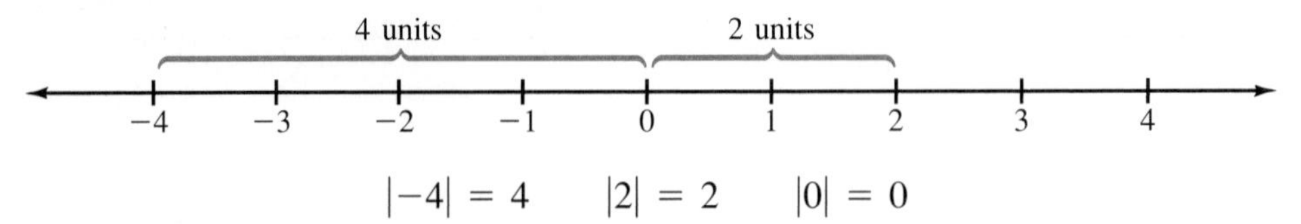

$|-4| = 4$ $|2| = 2$ $|0| = 0$

Absolute value is used in adding real numbers.

Addition of Real Numbers

- When both numbers are positive or both numbers are negative, add the absolute values and use the common sign as the sign of the sum.
- When one number is positive and the other negative, subtract the absolute values and use the sign of the number with the larger absolute value as the sign of the difference.

EXAMPLE 1 **Add:** **a.** $-7 + (-8)$ **b.** $4 + 12$ **c.** $3.6 + (-1.7)$ **d.** $\frac{1}{6} + \left(-\frac{5}{6}\right)$

a. $-7 + (-8) = -(|-7| + |-8|) = -(7 + 8) = -15$

b. $4 + 12 = (|4| + |12|) = 16$

c. $3.6 + (-1.7) = |3.6| - |-1.7| = 3.6 - 1.7 = 1.9$

d. $\frac{1}{6} + \left(-\frac{5}{6}\right) = -\left(\left|-\frac{5}{6}\right| - \left|\frac{1}{6}\right|\right) = -\left(\frac{5}{6} - \frac{1}{6}\right) = -\frac{4}{6}$ or $-\frac{2}{3}$

A real number and its **additive inverse** are the same distance from zero on a number line. For example, -6 and 6 are additive inverses.

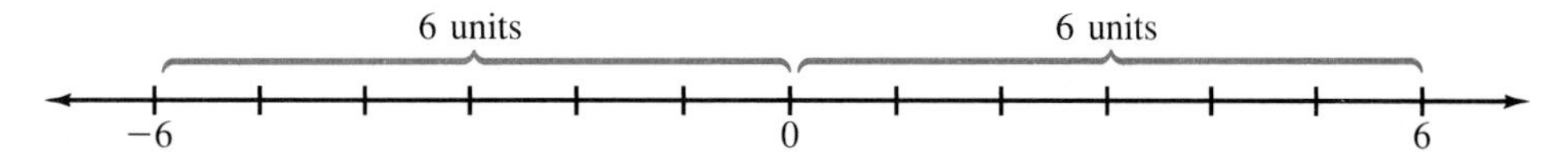

Additive Inverse Property

For every real number a there is a unique real additive inverse $-a$, such that $a + (-a) = 0$ and $(-a) + a = 0$.

The additive inverse of 6 is -6, since $6 + (-6) = 0$.

Subtraction of Real Numbers

For all real numbers a and b, $a - b = a + (-b)$. To subtract one real number from another, add the additive inverse of the number to be subtracted.

EXAMPLE 2 **Subtract:** **a.** $9 - 4$ **b.** $1.2 - 5$ **c.** $-18 - 7$ **d.** $-21 - (-6)$

a. $9 - 4 = 9 + (-4) = 5$

b. $1.2 - 5 = 1.2 + (-5) = -3.8$

c. $-18 - 7 = -18 + (-7) = -25$

d. $-21 - (-6) = -21 + 6 = -15$

Multiplication of Real Numbers

To multiply two real numbers, multiply their absolute values. If both real numbers are positive or both negative, the product is positive. If one number is positive and the other negative, the product is negative.

Multiplication Property of Zero

For every real number a, $0 \cdot a = 0$ and $a \cdot 0 = 0$.

Multiplication Property of −1

For every real number a, $(-1)a = -a$ and $a(-1) = -a$.

EXAMPLE 3 **Multiply:** **a.** $(-3)(-9)$ **b.** $-7(4)$ **c.** $0(-8)$

a. $(-3)(-9) = 27$ **b.** $-7(4) = -28$ **c.** $0(-8) = 0$

The **multiplicative inverse** of a nonzero real number is a unique real number called its **reciprocal.**

Multiplicative Inverse Property

For every nonzero real number a there is a unique multiplicative inverse $\frac{1}{a}$, such that $a \cdot \frac{1}{a} = 1$ and $\frac{1}{a} \cdot a = 1$.

The multiplicative inverse of 2 is $\frac{1}{2}$, since $2 \cdot \frac{1}{2} = 1$. The multiplicative inverse of -3 is $-\frac{1}{3}$, since $(-3) \cdot \left(-\frac{1}{3}\right) = 1$.

Division of Real Numbers

To divide real numbers, multiply by the multiplicative inverse of the divisor.

EXAMPLE 4 **Divide:** **a.** $-35 \div (-5)$ **b.** $-2.8 \div 7$

a. $-35 \div (-5) = (-35)\left(-\frac{1}{5}\right) = 7$ **b.** $-2.8 \div 7 = (-2.8)\left(\frac{1}{7}\right) = -0.4$

An absolute value symbol, like parentheses, indicates grouping. Therefore, perform operations within the symbol first and then take the absolute value.

EXAMPLE 5 **Simplify:** $|6 + (-3)| + 2|6 - (-3)| - |(-3) - 6|$

$$|6 + (-3)| + 2|6 - (-3)| - |(-3) - 6| = |3| + 2|9| - |-9| = 3 + 18 - 9 = 12$$

The expression 4^2 means 4×4. The real number 4 is the **base** and 2 is the **exponent.** The exponent indicates the number of times that the base is used as a factor. The value of 4^2 is 16.

EXAMPLE 6 **Simplify:** **a.** 2^3 **b.** $(-5)^2$

a. $2^3 = 2 \times 2 \times 2 = 8$ **b.** $(-5)^2 = (-5)(-5) = 25$

CLASS EXERCISES

1. What is the additive inverse of $\frac{a}{b}$, $b \neq 0$?

2. What is the multiplicative inverse of $-\frac{a}{b}$, $a \neq 0$, $b \neq 0$?

Find the additive inverse and the multiplicative inverse for each number.

3. -3 **4.** $\frac{1}{3}$ **5.** 7 **6.** -0.9

Perform the indicated operation.

7. $-4 + (-7)$ **8.** $-4 + (7)$ **9.** $-12 - (-3)$ **10.** $-42 - (3)$

11. $5(-4)$ **12.** $-7(-8)$ **13.** $-36 \div (4)$ **14.** $-12 \div (-2)$

Simplify.

15. $\left|-\frac{1}{3}\right|$ **16.** $-|-25|$ **17.** $0.2|-8|$ **18.** $-4|12|$

19. $3|7 - 10|$ **20.** $5|-6 + 10|$ **21.** $|-4| + |-3.54|$ **22.** $|5| - |-7|$

23. 6^2 **24.** $(-3)^2$ **25.** 2^5 **26.** $(-4)^3$

PRACTICE EXERCISES

Find the additive inverse and the multiplicative inverse for each number.

1. 23 **2.** -3.8 **3.** -5 **4.** π

5. $3\frac{2}{5}$ **6.** -4 **7.** $\frac{6}{7}$ **8.** $-\frac{9}{5}$

Perform the indicated operation.

9. $86 + (-28)$ **10.** $-11.2 + (-8.3)$ **11.** $78 - (-23)$ **12.** $-37 - (-29)$

13. $(-1.7)(7)$ **14.** $(-25)(-8)$ **15.** $73.2 \div (-1.2)$ **16.** $-144 \div (-9)$

17. $(-2)(-3)(0)(-2)$

18. $(-5)(8)(-4)(-2)$

19. $-10 + (14) + (-19) + (-20)$

20. $-20 + (-16) + (18) + (30)$

21. $\frac{3}{8} + \left(-\frac{1}{2}\right)$

22. $\left(-\frac{2}{3}\right) + \left(-\frac{1}{3}\right)$

23. $\left(-\frac{3}{5}\right)\left(-\frac{5}{3}\right)$

24. $\left(-\frac{7}{10}\right)\left(\frac{5}{14}\right)$

25. $\left(-\frac{5}{12}\right) \div (-2)$

26. $\left(\frac{9}{16}\right) \div \left(-\frac{1}{2}\right)$

Simplify.

27. $|-2| + |13| + |37| + |-12|$

28. $2|5| + 3|-7| - |-12|$

29. $|23 - 12| + |37 - 52| + |-65|$

30. $2|12 - 5| + 7|7 - 9| - 7|6 - 12|$

31. $(|-3| + |7| - |6|)(5|2| - |-7|)$

32. $(|7 - 8| + |7|)(3|-4| - |5 - 9|)$

33. $\left|\frac{-3}{2}\right| + \left|\frac{-7}{3}\right|$

34. $-3|5 - 10| - 4|5 - 12| + 7|4 + 14|$

35. $\frac{|3||-2| + |-7||3| - |15||-5|}{16}$

36. $\frac{|8 - 2|}{3} + \frac{|6 - 9|}{3} - \frac{|7 + 3|}{2}$

37. $(4 - 14)^2$

38. $5^2 \cdot 0$

39. $(3 \cdot 4)^2$

40. $(-7 - 2)^2$

41. $|-3|^2$

42. $|3 - 5|^3$

43. $(|5 - 6| + |-1|)^4$

44. $|4|^2 - |3|^3$

Indicate whether each statement is true or false. If false give values for *a* and *b* which make the statement false.

45. $|a| \cdot |b| = |ab|$

46. $|b - a| = |b| - |a|$

47. $|a + b| = |a| + |b|$

48. $|a| + |b| \geq |a + b|$

49. $|a| = a$

50. $|a| - |b| \leq |a - b|$

Applications

51. **Chemistry** Carbon dioxide (CO_2), which is commonly called dry ice, sublimates (changes from a gas to a solid) at −76°F. Ordinary ice melts at 32°F. How many degrees colder is a piece of dry ice than ordinary ice at its melting point?

52. **Meteorology** A weather balloon is sent aloft to determine temperatures at various altitudes. At 2500 ft the temperature is 4°C. At 5000 ft the temperature drops 18°C. What is the temperature at the higher altitude?

Mixed Review

Express each number as a fraction in simplest form.

53. 48%

54. 5.3

55. 0.5%

56. $0.16\frac{2}{3}$

1.3 Variables and Expressions

Objective: To evaluate numerical and algebraic expressions

In a softball league there are 10 teams. The league manager wants to schedule each team so that it plays every other team twice. In order to determine the total number of games necessary, the manager uses the *expression* $n(n - 1)$, where n stands for the number of teams in the league. The total number of games is therefore $10(10 - 1)$ or 90.

Capsule Review

Given the set $\left\{-3, -1.5, 0, 2, \sqrt{5}, \frac{5}{2}, \pi\right\}$, list the following subsets.

1. rational numbers **2.** whole numbers **3.** integers

4. irrational numbers **5.** real numbers **6.** natural numbers

A **numerical expression** is a combination of arithmetic operations and numbers. A numerical expression may also include symbols of grouping such as parentheses, brackets, absolute value symbols, and fraction bars. The following are numerical expressions:

$$12 + 2 \qquad 5 \div 2.75 \qquad \frac{7 \times 8^2}{6 - 2} \qquad 20 - [6 - (3 + 1)]$$

The parts separated by + or − symbols are called the *terms*. In $12 + 2$, 12 and 2 are the terms. In $20 - [6 - (3 + 1)]$, 20, $[6 - (3 + 1)]$, $(3 + 1)$, 6, 3, and 1 are terms. Every numerical expression has a unique value. To find the value of a numerical expression, use the *order of operations*.

Order of Operations

- When grouping symbols occur, simplify the expressions within grouping symbols first. Start with the innermost grouping symbols.
- Next, find the value of terms having exponents.
- Perform multiplication and/or division in order from left to right.
- Finally, perform addition and/or subtraction in order from left to right.

EXAMPLE 1 **Simplify:** **a.** $20 - [6 - (3 + 1)]$

b. $\frac{7 \times 8^2}{6 - 2}$ **c.** $\frac{-2 \times (-30) + 0.5 \times 20}{4^2 - 6}$

a. $20 - [6 - (3 + 1)] = 20 - [6 - (4)] = 20 - [2] = 18$

b. $\frac{7 \times 8^2}{6 - 2} = \frac{7 \times 64}{6 - 2} = \frac{448}{4} = 112$

c. $\frac{-2 \times (-30) + 0.5 \times 20}{4^2 - 6} = \frac{60 + 10}{16 - 6} = \frac{70}{10} = 7$

A **variable** is a letter which is used to represent a number or a set of numbers. Examples of variables are x, n, a, r, and t. An **algebraic expression** is a combination of arithmetic operations, numbers, and variables. An algebraic expression may include grouping symbols. The following are examples:

$$-2t \qquad 7x - 3y + 8 \qquad \frac{3a}{4b}$$

The parts separated by + or − symbols are terms. Notice that the algebraic expressions $-2t$ and $\frac{3a}{4b}$ each have one term and $7x - 3y + 8$ has three terms.

To **evaluate an algebraic expression,** substitute the given value(s) of the variable(s) then use the order of operations to find the value of the resulting numerical expression.

EXAMPLE 2 **Evaluate:** **a.** $3y + 2y^2 - 6(y - 4) + 8y^3$, for $y = 2$

b. $\frac{4w - 12}{w^2} + \frac{128 - 16w}{w^3}$, for $w = 4$

a. $3(2) + 2(2)^2 - 6(2 - 4) + 8(2)^3$ *Substitute 2 for y.*

$= 3(2) + 2(4) - 6(2 - 4) + 8(8)$ *Use the order of operations.*

$= 3(2) + 2(4) - 6(-2) + 8(8)$

$= 6 + 8 + 12 + 64$

$= 90$

b. $\frac{4(4) - 12}{4^2} + \frac{128 - 16(4)}{4^3}$ *Substitute 4 for w.*

$= \frac{4(4) - 12}{16} + \frac{128 - 16(4)}{64}$ *Use the order of operations.*

$= \frac{16 - 12}{16} + \frac{128 - 64}{64}$

$= \frac{4}{16} + \frac{64}{64} = \frac{1}{4} + 1 = 1\frac{1}{4}$

CLASS EXERCISES

Thinking Critically

1. How does the value of $\frac{1}{m}$ change as m increases or decreases?
2. Adapt your results from Exercise 1 to explain why division by zero is undefined.

Simplify each numerical expression.

3. $6 + 2 \times 8 - 12 + 9 \div 3$
4. $5 \times 3 - 8 \times 2 + 36 \times 4$
5. $2 + 5(3 \times 5 - 4 + 8 \times 2)$
6. $4 - 6(5 \times 7 - 9 \times 3)$
7. $\frac{18 - 3 \times 4}{3 \times 2} - \frac{5 \times 4 + 1}{7}$
8. $\frac{9 \times 6 + 18 \div 6 - 3 \times 3 - 15 \div 3}{3}$

Evaluate each expression using the value given.

9. $6a + 3a^2 - 2a^3$, for $a = -2$
10. $\frac{5d - 1}{2d} + \frac{6 - d}{d}$, for $d = 3$

PRACTICE EXERCISES

Use technology where appropriate.

Simplify each numerical expression.

1. $8 \times 3 + 4 \div 2$
2. $6 \times 2 + 3 \times 4 - 8 \div 2$
3. $0.25 \times 3 + 1 \div 4 - 0.54 + 1.26$
4. $9 \div 3 - 2 + 7 - 3 \times 4$
5. $3^4 + 2^3 - 5^2$
6. $2^4 - 3^2 + 4^3 - 5^2$

Evaluate each expression using the value given for each variable.

7. $4a + 7 + 3a - 2 + 2a$; $a = -5$
8. $5y - 3 + 4y - 1 - 3y$; $y = 3$
9. $12a^2 - 3a + 2$; $a = -5$
10. $36b - 4b^2 + 3b^3$; $b = -2$
11. $|x| + |2x| - |x - 1|$; $x = 2$
12. $|2z + 3| + |5 - 3z|$; $z = -3$
13. $3|2a + 5| + 2|3 - a|$; $a = 4$
14. $6|4b - 5| + 3|2 - 2b|$; $b = -1$
15. $\frac{3(2x + 1) - 2(x - 3)}{x + 6}$; $x = -3$
16. $\frac{5(2k - 3) - 3(k + 4)}{3k + 2}$; $k = -2$
17. $2y + 3 - y^2$; $y = 2\sqrt{3}$
18. $5c^3 - 6c^2 + 3c^4 - 2c$; $c = -5$
19. $\frac{\frac{1}{w} + \frac{w}{3} \times \frac{3}{w} + \frac{1}{w}}{w \times w - w}$; $w = 2$
20. $\frac{3(x + y) - 2(x - y)}{5x + y}$; $x = 3$, $y = 4$
21. $\frac{\frac{2}{y} + \frac{y}{4} - \frac{1}{2y}}{5n + 2}$; $y = 3$, $n = \frac{2}{3}$
22. $\frac{\frac{5}{x} + \frac{3}{x} - \frac{2}{7}}{\frac{6}{x} - \frac{4}{7} + \frac{5}{x}}$; $x = 7$

Insert one or more pairs of grouping symbols to make a true statement.

23. $3 \times 5 + 6 \times 3 + 1 = 132$

24. $5 + 3 \times 7 - 6 = 8$

25. $1 + 8 \times 2 - 3 \times 4 = -36$

26. $2 \times 2 + 3 \times 3 + 5 = 80$

Applications

27. Consumerism The expression $2F - 12$ is sometimes used to estimate a person's shoe size, where F is the length of the foot in inches. Measure your own foot and see if the expression gives your correct shoe size.

28. Medicine Various rules exist for determining the amount of medication for children when the adult dosage has been specified. One method is to use the expression $\left(\frac{A}{A + 12}\right)d$, where A is the age of the child in years and d is the adult dosage. A 10-year-old child is to take medication for which the correct adult dosage is 5 mg. What is the correct dosage for the child?

29. Business To determine cost a company uses the expression $an + b$, where each item costs $\$a$ to produce, n is the number of items to be produced, and b is the fixed cost. What is the cost for a company to make 10,000 items costing $0.38 apiece if the fixed cost is $885?

30. Computer Systems K is the letter programmers use to represent 1024 memory address locations. How many memory address locations are stored in a computer that has 256K of memory?

Mixed Review

Perform the indicated operation.

31. $35 - (-24)$ **32.** $4 + (-23)$ **33.** $(-18)(17)$ **34.** $-16 \div 32$

Replace each $\underline{?}$ with <, =, or > to make a true statement.

35. $0.25 \ \underline{?}\ \frac{1}{4}$ **36.** $6.03 \ \underline{?}\ 6.032$ **37.** $-\sqrt{3} \ \underline{?}\ -2$

38. What subset(s) of real numbers contain(s) -8?

39. Find the multiplicative inverse and the additive inverse of -7.

Developing Mathematical Power

Extension The order of operations is often implied by how the expression is written. Support this statement using the pairs of equivalent expressions.

40. $2 \div 9 + y \div 6; \frac{2}{9} + \frac{y}{6}$

41. $12 \times y - 5 \times x; 12y - 5x$

Properties of Real Numbers

Objective: To identify and use the properties of the real number system

Real numbers can be used to measure and describe real-world phenomena. You know that operations such as addition and multiplication can be performed on the set of real numbers. The set of real numbers and its operations have certain properties.

Capsule Review

Consider the following questions.

Is $4 + 3$ equal to $3 + 4$? Is $5 - 2$ equal to $2 - 5$?

If $13 + 3$ equals 16 and 4^2 equals 16, how is $13 + 3$ related to 4^2?

What is the sum of 7 and the additive inverse of 7?

The answers illustrate some of the properties of the real number system.

Simplify.

1. $\left(-\frac{2}{3}\right)\left(-\frac{3}{2}\right)$ **2.** $\left(\frac{7}{12}\right) - \left(\frac{5}{12}\right)$ **3.** $\left(\frac{5}{12}\right) - \left(\frac{7}{12}\right)$

4. $|-5| \cdot |-1|$ **5.** $\frac{1}{6}(12 - 18)$ **6.** $\frac{1}{8} + \left(\frac{1}{4} - \frac{1}{8}\right)$

The set of real numbers and the two operations of addition and multiplication form a mathematical system called a **field.** There are certain properties that every field must have.

For every a and b in the set of real numbers,

$a + b$ is a unique real number **Closure property of addition**
ab is a unique real number **Closure property of multiplication**

For example, $-2 + 3$, $-6 + (-7)$, $9 - (-5)$, $(8) - (-7)$, $(-3)(-6)$, and $(-4)(5)$ are all integers, but $9 \div 4$ is not an integer. The integers are closed under addition, subtraction, and multiplication, but they are not closed under division. Similarly, $(3)(5)$, $(19)(5)$, and $(3)(45)$ are all odd natural numbers, but $39 + 5$ is not an odd natural number. The odd natural numbers are closed under multiplication but not addition.

For every a and b that are real numbers,

$a + b = b + a$	**Commutative property of addition**
$ab = ba$	**Commutative property of multiplication**

For example, $-3 + 5 = 5 + (-3)$ and $(6)(-1.2) = (-1.2)(6)$.

However, $(-3) - (-5) \neq (-5) - (-3)$ and $12 \div 3 \neq 3 \div 12$.

The real numbers are not commutative with respect to the operations of subtraction and division.

For every a, b, and c that are real numbers,

$a + (b + c) = (a + b) + c$	**Associative property of addition**
$a(bc) = (ab)c$	**Associative property of multiplication**

For example, $4 + [(-1.2) + 0.6] = [4 + (-1.2)] + 0.6$.

The sum remains the same no matter what two numbers you add first.

$$(-0.3)[(2)(-4)] = [(-0.3)(2)](-4)$$

The product remains the same no matter what two numbers you multiply first.

There exists a unique real number 0 such that for every real number a, $a + 0 = 0 + a = a$. The number 0 is called the **additive identity.**

There exists a unique real number 1 such that for every real number a, $a(1) = 1(a) = a$. The number 1 is called the **multiplicative identity.**

For example, $-4.5 + 0 = 0 + (-4.5) = -4.5$
$(8.24)1 = 1(8.24) = 8.24$

For every a, b, and c that are real numbers,

$a(b + c) = a(b) + a(c)$ **Distributive property**

For example, $3(6.3 + 8.2) = 3(6.3) + 3(8.2)$.

The *additive inverse property* on page 9 and the *multiplicative inverse property* on page 10 are also properties of a field. Notice there are 11 field properties altogether. A set of elements together with one operation satisfying the properties of closure, associativity, identity, and inverse is called a *group*.

EXAMPLE 1 **Indicate which property is illustrated by each of the following:**

a. $8.24 + 0 = 8.24$ **b.** $4[(-12)6] = [4(-12)]6$

c. $3(6 + 8) = 3(6) + 3(8)$ **d.** $(-4)(5)$ is real

e. $(-2)\left(-\frac{1}{2}\right) = 1$ **f.** $-3 + 5 = 5 + (-3)$

a. additive identity **b.** associative of multiplication

c. distributive **d.** closure of multiplication

e. multiplicative inverse **f.** commutative of addition

Can other sets and other operations form a field? Use the field properties to determine whether a given set and operations form a field. A *counterexample* may be used to show a given set and operations is not a field.

EXAMPLE 2 **Determine if the set of integers and the operations of addition and multiplication form a field.**

The multiplicative inverse property is not satisfied by the set of integers.

For example, $3 \times \frac{1}{3} = 1$, but $\frac{1}{3}$ is not an integer *counterexample*

Therefore, the set of integers and the operations of addition and multiplication do not form a field.

The properties can also be used to determine whether a finite set and given operations form a field. Each element of the set must be checked.

EXAMPLE 3 **Determine if the set {0, 1, 2} and the operations of addition and multiplication form a field.**

$0 + 0 = 0$ *0 is in the set.*
$0 + 1 = 1$ *1 is in the set.*
$1 + 1 = 2$ *2 is in the set.*
$1 + 2 = 3$ *3 is not in the set.*

The closure property for addition is not satisfied. Hence, {0, 1, 2} and the operations of addition and multiplication do not form a field.

The following properties of equality are also true for real numbers.

For all real numbers a, b, and c

$a = a$	**Reflexive property**
If $a = b$, then $b = a$.	**Symmetric property**
If $a = b$ and $b = c$, then $a = c$.	**Transitive property**
If $a + b = c$ and $b = d$, then $a + d = c$.	**Substitution property**

Using the properties of real numbers, theorems involving operations with real numbers can be proved.

EXAMPLE 4 **Prove the theorem: If a, b, and c are real numbers and $a = b$, then $a + c = b + c$.**

1. a, b, and c are real numbers.	**1.** Given
2. $a + c$ is a real number.	**2.** Closure property of addition
3. $a + c = a + c$	**3.** Reflexive property
4. $a = b$	**4.** Given
5. $a + c = b + c$	**5.** Substitution property

Theorems which have been proved may be used in future proofs.

EXAMPLE 5 **Prove the theorem: If a, b, and x are real numbers and $x + a = b$, then $x = b - a$.**

1. a and b are real numbers.	**1.** Given
2. $x + a = b$	**2.** Given
3. $-a$ is a real number.	**3.** Additive inverse property
4. $(x + a)$ is a real number.	**4.** Closure property of addition
5. $(x + a) + (-a) = b + (-a)$	**5.** By theorem proved in Example 4
6. $x + [a + (-a)] = b + (-a)$	**6.** Associative property of addition
7. $x + 0 = b + (-a)$	**7.** Additive inverse property
8. $x = b + (-a)$	**8.** Additive identity property
9. $x = b - a$	**9.** Definition of subtraction

CLASS EXERCISES

Thinking Critically

1. Do the set $\{-1, 0, 1\}$ and the operations of addition and multiplication form a field? Explain.

2. Is the set $\{-1, 0, 1\}$ closed under multiplication? Is it closed under addition? If not, give a counterexample.

3. Is the set of even integers closed under addition? Is the set of even integers closed under multiplication? If not, give a counterexample.

Tell which property of real numbers is illustrated.

4. $|x| + |y| = |y| + |x|$

5. $(7 \cdot \pi) \in \{\text{real numbers}\}$

6. $(-2)[(-3)(-4)] = [(-2)(-3)](-4)$

7. $0 + \sqrt{3} = \sqrt{3}$

8. $\sqrt{5} + \sqrt{2}$ is a real number.

9. $x(x + 3) = x^2 + 3x$

PRACTICE EXERCISES

Name the property of real numbers illustrated by each of the following.

1. $92.5(1) = 92.5$

2. $\pi(a + b) = \pi a + \pi b$

3. $-7 + 4 = 4 + (-7)$

4. $14\sqrt{3}$ is a real number.

5. $29\pi = \pi \cdot 29$

6. $(2\sqrt{10}) \cdot \sqrt{3} = 2(\sqrt{10} \cdot \sqrt{3})$

7. $(-8) + [-(-8)] = 0$

8. $-\sqrt{5} + 0 = -\sqrt{5}$

9. $\left(\frac{1}{2} + \frac{1}{4}\right) + \left(-\frac{1}{4}\right) = \frac{1}{2} + \left[\frac{1}{4} + \left(-\frac{1}{4}\right)\right]$

10. $\frac{3}{5} \cdot \frac{5}{3} = 1$

11. $(-2)(-3) = (-3)(-2)$

12. $25(2x + 5y) = 50x + 125y$

Decide if each set below is closed under addition, subtraction, multiplication, and division. If not, give a counterexample.

13. {odd integers}

14. {even natural numbers}

15. {rational number}

16. {irrational numbers}

Which of the field properties *do not* hold for each set?

17. whole numbers

18. natural numbers

19. Is the set $\{0, 1\}$ closed under the operations of multiplication and addition? Which other field properties does it satisfy using the operations of addition and multiplication?

20. Is the set $\left\{2, 1, \frac{1}{2}\right\}$ closed under multiplication and division?

Name the property that justifies each step in each of the following series of statements.

21.
$$\begin{aligned} 3 + 9(2x + 1) &= 3 + (18x + 9) && \textbf{a. } \underline{?} \\ &= (3 + 18x) + 9 && \textbf{b. } \underline{?} \\ &= (18x + 3) + 9 && \textbf{c. } \underline{?} \\ &= 18x + (3 + 9) && \textbf{d. } \underline{?} \\ &= 18x + 12 && \textbf{e. } \underline{?} \end{aligned}$$

22.
$$\begin{aligned} (k \cdot 8)\frac{1}{8} + (-3k) &= k\left(8 \cdot \frac{1}{8}\right) + (-3k) && \textbf{a. } \underline{?} \\ &= k(1) + (-3k) && \textbf{b. } \underline{?} \\ &= k + (-3k) && \textbf{c. } \underline{?} \\ &= k[1 + (-3)] && \textbf{d. } \underline{?} \\ &= k(-2) && \textbf{e. } \underline{?} \\ &= -2k && \textbf{f. } \underline{?} \end{aligned}$$

23. Prove the theorem: If $a = b$, then $ca = cb$.

24. Prove the theorem: If $ab = ac$ and $a \neq 0$, then $b = c$.

25. Prove the theorem: $a(b - c) = ab - ac$.

26. Does the set $\{x: x$ is a rational number$\}$ form a field under the operations of addition and multiplication? Defend your answer.

27. Does the set $\{x: x$ is an irrational number$\}$ form a field under the operations of addition and multiplication? Defend your answer.

Applications

Clock Arithmetic is a specific case of *modular arithmetic* where 12 integer elements exist in the set. Using the face of a clock can help illustrate the operation of addition. For example, start at 8 o'clock, add 7 hours and the result is 3 o'clock. Therefore, $8 + 7 = 3$.

28. Find the sums in clock arithmetic: $3 + 2$; $7 + 9$; $6 + 6$.

29. Is addition in clock arithmetic commutative?

30. What number in clock arithmetic serves as the additive identity?

TEST YOURSELF

Identify the subset(s) of the set of real numbers to which each belongs:

1. $\sqrt{3}$ **2.** $-\frac{1}{2}$ **3.** 6 **4.** -8 1.1

Show that each decimal can be written as a quotient of two integers.

5. 0.175 **6.** $0.\overline{8}$ **7.** $0.\overline{39}$

Simplify.

8. $-3 + 12 \div 4 - 8 \times 2 + 5 \times 3$ **9.** $|6 - 3| + 2|2 - 8| - |-10|$ 1.2

10. $\dfrac{3(3 + 5) - 2(1 - 5)}{2 + 6 \div 3}$ **11.** $\frac{1}{2}\left(5 - \frac{27}{3}\right) + 30 \div 6$

Evaluate.

12. $x^2 - 3xy + y^2$; $x = 2$ and $y = 3$ **13.** $\dfrac{ab - 2a}{b + a}$; $a = 3$ and $b = -2$ 1.3

Identify the property of real numbers illustrated by each of the following.

14. $\dfrac{2}{7} \cdot \dfrac{7}{2} = 1$ **15.** $-\dfrac{4}{3} + 0 = -\dfrac{4}{3}$ 1.4

16. $9 + \sqrt{5}$ is a real number. **17.** $12(5x - 3y) = 60x - 36y$

1.5 Solving Equations in One Variable

Objective: To solve equations containing one variable

In basketball, before the 3-point line was introduced, all field goals counted as 2 points while all free throws counted as 1 point. In 1962 Wilt Chamberlain set a record by scoring 100 points in a game. If he scored exactly 28 points on free throws, how many field goals did he make?

This problem could be solved by using the equation $2x + 28 = 100$, where x is the number of field goals. Solving equations is a very important skill for solving problems.

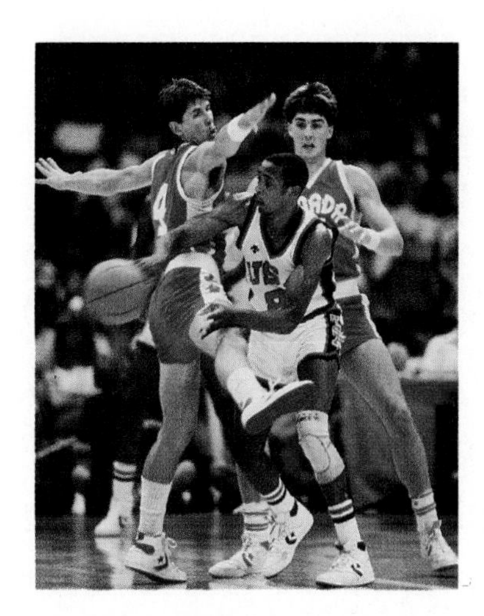

Capsule Review

Use the distributive property to combine like terms.

EXAMPLE **Simplify:** **a.** $5x - 9x + 3$ **b.** $2y + 7x + y - 1$

a. $5x - 9x + 3 = (5 - 9)x + 3$
$= -4x + 3$

b. $2y + 7x + y - 1 = 2y + y + 7x - 1$
$= (2 + 1)y + 7x - 1$
$= 3y + 7x - 1$

Simplify.

1. $x + 2y - 5y - 6x$

2. $10h + 12g - 8h - 4g$

3. $\frac{x}{6} + 3 - \frac{5x}{6}$

4. $\frac{x}{3} + \frac{y}{3} + \frac{2x}{3} - y$

5. $(x + y) - (x - y)$

6. $-(m + n) - 2(2m + 3n)$

An **equation** is a mathematical sentence stating that two quantities are equal. For example, $5 + 2 = 10 - 3$ is an equation. An equation that contains a variable is called an **open sentence.** The equation $2x + 28 = 100$ is an open sentence. The set of values which when substituted for x make the sentence true is the **solution set.** Unless otherwise stated, assume that equations are to be solved over the set of real numbers.

For all real numbers a, b, and c

If $a = b$, then $a + c = b + c$.	**Addition property**
If $a = b$, then $a - c = b - c$.	**Subtraction property**
If $a = b$, then $ac = bc$.	**Multiplication property**
If $a = b$, and $c \neq 0$, then $\frac{a}{c} = \frac{b}{c}$.	**Division property**

The solution can be checked by substituting it for the variable in the original equation. If this yields a true statement, the solution is correct.

EXAMPLE 1 **Solve for x: $2.4x + 35.6 = 122$**

$$2.4x + 35.6 = 122$$

$$2.4x = 86.4 \quad \text{Subtract 35.6 from both sides.}$$

$$x = 36 \quad \text{Divide both sides by 2.4.}$$

Check:

$$2.4x + 35.6 = 122$$

$$2.4(36) + 35.6 \stackrel{?}{=} 122 \quad \text{Substitute 36 for } x.$$

$$86.4 + 35.6 \stackrel{?}{=} 122$$

$$122 = 122 \checkmark$$

The solution set is $\{36\}$.

Most equations require many operations to find the solution. Be sure to keep the left and right members equal by operating on both sides each step.

EXAMPLE 2 **Solve for y: $13y + 48 = 8y - 47$**

$$13y + 48 = 8y - 47$$

$$5y + 48 = -47 \quad \text{Subtract } 8y \text{ from both sides.}$$

$$5y = -95 \quad \text{Subtract 48 from both sides.}$$

$$y = -19 \quad \text{Divide both sides by 5.}$$

Check:

$$13y + 48 = 8y - 47$$

$$13(-19) + 48 \stackrel{?}{=} 8(-19) - 47 \quad \text{Substitute } -19 \text{ for } y.$$

$$-247 + 48 \stackrel{?}{=} -152 - 47$$

$$-199 = -199 \checkmark$$

The solution set is $\{-19\}$.

When an equation is more complicated, simplify each member first.

EXAMPLE 3 **Solve for x: $3x - 7(2x - 13) = 3(-2x + 9)$**

$$3x - 7(2x - 13) = 3(-2x + 9)$$

$$3x - 14x + 91 = -6x + 27 \quad \text{Distributive property}$$

$$-11x + 91 = -6x + 27 \quad \text{Simplify.}$$

$$-5x + 91 = 27 \quad \text{Add } 6x \text{ to both sides.}$$

$$-5x = -64 \quad \text{Subtract 91 from both sides.}$$

$$x = \frac{64}{5} \text{ or } 12.8 \quad \text{Divide both sides by } -5.$$

The solution set is $\left\{\frac{64}{5}\right\}$ or $\{12.8\}$. *The check is left for the student.*

It is important to note that in some equations there is no solution.

EXAMPLE 4 **Solve for x: $6x - 4 = 2(3x + 8)$**

$6x - 4 = 2(3x + 8)$
$6x - 4 = 6x + 16$ *Distributive property*
$-4 = 16$ *Subtract 6x from both sides.*

Since no value of x will make the last line true, the solution set contains no elements and is called the **empty set, $\emptyset$**.

CLASS EXERCISES

Solve and check.

1. $14y - 3 = 25$

2. $3x - 4 + 8x + 3 = 32$

3. $8z + 12 = 5z - 21$

4. $6(t - 2) = 2(9 - 2t)$

5. $12 - 3(2w + 1) = 7w - 3(7 + w)$

6. $3(x + 2) = 2(2x - 1) - 12$

7. $6(x - 3) - 2(7 - x) = 8$

8. $7x - 6(11 - 2x) = 10$

9. $10x - 7 = 2(13 + 5x)$

10. $8x - 3(6 + x) = 5x + 4$

PRACTICE EXERCISES

Solve and check.

1. $4.2x + 6.4 = 40$

2. $6.5y + 3.5 = 49$

3. $2(y - 3) + 6 = 70$

4. $4(x + 2) - 8 = 80$

5. $2t - 3 = 9 - 4t$

6. $7w + 2 = 3w + 94$

7. $6g - (3 - 3g) = 24$

8. $4w - 2(1 - w) = -38$

9. $2(x + 3) - 2(x + 4) = 24$

10. $5(x - 2) - 4(x + 1) = 50$

11. $5x + \frac{1}{3} = 2x - \frac{3}{2}$

12. $4y - \frac{1}{10} = 3y + \frac{4}{5}$

13. $3(m - 2) - 5 = 8 - 2(m - 4)$

14. $7(a + 1) - 3a = 5 + 4(a - 1)$

Solve and check. Name the subset(s) of the real numbers to which each solution belongs.

15. $\frac{k}{3} + \frac{k}{6} = \frac{7}{2}$

16. $\frac{s}{4} + \frac{s}{2} + \frac{s}{3} = 13$

17. $2(3w + 2) - 12 = 3w - 11$

18. $3(8w - 2) = 46 - 2(12w + 1)$

19. $6x - 3(6 - 5x) + 3x = 10 - 4(2 - x)$

20. $\frac{1}{2}(6 + 4x) - \frac{1}{4}(8x - 12) = \frac{1}{2}(2x - 4)$

21. $5y - [7 - (2y - 1)] = 3(y - 5) + 4(y + 3)$

22. $3x + 1 + 2(1 + 2x^2) = 5(1 - x) + 4(x^2 + 1)$

Use the definition of absolute value to solve each equation.

23. $|x + 4| = 8$

24. $|x - 5| = 23$

25. $|2x - 7| = 3$

26. $2|x + 6| = 48$

Applications

27. **Physics** The height of a projectile in feet is given by the formula $h = -16t^2 + vt$, where t is time in seconds and v is the initial velocity of the projectile in ft/sec. Solve the formula for v.

28. **Geometry** The formula for the surface area of a cylinder is $S = 2\pi rh + 2\pi r^2$, where h is the height and r is the radius of the base. Solve for h.

29. **Automotive technology** The formula $p = \frac{sn}{6}$ is used to compute piston speed, where s is the length of the piston stroke in inches, n is the number of revolutions per minute, and p is the piston speed in ft/min. Solve the formula for n.

30. **Electricity** A lightbulb gets hot as electricity passes through the filament. The amount of electrical energy transformed to heat energy as current flows through a wire is given by the formula $w = RI^2t$, where w is heat energy in joules, R is the resistance of the wire in ohms, I is the current in amperes, and t is the time in seconds. Solve the formula for t.

Mixed Review

Write each decimal as a quotient of two integers.

31. 6.333

32. $-0.\overline{71}$

33. $6.1111\ldots$

Evaluate each expression using the value given for each variable.

34. $3|4x - 6| - 2x^2$; $x = -3$

35. $\frac{5r - r^2}{1 - 4r}$; $r = 4$

Name the property of real numbers illustrated by each of the following.

36. $16x + (-16x) = 0$

37. $4 + (x + 7) = 4 + (7 + x)$

38. $2.3 + \sqrt{2}$ is a real number.

39. $10(x + 5) = 10x + 50$

40. 5π is a real number.

41. $4(x - 9) = (x - 9)4$

1.6

Translating Word Sentences into Equations

Objective: To write equations for word sentences

A baseball player's batting average is determined by dividing the number of hits by the number of times at bat.

The preceding English sentence can be translated into the following mathematical sentence:

$$\text{batting average} = \frac{\text{hits}}{\text{times at bat}}$$

Using variables reduces the size of the sentence to

$$b = \frac{h}{t}, \text{ where } b = \text{batting average, } h = \text{hits, and } t = \text{times at bat}$$

Capsule Review

Recall the different symbols which are used when comparing real numbers.

is greater than	$>$	is less than or equal to	$\leq$
is less than	$<$	is greater than or equal to	$\geq$
is equal to	$=$		

Choose the comparison symbol which describes the relationship between each pair of numbers.

1. -9 and -8

2. 0.25 and $\frac{1}{4}$

3. 0 and -1

4. 0.5 and 0.3

5. 5 and $3\frac{1}{2} + 1\frac{1}{2}$

6. $-\frac{1}{2}$ and $\frac{1}{2}$

Most mathematical problems are stated in words. Algebraic symbols can be used to translate words into a mathematical expression or equation.

Word Phrase	Algebraic Symbol
8 more than a number	$x + 8$
a number decreased by 3	$n - 3$
n divided by 7	$\frac{n}{7}$
the difference between 4 and six times some number	$4 - 6x$
twice a number decreased by 4	$2x - 4$
9 more than five times a number	$5t + 9$

Word sentences can be translated phrase by phrase into algebraic sentences.

EXAMPLE 1 **Write an algebraic sentence for: A number increased by four is equal to nine.**

$$\underbrace{\text{a number}}_{x}\ \underbrace{\text{increased by}}_{+}\ \underbrace{\text{four}}_{4}\ \underbrace{\text{is equal to}}_{=}\ \underbrace{\text{nine}}_{9}$$

It is important to remember that an algebraic sentence may contain a comparison symbol other than an equal sign.

EXAMPLE 2 **Write an algebraic sentence for: Twice a number decreased by four is less than eight.**

$$\underbrace{\text{twice a number}}_{2n}\ \underbrace{\text{decreased by}}_{-}\ \underbrace{\text{four}}_{4}\ \underbrace{\text{is less than}}_{<}\ \underbrace{\text{eight}}_{8}$$

Notice the distinction between "six is less than a number" and "is six less than a number."

$$\underbrace{\text{six}}_{6}\ \underbrace{\text{is less than}}_{<}\ \underbrace{\text{a number}}_{n} \qquad \underbrace{\text{is}}_{=}\ \underbrace{\text{six less than}}_{n}\ \underbrace{\text{a number}}_{-6}$$

(In the second diagram, arrows cross to show that "six less than" and "a number" are reversed: $= n - 6$.)

The first expression involves the relation "less than," while the second involves the relation of equality.

Sometimes the phrase "the quantity" indicates that part of the expression is enclosed in grouping symbols such as parentheses.

EXAMPLE 3 **Write an algebraic sentence for: Twice the quantity of a number decreased by six is four more than the number.**

$$\underbrace{\text{Twice the quantity of}}_{2(}\ \underbrace{\text{a number}}_{y}\ \underbrace{\text{decreased by}}_{-}\ \underbrace{\text{six}}_{6)}\ \underbrace{\text{is}}_{=}\ \underbrace{\text{four more than}}_{y}\ \underbrace{\text{the number}}_{+4}$$

(Arrows cross to show that "four more than" and "the number" are reversed: $2(y - 6) = y + 4$.)

CLASS EXERCISES

Write an algebraic expression for each phrase using *x* for the variable.

1. five times a number decreased by four
2. ten decreased by five times a number
3. the square of a number decreased by ten
4. four more than the square root of a number

Write an algebraic sentence for each using *n* for the variable.

5. The sum of seven times a number and four times the same number is twenty-two.

6. Three times a number increased by two is greater than twice the number decreased by six.
7. The difference between the square of a number and the number is twelve.
8. The product of twice a number and ten is one hundred.

PRACTICE EXERCISES

Write an algebraic expression for each phrase using *n* for the variable.

1. eight more than twice a number
2. the sum of a number and its square
3. five times a number decreased by six

Write an algebraic sentence for each using *x* for the variable.

4. The sum of a number and five is 24.
5. Four times the quantity of a number increased by eight is equal to 20 more than the number.
6. Twice a number decreased by 10 is greater than 12.

Write an algebraic expression or sentence for each using *y* as the variable.

7. the difference between six times a number and 10
8. the square of a number increased by five more than the number
9. The square of a number decreased by twice the number is 24.
10. Five times a number is less than the number squared increased by 20.
11. The square root of a number is four less than the square of the number.
12. four more than seven times a number
13. The square of the quantity seven more than y is equal to 81.
14. Twelve times a number decreased by the square of a number is 10.

Write a word phrase for each of the following.

15. $9x - 3$ **16.** $7 - x^2$ **17.** $x + 10x$

18. $4 - 2x$ **19.** $4x - 12x^2$ **20.** $25x^2 - 36$

Write a word sentence for each of the following.

21. $8x + 3 = 12$ **22.** $4y > y + 15$ **23.** $z + 2 = 3z - 1$

24. $5x + 3 \geq 2x - 4$ **25.** $n^2 + 3n = 10$ **26.** $5x + 8 = 2x^2$

Write an algebraic expression or sentence for each using n for the variable.

27. the square root of three less than four times a number

28. the square of the sum of twice a number and three

29. Five times a number decreased by seven is equal to twice the number increased by 32.

30. The quotient of nine times a number and seven is equal to eight more than the number.

Write an algebraic sentence for each using x and y for the variables.

31. The sum of two numbers is greater than the difference of the same two numbers.

32. The square of a number decreased by a second number is equal to the product of the two numbers.

Write an algebraic expression for each of the following.

33. number of inches in x feet

34. number of feet in y miles

35. number of yards in z inches

36. number of kilograms in w grams

37. the next integer after the integer x

38. the next even integer after $n + 2$, if $n + 2$ is even

39. the sum of three consecutive odd integers if x is the first odd integer

40. the value in cents of x quarters, y dimes, and z nickels

Applications

Write a formula for each of the following.

41. Geometry The area A of a trapezoid is equal to one-half the product of the height h and the sum of the bases b_1 and b_2.

42. Chemistry A temperature on the Kelvin scale K is equal to the temperature on the Celsius scale C plus two hundred seventy-three.

43. Sports The earned run average ERA of a pitcher is equal to nine times the number of earned runs r the pitcher allows, divided by the number of innings i pitched.

Mixed Review

Place each number in its approximate location on a number line.

44. $-0.626262\ldots$

45. $-\sqrt{14}$

46. 2π

Simplify.

47. $23 + 5 \times 43 - 17$

48. $8 \times 1.25 - 1.25 + 1.25$

1.7

Problem Solving Strategy: Use a Mathematical Model

Mathematics is a useful and important subject, as it provides a way to solve problems in the real world. Real-world problems can be solved by constructing and analyzing mathematical models. A mathematical model represents the known parts of a problem. Therefore, when a mathematical model is set up and studied, the problem itself is being studied.

Many different kinds of models are used to solve problems mathematically. A drawing, a figure, a graph, or a table can be constructed as a model. A formula or a pattern can be developed to serve as a model. An equation or an inequality can be set up as a mathematical model.

Solving problems can be fun, but if you don't know where to begin, it can be frustrating. Problem solving skills, as with skills in athletics and music, can be improved greatly with patient and consistent practice.

Problem solving is a process, and consists of several steps which are applied sequentially. Each step consists of a set of skills.

Understand the Problem

Read the problem.
What are the given facts?
What is unknown, that is, *what are you asked to find?*
Review the definitions of all mathematical terms.

Plan Your Approach

Choose a strategy.
The strategy you choose is your plan of action for solving the problem. Many strategies will be developed in this book to help you plan your approach. In this lesson, the strategy is to use a mathematical model in the form of an equation.

Complete the Work

Apply the strategy.
Use the algebra you know to apply the strategy to solve the problem. Very often, this means using the appropriate algebraic operations. Keep an open mind. Change your strategy if it does not work.

Interpret the Results

State your answer.
Check your answer.
Does your answer make sense? Does it satisfy the conditions of the problem?

EXAMPLE 1 **The sum of three numbers is 188. The second number is 12 larger than the first, and the third is twice as large as the second. Find the three numbers.**

Understand the Problem

What are the given facts?

The sum of three numbers is 188. The second is 12 larger than the first. The third is twice as large as the second.

What are you asked to find?

Find the three unknown numbers.

Plan Your Approach

Choose a strategy.

The strategy to use is to translate the facts given in the problem into an equation. Then solve the equation to find the answer.

Assign variables to represent the unknown numbers.

Let n represent the first number.
Then $n + 12$ represents the second number.
$2(n + 12)$ represents the third number.

Write a word equation.

first number + second number + third number = the sum

Translate the word equation into an algebraic equation.

$n + n + 12 + 2(n + 12) = 188$

Complete the Work

Solve the equation.

$n + n + 12 + 2(n + 12) = 188$
$n + n + 12 + 2n + 24 = 188$ *Distributive property*
$4n + 36 = 188$ *Combine like terms.*
$4n = 152$ *Subtract 36 from both sides.*
$n = 38$ *Divide both sides by 4.*

The second number $= n + 12 = 38 + 12 = 50$.
The third number $= 2(n + 12) = 2(38 + 12) = 2(50) = 100$.

Interpret the Results

State your answer.

The numbers are 38, 50, 100.

Check your answer.

Is the sum of the numbers equal to 188?

$38 + 50 + 100 \stackrel{?}{=} 188$
$188 = 188$ ✓

EXAMPLE 2 **Two trains start at the same time and travel toward each other from cities 330 mi apart. How many hours will it take for them to meet if one train travels at 50 mi/h and the other travels at 60 mi/h?**

Understand the Problem

What are the given facts?

The trains are 330 mi apart. They are traveling toward each other. One train is moving at 50 mi/h and the other is moving at 60 mi/h.

What are you asked to find?

Find the time that the trains will meet.

Plan Your Approach

Choose a strategy.

The strategy is to write an equation and solve it.

Identify the important mathematical ideas.

distance = rate × time

Assign variables for the unknowns.

Let t represent the time from when the trains start to when they meet.

$50t$ represents the distance traveled by the first train.

$60t$ represents the distance traveled by the second train.

Write a word equation.

distance traveled by first train + distance traveled by second train = 330 mi

Translate the word equation into an algebraic equation.

$$50t + 60t = 330$$

Complete the Work

Solve the equation.

$$50t + 60t = 330$$
$$110t = 330$$
$$t = 3$$ Divide each side by 110.

Interpret the Results

State your answer.

It takes 3 hours for the trains to meet.

Check your answer.

Will the total distance traveled by both trains in 3 hours be 330 mi?

$$50t + 60t = 330$$
$$50(3) + 60(3) \stackrel{?}{=} 330$$ Substitute 3 for t.
$$150 + 180 \stackrel{?}{=} 330$$
$$330 = 330 \checkmark$$

EXAMPLE 3 **The measure of the supplement of an angle is 15 more than four times the measure of the angle. Find the measure of the angle and its supplement.**

Understand the Problem

What are the given facts?

The measure of an angle's supplement is 15 more than 4 times the measure of the angle.

What are you asked to find?

Find the measure of the angle and the measure of its supplement.

Plan Your Approach

Choose a strategy.

The strategy is to write an equation and solve it.

Identify the important mathematical ideas.

The sum of an angle and its supplement is 180°.

Assign variables for the unknowns.

Let x represent the measure of the angle.
Then $180 - x$ represents the measure of the supplement.

Write a word equation.

The measure of the supplement = 15 more than 4 times the measure of the angle.

Translate the word equation into an algebraic equation.

$180 - x = 15 + 4x$

Complete the Work

Solve the equation.

$180 - x = 15 + 4x$

$180 = 15 + 5x$ *Add x to each side.*

$165 = 5x$ *Subtract 15 from each side.*

$33 = x$ *Divide each side by 5.*

$180 - x = 180 - 33 = 147$ *Find the supplement.*

Interpret the Results

State your answer.

The measure of the angle is 33° and its supplement is 147°.

Check your answer.

Is the supplement 15 more than 4 times the angle?

$180 - x = 15 + 4x$

$147 \stackrel{?}{=} 15 + 4(33)$ *Substitute 33 for x.*

$147 \stackrel{?}{=} 15 + 132$

$147 = 147$ ✓

CLASS EXERCISES

Solve each of the following problems.

1. The sum of three consecutive even numbers is 90. Find the numbers.
2. Four times a number decreased by 16 is 86. Find the number.
3. The measure of the complement of an angle is 15 less than twice the measure of the angle. Find the measures of the angle and its complement.
4. The sum of four consecutive odd integers is 152. Find the four integers.
5. Two buses start at the same point and travel in opposite directions. If one bus travels at 40 mi/h and the other travels at 50 mi/h, in how many hours will the buses be 450 mi apart?

PRACTICE EXERCISES

1. Four times a number increased by 16 is 67. Find the number.
2. Six times a number decreased by 24 is 88. Find the number.
3. The measure of the supplement of an angle is 20 more than three times the original angle. Find the measures of the angle and its supplement.
4. The measure of an angle and its complement differ by 22. Find the angle and its complement.
5. The sum of four consecutive odd integers is 184. Find the four numbers.
6. Find four consecutive even integers such that the sum of the second and fourth is 76.
7. Five times a number decreased by 12 is 34 more than the original number. Find the number.
8. Five times a number decreased by 34 is equal to the number increased by 10. Find the number.
9. Two buses leave Houston at the same time and travel in opposite directions. One bus averages 55 mi/h and the other bus averages 45 mi/h. How long will it take them to be 400 mi apart?
10. Two planes leave an airport at noon, one flying east at a certain speed and the other flying west at twice the speed. If the planes are 2700 mi apart in 3 h, how fast was each plane flying?
11. Three times a number decreased by 22 is 87. Find the number and name the subset(s) of the real numbers to which the solution belongs.
12. 158 decreased by five times a number is equal to the number increased by 20. Find the number.

13. The sum of three integers is 242. The second number is three more than twice the first and the third number is nine less than five times the first. Find the numbers.

14. Find two rational numbers such that the second number is five more than three times the first number. The sum of the numbers is 19.

15. In an isosceles triangle the base is seven less than twice one of the legs. If the perimeter of the triangle is 65 cm, find the lengths of the three sides of the triangle.

16. In trapezoid $RSTV$ (with $\overline{RS} \parallel \overline{VT}$), the measure of $\angle R$ is 24° more than the measure of $\angle S$. The measure of $\angle R$ is four times the measure of $\angle V$. What is the measure of $\angle S$?

17. Find three consecutive even integers such that the sum of the first and third is 54 more than the second.

18. Find three consecutive odd integers such that the sum of the second and third is equal to the first integer increased by 79.

19. Find four consecutive even integers such that the product of the first and fourth is equal to the square of the second.

20. When you substitute a rational number into the algebraic expression $2x - 3(5x + 8) + x$, the value of the expression is -32. Find the rational number.

21. Adam made a trip to see his friend in college. The college is 295 mi from Adam's house. He averaged 50 mi/h for most of the trip but only 40 mi/h for the part where the road was under construction. If the trip took 6 h, how many miles of the road were under construction?

22. Eve opened her coin purse and found dimes, quarters, and nickels with a total value of $1.90. There are twice as many dimes as there are quarters and half as many nickels as quarters. How many coins of each type did Eve have in her purse?

Mixed Problem Solving Review

Translate each sentence into an algebraic equation or inequality. Use *x* for the variable.

1. Three times a number increased by six is 21.
2. The sum of a number and one-half the number is 15.
3. A number decreased by 9 is greater than -30.
4. Four times a number decreased by five is greater than or equal to the additive inverse of two and is less than the multiplicative identity.

Solve the problem.

5. The sum of three consecutive numbers is 36. Find the three numbers.
6. Two-thirds of a number less one-third of the same number is zero. Find the number.
7. The sum of two consecutive even integers is equal to seven times the first decreased by 18. Find the integers.
8. Two numbers differ by 13. When twice the smaller number is added to the larger, the result is 24 greater than the larger. Find the numbers.

PROJECT

The subject of mathematics contains many everyday words, such as *real* number, that have their own mathematical meanings. Use a good dictionary to look up the meaning of the word *real*. Is a number real in the sense that the word is defined in a dictionary? Explain your answer. Investigate the words *rational* and *irrational* in a similar manner.

TEST YOURSELF

Solve and check each of the following. 1.5

1. $23t + 35 = 219$
2. $6z + 24 = 2z - 48$
3. $4f + 6(2 - f) = 14$
4. $\frac{y}{2} + \frac{y}{4} = 12$

Write an algebraic expression or sentence for each using x as the variable. 1.6

5. The difference between five times a number and 12.
6. Three more than six times a number.
7. Twice the square of a number decreased by seven is equal to 15.
8. Four times the quantity of a number increased by nine is equal to 16 more than the number.

Solve. 1.7

9. One number is seven less than three times another number. The sum of the numbers is 61. Find the numbers.
10. Find three consecutive even integers such that the sum of the first and second is 48 more than the third.

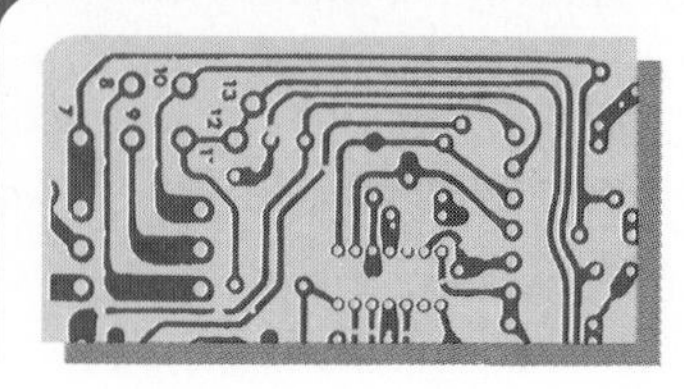

TECHNOLOGY: Expressions for Real Numbers

People have always been fascinated by number relationships. Leonhard Euler proved in 1773 that every integer is the sum of four or fewer squares.

For example:

$15 = 3^2 + 2^2 + 1^2 + 1^2$ $\quad$ $19 = 4^2 + 1^2 + 1^2 + 1^2$
$16 = 4^2$ $\quad$ $20 = 4^2 + 2^2$
$17 = 4^2 + 1^2$ $\quad$ $21 = 4^2 + 2^2 + 1^2$
$18 = 4^2 + 1^2 + 1^2$ $\quad$ $22 = 4^2 + 2^2 + 1^2 + 1^2$

Solve, using the x^2 key on your calculator.

1. Express 50, 83, 125, and 232 as the sum of four or fewer squares.
2. 325 is the smallest number that can be expressed as the sum of two squares in three different ways. Find the three expressions.

In 1770, Edward Waring found that every integer can be expressed as the sum of nine or fewer cubes.

Solve, using the x^3 feature or the exponent key on your calculator.

3. 1729 is the smallest number that can be expressed as the sum of two cubes in two different ways. Find the two expressions.
4. Express 9, 36, and 100 as the sum of nine or fewer cubes.
5. Describe any patterns created in the answers to Exercise 4. What would be the next two numbers in the pattern?

Although irrational numbers are numbers that neither repeat nor terminate, mathematicians have tried to find expressions with patterns in order to calculate approximations of irrational numbers. Of all the irrational numbers, none has intrigued mathematicians more than π.

More than 2000 years ago, Archimedes used the perimeter of a regular 96-gon to calculate that the value of π lies between $3\frac{10}{71}$ and $3\frac{10}{70}$, an accuracy of two decimal places. Perhaps the most familiar approximation of π—$\frac{22}{7}$—is the reduced version of Archimedes' boundary of $3\frac{10}{70}$. Five hundred years later, the Chinese philosopher and mathematician Tsu Ch'ung-Chi was able to calculate π to six decimal places by using the approximation $\frac{355}{113}$.

Later mathematicians were able to generate approximations of π by using formulas that involve infinite sequences of factors or terms. The accuracy of such approximations depends on how many factors or terms are used.

In the seventeenth century, John Wallis discovered this expression for π:

$$\pi = 2\left(\frac{2}{1} \times \frac{2}{3} \times \frac{4}{3} \times \frac{4}{5} \times \frac{6}{5} \times \frac{6}{7} \cdots\right)$$

6. Compare the calculator value of π with the value of this expression when you use (a) six factors; (b) twelve factors; (c) twenty factors.

In the 300 years since Wallis's formula, hundreds of methods of calculating π have been developed. The mathematicians Tamura and Kanada recently calculated the first sixteen million decimal places of π, using a method based on the work of the nineteenth-century mathematician Karl Gauss. The first 100 digits are shown here:

3.14159265358979323846264338327950288419716939937510
58209749445923078164062862089986280348253421170 67

Tamura and Kanada's method is based on a repeating sequence of commands; it is well-suited to the computer or to a calculator that has storage capability. The five storing steps are performed in order, followed by the display step. Then the steps are repeated; every repetition provides greater accuracy.

Before beginning the first trip through the loop, store the following initial values of the variables: $1 \rightarrow B$; $1 \rightarrow E$; $\frac{1}{\sqrt{2}} \rightarrow C$; $\frac{1}{4} \rightarrow D$

1. $B \rightarrow A$
2. $\frac{B + C}{2} \rightarrow B$
3. $\sqrt{CA} \rightarrow C$
4. $D - E(B - A)^2 \rightarrow D$
5. $2E \rightarrow E$

Display $\frac{(B + C)^2}{4D}$

The same method, in BASIC:

```
10 B=1:E=1:C=1/SQR(2):D=.25
20 FOR X=1 TO 5
30 A=B:B=(B+C)/2
40 C=SQR(C*A)
50 D=D-E*(B-A)^2:E=2*E
60 PRINT (B+C)^2/(4*D)
70 NEXT X
```

7. Examine the first approximation of π, comparing it with the value of π shown above. To how many decimal places is the approximation correct?

8. To how many decimal places is the second approximation correct?

9. Research the history of π to find other expressions for approximating π.

CHAPTER 1 SUMMARY AND REVIEW

Vocabulary

additive identity (18)
additive inverse (9)
algebraic expression (14)
associative property (18)
base (11)
closure property (17)
commutative property (18)
coordinate (4)
distributive property (18)
empty set (25)
equation (23)
evaluate an expression (14)
exponent (11)
field (17)
graph (4)
integers (2)
irrational numbers (3)
multiplication property of -1 (10)
multiplication property of zero (10)
multiplicative identity (18)
multiplicative inverse (10)
natural numbers (2)
numerical expression (13)
one-to-one correspondence (4)
open sentence (23)
order of operations (13)
rational numbers (3)
real numbers (3)
reciprocal (10)
reflexive property (19)
solution set (23)
substitution property (19)
symmetric property (19)
transitive property (19)
variable (14)
whole numbers (2)

Real Numbers The real numbers are the union of the rational numbers and the irrational numbers. 1.1

Name the subset(s) of the real numbers to which each belongs.

1. 13 **2.** -2 **3.** $-\sqrt{11}$ **4.** 35.444...

Are the following true or false?

5. $257.023 < 257.0023$
6. $-58 > -23$
7. 5.12121212 ... is a rational number.
8. -9 is an integer.

Operations with Real Numbers The operations with real numbers follow the same rules as the operations with integers. The absolute value of a real number is its distance from zero on the number line. 1.2

Simplify.

9. $(-12)(-4)$ **10.** $(36) \div (-12)$ **11.** $|15 - 23| - |34 - 28|$

Write the additive inverse and the multiplicative inverse.

12. -8 **13.** $\frac{5}{3}$

Variables and Expressions Mathematicians have agreed on a specified order in which to perform multiple operations within an expression. To evaluate an expression for a given number, substitute the number for the variable and follow the order of operations. 1.3

Simplify.

14. $8 + 2 \times 6 - 10 + 4 \times 2 \div 4$

15. $\frac{28 - 2 \times 4}{2 \times 2} - \frac{7 \times 2 + 1}{5}$

Evaluate each expression using the value given.

16. $5a - 2a^2 + 3a^3$; $a = 3$

17. $\frac{3d - 5}{d} - \frac{4 + d}{d - 2}$; $d = 5$

Properties of Real Numbers The real numbers along with the operations of addition and multiplication form a field. There are 11 properties associated with a field, called the *field properties*. 1.4

Tell which of the field properties of real numbers is illustrated.

18. $|b| + |a| = |a| + |b|$

19. $(8 \cdot 3\pi) \in$ (real numbers)

20. $(-5)[(-9)(-14)] = [(-5)(-9)](-14)$

21. $9 + 0 = 0 + 9 = 9$

Solving Equations in One Variable Most equations require the use of more than one operation to find the solution. 1.5

22. Solve for z: $12z + 36 = 8z - 48$

23. Solve for x: $2x - 5 + 4(6 - 3x) = 7(4 - x) - 3(x + 9)$

Translating Word Sentences into Equations Many mathematical problems are stated in words. These words must be translated into mathematical sentences before they can be solved. 1.6

Write an algebraic expression or a sentence for each using x as the variable.

24. four times a number decreased by eight

25. the square of a number increased by 20

26. Four times a number decreased by two is less than twice the number decreased by seven.

Writing an Equation Learning to write an equation using conditions stated in the problem is a problem solving skill. 1.7

27. Find three consecutive integers such that five times the second is 27 more than three times the third.

CHAPTER 1 TEST

Name the subset(s) of the real numbers to which each belongs.

1. -2

2. $\sqrt{3}$

Which property of real numbers is illustrated by each statement?

3. $a + (-1)a = 1 \cdot a + (-1)a$

4. $[a \cdot 0 + a \cdot 1] = a[0 + 1]$

5. $4 + [(-5) + 5] = 4 + 0$

6. $a \cdot \frac{1}{a} = 1$

Show that each decimal can be written as a quotient of two integers.

7. 0.6375

8. $0.\overline{312}$

Simplify each expression:

9. $(0.25 \times 4 + 1) \div 4 - 0.5 + 1.25$

10. $12 \div 3 - 2 + 7 - 4 \times 5^2$

11. Evaluate $\frac{4x + 15}{3} + \frac{2x - 7}{3}$ for $x = 2$.

Write a comparison statement for each pair using $<$, $=$, or $>$.

12. $-32 \ \underline{?}\ -24$

13. $56.56 \ \underline{?}\ 56.56565656$

14. Solve for n: $3n - 2(5n - 8) = 12n + 32$

Write an equation for each and solve.

15. Four times a number decreased by 12 is equal to the number increased by three.

16. The length of a rectangle is twice the width. Four times the length of a rectangle is equal to two times the width increased by 12.

17. The sum of three consecutive odd integers is 159. Find the integers.

True or False?

18. $|5| \cdot |-7| = |5(-7)|$

19. $-\sqrt{24}$ is not a real number.

20. $|4 - 9| = |4| - |9|$

21. $0.4 < \frac{2}{5}$

Challenge

Suppose the operation $*$ is defined as $a * b = (a + b)(a - b)$.
Is the operation $*$ commutative?
Does the set of integers have closure under $*$?

COLLEGE ENTRANCE EXAM REVIEW

Select the best choice for each question.

1. Which of the following is(are) true?

I. $3\sqrt{2}$ is a real number.

II. $\sqrt{81}$ is rational.

III. $\sqrt{a^2 + b^2} = a + b$

A. II only **B.** I, II only **C.** I, III only **D.** II, III only **E.** I, II, III

2. A man paid \$31.60 sales tax when he purchased a new mower for \$395. What is the sales tax rate in his community?

A. 6% **B.** $6\frac{1}{2}$% **C.** 7% **D.** $7\frac{1}{2}$% **E.** 8%

3. If p is a multiple of 3 and q is a multiple of 5, which of the following is(are) true?

I. $p + q$ is even.

II. pq is odd.

III. $5p + 3q$ is a multiple of 15.

A. None **B.** II only **C.** III only **D.** I, II only **E.** II, III only

4. Which is the best *estimate* of the fraction $\frac{68.4 \times 123 \times .45}{7.19 \times 11.82}$?

A. 500 **B.** 300 **C.** 50 **D.** 30 **E.** 5

5. If $9x - 6y = 11$, then $12x - 8y =$

A. $\frac{33}{8}$ **B.** $\frac{44}{9}$ **C.** $\frac{33}{4}$ **D.** 12 **E.** $\frac{44}{3}$

6. 200 is what percent of 40?

A. 500% **B.** 200% **C.** 50% **D.** 20% **E.** 2%

7. When the length of a rectangle is divided by 3 and the width is doubled, the area of the new rectangle is the area of the original one multiplied by

A. 6 **B.** 5 **C.** $\frac{3}{2}$ **D.** $\frac{2}{3}$ **E.** $\frac{1}{6}$

8. Square $OABC$ is drawn with vertex B on circle O. If the radius of the circle is 12 cm, find the length of $\widehat{BY}$.

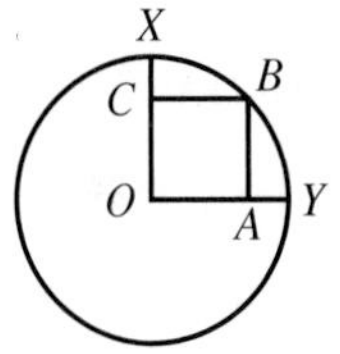

X
C
B
O
A
Y

A. $\frac{3}{2}\pi$ cm **B.** 2π cm **C.** $\frac{8}{3}\pi$ cm **D.** 3π cm **E.** 4π cm

9. In parallelogram $ABCD$, M is the midpoint of DC. The area of $\triangle ADM$ is what percent of the area of $ABCD$?

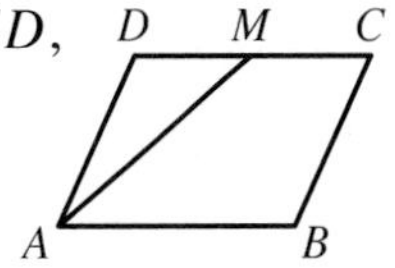

D
M
C
A
B

A. 20% **B.** $22\frac{1}{2}$% **C.** 25% **D.** $33\frac{1}{3}$% **E.** 40%

10. If 5 lines are drawn in a plane, what is the maximum possible number of points of intersection of the lines?

A. 12 **B.** 10 **C.** 8 **D.** 6 **E.** 4

11. Paul is 2 inches taller than Bob and Ron is 3 inches shorter than Paul. If Ron is 4 ft, 9 inches tall, how tall is Bob?

A. 5 ft 1 in. **B.** 5 ft **C.** 4 ft 11 in. **D.** 4 ft 10 in. **E.** 4 ft 8 in.

MIXED REVIEW

Order the set of numbers from least to greatest.

Example $\{-24, 23, -12, 19\}$ *Use a number line.*

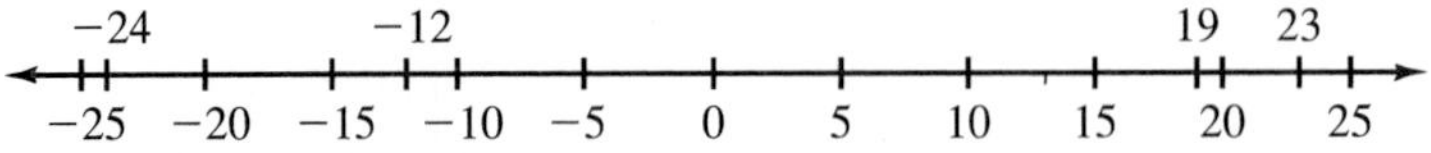

-24 is to the *left* of -12 $\quad -24$ *is less than* -12

$-24, -12, 19, 23$

1. $\{-18, -29, 31, 25\}$
2. $\{8, -9, -17, 5\}$
3. $\{-116, -128, -132, -143\}$
4. $\{1.5, -3.2, 0, -2.3\}$
5. $\{-1.4, -2.1, -3.5, -0.8\}$
6. $\{-5.3, -5.4, 5.2, 5.1\}$

Find the sum.

Examples $(-27) + 9$ $\qquad (-15) + 21 + (-8)$

$$\begin{array}{r}(-27)\\ +9\\ \hline -18\end{array} \qquad \text{Step 1} \quad \begin{array}{r}(-15)\\ +(-8)\\ \hline (-23)\end{array} \qquad \text{Step 2} \quad \begin{array}{r}(-23)\\ +21\\ \hline -2\end{array}$$

7. $(-16) + 7$
8. $45 + (-32)$
9. $(-7) + 37 + (-16)$
10. $15 + (-46) + 27$
11. $23 + (-19) + 5 + (-9)$
12. $(-12) + 9 + (-7) + 11$

Solve.

Example $3x + 5.4 = 32.4$

$3x = 27$ *Subtract 5.4 from both sides.*

$x = 9$ *Divide each side by 3.*

13. $4x + 3.8 = 27.8$
14. $y - \frac{1}{2} = 3\frac{1}{4}$
15. $\frac{x}{3} - 2 = 3$
16. $b + 0.5 = 4$
17. $5m - 1 = 0$
18. $\frac{1}{2}n = \frac{1}{8}$

Use the formula given to solve each word problem.

19. The width of a rectangular room is 9 ft. The distance around the room is 48 ft. What is the length? ($p = 2l + 2w$)
20. Jon drove 162 mi in 3 hours. What was his average rate of speed? ($d = rt$)
21. Aaron deposited \$2,500 in a certificate of deposit at 8% interest for 2 years. How much interest did he earn? ($I = prt$)

Equations and Inequalities

An age-old solution?

Developing Mathematical Power

Henry Dudeney (1847–1930) was a mathematician and one of the greatest inventors of puzzles in England. Test your wits against one of Dudeney's puzzles.

De Morgan and Another

Augustus De Morgan, a mathematician who died in 1871, used to boast that he was x *years old in the year* x^2. *Jasper Jenkins, wishing to improve on this, told me in 1925 that he was* $a^2 + b^2$ *in* $a^4 + b^4$*; that he was* 2m *in the year* $2m^2$*; and that he was* 3n *years old in the year* $3n^4$. *Can you give the years in which De Morgan and Jenkins were respectively born?*

Project

Solve Dudeney's puzzle. Then create your own mathematical puzzle and provide the solution.

On an average day, about 10,500 Americans are born.

2.1 Literal Equations and Formulas

Objective: To solve literal equations or formulas for one of the variables

If a temperature gauge shows readings in degrees Fahrenheit (F), the equivalent temperature in degrees Celsius (C) can be found using the formula

$$C = \frac{5}{9}(F - 32)$$

A **literal equation** is an equation that contains more than one letter or variable. A **formula** is a literal equation that states a general rule or principle. To solve literal equations, use the same methods that you use to solve equations containing just one variable.

Capsule Review

EXAMPLE **Use the properties of equality to solve the equation $7x + 5 = 3x - 10$ for x.**

$7x + 5 = 3x - 10$

$4x + 5 = -10$ *Subtract 3x from both sides.*

$4x = -15$ *Subtract 5 from both sides.*

$x = -\frac{15}{4}$ *Divide both sides by 4.*

Solve.

1. $5x - 8 = 32$

2. $4y - 6 = 2y + 8$

3. $3(2z + 1) = 35$

4. $5(3w - 2) - 7 = 23$

5. $t - 2(3 - 2t) = 2t + 9$

6. $8(s - 12) - 24 = 3(s + 2)$

The equation $ax = t$, where a, t, and x represent real numbers, is solved for t in terms of a and x. The equation can also be solved for x as shown at the right.

$ax = t$

$\frac{ax}{a} = \frac{t}{a}$ *Divide both sides by a.*

$x = \frac{t}{a}, a \neq 0$

Notice that you must exclude from the solution, values of a variable that make a denominator zero.

It is often necessary to use more than one property of equality to solve a literal equation. Example 1 illustrates cases in which it is convenient to use the addition or subtraction property before using the multiplication or division property.

EXAMPLE 1 **Solve for x and indicate any restrictions on the values of the variables.**

a. $ax + b = c$ **b.** $\frac{x}{r} - h = 4,\ r \neq 0$

c. $d - x = e$ **d.** $y - px - c = bk$

a. $ax + b = c$

$ax = c - b$ *Subtract b from both sides.*

$x = \frac{c - b}{a},\ a \neq 0$ *Divide both sides by a.*

b. $\frac{x}{r} - h = 4,\ r \neq 0$

$\frac{x}{r} = 4 + h$ *Add h to both sides.*

$x = r(4 + h)$ *Multiply both sides by r.*

c. $d - x = e$

$-x = e - d$ *Subtract d from both sides.*

$x = -e + d$ *Multiply both sides by −1.*

d. $y - px - c = bk$

$-px = bk - y + c$ *Subtract y and add c to both sides.*

$x = \frac{-bk + y - c}{p},\ p \neq 0$ *Divide both sides by −p.*

To solve the formula $C = \frac{5}{9}(F - 32)$ for F, the multiplication property is used before the addition property.

EXAMPLE 2 **Solve $C = \frac{5}{9}(F - 32)$ for F.**

$C = \frac{5}{9}(F - 32)$

$\frac{9}{5}C = F - 32$ *Multiply both sides by $\frac{9}{5}$.*

$\frac{9}{5}C + 32 = F$ *Add 32 to both sides.*

$F = \frac{9}{5}C + 32$ *Symmetric property*

When the variable for which an equation is being solved appears in more than one term, the distributive property can often be used to determine the coefficient of that variable.

EXAMPLE 3 **Solve $cx + dx = e$ for x, and indicate any restrictions on the values of the variables.**

$$cx + dx = e$$

$$x(c + d) = e \qquad \textit{Distributive property}$$

$$x = \frac{e}{c + d}, \; c \neq -d \qquad \textit{Divide both sides by } (c + d).$$

When the variable for which the equation is being solved appears on both sides of the equation, use the addition or subtraction property to get both terms on the same side.

EXAMPLE 4 **Solve $kt = \frac{t}{2} + r$ for t, and indicate any restrictions on the values of the variables.**

$$kt = \frac{t}{2} + r$$

$$2kt = t + 2r \qquad \textit{Multiply both sides by 2.}$$

$$2kt - t = 2r \qquad \textit{Subtract t from both sides.}$$

$$t(2k - 1) = 2r \qquad \textit{Distributive property}$$

$$t = \frac{2r}{2k - 1}, \; k \neq \frac{1}{2} \qquad \textit{Divide both sides by } (2k - 1).$$

Example 5 shows the procedure for solving a literal equation for a variable that appears in the denominator of a fraction.

EXAMPLE 5 **Solve $\frac{1}{x} + a = b, x \neq 0$ for x, and indicate any restrictions on the values of the variables.**

$$\frac{1}{x} + a = b, \; x \neq 0$$

$$\frac{1}{x} = b - a \qquad \textit{Subtract a from both sides.}$$

$$1 = x(b - a) \qquad \textit{Multiply both sides by x.}$$

$$\frac{1}{b - a} = x, \; b \neq a \qquad \textit{Divide both sides by } (b - a).$$

$$x = \frac{1}{b - a}, \; b \neq a \qquad \textit{Symmetric property}$$

CLASS EXERCISES

Tell what you would do first to solve each equation for x. Do not solve the equations.

1. $ax = f$
2. $cx + d = e$
3. $x - p = q$
4. $t - x = r$
5. $nx - s = c$
6. $a(x - 1) = 7a$
7. $ax + 9 = cx$
8. $\frac{x}{a} = c,\ a \neq 0$
9. $\frac{bx}{a} = d,\ a \neq 0$
10. $ax - bx = c$
11. $\frac{x - 2}{2} = m + n$
12. $\frac{2}{5}(x + 1) = g$
13. $\frac{2}{3}x = g$
14. $g\left(x + \frac{2}{3}\right) = 1$
15. $x + 1 = g + \frac{2}{3}$

PRACTICE EXERCISES

Solve for x and indicate any restrictions on the values of the variables.

1. $ax = c$
2. $bx = a$
3. $x + d = e$
4. $p + x = q$
5. $m - x = n$
6. $f - x = d$
7. $\frac{x}{b} = k,\ b \neq 0$
8. $\pi x = 7\pi$
9. $nx + p = r$
10. $rx + 8 = e$
11. $cx - d = e$
12. $bx - 3 = f$
13. $\frac{x}{a} + b = c,\ a \neq 0$
14. $\frac{x}{b} + 6 = d,\ b \neq 0$
15. $\frac{x}{a} - 5 = b,\ a \neq 0$
16. $\frac{x}{b} - c = d,\ b \neq 0$
17. $ax + bx = c$
18. $bx - cx = -c$

Solve each formula for the indicated variable. Indicate any restrictions on the values of the variables.

19. $A = \frac{1}{2}bh$, for h
20. $s = \frac{1}{2}gt^2$, for g
21. $V = lwh$, for w
22. $I = prt$, for r
23. $V = \pi r^2 h$, for h
24. $S = 2\pi rh$, for r

Solve for x and indicate any restrictions on the values of the variables.

25. $bx + a = dx + c$
26. $cx - b = ax + d$
27. $6bx + c = 2c - 3bx$
28. $8cx - e = 3e + cx$
29. $a(x - 3) + 8 = b(x - 1)$
30. $c(x + 2) - 5 = b(x - 3)$

31. $a(3tx - 2b) = c(dx - 2)$

32. $b(5px - 3c) = a(qx - 4)$

33. $\frac{1}{x} = c + \frac{1}{b}$, $x \neq 0$, $b \neq 0$

34. $\frac{1}{x} = c - \frac{1}{a}$, $x \neq 0$, $a \neq 0$

Solve each formula for the indicated variable. Indicate any restrictions on the values of the variables.

35. $R(r_1 + r_2) = r_1 \cdot r_2$, for r_2

36. $R(r_1 + r_2) = r_1 \cdot r_2$, for R

37. $h = vt - 5t^2$, for v

38. $v = s^2 + \frac{1}{2}sh$, for h

39. $A = \frac{1}{2}h(b_1 + b_2)$, for b_2

40. $A = p(1 + rt)$, for t

Solve for x and indicate any restrictions on the values of the variables.

41. $\frac{a}{b}(2x - 12) = \frac{c}{d}$, $b \neq 0$, $d \neq 0$

42. $\frac{3ax}{5} - 4c = \frac{ax}{5} + 2c$

43. $\frac{y - c}{x - a} = m$, $x \neq a$

44. $a - b = \frac{d}{x} + \frac{e}{x}$, $x \neq 0$

Solve each formula for the indicated variable. Indicate any restrictions on the values of the variables.

45. $S = 2\pi r^2 + 2\pi rh$, for h

46. $E = \frac{1}{2}mv^2$, for v

47. $V = \frac{4}{3}\pi r^3$, for r

48. $\frac{1}{R} = \frac{1}{a} + \frac{1}{b}$, $R, a, b \neq 0$, for a

Applications

49. **Investment** The relationship of the total amount (amount borrowed plus interest) A to the principal (amount borrowed) p, the rate of simple interest (expressed as a decimal) r, and the number of years t is given by the formula $A = p(1 + rt)$. Solve this formula for r.

50. **Manufacturing** The *break-even quantity* Q is the volume of production at which revenue is equal to total costs. The relationships among the break-even quantity and the sales price per unit P, the fixed costs F, and the variable costs V is given by the formula $PQ = F + VQ$. Solve this formula for Q.

51. **Communications** The relationship of the cost C of sending a telegram, to the minimum charge P, for 10 words or less, the additional charge per word a, and the number of words n is given by the formula $C = P + a(n - 10)$. Solve this formula for n.

Physics The relationship between the temperature t, in degrees Fahrenheit, and the speed of sound s, in feet per second, is given by the formula $s = 1055 + 1.1t$.

52. Solve the formula for t in terms of s.

53. Find the temperature, to the nearest degree Fahrenheit, at which the speed of sound is 1100 ft/s.

54. Express the temperature in Exercise 53 in degrees Celsius, to the nearest degree. Use the formula $C = \frac{5}{9}(F - 32)$.

BIOGRAPHY: Al-Khowarizimi

Mohammed ibn Musa Abu Dejefar al-Khowarizimi lived during the eighth and ninth centuries, when Arabia was a major center of scientific activity. He wrote *Hisat Al-jabr w'al-muqahalah,* which means "the science of reunion and the opposition." Through Latin translations, the word *al-jabr,* or *algebra,* became synonymous with the science of equations. Al-Khowarizimi's writings led to the discovery of a method for approximating the square root of a number x, using the formula

$$\sqrt{x} = p + \frac{x - p^2}{2p + 1}$$

where p is the square root of the nearest perfect square less than x.

EXAMPLE **Use the formula to approximate the square root of 13.**

Since 9 is the nearest perfect square that is less than 13, let $p = \sqrt{9} = 3$. Then

$$\begin{aligned} \sqrt{x} &= p + \frac{x - p^2}{2p + 1} \\ \sqrt{13} &= 3 + \frac{13 - 9}{2(3) + 1} \\ &= 3 + \frac{4}{7} \\ &\approx 3.571 \end{aligned}$$

Note that the answer does not differ much from the closer approximation, 3.606, obtained using a calculator.

Use the formula above to approximate each square root. Then use a calculator to find a closer approximation and compare the results.

1. $\sqrt{17}$ **2.** $\sqrt{23}$ **3.** $\sqrt{42}$

2.2

Solving Inequalities

Objective: To solve linear inequalities in one variable and graph their solution sets

A CAT (Computerized Axial Tomography) scan is a diagnostic procedure by which an image is formed using a computerized combination of many X-ray photographs. Suppose a technician is comparing lists of patients scheduled to undergo CAT scans on Monday and on Tuesday. The numbers of patients on the two lists can be compared by making just one of the following statements:

The number on Monday's list *is less than* the number on Tuesday's list.
The number on Monday's list *is equal to* the number on Tuesday's list.
The number on Monday's list *is greater than* the number on Tuesday's list.

Mathematical sentences that relate values that are unequal, such as $a < b$ and $b > a$, are called **inequalities.** Note that the statement *a is less than b* is equivalent to the statement *b is greater than a*. That is, $a < b$ is equivalent to $b > a$.

Capsule Review

When a is to the left of b on the real number line, then a is less than b.
When a is to the right of b on the real number line, then a is greater than b.

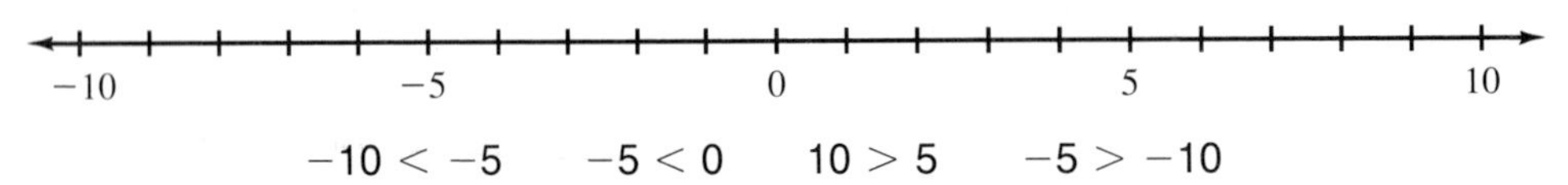

$-10 < -5$ $\quad -5 < 0$ $\quad 10 > 5$ $\quad -5 > -10$

Replace each ? with the symbol <, >, or = to make the sentence true.

1. -18 ? -32

2. 145 ? $(-9)(-5)$

3. 2 ? 3

4. $0.323435\ldots$? $0.343434\ldots$

5. $(3 - 24)$? $(4 - 36)$

6. -34 ? $-10 - 24$

7. 8 ? 4

8. $[5 - (-6)]$? -10

9. $5x$? $3x$, if $x > 0$

10. $5x$? $3x$, if $x < 0$

The three statements on page 52 illustrate the **trichotomy property,** also called the *comparison property*.

Trichotomy Property

If two real numbers, a and b, are chosen at random, then exactly one of the following statements is true:

a is less than b.	*a is equal to b.*	*a is greater than b.*
$a < b$	$a = b$	$a > b$

To solve an inequality such as $12 - 4x > 16$ or $5(x - 2) + 8 < 23$, you use the **properties of inequality,** which are also called the *properties of order*. Although these properties are similar to the properties of equality, you should note the differences in the cases of properties of multiplication and division. The properties of inequality are stated for the relation *is less than* ($<$) below, but they also apply to the relation *is greater than* ($>$).

Properties of Inequality

For all real numbers a, b, and c:

Transitive Property		If $a < b$ and $b < c$, then $a < c$.
	Example:	If $2 < 5$ and $5 < 9$, then $2 < 9$.
Addition Property		If $a < b$, then $a + c < b + c$.
	Example:	If $2 < 5$, then $2 + 9 < 5 + 9$.
Subtraction Property		If $a < b$, then $a - c < b - c$.
	Example:	If $2 < 5$, then $2 - 9 < 5 - 9$.
Multiplication Properties		If $a < b$ and $c > 0$, then $ac < bc$.
	Example:	If $2 < 5$, then $2(9) < 5(9)$.
		If $a < b$ and $c < 0$, then $ac > bc$.
	Example:	If $2 < 5$, then $2(-9) > 5(-9)$.
Division Properties		If $a < b$ and $c > 0$, then $\frac{a}{c} < \frac{b}{c}$.
	Example:	If $2 < 5$, then $\frac{2}{9} < \frac{5}{9}$.
		If $a < b$ and $c < 0$, then $\frac{a}{c} > \frac{b}{c}$.
	Example:	If $2 < 5$, then $\frac{2}{-9} > \frac{5}{-9}$.

EXAMPLE 1 **Solve and graph each inequality.**

a. $3x - 12 < 15$ **b.** $6 + 5(2 - x) < 41$

a. $3x - 12 < 15$

$3x < 27$ *Add 12 to both sides.*

$x < 9$ *Divide both sides by 3.*

All real numbers less than 9 are solutions. Using set-builder notation, the solution set is $\{x: x < 9\}$. The open circle at 9 shows that 9 is *not* included in the solution set.

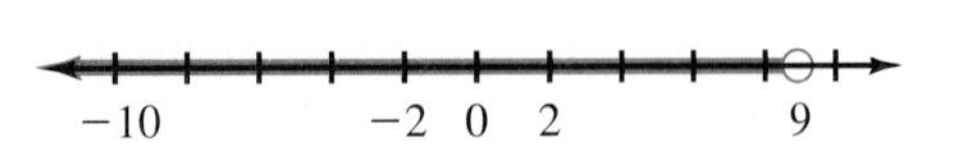

b. $6 + 5(2 - x) < 41$

$6 + 10 - 5x < 41$ *Distributive property*

$16 - 5x < 41$

$-5x < 25$ *Subtract 16 from both sides.*

$x > -5$ *Divide both sides by −5 and reverse the direction of the inequality symbol.*

Solution set: $\{x: x > -5\}$

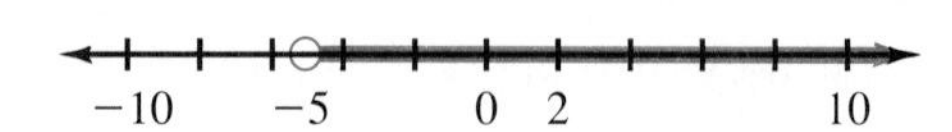

In Example 1, the inequalities $x < 9$ and $3x - 12 < 15$ have the same solution set. Similarly, $x > -5$ and $6 + 5(2 - x) < 41$ have the same solution set. Inequalities that have the same solution set are called **equivalent inequalities.**

The symbol $\geq$ (*is greater than or equal to*) or $\leq$ (*is less than or equal to*) may be used in an inequality. When graphing an inequality such as $x \geq 5$, use a solid dot to show that 5 is included.

EXAMPLE 2 **Solve and graph the inequality $4(5 - 3x) \geq 8x + 60$.**

$4(5 - 3x) \geq 8x + 60$

$20 - 12x \geq 8x + 60$ *Distributive property*

$20 - 20x \geq 60$ *Subtract 8x from both sides.*

$-20x \geq 40$ *Subtract 20 from both sides.*

$x \leq -2$ *Divide both sides by −20 and reverse the direction of the inequality symbol.*

Solution set: $\{x: x \leq -2\}$

The solid dot at -2 shows that -2 *is* in the solution set.

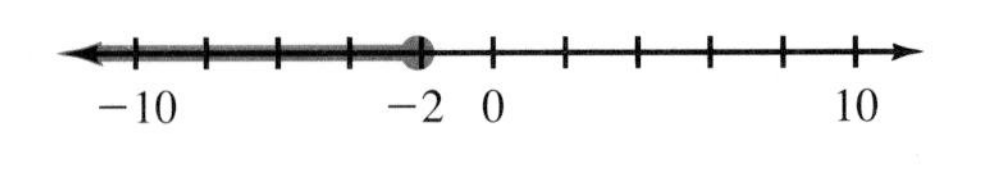

An inequality may not be true for any real number, in which case its solution set is the empty set, which is symbolized $\emptyset$. On the other hand, an inequality may be true for all real numbers, in which case its solution set is the set of all real numbers, $\{x: x \in \textit{real numbers}\}$. Example 3 illustrates each of these cases.

EXAMPLE 3 **Solve each inequality and graph the solution set if it is not the empty set.**

a. $2x - 3 \leq 2(x - 5)$ **b.** $7x + 6 > 7(x - 4)$

a.
$$\begin{aligned} 2x - 3 &\leq 2(x - 5) \\ 2x - 3 &\leq 2x - 10 \\ 2x &\leq 2x - 7 \\ 0 &\leq -7 \quad \textit{false} \end{aligned}$$

Solution set: $\emptyset$

b.
$$\begin{aligned} 7x + 6 &> 7(x - 4) \\ 7x + 6 &> 7x - 28 \\ 7x &> 7x - 34 \\ 0 &> -34 \quad \textit{true} \end{aligned}$$

Solution set: $\{x: x \in \textit{real numbers}\}$

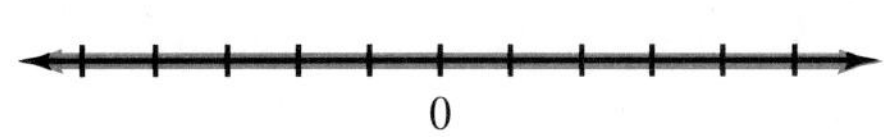

CLASS EXERCISES

Is the second inequality in each pair equivalent to the first? Give a reason for your answer.

1. $9x + 14 > 23;$ $9x > 9$

2. $-\frac{x}{3} < 6;$ $x > 18$

3. $-\frac{2}{5}x < 0;$ $x > 0$

4. $5x - 8 \geq 4;$ $5x \leq 12$

5. $5x < -15;$ $x > 3$

6. $-8x < 24;$ $x > -3$

Solve and graph.

7. $7x > 581$

8. $9z < 855$

9. $-6y < 474$

10. $-8y > 672$

11. $r + 13 \geq 34$

12. $r - 15 \leq -11$

13. $p - 5 < 27$

14. $p + 8 > 18$

15. $7 - x \geq 24$

PRACTICE EXERCISES

Solve and write the solution set using set-builder notation. Graph the solution set if it is not the empty set.

1. $3x - 6 > 27$

2. $3x + 10 \leq 25$

3. $5z - 6 > 14$

4. $8x - 15 > 73$

5. $-18 - 5y \geq 52$

6. $14 - 4y \geq 38$

7. $-5(4s + 1) < 23$

8. $57 - 4t \geq 13$

9. $12 < 2(3n + 1) + 22$

10. $4(t + 3) \leq 44$

11. $2(y - 3) + 7 < 21$

12. $4(x - 2) - 6 > 18$

13. $8(4z - 1) \geq 344$

14. $2(7 + 3x) \geq 86$

15. $5x + 7x - 8 > 16$

16. $3y + 14y - 5 \geq 354$

17. $9r - 12 < 6r + 36$

18. $8x + 15 > 15x - 24$

19. $3(6 - 5x) \leq 12x - 36$

20. $2(7 - 8x) \geq 14x - 46$

21. $9(n + 2) > 9(n - 3)$

22. $6x - 13 < 6(x - 2)$

23. $-6(2y - 10) < 180$

24. $-7(3x - 7) > 280$

25. $18 - 2(y + 6) < 76$

26. $21 - 3(7 - x) > 50$

27. $2 - 3z \geq 7(8 - 2z) + 12$

28. $17 - 2y \leq 5(7 - 3y) - 15$

29. $\frac{2}{3}(x - 12) \leq x + 8$

30. $\frac{3}{5}(x - 12) > x - 24$

31. $3[4x - (2x - 7)] < 2(3x - 5)$

32. $6[5y - (3y - 1)] \geq 4(3y - 7)$

Solve for x.

33. $3ax + 2b < 2ax - 8b$, $a < 0$

34. $7ex - 2c > 5ex + 6c$, $e < 0$

35. Prove that if $a > b$, then $a - b > 0$.

36. Prove that if $a < b$, then $a - b < 0$.

Applications

37. Personal Finance Josette is shopping for ballpoint pens. She has $5.79, and each pen costs 69¢. What is the greatest number of ballpoint pens she can buy? Assume that there is no sales tax.

38. Personal Finance Martin has $40.00 to buy tickets to a ball game, and the tickets cost $6.75 each. What is the greatest number of tickets he can buy?

39. Geometry The sum of the measures of any two sides of a triangle is always greater than the measure of the third side. In a triangle ABC, $BC = 4$ and $AC = 8 - AB$. What can be said about the measure of side AB?

40. Personal Finance Thelma is shopping for compact discs at a store that offers up to 20 discs at $9.99 each on the condition that the customer first buys 3 at $14.99 each. Thelma has a gift certificate for $100. How many compact discs can she buy?

41. Entertainment Chris is making dance tapes on 90-minute cassettes. He wants songs that are 3 to 5 minutes long. What is the maximum number of songs Chris can record on one tape? The minimum?

Developing Mathematical Power

42. Extension A train traveling x miles from New York to Boston makes 8 stops, including a 20 min stop in New Haven to change from electric to diesel locomotion. The train completes the trip in under 6 h. If the train averages 60 mi/h between stops, express (in terms of x) the average maximum number of minutes the train can wait at each of the remaining stops.

2.3

Conjunctions and Disjunctions

Objectives: To determine the truth or falsity of conjunctions and disjunctions
To graph the solution sets of conjunctions and disjunctions

In mathematics, as in English, sentences combined with the word *and* or the word *or* are called **compound sentences.** The truth of a compound sentence can be determined if the truth or falsity of the individual sentences can be determined.

Capsule Review

If it cannot be determined whether a sentence is true or false, the sentence is classified as an *open sentence*.

$x + 3 > 10$ Open sentence
$12 + 3 > 10$ True sentence
$14 < 3 + 10$ False sentence

Classify each sentence as true, false, or open.

1. $9 - 1 < 8$ **2.** $5 > 2 + 3$ **3.** $12 < 4 - x$

4. $40 \geq (10)(4)$ **5.** $x + 2 = 7$ **6.** $-(1 - 12) = 11$

7. $14 + 6 = 20$ **8.** $x + 23 < 10$ **9.** $24 + 3x = (8 + x)3$

10. $2n + 1 > 3n$ **11.** $4y - 1 = (2y - 1)2$ **12.** $2^4 < 4^2$

When two sentences are joined by the word *and,* the resulting sentence is called a **conjunction.** A conjunction is true only if *both* sentences are true.

EXAMPLE 1 **Classify each conjunction as true, false, or open.**

a. $5 + 8 = 13$ *and* Chicago is a state.
b. All squares have five sides and $3 - (-4) = -7$
c. $5 < 10$ *and* $x < 3$ **d.** $6 + 9 = 15$ *and* $-8 < 0$

a. $5 + 8 = 13$ is true, but *Chicago is a state* is false. The conjunction is false.

b. All squares have five sides is false, and $3 - (-4) = -7$ is false. The conjunction is false.

c. $5 < 10$ is true, but $x < 3$ is open. The truth or falsity of the conjunction cannot be established, so it is an open sentence.

d. $6 + 9 = 15$ and $-8 < 0$ are both true, so the conjunction is true.

When two sentences are joined by the word *or,* the resulting sentence is called a **disjunction.** A sentence such as $x \geq 3$ is also a disjunction because it can be written as the compound sentence $x > 3$ *or* $x = 3$. A disjunction is true only if *at least one* of the sentences is true.

EXAMPLE 2 **Classify each disjunction as true, false, or open.**

a. $0 < -5$ *or* $7 < 5$ **b.** $x < 5$ *or* $x > 7$

c. $7 < 10$ *or* $8 < 6$ **d.** $0 < 4$ *or* $-(-5) > 0$

a. $0 < -5$ and $7 < 5$ are both false, so the disjunction is false.

b. $x < 5$ and $x > 7$ are both open sentences, so the truth or falsity of the disjunction cannot be established. The disjunction is open.

c. $7 < 10$ is true and $8 < 6$ is false. The disjunction is true.

d. $0 < 4$ and $-(-5) > 0$ are both true, so the disjunction is true.

The shaded region in the **Venn diagram** below represents the intersection of sets A and B, where the universal set, U, is $\{1, 2, 3, 4, 5, 6, 7, 8, 9\}$. A and B are subsets of U such that A is the set of even numbers ($A = \{2, 4, 6, 8\}$) and B is the set of numbers less than seven ($B = \{1, 2, 3, 4, 5, 6\}$). The **intersection** of two sets A and B, written $A \cap B$, is the set of elements common to both A and B. Therefore, $A \cap B = \{2, 4, 6\}$.

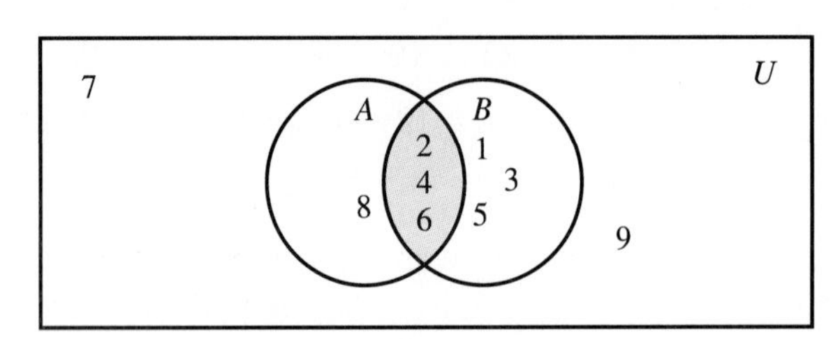

The solution set of a conjunction is the intersection of the solution sets of the individual sentences that form the conjunction.

EXAMPLE 3 **Graph the solution set of the conjunction $x > -4$ *and* $x < 5$.**

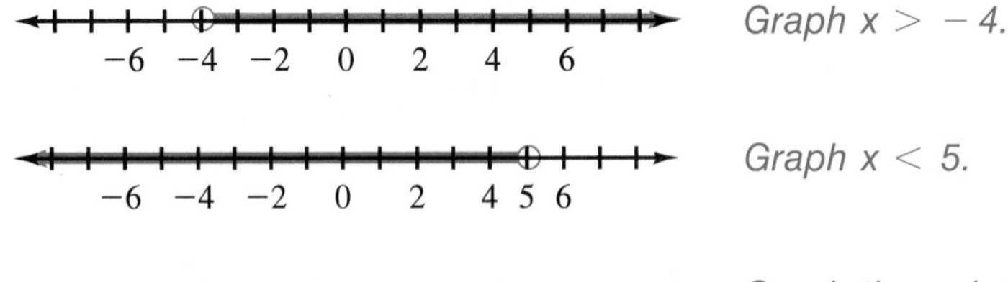

Graph $x > -4$.

Graph $x < 5$.

Graph the points common to the two graphs above. That is, graph $-4 < x < 5$.

The graph of $x > -4$ *and* $x < 5$ is the doubly shaded part of the line. Notice that only the numbers between -4 and 5 make the conjunction true. The solution set of the conjunction is $\{x: -4 < x < 5\}$, read "negative four is less than x and x is less than five," or "x is between negative four and five."

The sentences $x > -4$ *and* $x < 5$ and $-4 < x < 5$ are *equivalent sentences* because they have the same solution set. In general, a conjunction of the form $x > a$ *and* $x < b$ may be written compactly as $a < x < b$. Note that $a < x < b$ implies that $a < b$, by the transitive property.

The shaded region in the Venn diagram below represents the union of sets A and B discussed on page 58. The **union** of two sets A and B, written $A \cup B$, is the set of elements that are in either A or B or both. Therefore, $A \cup B = \{1, 2, 3, 4, 5, 6, 8\}$.

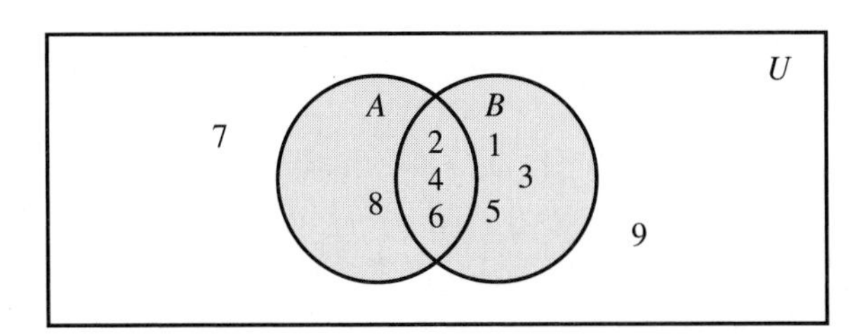

The solution set of a disjunction is the union of the solution sets of the individual sentences that form the disjunction.

EXAMPLE 4 **Graph the solution set of the disjunction $x < -2$ *or* $x > 4$.**

The solution set of the disjunction contains all numbers that are less than -2 and all numbers that are greater than 4.

−6 −4 −2 0 2 4 6

CLASS EXERCISES

State whether each sentence is true, false, or open. Give reasons for your answers.

1. $3 + 2 = 5$ *and* $5 - 7 = -2$

2. $-(5 - 10) > 0$ *or* $4 - 8 = 12$

3. $17 - x = 23$ *or* $1 + 2 = 3$

4. $9 - (-4) < 9 - 4$ *and* $-3 > 0$

5. Mars is a planet *and* $3 + 4 = 8$.

6. $0 < -5$ *or* $0 > -(2 - 3)$

7. $-6 > -7$ *or* *Hamlet* was written by Shakespeare.

8. This senator is from Texas *and* that senator is from California.

9. Ohio is smaller than Delaware *and* Rhode Island is smaller than Maine.

PRACTICE EXERCISES

Classify each sentence as a conjunction or a disjunction and state whether it is true, false, or open.

1. The sun rises in the east *and* the moon rises in the west.
2. A rectangle is a polygon *or* a triangle is a quadrilateral.
3. $-(-6) = 6$ *and* $-(-4) > -4$
4. $x + 3 > 5$ *and* $x - 5 < 2$
5. $-(-4) < 4$ *or* $-10 > 10 - 10$
6. $5 + 6 = 11$ *or* $9 - 2 = 11$
7. $17 > 12$ *or* $6 < 9$
8. $-9 - 10 < -8$ *or* $-2 < 6 - 34$

For each of the following sets, find $A \cap B$ and $A \cup B$.

9. $A = \{3, 4, 5, 6, 7, 8\}$; $B = \{2, 4, 6, 8, 10\}$
10. $A = \{a, e, i, o, u\}$; $B = \{h, i, d, e\}$
11. $A = \{3, 4, 5, 6, 7, 8\}$; $B = \{1, 2, 3, 4, 5, 6\}$
12. $A = \{0, 1, 2, 3, 4, 5, 6, 7, 8, 9, 10, 11, 12\}$; $B = \{0, 3, 6, 9, 12, 15\}$

Graph the solution set.

13. $x > 0$ *and* $x < 6$
14. $x > 4$ *and* $x < 10$
15. $x > -8$ *and* $x < -2$
16. $x < -3$ *or* $x > 3$
17. $x < -5$ *or* $x > 2$
18. $x < 2$ *or* $x > 6$
19. $x > -6$ *and* $x < 4$
20. $x > -8$ *and* $x < -1$
21. $x < 1$ *or* $x > 5$

State whether each expression represents an intersection or a union of sets and find the set that represents that intersection or union. Assume that the universal set is the set of integers *between* −10 and 10.

22. The set of multiples of 3 *and* the set of even integers
23. The set of odd integers *or* the set of multiples of 5
24. The set of odd integers *and* the set of even integers
25. The set of even integers *or* the set of odd integers
26. The set of nonnegative integers *and* the set of even integers
27. The set of nonpositive integers *or* the set of even integers
28. $\{x: x > 7\}$ *and* $\{x: x < 12\}$
29. $\{x: x < 4\}$ *or* $\{x: x > 8\}$
30. $\{x: x < 5\}$ *and* $\{x: x > 0\}$
31. $\{x: x < 4\}$ *and* $\{x: x < 0\}$
32. $\{x: x < 10\}$ *and* $\{x: x < 0\}$
33. $\{x: x > 8\}$ *and* $\{x: x < 6\}$

Graph the solution set of each of the following.

34. $\{x: x < 6\} \cap \{x: x \geq -3\}$

35. $\{x: x \geq 4\} \cap \{x: x < 7\}$

36. $\{x: x \geq 3\} \cup \{x: x < -4\}$

37. $\{x: x < 5\} \cup \{x: x \geq -2\}$

38. $\{x: x \leq -4\} \cap \{x: x \geq 7\}$

39. $\{x: x \geq -6\} \cup \{x: x \geq 3\}$

Applications

Logic If p represents one sentence and q represents another sentence, then "*p and q*" is a conjunction, "*p or q*" is a disjunction, $\sim p$ is the negation of p, and $\sim q$ is the negation of q.

Complete each *truth table* using T for true and F for false. Note that all possible combinations of true and false for two sentences are considered.

40.

p	q	p *and* q	p *or* q
T	T	T	?
T	F	F	T
F	T	?	?
F	F	?	?

41.

p	$\sim p$	p *and* $\sim p$	p *or* $\sim p$
T	?	?	?
F	?	?	?

42.

p	q	$\sim p$	$\sim q$	p *and* $\sim q$	$\sim p$ *and* q	$\sim p$ *or* q	p *or* $\sim q$
T	T	?	?	?	?	?	?
T	F	?	?	?	?	?	?
F	T	?	?	?	?	?	?
F	F	?	?	?	?	?	?

Mixed Review

Write an algebraic expression or sentence for each, using x as the variable.

43. Four times the absolute value of twice a number

44. Six less than the product of twelve and a number

45. The sum of three consecutive odd integers

46. The complement of an angle

Solve for x, indicating any restrictions on the variables.

47. $3x + 5a = dx + 8$

48. $24 = \dfrac{bx + c}{t}$

49. $\dfrac{3f - 5}{x} = 35c$

50. $\dfrac{19 + 2x}{3c} = c^2$

Solving Compound Sentences with Inequalities

Objective: To solve compound sentences involving inequalities and graph the solution sets

The solution set of a compound sentence may be the intersection or the union of the solution sets of the individual sentences. The intersection or the union of the solution sets can be graphed on a number line.

Capsule Review

This is the graph of the real numbers that satisfy the condition $x > 3$. The endpoint, 3, is *not* included.

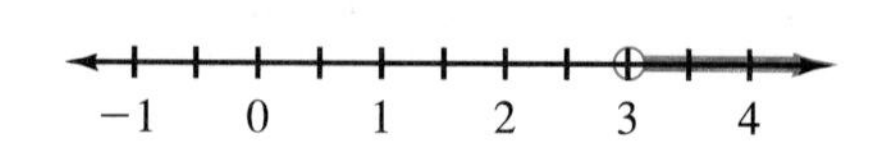

This is the graph of the real numbers that satisfy the condition $x \leq 2$. The endpoint, 2, *is* included.

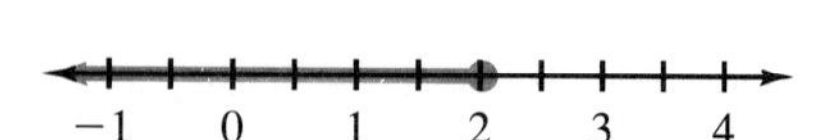

Graph the solution set.

1. $x \geq -4$ **2.** $x < 7$ **3.** $2x > 6$

4. $x - 20 > -14$ **5.** $2x + 3 \leq 9$ **6.** $3x - 1 > 7$

To solve a conjunction that is an open sentence in one variable, find the value(s) of the variable that make both sentences true.

EXAMPLE 1 **Solve and graph the solution set of the conjunction $3x - 1 > -28$ *and* $2x + 7 < 19$.**

$$\begin{aligned} 3x - 1 &> -28 \quad \textit{and} \quad 2x + 7 < 19 \\ 3x &> -27 \quad \textit{and} \quad 2x < 12 \\ x &> -9 \quad \textit{and} \quad x < 6 \end{aligned}$$

Therefore, the solution set is $\{x: x > -9 \textit{ and } x < 6\}$, or $\{x: -9 < x < 6\}$. That is, x is between -9 and 6.

The graph of the solution set is the graph of all values of x that make both $x > -9$ and $x < 6$ true.

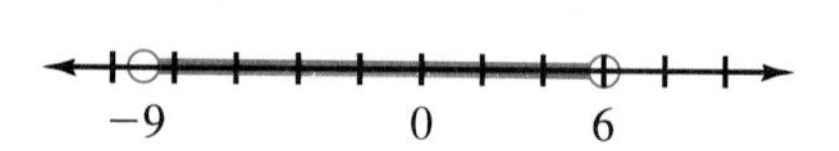

To solve a sentence such as $-7 < 2x - 3 < 9$, you may wish to write it first as a conjunction with the word *and*.

EXAMPLE 2 **Solve and graph the solution set of $-7 < 2x - 3 < 9$.**

$$\begin{aligned} 2x - 3 &> -7 \quad \textit{and} \quad & 2x - 3 &< 9 \\ 2x &> -4 \quad \textit{and} & 2x &< 12 \\ x &> -2 \quad \textit{and} & x &< 6 \end{aligned}$$

Solution set: $\{x: -2 < x < 6\}$

The solution set for the conjunction in Example 2 could also have been obtained, as shown below, without rewriting the original sentence.

$$-7 < 2x - 3 < 9$$

$$-7 + 3 < 2x - 3 + 3 < 9 + 3 \qquad \text{Add 3 to each of the three expressions.}$$

$$-4 < 2x < 12$$

$$\frac{-4}{2} < \frac{2x}{2} < \frac{12}{2} \qquad \text{Divide each expression by 2.}$$

$$-2 < x < 6$$

To solve a disjunction that is an open sentence in one variable, find value(s) of the variable that will make at least one of the sentences true.

EXAMPLE 3 **Solve and graph the solution set of the disjunction $4y - 2 \geq 14$ *or* $3y - 4 \leq -13$.**

$$\begin{aligned} 4y - 2 &\geq 14 \quad \textit{or} \quad & 3y - 4 &\leq -13 \\ 4y &\geq 16 \quad \textit{or} & 3y &\leq -9 \\ y &\geq 4 \quad \textit{or} & y &\leq -3 \end{aligned}$$

Solution set: $\{y: y \geq 4 \textit{ or } y \leq -3\}$

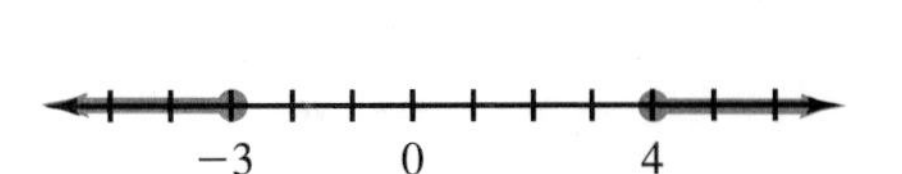

The intersection of the solution sets of the sentences that form a conjunction may be the same as the solution set of one of those sentences.

EXAMPLE 4 **Solve and graph the solution set of the conjunction $2 + 8z > 3z - 23$ *and* $4(1 - 2z) + 9 < 29$.**

$$\begin{aligned} 2 + 8z &> 3z - 23 \quad \textit{and} \quad & 4(1 - 2z) + 9 &< 29 \\ 8z &> 3z - 25 \quad \textit{and} & 4 - 8z + 9 &< 29 \\ 5z &> -25 \quad \textit{and} & -8z &< 16 \\ z &> -5 \quad \textit{and} & z &> -2 \end{aligned}$$

Any number greater than -2 is also greater than -5, so the solution set is $\{z: z > -2\}$.

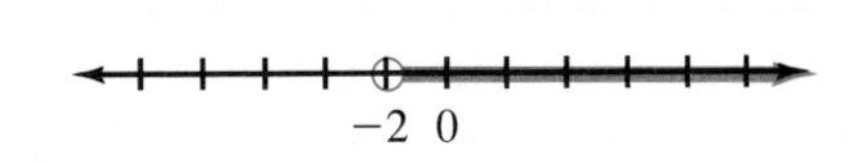

The solution set of a disjunction is the union of the solution sets of two sentences. Therefore, the solution set of a disjunction is not the empty set unless the solution set of each sentence in the disjunction is the empty set. However, the solution set of a disjunction may be the set of all real numbers.

EXAMPLE 5 **Solve and graph the solution set of the disjunction $3x \leq 15$ *or* $-2x \leq 10$.**

$3x \leq 15$ *or* $-2x \leq 10$
$x \leq 5$ *or* $x \geq -5$

Solution set: $\{x\colon x \in \text{real numbers}\}$

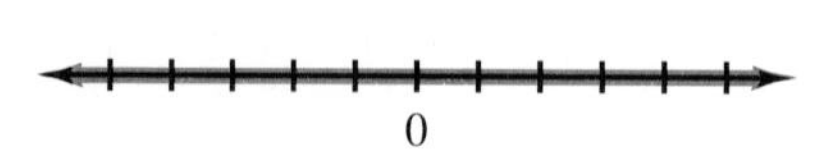

Since the solution set of a conjunction is the intersection of the solution sets of the two sentences, the solution set of the conjunction is sometimes the empty set.

EXAMPLE 6 **Solve the conjunction $4x > 24$ *and* $7x < -56$.**

$4x > 24$ *and* $7x < -56$
$x > 6$ *and* $x < -8$

Since there are no values of x that are both greater than 6 and less than -8, the solution set is the empty set, $\emptyset$.

CLASS EXERCISES

1. **Thinking Critically** Why is it incorrect to write the solution set for Example 3 (page 63) as $4 \leq y \leq -3$?

Which graph at the right is of the same general form as the graph of the solution set of the given compound sentence?

2. $4x < 144$ *and* $x + 3 > 5$
3. $3y - 6 \geq 21$ *or* $5y - 2 \leq 23$
4. $5t \leq 35$ *and* $2t \leq -14$
5. $-10 < 3x + 8 < 17$
6. $3x - 5 > 16$ *and* $8x + 2 > -14$
7. $2x - 8 < 10$ *or* $9x + 4 \geq 85$

a. 
b.
c. 
d.
e.

PRACTICE EXERCISES

Solve. Graph the solution set if it is not the empty set.

1. $2x > -10$ *and* $9x < 18$

2. $3x > -12$ *and* $8x < 16$

3. $6x > -24$ *and* $9x < 54$

4. $7x > -35$ *and* $5x < 30$

5. $-6 < 2x - 4 < 12$

6. $11 < 3x + 2 < 20$

7. $17 > 4x - 3 > -15$

8. $36 > 5x - 1 > -19$

9. $4y < 16$ *or* $12y > 144$

10. $3x > 2$ *or* $9x < 54$

11. $8x > -32$ *or* $6x < 48$

12. $9x < -27$ *or* $4x > 36$

13. $5y - 4 > 16$ *or* $3y + 2 < 17$

14. $6y + 3 < 15$ *or* $4y - 2 > 18$

15. $6x \leq 18$ *or* $-5x \leq 15$

16. $4x \leq 12$ *or* $-7x \leq 21$

17. $8x < -64$ *and* $5x > 25$

18. $15x > 30$ *and* $18x < -36$

19. $3 + 5z > 2z - 15$ *and* $3(1 - 3z) + 6 < 18$

20. $6 + 4z > 2z - 8$ *and* $5(1 - 2z) + 4 < 21$

21. $4w + 7 > 8 + 3w$ *or* $6(2w + 3) < 11w + 15$

22. $7(2w + 9) - 13w > 73$ *or* $16(w - 2) < 15(w - 2)$

23. $6(t - 4) - 5(t - 4) > 4$ *and* $9(2t - 1) - 7(2t - 3) < 32$

24. $4 - 2w \geq 18$ *or* $9 - 2w < 17$

25. $6 - 3w \geq 15$ *or* $7 - 4w < 19$

26. $3x + 4x - 7 < 2x + 18$ *and* $5x > 25$

27. $8 - 2z + 8z > 12$ *and* $3z < 18$

28. $3(2w - 7) + 7 \leq 10$ *or* $7 + w > -3$

29. $2(3w - 5) + 5 \leq 7$ *or* $8 + w > -4$

30. $5z - 4 + 3z > 6z - 24$ *and* $14z - 3 < 25$

31. $6z - 5 + 2z > 4z - 21$ *and* $13z - 3 < 23$

32. $2x - 3 > 9 - 4x$ *and* $6x - (3 - 3x) < 24$

33. $4x - 5 > 16 - 3x$ *and* $7x - (5 - 2x) < 31$

34. $0.3 - 0.7z > 0.4z - 3.0$ *and* $3(z - 2) - 6 \leq 15$

35. $0.5 - 0.8z > 0.3z - 5.0$ *and* $4(z - 3) - 12 \leq 8$

Solve for *t*, *x*, *w*, *y*, or *z*.

36. $\frac{y}{5} - 1 > 6$ *or* $\frac{y}{2} + 4 < 8$

37. $\frac{y}{6} + 1 > 0$ *or* $\frac{y}{3} + 5 < 9$

38. $\frac{x}{2} + \frac{x}{2} < \frac{5}{2}$ *and* $5x \geq 4x - 8$

39. $\frac{w}{3} + \frac{w}{6} > \frac{7}{2}$ *or* $8w + 7 < 63$

40. $\frac{x}{4} + \frac{x}{2} \leq -3$ *or* $\frac{x}{4} - \frac{x}{8} > 0$

41. $\frac{w}{6} - \frac{w}{3} > -1$ *and* $\frac{w}{4} + \frac{w}{6} \leq 5$

42. $3(z - 2) - 5 > 8 - 2(z - 4)$ *and* $1 - 2z > 10$

43. $2t + 3t + b > 3(b + t)$ *and* $7t - b < 13b$ *and* $b > 0$

Applications

44. **Consumerism** Mary knew that she had less than 4 dozen raisin bars and that there were more than 24 students in her class. Did she have enough so that each person could have two raisin bars?

45. **Geometry** The centers of two circles are 20 cm apart. If the radius of one circle is between 10 and 15 cm and the radius of the other circle is 8 cm, describe the possible number of intersections of the two circles.

TEST YOURSELF

Solve for the indicated variable. Indicate any restrictions on the values of the variables.

1. $at + c = b$, for t 2. $ax = bc + cs$, for x 3. $A = \pi r^2 h$, for h 2.1

Solve and write the solution set using set-builder notation. Graph the solution set if it is not the empty set.

4. $4(x - 2) - 8 > 12$ 5. $8 + 3(2 - x) < 32$ 2.2

Classify each sentence as a conjunction or a disjunction, and state whether it is true, false, or open.

6. $16 < 20$ *and* $3 < 0$ 7. $2 - (-1) \leq 3$ *or* $-7 \leq 0$ 2.3

State whether each expression represents an intersection or a union of sets and find the set that represents that intersection or union. Assume that the universal set is the set of integers *between* −8 and 8.

8. $\{x: x > 3\}$ *and* $\{x: x < 7\}$ 9. $\{x: x \geq 0\}$ *or* $\{x: x \geq -5\}$

Solve. Graph the solution set if it is not the empty set.

10. $2x + 1 \geq 3$ *or* $3x - 5 \geq 7$ 11. $-3 < 2y - 1 < 5$ 2.4

2.5

Problem Solving Strategy: Make a Drawing or a Table

An effective strategy to use in solving some problems is to make a simple drawing or sketch to illustrate the conditions of the problem. The use of a table can serve the same purpose, and drawings and tables are often used together. Once the conditions of a problem are made clear by a drawing or a table, it is easier to write an equation or an inequality to solve the problem.

EXAMPLE 1 The length of the rectangular swimming pool in the Nardis' yard is 2 ft more than twice its width. Find the dimensions of the pool if the perimeter is 124 ft.

Understand the Problem *What are the given facts?* The length of a rectangular pool is 2 ft greater than twice its width, and its perimeter is 124 ft.
What are you asked to find? The problem is to find the length and width of the pool.

Plan Your Approach *Choose a strategy.* A drawing can be used to illustrate the conditions of the problem. Assign variables to represent the unknown numbers. Let x represent the width of the pool in feet. Then $2x + 2$ represents the length in feet.

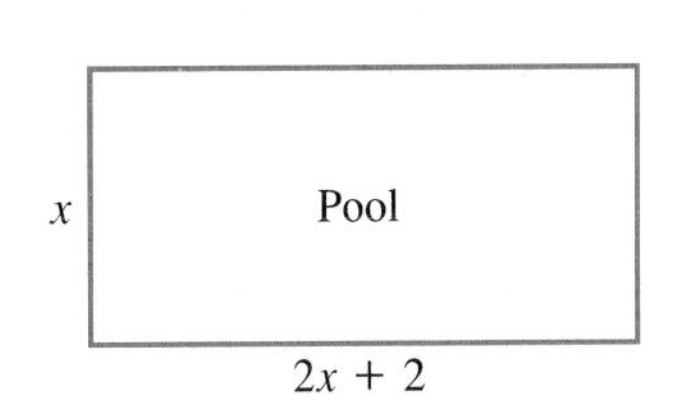

Write a word equation and translate it into an algebraic equation. The given data are related by the fact that the perimeter of a rectangle equals twice the sum of the length and width.

twice the sum of the length and width = the perimeter

$$2(2x + 2 + x) = 124$$

Complete the Work

$$2(3x + 2) = 124 \quad \textit{Solve the equation.}$$
$$6x + 4 = 124$$
$$x = 20 \quad \textit{width}$$
$$\text{Therefore,} \quad 2x + 2 = 42 \quad \textit{length}$$

Interpret the Results *State your answer.* The pool is 42 ft long and 20 ft wide.
Check your answer. Is the length 2 ft more than twice the width?

$$42 \stackrel{?}{=} 2(20) + 2$$
$$42 \stackrel{?}{=} 42 \checkmark$$

Do the dimensions give a perimeter of 124 ft?

$$2(42 + 20) \stackrel{?}{=} 124$$
$$124 = 124 \checkmark$$

Example 2 shows how to use a table along with a drawing to solve a problem involving motion.

EXAMPLE 2 Bonnie and her brother Larry left their home at the same time for their separate vacations. Bonnie drove east at an average rate of 75 km/h and Larry drove west at an average rate of 60 km/h. In how many hours will they be 810 km apart?

Understand the Problem *What are the given facts?* Bonnie and Larry left home at the same time. Bonnie drove east at an average rate of 75 km/h. Larry drove west at an average rate of 60 km/h.
What are you asked to find? The problem is to find how long it will take for Bonnie and Larry to be 810 km apart.

Plan Your Approach *Choose a strategy.* Use a drawing to illustrate the conditions of the problem. The drawing shows that Bonnie and Larry are traveling in opposite directions. The total distance, 810 km, is the sum of the distances they travel.

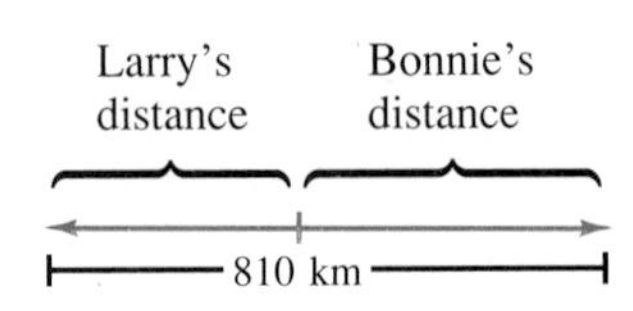

Organize the rates, times, and distances in a table. Let x represent the time, in hours, that it takes for Bonnie and Larry to be 810 km apart. Since they left at the same time, they each travel the same number of hours.

	rate ·	time =	distance
Bonnie	75	x	$75x$
Larry	60	x	$60x$

Write a word equation and translate it into an algebraic equation.

Bonnie's distance + Larry's distance = total distance

$$75x + 60x = 810$$

Complete the Work

$$135x = 810 \quad \text{Solve the equation.}$$
$$x = 6$$

Interpret the Results *State your answer.* Bonnie and Larry will be 810 km apart in 6 hours.
Check your answer. Find the distance each person drives in 6 hours and check that the sum of these distances is 810 km.

$$\text{distance} = \text{rate} \cdot \text{time}$$
$$\text{distance Bonnie drives} = 75(6) = 450$$
$$\text{distance Larry drives} = 60(6) = 360$$
$$\text{total distance} = 450 \text{ km} + 360 \text{ km} = 810 \text{ km} \checkmark$$

Example 3 does not require a drawing, but a chart is useful for organizing the given information.

EXAMPLE 3 At the end of one day, the change machine at Washburns' Launderette contained less than \$5.45 in nickels, dimes, and quarters. If there were 3 fewer dimes than twice the number of nickels and 2 more quarters than twice the number of nickels, what was the greatest possible number of nickels in the machine?

Understand the Problem *What are the given facts?* You are given certain relationships among the numbers of nickels, dimes, and quarters in a machine, as well as the fact that the total value of the coins is less than \$5.45.
What are you asked to find? You are asked to find the greatest possible number of nickels.

Plan Your Approach *Choose a strategy.* Use a table to organize the given facts. Let x represent the number of nickels. Then $2x - 3$ represents the number of dimes, and $2x + 2$ represents the number of quarters.

Coin	value of each ·	number =	value of coins
nickel	0.05	x	$0.05x$
dime	0.10	$2x - 3$	$0.10(2x - 3)$
quarter	0.25	$2x + 2$	$0.25(2x + 2)$

value of nickels + value of dimes + value of quarters < 5.45

Complete the Work

$$0.05x + 0.10(2x - 3) + 0.25(2x + 2) < 5.45$$
$$5x + 10(2x - 3) + 25(2x + 2) < 545$$
$$5x + 20x - 30 + 50x + 50 < 545$$
$$75x + 20 < 545$$
$$75x < 525$$
$$x < 7$$

Interpret the Results *State your answer.* The greatest possible number of nickels in the machine is 6.
Check your answer. Find the greatest possible number of each type of coin and the greatest possible total value.

There are at most 6 nickels, since $x < 7$.
Then there are at most $2(6) - 3$, or 9 dimes and $2(6) + 2$, or 14 quarters.

$6(\$.05) + 9(\$.10) + 14(\$.25) = \$4.70$, so the total value of the coins is at most \$4.70, which is less than \$5.45.

CLASS EXERCISES

Make a drawing and/or a chart and write an equation or an inequality for each problem.

1. The length of a rectangle is 3 times the width. The perimeter is 24 ft. Find the dimensions of the rectangle.
2. An express train and a freight train leave at the same time from two cities 270 mi apart and travel toward each other on parallel tracks. The rate of the express train is 50 mi/h and the rate of the freight train is 40 mi/h. In how many hours will the trains meet?
3. A passenger plane and a cargo plane leave at the same time from the same airport but travel in opposite directions. The passenger plane travels at twice the speed of the cargo plane. Find the speed of each plane if they are 2400 mi apart in 4 h.
4. The length of one side of a triangular shelf is 6 cm greater than the length of the shortest side. The length of the third side is 9 cm less than twice the length of the shortest side. What is the greatest possible length of the shortest side if the perimeter of the shelf is at most 45 cm?
5. The Calders sell sweaters and slacks. They make a profit of $25 on each sweater and $35 on each pair of slacks. One day they sold 20 more sweaters than pairs of slacks and made a profit of more than $1300. What is the least possible number of pairs of slacks they could have sold that day?

PRACTICE EXERCISES

1. The perimeter of Sportsland Park is 624 yd. If the length of the park is 8 yd more than 3 times the width, and the park is rectangular, find the dimensions of the park.
2. The length of a rectangular schoolyard is 5 m more than 3 times the width. Find the dimensions of the schoolyard if the perimeter is 166 m.
3. Two buses leave Houston at the same time and travel in opposite directions. One bus averages 50 mi/h and the other bus averages 45 mi/h. In how many hours will they be 570 mi apart?
4. Two planes leave at 9:00 AM from airports that are 2700 mi apart and fly toward each other at speeds of 200 mi/h and 250 mi/h. At what time will they pass each other?
5. Jim and Joe started on trips from San Francisco and traveled in opposite directions. Jim traveled 15 km/h faster than Joe. After 4 h, they were 420 km apart. How fast was each person traveling?

6. Nan and Peg started on trips from New York City and traveled in opposite directions. Nan traveled 10 mi/h faster than Peg. If they were 450 mi apart after 5 h, how fast was each person traveling?

7. The length of one side of a triangular flower bed is 3 ft less than twice the length of the shortest side, and the length of the third side is 3 ft greater than the length of the shortest side. If the perimeter is at least 36 ft, what is the least possible value for the length of the shortest side?

8. A model plane was flown in a triangular pattern from A to B, then from B to C, and then from C to A. The distance from B to C is 4 m less than twice the distance from A to B, and the distance from C to A is 2 m more than the distance from A to B. If the plane flew no more than 30 m, what is the greatest possible distance between A and B?

9. A coin bank contains more than $3 in quarters, dimes, and nickels. The number of dimes is twice the number of quarters and the number of nickels is 3 times the number of quarters. Find the least possible number of quarters in the bank.

10. Victor bought less than $8.40 worth of stamps. He bought 3 times as many 17¢ stamps as 25¢ stamps and 4 times as many 2¢ stamps as 25¢ stamps. Find the greatest possible number of 25¢ stamps he bought.

11. Cynthia invested 3 times as much money at 6% as at 5%. If the total of the simple interest for one year is at least $1150, what is the least possible amount Cynthia has invested at 5%?

12. The Chans invested twice as much money at 8% as at 6%. If the total of the simple interest for one year is at most $660, what is the greatest possible amount the Chans have invested at 6%?

13. Driving his car at 50 mi/h, José can cover a certain distance in 20 min less than when he drives at 45 mi/h. Find that distance.

14. Driving her car at 60 km/h, Rosita can cover a certain distance in 36 min less than when she drives at 50 km/h. Find that distance.

15. Michael drove to a friend's house at a rate of 40 mi/h. He came back by the same route, but at a rate of 45 mi/h. If the round-trip took less than 4 h, what is the greatest possible distance Michael could have traveled to visit his friend?

16. Denise drove to her parents' house at a rate of 70 km/h. She came back by the same route, but drove at a rate of 80 km/h. If the round-trip took no more than 3 h, what is the greatest possible distance she could have traveled to visit her parents?

17. The Leos invested part of $12,000 at 4% simple interest per year and the rest at 6% simple interest per year. The total interest for one year was at most $640. Find the least amount that could have been invested at 4%.

18. The Jensens invested part of $10,000 at 5% simple annual interest and the rest at 6% simple annual interest. The total interest was more than $550. What is the greatest amount that could have been invested at 5%?

19. In a triangle ABC, the ratio of the length of side AB to the length of side BC is $\frac{1}{2}$, and the ratio of the length of side AC to that of side AB is $\frac{5}{3}$. If the perimeter of the triangle is at most 28 cm, find the greatest possible lengths of the three sides.

20. A buyer of small appliances buys toasters for $21.50 each. The store sells the toasters at a 20% markup. What is the least possible number of toasters that must be sold to realize a profit of at least $1200?

Mixed Problem Solving Review

1. The sum of three numbers is 105. The second number is 10 larger than the first and the third is one-half the second. Find the three numbers.
2. Two trucks start at the same time and travel toward each other from cities 840 miles apart. How many hours will it take for them to meet if one travels at 65 mi/h and the other travels at 55 mi/h.

PROJECT

The following data was obtained from a survey of 200 people:

9 liked classical music, rock music, and light opera.
27 liked classical music and rock music.
33 liked rock music and light opera.
30 liked classical music and light opera.
72 liked classical music.
80 liked rock music.
93 liked light opera.

A Venn diagram can be useful for displaying data.

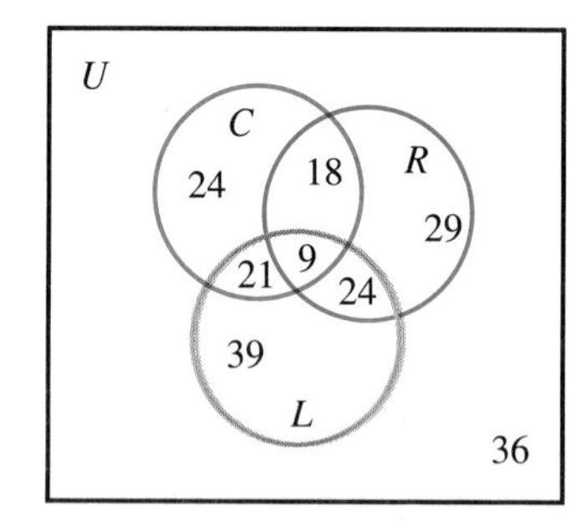

Draw three intersecting circles and label them C (classical), R (rock), and L (light opera). Write 9 in the center, where all three circular regions intersect. Since 27 people liked classical *and* rock music, write $27 - 9$, or 18, in the region where *only* regions C and R intersect. Use similar reasoning to complete the diagram. The number 36 is found by subtracting the sum of the numbers within the circles from 200.

Work together in small groups. Each group should design a survey and conduct interviews to obtain data. Use Venn diagrams to display the results.

2.6

Absolute Value Equations

Objective: To solve equations involving absolute value

Distance is expressed as a nonnegative number. In earlier math courses, you learned that the *absolute value* of a real number a, written $|a|$, is the distance between a and 0 on the number line.

The distance between 4 and 0 is 4.

$$|4| = 4$$

The distance between -4 and 0 is 4.

$$|-4| = 4$$

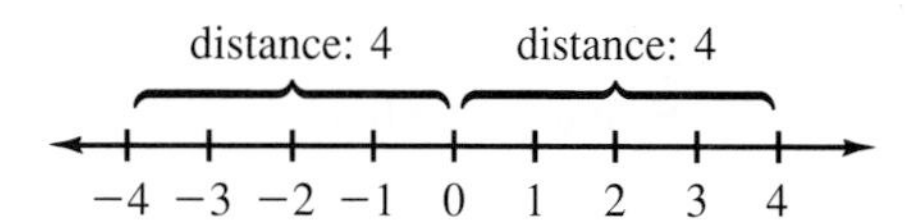

Absolute Value

For any real number a,

$$|a| = a, \text{ if } a \geq 0 \quad \text{and} \quad |a| = -a, \text{ if } a < 0$$

Capsule Review

Use the definition of absolute value to evaluate expressions.

EXAMPLE **Find the value of each expression.**

a. $|9 - 6|$ **b.** $|5x|$, if $x = -3$

a. $|9 - 6| = |3| = 3$ **b.** $|5(-3)| = |-15| = -(-15) = 15$

Find the value of each expression.

1. $|-10|$

2. $|3 - 8|$

3. $|15 + 34|$

4. $|-3x|$, if $x = -2$

5. $|4x + 8|$, if $x = -2$

6. $|1 + 8x|$, if $x = -\frac{1}{2}$

An equation in which the variable appears within absolute value symbols is called an **absolute value equation.** By the definition of absolute value, the equation $|x| = 8$ is equivalent to the disjunction $x = 8$ *or* $x = -8$. Similarly, $|4x| = 16$ is equivalent to the disjunction $4x = 16$ *or* $4x = -16$, or to the disjunction $x = 4$ *or* $x = -4$. To solve an absolute value equation, first write it as a disjunction.

EXAMPLE 1 **Solve and check: $|2y - 4| = 12$**

$$
\begin{aligned}
|2y - 4| &= 12 \\
2y - 4 = 12 \quad &\text{or} \quad 2y - 4 = -12 \\
2y = 16 \quad &\text{or} \quad 2y = -8 \\
y = 8 \quad &\text{or} \quad y = -4
\end{aligned}
$$

Check:

$$
\begin{aligned}
|2y - 4| &= 12 \\
|2(8) - 4| &\stackrel{?}{=} 12 \\
12 &= 12 \checkmark
\end{aligned}
\qquad
\begin{aligned}
|2y - 4| &= 12 \\
|2(-4) - 4| &\stackrel{?}{=} 12 \\
12 &= 12 \checkmark
\end{aligned}
$$

The solutions are 8 and -4.

You will find it easier to solve a more complicated absolute value equation if you first isolate the expression containing the absolute value symbols.

EXAMPLE 2 **Solve and check: $3|4w - 1| - 5 = 10$**

$$
\begin{aligned}
3|4w - 1| - 5 &= 10 \\
3|4w - 1| &= 15 \qquad \textit{Add 5 to both sides.} \\
|4w - 1| &= 5 \qquad \textit{Divide both sides by 3.}
\end{aligned}
$$

$$
\begin{aligned}
4w - 1 = 5 \quad &\text{or} \quad 4w - 1 = -5 \\
4w = 6 \quad &\text{or} \quad 4w = -4 \\
w = \frac{3}{2} \quad &\text{or} \quad w = -1
\end{aligned}
$$

Check in the original equation. The solutions are $\frac{3}{2}$ and -1.

An absolute value indicates a nonnegative value. Therefore, the equation $|2x + 7| = -2$ has no solution because $|2x + 7|$ cannot be negative. It is important to check possible solutions in the original equation, since one or more may not satisfy that equation.

EXAMPLE 3 **Solve and check: $|2x + 5| = 3x + 4$**

$$
\begin{aligned}
|2x + 5| &= 3x + 4 \\
2x + 5 = 3x + 4 \quad &\text{or} \quad 2x + 5 = -(3x + 4) \\
-x = -1 \quad &\text{or} \quad 2x + 5 = -3x - 4 \\
x = 1 \quad &\text{or} \quad 5x = -9 \\
x = 1 \quad &\text{or} \quad x = -\frac{9}{5}
\end{aligned}
$$

Check:

$$
\begin{aligned}
|2x + 5| &= 3x + 4 \\
|2(1) + 5| &\stackrel{?}{=} 3(1) + 4 \\
|7| &\stackrel{?}{=} 7 \\
7 &= 7 \checkmark
\end{aligned}
\qquad
\begin{aligned}
|2x + 5| &= 3x + 4 \\
\left|2\left(-\frac{9}{5}\right) + 5\right| &\stackrel{?}{=} 3\left(-\frac{9}{5}\right) + 4 \\
\left|\frac{7}{5}\right| &\stackrel{?}{=} -\frac{7}{5} \\
\frac{7}{5} &\neq -\frac{7}{5}
\end{aligned}
$$

$-\frac{9}{5}$ is not a solution.

The solution is 1.

CLASS EXERCISES

Express each absolute value equation as a disjunction using the word *or*.

1. $|x + 3| = 9$

2. $|3x - 5| = 10$

3. $|2x + 7| + 3 = 22$

4. $|3x - 6| - 7 = 14$

5. $|2x + 3| - 6 = 14$

6. $|6 - 5x| = 18$

Determine whether each sentence is true or false. If a sentence is false, show why it is false.

7. If $x \geq 0$, then $|x - 3| = 3 - x$

8. If $x \leq 0$, then $|x - 3| = |3 - x|$

9. If $x > 0$, then $|6 - x| = 6 - x$

10. If $x < 0$, then $|-x - 8| = x + 8$

11. If $x > 0$, then $|-x - 4| = x + 4$

12. If $x < 0$, then $|9 - x| = 9 + x$

PRACTICE EXERCISES

Solve and check. If an equation has no solution, so state.

1. $|3x| = 18$

2. $|5x| = 30$

3. $|-4x| = 32$

4. $|-9x| = 36$

5. $|x - 3| = 9$

6. $|x - 4| = 9$

7. $|x + 2| = 0$

8. $|x + 5| = 12$

9. $|5y - 8| = 12$

10. $|4y - 5| = 15$

11. $|3x + 2| = 7$

12. $|5y + 3| = 9$

13. $2|w + 6| = 10$

14. $3|x + 5| = 12$

15. $3|2w - 1| = 21$

16. $2|3w - 2| = 14$

17. $|3x + 4| = -3$

18. $|2x - 3| = -1$

19. $|3x + 5| = 5x + 2$

20. $|2x - 3| = 4x - 1$

21. $|x + 4| + 3 = 17$

22. $|y - 5| - 2 = 10$

23. $|6y - 2| + 4 = 32$

24. $|3x - 1| + 10 = 25$

25. $-|4 - 8x| = 12$

26. $-|2w - 6| = 10$

27. $7|3 - 2y| = 56$

28. $8|4 - 3y| = 48$

29. $4|3x + 4| = 4x + 8$

30. $6|2x + 5| = 6x + 24$

31. $5|6 - 5x| = 15x - 35$

32. $7|8 - 3x| = 21x - 49$

33. $\frac{1}{2}|3x + 5| = 6x + 4$

34. $\frac{1}{4}|4x + 7| = 8x + 16$

Solve for *x* and check. If an equation has no solution, so state. Assume that *a*, *b*, and *c* represent positive real numbers.

35. $|ax| - b = c$

36. $|cx - d| = ef$

37. $a|bx - c| = d$

38. $|3x - a| = x + 1$

39. $|x - 5| = 2x + 3a$

40. $2|x - 6| = 4x + b$

Applications

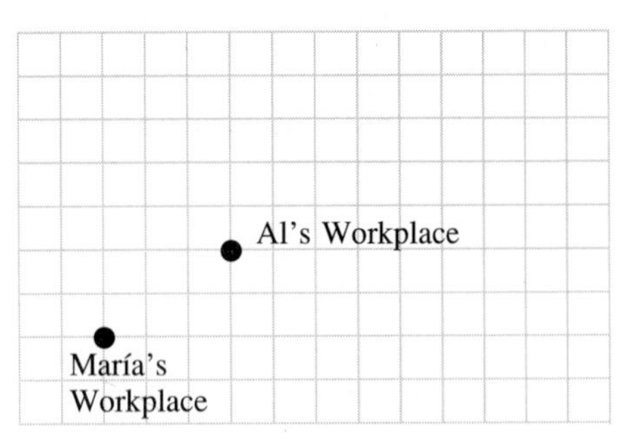

Geometry Taxicab Geometry is a geometry developed for measurement of distance in cities where the streets are in a grid pattern. Since taxicabs drive only on the streets, the absolute value function is used to measure the distance driven from one place to another.

Distance from A to B =

$$|\text{\# of North or South blocks}| + |\text{\# of East or West blocks}|$$

41. Locate all the places where María and her husband, Al, could find an apartment that is equidistant from both workplaces.

42. If Al's company moves to a new location 2 blocks north of the present location, where could María and Al live?

43. What would be a circle in Taxicab Geometry?

Mixed Review

44. The measure of an angle is 6 degrees less than twice its supplement. What is the measure of the angle?

Solve for x.

45. $3(2x - 1) = 9(3 - x)$

46. $4(3 - x) + 6 = \frac{x}{2}$

Solve each inequality and graph it on a number line.

47. $2x + 3(1 - x) > 5$

48. $2(4 + 5x) < 13x + 15$

Classify each sentence as true, false, or open.

49. $|13 - 15| = 0.5(5 - 1)$, and $-4.5 < -5$.

50. Vermont is in New England, or "c" is a vowel.

51. Clare and Alex are a happy couple, and $x = 5$.

Developing Mathematical Power

Thinking Critically Write each sentence with numbers and symbols. Is the sentence true?

52. The absolute value of the square of negative three is equal to the square of the absolute value of negative three.

53. The absolute value of the product of two and negative one is equal to the product of the absolute values of two and of negative one.

2.7 Absolute Value Inequalities

Objective: To solve inequalities involving absolute value and graph the solution sets

The methods that are used to solve compound sentences involving inequalities are also used to solve absolute value inequalities.

Capsule Review

To solve a conjunction or disjunction that is an open sentence in one variable, find the values of the variable that make both sentences true.

EXAMPLE **Solve and graph the solution set of the conjunction $2x + 5 > 1$ *and* $x - 1 < 3$.**

$2x + 5 > 1$ *and* $x - 1 < 3$
$2x > -4$ *and* $x < 4$
$x > -2$ *and* $x < 4$

Solution set: $\{x: -2 < x < 4\}$

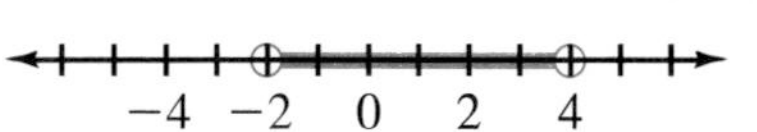

Solve and graph the solution set if it is not the empty set.

1. $2x > -4$ *and* $3x > 6$
2. $y + 2 \leq -1$ *or* $y - 1 \geq 1$
3. $w + 3 \geq 4$ *or* $2w - 3 \leq 1$
4. $z \leq 2$ *and* $3z + 4 \geq -8$
5. $4x - 3 > 17$ *and* $3x + 2 < 2$
6. $z + 5 < 9$ *or* $2z - 5 < -7$

The graph of $|a| < 5$ is shown at the right. The solution set is expressed as a conjunction, $\{a: -5 < a < 5\}$ *or* $\{a: a > -5$ *and* $a < 5\}$.

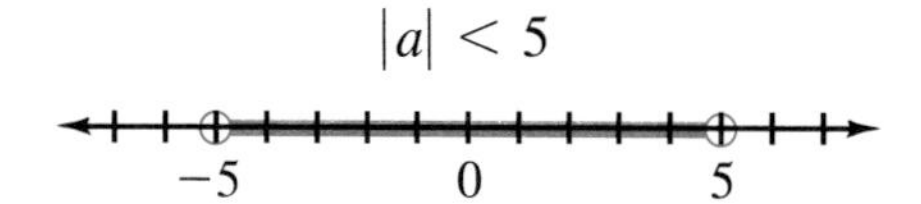

To solve an absolute value inequality with $<$ or $\leq$, express it as an equivalent conjunction, then solve the conjunction.

EXAMPLE 1 **Solve the inequality $|2x - 1| < 9$ and graph the solution set.**

$2x - 1 > -9$ *and* $2x - 1 < 9$ — *Write the conjunction equivalent to $|2x - 1| < 9$.*
$2x > -8$ *and* $2x < 10$
$x > -4$ *and* $x < 5$

Solution set: $\{x: -4 < x < 5\}$

−4 0 5

The graph of $|a| > 5$ is shown at the right. The solution set is expressed as a disjunction, $\{a: a < -5 \text{ or } a > 5\}$.

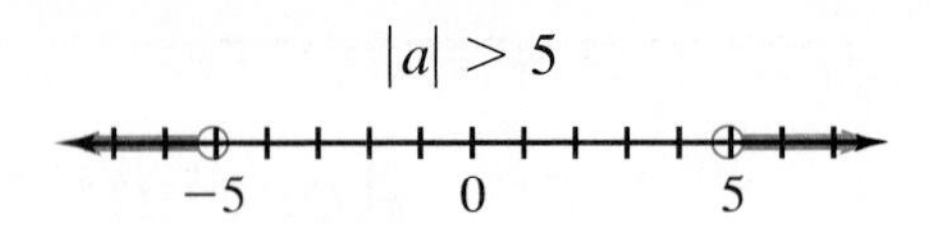

To solve an absolute value inequality with $>$ or $\geq$, express it as an equivalent disjunction and then solve the disjunction.

EXAMPLE 2 **Solve the inequality $|3x + 6| \geq 12$ and graph the solution set.**

$$\begin{aligned} 3x + 6 &\leq -12 \quad or \quad & 3x + 6 &\geq 12 \\ 3x &\leq -18 \quad or & 3x &\geq 6 \\ x &\leq -6 \quad or & x &\geq 2 \end{aligned}$$

Write the disjunction equivalent to $|3x + 6| \geq 12$.

Solution set: $\{x: x \leq -6 \text{ or } x \geq 2\}$

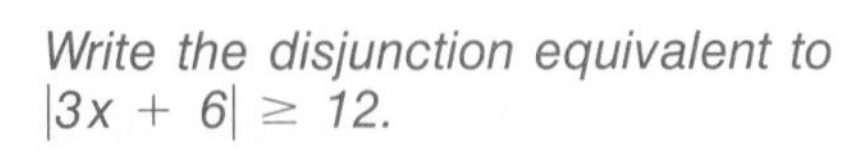

When an absolute value is combined with other operations, isolate the expression enclosed by the absolute value symbols before expressing the sentence as an equivalent conjunction or disjunction.

EXAMPLE 3 **Solve the inequality $3|2x + 6| - 9 < 15$ and graph the solution set.**

$$3|2x + 6| - 9 < 15$$
$$3|2x + 6| < 24 \quad \textit{Add 9 to both sides.}$$
$$|2x + 6| < 8 \quad \textit{Divide both sides by 3.}$$

$$\begin{aligned} 2x + 6 &> -8 \quad and \quad & 2x + 6 &< 8 \\ 2x &> -14 \quad and & 2x &< 2 \\ x &> -7 \quad and & x &< 1 \end{aligned}$$

Solution set: $\{x: -7 < x < 1\}$

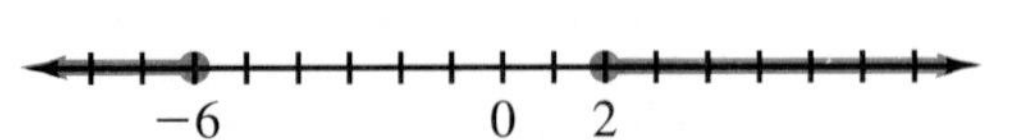

CLASS EXERCISES

Express each inequality as an equivalent conjunction or disjunction.

1. $|x| + 1 < 12$

2. $|x - 5| > 6$

3. $|3x| < 27$

4. $|x + 4| > 10$

5. $|7 - x| < 9$

6. $|x + 3| - 2 < 15$

7. $|2x + 1| \geq 9$

8. $|4x| + 7 \leq 35$

9. $2|x| - 3 \geq 5$

10. $|2x - 4| + 16 \leq 24$

11. $|2x + 4| - 6 < 0$

12. $|3x - 5| - 2 > 0$

13. $2|x + 3| \geq 10$

14. $6|x + 9| \leq 36$

15. $\frac{1}{7}|8x + 6| \leq 1$

16. $\frac{1}{13}|7x - 12| > 52$

17. $\frac{1}{16}|3x + 4| + 2 < 1$

18. $\frac{1}{9}|5x - 3| - 3 \geq 2$

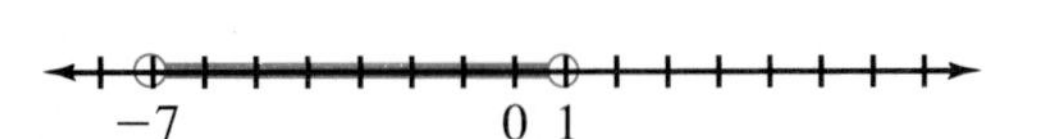

PRACTICE EXERCISES

Solve and graph the solution set.

1. $|x + 3| < 9$
2. $|x + 4| < 8$
3. $|x - 2| < 6$
4. $|x - 5| < 9$
5. $|2x - 3| < 7$
6. $|2x - 5| < 9$
7. $|y - 5| > 10$
8. $|y - 3| > 12$
9. $|5 - y| < 19$
10. $|4 - 2w| \geq 12$
11. $|2z + 3| - 6 < 14$
12. $|3x - 6| + 3 > 15$
13. $|3x - 4| + 5 \leq 27$
14. $|2x + 3| - 6 \geq 7$
15. $2|x + 4| \leq 22$
16. $2|w + 6| \leq 10$
17. $3|w - 9| > 27$
18. $3|2x - 1| \geq 21$
19. $|3z| - 4 \leq 8$
20. $|2x| + 8 > 12$
21. $|6y - 2| + 4 < 22$
22. $|5z + 3| - 7 > 34$
23. $4|2w + 3| - 7 < 9$
24. $3|5t - 1| + 9 > 23$
25. $|2y - 6| < 0$
26. $|3z + 15| > 0$
27. $|-2x + 1| > 2$
28. $|-2x - 1| < 2$
29. $\frac{1}{2}|x - 6| - 3 < 1$
30. $\frac{1}{4}|x - 3| + 2 > 1$
31. $\frac{1}{6}|2x - 1| + 2 \geq 5$
32. $\frac{1}{11}|2x - 4| + 10 \leq 11$
33. $\left|\frac{x - 3}{2}\right| + 2 < 6$
34. $\left|\frac{x + 5}{3}\right| - 3 > 6$
35. $4 \leq |x| \leq 7$
36. $5 < |x + 3| \leq 7$
37. $0 \leq |1 - x| < 8$

For Exercises 38–40, assume that each inequality is to be solved for real values of x. Also, assume that a, b, and c are real numbers.

38. What is the solution set for the equation $|ax + b| = c$, where $c < 0$?
39. What is the solution set for the inequality $|ax + b| > c$, where $c < 0$?
40. What is the solution set for the inequality $|ax + b| < c$, where $c < 0$?

Applications

Manufacturing The specifications for machined parts are given with *tolerance limits*. For example, if a part is to be 6.8 cm thick, with a tolerance of 0.01 cm, this means that the actual thickness must be at most 0.01 cm greater than or less than 6.8 cm. The part is acceptable if its thickness is at least 6.79 cm, but no more than 6.81 cm. If m represents the actual measured thickness of the part in centimeters, then the tolerance limit can be expressed as an absolute value inequality.

$$|m - 6.8| \leq 0.01$$

Express each tolerance limit as an absolute value inequality. Use the variable m for the actual measure of the part, in centimeters.

41. The length of a part is to be 9.6 cm, with a tolerance of 0.03 cm.

42. The diameter of a part is to be 1.38 cm, with a tolerance of 0.005 cm.

43. A machined part is to be 36.8 mm wide, with a tolerance of 0.05 mm. What is the greatest possible width that is acceptable?

44. A machined part is to be 9.51 mm thick, with a tolerance of 0.001 mm. What is the least possible thickness that is acceptable?

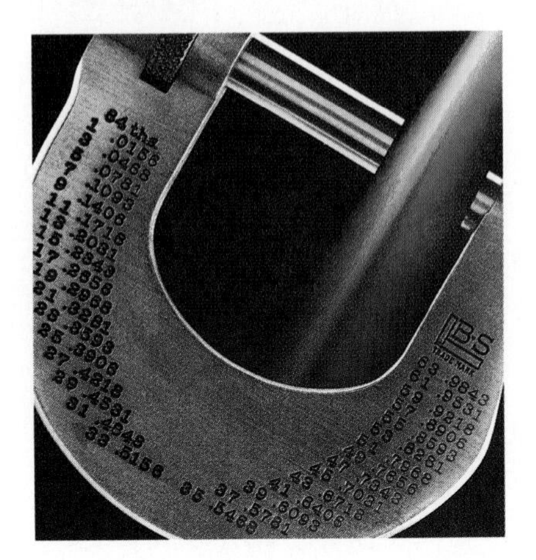

Mixed Review

Solve.

45. $|6x - 5| = 23$

46. $|9 - 2x| = 1$

Solve and graph the solution set.

47. $4(3 + 2x) \leq 28$

48. $12 - 5x > x + 19$

49. $3x + 2 \geq 20 - 3x$ or $14x - 9 < 2x - 9$

50. $3x + 1 < 22$ and $2x + 3 \leq 4 + 5x$

Name the property of real numbers illustrated by each of the following.

51. $4x + (6x + 3) = (4x + 6x) + 3$

52. $74x + 0 = 74x$

53. If $x = 32$ and $y = x + 7$, then $y = 32 + 7$.

54. If $c - 12 = 45$, then $c = 45 + 12$.

55. Two ants leave an anthill at noon, one crawling east (toward the garbage) at a certain speed and the other crawling west (toward the pizza) at twice that speed. If the ants are 180 inches apart in 30 seconds, how fast does each ant crawl?

2.8 Problem Solving: Using Equations and Inequalities

Equations and inequalities serve as mathematical models for many situations in everyday life.

EXAMPLE 1 Pat wants to fence in a rectangular plot for a garden. The garden is to be 5 ft longer than it is wide. If Pat can afford at most 146 ft of wire to build the fence, what is the greatest possible width for the garden?

Understand the Problem You are asked to find the greatest possible width of the garden, given the maximum perimeter and the relationship between the length and width. You will need the formula for the perimeter of a rectangle

$$P = 2(l + w)$$

Plan Your Approach Let w represent the width of the garden.

Then $w + 5$ represents the length.

The perimeter is less than or equal to 146 ft.

$w + 5$

w

Complete the Work

$$\begin{aligned} 2[(w + 5) + w] &\leq 146 \\ 2(2w + 5) &\leq 146 \\ 2w + 5 &\leq 73 \\ 2w &\leq 68 \\ w &\leq 34 \end{aligned}$$

Interpret the Results The greatest possible width of the garden is 34 ft.

To check, note that if $w \leq 34$, then $w + 5 \leq 39$. That is, the length is at most 39 ft. Use the formula to determine the maximum perimeter of the garden.

$$\begin{aligned} \text{maximum perimeter} &= 2(39 + 34) \qquad P = 2(l + w) \\ &= 146 \checkmark \end{aligned}$$

In a set of *consecutive integers,* each integer is 1 greater than the preceding integer. For example, -2, -1, 0, 1, and 2 are five consecutive integers. If n represents the first integer of a set of consecutive integers, then $n + 1$ represents the second integer, $n + 2$ the third integer, and so on.

EXAMPLE 2 Vinnie's father is 3 times as old as Vinnie; five years ago he was 4 times as old as Vinnie. How old is each of them now?

Understand the Problem You are asked to find their ages now. You need to organize your information about their current ages and their ages five years ago.

Plan Your Approach Let n represent Vinnie's age now.
Then $3n$ represents his father's age now.
Subtract 5 to find their ages 5 years ago:

$n - 5$ is Vinnie's age then, and
$3n - 5$ is his father's age then.

Complete the Work

$$\underbrace{\text{His father's age then}}_{3n-5}\ \text{was}\ \underbrace{\text{4 times}}_{4}\ \underbrace{\text{Vinnie's age then.}}_{(n-5)}$$

$$3n - 5 = 4(n - 5)$$
$$3n - 5 = 4n - 20$$
$$3n = 4n - 15$$
$$-n = -15$$
$$n = 15$$

Therefore, $3n = 45$

Interpret the Results Vinnie's age is 15 and his father's is 45.

To check, see if $3n - 5$ is equal to $4(n - 5)$.

$$45 - 5 \stackrel{?}{=} 4(15 - 5)$$
$$40 \stackrel{?}{=} 4(10)$$
$$40 = 40 \checkmark$$

To find the **average,** or **arithmetic mean,** of a set of numbers such as a set of scores, divide the sum of the scores by the number of scores in the set.

EXAMPLE 3 Lee's average on three math tests is between 90 and 93. Lee scored 15 fewer points on the second test than on the first test and 3 fewer points on the third test than on the first test. What scores might he have had on the first test?

Understand the Problem You are asked to find the possible scores for the first test. You will need to use the formula

$$\text{average} = \frac{\text{sum of scores}}{\text{number of scores}}$$

Plan Your Approach Let x represent Lee's score on the first test.
Then $x - 15$ represents his score on the second test,
and $x - 3$ represents his score on the third test.
The average of the test scores is between 90 and 93.

Complete the Work

$$90 < \text{average} < 93$$

$$90 < \frac{x + (x - 15) + (x - 3)}{3} < 93$$

$$90 < \frac{3x - 18}{3} < 93$$

$$90 < x - 6 < 93$$

$$96 < x < 99 \quad \textit{Add 6 to each expression.}$$

Interpret the Results

The possible scores for the first test are 97 and 98.

Check to show that a test score of either 97 or 98 gives an average score between 90 and 93.

CLASS EXERCISES

Write an equation or an inequality as a mathematical model for each problem.

1. Twice a number when increased by 24 is at least 3 times the original number.
2. What number when decreased by 24 is twice itself?
3. A rectangle is 10 ft longer than it is wide. If the perimeter is less than 84 ft, how long is the rectangle?
4. The sum of two consecutive integers is less than 33. What are the two integers?
5. Find the measures of two angles whose sum is between 85° and 90° if the measure of the second is 4 more than 3 times the measure of the first.
6. Find three consecutive even integers whose sum is greater than 86.
7. Robert is twice as old as his sister Alice. Their father is 4 times as old as Robert. If the sum of all their ages is at most 44, find the greatest possible age of each person.
8. That book is 3 times as old as it was 10 years ago.

PRACTICE EXERCISES

1. A rectangular tablecloth is to be 18 in. longer than it is wide. If Edith has at most 244 in. of lace trimming to sew around the cloth, what is the greatest possible width of the tablecloth?
2. A rectangular vegetable garden is to be 12 ft longer than it is wide. If David has at least 128 ft of fencing, what is the least measure that can be used for the width of the garden?

3. Find three consecutive integers whose sum is 126.

4. Find three consecutive integers whose sum is 159.

5. The sum of four consecutive odd integers is 216. Find the four integers.

6. The sum of four consecutive even integers is 204. Find the four integers.

7. Caridad is 6 years older than her sister. Four years ago she was 4 times as old as her sister. How old is Caridad now?

8. Sherman is 4 times as old as his pet rattlesnake. In 5 years, assuming both are still around, Sherman's age will be 3 years more than twice the snake's age. How old is each?

9. When James's great-great grandfather turned 80, he had lived through half the history of the United States. In what year was his 80th birthday?

10. The average age of the starting line-up on an NBA basketball team is 25 yr. If four of the five players are 21, 23, 24 and 31 yr old, find the age of the fifth player.

11. Julius bowled 121, 118, 132, 124, and 125 in five games. What must he bowl in the sixth game so that he will have an average of 125 for the six games?

12. Juliet would like her bowling average to be 133. If she bowled 135, 127, 119, 142, and 156 in five games, what must she bowl in the sixth game to achieve her goal?

13. Juanita's spelling average for three tests is between 85 and 90. If she scored 8 points more on the second test than on the first test and 5 points less on the third test than on the first test, what scores are possible for the first test?

14. Juan's math average for three quizzes is between 80 and 90. His second quiz score was 8 less than his first quiz score and his third quiz score was 2 more than his first quiz score. What scores are possible for Juan's first quiz?

15. Less than 40 ft of fencing were used to enclose a rectangular lot that has a length 12 ft less than 3 times the width. Find the set of all possible widths.

16. No more than 88 m of fencing were used to enclose a rectangular parking lot. Find the set of all possible widths if the length of the lot is 8 m more than 4 times the width.

17. Room 214 at City High School has twice as many chairs as Room 314, and Room 114 has 15 more chairs than Room 214. If the number of chairs in Room 114 is 12 less than the total number of chairs in Rooms 214 and 314, find the number of chairs in each room.

18. Three groups of people were hired to work during a special promotion campaign for a laundry detergent. The first group is twice as large as the second, and the third is as large as the first and second groups combined. There are 72 mi of streets in the area targeted for the promotion. How many blocks can be assigned to each group if there are 10 blocks to a mile? Assume that each person is given the same number of blocks to cover.

19. Jim's average score on three tests is 82. What is the least that he can score on his fourth test so that his average will be at least 86?

20. Carole had an average of 88 on four tests. After the fifth test, her average was between 82 and 85. What scores are possible for Carole's fifth test?

21. The sum of $\frac{1}{2}$ an integer and $\frac{2}{3}$ of the same integer is greater than 7. If $\frac{1}{3}$ of the integer is subtracted from $\frac{3}{4}$ of the same integer, the result is at most $4\frac{1}{2}$. Name the integers that satisfy these conditions.

22. Exactly 10% of the students in the junior class participate in varsity sports, and exactly 15% participate in intramurals. None participate in both. Exactly 10% of the students in the junior class are in the band, and exactly 5% are in the orchestra. Again, none participate in both. If fewer than 96 juniors participate in varsity or intramural sports, and more than 56 participate in the band or orchestra, how many juniors are there?

TEST YOURSELF

1. An express train and a freight train leave at the same time from the same station but travel in opposite directions. If the express travels at 55 mi/h and the freight travels at 45 mi/h, in how many hours will the trains be 500 mi apart? **2.5**

Solve and check.

2. $3|2x - 3| - 7 = 11$ **3.** $|3x - 7| = 5x - 4$ **2.6**

Solve and graph the solution set.

4. $|3x - 5| < 10$ **5.** $|4x + 2| \geq 10$ **6.** $3|2x - 5| + 6 < 9$ **2.7**

7. Find three consecutive integers whose sum is 156. **2.8**

8. Find all sets of three consecutive odd integers whose sum is at least 47 less than 4 times the third integer.

INTEGRATING ALGEBRA
Medical Resource Allocation

Did you know that one of the most important problems facing the medical profession is allocating resources? For example, if a hospital has only one bed available and two patients, a decision must be made as to which patient is in need of immediate attention and which patient is stable enough to be treated without hospitalization.

During World War I [1914–1918] a system of classifying military patients was designed to help allocate medical supplies. This system was given the name *triage* (*tri-* pronounced "tree," *-age* as in "barrage"), a French word meaning to *choose* or *sort*. Physicians divided patients into three categories: those too severely wounded to survive, those who would recover naturally with little treatment, and those who would survive only with immediate medical attention. The third group received the majority of the medical resources available.

Consider the following triage situation. The number of patients in each group will determine how medical resources should be allocated.

The staff in a clinic of 100 patients has diagnosed two diseases: 68 of the patients have disease A which requires an antidote and no other care for recovery; 52 of the patients have disease B which requires intensive therapy. Some of the 100 have both diseases and constitute an "overlap" group that can be comforted, but not cured. How many patients have both diseases? How many patients should receive intensive therapy?

This problem involves counting the number of elements in the regions that result when two sets share elements. In the *Venn diagram* to the right the number elements in $A \cup B$ represents the total number of patients in the clinic. Using counting notation, $n(A \cup B) = 100$.

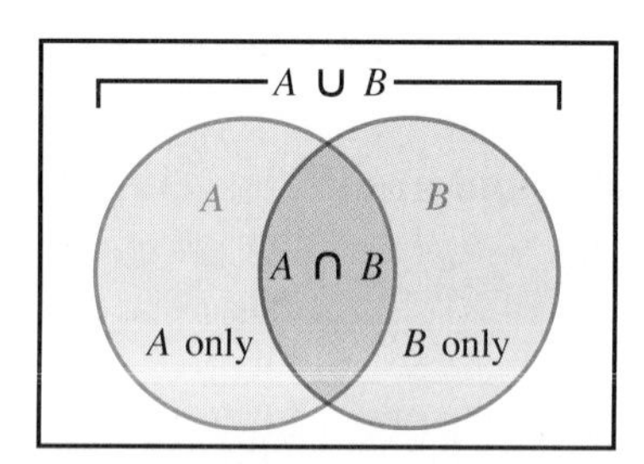

Finding $n(A \cup B)$ by counting all the elements in A and all the elements in B will yield an incorrect result because the "overlap" elements in $A \cap B$ will be counted twice. Therefore, by subtracting $n(A \cap B)$,

$$n(A \cup B) = n(A) + n(B) - n(A \cap B)$$

The "overlap" or intersection of the two sets represents the number of patients with both diseases. Solving the equation above for $n(A \cap B)$,

$$\begin{aligned} n(A \cap B) &= n(A) + n(B) - n(A \cup B) \\ &= 68 + 52 - 100 \\ &= 20 \end{aligned}$$

Therefore 20 patients have both diseases.

From the Venn diagram, the number of patients with disease B only can be determined in either of the following two ways:

$$\begin{aligned} n(B \text{ only}) &= n(B) - n(A \cap B) \\ &= 52 - 20 \\ &= 32 \end{aligned} \qquad \begin{aligned} n(B \text{ only}) &= n(A \cup B) - n(A) \\ &= 100 - 68 \\ &= 32 \end{aligned}$$

Hence, the clinic should provide intensive therapy for 32 patients and comfort 20. The remaining patients, having disease A only, should be administered an antidote and released.

EXERCISES

1. After a severe earthquake, a hospital has 77 patients: 49 patients have injury A and 38 have injury B. Patients with injury B will recover without treatment while those with injury A require medical care. Patients with both injuries can only be comforted. How many patients have both injuries? How many patients should be treated?
2. In an experimental group, 68 patients are given medication A and 44 are given medication B. Among these, 13 are administered both medications. How many subjects are in the experimental group?
3. In a two-disease hospital ward of 60 patients, three times as many patients have disease B than disease A. Four patients have both. How many have A? How many have B?
4. With three diseases among a group of patients, the problem of counting those with only one, those with two, and those with all three is more complicated, but the approach is the same. Using a Venn diagram, derive an expression for the total number of patients, $n(A \cup B \cup C)$, in terms of the number with each disease and in the "overlap" groups.

CHAPTER 2 SUMMARY AND REVIEW

Vocabulary

absolute value (73)
absolute value equation (73)
arithmetic mean (82)
average (82)
compound sentence (57)
conjunction (57)
disjunction (58)
equivalent inequalities (54)
formula (46)
inequality (52)
intersection of sets (58)
literal equation (46)
properties of inequality (53)
trichotomy property (53)
union of sets (59)
Venn diagram (58)

Literal Equations and Formulas Literal equations are equations that contain more than one letter or variable. A formula is a literal equation that states a general rule. 2.1

Solve for x and indicate any restrictions on the values of the variables.

1. $ax + b = c$
2. $dx + t = r - cx$
3. $3ax = 15c - 2ax$
4. Solve the formula $F = \frac{kWw}{d^2}$ for w.

Solving Inequalities If you multiply or divide both sides of an inequality by a negative number, you must reverse the order of the inequality. The order relationship is preserved under all other operations. 2.2

Solve and graph the solution set if it is not the empty set.

5. $2t + 3 > 17$
6. $4 - 3t \geq -14$
7. $2(3t - 8) < 4(t - 5)$
8. $-5(t + 3) \leq 3t + 2$

Conjunctions and Disjunctions A conjunction joins two sentences with the word *and*. A conjunction is true only when both sentences are true. The solution set of a conjunction is the intersection of the solution sets of the individual sentences that make up the conjunction. 2.3

A disjunction joins two sentences with the word *or*. A disjunction is true when either one or both of the sentences are true. The solution set of a disjunction is the union of the solution sets of the individual sentences.

Classify each sentence as a conjunction or a disjunction and state whether it is true, false, or open.

9. $-6 + 4 = -2$ *and* $6 - (-5) = 11$

10. $-8 - 4 < -13$ *or* $-2(-6) > 12$

11. $2x + 3 < 15$ *and* $-(-2 + 1) = 1$

12. $-6(3) > -5(4)$ *or* $3 - 6 > 4 - 5$

Solving Compound Sentences with Inequalities To solve a conjunction that is an open sentence in one variable, find the values of the variable that make both sentences true. To solve a disjunction that is an open sentence in one variable, find the values of the variable that make at least one of the sentences true. 2.4

Solve and graph the solution set if it is not the empty set.

13. $3y - 1 > 14$ *and* $2y \leq 16$

14. $2(y + 3) \leq 10$ *or* $3(y - 2) > 21$

15. $16 < 6 - 2y < 24$

16. Two school buses started from Middle School's parking lot and traveled in opposite directions. If one bus averaged 32 mi/h and the other bus averaged 28 mi/h, in how many hours will they be 100 mi apart? 2.5

17. A cashier started out with less than \$15 in nickels, dimes, and quarters in a cash register. Find the greatest number of nickels in the register if there were 5 fewer dimes than twice the number of nickels and twice as many quarters as nickels.

Absolute Value Equations To solve an absolute value equation, rewrite it as a disjunction and solve the disjunction. 2.6

Solve and check.

18. $|w| - 5 = 3$

19. $|2w + 4| = 10$

20. $3|w - 1| = 15$

Absolute Value Inequalities To solve an absolute value inequality, solve the equivalent conjunction or disjunction. 2.7

Solve and graph the solution set.

21. $|2z - 5| < 25$

22. $|4z - 3| \geq 13$

23. $-3|2z| + 3 \leq 7$

24. Find four consecutive odd integers such that the sum of the second and third integers is 48 less than the sum of all four integers. 2.8

25. A bowler scored a total of 620 in 5 games. What must the bowler score in the next game in order to average between 125 and 130 in 6 games?

CHAPTER 2 TEST

Solve for x and indicate any restrictions on the values of the variables.

1. $2cx - 4a = 8a$

2. $2ax + 4c = 8d - 2ax$

3. Solve the formula $A = 2\pi rh + 2\pi r^2$ for h.

Solve and graph the solution set if it is not the empty set.

4. $6(2x - 1) + 3 \geq 2(4 - x)$

5. $4(3t + 2) < 57 + 7(t - 2)$

Classify each sentence as a conjunction or a disjunction and state whether it is true, false, or open.

6. $7 - 2x = 5$ *or* $3x + 4 = 8$

7. $-(7 - 10) > 1$ *and* $-(-7 - 6) = 1$

Solve and graph the solution set if it is not the empty set.

8. $3y > 9$ *and* $2y - 6 < 10$

9. $2(z - 1) > 18$ *or* $-3z > 15$

Solve.

10. John traveled north from Miami at an average speed of 50 mi/h. Two hours after he left, Sharon left from Miami traveling in the same direction at an average speed of 55 mi/h. How long will it take Sharon to catch up with John?

Solve and check.

11. $|2y| - 3 = 7$

12. $|3z + 6| = 15$

Solve and graph the solution set.

13. $|2x - 5| \leq 11$

14. $|3 - 2t| > 9$

15. Find all possible sets of three consecutive even integers whose sum is between 168 and 180.

Challenge

Solve and graph the inequality $\frac{1}{x} < 2$.

COLLEGE ENTRANCE EXAM REVIEW

Select the best choice for each question.

1. When $\frac{a}{b} = \frac{c}{d}$ is a true statement, which of the following is not necessarily true?

A. $ac = bd$ **B.** $\frac{a}{c} = \frac{b}{d}$

C. $\frac{3d}{b} = \frac{3c}{a}$ **D.** $\frac{a+b}{c+d} = \frac{b}{d}$

E. $\frac{a+b}{a-b} = \frac{c+d}{c-d}$

2. The expression $|5x - 7| \leq 8$ is equivalent to

A. $5x - 7 \leq 8$
B. $-8 \leq 5x - 7 \leq 8$
C. $5x - 7 \geq 8$
D. $5x - 7 \leq 8$ *or* $5x - 7 \geq -8$
E. $5x - 7 \leq -8$ *and* $5x - 7 \geq 8$

3. This figure is formed by three arcs drawn using each vertex of the equilateral triangle as a center and a side of the triangle as a radius. If the shaded area is $18\pi - 27\sqrt{3}$, then a side of the triangle is

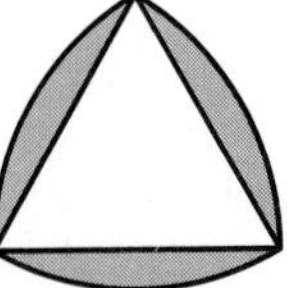

A. 2 **B.** 3 **C.** 4 **D.** 6 **E.** 9

4. Which of the following is *not* in the solution set of $x - 4y \leq 4$?

A. (0, 3) **B.** (7, 1)
C. (−1, 5) **D.** (−4, −1)
E. (3, −2)

5. What is the smallest value of y in the system

$$\begin{cases} x + 3y \geq 6 \\ 2x - 3y \leq 3? \end{cases}$$

A. −1 **B.** 0 **C.** 1 **D.** 2 **E.** 3

6. If an integer $N = \sqrt[3]{50{,}653}$, then N equals

A. 23 **B.** 27 **C.** 31
D. 37 **E.** 43

7. A square and a circle have equal areas. Express the radius of the circle in terms of s, a side of the square.

A. $r = \frac{s}{\pi}$ **B.** $r = \frac{s}{\sqrt{\pi}}$

C. $r = \sqrt{\frac{s}{\pi}}$ **D.** $r = s\pi$

E. $r = s\sqrt{\pi}$

8. Three circles of radii 2, 4, and 6 are tangent to each other externally. Find the area of the triangle formed by connecting their centers.

A. 4 **B.** 6 **C.** 12 **D.** 24
E. It cannot be determined from the information given.

9. What is $\frac{1}{3}\%$ of 690?

A. 230 **B.** 23 **C.** 2.3
D. 0.23 **E.** 0.023

10. Tom won a 1-mile race in the time of 4 minutes, 12 seconds. To the nearest tenth, what was his rate in miles per hour?

A. 7.0 **B.** 12.0 **C.** 13.4
D. 14.0 **E.** 14.3

11. Solve for y in the following:
$y(2y - 3) + 5(y - 1) = 2(y^2 + 5) - y$

A. 5 **B.** 3 **C.** 1
D. −1 **E.** −5

CUMULATIVE REVIEW

Solve and check.

1. $3m - 4 = 8 - m$
2. $|2x + 5| = 7$
3. $8y - \frac{1}{5} = 6y + \frac{3}{10}$
4. $3(t + 2) - 2(t - 4) = 24$
5. Write an algebraic sentence, using x for the variable: Three times a number decreased by five is greater than ten.
6. Solve for x and indicate any restrictions on the values of the variables: $rx + sx = s$

Solve.

7. Five times a number increased by thirteen is 17. Find the number.
8. Maureen's average on four math tests is 91. If her scores on the first three tests were 94, 83, and 95, what was her final test score?
9. Two trains leave Indianapolis at 10 AM and travel in opposite directions. If one train travels at 80 mi/h and the other at 95 mi/h, how far apart will they be at 2 PM?
10. The sum of three consecutive even integers is 72. Find the integers.

Simplify.

11. $\left(-\frac{33}{7}\right)\left(\frac{11}{3}\right)$
12. $4 \times 3 + 2 \times 8 - 12 \div 3$
13. $-2|5 - 7| - 3|4 - 11| + 6|3 + 5|$
14. $-108 - (-24)$
15. Evaluate $12a - 4a^2 + 7a^3$ using $a = -3$.
16. Convert to decimal form: $\frac{3}{11}$
17. Name the multiplicative inverse of $\left(-\frac{2}{5}\right)$.
18. Write a word phrase for $x^2 - 6$.
19. Graph the solution set: $x > 2$ and $x < 6$.
20. Express as the quotient of two integers: 1.275
21. Evaluate $3^x + 5^x - 6^x$ using $x = 2$.

Solve and graph the solution set.

22. $2|y - 6| \leq 2$
23. $-6x < 24$ or $5x < -35$

3 Functions and Graphs

An arm and a leg?

Developing Mathematical Power

A person's shoe size is not indicative of his or her height. Tall people can have small feet relative to their height. There is, however, a relationship between the length of a person's foot and the length of his or her forearm (wrist to elbow).

Project

Find the relationship between the length of a person's foot and forearm. Take a survey to collect data. Organize and analyze the data. Use graphs to present your interpretation of the relationship.

The Coordinate Plane

Objectives: To graph ordered pairs on a coordinate plane
To determine the coordinates of a point, given its graph

A graph is a pictorial representation of a point or a set of points. There are different systems of graphing that can be used. Each system has its own frame of reference.

Capsule Review

On a number line, numbers to the right of zero are *positive*. Numbers to the left of zero are *negative*.

The graph of $+2$, or 2, is two units to the right of zero.
The graph of -3 is three units to the left of zero.

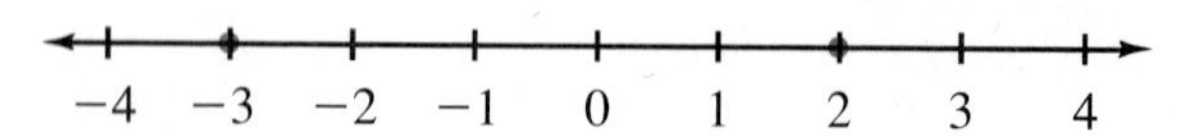

Graph each number on a number line.

1. $+4$ **2.** -7 **3.** 9 **4.** $-\frac{1}{2}$ **5.** $2\frac{3}{4}$ **6.** -3.5

One system used to graph points on a plane is the **rectangular coordinate system.** The frame of reference is two perpendicular lines called axes. The horizontal line is the ***x*-axis** and the vertical line is the ***y*-axis.** The point at which the two axes intersect is the **origin.** The arrowheads show the positive direction on each axis. The x- and y-axes separate the **coordinate plane** into four regions called **quadrants,** which are numbered in a counterclockwise direction, as shown.

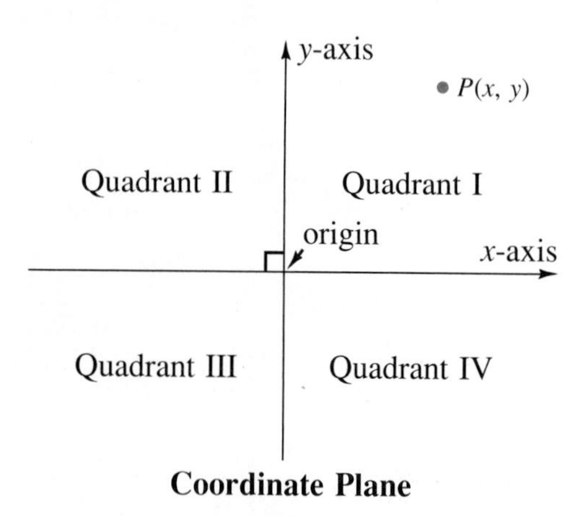

Coordinate Plane

There is a one-to-one correspondence between the points in the plane and the set of ordered pairs (x, y), where x and y are real numbers. The first number is the ***x*-coordinate,** or *abscissa,* which gives the horizontal distance of the point from the y-axis. The second number is the ***y*-coordinate,** or *ordinate,* which gives the vertical distance of the point from the x-axis.

To graph, or *plot,* the ordered pair (4, 3), start at the origin, move 4 units to the right and 3 units up, and place a dot at that point.

To graph the ordered pair (−4, −3), start at the origin, move 4 units to the left and 3 units down, and place a dot at that point.

Points on an axis are not in a quadrant. The coordinates of the origin are (0, 0).

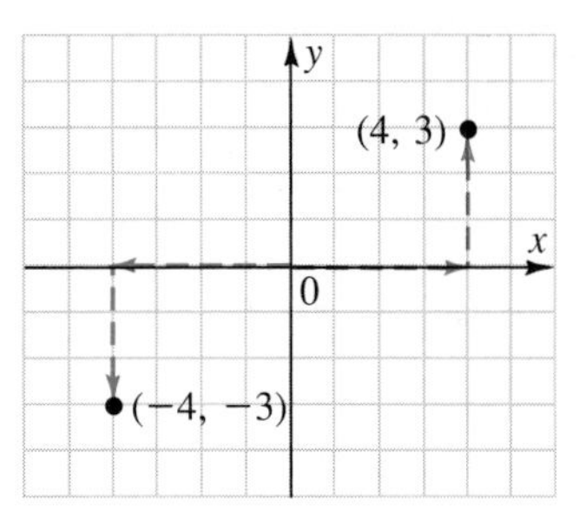

EXAMPLE 1 **Graph each ordered pair.**

a. (2, −3) **b.** (−4, 2)
c. (6, 2) **d.** (−5, −3)
e. (0, 3) **f.** (−3, 0)

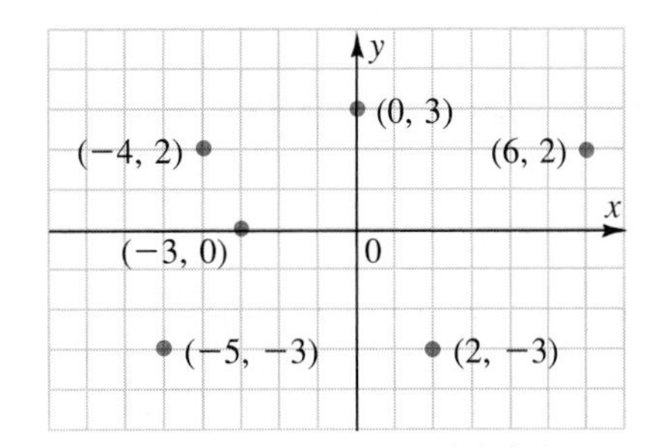

You can determine the coordinates of a point if you are given its graph.

EXAMPLE 2 **Give the coordinates of each point.**

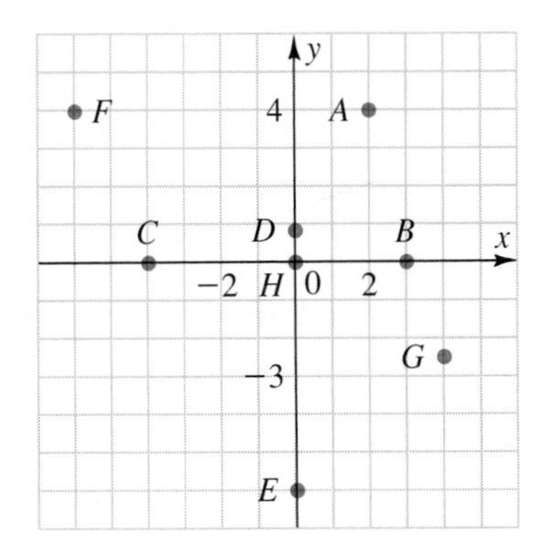

$A(2, 4)$
$B(3, 0)$
$C(-4, 0)$
$D(0, \frac{7}{8})$
$E(0, -6)$
$F(-6, 4)$
$G(4, -2\frac{1}{2})$
$H(0, 0)$

Notice that the *y*-coordinate of a point on the *x*-axis is always 0 and the *x*-coordinate of a point on the *y*-axis is always 0.

CLASS EXERCISES

Graph each ordered pair and name the quadrant or axis where each point lies.

1. (5, −2) **2.** (3, −4) **3.** (−4, −2) **4.** (−5, −7)

5. (5, 2) **6.** (3, 4) **7.** (5, 0) **8.** (0, −5)

Give the coordinates of each point.

9. A
10. B
11. C
12. D
13. E
14. F
15. G
16. H

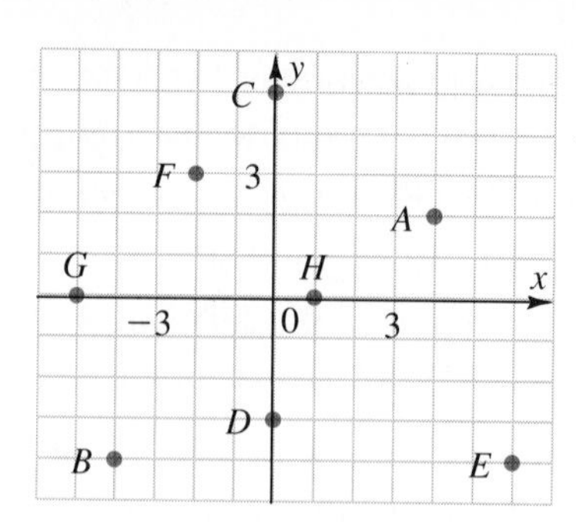

PRACTICE EXERCISES

Graph each ordered pair and name the quadrant or axis where each point lies.

1. (3, 4)
2. (−4, 2)
3. (−3, −1)
4. (4, −1)
5. (0, −2)
6. (0, 3)
7. (7, 0)
8. (−3, 0)
9. $\left(-\frac{1}{2}, -\frac{1}{2}\right)$
10. $\left(0, \frac{1}{2}\right)$
11. $\left(\frac{1}{3}, \frac{1}{2}\right)$
12. $\left(-\frac{1}{2}, 0\right)$

Give the coordinates of each point.

13. A
14. B
15. C
16. D
17. E
18. F
19. G
20. H

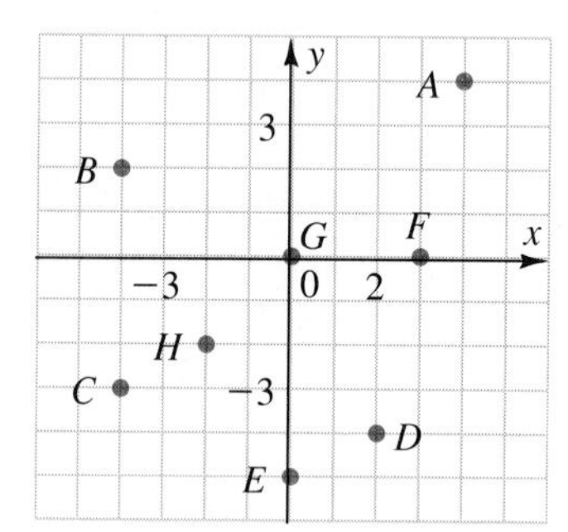

The ordered pair (x, y) represents a point in a coordinate plane. Name the quadrant, point, or axis that satisfies the given conditions.

21. $x > 0$ and $y < 0$
22. $x < 0$ and $y > 0$
23. $x > 0$ and $y > 0$
24. $x = 0$ and $y = 0$
25. $x < 0$ and $y < 0$
26. $y = 0$

Graph a point that satisfies the given conditions.

27. First coordinate is 4.
28. y-coordinate is −2.
29. x-coordinate is 0.
30. Coordinates are equal.

Find the coordinates of each point.

31. The distance from the y-axis is 3 units and the distance from the x-axis is 4 units. The point is in Quadrant I.

32. The distance from the y-axis is one unit and the distance from the x-axis is 2 units. The point is in Quadrant III.

33. The coordinates of three vertices of a rectangle are (1, 1), (1, 3), and (6, 1). Find the coordinates of the fourth vertex.

34. The coordinates of three vertices of a square are $(-2, 3)$, $(-2, -1)$, and (2, 3). Find the coordinates of the fourth vertex.

35. The coordinates of three vertices of a parallelogram are (0, 0), $(-6, 0)$, and $(-8, -3)$. Find the coordinates of the fourth vertex. *Hint:* There are three possible locations.

36. **Thinking Critically** The coordinates of two vertices of a square are $(0, a)$ and $(0, -a)$, $a \neq 0$. Find the coordinates of the other vertices. *Hint:* There are three possible locations.

Find the distance of each point from the x- and y-axes.

37. $(-2, 4)$ 38. $(-5, 2)$ 39. (0, 4) 40. $(-2, 0)$

Applications

41. **Architecture** A company wants to put lights in a parking lot. The architect treats the lot as a coordinate plane and uses the center of the lot as the origin. She puts lights at (4, 5), $(-4, -5)$, $(3, -2)$, and $(-1, 2)$. Graph these points to show where the lights should go.

42. **Engineering** An engineer is installing a sprinkler system in a large gymnasium. He treats the gym ceiling as a coordinate plane and lets the center be the origin. The sprinklers are to be placed at $(-3, 5)$, $(-2, 4)$, (6, 7), and $(5, -2)$. Graph these points to show where the sprinklers should go.

43. **Transportation** The stationmaster designates the train storage yard as the origin of a coordinate plane, and the locations of engines outside the yard are indicated by pins at points in the plane. At 6 PM there are engines at (0, 3), $(-4, 0)$, $(3, -4)$, and $(-6, 8)$. Graph these points.

Mixed Review

Write each of the following as the ratio of two integers.

44. 0.0023 45. 5 46. $12.3\overline{123}$

Solve and graph the solution set on a number line.

47. $14x + 3 < 18 - 5x$

48. $10 - 4x \geq 24$

49. $6x + 3 > 14x - 29 > 10 - 11x$

50. $2|5 - 2x| \leq 18$

3.2 Relations and Functions

Objectives: To determine the domain and range of relations and functions
To determine whether a relation is a function

Numerical information about two related variables can sometimes be written as a set of ordered pairs. In this form, the information can be used to develop a mathematical model to help solve business and scientific problems.

Capsule Review

Use substitution to evaluate an expression for given values of x.

EXAMPLE **Evaluate $x^2 - 4x$ for $x = 3$ and $x = -1$.**

$x^2 - 4x = (3)^2 - 4(3) = -3$ *Substitute 3 for x.*

$x^2 - 4x = (-1)^2 - 4(-1) = 5$ *Substitute −1 for x.*

Evaluate each expression for the given values of x.

1. $x + 2, x = -1, 2$
2. $x - 3, x = -2, 3$
3. $4x + 1, x = -3, 1$
4. $-2x + 3, x = 0, -1$
5. $2x^2 + 1, x = 5, 10$
6. $5x^2 - 1, x = 4, 7$

A **relation** is a set of ordered pairs, and a graph is one way to represent a relation.

EXAMPLE 1 **Graph the relation:**
$L = \{(-2, 4), (3, -2), (-1, 0), (1, 5)\}$

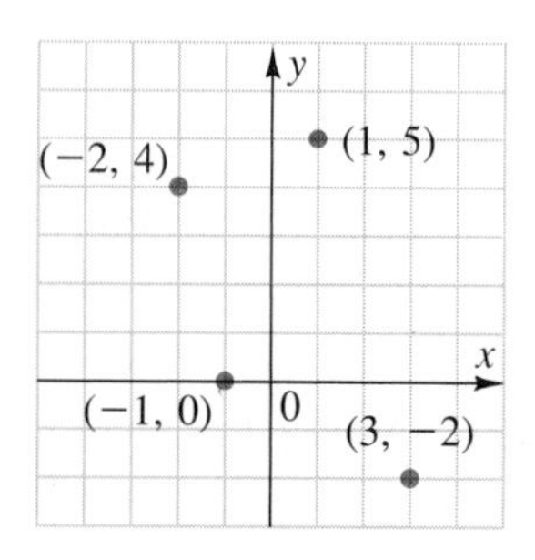

For this relation the set of values of x is

$$\{-2, 3, -1, 1\}$$

The corresponding set of values of y is

$$\{4, -2, 0, 5\}$$

The **domain** of a relation is the set of all the first coordinates of the ordered pairs. The **range** of a relation is the set of all the second coordinates of the ordered pairs.

Some relations may be described by rules. If $S = \{(1, 2), (2, 4), (3, 6), (4, 8)\}$, how is each y-coordinate related to its corresponding x-coordinate? Each value of y is twice the corresponding value of x.

Therefore, the relation can be expressed as a rule

$$S = \{(x, y): y = 2x, x = 1, 2, 3, 4\}$$

Read "the set of ordered pairs (x, y) such that $y = 2x$, for x equal to 1, 2, 3, and 4." The domain is $\{1, 2, 3, 4\}$ and the range is $\{2, 4, 6, 8\}$.

EXAMPLE 2 **Write a rule for the relation $\{(-1, 2), (0, 3), (1, 4), (2, 5)\}$. State the domain and range.**

$\{(x, y): y = x + 3, x = -1, 0, 1, 2\}$

The domain is $\{-1, 0, 1, 2\}$.

The range is $\{2, 3, 4, 5\}$.

You can sometimes determine the domain and the range from the graph of a relation.

EXAMPLE 3 **Write the ordered pairs for the relation graphed. State the domain and the range.**

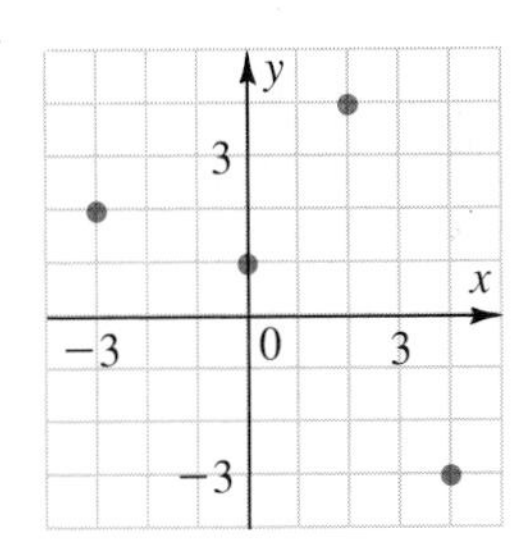

$\{(2, 4), (0, 1), (-3, 2), (4, -3)\}$

The domain is $\{2, 0, -3, 4\}$.

The range is $\{4, 1, 2, -3\}$.

Notice in Examples 1, 2, and 3 that each value of x corresponds to one and only one value of y. This is a special type of relation called a *function*. A **function** is a relation in which each element in the domain is paired with *one and only one* element in the range. One way to determine whether a relation is a function is to make a **mapping** diagram. Write the elements of the domain in one region and the elements of the range in another. Draw arrows to show how each element is mapped onto another.

EXAMPLE 4 **Make a mapping diagram for each relation and determine whether or not it is a function.**

a. $M = \{(-1, -2), (3, 6), (-5, -10)\}$
b. $K = \{(2, -3), (4, -1), (2, 3), (4, 1)\}$

a. $M = \{(-1, -2), (3, 6), (-5, -10)\}$

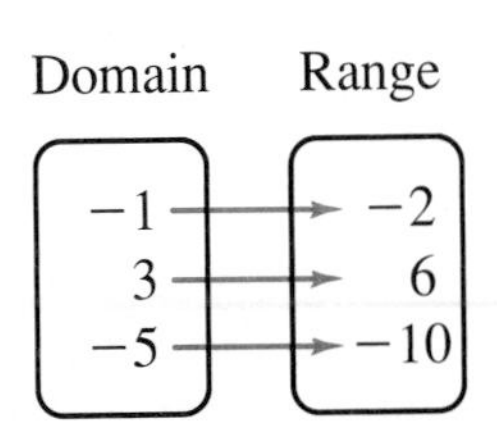

Each element of the domain maps onto one and only one element of the range.

The relation M is a function.

b. $K = \{(2, -3), (4, -1), (2, 3), (4, 1)\}$

Elements of the domain map onto more than one element of the range.

The relation K is *not* a function.

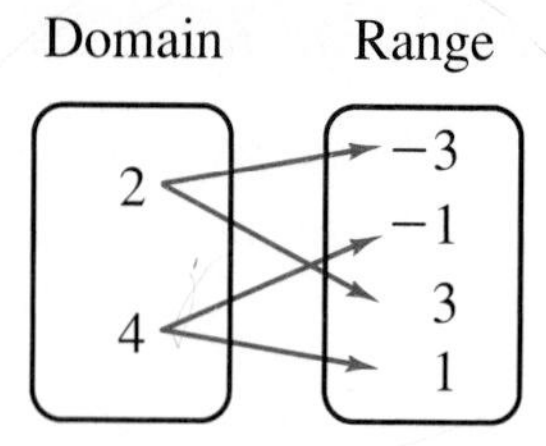

To determine if a relation represented by a graph is a function, use the **vertical line test.** Draw vertical lines through the graph, or simply picture them mentally. If no vertical line can be drawn that touches more than one point on the graph, the graph represents a function.

EXAMPLE 5 **Use the vertical line test to determine which of the graphs represents a function.**

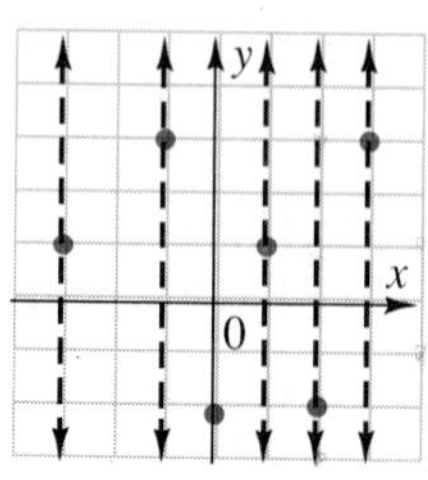

Function

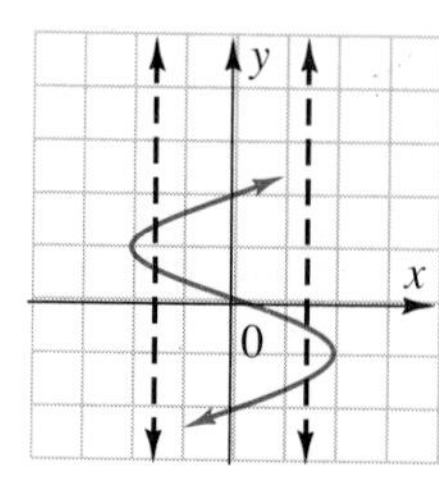

Not a function

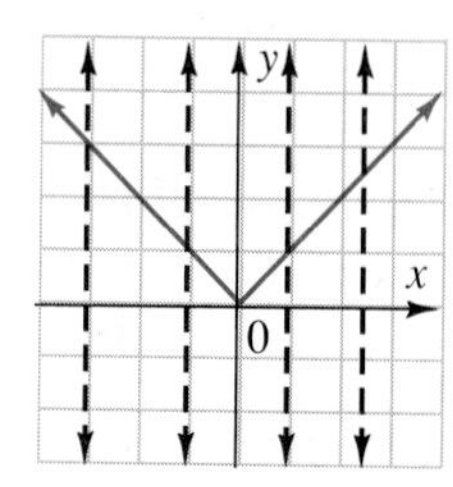

Function

CLASS EXERCISES

Graph each relation and state its domain and range.

1. $\{(-2, 1), (-1, 0), (0, 1), (1, 2)\}$

2. $\{(2, 4), (3, 5), (4, 6), (5, 7)\}$

Write a rule for each relation and state the domain and range.

3. $\{(0, 2), (1, 3), (2, 4)\}$

4. $\{(1, 3), (2, 6), (3, 9)\}$

Find the domain and range of each relation and state which are functions.

5. $\{(2, 6), (3, 6), (4, 5), (7, 8)\}$

6. $\{(2, -3), (2, 3), (4, 9), (5, 7)\}$

7.

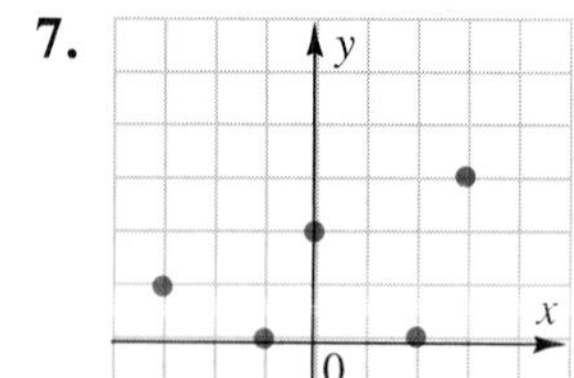

8.

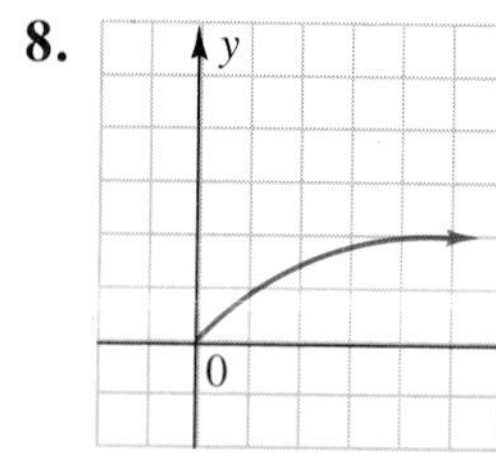

9. Domain Range

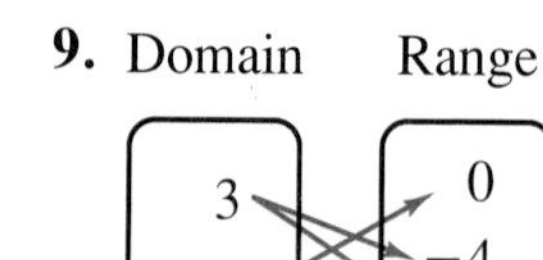

PRACTICE EXERCISES

Graph each relation and state its domain and range.

1. $\{(-1, 1), (-2, 2), (-3, 3), (-4, 4)\}$ **2.** $\{(6, 3), (4, 1), (2, -1), (0, -3)\}$

3. $\{(0, -2), (2, 0), (3, 1), (5, 3)\}$ **4.** $\{(-1, 3), (-2, 1), (-3, -3), (-4, -5)\}$

Write a rule for each relation and state its domain and range.

5. $\{(0, 4), (1, 5), (2, 6), (3, 7)\}$ **6.** $\{(2, 8), (3, 12), (4, 16)\}$

7. $\{(1, -2), (2, -1), (4, 1), (5, 2)\}$ **8.** $\{(1, 7), (2, 8), (3, 9), (4, 10)\}$

Write the ordered pairs for each relation graphed below. Find the domain and range.

9.

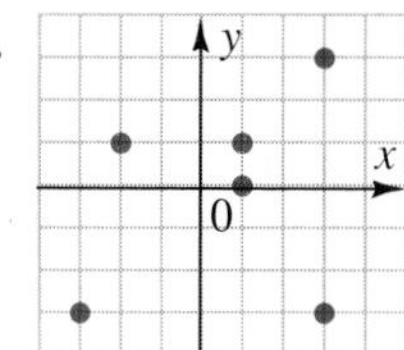

10.

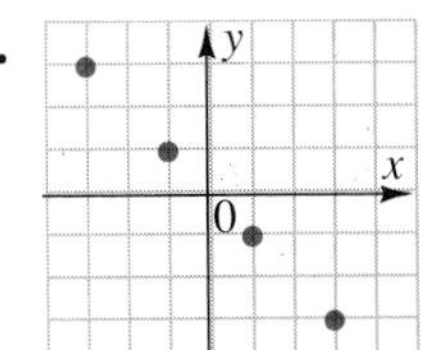

Determine which, if any, of the following relations are functions.

11. $M = \{(2, 4), (4, 8), (8, 16)\}$ **12.** $T = \{(-1, 2), (-2, 5), (-2, 7)\}$

13. Domain Range

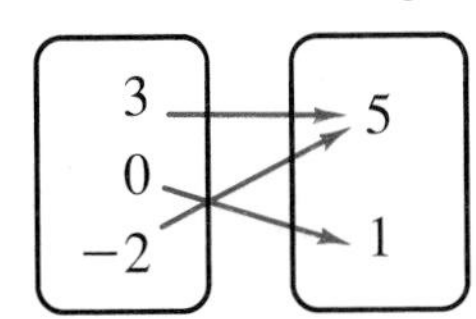

14. Domain Range

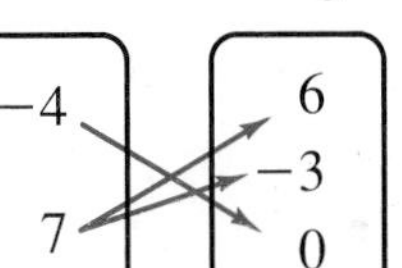

Use the vertical line test to determine which of the graphs are functions.

15.

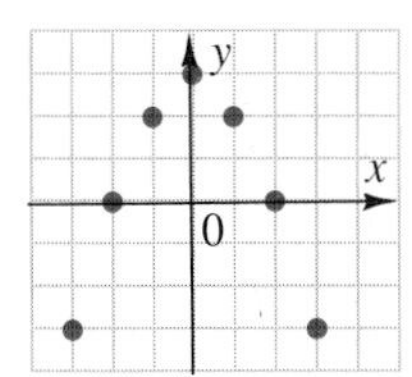

16.

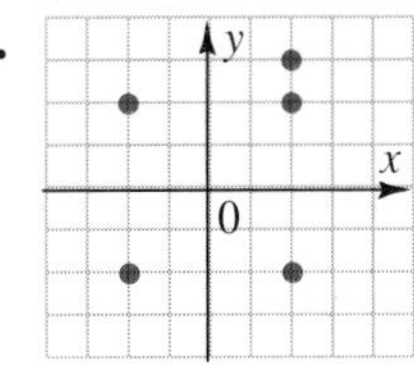

17.

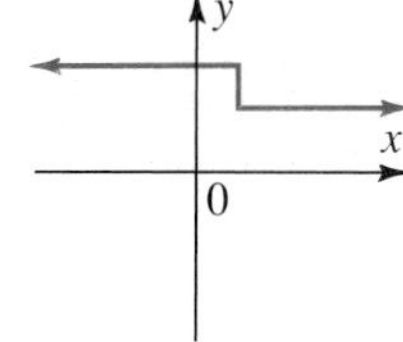

18.

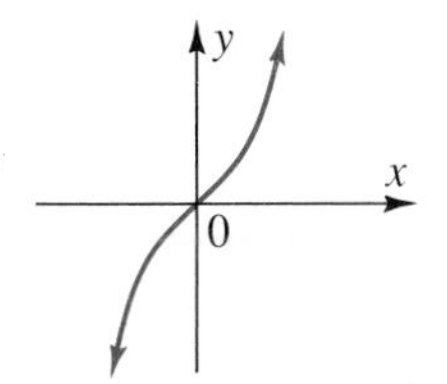

19.

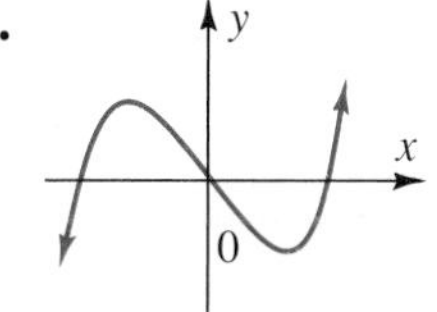

20.

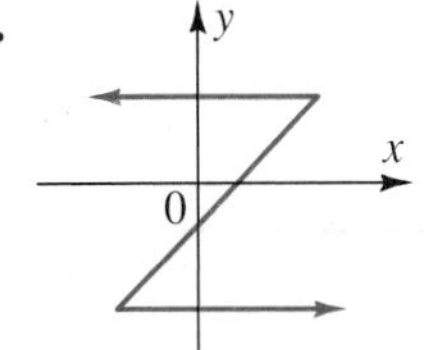

Write a rule for each relation and state its domain and range.

21. $\{(-2, -6), (-4, -12), (6, 18)\}$

22. $\{(1, -2), (2, -4), (-1, 2)\}$

23. $\{(-1, -1), (0, 1), (2, 5), (3, 7)\}$

24. $\{(2, 1), (1, -1), (-1, -5), (-2, -7)\}$

Make a mapping diagram for each relation and determine whether or not it is a function.

25. $\{(3, 5), (-7, 1), (2, 6), (4, 2)\}$

26. $\{(4, 6), (4, 7), (-2, -5), (3, -5)\}$

27. $\{(0, 1), (1, -3), (-2, -3), (3, -3)\}$

28. $\{(4, 0), (7, 0), (4, -1), (7, -1)\}$

Determine which, if any, of the following relations are functions.

29. $\{(x, y): y = 0\}$

30. $\{(x, y): x = 2\}$

31. Domain Range

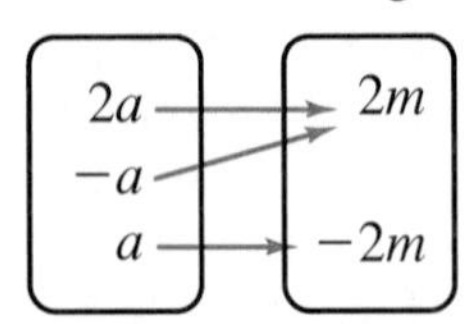

32. Domain Range

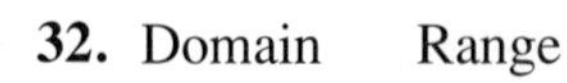

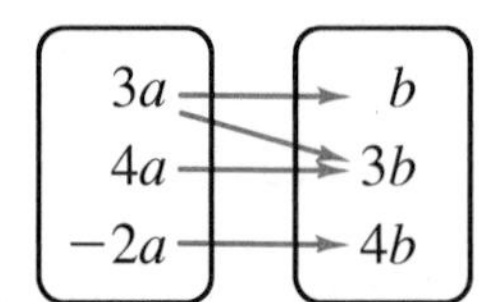

33.

34.

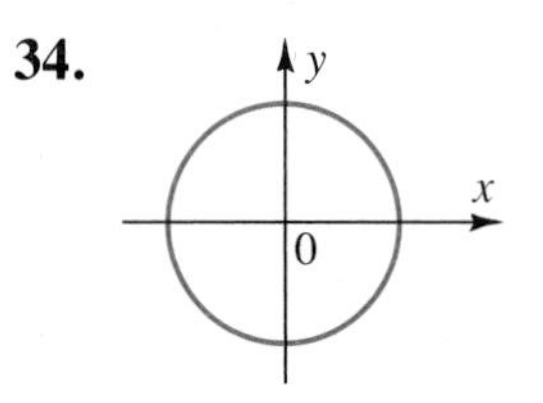

Write a rule for each relation.

35. $\{(-2, 2), (-1, 1), (0, 0), (1, 1), (2, 2)\}$

36. $\{(0, 0), (-1, 1), (1, 1), (-2, 4), (2, 4)\}$

37. $\{(0, 0), (-1, -1), (-2, -8), (-3, -27)\}$

38. $\{(-2, 8), (-1, 1), (0, 0), (1, 1), (2, 8)\}$

39. In the relation $\{(x, y): y = 3x + 4\}$, is y a function of x?

40. In the relation $\{(x, y): x = y^2\}$, is y a function of x?

Applications

41. Education A study found that a student's grades are related to the amount of time spent on homework. The relation was $\{(2, 85), (3, 90), (1, 76)\}$. Determine if this relation is a function.

42. Medicine There was found to be a relationship between the number of aspirins a person took and the number of days the fever and flu persisted. The relation was $\{(2, 3), (2, 5), (3, 4), (3, 5)\}$. Determine if this relation is a function.

43. Carpentry Carole is building drawers of different sizes for a cabinet. The bottoms of the drawers are rectangular, with the length, y, twice the width, x. If the widths she is using are 18 in., 20 in., and 22 in., write the rule for this relation. Determine if this relation is a function.

44. Construction John is building a patio of cement blocks. The length, y, of each block must be 2 in. greater than the width, x. He orders blocks with widths of 8 in., 14 in., and 29 in. Write the rule for this relation. Determine if this relation is a function.

45. Chemistry The time required for a certain chemical reaction to take place is related to the amount of catalyst present during the reaction. Suppose the domain of the relation is the number of grams of catalyst and the range is the number of seconds required for a fixed amount of the chemical to react. The following relation is the result of observing several reactions: $\{(2, 180), (2.5, 6), (2.7, 0.05), (2.9, 0.001), (3.0, 6), (3.1, 15), (3.2, 37), (3.3, 176)\}$. Is the relation a function? If the domain and range were interchanged, would the relation be a function?

BIOGRAPHY: Benjamin Banneker

Benjamin Banneker was born in Maryland in 1731, where his fascination with mathematics became evident at a very early age. While he worked on his parents' farm, he often entertained himself by making up and solving mathematical puzzles. As a student of astronomy, he published a series of almanacs for which he did all of the mathematical calculations.

Banneker, an accomplished surveyor, was one of three people chosen by George Washington to plan the city of Washington, D.C. When one man left, taking the plans with him, Banneker reconstructed the complex plans from memory. This assured the completion of the project.

Investigation

Almanacs played an important part in the lives of early Americans. What kind of information was provided in these publications? Give examples, and describe the mathematical knowledge required of the author and publisher of such material.

3.3

Graphing Equations

Objectives: To graph equations in two variables
To identify equations as linear

Linear equations can be used as mathematical models. In economics, linear equations are used to express relations such as those between price and supply, price and demand, and profit and loss.

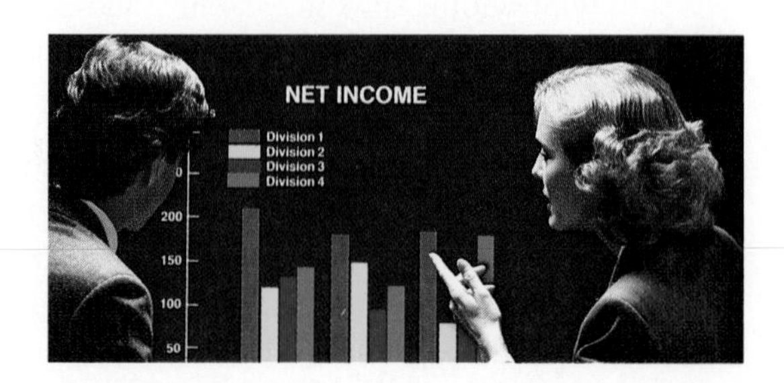

Capsule Review

Evaluate each expression for the given values of x.

1. $\frac{2}{3}x + 7, x = 1, 3, 5, 6$

2. $\frac{3}{5}x - 2, x = 1, 5, 6, 10$

For what values of x will the expression be an integer?

3. $\frac{2}{3}x + 7$

4. $\frac{3}{5}x - 2$

5. $3x + 1$

6. $\frac{1}{2}x - 8$

An equation such as $3x - 2y = 7$ has two variables. Its solutions are ordered pairs of numbers (x, y) that make the equation true.

EXAMPLE 1 **Determine whether each ordered pair is a solution of $3x - 2y = 7$.**
a. (3, 1) **b.** (2, 2) **c.** (−1, −5)

a. $3(3) - 2(1) \stackrel{?}{=} 7$
$9 - 2 \stackrel{?}{=} 7$
$7 = 7$ ✓

(3, 1) is a solution.

b. $3(2) - 2(2) \stackrel{?}{=} 7$
$6 - 4 \stackrel{?}{=} 7$
$2 \neq 7$

(2, 2) is *not* a solution.

c. $3(-1) - 2(-5) \stackrel{?}{=} 7$
$-3 + 10 \stackrel{?}{=} 7$
$7 = 7$ ✓

(−1, −5) is a solution.

Since the solutions of $3x - 2y = 7$ are ordered pairs, this equation defines a relation, $\{(x, y): 3x - 2y = 7\}$. The domain is the set of real numbers, since x can be any real number. The range is also the set of real numbers. This equation has an *infinite* number of solutions.

To graph an equation in two variables, first determine some ordered pairs that are solutions. Then plot enough points so that the shape of the graph is clear. Complete the graph by connecting the points, in order.

EXAMPLE 2 **Sketch the graph of $y = 2x + 3$.**

Make a table of values. Choose some values for x and find the corresponding values of y. Plot the point for each ordered pair and complete the graph by connecting the points in order.

x	$y = 2x + 3$	y	(x, y)
-2	$2(-2) + 3$	-1	$(-2, -1)$
-1	$2(-1) + 3$	1	$(-1, 1)$
0	$2(0) + 3$	3	$(0, 3)$
1	$2(1) + 3$	5	$(1, 5)$

The graph of $y = 2x + 3$ is a line. The arrowheads indicate that the line extends indefinitely in both directions. An equation whose graph is a line is called a **linear equation,** or **first-degree equation.** You can use a graphing calculator or a graphing utility to graph a linear equation solved for y.

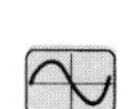

EXAMPLE 3 **Sketch the graph of $2x + 3y = 6$.**

$$2x + 3y = 6$$ *Solve for y in terms of x.*

$$3y = -2x + 6$$ *Add $-2x$ to each side.*

$$y = -\frac{2}{3}x + 2$$ *To simplify computations, choose multiples of 3 for values of x.*

x	$y = -\frac{2}{3}x + 2$	y	(x, y)
-3	$-\frac{2}{3}(-3) + 2$	4	$(-3, 4)$
0	$-\frac{2}{3}(0) + 2$	2	$(0, 2)$
3	$-\frac{2}{3}(3) + 2$	0	$(3, 0)$

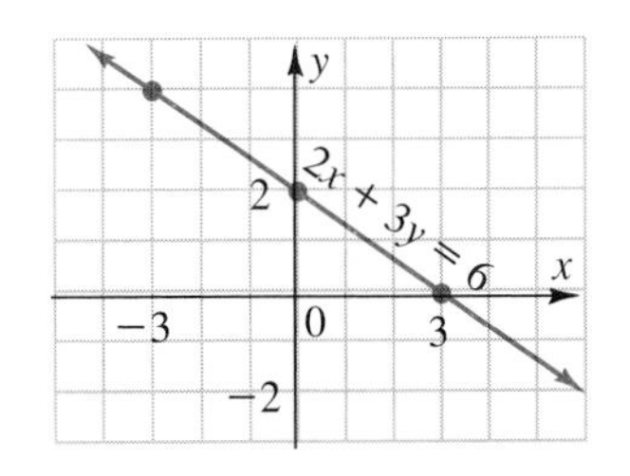

Since two points determine a line, it is necessary to plot only two points in order to graph a linear equation. However it is wise to plot a third point as a check. It is good practice to choose a positive, a negative, and zero value of x.

An equation is linear if it can be written in the **standard form** $Ax + By = C$, where A, B, and C are integers and A and B are not both zero. Thus, $x + 2y = 1$, $3x = 5$ and $x - y = 0$ are linear equations; $x^2 - 1 = 3$, $xy = 5$, and $\frac{1}{x} + y = 6$ are *not* linear.

EXAMPLE 4 **Sketch each graph.** **a.** $y = 2$ **b.** $x = -3$

a.

x	y	(x, y)
-2	2	$(-2, 2)$
0	2	$(0, 2)$
2	2	$(2, 2)$

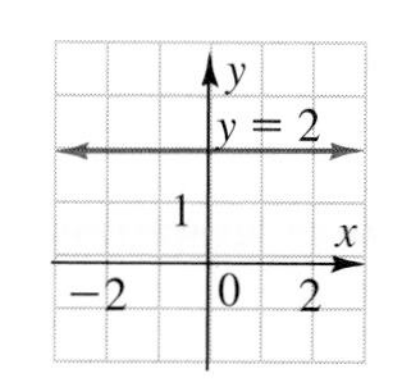

b.

x	y	(x, y)
-3	-2	$(-3, -2)$
-3	0	$(-3, 0)$
-3	2	$(-3, 2)$

The graph of a linear equation $y = k$ is a line parallel to the x-axis and k units from it. The graph of a linear equation $x = h$ is a line parallel to the y-axis and h units from it.

Study the graphs in Examples 2–4. Do the graphs of linear equations represent functions? The vertical line test shows that each of the graphs except the graph of $x = -3$ represents a function. In general, a linear equation $Ax + By = C$ represents a function except where $B = 0$.

To graph equations that are not linear, make a table of values and plot enough points to indicate the shape of the graph.

EXAMPLE 5 **Sketch the graph of $y = x^2$.**

x	y	(x, y)
-3	9	$(-3, 9)$
-2	4	$(-2, 4)$
-1	1	$(-1, 1)$
0	0	$(0, 0)$
1	1	$(1, 1)$
2	4	$(2, 4)$
3	9	$(3, 9)$

This graph is called a parabola.

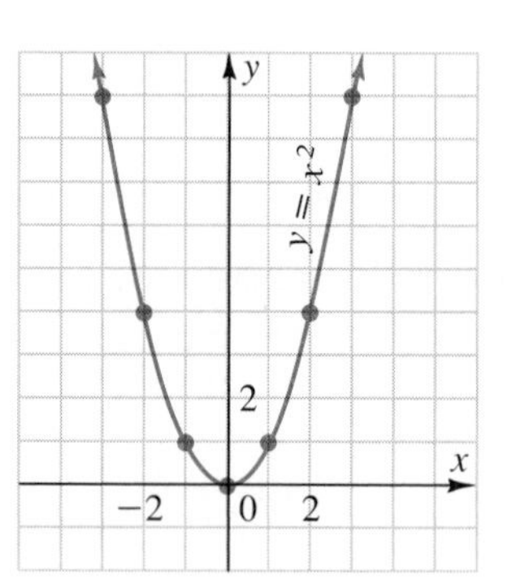

To Graph an Equation

- Make a table of values.
- Using graph paper, draw axes with arrowheads to show the positive directions. Label the axes with numbers and variables.
- Plot the ordered pairs from the table of values.
- Connect the points in order.
- Label the graph with the equation.

CLASS EXERCISES

1. **Thinking Critically** Most graphing calculators are designed to graph equations that are solved for y. What lines could not be graphed by this method?

Determine whether the ordered pair is a solution of the equation.

2. $y - x = 4$; $(2, 6)$

3. $x + y = -3$; $(-1, 2)$

4. $x - 3y = -1$; $(5, 2)$

5. $4y + x = 1$; $(-7, 2)$

6. $3x - 5y = -15$; $(0, 6)$

7. $3x - 3y = 9$; $(2, -1)$

Graph each equation.

8. $y = 2x$

9. $y = -3x$

10. $y = 3x - 2$

11. $y = -4x + 5$

12. $5x - 2y = -4$

13. $-2x + 5y = -10$

Determine which graphs represent *y* as a function of *x*.

14.

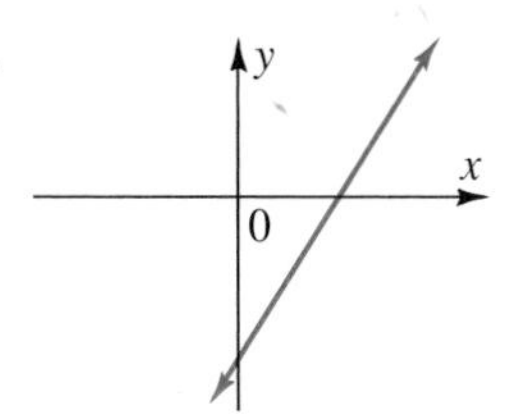

15.

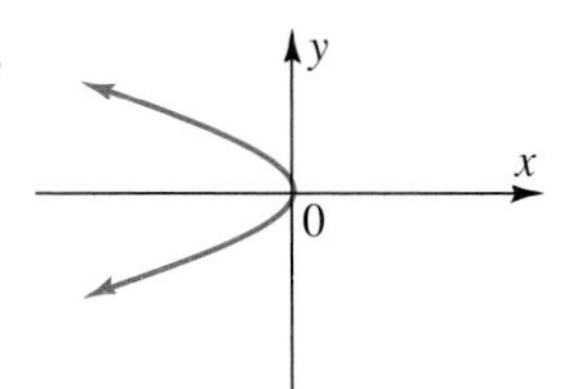

16.

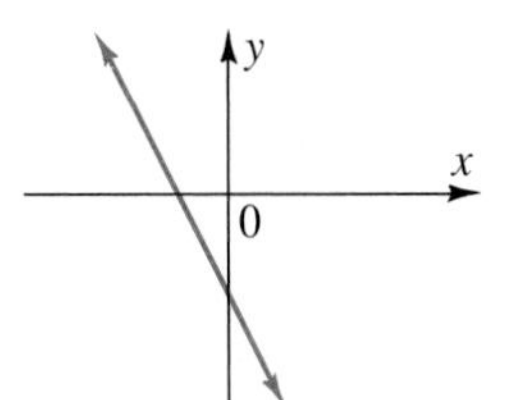

PRACTICE EXERCISES

Determine whether the ordered pair is a solution of the equation.

1. $x - y = 7$; $(2, -5)$

2. $x + y = 6$; $(-3, -9)$

3. $3y = 3x - 9$; $(2, -1)$

4. $-4x = 2y + 11$; $(-5, 7)$

5. $3x - 4y = 6$; $(-2, -3)$

6. $2x - 4y = 2$; $(1, -1)$

Graph each equation.

7. $y = 3x$

8. $y = -x$

9. $y = -2x + 5$

10. $y = -3x - 1$

11. $y = 4$

12. $x = -1$

13. $y - 3 = -2x$

14. $y + 4 = -3x$

15. $-2x + 3y = 9$

16. $-3x + 4y = -6$

17. $-x - 5y = -15$

18. $-x - 4y = -12$

Determine which graphs represent *y* as a function of *x*.

19.

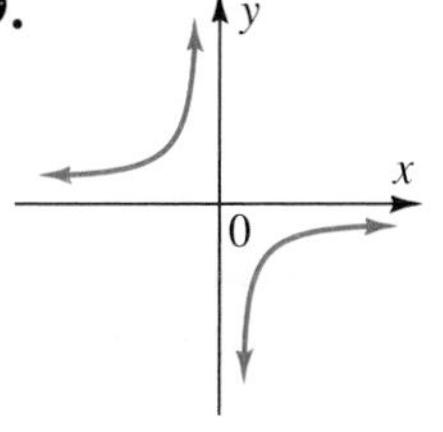

20.

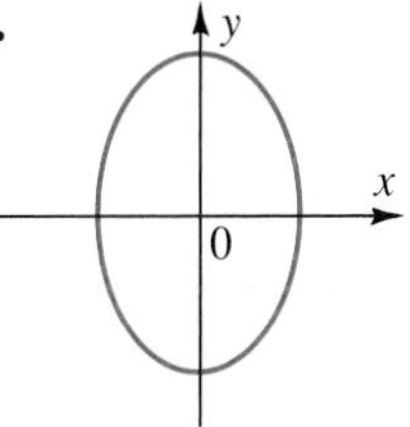

21.

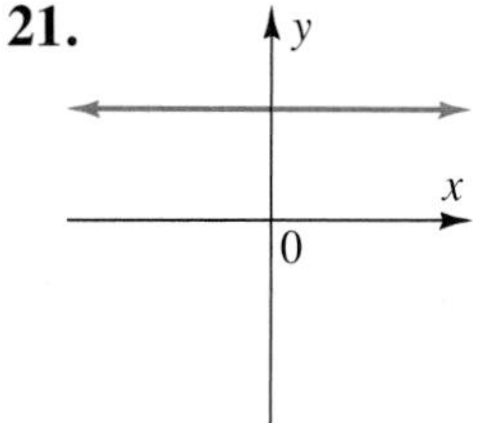

22.

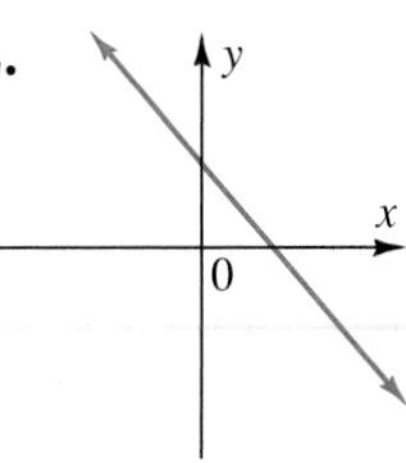

Determine whether the ordered pair is a solution of the equation.

23. $4x - 18 = 3y; \left(\frac{1}{2}, -\frac{16}{3}\right)$

24. $7 - 5y = 2x; \left(-\frac{1}{2}, \frac{6}{5}\right)$

25. $2x - 4y = -28; \left(\frac{3}{2}, \frac{31}{4}\right)$

26. $2x + 6y = -24; \left(\frac{2}{3}, -\frac{38}{9}\right)$

Graph each equation.

27. $3y - 2x = -12$

28. $2y + 3x = 4$

29. $4x + 5y = 20$

30. $4x - 3y = -6$

31. $y = 2x^2$

32. $y = x^2 - 1$

Determine which of the following are linear equations.

33. $3x - 5y = -15$

34. $xy = -2$

35. $y = -3x^2 + 1$

36. $15 - 5x = 3y$

37. $-2(x + y) = -4^2$

38. $x^2 + 3y = 5$

Graph each equation.

39. $\frac{2}{3}x + \frac{y}{3} = -\frac{1}{3}$

40. $\frac{3}{5}y - \frac{x}{5} = -\frac{6}{5}$

41. $y = |-3x|$

42. $y = |4x| - 1$

43. $y = -3x^2 + 1$

44. $y = x^2 - 2x - 8$

Applications

45. Earth science The number of blooms on a cactus is related to the number of days of sun it gets in a month by the equation $y = 7x - 1$, where x is the number of days of sun and y is the number of blooms. Graph this linear equation.

46. Cost analysis The cost of operating a car is related to the number of miles driven by the equation $y - 0.23x = 0$, where x is the number of miles driven and y is the cost. Graph this linear equation.

47. Biology The number of fish to be found in a pond is related to the number of plants in the pond by the equation $10x + 2y = -10$, where x is the number of plants and y is the number of fish. Graph this equation.

Developing Mathematical Power

Thinking Critically

48. In Example 5 on page 106, seven points were plotted. How could you have predicted that more than three points would be needed?

49. How many points would you have to plot in order to establish that an equation had a nonlinear graph?

3.4

Composition of Functions

Objectives: To find the composition of two functions
To evaluate composite functions

Functions are sometimes defined in terms of other functions. When this is done, it is important to determine the domain and range of each function.

Capsule Review

Recall that a relation is a set of ordered pairs. The set of first coordinates is the domain and the set of second coordinates is the range of the relation.

EXAMPLE **Write a rule for the relation $\{(-2, 8), (0, 4), (1, 2), (3, -2)\}$. State the domain and range.**

$\{(x, y): y = -2x + 4, x = -2, 0, 1, 3\}$
The domain is $\{-2, 0, 1, 3\}$.
The range is $\{8, 4, 2, -2\}$.

Write a rule for each relation and state the domain and range.

1. $\{(0, -5), (2, -3), (4, -1)\}$

2. $\{(-1, -6), (0, 0), (1, 6)\}$

3. $\{(-1, 3), (-2, 2), (3, 7)\}$

4. $\{(-3, 9), (0, 0), (2, -6)\}$

5. $\{(-1, -5), (0, -3), (1, -1)\}$

6. $\{(1, 1), (2, 0), (3, -1)\}$

The mapping diagram represents the linear function $y = 2x + 1$. Each number in the domain is mapped onto one and only one number in the range. It is common to represent functions by letters such as f, g, and h. In function notation, $y = 2x + 1$ can be expressed as $f(x) = 2x + 1$. The symbol $f(x)$, read "f of x," is used to represent the values of the function at x.

Domain, x	f	Range, y or $f(x)$
0	→	1
1	→	3
2	→	5
3	→	7
⋮		⋮

EXAMPLE 1 **Evaluate each function for the given value of x.**

a. If $f(x) = 3x + 4$, find $f(2)$.
b. If $g(x) = 2x^2 - 3x + 1$, find $g(-3)$.
c. If $h(x) = -4x - 7$, find $h(a + 1)$.

a. $f(x) = 3x + 4$
$f(2) = 3(2) + 4 = 10$ *Substitute 2 for x.*

b. $g(x) = 2x^2 - 3x + 1$
$g(-3) = 2(-3)^2 - 3(-3) + 1 = 28$ *Substitute −3 for x.*

c. $h(x) = -4x - 7$
$h(a + 1) = -4(a + 1) - 7 = -4a - 11$ *Substitute (a + 1) for x.*

$f(x) = |x|$ is called the **absolute value function.** The graph of this function is a pair of rays with the origin as their common endpoint.

EXAMPLE 2 **Sketch the graph of $f(x) = |x|$. Determine the domain and range.**

x	$f(x)$	(x, y)
−3	3	(−3, 3)
−1	1	(−1, 1)
0	0	(0, 0)
1	1	(1, 1)
3	3	(3, 3)

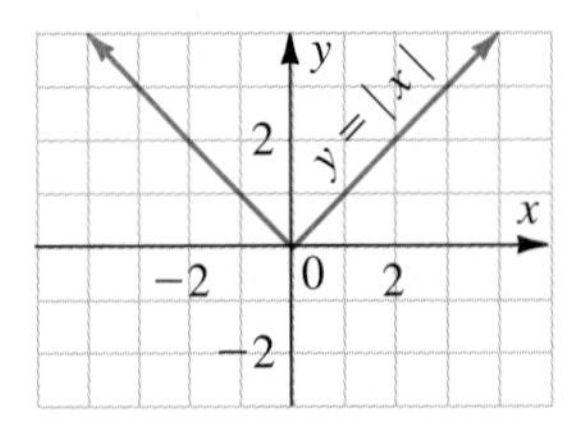

The domain is the set of all real numbers.
The range is the set of all nonnegative real numbers.

Given two functions $f(x)$ and $g(x)$, a **composite function** is written

$(g \circ f)(x) = g[f(x)]$ and is read "g of f of x."

The *composition* of two functions, $(g \circ f)(x)$, is the operation of applying $f(x)$ to values of x, and then $g(x)$ to values of $f(x)$.

EXAMPLE 3 **Given $f(x) = x^2$ and $g(x) = x + 3$, evaluate each composite function.**
a. $g[f(2)]$ **b.** $f[g(2)]$

a. $f(x) = x^2$ *First find f(2).*
$f(2) = 2^2 = 4$

$g(x) = x + 3$ *Then find g[f(2)].*
$g[f(2)] = g(4) = 4 + 3 = 7$

b. $g(x) = x + 3$ *First find g(2).*
$g(2) = 2 + 3 = 5$

$f(x) = x^2$ *Then find f[g(2)].*
$f[g(2)] = f(5) = 5^2 = 25$

Notice that in Example 3 $g[f(2)]$ and $f[g(2)]$ yielded two different values. Therefore, composition of functions is not commutative. That is, $g[f(x)]$ is not necessarily equal to $f[g(x)]$.

Calculators with retrieval capability are helpful when you are evaluating composite functions for more than one value.

EXAMPLE 4 **Given $f(x) = 2x + 5$ and $g(x) = 3x^2 + 1$, find:**
a. $(g \circ f)(-45)$ **b.** $(g \circ f)(21)$

a. The composite function is $(g \circ f)(x) = 3(2x + 5)^2 + 1$
Enter the composite function using −45 for x.
$(g \circ f)(-45) = 21{,}676$

b. Retrieve the last expression; change −45 to 21.
$(g \circ f)(21) = 6628$

```
3(2*⁻45+5)²+1
        21676
3(2*21+5)²+1
         6628
```

CLASS EXERCISES

Given $f(x) = 2x + 4$ and $g(x) = 3x$, find each value.

1. $f(1)$
2. $g(0)$
3. $f(-2)$
4. $g(-3)$
5. $g(4)$
6. $f(-5)$
7. $g(-2)$
8. $f(0)$

Given $f(x) = 2x + 1$ and $g(x) = 4x$, find each value.

9. $g[f(-3)]$
10. $f[g(-3)]$
11. $f[g(0)]$
12. $g[f(k)]$
13. $f[g(k)]$

PRACTICE EXERCISES

Use technology where appropriate.

Evaluate each pair of functions for $x = -1, 0, 2, a$.

1. $f(x) = -3x,\ g(x) = x + 2$
2. $f(x) = 4x,\ g(x) = x - 3$
3. $f(x) = x^2,\ g(x) = x + 5$
4. $f(x) = x^3,\ g(x) = x - 2$
5. $f(x) = 3x^3 + 1,\ g(x) = 2x$
6. $f(x) = -x^2,\ g(x) = -2x$
7. $f(x) = 2x^2 + 4x - 1,\ g(x) = 3x^2$
8. $f(x) = 3x^2 + 2x,\ g(x) = -x^2 + 2x$

Evaluate $g[f(1)]$, $g[f(-2)]$, and $g[f(0)]$ for each pair of functions.

9. $f(x) = -3x,\ g(x) = x + 2$
10. $f(x) = 4x,\ g(x) = x - 3$
11. $f(x) = 5x,\ g(x) = x + 5$
12. $f(x) = -4x,\ g(x) = x - 2$
13. $f(x) = 2x^2,\ g(x) = 2x$
14. $f(x) = -x^2,\ g(x) = -2x$
15. $f(x) = 2x^2,\ g(x) = 3x + 2$
16. $f(x) = 3x^2,\ g(x) = -x$
17. $f(x) = 2x,\ g(x) = -3x$
18. $f(x) = -3x,\ g(x) = 4x$
19. $f(x) = 2x + 1,\ g(x) = 4x + 2$
20. $f(x) = 3x + 5,\ g(x) = x - 1$
21. $f(x) = -3x + 5,\ g(x) = -2x + 1$
22. $f(x) = 5x - 2,\ g(x) = -x - 1$
23. $f(x) = x,\ g(x) = 2x$
24. $f(x) = 2x,\ g(x) = 3x$

Evaluate $f[g(x)]$ and $g[f(x)]$ for $x = -2, 0, 1, k$.

25. $f(x) = x^2,\ g(x) = x + 1$
26. $f(x) = 2x^2,\ g(x) = x - 2$
27. $f(x) = 2x^2,\ g(x) = 2x + 3$
28. $f(x) = 3x^2,\ g(x) = 4x - 1$
29. $f(x) = 3x^2,\ g(x) = -2x - 2$
30. $f(x) = -2x^2,\ g(x) = -2x + 5$
31. $f(x) = x^2 + 2x,\ g(x) = -3x + 2$
32. $f(x) = -x^2 - 4,\ g(x) = 2x + 3$

Find $(g \circ f)(x)$ and $(f \circ g)(x)$ for each pair of functions.

33. $f(x) = x^2,\ g(x) = 2x$
34. $f(x) = 2x^2,\ g(x) = 3x$
35. $f(x) = 3x^2,\ g(x) = -4x$
36. $f(x) = -x^2,\ g(x) = -2x$

Sketch the graph of each function. Determine the domain and range.

37. $f(x) = |x| + 1$ **38.** $f(x) = |x| - 1$ **39.** $f(x) = -|-x|$ **40.** $f(x) = 2 - |x|$

Given $f(x) = 2x + 3$ and $g(x) = x^2$, find each value.

41. $(f \circ g)(x^2)$

42. $(g \circ f)(x^2)$

43. $(g \circ g)(x)$

44. $(f \circ f)(x)$

45. $(f \circ g)(x + 1)$

46. $(g \circ f)(x + 1)$

47. If $f(x) = 2x - 3$, find:
$\dfrac{f(1 + h) - f(1)}{h}, h \neq 0$

48. If $f(x) = 4x - 1$, find:
$\dfrac{f(a + h) - f(a)}{h}, h \neq 0$

Applications

49. Cost Function A record sells for \$3 wholesale. The price the store pays is determined by the function $f(x) = x + 3$, where x is the wholesale price. The price the store charges is determined by $g(x) = x + 4$, where x is the price the store pays. Find the price the customer pays.

50. Cost Function Jane supplies balloons for parties. The manufacturer charges her \$0.50 above wholesale for each balloon. Jane then charges her customers \$1 above what she pays for each balloon. Write a function, $f(x)$, for the price Jane pays and a function, $g(x)$, for the price she charges her customers. Explain what $(g \circ f)(x)$ represents.

51. Biology The weight of the brain of an infant is given by $f(w) = 0.025w$, where w is the birth weight in pounds. The average weight of the brain of a 3-year-old can then be determined by $g(w) = w + 0.5$, where w is the brain weight at birth in pounds. Find the brain weight of a 3-year-old whose birth weight was 6.3 lb to the nearest hundredth.

Mixed Review

Solve. If an equation has no solution, write *no solution*.

52. $|x + 4| = 12$

53. $|3x - 7| = 16$

54. $|3x - 5| = 12x + 4$

55. $|4 - 5x| = 19x + 18$

56. $|2 + 7x| = -11$

57. $4 + |3x - 6| = 28$

58. The cheerleading coach set up a squad with twice as many girls as boys. But at tryouts, there were still 14 girls too many. If 48 girls in all were interested in cheerleading, how many boys made the squad?

59. The average of the heights of Abraham Lincoln and the Sears Tower in Chicago is 730 ft. If Abraham Lincoln's height was about 6 ft, about how tall is the Sears Tower?

Inverse Functions

Objectives: To find the inverse of a relation or a function
To determine if the inverse of a function is a function

Relationships between variables can be expressed in more than one way. The numbers and letters on the buttons or dial of a telephone are related. You can choose a letter and then find the corresponding number or you can start with a number and find the corresponding letters.

Capsule Review

To solve an equation for a given variable, isolate the variable on one side of the equals sign.

EXAMPLE **Solve for y: $3y - 1 = x$**

$$3y - 1 = x$$

$$3y = x + 1 \quad \textit{Add 1 to both sides.}$$

$$y = \frac{x}{3} + \frac{1}{3} \quad \textit{Divide both sides by 3.}$$

Solve each equation for y.

1. $2y = x$ **2.** $-3y + 3 = 6x$ **3.** $4y = x - 3$

4. $5x = 2y$ **5.** $7x - 5y = 20$ **6.** $y^2 = x$

At the right is a picture of two of the buttons on a pushbutton telephone. The letters and numbers form a relation.

EXAMPLE 1 **Write the ordered pairs of the relation formed by the letters (domain) and numbers (range) on the two telephone buttons. Draw a mapping diagram and determine if this relation is a function.**

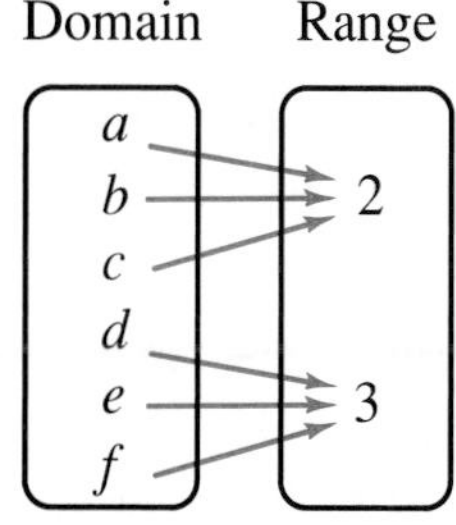

The set of ordered pairs of this relation is

$\{(a, 2), (b, 2), (c, 2), (d, 3), (e, 3), (f, 3)\}$.

Since each letter in the domain is mapped onto one and only one number in the range, this relation is a function.

Now look at the relationship between the numbers and letters on the telephone buttons in another way. Let the domain be the set of numbers and the range be the set of letters. This relation is called the *inverse* of the first relation. The **inverse** of a relation is the relation obtained by interchanging the first and second coordinates in every pair of the original relation. The domain of a relation is the range of its inverse relation. The range of a relation is the domain of its inverse relation.

EXAMPLE 2 **Write the ordered pairs of the inverse relation of the telephone buttons, letting the numbers be the domain and the letters the range. Use a mapping diagram to determine if this relation is a function.**

The set of ordered pairs of this relation is

$\{(2, a), (2, b), (2, c), (3, d), (3, e), (3, f)\}$.

Each number in the domain is mapped onto more than one letter in the range.

This relation is not a function.

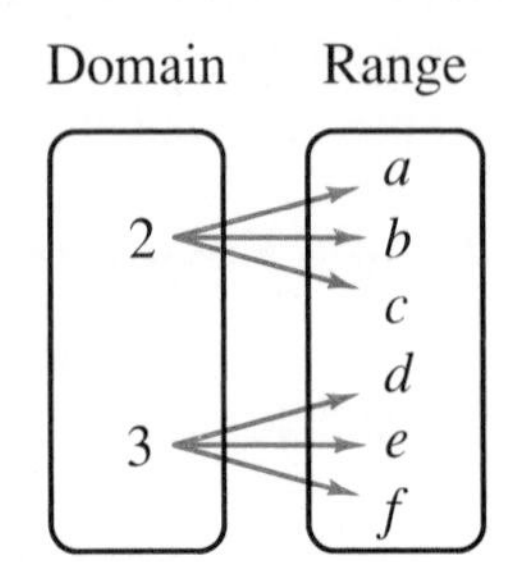

The relation in which the letters are mapped onto the numbers is a function, but the inverse relation in which the numbers are mapped onto the letters is *not* a function. In general, the inverse of a relation is a relation but the inverse of a function is not necessarily a function.

EXAMPLE 3 **Find the inverse of the function $\{(3, 4), (5, 6), (7, 8), (9, 10)\}$. State the domain and range of this inverse. Use a mapping diagram to determine if the inverse is a function.**

Interchange the first and second coordinates in each pair. The inverse of the function is $\{(4, 3), (6, 5), (8, 7), (10, 9)\}$.

Domain: $\{4, 6, 8, 10\}$ Range: $\{3, 5, 7, 9\}$

Since each element in the domain of the inverse maps onto one and only one element in the range, the inverse of the original function is also a function.

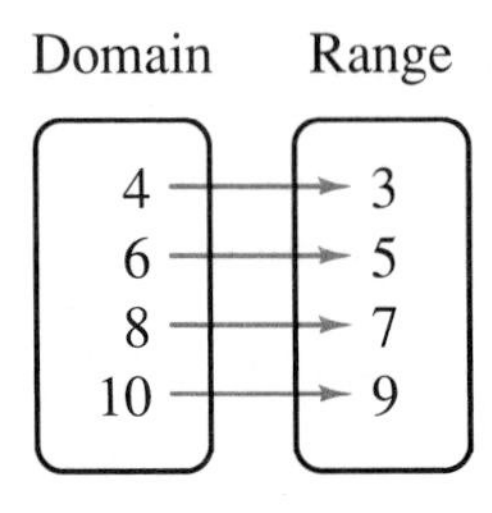

When the inverse of a function is a function, it is called the **inverse function.** If f is the function, then its inverse function is denoted by f^{-1}. Notice that if the inverse of a function is also a function, there must be a *one-to-one correspondence* between the elements of the domain and range.

The function $f(x) = 2x + 4$ can be written as the equation $y = 2x + 4$. To find the inverse of a function written as an equation, interchange the two variables x and y and solve for y.

EXAMPLE 4 **Find the inverse of the function $y = 2x + 4$. Sketch the graphs of f, f^{-1}, and $y = x$ on the same coordinate plane.**

Original function f: $y = 2x + 4$
Domain of f: $\{x: x \text{ is real}\}$ Range of f: $\{y: y \text{ is real}\}$

To find the inverse of the function, interchange x and y and solve for y.

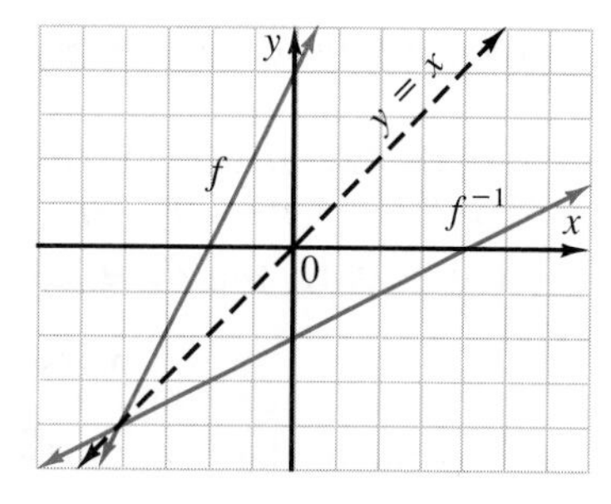

$$x = 2y + 4$$
$$x - 4 = 2y$$
$$\frac{x}{2} - 2 = y$$

Therefore the inverse function, f^{-1}, is

$$y = \frac{x}{2} - 2$$

Domain of f^{-1}: $\{x: x \text{ is real}\}$ Range of f^{-1}: $\{y: y \text{ is real}\}$

Notice that if the coordinate plane were folded along the line $y = x$, then the graphs of the function, f, and its inverse, f^{-1}, would coincide. That is, the graphs of f and f^{-1} are symmetric with respect to the line $y = x$. You can use a graphing calculator or a computer to draw the graph of f and f^{-1}.

EXAMPLE 5 **Find the inverse of the function $y = x^2$. Determine if the inverse of this function is also a function. Graph f, f^{-1}, and $y = x$ on the same coordinate plane.**

$x = y^2$ *Interchange x and y.*
$y = \pm\sqrt{x}$ *Solve for y.*

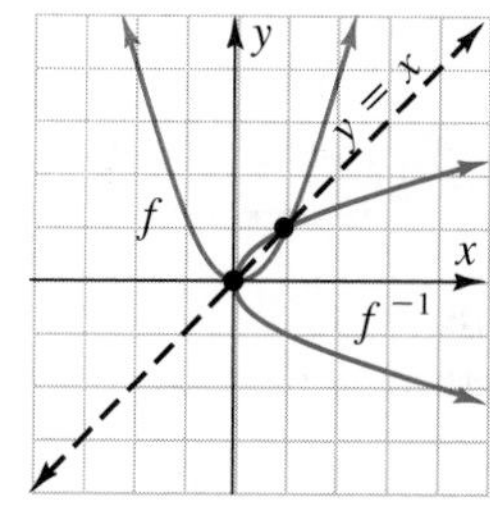

For each value of x $(x > 0)$, there are two values of y. Therefore, the inverse is not a function. The graphs of $y = x^2$ and $y = \pm\sqrt{x}$ are, however, symmetric with respect to the line $y = x$.

When the inverse of a function is not a function, it is sometimes possible to make it a function by restricting the values of the domain of the function. For example, if the domain of $y = x^2$ is restricted to the nonnegative real numbers, then the range of the inverse, $y = \sqrt{x}$, is also restricted to the nonnegative real numbers and $y = \sqrt{x}$ is a function.

CLASS EXERCISES

Find the inverse of each relation or function.

1. $\{(2, 5), (3, 7), (9, 8)\}$
2. $\{(-3, 4), (-2, 5), (-1, 0)\}$
3. $y = 3x + 1$
4. $y = 2x - 1$
5. $y = -3x + 4$
6. $y = -2x + 5$

Find the inverse of each function and determine whether the inverse is a function. Graph f, f^{-1}, and $y = x$ on the same coordinate plane.

7. $y = 3x - 1$ **8.** $y = -2x - 3$ **9.** $y = 2x^2$ **10.** $y = x^3$

PRACTICE EXERCISES

Find the inverse of each relation or function.

1. $\{(2, 4), (4, 8), (8, 16)\}$ **2.** $\{(-1, -2), (-2, -4), (-3, -6)\}$

3. $y = 4x - 1$ **4.** $y = -x + 2$ **5.** $y = 5x - 2$

6. $y = -5x + 6$ **7.** $y = -3x + 3$ **8.** $y = -2x - 5$

Find the inverse of each function and state its domain and range. Use a mapping diagram to determine if the inverse is also a function.

9. $\{(3, 3), (4, 4), (5, 5), (6, 6)\}$ **10.** $\{(-1, 2), (-2, 4), (-3, 6), (-4, 8)\}$

11. $\{(1, 3), (2, 3), (4, 5), (9, 5)\}$ **12.** $\{(0, 0), (1, 0), (2, 3), (3, 3)\}$

13. $\{(d, 3), (e, 3), (f, 3), (g, 4)\}$ **14.** $\{(j, 1), (k, 2), (l, 3), (m, 4)\}$

Find the inverse of each function. Graph f, f^{-1}, and $y = x$.

15. $y = 2x - 3$ **16.** $y = 7x - 3$ **17.** $y = 9x - 4$

18. $y = -x$ **19.** $y = -4x + 13$ **20.** $y + 7x = 2$

Find the inverse of each relation or function.

21. $y = 2x + 3, x = 1, 2, 3$ **22.** $y = -3x + 1, x = 0, 1, 2$

23. $y = 2x - 5, x = -1, -2, -3$ **24.** $y = -2x - 1, x = -2, -3, -4$

Find the inverse of each function and state its domain and range. Determine if the inverse is also a function.

25. $f(x) = 3x + 4$ **26.** $f(x) = 4x - 5$ **27.** $f(x) = -x + 7$

28. $f(x) = -2x + 3$ **29.** $f(x) = \frac{2}{3}x - 3$ **30.** $f(x) = -\frac{3}{4}x + 2$

31. $\{(x, y): y = 2x$, x is a positive integer less than $3\}$

32. $\{(x, y): y = 3x - 1$, x is a negative integer greater than $-5\}$

33. $\{(x, y): y = -3x + 4$, x is a positive integer less than $6\}$

34. $\{(x, y): y = -5x + 2$, x is a negative integer greater than $-7\}$

In Exercises 35–36, use $\{1, 2, 3, \ldots, 10\}$ as the domain.

35. Construct a function whose inverse is also a function.

36. Construct a function whose inverse is not a function.

37. The relation $\{(2, 3), (x, 4), (5, 6)\}$ is a function. What values, if any, may x *not* assume?

38. The inverse of the function $\{(2, -3), (3, y), (4, 5)\}$ is a function. What values, if any, may y *not* assume?

Determine the domain and range of each function so that the inverse is also a function.

39. $f(x) = 2x^2$

40. $f(x) = x^3$

Applications

41. Geometry The formula for the area of a square is $y = x^2$. Find the inverse of this function and determine if it is also a function.

42. Geometry The formula for the circumference of a circle is $y = 2\pi r$. Find the inverse of this function and determine if it is also a function.

43. Chemistry The formula for converting from Celsius to Fahrenheit temperatures is $y = \frac{9}{5}x + 32$. Find the inverse of this function and determine if the inverse of this function is also a function.

44. Banking For a fixed interest rate and time period the simple interest on a loan can be computed using the formula $y = 0.07x$. Find the inverse of this function and determine if the inverse is also a function.

TEST YOURSELF

Graph each ordered pair and name the quadrant in which it lies.

1. $(-2, 4)$ **2.** $(1, 3)$ **3.** $(-2, -3)$ 3.1

Determine which of the following relations are functions.

4. $L = \{(2, 3), (4, 5), (6, 9)\}$ **5.** $K = \{(2, 3), (2, -3), (5, 0), (5, -1)\}$ 3.2

Graph each linear equation. Determine which, if any, are not functions.

6. $x + 3y = 6$ **7.** $-4x + y = -8$ **8.** $x - 3 = 0$ **9.** $y = -2$ 3.3

Given $f(x) = 2x + 3$ and $g(x) = 4x$, evaluate each of the following.

10. $g[f(2)]$ **11.** $g[f(-3)]$ **12.** $f[g(-3)]$ 3.4

Find the inverse of each function.

13. $M = \{(4, 5), (3, 2), (1, 0)\}$ **14.** $y = 2x + 4$ 3.5

3.6

The Slope of a Line

Objectives: To find the slope of a line
To use the slope and y-intercept to write an equation of a line and to graph a linear equation

A builder can determine the R-factor (the amount of insulation provided) of various styrofoam sheets using the graph of the linear equation $y = 4.2x$, where y is the R-factor and x is the thickness of the styrofoam.

The graph of a linear equation is a line. The line may be parallel to an axis or it may cross both axes.

Capsule Review

The length of line segment AB, parallel to the x-axis, is $|x_2 - x_1|$.

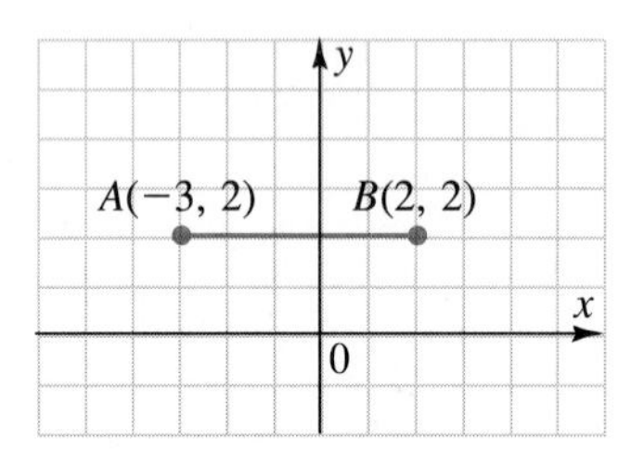

$AB = |x_2 - x_1| = |2 - (-3)| = 5$

The length of line segment CD, parallel to the y-axis, is $|y_2 - y_1|$.

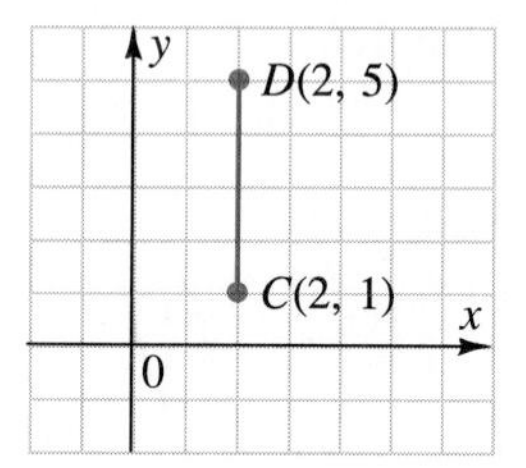

$CD = |y_2 - y_1| = |5 - 1| = 4$

Find the length of the line segment joining each pair of points.

1. (2, 4), (7, 4)
2. (−2, 9), (−2, −1)
3. (6, −1), (0, −1)
4. (3, 6), (3, −9)
5. (0, 5), (0, −3)
6. (7, 0), (−3, 0)

The steepness of a line is measured by the ratio of the change in the vertical distance, $y_2 - y_1$, to the change in the horizontal distance, $x_2 - x_1$, between any two points on the line. This measure is the **slope** of the line and is denoted by the symbol m.

$$m = \frac{\text{change in vertical distance}}{\text{change in horizontal distance}}$$

$$= \frac{y_2 - y_1}{x_2 - x_1}, \quad x_1 \neq x_2$$

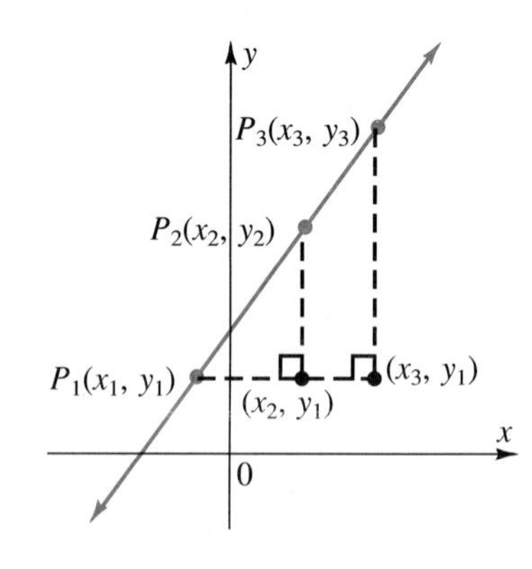

The triangles shown at the bottom of page 118 are similar, so their sides are proportional and $\frac{y_2 - y_1}{x_2 - x_1} = \frac{y_3 - y_1}{x_3 - x_1}$. Thus, the slope of a line is the same, regardless of which two points are chosen.

EXAMPLE 1 **Find the slope of the line passing through the given points:**
a. (1, 2), (2, 5) **b.** (−3, 5), (1, 2)

a. (1, 2), (2, 5)

$$m = \frac{y_2 - y_1}{x_2 - x_1} = \frac{5 - 2}{2 - 1} = \frac{3}{1} = 3$$

Notice the line rises to the right and the slope is *positive*.

b. (−3, 5), (1, 2)

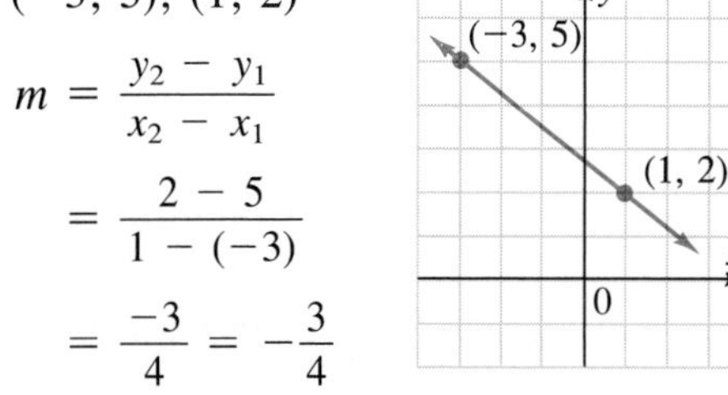

$$m = \frac{y_2 - y_1}{x_2 - x_1} = \frac{2 - 5}{1 - (-3)} = \frac{-3}{4} = -\frac{3}{4}$$

Notice the line falls to the right and the slope is *negative*.

Recall that the graph of $y = k$ is a horizontal line k units from the x-axis. The graph of $x = h$ is a vertical line h units from the y-axis.

EXAMPLE 2 **Find the slope of each line:** **a.** $y = 3$ **b.** $x = 4$

a.

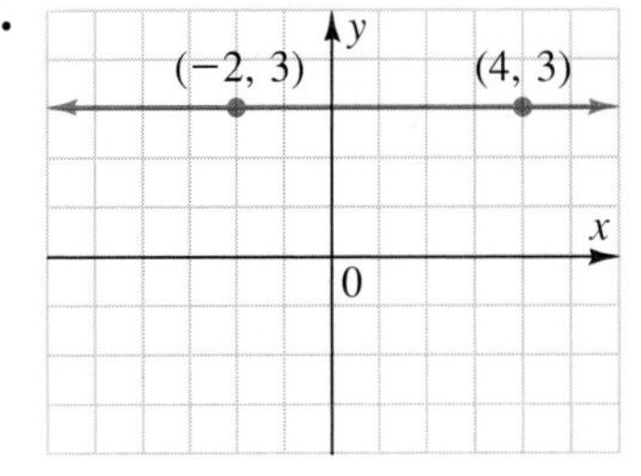

The y-coordinate of every point on the line is the same.

$$m = \frac{y_2 - y_1}{x_2 - x_1} = \frac{3 - 3}{4 - (-2)} = 0$$

The slope of a horizontal line is 0.

b.

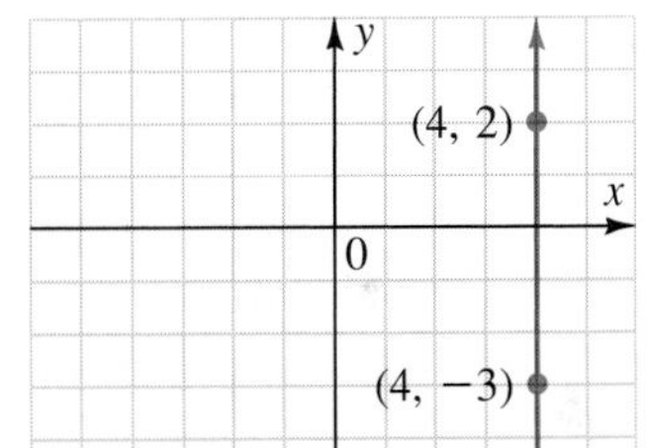

The x-coordinate of every point on the line is the same.

$$m = \frac{y_2 - y_1}{x_2 - x_1} = \frac{2 - (-3)}{4 - 4} = \frac{5}{0}$$

The slope of a vertical line is undefined since division by zero is not defined.

The y-intercept of a line is the y-coordinate of the point at which the line crosses the y-axis. The x-coordinate of this point is always 0. The x-intercept of a line is the x-coordinate of the point at which the line crosses the x-axis, and the y-coordinate of this point is always 0.

The graph of the line at the right has slope m and y-intercept b. Choose any point $P(x, y)$ on the line.

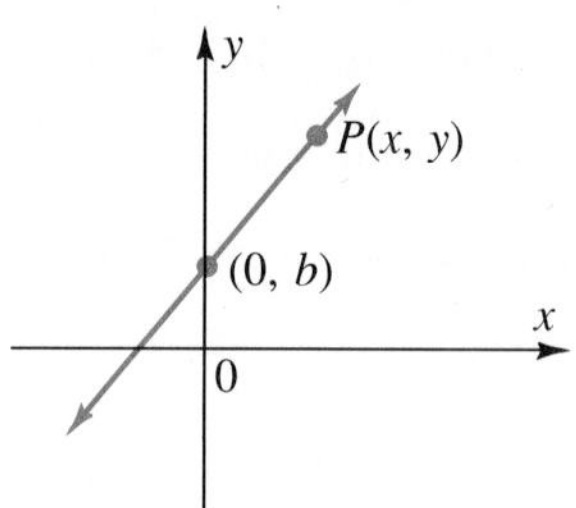

Then $$m = \frac{y - b}{x - 0}$$

$$y - b = mx$$
$$y = mx + b$$

The equation of a line with slope m and y-intercept b is

$$y = mx + b$$

slope (pointing to m), y-intercept (pointing to b)

This is called the **slope-intercept form** of a linear equation.

EXAMPLE 3 **Write the equation of a line with slope −2 and y-intercept 3.**

$y = mx + b$ *Use the slope-intercept form.*
$y = -2x + 3$ *Substitute −2 for m and 3 for b.*

You can determine the slope and the y-intercept of a line from a given equation by writing the equation in the slope-intercept form.

EXAMPLE 4 **Find the slope and y-intercept of the line $y - 3x = -4$.**

$y - 3x = -4$
$y = 3x - 4$ *Solve for y.*
$y = mx + b$ The slope is 3 and the y-intercept is −4.

The slope and y-intercept can be used to graph a line.

EXAMPLE 5 **Sketch the graph of $y = 3x - 4$.**

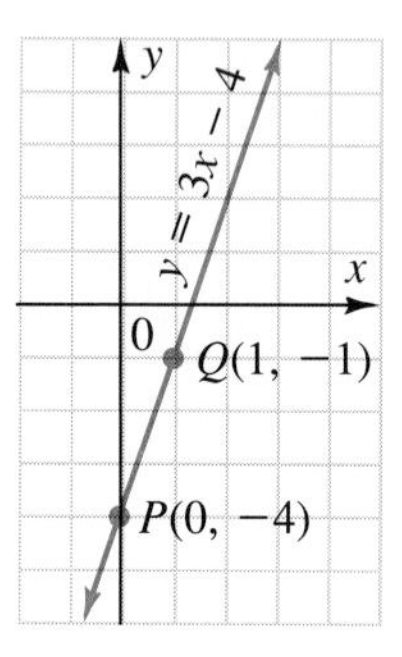

The y-intercept, b, is −4; plot the point $P(0, -4)$. Since $m = 3 = \frac{3}{1}$, move 3 units up and 1 unit to the right from $P(0, -4)$ to $Q(1, -1)$.

Draw the line through points P and Q.

CLASS EXERCISES

Thinking Critically

1. What is the slope of the x-axis? the y-axis?
2. When would a line with undefined slope intersect the y-axis?
3. When would a line with zero slope have the y-intercept 0?

Find the slope of the line passing through the given points.

4. (2, 5), (3, 7) **5.** (4, 3), (5, 8) **6.** (−1, 5), (−2, 3)

Write an equation of a line with the given slope and y-intercept.

7. $m = -2$, $b = 3$ **8.** $m = 0$, $b = 2$ **9.** $m = \frac{1}{2}$, $b = -2$

Find the slope and y-intercept for each line.

10. $3y = 3x - 6$ **11.** $2y + 3x = 4$ **12.** $3y - 2x = -3$

PRACTICE EXERCISES

Find the slope of the line passing through the given points.

1. (1, 3), (2, 4) **2.** (1, 5), (2, 6) **3.** (1, −3), (2, 1)

4. (−3, −6), (−5, 2) **5.** (−1, 3), (−4, −6) **6.** (−1, −6), (−6, −1)

Determine whether the slope of the line is positive, negative, zero, or undefined.

7.

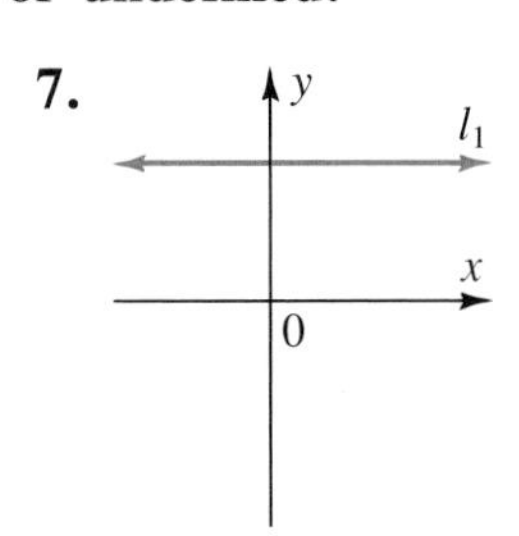

8.

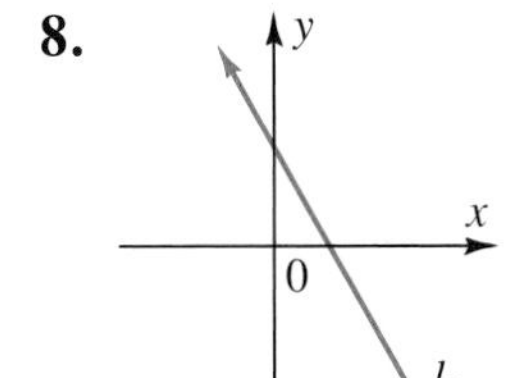

9.

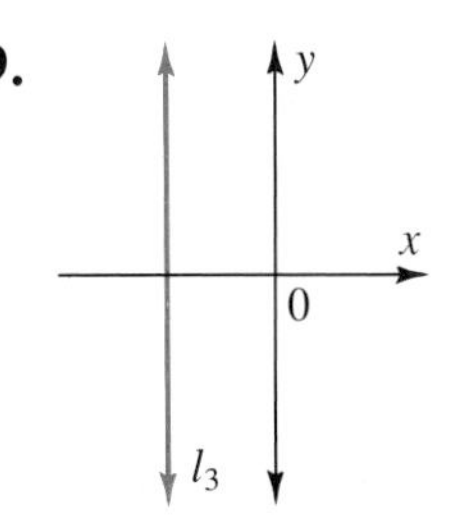

10.

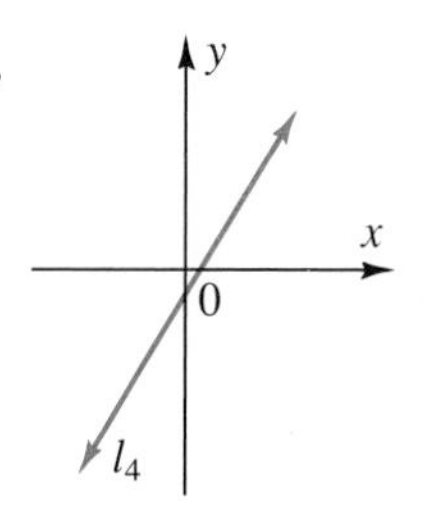

Write an equation of a line with the given slope and y-intercept.

11. $m = 1$, $b = 5$ **12.** $m = -2$, $b = -3$ **13.** $m = 1$, $b = -4$ **14.** $m = -2$, $b = 5$

Find the slope and y-intercept for each line.

15. $y = 3x - 4$ **16.** $y = -2x + 4$ **17.** $y - 4x = -1$ **18.** $y - 2x = -2$

Sketch the graph of each line.

19. $y = 2x - 1$ **20.** $y = -\frac{2x}{3} + 2$

Find the slope of the line passing through the given points.

21. (0, −3), (−2, −4) **22.** (−4, −2), (−2, −5) **23.** (−3, −4), (−4, −3)

24. (−1, 4), (−3, 2) **25.** (2, 5), (2, −4) **26.** (4, −5), (−1, −5)

Write an equation of a line with the given slope and *y*-intercept.

27. $m = -\frac{3}{2}, b = -1$

28. $m = -\frac{2}{5}, b = 3$

Find the slope and *y*-intercept for each line.

29. $2x + 3y = 6$

30. $-2x + 5y = -5$

31. $5x - 2y = -4$

32. $-3x + 4y = 4$

33. $3x - 3y = 0$

34. $4x - 2y = 0$

Determine whether the slope of each line is positive, negative, zero, or undefined.

35. The line has no y-intercept.

36. The line passes through the points (a, d) and (b, d), $a \neq b$.

37. The line passes through the points (a, b) and (a, c).

38. The line passes through the points (x, y) and $(x + a, y + a)$.

39. The line does not cross the x-axis.

40. The line passes through points $(-x, y)$ and $(x, -y)$, $x > 0$, $y > 0$.

Applications

41. Construction Find the slope of a road if a line representing the road passes through points (0, 0) and (100, 6).

42. Construction The R-factor, y, of styrofoam can be determined by the equation $y = 4.2x$, where x is the thickness. Find the slope and y-intercept of this line and sketch the graph.

Mixed Review

Write an algebraic expression for each of the following.

43. The product of twice some number and the square of a different number.

44. The sum of four consecutive odd integers.

45. The average of four numbers, each of which is twice the previous number.

46. Half the age of a woman whose age ten years ago was three times some number.

47. The product of some number and half its additive inverse.

48. The product of some number and twice its multiplicative inverse.

49. The absolute value of the sum of a number and the additive inverse of its square.

Parallel and Perpendicular Lines

Objectives: To determine if two lines are parallel or perpendicular
To write an equation of a line parallel to or perpendicular to a given line when the y-intercept is known

From the equation of one line it is possible to find a family of lines parallel to or perpendicular to the given line. When linear programming problems are solved graphically, families of parallel lines can be used to find maximum or minimum points.

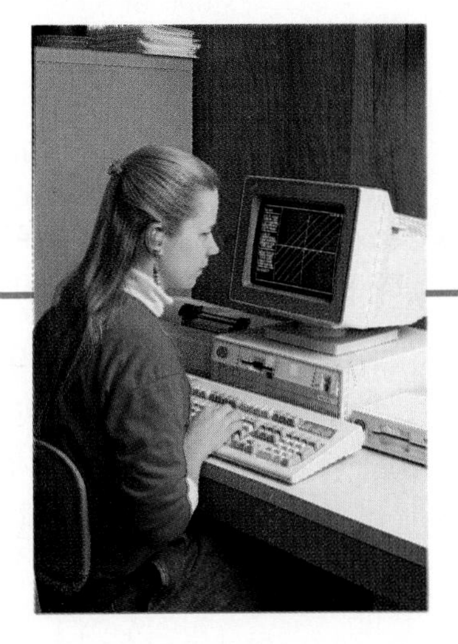

Capsule Review

If the product of two numbers is -1, the numbers are said to be *negative reciprocals.* In general, since $x \cdot -\frac{1}{x} = -1$, $-\frac{1}{x}$ is the negative reciprocal of x, $x \neq 0$.

Find the negative reciprocal of each number.

1. 4 **2.** $\frac{3}{2}$ **3.** -5

4. $-\frac{2}{3}$ **5.** $\frac{a}{b}$, $a \neq 0$, $b \neq 0$ **6.** $-\frac{1}{c}$, $c \neq 0$

Two distinct lines are parallel if and only if their slopes are equal or are both undefined. To see why this is so, note lines l_1 and l_2 in the figure. If you know the lines are parallel, then the two right triangles must be similar, so

$$\frac{y_2 - y_1}{x_2 - x_1} = \frac{y_4 - y_3}{x_4 - x_3}$$

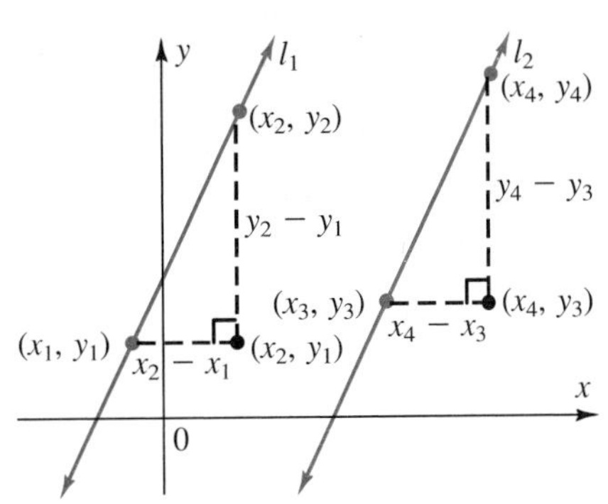

That is, the slopes of l_1 and l_2 are equal. Similarly, if you know that the slopes of l_1 and l_2 are equal, then

$$\frac{y_2 - y_1}{x_2 - x_1} = \frac{y_4 - y_3}{x_4 - x_3}$$

This implies that the triangles are similar, so lines l_1 and l_2 are parallel or coincident.

EXAMPLE 1 **Line k_1 passes through the points (4, 7) and (6, −1), and another line, k_2, passes through the points (−2, 2) and (0, −6). Determine if these two lines are parallel.**

slope of $k_1 = \frac{-1 - 7}{6 - 4} = -\frac{8}{2} = -4$ *Find the slope of line k_1.*

slope of $k_2 = \frac{-6 - 2}{0 - (-2)} = -\frac{8}{2} = -4$ *Find the slope of line k_2.*

Since the slopes of the lines are equal, the lines are parallel.

It can be shown that two nonvertical lines are perpendicular if and only if their slopes are negative reciprocals. Also, a vertical line and a horizontal line are perpendicular. You can use a graphing utility to draw the graph of two parallel or perpendicular lines.

EXAMPLE 2 **Line l_1 passes through the points (−1, −1) and (3, 1), and line l_2 passes through the points (2, −2) and (−2, 6). Determine if these two lines are perpendicular.**

slope of $l_1 = \frac{1 - (-1)}{3 - (-1)} = \frac{1}{2}$

slope of $l_2 = \frac{6 - (-2)}{-2 - 2} = -2$

Since $\frac{1}{2}(-2) = -1$, the slopes of the lines are negative reciprocals. Therefore, the lines are perpendicular.

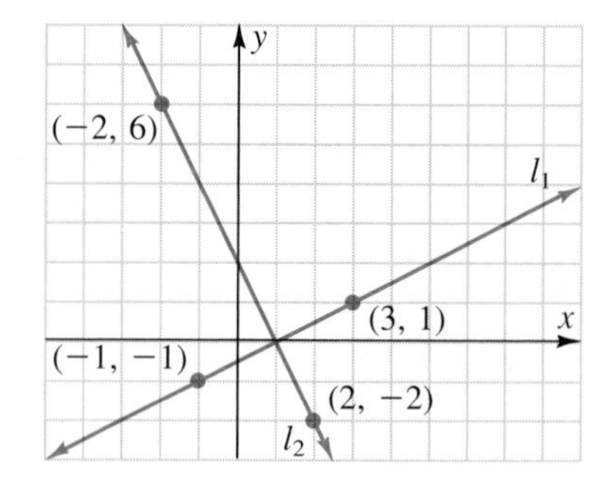

Given the y-intercept, you can determine an equation of a line that is parallel or perpendicular to another line.

EXAMPLE 3 **Given the line $2y = -4x + 8$, write an equation of a line**

a. parallel to the given line, with y-intercept 3.

b. perpendicular to the given line, with y-intercept −1.

$2y = -4x + 8$

$y = -2x + 4$ *Write the given line in slope-intercept form.*

a. $y = mx + b$

$y = -2x + 3$ *Parallel lines have equal slopes, so substitute −2 for m and 3 for b to find the equation of the new line.*

b. $y = mx + b$

$y = \frac{1}{2}x - 1$ *Slopes of perpendicular lines are negative reciprocals, so substitute $\frac{1}{2}$ for m and −1 for b to find the equation of the new line.*

CLASS EXERCISES

Determine if the lines passing through the given pairs of points are parallel, perpendicular, or neither. Assume the lines are distinct.

1. $(1, 5), (2, 7)$ and $(2, 5), (3, 7)$

2. $(3, 2), (-3, 0)$ and $(6, 2), (0, 0)$

3. $(-3, -6), (2, 4)$ and $(2, -1), (4, -2)$

4. $(2, 3), (4, 1)$ and $(1, 3), (2, 6)$

Write an equation of a line parallel to the given line and with the given y-intercept.

5. $y = 2x - 3, b = -1$

6. $2y = 4x - 8, b = 3$

7. $2y = x - 6, b = 2$

Write an equation of a line perpendicular to the given line and with the given y-intercept.

8. $y = 3x - 1, b = 5$

9. $y = -2x + 5, b = -3$

10. $2y = 3x - 2, b = 3$

PRACTICE EXERCISES

Use technology where appropriate.

Determine if the lines passing through the given pairs of points are parallel, perpendicular, or neither. Assume the lines are distinct.

1. $(1, 1), (2, 3)$ and $(2, 5), (3, 7)$

2. $(2, 3), (3, 5)$ and $(1, 5), (2, 7)$

3. $(3, 3), (6, 5)$ and $(2, -3), (4, -6)$

4. $(0, 2), (-4, -1)$ and $(3, -5), (6, -9)$

5. $(-6, 3), (-3, 1)$ and $(3, 0), (6, 0)$

6. $(2, -5), (4, -10)$ and $(1, 9), (2, -6)$

Write an equation of the line parallel to the given line and with the given y-intercept.

7. $y = 3x - 4, b = -3$

8. $y = 2x + 4, b = -2$

9. $y = -2x - 1, b = 3$

10. $y = -4x - 3, b = -5$

11. $2y = -2x + 4, b = 1$

12. $3y = -9x + 3, b = 7$

Write an equation of the line perpendicular to the given line and with the given y-intercept.

13. $y = -2x + 1, b = 2$

14. $y = -4x + 3, b = -2$

15. $2y = x - 3, b = 5$

16. $3y = x + 1, b = 1$

17. $3y = 2x + 3, b = -1$

18. $2y = -5x + 2, b = 6$

Determine if the given distinct lines are parallel, perpendicular, or neither.

19. $3y = x - 3, 2y = -6x - 12$

20. $4y = -2x + 4, y = 2x - 4$

21. $2y = 3x + 2, 4y = 6x$

22. $5y = -x, 10y = 2x - 10$

Write an equation of the line parallel to the given line with the given y-intercept.

23. $7y = 2x, b = 3$

24. $4y = -5x, b = -3$

25. $2y - 3x = 6, b = -4$

26. $3y - 4x = 3, b = 5$

Write an equation of the line perpendicular to the given line and with the given y-intercept.

27. $5y = 3x + 5,\ b = 4$

28. $3y = 4x - 3,\ b = -5$

29. $3y - 2x = 0,\ b = 2$

30. $2y - x = 0,\ b = -4$

Write an equation of the line perpendicular to the given line and passing through the given point.

31. $7x - 2y = -2,\ (7, 1)$

32. $4x - 3y = -4,\ (-4, 0)$

33. $5x - y - 1 = 0,\ \left(\frac{1}{2}, -\frac{1}{10}\right)$

34. $-2x - y - 1 = 0,\ \left(-\frac{1}{2}, -\frac{1}{4}\right)$

Use the definition of slope to determine if the following points are collinear.

35. $(-2, 6), (0, 2), (1, 0)$

36. $(3, -5), (-3, 3), (0, 2)$

Applications

37. **Geometry** Show that the triangle whose vertices have coordinates $(3, 5)$, $(-2, 6)$, and $(1, 3)$ is a right triangle.

38. **Geometry** Show that a quadrilateral whose vertices have coordinates $(1, 2)$, $(4, 5)$, $(3, 1)$, and $(0, -2)$ is a parallelogram.

39. **Geometry** Show that a quadrilateral whose vertices have coordinates $(2, 5)$, $(4, 8)$, $(7, 6)$, and $(5, 3)$ is a rectangle.

40. **Architecture** An architect is drawing a model of a house she is designing. The exterior is supposed to have decorative crossbeams placed perpendicular to each other. If the slope of one beam is $\frac{3}{4}$, find the slope of the beam perpendicular to it.

41. **Design** A designer is drawing a pattern for a new line of shirts. He uses graph paper to make the model. He wants two parallel stripes (lines). If the first stripe can be represented by the equation $3y = -6x + 9$, what must the slope of the second stripe be?

Developing Mathematical Power

42. **Logical Reasoning** Assume that this statement is true: All shims are shams. Which of the following statements must also be true?
    a. All shams are shims.
    b. If it is a sham, then it is a shim.
    c. If it is not a sham, then it is not a shim.
    d. If it is not a shim, then it is not a sham.

3.8

Equations of a Line

Objectives: To write an equation of a line, given the slope and a point on the line

To write an equation of a line, given two points

If the slope and y-intercept of a line are known, the equation of the line can be written using the slope-intercept form. When either two points or one point and the slope are given, it is helpful to develop a different form of an equation of a line.

Capsule Review

EXAMPLE **Solve for y:** $\frac{2}{3} = \frac{y+1}{x-3}$

$$\frac{2}{3} = \frac{y+1}{x-3}$$

$3(y+1) = 2(x-3)$ *Cross multiply.*

$3y + 3 = 2x - 6$ *Distributive property*

$3y = 2x - 9$

$y = \frac{2}{3}x - 3$

Solve for y.

1. $3 = \frac{y-1}{x+2}$
2. $\frac{5}{2} = \frac{y+1}{x-2}$
3. $\frac{3}{4} = \frac{y+2}{x+4}$
4. $\frac{2}{5} = \frac{y-1}{x-5}$
5. $\frac{y+4}{x-1} = -1$
6. $\frac{y-3}{x} = 4$

The line at the right has slope m and passes through the point $P_1(x_1, y_1)$. Let $P(x, y)$ be any other point on the line. Then the slope is

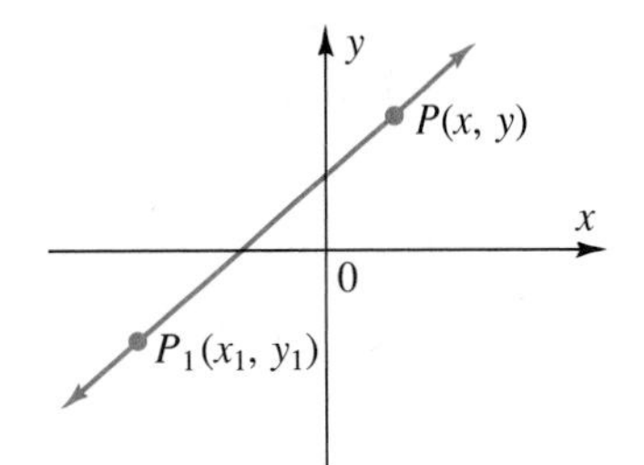

$$m = \frac{y - y_1}{x - x_1}$$

$$y - y_1 = m(x - x_1)$$

The **point-slope form** of an equation of a line is

$$y - y_1 = m(x - x_1)$$

where m is the slope and (x_1, y_1) are the coordinates of a given point on the line.

EXAMPLE 1 **Write in slope-intercept form the equation of the line that has slope $-\frac{2}{3}$ and that passes through the point (3, −1).**

$$y - y_1 = m(x - x_1) \qquad \textit{Use the point-slope form.}$$
$$y - (-1) = -\tfrac{2}{3}(x - 3) \qquad m = -\tfrac{2}{3},\ (x_1, y_1) = (3, -1)$$
$$3(y + 1) = -2(x - 3)$$
$$3y + 3 = -2x + 6 \qquad \textit{Distributive property}$$
$$y = -\tfrac{2}{3}x + 1 \qquad \textit{Slope-intercept form of the equation}$$

You can determine an equation of a line, given two points on the line.

EXAMPLE 2 **Write in slope-intercept form the equation of the line that passes through the points (2, 1) and (3, 4).**

$$m = \frac{y_2 - y_1}{x_2 - x_1} = \frac{4 - 1}{3 - 2} = \frac{3}{1} = 3 \qquad \textit{Find the slope.}$$
$$y - y_1 = m(x - x_1) \qquad \textit{Use the point-slope form.}$$
$$y - 1 = 3(x - 2) \qquad m = 3, \text{ and } (x_1, y_1) = (2, 1) \text{ or } (x_1, y_1) = (3, 4). \ (2, 1) \text{ is used here.}$$
$$y = 3x - 5 \qquad \textit{Slope-intercept form}$$

If (3, 4) had been used for (x_1, y_1) in Example 2, the final equation would have been the same, as you can easily verify.

In Example 3, you are asked to write a linear equation in standard form. Recall that the standard form of a linear equation is $Ax + By = C$.

EXAMPLE 3 **Write in standard form the equation of the line that passes through the points (−2, 4) and (−6, 1).**

$$m = \frac{y_2 - y_1}{x_2 - x_1} = \frac{1 - 4}{-6 - (-2)} = \frac{3}{4} \qquad \textit{Find the slope.}$$
$$y - y_1 = m(x - x_1) \qquad \textit{Use the point-slope form.}$$
$$y - 4 = \tfrac{3}{4}[x - (-2)] \qquad m = \tfrac{3}{4},\ (x_1, y_1) = (-2, 4)$$
$$4(y - 4) = 3(x + 2) \qquad \textit{Simplify.}$$
$$-3x + 4y = 22 \qquad \textit{Standard form}$$

CLASS EXERCISES

Write in slope-intercept form the equation of the line that has slope *m* and that passes through the given point.

1. $m = 2$, (1, 6) **2.** $m = 4$, (1, 5) **3.** $m = -2$, (−1, 1)

4. $m = \frac{1}{2}$, (2, 2) **5.** $m = \frac{2}{3}$, (3, 1) **6.** $m = -\frac{1}{2}$, (−2, 3)

Write in standard form the equation of the line that passes through the given points.

7. $(-1, -1), (2, 5)$
8. $(-3, -4), (6, 2)$
9. $(6, 13), (0, 5)$
10. $(-2, -4), (-4, -3)$
11. $(2, -2), (-4, 13)$
12. $(5, 5), (-5, 1)$

PRACTICE EXERCISES

Write in slope-intercept form the equation of the line that has slope *m* and that passes through the given point.

1. $m = -1, (-2, -3)$
2. $m = 1, (-2, 2)$
3. $m = 3, (2, 6)$
4. $m = -2, (2, -4)$
5. $m = -4, (1, -1)$
6. $m = -5, (-2, 8)$
7. $m = -3, (-2, -1)$
8. $m = 2, (3, 1)$
9. $m = 4, (0, 2)$

Write in standard form the equation of the line that passes through the given points.

10. $(1, -5), (2, -3)$
11. $(2, 6), (-3, 1)$
12. $(-1, 8), (1, 6)$
13. $(2, -5), (-3, 5)$
14. $(2, -1), (4, 1)$
15. $(-2, 4), (1, -5)$
16. $(1, -2), (3, 2)$
17. $(1, 6), (-2, -3)$
18. $(1, 4), (3, -8)$
19. $(-2, 3), (-4, 7)$
20. $(3, 7), (-2, -3)$
21. $(5, 4), (-5, -2)$
22. $(-5, 2), (7, 5)$
23. $(4, -5), (2, -5)$
24. $(3, 4), (3, -7)$

Write in slope-intercept form the equation of the line that has slope *m* and that passes through the given point.

25. $m = \frac{1}{2}, (2, 5)$
26. $m = \frac{3}{2}, (4, 3)$
27. $m = -\frac{3}{4}, (4, -6)$
28. $m = -\frac{2}{5}, (-5, 4)$
29. $m = -\frac{3}{5}, (0, 7)$
30. $m = -\frac{2}{7}, (0, 3)$

Write in standard form the equation of the line that passes through the given points.

31. $(9, 4), (0, 2)$
32. $(7, -6), (-7, 12)$
33. $(0, 5), (3, 6)$
34. $(-4, -1), (8, -4)$
35. $(2, -4), (0, 1)$
36. $(4, 4), (0, 1)$
37. $(0, -9), (9, -7)$
38. $(0, -7), (7, -4)$
39. $(13, -4), (1, 1)$
40. $(-3, 1), (3, 9)$
41. $(-3, 8), (0, 0)$
42. $(11, -5), (0, 0)$

Write in standard form the equation of the line that is parallel to the given line and passes through the given point.

43. $5y = 4x - 2, (5, -1)$
44. $2y = -3x + 4, (2, 2)$

Write in standard form the equation of the line that is perpendicular to the given line and passes through the given point.

45. $2y = x + 7$, $(3, -2)$

46. $4y = -x - 1$, $(1, -4)$

47. $2x = -9y$, $(0, 0)$

48. $x = 3$, $(2, 2)$

Applications

Business A store decides on a linear markup policy given by the following equation: $y = 1.2x + 8$

49. How much is the markup for a pair of skis that the store purchased for \$110? What is the cost to the customer?

50. Use a graphing calculator to graph the function. Interpret the graph for values of $x \leq 0$.

51. What would be a possible function to reflect a smaller markup? a larger?

52. Research markup policies from various stores to find out if the above linear model is reasonable.

53. If you know the original cost and the markup on two items in a store, what additional information would you need in order to predict the markup on other items?

Developing Mathematical Power

Did you know that the value of the United States dollar fluctuates daily in terms of foreign currencies? For example, the dollar was worth over 11.0 French francs in 1985, and it was below 5.8 francs in 1991. Hence, the dollar bought almost twice as much in France at times in 1985 as it did at times in 1991. Economists watch the Japanese yen, the German deutschmark, the British pound, and many other currencies. The foreign exchange rates can be found every weekday in the newspaper, usually in the business section.

54. Project Record the value of the dollar in three other currencies and plot each of them on a separate set of coordinate axes, with time on the horizontal axis and the value of the currency on the vertical axis. Continue to record data for each currency over a period of at least 3 months. Then form conclusions based on observations and answer the following questions. Which of the three currencies appears to be the most stable when compared to the dollar? Which currency is most volatile? Is the dollar getting stronger or weaker with respect to each of the other three currencies?

3.9

Problem Solving Strategy: Use Coordinate Geometry

Coordinate geometry combines concepts from algebra and geometry. The correspondence between the points in the coordinate plane and the set of ordered pairs (x, y) allows the solution sets of functions in two variables to be graphed in the plane. Then an equation of the function or of its graph can be used to solve a problem. For example, the relationship between the demand for a product and its price can be represented by a *demand curve* where x is the selling price and y is the demand in units.

EXAMPLE 1 A market research firm collected data on the sales of a new frozen gourmet dinner. Each day the firm recorded the number of dinners sold and the price. The data were graphed on a coordinate plane as shown below. In order to make predictions about sales, given different prices, the researchers want to know an equation of the graph. Find an equation of the graph and the number of dinners that will sell at a price of $5.25.

Understand the Problem *What are the given facts?* Two items of data are shown in the graph.

What are you asked to find? The problem is to find an equation of the graph and the number of dinners that will sell at a price of $5.25.

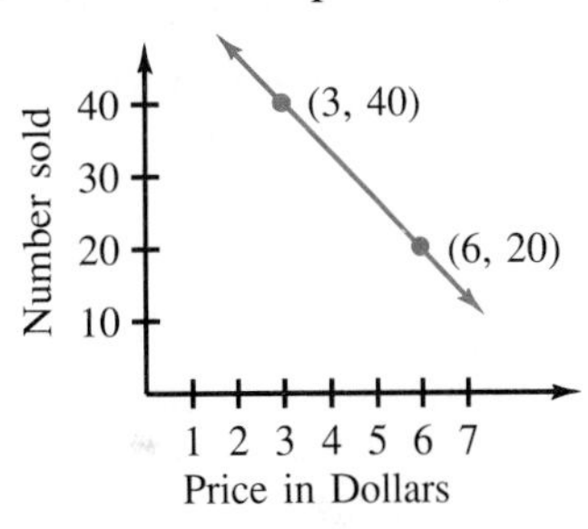

Plan Your Approach *Choose a strategy*. The problem can be solved by using coordinate geometry. The function is linear because its graph is a straight line. The equation of a line can be written if two points on the line are known. From the graph, two points are (3, 40) and (6, 20).

Let $(x_1, y_1) = (3, 40)$ and $(x_2, y_2) = (6, 20)$. Then the slope is

$$m = \frac{y_2 - y_1}{x_2 - x_2} = \frac{20 - 40}{6 - 3} = -\frac{20}{3}$$

Complete the Work

$$y_2 - y_1 = m(x - x_1)$$ *Use the point-slope form of a line.*

$$y - 40 = -\frac{20}{3}(x - 3)$$ *Substitute the known values into the point-slope form for the line.*

$$y = -\frac{20}{3}x + 20 + 40$$

$$y = -6\tfrac{2}{3}x + 60$$

Interpret the Results *State your answer*. An equation of the graph, in function form, is

$$f(x) = -6\tfrac{2}{3}x + 60$$

The function can be used to make predictions about how many dinners will sell at different prices. At a price of \$5.25,

$$f(5.25) = -6\tfrac{2}{3}(5.25) + 60 = 25$$

Thus, 25 dinners are expected to sell at a price of \$5.25.

EXAMPLE 2 A designer is making a triangular base for a sculpture to be placed in the local park. He determines the shape and location of the triangle using rectangular coordinates. The vertices of the triangle are (4, 5), (1, −1), and (0, 7). Show that the triangle is a right triangle.

Understand the Problem *What are the given facts?* A triangle has its vertices at (4, 5), (1, −1), and (0, 7).

What are you asked to find? Determine if the triangle has a right angle.

Two lines are perpendicular (form right angles) if their slopes are negative reciprocals.

Graph the points on the coordinate plane and draw the triangle.

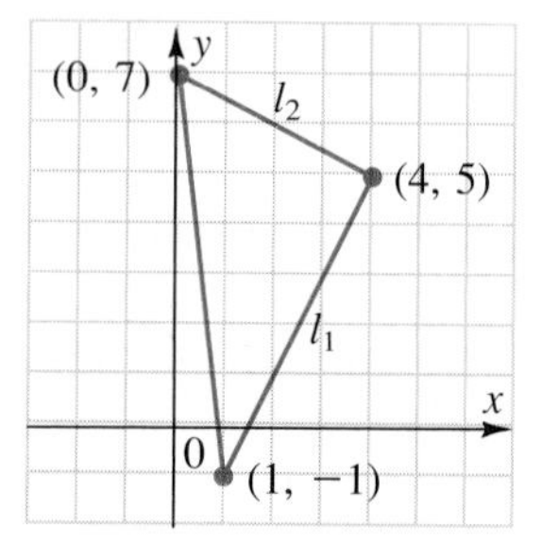

Plan Your Approach *Choose a strategy*. From the diagram lines l_1 and l_2 appear to be perpendicular. Therefore, determine the slope of each line and compare to see if they are negative reciprocals.

Let (1, −1) be (x_1, y_1) and (4, 5) be (x_2, y_2) on l_1.
Let (0, 7) be (x_1, y_1) and (4, 5) be (x_2, y_2) on l_2.

Complete the Work

$$m = \frac{y_2 - y_1}{x_2 - x_1} = \frac{5 - (-1)}{4 - 1} = \frac{6}{3} = 2 \qquad \textit{Find the slope of } l_1.$$

$$m = \frac{y_2 - y_1}{x_2 - x_1} = \frac{5 - 7}{4 - 0} = \frac{-2}{4} = -\frac{1}{2} \qquad \textit{Find the slope of } l_2.$$

Interpret the Results *State your answer*. The slopes of l_1 and l_2 are negative reciprocals. Therefore, l_1 and l_2 are perpendicular and the triangle is a right triangle.

Check your answer. Is the product of the slopes of l_1 and l_2 equal to −1?

$$(2)\left(-\tfrac{1}{2}\right) \stackrel{?}{=} -1$$
$$-1 = -1 \checkmark$$

CLASS EXERCISES

Use the function in Example 1 to find the number of dinners that are expected to sell at the following prices.

1. \$1 **2.** \$4 **3.** \$6.50 **4.** \$2.75 **5.** \$7

6. Market research shows that the demand curve for a new game is linear. It was found that there was a demand of 100 units at \$20 each and 200 units at \$10 each. Write the linear demand curve for this product and find the number of games that will sell at a price of \$5.

7. A quilt manufacturer received a design for a quilt. The design was a quadrilateral whose vertices, when plotted on the coordinate plane, had coordinates $(1, 7)$, $(2, 3)$, $(-2, 1)$, and $(-1, -3)$. Show that the quadrilateral is a parallelogram.

8. The initial cost of installing an alarm system is \$500. This covers the alarm box but does not include any door or window contacts. The cost of the installation with 4 contacts is \$750. The cost of the installation can be represented by a linear function. Find the cost with 10 contacts.

9. A company's earnings can be represented by a linear function. If the company sells 35 stoves, it earns \$17,000. If it sells 45 stoves, it earns \$23,000. If the company's earnings were \$32,000, how many stoves did they sell?

PRACTICE EXERCISES

1. A survey was done to determine the demand for a new novelty item. The demand curve was found to be linear. The demand was 10 units at \$4 each and 20 units at \$3 each. Write the linear demand curve for this item and find the number of units that could be sold at \$1.

2. The demand curve for a new brand of orange juice was found to be linear. The demand was 8 cases at \$50 and 10 cases at \$40. Write the linear demand curve for this product and find the number of cases that could be sold at \$20.

3. A manufacturer wanted to put out a line of cutting boards in the shape of a parallelogram. His designer sent him a sketch of a board plotted on the coordinate plane. The vertices of the board had coordinates $(1, -1)$, $(2, 1)$, $(4, 2)$, and $(3, 0)$. Show that this board is a parallelogram. (Show that the opposite sides are parallel.)

4. The base of a table was in the shape of a triangle. When plotted on the coordinate plane this triangle had coordinates $(4, 7)$, $(-4, 9)$, and $(-2, -17)$. Show that the triangle is a right triangle.

5. A company sells computers. Its sales can be represented by a linear function. When it sells 50 computers it earns $7000. When it sells 75 computers it earns $9000. How many computers did it sell if it earned $15,000? Assume that the sales function is linear.

6. The profits of a company that produces auto parts can be represented by a linear function. If the company sells 25 parts its profits are $80 and if the company sells 40 parts its profits are $125. Find the number of parts sold if the profits were $305. If the profits were $905.

7. An embankment has coordinates (4, 9) and (6, 17). Find an equation of a line that could be used to represent this embankment.

8. Railroad tracks are to be built parallel to an already existing road. The surveyor determines that two points on the road have coordinates (3, 5) and (23, 45). If the railroad tracks pass through the point (17, 19), find the linear equation representing the tracks.

9. A rectangular plot is surveyed. The equation representing one side of the plot is $y = -3x + 4$. Find the equation of an adjacent side, if the side passes through the point $(3, -5)$.

10. A firm charges $30 to install cable television plus a $5 charge for each additional television it is hooked up to. If the total cost of the hook-up is a linear function, find the cost of having a cable installation including 4 additional televisions.

11. A mover charges $250 for use of the truck and $75 per hour for labor. If the total cost of a move is a linear function, find the cost of a move requiring 4 hours of labor.

12. A row of trees is to be planted in a straight line up a hill. A surveyor determines that the coordinates of two points along the row are to be (40, 50) and (70, 70). The trees are to be planted 50 ft apart measured horizontally. Give the coordinates of the second tree, if the first tree is to be 40 ft from the base of the hill measured horizontally.

13. A ski lift is to be built. The builder wants the lines to be parallel to the sides of the mountain with its lowest point at (0, 40). A surveyor tells the builder that the coordinates of two points on the side of the mountain are (400, 800) and (600, 10,000). Write a linear function to represent the ski lift.

14. The undepreciated portion of the cost of a house is measured by the linear function $y = 60{,}000 - 2000x$, where x is the age of the house in years. Find the undepreciated portion of the cost if the house is 6 months old. Find the undepreciated portion of the cost if the house is 9 months old.

15. The undepreciated portion of the cost of a piece of equipment is measured by the linear function $y = 5000 - 700x$, where x is the age of the equipment in years. Find the undepreciated portion of the cost if the equipment is 18 months old. 30 months old.

16. A brick patio is built in the shape of a rhombus. An equation of a line representing the diagonal of this rhombus is $y = -\frac{2}{3}x + 7$. Find the equation of the other diagonal if it passes through the point (3, 5).

17. A road is to be constructed perpendicular to an established road. A surveyor determines that the original road passes through the points (4, 9) and (2, 13). The roads must intersect at (4, 9). Write the equation of the second road.

18. A company could sell 20 talking bears for $200 each and 70 talking bears at $150 each. If the demand curve is linear, determine the demand if the selling price was $70.

19. The coordinates of the vertices of one side of a rectangular field are (5, 1) and (7, 9). A fence post is erected at the point (6, 4) on the opposite side. Write a linear equation representing the opposite side, and determine the x-coordinate of a post on that side whose y-coordinate is 3.

20. Two new roads are to be built perpendicular to an existing road whose equation is $2y - 3x = 6$. The engineer has to decide where each new road should cross the existing road. He decides that the first road should cross at (2, 6) and the second road should cross at a point whose x-coordinate is twice the x-coordinate of the first new road and whose y-coordinate is 3 more than the y-coordinate of the first new road. Write a linear equation to represent each new road.

21. A company's profit on a new line of jeans is represented by the linear function $f(x) = \frac{3}{4}x - 243$, where x is the number of pairs of jeans sold. Find the number of pairs of jeans the company must sell to break even (profit is $0).

22. The pitch of a beam of a roof is $\frac{1}{3}$. If the edge of the beam is measured 12 ft from a vertical tree and 7 ft from the ground, write an equation representing a line drawn along this beam.

Mixed Problem Solving Review

Solve each problem.

1. The measure of the supplement of an angle is sixty more than five times the measure of the angle. Find the measure of the angle and its supplement.

2. Find four consecutive even integers such that the sum of the second and fourth is 28.

3. The length of a rectangle is 5 times its width. The perimeter is 120 ft. Find the dimensions of the rectangle.

4. Alex bowled 119, 131, 127, 146, and 152 in 5 games. What must he bowl in the sixth game to have an average of 135?

PROJECT

The concept of *function* is one of the most important ideas in mathematics. Many of the topics covered in this textbook involve the study of different kinds of functions. A linear function has a graph that is a straight line. A straight line is an example of a *continuous curve,* that is, a curve with no breaks or gaps. Write an explanation of why there are no breaks or gaps in the graph of a linear function.

TEST YOURSELF

Find the slope of the line passing through the given points.

1. (7, 2), (6, 1) **2.** (−3, 5), (1, −3) **3.** (2, 4), (−5, 4) 3.6

Write an equation of the line with the given slope and y-intercept.

4. $m = -3$, $b = 2$ **5.** $m = \frac{2}{3}$, $b = -1$ **6.** $m = 0$, $b = -4$ 3.6

7. Write in slope-intercept form the equation of the line parallel to the line $3y = 6x - 4$ and with y-intercept 5. 3.7

8. Write in slope-intercept form the equation of the line perpendicular to the line $4y = -8x - 1$ and with y-intercept -3. 3.7

9. Write in slope-intercept form the equation of the line that has slope $\frac{4}{3}$ and that passes through the point (6, 3). 3.8

10. Write in standard form the equation of the line that passes through the points (4, −1) and (5, −2). 3.8

11. Write in standard form the equation of the line that passes through the points (−4, 1) and (2, −2).

12. A straight road is to be built parallel to a given straight road. The engineer makes a model of the existing road on a grid and determines that its equation is $y = -\frac{1}{5}x - 4$. If the new road is to pass through the point (5, 8), write an equation in standard form for the new road. 3.9

INTEGRATING ALGEBRA
Economic Models

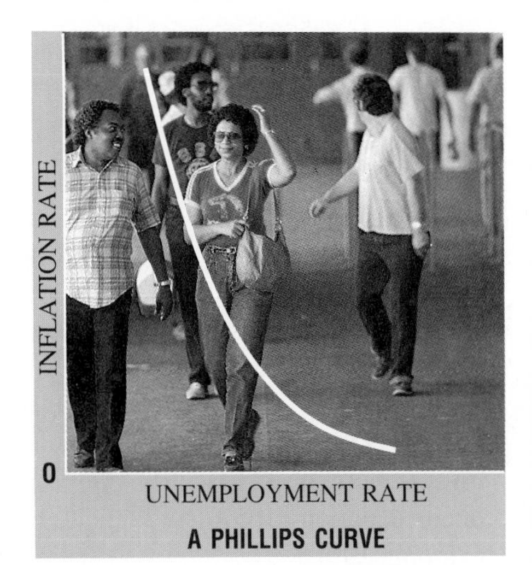

A PHILLIPS CURVE

The study of economics is an ever changing science. Economic models based on past economic trends are frequently used to predict future economic activity. Some models, however, soon become inadequate because economic conditions change in unexpected ways. In the United States the government responds to economic indicators in an effort to stabilize the economy and to foster consistent economic growth. The need for reliable economic models is, therefore, of significant importance for the economic well-being of the country. One economic model, the Phillips curve, suggests a tradeoff between unemployment and inflation. According to this model, less unemployment can be obtained by incurring higher inflation, or conversely, inflation can be reduced by allowing more unemployment. A sketch of this model appears above.

The table below and to the right contains unemployment (U) and inflation (I) rates in percents for the years 1976 to 1990 in the United States.

For the following exercises graph U, the unemployment rate, on the horizontal axis and I, the inflation rate, on the vertical axis.

1. Graph all 15 points and connect them in order of year.

2. Does the curve obtained in the exercise above support the Phillips curve model? Do any parts of the curve follow the model?

3. On another set of axes, graph the data for the years 1976 to 1979 and connect the points in order of year.

4. On a third set of axes, graph the data for the years 1986 to 1990 and connect the points in order of year.

5. Based on the results of the previous exercises, is the Phillips curve a reasonable economic model? Explain why or why not.

Year	U	I
1976	7.7	5.7
1977	7.0	6.5
1978	6.0	7.6
1979	5.8	11.3
1980	7.1	13.5
1981	7.6	10.4
1982	9.7	6.2
1983	9.6	3.2
1984	7.5	4.3
1985	7.2	3.5
1986	7.0	1.9
1987	6.2	3.7
1988	5.4	4.1
1989	5.2	4.7
1990	5.2	5.4

CHAPTER 3 SUMMARY AND REVIEW

Vocabulary

absolute value function (110)
coordinate plane (94)
composite function (110)
domain (98)
first-degree equation (105)
function (99)
inverse (114)
inverse function (114)
linear equation (105)
mapping (99)
origin (94)
point-slope form of a linear equation (127)
quadrants (94)
range (98)
rectangular coordinate system (94)
relation (98)
slope (118)
slope-intercept form of a linear equation (120)
standard form of a linear equation (105)
vertical line test (100)
x-axis (94)
x-coordinate (94)
y-axis (94)
y-coordinate (94)

Rectangular Coordinate System Each point on the coordinate plane can be represented by two real numbers written as an ordered pair, (x, y). 3.1

Graph each ordered pair and name the quadrant in which it lies.

1. $(2, 5)$ **2.** $(-6, 3)$ **3.** $(-1, -4)$ **4.** $(7, -2)$

Relations and Functions A relation is a set of ordered pairs. A function is a relation in which each element in the domain is paired with one and only one element in the range. 3.2

Graph each relation and state its domain and range.

5. $\{(-1, 2), (0, 3), (1, 4)\}$ **6.** $\{(-4, 5), (-2, 6), (-4, 7)\}$

Determine which, if any, of the following relations are functions.

7. $L = \{(2, 4), (3, 5), (4, 6)\}$ **8.** $M = \{(2, 4), (2, -4), (3, 5), (3, -5)\}$

Graphing Equations The graph of a linear equation of the form $Ax + By = C$, with A, B, and C real numbers and both A and B not equal to 0, is a line. 3.3

Graph each equation.

9. $y = -3x + 2$ **10.** $2x + y = 5$ **11.** $-3x + 2y = 6$

Determine which, if any, of the following are linear equations.

12. $x - 3y = -3$ **13.** $x = 2y^2$ **14.** $-2x - 2y = -4$

Composition of Functions Given two functions $f(x)$ and $g(x)$, a composite function of $f(x)$ and $g(x)$ is $(g \circ f)(x) = g[f(x)]$. 3.4

For each function, find $f(x)$ for $x = -1, 2, 5$.

15. $f(x) = 3x - 2$ **16.** $f(x) = -x - 2$ **17.** $f(x) = x^2$

Evaluate $g[f(2)]$ and $f[g(-3)]$ for each pair of functions.

18. $f(x) = -2x$, $g(x) = x + 3$ **19.** $f(x) = 3x$, $g(x) = -x + 1$

Inverse Functions To find the inverse of a function written as an equation, interchange the two variables x and y and solve for y. If the inverse is also a function, it is denoted by f^{-1}. 3.5

Find the inverse of each relation or function. Determine if the inverse is also a function.

20. $y = 3x + 1$ **21.** $y = -2x + 4$ **22.** $y = x^3$

Slope of a Line If $P(x_1, y_1)$ and $Q(x_2, y_2)$ are on a line, then the slope, m, of the line is given by $m = \frac{y_2 - y_1}{x_2 - x_1}$, $x_1 \neq x_2$. Slope-intercept form of a line: $y = mx + b$, where m is the slope and b is the y-intercept. 3.6

Find the slope of the line that passes through the given points. 3.6

23. (1, 4), (2, 7) **24.** (−2, 4), (5, −10) **25.** (−3, 6), (−4, −1)

Write an equation of the line with the given slope and y-intercept.

26. $m = 2$, $b = -3$ **27.** $m = -4$, $b = 5$ **28.** $m = -2$, $b = -7$

Parallel and Perpendicular Lines Two lines are parallel if and only if their slopes are equal or both undefined. Two nonvertical lines are perpendicular if and only if their slopes are negative reciprocals. 3.7

Write an equation of the line parallel to the given line and with the given y-intercept.

29. $y = 2x - 3$, $b = -2$ **30.** $y = 3x + 5$, $b = 4$

Write an equation of the line perpendicular to the given line and with the given y-intercept.

31. $y = -3x + 2$, $b = 4$ **32.** $y = 2x - 5$, $b = -2$

Equation of a Line Point-slope form: $y - y_1 = m(x - x_1)$, where m is the slope and (x_1, y_1) is a point on the line. 3.8

Write in slope-intercept form the equation of each line.

33. $m = 2$, passes through (3, 4) **34.** passes through (−2, 5) and (4, 6)

35. One side of a parallelogram has slope −4. If the opposite side passes through point (−2, −3), write the equation of the line containing that side. 3.9

CHAPTER 3 TEST

Graph each ordered pair and name the quadrant in which it lies.

1. $(-3, 5)$ **2.** $(4, -2)$ **3.** $(-1, -6)$

Determine which, if any, of the following relations are functions.

4. $M = \{(1, -3), (1, 3), (2, 5), (2, -5)\}$ **5.** $y = -5x + 7$

Graph each linear equation. Determine which, if any, are not functions.

6. $-4x + y = -6$ **7.** $2x + y = -8$ **8.** $x = 5$

For each function, find $f(x)$ for $x = -3, 0, 2$.

9. $f(x) = 4x - 2$ **10.** $f(x) = 3x^2$

Evaluate $g[f(-2)]$ and $f[g(3)]$ for each pair of functions.

11. $f(x) = 3x,\ g(x) = 2x + 3$ **12.** $f(x) = -x,\ g(x) = x + 5$

Find the inverse of each relation or function. Determine if the inverse is also a function.

13. $2x + 3y = 6$ **14.** $y = -2x - 8$ **15.** $y = x^2 + 1$

Write in slope-intercept form the equation of each line.

16. $m = -3$, passes through $(-1, 6)$ **17.** passes through $(0, -4)$ and $(1, 2)$

Solve.

18. A road is built on an incline. Find the slope of the road if a line representing the road passes through the points (1, 10) and (20, 11).

Challenge

The cost of installing a boiler is $520. Each radiator costs $70.50. Write an equation of the linear function representing the cost of installing a heating system consisting of one boiler and at least one radiator. Given a budget of $1050, determine the greatest number of radiators that can be installed with the heating system.

COLLEGE ENTRANCE EXAM REVIEW

Solve. Grid in your response on the answer sheet.

1. Simplify: $\dfrac{\frac{1}{2} + \frac{4}{5}}{1 - \frac{1}{3}}$

2. Find 0.3% of 60.

3. What is the slope of the line passing through the points $A(6, -2)$ and $B(4, 8)$?

Use this figure for 4 and 5.

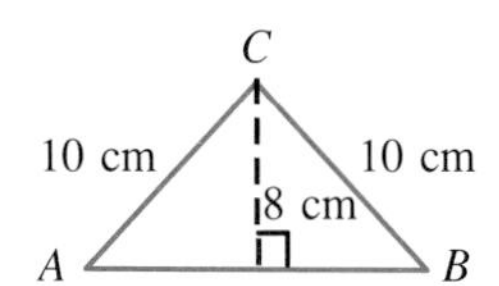

4. Find the area of $\triangle ABC$ in cm^2.

5. What is the perimeter of $\triangle ABC$, in cm?

Use this information for 6 and 7.

$$f(x) = 2x^2 - x + 1$$
$$g(x) = \sqrt{x + 2}$$

6. What is the value of $g(0)$ to the nearest tenth?

7. Find the value of $f(g(1))$ to the nearest tenth.

Use this diagram for 8–10.

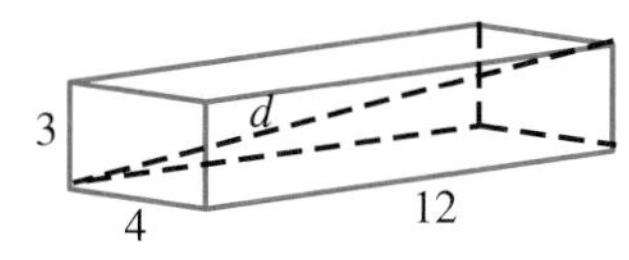

8. What is the volume of the rectangular solid?

9. Find the length of diagonal d.

10. What is the total surface area of the rectangular solid?

11. Written as a decimal, what is the value of $\frac{31}{1000} + \frac{2}{10} + \frac{8}{100}$?

12. Find the second of four consecutive even integers such that the product of the first and fourth is equal to the square of the second.

13. In a 3-drawer chest, Sadie has 25 percent of her sweaters in drawer A, 30 percent in drawer B, and the rest in drawer C. What is the smallest number of sweaters she could have in drawer C?

14. If $1 \leq a < b \leq 9$ and a and b are integers, what is the least possible value of $\frac{a + b}{ab}$?

15. If a car averages 55 mi/h on a certain trip, it will arrive at its destination 2 h early. If the car averages 35 mi/h, it will arrive 2 h late. What is the length, in miles, of the trip?

16. In March, Alvin spent $\frac{1}{4}$ of the money in his savings account. The next month he spent $\frac{1}{5}$ of the remainder. If he then had $300 left, how much was in his savings account originally, in dollars?

MIXED REVIEW

Express each fraction as a decimal.

Example $\frac{7}{8} \longrightarrow 8\overline{)7.000}$ with quotient 0.875 $\qquad \frac{7}{8} = 0.875$

1. $\frac{1}{2}$ **2.** $\frac{3}{8}$ **3.** $\frac{2}{5}$ **4.** $\frac{3}{4}$

5. $\frac{5}{8}$ **6.** $\frac{7}{10}$ **7.** $\frac{9}{20}$ **8.** $\frac{12}{25}$

Solve each proportion.

Example $\frac{3}{8} = \frac{x}{20}$ $\qquad \frac{3}{8} \times \frac{x}{20}$ *The product of the extremes equals the product of the means.*

$$\begin{aligned} \text{extremes} \quad & \quad \text{means} \\ 3 \cdot 20 &= 8x \\ 60 &= 8x \quad \textit{Solve for x.} \\ x &= 7.5, \text{ or } 7\tfrac{1}{2} \end{aligned}$$

9. $\frac{7}{12} = \frac{28}{x}$ **10.** $\frac{x}{9} = \frac{3}{18}$ **11.** $\frac{6}{x} = \frac{15}{12.5}$ **12.** $\frac{1.2}{3.8} = \frac{x}{19}$

Solve each equation or inequality.

Example
$$\begin{aligned} 5 - 3x &> 20 \\ -3x &> 15 \quad \textit{Subtract 5 from both sides.} \\ x &< -5 \quad \textit{Divide each side by } -3. \end{aligned}$$

13. $\frac{1}{2}x + 9 = 3$ **14.** $5x + 2 < 1$ **15.** $3 - 4x = -5$

16. $8 - 5x \geq 13$ **17.** $3x + 4 = -2$ **18.** $9 - \frac{1}{2}x \leq 0$

Make a table to solve each word problem.

19. It cost \$25 a day and \$0.18 a mile to rent a car. Jennifer rented a car for 3 days and drove 400 miles. How much did she pay for the car rental?

20. Brian earns \$4.50 an hour and $1\frac{1}{2}$ times as much for overtime. If Brian worked 15 regular hours and 4 overtime hours, how much did he earn?

21. Twelve ounces of orange juice costs \$1.44. Sixteen ounces of orange juice costs \$1.76. Which size is the better buy?

Systems of Equations and Inequalities

Need a lift?

Developing Mathematical Power

In a hydraulic lift, the ratio of the force from the large piston to the force on the small piston is called the *mechanical advantage* (MA).

The Easy-Does-It Engineering company designs hydraulic lifts. The company has two new designs. One lift has a mechanical advantage of 49 and the other has a mechanical advantage of 46.

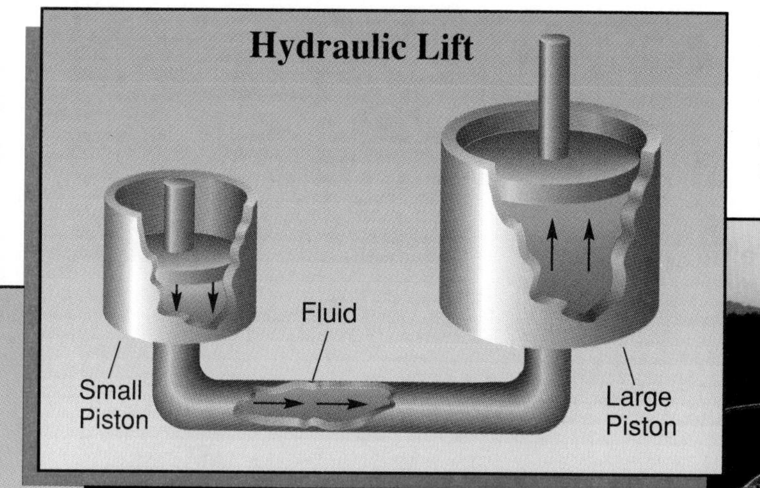

Project

Suppose you are a sales representative for Easy-Does-It. How would you use the different mechanical advantages of the hydraulic lifts as a selling point? Create a marketing plan for each lift.

4.1 Direct Variation

Objectives: To determine when a function is a direct variation
To solve problems involving direct variation

Roast beef is on sale for $3.00 per pound. The price y of a roast depends on its weight x. The ratio $\frac{y}{x}$ describes the relationship between the price of the roast and its weight.

Capsule Review

A *proportion*, $\frac{a}{b} = \frac{c}{d}$, $b \neq 0$ and $d \neq 0$, is an equation showing that two ratios are equal.

EXAMPLE **Is $\frac{6}{9} = \frac{18}{27}$ a true proportion?**

$\frac{6}{9} = \frac{2}{3}$ and $\frac{18}{27} = \frac{2}{3}$ $\qquad$ $\frac{6}{9} = \frac{18}{27}$ is a true proportion.

Tell whether each equation is a true proportion.

1. $\frac{1}{4} = \frac{2}{8}$ **2.** $\frac{2}{5} = \frac{6}{15}$ **3.** $\frac{9}{24} = \frac{12}{36}$ **4.** $\frac{20}{24} = \frac{30}{36}$

The table at the right shows the price of roasts of different weights. Notice that the ratio $\frac{y}{x}$ is the constant $\frac{3}{1}$, or 3. The function defined by the table can also be defined by an equation.

$$\frac{y}{x} = 3, \text{ or } y = 3x$$

Weight (in pounds) x	Price (in dollars) y	Ratio $\frac{y}{x}$
2	6	$\frac{6}{2}$, or $\frac{3}{1}$
4	12	$\frac{12}{4}$, or $\frac{3}{1}$
6	18	$\frac{18}{6}$, or $\frac{3}{1}$
8	24	$\frac{24}{8}$, or $\frac{3}{1}$

A linear function defined by an equation that can be written in the form $y = kx$, where $k \neq 0$, is called a **direct variation.** It is said that ***y* varies directly as *x*,** and k is the **constant of variation.**

EXAMPLE 1 **For each function, determine whether y varies directly as x. If so, find the constant of variation and write the equation.**

a.

x	y
2	8
3	12
5	20

$\frac{y}{x} = \frac{8}{2} = \frac{12}{3} = \frac{20}{5} = 4$, so y varies directly as x.

The constant of variation is 4.

The equation is $y = 4x$.

b.

x	y
1	4
2	7
5	10

Since $\frac{4}{1}, \frac{7}{2}$, and $\frac{10}{5}$ are not equal, the ratio $\frac{y}{x}$ is not a constant.

y does *not* vary directly as x.

You can use a calculator to compute the ratio $\frac{y}{x}$ for each ordered pair. If all of the ratios are not equal, the function is not a direct variation.

EXAMPLE 2 **For each function, determine whether y varies directly as x. If so, find the constant of variation.**

a. $y = \frac{2}{3}x$ **b.** $y = 2x + 3$

a. $y = \frac{2}{3}x$ Since the equation is of the form $y = kx$, y varies directly as x. The constant of variation is $\frac{2}{3}$.

b. $y = 2x + 3$ Since the equation cannot be written in the form $y = kx$, y does *not* vary directly as x.

An equation of the form $y = kx$ can be used to solve many direct variation problems.

EXAMPLE 3 ***y* varies directly as *x*, and *y* is 15 when *x* is 3. Find the constant of variation and find *y* when *x* is 6.**

First find the constant of variation.

$\frac{y}{x} = \frac{15}{3} = 5$ Therefore, $k = 5$.

Then write and solve an equation.

$y = kx$

$y = 5(6)$ *Substitute 5 for k and 6 for x.*

$y = 30$

y is 30 when x is 6.

Use the trace feature of your graphing utility to check answers.

If y varies directly as x, then for any two ordered pairs of nonzero numbers (x_1, y_1) and (x_2, y_2), $\frac{y_1}{x_1} = k$ and $\frac{y_2}{x_2} = k$, where $k \neq 0$. Since the ratios are equal, they form a proportion

$$\frac{y_1}{x_1} = \frac{y_2}{x_2} \qquad \textit{Read: } y_1 \text{ is to } x_1 \text{ as } y_2 \text{ is to } x_2.$$

Because of this, y is said to be *directly proportional to* x, and the constant of variation k is also called the **constant of proportionality.**

The proportion shown above can also be written this way:

$$y_1 : x_1 = y_2 : x_2$$

The first and fourth terms, y_1 and x_2, are called the **extremes,** and the second and third terms, x_1 and y_2, are called the **means.** If each side of the proportion $\frac{y_1}{x_1} = \frac{y_2}{x_2}$ is multiplied by $x_1 x_2$, the resulting equivalent equation is $y_1 x_2 = x_1 y_2$. Thus, the following statement is true for any proportion:

The product of the means equals the product of the extremes.

You can use proportions to solve problems involving direct variation. This can save you time if the problem does not ask for the constant of proportionality.

EXAMPLE 4 **If y varies directly as x, and $y = -51$ when $x = 27$, find x when $y = -17$.**

Let $(x_1, y_1) = (27, -51)$ and let $(x_2, y_2) = (x_2, -17)$.

$\frac{y_1}{x_1} = \frac{y_2}{x_2}$ — *Write a proportion.*

$\frac{-51}{27} = \frac{-17}{x_2}$ — *Substitute.*

$-51x_2 = 27(-17)$ — *The product of the means equals the product of the extremes.*

$x_2 = \frac{27(-17)}{-51}$ — *Calculation-ready form*

$x_2 = 9$ — x is 9 when y is -17.

In general, an equation shows a direct variation if y is equal to the product of a nonzero constant and a function of x. The following examples illustrate direct variation:

$\frac{y}{x^2} = \frac{5}{1}$, or $y = 5x^2$ — y varies directly as the square of x.

$\frac{y}{x^3} = \frac{2}{5}$, or $y = \frac{2}{5}x^3$ — y varies directly as the cube of x.

EXAMPLE 5 **If y varies directly as the square of x, and $y = 50$ when $x = 2$, find y when $x = 6$.**

Let $(x_1, y_1) = (6, y_1)$ and let $(x_2, y_2) = (2, 50)$.

$$\frac{y_1}{(x_1)^2} = \frac{y_2}{(x_2)^2} \qquad \textit{Write a proportion.}$$

$$\frac{y_1}{6^2} = \frac{50}{2^2} \qquad \textit{Substitute.}$$

$$\frac{y_1}{36} = \frac{50}{4}$$

$$y_1 = \frac{36(50)}{4}$$

$$y_1 = 450 \qquad y \text{ is } 450 \text{ when } x \text{ is } 6.$$

CLASS EXERCISES

For each function, determine whether y varies directly as x. If so, find the constant of variation and write the equation.

1.

x	y
2	4
4	8
16	32

2.

x	y
2	−6
4	−12
5	−15

3.

x	y
11	22
16	32
7	42

4.

x	y
27	9
30	10
60	20

For each function, determine whether y varies directly as x. If so, find the constant of variation.

5. $y = 4x - 3$ **6.** $y = -5x$ **7.** $y - 6x = 0$ **8.** $y + 3 = -3x$

In Exercises 9–10, use k as the constant of variation and write an equation for the direct variation.

9. The amount of commission C varies directly as the amount of sales S.

10. The amount earned E varies directly as the number of hours worked h.

PRACTICE EXERCISES

Use technology where appropriate.

For each function, determine whether y varies directly as x. If so, find the constant of variation and write the equation.

1.

x	y
2	14
3	21
5	35

2.

x	y
3	9
4	12
7	21

3.

x	y
−2	4
−3	6
−5	10

4.

x	y
1	−2
3	−8
5	14

5.

x	y
9	6
12	8
15	10

6.

x	y
4	1
6	2
8	3

7.

x	y
23	24
55	56
66	67

8.

x	y
2	2.6
3	3.9
4	5.2

For each function, determine whether y varies directly as x. If so, find the constant of variation.

9. $y = 12x$

10. $y = 6x$

11. $y = -2x$

12. $y = 4x + 1$

13. $y = -2x + 1$

14. $y + 2x = 1$

15. $y = x$

16. $y = 3x - 5$

17. $3y = 4x$

18. $2y - 5x = 0$

19. $\frac{y}{x} = 6$

20. $\frac{y}{x} - 8 = 3$

In Exercises 21–26, y varies directly as x.

21. If $y = 4$ when $x = 3$, find y when $x = 6$.

22. If $y = 7$ when $x = 2$, find y when $x = 8$.

23. If $y = 4$ when $x = -2$, find x when $y = 6$.

24. If $y = 6$ when $x = 2$, find x when $y = 12$.

25. If $y = 7$ when $x = 2$, find y when $x = 3$.

26. If $y = 5$ when $x = -3$, find y when $x = -1$.

For each function, determine whether y varies directly as x. If so, find the constant of variation.

27. $4y - 2x = 4$

28. $y = 2\pi x$

29. $y = x + \pi$

30. $y = 2x^2$

31. $y - 4x^3 = 0$

32. $\frac{3y}{5} = 4x$

In Exercises 33–40, y varies directly as x.

33. If $y = 3$ when $x = 2$, find y when $x = 22$.

34. If $y = 4$ when $x = 1$, find y when $x = 36$.

35. If $y = 7$ when $x = 3$, find x when $y = 21$.

36. If $y = 25$ when $x = 15$, find x when $y = 10$.

37. If $y = 30$ when $x = -3$, find y when $x = -9$.

38. If $y = -20$ when $x = 2$, find y when $x = 14$.

39. If $y = 0.9$ when $x = 4.8$, find y when $x = 6.4$.

40. If $y = -7.2$ when $x = 2.6$, find y when $x = 3.9$.

In Exercises 41–44, y varies directly as x^2.

41. If $y = 4$ when $x = 6$, find y when $x = 9$.

42. If $y = 15$ when $x = 5$, find y when $x = 10$.

43. If $y = 16$ when $x = 4$, find x when $y = 25$.

44. If $y = 9$ when $x = -3$, find x when $y = 81$.

In Exercises 45–48, y varies directly as x.

45. If x is doubled, what happens to y?

46. If x is halved, what happens to y?

47. If x is divided by 7, what happens to y?

48. If x is multiplied by 10, what happens to y?

In Exercises 49–52, y varies directly as x^2.

49. If x is doubled, what happens to y?

50. If x is halved, what happens to y?

51. If x is multiplied by 3, what happens to y?

52. If x is divided by 4, what happens to y?

Applications

53. Transportation A train travels at a constant rate. If it travels 100 mi in 2 h, how far will it travel in 7 h?

54. Electricity In a circuit, the voltage varies directly as the current. If the voltage is 75 volts when the current is 15 amps, find the voltage when the current is 10 amps.

55. Real Estate The amount of commission is directly proportional to the amount of sales. A realtor received a commission of $18,000 on the sale of a $225,000 house. How much would the commission be on a $130,000 house?

56. Consumerism The Plant Stop is selling marigolds in units of one tray, three trays, or flats of eight trays, with different rates for each amount. One tray sells for $1.49, a batch of three trays sells for $3.99, and a flat of eight trays sells for $11.00. Use three direct variations to determine what each unit would cost at the first rate, at the second rate, and at the third rate. Which is the best buy?

Developing Mathematical Power

57. Thinking Critically What characteristic(s) is (are) common to the graphs of all direct variations?

58. Thinking Critically Direct variations have been described as linear functions. Yet $y = 5x^2$, which can show direct variation (y varying directly as the square of x), has a graph that is not a line. How could the axes of the graph of $y = 5x^2$ be labeled to demonstrate the linear nature of this direct variation?

4.2

Problem Solving Strategy: Draw a Graph

A good strategy to use in solving problems involving mathematical functions is to draw graphs of the functions. A graph of a function is the set of points (x, y) in a coordinate plane for which the rule of the function assigns y to x. A graph may or may not give the precise solution to a problem. However, a graph is usually very helpful in understanding the conditions and solution of a problem.

EXAMPLE The equations of two linear functions are $x - y = 3$ and $2x + y = 6$. Do these two equations have a solution in common? If so, what is it?

Understand the Problem *What are the given facts?* The equations of two linear functions are $x - y = 3$ and $2x + y = 6$.
What are you asked to find? Do the equations have a common solution? If so, what is it?

Plan Your Approach *Choose a strategy.* The graph of a linear function is a line. Therefore, the equations $x - y = 3$ and $2x + y = 6$ can each be represented by a line.

The strategy to use in order to answer the question is to draw the graphs of the two equations. First, write each equation in slope-intercept form.

$$\begin{aligned} x - y &= 3 \\ -y &= -x + 3 \quad \text{Solve for } y. \\ y &= x - 3 \end{aligned}$$

$$\begin{aligned} 2x + y &= 6 \\ y &= -2x + 6 \quad \text{Solve for } y. \end{aligned}$$

The slope is 1.
The y-intercept is -3.

The slope is -2.
The y-intercept is 6.

Think of the possible relationships of two lines in a plane.

- The two lines are *parallel* if the slopes are the same and the y-intercepts are different.
- The two lines *coincide* if the slopes and y-intercepts are the same. If this is the case, the two equations are equivalent.
- The two lines *intersect* in one point if the slopes are different. If the y-intercepts are the same, then this is the point of intersection.

Since the two lines $x - y = 3$ and $2x + y = 6$ have different slopes and y-intercepts, they intersect in one point, which is the common solution.

Complete the Work Graph each equation on the same set of axes. Read the coordinates of the point of intersection.

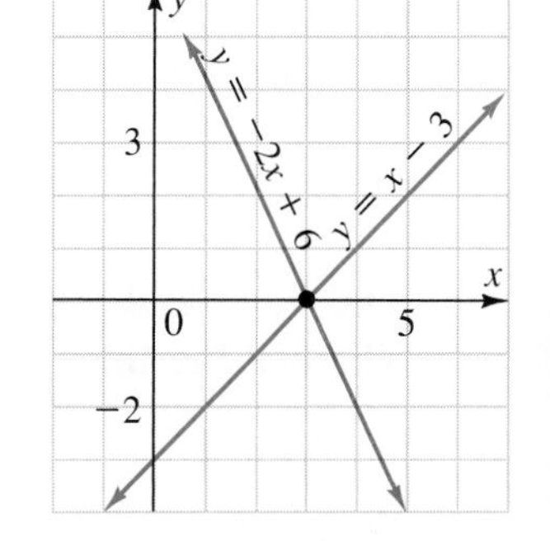

Interpret the Results *State your answer.* The coordinates of the point of intersection are (3, 0). Thus, (3, 0) is a solution to both equations $x - y = 3$ and $2x + y = 6$.

Check your answer. Substitute (3, 0) into each equation.

$x - y = 3$	$2x + y = 6$
$3 - 0 \stackrel{?}{=} 3$	$2(3) + 0 \stackrel{?}{=} 6$
$3 = 3$ ✓	$6 = 6$ ✓

CLASS EXERCISES

1. How many variables does the equation $x - y = 3$ have?
2. What is a solution to an equation such as $2x + y = 6$ called?
3. Is it possible to list all the solutions to $x - y = 3$ or $2x + y = 6$? Explain your answer.
4. Are all the solutions to $x - y = 3$ pairs of whole numbers?
5. Do any of the solutions to $2x + y = 6$ contain pairs of rational numbers?
6. How can you determine if two equations in two variables are equivalent equations?
7. If two lines are parallel, what is true about their slopes?
8. If two lines are perpendicular, what is true about their slopes?
9. What is the x-coordinate of a line at the point it intersects the y-axis?
10. What is the y-coordinate of a line at the point it intersects the x-axis?
11. If two lines are parallel, is there an ordered pair of numbers that is a solution to both of the equations?
12. Is every point on the graph of the line $x - y = 3$ a solution of the equation?

PRACTICE EXERCISES

Find the slope and y-intercept for each linear function.

1. $y - 3x = 7$

2. $y + 4x = 5$

3. $y - \frac{1}{2}x = -3$

4. $2x + 3y = 9$

5. $-6x + 2y = 9$

6. $3y + 2x = 4$

7. $\frac{1}{2}x + y = 1$

8. $8x + 9y = 16$

9. $\frac{2}{3}x - \frac{1}{2}y = 4$

The slope and y-intercept of different pairs of lines are given below. Tell whether the lines are parallel, perpendicular, intersect, or coincide.

	Line 1		Line 2	
10.	slope:	-2	slope:	-2
	y-intercept:	4	y-intercept:	0
11.	slope:	$\frac{3}{4}$	slope:	4
	y-intercept:	2	y-intercept:	-1
12.	slope:	5	slope:	5
	y-intercept:	-4	y-intercept:	-4
13.	slope:	$-\frac{2}{3}$	slope:	$\frac{3}{2}$
	y-intercept:	4	y-intercept:	-3

14. If two lines coincide, how many common solutions do they have?

15. If two lines are parallel, how many common solutions do they have?

16. If two lines intersect, how many common solutions do they have?

Tell whether or not the following linear functions have a common solution. If so, state the number of solutions.

17. $4x + 3y = 5$ $\quad$ $y + 8x = 5$

18. $-x + 2y = 3$ $\quad$ $6y - 3x = 9$

19. $7y + 14x = 21$ $\quad$ $y = -2x + 3$

20. Can graphs of linear functions be used to find the exact solution to any pair of functions? Explain your answer.

21. Draw a circle and a line to show their possible relationships. What are the possible number of solutions that exist for a linear function and a circle?

Mixed Problem Solving Review

1. The measure of the complement of an angle is fifteen less than four times the measure of the angle. Find the measure of the angle and its complement.
2. Two people leave the same place together and drive in opposite directions at the rates of 60 mi/h and 40 mi/h. In how many hours will they be 400 miles apart?
3. A rectangular plot of land has a perimeter of 680 ft. The land is 60 ft longer than it is wide. Find the length and width of the land.
4. The sum of four consecutive even integers is 172. Find the four integers.
5. The demand curve for a new brand of cereal was found to be linear. The demand was 10 cases at $60 and 15 cases at $55. Write the function for this product and find the price at which 25 cases could be sold.
6. The Quickstar Baseball Company manufactures two grades of baseballs, the Speedee, which sells for $2.10, and the Quickfire, which sells for $3.50. The company receives $843.50 for an order of 305 balls. How many of each ball were ordered?
7. Enid scored 87, 93, 100, and 85 on four tests. What must she score on her fifth test to have an average of 90?
8. A coin bank contains $12.45 in quarters and dimes. If there are 16 more dimes than quarters, how many are there of each?
9. Mike spent $108 clothes shopping. He bought one pair of jeans for $5x$ dollars, a shirt for $4x$ dollars, and 3 t-shirts at x dollars each. How much did each item cost?
10. Value Hardware sells 3-in. finishing nails for 0.99/lb, and 2-in. finishing nails sell for $1.39/lb. Tony needs 5 lb of nails. If his bill totals $6.15, how many pounds of each size did he buy?

PROJECT

Algebra and geometry are two different branches of mathematics. Algebra is concerned with variables, equations, and inequalities. Geometry deals with points, lines, triangles, and other figures. But you have now learned that certain equations in two variables can be graphed as lines and that lines can thus be represented by algebraic equations. Do library research to determine who the mathematician was that united algebra and geometry. Write a brief report about this person and what he did.

4.3

Solving Linear Systems Using Graphs

Objectives: To find the solution of a linear system of equations in two variables graphically
To determine whether a linear system of equations is consistent or inconsistent, dependent or independent

If a linear equation has two variables, there is generally an infinite number of ordered pairs of numbers that make it true. Each ordered pair is a solution of the equation and represents a point on its graph. The set of all such ordered pairs is the solution set of the equation, and the graph of the solution set is a line.

Capsule Review

The slope-intercept method can be used to graph a linear equation.

EXAMPLE **Graph: $x + 2y = -4$**

Solve for y.

$x + 2y = -4$

$2y = -x - 4$

$y = -\frac{1}{2}x - 2$

Slope: $-\frac{1}{2}$

y-intercept: -2

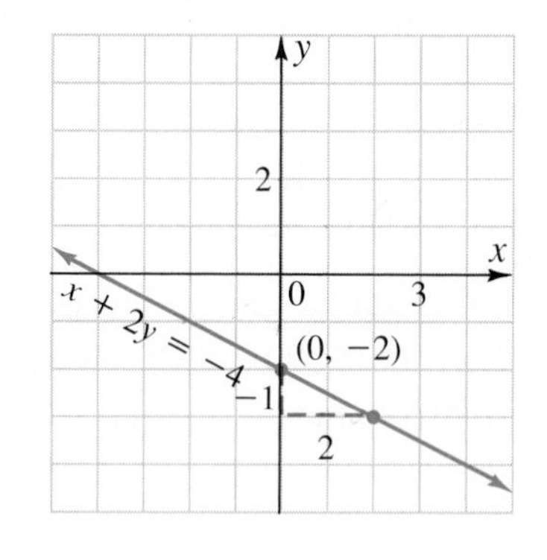

Graph each line.

1. $y = 2x - 1$ **2.** $y - 3x = 4$ **3.** $2y + 3x = 6$ **4.** $-2y + x = 0$

To solve some mathematical problems it is necessary to consider a *common solution* of two equations in two variables. A **system of equations** is a set of two or more equations. A system consisting of linear equations is called a **linear system.**

To find a common solution for a linear system of two equations in two variables, find an ordered pair of numbers that satisfies both equations. One way to do this is to graph both equations on one set of axes and find the point of intersection of the graphs, if it exists.

EXAMPLE 1 **Solve the system graphically:** $\begin{cases} y - x = 3 \\ y + 2x = -3 \end{cases}$

Write each equation in slope-intercept form.

$$y - x = 3 \qquad\qquad y + 2x = -3$$
$$y = x + 3 \qquad\qquad y = -2x - 3$$

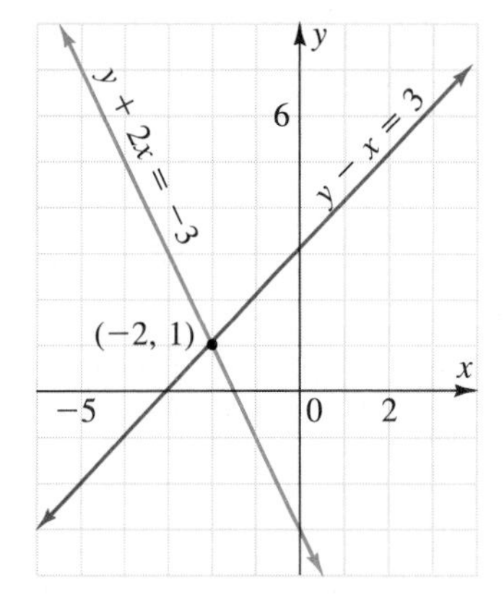

Graph each equation on the same set of axes. Read the coordinates of the point of intersection, if it exists. In this example, that point is $(-2, 1)$.

Check $(-2, 1)$ in each equation.

$$y - x = 3 \qquad\qquad y + 2x = -3$$
$$1 - (-2) \stackrel{?}{=} 3 \qquad\qquad 1 + 2(-2) \stackrel{?}{=} -3$$
$$3 = 3 \checkmark \qquad\qquad -3 = -3 \checkmark$$

The solution is $(-2, 1)$.

A linear system that has *at least one* solution is called a **consistent system.** If a linear system has *exactly one* solution, it is said to be an **independent system.** Therefore, the system in Example 1 is consistent and independent. A graphing utility can be used to verify the solution of the system.

EXAMPLE 2 **Solve the system graphically:** $\begin{cases} 2y - x = 4 \\ 4y = 2x + 12 \end{cases}$

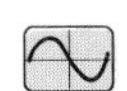

Write each equation in slope-intercept form.

$$2y - x = 4 \qquad\qquad 4y = 2x + 12$$
$$2y = x + 4 \qquad\qquad y = \frac{1}{2}x + 3$$
$$y = \frac{1}{2}x + 2$$

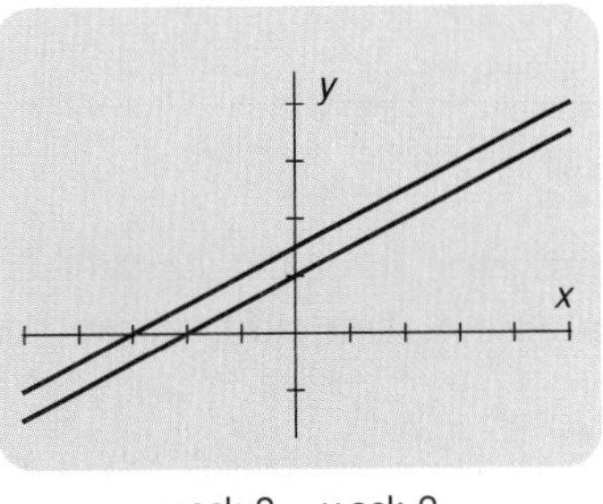

x scl: 2 y scl: 2

Graph the equations. This graph is shown as it might appear on a graphing utility.

The lines have the same slope but different y-intercepts, so they are parallel. Since the lines have no point of intersection, the system has *no* solution.

A linear system of equations that has *no* solution is said to be an **inconsistent system.** Therefore, the system in Example 2 is inconsistent. Some linear systems, such as the one in the next example, have an *infinite number* of solutions. Such a system is called a **dependent system.** A dependent system has *at least one* solution, so it is also consistent. What difficulties exist in solving a system of linear equations graphically? Explain.

EXAMPLE 3 **Solve the system graphically:** $\begin{cases} 2y = 4x - 4 \\ -8x + 4y = -8 \end{cases}$

Write each equation in slope-intercept form.

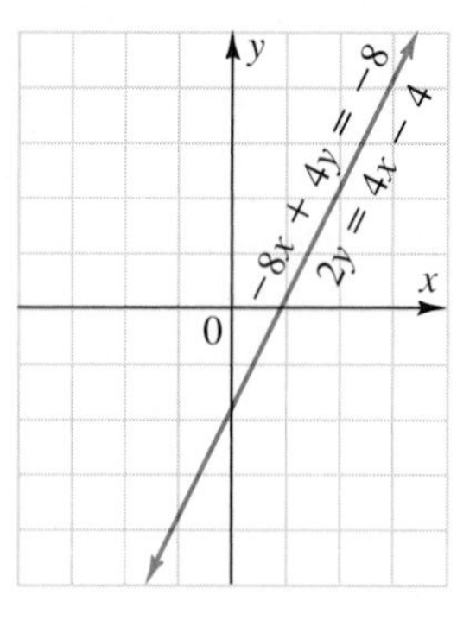

$$\begin{aligned} 2y &= 4x - 4 \\ y &= 2x - 2 \end{aligned} \qquad \begin{aligned} -8x + 4y &= -8 \\ 4y &= 8x - 8 \\ y &= 2x - 2 \end{aligned}$$

Graph the equations.

Since the slopes and y-intercepts are identical, the lines coincide. Every point on the line is a common solution of the system. The solution set of the system is $\{(x, y): y = 2x - 2\}$.

Graphical Solutions of Linear Systems of Equations in Two Variables

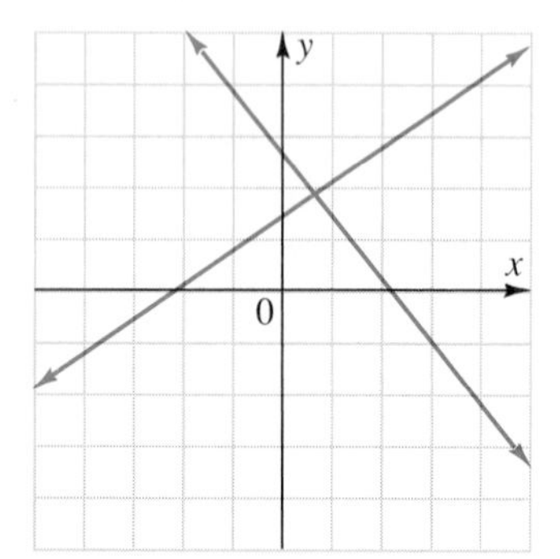

Intersecting lines
One solution
Consistent and independent

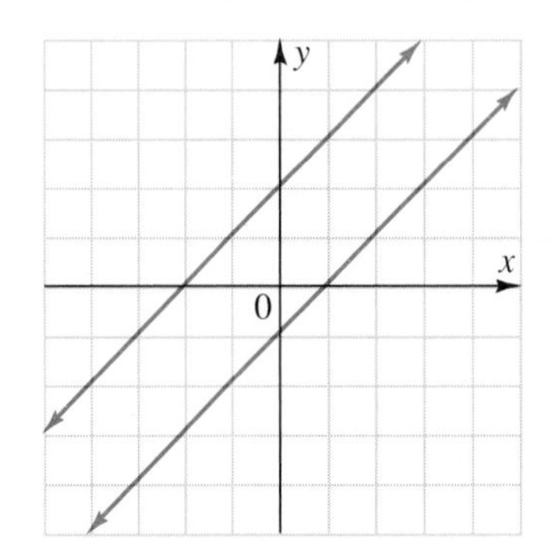

Parallel lines
No solution
Inconsistent

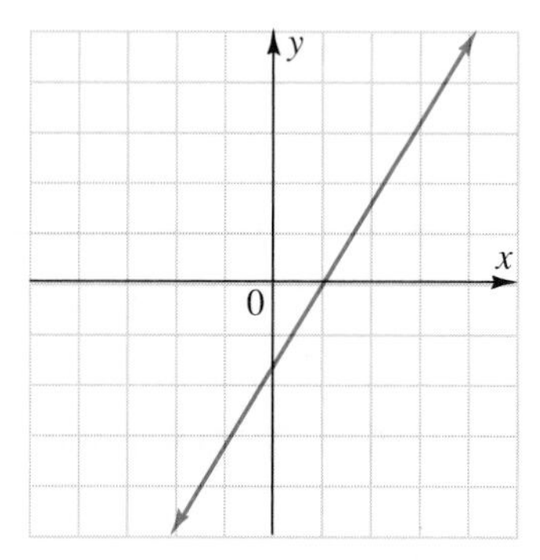

Identical lines
Infinite number of solutions
Consistent and dependent

CLASS EXERCISES

For each linear system of equations, predict whether the lines will intersect, coincide, or be parallel.

1. $\begin{cases} y = x + 2 \\ y = -2x + 3 \end{cases}$

2. $\begin{cases} y = -3x + 3 \\ y = 2x - 7 \end{cases}$

3. $\begin{cases} y = -4x + 9 \\ y = 3x - 5 \end{cases}$

4. $\begin{cases} y = x - 2 \\ y = -2x + 7 \end{cases}$

5. $\begin{cases} y = \frac{3}{4}x - 8 \\ y = \frac{3}{4}x + 8 \end{cases}$

6. $\begin{cases} y = \frac{1}{3}x + 3 \\ y = \frac{1}{3}x - 3 \end{cases}$

7. $\begin{cases} x - 3y = -6 \\ y = -2x + 6 \end{cases}$

8. $\begin{cases} 2x - 4y = 4 \\ -x + 2y - 2 = 0 \end{cases}$

9. $\begin{cases} 5x - 7y = -7 \\ y = \frac{5}{7}x + 1 \end{cases}$

PRACTICE EXERCISES

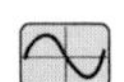 Use technology where appropriate.

Solve each linear system graphically. Determine whether the system is consistent and independent, inconsistent, or consistent and dependent.

1. $\begin{cases} x + y = 2 \\ y = 2x - 1 \end{cases}$

2. $\begin{cases} x + y = 3 \\ y = 3x - 1 \end{cases}$

3. $\begin{cases} 3x + y = 5 \\ x - y = 7 \end{cases}$

4. $\begin{cases} 3y - 4x = 0 \\ y - x = 1 \end{cases}$

5. $\begin{cases} 2x + 4y = 12 \\ x + y = 2 \end{cases}$

6. $\begin{cases} 2x - 2y = 4 \\ y - x = 6 \end{cases}$

7. $\begin{cases} 2x + y = 5 \\ y - x = -1 \end{cases}$

8. $\begin{cases} x + y = 3 \\ 2x + y = 10 \end{cases}$

9. $\begin{cases} x + 3y = 6 \\ -6y - 2x + 12 = 0 \end{cases}$

10. $\begin{cases} 2x - y = 4 \\ 2y - 4x + 8 = 0 \end{cases}$

11. $\begin{cases} -2x = y - 1 \\ y + x = 4 \end{cases}$

12. $\begin{cases} y = 1 - x \\ 2x + y = 9 \end{cases}$

13. $\begin{cases} 3x - 4y = 13 \\ 2x + y = 5 \end{cases}$

14. $\begin{cases} 2x - y = 0 \\ 2x + 4y = -10 \end{cases}$

15. $\begin{cases} 2x = y - 7 \\ 4x - 2y - 14 = 0 \end{cases}$

16. $\begin{cases} 3y = 1 - 4x \\ 8x + 6y - 12 = 0 \end{cases}$

17. $\begin{cases} y = 2 \\ y - 4 = x \end{cases}$

18. $\begin{cases} y = x \\ y - 7x = 0 \end{cases}$

19. $\begin{cases} y = 3 \\ 7x = 5y - 1 \end{cases}$

20. $\begin{cases} x = 2 \\ 2y = 4x + 2 \end{cases}$

21. $\begin{cases} 4y - 2x = 6 \\ 8y = 4x - 12 \end{cases}$

22. $\begin{cases} x - y = 1 \\ -7x + 7y = -7 \end{cases}$

23. $\begin{cases} 2y = -3x \\ 2y - 3x = 0 \end{cases}$

24. $\begin{cases} x - 2y + 1 = 0 \\ 2x + 4y - 6 = 0 \end{cases}$

25. $\begin{cases} x = 2 \\ y = 3 \end{cases}$

26. $\begin{cases} x = 4 \\ y = -7 \end{cases}$

27. $\begin{cases} y - x = 0 \\ y = -x \end{cases}$

28. $\begin{cases} 2y - x = 0 \\ \frac{1}{2}x - y = 0 \end{cases}$

29. $\begin{cases} 0.2x + 0.5y = 0.1 \\ 0.4x - 0.2y = -1 \end{cases}$

30. $\begin{cases} 0.3x - 0.2y = 1.1 \\ 0.1x - 0.4y = 0.7 \end{cases}$

Without graphing, determine whether each linear system is consistent and independent, inconsistent, or consistent and dependent. Check by graphing.

31. $\begin{cases} 2x - 3y = 4 \\ 3x - 4y = 7 \end{cases}$

32. $\begin{cases} 2x + 3y = -5 \\ x - 2y = 7 \end{cases}$

33. $\begin{cases} 3x - 2y = 8 \\ y = \dfrac{6x - 5}{4} \end{cases}$

34. $\begin{cases} 3x = -5y + 4 \\ y = \dfrac{-3x + 6}{5} \end{cases}$

35. $\begin{cases} \dfrac{2}{3} = \dfrac{x - y}{5} \\ \dfrac{1}{2} = \dfrac{2x + y}{4} \end{cases}$

36. $\begin{cases} \dfrac{3}{5} = \dfrac{x + 2y}{2} \\ \dfrac{4y + 2x}{4} = \dfrac{6}{10} \end{cases}$

37. Find and graph two linear equations that form a consistent, dependent system.

38. Is it possible to have an inconsistent linear system in which both lines have the same y-intercept? Explain.

Applications

Market Research The pricing of products, especially in fluctuating markets, is a large component of business market research and development. Generally, the lower the price, the more people will buy a product. However, when the demand exceeds the supply, the price begins to rise. The question about any product is this: At what price, called the *equilibrium price,* does *supply equal demand*?

For example, the sales of T-shirts at a tourist attraction can be represented by these equations:

$y = -0.3x + 10$ the demand equation
$y = 0.5x + 2$ the supply equation

where $y =$ the price of a T-shirt, in dollars, and
$x =$ the number of T-shirts, in hundreds.

If you graph these equations on a graphing utility, you can use the trace feature to find the equilibrium price, which is $7.00.

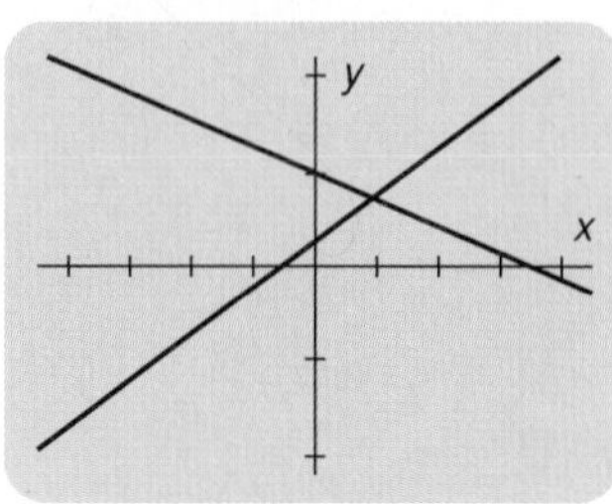

x scl: 10 y scl: 10

Since the equilibrium price will always be positive, you may wish to explore the range or the zoom feature on the calculator in order to show only the first quadrant of the graph.

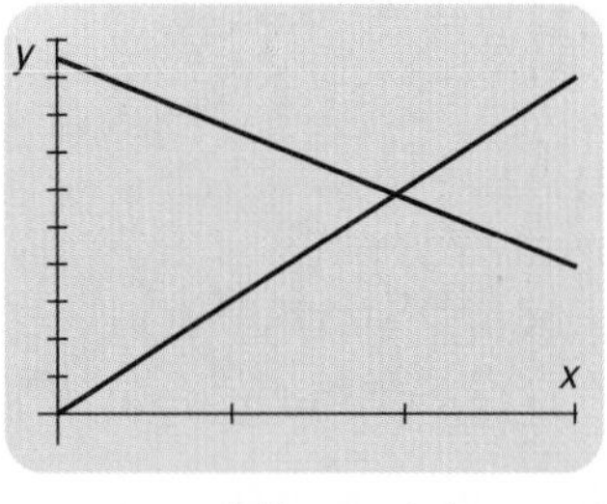

x scl: 5 y scl: 1

39. How would you interpret the graph of the two equations for values of $x < 10$? for values of $x > 10$?

40. The fruit market is subject to great fluctuation. Suppose the demand for strawberries in a region is given by $y = -0.2x + 6$ and the supply for strawberries is given by $y = 0.8x$ where y represents the price of a gallon of strawberries in dollars. Find the equilibrium price.

41. Gather the data for the supply and demand of different products at a local store. Try describing the supply-and-demand data in a linear model and find the equilibrium price. Does that price match the store's price for the product? Are there products for which the data does not easily fit into a linear model? Why?

Mixed Review

42. Find an equation for the line passing through each pair of points.
a. (0, 5), (3, 7) **b.** (4, −2), (−2, 10) **c.** (5, 12), (−3, 4)

4.4

Solving Linear Systems Using the Addition Method

Objective: To solve a linear system in two variables by addition

Ted Johnson bought antifreeze and oil for his truck—7 quarts in all. If you know the price of a quart of each and the difference between the total amounts paid for each, you can find the amount of antifreeze and the amount of oil he bought by writing a system of equations. One method for solving a system of linear equations uses additive inverses and the addition property of equations.

Capsule Review

For any number a and its *additive inverse* $-a$, $a + (-a) = 0$.

Find the additive inverse of each term.

1. 5 **2.** -18 **3.** 1 **4.** $-6x$ **5.** $7y$ **6.** $-x$

For a linear system of equations, if the coefficients of one variable are additive inverses of each other, that variable can be eliminated by adding the equations. To add two equations, add left side to left side and right side to right side. The result is an equation in one variable.

EXAMPLE 1 Antifreeze costs \$2 per quart, and oil costs \$1 per quart. Ted bought 7 quarts in all, and he paid \$5 more for the antifreeze than for the oil. How many quarts of each did he buy?

Let x = number of quarts of antifreeze.
Let y = number of quarts of oil.

$\begin{cases} x + y = 7 \\ 2x - y = 5 \end{cases}$ *Write a system of linear equations.*

$3x = 12$ *Add the equations.*
$x = 4$ *Solve for x.*

$x + y = 7$ *Use one of the original equations.*
$4 + y = 7$ *Substitute 4 for x.*
$y = 3$ *Solve for y.*

Ted bought 4 qt of antifreeze and 3 qt of oil.

$2x - y = 5$ *To check, use the other original equation, since the solution must make both equations true.*
$2(4) - 3 \stackrel{?}{=} 5$
$5 = 5$ ✓

If the coefficients of one variable are equal, multiply both sides of one equation by -1 to make those coefficients additive inverses. The multiplication property of equations guarantees that the resulting system of equations is equivalent to the original system. **Equivalent systems** are systems that have the same solutions.

EXAMPLE 2 **Solve the system by addition:** $\begin{cases} 3x + y = 2 \\ 3x + 2y = 7 \end{cases}$

$$\begin{cases} 3x + y = 2 \\ 3x + 2y = 7 \end{cases}$$

$$\begin{aligned} 3x + y &= 2 \\ -3x - 2y &= -7 \end{aligned}$$ *Multiply both sides of the second equation by -1.*

$$-y = -5$$ *Add the equations.*

$$y = 5$$

$$3x + y = 2$$ *Use one original equation.*

$$3x + 5 = 2$$ *Substitute 5 for y.*

$$3x = -3$$ *Solve for x.*

$$x = -1$$

Check: $3x + 2y = 7$ *Use the other original equation.*

$3(-1) + 2(5) \stackrel{?}{=} 7$ *Substitute -1 for x and 5 for y.*

$7 = 7$ ✓ The solution is $(-1, 5)$.

It is sometimes necessary to multiply one or both equations by a nonzero real number other than -1 in order to make the coefficients of one of the variables additive inverses of each other.

EXAMPLE 3 **Solve the system by addition:** $\begin{cases} 3x + 4y = -10 \\ 5x - 2y = 18 \end{cases}$

$$\begin{cases} 3x + 4y = -10 \\ 5x - 2y = 18 \end{cases}$$

$$\begin{aligned} 3x + 4y &= -10 \\ 10x - 4y &= 36 \end{aligned}$$ *Multiply both sides of the second equation by 2.*

$$13x = 26$$ *Add.*

$$x = 2$$

$$3x + 4y = -10$$

$$3(2) + 4y = -10$$ *Substitute 2 for x.*

$$4y = -16$$

$$y = -4$$

Check: $5x - 2y = 18$ *Use the other equation.*

$5(2) - 2(-4) \stackrel{?}{=} 18$

$18 = 18$ ✓ The solution is $(2, -4)$.

It makes no difference which variable is eliminated when the equations are added. The solution will be the same. But before you add, it may be necessary to rewrite the equations so that the variables are all on one side.

EXAMPLE 4 **Solve the system by addition:** $\begin{cases} 5x = -3y - 2 \\ 2y = -3x - 1 \end{cases}$

$$\begin{cases} 5x = -3y - 2 \\ 2y = -3x - 1 \end{cases}$$

$$\begin{cases} 5x + 3y = -2 \\ 3x + 2y = -1 \end{cases}$$ *Write the equations with the variables on one side.*

$10x + 6y = -4$ *Multiply both sides by 2.*

$-9x - 6y = 3$ *Multiply both sides by −3.*

$x = -1$ *Add.*

$2y = -3x - 1$

$2y = -3(-1) - 1$ *Substitute −1 for x.*

$2y = 2$

$y = 1$

Check: $5x = -3y - 2$

$5(-1) \stackrel{?}{=} -3(1) - 2$

$-5 = -5$ ✓ The solution is $(-1, 1)$.

The linear systems of equations in Examples 1–4 are consistent and independent. When the addition method is used with an inconsistent system, both variables are eliminated and the resulting equation is a contradiction, $0 = c$, where c is a nonzero constant. An inconsistent system has *no* solution.

EXAMPLE 5 **Solve the system by addition:** $\begin{cases} 2x - 4y = 8 \\ 6x - 12y = 10 \end{cases}$

$$\begin{cases} 2x - 4y = 8 \\ 6x - 12y = 10 \end{cases}$$

$-6x + 12y = -24$ *Multiply each side by −3.*

$6x - 12y = 10$

$0 = -14$ *Add. The result is a contradiction.*

No ordered pair (x, y) can satisfy both equations. There is no solution.

When the addition method is used with a dependent system, both variables are eliminated and the result is the true equation $0 = 0$. A dependent system has an infinite number of solutions. Since the equations in the system are equivalent, the solution set of the system is all ordered pairs (x, y) that satisfy either equation.

EXAMPLE 6 **Solve the system by addition:** $\begin{cases} -8x + 12y = 24 \\ 2x - 3y = -6 \end{cases}$

$$\begin{cases} -8x + 12y = 24 \\ 2x - 3y = -6 \end{cases}$$

$$\begin{array}{rl} -8x + 12y = 24 & \\ 8x - 12y = -24 & \textit{Multiply by 4.} \\ \hline 0 = 0 & \textit{Add.} \end{array}$$

The solution is $\{(x, y): 2x - 3y = -6\}$. Thus there are infinitely many solutions, two of which are (0, 2) and (3, 4).

CLASS EXERCISES

By what number would you multiply the first equation in order to eliminate the variable *x* when the equations are added?

1. $\begin{cases} x + 3y = 10 \\ x + y = 6 \end{cases}$ **2.** $\begin{cases} x + 2y = -2 \\ 3x + y = 9 \end{cases}$ **3.** $\begin{cases} -2x + 3y = 13 \\ 8x + 7y = 5 \end{cases}$

By what numbers would you multiply the equations in order to eliminate the variable *y* when the two equations are added?

4. $\begin{cases} 9x + 5y = 1 \\ -2x + 3y = 8 \end{cases}$ **5.** $\begin{cases} -5x + 6y = 8 \\ 11x + 8y = 46 \end{cases}$ **6.** $\begin{cases} -4x + 3y = 12 \\ 5x - 2y = -8 \end{cases}$

Are the given linear systems of equations equivalent? Give a reason for your answer.

7. $\begin{cases} x + y = 5 \\ x - y = 1 \end{cases}$ $\begin{cases} x = 2 \\ y = 3 \end{cases}$ **8.** $\begin{cases} 3x - 2y = 8 \\ x + y = -1 \end{cases}$ $\begin{cases} x = 2 \\ y = -1 \end{cases}$

9. $\begin{cases} 3x + 2y = 12 \\ 6x + 5y = 27 \end{cases}$ $\begin{cases} x - 2 = 0 \\ y - 3 = 0 \end{cases}$ **10.** $\begin{cases} 6x + 8y = 24 \\ 3x - 5y = 30 \end{cases}$ $\begin{cases} x = \frac{20}{3} \\ y = -2 \end{cases}$

PRACTICE EXERCISES

Solve each linear system by addition. If the system is inconsistent, write *no solution*.

1. $\begin{cases} x + y = 12 \\ x - y = 2 \end{cases}$ **2.** $\begin{cases} x + y = 13 \\ x - y = 5 \end{cases}$ **3.** $\begin{cases} -x + 2y = -1 \\ x - 3y = -1 \end{cases}$

4. $\begin{cases} 3x + 4y = 9 \\ -3x - 2y = -3 \end{cases}$ **5.** $\begin{cases} x + 2y = 10 \\ x + y = 6 \end{cases}$ **6.** $\begin{cases} x + 3y = 11 \\ x + 4y = 14 \end{cases}$

7. $\begin{cases} 4x + 2y = 4 \\ 6x + 2y = 8 \end{cases}$ **8.** $\begin{cases} 5x + 3y = 30 \\ 3x + 3y = 18 \end{cases}$ **9.** $\begin{cases} x + 3y = 11 \\ 3x - y = 3 \end{cases}$

10. $\begin{cases} 3x + 2y = 9 \\ -x + 3y = 8 \end{cases}$

11. $\begin{cases} x - 3y = 1 \\ 6x - y = 6 \end{cases}$

12. $\begin{cases} x + 2y = 9 \\ 2x - y = 8 \end{cases}$

13. $\begin{cases} 3x + y = -3 \\ x + 4y = 10 \end{cases}$

14. $\begin{cases} 5x - 2y = 3 \\ 2x - y = 0 \end{cases}$

15. $\begin{cases} 4x - 6y = -26 \\ -2x + 3y = 13 \end{cases}$

16. $\begin{cases} 5x - 2y = -19 \\ 2x + 3y = 0 \end{cases}$

17. $\begin{cases} 2x - 3y = 6 \\ 6x - 9y = 9 \end{cases}$

18. $\begin{cases} 3x + 2y = 10 \\ 6x + 4y = 15 \end{cases}$

19. $\begin{cases} 2x = 8y + 24 \\ 3x + 5y = 2 \end{cases}$

20. $\begin{cases} 2x = -9y - 12 \\ 4x + 3y = 6 \end{cases}$

21. $\begin{cases} 5x - 7y = 2 \\ 3x = 4y \end{cases}$

22. $\begin{cases} 0.03x - 0.02y = 0.03 \\ 0.8x - 0.4y = 0.4 \end{cases}$

23. $\begin{cases} 0.01x + 0.02y = 0.14 \\ 0.3x + 0.4y = 1.8 \end{cases}$

24. $\begin{cases} 0.03x - 0.02y = 0.03 \\ 0.09x - 0.06y = 0.08 \end{cases}$

25. $\begin{cases} 0.04x + 0.06y = 0.12 \\ 0.16x + 0.24y = 0.36 \end{cases}$

26. $\begin{cases} 0.03x + 0.04y = 0.01 \\ 0.09x + 0.12y = 0.3 \end{cases}$

27. $\begin{cases} 1.5x - 2.3y = 3.4 \\ 0.15x - 0.23y = 0.34 \end{cases}$

28. $\begin{cases} 5x - 4y = 47 + 5y \\ 3x + 2y = 18 - 3x \end{cases}$

29. $\begin{cases} 10x + 10y = 40 - 5x \\ 9x + 2y = 32 - 6y \end{cases}$

30. $\begin{cases} 2(2x + 3y) = 0 \\ 7x = 3(2y + 3) + 2 \end{cases}$

31. $\begin{cases} 4(3x - y) = 0 \\ 3(x + 3) = 10y \end{cases}$

32. $\begin{cases} x - \dfrac{3y}{2} = -\dfrac{1}{2} \\ x + \dfrac{4y}{3} = 8 \end{cases}$

33. $\begin{cases} x + \dfrac{y}{2} = \dfrac{3}{2} \\ x + \dfrac{y}{4} = \dfrac{13}{4} \end{cases}$

34. $\begin{cases} \dfrac{3y - 1}{2} = \dfrac{3x + 8}{5} \\ \dfrac{x + y}{2} = \dfrac{6 + x - y}{2} \end{cases}$

35. $\begin{cases} 2y - 1 = \dfrac{3x - y}{5} \\ \dfrac{3x - 4}{8} = \dfrac{y}{4} \end{cases}$

Applications

36. **Geometry** One side of an angle lies on a line with the equation $2x - y = 3$. The other side of the angle lies on a line with the equation $y - 3x = -4$. Solve the system by the addition method to find the coordinates of the vertex of the angle.

37. **Finance** There are 8 more quarters than dimes in a parking meter. Three times the number of dimes is 1 less than twice the number of quarters. Write a linear system of equations in two variables. Solve it by the addition method to find the number of dimes and the number of quarters.

38. **Consumerism** Eric bought 3 rolls of film and 1 battery for \$11. The next day he purchased 2 rolls of film and 3 batteries for \$12. Write a linear system of equations in two variables. Solve it by the addition method to find the cost of a roll of film and the cost of a battery.

Developing Mathematical Power

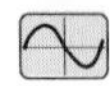 **Thinking Critically** Use a graphing utility to create a graph of the system

$$\begin{cases} y = 3x - 7 \\ y = -x - 31 \end{cases}$$

39. Without first multiplying anything, add the two equations. What can you predict about the graph of the result? Check your answer on the graphing utility.

40. Use the addition method to eliminate the x-term from the original system. What can you predict about the result when it is graphed with the two original equations? Check your answer on the graphing utility.

BIOGRAPHY: Gottfried Wilhelm Leibniz

Leibniz, a brilliant German mathematician, is considered to have been one of the greatest thinkers of his time. Leibniz is credited with many significant contributions in the fields of philosophy, theology, history, law, and literature, as well as mathematics. In the latter area, he is credited with the development of *determinants* in connection with his study of linear systems of equations. (Determinants will be introduced in the next chapter of this book.) Seki Kowa of Japan had actually invented determinants ten years earlier, but it was Leibniz who popularized their use.

Leibniz is also well known for his invention of the calculus and the fundamental theorem of calculus. Although it is a recognized fact that both he and one of his contemporaries, Sir Isaac Newton, invented the calculus, they did so independently. Leibniz was the first to publish his results, and his system of notation remains more popular than the system devised by Newton.

Investigation

Read about the life and work of Leibniz (1646–1716). Then list two or three of his important contributions to mathematics, other than those mentioned in the paragraph above.

4.5

Solving Linear Systems Using the Substitution Method

Objective: To solve a linear system in two variables by substitution

The substitution property of equality provides a second method of reducing a system of two equations in two variables to one equation in one variable.

Capsule Review

EXAMPLE **Solve $3x + 2y = 6$ for x, if $y = x - 2$.**

$3x + 2y = 6$	
$3x + 2(x - 2) = 6$	*Substitute $x - 2$ for y.*
$3x + 2x - 4 = 6$	*Distributive property*
$5x - 4 = 6$	
$5x = 10$	
$x = 2$	

Solve each equation for y, if $x = 2y - 1$.

1. $x + 2y = 3$ **2.** $y - x = 4$ **3.** $2y + 3x = -5$ **4.** $y - 2x = 8$

To solve a system of equations by substitution, use one equation to find one variable in terms of the other. Then substitute for that variable in the other equation. In Example 1, the first equation gives y in terms of x.

EXAMPLE 1 **Solve the system by substitution:** $\begin{cases} y = x + 1 \\ 2x + y = 7 \end{cases}$

$2x + y = 7$	*Start with the second equation.*
$2x + (x + 1) = 7$	*Substitute $x + 1$ for y.*
$3x + 1 = 7$	
$3x = 6$	
$x = 2$	
$y = x + 1$	*Use an original equation to find y.*
$y = 2 + 1$	*Substitute 2 for x.*
$y = 3$	
Check: $2x + y = 7$	*Use the other original equation.*
$2(2) + 3 \stackrel{?}{=} 7$	*Substitute 2 for x and 3 for y.*
$7 = 7$ ✓	The solution is (2, 3).

Solutions can also be checked with a graphing utility.

To solve a linear system of equations in two variables by the substitution method

1. Solve either equation for one variable in terms of the other.
2. Substitute for that variable in the other equation and solve.
3. Substitute in either of the original equations to find the value of the other variable.
4. Substitute the ordered pair in the *other* original equation to check.

EXAMPLE 2 **Solve the system by substitution:** $\begin{cases} 2x + 3y = -7 \\ x - 4y = 13 \end{cases}$

$$x - 4y = 13$$ *Start with the second equation.*
$$x = 4y + 13$$ *Solve for x.*

$$2x + 3y = -7$$ *Use the other equation.*
$$2(4y + 13) + 3y = -7$$ *Substitute 4y + 13 for x.*
$$8y + 26 + 3y = -7$$
$$11y = -33$$
$$y = -3$$

$$x - 4y = 13$$ *Use an original equation.*
$$x - 4(-3) = 13$$ *Substitute −3 for y.*
$$x + 12 = 13$$
$$x = 1$$

Check:
$$2x + 3y = -7$$ *Use the other original equation.*
$$2(1) + 3(-3) \stackrel{?}{=} -7$$ *Substitute 1 for x and −3 for y.*
$$-7 = -7 \checkmark$$ The solution is $(1, -3)$

The next example illustrates the result that is obtained when a dependent system is solved by the substitution method.

EXAMPLE 3 **Solve the system by substitution:** $\begin{cases} 3x + y = 4 \\ -6x - 2y = -8 \end{cases}$

$$3x + y = 4$$ *Use the first equation.*
$$y = -3x + 4$$ *Solve for y.*

$$-6x - 2y = -8$$ *Use the second equation.*
$$-6x - 2(-3x + 4) = -8$$ *Substitute −3x + 4 for y.*
$$-6x + 6x - 8 = -8$$
$$0 = 0$$

The equation $0 = 0$ is always true, the number of solutions is *infinite*, and this system is dependent. The solution is $\{(x, y): 3x + y = 4\}$.

In Examples 1–3, 1 is the coefficient of at least one term containing a variable. The substitution method can also be used when no term containing a variable has the coefficient 1.

EXAMPLE 4 **Solve the system by substitution:** $\begin{cases} \mathbf{2x + 2y = 6} \\ \mathbf{5x = -5y + 10} \end{cases}$

$5x = -5y + 10$ *Use the second equation.*

$x = -y + 2$ *Solve for x.*

$2x + 2y = 6$

$2(-y + 2) + 2y = 6$ *Substitute $-y + 2$ for x.*

$-2y + 4 + 2y = 6$

$0 = 2$ Contradiction

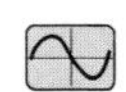

Since the equation $0 = 2$ is a contradiction, there is no solution, and the system is inconsistent. You can use a graphing utility to verify that the lines are parallel.

CLASS EXERCISES

Solve each equation for the indicated variable.

1. $x + 2y = 14$; x **2.** $2x + y = 10$; y **3.** $x + 2y = -1$; x

4. $3x - y = 2$; y **5.** $3x + 4y = 10$; x **6.** $2x + 3y = 4$; y

7. $2x - 3y = -8$; x **8.** $x - 3y = 0$; y **9.** $4x - 5y = 6$; y

Tell which method—addition or substitution—you would choose to solve each linear system. Give a reason for your answer.

10. $\begin{cases} 2x + 5y = 6 \\ 3x - 5y = 9 \end{cases}$ **11.** $\begin{cases} 13x - 3y = -16 \\ x + 5y = 4 \end{cases}$ **12.** $\begin{cases} 6x + 3y = 17 \\ 5x + 6y = 13 \end{cases}$

PRACTICE EXERCISES

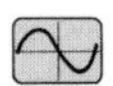 Use technology where appropriate.

Solve each system of equations by substitution. If the system is inconsistent, write *no solution*.

1. $\begin{cases} y = x + 2 \\ 2x - y = 1 \end{cases}$ **2.** $\begin{cases} x = y - 4 \\ 2x - y = 6 \end{cases}$ **3.** $\begin{cases} y = x - 3 \\ x - 2y = -10 \end{cases}$

4. $\begin{cases} x = y + 1 \\ 4x - y = 19 \end{cases}$ **5.** $\begin{cases} x = 4y - 5 \\ 5x - y = -6 \end{cases}$ **6.** $\begin{cases} y = 2x + 3 \\ 4x - 3y = 1 \end{cases}$

7. $\begin{cases} y - x = -5 \\ 2x + 4y = 10 \end{cases}$ **8.** $\begin{cases} x + y = -12 \\ 2x - 3y = 6 \end{cases}$ **9.** $\begin{cases} y - 3x = 0 \\ 4x + 3y = 26 \end{cases}$

10. $\begin{cases} x - 2y = 0 \\ 5x + 2y = 24 \end{cases}$

11. $\begin{cases} 2x + y = 7 \\ 4x - 2y = -6 \end{cases}$

12. $\begin{cases} x + 3y = 16 \\ 7x + 4y = 10 \end{cases}$

13. $\begin{cases} 4x - y = 0 \\ 12x - 6y = 24 \end{cases}$

14. $\begin{cases} x - 2y = 0 \\ 8x + 4y = 20 \end{cases}$

15. $\begin{cases} x = 3y + 1 \\ 2x - 6y = 2 \end{cases}$

16. $\begin{cases} 2x + 4y = 0 \\ 5x = -10y \end{cases}$

17. $\begin{cases} y = 2x - 3 \\ -2x + y = 5 \end{cases}$

18. $\begin{cases} 3x - 6y = 7 \\ 3x = 6y \end{cases}$

19. $\begin{cases} 2x + y = 4 \\ 3x - y = 6 \end{cases}$

20. $\begin{cases} 2x + 3y = -5 \\ x - 3y = 2 \end{cases}$

21. $\begin{cases} 3x + y = -4 \\ x + 3y = -12 \end{cases}$

22. $\begin{cases} 5y = x \\ 2x - 3y = 7 \end{cases}$

23. $\begin{cases} 3x + 15y = 3 \\ 2x - 5y = -28 \end{cases}$

24. $\begin{cases} -5x + 20y = 15 \\ 2x - 3y = 4 \end{cases}$

25. $\begin{cases} 3y = 2x \\ 2x + 9y = 24 \end{cases}$

26. $\begin{cases} 4y = -5x \\ 5x + 8y = 20 \end{cases}$

27. $\begin{cases} 8y = 4x \\ 7x + 2y = -8 \end{cases}$

28. $\begin{cases} 15x = -5y \\ 3x - 2y = -15 \end{cases}$

29. $\begin{cases} y - 7 = 0 \\ -7x + 3y = 0 \end{cases}$

30. $\begin{cases} x - 4 = 0 \\ 9x - 2y = 0 \end{cases}$

31. $\begin{cases} 4x - 5y = 0 \\ 8x = 10y + 15 \end{cases}$

32. $\begin{cases} 3x + 7y = 0 \\ 15x = -35y \end{cases}$

33. $\begin{cases} 0.4x - 0.1y = 0.6 \\ 2x + 3y = 10 \end{cases}$

34. $\begin{cases} 0.03x = 0.02y \\ 9x + 3y = -27 \end{cases}$

35. $\begin{cases} \dfrac{x}{4} - \dfrac{y}{4} = -1 \\ -3x + 7y = 8 \end{cases}$

36. $\begin{cases} \dfrac{x}{6} - \dfrac{y}{2} = \dfrac{1}{3} \\ x + 2y = -3 \end{cases}$

37. $\begin{cases} 2x + 3y = x + 2y + 2 \\ 5x - 3x + 2y = 22 + y \end{cases}$

38. $\begin{cases} 2(x + 2y) = 6 \\ 3(x + 2y - 3) = 0 \end{cases}$

39. $\begin{cases} 2(x + y) = 4x + 1 \\ 3(x - y) = x + y - 3 \end{cases}$

40. $\begin{cases} 3x + 6y + 5 = 9x - 3y - 5 \\ x - 2y = 2(x + y + 1) \end{cases}$

41. $\begin{cases} x + 4y = -\dfrac{1}{2} + 4x \\ 2x + 3y = \dfrac{x}{2} + 2y + 1 \end{cases}$

42. $\begin{cases} 2x + y = \dfrac{y}{4} - 4x \\ x + 4 + \dfrac{y}{2} = 2 - \dfrac{3y}{4} - x \end{cases}$

Find the value of *a* that makes each of the following a dependent system.

43. $\begin{cases} y = 3x + a \\ 3x - y = 2 \end{cases}$

44. $\begin{cases} 3y = 2x \\ 6y - a - 4x = 0 \end{cases}$

45. $\begin{cases} y = \dfrac{x}{2} + 4 \\ 2y - x = a \end{cases}$

Find the value of *m* that makes each of the following an inconsistent system.

46. $\begin{cases} y = mx \\ 2y - 3x = 3 \end{cases}$

47. $\begin{cases} y = mx + 3 \\ -2y + 5x = 7 \end{cases}$

48. $\begin{cases} y = mx - \dfrac{2}{3} \\ 4x + 5y = 5 \end{cases}$

Applications

49. Sports Noreen ran 4 mi farther than Heather. Together they ran a distance of 20 mi. How far did each girl run?

50. Finance Craig has three times as many quarters as dimes. If the sum of the number of dimes and twice the number of quarters is 21, how many of each type of coin does he have?

51. Politics In a mayoral election, the incumbent received 25% more votes than her opponent. Altogether, 5175 votes were cast for the two candidates. How many votes did the incumbent mayor receive?

Developing Mathematical Power

52. Thinking Critically Given the equations $ax + by = c$ and $dx + ey = f$, use substitution to solve for x and for y. How might your results provide you with a useful shortcut in solving systems of equations?

TEST YOURSELF

For each function, determine whether y varies directly as x. If so, give the constant of variation.

1. $\frac{y}{x} = 5$

2. $2y - 3x = 6$

4.1

In Exercises 3–4, y varies directly as x.

3. If $y = 4$ when $x = 7$, find y when $x = -14$.

4. If $y = 9$ when $x = 15$, find x when $y = 6$.

Solve each linear system graphically. Determine whether the system is consistent and independent, inconsistent, or consistent and dependent.

5. $\begin{cases} 2y - 4x = 12 \\ y + 2x = -2 \end{cases}$

6. $\begin{cases} 2y - 4x = 8 \\ 4y = 8x + 8 \end{cases}$

4.2, 4.3

Solve each linear system by the addition method.

7. $\begin{cases} 2x + 3y = 7 \\ x + 4y = 6 \end{cases}$

8. $\begin{cases} 6x - 3y = 3 \\ 5x - 5y = 10 \end{cases}$

4.4

Solve each linear system by the substitution method.

9. $\begin{cases} x = y - 3 \\ x + 2y = 12 \end{cases}$

10. $\begin{cases} y - 4x = 0 \\ 2x + 3y = 28 \end{cases}$

4.5

4.6

Problem Solving: Using Linear Systems of Equations

Many problems in business and science require the solution of a linear system of equations. Organizing the given information in a table may make it easier to understand the problem.

EXAMPLE 1 A travel agent offers two package vacation plans. The first plan costs \$360 and includes 3 days at a hotel and a rental car for 2 days. The second plan costs \$500 and includes 4 days at a hotel and a rental car for 3 days. The daily charge for the room is the same under each plan, as is the daily charge for the car. Find the cost per day for the room and for the car.

Understand the Problem Organize the information in a table.

	cost	days in hotel	days for car
Plan 1	\$360	3	2
Plan 2	\$500	4	3

Plan Your Approach Let r represent the cost of the room per day, in dollars. Let c represent the cost of the car per day, in dollars.

cost of room + cost of car = cost of plan

$$\begin{cases} 3r + 2c = 360 \\ 4r + 3c = 500 \end{cases}$$

Complete the Work Use the addition method to solve the system.

$$\begin{aligned} 9r + 6c &= 1080 && \textit{Multiply first equation by 3.} \\ -8r - 6c &= -1000 && \textit{Multiply second equation by } -2. \\ \hline r &= 80 && \textit{Add.} \end{aligned}$$

$$\begin{aligned} 3r + 2c &= 360 \\ 3(80) + 2c &= 360 && \textit{Substitute 80 for r.} \\ 240 + 2c &= 360 \\ 2c &= 120 \\ c &= 60 \end{aligned}$$

Interpret the Results The room costs \$80 per day and the car costs \$60 per day.

Is \$360 the cost of Plan 1?

room cost + car cost = \$360

$3(80) + 2(60) \stackrel{?}{=} 360$

$360 = 360$ ✓

Is \$500 the cost of Plan 2?

room cost + car cost = \$500

$4(80) + 3(60) \stackrel{?}{=} 500$

$500 = 500$ ✓

The time it takes an aircraft to travel a specified distance depends on whether it is flying *with* or *against* the wind. **Air speed** is the speed of an aircraft in still (calm) air. A **tail wind** is a wind blowing in the same direction in which the aircraft is traveling. A **head wind** is a wind blowing in the opposite direction from that in which the aircraft is traveling. **Ground speed** is the speed of the aircraft relative to the ground.

With a tail wind: ground speed = air speed + wind speed

With a head wind: ground speed = air speed − wind speed

For a boat traveling in a stream, current has an effect similar to that of wind on an aircraft. A boat traveling **upstream** is traveling *against* the current. A boat traveling **downstream** is traveling *with* the current.

EXAMPLE 2 Flying with a tail wind, a plane flew 4200 mi in 6 h. With a head wind, the plane took 7 h to make the return flight. Find the plane's air speed and the wind speed. Assume both are constant.

Understand the Problem You are asked to find the air speed of the plane and the speed of the wind. You will need to use the formula

$$\text{rate} \cdot \text{time} = \text{distance}$$

Plan Your Approach Let a represent the plane's air speed in miles per hour. Let w represent the wind speed in miles per hour.

	rate	· time	= distance
With tail wind	$a + w$	6 h	4200 mi
With head wind	$a - w$	7 h	4200 mi

Use the formula to write a system of equations.

$$\text{rate} \cdot \text{time} = \text{distance}$$
$$\begin{cases} (a + w)6 = 4200 \\ (a - w)7 = 4200 \end{cases}$$

Complete the Work

$$a + w = 700 \quad \textit{Multiply first equation by } \tfrac{1}{6}.$$
$$a - w = 600 \quad \textit{Multiply second equation by } \tfrac{1}{7}.$$
$$2a = 1300 \quad \textit{Add.}$$
$$a = 650$$

$$(a + w)6 = 4200 \quad \textit{Use the other equation.}$$
$$(650 + w)6 = 4200 \quad \textit{Substitute 650 for a.}$$
$$650 + w = 700 \quad \textit{Divide both sides by 6.}$$
$$w = 50$$

Interpret the Results The plane's air speed is 650 mi/h and the wind speed is 50 mi/h.

Does the plane travel 4200 mi with a tail wind?

$$(650 + 50)6 \stackrel{?}{=} 4200$$
$$4200 = 4200 \checkmark$$

Does the plane travel 4200 mi with a head wind?

$$(650 - 50)7 \stackrel{?}{=} 4200$$
$$4200 = 4200 \checkmark$$

A system of equations can be used to solve problems involving two-digit numbers. Any two-digit number can be written in the form $10t + u$, where t is the tens digit and u is the units digit. For example,

$$74 = 10(7) + 4 \qquad 30 = 10(3) + 0$$

When the digits of a two-digit number are reversed, the new number is of the form $10u + t$. For example, 74 becomes 47 when its digits are reversed.

EXAMPLE 3 The sum of the digits of a two-digit number is 13. If the digits are reversed, the new number is 27 more than the original number. What is the original number?

Understand the Problem You are given the relationship of the digits in a two-digit number and the relationship of the given number to the number formed by reversing its digits. You are asked to find the original number.

Plan Your Approach Let t represent the tens digit. Let u represent the units digit. Let $10t + u$ represent the original number and $10u + t$ represent the new number.

$$\text{tens digit} + \text{units digit} = 13$$
$$t + u = 13$$

$$\text{new number} = 27 + \text{original number}$$
$$10u + t = 27 + 10t + u$$
$$-9t + 9u = 27 \qquad \textit{Simplify.}$$

Complete the Work Use the addition method to solve the system of equations.

$$\begin{cases} t + u = 13 \\ -9t + 9u = 27 \end{cases}$$

$$t + u = 13$$
$$-t + u = 3 \qquad \textit{Multiply second equation by } \tfrac{1}{9}.$$
$$2u = 16 \qquad \textit{Add.}$$
$$u = 8$$

$$t + u = 13$$
$$t + 8 = 13$$
$$t = 5$$

Original number: $10t + u = 10(5) + 8 = 58$

New number: $10u + t = 10(8) + 5 = 85$

Interpret the Results

The original number is 58.

Is the sum of the digits 13?

$5 + 8 \stackrel{?}{=} 13$

$13 = 13$ ✓

Is the new number 27 more than the original number?

$85 \stackrel{?}{=} 27 + 58$

$85 = 85$ ✓

CLASS EXERCISES

For each problem, select two variables and state what each one represents. Then write a system of equations that could be used to solve the problem.

1. A company ordered two types of parts, brass and steel. A shipment containing 3 brass and 10 steel parts costs $48. A second shipment containing 7 brass and 4 steel parts costs $54. Find the cost of a brass part and the cost of a steel part.

2. Two coin banks contain only dimes and quarters. There are 25 coins in the first bank. The value of the coins in the second bank is $5, and it contains 5 fewer dimes and 3 more quarters than the first bank. Find the number of dimes and the number of quarters in the first bank.

3. A ferry traveled 6 mi downstream in 1.5 h. On the return trip, the ferry traveled 5.25 mi upstream in the same amount of time. What is the ferry's rate in still water, and what is the rate of the current?

4. The sum of the digits of a two-digit number is 12. If the digits are reversed, the new number is 15 more than twice the original number. What is the original number?

PRACTICE EXERCISES

1. A music store receives shipments of recorders and harmonicas. A shipment of 5 recorders and 4 harmonicas costs $62. A shipment of 10 recorders and 3 harmonicas costs $84. Find the cost of a recorder and the cost of a harmonica.

2. A company manufactures two types of fishing rods, standard and folding. The cost of manufacturing 1 standard and 2 folding rods is $35. If 5 standard and 7 folding rods were manufactured, the cost would be $145. How much does it cost to manufacture each type of rod?

3. Celia wants to buy pets for each of her five grandchildren. If she buys 3 poodles and 2 Siamese cats, she will spend $1200. If she buys 2 poodles and 3 Siamese cats, she will spend $1050. Find the cost of a poodle and the cost of a Siamese cat.

4. Jennifer received a $25 gift certificate from the stereo store. She can use it to buy 2 albums and 5 singles or 1 album and 15 singles. Find the cost of an album and the cost of a single.

5. There are twice as many boys as girls in Ms. Taylor's class. When three new girls join her class, she will have 24 students. How many boys and girls were originally in her class?

6. A seamstress purchased 100 yards of fabric. She can use it to make 20 skirts and 8 dresses. If she makes 14 dresses, however, she will have enough fabric to make only 10 skirts. How many yards of fabric does it take to make a skirt, and how many yards does it take to make a dress?

7. Hal makes tables and chairs. Each chair requires 4 ft of oak and 3 ft of pine, while each table requires 8 ft of oak and 2 ft of pine. Hal has 52 ft of oak and 23 ft of pine. How many chairs and tables can he build?

8. Sheridan's monthly phone bill includes a fixed charge, which is the same each month, and an additional charge for toll calls. One month the bill was $22, which included charges for 5 toll calls. The following month the bill was $26, which included charges for 7 toll calls. Find the fixed charge and the average charge per toll call.

9. The local cable television company offers two deals. Basic cable service with one movie channel costs $35 per month. Basic service with two movie channels costs $45 per month. Find the charge for the basic cable service and the charge for each movie channel.

10. Nathan's monthly electric bill includes a fixed service charge and an additional charge for the number of kilowatt-hours (kwh) of electricity used. One bill, including a charge for 595 kwh of electricity, is $61.72. Another bill, including a charge for 650 kwh, is $67.11. Find the fixed charge and the cost per kilowatt-hour.

11. With a tail wind, a plane traveled 800 mi in 5 h. With a head wind, the plane traveled the same distance in 8 h. Find the plane's air speed and the speed of the wind. Assume that both were constant.

12. A boat's crew rowed 16 km downstream in 2 h, but the return trip upstream took 4 h. What was the rate of the current and the boat's rate in still water? Assume that both were constant.

13. With a head wind, a plane flew 3600 km in 10 h. The return flight over the same route took only 9 h. Find the wind speed and the plane's air speed. Assume that both were constant.

14. With a tail wind, a plane flew 1740 km in 6 h. With a head wind, it flew only 570 km in 3 h. Find the plane's air speed and the wind speed. Assume that both were constant.

15. The units digit of a two-digit number is twice the tens digit. If the digits are reversed, the new number is 9 less than twice the original number. What is the original number?

16. The sum of the digits of a two-digit number is 10. If the digits are reversed, the new number is 36 less than the original number. Find the original number.

17. A two-digit number is 3 times the sum of its digits. The number is also 45 less than the number formed by reversing the digits of the original number. What is the original number?

18. A coin bank contains only dimes and nickels. The bank contains 46 coins. If 5 dimes and 2 nickels were removed, the total value of the coins would be $3.40. Find the number of dimes and the number of nickels in the bank.

19. Lou has only quarters and dimes in his coin bank. There are currently 26 coins in the bank. If he adds 7 quarters and 3 dimes, the total value of the coins will be $7.35. Find the number of quarters and the number of dimes Lou originally had in the bank.

20. Marci has a total of 50 stamps, consisting of 25¢ stamps and 15¢ stamps. If she doubles the number of 25¢ stamps and halves the number of 15¢ stamps, they will be worth $16.50. How many 25¢ stamps and how many 15¢ stamps were there in the original 50?

21. In a new development, 50 one- and two-bedroom condominiums were sold. Each one-bedroom condominium sold for $185,900 and each two-bedroom condominium sold for $195,900. If sales totaled $9,515,000, how many of each type of unit were sold?

22. A resort hotel has 200 rooms. Those with kitchen facilities rent for $100 per night and those without kitchen facilities rent for $80 per night. On a night when the hotel was completely occupied, revenues were $17,000. How many of each type of room does the hotel have?

23. When she rows with the current, Mindy can row 24 mi in 3 h. Against the same current, Mindy can row only $\frac{2}{3}$ of this distance in 4 h. Find the rate of the current and Mindy's rowing rate in still water. Assume that both rates are constant.

24. When he runs against the wind, Ricky can run 16 km in 2 h. With the wind, he can run $\frac{5}{4}$ of this distance in 2 h. Find Ricky's running rate and the rate of the wind. Assume that both are constant.

25. The sum of the digits of a two-digit number is 12. If 6 is subtracted from the number, the result is $\frac{1}{2}$ the number formed by reversing the digits of the original number. What is the original number?

26. Craig can get to his office, which is 14 mi away, by riding a bus for 30 min and walking for 30 min. When he wants more exercise, he gets there by riding the bus for 25 min and walking for 1 h. What are his average walking rate and the average speed of the bus?

27. The Cranshaws invested $20,000, part at 8% per year and the remainder at 6% per year. After 18 months, they had earned a total of $2160 in simple interest. Find the amount invested at each rate.

Mixed Review

28. Evaluate $m(x) = 3x^2 + 4x - 1$ for **a.** $x = -3$ **b.** $x = 5$ **c.** $x = t$.

29. Find the inverse of $y = 2x + 3$. Is it a function?

30. Name the quadrant containing each point:

$$A(3, 5);\ B(-2, 6);\ C(2, -5);\ D(-4, -1)$$

31. On a number line, graph $\{x: x \geq 2\} \cup \{x: x < 0\}$.

32. Is $(-3, 4)$ a solution of $y = 2x^2 - 6x$?

33. Given $f(x) = 3x - 4$, graph $f(x)$, $f^{-1}(x)$, and $y = x$ on one set of axes.

34. What is the slope of a line passing through $(2, 4)$ and $(0, -2)$?

35. What is the slope of a line perpendicular to the graph of $2x + y = 5$?

36. The coordinates of three vertices of a parallelogram are $(3, 4)$, $(13, 0)$, and $(10, -4)$. Find the coordinates of the fourth vertex.

37. What property of equality guarantees that $2(3x + 0) = 2(3x)$?

Developing Mathematical Power

Thinking Critically Elizabeth and Jean covered x miles on an automobile trip. At their destination, they replaced their tires with larger snow tires. On the return trip over the same route, they clocked the distance as $(x - 10)$ mi.

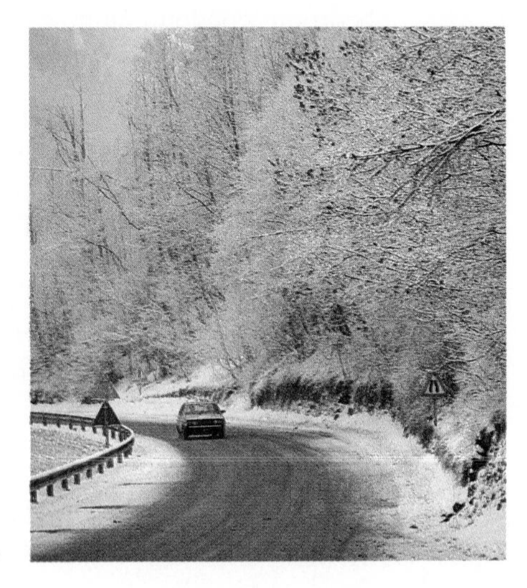

38. If the radius of each original tire was 14 in., express the difference between the radius of the snow tire and the radius of the original tire.

39. Explain why the gas mileage will increase (or decrease) as a result of this tire change.

Solving Linear Systems of Inequalities

Objectives: To graph linear inequalities in two variables
To solve linear systems of inequalities in two variables graphically

Solving a linear system of equations graphically involves finding the set of points that satisfy all equations in the system. Solving a linear system of inequalities graphically involves finding the set of points that satisfy all the inequalities in the system.

Capsule Review

When you multiply or divide both sides of an inequality by a positive number, the inequality sign remains the same. When you multiply or divide by a negative number, the inequality sign is reversed.

EXAMPLE **Solve for y:** **a.** $3y > 2x$ **b.** $-2y + 4x \leq 8$

a. $3y > 2x$

$y > \frac{2}{3}x$ *Divide both sides by 3.*

b. $-2y + 4x \leq 8$

$-2y \leq -4x + 8$ *Add $-4x$ to both sides.*

$y \geq 2x - 4$ *Divide both sides by -2.*

Solve each inequality for y.

1. $-2y \leq 4x$ **2.** $-y + x < 0$ **3.** $4y + 5x \geq 20$ **4.** $-3y - 2x \leq 6$

An inequality has a *corresponding equation*. For example, the equation

$$y = -2x - 5$$

is the corresponding equation for inequalities such as

$$y < -2x - 5 \quad \text{and} \quad y \geq -2x - 5$$

The line that is the graph of the corresponding equation is the **boundary** of the **half-plane** that is the graph of the inequality.

If the symbol of inequality is $\leq$ or $\geq$, the graph of the inequality is a **closed half-plane,** which consists of the boundary and all points in the plane that are on one side of the boundary. To graph such an inequality, first graph the corresponding equation as a *solid line*.

EXAMPLE 1 **Graph:** $y - 2x \leq 1$

Solve the inequality for y.

$$y - 2x \leq 1$$
$$y \leq 2x + 1$$

Graph $y = 2x + 1$, using a *solid line*. Then test a point in each half-plane to see which half-plane is the graph of the inequality.

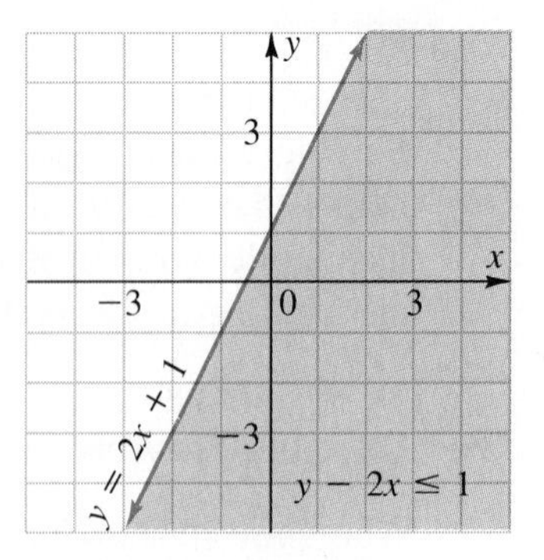

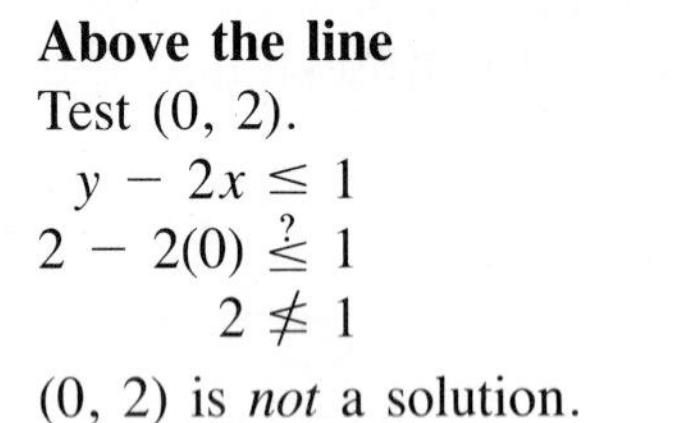

Above the line
Test (0, 2).

$$y - 2x \leq 1$$
$$2 - 2(0) \overset{?}{\leq} 1$$
$$2 \nleq 1$$

(0, 2) is *not* a solution.

Below the line
Test (0, 0).

$$y - 2x \leq 1$$
$$0 - 2(0) \overset{?}{\leq} 1$$
$$0 \leq 1 \checkmark$$

(0, 0) is a solution.

Shade the half-plane below the boundary. The graph of $y - 2x \leq 1$ is the line $y = 2x + 1$ and all points in the plane below it.

If the inequality symbol is $<$ or $>$, the boundary is not included in the graph of the inequality and it is drawn as a *dashed line*. The graph of an inequality that does not include the boundary is called an **open half-plane.**

EXAMPLE 2 **Graph:** $-2y < -6x + 2$

$$-2y < -6x + 2 \qquad \textit{Solve for y.}$$
$$y > 3x - 1$$

Graph $y = 3x - 1$, using a *dashed line*. Shade the half-plane above the line. Check by testing a point in the shaded half-plane.

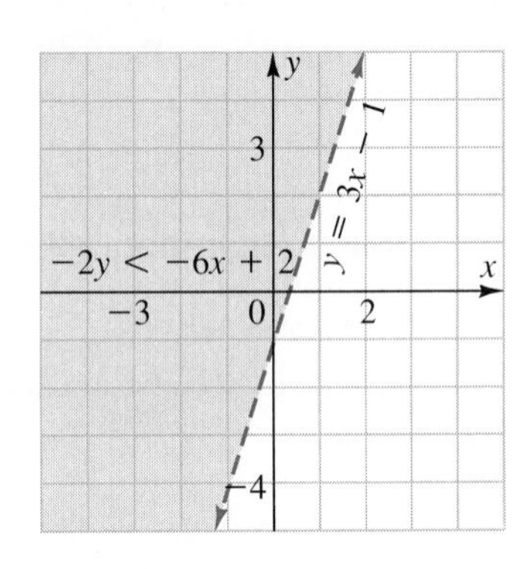

Test (0, 0).

$$-2y < -6x + 2$$
$$-2(0) \overset{?}{<} -6(0) + 2$$
$$0 < 2 \checkmark$$

The graph is the open half-plane above the line $y = 3x - 1$.

The graph of an inequality of the form

$y \leq mx + b$ is the closed half-plane *below* the line $y = mx + b$
$y < mx + b$ is the open half-plane *below* the line $y = mx + b$
$y \geq mx + b$ is the closed half-plane *above* the line $y = mx + b$
$y > mx + b$ is the open half-plane *above* the line $y = mx + b$

If $|x|$ or $|y|$ is isolated on the left side of an absolute-value inequality with the symbol $<$ or $\leq$, then the graph is a single region. However, for the symbols $>$ and $\geq$, the graph consists of points in two different half-planes.

EXAMPLE 3 **Graph on the coordinate plane: a.** $|y| \leq 3$ **b.** $|x| > 2$

a. $|y| \leq 3$

$-3 \leq y \leq 3$

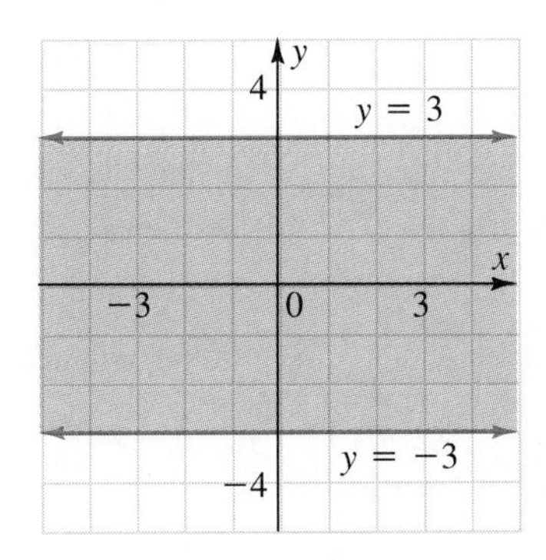

The graph consists of the solid lines $y = 3$ and $y = -3$ and the region between these lines.

b. $|x| > 2$

$x < -2$ or $x > 2$

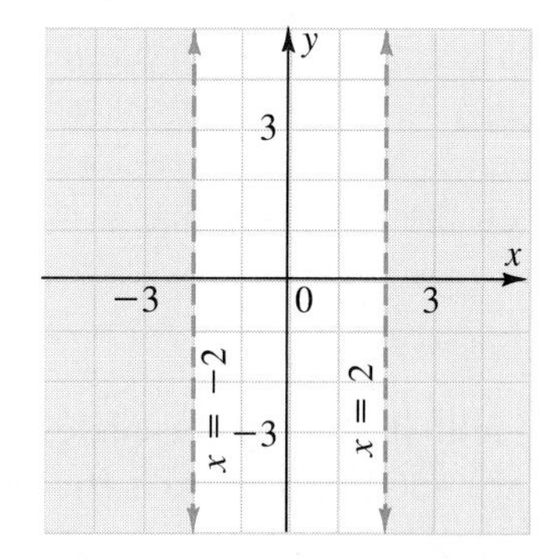

The graph consists of all points to the left of $x = -2$ and all points to the right of $x = 2$.

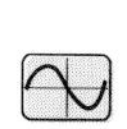

To solve a **system of inequalities** graphically, graph the inequalities on the same coordinate plane. If the graphs overlap, all points in the overlapping region have coordinates that are solutions to the system. Many graphing utilities allow you to graph systems of inequalities.

EXAMPLE 4 **Solve the system graphically:** $\begin{cases} y \leq x - 3 \\ -2y \leq 2x + 4 \end{cases}$

Graph both inequalities on one coordinate plane.

$$y \leq x - 3 \qquad -2y \leq 2x + 4$$
$$y \geq -x - 2$$

Choose a point in the double-shaded region and check it in both of the original inequalities. Test (4, 0).

$$y \leq x - 3 \qquad -2y \leq 2x + 4$$
$$0 \stackrel{?}{\leq} 4 - 3 \qquad -2(0) \stackrel{?}{\leq} 2(4) + 4$$
$$0 \leq 1 \checkmark \qquad 0 \leq 12 \checkmark$$

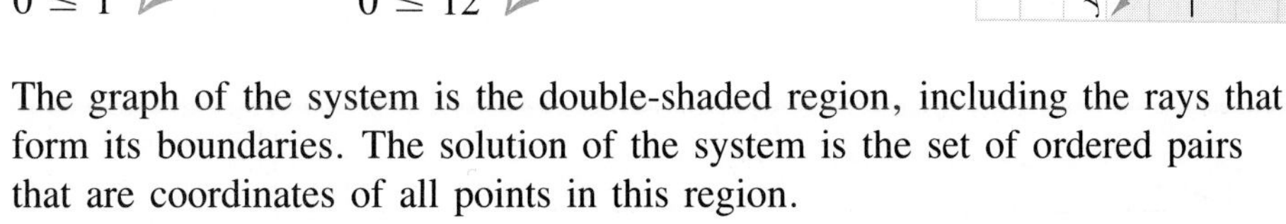

The graph of the system is the double-shaded region, including the rays that form its boundaries. The solution of the system is the set of ordered pairs that are coordinates of all points in this region.

EXAMPLE 5 **Solve the system graphically:** $\begin{cases} 3y + 9x < 3 \\ |y| \geq 2 \end{cases}$

$$3y + 9x < 3 \qquad |y| \geq 2$$
$$3y < -9x + 3 \qquad y \geq 2 \text{ or } y \leq -2$$
$$y < -3x + 1$$

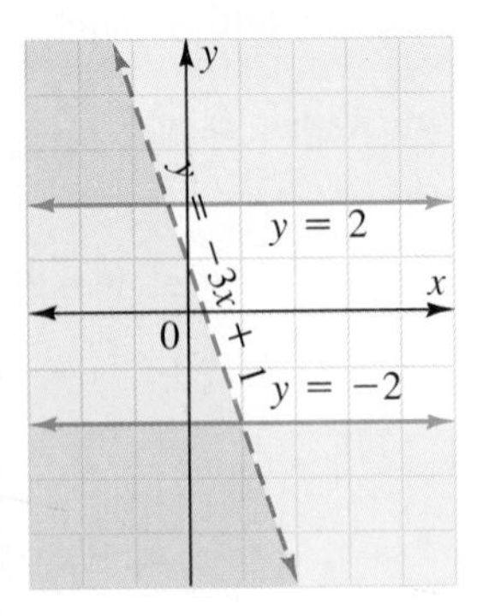

The solution of the system is the set of ordered pairs that are coordinates of all points in the double-shaded regions. This includes points in the *half-planes* consisting of all points on or above $y = 2$ to the left of $y = -3x + 1$ and all points on or below the line $y = -2$ to the left of $y = -3x + 1$. Points on the boundary line $y = -3x + 1$ are not included.

CLASS EXERCISES

Tell which, if any, of the ordered pairs (2, −1), (−1, 2), (−3, −2), and (1, 2) are solutions of each inequality or system of inequalities.

1. $y \leq 2x + 3$
2. $y \geq -3x + 2$
3. $-2y < 4x$
4. $3y > 6x$
5. $|y| \leq 3$
6. $|x| \geq -2$
7. $\begin{cases} y \leq 2x + 1 \\ y \geq -x + 3 \end{cases}$
8. $\begin{cases} y \geq 3x - 1 \\ y \geq -2x + 2 \end{cases}$
9. $\begin{cases} 4y > 8x \\ |x| \leq 3 \end{cases}$

PRACTICE EXERCISES

Use technology where appropriate.

Graph each inequality on the coordinate plane.

1. $y \leq 4x - 1$
2. $y \geq 3x + 4$
3. $3y \geq 6x + 3$
4. $2y \leq 4x - 2$
5. $-3y < -12x + 3$
6. $-4y > -4x + 8$
7. $4y - 2x \leq 4$
8. $-5y + 2x \geq -5$
9. $|x| \leq 4$

Solve each system of inequalities by graphing on the coordinate plane.

10. $\begin{cases} y \leq 2x + 2 \\ y < -x + 1 \end{cases}$
11. $\begin{cases} y \geq -3x + 3 \\ y > x + 2 \end{cases}$
12. $\begin{cases} y > -4x + 1 \\ y \leq 2x - 2 \end{cases}$
13. $\begin{cases} y \leq 3x - 6 \\ y > -4x + 2 \end{cases}$
14. $\begin{cases} -2y < 4x + 2 \\ y > 2x + 1 \end{cases}$
15. $\begin{cases} 3y \geq 6x - 3 \\ y > -x - 5 \end{cases}$
16. $\begin{cases} -4y + 8x \geq 0 \\ y - 3x < 2 \end{cases}$
17. $\begin{cases} -x - y \leq 2 \\ y - 2x > 1 \end{cases}$
18. $\begin{cases} -2x - y > -1 \\ -y + 3x < 1 \end{cases}$
19. $\begin{cases} 2y \leq x - 4 \\ |y| \leq 1 \end{cases}$
20. $\begin{cases} 3y > x + 3 \\ |y| \geq 2 \end{cases}$
21. $\begin{cases} -3y < 6x \\ |x| > 4 \end{cases}$
22. $\begin{cases} 2y < -4x \\ |x| < 2 \end{cases}$
23. $\begin{cases} |x| \leq 2 \\ 1 \leq y \leq 3 \end{cases}$
24. $\begin{cases} |y| > 4 \\ -4 < x < -2 \end{cases}$

25. $\begin{cases} |y + 2| \le 6 \\ |x - 3| \ge 4 \end{cases}$

26. $\begin{cases} |y - 3| \le 4 \\ |x + 1| \le 5 \end{cases}$

27. $\begin{cases} |y| \le x \\ |y| \le 3 \end{cases}$

28. $\begin{cases} y \le -2x + 4 \\ x > -3 \\ y \ge 1 \end{cases}$

29. $\begin{cases} 2y - 4x \le 0 \\ x \ge 0 \\ y \ge 0 \end{cases}$

30. $\begin{cases} |y + x| \le 3 \\ |x| \ge 0 \\ |y| \ge 0 \end{cases}$

Applications

Education An entrance exam has two parts, mathematics and verbal. Each part is scored on 800 points. The formulas at the right are used, where x is the mathematics score and y is the verbal score. Graph the system of inequalities and use the graph to answer the questions.

For admission	$x + y \ge 550$
For partial scholarship	$3x + 4y \ge 3000$
For a full scholarship	$2x + 3y \ge 3000$

31. What would be the status of a student who scored 350 points on each test?

32. What scholarship would be awarded to a student who scored 550 points on each test?

33. What scholarship would be awarded to a student who scored 600 points on each test?

34. Which score does the admissions office consider more important when awarding a scholarship?

Developing Mathematical Power

Thinking Critically This graph shows the cost of renting the same model car for one day from two different firms. Firm A charges \$10 a day plus 50¢ per mile. Firm B charges \$20 a day plus 25¢ per mile.

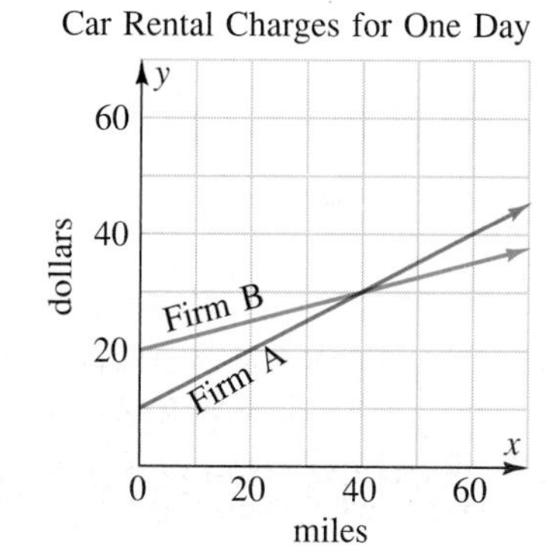

35. Which firm offers the better deal for one day
 a. if you plan to drive more than 40 mi?
 b. if you plan to drive less than 40 mi?

36. Suppose you plan to rent the car for two days. Write and then graph a system of equations relating the distance x in miles and the charges y in dollars for the two firms.
 a. For what number of miles are the two firms' charges the same?
 b. If you plan to drive 40 miles, which firm offers the better deal?

37. If you plan to rent the car for three days, for what number of miles will the two firms' charges be the same? (Study the graph above and the graph you drew for Exercise 2. Then picture the graph for three days.)

4.8 Problem Solving: Linear Programming

Decision makers in many fields, including business and science, try to find ways to allocate resources in order to maximize profits or productivity and to minimize cost. The quantity to be maximized or minimized is represented by a function called the **objective function.** The available resources, and the restrictions placed on them, can be represented by a system of inequalities called **constraints. Linear programming** problems involve maximizing or minimizing an objective function by subjecting it to linear constraints.

The graph of the linear system of constraints is called the **feasible region.** Any point in this region satisfies each of the constraints. However, mathematicians have proved that the following statement is true:

> If there is a maximum or minimum value of the objective function, it will occur at a *vertex* of the feasible region.

EXAMPLE 1 **Find the maximum and minimum values of the objective function $P = 2x + y$ under the following constraints:**

$$\begin{cases} y \geq 2x - 2 \\ y \leq -x + 4 \\ x \geq 0 \\ y \geq 0 \end{cases}$$

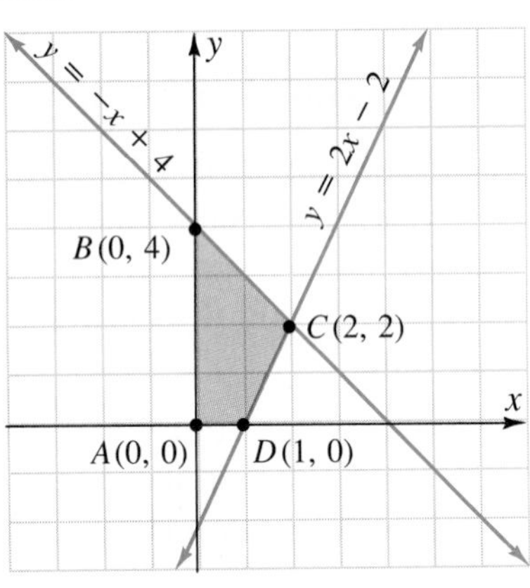

Graph the system of linear inequalities. Notice that the constraints

$$x \geq 0 \text{ and } y \geq 0$$

restrict the graph to Quadrant I. The shaded region, or feasible region, represents the set of all possible solutions.

Label the vertices and read their coordinates from the graph. Then evaluate the objective function at each vertex.

Vertex	$P = 2x + y$
A (0, 0)	$P = 2(0) + 0 = 0$
B (0, 4)	$P = 2(0) + 4 = 4$
C (2, 2)	$P = 2(2) + 2 = 6$
D (1, 0)	$P = 2(1) + 0 = 2$

The maximum value of the objective function is 6, and it occurs at (2, 2).

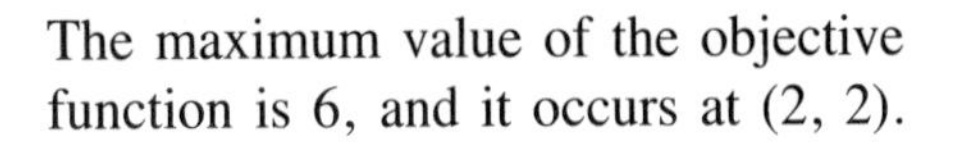
The minimum value of the objective function is 0, and it occurs at (0, 0).

If your graphing utility has no shading capability or the instructions for shading are awkward, you may choose to graph only the boundaries of the feasible region. Then use the trace feature to determine the coordinates of the vertices of the region.

The following example shows a simple application of linear programming.

EXAMPLE 2 Jim Olsen makes and sells gourmet food items. He makes two types of salad dressing, garlic and tofu. Each gallon of garlic dressing requires 2 qt of oil and 2 qt of vinegar. Each gallon of tofu dressing requires 3 qt of oil and 1 qt of vinegar. Jim makes a \$3 profit on each gallon of garlic dressing and a \$2 profit on each gallon of tofu dressing. He has 18 qt of oil and 10 qt of vinegar on hand. How many gallons of each type of dressing should he make to maximize his profits?

Understand the Problem You are given the amounts of oil and vinegar required to make a gallon of each type of dressing. You are given the amounts of oil and vinegar available and the profit on each gallon of dressing.

You are asked to find the number of gallons of each type of dressing required to maximize Jim's profits.

Plan Your Approach Let x represent the number of gallons of garlic dressing.
Let y represent the number of gallons of tofu dressing.

It may help to organize the information in a table.

	Gallons of dressing	Quarts of oil	Quarts of vinegar	Profit (in dollars)
Garlic dressing	x	$2x$	$2x$	$3x$
Tofu dressing	y	$3y$	y	$2y$
Totals		18	10	P

Write the objective function.

$$\text{total profit} = \text{profit on garlic} + \text{profit on tofu}$$
$$P = 3x + 2y$$

Then write the constraints as inequalities.

The total number of quarts of oil used is no more than 18: $2x + 3y \leq 18$

The total number of quarts of vinegar used is no more than 10: $2x + y \leq 10$

The number of gallons of garlic dressing made is greater than or equal to 0: $x \geq 0$

The number of gallons of tofu dressing made is greater than or equal to 0: $y \geq 0$

Complete the Work Graph the following system.

$$\begin{cases} 2x + 3y \leq 18 \\ 2x + y \leq 10 \\ x \geq 0 \\ y \geq 0 \end{cases}$$

Shade the feasible region and label the vertices.

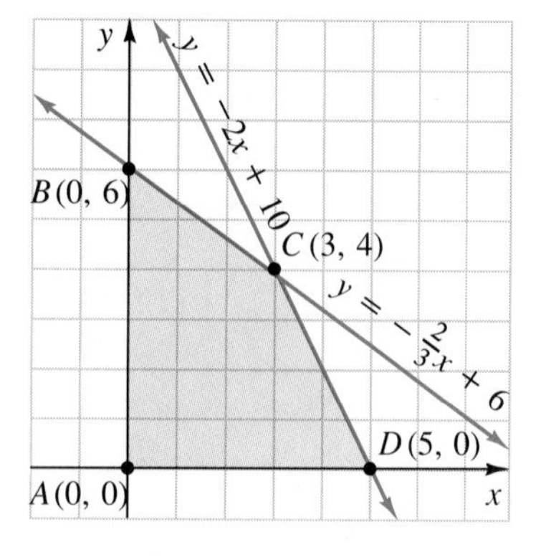

Interpret the Results The maximum value occurs at a vertex of the feasible region. Evaluate the objective function at each vertex. The maximum, 17, occurs at (3, 4).

To maximize profits, Jim should make 3 gal of garlic dressing and 4 gal of tofu dressing.

Vertex	$P = 3x + 2y$
A (0, 0)	$P = 3(0) + 2(0) = 0$
B (0, 6)	$P = 3(0) + 2(6) = 12$
C (3, 4)	$P = 3(3) + 2(4) = 17$
D (5, 0)	$P = 3(5) + 2(0) = 15$

CLASS EXERCISES

Find the maximum and minimum value of each objective function for the given coordinates of vertices of the feasible region.

1. $P = 2x + 3y$ (0, 0), (0, 10), (3, 5)
2. $P = 3x + y$ (0, 3), (2, 6), (8, 3), (4, 0)
3. $P = x + 4y$ (6, 6), (6, 10), (8, 8), (8, 6)
4. $P = 2x + 3y$ (20, 0), (20, 50), (30, 60), (40, 50)
5. $P = x + y$ (10, 10), (10, 30), (50, 30), (40, 20), (30, 20)

PRACTICE EXERCISES

Use technology where appropriate.

1. If profit is represented by $P = x + 3y$, find the maximum profit under these constraints:

$$\begin{cases} x + y \leq 5 \\ x + 2y \leq 8 \\ x \geq 0 \\ y \geq 0 \end{cases}$$

2. If profit is represented by $P = 4x + y$, find the maximum profit under these constraints:

$$\begin{cases} x + y \leq 6 \\ 2x + y \leq 10 \\ x \geq 0 \\ y \geq 0 \end{cases}$$

3. If cost is represented by $C = 2x + 2y$, find the minimum cost under these constraints:

$$\begin{cases} 2x + y \le 6 \\ x \ge 0 \\ y \ge 2 \end{cases}$$

4. If cost is represented by $C = x + 3y$, find the minimum cost under these constraints:

$$\begin{cases} x + 2y \le 8 \\ x \ge 2 \\ y \ge 0 \end{cases}$$

5. If profit is represented by $P = 3x + 4y$, find the maximum profit under these constraints:

$$\begin{cases} x + y \le 3 \\ x \ge 0 \\ y \le 2 \end{cases}$$

6. If cost is represented by $C = 2x + 3y$, find the minimum cost under these constraints:

$$\begin{cases} x + y \le 5 \\ x \ge 2 \\ y \ge 1 \end{cases}$$

Solve each linear programming problem.

7. Lois makes banana bread and nut bread to sell at a bazaar. A loaf of banana bread requires 2 c flour and 2 eggs. A loaf of nut bread takes 3 c flour and 1 egg. Lois has 12 c flour and 8 eggs on hand. She makes $2 profit per loaf of banana bread and $2 per loaf of nut bread. To maximize profits, how many loaves of each type should she bake?

8. A tray of corn muffins requires 4 c milk and 3 c wheat flour. A tray of bran muffins takes 2 c milk and 3 c wheat flour. There are 16 c milk and 15 c wheat flour available, and the baker makes $3 profit per tray of corn muffins and $2 per tray of bran muffins. How many trays of each should he make in order to maximize profits?

9. Juan makes two types of wood clocks to sell at local stores. It takes him 2 h to assemble a pine clock, which requires 1 oz of varnish. It takes 2 h to assemble an oak clock, which takes 4 oz of varnish. Juan has 16 oz of varnish in stock, and he can work 20 hours. If he makes $3 profit on each pine clock and $4 on each oak clock, how many of each type should he make to maximize his profits?

10. Kay grows and sells tomatoes and green beans. It costs $1 to grow a bushel of tomatoes, and it takes 1 yd^2 of land. It costs $3 to grow a bushel of beans, and it takes 6 yd^2 of land. Kay's budget is $15, and she has 24 yd^2 of land available. If she makes $1 profit on each bushel of tomatoes and $4 profit on each bushel of beans, how many bushels of each should she grow in order to maximize profits?

11. A biologist needs at least 40 fish for her experiment. She cannot use more than 25 perch or more than 30 bass. Each perch costs $5, and each bass costs $3. How many of each fish should she use in order to minimize the cost?

12. A restaurant provides individual servings of ketchup and mustard free to all customers. One single-serving container of ketchup costs the restaurant 2¢; one of mustard costs 4¢. A ketchup container takes up 3 cm^2 of counter space, and a mustard container takes 2 cm^2 of space. Only 600 cm^2 of counter space is available for these condiments. The restaurant must have at least 50 servings of ketchup and 150 servings of mustard available for the customers. How many servings of each should be provided to minimize the cost?

13. A company makes whole wheat crackers and sesame crackers. The crackers are sold by the box. Each box contains 5 packets of whole wheat crackers or 3 packets of sesame crackers. The company cannot produce more than 150 packets of crackers per minute, but at least 15 boxes of whole wheat crackers and at least 20 boxes of sesame crackers must be produced per minute. If the profit per box of whole wheat crackers is 10 cents and the profit per box of sesame crackers is 5 cents, how many boxes of each type should be produced per minute in order to maximize profits?

14. A biologist is developing two new strains of bacteria. Each sample of Type I bacteria produces 4 new viable bacteria, and each sample of Type II produces 3 new viable bacteria. Altogether, at least 240 new viable bacteria must be produced. At least 30, but not more than 60, of the original samples must be Type I. Not more than 70 of the samples are to be Type II. A sample of Type I costs \$5 and a sample of Type II costs \$7. If both types are to be used, how many samples of each should be used to minimize the cost?

TEST YOURSELF

4.6

1. With a tail wind, a plane travels 600 mi in 4 h. With a head wind, it travels only 560 mi in 4 h. Find the plane's air speed and the speed of the wind. Assume that both are constant.

2. The tens digit of a two-digit number is 3 less than the units digit. If the digits are reversed, the sum of the new number and the original number is 1 less than 4 times the original number. What is the original number?

4.7, 4.8

3. Solve this linear system of inequalities graphically:

$$\begin{cases} 3x - 2y \geq 6 \\ |x| \leq 4 \end{cases}$$

4. If profit is represented by $P = 3x + y$, find the maximum profit under these constraints:

$$\begin{cases} x + y \leq 7 \\ x + 2y \leq 8 \\ x \geq 0 \\ y \geq 0 \end{cases}$$

INTEGRATING ALGEBRA
Break-Even Point

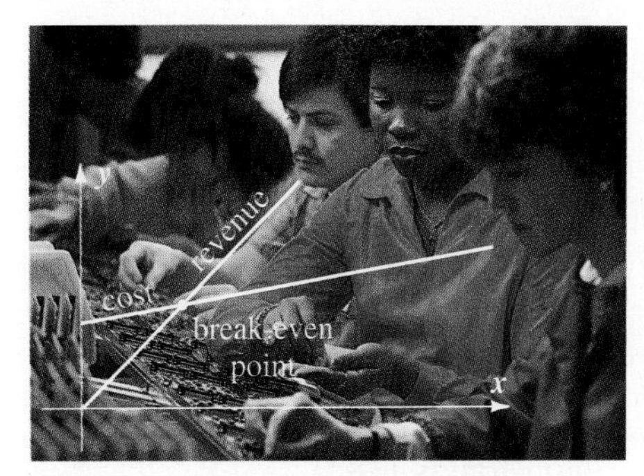

In order for any business venture to be successful, the *revenues* must exceed the *costs*. **Revenue** is the term for money that is earned from sales while **cost** is the term for expenses that are paid. Costs include *overhead,* such as rent, telephone, and office furniture and *production costs,* such as workers' salaries and materials. Because of costs, business may experience a period after opening in which it loses money, until the revenues "catch up" with the costs. The point at which the revenues and costs are equal is called the **break-even point.** If the cost function and the revenue function for an operation are linear, a system of linear equations can be used to find the break-even point.

EXAMPLE Producer Jake Murray invests in a new play. The cost includes an overhead of \$30,000, plus production costs of \$2500 per performance. If a sold-out performance brings in \$3125, how many sold-out performances must be played to break even?

Let x be the number of sold-out performances. Then

$$\text{Cost} = 30{,}000 + 2500x \qquad \text{Revenue} = 3125x$$

$$\begin{aligned} \text{Cost} &= \text{Revenue} \qquad \textit{To break even} \\ 30{,}000 + 2500x &= 3125x \\ 30{,}000 &= 625x \\ 48 &= x \end{aligned}$$

The play must run 48 sold-out performances to break even.

EXERCISES

1. Charlie and Spike open a lemonade stand. Their expenses are \$8.40 for supplies, plus 1¢ per glass. If they charge 15¢ per glass of lemonade, how many glasses must they sell to break even?
2. An entrepreneur sells ant farms for \$18 each. His expenses include a fixed overhead of \$290 plus production costs of \$4 per ant farm. How many ant farms must he sell before he starts to earn a profit?
3. Annie paid \$30,000 in college tuition and then started a business writing greeting cards. If her supplies cost 2¢ per card and she sells each card for 50¢, how many cards must she sell to cover her tuition and her costs?

CHAPTER 4 SUMMARY AND REVIEW

Vocabulary

boundary of a half-plane (177)
closed half-plane (177)
consistent system of equations (155)
constant of proportionality (146)
constant of variation (144)
constraints (182)
dependent system of equations (155)
direct variation (144)
equivalent systems of equations (160)
extremes of a proportion (146)
feasible region (182)
half-plane (177)
inconsistent system of equations (155)
independent system of equations (155)
linear programming (182)
linear system of equations (154)
means of a proportion (146)
objective function (182)
open half-plane (178)
system of equations (154)
system of inequalities (179)
y varies directly as x (144)

Direct Variation If $y = kx$, $k \neq 0$, then y varies directly as x. k is the constant of variation. 4.1

For each function, determine whether y varies directly as x. If so, give the constant of variation.

1. $y = 9x$
2. $x + y = 1$
3. $y = -3x + 4$
4. $3y - 2x = 0$

In Exercises 5–6, y varies directly as x.

5. If $y = 12$ when $x = 2$, find y when $x = 6$.
6. If $y = 4$ when $x = 8$, find x when $y = 13$.

Solving a Linear System of Equations Graphically Graph both equations on the same coordinate plane and find the point of intersection, if it exists. A linear system of equations may have no solutions, one solution, or an infinite number of solutions. 4.2–4.3

Solve each system graphically. Determine whether the system is consistent and independent, inconsistent, or consistent and dependent.

7. $\begin{cases} 4x + 3y = 6 \\ -2x + 3y = -12 \end{cases}$
8. $\begin{cases} 4x + 2y = -4 \\ 12x + 6y = 12 \end{cases}$

Solving a Linear System by Addition Multiply one or both of the equations by nonzero constants so that the coefficients of one variable are additive inverses. Add the equations and solve for one variable. Then substitute in an original equation and solve for the other variable. 4.4

Solve each system by addition.

9. $\begin{cases} 4x - y = 17 \\ 3x + 2y = -1 \end{cases}$

10. $\begin{cases} 4x + 5y = -12 \\ 3x - 4y = 22 \end{cases}$

Solving a Linear System by Substitution Solve either equation for one variable in terms of the other. Substitute in the other equation and solve. Then substitute that value in an original equation to find the other variable. 4.5

Solve each system by substitution.

11. $\begin{cases} 5x - 4y = 6 \\ y = 3x + 2 \end{cases}$

12. $\begin{cases} 4x + 7y = 28 \\ y = 2x - 14 \end{cases}$

Problem Solving Linear systems of equations can be used to solve many problems involving two unknown quantities. 4.6

Solve.

13. Jennifer bought 4 cartons of milk and 6 cartons of juice for \$19.32. Alex paid \$15.67 for 3 cartons of the same brand of milk and 5 cartons of the same juice. What is the cost per carton for each?

14. With a head wind, a plane flew 6160 mi in 14 h. With a tail wind of the same strength, the return trip took only 11 h. Find the plane's speed and the wind speed.

Solving a System of Inequalities Find the intersection of their graphs. 4.7

Solve each linear system of inequalities graphically.

15. $\begin{cases} 2y \leq -3x + 12 \\ y \geq x - 2 \end{cases}$

16. $\begin{cases} y - 2x \geq 5 \\ 2 \leq y \leq 9 \end{cases}$

Linear Programming First graph the constraints on the same coordinate plane and shade the feasible region. Find the coordinates of the vertices of the feasible region. Evaluate the objective function at each vertex in order to find its maximum or minimum value. 4.8

Solve.

17. If profit is represented by $P = 4x + 3y$, find the maximum profit under the constraints shown at the right.

$\begin{cases} x + y \leq 3 \\ 2x + y \leq 4 \\ x \geq 0 \\ y \geq 0 \end{cases}$

CHAPTER 4 TEST

For each function, determine whether y varies directly as x. If so, give the constant of variation.

1. $3y + 6x = 0$

2. $2y - x = 5$

3. $\frac{y}{x} - 4 = 0$

In Exercises 4–5, y varies directly as x.

4. If $y = 24$ when $x = 12$, find y when $x = 6$.

5. If $y = 10$ when $x = -2$, find x when $y = 55$.

Solve each linear system by the given method. Determine if the system is consistent and independent, inconsistent, or consistent and dependent.

6. Solve graphically.
$$\begin{cases} 5x - 6y = 30 \\ -10x = -12y - 60 \end{cases}$$

7. Solve by addition.
$$\begin{cases} 5x + 6y = -19 \\ 3x + 4y = -13 \end{cases}$$

8. Solve by substitution.
$$\begin{cases} 4x - y = -2 \\ 3y = 4x + 2 \end{cases}$$

Solve.

9. Pat rowed 40 km downstream in 2 h. Her rowing rate was the same going back upstream, but the return trip took her 5 h. Find her rowing rate in still water and the rate of the current.

10. The units digit of a two-digit number is 4 less than twice the tens digit. When twice the number is subtracted from 3 times the number formed by reversing the digits of the original number, the result is 83. What is the original number?

11. Solve the system of inequalities graphically:
$$\begin{cases} -2x + y \le 8 \\ 2x + y \ge 0 \end{cases}$$

12. If cost is represented by $C = 5x + 2y$, find the minimum cost under these constraints:
$$\begin{cases} 2x + 2y \ge 18 \\ y \le 9 \\ x \le 3 \end{cases}$$

Challenge

The perimeter of a regular polygon varies directly as the length of a side. The perimeter of a certain regular polygon is 720 cm. If the length of each side were decreased by 40 cm, the perimeter would be $\frac{1}{3}$ less. How many sides does the polygon have?

COLLEGE ENTRANCE EXAM REVIEW

Select the best choice for each question.

1. If $3x = 4y - 15$, then $12y - 9x =$
A. 60 **B.** -60 **C.** 45 **D.** -45
E. It cannot be determined from the information given.

2. If d, e, and f are consecutive integers, which of the following is false?
A. $d + f$ is even **B.** def is even
C. $e + f$ is odd **D.** ef is odd
E. $\frac{def}{3}$ is an integer

3. In the figure below, $p \parallel q$ and $110 < x < 150$. Which is true of y?

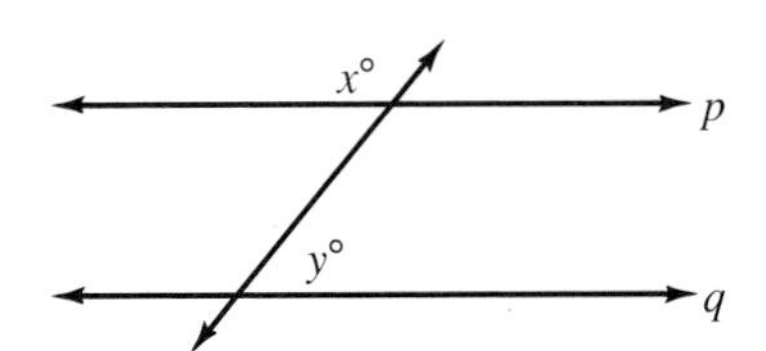

A. $30 < y < 70$ **B.** $40 < y < 70$
C. $30 < y < 40$ **D.** $9 < y < 40$
E. $0 < y < 30$

4. In a class there are $\frac{2}{3}$ as many boys as girls. If 30% of the boys ride the school bus and 40% of the girls ride it, what percent of the class rides the bus?
A. 35% **B.** 36% **C.** 38% **D.** 39%
E. It cannot be determined from the information given.

5. The solution for the system of equations $\begin{cases} 2x - y = 14 \\ x + 3y = 14 \end{cases}$ is
A. (10, 6) **B.** (9, 4) **C.** (8, 2)
D. (7, 0) **E.** (5, 3)

6. An artist plans to have the length of each wire in his mobile vary directly with the weight it supports. If a 10 cm wire holds a 36-g weight, how many centimeters of wire would hold a 54-g weight?
A. 9 **B.** 12 **C.** 15 **D.** 18 **E.** 20

7. When $0 < n < 1$, which of these *must* be true about n?

I. $n^2 > 1$
II. $n^2 > n^3$
III. $\sqrt{n} > n$

A. III only **B.** I and II only
C. I and III only **D.** II and III only
E. I, II, and III

8. If $\frac{a}{b} = \frac{5}{6}$, then $\frac{b}{2a}$ equals

A. $\frac{3}{5}$ **B.** $\frac{5}{3}$ **C.** $\frac{5}{12}$ **D.** $\frac{12}{5}$ **E.** $\frac{3}{10}$

9. Suppose the symbol $p(n)$ means the sum of all the prime numbers that are factors of n. (Remember that 1 is not prime.) For example

$p(18) = 2 + 3 = 5$
$p(7) = 7$
$p(66) = 2 + 3 + 11 = 16$

Find $p(70)$.
A. 7 **B.** 9 **C.** 12 **D.** 14 **E.** 17

10. Find the coordinates of the point with the maximum value of y in the solution of the system:

$$\begin{cases} 4x + y \le 16 \\ 2x - 3y \ge -6 \end{cases}$$

A. (1, 12) **B.** (2, 7) **C.** (6, 6)
D. (5, 5) **E.** (3, 4)

CUMULATIVE REVIEW (CHAPTERS 1–4)

Graph.

1. $y = -3$
2. $y = -x + 4$
3. $3y + 6 = 9x$
4. $3x + 2y = 6$
5. Write in standard form the equation of the line that passes through (2, 3) and (1, −4).
6. Solve the linear system graphically. Determine whether the system is consistent and independent, inconsistent, or consistent and dependent.

$$\begin{cases} y = 3 - x \\ 3x - y = 5 \end{cases}$$

Solve and graph the solution set.

7. $-3|x + 4| \geq 6$
8. $4x > 24$ or $-3x > 9$
9. Find three consecutive odd integers, if the sum of the smallest and the largest is 70.

Graph each ordered pair and name the quadrant or axis where each point lies.

10. (0, 3)
11. (−1, −4)
12. (2, 5)
13. (−3, 2)
14. (1, −2)
15. (−2, 0)
16. Evaluate $4c^4 - 3c^3 + c^2$, using $c = -1$.
17. Write an algebraic sentence for the following, using y for the variable: Four times a number increased by six is equal to thirty-two.
18. Graph the solution set: $x \leq -2$ or $x > 3$.
19. A mountain climber planning an expedition is concerned with two types of synthetic food. One food contains 100 calories per ounce, 24 units of protein per ounce, and 4 units of fat per ounce. A second food contains 125 calories per ounce, 20 units of protein per ounce, and 10 units of fat per ounce. Suppose that the minimum daily requirements for an average man are calories: 2000; protein: 400; and fat: 100. Which food or food combination should be used to meet the minimum daily requirements and minimize the total weight?
20. Name the additive inverse of $\frac{2}{3}$.

Solve and check.

21. $|4x - 3| = 5$

22. $5y - 2 = 3y + 8$

23. Graph: $y = -\frac{1}{3}x + 1$

24. Solve the system of equations by substitution:

$$\begin{cases} x + y = -10 \\ 3x - 6y = -12 \end{cases}$$

25. The measure of an angle is twenty-four greater than two times its complement. Find the angle and its complement.

26. Graph: $-3y \geq -12x - 6$

27. Write a rule for the relation and state its domain and range: $\{(0, -1), (1, 1), (2, 3), (3, 5)\}$

Solve each system by graphing.

28. $\begin{cases} -x + y \geq 4 \\ 3x - y \leq 6 \end{cases}$

29. $\begin{cases} y > x - 5 \\ y \leq 2x + 3 \end{cases}$

30. Write an equation of the line parallel to $y = 2x + 3$ with y-intercept $b = -2$.

Evaluate each pair of functions for $x = -2, -1, 0, 1, c$.

31. $f(x) = -2x$, $g(x) = 2x^2 + 1$

32. $f(x) = x^3 - 1$, $g(x) = 4x - 3$

33. Find the slope and y-intercept of $3x + 2y = 4$.

34. Write the equation of the line having slope $m = 2$ and passing through $(-3, 1)$.

35. Determine whether y varies directly as x. If so, give the constant of variation. $y + 3x = 0$

36. Graph the relation and state its domain and range. $\{(-3, 1), (4, 3), (0, 2), (-1, 4), (3, 0)\}$

37. The demand curve for a new brand of floppy disk was found to be linear. The demand was 50 disks at \$2 each and 100 disks at \$1 each. Write the equation for the linear demand curve for this product and determine at what price the demand would be 150.

38. Solve the linear system by the addition method.

$$\begin{cases} y = 4 - x \\ 3x + y = 6 \end{cases}$$

39. Find the slope and y-intercept of $y = -x - 3$.

40. Evaluate $g(f(2))$ given $f(x) = 3x^2$ and $g(x) = x + 3$.

41. Find the inverse of $y = -2x + 4$.

42. y varies directly as x. If $y = 5$ when $x = 3$, find y when $x = -1$.

43. Mark ordered 10 disks and 1 printer ribbon, costing \$6.50. A second order from the same company consisted of 15 disks and 2 printer ribbons. It totalled \$11.75. What was the cost of each disk and each printer ribbon?

44. Find $(g \circ f)(x)$, given $f(x) = 2x^3$ and $g(x) = 5x$.

45. Determine whether the lines are parallel, perpendicular, or neither.
$3y = 2x - 6$, $4x - 6y = 3$

46. Write an equation of a line with slope $m = 3$ and y-intercept $b = -2$.

47. Make a mapping diagram and determine whether or not it is a function.
$\{(4, 2), (1, 5), (0, -3), (-1, -2)\}$

48. Find the inverse of $\{(3, 4), (2, 5), (-1, 3), (0, 4)\}$.

49. Write an equation of the line perpendicular to $2y = 3x - 6$ with y-intercept $b = -1$.

Solve each system of equations by addition. If inconsistent, write "no solution."

50. $\begin{cases} x + 3y = 7 \\ 2x - y = 7 \end{cases}$

51. $\begin{cases} 0.05x - 0.03y = -0.01 \\ 0.04x + 0.01y = 0.06 \end{cases}$

52. Find the slope of the line passing through $(-2, 3)$ and $(5, -4)$.

Solve each equation for x and indicate any restrictions on the values of the variables.

53. $ax + bx = r$

54. $m - x = 7$

55. $\frac{x}{c} - g = h;\ c \neq 0$

56. $\frac{s + t}{x - y} = 2;\ x \neq y$

State which property of real numbers is illustrated by each of the following.

57. $x(3x + 5) = 3x^2 + 5x$

58. $\pi \cdot 1 = \pi$

59. $\left(-\frac{11}{15}\right)\left(-\frac{15}{11}\right) = 1$

60. $\sqrt{6} + \sqrt{2}$ is irrational

Find the inverse of each relation.

61. $y = 2x - 10$

62. $2y = 6x + 14$

63. Show that the coordinates $(-2, -2)$, $(-4, 1)$, $(7, 4)$, and $(5, 7)$ are the vertices of a rectangle.

Matrices and Determinants

Help this band get away!

Developing Mathematical Power

The Metro High School marching band needs to raise $3000 to go to a music festival. The students have 12 weeks to raise that amount. Local merchants have donated merchandise for students to sell. The school will allow the students to use the school store and a computer for the duration of the sale.

DONATIONS

- 300 pens
- 100 packages of loose-leaf paper
- 125 T-shirts
- 35 cans of tennis balls
- 25 packs of golf balls
- 100 free-dinner coupons
- 1500 pencils
- 40 CDs
- 75 caps
- 5 tennis rackets
- 4 basketballs

Project

Suppose you are in charge of the sale. How would you organize it? Keep track of the inventory and revenue so that current information is easily obtained. Schedule students to work in the store and keep track of their schedules. Present a final report of the inventory and revenues to the student body.

5.1

Matrix Addition

Objective: To find the sum and difference of two matrices and the product of a matrix and a scalar

One way to organize numerical data is to write the numbers in rows and columns, forming a *matrix*. This format is especially useful for entering information into a computer. More information about the data can be obtained by performing mathematical operations on matrices. You will find that the properties of matrix addition and scalar multiplication are similar to the properties of the real number system.

Capsule Review

Properties of the Real Number System

If a, b, and c are real numbers, then

$a + b$ and ab are real numbers.	Closure properties
$a + b = b + a$ and $ab = ba$	Commutative properties
$(a + b) + c = a + (b + c)$ and $(ab)c = a(bc)$	Associative properties
$a + 0 = 0 + a = a$	Additive identity property
$a \cdot 1 = 1 \cdot a = a$	Multiplicative identity property
$a + (-a) = 0$	Additive inverse property
$a \cdot \frac{1}{a} = 1, a \neq 0$	Multiplicative inverse property
$a(b + c) = ab + ac$	Distributive property
$a \cdot 0 = 0 \cdot a = 0$	Multiplicative property of zero

Name the property illustrated by each equation.

1. $3 + 5 = 5 + 3$

2. $\sqrt{7} + 0 = \sqrt{7}$

3. $12 + 9$ is a real number.

4. $2(3 + 5) = 2(3) + 2(5)$

5. $\pi + (-\pi) = 0$

6. $1.5 \cdot 1 = 1.5$

7. $8 \cdot \frac{1}{8} = 1$

8. $(9 + 4) + 2 = 9 + (4 + 2)$

A **matrix** (plural: *matrices*) is a rectangular array of numbers written within brackets. A matrix is represented by a capital letter and is classified by its dimensions, which are determined by the number of rows (horizontal) and columns (vertical). Matrix A shown below has 2 rows and 3 columns, so it is a 2×3 (read "two by three") matrix and may be designated $A_{2\times 3}$.

$$A = \begin{bmatrix} 2 & 3 & 4 \\ 6 & 7 & 0 \end{bmatrix}$$

A matrix with m rows and n columns is an $m \times n$ matrix and may be written

$$A_{m\times n}$$

rows / columns

Each number in a matrix is called an **element** of the matrix. The elements of matrix A are 2, 3, 4, 6, 7, and 0. An element of a matrix is identified by its position, given by the row and column numbers.

EXAMPLE 1 **State the dimensions of each matrix and name the indicated element.**

a. $M = \begin{bmatrix} 4 & 6 & 5 \\ 2 & -3 & -7 \\ 1 & 0 & 9 \end{bmatrix}$; 1st row, 1st column

b. $N = \begin{bmatrix} -4 & 1 & -3 \end{bmatrix}$; 1st row, 3rd column

c. $P = \begin{bmatrix} 1 \\ 2 \\ 0 \end{bmatrix}$; 2nd row, 1st column

a. M has 3 rows and 3 columns and is therefore a 3×3 matrix. The element in the first row, first column is 4.

b. N has 1 row and 3 columns and is therefore a 1×3 matrix. The element in the first row, third column is -3.

c. P has 3 rows and 1 column and is therefore a 3×1 matrix. The element in the second row, first column is 2.

A matrix that has the same number of rows and columns, such as matrix M in Example 1, is called a **square matrix.** A matrix that has only one row, such as matrix N, is called a **row matrix.** A matrix that has only one column, such as matrix P, is called a **column matrix.**

Elements that occupy corresponding positions in two matrices with the same dimensions are called *corresponding elements*. Two matrices are **equal matrices** if and only if they have the same dimensions and all of their corresponding elements are equal.

Matrix addition is defined only for matrices with the same dimensions.

> **Matrix Addition** If two matrices, A and B, have the same dimensions, then their sum, $A + B$, is a matrix whose elements are the sums of the corresponding elements of A and B.

EXAMPLE 2 **If $A = \begin{bmatrix} 1 & 2 & 4 \\ -3 & 1 & 0 \end{bmatrix}$ and $B = \begin{bmatrix} 5 & 1 & -3 \\ 3 & 6 & 1 \end{bmatrix}$ find each sum:**

a. $A + B$ **b.** $B + A$

a. $A + B = \begin{bmatrix} 1 & 2 & 4 \\ -3 & 1 & 0 \end{bmatrix} + \begin{bmatrix} 5 & 1 & -3 \\ 3 & 6 & 1 \end{bmatrix} = \begin{bmatrix} 1+5 & 2+1 & 4-3 \\ -3+3 & 1+6 & 0+1 \end{bmatrix} = \begin{bmatrix} 6 & 3 & 1 \\ 0 & 7 & 1 \end{bmatrix}$

b. $B + A = \begin{bmatrix} 5 & 1 & -3 \\ 3 & 6 & 1 \end{bmatrix} + \begin{bmatrix} 1 & 2 & 4 \\ -3 & 1 & 0 \end{bmatrix} = \begin{bmatrix} 5+1 & 1+2 & -3+4 \\ 3-3 & 6+1 & 1+0 \end{bmatrix} = \begin{bmatrix} 6 & 3 & 1 \\ 0 & 7 & 1 \end{bmatrix}$

Example 2 illustrates the fact that matrix addition is commutative.

The *additive identity* matrix for the set of all $m \times n$ matrices is the **zero matrix** O, or $O_{m \times n}$. For example $O_{2\times 2} = \begin{bmatrix} 0 & 0 \\ 0 & 0 \end{bmatrix}$ and $O_{2\times 3} = \begin{bmatrix} 0 & 0 & 0 \\ 0 & 0 & 0 \end{bmatrix}$.

The *opposite,* or *additive inverse,* of an $m \times n$ matrix A is $-A$. $-A$ is the $m \times n$ matrix whose elements are the opposites of the corresponding elements of A. For example, if $A = \begin{bmatrix} 6 & 3 & 1 \\ 0 & 7 & -2 \end{bmatrix}$, then $-A = \begin{bmatrix} -6 & -3 & -1 \\ 0 & -7 & 2 \end{bmatrix}$.

The properties of matrix addition are summarized below. You will be asked to prove some of these properties in Exercises 33–40.

Properties of Matrix Addition

If A, B, and C are $m \times n$ matrices, then

$A + B$ is an $m \times n$ matrix.	Closure property
$A + B = B + A$	Commutative property
$(A + B) + C = A + (B + C)$	Associative property
There exists a unique $m \times n$ matrix O such that $O + A = A + O = A$.	Additive identity property
For each A, there exists a unique opposite, $-A$, such that $A + (-A) = O$.	Additive inverse property

Matrix subtraction can be defined using the additive inverse in a manner similar to subtraction for real numbers.

Matrix Subtraction If two matrices, A and B, have the same dimensions, then $A - B = A + (-B)$.

EXAMPLE 3 **If $A = \begin{bmatrix} 3 & 2 & 4 \\ -1 & 4 & 0 \end{bmatrix}$ and $B = \begin{bmatrix} 1 & 4 & 3 \\ -2 & 2 & 4 \end{bmatrix}$, find $A - B$.**

$$A - B = A + (-B) = \begin{bmatrix} 3 & 2 & 4 \\ -1 & 4 & 0 \end{bmatrix} + \begin{bmatrix} -1 & -4 & -3 \\ 2 & -2 & -4 \end{bmatrix}$$
$$= \begin{bmatrix} 2 & -2 & 1 \\ 1 & 2 & -4 \end{bmatrix}$$

Note that $A - B$ can also be computed by subtracting corresponding elements, just as $A + B$ is computed by adding corresponding elements. That is,

$$A - B = \begin{bmatrix} 3 & 2 & 4 \\ -1 & 4 & 0 \end{bmatrix} - \begin{bmatrix} 1 & 4 & 3 \\ -2 & 2 & 4 \end{bmatrix}$$
$$= \begin{bmatrix} 3 - 1 & 2 - 4 & 4 - 3 \\ -1 - (-2) & 4 - 2 & 0 - 4 \end{bmatrix} = \begin{bmatrix} 2 & -2 & 1 \\ 1 & 2 & -4 \end{bmatrix}$$

In matrix algebra, any real number is called a **scalar.** The product of a scalar c and a matrix A, called a **scalar product,** is a matrix of the same dimensions as A whose elements are the products of c and the corresponding elements of A. The scalar product is denoted by cA. For example,

$$-2\begin{bmatrix} 3 & -4 \\ 5 & 8 \end{bmatrix} = \begin{bmatrix} -6 & 8 \\ -10 & -16 \end{bmatrix}$$

EXAMPLE 4 **If $A = \begin{bmatrix} 2 & 3 & -7 \\ 1 & 4 & 5 \end{bmatrix}$ and $B = \begin{bmatrix} 3 & 0 & 6 \\ -1 & 8 & 2 \end{bmatrix}$ find the following matrices:**

a. $2A$ **b.** $3A + 2B$ **c.** $5A - 3B$

a. $2A = 2\begin{bmatrix} 2 & 3 & -7 \\ 1 & 4 & 5 \end{bmatrix} = \begin{bmatrix} 2(2) & 2(3) & 2(-7) \\ 2(1) & 2(4) & 2(5) \end{bmatrix} = \begin{bmatrix} 4 & 6 & -14 \\ 2 & 8 & 10 \end{bmatrix}$

b. $3A + 2B = 3\begin{bmatrix} 2 & 3 & -7 \\ 1 & 4 & 5 \end{bmatrix} + 2\begin{bmatrix} 3 & 0 & 6 \\ -1 & 8 & 2 \end{bmatrix}$

$$= \begin{bmatrix} 6 & 9 & -21 \\ 3 & 12 & 15 \end{bmatrix} + \begin{bmatrix} 6 & 0 & 12 \\ -2 & 16 & 4 \end{bmatrix} = \begin{bmatrix} 12 & 9 & -9 \\ 1 & 28 & 19 \end{bmatrix}$$

c. $5A - 3B = 5\begin{bmatrix} 2 & 3 & -7 \\ 1 & 4 & 5 \end{bmatrix} - 3\begin{bmatrix} 3 & 0 & 6 \\ -1 & 8 & 2 \end{bmatrix}$

$$= \begin{bmatrix} 10 & 15 & -35 \\ 5 & 20 & 25 \end{bmatrix} - \begin{bmatrix} 9 & 0 & 18 \\ -3 & 24 & 6 \end{bmatrix} = \begin{bmatrix} 1 & 15 & -53 \\ 8 & -4 & 19 \end{bmatrix}$$

Properties of Scalar Multiplication

If A, B, and O are $m \times n$ matrices and c and d are scalars, then

cA is an $m \times n$ matrix.	Closure property
$(cd)A = c(dA)$	Associative property
$c(A + B) = cA + cB$ and $(c + d)A = cA + dA$	Distributive properties
$1 \cdot A = A$	Multiplicative identity property
$0A = O$ and $cO = O$	Multiplicative properties of zero

You can use the properties of scalar multiplication and the definitions of matrix addition and subtraction to solve simple matrix equations. In a **matrix equation,** the variable represents a matrix.

EXAMPLE 5 **Solve:** $\mathbf{4X + 2\begin{bmatrix} 3 & 4 \\ -2 & 1 \end{bmatrix} = \begin{bmatrix} 10 & 0 \\ 4 & 2 \end{bmatrix}}$

$$4X + \begin{bmatrix} 6 & 8 \\ -4 & 2 \end{bmatrix} = \begin{bmatrix} 10 & 0 \\ 4 & 2 \end{bmatrix}$$ *Scalar multiplication*

$$4X = \begin{bmatrix} 10 & 0 \\ 4 & 2 \end{bmatrix} - \begin{bmatrix} 6 & 8 \\ -4 & 2 \end{bmatrix}$$

$$4X = \begin{bmatrix} 4 & -8 \\ 8 & 0 \end{bmatrix}$$ *Matrix subtraction*

$$X = \frac{1}{4}\begin{bmatrix} 4 & -8 \\ 8 & 0 \end{bmatrix}$$ *Multiply both sides by $\frac{1}{4}$.*

$$X = \begin{bmatrix} 1 & -2 \\ 2 & 0 \end{bmatrix}$$ Substitute the value of X into the original equation to check.

Example 6 illustrates the use of scalar multiplication, matrix addition, and the definition of equal matrices, to solve an equation for unknown values of elements of matrices.

EXAMPLE 6 **If $6[x \quad y] = [x \quad 4] + [5 \quad 4y]$, find the values of x and y.**

$6[x \quad y] = [x \quad 4] + [5 \quad 4y]$

$[6x \quad 6y] = [x + 5 \quad 4 + 4y]$ *Scalar multiplication; matrix addition*

$6x = x + 5$ and $6y = 4 + 4y$ *Definition of equal matrices*

$x = 1$ and $y = 2$

Replace x with 1 and y with 2 in the original equation to check.

A computer graphing utility can simplify matrix operations. Most graphing calculators can also perform limited matrix operations.

CLASS EXERCISES

1. **Thinking Critically** Explain why matrix addition is defined only for matrices with the same dimensions.

State the dimensions of each matrix and identify any square, row, or column matrices.

2. $\begin{bmatrix} 2 & 3 & 1 \\ 4 & 6 & 1 \end{bmatrix}$

3. $\begin{bmatrix} 1 \\ 2 \end{bmatrix}$

4. $\begin{bmatrix} 7 & 4 \\ 5 & 6 \end{bmatrix}$

5. $[5 \quad 9]$

Determine whether the matrices in each pair are equal.

6. $\begin{bmatrix} 4 \\ 6 \\ 8 \end{bmatrix}, \begin{bmatrix} \frac{8}{2} & \frac{18}{3} & \frac{16}{2} \end{bmatrix}$

7. $\begin{bmatrix} -2 & 3 \\ 5 & 0 \end{bmatrix}, \begin{bmatrix} -\frac{8}{4} & 6-3 \\ \frac{15}{3} & 4-4 \end{bmatrix}$

8. Solve the matrix equation.

$$2X = \begin{bmatrix} 4 & 12 \\ 1 & -4 \end{bmatrix} + \begin{bmatrix} -2 & 0 \\ 3 & 4 \end{bmatrix}$$

9. Find the value of each variable.

$$\begin{bmatrix} 2 & 2 \\ -1 & 6 \end{bmatrix} - \begin{bmatrix} 4 & -1 \\ 0 & 5 \end{bmatrix} = \begin{bmatrix} x & y \\ -1 & z \end{bmatrix}$$

PRACTICE EXERCISES

 Use technology where appropriate.

Perform the indicated operations, if they are defined, for matrices A, B, C, and D shown below. If an operation is not defined for the given matrices, label it *undefined*.

$$A = \begin{bmatrix} 3 & 4 \\ 6 & -2 \\ 1 & 0 \end{bmatrix} \quad B = \begin{bmatrix} -3 & 1 \\ 2 & -4 \\ -1 & 5 \end{bmatrix} \quad C = \begin{bmatrix} 1 & 2 \\ -3 & 1 \end{bmatrix} \quad D = \begin{bmatrix} 5 & 1 \\ 0 & 2 \end{bmatrix}$$

1. $A + B$
2. $B + A$
3. $A + C$
4. $B + D$
5. $C + D$
6. $D + C$
7. $B - A$
8. $A - B$
9. $C - D$
10. $D - C$
11. $3A$
12. $4B$
13. $3A + 2B$
14. $4C + 3D$
15. $2A - 5B$
16. $2C - 5D$

Solve each matrix equation.

17. $X - \begin{bmatrix} 1 & 4 \\ -2 & 3 \end{bmatrix} = \begin{bmatrix} 5 & -2 \\ 1 & 0 \end{bmatrix}$

18. $X + \begin{bmatrix} 6 & 1 \\ -2 & 3 \end{bmatrix} = \begin{bmatrix} 2 & 0 \\ -3 & 1 \end{bmatrix}$

19. $2X - \begin{bmatrix} 6 & 2 & 8 \\ 0 & 4 & 2 \end{bmatrix} = 2\begin{bmatrix} 7 & -3 & -1 \\ 5 & 4 & -2 \end{bmatrix}$

20. $3X - 3\begin{bmatrix} 1 & -2 & 0 \\ 3 & -5 & 7 \end{bmatrix} = 2\begin{bmatrix} 3 & 9 & -3 \\ 0 & 6 & 12 \end{bmatrix}$

Find the value of each variable.

21. $\begin{bmatrix} x+2 & 4 \\ 6 & -y \end{bmatrix} + \begin{bmatrix} 3 & 2 \\ 5 & -1 \end{bmatrix} = \begin{bmatrix} 7 & 6 \\ 11 & -5 \end{bmatrix}$

22. $\begin{bmatrix} x+3 & y-2 \\ z+3 & w+4 \end{bmatrix} + \begin{bmatrix} 2 & 5 \\ -2 & 4 \end{bmatrix} = \begin{bmatrix} 6 & 1 \\ 4 & 8 \end{bmatrix}$

23. $\begin{bmatrix} x & 3 \\ x & -2 \end{bmatrix} + \begin{bmatrix} y & 6 \\ -y & 3 \end{bmatrix} = \begin{bmatrix} 6 & 9 \\ 4 & 1 \end{bmatrix}$

24. $\begin{bmatrix} 2x & 4 \\ 3 & 3x \end{bmatrix} + \begin{bmatrix} 4y & 6 \\ 4 & y \end{bmatrix} = \begin{bmatrix} 2 & 10 \\ 7 & -7 \end{bmatrix}$

For Exercises 25–30, use matrices E, F, G, and H shown below.

$$E = \begin{bmatrix} 3 \\ 4 \\ 7 \end{bmatrix} \qquad F = \begin{bmatrix} 5 & 9 & 2 \\ 6 & 1 & -1 \\ 0 & 2 & 3 \end{bmatrix} \qquad G = \begin{bmatrix} -2 \\ 0 \\ 5 \end{bmatrix} \qquad H = [-5 \quad 10 \quad 15]$$

25. Find the sum of the column matrices.

26. Find the scalar product of 3 and the square matrix.

27. Find $\frac{1}{5}$ of the row matrix.

28. Write the zero matrix for matrix F.

29. Write the additive inverse for matrix H.

30. Find the sum of matrix E and the additive inverse for matrix G.

Find the value of each variable.

31. $x[3 \quad 5] + y[2 \quad 5] = [4 \quad 5]$

32. $x\begin{bmatrix} 4 \\ 1 \end{bmatrix} + y\begin{bmatrix} 2 \\ 3 \end{bmatrix} = \begin{bmatrix} -8 \\ 3 \end{bmatrix}$

33. Prove that matrix addition is commutative for 2×2 matrices.

34. Prove that matrix addition is associative for 2×2 matrices.

For Exercises 35–40, use the notation illustrated at the right, where the element in the ith row, jth column of matrix A is denoted a_{ij}.

$$A = \begin{bmatrix} a_{11} & a_{12} & \cdots & a_{1n} \\ a_{21} & a_{22} & \cdots & a_{2n} \\ \vdots & \vdots & & \vdots \\ a_{m1} & a_{m2} & \cdots & a_{mn} \end{bmatrix}$$

35. Prove that matrix addition is commutative for all $m \times n$ matrices.

36. Prove that matrix addition is associative for all $m \times n$ matrices.

37. Prove that $0 \cdot A = O$, where O and A have the same dimensions.

38. Prove that for any scalar c, $c(A + B) = cA + cB$.

39. Prove that for any matrix A and scalars c and d, $(cd)A = c(dA)$.

40. Prove that for any matrix A and scalars c and d, $(c + d)A = cA + dA$.

Applications

Business A shoe store has three outlets. The inventory for each is listed in matrix form, where the elements represent the number of pairs of each type of footwear.

$$\begin{array}{lccc} & \text{Outlet } A & \text{Outlet } B & \text{Outlet } C \\ \begin{array}{l}\text{Boots}\\ \text{Shoes}\\ \text{Sneakers}\end{array} & \begin{bmatrix}300\\400\\600\end{bmatrix} & \begin{bmatrix}800\\900\\1000\end{bmatrix} & \begin{bmatrix}400\\700\\300\end{bmatrix}\end{array}$$

41. Enter matrices A, B, and C into a calculator or computer graphing utility. What is the total inventory for outlets A and B? for all three outlets?

42. Write matrices to show how much greater the inventory is at outlet B than at A, and how much greater it is at outlet B than at C.

Developing Mathematical Power

Extension It is important to be able to identify the location of elements in a matrix. One way to do this is to use variables with subscripts that give the row and column of each element. Matrices A and B shown below illustrate this procedure.

$$A = \begin{bmatrix} a_{11} & a_{12} & a_{13} \\ a_{21} & a_{22} & a_{23} \\ a_{31} & a_{32} & a_{33} \end{bmatrix} \qquad B = \begin{bmatrix} b_{11} & b_{12} & b_{13} \\ b_{21} & b_{22} & b_{23} \\ b_{31} & b_{32} & b_{33} \end{bmatrix}$$

43. Name the row and column of the element a_{22} in matrix A.

44. Name the row and column of the element b_{31} in matrix B.

45. For the element a_{13} in matrix A, name the corresponding element in matrix B.

Mixed Review

46. The sum of the first and last of four consecutive odd integers is 48. What are the four integers?

47. If y varies directly as x, and $y = 6$ when $x = 4$, find y when $x = 9$.

48. If y varies directly as x^2, and $y = 8$ when $x = 4$, find y when $x = 9$.

49. What subset(s) of real numbers contain(s) 6? $\sqrt{2}$?

5.2

Matrix Multiplication

Objective: To find the product of two matrices

Matrices are used in business to keep track of such data as inventories, prices, and sales. Matrix multiplication can be used to help analyze the data and provide more information.

Capsule Review

In order to determine whether or not the product of two matrices is defined, it is necessary to recognize whether the number of columns in the first matrix is equal to the number of rows in the second matrix.

Is the number of *columns* in the first matrix equal to the number of *rows* in the second matrix?

$$P = \begin{bmatrix} 0 & -7 & 4 \\ 2 & 0 & -3 \\ -1 & 1 & 8 \end{bmatrix} \quad Q = \begin{bmatrix} 1 & -9 \\ 5 & -1 \\ 4 & 5 \end{bmatrix}$$

$$R = \begin{bmatrix} 2 & -2 & 2 \\ -2 & 2 & -2 \end{bmatrix} \quad S = \begin{bmatrix} 1 & 1 \\ 1 & 1 \end{bmatrix}$$

$$T = [8 \quad -1 \quad 2] \quad U = \begin{bmatrix} 150 \\ -98 \end{bmatrix}$$

1. $P; Q$ **2.** $Q; P$ **3.** $Q; R$

4. $R; Q$ **5.** $S; U$ **6.** $U; S$

7. $T; P$ **8.** $P; T$ **9.** $S; R$

10. $S; S$ **11.** $R; R$ **12.** $T; R$

A record store sells tapes, records, and compact discs. The prices of the items are displayed in a 1×3 row matrix. The number of each item sold per day is displayed in a 3×1 column matrix.

Price

Tapes	Records	Discs
[$8	$6	$13]

Number of Items Sold

Tapes	9
Records	30
Discs	20

To find the gross income for the day, it would make sense to multiply as follows

$$\begin{aligned} &(\$8)(9) + (\$6)(30) + (\$13)(20) \\ &= \$72 + \$180 + \$260 \\ &= \$512 \end{aligned}$$

This type of pattern is, in fact, the basis for the following definition of matrix multiplication.

Matrix Multiplication The product of two matrices, $A_{m \times p}$ and $B_{p \times n}$, is a matrix with dimensions $m \times n$. Any element in the ith row and jth column of this product matrix is the sum of the products of the corresponding elements of the ith row of A and jth column of B.

Note that the product of two matrices A and B, written AB, is defined only if the number of *columns* of A is equal to the number of *rows* of B. To illustrate, let $A = \begin{bmatrix} 3 & 4 \\ 6 & 2 \end{bmatrix}$ and $B = \begin{bmatrix} 1 & 4 & 1 \\ 3 & 5 & 2 \end{bmatrix}$. Then, the product AB exists, since the number of columns in A equals the number of rows in B.

$$AB = \begin{bmatrix} 3 & 4 \\ 6 & 2 \end{bmatrix} \begin{bmatrix} 1 & 4 & 1 \\ 3 & 5 & 2 \end{bmatrix}$$

To obtain the element in the first row, first column of the product matrix, multiply the first element in the first row of A and the first element in the first column of B. Then multiply the second element in the first row of A and the second element in the first column of B, and add.

$$\begin{bmatrix} 3 & 4 \\ 6 & 2 \end{bmatrix} \begin{bmatrix} 1 & 4 & 1 \\ 3 & 5 & 2 \end{bmatrix} \longrightarrow \begin{bmatrix} 3(1) + 4(3) & & \\ & & \end{bmatrix}$$

To find the element in the first row, second column of the product matrix, multiply the elements in the first row of A and the second column of B, and add. The other elements in the product matrix are found in a similar manner.

$$\begin{bmatrix} 3 & 4 \\ 6 & 2 \end{bmatrix} \begin{bmatrix} 1 & 4 & 1 \\ 3 & 5 & 2 \end{bmatrix} \longrightarrow \begin{bmatrix} 15 & 3(4) + 4(5) & \\ & & \end{bmatrix}$$

$$\begin{bmatrix} 3 & 4 \\ 6 & 2 \end{bmatrix} \begin{bmatrix} 1 & 4 & 1 \\ 3 & 5 & 2 \end{bmatrix} \longrightarrow \begin{bmatrix} 15 & 32 & 3(1) + 4(2) \\ & & \end{bmatrix}$$

$$\begin{bmatrix} 3 & 4 \\ 6 & 2 \end{bmatrix} \begin{bmatrix} 1 & 4 & 1 \\ 3 & 5 & 2 \end{bmatrix} \longrightarrow \begin{bmatrix} 15 & 32 & 11 \\ 6(1) + 2(3) & & \end{bmatrix}$$

$$\begin{bmatrix} 3 & 4 \\ 6 & 2 \end{bmatrix} \begin{bmatrix} 1 & 4 & 1 \\ 3 & 5 & 2 \end{bmatrix} \longrightarrow \begin{bmatrix} 15 & 32 & 11 \\ 12 & 6(4) + 2(5) & \end{bmatrix}$$

$$\begin{bmatrix} 3 & 4 \\ 6 & 2 \end{bmatrix} \begin{bmatrix} 1 & 4 & 1 \\ 3 & 5 & 2 \end{bmatrix} \longrightarrow \begin{bmatrix} 15 & 32 & 11 \\ 12 & 34 & 6(1) + 2(2) \end{bmatrix}$$

$$AB = \begin{bmatrix} 3 & 4 \\ 6 & 2 \end{bmatrix} \begin{bmatrix} 1 & 4 & 1 \\ 3 & 5 & 2 \end{bmatrix} = \begin{bmatrix} 15 & 32 & 11 \\ 12 & 34 & 10 \end{bmatrix}$$

You can use the subscript of the element you are solving for to remember which row and column to multiply. When you multiply the first row of the first matrix and third column of the second matrix, you obtain element a_{13}.

EXAMPLE 1 **If** $A = \begin{bmatrix} 2 & 3 \\ 6 & 9 \end{bmatrix}$ **and** $B = \begin{bmatrix} -3 & 6 \\ 2 & -4 \end{bmatrix}$**, find:** **a.** AB **b.** BA

a. $AB = \begin{bmatrix} 2 & 3 \\ 6 & 9 \end{bmatrix}\begin{bmatrix} -3 & 6 \\ 2 & -4 \end{bmatrix} = \begin{bmatrix} 2(-3) + 3(2) & 2(6) + 3(-4) \\ 6(-3) + 9(2) & 6(6) + 9(-4) \end{bmatrix} = \begin{bmatrix} 0 & 0 \\ 0 & 0 \end{bmatrix}$

b. $BA = \begin{bmatrix} -3 & 6 \\ 2 & -4 \end{bmatrix}\begin{bmatrix} 2 & 3 \\ 6 & 9 \end{bmatrix} = \begin{bmatrix} -3(2) + 6(6) & -3(3) + 6(9) \\ 2(2) + (-4)(6) & 2(3) + (-4)(9) \end{bmatrix} = \begin{bmatrix} 30 & 45 \\ -20 & -30 \end{bmatrix}$

If the product of two real numbers is zero, then either one or both of the factors must be zero. This is known as the *zero-product property for real numbers*. However, part *a* of Example 1 illustrates the fact that the product of two matrices, A and B, can be a zero matrix even if neither A nor B is a zero matrix. That is, the zero-product property does *not* hold for matrix multiplication. Example 1 also shows that AB is not equal to BA. That is, matrix multiplication is *not* necessarily commutative.

Raising a matrix to a power is defined only for $n \times n$ (square) matrices. In general, for any square matrix A:

$$A^m = \underbrace{A \cdot A \cdot A \cdot \ldots \cdot A}_{m \text{ factors}}$$

EXAMPLE 2 **If** $A = \begin{bmatrix} 3 & 1 & -1 \\ -2 & 0 & 2 \end{bmatrix}$ **and** $B = \begin{bmatrix} 0 & -5 & -1 \\ 1 & 3 & 0 \\ 2 & 4 & -2 \end{bmatrix}$**, find the following matrix products, if they are defined:** **a.** B^2 **b.** A^2B

a. $B^2 = \begin{bmatrix} 0 & -5 & -1 \\ 1 & 3 & 0 \\ 2 & 4 & -2 \end{bmatrix}\begin{bmatrix} 0 & -5 & -1 \\ 1 & 3 & 0 \\ 2 & 4 & -2 \end{bmatrix}$

$$= \begin{bmatrix} 0(0) + (-5)1 + (-1)2 & 0(-5) + (-5)3 + (-1)4 & 0(-1) + (-5)0 + (-1)(-2) \\ 1(0) + 3(1) + 0(2) & 1(-5) + 3(3) + 0(4) & 1(-1) + 3(0) + 0(-2) \\ 2(0) + 4(1) + (-2)2 & 2(-5) + 4(3) + (-2)4 & 2(-1) + 4(0) + (-2)(-2) \end{bmatrix}$$

$$= \begin{bmatrix} -7 & -19 & 2 \\ 3 & 4 & -1 \\ 0 & -6 & 2 \end{bmatrix}$$

b. As A is not a square matrix, A^2 is not defined; thus A^2B is not defined.

Because matrix multiplication can require a great many calculations, the potential for mistakes in computation is great. Graphing utilities often provide faster and more accurate results.

The number 1 is the identity element for the multiplication of real numbers. If only the set of $n \times n$ matrices is considered, then the *multiplicative identity matrix* is an $n \times n$ matrix denoted $I_{n\times n}$ and having this form:

$$I_2 = \begin{bmatrix} 1 & 0 \\ 0 & 1 \end{bmatrix} \qquad I_3 = \begin{bmatrix} 1 & 0 & 0 \\ 0 & 1 & 0 \\ 0 & 0 & 1 \end{bmatrix} \qquad I_n = \begin{bmatrix} 1 & 0 & 0 & 0 & \cdots & 0 \\ 0 & 1 & 0 & 0 & \cdots & 0 \\ 0 & 0 & 1 & 0 & \cdots & 0 \\ 0 & 0 & 0 & 1 & \cdots & 0 \\ \vdots & \vdots & \vdots & \vdots & & \vdots \\ 0 & 0 & 0 & 0 & \cdots & 1 \end{bmatrix}$$

The identity matrix for the set of $n \times n$ matrices is the $n \times n$ square matrix with 1's along the *main diagonal* (from the upper left corner to the lower right corner) and 0's in all other positions.

It is true that if $m \neq n$, then $I_{m \times m}A_{m \times n} = A_{m \times n}$. This can be illustrated using matrix A from Example 2.

$$\begin{bmatrix} 1 & 0 \\ 0 & 1 \end{bmatrix}\begin{bmatrix} 3 & 1 & -1 \\ -2 & 0 & 2 \end{bmatrix} = \begin{bmatrix} 3 & 1 & -1 \\ -2 & 0 & 2 \end{bmatrix} \qquad I_{2 \times 2}A_{2 \times 3} = A_{2 \times 3}$$

If $I_{2 \times 2}$ were the multiplicative identity matrix for $A_{2 \times 3}$, then $A_{2 \times 3}I_{2 \times 2}$ would also have to be equal to $A_{2 \times 3}$. However, the product $A_{2 \times 3}I_{2 \times 2}$ is not defined. Why? In general, if A is an $m \times n$ matrix with $m \neq n$, then A does *not* have a multiplicative identity matrix.

Properties of Matrix Multiplication

If A, B, and C are $n \times n$ matrices, then

AB is an $n \times n$ matrix.	Closure property
$(AB)C = A(BC)$	Associative property
$A(B + C) = AB + AC$ $(B + C)A = BA + CA$	Distributive properties
$IA = AI = A$, where I has the same dimensions as A.	Multiplicative identity property
$OA = AO = O$, where O has the same dimensions as A.	Multiplicative property of zero
A^{-1} is the multiplicative inverse of A if A^{-1} is defined and $AA^{-1} = A^{-1}A = I$.	Multiplicative inverse property

The last property listed, the multiplicative inverse property for matrices, will be discussed on pages 227 and 228.

CLASS EXERCISES

State the dimensions of the product matrix, AB, if it is defined.

1. $A_{2 \times 3}B_{3 \times 2}$ **2.** $A_{4 \times 1}B_{1 \times 2}$ **3.** $A_{3 \times 3}B_{3 \times 1}$ **4.** $A_{1 \times 2}B_{3 \times 4}$

For Class Exercises 5–16, use the matrices given below.

$$A = [3 \quad 2 \quad -1] \qquad B = \begin{bmatrix} 2 & 3 & 0 \\ 1 & -4 & -2 \\ 0 & 1 & 1 \end{bmatrix} \qquad C = \begin{bmatrix} -2 & 3 \\ 6 & -1 \\ 0 & 2 \end{bmatrix} \qquad I = \begin{bmatrix} 1 & 0 & 0 \\ 0 & 1 & 0 \\ 0 & 0 & 1 \end{bmatrix}$$

Determine which of the following products are defined, and state the dimensions of such products. If a product is not defined, state why.

5. AB **6.** BA **7.** BC **8.** CB

9. AC **10.** CA **11.** $A(B + C)$ **12.** $(B + I)C$

13. IB **14.** BI **15.** B^2 **16.** C^2

PRACTICE EXERCISES

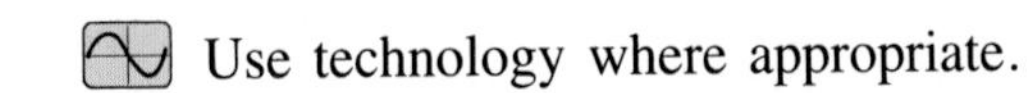
Use technology where appropriate.

For Exercises 1–16, use matrices *D*, *E*, *F*, and *I*, shown below. Perform the indicated operations if they are defined. If an operation is not defined, label it "undefined."

$$D = \begin{bmatrix} 1 & 2 & -1 \\ 0 & 3 & 1 \\ 2 & -1 & -2 \end{bmatrix} \qquad E = \begin{bmatrix} 2 & -5 & 0 \\ 1 & 0 & -2 \\ 3 & 1 & 1 \end{bmatrix} \qquad F = \begin{bmatrix} -3 & 2 \\ -5 & 1 \\ 2 & 4 \end{bmatrix} \qquad I = \begin{bmatrix} 1 & 0 & 0 \\ 0 & 1 & 0 \\ 0 & 0 & 1 \end{bmatrix}$$

1. DE **2.** ED **3.** DF **4.** EF

5. FD **6.** FE **7.** $(DE)F$ **8.** $D(EF)$

9. ID **10.** DI **11.** $(D + E)F$ **12.** $DF + EF$

13. D^2 **14.** E^2 **15.** D^2E **16.** E^2D

For Exercises 17–24, use matrices *P*, *Q*, *R*, *S*, and *I*, shown below. Determine if the expressions in each pair are equal.

$$P = \begin{bmatrix} 3 & 4 \\ 1 & 2 \end{bmatrix} \qquad Q = \begin{bmatrix} -1 & 0 \\ 3 & -2 \end{bmatrix} \qquad R = \begin{bmatrix} 1 & 4 \\ -2 & 1 \end{bmatrix} \qquad S = \begin{bmatrix} 0 & 1 \\ 2 & 0 \end{bmatrix} \qquad I = \begin{bmatrix} 1 & 0 \\ 0 & 1 \end{bmatrix}$$

17. $(P + Q)R$ and $PR + QR$ **18.** $(P - Q)R$ and $RP - RQ$

19. $(IP)Q$ and $(PI)Q$ **20.** $(P + Q)I$ and $PI + QI$

21. $(P + Q)^2$ and $P^2 + PQ + QP + Q^2$

22. $(P + Q)^2$ and $P^2 + 2PQ + Q^2$

23. $(P + Q)(R + S)$ and $(P + Q)R + (P + Q)S$

24. $(P + Q)(R + S)$ and $PR + PS + QR + QS$

Solve for x and y.

25. $\begin{bmatrix} 0 & 2 \\ 3 & 0 \end{bmatrix}\begin{bmatrix} x & 4 \\ -1 & 3y \end{bmatrix} = \begin{bmatrix} -2 & 6 \\ 12 & 12 \end{bmatrix}$

26. $\begin{bmatrix} 2x & 1 \\ 2 & 0 \end{bmatrix}\begin{bmatrix} 0 & 3 \\ 2x & -y \end{bmatrix} = \begin{bmatrix} -4 & -9 \\ 0 & 6 \end{bmatrix}$

27. $\begin{bmatrix} x & y \\ 5 & 6 \end{bmatrix}\begin{bmatrix} 1 & 2 \\ 3 & 4 \end{bmatrix} = \begin{bmatrix} 5 & 8 \\ 23 & 34 \end{bmatrix}$

28. $\begin{bmatrix} 2 & 1 \\ x & 2y \end{bmatrix}\begin{bmatrix} 1 & 4 \\ -2 & 3 \end{bmatrix} = \begin{bmatrix} 0 & 11 \\ -13 & 14 \end{bmatrix}$

Prove each of the following statements for all 2×2 matrices, A, B, and C, where O and I are the 2×2 zero and identity matrices, respectively.

29. $(AB)C = A(BC)$

30. $AI = IA = A$

31. $(A + B)C = AC + BC$

32. $OA = AO$

Applications

Business A hardware company shows prices in a 1×3 matrix and daily sales at its three stores in a 3×3 matrix.

Prices

Hammers	Flashlights	Lanterns
[\$3	\$5	\$7]

Number of Items Sold

	Store A	Store B	Store C
Hammers	10	9	8
Flashlights	3	14	6
Lanterns	2	5	7

33. Use a graphing utility to find the product of the two matrices. Explain what this product represents.

34. How would you find the total gross revenue from all three stores?

35. Find the total gross revenue from the flashlights sold at all three stores.

Developing Mathematical Power

Thinking Critically Numbers arranged in rows and columns have fascinated mathematicians for centuries. While matrices are examples of such arrangements, the rows and columns of numbers shown at the right illustrate a different arrangement. The problem below is similar to one that appeared in an Annual High School Examination sponsored by The Mathematical Association of America. See how long it takes you to find the answer!

	2	3	4	5
9	8	7	6	
	10	11	12	13
17	16	15	14	
	18	19	20	21
25	24	23	22	
	.	.	.	.
.	.	.	.	
	.	.	.	.

36. All integers greater than 1 are arranged in five columns, as shown at the right above. In which column will the number 1000 be located?

5.3

Problem Solving Strategy: Use Matrices

Matrices are used as a mathematical model to solve problems in many diversified fields. Problems in engineering, chemistry, physics, psychology, economics, statistics, and education have been solved by using matrices. If a problem contains a great deal of data, then the strategy of using matrices should be considered. In order to illustrate the strategy, some simple examples are presented in this lesson.

EXAMPLE 1 A travel agent offers three different travel packages to Orlando, Florida. Package I consists of 4 nights at a hotel, 3 passes to local attractions, and 5 meals. Package II consists of 3 nights at a hotel, 4 passes, and 7 meals. Package III consists of 5 nights at a hotel, 4 passes, and no meals. The agent can book a hotel room for \$50 per night, get passes for \$10 each and provide a meal at a given restaurant for an average of \$15. She wants to run an ad featuring the least expensive package. Which plan should she advertise?

Understand the Problem

What are the given facts? You are given the following data.

- the number of nights at a hotel
- the number of passes
- the number of meals for each of three travel packages
- the cost of one night at a hotel, one pass, and one meal

What are you asked to find? The problem is to find the least expensive travel package.

Plan Your Approach

Choose a strategy. The problem contains many numbers which can be organized by using a matrix. In order to find the least expensive travel package, the cost of *each* package must be found. The cost of a package is the sum of the products of the cost of each item times the number of items.

To find the cost of each plan, write the information in matrix form. The *prices* matrix is a 1×3 matrix. The *number of items* matrix is a 3×3 matrix. Note that the column headings for the *prices* matrix must be written in the same order as the row headings in the *number of items* matrix.

Prices

Hotel	Pass	Meal
[\$50	\$10	\$15]

Number of Items

	Plan I	Plan II	Plan III
Nights at Hotel	4	3	5
Passes	3	4	4
Meals	5	7	0

The product of the *prices* matrix times the *number of items* matrix is a matrix that shows the cost of each plan.

prices matrix × *number of items* matrix = *cost* matrix

Complete the Work

$$\begin{bmatrix} 50 & 10 & 15 \end{bmatrix} \begin{bmatrix} 4 & 3 & 5 \\ 3 & 4 & 4 \\ 5 & 7 & 0 \end{bmatrix} = \begin{matrix} \text{Plan I} & \text{Plan II} & \text{Plan III} \\ [\$305 & \$295 & \$290] \end{matrix}$$

Interpret the Results The agent should advertise Plan III, since it is the least expensive.

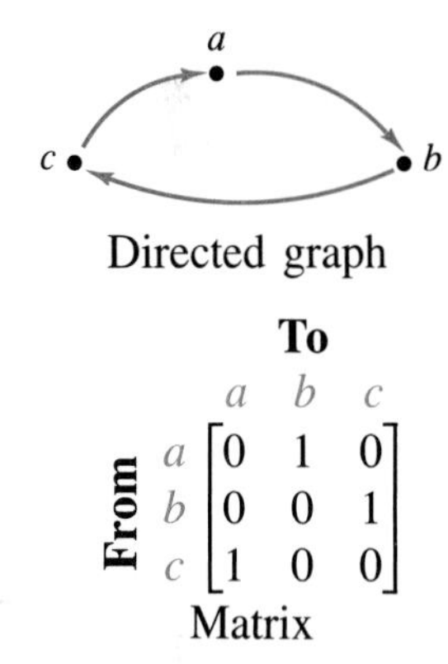

Directed graph

$$\begin{bmatrix} 0 & 1 & 0 \\ 0 & 0 & 1 \\ 1 & 0 & 0 \end{bmatrix}$$

Matrix

Communications systems can be analyzed using matrices. If person a can speak to person b, person b can speak to person c, and person c can speak to person a, then person a can communicate with person c only in two steps. A **directed graph** is used to represent this geometrically. A directed graph can be interpreted as a matrix in this way: Use 1 to represent the ability to communicate directly and 0 to represent the inability to do so. It can be shown that the square of such a matrix gives the number of ways the people can communicate with each other in two steps.

EXAMPLE 2 Tom, Jan, Kay, and Pat all have modems for their computers. Their ability to speak to each other's computers is given by the directed graph shown. Find the number of different ways that Kay can communicate with Jan in two steps.

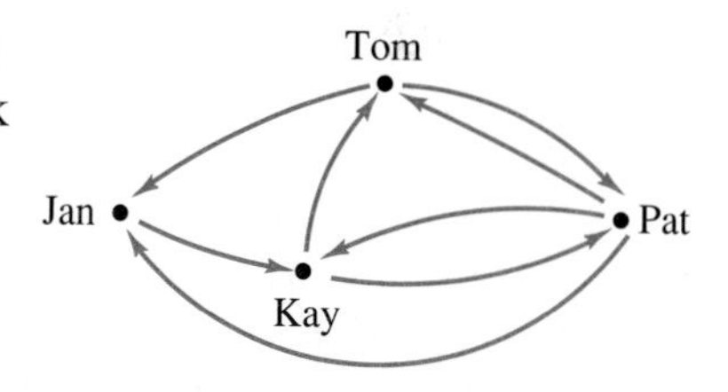

Understand the Problem You are given a graph showing how four persons communicate with each other using their computer modems. You are asked to find the number of ways that one person, Kay, can communicate with another, Jan, in two steps.

Plan Your Approach Write the information in matrix form. Square the matrix to obtain the number of ways the four persons can communicate with each other in two steps. Read the desired information from that matrix.

Complete the Work Write the matrix and label it matrix C. Assume that a person cannot communicate *directly* with himself using his own modem.

$$\textbf{From}\;\begin{array}{c} \\ \text{Tom} \\ \text{Jan} \\ \text{Kay} \\ \text{Pat} \end{array}\begin{array}{c} \textbf{To} \\ \begin{array}{cccc} \text{Tom} & \text{Jan} & \text{Kay} & \text{Pat} \end{array} \\ \begin{bmatrix} 0 & 1 & 0 & 1 \\ 0 & 0 & 1 & 0 \\ 1 & 0 & 0 & 1 \\ 1 & 1 & 1 & 0 \end{bmatrix} \end{array} = C$$

Next, find C^2.

$$C^2 = \begin{bmatrix} 0 & 1 & 0 & 1 \\ 0 & 0 & 1 & 0 \\ 1 & 0 & 0 & 1 \\ 1 & 1 & 1 & 0 \end{bmatrix}\begin{bmatrix} 0 & 1 & 0 & 1 \\ 0 & 0 & 1 & 0 \\ 1 & 0 & 0 & 1 \\ 1 & 1 & 1 & 0 \end{bmatrix} = \begin{array}{c} \text{Tom} \\ \text{Jan} \\ \text{Kay} \\ \text{Pat} \end{array}\begin{array}{c} \begin{array}{cccc} \text{Tom} & \text{Jan} & \text{Kay} & \text{Pat} \end{array} \\ \begin{bmatrix} 1 & 1 & 2 & 0 \\ 1 & 0 & 0 & 1 \\ 1 & 2 & 1 & 1 \\ 1 & 1 & 1 & 2 \end{bmatrix} \end{array}$$

Interpret the Results Kay can communicate with Jan in two steps in 2 different ways, since the element in the third row, second column of the matrix C^2 is 2.

To show that this element of C^2 does, indeed, give the correct number of two-step communication paths from Kay to Jan, refer to the directed graph on page 211. Kay can communicate with Tom, who can relay the message to Jan. Or, Kay can communicate with Pat, who can relay the message to Jan.

CLASS EXERCISES

A manufacturer uses cotton, wool, polyester, and rayon yarns to make two different blended fabrics. Each bolt of fabric a requires 2 units of cotton, 6 units of wool, 4 units of polyester, and 4 units of rayon. Fabric b contains no cotton, but one bolt of fabric b requires 7 units of wool, 3 units of polyester, and 6 units of rayon. The cotton yarn costs $4 per unit, the wool $6 per unit, the polyester $2 per unit, and the rayon $5 per unit.

1. Write a cost matrix.
2. Write a matrix showing the number of units of each type of yarn needed for each type of fabric.
3. Use matrix multiplication to write a matrix showing the total cost of the yarn for one bolt of each type of fabric.
4. What is the total cost of the yarn for one bolt of fabric a?
5. What is the total cost of the yarn for two bolts, one of each fabric?

The directed graphs below are communications graphs for executives within two companies.

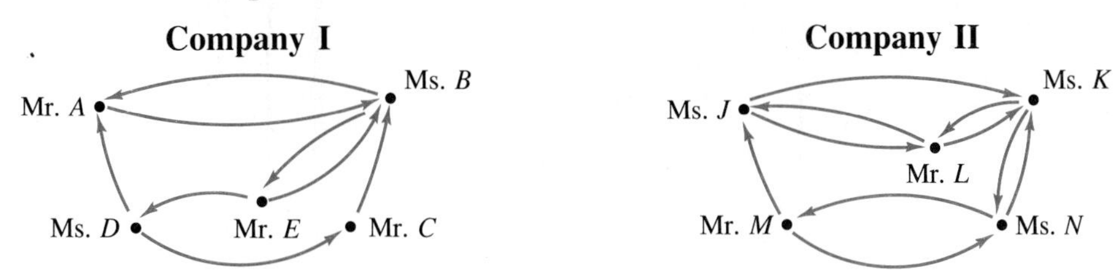

6. Write a direct communication matrix for each company.

7. Write a matrix for each company showing the number of ways the executives can communicate in two steps.

8. In how many ways can Mr. *E* communicate with Mr. *C* in two steps? Mr. *M* with Ms. *K*?

PRACTICE EXERCISES

Four elective subjects are available to the students in a high school: woodworking, cooking, auto repair, and computer repair. For this electives program, the school must budget \$5 per student in woodworking, \$6 per student in cooking, \$10 per student in auto repair, and \$7 per student in computer repair. The numbers of tenth, eleventh, and twelfth grade students assigned to these classes are shown in the matrix below.

	Tenth Grade	Eleventh Grade	Twelfth Grade
Woodworking	6	11	9
Cooking	8	15	10
Auto Repair	10	19	3
Computer Repair	14	6	9

1. Express the cost information for the electives program in matrix form.

2. Express the cost of providing the electives program for all grades in matrix form.

3. Find the total cost of providing the electives program.

4. Find the cost of providing auto repair classes for all three grades.

5. Assume that the number of students assigned to each of these subjects is tripled in all grades. Write a matrix representing the number of students, by grade, in the electives program.

6. Assume that the number of students assigned to each of these subjects is doubled in all grades. Write a matrix showing the cost of providing the electives program for each grade.

Firm *X*

Direct communications lines for executives a, b, c, and d

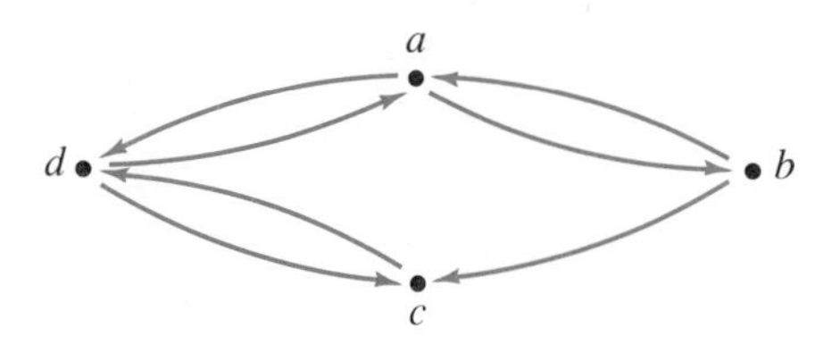

Firm *Y*

Direct communications lines for executives e, f, g, and h

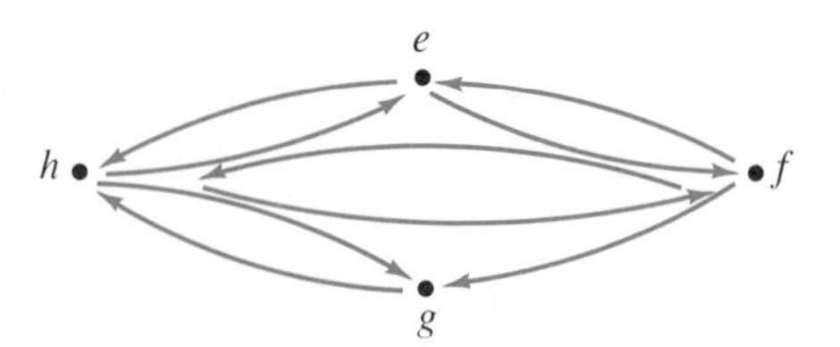

7. Write a matrix for Firm X.

8. Write a matrix for Firm Y.

9. Write a matrix that shows the number of ways the executives in Firm X can communicate in two steps. How many ways can a talk to c in two steps? d to c?

10. Write a matrix that shows the number of ways the executives in Firm Y can communicate in two steps. How many ways can f talk to g in two steps? g to e?

The All-Tools hardware store keeps its inventory in computer form. It has *good, better,* and *best* tools. The matrix on the left below represents the cost of each type of tool. The matrix on the right represents the number of each type of tool in the inventory.

Cost

	Good	Better	Best
Pliers	$3	$4	$ 7
Screwdrivers	$2	$5	$11
Hammers	$3	$4	$ 8

Current Inventory

	Pliers	Screwdrivers	Hammers
Good	4	5	3
Better	7	6	12
Best	9	6	10

11. Find the value of the pliers in the current inventory. Find the value of the hammers in the current inventory.

The All-Tools company sales for two days are represented by the matrices shown below.

Monday's Sales

	Pliers	Screwdrivers	Hammers
Good	1	0	2
Better	2	1	3
Best	1	5	2

Tuesday's Sales

Pliers	Screwdrivers	Hammers
0	3	1
1	0	0
1	0	2

12. Find the new matrix of current inventory after Monday's sales.

13. Find the new matrix of current inventory after Monday's and Tuesday's sales.

14. Find the value of the hammers in the current inventory after Monday's and Tuesday's sales.

Three researchers, *a*, *b*, and *c*, in different cities are working on a project. They communicate by phone if they have each other's phone numbers. They communicate by mail if they have each other's addresses. If they know both the phone number and the address, they may communicate either or both ways. The communications matrices, *M* (by mail) and *P* (by phone), are given below.

$$\textbf{From}\ \begin{matrix} & \textbf{To} \\ & \begin{matrix} a & b & c \end{matrix} \\ \begin{matrix} a \\ b \\ c \end{matrix} & \begin{bmatrix} 0 & 1 & 0 \\ 1 & 0 & 1 \\ 0 & 1 & 0 \end{bmatrix} \end{matrix} = M \qquad \textbf{From}\ \begin{matrix} & \textbf{To} \\ & \begin{matrix} a & b & c \end{matrix} \\ \begin{matrix} a \\ b \\ c \end{matrix} & \begin{bmatrix} 0 & 0 & 0 \\ 1 & 0 & 1 \\ 0 & 1 & 0 \end{bmatrix} \end{matrix} = P$$

15. Find M^3 and P^3. How can these matrices be interpreted?

16. Draw a directed communications graph for $(M + P)$.

Mixed Problem Solving Review

1. What number, decreased by 36, is three times itself?

2. A company sells computers. Its sales can be represented by a linear function. When it sells 25 computers, the company earns \$5000. On sales of 75 computers the company earns \$15,000. How much would the company earn on the sales of 225 computers.

3. An airplane flew 3000 mi in 5 h with a tailwind. With a headwind on the return flight, the plane took 6 h. Find the plane's air speed and the wind speed. Assume both the air speed and the wind speed are constant.

4. The sum of the digits of a two-digit number is 12. If the digits are reversed, the new number is 54 more than the original number. What is the original number?

PROJECT

The subject of linear algebra is the branch of mathematics that has been developed to study the problem of solving systems of linear equations. Linear algebra begins with a study of the real number system, then considers the topics of systems of linear equations, matrices, determinants, and vectors. Working with a classmate, look up the subject of linear algebra and, in particular, the definition of a vector. Does a concept of vector addition or vector multiplication exist? If so, explain what they are in writing.

5.4

Systems of Equations in Three Variables

Objective: To solve a system of linear equations in three variables

Any equation that can be written in the form $ax + by = c$, where a, b, and c are constants (a and b not both zero) is a *linear equation in two variables.* The solution to such an equation is an *ordered pair* (x, y). Similarly, any equation that can be written in the form $ax + by + cz = d$, where a, b, c, and d are constants (a, b, and c not all zero) is a *linear equation in three variables.* The solution to such an equation is an **ordered triple** (x, y, z). Solving a system of equations in three variables is similar to solving a system of equations in two variables.

Capsule Review

EXAMPLE **Solve the system:** $\begin{cases} 2x + y = 3 \\ 3x - 2y = 8 \end{cases}$

$4x + 2y = 6$ *Multiply both sides of the first equation by 2.*
$3x - 2y = 8$

$7x = 14$ *Add the equations.*
$x = 2$ *Solve for x.*

$2x + y = 3$ *Use one of the original equations.*
$2(2) + y = 3$ *Substitute 2 for x.*
$y = -1$ *Solve for y.*

Check in both equations to show that the solution is $(2, -1)$.

Solve each linear system.

1. $\begin{cases} x + y = 7 \\ 2x - y = 2 \end{cases}$ **2.** $\begin{cases} 2x + 2y = 4 \\ 3x - 2y = 11 \end{cases}$ **3.** $\begin{cases} 2x + 3y = 19 \\ x - 2y = 6 \end{cases}$

4. $\begin{cases} x - 3y = 4 \\ 2x + y = 15 \end{cases}$ **5.** $\begin{cases} 2x - 3y = 4 \\ 3x + 2y = 6 \end{cases}$ **6.** $\begin{cases} 3x - 4y = -1 \\ 2x + 5y = 7 \end{cases}$

The position of a point in space can be expressed as an ordered triple (x, y, z), and the graph of a linear equation in three variables is a *plane.* A system of three linear equations in three variables has one real solution if all three planes represented by the equations intersect in exactly one point. If the planes meet in an infinite number of points, the system has an infinite number of real solutions. If there is no point common to all three planes, the system has no real solutions.

Consistent, Independent System
(One Solution)
3 planes intersect in exactly one point.

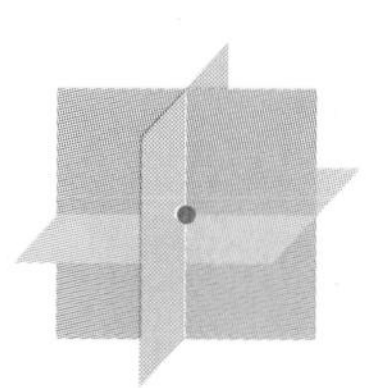

Consistent, Dependent Systems
(Infinite Number of Solutions)

3 planes intersect in one line.

3 planes coincide.

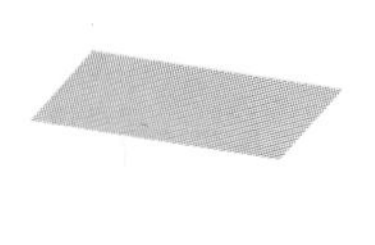

Inconsistent Systems
(No Solutions)

Planes intersect in 2 parallel lines.

Planes intersect in 3 parallel lines.

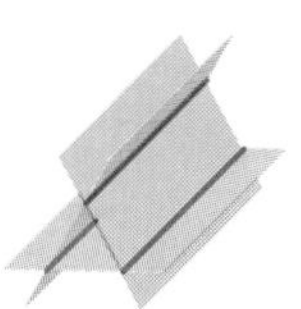

2 planes coincide; third is parallel.

3 planes are parallel.

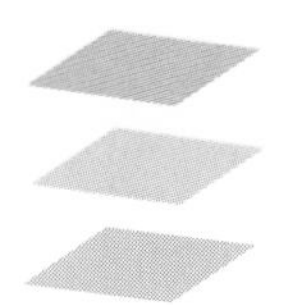

To solve a system of three linear equations in three variables by the addition method, eliminate one variable from two equations. Repeat the process by eliminating the same variable from a different pair of equations. A system of linear equations in two variables is now obtained, and it can be solved by any of the methods you have learned.

EXAMPLE 1 **Solve the system by addition:** $\begin{cases} 2x - y + 3z = -4 \\ x + 2y - 5z = 11 \\ x + 3y - 2z = 5 \end{cases}$

Add the first two equations to eliminate y.

$$\begin{array}{rl} 4x - 2y + 6z = -8 & \textit{Multiply both sides of the first equation by 2.} \\ x + 2y - 5z = 11 & \\ \hline 5x \qquad + z = 3 & \textit{Add.} \end{array}$$

Add the first and third equations to eliminate y.

$$\begin{array}{rl} 6x - 3y + 9z = -12 & \textit{Multiply both sides of the first equation by 3.} \\ x + 3y - 2z = 5 & \\ \hline 7x \qquad + 7z = -7 & \textit{Add.} \\ x \qquad + z = -1 & \textit{Simplify.} \end{array}$$

You now have a system of two equations in two variables: $\begin{cases} 5x + z = 3 \\ x + z = -1 \end{cases}$

$$\begin{aligned} 5x + z &= 3 \\ -x - z &= 1 \\ \hline 4x \quad &= 4 \end{aligned}$$

Multiply both sides of the second equation by −1. Then, add the equations to eliminate z.

$$x = 1$$

Solve for x.

$$\begin{aligned} x + z &= -1 \\ 1 + z &= -1 \end{aligned}$$

Substitute 1 for x in x + z = −1 or in 5x + z = 3.

$$z = -2$$

Solve for z.

$$\begin{aligned} 2x - y + 3z &= -4 \\ 2(1) - y + 3(-2) &= -4 \\ 2 - y - 6 &= -4 \end{aligned}$$

Substitute 1 for x and −2 for z in one of the original equations.

$$y = 0$$

Solve for y.

Check to show that the solution is $(1, 0, -2)$. Substitute 1 for x, 0 for y, and -2 for z in each of the original equations.

$$\begin{aligned} 2x - y + 3z &= -4 \\ 2(1) - 0 + 3(-2) &\stackrel{?}{=} -4 \\ -4 &= -4 \checkmark \end{aligned} \qquad \begin{aligned} x + 2y - 5z &= 11 \\ 1 + 2(0) - 5(-2) &\stackrel{?}{=} 11 \\ 11 &= 11 \checkmark \end{aligned} \qquad \begin{aligned} x + 3y - 2z &= 5 \\ 1 + 3(0) - 2(-2) &\stackrel{?}{=} 5 \\ 5 &= 5 \checkmark \end{aligned}$$

Not all of the equations in a system may contain every variable.

EXAMPLE 2 **Solve the system:** $\begin{cases} 2x - y = 5 \\ 3y + z = 0 \\ x + 4z = 14 \end{cases}$

$$\begin{aligned} -12y - 4z &= 0 \\ x \qquad + 4z &= 14 \\ \hline x - 12y \qquad &= 14 \end{aligned}$$

Multiply both sides of the second equation by −4. Then, add this equation to the third equation to eliminate z.

$$\begin{aligned} -2x + 24y &= -28 \\ 2x - \quad y &= \quad 5 \\ \hline 23y &= -23 \\ y &= -1 \end{aligned}$$

Multiply both sides of x − 12y = 14 by −2. Then, add this equation to the first original equation to eliminate x.

$$\begin{aligned} 2x - y &= 5 \\ 2x - (-1) &= 5 \\ x &= 2 \end{aligned}$$

Substitute −1 for y in the first original equation and solve for x.

$$\begin{aligned} 3y + z &= 0 \\ 3(-1) + z &= 0 \\ z &= 3 \end{aligned}$$

Substitute −1 for y in the second original equation to solve for z.

Check in the three original equations. The solution is $(2, -1, 3)$.

If a pair of equations in a system reduces to the contradiction $0 = c$, where $c \neq 0$, then the system is inconsistent and there is no solution. If a pair of equations reduces to the true equation $0 = 0$, (but no pair reduces to $0 = c$, $c \neq 0$), then the system is dependent and there are an infinite number of solutions.

CLASS EXERCISES

Solve each system. State whether the system is consistent and independent, consistent and dependent, or inconsistent.

1. $\begin{cases} x + 2z = 5 \\ 2x - y + z = 3 \\ x + y - z = 0 \end{cases}$

2. $\begin{cases} 2x + y - z = 2 \\ x - 2y + z = 5 \\ -x + 2y - z = 1 \end{cases}$

3. $\begin{cases} x + 2y - 3z = -1 \\ 2x + 4y - z = 0 \\ -2x - 4y + 6z = 2 \end{cases}$

PRACTICE EXERCISES

Solve each system of equations and state whether the system is consistent and independent, consistent and dependent, or inconsistent.

1. $\begin{cases} x - y + 2z = 7 \\ 2x + y + z = 8 \\ x - z = 5 \end{cases}$

2. $\begin{cases} x + y + z = 2 \\ 2x + y - z = -1 \\ x + 2z = 5 \end{cases}$

3. $\begin{cases} x + y = -1 \\ y + z = 4 \\ x + z = 1 \end{cases}$

4. $\begin{cases} x - y = 3 \\ y + z = 4 \\ x - z = 3 \end{cases}$

5. $\begin{cases} 2x - 3y + z = 6 \\ x - y + z = 2 \\ x - y - 2z = 8 \end{cases}$

6. $\begin{cases} 3x - y + z = 3 \\ x + y + 2z = 4 \\ x + 2y + z = 4 \end{cases}$

7. $\begin{cases} 3x + 2y = 7 \\ -y - 3z = 2 \\ x + y + z = 2 \end{cases}$

8. $\begin{cases} 5y + 4z = 6 \\ x + y + z = 3 \\ 4x - y = 8 \end{cases}$

9. $\begin{cases} 2x + 3y - z = 5 \\ x + y + z = 9 \\ x - 2y - z = -7 \end{cases}$

10. $\begin{cases} 5x - y + z = 4 \\ x + 2y - z = 5 \\ 2x + 3y - 3z = 5 \end{cases}$

11. $\begin{cases} 2x + 4y = 6 - 3z \\ x - 3y = -7 + 2z \\ x - 2y + z = -5 \end{cases}$

12. $\begin{cases} 4x + z = y - 5 \\ -x + y = z + 5 \\ 2x - z - 1 = y \end{cases}$

13. $\begin{cases} x + 4y - z = 6 \\ x + 3y = 8 - z \\ 2x - y - z = 1 \end{cases}$

14. $\begin{cases} x = -2y + z - 1 \\ 4x - 2y + z = 6 \\ -3z = -5x - 2y + 8 \end{cases}$

15. $\begin{cases} x - y + 2z = 5 \\ x + y + 2z = 5 \\ 2x + 4z = 10 \end{cases}$

16. $\begin{cases} \frac{x}{2} + \frac{y}{4} = 10 \\ \frac{3x}{10} + \frac{z}{10} = 6 \\ x - \frac{y}{20} = 9 \end{cases}$

17. $\begin{cases} \frac{3x}{2} + \frac{y}{4} - z = -2 \\ \frac{y}{4} - \frac{z}{6} = 0 \\ x - \frac{z}{2} = -1 \end{cases}$

18. $\begin{cases} \frac{x}{4} + \frac{3z}{2} = 3 \\ y - \frac{z}{2} = \frac{5}{2} \\ -\frac{x}{3} + \frac{y}{2} = 1 \end{cases}$

Solve.

19. $\begin{cases} w + x + 2z = -3 \\ -w + x + 2y = 3 \\ w + 2x - y + z = -3 \\ x + 2y - z = 6 \end{cases}$

20. $\begin{cases} x + 3y - z = 4 \\ -w + 2x + 3z = 0 \\ -2w + x + 2y = -1 \\ w + x + y + z = 4 \end{cases}$

Applications

21. Consumerism Assorted nuts, bolts, and washers are packaged and sold. A package containing 1 nut, 1 bolt, and 1 washer costs 20¢. A package containing 2 nuts, 2 bolts, and 4 washers costs 50¢. A package containing 5 nuts and 4 bolts costs 70¢. Find the price of a nut, a bolt, and a washer.

22. Consumerism Travel packs contain toothpaste, combs, and toothbrushes. A pack containing 2 combs and 1 toothbrush costs $1.00. A pack containing 5 combs and 5 toothbrushes costs $3.75. A pack containing 2 toothbrushes and 2 tubes of toothpaste costs $2.50. Find the price of a tube of toothpaste, a comb, and a toothbrush.

23. Geometry x, y, and z, which are related by this system of equations, represent the number of sides of three different polygons. Name the polygons.

$$\begin{cases} x + y + z = 14 \\ 2x - z = 0 \\ y + 2z = 17 \end{cases}$$

TEST YOURSELF

For the matrices A, B, and C below, perform the indicated operations if they are defined. If an operation is not defined, label it "undefined."

$$A = \begin{bmatrix} 2 & 3 & 0 \\ 3 & 4 & 1 \end{bmatrix} \qquad B = \begin{bmatrix} 1 & 0 & 4 \\ 2 & 1 & 0 \\ 0 & 3 & 2 \end{bmatrix} \qquad C = \begin{bmatrix} 5 & 2 & 1 \\ 1 & 1 & 0 \\ -1 & -2 & 0 \end{bmatrix}$$

1. $B + C$ **2.** $2A$ **3.** AB **4.** BC **5.** $(3C)(A)$ **6.** $A(B + C)$ **5.1–5.2**

7. On Thursday, 60 toasters, 75 irons, 40 mixers, and 25 blenders were sold. On Friday, 85 toasters, 52 irons, 55 mixers, and 32 blenders were sold. If the initial inventory consisted of 350 toasters, 200 irons, 235 mixers, and 115 blenders, express in matrix form the number of appliances remaining in stock after Friday's sales. **5.3**

Solve.

8. $\begin{cases} x + 4y + 3z = 3 \\ 2x - 5y - z = 5 \\ 3x + 2y - 2z = -3 \end{cases}$

9. $\begin{cases} x + y + z = -1 \\ y + 3z = -5 \\ x + z = -2 \end{cases}$ **5.4**

5.5

Augmented Matrix Solutions

Objective: To solve a system of linear equations in three variables using the augmented matrix method

The solution of systems of linear equations can be simplified by using matrices. The matrix of the coefficients of the variables in a system of two or three linear equations is a square matrix. Matrix methods are especially suitable for computer solution of large systems of equations with many variables.

Capsule Review

$$A = \begin{bmatrix} 1 & 2 \\ 3 & -5 \end{bmatrix} \qquad B = \begin{bmatrix} 2 & 4 & 6 \\ -1 & 0 & 1 \\ 3 & 0 & 7 \end{bmatrix}$$

A and *B* are square matrices; that is, the number of rows in each matrix is equal to the number of columns. The element in the second row, first column of *A* is 3. The element in the second row, second column of *B* is 0. The *main diagonal* of a square matrix consists of the elements from the upper left corner to the lower right corner.

For matrices *A* and *B*, find the indicated elements.

1. First row, second column of *A*

2. First row, third column of *B*

3. Second row, second column of *A*

4. Third row, second column of *B*

5. Main diagonal of *A*

6. Main diagonal of *B*

7. Second row, third column of *B*

8. Third row, first column of *B*

A system of linear equations can be written in matrix form as $AX = C$, where A is the matrix of coefficients, X is a column matrix of the variables, and C is a column matrix of the constants. For example,

$$\overset{A}{\begin{bmatrix} 2 & -1 & 3 \\ 1 & 0 & 1 \\ 3 & 5 & 0 \end{bmatrix}} \overset{X}{\begin{bmatrix} x \\ y \\ z \end{bmatrix}} \overset{=}{=} \overset{C}{\begin{bmatrix} 4 \\ 2 \\ 8 \end{bmatrix}} \quad \text{represents the system} \quad \begin{cases} 2x - y + 3z = 4 \\ x + z = 2 \\ 3x + 5y = 8 \end{cases}$$

Notice that when a variable is missing in an equation, its coefficient in matrix A is 0. To verify that the matrix equation represents the system on its right, find the product AX and use the definition of equal matrices.

$$\begin{bmatrix} 2 & -1 & 3 \\ 1 & 0 & 1 \\ 3 & 5 & 0 \end{bmatrix}\begin{bmatrix} x \\ y \\ z \end{bmatrix} = \begin{bmatrix} 4 \\ 2 \\ 8 \end{bmatrix} \longrightarrow \begin{bmatrix} 2x - 1y + 3z \\ 1x + 0y + 1z \\ 3x + 5y + 0z \end{bmatrix} = \begin{bmatrix} 4 \\ 2 \\ 8 \end{bmatrix} \longrightarrow \begin{cases} 2x - y + 3z = 4 \\ x + z = 2 \\ 3x + 5y = 8 \end{cases}$$

An **augmented matrix** is formed by writing the constants of the equation as a column and attaching this column to the matrix of coefficients. The augmented matrix for the system shown above is

$$\left[\begin{array}{ccc|c} 2 & -1 & 3 & 4 \\ 1 & 0 & 1 & 2 \\ 3 & 5 & 0 & 8 \end{array}\right]$$

The vertical line segment separates the coefficients from the constants.

To solve the system, transform the augmented matrix into a simpler *equivalent matrix,* which has the same solution, by using one or more of the following **row operations**

- Interchange any two rows.
- Multiply the elements of any row by a nonzero constant.
- Add a multiple of the elements of one row to the corresponding elements of another row.

The *augmented matrix method* provides a systematic way to transform an augmented matrix so that the elements on the main diagonal of the coefficient matrix are 1's and all other elements in the coefficient matrix are 0's. The values of the variables can then be read directly from the matrix. The final form of the augmented matrix for a system of three linear equations is

$$\left[\begin{array}{ccc|c} 1 & 0 & 0 & a \\ 0 & 1 & 0 & b \\ 0 & 0 & 1 & c \end{array}\right] \qquad x = a,\ y = b,\ z = c$$

To Solve a System of Three Linear Equations Using Augmented Matrices

Step 1 Using a row operation, if necessary, make the element in the first row, first column 1.

Step 2 Add multiples of the first row to the second and third rows to make the other elements in the first column 0's.

Step 3 Using a row operation, if necessary, make the element in the second row, second column 1.

Step 4 Add multiples of the second row to the first and third rows to make the other elements in the second column 0's.

Step 5 Using a row operation, if necessary, make the element in the third row, third column 1.

Step 6 Add multiples of the third row to the first and second rows to make the other elements in the third column 0's.

Step 7 Read the values of the variables from the matrix.

EXAMPLE **Solve the system of equations at the right, using the augmented matrix method.**

$$\begin{cases} 2x + 2y + z = 1 \\ x + 2y - z = -3 \\ -x + 2y + 2z = -4 \end{cases}$$

$$\left[\begin{array}{ccc|c} 2 & 2 & 1 & 1 \\ 1 & 2 & -1 & -3 \\ -1 & 2 & 2 & -4 \end{array}\right]$$

Write the augmented matrix. If your graphing utility is able to perform row operations, use it to save time and paper.

$$\left[\begin{array}{ccc|c} 1 & 2 & -1 & -3 \\ 2 & 2 & 1 & 1 \\ -1 & 2 & 2 & -4 \end{array}\right]$$

Step 1 *To get 1 in the first row, first column, interchange rows 1 and 2.*

$$\left[\begin{array}{ccc|c} 1 & 2 & -1 & -3 \\ 0 & -2 & 3 & 7 \\ -1 & 2 & 2 & -4 \end{array}\right]$$

Step 2 *To get 0 in the second row, first column, multiply the elements in row 1 by -2 and add the results to the corresponding elements in row 2.*

$$\left[\begin{array}{ccc|c} 1 & 2 & -1 & -3 \\ 0 & -2 & 3 & 7 \\ 0 & 4 & 1 & -7 \end{array}\right]$$

To get 0 in the third row, first column, add the elements in row 1 to the corresponding elements in row 3.

$$\left[\begin{array}{ccc|c} 1 & 2 & -1 & -3 \\ 0 & 1 & -\frac{3}{2} & -\frac{7}{2} \\ 0 & 4 & 1 & -7 \end{array}\right]$$

Step 3 *To get 1 in the second row, second column, multiply the elements in row 2 by $-\frac{1}{2}$.*

$$\left[\begin{array}{ccc|c} 1 & 0 & 2 & 4 \\ 0 & 1 & -\frac{3}{2} & -\frac{7}{2} \\ 0 & 4 & 1 & -7 \end{array}\right]$$

Step 4 *To get 0 in the first row, second column, multiply the elements in row 2 by -2 and add to the corresponding elements in row 1.*

$$\left[\begin{array}{ccc|c} 1 & 0 & 2 & 4 \\ 0 & 1 & -\frac{3}{2} & -\frac{7}{2} \\ 0 & 0 & 7 & 7 \end{array}\right]$$

To get 0 in the third row, second column, multiply the elements in row 2 by -4 and add to the corresponding elements in row 3.

$$\left[\begin{array}{ccc|c} 1 & 0 & 2 & 4 \\ 0 & 1 & -\frac{3}{2} & -\frac{7}{2} \\ 0 & 0 & 1 & 1 \end{array}\right]$$

Step 5 *To get 1 in the third row, third column, multiply the elements in row 3 by $\frac{1}{7}$.*

$$\left[\begin{array}{ccc|c} 1 & 0 & 0 & 2 \\ 0 & 1 & -\frac{3}{2} & -\frac{7}{2} \\ 0 & 0 & 1 & 1 \end{array}\right]$$

Step 6 *To get 0 in the first row, third column, multiply the elements in row 3 by -2 and add to the corresponding elements in row 1.*

$$\left[\begin{array}{ccc|c} 1 & 0 & 0 & 2 \\ 0 & 1 & 0 & -2 \\ 0 & 0 & 1 & 1 \end{array}\right]$$

To get 0 in the second row, third column, multiply the elements in row 3 by $\frac{3}{2}$ and add to the corresponding elements in row 2.

At this point, check to see that the elements on the main diagonal of the coefficient matrix are 1's and all other elements in the coefficient matrix are 0's.

$x = 2, y = -2, z = 1$ **Step 7** *Read the values of x, y, and z from the matrix.*

Solution: $(2, -2, 1)$

Substitute for x, y, and z in the three original equations to check.

If a row of the matrix reduces to the form $000|c$, where $c \neq 0$, this implies that the product of zero and one of the variables is not zero. This is false for any value of the variable, so the system is inconsistent and there is no solution. If a row reduces to the form $000|0$ (but no row reduces to $000|c$, $c \neq 0$), this implies that the product of zero and each of the variables is equal to zero. This is true for all real values of the variables, so the system is dependent and there are an infinite number of solutions.

The approach outlined above is also called the *Gaussian elimination method,* and it can also be applied to systems of linear equations in two variables, or to systems with four or more variables.

CLASS EXERCISES

1. **Thinking Critically** Explain how to solve the system of equations at the right, using the augmented matrix method. $\begin{cases} x + 2y = 7 \\ 2x - 5y = -4 \end{cases}$

Give the augmented matrix for each system of equations. Do not solve.

2. $\begin{cases} x + 4y - z = 4 \\ x - 2y + z = -2 \\ 5x - 3y + 8z = 13 \end{cases}$

3. $\begin{cases} 3x - 7y + 3z = -3 \\ x + y + 2z = -3 \\ 2x - 3y + 5z = -8 \end{cases}$

4. $\begin{cases} 2x + y + z = 1 \\ x + y + z = 2 \\ x - y + z = -2 \end{cases}$

5. $\begin{cases} 3x + z = 4 \\ x + y + z = 0 \\ 2x + y - z = 6 \end{cases}$

6. $\begin{cases} x + z = 3 \\ y + z = 0 \\ x - 2y = -4 \end{cases}$

7. $\begin{cases} y - z = 7 \\ x - y = 2 \\ x + 3z = -11 \end{cases}$

PRACTICE EXERCISES

Use technology where appropriate.

Solve each system of equations by using the augmented matrix method. Determine whether the system is consistent and independent, consistent and dependent, or inconsistent.

1. $\begin{cases} x + y + z = 2 \\ 2y - 2z = 2 \\ x - 3z = 1 \end{cases}$

2. $\begin{cases} x - y + z = 3 \\ x + 3z = 6 \\ y - 2z = -1 \end{cases}$

3. $\begin{cases} x + y - z = 3 \\ 3x + 4y - z = 1 \\ 6x + 8y - 2z = 5 \end{cases}$

4. $\begin{cases} x + y - z = 1 \\ 3x + 3y + z = 7 \\ 2x + 2y - 2z = 2 \end{cases}$

5. $\begin{cases} x + y = 1 \\ y + z = 2 \\ x - z = -1 \end{cases}$

6. $\begin{cases} x + z = 2 \\ y - z = 1 \\ x + y = 3 \end{cases}$

7. $\begin{cases} x + y + z = 1 \\ y - 3z = 4 \\ x - z = 2 \end{cases}$

8. $\begin{cases} x + y + z = 0 \\ y + 4z = -6 \\ 2x - 2z = 4 \end{cases}$

9. $\begin{cases} 2x + y = 8 \\ x + z = 5 \\ y - z = -1 \end{cases}$

10. $\begin{cases} 3x - y = 3 \\ y - z = 3 \\ x + 4z = 2 \end{cases}$

11. $\begin{cases} 2x + y + z = 7 \\ y - 2z = -1 \\ x + 3z = 7 \end{cases}$

12. $\begin{cases} 2x + 2y + z = 12 \\ y + 3z = 7 \\ x - z = 2 \end{cases}$

Write the system of equations represented by each augmented matrix. Do not solve.

13. $\left[\begin{array}{cc|c} 3 & 6 & 2 \\ 2 & -1 & 3 \end{array}\right]$

14. $\left[\begin{array}{ccc|c} -2 & 3 & 5 & 9 \\ -1 & 0 & 3 & 0 \\ 0 & 4 & -2 & 6 \end{array}\right]$

Solve.

15. $\begin{cases} 2x - 3y + 2z = 10 \\ x + 3y + 4z = 14 \\ 3x - y + z = 9 \end{cases}$

16. $\begin{cases} 4x - y + z = 3 \\ x + 2y + z = 0 \\ 3x + 7y - 3z = 6 \end{cases}$

17. $\begin{cases} x + 2y + z = 4 \\ 3x + 6y + 3z = 2 \\ x - y + z = 3 \end{cases}$

18. $\begin{cases} x + y + z = 5 \\ 3x - y - z = 2 \\ -6x + 2y + 2z = 10 \end{cases}$

19. $\begin{cases} x + y - z = 5 \\ 2x + 2y + 2z = 10 \\ 3x + 3y + z = 15 \end{cases}$

20. $\begin{cases} x - 2y + z = 1 \\ 3x + 6y + 3z = 3 \\ 5x + 11y + 5z = 5 \end{cases}$

21. $\begin{cases} 2x + y = z + 1 \\ y = 2x - z + 7 \\ z - y = x \end{cases}$

22. $\begin{cases} 2x + y = z \\ y = -z + 2x \\ -3x + 4z = -y \end{cases}$

23. $\begin{cases} w + x + y + z = 3 \\ -w + x - 2y + z = -2 \\ 2x - y + z = 1 \\ w + y - z = 2 \end{cases}$

24. $\begin{cases} 2x + 2y + z = 4 \\ w + y - z = -2 \\ w + x + y + z = 3 \\ -4w + z = 2 \end{cases}$

Application

Nutrition While stranded on an island, the crew of a sailboat has access to only three sources of food. Food *B* is high in carbohydrate, food *C* is high in protein, and food *A* contains some of all three nutrients. One of the crew members designs a daily diet to supply each person with 120 g of fat, 220 g of carbohydrate, and 160 g of protein. The nutritional information per portion of each of the three food sources is given below.

	Food *A*	Food *B*	Food *C*
Fat	10 g	4 g	12 g
Carbohydrate	11 g	77 g	0 g
Protein	8 g	2 g	32 g

25. Set up a system of three equations in three variables to determine the number of portions of each food source each person must have to meet the dietary requirements.

26. Solve the system of equations set up in the problem above by the Gaussian elimination method. Use a calculator and round each result to the nearest tenth.

27. If only food A were available, how many portions would be required to provide the daily nutrition of one crew member?

28. Too much of a nutrient can be as dangerous to a person's health as too little. If only food B were available, how many times the prescribed amount of carbohydrate would a crew member receive in order to consume an adequate amount of protein?

29. Suppose that food C is gone and food A must be limited to 2 portions per person per day. How many portions of food B does each crew member need to obtain the prescribed amounts of all three nutrients?

HISTORICAL NOTE

The invention of matrices and their algebra dates back to a memoir of Arthur Cayley written in 1858. Cayley was a British mathematician who worked extensively in both algebra and geometry. He found that arrays of the type shown below were useful in representing certain types of linear transformations that he was studying.

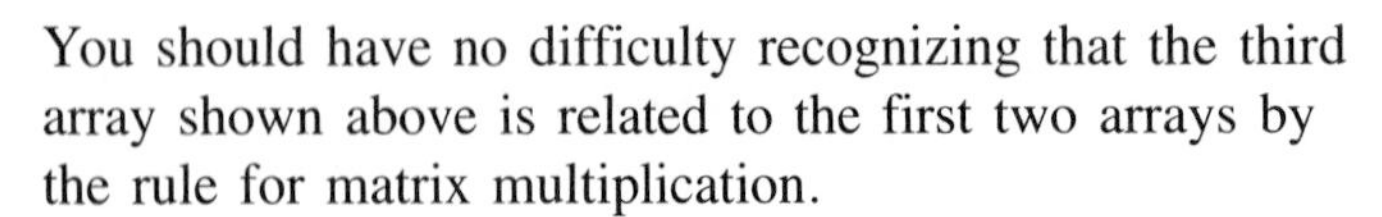

$$\begin{matrix} p & q \\ r & s \end{matrix} \qquad \begin{matrix} P & Q \\ R & S \end{matrix} \qquad \begin{matrix} pP + qR & pQ + qS \\ rP + sR & rQ + sS \end{matrix}$$

You should have no difficulty recognizing that the third array shown above is related to the first two arrays by the rule for matrix multiplication.

It often happens that the mathematical tools necessary for certain scientific applications are invented decades before the need for these tools arises. It was 67 years after Cayley invented the algebra of matrices that the nuclear physicist Werner Heisenberg recognized it as the tool he needed for his revolutionary work in quantum mechanics.

Investigation

It is interesting that matrix multiplication is not commutative. Work together with your classmates to see whether you can develop another system in which multiplication, as you define it, is not commutative.

5.6

Inverse of a Matrix

Objective: To find the multiplicative inverse of a matrix and to use it to solve a system of equations

A system of linear equations can be represented by the matrix equation $AX = C$, where A is the matrix of coefficients, X is a column matrix of the variables, and C is a column matrix of the constants. You have learned to solve systems of linear equations using augmented matrices of coefficients. In this lesson, you will learn how to solve certain systems using multiplicative inverses of matrices. First, it will be helpful to review the concept of a multiplicative identity matrix.

Capsule Review

For any real number a, 1 is the *multiplicative identity* of a, since $a \cdot 1 = 1 \cdot a = a$. The *multiplicative identity matrix* is defined only for the set of $n \times n$ square matrices. It is an $n \times n$ matrix, I, or $I_{n \times n}$, whose elements on the main diagonal are 1's and whose other elements are 0's.

$$I_{2\times 2} = \begin{bmatrix} 1 & 0 \\ 0 & 1 \end{bmatrix} \qquad I_{3\times 3} = \begin{bmatrix} 1 & 0 & 0 \\ 0 & 1 & 0 \\ 0 & 0 & 1 \end{bmatrix}$$

If A is a square matrix, then $AI = IA = A$.

Write the multiplicative identity matrix for each given matrix, if it exists, and find the product of the matrix and its multiplicative identity. If a matrix does not have a multiplicative identity, so state.

1. $\begin{bmatrix} 1 & 3 \\ 2 & 0 \end{bmatrix}$ **2.** $\begin{bmatrix} 1 & 3 & 2 \\ 2 & 4 & 1 \end{bmatrix}$ **3.** $\begin{bmatrix} 1 & -2 \\ 3 & 0 \end{bmatrix}$

4. $\begin{bmatrix} 3 & 1 & 4 \\ -2 & 1 & 0 \\ 2 & 4 & 0 \end{bmatrix}$ **5.** $\begin{bmatrix} 0 & 1 & 2 \\ -2 & 1 & 1 \\ -1 & 0 & 5 \end{bmatrix}$ **6.** $\begin{bmatrix} 1 & 3 \\ -2 & -4 \\ 0 & 2 \end{bmatrix}$

If for any real number a, there exists a real number b such that $ab = ba = 1$, then b is the *multiplicative inverse* of a. Some square matrices also have multiplicative inverses.

If A and B are $n \times n$ matrices, and $AB = BA = I$, then B is the multiplicative inverse of A and B is denoted by A^{-1}.

EXAMPLE 1 **Show that B is the multiplicative inverse of A.**

$$A = \begin{bmatrix} 2 & 3 \\ 1 & 2 \end{bmatrix} \quad B = \begin{bmatrix} 2 & -3 \\ -1 & 2 \end{bmatrix}$$

$$AB = \begin{bmatrix} 2 & 3 \\ 1 & 2 \end{bmatrix}\begin{bmatrix} 2 & -3 \\ -1 & 2 \end{bmatrix} = \begin{bmatrix} 1 & 0 \\ 0 & 1 \end{bmatrix} \qquad BA = \begin{bmatrix} 2 & -3 \\ -1 & 2 \end{bmatrix}\begin{bmatrix} 2 & 3 \\ 1 & 2 \end{bmatrix} = \begin{bmatrix} 1 & 0 \\ 0 & 1 \end{bmatrix}$$

$AB = I_{2\times2}$ and $BA = I_{2\times2}$, so $AB = BA = I_{2\times2}$. Therefore, B is the multiplicative inverse of A. A is also the multiplicative inverse of B.

Only square matrices can have multiplicative inverses. To find the inverse of a square matrix, such as matrix A shown below

$$A = \begin{bmatrix} a_1 & b_1 & c_1 \\ a_2 & b_2 & c_2 \\ a_3 & b_3 & c_3 \end{bmatrix}$$

write an augmented matrix $[A|I]$

$$[A|I] = \left[\begin{array}{ccc|ccc} a_1 & b_1 & c_1 & 1 & 0 & 0 \\ a_2 & b_2 & c_2 & 0 & 1 & 0 \\ a_3 & b_3 & c_3 & 0 & 0 & 1 \end{array}\right]$$

Using row operations, try to obtain an equivalent matrix $[I|B]$.

$$[I|B] = \left[\begin{array}{ccc|ccc} 1 & 0 & 0 & d_1 & e_1 & f_1 \\ 0 & 1 & 0 & d_2 & e_2 & f_2 \\ 0 & 0 & 1 & d_3 & e_3 & f_3 \end{array}\right]$$

It can be shown that if $[I|B]$ exists, then B is the multiplicative inverse of A. If one or more rows to the left of the vertical line segment contain all zero elements, then A has no multiplicative inverse.

EXAMPLE 2 **Find the multiplicative inverse of matrix A, if it exists.**

$$A = \begin{bmatrix} 0 & -1 & 2 \\ 1 & 1 & -1 \\ 2 & 0 & 3 \end{bmatrix}$$

$$\left[\begin{array}{ccc|ccc} 0 & -1 & 2 & 1 & 0 & 0 \\ 1 & 1 & -1 & 0 & 1 & 0 \\ 2 & 0 & 3 & 0 & 0 & 1 \end{array}\right]$$

Write the augmented matrix [A|I].

$$\left[\begin{array}{ccc|ccc} 1 & 1 & -1 & 0 & 1 & 0 \\ 0 & -1 & 2 & 1 & 0 & 0 \\ 2 & 0 & 3 & 0 & 0 & 1 \end{array}\right]$$

To get 1 in the first row, first column and 0 in the second row, first column, interchange the first and second rows.

$$\left[\begin{array}{ccc|ccc} 1 & 1 & -1 & 0 & 1 & 0 \\ 0 & -1 & 2 & 1 & 0 & 0 \\ 0 & -2 & 5 & 0 & -2 & 1 \end{array}\right]$$

To get 0 in the third row, first column, multiply the elements in row 1 by −2 and add the results to the corresponding elements in row 3.

$$\left[\begin{array}{rrr|rrr} 1 & 1 & -1 & 0 & 1 & 0 \\ 0 & 1 & -2 & -1 & 0 & 0 \\ 0 & -2 & 5 & 0 & -2 & 1 \end{array}\right]$$

To get 1 in the second row, second column, multiply row 2 by −1.

$$\left[\begin{array}{rrr|rrr} 1 & 0 & 1 & 1 & 1 & 0 \\ 0 & 1 & -2 & -1 & 0 & 0 \\ 0 & -2 & 5 & 0 & -2 & 1 \end{array}\right]$$

To get 0 in the first row, second column, multiply row 2 by −1 and add to row 1.

$$\left[\begin{array}{rrr|rrr} 1 & 0 & 1 & 1 & 1 & 0 \\ 0 & 1 & -2 & -1 & 0 & 0 \\ 0 & 0 & 1 & -2 & -2 & 1 \end{array}\right]$$

To get 0 in the third row, second column, multiply row 2 by 2 and add to row 3.

$$\left[\begin{array}{rrr|rrr} 1 & 0 & 0 & 3 & 3 & -1 \\ 0 & 1 & -2 & -1 & 0 & 0 \\ 0 & 0 & 1 & -2 & -2 & 1 \end{array}\right]$$

To get 0 in the first row, third column, multiply row 3 by −1 and add to row 1.

$$\left[\begin{array}{rrr|rrr} 1 & 0 & 0 & 3 & 3 & -1 \\ 0 & 1 & 0 & -5 & -4 & 2 \\ 0 & 0 & 1 & -2 & -2 & 1 \end{array}\right]$$

To get 0 in the second row, third column, multiply row 3 by 2 and add to row 2.

$$A^{-1} = \begin{bmatrix} 3 & 3 & -1 \\ -5 & -4 & 2 \\ -2 & -2 & 1 \end{bmatrix}$$

Check by showing that $AA^{-1} = A^{-1}A = I$.

If the inverse of a matrix exists, it can be used to solve a matrix equation of the form $AX = C$. The multiplication property of equality can be extended to include square matrices. Thus, if $M = N$, then $LM = LN$ and $ML = NL$.

$AX = C$	
$A^{-1}(AX) = A^{-1}C$	*Multiply both sides by A^{-1}.*
$(A^{-1}A)X = A^{-1}C$	*Associative property*
$IX = A^{-1}C$	*Definition of multiplicative inverse*
$X = A^{-1}C$	*Definition of multiplicative identity*

EXAMPLE 3 **Solve the system of equations at the right using the multiplicative inverse of a matrix.**

$$\begin{cases} 2x + 3y = 11 \\ x + 2y = 6 \end{cases}$$

$$\overset{A}{\begin{bmatrix} 2 & 3 \\ 1 & 2 \end{bmatrix}} \overset{X}{\begin{bmatrix} x \\ y \end{bmatrix}} \overset{=}{=} \overset{C}{\begin{bmatrix} 11 \\ 6 \end{bmatrix}}$$

Write the matrix equation $AX = C$.

$$A^{-1} = \begin{bmatrix} 2 & -3 \\ -1 & 2 \end{bmatrix}$$

Find A^{-1}. (See Example 1.)

Then, $X = A^{-1}C = \begin{bmatrix} 2 & -3 \\ -1 & 2 \end{bmatrix}\begin{bmatrix} 11 \\ 6 \end{bmatrix} = \begin{bmatrix} 4 \\ 1 \end{bmatrix}$ $X = \begin{bmatrix} x \\ y \end{bmatrix} = \begin{bmatrix} 4 \\ 1 \end{bmatrix}$

Check in the original equations to show that the solution is (4, 1).

A system of linear equations in three variables can also be solved using the inverse of a matrix, provided the inverse exists.

EXAMPLE 4 **Solve the system of equations at the right. Use the multiplicative inverse of a matrix.**

$$\begin{cases} -y + 2z = 4 \\ x + y - z = 0 \\ 2x + 3z = 11 \end{cases}$$

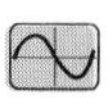

$$\underset{A}{\begin{bmatrix} 0 & -1 & 2 \\ 1 & 1 & -1 \\ 2 & 0 & 3 \end{bmatrix}} \underset{X}{\begin{bmatrix} x \\ y \\ z \end{bmatrix}} = \underset{C}{\begin{bmatrix} 4 \\ 0 \\ 11 \end{bmatrix}}$$

Write the matrix equation $AX = C$.

$$A^{-1} = \begin{bmatrix} 3 & 3 & -1 \\ -5 & -4 & 2 \\ -2 & -2 & 1 \end{bmatrix}$$

Use the inverse function on your graphing utility to find A^{-1}, or use the method from Example 2.

Then, $X = A^{-1}C = \begin{bmatrix} 3 & 3 & -1 \\ -5 & -4 & 2 \\ -2 & -2 & 1 \end{bmatrix} \begin{bmatrix} 4 \\ 0 \\ 11 \end{bmatrix} = \begin{bmatrix} 1 \\ 2 \\ 3 \end{bmatrix}$ $\quad X = \begin{bmatrix} x \\ y \\ z \end{bmatrix} = \begin{bmatrix} 1 \\ 2 \\ 3 \end{bmatrix}$

Check to show that the solution to the system is (1, 2, 3).

CLASS EXERCISES

1. **Thinking Critically** Explain why $A_{2\times 3}$ does not have a multiplicative inverse.

Find the multiplicative inverse of each matrix, if it exists. If it does not exist, so state.

2. $\begin{bmatrix} 1 & 2 \\ 1 & 3 \end{bmatrix}$ **3.** $\begin{bmatrix} 3 & 0 \\ 6 & 0 \end{bmatrix}$ **4.** $\begin{bmatrix} 2 & 2 & 2 \\ 2 & -2 & 2 \\ 2 & 2 & -2 \end{bmatrix}$ **5.** $\begin{bmatrix} 1 & 0 & 3 \\ 1 & 1 & 0 \\ -1 & 0 & -1 \end{bmatrix}$

Write the matrix equation for each system of equations. Do not solve.

6. $\begin{cases} x + 2y = 11 \\ 2x + 3y = 18 \end{cases}$ **7.** $\begin{cases} x + y + z = 4 \\ 4x + 5y = 4 \\ y - 3z = -9 \end{cases}$ **8.** $\begin{cases} x + y + z = 4 \\ 4x + 5y = 3 \\ y - 3z = -10 \end{cases}$

PRACTICE EXERCISES

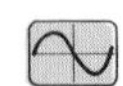 Use technology where appropriate.

Find the multiplicative inverse of each matrix, if it exists. If it does not exist, so state.

1. $\begin{bmatrix} 2 & -1 \\ 1 & 0 \end{bmatrix}$ **2.** $\begin{bmatrix} 2 & 3 \\ 1 & 1 \end{bmatrix}$ **3.** $\begin{bmatrix} 1 & 4 \\ 1 & 3 \end{bmatrix}$ **4.** $\begin{bmatrix} 4 & 7 \\ 3 & 5 \end{bmatrix}$

5. $\begin{bmatrix} 5 & 2 \\ -8 & -3 \end{bmatrix}$ **6.** $\begin{bmatrix} -3 & 11 \\ 2 & -7 \end{bmatrix}$ **7.** $\begin{bmatrix} 2 & 0 \\ 0 & 2 \end{bmatrix}$ **8.** $\begin{bmatrix} 0 & 3 \\ 3 & 0 \end{bmatrix}$

Solve each system of equations by using the multiplicative inverse of a matrix.

9. $\begin{cases} x + 3y = 5 \\ x + 4y = 6 \end{cases}$ **10.** $\begin{cases} x + 5y = -4 \\ x + 6y = -5 \end{cases}$ **11.** $\begin{cases} 2x + 3y = 12 \\ x + 2y = 7 \end{cases}$

12. $\begin{cases} -3x + 4y = 2 \\ x - y = -1 \end{cases}$ **13.** $\begin{cases} x + 2y = 10 \\ 3x + 5y = 26 \end{cases}$ **14.** $\begin{cases} x - 3y = -1 \\ -6x + 19y = 6 \end{cases}$

Find the multiplicative inverse of each matrix, if it exists. If it does not exist, so state.

15. $\begin{bmatrix} -1 & 3 \\ 2 & 0 \end{bmatrix}$ **16.** $\begin{bmatrix} 1 & 2 \\ 2 & 1 \end{bmatrix}$ **17.** $\begin{bmatrix} 1 & -3 & 0 \\ 0 & 3 & 1 \\ 2 & -1 & 2 \end{bmatrix}$

18. $\begin{bmatrix} 1 & 2 & 0 \\ 0 & 2 & -2 \\ 1 & 0 & 2 \end{bmatrix}$ **19.** $\begin{bmatrix} 2 & -1 & 3 \\ 1 & 0 & 1 \\ 3 & 5 & 0 \end{bmatrix}$ **20.** $\begin{bmatrix} 0 & 1 & 1 \\ -1 & 0 & 3 \\ 1 & 0 & 2 \end{bmatrix}$

Solve each system of equations by using the multiplicative inverse of a matrix.

21. $\begin{cases} x - 3z = 0 \\ y + 3z = 3 \\ 2x + 2y - z = 5 \end{cases}$ **22.** $\begin{cases} 9y + 2z = 18 \\ 3x + 2y + z = 1 \\ x - y = -3 \end{cases}$

23. $\begin{cases} x - 3z = -2 \\ y + 3z = 4 \\ 2x + 2y - z = 3 \end{cases}$
(See Exercise 21 for A^{-1}.)

24. $\begin{cases} 9y + 2z = 14 \\ 3x + 2y + z = 5 \\ x - y = -1 \end{cases}$
(See Exercise 22 for A^{-1}.)

25. $\begin{cases} x = 5 - y \\ 3y = z \\ x + z = 7 \end{cases}$ **26.** $\begin{cases} -x = -4 - z \\ 2y = z - 1 \\ x = 6 - y - z \end{cases}$

27. $\begin{cases} -2w + x + y = 0 \\ -w + 2x - y + z = 1 \\ -2w + 3x + 3y + 2z = 6 \\ w + x + 2y + z = 5 \end{cases}$ **28.** $\begin{cases} -2w + x + y = -2 \\ -w + 2x - y + z = -4 \\ -2w + 3x + 3y + 2z = 2 \\ w + x + 2y + z = 6 \end{cases}$

Applications

29. Geometry The coordinates (x, y) of a point in a plane are the solution of this system of equations. Find the coordinates of the point. $\begin{cases} 2x + 3y = 13 \\ 5x + 7y = 31 \end{cases}$

30. **Geometry** The dimensions, x and y, of a rectangle are related by this system of equations. Find the dimensions.

$$\begin{cases} -x + 5y = 11 \\ x - 6y = -14 \end{cases}$$

31. **Consumerism** A plumbing supply store sells packages containing large and small O-rings. A package of 2 large and 3 small rings costs $0.78. A package of 3 large and 4 small rings costs $1.10. Find the cost of a large and a small O-ring.

Developing Mathematical Power

32. **Thinking Critically** Analyze the strategy for finding the inverse of a matrix. Then write a 3×3 matrix that has no inverse.

CAREER

A computer programmer analyzes a request for information, locates suitable data, and plans instructions for the computer to fulfill the request. The programmer prepares a flow chart showing the sequence of steps that the computer must make. These minutely detailed steps are translated into a language that the computer can read. When the program is completed, the programmer tests it by having the computer perform on simulated data. If the program does not work as expected, it is examined for errors. Elimination of these errors is called *debugging*.

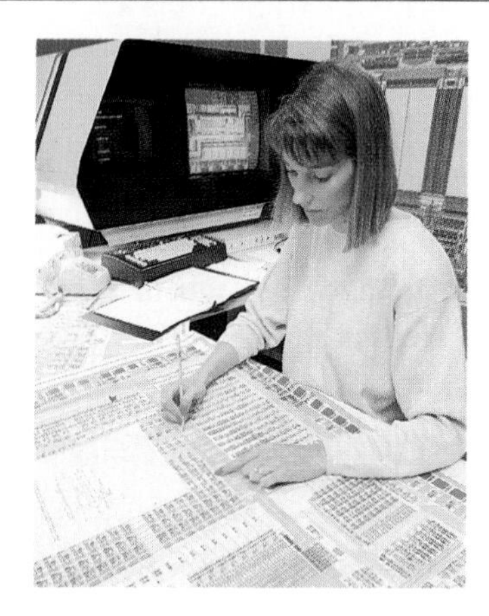

If you

- like mathematics
- pay attention to the smallest details
- and enjoy solving complicated problems involving logic and reasoning

then you may be interested in becoming a computer programmer.

Investigation

The first known mechanical computer is the abacus, which was in use as early as the sixth century BC. The first electronic computer was the Electronic Numerical Integrator And Computer (ENIAC), completed in 1946 at the University of Pennsylvania. ENIAC contained some 18,000 vacuum tubes and diodes and weighed 30 tons. Research the differences between the capabilities of the first computers and those of today. What new capabilities are expected in computers of the future?

5.7

Determinants and Cramer's Rule

Objectives: To evaluate the determinant of a 2×2 or a 3×3 matrix
To solve systems of equations using Cramer's rule

The matrix of coefficients of a system of two linear equations in two variables or three linear equations in three variables is a square matrix. A numerical value can be associated with any square matrix, and this value can then be used to solve the system of equations.

Capsule Review

For the system of two equations in two variables $\begin{cases} x + 2y = -3, \\ 3x - 4y = 5 \end{cases}$

the matrix of coefficients is $\begin{bmatrix} 1 & 2 \\ 3 & -4 \end{bmatrix}$.

Write the matrix of coefficients for each system of equations.

1. $\begin{cases} 2x + 3y = 5 \\ x + 2y = 6 \end{cases}$

2. $\begin{cases} 3x - y = 4 \\ 2x + 3y = 7 \end{cases}$

3. $\begin{cases} 3x + 5y = 7 \\ 2x - 6y = 9 \end{cases}$

4. $\begin{cases} x + 2y + z = 4 \\ 2x + 3y + 5z = 2 \\ 5x - 2y + 3z = -5 \end{cases}$

5. $\begin{cases} x + z = 7 \\ 2y - z = 3 \\ x + y + z = 4 \end{cases}$

6. $\begin{cases} 3x - y + z = 2 \\ 5x + 2y - 3z = 1 \\ 2x - 3y + 4z = 3 \end{cases}$

Associated with every square matrix is a real number called its **determinant.**

The determinant of a 2×2 matrix $\begin{bmatrix} a_1 & b_1 \\ a_2 & b_2 \end{bmatrix}$ is defined to be

$$\begin{vmatrix} a_1 & b_1 \\ a_2 & b_2 \end{vmatrix} = a_1 b_2 - a_2 b_1$$

The elements of a determinant are enclosed within vertical bars, not brackets. The value of a 2×2 determinant is the product of the elements on the main diagonal minus the product of the elements on the other diagonal.

The reason that the determinant is defined in this way will become apparent on pages 235 and 236, where the relationship between determinants and the solution of a general system of two equations in two variables is discussed.

EXAMPLE 1 **Evaluate the determinant:** $D = \begin{vmatrix} 8 & 7 \\ 2 & 3 \end{vmatrix}$

$$D = \begin{vmatrix} 8 & 7 \\ 2 & 3 \end{vmatrix} = 8(3) - 2(7) = 10 \qquad a_1b_2 - a_2b_1$$

The determinant of a 3×3 matrix $\begin{bmatrix} a_1 & b_1 & c_1 \\ a_2 & b_2 & c_2 \\ a_3 & b_3 & c_3 \end{bmatrix}$ is defined to be

$$\begin{vmatrix} a_1 & b_1 & c_1 \\ a_2 & b_2 & c_2 \\ a_3 & b_3 & c_3 \end{vmatrix} = a_1b_2c_3 + a_2b_3c_1 + a_3b_1c_2 - a_1b_3c_2 - a_2b_1c_3 - a_3b_2c_1$$

To evaluate a 3×3 determinant, repeat the first two columns, writing them to the right of the third column. Then subtract the sum of the products along the three right-to-left diagonals from the sum of the products along the three left-to-right diagonals.

$$\begin{matrix} a_1 & b_1 & c_1 & a_1 & b_1 \\ a_2 & b_2 & c_2 & a_2 & b_2 \\ a_3 & b_3 & c_3 & a_3 & b_3 \end{matrix}$$

Left-to-right: $(a_1b_2c_3 + b_1c_2a_3 + c_1a_2b_3)$

Right-to-left: $-(b_1a_2c_3 + a_1c_2b_3 + c_1b_2a_3)$

It is also possible to use 2×2 determinants to evaluate a 3×3 determinant, as shown below.

$$\begin{aligned} \begin{vmatrix} a_1 & b_1 & c_1 \\ a_2 & b_2 & c_2 \\ a_3 & b_3 & c_3 \end{vmatrix} &= a_1b_2c_3 + a_2b_3c_1 + a_3b_1c_2 - a_1b_3c_2 - a_2b_1c_3 - a_3b_2c_1 \\ &= a_1(b_2c_3 - b_3c_2) - b_1(a_2c_3 - a_3c_2) + c_1(a_2b_3 - a_3b_2) \\ &= a_1\begin{vmatrix} b_2 & c_2 \\ b_3 & c_3 \end{vmatrix} - b_1\begin{vmatrix} a_2 & c_2 \\ a_3 & c_3 \end{vmatrix} + c_1\begin{vmatrix} a_2 & b_2 \\ a_3 & b_3 \end{vmatrix} \end{aligned}$$

The 2×2 determinants shown above are called the **minors** of the elements a_1, b_1, and c_1. In general, the minor of an element in a 3×3 determinant is the 2×2 determinant that is found by eliminating the row and column that contain that element. Thus, the minor of a_1 in the 3×3 determinant above is

$$\begin{vmatrix} b_2 & c_2 \\ b_3 & c_3 \end{vmatrix} \qquad \begin{vmatrix} a_1 & b_1 & c_1 \\ a_2 & b_2 & c_2 \\ a_3 & b_3 & c_3 \end{vmatrix}$$

The minor of b_3 is $\begin{vmatrix} a_1 & c_1 \\ a_2 & c_2 \end{vmatrix} \qquad \begin{vmatrix} a_1 & b_1 & c_1 \\ a_2 & b_2 & c_2 \\ a_3 & b_3 & c_3 \end{vmatrix}$

The evaluation of a determinant using minors is called *expansion by minors*.

EXAMPLE 2 **Evaluate determinant D by expanding it using elements in the first column and their minors.**

$$D = \begin{vmatrix} \mathbf{1} & \mathbf{2} & \mathbf{1} \\ \mathbf{3} & \mathbf{1} & \mathbf{0} \\ \mathbf{2} & \mathbf{5} & \mathbf{2} \end{vmatrix}$$

$$\begin{aligned} D &= 1\begin{vmatrix} 1 & 0 \\ 5 & 2 \end{vmatrix} - 3\begin{vmatrix} 2 & 1 \\ 5 & 2 \end{vmatrix} + 2\begin{vmatrix} 2 & 1 \\ 1 & 0 \end{vmatrix} \\ &= 1[1(2) - 5(0)] - 3[2(2) - 5(1)] + 2[2(0) - 1(1)] \\ &= 1[2 - 0] - 3[4 - 5] + 2[0 - 1] \\ &= 2 + 3 - 2 = 3 \end{aligned}$$

A 3×3 determinant can be evaluated using the elements in any row or column, along with their minors. The diagram at the right shows the pattern of the signs for the expansion of a 3×3 determinant by minors.

$$\begin{vmatrix} + & - & + \\ - & + & - \\ + & - & + \end{vmatrix}$$

The value of the determinant is the sum of the products of the elements in the chosen row or column, each with the appropriate sign, and their minors. To illustrate, determinant D in Example 2 could have been expanded using elements in the second row and their minors

$$\begin{aligned} D &= -3\begin{vmatrix} 2 & 1 \\ 5 & 2 \end{vmatrix} + 1\begin{vmatrix} 1 & 1 \\ 2 & 2 \end{vmatrix} - 0\begin{vmatrix} 1 & 2 \\ 2 & 5 \end{vmatrix} \\ &= -3[2(2) - 5(1)] + 1[1(2) - 2(1)] - 0 = 3 \end{aligned}$$

Expansion by minors can be used with any determinant of order three or greater.

Determinants can be used to solve systems of equations. To see how this can be done, first solve a general system of two equations in two variables.

$$\begin{cases} a_1x + b_1y = c_1 \\ a_2x + b_2y = c_2 \end{cases}$$

$$\begin{aligned} a_1b_2x + b_1b_2y &= c_1b_2 \\ -a_2b_1x - b_2b_1y &= -c_2b_1 \end{aligned}$$

Multiply both sides of the first equation by b_2 and both sides of the second equation by $-b_1$.

$$a_1b_2x - a_2b_1x = c_1b_2 - c_2b_1$$

Add.

$$x = \frac{c_1b_2 - c_2b_1}{a_1b_2 - a_2b_1}, \ a_1b_2 - a_2b_1 \neq 0$$

Solve for x.

$$\begin{aligned} -a_1a_2x - a_2b_1y &= -a_2c_1 \\ a_1a_2x + a_1b_2y &= a_1c_2 \end{aligned}$$

Multiply both sides of the first equation by $-a_2$ and both sides of the second equation by a_1.

$$a_1b_2y - a_2b_1y = a_1c_2 - a_2c_1$$

Add.

$$y = \frac{a_1c_2 - a_2c_1}{a_1b_2 - a_2b_1}, \ a_1b_2 - a_2b_1 \neq 0$$

Solve for y.

The determinant of the coefficient matrix of the system of equations is

$$D = \begin{vmatrix} a_1 & b_1 \\ a_2 & b_2 \end{vmatrix} = a_1 b_2 - a_2 b_1$$

The determinant of the coefficient matrix, where the coefficients of x are replaced by the constants, is

$$D_x = \begin{vmatrix} c_1 & b_1 \\ c_2 & b_2 \end{vmatrix} = c_1 b_2 - c_2 b_1$$

Therefore, if $D \neq 0$, then $x = \dfrac{D_x}{D}$. $\quad x = \dfrac{c_1 b_2 - c_2 b_1}{a_1 b_2 - a_2 b_1},\ a_1 b_2 - a_2 b_1 \neq 0$

Similarly, the determinant of the coefficient matrix, where the coefficients of y are replaced by the constants, is

$$D_y = \begin{vmatrix} a_1 & c_1 \\ a_2 & c_2 \end{vmatrix} = a_1 c_2 - a_2 c_1$$

Therefore, if $D \neq 0$, then, $y = \dfrac{D_y}{D}$. $\quad y = \dfrac{a_1 c_2 - a_2 c_1}{a_1 b_2 - a_2 b_1},\ a_1 b_2 - a_2 b_1 \neq 0$

This method of solving systems of equations using determinants is called **Cramer's rule.**

EXAMPLE 3 **Solve the system using Cramer's Rule:** $\begin{cases} \mathbf{2x - 5y = 11} \\ \mathbf{3x + 4y = \ \ 5} \end{cases}$

$$D = \begin{vmatrix} a_1 & b_1 \\ a_2 & b_2 \end{vmatrix} = \begin{vmatrix} 2 & -5 \\ 3 & 4 \end{vmatrix} = 8 - (-15) = 23 \qquad \text{Find } D.$$

$$D_x = \begin{vmatrix} c_1 & b_1 \\ c_2 & b_2 \end{vmatrix} = \begin{vmatrix} 11 & -5 \\ 5 & 4 \end{vmatrix} = 44 - (-25) = 69 \qquad \text{Find } D_x.$$

$$D_y = \begin{vmatrix} a_1 & c_1 \\ a_2 & c_2 \end{vmatrix} = \begin{vmatrix} 2 & 11 \\ 3 & 5 \end{vmatrix} = 10 - 33 = -23 \qquad \text{Find } D_y.$$

$$x = \frac{D_x}{D} = \frac{69}{23} = 3 \quad \text{and} \quad y = \frac{D_y}{D} = -\frac{23}{23} = -1 \qquad \text{Find } x \text{ and } y.$$

Check to show that the solution to the system is $(3, -1)$.

Cramer's rule can also be applied to a system of three equations in three variables. The required determinants are shown below.

$$\begin{cases} a_1 x + b_1 y + c_1 z = d_1 \\ a_2 x + b_2 y + c_2 z = d_2 \\ a_3 x + b_3 y + c_3 z = d_3 \end{cases}$$

$$D = \begin{vmatrix} a_1 & b_1 & c_1 \\ a_2 & b_2 & c_2 \\ a_3 & b_3 & c_3 \end{vmatrix} \quad D_x = \begin{vmatrix} d_1 & b_1 & c_1 \\ d_2 & b_2 & c_2 \\ d_3 & b_3 & c_3 \end{vmatrix} \quad D_y = \begin{vmatrix} a_1 & d_1 & c_1 \\ a_2 & d_2 & c_2 \\ a_3 & d_3 & c_3 \end{vmatrix} \quad D_z = \begin{vmatrix} a_1 & b_1 & d_1 \\ a_2 & b_2 & d_2 \\ a_3 & b_3 & d_3 \end{vmatrix}$$

$$x = \frac{D_x}{D} \qquad y = \frac{D_y}{D} \qquad z = \frac{D_z}{D} \qquad D \neq 0$$

CLASS EXERCISES

Evaluate each determinant using the elements in the indicated row or column and their minors.

1. $\begin{vmatrix} 1 & 2 & 5 \\ 3 & 1 & 0 \\ 1 & 2 & 1 \end{vmatrix}$ First column

2. $\begin{vmatrix} 1 & 4 & 0 \\ 2 & 3 & 5 \\ 0 & 1 & 0 \end{vmatrix}$ Third column

3. $\begin{vmatrix} 2 & 4 & 1 \\ 3 & 0 & 1 \\ 1 & 2 & 1 \end{vmatrix}$ Second row

4. $\begin{vmatrix} 2 & 3 & 0 \\ 1 & 2 & 5 \\ 7 & 0 & 1 \end{vmatrix}$ Third row

Give the determinants you would use to solve for the variables in each system of equations. ***Do not solve the systems.***

5. $\begin{cases} 2x + 4y = 10 \\ 3x + 5y = 14 \end{cases}$

6. $\begin{cases} x + 2y = 5 \\ x + y + z = 4 \\ 3y - 4z = 2 \end{cases}$

7. $\begin{cases} 3x + y + z = 1 \\ 5x - 4y + 3z = -1 \\ -2x + y + 4z = -10 \end{cases}$

PRACTICE EXERCISES

Evaluate each determinant.

1. $\begin{vmatrix} 5 & 2 \\ 1 & 3 \end{vmatrix}$

2. $\begin{vmatrix} 2 & -1 \\ 5 & -4 \end{vmatrix}$

3. $\begin{vmatrix} -4 & 3 \\ 2 & 0 \end{vmatrix}$

4. $\begin{vmatrix} 7 & 2 \\ 0 & -3 \end{vmatrix}$

5. $\begin{vmatrix} 6 & 2 \\ -6 & -2 \end{vmatrix}$

Evaluate each determinant using minors.

6. $\begin{vmatrix} 5 & 1 & 0 \\ 0 & 2 & -1 \\ -2 & -3 & 1 \end{vmatrix}$

7. $\begin{vmatrix} 4 & 6 & -1 \\ 2 & 3 & 2 \\ 1 & -1 & 1 \end{vmatrix}$

8. $\begin{vmatrix} -3 & 2 & -1 \\ 2 & 5 & 2 \\ 1 & -2 & 0 \end{vmatrix}$

9. $\begin{vmatrix} 0 & 2 & -3 \\ 1 & 2 & 4 \\ -2 & 0 & 1 \end{vmatrix}$

Solve each system of equations using Cramer's rule.

10. $\begin{cases} 3x + y = 5 \\ 2x + 3y = 8 \end{cases}$

11. $\begin{cases} 2x + 4y = 10 \\ 3x + 5y = 14 \end{cases}$

12. $\begin{cases} 5x - 3y = -19 \\ 3x + 4y = 6 \end{cases}$

13. $\begin{cases} 2x + 7y = 1 \\ 3x - 4y = 16 \end{cases}$

14. $\begin{cases} 3x + 2y = 24 \\ 4x - 3y = -2 \end{cases}$

15. $\begin{cases} 2x + 5y = 25 \\ -4x + 7y = 1 \end{cases}$

16. $\begin{cases} 0.5x + 1.5y = 7 \\ 2.5x - 3.5y = -9 \end{cases}$

17. $\begin{cases} 0.5x + 0.2y = 1.1 \\ 0.4x - 0.7y = 2.6 \end{cases}$

18. $\begin{cases} -1.2x - 0.3y = 2.1 \\ -0.2x + 0.8y = 4.6 \end{cases}$

19. $\begin{cases} \dfrac{x}{5} - \dfrac{2y}{5} = 4 \\ \dfrac{2x}{5} - \dfrac{3y}{15} = 5 \end{cases}$

20. $\begin{cases} \dfrac{x}{2} + \dfrac{y}{4} = 4 \\ \dfrac{x}{4} - \dfrac{3y}{8} = -2 \end{cases}$

21. $\begin{cases} y + 4z = 5 \\ x + y + z = 8 \\ 2x - 5y = 7 \end{cases}$

22. $\begin{cases} 2x + 3y + z = 5 \\ x + y - 2z = -2 \\ -3x + z = -7 \end{cases}$

23. $\begin{cases} -3x + 4y - z = 3 \\ x + 2y - 3z = 9 \\ y - 5z = -1 \end{cases}$

24. $\begin{cases} x + 2y - z = -5 \\ 2x + 3y + 2z = 1 \\ 3x + 3y - 4z = -8 \end{cases}$

25. $\begin{cases} 2x + 3y + 5z = 12 \\ 4x + 2y + 4z = -2 \\ 5x + 4y + 7z = 7 \end{cases}$

Prove the following statements for all 3 × 3 determinants.

26. If all elements of one column are 0's, then the determinant equals 0.

27. If the only nonzero elements lie on the main diagonal, then the determinant is equal to the product of the elements on that diagonal.

Solve using Cramer's rule. ***Hint:*** **Start by letting $m = \frac{1}{x}$ and $n = \frac{1}{y}$.**

28. $\begin{cases} \frac{4}{x} + \frac{1}{y} = 1 \\ \frac{8}{x} + \frac{4}{y} = 3 \end{cases}$

29. $\begin{cases} \frac{4}{x} - \frac{2}{y} = 1 \\ \frac{10}{x} + \frac{20}{y} = 0 \end{cases}$

30. Prove that if two columns of a 3 × 3 determinant are identical, then the determinant equals 0.

31. Let $M = \begin{bmatrix} a & b \\ c & d \end{bmatrix}$ and $N = \begin{bmatrix} e & f \\ g & h \end{bmatrix}$. Prove that the product of the determinants of M and N equals the determinant of the matrix product MN.

Applications

32. Business A manufacturer sells pencils and erasers in packages. A package of 5 erasers and 2 pencils costs 23¢. A package of 7 erasers and 5 pencils costs 41¢. Find the price of 1 eraser and 1 pencil.

33. Consumerism A store has a sale on almonds, pecans, and pistachios. One lb of almonds, 1 lb of pecans, and 1 lb of pistachios cost $12. Two lb of almonds and 3 lb of pecans cost $16. Three lb of pecans and 2 lb of pistachios cost $24. Find the price of a pound of each type of nut.

Mixed Review

34. Solve the system by addition.

$$\begin{aligned} 2x + 3y &= 11 \\ 5x - 6y &= -40 \end{aligned}$$

35. Solve the system by substitution.

$$\begin{aligned} 2x - y &= 7 \\ 4x + 3y &= 9 \end{aligned}$$

36. Solve the system by graphing.

$$\begin{aligned} 3x - 2y &= 2 \\ x + 4y &= 10 \end{aligned}$$

37. Solve the system by using the multiplicative inverse of a matrix.

$$\begin{aligned} x + 2y &= 2 \\ 2x - 7y &= 37 \end{aligned}$$

5.8

Problem Solving: Systems of Equations

A mathematical model that results in a system of linear equations gives you many options for finding a solution. You may choose to solve the system by using the algebraic method of adding equations to eliminate a variable, or you may choose to use an augmented matrix, the multiplicative inverse of a matrix, or determinants and Cramer's rule. The tools you have available, such as a calculator or a computer, may also influence your choice.

EXAMPLE 1 A ceramic tile designer has some odd lots of tiles in boxes. A particular box contains square, pentagonal, and hexagonal tiles. Altogether there are 100 tiles, which have a total of 510 sides. There are 20 more pentagonal tiles than hexagonal tiles in the box. Find the number of each type of tile.

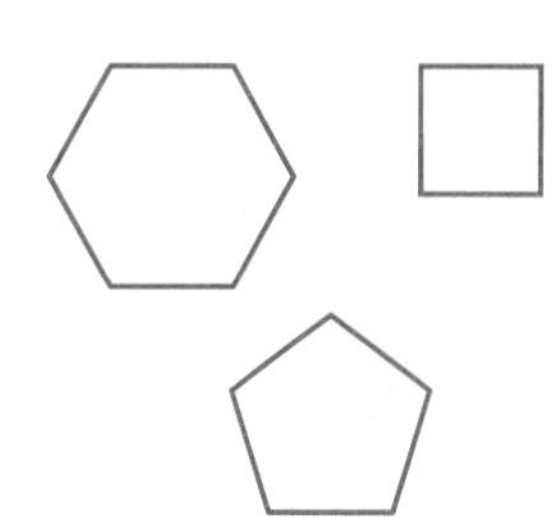

Understand the Problem You are asked to find the number of tiles of each shape.

Plan Your Approach Let x represent the number of square tiles.

Then the total number of sides of the square tiles is $4x$.

Let y represent the number of pentagonal tiles.

Then the total number of sides of the pentagonal tiles is $5y$.

Let z represent the number of hexagonal tiles.

Then the total number of sides of the hexagonal tiles is $6z$.

Write a system of equations.

$$\begin{cases} x + y + z = 100 \\ 4x + 5y + 6z = 510 \\ y = z + 20 \end{cases}$$

The total number of tiles is 100.
The total number of sides is 510.
There are 20 more pentagonal tiles than hexagonal tiles.

Complete the Work

$$\begin{cases} x + y + z = 100 \\ 4x + 5y + 6z = 510 \\ y - z = 20 \end{cases}$$

Rewrite the third equation with the variables on the same side.

Use any method to solve the system of equations. The use of the inverse of a matrix is illustrated on the next page.

$$\begin{array}{ccc} A & X & = & C \end{array}$$

$$\begin{bmatrix} 1 & 1 & 1 \\ 4 & 5 & 6 \\ 0 & 1 & -1 \end{bmatrix} \begin{bmatrix} x \\ y \\ z \end{bmatrix} = \begin{bmatrix} 100 \\ 510 \\ 20 \end{bmatrix}$$

Write the matrix equation.

$$\left[\begin{array}{ccc|ccc} 1 & 1 & 1 & 1 & 0 & 0 \\ 4 & 5 & 6 & 0 & 1 & 0 \\ 0 & 1 & -1 & 0 & 0 & 1 \end{array}\right]$$

To find A^{-1}, write the augmented matrix for the matrix of coefficients.

$$\left[\begin{array}{ccc|ccc} 1 & 0 & 0 & \frac{11}{3} & -\frac{2}{3} & -\frac{1}{3} \\ 0 & 1 & 0 & -\frac{4}{3} & \frac{1}{3} & \frac{2}{3} \\ 0 & 0 & 1 & -\frac{4}{3} & \frac{1}{3} & -\frac{1}{3} \end{array}\right]$$

Use row operations.

$$A^{-1} = \begin{bmatrix} \frac{11}{3} & -\frac{2}{3} & -\frac{1}{3} \\ -\frac{4}{3} & \frac{1}{3} & \frac{2}{3} \\ -\frac{4}{3} & \frac{1}{3} & -\frac{1}{3} \end{bmatrix} = \frac{1}{3}\begin{bmatrix} 11 & -2 & -1 \\ -4 & 1 & 2 \\ -4 & 1 & -1 \end{bmatrix}$$

$$X = \frac{1}{3}\begin{bmatrix} 11 & -2 & -1 \\ -4 & 1 & 2 \\ -4 & 1 & -1 \end{bmatrix}\begin{bmatrix} 100 \\ 510 \\ 20 \end{bmatrix} = \frac{1}{3}\begin{bmatrix} 60 \\ 150 \\ 90 \end{bmatrix} = \begin{bmatrix} 20 \\ 50 \\ 30 \end{bmatrix}$$

$X = A^{-1}C$

Interpret the Results There are 20 square, 50 pentagonal, and 30 hexagonal tiles in the box.

Is the total number of tiles 100?

$$\begin{aligned} x + y + z &= 100 \\ 20 + 50 + 30 &\stackrel{?}{=} 100 \\ 100 &= 100 \checkmark \end{aligned}$$

Is the total number of sides 510?

$$\begin{aligned} 4x + 5y + 6z &= 510 \\ 4(20) + 5(50) + 6(30) &\stackrel{?}{=} 510 \\ 510 &= 510 \checkmark \end{aligned}$$

Is the number of pentagonal tiles 20 more than the number of hexagonal tiles?

$$\begin{aligned} y &= z + 20 \\ 50 &\stackrel{?}{=} 30 + 20 \\ 50 &= 50 \checkmark \end{aligned}$$

The next example involves finding the amounts of money invested in several accounts at different simple interest rates.

EXAMPLE 2 Caryn earned $5000 working part time as a cashier in a department store. She put her earnings into three different bank accounts. She put some of the money in a checking account paying 5% annual interest, some in a savings account paying 6%, and the rest in a 1-year certificate of deposit paying 7%. The accounts earned simple interest. No additional money was deposited into or withdrawn from the accounts during the year, and the total interest at the end of the year was $325. If she invested 3 times as much at 6% as at 5%, find the amount she invested at each rate.

Understand the Problem

Simple interest is found by multiplying the principal, the annual rate in decimal form, and the time in years. The formula is

$$I = prt$$

Plan Your Approach

Let x represent the number of dollars invested at 5%.
Then the interest on x dollars for 1 year at $0.05x$.
Let y represent the number of dollars invested at 6%.
Then the interest on y dollars for 1 year was $0.06y$.
Let z represent the number of dollars invested at 7%.
Then the interest on z dollars for 1 year was $0.07z$.

$$\begin{cases} x + y + z = 5000 \\ 0.05x + 0.06y + 0.07z = 325 \\ y = 3x \end{cases}$$

Write a system of equations.

Complete the Work

$$\begin{aligned} x + y + z &= 5000 \\ 5x + 6y + 7z &= 32{,}500 \\ -3x + y + &= 0 \end{aligned}$$

Multiply each side of the second equation by 100.

Rewrite the third equation.

Use any method to solve the system of equations. The use of an augmented matrix is shown below.

$$\left[\begin{array}{rrr|r} 1 & 1 & 1 & 5000 \\ 5 & 6 & 7 & 32{,}500 \\ -3 & 1 & 0 & 0 \end{array}\right]$$

Write the augmented matrix.

$$\left[\begin{array}{rrr|r} 1 & 0 & 0 & 500 \\ 0 & 1 & 0 & 1500 \\ 0 & 0 & 1 & 3000 \end{array}\right]$$

Use row operations to transform the augmented matrix.

Read the values of the variables from the transformed matrix.

$$x = 500 \qquad y = 1500 \qquad z = 3000$$

Interpret the Results

Caryn invested $500 at 5%, $1500 at 6%, and $3000 at 7%.

Is the total amount invested $5000?

$$\begin{aligned} x + y + z &= 5000 \\ 500 + 1500 + 3000 &\stackrel{?}{=} 5000 \\ 5000 &= 5000 \ \checkmark \end{aligned}$$

Is the total interest \$325?

$$
\begin{aligned}
0.05x + 0.06y + 0.07z &= 325 \\
0.05(500) + 0.06(1500) + 0.07(3000) &\stackrel{?}{=} 325 \\
325 &= 325 \checkmark
\end{aligned}
$$

Is the amount invested at 6% three times the amount at 5%?

$$
\begin{aligned}
y &= 3x \\
1500 &\stackrel{?}{=} 3(500) \\
1500 &= 1500 \checkmark
\end{aligned}
$$

CLASS EXERCISES

Use a variable to represent each unknown quantity and write a system of equations that can be used to solve the problem. Do not solve.

1. A coin bank holds nickels, dimes, and quarters. There are 45 coins in the bank and the value of the coins is \$4.75. If there are 5 more nickels than quarters, find the number of each type of coin in the bank.
2. John invested \$6500 in three different mutual funds for 1 year. He earned a total of \$560 in simple interest on the three investments. The first fund paid 5% interest, the second paid 8% interest, and the third paid 10% interest. If the sum of the first two investments was \$500 less than the amount of the third investment, find the amount he invested at each rate.
3. The sum of the digits of a three-digit number is 12. Five times the units digit plus 6 times the tens digit is 28. If 2 times the tens digit is subtracted from 3 times the hundreds digit, the result is 15. Find the number.
4. A rectangle $ABCD$ is drawn with one diagonal from A to C. The perimeter of triangle ABC is 24 in. and the perimeter of the rectangle is 28 in. If the length of the diagonal of the rectangle is 2 in. greater than the length of the longer side, find the dimensions of the rectangle and the length of the diagonal.

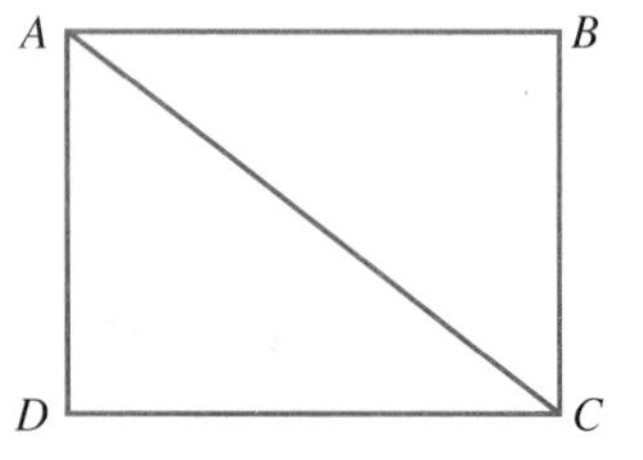

PRACTICE EXERCISES

1. Derrick invested \$10,000 for one year in three different investments. The investments paid simple annual interest of 5%, 6%, and 7%, respectively, and he received a total of \$610 in interest for the year. He invested \$3000 more at 5% than at 6%. Find the amount invested at each rate.

2. Carmel put her \$5000 earnings into three different accounts. The accounts paid simple annual interest of 8%, 10%, and 7%, respectively. The total interest at the end of one year was \$405. If Carmel invested \$500 more at 10% than at 8%, find the amount she invested in each account.

3. A box of wood pieces contains wood cut into triangular, square, and pentagonal shapes. There are 80 pieces of wood in the box, and the pieces have a total of 290 sides. If there are 10 more triangular pieces than square pieces, find the number of pieces of wood of each shape in the box.

4. A manufacturer of educational supplies makes plastic shapes to hang in classrooms. A package of shapes contains squares, pentagons, and hexagons. There are 24 shapes in the package with a total of 112 sides. If there are 10 more squares than pentagons, find the number of each shape in the package.

5. A change machine contains nickels, dimes, and quarters. There are 75 coins in the machine, and the value of the coins is \$7.25. If there are 5 times as many nickels as dimes, find the number of coins of each type in the machine.

6. A bin in a nut store contains 100 lb of a mixture of almonds, peanuts, and raisins. Almonds sell for \$1.89/lb, peanuts for \$1.58/lb, and raisins for \$1.39/lb. If the mixture contains twice as many pounds of peanuts as almonds, and if the total value of the almonds and the raisins in the mixture is \$93.40, how many pounds of each item does the mixture contain?

7. The sum of the digits of a three-digit number is 18. Three times the tens digit minus 5 times the units digit is 17. If 4 times the units digit is added to twice the hundreds digit, the result is 22. Find the three-digit number.

8. The sum of the digits of a three-digit number is 17. Four times the hundreds digit minus 5 times the tens digit is 12. If 7 times the units digit is added to 3 times the tens digit, the result is 47. Find the number.

9. A rectangular box is twice as long as it is wide and twice as wide as it is high. The sum of its length, width, and height is 35 in. What are the dimensions of the box?

10. The length of a rectangular shed is twice its height, and the height of the shed is 1 ft greater than its width. If the base of the shed has a perimeter of 40 ft, find the dimensions of the shed.

11. A merchant keeps an account of his inventory in matrix form. The first matrix shows the inventory of flour, measured in pounds, as of Monday night. The second matrix shows Tuesday's sales, measured in pounds. The value of the inventory at store A after Tuesday's sales is \$8. The value of the inventory at store B after Tuesday's sales is \$17. The value of the inventory at store C after Tuesday's sales is \$19. Find the price of a pound of each type of flour: wheat, corn, and oat.

Inventory—Monday Night **Tuesday's Sales**

$$\begin{array}{c} \\ A \\ B \\ C \end{array}\begin{array}{c} \begin{array}{ccc} W & C & O \end{array} \\ \begin{bmatrix} 4 & 3 & 6 \\ 5 & 8 & 5 \\ 8 & 5 & 4 \end{bmatrix} \end{array} \qquad \begin{array}{c} \begin{array}{ccc} W & C & O \end{array} \\ \begin{bmatrix} 1 & 1 & 5 \\ 0 & 4 & 2 \\ 3 & 1 & 0 \end{bmatrix} \end{array}$$

12. Jana, Natalie, and Donna work after school and weekends for a coat manufacturer. They get paid a different rate for afternoons, evenings, and weekends. The number of hours they worked during one week is given in matrix form.

$$\begin{array}{c} \\ \text{Jana} \\ \text{Natalie} \\ \text{Donna} \end{array}\begin{array}{c} \begin{array}{ccc} \text{Afternoons} & \text{Evenings} & \text{Weekends} \end{array} \\ \begin{bmatrix} 5 & 2 & 3 \\ 1 & 2 & 6 \\ 2 & 2 & 3 \end{bmatrix} \end{array}$$

If Jana had worked twice the number of hours for the week, her salary would have been \$98. If Natalie had worked 2 more hours in the evening, her salary would have been \$62. If Donna had worked 1 more hour on the weekend, her salary would have been \$43. Find the rate of pay for the afternoon, evening, and weekend.

13. A store sells three items for the regular price on Monday. The total of the regular prices of the three items is \$48. On Tuesday, the store sells item A for $\frac{1}{2}$ the regular price, item B for the regular price, and item C for $\frac{1}{3}$ the regular price. The cost of buying the three items is \$31. On Wednesday, the store sells item A for $\frac{1}{2}$ the regular price, item B for $\frac{1}{4}$ the regular price, and item C for the regular price. The cost of buying the three items on Wednesday is \$28. Find the regular price of each item.

14. During the day, a truck driver encounters light, moderate, and heavy traffic. If he travels $\frac{1}{3}$ hour in light traffic, $\frac{1}{2}$ hour in moderate traffic, and $\frac{1}{5}$ hour in heavy traffic, he can cover 21 mi. If he travels $\frac{1}{5}$ hour in light, $\frac{1}{4}$ hour in moderate, and 1 h in heavy traffic, he can cover 16 mi. The sum of his speeds in moderate and heavy traffic is 5 mi/h less than his speed in light traffic. Find his speed in each type of traffic.

15. Mr. & Mrs. Koch put \$6000 into three different investments paying 9.5%, 8.5%, and 6.25%, respectively. The total simple annual interest at the end of 2 years was \$970. The average of the amounts invested at 9.5% and 8.5% was equal to the amount invested at 6.25%. How much was invested at each rate?

16. $ABCD$ is an isosceles trapezoid. Segments BF and CE are both perpendicular to base AD. The ratio of AF to AD is 1:4. The perimeter of triangle ABF is 24 cm and the perimeter of trapezoid $ABCE$ is 48 cm. If the perimeter of rectangle $BCEF$ is 40 cm, find the perimeter of trapezoid $ABCD$.

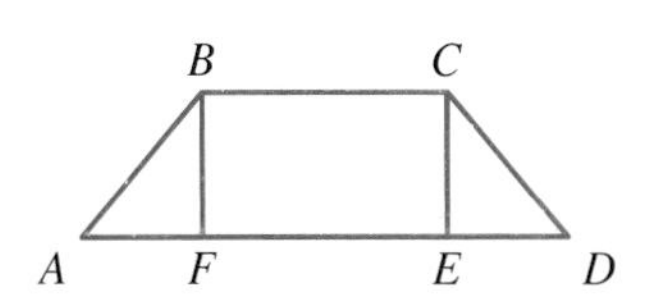

TEST YOURSELF

Solve each system of equations using the augmented matrix method. 5.5

1. $\begin{cases} 2x - 3y - 2z = 6 \\ x + 2y + 3z = 8 \\ 3x - y - 2z = 5 \end{cases}$

2. $\begin{cases} x + z = 6 \\ y + z = 0 \\ x + 7y = -2 \end{cases}$

Solve each system of equations using the inverse of a matrix. 5.6

3. $\begin{cases} -3x + 4y = 7 \\ x - y = -1 \end{cases}$

4. $\begin{cases} x + 3y = 7 \\ 2x + 5y = 11 \end{cases}$

Evaluate. 5.7

5. $\begin{vmatrix} 6 & -4 \\ 2 & 3 \end{vmatrix}$

6. $\begin{vmatrix} 2 & 3 & -1 \\ 4 & 1 & 1 \\ 6 & -1 & 2 \end{vmatrix}$

7. $\begin{vmatrix} 7 & 2 & 6 \\ 3 & 4 & 0 \\ 1 & 5 & 8 \end{vmatrix}$

8. Solve the system using Cramer's rule:

$$\begin{cases} 2x + 3y = -9 \\ 3x - 5y = 34 \end{cases}$$

9. The sum of the digits of a three-digit number is 12. Four times the hundreds digit minus 3 times the units digit is 6. Six times the units digit plus 3 times the tens digit is 33. Find the number. 5.8

10. A certain coin bank accepts pennies, nickels, and dimes. There are 156 coins in the bank, and the value of the coins is \$6.93. If there are 12 more dimes than nickels, find the number of coins of each type in the bank.

INTEGRATING ALGEBRA
The Airline Problem

One advantage of arranging data in rectangular arrays such as matrices, tables, and charts, is that patterns in the data become easier to discern. Matrices and matrix operations can be used to help solve the so-called *Airline Problem*, namely how to minimize flight time and ticket cost for passengers, while maximizing flight availability. An example to illustrate the problem appears below.

EXAMPLE A salesperson plans a business trip from New York to San Francisco, with stops in Toronto, Chicago, and Denver, in any order. What is the least expensive route? A table of city-to-city ticket prices (in dollars) is given below.

	N	*T*	*C*	*D*	*S*
N	0	100	120	400	600
T	100	0	80	250	500
C	120	80	0	200	400
D	400	250	200	0	150
S	600	500	400	150	0

Construct a flow diagram illustrating all possible routes. Each route will consist of 4 flights. There are 6 possible routes

$$NTCDS,\ NTDCS,\ NCDTS,\ NCTDS,\ NDTCS,\ NDCTS.$$

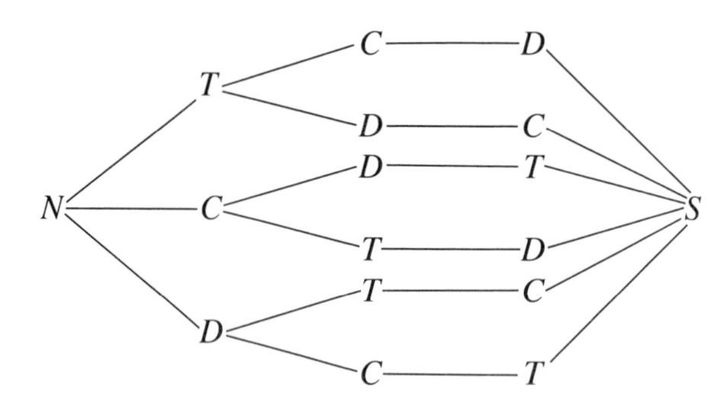

Use the flow diagram and the ticket price table to make a route-cost chart.

	I		II		III		IV	
N	100	T	80	C	200	D	150	S
N	100	T	250	D	200	C	400	S
N	120	C	200	D	250	T	500	S
N	120	C	80	T	250	D	150	S
N	400	D	250	T	80	C	400	S
N	400	D	200	C	80	T	500	S

It would be easy enough to simply add each of the six rows to find the cheapest cost, but it would be time consuming and tedious, particularly if such a business trip included many more cities. There is an easier way.

Note that in column I and column III each price repeats immediately. Columns II and IV repeat their first three entries as their last three. Without these repetitions, two 3×1 matrices can be formed in the following way

$$\text{Let } A = \begin{bmatrix} 100 + 200 \\ 120 + 250 \\ 400 + 80 \end{bmatrix} = \begin{bmatrix} 300 \\ 370 \\ 480 \end{bmatrix} \quad \text{and} \quad \text{let } B = \begin{bmatrix} 80 + 150 \\ 250 + 400 \\ 200 + 500 \end{bmatrix} = \begin{bmatrix} 230 \\ 650 \\ 700 \end{bmatrix}$$

(In A, the first addends are from column I and the second from column III; in B, the first addends are from column II and the second from column IV.)

Now construct two 6×1 matrices, A' and B', by reincorporating the repetitions in the route-cost chart. $A' + B'$ will yield the costs of each route.

$$A' = \begin{bmatrix} 300 \\ 300 \\ 370 \\ 370 \\ 480 \\ 480 \end{bmatrix}, \quad B' = \begin{bmatrix} 230 \\ 650 \\ 700 \\ 230 \\ 650 \\ 700 \end{bmatrix}, \quad \text{and} \quad \text{cost} = A' + B' = \begin{bmatrix} 530 \\ 950 \\ 1070 \\ 600 \\ 1130 \\ 1180 \end{bmatrix} \begin{matrix} NTCDS \\ NTDCS \\ NCDTS \\ NCTDS \\ NDTCS \\ NDCTS \end{matrix}$$

The *NTCDS* route, costing \$530, is the least expensive.

Note in the flow diagram that the order of the cities in the column after New York (*TCD*) is in the opposite order of the cities in the column before San Francisco (*DCT* twice). Observing this procedure will simplify the arrangement of the ticket prices into matrices.

EXERCISES

1. What would be the least expensive route from San Francisco to New York with stops in each of the other three cities? How much would it cost?
2. What would be the least expensive route from Chicago to Denver with stops in each of the three other cities? How much would it cost?

CHAPTER 5 SUMMARY AND REVIEW

Vocabulary

augmented matrix (222)
column matrix (197)
Cramer's rule (236)
determinant (233)
directed graph (211)
element of a matrix (197)
equal matrices (197)
matrix (197)
matrix equation (200)
minors (234)
ordered triple (216)
row matrix (197)
row operations (222)
scalar (199)
scalar product (199)
square matrix (197)
zero matrix (198)

Matrix Addition To add matrices with the same dimensions, add corresponding elements. To subtract matrices with the same dimensions, add the first matrix to the additive inverse of the second matrix. To multiply a matrix by a scalar, multiply each element in the matrix by the scalar. 5.1

For matrices A, B, C, and D, perform the indicated operations if they are defined. If an operation is not defined, label it "undefined."

$$A = \begin{bmatrix} 1 & 4 \\ 3 & 0 \\ -2 & 6 \end{bmatrix} \quad B = \begin{bmatrix} -2 & 1 \\ 3 & -2 \\ 1 & 0 \end{bmatrix} \quad C = \begin{bmatrix} 7 & 3 \\ 5 & 2 \end{bmatrix} \quad D = \begin{bmatrix} 4 & -3 \\ 3 & -2 \end{bmatrix}$$

1. $A + B$ **2.** $A - B$ **3.** $A + C$

4. $3A$ **5.** $5B$ **6.** $-2C + D$

Matrix Multiplication The product $A_{m \times p}B_{p \times n}$ is a matrix with dimensions $m \times n$ whose element in the ith row and jth column is the sum of the products of the elements of the ith row of A and the jth column of B. 5.2

Using the matrices for Exercises 1–6, perform the indicated operations if they are defined. If an operation is not defined, label it "undefined."

7. CD **8.** AC **9.** DC

10. C^2 **11.** A^2 **12.** $2CD - 4AC$

13. This directed graph shows the ways three people can communicate in one step using an intercom system. Write a matrix to show the information in the graph. Then, write a matrix to show the numbers of ways the people can communicate in two steps. 5.3

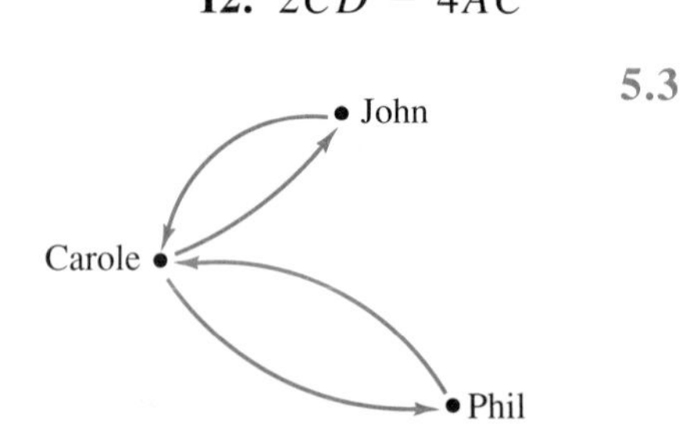

Systems of Equations in Three Variables To solve such a linear system by addition, eliminate one variable from two equations, eliminate the same variable from a different pair of equations, and solve the resulting system of two equations. 5.4

Solve and check.

14. $\begin{cases} x + 2y + z = 8 \\ y - 2z = 1 \\ 3x + 3z = 6 \end{cases}$

15. $\begin{cases} x + 4y - 2z = 7 \\ 5y + 3z = 10 \\ 3x + 2y = 1 \end{cases}$

Augmented Matrix Solutions A system of linear equations in three variables may be written as a matrix equation $AX = C$, where A is the matrix of coefficients, X is a column matrix of variables, and C is a column matrix of constants. To solve the system, write the augmented matrix $[A|C]$. Use row transformations to find an equivalent matrix of the form at the right, where $x = a$, $y = b$, and $z = c$. 5.5

$$\left[\begin{array}{ccc|c} 1 & 0 & 0 & a \\ 0 & 1 & 0 & b \\ 0 & 0 & 1 & c \end{array}\right]$$

Solve using the augmented matrix method.

16. $\begin{cases} x - 2y + 3z = 10 \\ 2x + y = 9 \\ 7y + z = -6 \end{cases}$

17. $\begin{cases} x + 3y - 4z = 11 \\ y - 5z = -4 \\ x + 7z = 15 \end{cases}$

Inverse of a Matrix To solve $AX = C$, first find A^{-1}. Then, $X = A^{-1}C$. 5.6

Solve using the inverse of a matrix.

18. $\begin{cases} 3x + 4y = -9 \\ 2x + 3y = -8 \end{cases}$

19. $\begin{cases} x + 3z = 7 \\ y - 3z = -8 \\ 2x + 2y - z = -4 \end{cases}$

Determinants and Cramer's Rule For a system $\begin{cases} a_1x + b_1y = c_1 \\ a_2x + b_2y = c_2 \end{cases}$ 5.7

$$x = \frac{D_x}{D} \text{ and } y = \frac{D_y}{D}\ (D \neq 0), \text{ where } D = \begin{vmatrix} a_1 & b_1 \\ a_2 & b_2 \end{vmatrix}, D_x = \begin{vmatrix} c_1 & b_1 \\ c_2 & b_2 \end{vmatrix}, D_y = \begin{vmatrix} a_1 & c_1 \\ a_2 & c_2 \end{vmatrix}$$

Solve using Cramer's rule.

20. $\begin{cases} 3x + 2y = -2 \\ 4x - 3y = 20 \end{cases}$

21. $\begin{cases} x + y + z = 2 \\ 2x + y = 7 \\ 3y + 4z = -5 \end{cases}$

22. The sum of the digits of a three-digit number is 16. The sum of the tens and units digits is 2 more than the hundreds digit. Three times the hundreds digit plus 4 times the units digit is 25. Find the number. 5.8

CHAPTER 5 TEST

For matrices A, B, C, and D below, perform the indicated operations if they are defined. If an operation is not defined, label it "undefined."

$$A = \begin{bmatrix} 2 & 1 & 0 \\ 4 & 3 & -1 \end{bmatrix} \quad B = \begin{bmatrix} 5 & 1 & 3 \\ 0 & -1 & 2 \end{bmatrix} \quad C = \begin{bmatrix} 4 & 5 \\ 1 & 1 \end{bmatrix} \quad D = \begin{bmatrix} 1 & 2 \\ 3 & 5 \end{bmatrix}$$

1. $C - D$ **2.** $2A + B$ **3.** CA **4.** AB

5. The graph at the right shows the number of ways three people can communicate with each other in one step. Write a matrix to show the information in the graph. Then, write a matrix showing the numbers of ways the three people can communicate with each other in two steps.

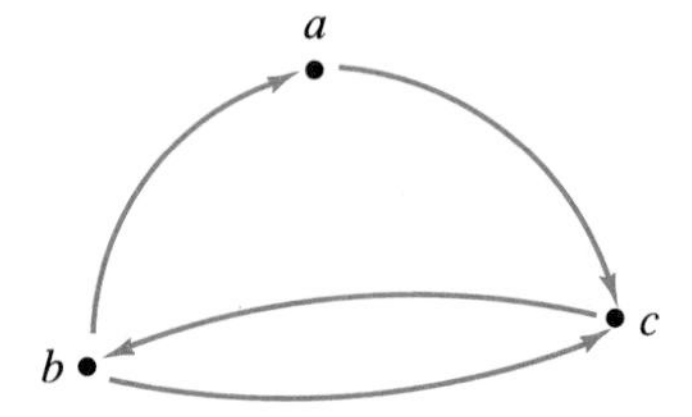

6. Solve and check: $\begin{cases} x + y + 2z = 3 \\ y - 4z = 7 \\ 3x + 2y = 12 \end{cases}$

7. Solve using an augmented matrix: $\begin{cases} 3x + y = 13 \\ x - 5z = 4 \\ x + y + z = 5 \end{cases}$

8. Solve using the inverse of a matrix: $\begin{cases} 2x + 5y = 7 \\ 3x + 7y = 11 \end{cases}$

9. Solve using Cramer's rule: $\begin{cases} 5x - 2y = -24 \\ 3x + 5y = -2 \end{cases}$

10. There are nickels, dimes, and quarters in a change machine. In all, there are 70 coins with a total value of $10.85. If there are 4 times as many nickels as dimes, find the number of each type of coin in the machine.

Challenge

The sum of the digits of a three-digit number is 18. If $\frac{1}{2}$ the hundreds digit is added to $\frac{1}{3}$ the units digit, the result is 2 more than the tens digit. The average of the hundreds and the units digits is 1 more than the average of the three digits. Find the number.

COLLEGE ENTRANCE EXAM REVIEW

Solve. Grid in your response on the answer sheet.

1. Evaluate: $\begin{vmatrix} 5 & 1 \\ -2 & -1 \end{vmatrix}$

2. If f and g are inverse functions and $f(x) = 2x - 6$, then find $g(-2)$.

3. If $(5^3)(2^5) = 4(10^k)$, then what is the value of k?

4. Evaluate: $\begin{vmatrix} 1 & 0 & -2 \\ 3 & 1 & 4 \\ 0 & -1 & 2 \end{vmatrix}$

5. When using Cramer's Rule to solve the system

$$\begin{cases} 3x - 7y = 11 \\ 5x + 2y = -8 \end{cases}$$

what is the value of the determinant that would be used as denominator of both x and y?

6. What is the value of x in the solution of the system of equations

$$\begin{cases} x + y - z = 3 \\ y + z = 1 \\ z = 4? \end{cases}$$

7. Starting at X, Greg traveled $\frac{2}{5}$ of the distance from X to Y by car, then traveled 15 mi by bus, and then walked 3 mi to reach Y. For what fractional part of his trip from X to Y was Greg traveling by car or by bus?

8. Find the positive value for x in the equation

$$\frac{x - 1}{x - 2} + \frac{x}{3} = \frac{3x + 2}{3x - 6} - \frac{1}{2 - x}.$$

9. Rachael has three times as many quarters as dimes. If the sum of the number of dimes and twice the number of quarters is 21, how many quarters does she have?

10. A total of \$24 is to be divided among three girls in amounts proportional to their ages. If the girls' ages are 9, 12, and 15 years, how much money in dollars will the youngest girl receive?

11. Properties of the whole number n:

    I. n is a multiple of 6.
    II. n is between 30 and 70.
    III. $n - 1$ and $n + 1$ are prime numbers.

    What is the value of n?

12. The perimeter of a triangle is 90 cm. The measure of the longest side is 6 less than the sum of the measures of the other two sides. Four times the measure of the shortest side equals the sum of the measures of the other two sides. What is the dimension in cm of the shortest side of the triangle?

MIXED REVIEW

Simplify.

Example 3^4 $\quad 3^4$ means $\overbrace{3 \times 3 \times 3 \times 3}^{\text{4 factors}}$

$3^4 = 81$

1. 2^4 **2.** 4^3 **3.** 7^2 **4.** 5^3 **5.** 10^4 **6.** 3^1

Simplify.

Example $3y + 2x - 9 - 4 - 3x + 12$

$(3y - y) + (2x - 3x) + (-9 + 12)$ *Group like terms together.*

$2y + (-x) + 3$ *Combine like terms.*

$2y - x + 3$

7. $a - b + 2b - 3a + 7$

8. $5n + 3 - 2m - 4 + 3m - 5n$

9. $3s - 5 + 2t - (s + 3t) + 8$

10. $12 - (2p - 3q) + 5p - 2q + 6$

11. $5c - 3d + 2c - (8 + 3c)$

12. $9w - 2u + 7 - (3u + 2) + 5u$

Use the distributive property to rewrite each expression.

Example $12y + 4x - 9$

$4 \cdot 3y + 4 \cdot x - 9$ *Look for a common factor.*

$4(3y + x) - 9$ *Use the distributive property.*

13. $15a - 9b + 21$

14. $6m + 24n - 30$

15. $28r - 35s + 42$

16. $9x - 12y + 16$

17. $5p + 24q - 6$

18. $8c + 6d - 32$

Solve.

19. The area of a rectangular rug is 84 ft^2. Find its width if the length is 12 ft. $(A = lw)$

20. The sum of two consecutive integers is 25. Find the integers.

21. The sum of three consecutive odd integers is 63. Find the integers.

22. The base of a triangular banner is 30 cm. If the area of the banner is 300 cm^2, what is the height? $\left(A = \frac{1}{2}bh\right)$

6 Polynomials

The magic carpet?

Developing Mathematical Power

The Charles Carpet Company wants to offer customers custom-designed and color-bordered rugs. Each Charles Carpet store will have a special computer. A sales representative will keyboard the length (l) and width (w) of the rug, the width of the border (b), and the pattern and color codes for the center section and border. The computer will display the rug design and compute the cost based on the number of square feet in each part as well as the number of linear feet of seaming needed to join the two parts.

Project

You are part of the team that is writing the computer program for the custom-designed rugs. Prepare a list of all the questions the customer will have to answer and all computations the computer will have to perform.

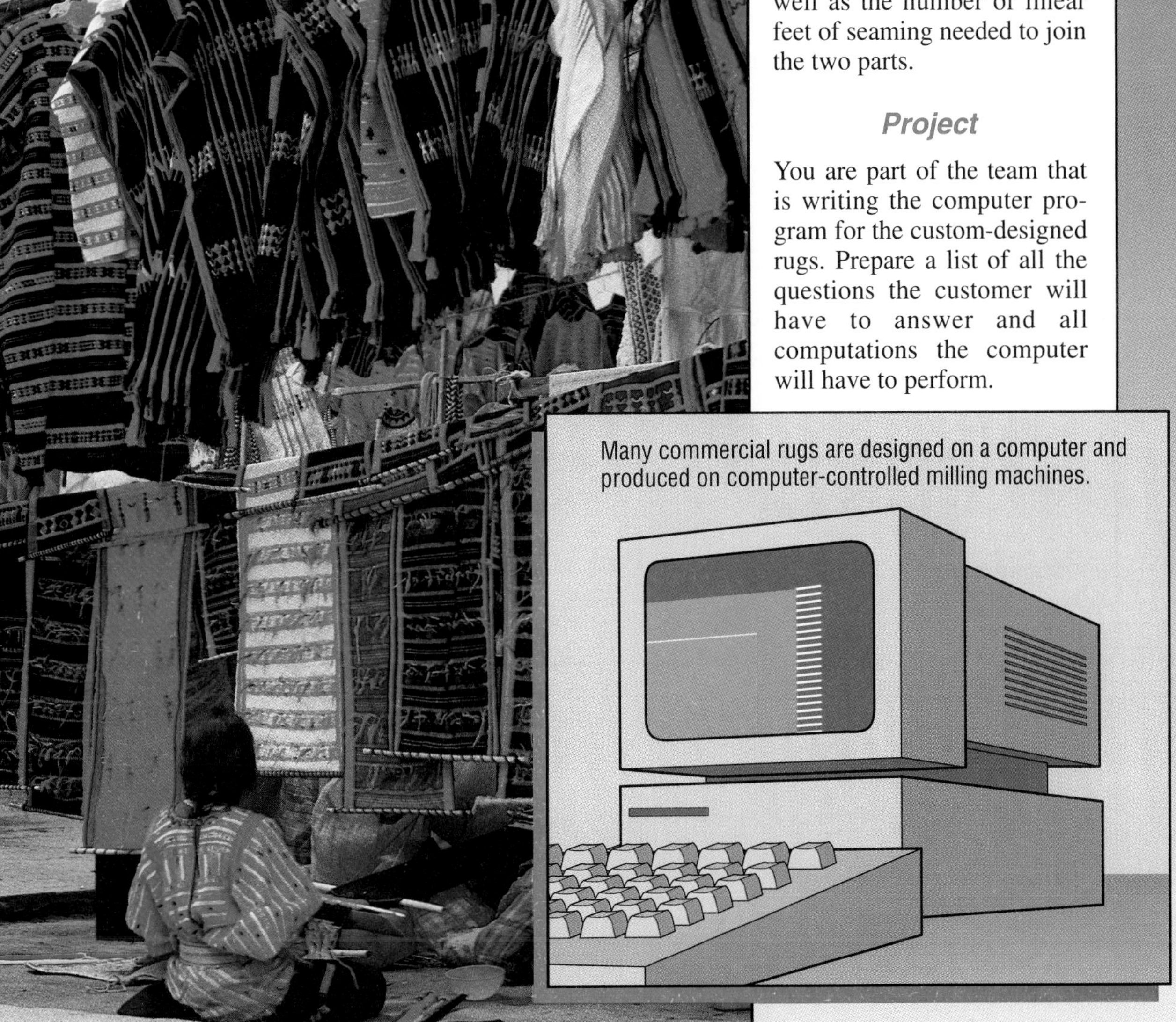

Many commercial rugs are designed on a computer and produced on computer-controlled milling machines.

6.1 Exponents and Monomials

Objectives: To review the properties of exponents
To define monomials

Mathematical notation can often be used to simplify English sentences. For example, the following sentence can be written more precisely using mathematical notation. Recall that an exponent shows repeated use of the same factor.

The area of a square launching pad is equal to the length of a side, multiplied by itself.

$$A = s^2$$

Capsule Review

In the expression s^2, s is the *base,* 2 is the *exponent.*

s^2 means $\underbrace{s \cdot s}_{\text{2 factors}}$ The base, s, is used 2 times as a factor.

$(-4x)^3$ means $\underbrace{(-4x)(-4x)(-4x)}_{\text{3 factors}}$ The base, $-4x$, is used 3 times as a factor.

Name the base and the exponent.

1. x^4 **2.** $(5x)^3$ **3.** $(-7a)^5$ **4.** $(-y)^2$ **5.** $(2x)^4$ **6.** $4x^3$

Exponent For any real number r and any positive integer n, r^n means that r is used as a factor n times.

The definition of exponent can be used to simplify many expressions.

$$x^3 \cdot x^4 = \underbrace{x \cdot x \cdot x}_{\text{3 factors}} \cdot \underbrace{x \cdot x \cdot x \cdot x}_{\text{4 factors}} = x^7$$

The exponent 7 is the sum of the exponents 3 and 4.

$$\frac{x^5}{x^2} = \frac{x \cdot x \cdot x \cdot x \cdot x}{x \cdot x} = x \cdot x \cdot x = x^3$$

The exponent 3 is the difference between the exponents 5 and 2.

$$r^m \cdot r^n = r^{m+n}$$

where r is a real number and m and n are positive integers.

$$\frac{r^m}{r^n} = r^{m-n}$$

where r is a nonzero real number and m and n are positive integers, with $m > n$.

EXAMPLE 1 **Simplify:** **a.** $y^4 \cdot y^7$ **b.** $\dfrac{x^6}{x}$

a. $y^4 \cdot y^7 = y^{4+7}$
$= y^{11}$

b. $\dfrac{x^6}{x} = x^{6-1}$
$= x^5$

The definition of exponent and the first property shown above are used to establish another useful property.

$(x^2)^3 = x^2 \cdot x^2 \cdot x^2$
$= x^{2+2+2} = x^6$ *The exponent 6 is the product of the exponents 2 and 3.*

The general statement of the property is shown below.

$$(r^m)^n = r^{mn}$$

where r is a real number and m and n are positive integers.

Other types of expressions can also be simplified. Note the use of the associative and commutative properties in the example on the left below.

$$\begin{aligned}(2y)^3 &= (2y)(2y)(2y)\\ &= (2 \cdot 2 \cdot 2)(y \cdot y \cdot y)\\ &= 2^3y^3, \text{ or } 8y^3\end{aligned}$$

$$\begin{aligned}\left(\frac{x}{3}\right)^2 &= \frac{x}{3} \cdot \frac{x}{3}\\ &= \frac{x \cdot x}{3 \cdot 3}\\ &= \frac{x^2}{3^2}, \text{ or } \frac{x^2}{9}\end{aligned}$$

These examples suggest the following properties.

$$(rs)^m = r^m s^m$$

where r and s are real numbers and m is a positive integer.

$$\left(\frac{r}{s}\right)^m = \frac{r^m}{s^m}$$

where r and s are real numbers, s is not equal to zero, and m is a positive integer.

EXAMPLE 2 **Simplify:** **a.** $(y^3)^5$ **b.** $(-3x)^4$ **c.** $\left(\frac{t}{5}\right)^3$

a. $(y^3)^5 = y^{3\cdot5} = y^{15}$

b. $(-3x)^4 = (-3)^4x^4 = 81x^4$

c. $\left(\frac{t}{5}\right)^3 = \frac{t^3}{5^3} = \frac{t^3}{125}$

To *simplify* an expression like those in this lesson, write an equivalent expression without parentheses, and write the coefficient in simplest form.

EXAMPLE 3 **Simplify:** $\mathbf{(-3x^2a^3)(2x^3a)}$

$$\begin{aligned}(-3x^2a^3)(2x^3a) &= (-3 \cdot 2)(x^2 \cdot x^3)(a^3 \cdot a) \\ &= -6x^{2+3}a^{3+1} \\ &= -6x^5a^4\end{aligned}$$

Sometimes it is necessary to use more than one of the properties of exponents to simplify an expression.

EXAMPLE 4 **Simplify:** $\left(\frac{-3x^4}{2}\right)^3$

$$\begin{aligned}\left(\frac{-3x^4}{2}\right)^3 &= \frac{(-3x^4)^3}{2^3} \\ &= \frac{(-3)^3(x^4)^3}{8} \\ &= \frac{-27x^{12}}{8}\end{aligned}$$

Some expressions are *monomials*. A **monomial** is a real number, a variable, or a product of a real number and one or more variables. The expressions 13, x, $27y^2$, and $-18v^2w^3$ are monomials. The expressions $\sqrt{3x}$, 7^x, and $\frac{5x}{y}$ are *not* monomials. A monomial cannot have a variable under a radical sign, in an exponent, or in a denominator.

The **coefficient** or **numerical coefficient** of a monomial is the real-number factor, and the **degree** of a monomial is the sum of the exponents of its variables. The degree of a nonzero real number is zero. The *zero monomial,* 0, has no degree.

EXAMPLE 5 **State the degree and the coefficient of each monomial.**

Monomial	$-13x$	$0.5x^2$	23	$2a^3b^2$	xyz
Degree	1	2	0	5	3
Coefficient	-13	0.5	23	2	1

CLASS EXERCISES

Match each expression on the left with an equivalent expression on the right.

1. $2(x^5)^3$ **a.** $8x^{15}$

2. $\left(\frac{x^5}{2}\right)^3$ **b.** x^{10}

3. $(2x^5)^3$ **c.** $2x^8$

4. $\frac{2x^5}{x^3}$ **d.** $\frac{x^{15}}{8}$

5. $\left(\frac{2x}{5}\right)^3$ **e.** $\frac{8x^3}{125}$

6. $2x^5 \cdot x^3$ **f.** $2x^{15}$

7. $x^5(2x)^3$ **g.** $2x^2$

8. $x^2 \cdot x^3 \cdot x^5$ **h.** $8x^8$

Tell whether each expression is a monomial. If it is not a monomial, explain why it is not. For each monomial, give its degree and coefficient.

9. $7x^5$ **10.** $\frac{3}{x}$ **11.** $\frac{1}{2}a^4b^2$ **12.** πr^2

PRACTICE EXERCISES

Simplify.

1. $x^6 \cdot x^7$ **2.** $s^3 \cdot s^2$ **3.** $y^2(y^4)(y^5)$ **4.** $p(p^3)(p^5)$

5. $\frac{x^4}{x^3}$ **6.** $\frac{y^8}{y^3}$ **7.** $\frac{t^7}{t}$ **8.** $\frac{z^{10}}{z^6}$

9. $(m^5)^7$ **10.** $(n^6)^3$ **11.** $(x^2)^4$ **12.** $(y^3)^3$

13. $(3x)^4$ **14.** $(-2y)^3$ **15.** $(3ab)^2$ **16.** $(-pq)^6$

17. $\left(\frac{x}{4}\right)^2$ **18.** $\left(\frac{y}{2}\right)^3$ **19.** $\left(\frac{2t}{3}\right)^2$ **20.** $\left(\frac{3x^2}{2}\right)^4$

21. $-(x^4y^3)^2$ **22.** $(-x^4y^3)^2$ **23.** $(-6m^2n^2)(3mn)$ **24.** $(8s^2t^3)(3st^4)$

Give the degree and the coefficient of each monomial.

25. $3x^2$ **26.** 0.7 **27.** $-3abc$ **28.** $\sqrt{6}a^4b$

Simplify.

29. $\frac{1}{4}a^2 \cdot \frac{1}{2}ab$ **30.** $\frac{13}{2}s \cdot \frac{s}{2}$ **31.** $(a^5b^4)^3(a^2b)$ **32.** $(xy^2)^2(x^2y)^3$

33. $(0.4xy)^3(0.2x^2y^4z)$ **34.** $(-3c^2d)^3(d^2)^4$ **35.** $(-0.5a^2b^2)^3(8a^4b^5)$

36. $(ef)^2(e^3f^4)$

37. $(2xy)^2(3x)^3(-2z)^3$

38. $(-0.2mn^2p)^3(-8m^2p^3)^2$

39. $\left(\frac{2}{3}a^2b\right)\left(\frac{1}{4}a^3b\right)(3ab^2)$

40. $\left(\frac{2m^7n^3}{3}\right)^2$

41. $\left(\frac{3r^2s}{4}\right)^3$

42. $(-5xyz^3)^3(-2x^2yz)^3$

43. $(-6a^4b)^2(-3ab^2)^2$

44. $(-xy^4)(-2x^4y)^3$

Tell whether each expression is a monomial. If it is not a monomial, explain why it is not.

45. $\sqrt{2}xy^m$

46. $\sqrt{5xy^2}$

47. $\frac{x^2}{y^3}$

48. $-\sqrt{3}x^2y$

Simplify. Assume that all exponents are positive integers.

49. $[(x^2)^3]^4$

50. $[(y^3)^3]^3$

51. $(a^2)^k$

52. $x^ay^{3b}(x^{2a}y^b)$

53. $(s^at^bm^c)^n$

54. $2^{2m}(4^m)$

55. $(x^2)^n(x^n)^2$

56. $2(x^3)^m(2x^3)^m$

Applications

57. Geometry A triangle has a base of $6xy$ and a height equal to the square of the base. Express the area as a monomial.

58. Manufacturing Express as a monomial the volume of a box with a square base if the length of a side is $2ab$ and the height is $3b^2$.

Developing Mathematical Power

Number Theory A *perfect number* is a positive integer that is equal to the sum of its positive integral divisors, excluding itself. A *prime number* is an integer greater than 1 whose only positive integral factors are itself and 1. Euclid described an important fact about prime numbers and perfect numbers.

> If n is a prime number and $2^n - 1$ is a prime number, then $(2^n - 1)(2^{n-1})$ is a perfect number.

For example, if $n = 2$, then

$$\begin{aligned}(2^n - 1)(2^{n-1}) &= (2^2 - 1)(2^{2-1})\\ &= (4 - 1)(2) = 6\end{aligned}$$

Note that 6 is a perfect number because the sum of its divisors, excluding itself, is 6. That is, $1 + 2 + 3$ is equal to 6.

Evaluate $(2^n - 1)(2^{n-1})$ for the given value of n. Then show that the resulting number is a perfect number.

59. $n = 3$

60. $n = 5$

6.2

Polynomials: Classification, Addition, and Subtraction

Objective: To classify, add, and subtract polynomials

The practice you have had in adding monomials should make addition of other polynomials easier. To add and subtract polynomials, you must recall the difference between like and unlike terms.

Capsule Review

Like terms	Unlike terms
$3, -15, 7.2$	$8, x$
$a, 8a - \frac{2}{3}a$	$9x^2, \frac{3}{4}x^3$
$2xy^2, 5xy^2$	$3xy^2, -12x^2y$

Like terms, or *similar terms,* are monomials that contain the same variables with the same exponents. Like terms differ only in their coefficients.

Determine which of the following are pairs of like terms.

1. $24a, 24a^2$ **2.** $-0.5x, \frac{2}{3}x$ **3.** $-4, \frac{5}{2}$

4. $-3x^2y^3, 4x^3y^2$ **5.** $x^4y^2t, -11x^4y^2t$ **6.** $1.3x^5y, -3.7xy^5$

A **polynomial** is a monomial or a sum of monomials. Examples are:

$$4x^2 + 3y \qquad -5x^2 + 6x + 2 \qquad \frac{1}{2}xy \qquad 7a - 5b$$

Each monomial in a polynomial is a **term of the polynomial.** A polynomial of one term is a monomial. A polynomial of two terms is a **binomial,** and a polynomial of three terms is a **trinomial.** Polynomials of more than three terms are not usually given special names. Such a polynomial is called a polynomial of four terms, a polynomial of five terms, and so forth.

EXAMPLE 1 **Classify each polynomial by the number of terms.**

Polynomial	Classification
$3x + 4y$	Binomial
$8x^3 - 9y^4 - 7$	Trinomial
$7xy^3$	Monomial
$5y^2z + 10xy - 9xz^6 + 6$	Polynomial of four terms

You know that the degree of a monomial is the sum of the exponents of its variables. The **degree of a polynomial** is the degree of the term of highest degree.

EXAMPLE 2 **Give the degree of each polynomial.**

Polynomial	$4x$	$x^2 - xy^2$	$x^3y^3 + x^2y^5$	18
Degree	1	3	7	0

A polynomial is in *simplest form* when all like terms are combined. Any work with polynomials involving two or more terms can be made easier if the terms of each polynomial are arranged in descending or ascending order by the exponents of a single variable. These polynomials are in simplest form.

$4z^4 - 10z^3 + 8z^2 - 17z + 100$	Descending order of z
$6 - x + x^2 - 3x^3 + 5x^4$	Ascending order of x
$a^3 + 2a^2b + 3ab^2 + b^3$	Descending order of a; ascending order of b

EXAMPLE 3 **Simplify:** $\mathbf{4x^3 + 6x + 2x^2 + x^2}$

$$4x^3 + 6x + 2x^2 + x^2 = 4x^3 + 6x + (2 + 1)x^2 \quad \textit{Distributive property}$$
$$= 4x^3 + 6x + 3x^2$$
$$= 4x^3 + 3x^2 + 6x \quad \textit{Descending order of x}$$

In order to group like terms, it is sometimes necessary to apply both the commutative and associative properties.

EXAMPLE 4 **Simplify:** $\mathbf{3ab^2 - 5c^2d + 7 - ab^2 + 8c^2d}$

$$3ab^2 - 5c^2d + 7 - ab^2 + 8c^2d$$
$$= (3ab^2 - ab^2) + (8c^2d - 5c^2d) + 7 \quad \textit{Group like terms.}$$
$$= (3 - 1)ab^2 + (8 - 5)c^2d + 7 \quad \textit{Distributive property}$$
$$= 2ab^2 + 3c^2d + 7$$

The sum of two polynomials is a polynomial. To add polynomials, combine like terms and write the polynomial in simplest form.

EXAMPLE 5 **Add:** $\mathbf{(9 + 7x + 3x^2) + (5x^2 - 8x - 5)}$

$$(9 + 7x + 3x^2) + (5x^2 - 8x - 5)$$
$$= (3x^2 + 5x^2) + (7x - 8x) + (9 - 5) \quad \textit{Group like terms.}$$
$$= 8x^2 - x + 4$$

A vertical arrangement may be used to add polynomials. In a vertical arrangement, like terms are written in the same column.

$$\begin{array}{r} 3x^2 + 7x + 9 \\ + \; 5x^2 - 8x - 5 \\ \hline 8x^2 - x + 4 \end{array}$$

EXAMPLE 6 **Add: $(6 - 2x + 4x^2) + (7x^2 + 5x - 1)$. Use a calculator to evaluate each of the addends and the resulting sum for $x = 7$.**

$$(6 - 2x + 4x^2) + (7x^2 + 5x - 1)$$
$$= (4x^2 + 7x^2) + (-2x + 5x) + (6 - 1)$$
$$= 11x^2 + 3x + 5$$

For $x = 7$,

$6 - 2 \cdot 7 + 4 \cdot 7^2 = 188$ $\qquad$ $11 \cdot 7^2 + 3 \cdot 7 + 5 = 565$

$7 \cdot 7^2 + 5 \cdot 7 - 1 = 377$

$188 + 377 = 565$

Subtraction of polynomials is almost identical to addition of polynomials since subtraction is defined as addition of the opposite or additive inverse. The additive inverse of 5 is -5 and the additive inverse of $-3x$ is $3x$. To find the additive inverse of a polynomial of more than one term, replace every term by its additive inverse.

EXAMPLE 7 **Find the additive inverse of $2x^2 - 8x + 23$.**

$$-(2x^2 - 8x + 23) = -2x^2 + 8x - 23$$

To subtract one polynomial from another, add the additive inverse of the polynomial being subtracted.

EXAMPLE 8 **Subtract $9x^3 - 14x^2 + 8x - 17$ from $11x^3 - 21x^2 + 5x - 13$.**

$$(11x^3 - 21x^2 + 5x - 13) - (9x^3 - 14x^2 + 8x - 17)$$
$$= (11x^3 - 21x^2 + 5x - 13) + (-9x^3 + 14x^2 - 8x + 17)$$ *Add the additive inverse.*
$$= 11x^3 - 9x^3 - 21x^2 + 14x^2 + 5x - 8x - 13 + 17$$
$$= 2x^3 - 7x^2 - 3x + 4$$

The following scheme may be used: First, write like terms under one another, leaving a space if a term is missing. Then, subtract.

EXAMPLE 9 **Subtract: $(13x^3 + 18 - 29x^2) - (25x^3 - 18x^2 + 7x - 12)$**

$$\begin{array}{rrrr} 13x^3 - 29x^2 & & + 18 \\ -\ 25x^3 - 18x^2 & + 7x & - 12 \\ \hline -12x^3 - 11x^2 & - 7x & + 30 \end{array}$$

Arrange terms in descending order of x.
Subtract by adding the additive inverse, $-25x^3 + 18x^2 - 7x + 12$.

You should remember two important points when adding or subtracting polynomials. First, you always add or subtract coefficients of like terms. Second, the exponents never change in value when adding or subtracting.

CLASS EXERCISES

Tell whether each expression is a polynomial.

1. $3xy - 5$
2. $\frac{1}{4}a^2 + a$
3. $4x^2 + 3x - \frac{7}{x}$

Classify each polynomial as a monomial, a binomial, or a trinomial. Then give the degree of the polynomial.

4. $0.8x^5 - y$
5. $\sqrt{11}x^2y^3$
6. $x^3 - 3x^2y + 3xy^3$

Arrange each polynomial in the indicated order.

7. $3x + 3x^3 - 2x^4 + 9x^2 - 8$ in descending order of x
8. $y - 7 + y^2 - 8y^4$ in ascending order of y

Simplify.

9. $3xy - 2yz - 5xy$
10. $6c^2 - 4c + 7 - 8c^2$

Add.

11. $(6x + 12) + (9x + 15)$
12. $(7x^2 + 8x - 5) + (9x^2 - 9x)$

Subtract.

13. $(12x - 22) - (19x + 13)$
14. $(5x^2 - 6x + 8) - (3x^2 - 9)$

PRACTICE EXERCISES

Classify each polynomial as a monomial, a binomial, or a trinomial. Then give the degree of the polynomial.

1. $-0.7a^{15}$
2. $4x^3 - \sqrt{13}x + 1$
3. -7
4. $6x^2y^2 - 4x^5$

Simplify.

5. $a^2b + 4ab^2 - 5a^2b - ab^2$
6. $12 + 3m - 5n - m + 4n$
7. $3x^2 - 5x - x^2 + x + 4x$
8. $2y^2 + 3xy + y^2 - x - 6xy$
9. $n^3 + 3n^2 - n - 3 - 3n^3$
10. $15 - y^2 - 10y - 8 + 8y$

Add.

11. $\begin{array}{r} 5x^2 - 8x - 13 \\ \underline{9x^2 + 7x - 26} \end{array}$

12. $\begin{array}{rrrr} 7x^3 & - 44x^2 & - 23x & + 37 \\ \underline{3x^3} & & \underline{+ 5x} & \underline{- 25} \end{array}$

13. $(4x - 7) + (9x + 8)$

14. $(13x + 16) + (-17x + 19)$

Subtract.

15. $\begin{array}{r} 17x^2 + 5x - 4 \\ 9x^2 + 8x + 10 \\ \hline \end{array}$

16. $\begin{array}{r} 3x^2 - 7x + 12 \\ 15x^2 - 8x + 22 \\ \hline \end{array}$

17. $(-11x - 14) - (12x + 13)$

18. $(46x^2 + 13x - 24) - (32x^2 + 16)$

Add or subtract as indicated.

19. $(3a - 2b) + (6b - 2a)$

20. $(4x - 5y) - (4x + 7y)$

21. $(2c^2 + 9) - (3c^2 - 7)$

22. $(-8d - 7) + (-d - 6)$

23. $(3x^2 - 6y - 1) + (5x^2 + 1)$

24. $(-a^2 - 3) - (3a - a^2 - 5)$

25. $(6x^2y - 5xy^2 + 7xy - 7) + (-19x^2y + xy^2 - 11xy + 14)$

26. $(8x^2y + 12xy^2 - 15xy + 21) + (27x^2y + 13xy^2 - 21xy - 17)$

27. $(7x^3 + 9x^2 - 8x + 11) - (5x^3 - 13x - 16)$

28. $(-3x^3 + 7x^2 - 8) - (-5x^3 + 9x^2 - 8x + 19)$

29. $(-12x^3 + 5x - 23) - (4x^4 + 31 - 9x^3)$

30. $(30x^3 - 49x^2 + 7x) + (50x^3 - 75 - 60x^2)$

31. $(0.348x^2 - 3.316) + (-7.829x^2 - 3.957x - 6.387)$

32. $(3.521x^2 - 6.309x) + (-6.217x^2 - 4.208x - 8.492)$

33. $(3a^2 - ab - 7) + (5a^2 + ab + 8) - (-2a^2 + 3ab - 9)$

34. $(a^2 - 2ab + b^2) - (3a^2 - 2ab + b^2) + (4a^2 + 7b^2)$

35. Subtract $5x^3 - 3x^2 + 5x - 1$ from $4x^2 - 7$.

36. From $(3cd - ef)$, subtract $(-4ef + 1)$.

37. By how much does $(3y^2 - 5y - 12)$ exceed $(2y^2 - 3y + 8)$?

38. Subtract $(10x^3 + 5 - 4x^2 - x^4)$ from $(3x - 4x^2 - 5x^3 + 2)$.

First use the distributive property to simplify each product. Then, add or subtract as indicated.

39. $2(3a - 2b) - 4(c - 2d) + 3(6c - 2a)$

40. $-3x(4x + 5y + 6y^2) + 9y(5x + 4xy - 7x^2)$

41. State the closure property for addition of polynomials. Illustrate this property using $2x^2 + 5x + 2$ and $x^2 - x + 8$.

42. Does the set of polynomials have an additive identity element? If so, what is it?

43. State the associative property for addition of polynomials. Illustrate this property using $5x^2 + 7x + 4$, $8x - 9$, and $4x^2 - 8x - 7$.

44. State the commutative property for addition of polynomials. Illustrate using $9x^2 - 11$ and $-13x^2 + 21$.

45. What is the additive inverse of $-x^3 + 3x^2 + 7x - 9$?

46. What properties must be true for the set of polynomials to be a group under addition? Are those properties true?

47. Is the set of polynomials commutative under subtraction? Illustrate with an example.

48. Is the set of polynomials associative under subtraction? Illustrate with an example.

Applications

49. Business Tara's salary is $15,000 per year. Let x represent her annual percent increase expressed as a decimal. Express Tara's next year's salary as a polynomial.

50. Business For Exercise 49, what polynomial would represent Tara's salary in two years?

51. Manufacturing Express the surface area of a rectangular box with height h, width w, and length l as a polynomial.

52. Travel Tickets to Adventureland cost $25 each for the first 10 members of a group and $22 for each additional member. Express the cost of n tickets as a polynomial. Assume that $n \geq 10$.

Mixed Review

Solve each system of inequalities by graphing on the coordinate plane.

53. $\begin{cases} 2x - 4y \leq 0 \\ y + 5x < 3 \end{cases}$

54. $\begin{cases} -x - y \leq 5 \\ x + y > 3y \end{cases}$

55. $\begin{cases} y > -2x \\ |x| > 6 \end{cases}$

For Exercises 56–61, use matrices *C*, *N*, and *R* shown below. Perform the indicated operations.

$$C = \begin{bmatrix} 5 & 0 \\ 2 & 4 \\ -3 & 1 \end{bmatrix} \quad N = \begin{bmatrix} 1 & 1 & 0 \\ 3 & 0 & -2 \\ 4 & 3 & -5 \end{bmatrix} \quad R = \begin{bmatrix} 0 & 1 \\ 2 & 0 \\ 3 & -12 \end{bmatrix}$$

56. $C + R$

57. $4N$

58. $\frac{1}{2}R$

59. $3R - C$

60. $8N^{-1}$

61. NR

6.3

Multiplication of Polynomials

Objective: To multiply polynomials, including special products

The product of two polynomials is a polynomial and it consists of all terms obtained by multiplying each term of one polynomial by every term of the other polynomial. In multiplying polynomials, the distributive property and the properties of exponents are used extensively.

Capsule Review

The distributive property is used to multiply a polynomial by a monomial. The monomial is distributed over each of the terms of the polynomial.

EXAMPLE **Multiply:** $3x(2x^2 - 4x + 7)$

$3x(2x^2 - 4x + 7) = 3x(2x^2) + 3x(-4x) + 3x(7)$ *Distributive property*
$= 6x^3 - 12x^2 + 21x$ *Simplify.*

Multiply.

1. $3a(2a^2 + 3ab + 6b)$

2. $5x^2(3x^2 - 4x + 8)$

3. $5a(4a - 6a - [-3a])$

4. $(-5x)^2(6x^3 - 2x^2)$

5. $(-4a^3)(-7b + 2ab)$

6. $(-6x^2y^3)(-3x^3y^2)$

7. $(-3s^4t^3)(2t^2 - 4tv - 3v^2)$

8. $(x^2y)^2(2x^2 - 7y^2)$

9. $4x^2(3x^3 + 2x^4 - 8x^5 + 9x^6)$

10. $7a^3(3a^2 - 2ab^2)$

To multiply a binomial by another binomial, use the distributive property twice: once to distribute the second binomial as a whole over every term of the first binomial, and then to distribute each term of the first binomial over every term of the second binomial.

EXAMPLE 1 **Multiply: $(4x - 3)(2x + 5)$**

$(4x - 3)(2x + 5)$
$= 4x(2x + 5) - 3(2x + 5)$ *Distributive property*
$= (4x)(2x) + (4x)(5) + (-3)(2x) + (-3)(5)$ *Distributive property*
$= 8x^2 + 20x - 6x - 15$
$= 8x^2 + 14x - 15$ *Combine like terms.*

In Example 1, notice that the terms in the product are the products of each term in the first binomial multiplied by each term in the second binomial. There is a simple memory aid (called a *mnemonic device*) to simplify the process of multiplying two binomials.

FOIL The *product of two binomials* is the sum of the products of the **F**irst terms, the **O**uter terms, the **I**nner terms, and the **L**ast terms.

$$\begin{aligned}(4x - 3)(2x + 5) &= \overset{\text{F}}{8x^2} + \overset{\text{O}}{20x} - \overset{\text{I}}{6x} - \overset{\text{L}}{15} \\ &= 8x^2 + 14x - 15 \qquad \textit{Combine like terms.}\end{aligned}$$

In Example 1, the product of the two binomials is a trinomial. In the next example, the product is a polynomial of four terms.

EXAMPLE 2 **Multiply: $(4x + y)(a - 2b)$**

$$(4x + y)(a - 2b) = \overset{\text{F}}{4ax} - \overset{\text{O}}{8bx} + \overset{\text{I}}{ay} - \overset{\text{L}}{2by}$$

When the two binomial factors are identical, the product is a trinomial.

EXAMPLE 3 **Multiply: $(3x - 5)(3x - 5)$**

$$\begin{aligned}(3x - 5)(3x - 5) &= \overset{\text{F}}{9x^2} - \overset{\text{O}}{15x} - \overset{\text{I}}{15x} + \overset{\text{L}}{25} \\ &= 9x^2 - 30x + 25 \qquad \textit{Combine like terms.}\end{aligned}$$

Since $(3x - 5)(3x - 5) = (3x - 5)^2$, the product is called a **perfect square trinomial.** The terms of this trinomial are the square of the first term of the binomial plus twice the product of the terms plus the square of the last term of the binomial.

Special Product: Perfect Square Trinomial

$$(a + b)^2 = a^2 + 2ab + b^2$$
$$(a - b)^2 = a^2 - 2ab + b^2$$

Two binomials whose first terms are equal and whose last terms are opposites are called **conjugates.** For example, $3x - 5$ and $3x + 5$ are conjugates. The product of two conjugate binomials is a special product with just two terms. It is called the **difference of two squares.**

EXAMPLE 4 **Multiply: $(8x + 3)(8x - 3)$**

$$\begin{aligned} & \quad\ \ \text{F} \qquad\ \ \text{O} \qquad \text{I} \qquad \text{L} \\ (8x + 3)(8x - 3) &= 64x^2 - 24x + 24x - 9 \\ &= 64x^2 - 9 \end{aligned}$$

Special Product: Difference of Two Squares

$$(a + b)(a - b) = a^2 - b^2$$

To multiply polynomials when one or both polynomial factors contain more than two terms, use the distributive property repeatedly.

EXAMPLE 5 **Multiply: $(3x - 2)(5x^2 - 4x - 3)$**

$$\begin{aligned} &(3x - 2)(5x^2 - 4x - 3) \\ &\quad = 3x(5x^2 - 4x - 3) - 2(5x^2 - 4x - 3) && \textit{Distributive property} \\ &\quad = 15x^3 - 12x^2 - 9x - 10x^2 + 8x + 6 && \textit{Distributive property} \\ &\quad = 15x^3 - 22x^2 - x + 6 \end{aligned}$$

You can also use a vertical arrangement.

$$\begin{array}{rrrrl} & 5x^2 & -\ 4x & -\ 3 & \\ \times & & 3x & -\ 2 & \\ \hline 15x^3 & -\ 12x^2 & -\ 9x & & \leftarrow \textit{Multiply: } 3x(5x^2 - 4x - 3) \\ & -\ 10x^2 & +\ 8x & +\ 6 & \leftarrow \textit{Multiply: } -2(5x^2 - 4x - 3) \\ \hline 15x^3 & -\ 22x^2 & -\ x & +\ 6 & \textit{Add.} \end{array}$$

CLASS EXERCISES

Predict the number of terms that will be in the product when it is in simplest form.

1. $(4x - 3)(4x - 3)$
2. $(2a + 3)(2a - 3)$
3. $(x + y)(a + b)$
4. $(5x + 1)^2$
5. $(w + l)(r + h)$
6. $(x + 2y)(y - 2x)$

Multiply.

7. $3y(2y^2 - 5)$
8. $-4a^2(3ab - 2ab^2 + 5)$
9. $(x + 3)(x - 4)$
10. $(2x - 3y)(x - 2y)$
11. $(4c - 5d)(2x + 3y)$
12. $(4x + 5)^2$
13. $(2m - n)(2m + n)$
14. $(4d + 7)(4d - 7)$
15. $(x - 3)(3 - x)$

PRACTICE EXERCISES

Multiply.

1. $x(4 - 3x)$
2. $-5(x^2 - 1)$
3. $-3x^2(x^2 + x - 1)$
4. $0.5d(a + b + c)$
5. $(x + 3)(2x - 4)$
6. $(4x + 7)(4x - 1)$
7. $(11 - x)(x + 3)$
8. $(2x - 1)(1 - x)$
9. $(x + 5)(x - 8)$
10. $(x - 9)(2x - 5)$
11. $(5x - 17)(x + 2)$
12. $(x + 2)(13 - 4x)$
13. $(3x - 11)(x - 6)$
14. $(4x - 5)(3x - 8)$
15. $(2a + 3b)(a - 7b)$
16. $(4x + y)(2x - y)$
17. $(3x - y)(2x + y)$
18. $(2x + 3y)(2x - 5y)$
19. $(x^2 - y^2)(x - y)$
20. $(xy - 2)(xy + 4)$
21. $(2a - c)(3b + 5d)$
22. $(x + y)(3m - 5n)$
23. $(x - 2)^2$
24. $(c + 7)^2$
25. $(2w + 3t)^2$
26. $(z - 2y)^2$
27. $(c + 9)(c - 9)$
28. $(a - 5)(5 + a)$
29. $(3a + 4)(3a - 4)$
30. $(4x - y)(4x + y)$
31. $(2a - 5)(4a^2 + 7a - 3)$
32. $(d + 3)(d^2 - 3d + 8)$
33. $(2x + 3)(4x^2 - 6x + 9)$
34. $(x^2 - xy + y^2)(x - y)$
35. $(a + b)(a^2 - ab + b^2)$
36. $(5 - 2a)(25 + 10a + 4a^2)$
37. $(x - y)(x^2 + xy + y^2)$
38. $(y^2 - 1)(y^4 + y^2 + 1)$
39. $(x + y)^2(x - y)$
40. $(a - b)^2(a + b)$
41. $(3 + 2y)^2(4 - 3y)$
42. $(3a + 2b)(2a - b)^2$
43. $(2d - 1)(d + 4)(3d - 5)$
44. $(2s + 3)(4s - 1)(3s + 7)$
45. $(2x^2 + 3x)(x + 4)(x^2 - 8)$
46. $(9c - d)(2d - c)(3c + d)$
47. $(3a + 2)^2(a - 4)^2$
48. $(2x + 5)^3$
49. $(2a - 1)^4$
50. $0.253(1.86x + 3.84)$

Multiply. Assume all exponents are positive integers.

51. $(y^a + 7)(y^a - 7)$
52. $(2x^a - 3y^b)(x^a - y^b)$
53. $(4x^{2a} - 1)(4x^{2a} + 7)$
54. $(3x^{2a} + 1)^2$
55. $[(x + y) - z]^2$
56. $[(x - y) + z]^2$
57. $(a + b + c)(a - b - c)$
58. $(x^{2a+1} - y^{3b})(x^a - y^{2b-1})$
59. $(w^c + z^e)(w^{2c} - w^c z^e + z^{2e})$
60. $(r + s)(r - s)(r^2 + s^2)$

Applications

61. Geometry The length of a side of a square is $x + 3$ cm. Express the perimeter as a polynomial.

62. Geometry The length of a rectangular lot is $x + 4$ m and its width is $x - 2$ m. Express the area as a polynomial.

63. Geometry A rectangular box is $2x + 1$ units long, $x - 1$ units wide, and x units high. Express the volume of the box as a polynomial.

64. Geometry In the figure on the right, express the area of one of the circles as a polynomial.

65. Geometry In the figure on the right, express the area of the square as a polynomial.

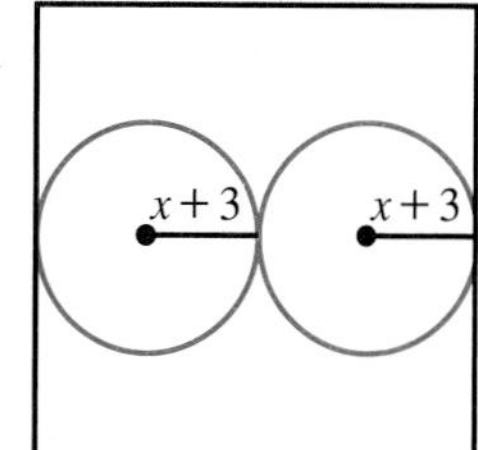

Developing Mathematical Power

Investigation Did you know that the product of two polynomials can be represented by a rectangle? The length and width of the rectangle represent the factors, and their product is the area of the rectangle. The large rectangular region can be separated into smaller rectangular regions, and the sum of the areas of these rectangles is equal to the area of the large rectangle.

The length of the large rectangle is $x + 2$ units, and the width is $x + 1$ units. So, its area is

$$(x + 2)(x + 1) \text{ square units}$$

The sum of the areas of the smaller rectangles is $x^2 + 3x + 2$ square units.

Therefore

$$(x + 2)(x + 1) = x^2 + 3x + 2$$

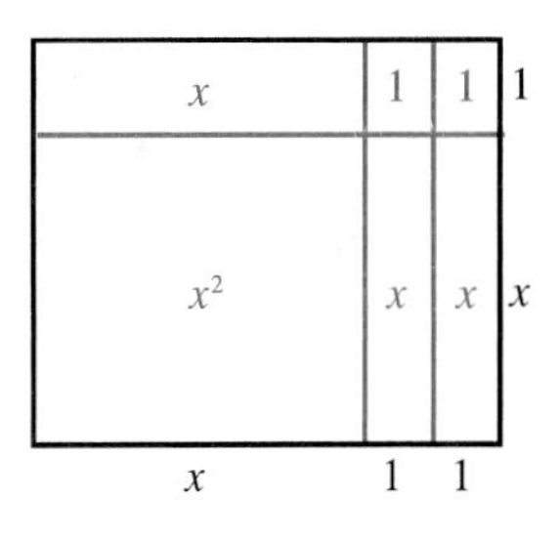

For each figure, write an equation showing that the area of the large rectangle is equal to the sum of the areas of the smaller rectangles.

66.

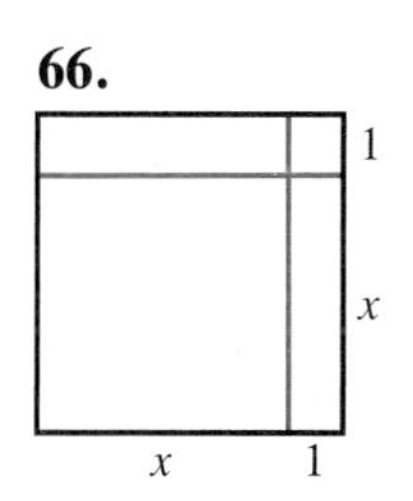

67.

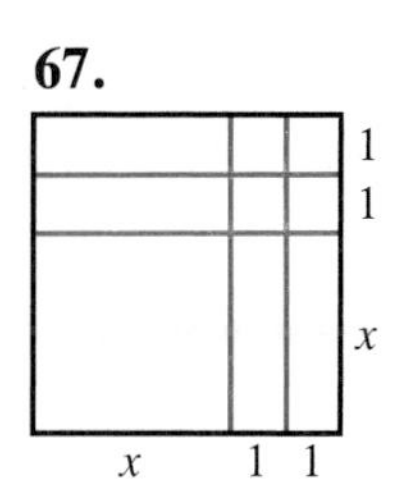

68.

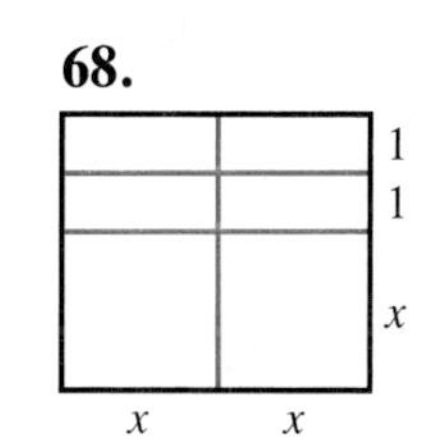

69.

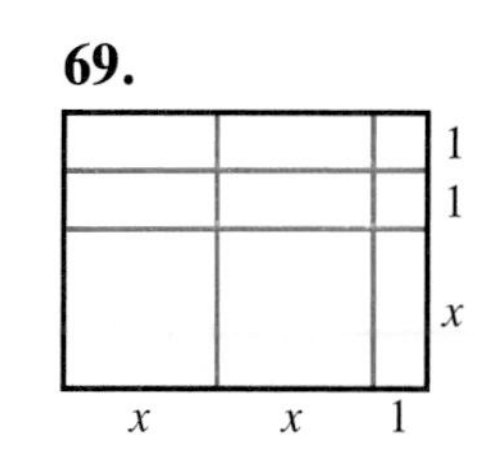

6.4

Factoring Polynomials

Objectives: To factor a polynomial with terms containing common factors
To factor a perfect square trinomial, a difference of two squares, and a sum or difference of two cubes

The *greatest common factor* (*GCF*) of two or more integers is the greatest integer that is a factor of each of the given numbers.

Capsule Review

A *prime number* is an integer greater than 1 whose only positive integral factors are itself and 1. To find the GCF of two or more integers, first write their *prime factorizations*. That is, write each integer as the product of prime factors. The GCF is the product of the prime factors that appear in all of the factorizations.

EXAMPLE **Find the GCF of 24, 36, and 60.**

$24 = 2 \cdot 2 \cdot 2 \cdot 3$
$36 = 2 \cdot 2 \cdot 3 \cdot 3$
$60 = 2 \cdot 2 \cdot 3 \cdot 5$
GCF: $2 \cdot 2 \cdot 3$, or 12

Find the GCF of the numbers in each group.

1. 8, 20 **2.** 18, 25 **3.** 6, 9, 15 **4.** 12, 30, 42

The **greatest common factor (GCF)** of two or more monomials is the common factor that has the greatest coefficient and the greatest degree.

EXAMPLE 1 **Find the GCF of $4x^3y^2$, $12xy^2$, and $20x^2y^3$.**

$4x^3y^2 = 2 \cdot 2 \cdot x \cdot x \cdot x \cdot y \cdot y$
$12xy^2 = 2 \cdot 2 \cdot 3 \cdot x \cdot y \cdot y$
$20x^2y^3 = 2 \cdot 2 \cdot 5 \cdot x \cdot x \cdot y \cdot y \cdot y$
GCF: $2 \cdot 2 \cdot x \cdot y \cdot y$, or $4xy^2$

A polynomial may also be factored. When the terms of a polynomial have a GCF greater than 1, the polynomial is factored as the product of that GCF and a polynomial.

EXAMPLE 2 **Factor:** $9mn^2 - 24mn + 15m$

$$9mn^2 - 24mn + 15m = 3m(3n^2) - 3m(8n) + 3m(5) \quad \text{The GCF is } 3m.$$
$$= 3m(3n^2 - 8n + 5)$$

The trinomial $x^2 + 8x + 16$ can be written as $(x)^2 + 2(x \cdot 4) + (4)^2$, which is in the form $a^2 + 2ab + b^2$. Thus, $x^2 + 8x + 16$ is a perfect square trinomial. You know that a perfect square trinomial is obtained by multiplying two identical binomials, or *squaring a binomial.* A perfect square trinomial then has two identical binomial factors. To find these factors, reverse the formula for the special product.

Factoring a Perfect Square Trinomial

$$a^2 + 2ab + b^2 = (a + b)^2$$
$$a^2 - 2ab + b^2 = (a - b)^2$$

EXAMPLE 3 **Factor:** **a.** $9x^2 + 42x + 49$ **b.** $64x^2 - 16xy + y^2$

a. $9x^2 + 42x + 49 = (3x)^2 + 2(3x \cdot 7) + (7)^2$ $\quad a = 3x, b = 7$

$= (3x + 7)^2$ $\quad (a + b)^2$

b. $64x^2 - 16xy + y^2 = (8x)^2 - 2(8x \cdot y) + (y)^2$ $\quad a = 8x, b = y$

$= (8x - y)^2$ $\quad (a - b)^2$

If the expression to be factored is a binomial, check to see if it is the difference of two squares. The difference of two squares is obtained by multiplying a pair of conjugates. Thus, a difference of two squares has a pair of conjugates as its factors. To factor a difference of squares, reverse the special product formula.

Factoring the Difference of Two Squares

$$a^2 - b^2 = (a + b)(a - b)$$

EXAMPLE 4 **Factor:** **a.** $16x^2 - 9$ **b.** $25s^2 - t^2$

a. $16x^2 - 9 = (4x)^2 - (3)^2$ $\quad a = 4x, b = 3$

$= (4x + 3)(4x - 3)$ $\quad (a + b)(a - b)$

b. $25s^2 - t^2 = (5s)^2 - (t)^2$ $\quad a = 5s, b = t$

$= (5s + t)(5s - t)$ $\quad (a + b)(a - b)$

The sum or difference of two cubes can be factored as the product of a binomial and a trinomial. By multiplying the factors, you can show that the formulas below are correct.

> **Factoring a Sum or Difference of Two Cubes**
>
> $$a^3 + b^3 = (a + b)(a^2 - ab + b^2)$$
> $$a^3 - b^3 = (a - b)(a^2 + ab + b^2)$$

EXAMPLE 5 **Factor:** **a.** $64x^3 + 27$ **b.** $125x^3 - y^6$

a.
$$\begin{aligned} 64^3 + 27 &= (4x)^3 + (3)^3 \\ &= (4x + 3)(16x^2 - 12x + 9) \\ & \quad (\ a + b)(\ a^2 - \ ab + b^2) \end{aligned}$$

b.
$$\begin{aligned} 125x^3 - y^6 &= (5x)^3 - (y^2)^3 \\ &= (5x - y^2)(25x^2 + 5xy^2 + y^4) \\ & \quad (\ a - b\ )(\ a^2 + \ ab \ + b^2) \end{aligned}$$

When you factor a polynomial, you should factor it so that all terms have integral coefficients. In some cases you may need to factor out a rational number first.

EXAMPLE 6 **Factor:** **a.** $x^2 + \frac{2}{3}x + \frac{1}{9}$ **b.** $0.25y^2 - 0.04$

a.
$$\begin{aligned} x^2 + \tfrac{2}{3}x + \tfrac{1}{9} &= \tfrac{1}{9}(9x^2 + 6x + 1) \\ &= \tfrac{1}{9}(3x + 1)^2 \end{aligned}$$

b.
$$\begin{aligned} 0.25y^2 - 0.04 &= 0.01(25y^2 - 4) \\ &= 0.01(5y + 2)(5y - 2) \end{aligned}$$

CLASS EXERCISES

Match each polynomial on the left with its factored form on the right.

1. $64x^3 + y^3$

2. $64x^2 + 16xy + 8y^2$

3. $64x^2 - y^2$

4. $x^3 - 64$

5. $64x^2 - 16xy + y^2$

a. $(x - 4)(x^2 + 4x + 16)$

b. $(8x - y)^2$

c. $(8x + y)(8x - y)$

d. $8(8x^2 + 2xy + y^2)$

e. $(4x + y)(16x^2 - 4xy + y^2)$

PRACTICE EXERCISES

Factor.

1. $3a^2 + 3a$
2. $2x^4 + 14x^3$
3. $25a^2 - 35a$
4. $4x^3 - 8x^2 + 16x$
5. $30w^2 - 12hw + 6h^2w^2$
6. $4a^2 - 4ac + 4c^2$
7. $12a^3 + 6a^2b + 36ab$
8. $28t^3s + 8t^4s$
9. $x^2 + 4x + 4$
10. $x^2 + 2x + 1$
11. $x^2 - 14x + 49$
12. $x^2 - 18x + 81$
13. $9x^2 + 48x + 64$
14. $x^2 - 2xy + y^2$
15. $25x^2 - 40xy + 16y^2$
16. $81x^2 + 36xy + 4y^2$
17. $4a^2 - 1$
18. $9x^2 - 1$
19. $c^2 - 4d^2$
20. $16a^2 - 9b^2$
21. $4a^2 - 49c^2$
22. $y^2 - 81z^2$
23. $c^3 + 8$
24. $d^3 + 64$
25. $8s^3 + 343t^3$
26. $1000f^3 + 27g^3$
27. $m^3 - 125$
28. $f^3 - 216$
29. $27x^3 - y^3$
30. $8x^3 - 27y^3$
31. $10axc + 20cay - 30acz$
32. $12s^6 + 8s^7 - 32s^8 + 36s^9$
33. $a^4 - 4$
34. $b^6 - 8$
35. $4f^4 - 20e^2f^2 + 25e^4$
36. $x^6 + 2x^3y^2 + y^4$
37. $36x^6 - 25y^4$
38. $100 - 81t^6$
39. $y^6 - 64z^8$
40. $8x^6 - 27y^6$
41. $64x^6y^3 - 27z^3$
42. $x^9 - y^3z^3$
43. $125x^9 + 729y^6$
44. $1331s^{12} - 1728t^6$

Factor. Remember to factor so that the terms have integral coefficients.

45. $x^2 + x + \frac{1}{4}$
46. $\frac{1}{4}x^2 + \frac{1}{2}x + \frac{1}{4}$
47. $x^2 - \frac{1}{9}$
48. $x^3 - \frac{1}{8}$
49. $0.16y^2 - 0.01$
50. $0.04y^2 + 0.12y + 0.09$

Factor. Assume all exponents are positive integers.

51. $(a + 2b)^2 - d^2$
52. $z^2 - (3x + 2y)^2$
53. $2x^{n+3} + 3x^n$
54. $3a^n + 5a^{n+1}$
55. $x^{2a} - y^{2b}$
56. $r^{4a}s^{2b} + 2r^{2a}s^bt^{3a} + t^{6a}$

57. $4x^{2a}y^{8b} - 20x^{a}y^{4b}z^{c} + 25z^{2c}$

58. $25x^{4m+2} + 30x^{2m+1}y^{m-2} + 9y^{2m-4}$

59. $16x^{4m} - 88x^{2m}y^{m+3} + 121y^{2m+6}$

60. $x^{3c+18} - 27y^{9d}$

Applications

61. Manufacturing The area of a square mat is $25x^2 - 10x + 1$ cm^2. Express the length of a side of the mat in terms of x.

62. Storage The volume of a cubic storage bin is $8x^3 + 125$ m^3. Express the volume in factored form.

63. Manufacturing The volume of a cubic box is $27x^3 - 64$. Express the volume in factored form.

64. Manufacturing A small square of plastic is to be cut from a square plastic cover. Express the area of the shaded portion in factored form.

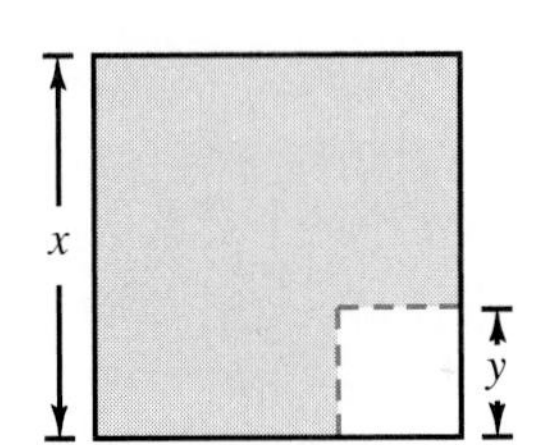

65. Manufacturing A small square is to be cut from a square of cardboard. Express the area of the shaded portion in factored form.

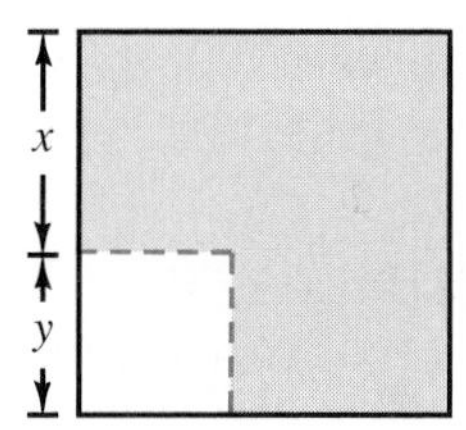

TEST YOURSELF

Identify the degree of each polynomial.

6.1

1. $5xy^2z^3$

2. $4x^3 - 3y^2 + 7x^4y$

Perform the indicated operations.

3. $(7ab^3c^2)(5a^2bc^3)$

4. $(-2m^2n^3)^2(3mn^2)^3$

6.2

5. $(3a^2 - 2b) + (8b - 7a^2)$

6. $(5t^2 - 3v^3 + 8w^4) - (2t^2 + 5v^3)$

6.3

7. $(3x - 2)(2x + 5)$

8. $(2x + 3y)^2$

9. $(a - 3b)(9a + 3b)$

10. $(w + 2a)(w^2 - 2aw + 4a^2)$

Factor.

6.4

11. $5x^2 - 15x + 45$

12. $9a^2 - b^2$

13. $9x^2 - 24xy + 16y^2$

14. $27x^3 + 8$

Factoring Quadratic Trinomials

Objective: To factor a quadratic trinomial

A polynomial of the form $ax^2 + bx + c$, $a, b, c \neq 0$, is a **quadratic trinomial** because it contains a second-degree term, ax^2, called the **quadratic term.** Examples of quadratic trinomials are

$$x^2 + 12x + 35 \qquad 2x^2 - 7x - 15 \qquad 4x^2 - xy - 5y^2$$

Capsule Review

The product of two binomials is a quadratic trinomial if the first term of each binomial factor contains the same literal factor(s).

EXAMPLE

$$(x + 2)(x + 3) = x^2 + 3x + 2x + 6$$
$$= x^2 + 5x + 6$$

$$(x + 2)(x - 3) = x^2 - 3x + 2x - 6$$
$$= x^2 - x - 6$$

Find each product. Observe how the signs of the numerical terms in the binomials influence the signs of the terms in the trinomials.

1. $(x + 4)(x + 3)$ **2.** $(x - 4)(x - 3)$ **3.** $(x + 4)(x - 3)$ **4.** $(x - 4)(x + 3)$

5. $(x + 5)(x + 6)$ **6.** $(x - 5)(x - 6)$ **7.** $(x + 5)(x - 6)$ **8.** $(x - 5)(x + 6)$

Sometimes a quadratic trinomial can be factored as the product of two binomials. If the coefficient of the second-degree term is 1 and the last term is *positive,* the factors are of the form

$$(d + m)(d + n) \quad \text{or} \quad (d - m)(d - n)$$

Look for integers whose product is mn and whose sum is $m + n$ or $-m + (-n)$.

EXAMPLE 1 **Factor:** **a.** $h^2 + 8h + 7$ **b.** $f^2 - 17f + 72$

F O + I L

a. $h^2 + 8h + 7 = (h + \underline{?})(h + \underline{?})$
$= (h + 7)(h + 1)$ $\quad 7(1) = 7;\ 1h + 7h = 8h$

b. $f^2 - 17f + 72 = (f - \underline{?})(f - \underline{?})$
$= (f - 9)(f - 8)$ $\quad -9(-8) = 72;\ -8f + (-9f) = -17f$

When the coefficient of the second-degree term is 1 and the last term is *negative,* the factors will be of the form $(d + m)(d - n)$. Look for integers whose product is $-mn$ and whose sum is $m + (-n)$.

EXAMPLE 2 **Factor:** **a.** $x^2 - x - 12$ **b.** $x^2 + x - 12$

a. $x^2 - x - 12 = (x + 3)(x - 4)$ *$3(-4) = -12; (-4x) + 3x = -x$*

b. $x^2 + x - 12 = (x + 4)(x - 3)$ *$4(-3) = -12; (-3x) + 4x = x$*

When the coefficient of the second-degree term is *not* 1, try possible pairs of binomial factors until you find the pair that gives the correct middle term. If the last term of the trinomial is *positive,* its factors have the same sign.

EXAMPLE 3 **Factor:** $2x^2 + 11x + 12$

The factors of the first term $2x^2$, are $2x$ and x.
The factors of the last term, 12, could be

1 and 12	2 and 6	3 and 4
−1 and −12	−2 and −6	−3 and −4

Since the last term (the **constant**) and the middle term (the **linear term**) are both positive, try only *positive* factors of 12.

Possible Factors	Middle Term (O + I)	
$(2x + 12)(x + 1)$	$2x + 12x = 14x$	
$(2x + 1)(x + 12)$	$24x + x = 25x$	
$(2x + 6)(x + 2)$	$4x + 6x = 10x$	
$(2x + 2)(x + 6)$	$12x + 2x = 14x$	
$(2x + 4)(x + 3)$	$6x + 4x = 10x$	
$(2x + 3)(x + 4)$	$8x + 3x = 11x$	*Correct middle term*

Thus, $2x^2 + 11x + 12 = (2x + 3)(x + 4)$

If the last term of the trinomial is *negative,* its factors have different signs. The inner or outer product—whichever has the greater absolute value—has the same sign as the middle term.

EXAMPLE 4 **Factor:** $3y^2 - 4y - 15$

Possible Factors	Middle Term (O + I)	*Try factors that give a negative middle term.*
$(3y - 15)(y + 1)$	$3y - 15y = -12y$	
$(3y + 1)(y - 15)$	$-45y + y = -44y$	
$(3y + 5)(y - 3)$	$-9y + 5y = -4y$	*Correct middle term*
$(3y + 3)(y - 5)$	$-15y + 3y = -12y$	

Thus, $3y^2 - 4y - 15 = (3y + 5)(y - 3)$

Use the same procedures to factor quadratic trinomials that contain more than one variable.

EXAMPLE 5 **Factor:** $6x^2 + 7xy - 3y^2$

Possible Factors	Middle Term (O + I)	
$(6x - 3y)(x + y)$	$6xy - 3xy = 3xy$	*Try factors that give a positive middle term.*
$(6x - y)(x + 3y)$	$18xy - xy = 17xy$	
$(3x - y)(2x + 3y)$	$9xy - 2xy = 7xy$	*Correct middle term*
$(2x - y)(3x + 3y)$	$6xy - 3xy = 3xy$	

Thus, $6x^2 + 7xy - 3y^2 = (3x - y)(2x + 3y)$

A trinomial such as $ax^4 + bx^2 + c$ can be written in *quadratic form* as

$$a(x^2)^2 + b(x^2) + c$$

Some trinomials in quadratic form can be factored as the product of two binomials.

EXAMPLE 6 **Factor:** $2x^4 + 9x^2 - 18$

Write the expression in quadratic form: $2(x^2)^2 + 9(x^2) - 18$

Possible Factors	Middle Term (O + I)	
$(2x^2 + 18)(x^2 - 1)$	$-2x^2 + 18x^2 = 16x^2$	
$(2x^2 - 1)(x^2 + 18)$	$36x^2 - x^2 = 35x^2$	
$(2x^2 + 9)(x^2 - 2)$	$-4x^2 + 9x^2 = 5x^2$	
$(2x^2 - 2)(x^2 + 9)$	$18x^2 - 2x^2 = 16x^2$	
$(2x^2 - 3)(x^2 + 6)$	$12x^2 - 3x^2 = 9x^2$	*Correct middle term*
$(2x^2 + 6)(x^2 - 3)$	$-6x^2 + 6x^2 = 0$	
$(2x^2 - 6)(x^2 + 3)$	$6x^2 - 6x^2 = 0$	

Thus, $2x^4 + 9x^2 - 18 = (2x^2 - 3)(x^2 + 6)$

CLASS EXERCISES

One factor of each trinomial is given. Give the other factor.

1. $x^2 - 13x + 36, (x - 4)$

2. $2x^2 + 13x - 24, (2x - 3)$

3. $4a^2 + 4ab - 3b^2, (2a + 3b)$

4. $3s^2 + 25s + 42, (s + 6)$

5. $3x^2 + 8x - 35, (x + 5)$

6. $3c^2 - cd - 4d^2, (3c - 4d)$

Tell whether each trinomial can be factored into two identical binomials. If so, give those factors. If not, give a reason for your answer.

7. $x^2 + x + 1$

8. $x^2 - 2x + 1$

9. $x^2 + x + 2$

PRACTICE EXERCISES

Factor.

1. $x^2 + 5x + 6$
2. $x^2 - 6x + 8$
3. $t^2 - 12t + 27$
4. $r^2 - 11r + 18$
5. $x^2 + 10x + 16$
6. $y^2 + 15y + 36$
7. $x^2 - 5x - 14$
8. $x^2 + x - 20$
9. $x^2 - 3x - 40$
10. $c^2 + 2c - 63$
11. $d^2 + 10d - 75$
12. $f^2 - 7f - 44$
13. $3y^2 + 31y + 36$
14. $2y^2 - 19y + 24$
15. $5r^2 + 23r + 26$
16. $2m^2 - 11m + 15$
17. $5t^2 + 28t + 32$
18. $2p^2 - 27p + 36$
19. $3y^2 + 7y - 20$
20. $5y^2 + 12y - 32$
21. $7z^2 - 8z - 12$
22. $2z^2 + z - 28$
23. $6c^2 + 11c + 4$
24. $28a^2 + 13a - 6$
25. $4a^2 + ab - 15b^2$
26. $14x^2 + 11xy + 2y^2$
27. $3x^2 - 8xy - 16y^2$
28. $x^2 - 9x - 13$
29. $x^4 - 8x^2 + 15$
30. $2x^4 + x^2 - 6$
31. $12y^2 - 16y - 35$
32. $12x^2 - 11x - 56$
33. $15c^2 - 34c - 16$
34. $21x^2 + 88x + 60$
35. $24m^2 + 31mp - 15p^2$
36. $36r^2 + 91rs - 22s^2$
37. $21x^2y^2 - 59xy + 40$
38. $49m^2n^2 + 7mn - 72$
39. $12x^4 + x^2y^2 - 20y^4$
40. $4x^4 + 5x^2y^2 - 6y^4$
41. $4x^2 + 8x + 9$
42. $5x^2 - 3x + 17$

43. $x^8 - 2x^4y^4 - 15y^8$
44. $6x^6 - x^3y^3 - 35y^6$
45. $12x^4 - 11x^2y - y^2$
46. $15x^4 + 29x^2y - 2y^2$
47. $12x^2y^2 + 5xyz - 3z^2$
48. $6a^2b^2 + 11abc + 4c^2$
49. $4a^2b^2c^2 - 25abc + 6$
50. $6x^2y^2z^2 - xyz - 5$

Factor. Assume all exponents are positive integers.

51. $x^{2n} - 8x^n + 15$
52. $x^{2n} + 3x^n - 28$
53. $2x^{6a} - 11x^{3a} + 14$
54. $6x^{2f} + 19x^f + 15$
55. $6x^{4m} - 5x^{2m} - 50$
56. $12y^{2a+2} - 7y^{a+1} - 12$
57. $24x^{2n+4} + 17x^{n+2} - 20$
58. $30x^{4a} + 11x^{2a}y^{b+1} - 30y^{2b+2}$
59. $15y^{6b+2} + 19y^{3b+1}x^c - 5x^{2c}$
60. $42x^{4a+10} + 17x^{2a+5}y^{a-3} - 15y^{2a-6}$

Applications

Give each answer in terms of *x*.

61. Interior Design A rectangular rug has an area of $(x^2 - 17x + 30)$ m^2. The length of the rug is $x - 2$ m. What is the width?

62. **Agriculture** The area of a rectangular field is $x^2 - 120x + 3500$ ft^2. If the width is $x - 50$ ft, what is the length?

63. **Textiles** The area of a rectangular cloth is $6x^2 - 19x - 85$ cm^2. If the length is $2x + 5$ cm, find the width.

64. **Real Estate** The area of a rectangular lot is $5x^2 - 3x - 2$ ft^2. If the length is $5x + 2$ ft, what is the perimeter of the lot?

HISTORICAL NOTE

Early Quadratic Trinomials

As early as 2000 BC, the Babylonians had a well-developed algebra. It was mostly verbal rather than symbolic, and it was closely related to practical measurement. The development of a more sophisticated algebraic symbolism, such as the one you use today, lagged well behind the development of the underlying concepts. In fact, it was not until the seventeenth century that mathematicians regularly wrote numerals to the upper right of a base to represent the number of times the base is to be used as a factor. Thus, such terms as x^2, x^3, and $7x^2$ came into existence.

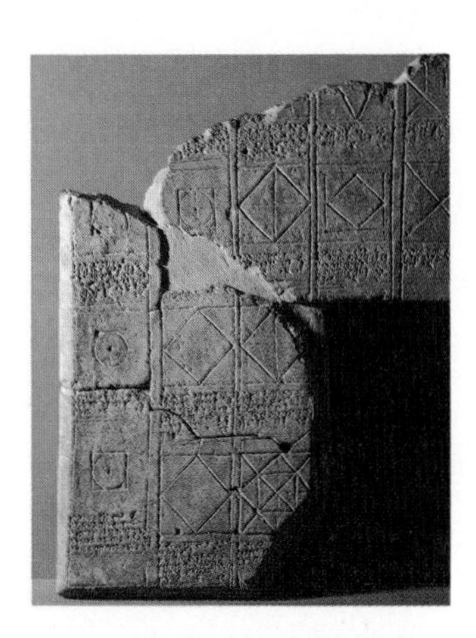

Four thousand years ago, the Babylonians might have said: Find the side of a square if the area, less the side, is twelve. Today, you would find the answer by writing and solving the *quadratic trinomial equation* $x^2 - x = 12$, or $x^2 - x - 12 = 0$, where x represents the length of a side of the square.

Given below are examples of problems that might have been posed by the early Babylonians. Write a quadratic trinomial equation that could be used to solve each problem.

1. Find the side of a square if the area, less the side, is twenty.
2. Find the side of a square if the area, added to twice the side, is thirty-five.
3. Find the side of a square if the area is the sum of fifty-six and ten times the side.
4. Find the side of a square if the area is the sum of three times the side and four.
5. Find the side of a square if ten times the area is the sum of ten and twenty-one times the side.

6.6

Factoring Polynomials Completely

Objectives: To factor a polynomial by grouping terms
To factor a polynomial completely

Some polynomials can be factored by rearranging and grouping terms or by using a combination of the methods presented in the two preceding lessons.

Capsule Review

You have used the following patterns to factor polynomials.

A. $am + bm + cm = m(a + b + c)$

B. $a^2 + 2ab + b^2 = (a + b)^2$

C. $a^2 - 2ab + b^2 = (a - b)^2$

D. $a^2 - b^2 = (a + b)(a - b)$

E. $a^3 + b^3 = (a + b)(a^2 - ab + b^2)$

F. $a^3 - b^3 = (a - b)(a^2 + ab + b^2)$

Write the letter for the factoring pattern that could be used to factor each polynomial.

1. $4x^2 - y^2$ **2.** $4x^2 - 8y^2$ **3.** $4x^2 + 4xy + y^2$

4. $x^3 - 8$ **5.** $x^3 - 2x^2 + 8x$ **6.** $1 + 8x^3$

7. $4x^2 - 4x + 1$ **8.** $x^6 - y^3$ **9.** $4 - y^6$

10. $x^2 + 10x + 25$ **11.** $27x^3 + 8y^3$ **12.** $9x^2 - 12x + 4$

If a polynomial does not fit one of the factoring patterns you have been using, you may be able to factor it by rearranging and grouping its terms. This method, called **factoring by grouping,** may be especially helpful when a polynomial with four or more terms is to be factored.

EXAMPLE 1 **Factor:** $2ax - bx + 2ay - by$

$2ax - bx + 2ay - by = (2ax - bx) + (2ay - by)$ *Group terms.*

$= x(2a - b) + y(2a - b)$ *Factor each binomial.*

$= (2a - b)(x + y)$ *Factor again. The GCF is the binomial $2a - b$.*

In Example 1, the greatest common factor was used in the factoring process. Sometimes other methods of factoring can be used.

EXAMPLE 2 **Factor: $8x^3 - 4x^2 + y^2 - y^3$**

$8x^3 - 4x^2 + y^2 - y^3$

$= (8x^3 - y^3) - (4x^2 - y^2)$ — *Rearrange and group terms.*

$= (2x - y)(4x^2 + 2xy + y^2) - (2x - y)(2x + y)$ — *Factor each binomial.*

$= (2x - y)[4x^2 + 2xy + y^2 - (2x + y)]$ — *Factor again. GCF is $2x - y$.*

$= (2x - y)(4x^2 + 2xy - 2x - y + y^2)$ — *Descending exponents of x*

Sometimes you may need to group three terms together before factoring.

EXAMPLE 3 **Factor: $a^2 - 2ay + y^2 + ca - cy$**

$a^2 - 2ay + y^2 + ca - cy = (a^2 - 2ay + y^2) + (ca - cy)$ — *Group terms.*

$= (a - y)^2 + c(a - y)$ — *Factor.*

$= (a - y)(a - y + c)$ — *Factor again.*

Grouping can be used to factor some polynomials that do not have a binomial factor.

EXAMPLE 4 **Factor: $25a^2 - 20ab + 4b^2 - 9c^2$**

$25a^2 - 20ab + 4b^2 - 9c^2$

$= (25a^2 - 20ab + 4b^2) - 9c^2$ — *Group terms.*

$= (5a - 2b)^2 - 9c^2$ — *Factor.*

$= [(5a - 2b) + 3c][(5a - 2b) - 3c]$ — *Factor the difference of two squares.*

$= (5a - 2b + 3c)(5a - 2b - 3c)$

A **prime polynomial** is a polynomial that cannot be factored. A polynomial is *factored completely* if each of its factors is either a monomial or a prime polynomial.

EXAMPLE 5 **Factor completely:**

a. $24a^2x^2 - 150a^2$ **b.** $147by^2 - 168by + 48b$

c. $-x^6 - 64$ **d.** $8ax^2 + 2axy - 15ay^2$

a. $24a^2x^2 - 150a^2 = 6a^2(4x^2 - 25)$ — *The GCF is $6a^2$.*

$= 6a^2(2x + 5)(2x - 5)$ — *Difference of two squares*

b. $147by^2 - 168by + 48b = 3b(49y^2 - 56y + 16)$

$= 3b(7y - 4)^2$ — *Perfect square trinomial*

c. $-x^6 - 64 = -1(x^6 + 64)$ — *Factor out -1.*

$= -1(x^2 + 4)(x^4 - 4x^2 + 16)$ — *Sum of two cubes*

d. $8ax^2 + 2axy - 15ay^2 = a(8x^2 + 2xy - 15y^2)$

$= a(2x + 3y)(4x - 5y)$

Sometimes you will find it possible to factor more than twice.

EXAMPLE 6 **Factor completely:**
a. $16x^4 + 20x^2 - 126$ **b.** $-225c^4 + 34c^2 - 1$

a. $16x^4 + 20x^2 - 126 = 2(8x^4 + 10x^2 - 63)$
$= 2(2x^2 + 7)(4x^2 - 9)$
$= 2(2x^2 + 7)(2x + 3)(2x - 3)$

b. $-225c^4 + 34c^2 - 1 = -1(225c^4 - 34c^2 + 1)$
$= -1(25c^2 - 1)(9c^2 - 1)$
$= -1(5c + 1)(5c - 1)(3c + 1)(3c - 1)$

To factor a polynomial completely

Factor using the GCF, if possible. Then factor the resulting polynomial until each of its factors is a prime polynomial.

CLASS EXERCISES

Tell whether each polynomial is *prime* or *not prime*. If it is not prime, factor it.

1. $y^3 + y^2 + y$
2. $y^2 + 8y + 16$
3. $m^2 + 1$
4. $m^3 + 1$
5. $9x^2 - 12x + 4 - y^2$
6. $r^2 + r$

Tell whether each polynomial is factored completely. If not, complete the factoring.

7. $6x^2 + 10x - 4 = (2x + 4)(3x - 1)$

8. $3cx - 4bx + 3cy - 4by = 3c(x + y) - 4b(x + y)$

9. $a^2 - 2ab + 1 - b^2 = (a - 1)^2 - b^2$

10. $12x^2 - 20xy - 8y^2 = 4(3x^2 - 5xy - 2y^2)$

11. $x^4 + 18x^2 + 81 = (x^2 + 9)^2$

PRACTICE EXERCISES

Factor by grouping.

1. $x^2 - ax + bx - ab$
2. $br + b + cr + c$
3. $2ms + 3mt - 4ns - 6nt$
4. $3cx - bx + 3cy - by$
5. $f^2 + 2fg + mf + mg + g^2$
6. $4r^2 - 12rs - 6s + 4r + 9s^2$
7. $a^2 - 4ab + 4b^2 - a + 2b$
8. $t^2 - s^2 + rs - rt$

9. $a^3 + b^3 + a^2 - b^2$

10. $a^3 + b^2 - b^3 - a^2$

11. $p^3 - 4q^2 + p^2 - 8q^3$

12. $w^3 + 27z^3 - w^2 + 9z^2$

13. $25x^2 - 10x + 1 - 4y^2$

14. $x^2 + 2xy + y^2 - 4z^2$

15. $a^2 + 4b^2 - 25c^2 + 4ab$

16. $x^3 + x^2 - 4x - 4$

Factor completely.

17. $9x^2 - 36$

18. $18a^2 - 8b^2$

19. $12m^2 - 75n^2$

20. $64c^2 - 16d^2$

21. $12x^2 + 36x + 27$

22. $16x^2 - 80x + 100$

23. $2a^2b - 16ab + 32b$

24. $-x^3 - 8$

25. $-x^3 + 27$

26. $ax^2 - 8ax + 12a$

27. $3x^2 - 24x - 27$

28. $18bm^2 + 24bm - 10b$

29. $4f^2 - 20f + 24$

30. $3x^2 + 24x + 45$

31. $16x^4 + 4x^2 - 2$

32. $4ax^4 - 22ax^2 + 10a$

33. $-x^2 + 5x - 4$

34. $-36x^4 + 25x^2 - 4$

35. $5x^3 - 6y^3 + 6x^2y - 5xy^2$

36. $3p^3 - 6q^3 + 6p^2q - 3pq^2$

37. $y^5 - 16y^3 + 8y^2 - 128$

38. $ax^2 - a + bx^2 - b$

39. $a^2 - 4ab + 4b^2 - c^2 - 2cd - d^2$

40. $8x^3 + 4x^2 + 4xz + z^2 + z^3$

41. $k^4 - 1$

42. $k^8 - 16$

43. $x^2 - 13xy + 36y^2$

44. $100x^4 - 41x^2y^2 + 4y^4$

45. $-36c^4 + 289c^2 - 400$

46. $-5p^4 + 2000$

47. $-6q^5 - 6000q^2$

48. $20x^4 - 45x^2 - 500$

49. $9y^8 - 26y^4 + 16$

50. $x^4 - 18x^2 + 81$

51. $-81p^3 + 375$

52. $-5y^3 + 110y^2z - 605yz^2$

53. $16x^4 - 200x^2 + 625$

54. $60x^2 - 64xy - 60y^2$

Factor completely. First factor out a rational number. Then factor the binomial or trinomial so the terms have integral coefficients.

55. $\frac{1}{2}x^2 - \frac{1}{2}$

56. $\frac{x^3}{3} - \frac{x^2}{3} - \frac{2x}{3}$

57. $1.8x^2 + 0.3x - 10.5$

58. $0.4y^3 - 0.4y^2 + 0.1y$

Factor completely. Assume all exponents are positive integers.

59. $6x^{4a} - 54y^{4b}$

60. $x^{7c+21} + 5x^7$

61. $-60y^{4a} - 55y^{2a}z^{b} + 75z^{2b}$

62. $m^{p+6} - m^{p+4}$

63. $6(x + y)^2 - 5z(x + y) + z^2$

64. $3(2a - b)^2 + 17(2a - b) + 10$

65. $a^2 - 2a + 1 - 4b^2 - 12b - 9$

66. $3ab^2x^{10n} - 24ab^2x^{5n}y^{3n} + 48ab^2y^{6n}$

67. $(x + y + 1)^2 - 2(x + y + 1)(x - y + 3) + (x - y + 3)^2$

68. $(x - y + 2)^2 - 2(x - y + 2)(x + y + 2) + (x + y + 2)^2$

Applications

69. **Geometry** The area of the ring below is $\pi M^2 - \pi m^2$ square units. Express the area in completely factored form.

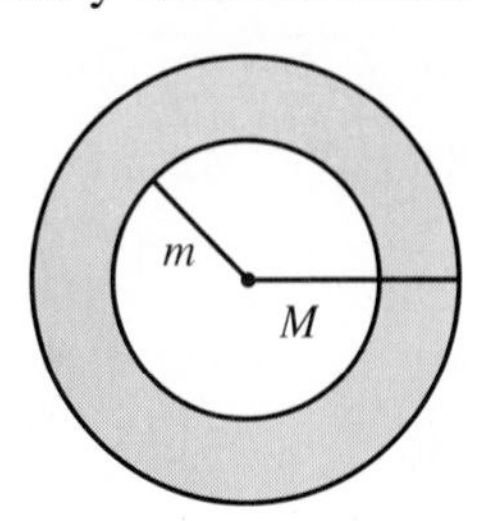

70. **Geometry** The volume of the shaded portion of the cylinder below is $(\pi R^2h - \pi r^2h)$ cm^3. Express the volume in factored form.

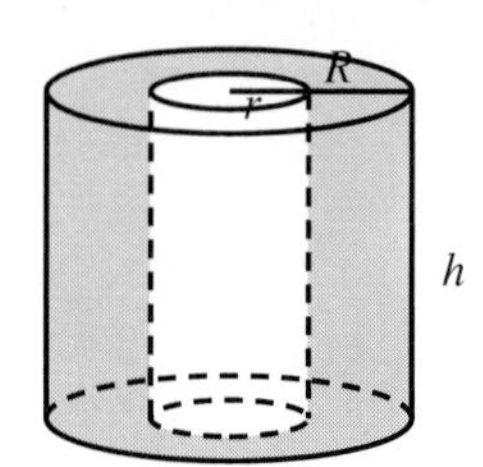

Mixed Review

71. If $f(x) = x^2 + 3$ and $g(x) = x^2 - 3$, find $f[g(4)]$.

72. If $h(x) = 4x - 1$, find $h^{-1}(x)$.

73. Write an equation of the line that passes through $(-2, 3)$ and is parallel to the graph of $3x + 4y = 5$.

Solve for x and indicate any restrictions on the variables.

74. $a^2(x + 2) - 4 = c(x - 5)$

75. $2x + 1 = \frac{5xy}{3} - t$

76. Solve and graph the solution set on a number line: $|4x + 3| - 6 < 15$

77. Solve for X:

$$4X - 2\begin{bmatrix} 1 & 9 \\ -2 & 4 \\ 5 & 10 \end{bmatrix} = \begin{bmatrix} 6 & -14 \\ 24 & 28 \\ 2 & -20 \end{bmatrix}$$

78. Find a zero matrix for $\begin{bmatrix} 3 & 4 & 5 \\ -1 & -2 & -3 \end{bmatrix}$

Evaluate each determinant.

79. $\begin{vmatrix} 3 & 1 \\ 2 & 0 \end{vmatrix}$

80. $\begin{vmatrix} -5 & 4 \\ 3 & 6 \end{vmatrix}$

81. $\begin{vmatrix} 2 & 1 & 0 \\ 0 & 5 & 4 \\ 3 & 2 & 6 \end{vmatrix}$

6.7

Solving Polynomial Equations

Objective: To solve polynomial equations by factoring

The width of the huge Vehicle Assembly Building at the Kennedy Space Center in Florida is 60 m less than the length, and the area of the rectangular base of the building is 34,444 m^2. If l represents the length, then $l - 60$ represents the width. The product of the length and the width is equal to the area.

$$l(l - 60) = 34{,}444$$

or, $$l^2 - 60l - 34{,}444 = 0$$

An equation that can be written with a polynomial as one side and 0 as the other side is a **polynomial equation.** Solving polynomial equations involves solving linear equations.

Capsule Review

To solve a linear equation, first use the addition or subtraction property of equations, then use the multiplication or division property.

EXAMPLE **Solve:** $\mathbf{2x + 5 = 0}$

$$2x + 5 = 0$$
$$2x = -5$$
$$x = -\frac{5}{2}$$

Check:

$$2x + 5 = 0$$
$$2\left(-\frac{5}{2}\right) + 5 \stackrel{?}{=} 0$$
$$-5 + 5 \stackrel{?}{=} 0$$
$$0 = 0 \checkmark$$

Solve and check.

1. $x + 7 = 0$ **2.** $x - 3 = 0$ **3.** $3x - 5 = 0$ **4.** $3x + 1 = 0$

5. $3x + 2 = 0$ **6.** $2x + 8 = 0$ **7.** $4x - 120 = 0$ **8.** $5x + 216 = 0$

Polynomial equations with 0 as one side can often be solved by factoring the polynomial into linear factors and then using the Zero-product property.

Zero-Product Property

For all real numbers a and b, $ab = 0$ if and only if $a = 0$ or $b = 0$.

A polynomial equation of the form $ax^2 + bx + c = 0$, $a \neq 0$, is a **quadratic,** or **second-degree, equation.** Any value of the variable that satisfies the equation is a *solution* of the polynomial equation.

EXAMPLE 1 **Solve and check: $3x^2 - 13x - 10 = 0$**

$$3x^2 - 13x - 10 = 0$$

$$(3x + 2)(x - 5) = 0 \qquad \textit{Factor.}$$

$$3x + 2 = 0 \quad \text{or} \quad x - 5 = 0 \qquad \textit{Zero-product property}$$

$$x = -\frac{2}{3} \quad \text{or} \quad x = 5 \qquad \textit{Solve each equation.}$$

Check:

$$3x^2 - 13x - 10 = 0$$

$$3\left(-\frac{2}{3}\right)^2 - 13\left(-\frac{2}{3}\right) - 10 \stackrel{?}{=} 0$$

$$\frac{4}{3} + \frac{26}{3} - \frac{30}{3} \stackrel{?}{=} 0$$

$$0 = 0 \checkmark$$

$$3x^2 - 13x - 10 = 0$$

$$3(5)^2 - 13(5) - 10 \stackrel{?}{=} 0$$

$$75 - 65 - 10 \stackrel{?}{=} 0$$

$$0 = 0 \checkmark$$

The solutions are $-\frac{2}{3}$ and 5.

If one side of a polynomial equation is not 0, you may be able to rewrite the equation in that form and then factor the resulting polynomial. If the GCF of the terms has a coefficient other than 1, divide both sides of the equation by that number before factoring.

EXAMPLE 2 **Solve and check: $6y^2 = 48y$**

$$6y^2 = 48y$$

$$6y^2 - 48y = 0 \qquad \textit{Rewrite as an expression equal to 0.}$$

$$y^2 - 8y = 0 \qquad \textit{The GCF of } 6y^2 \textit{ and } 48y \textit{ is } 6y, \textit{ so divide both sides by 6.}$$

$$y(y - 8) = 0 \qquad \textit{Factor.}$$

$$y = 0 \quad \text{or} \quad y - 8 = 0 \qquad \textit{Zero-product property}$$

$$y = 0 \quad \text{or} \quad y = 8$$

Check:

$$6y^2 = 48y \qquad\qquad 6y^2 = 48y$$

$$6(0)^2 \stackrel{?}{=} 48(0) \qquad\qquad 6(8)^2 \stackrel{?}{=} 48(8)$$

$$0 = 0 \checkmark \qquad\qquad 384 = 384 \checkmark$$

Check your work with a calculator. In this case, the solutions are 0 and 8.

The polynomial equation in the next example is a **cubic,** or **third-degree, equation** since the highest degree of a term is 3. Polynomial equations of degree four and higher are usually not given special names. Note that the zero-product property can be extended to any number of factors, since a product is zero if and only if one of its factors is zero.

EXAMPLE 3 **Solve and check:** $x^3 + x^2 - 4x - 4 = 0$

$$
\begin{aligned}
x^3 + x^2 - 4x - 4 &= 0 \\
(x^3 + x^2) - (4x + 4) &= 0 && \textit{Factor by grouping.} \\
x^2(x + 1) - 4(x + 1) &= 0 \\
(x + 1)(x^2 - 4) &= 0 \\
(x + 1)(x + 2)(x - 2) &= 0 && \textit{Factor again.}
\end{aligned}
$$

$x + 1 = 0 \quad \text{or} \quad x + 2 = 0 \quad \text{or} \quad x - 2 = 0$ *Zero-product property*

$x = -1 \quad \text{or} \quad x = -2 \quad \text{or} \quad x = 2$

Check:

$$
\begin{aligned}
x^3 + x^2 - 4x - 4 &= 0 \\
(-1)^3 + (-1)^2 - 4(-1) - 4 &\stackrel{?}{=} 0 \\
-1 + 1 + 4 - 4 &\stackrel{?}{=} 0 \\
0 &= 0 \checkmark
\end{aligned}
$$

$$
\begin{aligned}
x^3 + x^2 - 4x - 4 &= 0 \\
(-2)^3 + (-2)^2 - 4(-2) - 4 &\stackrel{?}{=} 0 \\
-8 + 4 + 8 - 4 &\stackrel{?}{=} 0 \\
0 &= 0 \checkmark
\end{aligned}
$$

$$
\begin{aligned}
x^3 + x^2 - 4x - 4 &= 0 \\
(2)^3 + (2)^2 - 4(2) - 4 &\stackrel{?}{=} 0 \\
8 + 4 - 8 - 4 &\stackrel{?}{=} 0 \\
0 &= 0 \checkmark
\end{aligned}
$$

The solutions are -1, -2, and 2.

Sometimes a factor is repeated in the factorization of a polynomial. For example, $x^2 - 4x + 4 = 0$ is equivalent to $(x - 2)^2 = 0$. In this case, 2 is a *double solution* of the equation $x^2 - 4x + 4 = 0$. Solutions that are obtained from repeated factors are called **multiple solutions.** For $(x - 5)^3 = 0$, 5 is a *triple solution,* or a solution with **multiplicity** 3.

EXAMPLE 4 **Solve and check:** $12x^3 - 60x^2 + 75x = 0$

$$
\begin{aligned}
12x^3 - 60x^2 + 75x &= 0 \\
3x(4x^2 - 20x + 25) &= 0 \\
3x(2x - 5)^2 &= 0
\end{aligned}
$$

$3x = 0 \quad \text{or} \quad 2x - 5 = 0 \quad \text{or} \quad 2x - 5 = 0$

$x = 0 \quad \text{or} \quad x = \frac{5}{2} \quad \text{or} \quad x = \frac{5}{2}$

Check to show that the solutions are 0 and $\frac{5}{2}$; $\frac{5}{2}$ is a double solution.

CLASS EXERCISES

Examine the following equations.

a. $-2x(x + 5) = 0$ **b.** $(x + 1)(2x - 3) = 0$ **c.** $x(x + 1)(2x - 1) = 0$

d. $x^2(x - 2) = 0$ **e.** $(2x - 1)^2(x - 1) = 0$ **f.** $(x - 1)(2x + 1) = 0$

1. Which of the above equations have 0 as a solution? Explain.
2. Which of the equations have -1 as a solution? Explain.
3. Which of the equations have $\frac{1}{2}$ as a solution? Explain.

Solve and check.

4. $m^2 - 3m + 2 = 0$
5. $g^2 + 7g = 0$
6. $x^2 - 4 = 0$
7. $y^2 - 6y + 8 = 0$
8. $q^2 - 2q + 1 = 0$
9. $2x^2 + 9x - 5 = 0$
10. Find the dimensions of the base of the Vehicle Assembly Building at the Kennedy Space Center. Use the equation on page 285. *Hint:* 158 is a factor of 34,444.

PRACTICE EXERCISES

 Use technology where appropriate.

Solve and check.

1. $x^2 - 7x + 12 = 0$
2. $x^2 + 5x = 0$
3. $x^2 - 8x + 15 = 0$
4. $x^2 + 9x + 20 = 0$
5. $2x^2 - 3x + 1 = 0$
6. $x^2 - 8x + 12 = 0$
7. $x^2 - 6x = 0$
8. $x^2 - 8x = 0$
9. $4x^2 - 8x = 0$
10. $3x^2 - 6x = 0$
11. $-8p^2 = 40p$
12. $12g^2 = -16g$
13. $x^2 + 10x = -25$
14. $x^2 - 8x + 16 = 0$
15. $4x^2 + 9 = 12x$
16. $p^2 - 4p + 4 = 0$
17. $x^2 - 25 = 0$
18. $9y^2 = 6y - 1$
19. $x^3 - 3x^2 - x + 3 = 0$
20. $x^3 + 4x^2 - 4x - 16 = 0$
21. $6x^2 + 11x - 10 = 0$
22. $8m^2 + 10m - 25 = 0$
23. $9x^3 - 18x^2 - 4x + 8 = 0$
24. $4x^3 - 8x^2 - x + 2 = 0$
25. $c^4 - 2c^2 + 1 = 0$
26. $4p^4 - 12p^2 + 9 = 0$
27. $(y - 3)^3 - (y - 3)^2 = 0$
28. $(y - 5)^3 - (y - 5)^2 = 0$
29. $(2x - 1)^2 + 3(2x - 1) = 0$
30. $(3x - 1)^2 + 4(3x - 1) = 0$
31. $28q = 6q^2$
32. $36p^2 = 8p$
33. $2t^4 - 5t^2 + 3 = 0$
34. $15t^4 - 34t^2 + 15 = 0$

Solve for x.

35. $x^2 - 4x + 4 - b^2 = 0$

36. $x^2 - 10x + 25 - a^2 = 0$

37. $ax - bx + 3a - 3b = 0$

38. $25x^2 - 10x + 1 = 9a^2b^2$

39. $(x + a)^2 - 8(x + a) + 15 = 0$

40. $(x - 2c)^2 - 10(x - 2c) + 21 = 0$

Applications

41. **Manufacturing** A rectangular piece of cardboard has an area of 108 cm^2. The length of the cardboard is 3 times its width. Find the length and width of the cardboard.

42. **Sports** Shaun set a new state record by throwing a baseball at a speed of 98 mi/h. If he had thrown it straight up, the height in feet, f, at s seconds would be given by the equation $f = 144s - 16s^2$. After how many seconds would the ball have a height of 288 ft? Explain your answer.

43. **Construction** Marisel is building a storage shed behind her father's bodega and wants to form a triangular brace, using a 75-in. board as the hypotenuse. Given that the vertical leg is to be 51 in. longer than the horizontal leg, use the Pythagorean theorem to find the length of each.

BIOGRAPHY: Julia Robinson

Julia Robinson was born in Missouri in 1919 and raised in southern California. Mathematics was her favorite subject; she was excited by number theory and kept her sister up at night recounting theorems. Reading E. T. Bell's *Men of Mathematics* gave her an insight into what mathematicians do. While studying at Berkeley, she married her number theory professor. Since members of the same family could not teach in the same department at Berkeley, she became a mathematician. She was fascinated with Hilbert's Tenth Problem: to find an effective method for determining if a given diophantine equation is solvable in integers. Indeed, every year when she blew out the candles on her birthday cake, she wished for its solution! In 1970, she got her wish. Julia Robinson, Martin Davis, and Yuri Matijasevic are credited with the solution of this problem. Julia Robinson was the first woman mathematician elected to the National Academy of Sciences. She was also elected President of the American Mathematical Society in 1982.

Research Julia Robinson and David Hilbert.

6.8 Problem Solving Strategy: Solve a Simpler Problem

Many mathematics problems can be solved by employing the strategy of *solving a simpler problem*. The solution to the simpler problem is then used to solve the original, more difficult problem.

EXAMPLE The Bergen County Recreational League has n teams in it. Each team plays every other team twice, once at home and once away. How many games are played altogether?

Understand the Problem *What are the given facts?* There are n teams. Each team plays every other team, once at home and once away.

What are you asked to find? How many games are played altogether?

Plan Your Approach *Choose a strategy*. Use the strategy of solving a simpler problem. Choose some specific values for n, say 2, 3, and 4, and solve these problems. Suppose there are 2 teams. How many games are played?

Team 1 plays Team 2 at home and away. The home game for Team 1 is the away game for Team 2 and vice versa.
Answer: There are 2 games played.

Suppose there are 3 teams. How many games are played?

Team 1 plays Team 2 at home and away: 2 games.
Team 2 plays Team 3 at home and away: 2 games.
Team 1 plays Team 3 at home and away: 2 games.
Answer: For 3 teams, a total of 6 games are played.

Suppose there are 4 teams. How many games are played?

Team 1 plays 2 games each with Teams 2, 3, and 4: 6 games.
Team 2 plays 2 games each with Teams 3 and 4: 4 games.
Team 3 plays 2 games with Team 4: 2 games.
Answer: For 4 teams, a total of 12 games are played.

Complete the Work The strategy of solving simpler problems, using a specific number of teams can be continued until a pattern emerges.

Number of teams:	2	3	4	5	n
Number of games:	$2 = 2 \cdot 1$	$6 = 3 \cdot 2$	$12 = 4 \cdot 3$	$20 = 5 \cdot 4$	$n(n - 1)$

Interpret the Results *State your answer*. There are $n(n - 1)$ games played.

CLASS EXERCISES

1. Can more than one problem solving strategy be used to solve some problems?
2. Name two methods of simplifying a system of two linear equations in two variables.
3. What would you do if a particular strategy to solve a problem does not work?
4. Do you think the strategy of solving a simpler problem can be used effectively to solve many mathematics problems?
5. Do you think you will become a better problem solver as you learn more problem solving strategies?
6. Name three problem solving strategies that were used in the Example.
7. Is it always clear which strategy to use first in solving a mathematics problem?

PRACTICE EXERCISES

Classify each statement as true or false.

1. An equation in two variables expresses a relationship between two unknown numbers.
2. An equation in two variables has only one solution.
3. Equations can be used as mathematical models to solve real-world problems.
4. The strategy of solving a simpler problem can be used to solve word problems as well as other kinds of mathematics problems.
5. Patterns are not useful in solving problems.
6. Only one strategy can be used to solve a given problem.

Find the number of games played for the given number of teams in the example.

7. $n = 6$ **8.** $n = 7$ **9.** $n = 8$ **10.** $n = 12$

Polynomial equations can be solved by using the strategy of solving a simpler problem. First, the polynomial equation is changed to simpler linear equations. Then the linear equations are solved. Solve each of the following polynomial equations by factoring and using the zero-product property.

11. $y^2 - 2y - 35 = 0$ **12.** $x^2 - 5x + 4 = 0$ **13.** $t^2 - 4t = -4$

14. $a^2 - 2a - 15 = 0$ **15.** $b^2 + 4b = 12$ **16.** $x^2 + 10x + 25 = 0$

Solve each problem.

17. A business has 67 employees. If there are 9 more women than men, how many women and men work for the company?

18. A backyard is rectangular in shape. The width is 40 ft less than the length. If the area of the yard is 25,200 ft^2, find the length and width of the yard.

19. Find the sum of the first n whole numbers.

Mixed Problem Solving Review

1. Find two rational numbers such that the second number is six more than three times the first. The sum of the numbers is $19\frac{1}{3}$.

2. If the sum of the base angles of an isosceles triangle is one-half the vertex angle, find the measure of each angle.

3. A private detective charges a $400 retainer fee plus $125 per day for his services. If the cost of hiring him is a linear function, find the total cost of a job requiring 5 days to complete.

4. A coin bank contains only nickels and dimes. The bank contains 34 coins. If 2 nickels and 2 dimes are removed, the total value of the coins would be $2. Find the number of nickels and the number of dimes in the bank.

5. The sum of the digits of a three-digit number is 12. Three times the hundreds digit minus twice the tens digit is 5. If six times the units digit is added to four times the tens digit, the result is 50. Find the number.

6. Two motorists leave a rest stop at 3:00 PM and travel in opposite directions. One is traveling at 55 mi/h and the other is driving at 45 mi/h. At what time will they be 600 mi apart?

PROJECT

An interesting mathematical fact is the following: If you multiply any four consecutive integers and then add 1, the result is always a perfect square. For example

$$1 \times 2 \times 3 \times 4 + 1 = 25 = 5^2$$

Check other examples of your own using a calculator if necessary. Write a polynomial expression to represent the product of four consecutive integers. Use the distributive property repeatedly to multiply all the terms of the polynomial. What is the degree of the polynomial?

6.9

Problem Solving: Using Polynomial Equations

Polynomial equations serve as mathematical models for many different types of word problems. Since a polynomial equation often has more than one solution, there may be more than one solution to the problem itself.

EXAMPLE 1 Find three consecutive even integers if the product of the second and third integers is 200 more than 10 times the first.

Understand the Problem Consecutive even integers differ by 2.

Plan Your Approach

Let x = first even integer
Then $x + 2$ = second even integer
and $x + 4$ = third even integer

Complete the Work

product of second and third = 200 more than 10 times first

$$(x + 2)(x + 4) = 200 + 10x$$
$$x^2 + 6x + 8 = 200 + 10x \quad \textit{Multiply.}$$
$$x^2 - 4x - 192 = 0$$
$$(x - 16)(x + 12) = 0 \quad \textit{Factor.}$$
$$x - 16 = 0 \quad \text{or} \quad x + 12 = 0$$

Therefore, $x = 16$ or $x = -12$
and $x + 2 = 18$ or $x + 2 = -10$
and $x + 4 = 20$ or $x + 4 = -8$

Interpret the Results The numbers are 16, 18, and 20, or −12, −10, and −8. Since each number is a multiple of 2, they are all even integers. Is the product of the second and third integer 200 more than 10 times the first integer?

$$(18)(20) \stackrel{?}{=} 200 + 10(16)$$
$$360 \stackrel{?}{=} 200 + 160$$
$$360 = 360 \ \checkmark$$

$$(-10)(-8) \stackrel{?}{=} 200 + 10(-12)$$
$$80 \stackrel{?}{=} 200 - 120$$
$$80 = 80 \ \checkmark$$

When a polynomial equation is used to solve a problem, a solution to the equation may not be a solution to the problem itself. This is why it is important to check all possible solutions.

EXAMPLE 2 The height of the sail on a boat is 7 ft less than 3 times the length of its base. If the area of the sail is 68 ft^2, find its height and the length of its base.

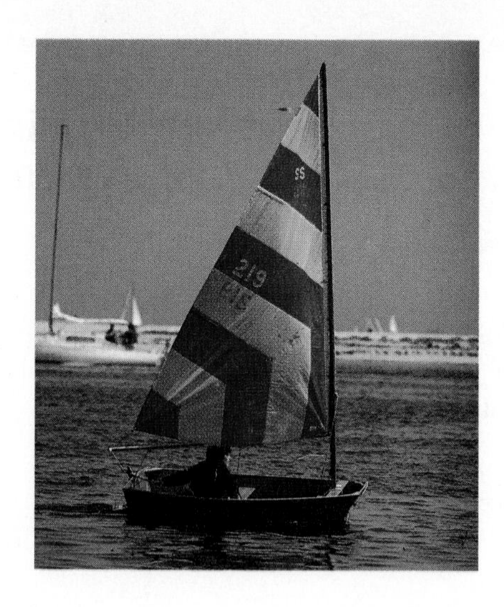

Understand the Problem A triangle is a model for the sail, and the formula for the area of a triangle is

$$A = \frac{1}{2}bh$$

Plan Your Approach Let x = length of base, in feet
Then $3x - 7$ = height, in feet

$$\frac{1}{2}bh = A$$

$$\frac{1}{2}x(3x - 7) = 68$$

Complete the Work

$$x(3x - 7) = 136 \quad \textit{Multiply each side by 2.}$$

$$3x^2 - 7x - 136 = 0$$

$$(x - 8)(3x + 17) = 0$$

$$x - 8 = 0 \quad \text{or} \quad 3x + 17 = 0$$

$$x = 8 \quad \text{or} \quad x = -\frac{17}{3}$$

Interpret the Results Since the length of the base of the sail cannot be a negative number of feet, the answer $-\frac{17}{3}$ is rejected.
Evaluate $3x - 7$ when $x = 8$.

$$3x - 7 = 3(8) - 7$$
$$= 17$$

The height of the sail is 17 ft and the length of its base is 8 ft. Is the height 7 ft less than 3 times the length of the base?

$$17 \stackrel{?}{=} 3(8) - 7$$
$$17 = 17 \checkmark$$

Is the area 68 ft^2?

$$\frac{1}{2} \cdot 8 \cdot 17 \stackrel{?}{=} 68$$
$$68 = 68 \checkmark$$

The approximate height of an object that is projected upward from ground level can be determined by using the following formulas.

$$h = vt - 4.9t^2$$

where t is the time in **seconds,** h is the height in **meters,** and v is the initial upward velocity in **meters per second.**

$$h = vt - 16t^2$$

where t is the time in **seconds,** h is the height in **feet,** and v is the initial upward velocity in **feet per second.**

EXAMPLE 3 A rocket is fired upward from ground level with an initial velocity of 68.6 m/s. When will it reach a height of 196 m?

Understand the Problem The data is given in meters and in meters per second, so use the formula $h = vt - 4.9t^2$.

Plan Your Approach In the formula, t represents the time it takes the rocket to reach a height of 196 m.

$$h = vt - 4.9t^2$$
$$196 = 68.6t - 4.9t^2$$

Complete the Work

$$4.9t^2 - 68.6t + 196 = 0$$
$$49t^2 - 686t + 1960 = 0$$
$$t^2 - 14t + 40 = 0 \quad \text{Divide both sides by 49.}$$
$$(t - 4)(t - 10) = 0$$
$$t = 4 \quad \text{or} \quad t = 10$$

Interpret the Results The rocket reaches a height of 196 m in 4 s and in 10 s. There are two answers, since the rocket reaches this height once on the way up and once on the way down.

After 4 s, is the rocket at 196 m?
$196 \stackrel{?}{=} 68.6(4) - 4.9(4)^2$
$196 \stackrel{?}{=} 274.4 - 78.4$
$196 = 196$ ✓

After 10 s, is the rocket at 196 m?
$196 \stackrel{?}{=} 68.6(10) - 4.9(10)^2$
$196 \stackrel{?}{=} 686 - 490$
$196 = 196$ ✓

CLASS EXERCISES

Write an equation for each problem.

1. Find three consecutive even integers if the product of the second and third integer is 78 more than 9 times the first.

2. An object is hurled upward from the ground at an initial velocity of 128 ft/s. When will the object reach a height of 192 ft?

3. The width of a rectangular garden is 4 ft less than the length. If the area of the rectangle is 96 ft^2, what are the dimensions of the rectangle?

4. A rectangular pool 60 ft long by 40 ft wide is surrounded by a walk of uniform width. Find the width of the walk if its area is 416 ft^2.

PRACTICE EXERCISES

1. Find three consecutive odd integers if the product of the first and second integer is 7 less than 10 times the third integer.
2. The length of the sides of a triangular flower bed can be expressed as integral numbers of feet, and the three integers are consecutive. The product of the shortest and longest length is 27 ft more than 12 times the remaining length. Find the length of each side of the flower bed.
3. The product of two consecutive multiples of 5 is 500. Find the numbers.
4. Find two numbers that have a sum of 25 and a product of 156.
5. A wall of a tent is shaped like a triangle and has an area of 48 ft^2. The base of the wall is 2 ft shorter than 3 times the height. Find the length of the base and the height of the wall.
6. A triangular frame has an area of 36 m^2. The height of the frame is 1 m longer than the base. Find the height and the length of the base.
7. A rectangular garden is 10 ft wide and 12 ft long. A walk of uniform width surrounds the garden. If the total area of the garden and the walk is 168 ft^2, what is the width of the walk?
8. A cement walk of uniform width surrounds a rectangular swimming pool that is 10 m wide and 50 m long. Find the width of the walk if its area is 864 m^2.
9. The area of a rectangular card is 128 cm^2. The length of the card is twice its width. What are the dimensions of the card?
10. The area of a rectangular piece of sheet metal is 216 ft^2. The width of the sheet is 6 ft less than the length. What are the dimensions of the piece of metal?
11. A rocket is launched from ground level with an initial velocity of 224 ft/s. When will the rocket reach a height of 528 ft?
12. From ground level, a ball is thrown upward with an initial velocity of 83.3 m/s. When will the ball reach a height of 294 m?
13. The sum of the number of A's and B's on a math test is 15. There are more B's than A's, and the sum of the squares of the two numbers is 113. Find the number of A's and the number of B's.

14. The units digit of a two-digit number is twice the tens digit. If the square of the units digit equals the number itself, what is the number?

15. The lengths of the three sides of a right triangle are consecutive even integers. Find the length of each side of the triangle. (Recall the Pythagorean Theorem: $a^2 + b^2 = c^2$, where a and b are the lengths of the two legs and c is the length of the hypotenuse.)

16. A landscaper cut the grass of a rectangular lawn in a strip of uniform width along the edge of the lawn. This left one-fifth of the grass uncut. What is the width of the strip if the lawn is 50 ft wide and 60 ft long?

17. A rocket is fired upward at an initial velocity of 1008 ft/s from a tower that is 120 ft high. When will the rocket reach a height of 3896 ft above ground level?

18. A rocket is launched vertically and it runs out of fuel at an altitude of 1000 m above ground level, when its velocity is 78.4 m/s. How long will it be before the rocket again is at an altitude of 1000 m above ground level?

19. The length of a rectangular table is 1 ft more than twice the length of a side of a square rug and the width of the table is 3 ft less than the length of a side of the rug. If the area of the table is 81 ft^2 greater than the area of the rug, what is the area of the rug?

20. The product of 25 less than the square of a number and 3 more than the number is zero. Find the number(s) that meets these requirements.

TEST YOURSELF

Factor completely.

1. $3x^2 - 10x - 8$ **2.** $12x^2 - 3y^2$ 6.5–6.6

3. $x^3 + x^2y - x - y$ **4.** $x^3 - xy^2$

Solve and check.

5. $x^2 + 3x - 18 = 0$ **6.** $4x^2 + 1 = 4x$ **7.** $x^3 = 4x$ 6.7

8. A rectangle is 8 cm longer than it is wide, and its area is 153 cm^2. Find the dimensions of the rectangle. 6.8

9. The square of a positive integer is 6 more than twice the sum of the next two consecutive integers. What are the integers?

INTEGRATING ALGEBRA
Data Analysis and Polynomial Functions

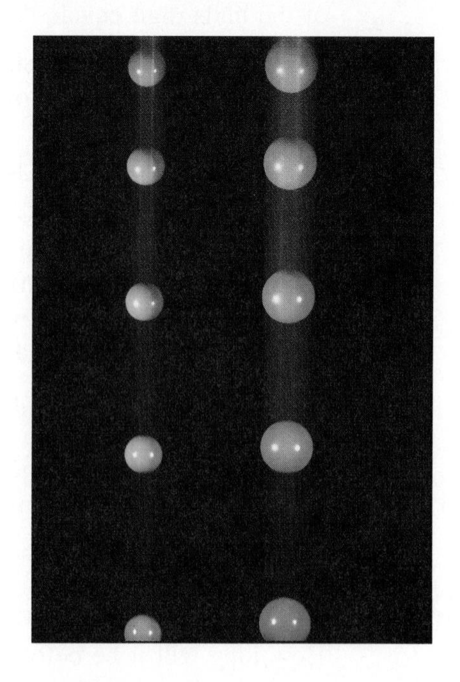

Did you know that a set of data in two variables can sometimes be described by a polynomial function? Consider the stroboscopic multiple-exposure photograph of an object in free fall to the right. Is there a relationship between elapsed time and the distance the ball has traveled? If so, what kind of function best describes this relationship? The *method of finite differences* enables you to determine whether or not the relationship between paired data can be described by a polynomial function.

The method of finite differences can be used only when the x-values of the paired data increase in constant increments. When that occurs, then a difference table can be generated, where the 1st-order differences are formed by finding the differences between successive y-values. The 2nd-order differences record the differences between successive 1st-order differences and so on, until all the differences in a row equal 0 or until it is obvious that the differences will never all be zero.

EXAMPLE Given the set of paired data below, construct a difference table.

$$\{(x, y): (1, -1), (2, 2), (3, 9), (4, 20), (5, 35), (6, 54)\}$$

Since the x-values are 1, 2, 3, 4, 5, and 6, they increase in constant increments, and a difference table can be constructed.

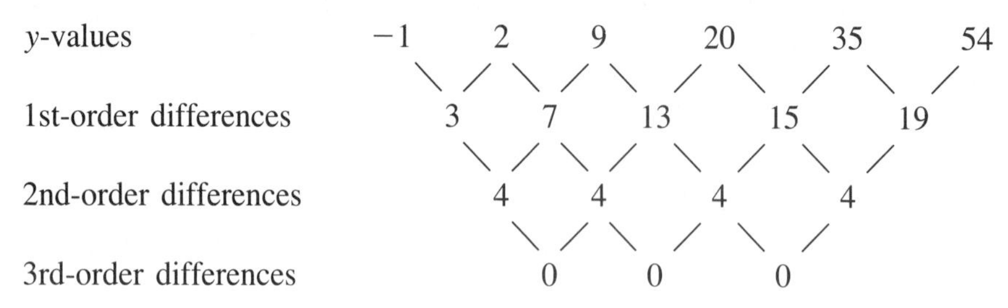

y-values	−1		2		9		20		35		54
1st-order differences		3		7		13		15		19	
2nd-order differences			4		4		4		4		
3rd-order differences				0		0		0			

Note that each 3rd-order difference is equal to zero.

The results of a difference table are interpreted by the following rule: if the $(n + 1)$th order differences are zero, then a polynomial of degree n can be found to relate the x and y values of the data set. Otherwise, the relationship between x and y cannot be described by a polynomial function.

Use the following guidelines to determine the form of the polynomial function.

If the 1st order differences are 0, then $y = c$, with c constant.
If the 2nd order differences are 0, then $y = ax + b$, with a and b constant.
If the 3rd order differences are 0, then $y = ax^2 + bx + c$, with a, b, and c constant.
If the 4th order differences are 0, then $y = ax^3 + bx^2 + cx + d$, with a, b, c, and d constant.

Since the 3rd-order differences in the example above are all equal to zero, a 2nd-order polynomial of the form $y = ax^2 + bx + c$, with a, b, and c constant, can be used to describe the relationship between x and y.

EXERCISES

For each of the following sets of data, construct a difference table. Determine which form of polynomial function can be used to describe the relationship, if possible.

1. (15, 25), (20, 35), (25, 45), (30, 55), (35, 65), (40, 75)

2. (1, 5), (3, 5), (5, 13), (7, 29), (9, 53), (11, 85)

3. (1, −10), (2, −11), (3, −6), (4, 11), (5, 46), (6, 105)

4. (2, 5), (4, 41), (6, 65), (8, 192), (10, 293)

5. Show that the quadratic polynomial $y = 2x^2 - 3x$ describes the data in the example and use this polynomial to generate the next three pairs of data.

6. Determine if a polynomial function can be used to describe the data relating the speed of a car x to the approximate distance in feet that a car must travel before stopping y. If so, give the form of the polynomial.

Speed of Car (mi/h)	Stopping Distance (feet)
20	36
25	50
30	67
35	87
40	110
45	136

CHAPTER 6 SUMMARY AND REVIEW

Vocabulary

binomial (259)
coefficient (256)
conjugates (266)
constant (276)
cubic equation (287)
degree of monomial (256)
degree of polynomial (260)
difference of two cubes (272)
difference of two squares (266)
exponent (254)
factoring by grouping (280)
FOIL method (266)
greatest common factor (GCF) (270)
linear term (276)
monomial (256)
multiple solutions (287)
multiplicity (287)
perfect square trinomial (266)
polynomial (259)
polynomial equation (285)
prime polynomial (281)
quadratic equation (286)
quadratic term (275)
quadratic trinomial (275)
sum of two cubes (272)
term of a polynomial (259)
trinomial (259)
zero-product property (286)

Properties of Exponents If r and s are real numbers and m and n are positive integers, then **6.1–6.2**

$$r^m \cdot r^n = r^{m+n} \qquad (r^m)^n = r^{mn} \qquad (rs)^m = r^m s^m$$

If r and s are real numbers, $s \neq 0$, and m and n are positive integers, with $m > n$, then

$$\frac{r^m}{r^n} = r^{m-n} \qquad \left(\frac{r}{s}\right)^m = \frac{r^m}{s^m}$$

Degree of a Monomial and a Polynomial The degree of a monomial is the sum of the exponents of its variables. The degree of a polynomial is the degree of the term of highest degree.

Simplify.

1. $(5a^2b^3)(-3ab^3c^2)$ **2.** $(2xy^3z)(4x^2y^4z^3)$ **3.** $(-2mn^2)^3$

Give the degree of each polynomial.

4. $9xyz$ **5.** $5x^3y^2 + 3xy^6 + y^{10}$ **6.** $6x^2 - xy - y^2$

Adding and Subtracting Polynomials To find the sum of two or more polynomials, add their like terms. To subtract one polynomial from another, add the additive inverse of the polynomial being subtracted.

Add or subtract as indicated.

7. $(3a + 2b - 5c) + (4b - 2a + 6c)$
8. $(5x - 6x^2 - 10) - (3x^2 - 2x + 23)$
9. $(2a + b) - (b - 3a) + 5a$
10. $(4y^2 - 7 + 9y) + (6y - 3y^2 - 14)$

Multiplying Two Binomials To multiply two binomials use the FOIL method, and look at special-product patterns: 6.3

Perfect square trinomial	$(a + b)^2 = a^2 + 2ab + b^2$
	$(a - b)^2 = a^2 - 2ab + b^2$
Difference of two squares	$(a + b)(a - b) = a^2 - b^2$

Multiply.

11. $(5x - y)(3x + 2y)$
12. $(3s - 2t)^2$
13. $(x + 4y)(x - 4y)$
14. $(s + t)(s^2 - st + t^2)$
15. $(3y - z)(9y^2 + 3yz + z^2)$
16. $(x - 2y)(x + y)(3x - y)$

Factoring Patterns Look first for a greatest common factor (GCF). Is the polynomial one of the special types? 6.4–6.6

$$a^2 - b^2 = (a - b)(a + b)$$
$$a^2 + 2ab + b^2 = (a + b)^2$$
$$a^2 - 2ab + b^2 = (a - b)^2$$
$$a^3 + b^3 = (a + b)(a^2 - ab + b^2)$$
$$a^3 - b^3 = (a - b)(a^2 + ab + b^2)$$

Is the polynomial a quadratic trinomial $ax^2 + bx + c$? Can the polynomial be grouped and then factored?

Factor completely.

17. $a^2b^2 - 25c^2$
18. $q^3 + 343p^3$
19. $36p^2 + 60pq + 25q^2$
20. $h^2 - 4h - 32$
21. $2a^2 - 11a + 9$
22. $3y^2 + 8xy + 4x^2$
23. $x^3 + x^2 + x + 1$
24. $4b^2 + 16a^2$
25. $6x^4 - 39x^2 + 60$

Solving Polynomial Equations To solve a polynomial equation by factoring, use the zero-product property: 6.7

For all real numbers a and b, $ab = 0$, if and only if $a = 0$ or $b = 0$.

26. $x^2 - 5x + 6 = 0$
27. $6x^2 = x + 1$
28. $x^3 - 4x = 0$

29. Find three consecutive even integers if the product of the first and third integer is 192. 6.9

30. The base of a triangle is 3 times the length of the altitude to that base. The area of the triangle is 96 cm^2. Find the length of the base.

CHAPTER 6 TEST

Give the degree of each polynomial.

1. $9xy^3z^6$

2. $3x^3 - 2x^2y^2 + y^2$

Perform the indicated operations.

3. $(3a^2 - 4a + 6) + (4a^2 + 2a - 8)$

4. $(3x^2 + 7) - (x^2 - 4x + 2)$

5. $(4ab^3c^2)(5ab^2c)$

6. $(6x^4y^3)^2$

7. $(a + 3b)(a - 3b)$

8. $(5x - 2)(x + 3)$

9. $(5x - 7)^2$

10. $(2y - 3)(4y^2 + 6y + 9)$

Factor completely.

11. $2x^2 + 3x - 5$

12. $9c^2d^2 - 25f^2$

13. $5x^2 + 35y^2$

14. $8x^3 + y^3$

15. $4x^3 + 8x^2 - 12x$

16. $25x^2 - 70xy + 49y^2$

17. $am - 3a + 2m - 6$

18. $3ax^2 - 27a + 2x^2 - 18$

Solve and check.

19. $6x^2 - x - 15 = 0$

20. $3x^2 + 8x = 3$

21. $x^2 + 12x + 36 = 0$

22. $2x^3 + 3x^2 - 18x - 27 = 0$

23. The product of two consecutive positive integers is 210. Find the integers.

24. Find three consecutive even integers if the product of the second and third is eighty-eight more than four times the first.

25. The length of a rectangular deck is 8 ft greater than its width. If the width were doubled, the area of the deck would be 84 ft^2 greater than it is now. Find the dimensions of the deck.

26. A rocket is fired upward from the ground at an initial velocity of 176 ft/s. When will the object reach a height of 160 ft?

Challenge

Factor completely.

1. $a^2 + b^2 - 2ab + 3b - 3a - 4$

2. $x^2 - 4xy + 4y^2 - z^2 + 8z - 16$

COLLEGE ENTRANCE EXAM REVIEW

In each item you are to compare a quantity in Column 1 with a quantity in Column 2. Write the letter of the correct answer from these choices:

A. The quantity in Column 1 is greater than the quantity in Column 2.
B. The quantity in Column 2 is greater than the quantity in Column 1.
C. The quantity in Column 1 is equal to the quantity in Column 2.
D. The relationship cannot be determined from the information given.

Notes: Information centered over both columns refers to one or both of the quantities being compared. A symbol that appears in both columns has the same meaning in each column. All variables represent real numbers. Most figures are not drawn to scale.

	Column 1	Column 2
1.	$\frac{x + y}{xy}$	$\frac{1}{x} + \frac{1}{y}$

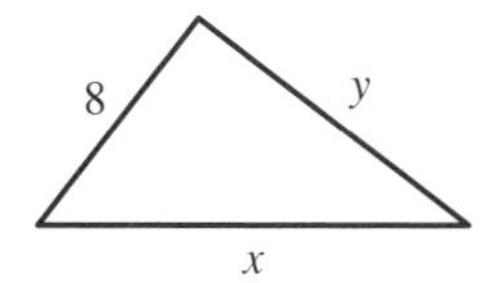

	Column 1	Column 2
2.	y	$x - 8$

$x < 0$

	Column 1	Column 2
3.	$x^3 + x$	x^2

$2x + y = 14$
$3x - y = 6$

	Column 1	Column 2
4.	x	y

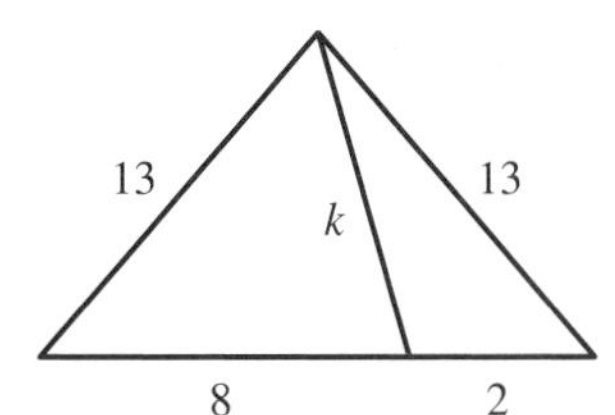

	Column 1	Column 2
5.	k	12
6.	$(x - 3)(2x + 5)$	$2x^2 - x - 15$
7.	$4p^2 - 20p + 25$	$(4p - 5)(p - 5)$

$3x + 4 < 10$

	Column 1	Column 2
8.	x	1
9.	$(-1^2)^3$	$((-1)^2)^3$

$\begin{vmatrix} 3 & -4 \\ 5 & -6 \end{vmatrix} = d$

	Column 1	Column 2
10.	d	2

$pq < 0$
$q < p$

	Column 1	Column 2
11.	p	0

Use this diagram for 12–14.

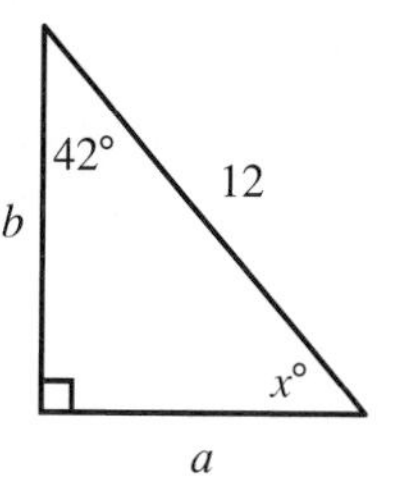

	Column 1	Column 2
12.	x	42
13.	a	b
14.	b	$6\sqrt{2}$

CUMULATIVE REVIEW

1. Graph the solution set: $x > -1$ and $x < 3$

2. Solve the linear system graphically. Determine whether the system is consistent and independent, inconsistent, or consistent and dependent. $\begin{cases} 2x - 2y = 8 \\ y = 4 - 3x \end{cases}$

3. Solve using Cramer's rule: $\begin{cases} 3x - y = -1 \\ 2x + 3y = 14 \end{cases}$

Solve and check.

4. $2x - 3 = 9 - x$

5. $|3x - 2| = 7$

6. $x^2 - 4x + 3 = 0$

7. Given $f(x) = -2x$, $g(x) = x + 3$. Evaluate $g[f(-3)]$ and $f[g(2)]$.

8. y varies directly as x. If $y = 3$ when $x = 2$, find y when $x = 4$.

9. Four times a number decreased by eleven is 45. Find the number.

10. Express in standard form the equation of the line having slope -3 and passing through $(-1, 4)$.

11. Express as the quotient of two integers: $0.\overline{21}$

12. Given $M = \begin{bmatrix} 2 & 0 & -1 \\ 3 & -2 & 4 \\ -4 & -2 & 0 \end{bmatrix}$. Find M^2.

13. Solve the linear system by addition. Determine whether the system is consistent and independent, inconsistent, or consistent and dependent. $\begin{cases} 4x - y = 6 \\ 2x + 3y = 7 \end{cases}$

14. Write an algebraic sentence, using x for the variable. Five times a number increased by eleven is less than twenty-three.

15. Solve and check: $\begin{cases} 2x - 3y + z = 13 \\ x - z = -1 \\ 3x + 4y = 5 \end{cases}$

Factor.

16. $16x^2 - 9$

17. $a^2 - 4a - 32$

18. $m^3 - 125$

19. $3c^2 - 27$

20. Solve and graph the solution set: $|6 - 3x| > 12$

21. Solve using an augmented matrix: $\begin{cases} 3x - y - z = 1 \\ x - 3z = 6 \\ 2x - 3y = -3 \end{cases}$

7 Rational Expressions

A revolutionary law!

Developing Mathematical Power

For twenty-two years, Johannes Kepler (1571–1630), a German astronomer and mathematician, searched for rules that describe the motion of the planets around the sun. Using observations of the motions of Mars and Earth made by Danish astronomer Tycho Brahe (1546–1601), Kepler discovered three laws of planetary motion.

Kepler's First Law (1609) states that the shape of each planet's orbit is an ellipse with the sun at one focus. *Kepler's Second Law* (1609) proves that the planets move faster when they are closer to the sun. *Kepler's Third Law* (1619) states that the square of the period of a planet's orbit (time of one complete revolution around the sun) is proportional to the cube of its mean distance from the sun.

Project

Prove that Kepler's Third Law is true for each planet in the solar system.

7.1

Negative Exponents and Scientific Notation

Objectives: To define zero and negative exponents
To write numbers in scientific notation

Computers operate at such great speeds that the time required to perform certain operations can be measured in nanoseconds. One *nanosecond* is equivalent to 0.000000001 second.

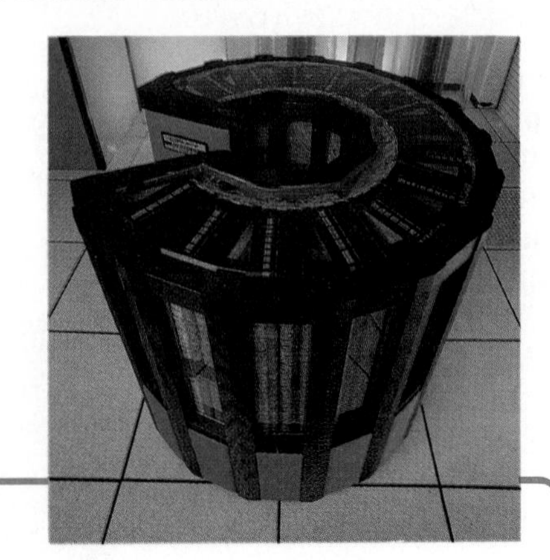

Computations with very large and very small numbers can often be simplified by writing the numbers using integral exponents.

Capsule Review

Property of Exponents	Example
$a^m \cdot a^n = a^{m+n}$	$x^3 \cdot x^4 = x^{3+4} = x^7$
$(a^m)^n = a^{mn}$	$(x^5)^4 = x^{5 \cdot 4} = x^{20}$
$(a^m b^n)^p = a^{mp} b^{np}$	$(x^3y^2)^4 = x^{3 \cdot 4}y^{2 \cdot 4} = x^{12}y^8$
$\frac{a^m}{a^n} = a^{m-n}$, $a \neq 0$ and $m > n$	$\frac{x^8}{x^3} = x^{8-3} = x^5$

Simplify.

1. $x \cdot x^7$ **2.** $(x^6)^3$ **3.** $(x^2y^5)^3$ **4.** $\frac{x^9}{x^6}$ **5.** $\frac{(x^5)^3}{x^8}$

You know that if $a \neq 0$, the property $\frac{a^m}{a^n} = a^{m-n}$ is true when $m > n$. Why the restriction that $a \neq 0$? Assume that this property is also true when $m = n$. Then if $x \neq 0$, $\frac{x^7}{x^7} = x^{7-7} = x^0$. But by using the definition of exponent, $\frac{x^7}{x^7}$, $x \neq 0$, may be simplified as follows:

$$\frac{x^7}{x^7} = \frac{x \cdot x \cdot x \cdot x \cdot x \cdot x \cdot x}{x \cdot x \cdot x \cdot x \cdot x \cdot x \cdot x} = 1$$

Since $\frac{x^7}{x^7} = 1$ and $\frac{x^7}{x^7} = x^0$, then x^0 is equal to 1.

Zero Exponent $a^0 = 1, a \neq 0$

In this book, you may assume that when the exponent is zero and the base is an algebraic expression, that expression is not equal to zero.

EXAMPLE 1 **Simplify:** **a.** $(2x)^0$ **b.** $2x^0$ **c.** $(4 + x)^0$ **d.** $4 + x^0$

a. $(2x)^0 = 1$ **b.** $2x^0 = 2 \cdot 1 = 2$

c. $(4 + x)^0 = 1$ **d.** $4 + x^0 = 4 + 1 = 5$

Now assume that the property $\frac{a^m}{a^n} = a^{m-n}$, $a \neq 0$, is also true when $m < n$. Then, if $x \neq 0$, $\frac{x^4}{x^6} = x^{4-6} = x^{-2}$. By the definition of exponent, if $x \neq 0$, $\frac{x^4}{x^6} = \frac{1}{x^2}$. Since $\frac{x^4}{x^6} = \frac{1}{x^2}$ and $\frac{x^4}{x^6} = x^{-2}$, then $\frac{1}{x^2} = x^{-2}$.

Negative Exponent If n is an integer, then $a^{-n} = \frac{1}{a^n}$, $a \neq 0$.

The properties of positive integral exponents can be extended to include nonpositive integral exponents. Throughout the remainder of this book you may assume that in any given algebraic expression, denominators are not equal to zero and bases with negative exponents are not equal to zero.

EXAMPLE 2 **Simplify and write with positive exponents:**

a. $(x^{-3})(x^4)$ **b.** $(x^{-2})^3$ **c.** $(x^{-4}y^2)^2$ **d.** $\frac{x^{-3}}{x^{-2}}$

a. $(x^{-3})(x^4) = x^{-3+4} = x^1$, or x

b. $(x^{-2})^3 = x^{(-2)(3)} = x^{-6} = \frac{1}{x^6}$

c. $(x^{-4}y^2)^2 = x^{(-4)(2)}y^{(2)(2)} = x^{-8}y^4 = \frac{y^4}{x^8}$

d. $\frac{x^{-3}}{x^{-2}} = x^{-3-(-2)} = x^{-1} = \frac{1}{x}$

> A positive number expressed in the form
>
> $a \times 10^b$, where $1 \leq a < 10$ and b is an integer
>
> is expressed in **scientific notation.**

Nanoseconds can be expressed in scientific notation.

$$1 \text{ ns} = 0.000000001 \text{ s} = 1 \times 10^{-9} \text{ s}$$

Five seconds are equivalent to 5,000,000,000 nanoseconds. This number can be expressed in scientific notation as 5×10^9.

When a decimal number is expressed in scientific notation, the number of places the decimal point is moved is the absolute value of the exponent.

EXAMPLE 3 **Express each number in scientific notation:** **a.** 4,380,000 **b.** 0.006

a. 4,380,000 ⟶ 4.38

Move the decimal point to form the number a such that $1 \leq a < 10$.

$4{,}380{,}000 = 4.38 \times 10^6$

*The decimal point was moved 6 places to the **left**. Since $4{,}380{,}000 > 4.38$, the exponent b is **positive.***

b. 0.006 ⟶ 6

Move the decimal point to form the number a such that $1 \leq a < 10$.

$0.006 = 6 \times 10^{-3}$

*The decimal point was moved 3 places to the **right**. Since $0.006 < 6$, the exponent b is **negative.***

When a number in scientific notation is expressed in decimal form, the absolute value of the exponent is the number of places the decimal point is moved. You can use the EXP (or EE) key on a calculator to enter numbers in scientific notation.

EXAMPLE 4 **Express each number in decimal form:**
a. 2.50×10^7 **b.** 3.04×10^{-5}

a. $2.50 \times 10^7 \longrightarrow 2.5000000$
$2.50 \times 10^7 = 25{,}000{,}000$

*Multiplication by 10^7 yields a **larger** number, so move the decimal point to the **right** 7 places.*

b. $3.04 \times 10^{-5} \longrightarrow 00003.04$
$3.04 \times 10^{-5} = 0.0000304$

*Multiplication by 10^{-5} yields a **smaller** number, so move the decimal point to the **left** 5 places.*

Any nonzero digit, or any zero which serves a purpose other than to locate the decimal point, is a **significant digit.** In the numbers below, the significant digits are shown in color.

0.00675	Three significant digits
67.50	Four significant digits
60,705	Five significant digits

In any approximate number, such as a measurement, the greater the number of significant digits the greater the *accuracy* of the approximation. But it may not be clear which, if any, of the zeros are significant as in 16,000 km, for example. Scientific notation eliminates this problem, since all digits in the first factor are significant. Expressing 16,000 km as

1.6×10^4 km $\quad 1.60 \times 10^4$ km $\quad 1.600 \times 10^4$ km $\quad$ or $\quad 1.6000 \times 10^4$ km

indicates that none, one, two, or three of the zeros are significant and 1.6000×10^4 km is the most accurate.

EXAMPLE 5 Express in scientific notation and perform the indicated operations. Assume that the final zeros in an integer are not significant. Write the answer in decimal form with the same number of significant digits as in the *least* accurate number given. Check by estimating.

a. $9000 \times 3{,}140{,}000$ **b.** $\dfrac{0.000426}{0.357}$

a.
$$\begin{aligned} 9000 \times 3{,}140{,}000 &= 9 \times 10^3 \times 3.14 \times 10^6 \\ &= 9 \times 3.14 \times 10^{3+6} \\ &= 28.26 \times 10^9 \\ &= 28{,}260{,}000{,}000 \\ &\approx 30{,}000{,}000{,}000 \end{aligned}$$

The least accurate number given has 1 significant digit.

Check: $9000 \times 3{,}140{,}000 \approx 9000 \times 3{,}000{,}000$, or $27{,}000{,}000{,}000$
The answer 30,000,000,000 is reasonable.

b.
$$\begin{aligned} \frac{0.000426}{0.357} &= \frac{4.26 \times 10^{-4}}{3.57 \times 10^{-1}} \\ &\approx (4.26 \div 3.57) \times 10^{-4-(-1)} \\ &\approx 1.193 \times 10^{-3} \\ &\approx 0.00119 \end{aligned}$$

The least accurate number given has 3 significant digits.

Check: $\dfrac{0.000426}{0.357} \approx \dfrac{0.0004}{0.4}$, or 0.001 The answer 0.00119 is reasonable.

CLASS EXERCISES

Tell how many significant digits there are in each number. Give a reason for your answer. If an integer ends in zeros, assume that the zeros are not significant.

1. 0.107 **2.** 0.000060 **3.** 6.07 **4.** 5000 **5.** 0.003600

Simplify.

6. 5^0 **7.** 36^0 **8.** $(4y)^0$ **9.** $3m^0$

10. $(2 + z)^0$ **11.** $(3 - y)^0$ **12.** $6^0 + t$ **13.** $18 + p^0$

Simplify and write with positive exponents.

14. x^{-5} **15.** $(y^{-2})(y^{-4})$ **16.** $(x^2y^{-3})^4$ **17.** $(2x^{-5}y^4)^3$

18. $\dfrac{x^{-4}}{x^{-3}}$ **19.** $\dfrac{x^{-6}}{x^{-7}}$ **20.** $\dfrac{x^4x^{-2}}{x^{-5}}$ **21.** $\dfrac{6x^7y^{-5}}{3x^{-1}}$

Express each number in scientific notation. If an integer ends in zeros, assume that the zeros are not significant.

22. 94.2 **23.** 0.009800 **24.** 802,000,000 **25.** 0.0000465

Express each number in decimal form.

26. 1.36×10^7 **27.** 8.2×10^{-4} **28.** 6.44×10^{-3} **29.** 7×10^9

PRACTICE EXERCISES

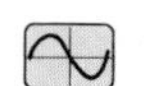 Use technology where appropriate.

Simplify.

1. 43^0 **2.** 150^0 **3.** $(5a)^0$ **4.** $(17b)^0$

5. $5a^0$ **6.** $(7 + q)^0$ **7.** $8 + p^0$ **8.** $a^0 - b^0$

Simplify and write with positive exponents.

9. x^{-2} **10.** y^{-6} **11.** $x \cdot x^{-2}$ **12.** $y^{-3} \cdot y$

13. xy^{-1} **14.** $x^{-2}y^3$ **15.** $(x^{-4})^2$ **16.** $(y^{-3})^3$

17. $(x^2)^{-3}$ **18.** $(x^5)^{-2}$ **19.** $\dfrac{x^5}{x^6}$ **20.** $\dfrac{x^{-3}}{x^{-4}}$

Express in scientific notation. If an integer ends in zeros, assume that the zeros are not significant.

21. 32,050 **22.** 640,000,000 **23.** 0.0045 **24.** 0.0007430

Express in decimal form.

25. 2.5×10^2 **26.** 8.71×10^4 **27.** 5.41×10^{-4} **28.** 4.39×10^{-5}

Simplify and write with positive exponents.

29. $(3x^{-2}y)^2$ **30.** $(2x^{-1}y^{-3})^4$ **31.** $(xy^{-1})^0$

32. $(x^2y^{-3})^0$ **33.** $\dfrac{x^{-2}y^{-1}}{y}$ **34.** $\dfrac{x^{-4}y^{-3}}{y^2}$

35. $\dfrac{2x^{-4}y^{-3}}{4x^2y^{-2}}$ **36.** $\dfrac{3x^{-4}y^{-3}}{9xy^3}$ **37.** $(-2x^{-1})^3(-3x^{-2})^2$

38. $(4x^{-2})^2(x^{-2})^4$ **39.** $(2x^{-1}y^{-3})^3$ **40.** $(-3x^4y^{-1})^2$

Write with positive exponents and evaluate.

41. $(2^{-3}4)^2$ **42.** $(2^{-4})^{-2}(2^{-2})^3$ **43.** $\dfrac{(3^{-2})^{-1}}{(3^{-2})^2}$

Express in scientific notation and perform the indicated operations. Write the answer in decimal form with the same number of significant digits as in the *least* accurate number given. Check by estimating.

44. $0.00034 \times 9{,}000{,}000$ **45.** $0.00064 \div 0.092$ **46.** $(0.04)^3$

47. $\dfrac{9{,}060{,}000 \times 0.00447}{0.00042 \times 0.0034}$ **48.** $\dfrac{25{,}080 + 3460}{32{,}815 - 8460}$ **49.** $\dfrac{301{,}000 \times 56{,}000}{28{,}068 - 9731}$

Simplify and write with positive exponents.

50. $(x^{-2}y^{-3})^3(x^2y^4)^{-3}$ **51.** $(x^{-4}y^3)^2(x^{-3}y^{-4})^3$ **52.** $[(x^{-4}y^{-1})^2]^3$

53. $(x^{-2})^3 + (y^{-1})^2$ **54.** $(x^2y^{-2})^3 + (x^3y^{-3})^2$ **55.** $[(x^{-5}y^{-16})^2]^0$

Applications

Technology Use the exponent key on your calculator to verify computation with negative exponents. For example, compute 2^{-5} and store it in memory. Then compute $\frac{1}{2^5}$ and compare this result with the one in memory. They should be the same. Verify the following statements.

56. $3^{-4} \cdot 3^{-2} = 3^{-6}$ **57.** $\frac{1}{4^3} = 4^{-3}$ **58.** $\frac{1}{17^{-2}} = 17^2$ **59.** $(2^{-4})^3 = 2^{-12}$

BIOGRAPHY: Grace Murray Hopper

A born rebel, Grace Murray Hopper kept a clock that ran counterclockwise in her office, "to remind people that things don't always have to be done the conventional way."

During World War II she enlisted in the U.S. Navy, where she developed operating programs for the Mark I computer. Her insatiable curiosity and extreme practicality led her to design the first compiler. Later she created the first higher order computer language, COBOL.

It was Hopper's computer team that coined the term "bug" when, in 1945, someone pulled a two-inch moth from the circuit of a malfunctioning Mark I. Grace Hopper has been termed the "Grand Old Lady of Software," "a computer pioneer," and the "mother of electronic computer automatic programming." Born in New York City, she studied at Vassar College and earned a Ph.D. in mathematics at Yale University. As a retiree recalled to service, Hopper received an unprecedented promotion by a special act of Congress.

7.2

Simplifying Rational Expressions

Objective: To simplify rational expressions

The quotient of two integers is a rational number that can be simplified by dividing numerator and denominator by their greatest common factor (GCF). Examples of rational numbers are:

$$\frac{12}{28} \qquad \frac{0}{6} \qquad -\frac{13}{39} \qquad \frac{17}{1} \qquad 35 \qquad \frac{216}{100}$$

The quotient of two polynomials is called a *rational algebraic expression* or simply a *rational expression*. Examples of rational expressions are:

$$\frac{3x^2y}{15xy^3} \qquad \frac{2x+5}{y-3} \qquad \frac{2x^2+3x+1}{y+1} \qquad x+9 \qquad \frac{13}{25}$$

Simplifying rational expressions requires knowing how to factor polynomials.

Capsule Review

These examples illustrate several ways to factor polynomials.

Use the GCF of the terms.	$4x^2 - 8x = 4x(x - 2)$
Factor a quadratic trinomial.	$6x^2 + 7x - 3 = (2x + 3)(3x - 1)$
Factor the difference of two squares.	$4x^2 - 25 = (2x + 5)(2x - 5)$
Factor a perfect square.	$9x^2 + 12x + 4 = (3x + 2)(3x + 2)$

Factor.

1. $2x^2 - 3x + 1$ **2.** $4x^2 - 9$ **3.** $5x^2 + 6x + 1$

4. $10x^2 - 10$ **5.** $x^2 + 12x + 36$ **6.** $4x^2 - 40x + 100$

When the denominator of a rational expression is zero, the expression is undefined.

EXAMPLE 1 **For what values of the variables is each expression undefined?**

a. $\frac{4x^3y}{3t^2z}$ **b.** $\frac{3x-7}{y-2}$ **c.** $\frac{4x^2+3x-5}{(x+6)(2x-1)}$

a. $\frac{4x^3y}{3t^2z}$ $t = 0,\ z = 0$ **b.** $\frac{3x-7}{y-2}$ $y = 2$ **c.** $\frac{4x^2+3x-5}{(x+6)(2x-1)}$ $x = -6, \frac{1}{2}$

From now on, assume that the domains of the variables are restricted so that denominators will be nonzero.

> A **rational expression** is an expression that can be written in the form $\frac{P}{Q}$, where P and Q are polynomials and $Q \neq 0$.

Rational expressions can be simplified by following the same procedure used to simplify rational numbers. A rational expression is in *simplest form* when the greatest common factor of the numerator and the denominator is 1. Two rational expressions are **equivalent** if they have the same simplest form.

EXAMPLE 2 **Simplify:** **a.** $\frac{-27x^3y}{9x^4y}$ **b.** $\frac{-45x^6y^4}{15x^7y^4}$

a. $\frac{-27x^3y}{9x^4y} = \frac{-3 \cdot 9x^3y}{x \cdot 9x^3y}$ *Factor the numerator and denominator.*

$= \frac{-3}{x} \cdot 1$ *$\frac{9x^3y}{9x^3y} = 1$*

$= \frac{-3}{x}$ *The GCF is 1. The expression is in simplest form.*

b. $\frac{-45x^6y^4}{15x^7y^4} = \frac{-3 \cdot 15x^6y^4}{x \cdot 15x^6y^4}$ *Factor the numerator and denominator.*

$= \frac{-3}{x} \cdot 1$ *$\frac{15x^6y^4}{15x^6y^4} = 1$*

$= \frac{-3}{x}$ *The GCF is 1. The expression is in simplest form.*

$\frac{-27x^3y}{9x^4y}$, $\frac{-45x^6y^4}{15x^7y^4}$, and $\frac{-3}{x}$ are *equivalent* expressions.

When you simplify rational expressions in which factors of the form $a - b$ and $b - a$ occur, use the fact that $b - a = -1(a - b)$.

EXAMPLE 3 **Simplify:** $\frac{6 - 3x}{x^2 - 5x + 6}$

$\frac{6 - 3x}{x^2 - 5x + 6} = \frac{3(2 - x)}{(x - 3)(x - 2)}$ *Factor.*

$= \frac{3(-1)(x - 2)}{(x - 3)(x - 2)}$ *Replace $(2 - x)$ with $(-1)(x - 2)$.*

$= \frac{-3}{x - 3}$ *The GCF is 1. The expression is in simplest form.*

Sometimes you may have to factor more than once in order to *factor completely*.

EXAMPLE 4 **Simplify:** $\dfrac{3x^3 + 3x^2 - 6x}{x^3 + 2x^2 - 3x}$

$$\frac{3x^3 + 3x^2 - 6x}{x^3 + 2x^2 - 3x} = \frac{3x(x^2 + x - 2)}{x(x^2 + 2x - 3)} \quad \text{Factor.}$$

$$= \frac{3x(x + 2)(x - 1)}{x(x + 3)(x - 1)} \quad \text{Factor again.}$$

$$= \frac{3(x + 2)}{x + 3}$$

CLASS EXERCISES

For what values of the variables is each expression undefined?

1. $\dfrac{7x^2y}{3xyz}$
2. $\dfrac{3x(x - 5)}{2x(x + 2)}$
3. $\dfrac{2x^2 + 3xy - 2y^2}{6x^2 - x - 1}$

Simplify.

4. $\dfrac{10x^2y}{15xy^2}$
5. $\dfrac{6x - 12}{18x - 36}$
6. $\dfrac{4y - 8}{8 - 4y}$
7. $\dfrac{y^2 - 1}{2y + 2}$
8. $\dfrac{x^2 + 5x + 6}{x^2 + 6x + 9}$
9. $\dfrac{y^2 - 5y + 6}{y^2 - 4}$
10. $\dfrac{4x^2 - 9}{4x + 6}$
11. $\dfrac{2x - 4}{x^2 + 3x - 10}$
12. $\dfrac{2x^2 - 3x - 2}{x^2 - 5x + 6}$

PRACTICE EXERCISES

Use technology where appropriate.

Simplify.

1. $\dfrac{15x^4y^5}{5x^3y^2}$
2. $\dfrac{32x^5y^3}{40x^3y^2}$
3. $\dfrac{3x^3y^4}{6x^2y^5}$
4. $\dfrac{-6x^2y^4}{9xy^3}$
5. $\dfrac{12x^3y^6}{8x^4y^7}$
6. $\dfrac{-28x^5y}{35x^6y^2}$
7. $\dfrac{5x - 15}{2x - 6}$
8. $\dfrac{4y - 8}{8y - 16}$
9. $\dfrac{7x - 28}{3x - 12}$
10. $\dfrac{3x - 6}{6 - 3x}$
11. $\dfrac{4x - 8}{4 - 2x}$
12. $\dfrac{7x - 56}{40 - 5x}$
13. $\dfrac{2y}{y^2 - 2y}$
14. $\dfrac{3x^2}{x^2 - 5x}$
15. $\dfrac{2x + 10}{x^2 + 10x + 25}$

16. $\frac{3x^2 - 2x}{6x - 4}$

17. $\frac{2x^2 - 72}{x^2 + 3x - 18}$

18. $\frac{3x^2 - 27}{x^2 - x - 12}$

19. $\frac{5x^2 - 20}{x^2 - 5x - 14}$

20. $\frac{x^2 + 2x - 3}{x^2 + 6x + 9}$

21. $\frac{y^2 + 2y - 15}{y^2 + y - 12}$

22. $\frac{y^2 + 4y - 21}{y^2 + 2y - 35}$

23. $\frac{x^2 + 4x + 4}{x^2 - 4}$

24. $\frac{x^2 - 25}{x^2 + 10x + 25}$

25. $\frac{y^2 - 49}{y^2 - 14y + 49}$

26. $\frac{x^2 - 6x + 9}{x^2 + x - 12}$

27. $\frac{x^2 + 8x + 16}{x^2 - 2x - 24}$

For what values of the variables is each expression undefined?

28. $\frac{5y - 45}{y^2 - 6y}$

29. $\frac{5x^2 - 20x}{3x^2 - 18x + 24}$

30. $\frac{4y^2 + 4y - 24}{3y^2 - 27}$

Simplify.

31. $\frac{6x^2 - 7x + 2}{6x^2 + 5x - 6}$

32. $\frac{8x^2 - 10x + 3}{6x^2 + 3x - 3}$

33. $\frac{x^2 + 6xy + 5y^2}{x^2 - y^2}$

34. $\frac{x^2 - 3xy - 4y^2}{y^2 - x^2}$

35. $\frac{2x^2 + 8xy + 8y^2}{3x^2 + 9xy + 6y^2}$

36. $\frac{2x^2 - 10xy + 12y^2}{2x^2 - 18y^2}$

37. $\frac{2x^2 - 8x - 42}{-x^2 - 6x - 9}$

38. $\frac{-2y^2 - 8y + 24}{2y^2 - 8y + 8}$

39. $\frac{5y^3 - 45y}{6 - y - y^2}$

40. $\frac{-8 + 6x - x^2}{2x^3 - 8x}$

41. $\frac{xy^3 - 9xy}{12xy^2 + 12xy - 144x}$

42. $\frac{x^2y^2 + 3xy^2 + 2y^2}{-5x^2y^4 + 20y^4}$

43. $\frac{y^3x^2 - 16y^3}{x^2y^3 - 8xy^3 + 16y^3}$

44. $\frac{y^2z^2 - 3yz^2 - 10z^2}{4z^2 - z^2y^2}$

45. $\frac{x^2y^2 + 6xy + 5}{1 - x^2y^2}$

46. $\frac{x^{4a}y^{5b}}{x^{5a}y^{6b}}$

47. $\frac{3x^{5a}y^{3b}}{27x^{2a}y^{b}}$

48. $\frac{x^{n+2} - x^ny^2}{y^{n+2} - y^nx^2}$

49. $\frac{(x^{2n} - y^{3n})^3}{x^{4n} - 2x^{2n}y^{3n} + y^{6n}}$

50. $\frac{x^{2n} + 8x^ny^n + 16y^{2n}}{(x^n + 4y^n)^5}$

51. $\frac{(x^n - 2y^{2n})^3}{x^{2n} - 4y^{4n}}$

52. $\frac{x^{2n} + 4x^n + 3}{x^{2n} + 6x^n + 5}$

53. $\frac{x^{2n} + 6x^n + 8}{x^{2n} - 3x^n - 10}$

54. $\frac{x^{4n} + 2x^{2n} - 15}{x^{4n} - 8x^{2n} + 15}$

Applications

55. **Manufacturing** A machine has a work output of $14ax^4$ J (joules) and a work input of $7a^2x^5$ J. What is the efficiency of the machine?

$$\text{efficiency in }\% = \frac{\text{work output in joules}}{\text{work input in joules}}$$

56. Geometry The perimeter of a hexagon is $6a^2s$ ft. The perimeter of a triangle is $3a^3b^2s$ ft. Find the ratio of the perimeter of the hexagon to the perimeter of the triangle.

57. Transportation A bus travels d miles in a hours. The return trip takes b hours. Express the average rate for the round trip in terms of d, a, and b.

Mixed Review

Use the substitution method to solve the following systems.

58. $\begin{cases} 2x - 3y = -10 \\ -x - 4y = 5 \end{cases}$ **59.** $\begin{cases} 4x - 3y = 28 \\ 3x + 2y = -30 \end{cases}$ **60.** $\begin{cases} 14x - 9y = 1 \\ 5x + 4y = 22 \end{cases}$

Developing Mathematical Power

61. Thinking Critically Given the equation $\dfrac{(ax + by)^m}{(ax + by)^n} = \dfrac{1}{ax + by}$, write an equation that defines the relationship between m and n.

Technology The graphing calculator has three different formats for displaying numbers: Normal, Scientific, and Engineering. For these formats, the calculator uses an exponential form called **floating-point notation,** in which the number after E corresponds to the exponent seen when a number is written in scientific notation.

For example: 12345 * 67891 = Norm 838114395
Sci 8.38114395E8
Eng 838.114395E6

Normal mode displays numbers as they are usually written, with digits to the left and/or right of the decimal. Scientific mode displays numbers with exactly one digit to the left of the decimal. Engineering mode displays numbers with 1, 2, or 3 digits to the left of the decimal.

If you are working in Normal mode and the result is too large or too small, the calculator will automatically switch to Scientific mode for that result. If a result is too large or too small for Scientific mode, your calculator may display an error message.

62. What is the range of results that can be expressed by the calculator in Normal mode? in Scientific mode?

63. Use Engineering mode to express several numbers. What do all of the "exponents" have in common?

64. Describe situations in which Scientific mode is preferable to Normal mode.

7.3 Multiplying and Dividing Rational Expressions

Objective: To multiply and divide rational expressions

Rational numbers are special types of rational expressions. The methods for multiplying and dividing rational numbers can be applied to multiplying and dividing other rational expressions.

Capsule Review

Study these products and quotients of rational numbers. Recall that if $a \neq 0$ and $b \neq 0$, the rational numbers $\frac{a}{b}$ and $\frac{b}{a}$ are *reciprocals*, since $\frac{a}{b} \cdot \frac{b}{a} = 1$.

Products

$$\frac{1}{2} \cdot \frac{1}{4} = \frac{1 \cdot 1}{2 \cdot 4} = \frac{1}{8}$$

$$\frac{2}{3} \cdot \frac{3}{8} = \frac{\overset{1}{\cancel{2}} \cdot \overset{1}{\cancel{3}}}{\underset{1}{\cancel{3}} \cdot \underset{4}{\cancel{8}}} = \frac{1}{4}$$

$$\frac{6}{5} \cdot \frac{4}{2} = \frac{\overset{3}{\cancel{6}} \cdot 4}{5 \cdot \underset{1}{\cancel{2}}} = \frac{12}{5}$$

Quotients

$$\frac{3}{4} \div 2 = \frac{3}{4} \cdot \frac{1}{2} = \frac{3 \cdot 1}{4 \cdot 2} = \frac{3}{8}$$

$$\frac{5}{6} \div \frac{1}{3} = \frac{5}{6} \cdot \frac{3}{1} = \frac{5 \cdot \overset{1}{\cancel{3}}}{\underset{2}{\cancel{6}} \cdot 1} = \frac{5}{2}$$

$$\frac{4}{5} \div \frac{12}{10} = \frac{4}{5} \cdot \frac{10}{12} = \frac{\overset{1}{\cancel{4}} \cdot \overset{2}{\cancel{10}}}{\underset{1}{\cancel{5}} \cdot \underset{3}{\cancel{12}}} = \frac{2}{3}$$

Multiply or divide as indicated.

1. $\frac{3}{8} \cdot \frac{5}{6}$
2. $\frac{1}{2} \cdot \frac{4}{6}$
3. $\frac{8}{3} \cdot \frac{2}{16}$
4. $\frac{2}{5} \cdot \frac{3}{7}$
5. $\frac{5}{8} \div 4$
6. $\frac{3}{4} \div \frac{1}{2}$
7. $\frac{9}{16} \div \frac{3}{4}$
8. $\frac{5}{4} \div \frac{15}{8}$

The product of two rational expressions is a rational expression whose numerator is the product of the numerators and whose denominator is the product of the denominators of the given expressions.

If $\frac{P}{Q}$ and $\frac{R}{S}$ are rational expressions, then $\frac{P}{Q} \cdot \frac{R}{S} = \frac{P \cdot R}{Q \cdot S}$.

EXAMPLE 1 **Multiply:** $\dfrac{4x^3}{3y^4} \cdot \dfrac{9y^2}{16x^2}$

$$\frac{4x^3}{3y^4} \cdot \frac{9y^2}{16x^2} = \frac{4x^3 \cdot 9y^2}{3y^4 \cdot 16x^2}$$ *Definition of multiplication*

$$= \frac{\overset{1\ x}{\cancel{4}\cancel{x^3}} \cdot \overset{3\ 1}{\cancel{9}\cancel{y^2}}}{\underset{1\ y^2}{\cancel{3}\cancel{y^4}} \cdot \underset{4\ 1}{\cancel{16}\cancel{x^2}}}$$ *Divide out common factors.*

$$= \frac{x \cdot 3}{y^2 \cdot 4}$$ *Multiply remaining factors in the numerator and in the denominator.*

$$= \frac{3x}{4y^2}$$

When you multiply rational expressions, place parentheses around each polynomial. Factor completely before simplifying the product.

EXAMPLE 2 **Multiply:** $\dfrac{3x^2 - 9x}{x - 2} \cdot \dfrac{4x - 8}{x^2 - 9}$

$$\frac{3x^2 - 9x}{x - 2} \cdot \frac{4x - 8}{x^2 - 9} = \frac{(3x^2 - 9x)(4x - 8)}{(x - 2)(x^2 - 9)}$$ *Place parentheses around each polynomial.*

$$= \frac{3x(x - 3)(4)(x - 2)}{(x - 2)(x + 3)(x - 3)}$$ *Factor numerator and denominator completely.*

$$= \frac{3x(x - 3)(4)(x - 2)}{(x - 2)(x + 3)(x - 3)}$$ *Divide out common factors.*

$$= \frac{12x}{x + 3}$$

To divide with rational expressions, multiply the dividend by the reciprocal (multiplicative inverse) of the divisor.

> If $R \neq 0$ and $S \neq 0$, the rational expressions $\dfrac{R}{S}$ and $\dfrac{S}{R}$ are reciprocals since $\dfrac{R}{S} \cdot \dfrac{S}{R} = 1$.
>
> If $\dfrac{P}{Q}$ and $\dfrac{R}{S}$ are rational expressions, and $R \neq 0$, then $\dfrac{P}{Q} \div \dfrac{R}{S} = \dfrac{P}{Q} \cdot \dfrac{S}{R}$.

EXAMPLE 3 **Divide:** $\dfrac{x^2 + 2x - 15}{2x - 8} \div \dfrac{x^2 + x - 12}{x^2 - 16}$

$$\frac{x^2 + 2x - 15}{2x - 8} \div \frac{x^2 + x - 12}{x^2 - 16} = \frac{x^2 + 2x - 15}{2x - 8} \cdot \frac{x^2 - 16}{x^2 + x - 12}$$

Find reciprocal of divisor and multiply.

$$= \frac{(x + 5)(x - 3)(x + 4)(x - 4)}{2(x - 4)(x - 3)(x + 4)}$$

Factor.

$$= \frac{x + 5}{2}$$

CLASS EXERCISES

Name the reciprocal of each rational expression.

1. $x - 2$

2. $\dfrac{1}{x + 3}$

3. $\dfrac{x^2 - 1}{2x}$

4. $\dfrac{3x}{5x - 6}$

Multiply or divide as indicated.

5. $\dfrac{6x^2}{y} \cdot \dfrac{y^3}{12x^4}$

6. $\dfrac{7ax^3}{8by^2} \div \dfrac{14ax^4}{4by}$

7. $\dfrac{3x - 6}{5x - 20} \cdot \dfrac{2x - 8}{5x - 10}$

8. $\dfrac{14x + 7}{4x - 6} \cdot \dfrac{8x - 12}{42x + 21}$

9. $\dfrac{3x - 6}{12x + 24} \div \dfrac{x^2 - 5x + 6}{3x^2 - 12}$

10. $\dfrac{5x + 15}{10x - 10} \div \dfrac{x^2 + 6x + 9}{3x^2 - 3}$

PRACTICE EXERCISES

Multiply or divide as indicated.

1. $\dfrac{4x^2}{5y} \cdot \dfrac{7y}{12x^4}$

2. $\dfrac{2x^4}{10y^2} \cdot \dfrac{5y^3}{4x^3}$

3. $\dfrac{7x}{4y^3} \div \dfrac{21x^3}{8y}$

4. $\dfrac{3x^3}{5y^2} \div \dfrac{6x^5}{5y^3}$

5. $\dfrac{8y - 4}{10y - 5} \cdot \dfrac{5y - 15}{3y - 9}$

6. $\dfrac{2x + 12}{3x - 9} \cdot \dfrac{2x - 6}{3x + 18}$

7. $\dfrac{6x + 6y}{x - y} \div \dfrac{18}{5x - 5y}$

8. $\dfrac{3y - 12}{2y + 4} \div \dfrac{6y - 24}{4y + 8}$

9. $\dfrac{x^2}{x^2 + 2x + 1} \div \dfrac{3x}{x^2 - 1}$

10. $\dfrac{y^2 - 5y + 6}{y^3} \div \dfrac{y^2 + 3y - 10}{4y^2}$

11. $\dfrac{x^2 - 4}{x^2 - 1} \cdot \dfrac{x + 1}{x^2 + 2x}$

12. $\dfrac{y^2 - 25}{y^2 - 16} \cdot \dfrac{y^2 - 4y}{2y + 10}$

13. $\dfrac{x^2 - 49}{x^2y^3} \div \dfrac{x^2 - 14x + 49}{3x^2y^3}$

14. $\dfrac{x^2 - 7x + 10}{xy^6} \div \dfrac{x^2 - 11x + 30}{x^2y^5}$

15. $\dfrac{x^2 - 5x + 6}{x^2 - 4} \cdot \dfrac{x^2 + 3x + 2}{x^2 - 2x - 3}$

16. $\dfrac{y^2 + 2y - 8}{y^2 - 16} \cdot \dfrac{y^2 - 8y + 16}{y^2 + 3y - 10}$

17. $\dfrac{x^2 + 9x + 14}{x^2 - 3x - 10} \div \dfrac{x^2 + 14x + 49}{x^2 + 2x - 35}$

18. $\dfrac{y^2 - 5y - 36}{y^2 + y - 12} \div \dfrac{y^2 - 4y - 45}{y^2 + 2y - 15}$

19. $\dfrac{x^2 + 2x}{3x^2 + 5x - 2} \cdot \dfrac{6x^2 + 13x - 5}{2x^2 + 5x}$

20. $\dfrac{y^2 - 9y + 14}{y^2 - 8y + 7} \cdot \dfrac{y^2 + 3y - 4}{y^2 + 3y - 10}$

21. $\dfrac{y^2 - 25}{(y + 5)^2} \div \dfrac{2y - 10}{4y + 20}$

22. $\dfrac{y^2 - 36}{(y + 3)^2} \div \dfrac{5y - 30}{2y + 6}$

23. $\dfrac{x^2 + x - 20}{2x^2 - 32} \div \dfrac{2x^2 - 50}{2x^2 + 8x}$

24. $\dfrac{x^2 + 6x + 9}{2x^2 - 18} \div \dfrac{6x + 18}{3x^2 - 27}$

25. $\dfrac{2x^2 + 9x + 10}{x^2 + 5x + 6} \cdot \dfrac{x^2 + 7x + 12}{4x^2 + 6x - 10}$

26. $\dfrac{3y^2 - 6y + 3}{2y^2 - 7y + 6} \cdot \dfrac{2y^2 - y - 3}{5y^2 - 5}$

27. $\dfrac{2x^2 - 6x}{x^2 + 18x + 81} \cdot \dfrac{9x + 81}{x^2 - 9}$

28. $\dfrac{3y^2 + 3y - 6}{2y^2 - 98} \cdot \dfrac{y^2 - 13y + 42}{5y^2 + 10y}$

29. $\dfrac{2x^2 - 3x - 20}{2x^2 - 7x - 30} \cdot \dfrac{4x^2 + 12x + 9}{2x^2 - 5x - 12}$

30. $\dfrac{2y^2 + 5y + 2}{4y^2 - 1} \cdot \dfrac{2y^2 + y - 1}{y^2 + y - 2}$

31. $\dfrac{6x^2 - x - 2}{12x^2 + 5x - 2} \div \dfrac{4x^2 - 1}{8x^2 - 6x + 1}$

32. $\dfrac{3y^2 + 11y + 6}{4y^2 + 16y + 7} \div \dfrac{3y^2 - y - 2}{2y^2 - y - 28}$

33. $\dfrac{6y^2 + 13y + 6}{4y^2 - 9} \div \dfrac{6y^2 + y - 2}{4y^2 - 1}$

34. $\dfrac{3y^2 + 14y + 8}{2y^2 + 9y - 5} \div \dfrac{3y^2 + 20y + 12}{2y^2 + 11y - 6}$

35. $\dfrac{x^2 + xy - 12y^2}{x^2 - 5xy - 36y^2} \div \dfrac{x^2 + 2xy - 15y^2}{x^2 - 4xy - 45y^2}$

36. $\dfrac{x^2 - 3xy - 10y^2}{x^2 + 3xy + 2y^2} \div \dfrac{x^2 + 2xy - 15y^2}{x^2 - 2xy - 3y^2}$

37. $\dfrac{2y^2 - y}{2y^2 - y - 1} \cdot \dfrac{4y^2 - 1}{y - y^2} \div \dfrac{(2y - 1)^2}{2y - 2}$

38. $\dfrac{x^3 - 25x}{x^2 - 6x + 5} \cdot \dfrac{2x^2 - 2}{4x^2} \div \dfrac{x^2 + 5x}{7x + 7}$

39. $\dfrac{y^2 + y - 2}{x^2 - 4} \div \left(\dfrac{4 - y^2}{1 - x^2} \div \dfrac{10y - 20}{5y - 5}\right)$

40. $\dfrac{y^3 - 4y}{5y^2 - 20} \div \left(\dfrac{2x - 6}{2y^2 - 8} \div \dfrac{5x - 15}{(y + 4)^2}\right)$

41. $\dfrac{x^{2a} + x^a - 2}{x^a y^{b+2}} \cdot \dfrac{x^{a+4}y^{b+6}}{x^{2a} - 1}$

42. $\dfrac{x^{4a} - y^{6a}}{x^{a+2} - x^a} \div \dfrac{x^{4a} - 2x^{2a}y^{3a} + y^{6a}}{x^2 - 1}$

43. $\dfrac{ax + ay + bx + by}{ax - ay - bx + by} \div \dfrac{ax - ay + bx - by}{ax + ay - bx - by}$

44. $\left(\dfrac{x^3 + y^3}{x^2 - 4xy + 3y^2} \div \dfrac{x^2 - 2xy - 3y^2}{x^2 + xy - 2y^2}\right) \cdot \left(\dfrac{x + y}{x - 3y}\right)^{-1}$

45. $\dfrac{x^2 + 7xy + 10y^2}{x^2 + 5xy + 4y^2} \cdot (x + 3y) \cdot \left(\dfrac{x^2 + 10xy + 25y^2}{x + 4y}\right)^{-1}$

Applications

46. Physics A space probe with a mass of $\frac{6x^2 - x - 2}{12x^2 + 5x - 2}$ kg lands on a distant planet where its weight is $\frac{4x^2 - 1}{8x^2 - 6x + 1}$ N (newtons). Find the free fall acceleration g in m/s^2 on this planet, if free fall acceleration $g = \frac{\text{weight in N}}{\text{mass in kg}}$ and $1 \text{ N} = 1\frac{\text{kg} \cdot \text{m}}{\text{s}^2}$.

47. Metallurgy In a laboratory experiment you are given a sample of an unknown metal with a mass of $\frac{x^2 + 2x - 15}{2x - 8}$ g. You determine that the volume is $\frac{x^2 + x - 12}{x^2 - 16}$ cm^3. Is the sample made of solid brass, which has a density of $\frac{x + 5}{2}$ g/cm^3? (density = mass ÷ volume)

Developing Mathematical Power

Thinking Critically Look for similarities and differences in these two statements.

A rational number is the quotient of two integers.
A rational expression is the quotient of two polynomials.

When you compare them, you see that they are the same with the exception of the underlined terms. When you study the operations involving rational expressions, you are encouraged to think of the operations involving rational numbers, operations already familiar to you. You are encouraged to view the new process as one similar to a familiar process.

Understanding a new concept often involves **comparing and contrasting**. This skill creates a link between a familiar concept and an unfamiliar concept, or between a simple idea and a more complex idea. It is looking at a process in one context and transferring the process to a new or more complex context.

Describe what is similar about simplifying both expressions.

48. $\frac{42}{70} = \frac{3}{5}$

49. $\frac{x(x - 1)}{5(x + 1)} \cdot \frac{4x + 4}{x^2 - x} = \frac{4}{5}$

Describe what common process is used to find both sums.

50. $\frac{7}{18} + \frac{4}{27} = \frac{29}{54}$

51. $\frac{5}{2x^2y} + \frac{7}{6xy^2} = \frac{15y + 7x}{6x^2y^2}$

7.4

Adding and Subtracting Rational Expressions

Objective: To add and subtract rational expressions

As with multiplication and division, rules for adding and subtracting rational numbers can be applied to other rational expressions.

Capsule Review

To add or subtract rational numbers with different denominators, find the least common denominator (LCD) and rewrite each number with that LCD. Then add or subtract.

EXAMPLE **Add:** $\frac{7}{18} + \frac{4}{27}$

$\frac{7}{18} + \frac{4}{27} = \frac{7}{2 \cdot 3 \cdot 3} + \frac{4}{3 \cdot 3 \cdot 3}$ *Factor each denominator.*

$= \frac{7 \cdot 3}{2 \cdot 3 \cdot 3 \cdot 3} + \frac{4 \cdot 2}{3 \cdot 3 \cdot 3 \cdot 2}$ *The LCD is 2 · 3 · 3 · 3, or 54.*

$= \frac{21}{54} + \frac{8}{54} = \frac{29}{54}$

Add or subtract as indicated.

1. $\frac{5}{19} + \frac{7}{38}$

2. $\frac{2}{15} + \frac{3}{25}$

3. $\frac{7}{24} - \frac{5}{36}$

4. $\frac{11}{12} - \frac{7}{45}$

5. $\frac{3}{21} + \frac{11}{42} - \frac{9}{14}$

6. $\frac{2}{9} + \frac{7}{12} - \frac{13}{21}$

To add or subtract rational expressions with the same denominator, add or subtract only the numerators.

If $\frac{P}{Q}$ and $\frac{R}{Q}$ are rational expressions, then

$$\frac{P}{Q} + \frac{R}{Q} = \frac{P + R}{Q} \quad \text{and} \quad \frac{P}{Q} - \frac{R}{Q} = \frac{P - R}{Q}.$$

EXAMPLE 1 **Add or subtract as indicated:**

a. $\dfrac{x-2}{3x^3} + \dfrac{2x+4}{3x^3}$ **b.** $\dfrac{3x}{4x-3} - \dfrac{2x+1}{4x-3}$

a. $$\frac{x-2}{3x^3} + \frac{2x+4}{3x^3} = \frac{(x-2)+(2x+4)}{3x^3} = \frac{x-2+2x+4}{3x^3} = \frac{3x+2}{3x^3}$$

b. $$\frac{3x}{4x-3} - \frac{2x+1}{4x-3} = \frac{3x-(2x+1)}{4x-3} = \frac{3x-2x-1}{4x-3} = \frac{x-1}{4x-3}$$

To add or subtract rational expressions with different denominators, find the least common denominator (LCD) and write equivalent rational expressions with the LCD as the denominator. Then add or subtract.

EXAMPLE 2 **Add:** $\dfrac{5}{2x^2y} + \dfrac{7}{6xy^2}$

$$\frac{5}{2x^2y} + \frac{7}{6xy^2} = \frac{5 \cdot 3y}{2x^2y \cdot 3y} + \frac{7 \cdot x}{6xy^2 \cdot x} = \frac{15y}{6x^2y^2} + \frac{7x}{6x^2y^2} = \frac{15y+7x}{6x^2y^2}$$

Notice that $2x^2y = 2 \cdot x^2 \cdot y$ and $6xy^2 = 2 \cdot 3 \cdot x \cdot y^2$, so the LCD is $2 \cdot 3 \cdot x^2 \cdot y^2$, or $6x^2y^2$.

When the denominators of the rational expressions are polynomials of more than one term, begin by factoring the denominators.

EXAMPLE 3 **Subtract:** $\dfrac{x+1}{2x-2} - \dfrac{2x}{x^2+2x-3}$

$$\frac{x+1}{2x-2} - \frac{2x}{x^2+2x-3}$$

$$= \frac{x+1}{2(x-1)} - \frac{2x}{(x-1)(x+3)}$$

$$= \frac{(x+1)(x+3)}{2(x-1)(x+3)} - \frac{2x(2)}{(x-1)(x+3)(2)}$$

The LCD is $2(x-1)(x+3)$.

$$= \frac{(x+1)(x+3)-2x(2)}{2(x-1)(x+3)}$$

$$= \frac{x^2+4x+3-4x}{2(x-1)(x+3)}$$

$$= \frac{x^2+3}{2(x-1)(x+3)}, \text{ or } \frac{x^2+3}{2x^2+4x-6}$$

Remember to simplify the rational expression, if possible, after you add or subtract.

EXAMPLE 4 **Simplify:** $\dfrac{1}{x} + \dfrac{4}{x^2 - 4} - \dfrac{2}{x^2 - 2x}$

$$\frac{1}{x} + \frac{4}{x^2 - 4} - \frac{2}{x^2 - 2x}$$

$$= \frac{1}{x} + \frac{4}{(x + 2)(x - 2)} - \frac{2}{x(x - 2)}$$

$$= \frac{1(x + 2)(x - 2)}{x(x + 2)(x - 2)} + \frac{4x}{(x + 2)(x - 2)x} - \frac{2(x + 2)}{x(x - 2)(x + 2)}$$

$$= \frac{x^2 - 4 + 4x - 2x - 4}{x(x + 2)(x - 2)}$$

$$= \frac{x^2 + 2x - 8}{x(x + 2)(x - 2)}$$

$$= \frac{(x - 2)(x + 4)}{x(x + 2)(x - 2)}$$ *Factor the numerator.*

$$= \frac{x + 4}{x(x + 2)}, \text{ or } \frac{x + 4}{x^2 + 2x}$$

CLASS EXERCISES

Name the least common denominator of the rational expressions.

1. $\dfrac{2}{3x}, \dfrac{5}{6x^2}$

2. $\dfrac{1}{a^2b}, \dfrac{3}{ab^2}$

3. $\dfrac{5}{x}, \dfrac{2}{x + 3}$

Add or subtract as indicated.

4. $\dfrac{x}{2} + \dfrac{3x}{2}$

5. $\dfrac{5}{x} - \dfrac{2}{x}$

6. $\dfrac{x + 2}{5} + \dfrac{2x - 4}{5}$

7. $\dfrac{3x - 5}{y} - \dfrac{2x + 2}{y}$

8. $\dfrac{5}{x} - \dfrac{7}{y}$

9. $\dfrac{2}{x} + \dfrac{5}{3y}$

10. $\dfrac{2x + 1}{xy} + \dfrac{4x - 2}{xy}$

11. $\dfrac{5x + 2}{xy^2} - \dfrac{2x - 4}{4xy}$

12. $\dfrac{5x}{x^2 - 16} + \dfrac{2}{x + 4}$

13. $\dfrac{-3x}{x^2 - 9} + \dfrac{4}{2x + 6}$

14. $\dfrac{5x}{x^2 - x - 6} - \dfrac{4}{x^2 + 4x + 4}$

PRACTICE EXERCISES

Name the least common denominator of the rational expressions.

1. $\frac{7}{9y}, \frac{11}{18y}$

2. $\frac{1}{xy^3}, \frac{3}{5x^2y}$

3. $\frac{7}{x-1}, \frac{4}{x-2}$

4. $\frac{5}{8y}, \frac{1}{y(y+2)}$

5. $\frac{1}{x+3}, \frac{2}{(x+3)^2}$

6. $\frac{3}{x}, \frac{4}{x-4}$

Add or subtract as indicated.

7. $\frac{x}{5} + \frac{2x}{5}$

8. $\frac{2x}{7} - \frac{3x}{7}$

9. $\frac{x+5}{4y} + \frac{2x+6}{4y}$

10. $\frac{x+9}{8y^2} - \frac{3x}{8y^2}$

11. $\frac{5}{x^2+2} - \frac{2}{x^2+2}$

12. $\frac{x+2}{x^2+7} + \frac{2x-4}{x^2+7}$

13. $\frac{8}{3x^3y} + \frac{4}{9xy^3}$

14. $\frac{7}{4xy} + \frac{5}{12x^2y^2}$

15. $\frac{9}{5x^2y} - \frac{3}{10xy^2}$

16. $\frac{8}{3xy^4} - \frac{7}{9x^2y}$

17. $\frac{y}{2y+4} + \frac{3}{y+2}$

18. $\frac{x}{6x-9} + \frac{5}{2x-3}$

19. $\frac{3y+1}{4y+4} - \frac{2y+7}{2y+2}$

20. $\frac{x+2}{5x-10} - \frac{3x+5}{2x-4}$

21. $\frac{2}{3x+9} + \frac{4}{2x+6}$

22. $\frac{7}{5y+25} - \frac{4}{3y+15}$

23. $\frac{y}{2y+4} + \frac{5}{y^2+2y}$

24. $\frac{x}{3x+9} + \frac{8}{x^2+3x}$

25. $\frac{5x}{2y+4} - \frac{6}{y^2+2y}$

26. $\frac{3y}{y^2-25} - \frac{8}{y-5}$

27. $\frac{4}{x^2-9} + \frac{7}{x+3}$

28. $\frac{5}{x^2-36} + \frac{9}{x^2+5x-6}$

29. $\frac{2x}{x^2-x-2} - \frac{5x}{x^2-3x+2}$

30. $\frac{y}{y^2-y-6} - \frac{y+2}{y^2+5y+6}$

31. $\frac{-x}{x^2-2x-3} - \frac{2x}{2x^2-2}$

32. $\frac{2y}{y^2+8y+7} - \frac{4y}{5y^2-5}$

33. $\frac{y+4}{y^2-2y} + \frac{2-y}{3y^2-6y}$

34. $\frac{2x+1}{x^2-2x} + \frac{3-x}{5x^2-20}$

35. $3x + \frac{x^2+5x}{x^2-2}$

36. $4y - \frac{y+2}{y^2+3y}$

37. $\frac{4x}{x^2-2x} + \frac{2}{3x+6}$

38. $\frac{5y}{y^2-7y} - \frac{4}{2y-14} + \frac{9}{y}$

39. $\frac{3x}{x^2-4} + \frac{5x}{x^2+x-2} - \frac{3}{x^2-4x+4}$

40. $\frac{3x-y}{x^2-9xy+20y^2} + \frac{2y}{x^2-25y^2}$

41. $\frac{x+2y}{x^2+4xy+4y^2} - \frac{2x}{x^2-4y^2}$

42. $\dfrac{x-1}{x^2+2x+1}+\dfrac{x}{x^2-1}-\dfrac{3}{2x-2}$

43. $\dfrac{7y-1}{y^2-9}-\dfrac{2y}{y^2+6y+9}+\dfrac{3}{2y^2+6y}$

44. $\dfrac{4x-3}{x^2-25}+\dfrac{2x^2+1}{x^2+x-20}-\dfrac{5x^2}{x^2-16}$

45. $\dfrac{2x+9}{3x-x^2}+\dfrac{x+5}{x^2-49}-\dfrac{2x^2+x}{x^2+4x-21}$

46. $x+1+\dfrac{x^2-4xy}{2y-6x}-5y$

47. $\dfrac{x}{x^3+1}-\dfrac{2}{x+1}+\dfrac{3}{x^2-x+1}$

48. $\left(\dfrac{1}{x-2}+\dfrac{1}{x+5}\right)\left(\dfrac{2x+10}{2x^2+3x}\right)$

49. $\left(\dfrac{1}{x+5}-\dfrac{2}{x+1}\right)\left(\dfrac{3x+15}{3x^2+4x}\right)$

50. $\left(\dfrac{2}{x^2-4}-\dfrac{1}{x-2}\right)\div\left(\dfrac{2x+5}{x+2}\right)$

51. $\left(\dfrac{3}{x^2-36}-\dfrac{1}{x+6}\right)\div\left(\dfrac{x-9}{x-6}\right)$

Applications

52. **Navigation** The time a boat takes to go upstream is $\dfrac{2}{x-1}$ h. The time it takes the boat to go downstream is $\dfrac{5}{x+1}$ h. Find the total time.

53. **Physics** The distance a force can move an object is $\dfrac{6}{x^2-5x}$ ft. The distance a second force can move an object is $\dfrac{3x}{x^2-10x+25}$ ft. Find the difference of the first distance from the second distance.

54. **Sports** The pitcher of a major league baseball team throws his fast ball with a speed of $\dfrac{5x+2}{xy^2}$ km/h. His knuckle ball has an initial speed of $\dfrac{2x-4}{4xy}$ km/h. How much faster is his fast ball?

TEST YOURSELF

Simplify and express with positive exponents. 7.1

1. $(x^2)(x^{-5})$
2. $(x^{-4})^{-5}$
3. $\dfrac{x^3}{x^6}$
4. $(2x^{-3}y^{-1})^3$

Express in scientific notation.

5. 0.0000456
6. 279,000,000

Express in decimal form.

7. 3.78×10^5
8. 9.41×10^{-5}

Simplify. 7.2–7.4

9. $\dfrac{x^2+x-12}{2x^2+4x-30}$

10. $\dfrac{4x^2+20x+25}{x+2}\cdot\dfrac{3x+6}{4x^2-25}$

11. $\dfrac{x^2+2x-35}{x^2-25}\div\dfrac{x^2+14x+49}{2x+10}$

12. $\dfrac{x}{x^2+7x}+\dfrac{2x-4}{x^2-49}$

7.5

Complex Rational Expressions

Objective: To simplify complex rational expressions

An expression that has a rational expression in its numerator or in its denominator is a **complex rational expression.** Some examples are:

$$\frac{\frac{1}{x}}{y} \qquad \frac{2 + \frac{3}{4y}}{1 - \frac{1}{2y}} \qquad \frac{1 + \frac{1}{2}}{\frac{2}{3}} \qquad \frac{\frac{x-2}{x} - \frac{2}{x+1}}{\frac{3}{x-1} - \frac{1}{x+1}}$$

When simplifying complex rational expressions, you will need to use the distributive property.

Capsule Review

To multiply a binomial by a monomial, use the distributive property.

EXAMPLE **Multiply:** **a.** $4x(x + 3y)$ **b.** $14x^4y\left(\frac{3y}{14x^4} + \frac{2y^2}{7x^3y}\right)$

a. $4x(x + 3y)$

$= 4x \cdot x + 4x \cdot 3y$

$= 4x^2 + 12xy$

b. $14x^4y\left(\frac{3y}{14x^4} + \frac{2y^2}{7x^3y}\right)$

$= 14x^4y \cdot \frac{3y}{14x^4} + 14x^4y \cdot \frac{2y^2}{7x^3y}$

$= y \cdot 3y + 2x \cdot 2y^2$

$= 3y^2 + 4xy^2$

Multiply.

1. $30\left(\frac{x}{5} + \frac{y}{15}\right)$

2. $8x^2y^2\left(\frac{x}{8y^2} - \frac{5y}{2x^2}\right)$

3. $3(x + 3)\left(\frac{2}{3} + \frac{3y}{x + 3}\right)$

4. $(x + 2)(x - 1)\left(\frac{5}{x - 1} + \frac{6}{x + 2}\right)$

To simplify a complex rational expression, find the LCD of all the rational expressions in the numerator and denominator. Let this LCD be called P. Then multiply the complex rational expression by 1 in the form $\frac{P}{P}$.

EXAMPLE 1 **Simplify:** $\dfrac{\dfrac{1}{x} + 3}{4 + \dfrac{5}{y}}$

$$\frac{\frac{1}{x} + 3}{4 + \frac{5}{y}} = \frac{xy}{xy} \cdot \frac{\left(\frac{1}{x} + 3\right)}{\left(4 + \frac{5}{y}\right)}$$ *The LCD is xy, so multiply by $\frac{xy}{xy}$.*

$$= \frac{xy\left(\frac{1}{x} + 3\right)}{xy\left(4 + \frac{5}{y}\right)}$$

$$= \frac{xy \cdot \frac{1}{x} + xy \cdot 3}{xy \cdot 4 + xy \cdot \frac{5}{y}}$$ *Distributive property*

$$= \frac{y + 3xy}{4xy + 5x}$$

Remember to express every answer in simplest form. You may need to factor and divide out any common factors.

EXAMPLE 2 **Simplify:** $\dfrac{\dfrac{1}{3x^2} - \dfrac{25}{3y^2}}{\dfrac{1}{x^2y} + \dfrac{5}{xy^2}}$

$$\frac{\frac{1}{3x^2} - \frac{25}{3y^2}}{\frac{1}{x^2y} + \frac{5}{xy^2}} = \frac{3x^2y^2}{3x^2y^2} \cdot \frac{\left(\frac{1}{3x^2} - \frac{25}{3y^2}\right)}{\left(\frac{1}{x^2y} + \frac{5}{xy^2}\right)}$$ *The LCD is $3x^2y^2$.*

$$= \frac{3x^2y^2 \cdot \frac{1}{3x^2} - 3x^2y^2 \cdot \frac{25}{3y^2}}{3x^2y^2 \cdot \frac{1}{x^2y} + 3x^2y^2 \cdot \frac{5}{xy^2}}$$ *Distributive property*

$$= \frac{y^2 - 25x^2}{3y + 15x}$$

$$= \frac{(y + 5x)(y - 5x)}{3(y + 5x)}$$ *Factor.*

$$= \frac{y - 5x}{3}$$

When the denominators are polynomials of more than one term, you may need to factor to find the LCD.

EXAMPLE 3 **Simplify:** $$\frac{\frac{5x}{x^2-9}+\frac{7}{x+3}}{\frac{3}{x-3}-\frac{4}{2x+6}}$$

$$\frac{\frac{5x}{x^2-9}+\frac{7}{x+3}}{\frac{3}{x-3}-\frac{4}{2x+6}} = \frac{\frac{5x}{(x+3)(x-3)}+\frac{7}{x+3}}{\frac{3}{x-3}-\frac{4}{2(x+3)}}$$

The LCD is 2(x + 3)(x − 3).

$$= \frac{2(x+3)(x-3)}{2(x+3)(x-3)} \cdot \frac{\left(\frac{5x}{(x+3)(x-3)}+\frac{7}{x+3}\right)}{\left(\frac{3}{x-3}-\frac{4}{2(x+3)}\right)}$$

$$= \frac{2(5x)+2(x-3)(7)}{2(x+3)(3)-(x-3)(4)}$$

Distributive property

$$= \frac{24x-42}{2x+30}$$

$$= \frac{6(4x-7)}{2(x+15)} = \frac{3(4x-7)}{x+15}, \text{ or } \frac{12x-21}{x+15}$$

CLASS EXERCISES

Simplify.

1. $\frac{\frac{1}{x}}{\frac{2}{y}}$

2. $\frac{1-\frac{1}{4}}{2-\frac{3}{5}}$

3. $\frac{\frac{2}{x}+y}{3}$

4. $\frac{\frac{2}{3}}{\frac{3}{b}}$

5. $\frac{1}{1+\frac{x}{y}}$

6. $\frac{3}{\frac{2}{x}+y}$

7. $\frac{\frac{2}{x+y}}{\frac{5}{x+y}}$

8. $\frac{\frac{3}{x-4}}{1-\frac{2}{x-4}}$

9. $\frac{1+\frac{2}{x}+\frac{3}{x^2}}{\frac{4}{x}+2}$

10. $\frac{\frac{2}{x+2}+\frac{4}{x-3}}{\frac{1}{x-3}+\frac{3}{x+2}}$

11. $\frac{\frac{5}{2x-6}-\frac{2}{x+3}}{\frac{7}{3x-9}+\frac{2}{x+3}}$

12. $\frac{\frac{1}{x^2-1}+\frac{x}{x+1}}{\frac{x}{x-1}}$

PRACTICE EXERCISES

Use technology where appropriate.

Simplify.

1. $\dfrac{\frac{1}{x}}{\frac{2}{x}}$

2. $\dfrac{\frac{4}{y}}{\frac{3}{y}}$

3. $\dfrac{\frac{3}{x}}{\frac{2}{y}}$

4. $\dfrac{\frac{3}{z}}{\frac{1}{y}}$

5. $\dfrac{4}{2 + \frac{x}{y}}$

6. $\dfrac{-3}{\frac{5}{x} + y}$

7. $\dfrac{\frac{2}{x} + \frac{3}{y}}{\frac{-5}{x} + \frac{7}{y}}$

8. $\dfrac{\frac{5}{x} - \frac{2}{y}}{\frac{-4}{x} - \frac{6}{y}}$

9. $\dfrac{1 + \frac{2}{x}}{2 + \frac{3}{2x}}$

10. $\dfrac{\frac{1}{xy} - \frac{1}{y^2}}{\frac{1}{x^2y} - \frac{1}{xy^2}}$

11. $\dfrac{\frac{2}{x + 4} + 2}{1 + \frac{3}{x + 4}}$

12. $\dfrac{\frac{3}{x - 2} - 5}{2 - \frac{4}{x - 2}}$

13. $\dfrac{\frac{5}{x^2} + 6}{-8 + \frac{1}{x}}$

14. $\dfrac{\frac{3}{x} - 5}{4 - \frac{9}{x^2}}$

15. $\dfrac{\frac{2x}{x^2 - 9}}{\frac{6}{x - 3}}$

16. $\dfrac{\frac{2y}{y^2 - 16}}{\frac{3}{y + 4}}$

17. $\dfrac{\frac{4}{2x - 10}}{\frac{7x}{x^2 - 25}}$

18. $\dfrac{\frac{2}{3y - 6}}{\frac{5y}{y^2 - 4}}$

19. $\dfrac{\frac{4y}{y^2 - 4} - \frac{5}{y - 2}}{\frac{2}{y - 2} + \frac{3}{y + 2}}$

20. $\dfrac{\frac{2}{x + 3} + \frac{5x}{x^2 - 9}}{\frac{4}{x + 3} + \frac{2}{x - 3}}$

21. $\dfrac{\frac{2x}{x^2 + 4x + 3}}{\frac{1}{x + 3} + \frac{2}{x + 1}}$

22. $\dfrac{\frac{5y}{y^2 - 5y + 6}}{\frac{3}{y - 3} + \frac{2}{y - 2}}$

23. $\dfrac{\frac{7y}{y^2 + 14y + 49}}{\frac{3}{y + 7} + \frac{2y}{(y + 7)^2}}$

24. $\dfrac{\frac{4x}{x^2 + 6x + 9}}{\frac{-5}{(x + 3)^2} + \frac{7}{x + 3}}$

25. $\dfrac{x - 3 + \frac{5}{x + 1}}{x + 5 + \frac{2}{2x + 2}}$

26. $\dfrac{2 + y + \frac{2}{y - 3}}{5 - y - \frac{8}{4y - 12}}$

27. $\dfrac{2x + 4 + \frac{12}{x + 5}}{3x - 1 + \frac{10}{(x + 5)^2}}$

28. $\dfrac{4y - 3 + \dfrac{11}{y - 4}}{2 + 2y - \dfrac{9}{(y - 4)^2}}$

29. $\dfrac{\dfrac{-3}{x} + 4 + \dfrac{20}{x - 2}}{-x - \dfrac{7}{x} + \dfrac{10}{(x - 2)^2}}$

30. $\dfrac{7y - \dfrac{4}{y} + \dfrac{15}{y + 1}}{\dfrac{-2}{y} + 4y - \dfrac{6}{(y + 1)^2}}$

31. $\dfrac{\dfrac{2x}{x^2 - 25} + \dfrac{1}{3x - 15}}{\dfrac{5}{x - 5} + \dfrac{3}{4x - 20}}$

32. $\dfrac{\dfrac{7y}{2y - 2} + \dfrac{y}{y^2 - 1}}{\dfrac{-1}{3y + 3} + \dfrac{4}{y + 1}}$

33. $\dfrac{\dfrac{3 + x}{2x^2 - 32} + \dfrac{5}{2x + 8}}{\dfrac{-7}{x - 4} + \dfrac{3}{x + 4}}$

34. $\dfrac{\dfrac{2x}{(x + 3)^3} + \dfrac{1}{(x + 3)}}{\dfrac{5}{(x + 3)^2} + \dfrac{9}{(x + 3)^2}}$

35. $\dfrac{\dfrac{2}{(2y - 4)^4} - \dfrac{1}{(y - 2)^5}}{\dfrac{-4y}{(y - 2)^5} + \dfrac{6}{(y - 2)^4}}$

36. $\dfrac{\dfrac{4a}{(2x - 2y)^3} + \dfrac{3b}{(x - y)^4}}{\dfrac{2c}{(2x - 2y)^4}}$

37. $\dfrac{\dfrac{-y}{(3y - 6)^3} - \dfrac{2y}{(y - 2)^2}}{\dfrac{-1}{(y - 2)^2} + \dfrac{5}{(2y - 4)^2}}$

38. $\dfrac{\dfrac{1}{x + 1}}{x - \dfrac{1}{x + \dfrac{1}{x}}}$

39. $x + \dfrac{1}{x + \dfrac{1}{x + \dfrac{1}{x}}}$

Applications

Technology A complex fraction of this form is called a *continued fraction,* and many computers represent irrational numbers in this way.

$$1 + \cfrac{1}{a_1 + \cfrac{1}{a_2 + \cfrac{1}{a_3}}} \cdots$$

The Store function on a graphing calculator can be used to evaluate these fractions. For example:

$$\sqrt{2} = 1 + \cfrac{1}{2 + \cfrac{1}{2 + \cfrac{1}{2}}} \cdots$$

1. Store $\frac{1}{2} \rightarrow A$.
2. Store $2 + A \rightarrow B$.
3. Store $B^{-1} \rightarrow A$.
4. Repeat steps 2–3 several times.
5. Evaluate $1 + B^{-1}$.

40. How many times must you perform these steps in order to obtain an approximation of $\sqrt{2}$ that is accurate to 3 decimal places? to 5 places?

41. What complex fraction did you use to approximate $\sqrt{2}$ to 3 places?

42. What number is approximated by substituting 1 for every a_n?

Developing Mathematical Power

43. **Thinking Critically** Can $\frac{2}{3}$ be considered a complex rational expression? Explain.

7.6

Solving Rational Equations

Objective: To solve rational equations

Equations are mathematical models for many scientific and business problems. Many such problems require solving equations containing rational expressions. Such equations are called **rational equations.** Solving a rational equation may lead to solving a quadratic equation as in Example 4.

Capsule Review

Remember you can use the zero-product property to solve a quadratic equation.

EXAMPLE **Solve:** $x^2 - 4x = -3$

$$x^2 - 4x = -3$$
$$x^2 - 4x + 3 = 0$$
$$(x - 3)(x - 1) = 0 \quad \textit{Factor.}$$
$$x - 3 = 0 \text{ or } x - 1 = 0 \quad \textit{Zero-product property}$$
$$x = 3 \text{ or } x = 1$$

Check: For $x = 3$, $x^2 - 4x = -3$; $3^2 - 4(3) \stackrel{?}{=} -3$; $9 - 12 \stackrel{?}{=} -3$; $-3 = -3$ ✓

For $x = 1$, $x^2 - 4x = -3$; $1^2 - 4(1) \stackrel{?}{=} -3$; $1 - 4 \stackrel{?}{=} -3$; $-3 = -3$ ✓

The solutions are 3 and 1.

Solve.

1. $x^2 + 5x - 6 = 0$

2. $x^2 - 36 = 0$

3. $x^2 + 14x = -49$

4. $x^2 = 6x$

5. $x^2 + 15 = 8x$

6. $x^2 - 3x = 40$

If both sides of an equation containing fractions are multiplied by the LCD of the fractions, an equivalent equation with integral coefficients is obtained. Also, if both sides of an equation containing rational expressions are multiplied by the LCD of these expressions, an equation without denominators is obtained. However, multiplication by a polynomial does not necessarily yield an equation equivalent to the original equation. Therefore, be sure that any possible solution is in the replacement set of the original equation. If a value of the variable yields a zero denominator, that value is *not* in the replacement set for the equation.

Compare and contrast the solutions of $\frac{1}{5x} = \frac{1}{9x}$ and $5x = 9x$. Are the equations equivalent? Why or why not? Multiply both sides of the equation $\frac{1}{5x} = \frac{1}{9x}$ by the LCD. What happened? Interpret your results.

EXAMPLE 1 **Solve and check:** $\mathbf{\frac{1}{4x} - \frac{3}{4} = \frac{7}{x}}$

$$\frac{1}{4x} - \frac{3}{4} = \frac{7}{x}$$

If $x = 0$, the denominators $4x$ and x are 0. So, the replacement set is all real numbers except 0.

$$4x\left(\frac{1}{4x} - \frac{3}{4}\right) = 4x\left(\frac{7}{x}\right)$$

Multiply both sides by the LCD, $4x$.

$$4x\left(\frac{1}{4x}\right) - 4x\left(\frac{3}{4}\right) = 4x\left(\frac{7}{x}\right)$$

Distributive property

$$1 - 3x = 28$$
$$3x = -27$$
$$x = -9$$

Check:

$$\frac{1}{4x} - \frac{3}{4} = \frac{7}{x}$$

$$\frac{1}{4(-9)} - \frac{3}{4} \stackrel{?}{=} \frac{7}{-9}$$

-9 is a member of the replacement set.

$$-\frac{1}{36} - \frac{27}{36} \stackrel{?}{=} -\frac{28}{36}$$

$$-\frac{28}{36} = -\frac{28}{36} \checkmark$$

The solution is -9.

If the denominators in a rational equation are polynomials of more than one term, begin by factoring them, if possible.

EXAMPLE 2 **Solve and check:** $\mathbf{\frac{5}{2x - 2} = \frac{15}{x^2 - 1}}$

$$\frac{5}{2x - 2} = \frac{15}{x^2 - 1}$$

If $x = 1$, then $2x - 2 = 0$, and $x^2 - 1 = 0$. Also, if $x = -1$, $x^2 - 1 = 0$.

$$\frac{5}{2(x - 1)} = \frac{15}{(x + 1)(x - 1)}$$

So, the replacement set is all real numbers except 1 and -1.

$$2(x + 1)(x - 1)\frac{5}{2(x - 1)} = 2(x + 1)(x - 1)\frac{15}{(x + 1)(x - 1)}$$

The LCD is $2(x + 1)(x - 1)$.

$$5x + 5 = 30$$
$$5x = 25$$
$$x = 5$$

Check: $\frac{5}{2x - 2} = \frac{15}{x^2 - 1}$

$\frac{5}{2(5) - 2} \stackrel{?}{=} \frac{15}{5^2 - 1}$ *5 is a member of the replacement set.*

$\frac{5}{8} \stackrel{?}{=} \frac{15}{24}$

$\frac{5}{8} = \frac{5}{8}$ ✓ The solution is 5.

Sometimes an equation has no solution, as illustrated by the next example.

EXAMPLE 3 **Solve and check:** $\frac{x}{x - 2} = \frac{1}{2} + \frac{2}{x - 2}$

$\frac{x}{x - 2} = \frac{1}{2} + \frac{2}{x - 2}$ *If $x = 2$, then $x - 2 = 0$. So, the replacement set is all real numbers except 2.*

$2(x - 2)\left(\frac{x}{x - 2}\right) = 2(x - 2)\left(\frac{1}{2} + \frac{2}{x - 2}\right)$ *The LCD is $2(x - 2)$.*

$2x = x - 2 + 4$ Since 2 is not a member of the replacement set,
$x = 2$ the equation has no solution.

When multiplying both sides of an equation by an LCD containing a variable, you may introduce *extraneous* solutions. Be sure to check the replacement set.

EXAMPLE 4 **Solve and check:** $\frac{x}{x - 3} - \frac{7x - 6}{x^2 - x - 6} = \frac{2}{x + 2}$

$\frac{x}{x - 3} - \frac{7x - 6}{x^2 - x - 6} = \frac{2}{x + 2}$

$\frac{x}{x - 3} - \frac{7x - 6}{(x - 3)(x + 2)} = \frac{2}{x + 2}$ *The replacement set is all real numbers except 3 and -2.*

$(x - 3)(x + 2)\left(\frac{x}{x - 3} - \frac{7x - 6}{(x - 3)(x + 2)}\right) = (x - 3)(x + 2)\left(\frac{2}{x + 2}\right)$

$x^2 + 2x - 7x + 6 = 2x - 6$

$x^2 - 7x + 12 = 0$ *Quadratic equation*

$(x - 3)(x - 4) = 0$ *Zero-product property*

$x - 3 = 0$ or $x - 4 = 0$

$x = 3$ or $x = 4$

Only 4 can be a solution, since 3 is not in the replacement set for x.

Check: For $x = 4$, $\frac{x}{x - 3} - \frac{7x - 6}{x^2 - x - 6} = \frac{2}{x + 2}$

$$\frac{4}{4 - 3} - \frac{7(4) - 6}{4^2 - 4 - 6} \stackrel{?}{=} \frac{2}{4 + 2}$$

$$4 - \frac{22}{6} \stackrel{?}{=} \frac{2}{6}$$

$$\frac{12}{3} - \frac{11}{3} \stackrel{?}{=} \frac{1}{3}$$

$$\frac{1}{3} = \frac{1}{3} \checkmark$$

The solution is 4.

> **Reminder** Be sure to always check that any possible solution is in the replacement set of the *original* equation.

CLASS EXERCISES

Without solving, determine the replacement set for *x*. Give a reason for your answer.

1. $\frac{1}{3x} = \frac{5}{6}$
2. $\frac{x}{2} + \frac{1}{3} = \frac{1}{6}$
3. $\frac{3}{x - 5} = \frac{2}{x - 5}$
4. $\frac{5}{x^2 - x} + \frac{3}{x - 1} = 6$
5. $\frac{3x - 5}{x^2 - 6x + 8} + \frac{x}{x - 2} = \frac{3}{x - 4}$

Find the replacement set. Then solve and check.

6. $\frac{1}{2x} = \frac{1}{8}$
7. $\frac{x}{3} + \frac{1}{2} = \frac{1}{6}$
8. $\frac{y + 1}{8} = \frac{1}{4}$
9. $\frac{x + 1}{9} = \frac{2}{3}$
10. $\frac{y}{5} + 2 = \frac{y}{3}$
11. $\frac{5x}{2} - \frac{1}{2} = x + 1$
12. $\frac{3}{x - 3} = \frac{x}{x - 3} - 2$
13. $\frac{x}{x - 1} + \frac{2}{x} = \frac{3x - 2}{x^2 - x}$

PRACTICE EXERCISES

Find the replacement set. Then solve and check.

1. $\frac{1}{5x} = \frac{1}{10}$
2. $\frac{1}{7y} = \frac{1}{21}$
3. $\frac{x}{4} + \frac{x}{5} = 9$
4. $\frac{x}{5} + \frac{x}{3} = 16$
5. $\frac{x}{3} + \frac{1}{6} = \frac{x}{4} + \frac{1}{4}$
6. $\frac{2y}{9} - \frac{5}{6} = \frac{y}{9} - \frac{1}{2}$

7. $\frac{x}{3} + \frac{x}{2} = 5$

8. $\frac{y}{5} + \frac{y}{2} = 7$

9. $\frac{x}{4} - 1 = \frac{x}{8}$

10. $\frac{x - 4}{6} = \frac{x - 4}{2}$

11. $\frac{y + 4}{5} = \frac{y - 2}{3}$

12. $\frac{10}{6x + 7} = \frac{6}{2x + 9}$

13. $\frac{4}{y - 3} = \frac{6}{y + 3}$

14. $\frac{2x}{3} - \frac{1}{2} = \frac{2x + 5}{6}$

15. $\frac{3x - 2}{12} - \frac{1}{6} = \frac{1}{6}$

16. $\frac{-6}{2 - 4x} = \frac{10}{6x - 8}$

17. $\frac{2}{3x - 5} = \frac{4}{x - 15}$

18. $\frac{1}{x + 12} = \frac{3x}{3x^2 + 36}$

19. $\frac{7y}{y^2 - 4} + \frac{5}{y - 2} = \frac{2y}{y^2 - 4}$

20. $\frac{4x}{x^2 - 16} + \frac{5}{x + 4} = \frac{5x}{x^2 - 16}$

21. $\frac{7x - 2}{11} - 1 = \frac{2x - 7}{3}$

22. $\frac{3x - 6}{x} + 1 = \frac{10}{x} + \frac{4}{x}$

23. $2 - \frac{1}{x + 1} = \frac{x}{x + 1}$

24. $3 + \frac{4}{y - 4} = \frac{y}{y - 4}$

25. $\frac{5}{x^2 - 7x + 12} - \frac{2}{3 - x} = \frac{5}{x - 4}$

26. $\frac{3}{x + 5} + \frac{2}{5 - x} = \frac{-4}{x^2 - 25}$

27. $\frac{8}{x^2 - x} + \frac{8}{x} = \frac{6}{2x - 2}$

28. $\frac{5}{x + 2} = \frac{-1}{x^2 + 7x + 10} + \frac{3}{-x - 5}$

29. $\frac{7x + 3}{x^2 - 8x + 15} + \frac{3x}{x - 5} = \frac{-1}{x - 3}$

30. $\frac{10}{2y + 8} - \frac{7y + 8}{y^2 - 16} = \frac{-8}{2y - 8}$

31. $\frac{x - 5}{x^2 - 4x - 5} - \frac{2x - 5}{x^2 - x - 2} = 0$

32. $\frac{x + 4}{x^2 + 4x} - \frac{x}{x^2 + 6x} = 0$

33. $\frac{2}{x + 3} - \frac{3}{4 - x} = \frac{2x - 2}{x^2 - x - 12}$

34. $\frac{4}{x + 4} = \frac{-5}{x^2 - 16} + 1$

Find the replacement set. Then solve for *x* or *y*.

35. $\left(\frac{x - 2}{x + 5}\right)^2 \cdot \left(\frac{x + 5}{2x - 4}\right)^3 = 1$

36. $\left(\frac{x + 1}{x + 7}\right)^2 \div \left(\frac{x + 1}{x + 7}\right)^4 = 0$

37. $(-1)^4 + \frac{y + 1}{2y} = \frac{3}{4y^5} \div \frac{1}{y^5}$

38. $\left(\frac{x}{(x - 2)^2} - \frac{3x}{(x - 2)^3}\right)(x - 2)^3 = 6$

39. $\frac{2}{5}\left(\frac{x}{b} + 1\right) = \frac{3}{8}\left(\frac{x}{b} - 1\right)$

40. $\frac{y}{y - b} - \frac{y - b}{y} = \frac{b}{y}$

Applications

41. **Personal Finance** Alan invested \$40,000 more than Scott. If each had invested \$30,000 more, the ratio of their investments would have been 3:5. How much did each invest?

42. Carpentry The dimensions of a portrait are $10\frac{1}{2}$ in. by 14 in. A carpenter wishes to build a frame for it of uniform width so that the outside of the framed portrait is in the ratio 5:6. Find the width of the frame.

43. Masonry A rectangular pool is 35 ft by 62 ft. A deck of uniform width is to be built around the pool so that the outside edges of the deck are in the ratio 2:3. Find the width of the deck.

44. Money Emily's savings account contains $72 less than Nicole's. If Nicole withdraws $16 from her account, the ratio of their accounts will be 7:15. Find the amount in each account.

Mixed Review

Find the additive and multiplicative inverses of each.

45. 8

46. $-\frac{4}{3}$

47. $\frac{c^2}{x}$

48. $\begin{bmatrix} 1 & 0 \\ -2 & -1 \end{bmatrix}$

Use x and y to write an algebraic equation for the following.

49. Twice the square root of one number is three more than the complement of another number.

50. The product of one number and its additive inverse is equal to the sum of another number and its multiplicative inverse.

51. Solve: $2|3x - 12| = x - 4$

Determine whether (2, −4) is a solution for each equation.

52. $2x + 3y = -12$

53. $y^2 = x^2 + 6x$

54. $\frac{xy}{2x^2} = \frac{x - y}{y - x}$

55. Find the slope of the line passing through (2, 1) and (−5, 3).

For each function, determine whether y varies directly as x. If so, find the constant of variation.

56. $y = \frac{3\pi x}{4}$

57. $y + 2 = 4x$

58. $3y = \frac{2x}{3}$

Developing Mathematical Power

59. Thinking Critically Write three different rational equations whose replacement set is the set of all real numbers except a.

7.7

Problem Solving Strategy: Classify a Problem

An important first step in solving a problem is classifying the problem. To classify means to try to relate it to a type of problem you know how to solve. If the problem fits into a class of problems you can solve, then apply the known problem solving strategy to the new problem. Very often, the appropriate strategy is to write a mathematical model. Here are some ways to classify a problem:

- The problem can be solved by doing some arithmetic calculations.
- The problem can be solved by applying a formula.
- The problem can be solved by using an algebraic model such as
    - a linear equation or inequality
    - an absolute value equation
    - a system of equations or inequalities in two variables
    - a matrix or determinant
    - a polynomial equation
    - a rational equation
- The problem can be solved by using geometric or coordinate techniques.

EXAMPLE Ted drove from his home to Dallas at an average speed of 40 mi/h. He returned home over the same route at an average speed of 50 mi/h. What is his average speed for the entire trip?

Understand the Problem *What are the given facts?* Ted drove to Dallas at an average speed of 40 mi/h and returned home at an average speed of 50 mi/h.
What are you asked to find? What is the average speed for the entire trip?

Plan Your Approach *Choose a strategy*. This is a problem involving rates. Rate problems usually can be solved by using a rational equation to model the conditions of the problem. The facts of the problem can be organized into a table.

Let d represent the distance between Ted's house and Dallas.

	distance	÷ rate	= time
going	d	40	$\frac{d}{40}$
returning	d	50	$\frac{d}{50}$

The average rate a for the entire trip is the total distance divided by the total time.

Total distance: $d + d = 2d$ Total time: $\frac{d}{40} + \frac{d}{50}$

$$a = \frac{2d}{\frac{d}{40} + \frac{d}{50}}$$

Write an equation.
Simplify the right side of the equation to find a.

Complete the Work

$$a = \frac{2d}{\frac{d}{40} + \frac{d}{50}} \cdot \frac{200}{200}$$

$$a = \frac{400d}{5d + 4d}$$

$$a = \frac{400d}{9d}$$

$$a = \frac{400}{9} = 44.44 \text{ mi/h}$$

Interpret the Results

State your answer. The average rate for the entire trip is 44.44 mi/h.

Note that the average rate is not the average of the two rates (45 mi/h). Imagine a trip of 100 miles each way. The rate is not 45 mi/h because Ted goes for a longer time at 40 mi/h than at 50 mi/h.

CLASS EXERCISES

Classify each statement as true or false.

1. Experience in solving many different kinds of problems is important for becoming a good problem solver.

2. All problems can be solved by using arithmetic or algebraic skills.

3. Writing a mathematical model is often a good strategy.

4. A good problem solver tries to relate a new problem to one already solved.

5. Matrices and determinants are not useful mathematical models.

6. The coordinate plane can be used as a problem solving tool for certain classes of problems.

7. Problem solving strategies help to plan an approach to solving a problem.

8. A problem that cannot be classified cannot be solved.

9. A good problem solver understands mathematical concepts and can use algebraic skills effectively.

10. Solving problems is an unimportant part of learning algebra.

PRACTICE EXERCISES

Give an example of each mathematical model listed below.

1. A multiplication example using only whole numbers.
2. A formula to find the area of a circle.
3. A linear equation whose solution is a negative integer.
4. An inequality whose solution set is greater than or equal to zero.
5. An absolute value equation.
6. A system of equations in two variables whose lines are perpendicular.
7. A matrix equation that will solve a system of two equations in two variables.
8. A polynomial equation that can be solved by factoring.
9. A rational equation with the variable y^2 in the denominator.
10. An equation of a line that is parallel to the x-axis.

Classify each problem by naming the type of mathematical model you would use to solve it.

11. Joyce and Ann each have some quarters. Joyce has 7 more quarters than Ann. The total value of the quarters is $3.25. How many quarters does each have?
12. Billy is paid $290 a week plus a commission equal to 2% of his sales. What must his sales be if he is to earn a weekly income of no less than $500 per week?
13. Felipe works in a supermarket. He wants to stack 12 dozen cans of soup with an equal number on each of 3 shelves. How many dozen cans should he stack on each shelf?
14. Find three consecutive even integers if the product of the first and second is 12 less than 6 times the third integer.
15. Karen rode her bike from home to school at an average speed of 20 mi/h. To return home from school, she rode at an average speed of 8 mi/h. Find her average speed for the entire trip.
16. The sum of the digits of a two-digit number is 11. If the digits are reversed, the new number is 7 times more than twice the original number. What is the original number?
17. Briana is 4 years older than her sister. Briana's age squared is equal to eleven times her sister's age plus 26. How old is Briana?

Solve each problem.

18. Robert drove his car to New York City from his office at an average speed of 60 mi/h. He returned to the office at an average speed of 40 mi/h. Find his average speed for the entire trip.

19. The length of a rectangular yard is 25 ft more than its width. If the width were doubled, the area of the yard would be 1800 ft^2. Find the dimensions of the yard.

Mixed Problem Solving Review

Solve each problem.

1. Juan traveled west from St. Louis at an average speed of 50 mi/h. Two hours after he left, Gail left St. Louis and traveled over Juan's route at an average speed of 60 mi/h. How long will it take Gail to catch up with Juan?

2. A small business firm determined that its earnings for the past five years could be represented by a linear function. In 1987, the company sold 15,000 products and earned $250,000. In 1988, sales were 17,000 units and earnings were $283,390. Find the earnings in 1989 on sales of 20,000 units.

3. A rectangular box is twice as long as it is wide and three times as wide as it is high. The sum of its length, width and height is 26 inches. Find the dimensions of the box.

4. The product of two consecutive multiples of 7 is 588. Find the numbers.

5. The height of a triangular sail on a boat is 2 ft less than 4 times the length of its base. If the area of the sail is 45 ft^2, find the length of its base and height.

6. If each of the base angles of an isosceles triangle is four times the vertex angle, find the measure of each angle.

7. A mover charges $150 for use of a truck and $85 per hour for labor. If the total cost of a move is a linear function, find the cost of a move requiring 6 hours of labor.

PROJECT

Working with a classmate, review all the word problems presented in Chapters 1 through 6. Make a list of any classes of problems that were not discussed in this lesson. Identify the type of mathematical model needed to solve each class.

7.8 Problem Solving: Using Rational Equations

Rational equations often serve as mathematical models for problem-solving situations involving rates. Organizing the information in a table may help you understand the problem and write an equation to solve it.

EXAMPLE 1 Carlos can travel 40 mi on his motorbike in the same time it takes Yuri to travel 15 mi on his bicycle. If Yuri rides his bike 20 mi/h slower than Carlos rides his motorbike, find the rate for each bike.

Understand the Problem A motorbike travels 40 mi. A bicycle travels 15 mi in the same time. The rate of the bicycle is 20 mi/h less than the rate of the motorbike. The time is found by dividing the distance by the rate.

Plan Your Approach Let x represent the rate of the motorbike in mi/h.
Then $x - 20$ represents the rate of the bicycle in mi/h.
Organize the information in a table.

	distance ÷	rate =	time
motorbike	40	x	$\frac{40}{x}$
bicycle	15	$x - 20$	$\frac{15}{x-20}$

Time for motorbike = Time for bicycle *Write a word equation.*

$$\frac{40}{x} = \frac{15}{x-20}$$ *Translate the word equation into an algebraic equation.*

Complete the Work

$$x(x-20)\frac{40}{x} = x(x-20)\frac{15}{x-20}$$ *LCD is $x(x - 20)$.*

$$40x - 800 = 15x$$
$$25x = 800$$
$$x = 32$$

Therefore, $x - 20 = 12$

Interpret the Results The rate of the motorbike is 32 mi/h. The rate of the bicycle is 12 mi/h. Is the time for the motorbike equal to the time for the bicycle?

$$\text{Time for motorbike} = \text{Time for bicycle}$$

$$\frac{40}{32} \stackrel{?}{=} \frac{15}{12}$$

$$\frac{5}{4} = \frac{5}{4} \checkmark$$

EXAMPLE 2 Shelley can paint the outside of a house in 8 h. Karen can do it in 4 h. How long will it take them to do the job if they work together?

Understand the Problem Shelley can do the job in 8 h and Karen can do it in 4 h. You are asked to find how long it will take them to do the job if they work together. If a job can be done in n hours, then

the part of the job that can be done in 1 hour is $\frac{1}{n}$

the part of the job that can be done in x hours is $x \cdot \frac{1}{n}$, or $\frac{x}{n}$

Plan Your Approach Let x represent the number of hours it takes to complete the job if they work together.

	part done in 1 hour	no. of hours	= part done in x hours
Shelley	$\frac{1}{8}$	x	$\frac{x}{8}$
Karen	$\frac{1}{4}$	x	$\frac{x}{4}$

Part done by Shelley plus part done by Karen is 1.

$$\frac{x}{8} + \frac{x}{4} = 1$$ Why does the sum of the parts of the job equal 1?

Complete the Work

$$8\left(\frac{x}{8} + \frac{x}{4}\right) = 8(1)$$ *LCD is 8.*

$$x + 2x = 8$$

$$3x = 8$$

$$x = \frac{8}{3}, \text{ or } 2\frac{2}{3}$$

Interpret the Results It would take Shelley and Karen $2\frac{2}{3}$ h to put one coat of paint on the house if they work together.

Does the sum of the parts of the job done by each person in $2\frac{2}{3}$ h equal 1?

$$\frac{\frac{8}{3}}{8} + \frac{\frac{8}{3}}{4} \stackrel{?}{=} 1$$

$$\frac{8}{24} + \frac{8}{12} \stackrel{?}{=} 1$$

$$1 = 1 \checkmark$$

Problems involving percents are also considered rate problems. You know that percent means "per hundred." Thus, 5% means 0.05, or $\frac{5}{100}$.

EXAMPLE 3 A pharmacist has 8 L of a 6% alcohol solution. How much water should be added to obtain a 4% alcohol solution?

Understand the Problem You are asked to find the amount of water that should be added to the 6% alcohol solution to obtain a 4% alcohol solution. The number of liters of solution, multiplied by the percent of alcohol, gives the number of liters of alcohol.

Plan Your Approach Let x represent the number of liters of water to be added. Then $8 + x$ represents the number of liters of 4% alcohol solution.

	L of solution ·	% alcohol =	L of alcohol
6% solution	8	6%	0.06(8)
4% solution	$8 + x$	4%	$0.04(8 + x)$

L alcohol in 6% solution = L alcohol in 4% solution

Complete the Work

$$0.06(8) = 0.04(8 + x)$$
$$6(8) = 4(8 + x) \quad \textit{Multiply both sides by 100.}$$
$$48 = 32 + 4x$$
$$16 = 4x$$
$$4 = x$$

Interpret the Results To obtain a 4% alcohol solution from the 6% alcohol solution, 4 L of water should be added. Do the original and final solutions contain the same number of liters of alcohol?

L alcohol in 6% solution = L alcohol in 4% solution

$$0.06(8) \stackrel{?}{=} 0.04(8 + x)$$
$$0.48 \stackrel{?}{=} 0.04(8 + 4)$$
$$0.48 \stackrel{?}{=} 0.04(12)$$
$$0.48 = 0.48 \checkmark$$

Problem Solving Reminders

- A table may help you to organize the information and understand the problem.
- Write a word equation before you write the algebraic equation.
- Be sure that your solution satisfies the conditions of the problem.

CLASS EXERCISES

Make a table and write an equation for each problem.

1. A passenger train travels 392 mi in the same time that it takes a freight train to travel 322 mi. If the passenger train travels 20 mi/h faster than the freight train, find the rate of each train.

2. Ann can sort a bag of mail in 3 h. Paul can sort the same bag in 2 h. If they work together, how long will it take them to sort the mail?

3. One pump can fill a tank with oil in 4 h. A second pump can fill the same tank in 3 h. If both pumps are used at the same time, how long will it take to fill the tank?

4. A chemist has 250 mL of a 5% acid solution. How much water should be added to this solution to reduce it to a 3% acid solution?

PRACTICE EXERCISES

1. It takes a person the same time to walk 10 mi as it takes a car to travel 15 mi in heavy traffic. If the car is traveling 3 mi/h faster than the person is walking, how fast is each traveling?

2. It takes a train the same time to travel 450 km as it takes a bus to travel 300 km. If the bus is traveling 30 km/h slower than the train, find the rate at which each is traveling.

3. Gilda can jog 5 mi downhill in the same time that it takes her to jog 3 mi uphill. Find her jogging rate for each way if she jogs downhill 4 mi/h faster than she jogs uphill.

4. Tony can type 100 words in the same amount of time that it takes Fred to type 75 words. If Fred's typing rate is 8 words per minute less than Tony's, find each person's typing rate.

5. Diana can mow the lawn in 20 min. Joan can mow the lawn in 30 min. If they work together, how long will it take them to mow the lawn?

6. Arthur can edit 15 essays in 10 h. Debra needs 15 h to do the same job. If they work together, how long will it take them to edit the essays?

7. Mr. Gomez must leave in 10 min to drive to his office, but he is snowed in. Sam claims he can shovel the driveway in 20 min and Paul claims he can do it in 15 min. If they shovel together, will they be able to clear the driveway before Mr. Gomez has to leave?

8. Fran can wash the car in 30 min. Gail can wash the car in 40 min. Working together, can they wash the car in less than 16 min?

9. A small tank contains 500 L of a 5% salt solution. How much water should be added to obtain a 2% salt solution?

10. A big barrel contains 300 L of a 6% solution of lawn fertilizer. How much water should be added to obtain a 5% solution?

11. A pharmacist has 60 mL of an 8% alcohol solution. How much water should be evaporated so that the solution that remains is 10% alcohol?

12. A chemist has 80 mL of a 6% acid solution. How much water should be evaporated so that the solution that remains is 8% acid?

13. Machine A can do a job in 10 h, machine B can do the job in 15 h, and machine C can do it in 20 h. If the three machines are being used at the same time, how long will it take to complete the job?

14. A man drove his car 132 mi before the water pump broke. Then the car was pushed 1 mi to a gas station. The man could drive the car 12 times faster than it could be pushed. If the total trip took 3 h, find how fast the car was pushed.

15. Nancy does one-quarter of a job in 3 h. If Carl works 2 hours more than Nancy, he can finish one-sixth of the job. How long does it take to do the job if they work together?

16. The members of a math club invested \$150 more at 9% than at 10%, but the income from each investment was the same. How much was invested at each rate?

17. Jan invested \$3600, part at 8% and the rest at 7%. If the total income from the investments was \$274, how much did she invest at each rate?

18. Three machines are twisting pretzels. Two are old and one is new. The new machine can twist pretzels twice as fast as an old machine can. An old machine can twist the daily quota of pretzels in 20 h. How long will it take to make the quota if the machines are operated at the same time?

19. A generator in an electrical power plant produces 10 MW (megawatts) of electrical power in one hour. A second generator produces $2\frac{1}{2}$ times as much power as the first, and a third generator produces 50% of the power of the first two combined. How long does it take the three generators to produce 210 MW of power?

20. It normally takes 2 h to fill a swimming pool. The pool has developed a slow leak. If the pool were full, it would take 10 h for all the water to leak out. If the pool is empty, how long will it take to fill it?

21. How much of a 25% acid solution should be mixed with 5 L of a 15% acid solution to obtain a 20% acid solution?

22. How much pure alcohol must be added to 4 L of a 10% alcohol solution to make a solution that is 20% alcohol?

23. George jogged downhill at 6 mi/h and then jogged back up at 4 mi/h. If the total jogging time was $1\frac{1}{4}$ h, how far did he jog in all?

24. A canoe can travel 15 mi/h in still water. Going with the current it can travel 20 mi in the same time that it takes to travel 10 mi against the current. Find the rate of the current.

25. An experienced carpenter can panel a room 3 times faster than an apprentice can. Working together, they can panel the room in 6 h. How long would it take each one working alone to do the job?

26. An automatic pitching machine can pitch all its balls in $1\frac{1}{4}$ h. One attendant working $3\frac{1}{2}$ h can retrieve all the balls pitched by one machine. How many attendants working at the same rate should be hired so that the balls from 10 machines are all retrieved in 8 h?

27. The Petriks earned $120 simple interest on a certain amount of money for one year. If the rate had been 2% higher, they could have invested $300 less and still earned the same amount of interest. How much did they invest and what was the rate of interest?

TEST YOURSELF

Simplify.

1. $\dfrac{2 - \frac{3}{4}}{\frac{3}{8}}$ **2.** $\dfrac{1 + \frac{3}{x}}{2 + \frac{2}{3x}}$ **3.** $\dfrac{\frac{1}{x+2} - \frac{2}{x^2-4}}{\frac{3}{x-2} + \frac{4}{x+2}}$ 7.5

Find the replacement set. Then solve and check.

4. $\frac{5}{2x} - \frac{2}{3} = \frac{1}{x} + \frac{5}{6}$ **5.** $\frac{2}{x} - \frac{1}{x-1} = \frac{1}{x+1}$ **6.** $\frac{3}{2y-6} = \frac{5}{3y+9}$ 7.6

7. Joe can wash the car in 45 min. Jim needs 30 min to wash it. If they work together, how long will it take them to wash the car? 7.8

8. A chemist wants to make 700 mL of a 6% nitric acid solution by using 5% and 12% nitric acid solutions. How much of each should she use?

INTEGRATING ALGEBRA
Thermal Expansion of Solids

Did you know that solid materials expand when temperature increases and contract when temperature decreases? Thermal expansion in solids is an important factor in many practical problems. Provisions for expansion at joints must be made in the construction of buildings, bridges, pavements, and other structures. In laying railroad tracks, a gap is left between the rails to allow for expansion of the rails on a warm day.

The change in any linear dimension of a solid is called **linear expansion.** The **coefficient of linear expansion** is the change in length per unit length per degree change in temperature. The coefficient of linear expansion k is different for each substance, and physicists use the following formula to determine what it is.

$$k = \frac{\Delta l}{l(T - t)}$$

where
l = original length of the solid
Δl = change in length of the solid
T = highest temperature in degrees
t = lowest temperature in degrees

Use this formula to find the change in length of a solid (Δl), or the highest temperature value (T) required to produce a specific change in length. You must know the coefficient of the linear expansion (k) for the given substance.

EXAMPLE 1 Suppose a steel tape is accurate at 40°C and is used to measure a mile when the temperature is 10°C. If the coefficient of expansion for the tape is 12×10^{-6}, how many feet and inches should be added to 5280 ft to get a true reading?

$$k = \frac{\Delta l}{l(T - t)}$$

$$(12 \times 10^{-6}) = \frac{\Delta l}{5280(40 - 10)}$$ *Substitute 12×10^{-6} for k, 5280 for Δl, 40 for T, and 10 for t.*

$$\Delta l = (5280)(30)(12 \times 10^{-6}) \approx 1.9$$ *Solve for Δl.*

The tape will contract about 1.9 ft at 10°C. Add about 1.9 ft, or 1 ft 11 in. to 5280 ft to get a true reading. The tape should read about 5281 ft 11 in.

EXAMPLE 2 A platinum rod and a brass rod are each 1 m long at 0°C. They are heated until the brass rod is 2 mm longer than the platinum rod. What temperature was used to accomplish this? ($k = 8.9 \times 10^{-6}$ for platinum, and $k = 1.89 \times 10^{-5}$ for brass)

$$k = \frac{\Delta l}{l(T - t)}$$

$$k = \frac{\Delta l}{lT}$$ *Substitute 0°C for t.*

$$T = \frac{\Delta l}{lk}$$ *Solve for T.*

T for platinum $=$ T for brass *Both rods are heated to the same temperature.*

$$\frac{x}{(1)(0.0000089)} = \frac{x + 2}{(1)(0.0000189)}$$ *Let $x = \Delta l$ for platinum, then $x + 2 = \Delta l$ for brass.*

$$0.0000189x = 0.0000089x + 0.0000178$$

$$x = \frac{0.0000178}{0.00001} = 1.78$$ *The platinum changed by 1.78 mm.*

$$T = \frac{1.78}{1000(0.0000089)}$$ *Substitute 1.78 for x and use 1000 mm in place of 1 m for the original length.*

$T = 200$ The temperature used was 200°C.

EXERCISES

1. Steel railroad rails are put into place when the temperature is 32°F. Each rail is 39 ft long. How much gap should be left between one rail and the next if the two rails should just touch when the temperature rises to 98°F?
2. The World Trade Center with its steel framework is 415 m high at 10°C. How tall is it on a hot summer day when the temperature is 40°C?
3. A steel tape is accurate at 30°C. The tape is used to measure 4 km at −15°C. How many meters and centimeters should be read on the tape to get a true 4 km reading at this temperature?
4. Suppose an aluminum tape is used instead of steel in Exercise 3. If $k = 24 \times 10^{-6}$ for aluminum, what should the reading be?
5. A platinum and a gold rod are each 3 m long at 32°F. The rods are heated until the gold rod is 3.5 mm longer than the platinum rod. What temperature was used? ($k = 14 \times 10^{-6}$ for gold)
6. A highway is made of concrete slabs placed end to end. Each slab is 25 m long. How wide an expansion gap should be left at 20°C if the slabs are just to touch at 90°C? ($k = 1.0 \times 10^{-5}$ for concrete)

CHAPTER 7 SUMMARY AND REVIEW

Vocabulary

complex rational expression (327)
equivalent rational expressions (313)
negative exponent (307)
rational equation (332)
rational expression (313)
scientific notation (307)
significant digits (308)
zero exponent (307)

Properties of Exponents If a is a nonzero real number and n is an integer, then $a^0 = 1$ and $a^{-n} = \frac{1}{a^n}$. **7.1**

Simplify and write with positive exponents.

1. $(x^{-3})(x^{-4})$ **2.** $(x^{-5}y^{-1})^2$ **3.** $(x^4 - x^{-3})^0$ **4.** $\frac{x^{-6}}{x^{-2}}$

Changing from Decimal Form to Scientific Notation To change a number from decimal form to scientific notation, write the number in the form

$$a \times 10^b \text{ where } 1 \leq a < 10, \text{ and } b \text{ is an integer.}$$

The digits of a should be significant.

Write in scientific notation. If an integer ends in zeros, assume that the zeros are not significant.

5. 65,000,000,000 **6.** 0.00000723 **7.** 5,709,000 **8.** 0.002400

Changing from Scientific Notation to Decimal Form To change a number from scientific notation to decimal form, use the absolute value of the exponent to determine the number of places to move the decimal point.

Write in decimal form.

9. 3.4×10^7 **10.** 6.88×10^{-4} **11.** 9.0×10^{-3} **12.** 6.02×10^5

Simplifying Rational Expressions To simplify a rational expression, factor the numerator and denominator completely and then divide both by the common factors. **7.2**

Simplify.

13. $\frac{21x^4y^3}{-7x^5y^2}$ **14.** $\frac{3x - 15}{5 - x}$ **15.** $\frac{2x^2 - 6x - 36}{x^2 - 36}$

Multiplying and Dividing Rational Expressions To multiply and divide rational expressions use the product and quotient rules. 7.3

Multiply or divide as indicated.

16. $\frac{-24x^4y^5}{-14x^2z} \cdot \frac{7z^3}{3x^5}$ **17.** $\frac{x}{2x-6} \div \frac{2x^2}{x^2-9}$ **18.** $\frac{x^2-5x-14}{x^2+5x+6} \div \frac{x^2-49}{5x+15}$

Adding and Subtracting Rational Expressions To add or subtract with the same denominators, add or subtract numerators and keep the common denominator. To add or subtract with different denominators, find the least common denominator (LCD) and write the equivalent rational expressions with the LCD as denominator. Then add or subtract. 7.4

Add or subtract as indicated.

19. $\frac{4x-6}{x-3} - \frac{5x+2}{x-3}$ **20.** $\frac{5x}{x^2-4} + \frac{7}{9x+18}$ **21.** $\frac{2x+3}{x+4} + \frac{3-7x}{x-5}$

Simplifying Complex Rational Expressions To simplify a complex rational expression, multiply the numerator and denominator by the LCD of all the rational expressions in the complex rational expression. 7.5

Simplify.

22. $\frac{\frac{1}{x^2}}{\frac{2}{x}}$ **23.** $\frac{\frac{4}{x-1} - \frac{3}{x+1}}{\frac{5}{x-1} + \frac{2}{x+1}}$ **24.** $\frac{\frac{2}{x-7} + \frac{x}{x^2-49}}{\frac{5}{x+7}}$

Solving Rational Equations To solve a rational equation, multiply both sides of the equation by the LCD of the rational expressions, then solve. 7.6

Find the replacement set. Then solve and check.

25. $\frac{5}{7x} = \frac{2}{28}$ **26.** $\frac{9}{x+2} = \frac{6}{x-5}$ **27.** $\frac{x}{x-7} = \frac{x}{x-7} + 11$

28. The numerator of a fraction is 3 less than the denominator. If the fraction is equivalent to $\frac{9}{10}$, find the fraction. 7.7

29. Twice an integer minus 8 times its reciprocal is 15. Find the integer.

30. Marge can file monthly reports in 6 h and Brian can do it in 8 h. How long will it take them to file the reports if they work together?

CHAPTER 7 TEST

Simplify and write with positive exponents.

1. $(x^{-7})(x^3)$
2. $(x^5y^{-3})^{-2}$
3. $(x^5x^{-3})^0$

Write in scientific notation. If an integer ends in zeros, assume that the zeros are not significant.

4. 735,000,000,000
5. 0.008970

Write in decimal form.

6. 2.41×10^{-7}
7. 5.3×10^4

Simplify.

8. $\dfrac{14x^7y^3z}{-7x^2y^5z^3}$
9. $\dfrac{3x^2 - 9x - 12}{x^2 + x - 20}$
10. $\dfrac{22x^2y}{14z} \div \dfrac{2xy}{7z}$
11. $\dfrac{x^2 + 8x - 9}{x^2 + 2x - 3} \cdot \dfrac{x^2 + 3x}{x^2 + 13x + 36}$
12. $\dfrac{5x - 1}{x^2 + 1} + \dfrac{4x}{x^2 + 1}$
13. $\dfrac{3}{x^2 - x} - \dfrac{2x}{3x - 3}$
14. $\dfrac{\frac{3}{x}}{\frac{5}{4x}}$
15. $\dfrac{\frac{4}{x + 2} + \frac{3}{x + 1}}{\frac{2}{x + 2} + \frac{8}{x + 1}}$

Find the replacement set. Then solve and check.

16. $\dfrac{6}{5x} = \dfrac{3}{20}$
17. $\dfrac{1}{x + 2} = \dfrac{4x}{x - 1} - 1$
18. A cyclist can travel 4 mi/h faster going downhill than going uphill. If it takes the same amount of time to go 3 mi downhill as it does to go 2 mi uphill, find the downhill rate.

Challenge

Josie works from 7 AM to noon every day. She can assemble a computer in 12 h and Eric can do it in 16 h. Josie starts to assemble a computer at 7 AM and Eric helps her from 10:00 AM to noon and then completes the task himself. How many hours will Eric have to work alone?

COLLEGE ENTRANCE EXAM REVIEW

Select the best choice for each question.

1. When the graph of $3x - 2y \geq 6$ is drawn, which of the following points is NOT included?
A. $(2, -2)$ **B.** $(2, 0)$ **C.** $(1, 2)$
D. $(0, -4)$ **E.** $(3, 1)$

2. The expression $\frac{x}{(x^2 + 4)(x - 3)}$ represents a real number for all real values of x except
A. $3, 2, 0, -2$ **B.** $3, 2, -2$
C. $2, -2$ **D.** 0 **E.** 3

3. In the figure below, $ABCD$ is a parallelogram with angle measures as marked. Find $x + y$.

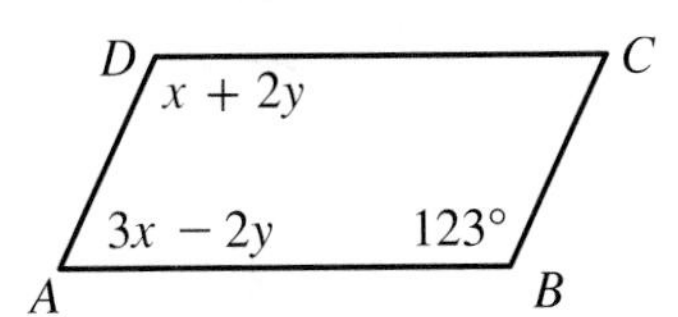

A. 84 **B.** 79 **C.** 57
D. 45 **E.** 39

4. If $x = 1 - \frac{1}{1 - \frac{1}{3}}$, then $x =$

A. $\frac{3}{2}$ **B.** $-\frac{3}{2}$ **C.** $\frac{1}{2}$
D. $-\frac{1}{2}$ **E.** $\frac{2}{3}$

5. One day Tim, Paul, Peter, and Al decided to pool their money. Tim had \$4.81, Paul had \$5.42, and Peter had \$3.94. Al said, "With mine we have an average of \$5.00 each." How much did Al have?
A. \$6.03 **B.** \$5.83 **C.** \$5.17
D. \$5.00 **E.** \$4.72

For questions 6–8 the operation $\oplus$ is defined as: $a \oplus b = a^2b - ab^2$

Example: $3 \oplus 2 = 9 \cdot 2 - 3 \cdot 4 = 6$

6. $5 \oplus (-2) =$
A. -70 **B.** -30 **C.** 30
D. 70 **E.** 100

7. $0.2 \oplus 0.1 =$
A. 0.2 **B.** 0.01 **C.** 0.02
D. 0.001 **E.** 0.002

8. $(3 \oplus 5) - (5 \oplus 3) =$
A. 60 **B.** -60 **C.** 30
D. -30 **E.** 0

9. The figure below has 4 squares placed side by side. How many rectangles are in the figure?

A. 11 **B.** 10 **C.** 9
D. 8 **E.** 6

10. When m, n, and p are all odd integers, which of these three statements is (are) true?

I. $m^2 + n^3 + p^4$ is even
II. $mn + p$ is even
III. $mnp + 1$ is even

A. I only
B. III only
C. I and II only
D. II and III only
E. I, II, and III

MIXED REVIEW

Compute.

Examples $3\frac{1}{5} \div 2\frac{2}{3}$

$3\frac{1}{5} \div 2\frac{2}{3} = \frac{16}{5} \div \frac{8}{3}$ *Change to fractions.*

$= \frac{\overset{2}{\cancel{16}}}{5} \times \frac{3}{\underset{1}{\cancel{8}}} = \frac{6}{5}$ or $1\frac{1}{5}$

$$\begin{array}{rr} 10\frac{1}{2} \rightarrow & 9\frac{12}{8} \\ -7\frac{5}{8} & -7\frac{5}{8} \\ \hline & 2\frac{7}{8} \end{array}$$

1. $1\frac{2}{3} \times 2\frac{3}{5}$ **2.** $3\frac{3}{4} \times 2\frac{2}{3}$ **3.** $1\frac{7}{8} \div 5$ **4.** $2\frac{2}{5} \div 4\frac{1}{2}$

5. $5\frac{7}{8} + 4\frac{5}{6}$ **6.** $10\frac{2}{3} + 8\frac{4}{5}$ **7.** $8 - 3\frac{5}{6}$ **8.** $12\frac{5}{6} - 9\frac{7}{8}$

Simplify.

Examples $\sqrt{0.25} = \sqrt{(0.5)^2} = 0.5$ $\sqrt{\frac{9}{16}} = \sqrt{\left(\frac{3}{4}\right)^2} = \frac{3}{4}$

9. $\sqrt{49}$ **10.** $\sqrt{64}$ **11.** $\sqrt{0.16}$ **12.** $\sqrt{0.09}$

13. $\sqrt{\frac{4}{9}}$ **14.** $\sqrt{\frac{25}{81}}$ **15.** $\sqrt{x^2}$ **16.** $\sqrt{y^2}$

Multiply or divide.

Example $(2a^2b^3)(3ab^2) = (2 \cdot 3)(a^2 \cdot a)(b^3 \cdot b^2)$
$= 6(a^{2+1})(b^{3+2})$ *Add the exponents.*
$= 6a^3b^5$

17. $(7m^4)(-m^2)$ **18.** $(8d^4)(2d^3)(d^2)$

19. $(4x^2)(3xy^2)$ **20.** $(-5c^3)(3c^3d)$

21. $6a^3b^5 \div 2a^2b^3$ **22.** $12x^7 \div 3x^5$

23. $8x^6y^7 \div (-4xy)$ **24.** $9p^5q^3 \div 3p^2q^3$

Solve.

25. The length of $\overline{AB}$ is 15 cm. $\overline{CD}$ bisects $\overline{AB}$ at E. What is the length of $\overline{AE}$?

26. In triangle ABC, $\angle C$ is a right angle. Find CB, if $AB = 15$ and $AC = 9$.

Irrational and Complex Numbers

A calculated breakup?

Developing Mathematical Power

When you think of a pendulum, you probably picture a grandfather clock. The wrecker's ball is a special type of pendulum.

In the 16th century, Galileo discovered that the period of oscillation, the time it takes a pendulum to swing from one side to the other, depends on the pendulum's length. He also discovered that the oscillation period can be changed by altering the pendulum's length and that the period does not depend on the weight at the end of the pendulum.

Project

Devise an experiment to demonstrate Galileo's discoveries. Then substitute your values for L in the formula below and compare the calculations with the results of your experiment.

Period of Oscillation of a Simple Pendulum

$$T = 2\pi \sqrt{\frac{L}{g}}$$

where T is the period of oscillation, L is the length (in feet), and g is the force of gravity, 32 ft/s^2.

8.1

Roots and Radicals

Objectives: To simplify an nth root
To solve equations of the form $x^n = k$

A *perfect square* is the product of two equal rational numbers or expressions.

Capsule Review

16 is a perfect square.	$16 = 4^2$ or $(-4)^2$
$\frac{1}{36}$ is a perfect square.	$\frac{1}{36} = \left(\frac{1}{6}\right)^2$ or $\left(-\frac{1}{6}\right)^2$
21 is *not* a perfect square.	There is no rational number k such that $k^2 = 21$.
x^2 is a perfect square.	$x^2 = (x)^2$ or $(-x)^2$
y^4 is a perfect square.	$y^4 = (y^2)^2$ or $(-y^2)^2$
x^8y^6 is a perfect square.	$x^8y^6 = (x^4y^3)^2$ or $(-x^4y^3)^2$

Write each perfect square as a number or a term with the exponent 2.

1. 25 **2.** 0.09 **3.** $\frac{4}{49}$ **4.** x^{10} **5.** x^4y^2 **6.** $169x^6y^{12}$

The solutions of a quadratic equation of the form $x^2 = k$ are the *positive square root* of k, $+\sqrt{k}$, and the *negative square root* of k, $-\sqrt{k}$. These solutions are often written as $\pm\sqrt{k}$, read "the positive or negative square roots of k." If $k < 0$, then k has no real square roots, since the square of a real number is never negative. It follows that if $k < 0$, then $x^2 = k$ has no real-number solutions.

EXAMPLE 1 **Find the real-number solutions, if any, for each equation.**
a. $y^2 - 8 = 41$ **b.** $3m^2 + 48 = 0$

a. $y^2 - 8 = 41$
$y^2 = 49$
$y = \pm 7$

Check: $y^2 - 8 = 41$
$7^2 - 8 \stackrel{?}{=} 41$
$41 = 41$ ✓

$y^2 - 8 = 41$
$(-7)^2 - 8 \stackrel{?}{=} 41$
$41 = 41$ ✓

The solutions are 7 and -7.

b. $3m^2 + 48 = 0$
$m^2 = -16$

There is no real-number solution because the square of a real number is never negative.

The positive square root of a number is its **principal square root,** which is usually written without the + symbol. Therefore, $\sqrt{k}$ indicates the principal square root of k and $-\sqrt{k}$ indicates the negative square root of k.

EXAMPLE 2 **Simplify each principal square root if it is a real number.**

a. $\sqrt{16}$ **b.** $\sqrt{-16}$ **c.** $\sqrt{0.16}$

a. $\sqrt{16} = 4$ *The principal square root is the positive square root, and $16 = (4)(4)$.*

b. $\sqrt{-16}$ is *not* a real number. *There is no real number whose square is -16.*

c. $\sqrt{0.16} = 0.4$ *The principal square root is the positive square root, and $0.16 = (0.4)(0.4)$.*

It can be shown that an equation of the form $x^3 = k$ has exactly one real-number solution for all real numbers k. This solution is called the **principal cube root** of k, written $\sqrt[3]{k}$. Any real number k, whether positive or negative, has exactly one real-number cube root.

EXAMPLE 3 **Simplify each principal cube root.**

a. $\sqrt[3]{64}$ **b.** $\sqrt[3]{-64}$ **c.** $\sqrt[3]{-0.064}$

a. $\sqrt[3]{64} = 4$ *$(4)(4)(4) = 64$*

b. $\sqrt[3]{-64} = -4$ *$(-4)(-4)(-4) = -64$*

c. $\sqrt[3]{-0.064} = -0.4$ *$(-0.4)(-0.4)(-0.4) = -0.064$*

For an equation of the nth degree, such as $x^n = k$, the ***n*th root** of k may be a real-number solution. The symbol for the **principal *n*th root** of k, $\sqrt[n]{k}$, is called a **radical expression,** or simply a **radical.** A radical expression consists of the following parts:

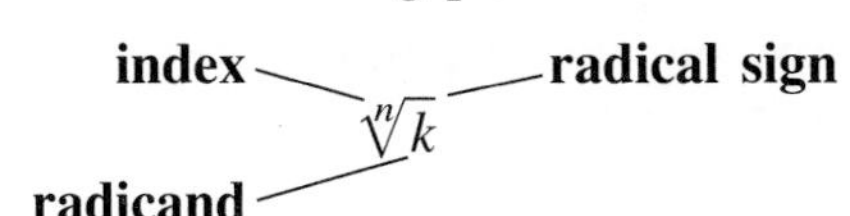

When the index is 2, it is usually not written.

For any real number k and any integer n, $n > 1$, the principal nth root of k can be described as follows:

If $k > 0$, then $\sqrt[n]{k}$ is a positive real number.
If $k = 0$, then $\sqrt[n]{k}$ is zero.
If $k < 0$ and n is odd, then $\sqrt[n]{k}$ is a negative real number.
If $k < 0$ and n is even, then $\sqrt[n]{k}$ is not a real number.
(This case will be discussed in Lesson 8.7.)

EXAMPLE 4 **Find the principal *n*th root if it is a real number.**

a. $\sqrt[3]{-8}$ **b.** $\sqrt[4]{-16}$ **c.** $\sqrt{49}$

a. $\sqrt[3]{-8} = -2$ *$-8 < 0$ and n is odd.*

b. $\sqrt[4]{-16}$ is *not* a real number. *$-16 < 0$ and n is even.*

c. $\sqrt{49} = 7$ *$49 > 0$*

The following equivalent expressions are useful when you simplify a radical expression involving exponents.

$\sqrt[n]{k^n} = k$ when n is odd. $\sqrt[3]{6^3} = 6$ $\sqrt[5]{(-2)^5} = -2$ $\sqrt[7]{x^7} = x$

$\sqrt[n]{k^n} = |k|$ when n is even. $\sqrt{(-7)^2} = |-7| = 7$ $\sqrt[4]{5^4} = |5| = 5$

The principal nth root of k^n is nonnegative when n is even. To avoid the use of absolute value signs, assume the following, unless stated otherwise: all variable expressions in a radicand with an even index represent nonnegative real numbers.

EXAMPLE 5 **Simplify:** **a.** $\sqrt{4x^2y^4}$ **b.** $\sqrt[3]{-27c^6}$

a. $\sqrt{4x^2y^4} = \sqrt{2^2x^2(y^2)^2} = \sqrt{(2xy^2)^2} = 2xy^2$

b. $\sqrt[3]{-27c^6} = \sqrt[3]{(-3)^3(c^2)^3} = \sqrt[3]{(-3c^2)^3} = -3c^2$

An equation of the form $x^n = k$, where n is a positive integer, may have no real solutions or it may have one or two real solutions.

When n is even and $k < 0$, $x^n = k$ has *no* real solutions.
When n is odd and k is any real number, $x^n = k$ has *exactly one* real solution, $\sqrt[n]{k}$.
When n is even and $k > 0$, $x^n = k$ has *two* real solutions, $\sqrt[n]{k}$ and $-\sqrt[n]{k}$.

EXAMPLE 6 **Find the real solutions, if any, for each equation.**

a. $3x^3 = -24$ **b.** $x^4 - 625 = 0$

a.
$$3x^3 = -24$$
$$x^3 = -8$$
$$x = -2$$

Check:
$$3x^3 = -24$$
$$3(-2)^3 \stackrel{?}{=} -24$$
$$3(-8) \stackrel{?}{=} -24$$
$$-24 = -24 \checkmark$$

The solution is -2.

b.
$$x^4 - 625 = 0$$
$$x^4 = 625$$
$$x = \pm 5$$

Check:
$$x^4 - 625 = 0$$
$$(\pm 5)^4 - 625 \stackrel{?}{=} 0$$
$$625 - 625 \stackrel{?}{=} 0$$
$$0 = 0 \checkmark$$

The solutions are 5 and -5.

CLASS EXERCISES

Determine whether each statement is true or false. If a statement is false, correct it to make it true.

1. The index in $\sqrt{3}$ is 3.
2. The radicand in $\sqrt[5]{32}$ is $\sqrt[5]{32}$.
3. The index in $\sqrt{5x^4}$ is 2.
4. The radicand in $\sqrt[4]{x^8y^4}$ is x^8y^4.
5. $\sqrt[6]{-1}$ is a real number.
6. $\sqrt[3]{10x^2}$ is a radical expression.
7. $\sqrt[5]{0}$ is not a real number.
8. $\sqrt[4]{-16x^8}$ is not a real number.

Predict the number of real solutions for each equation.

9. $3x^2 + 10 = 8$
10. $5x^3 - 6 = -6$
11. $x^4 = 1$

PRACTICE EXERCISES

Find the real solutions, if any, for each equation.

1. $x^2 = 100$
2. $3x^2 = 75$
3. $m^2 - 36 = 0$
4. $m^2 - 4 = 12$
5. $y^2 + 12 = 3$
6. $5y^2 = -30$
7. $4x^3 = 32$
8. $x^4 + 81 = 0$
9. $5x^3 + 16 = 16$
10. $2x^2 - 50 = 0$
11. $y^3 = -216$
12. $2x^4 - 32 = 0$

Simplify each radical expression if it is a real number.

13. $\sqrt{36}$
14. $-\sqrt{36}$
15. $\pm\sqrt{36}$
16. $\sqrt{0.36}$
17. $\sqrt{-1}$
18. $\sqrt{100}$
19. $-\sqrt{121}$
20. $\sqrt{-81}$
21. $\sqrt[3]{64}$
22. $\sqrt[3]{-64}$
23. $\sqrt{16x^2}$
24. $\sqrt{0.25x^6}$
25. $\sqrt{x^8y^{18}}$
26. $\sqrt{x^{10}y^{100}}$
27. $\sqrt{x^{80}y^{50}}$
28. $\sqrt{64b^{48}}$
29. $\sqrt{121a^{90}}$
30. $-\sqrt{81c^{48}d^{64}}$
31. $\sqrt{64x^{36}y^{96}}$
32. $\sqrt[3]{-64a^{81}}$
33. $\sqrt[5]{32y^{25}}$
34. $\sqrt[7]{x^{14}y^{35}}$
35. $\sqrt{0.0064x^{40}}$
36. $-\sqrt[3]{0.000027y^{33}}$
37. $\sqrt{(x + 3)^2}$
38. $\sqrt{(x + 1)^4}$
39. $\sqrt[3]{(x - 5)^6}$
40. $\sqrt[3]{(x - 2)^9}$
41. $\sqrt[3]{(x^2 - 8x + 16)^9}$
42. $-\sqrt[4]{(x^2 + 2x + 1)^8}$
43. $\pm\sqrt[11]{-m^{33a}}$
44. $\sqrt{25p^{4c-2}}$
45. $\sqrt[6]{64q^{12a+54}}$
46. $\sqrt{x^{4a+16}}$
47. $\sqrt[3]{s^{3n}}$
48. $\sqrt[n]{r^n}$
49. $\sqrt[n]{(p + q)^n}$

Applications

50. Geometry The radius of a sphere is $\sqrt[3]{125}$ cm. Express the radius in simplest form.

51. Number Theory A solution of a certain fifth-degree equation is $\sqrt[5]{a^{10}}$. Write the solution in simplest form.

52. Geometry The length of a side of a square room is $\sqrt{36a^4b^8}$ ft. Express the length in simplest form.

53. Physics The velocity of a bowling ball with a given kinetic energy is $\sqrt{0.49}$ m/s. Express the velocity in simplest form.

54. Electricity The current loss I_L in a wire is $I_L = K\frac{r^2}{l}$, where r is the radius of the wire in millimeters and l is its length in meters. If a 2 m wire has a current loss of $18K$, what is its radius?

Developing Mathematical Power

Thinking Critically Euclid proved in the tenth book of his *Elements* that $\sqrt{2}$ is an irrational number. He used an indirect proof. That is, he assumed that $\sqrt{2}$ was rational. Then he showed that this assumption led to a contradiction.

Assume that $\sqrt{2}$ is a rational number. Then $\sqrt{2} = \frac{a}{b}$, where a and b are integers, $b \neq 0$. Also, assume that a and b are *relatively prime*. That is, they have no common integral factor other than 1.

$$\text{If} \quad \sqrt{2} = \frac{a}{b}$$

$$\text{then} \quad 2 = \frac{a^2}{b^2} \qquad \text{Square both sides.}$$

$$\text{and} \quad 2b^2 = a^2$$

Since $2b^2$ is an even number, a^2 is even. Since a^2 is even, a is even. Therefore, for some integer c

$$a = 2c$$

$$a^2 = 4c^2 \qquad \text{Square both sides.}$$

$$2b^2 = 4c^2 \qquad \text{Recall that } a^2 = 2b^2.$$

$$b^2 = 2c^2$$

Since $2c^2$ is an even number, b^2 is even. Since b^2 is even, b is even. However, two even numbers cannot be relatively prime, so a and b are not relatively prime. This contradicts the original assumption, so it is not true that $\sqrt{2}$ is rational. Thus, $\sqrt{2}$ is irrational.

55. Use an indirect proof to show that $\sqrt{3}$ is irrational.

8.2 Multiplying and Dividing Radicals

Objectives: To multiply and divide radical expressions
To write radical expressions in simplest form

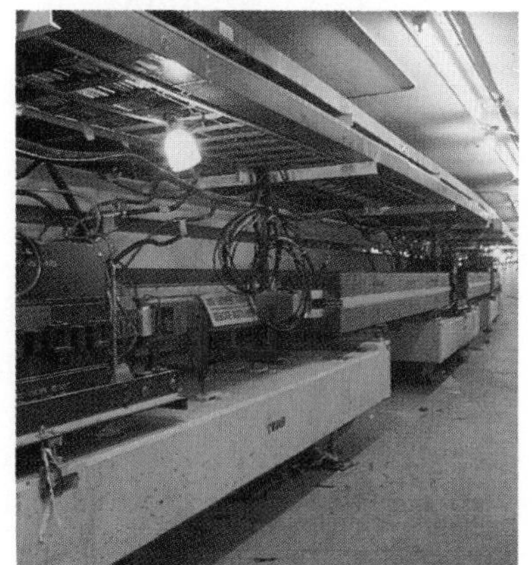

A linear accelerator causes particles, such as protons, to move at increasing velocities. The velocity, v, of a particle can be represented by the formula

$$v = \sqrt{\frac{2K}{m}}$$

In the formula, K represents the kinetic energy of the particle, or the energy resulting from its motion, measured in joules; m represents the mass of the particle, in milligrams.

The expression on the right side of the formula is not in simplest form. In order to simplify radical expressions and their products and quotients, it is often helpful to find appropriate factors of the radicand.

Capsule Review

Find the missing factor.

EXAMPLE $24a^5b^4 = 8a^3b^3(\underline{?})$
$= 8a^3b^3(3a^2b)$

1. $150 = 25(\underline{?})$ **2.** $54 = (\underline{?})(2)$ **3.** $48 = 16(\underline{?})$

4. $x^5y^2 = x^4y^2(\underline{?})$ **5.** $3a^3b^4 = a^3b^3(\underline{?})$ **6.** $75a^7b^8 = 25a^6b^6(\underline{?})$

Radical expressions such as $\sqrt{0.04}$ and $\sqrt[3]{-27}$ represent rational numbers, since $\sqrt{0.04} = 0.2$ and $\sqrt[3]{-27} = -3$. Radical expressions such as $\sqrt{5}$ and $\sqrt[3]{4}$ represent real numbers that are *irrational,* or not rational. What distinction can you make between the radicands of the rational and irrational numbers represented above?

To multiply radicals, consider the following:

$$\sqrt{16} \cdot \sqrt{9} = 4 \cdot 3 = 12 \quad \text{and} \quad \sqrt{16 \cdot 9} = \sqrt{144} = 12$$
$$\text{Therefore, } \sqrt{16} \cdot \sqrt{9} = \sqrt{16 \cdot 9}$$

In general, the product of the principal nth roots of two positive numbers is equal to the principal nth root of their product.

For all real numbers a and b, where $a \geq 0$ and $b \geq 0$, and for any integer n, where $n > 1$

$$\sqrt[n]{a} \cdot \sqrt[n]{b} = \sqrt[n]{ab}$$

EXAMPLE 1 **Multiply:** **a.** $\sqrt{5} \cdot \sqrt{7}$ **b.** $\sqrt[3]{3x} \cdot \sqrt[3]{11y}$

a. $\sqrt{5} \cdot \sqrt{7} = \sqrt{5 \cdot 7} = \sqrt{35}$

b. $\sqrt[3]{3x} \cdot \sqrt[3]{11y} = \sqrt[3]{3x \cdot 11y} = \sqrt[3]{33xy}$

To simplify a radical expression, begin by factoring the radicand. If you are simplifying a square root, factor the radicand into as many perfect squares as possible. Then take the square root of each perfect square and leave the remaining factors under the radical sign. Similarly, to simplify a cube root, factor the radicand into as many perfect cubes as possible. Then take the cube root of each perfect cube.

EXAMPLE 2 **Simplify:** **a.** $\sqrt{72x^3}$ **b.** $\sqrt[3]{80n^5}$

a.
$$\begin{aligned} \sqrt{72x^3} &= \sqrt{36 \cdot 2 \cdot x^2 \cdot x} && \text{Factor into perfect squares.} \\ &= \sqrt{36 \cdot x^2 \cdot 2 \cdot x} \\ &= \sqrt{36 \cdot x^2} \cdot \sqrt{2 \cdot x} && \text{Then use } \sqrt{ab} = \sqrt{a} \cdot \sqrt{b}. \\ &= 6x\sqrt{2x} \end{aligned}$$

b.
$$\begin{aligned} \sqrt[3]{80n^5} &= \sqrt[3]{8 \cdot 10 \cdot n^3 \cdot n^2} && \text{Factor into perfect cubes.} \\ &= \sqrt[3]{8 \cdot n^3 \cdot 10 \cdot n^2} \\ &= \sqrt[3]{8n^3} \cdot \sqrt[3]{10n^2} && \text{Then use } \sqrt[3]{ab} = \sqrt[3]{a} \cdot \sqrt[3]{b}. \\ &= 2n\sqrt[3]{10n^2} \end{aligned}$$

The product of two radicals should be simplified, when possible.

EXAMPLE 3 **Simplify:** **a.** $\sqrt{8x^2} \cdot \sqrt{12xy^2}$ **b.** $\sqrt[3]{54x^2y^3} \cdot \sqrt[3]{5x^3y^4}$

a.
$$\begin{aligned} \sqrt{8x^2} \cdot \sqrt{12xy^2} &= \sqrt{8x^2 \cdot 12xy^2} && \sqrt{a} \cdot \sqrt{b} = \sqrt{ab} \\ &= \sqrt{4 \cdot 2 \cdot x^2 \cdot 4 \cdot 3 \cdot x \cdot y^2} && \text{Factor into perfect squares.} \\ &= \sqrt{4 \cdot 4 \cdot x^2 \cdot y^2} \cdot \sqrt{2 \cdot 3 \cdot x} \\ &= 4xy\sqrt{6x} \end{aligned}$$

b.
$$\begin{aligned} \sqrt[3]{54x^2y^3} \cdot \sqrt[3]{5x^3y^4} &= \sqrt[3]{54x^2y^3 \cdot 5x^3y^4} && \sqrt[3]{a} \cdot \sqrt[3]{b} = \sqrt[3]{ab} \\ &= \sqrt[3]{27 \cdot 2 \cdot x^2 \cdot y^3 \cdot 5 \cdot x^3 \cdot y^3 \cdot y} && \text{Factor into perfect cubes.} \\ &= \sqrt[3]{27 \cdot x^3 \cdot y^3 \cdot y^3} \cdot \sqrt[3]{2 \cdot 5 \cdot x^2 \cdot y} \\ &= 3xy^2\sqrt[3]{10x^2y} \end{aligned}$$

When you multiply radical expressions, it is often helpful to use the commutative and associative properties to group like factors.

EXAMPLE 4 **Simplify:** **a.** $-5\sqrt{6x} \cdot 3\sqrt{6x^2}$ **b.** $3\sqrt[3]{4x^2} \cdot 7\sqrt[3]{12x^4}$

a. $-5\sqrt{6x} \cdot 3\sqrt{6x^2} = (-5 \cdot 3)(\sqrt{6x} \cdot \sqrt{6x^2})$ *Group like factors.*

$= -15\sqrt{6 \cdot 6 \cdot x \cdot x^2}$

$= -15 \cdot 6x\sqrt{x}$

$= -90x\sqrt{x}$

b. $3\sqrt[3]{4x^2} \cdot 7\sqrt[3]{12x^4} = (3 \cdot 7)(\sqrt[3]{4x^2} \cdot \sqrt[3]{12x^4})$ *Group like factors.*

$= 21\sqrt[3]{48x^6}$

$= 21\sqrt[3]{8 \cdot 6 \cdot x^6}$

$= 21\sqrt[3]{8x^6} \cdot \sqrt[3]{6}$

$= 21 \cdot 2x^2\sqrt[3]{6}$

$= 42x^2\sqrt[3]{6}$

To see how to divide with radicals, consider the following:

$$\frac{\sqrt{36}}{\sqrt{25}} = \frac{6}{5} \quad \text{and} \quad \sqrt{\frac{36}{25}} = \frac{6}{5} \qquad \text{Therefore,} \quad \frac{\sqrt{36}}{\sqrt{25}} = \sqrt{\frac{36}{25}}$$

In general, the quotient of the principal nth roots of two positive numbers is equal to the principal nth root of their quotient.

For all real numbers a and b, where $a \geq 0$ and $b > 0$, and for any integer n, where $n > 1$

$$\frac{\sqrt[n]{a}}{\sqrt[n]{b}} = \sqrt[n]{\frac{a}{b}}$$

In the exercises and examples in this chapter, assume that no denominator is equal to zero.

EXAMPLE 5 **Simplify:** **a.** $\frac{\sqrt{125}}{\sqrt{5}}$ **b.** $\frac{\sqrt[3]{162x^5}}{\sqrt[3]{3x^2}}$

a. $\frac{\sqrt{125}}{\sqrt{5}} = \sqrt{\frac{125}{5}}$ $\quad \frac{\sqrt{a}}{\sqrt{b}} = \sqrt{\frac{a}{b}}$

$= \sqrt{25}$ *Divide.*

$= 5$

b. $\frac{\sqrt[3]{162x^5}}{\sqrt[3]{3x^2}} = \sqrt[3]{\frac{162x^5}{3x^2}}$ $\quad \frac{\sqrt[n]{a}}{\sqrt[n]{b}} = \sqrt[n]{\frac{a}{b}}$

$= \sqrt[3]{54x^3}$ *Divide.*

$= \sqrt[3]{27 \cdot 2 \cdot x^3}$

$= 3x\sqrt[3]{2}$

It is customary to simplify a radical expression in such a way that there are no fractions in a radicand and no radicals in a denominator. This procedure is called **rationalizing the denominator.** To rationalize the denominator, multiply the radicand by 1 in the form $\frac{p}{p}$, where p is an expression that will make the denominator a perfect nth power.

EXAMPLE 6 **Simplify:** **a.** $\frac{\sqrt{2x^3}}{\sqrt{10xy}}$ **b.** $\frac{\sqrt[3]{4}}{\sqrt[3]{6x}}$

a.
$$\begin{aligned}\frac{\sqrt{2x^3}}{\sqrt{10xy}} &= \sqrt{\frac{2x^3}{10xy}}\\ &= \sqrt{\frac{x^2}{5y}}\\ &= \sqrt{\frac{x^2}{5y}\cdot\frac{5y}{5y}} && \text{Make the denominator a perfect nth power.}\\ &= \sqrt{\frac{5x^2y}{25y^2}}\\ &= \frac{x\sqrt{5y}}{5y} && \text{The denominator is rationalized.}\end{aligned}$$

b.
$$\begin{aligned}\frac{\sqrt[3]{4}}{\sqrt[3]{6x}} &= \sqrt[3]{\frac{4}{6x}}\\ &= \sqrt[3]{\frac{2}{3x}}\\ &= \sqrt[3]{\frac{2}{3x}\cdot\frac{3^2x^2}{3^2x^2}}\\ &= \sqrt[3]{\frac{18x^2}{27x^3}}\\ &= \frac{\sqrt[3]{18x^2}}{3x}\end{aligned}$$

The formula for the velocity of a particle can now be simplified as follows:

$$v = \sqrt{\frac{2K}{m}} = \sqrt{\frac{2K}{m}\cdot\frac{m}{m}} = \frac{\sqrt{2Km}}{m}$$

To summarize, an nth root is in **simplest radical form** when:

the radicand contains no perfect nth power, other than 1,
the radicand contains no fractions, and
the denominator contains no radicals

EXAMPLE 7 **Simplify:** **a.** $4(\sqrt{3} - 2\sqrt{2})$ **b.** $\frac{\sqrt{2} + \sqrt{5}}{\sqrt{8}}$

a.
$$\begin{aligned}4(\sqrt{3} - 2\sqrt{2}) &= 4(\sqrt{3}) + 4(-2\sqrt{2}) && \text{Distributive property}\\ &= 4\sqrt{3} - 8\sqrt{2}\end{aligned}$$

b.
$$\begin{aligned}\frac{\sqrt{2}+\sqrt{5}}{\sqrt{8}} &= \frac{\sqrt{2}+\sqrt{5}}{\sqrt{8}}\cdot\frac{\sqrt{2}}{\sqrt{2}} && \text{Rationalize the denominator.}\\ &= \frac{(\sqrt{2}\cdot\sqrt{2}) + (\sqrt{5}\cdot\sqrt{2})}{\sqrt{8}\cdot\sqrt{2}} && \text{Distributive property}\\ &= \frac{\sqrt{4}+\sqrt{10}}{\sqrt{16}} = \frac{2+\sqrt{10}}{4}\end{aligned}$$

In Example 7b, the expression could be multiplied by $\frac{\sqrt{8}}{\sqrt{8}}$ instead of $\frac{\sqrt{2}}{\sqrt{2}}$. Then the result, $\frac{4 + 2\sqrt{10}}{8}$, would have to be simplified by dividing the numerator and the denominator by 2. The choice of the least number that will make the denominator a perfect square can eliminate steps.

CLASS EXERCISES

By what number would you multiply to rationalize the denominator?

1. $\sqrt{\frac{1}{5}}$
2. $\sqrt{\frac{5}{12}}$
3. $\sqrt[3]{\frac{1}{2}}$
4. $\sqrt[3]{\frac{2}{9}}$

Tell whether or not each radical is in simplest form. If the expression is not in simplest form, tell why it is not.

5. $\sqrt{20x^3}$
6. $\sqrt[3]{81x}$
7. $\sqrt{\frac{6}{2}}$
8. $\frac{\sqrt{2}}{5}$

PRACTICE EXERCISES

Simplify.

1. $\sqrt{3} \cdot \sqrt{3}$
2. $\sqrt{2} \cdot \sqrt{2}$
3. $\sqrt{11} \cdot \sqrt{11}$
4. $\sqrt{6} \cdot \sqrt{12}$
5. $\sqrt{5} \cdot \sqrt{40}$
6. $\sqrt[3]{3} \cdot \sqrt[3]{18}$
7. $\sqrt[3]{10y^3} \cdot \sqrt[3]{25y^3}$
8. $\sqrt{3x} \cdot \sqrt{3x}$
9. $\sqrt{7xy} \cdot \sqrt{7xy}$
10. $\sqrt{3x} \cdot \sqrt{5x}$
11. $3\sqrt{7x^3} \cdot 2\sqrt{21x^3y^2}$
12. $4\sqrt{2x} \cdot 5\sqrt{6xy^2}$
13. $4\sqrt[3]{5y^3} \cdot 2\sqrt[3]{50y^4}$
14. $-\sqrt[3]{2x^2y^2} \cdot 2\sqrt[3]{16x^5y}$
15. $\frac{\sqrt{500}}{\sqrt{5}}$
16. $\frac{\sqrt{32}}{\sqrt{2}}$
17. $\frac{\sqrt{96}}{\sqrt{8}}$
18. $\frac{\sqrt{48x^3y^4}}{\sqrt{3xy^2}}$
19. $\frac{\sqrt{56x^5y^5}}{\sqrt{7xy}}$
20. $\frac{\sqrt{36x^3}}{\sqrt{12x}}$
21. $\frac{\sqrt{x}}{\sqrt{2}}$
22. $\frac{\sqrt{5}}{\sqrt{8x}}$
23. $\frac{\sqrt{2}}{\sqrt{5}}$
24. $\frac{\sqrt{3x}}{\sqrt{6}}$
25. $3(\sqrt{2} - 3\sqrt{5})$
26. $-2(\sqrt[3]{6} + \sqrt[3]{2})$
27. $\frac{1 + \sqrt{2}}{\sqrt{2}}$
28. $\frac{3 + \sqrt{5}}{\sqrt{5}}$
29. $\frac{2 - \sqrt{7}}{\sqrt{7}}$
30. $\frac{\sqrt{3} - \sqrt{2}}{\sqrt{8}}$
31. $\sqrt{8y^5} \cdot \sqrt{40y^2}$
32. $\sqrt{7x^5} \cdot \sqrt{42xy^9}$
33. $\sqrt[3]{6} \cdot \sqrt[3]{16}$

34. $\sqrt[3]{4} \cdot \sqrt[3]{80}$

35. $\sqrt{x^5y^5} \cdot 3\sqrt{2x^7y^6}$

36. $5\sqrt{2xy^6} \cdot 2\sqrt{2x^3y}$

37. $\sqrt{2}(\sqrt{50} + 7)$

38. $\sqrt{3}(5 + \sqrt{21})$

39. $\sqrt{5}(\sqrt{5} + \sqrt{15})$

40. $\sqrt{8}(\sqrt{24} + 3\sqrt{8})$

41. $\sqrt[3]{2x} \cdot \sqrt[3]{4} \cdot \sqrt[3]{2x^2}$

42. $\sqrt[3]{3x^2} \cdot \sqrt[3]{x^2} \cdot \sqrt[3]{9x^3}$

43. $\dfrac{15\sqrt{60x^5}}{3\sqrt{12x}}$

44. $\dfrac{\sqrt{3xy^2}}{\sqrt{5xy^3}}$

45. $\dfrac{\sqrt{5x^4y}}{\sqrt{2x^2y^3}}$

46. $\dfrac{5\sqrt{2}}{3\sqrt{7x}}$

47. $\dfrac{-6\sqrt{7}}{-5\sqrt{3y^3}}$

48. $\dfrac{4\sqrt{2xy}}{9\sqrt{5x^2y}}$

49. $\dfrac{1}{\sqrt[3]{9x}}$

50. $\dfrac{10}{\sqrt[3]{5x^2}}$

51. $\dfrac{\sqrt[3]{14}}{\sqrt[3]{7x^2y}}$

52. $\dfrac{3\sqrt{11x^3y}}{-2\sqrt{12x^4y}}$

53. $\dfrac{4 + \sqrt{5}}{\sqrt{72}}$

54. $\dfrac{7 + \sqrt{6}}{\sqrt{84}}$

55. $\dfrac{(\sqrt{4x})^2(\sqrt[3]{3x})^3}{\sqrt{5}}$

56. $\dfrac{(\sqrt[3]{7x^3y})^3(x)^2}{\sqrt{32x}}$

57. $\sqrt{\sqrt{16x^4y^4}}$

58. $\sqrt[3]{\sqrt{64x^6y^{12}}}$

59. $\sqrt{(\sqrt[4]{16x^8y^{20}})^2}$

60. $\sqrt{\sqrt[3]{8000}}$

61. $\sqrt[3]{x^{-1}y^{-2}}$

62. $\sqrt[5]{x^{-4}y}$

63. $\sqrt[6]{\dfrac{y^{-3}}{x^{-4}}}$

Applications

64. **Geometry** A rectangular shelf is $\sqrt{440}$ cm by $\sqrt{20}$ cm. Find its area.

65. **Geometry** The height of a triangular model is $\sqrt{54a^5b^3}$ ft and the base is $\sqrt{24ab}$ ft. Write the ratio of the height to the base in simplest form.

66. **Physics** The time, t, it takes a freely falling object to hit the ground is $\sqrt{18a^5}$ seconds. If it was dropped from rest, find the height, h, in ft in terms of a using the formula $h = 16t^2$.

Mixed Review

Use the vertical line test to determine whether the relation is a function.

67.

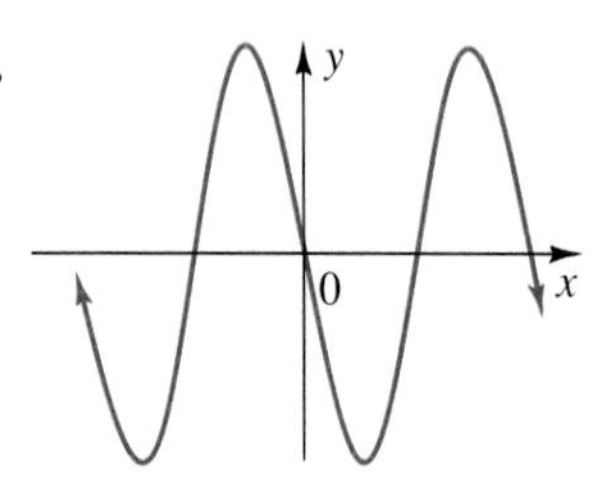

68.

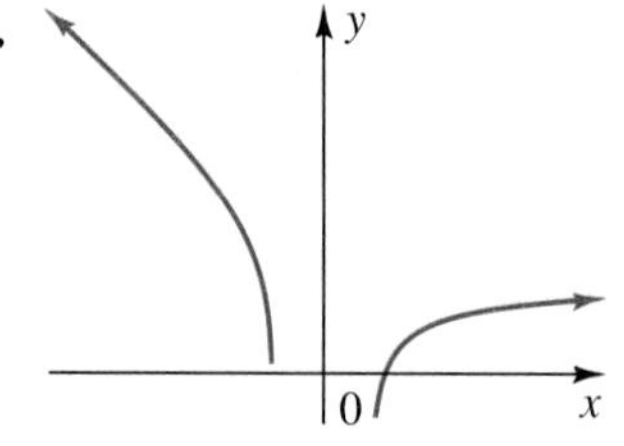

69.

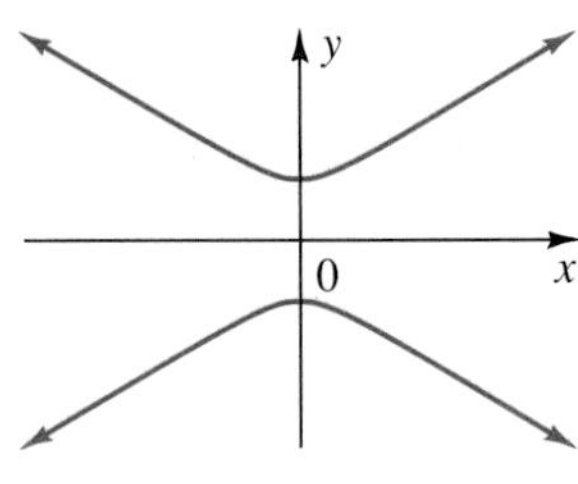

Developing Mathematical Power

70. **Thinking Critically** When a given radical expression $\sqrt{x^ay^b}$ is simplified, the result is $\dfrac{1}{x^cy^{3d}}$, where c and d are positive integers. Express a in terms of c, and b in terms of d.

8.3

Adding and Subtracting Radicals

Objectives: To add and subtract radical expressions
To multiply and divide binomials that contain radicals

You can add or subtract expressions involving radicals using the same procedures you would use to add or subtract polynomials. Similarly, you can multiply binomials that contain radicals in the same way you would multiply two binomials.

Capsule Review

To multiply a binomial by a binomial, use the FOIL method. Then combine like terms.

EXAMPLE $(3x + 2)(2x + 5) = 6x^2 + 15x + 4x + 10$ (F O I L)
$= 6x^2 + 19x + 10$ *Combine like terms.*

Multiply.

1. $(5x + 4)(3x - 2)$ **2.** $(-8x + 5)(3x - 7)$ **3.** $(x + 4)(x - 4)$

4. $(4x + 5)(4x - 5)$ **5.** $(x + 5)^2$ **6.** $(2x - 9)^2$

Like radicals are radical expressions that have the same index and the same radicand. To add or subtract like radicals, use the distributive property.

EXAMPLE 1 **Simplify:** **a.** $6\sqrt{2} + 5\sqrt{2}$ **b.** $5\sqrt[3]{7} - 3\sqrt[3]{7}$ **c.** $4\sqrt{2} + 5\sqrt{3}$

a. $6\sqrt{2} + 5\sqrt{2} = (6 + 5)\sqrt{2}$ *Factor.*
$= 11\sqrt{2}$

b. $5\sqrt[3]{7} - 3\sqrt[3]{7} = (5 - 3)\sqrt[3]{7}$
$= 2\sqrt[3]{7}$

c. $4\sqrt{2} + 5\sqrt{3}$

The expression is in simplest form. The radicals are not like radicals, so they cannot be combined in one term.

Be sure to simplify radicals before adding or subtracting, so that you find all like radicals.

EXAMPLE 2 **Simplify:** $6\sqrt{18} + 4\sqrt{8} - 3\sqrt{72}$

$$\begin{aligned} 6\sqrt{18} + 4\sqrt{8} - 3\sqrt{72} &= 6\sqrt{9 \cdot 2} + 4\sqrt{4 \cdot 2} - 3\sqrt{36 \cdot 2} && \text{Simplify each radical.} \\ &= 6 \cdot 3\sqrt{2} + 4 \cdot 2\sqrt{2} - 3 \cdot 6\sqrt{2} \\ &= 18\sqrt{2} + 8\sqrt{2} - 18\sqrt{2} \\ &= (18 + 8 - 18)\sqrt{2} && \text{Use the distributive property.} \\ &= 8\sqrt{2} \end{aligned}$$

If a is a rational number and $\sqrt{b}$ is an irrational number, then their sum, $a + \sqrt{b}$, is an irrational number. Example 3 shows that the product of two binomials that contain irrational numbers may be an irrational number.

EXAMPLE 3 **Simplify:** **a.** $(3 + 2\sqrt{5})(2 + 4\sqrt{5})$ **b.** $(\sqrt{2} - \sqrt{3})^2$

a.
$$\begin{aligned} (3 + 2\sqrt{5})(2 + 4\sqrt{5}) &= 6 + 12\sqrt{5} + 4\sqrt{5} + 8(\sqrt{5})^2 && \text{Use FOIL.} \\ &= 6 + (12 + 4)\sqrt{5} + 40 && \text{Combine like radicals.} \\ &= 46 + 16\sqrt{5} \end{aligned}$$

b.
$$\begin{aligned} (\sqrt{2} - \sqrt{3})^2 &= (\sqrt{2})^2 - 2 \cdot \sqrt{2} \cdot \sqrt{3} + (\sqrt{3})^2 && (a - b)^2 = a^2 - 2ab + b^2 \\ &= 2 - 2\sqrt{6} + 3 \\ &= 5 - 2\sqrt{6} \end{aligned}$$

The product of two binomials that contain irrational numbers is not always irrational. Expressions like $\sqrt{a} + \sqrt{b}$ and $\sqrt{a} - \sqrt{b}$, which differ only in the sign of the second term, are called **conjugates.** The product of two conjugates is the difference of two squares, since for all positive real numbers a and b, $(a + b)(a - b) = a^2 - b^2$. Thus,

$$(\sqrt{a} + \sqrt{b})(\sqrt{a} - \sqrt{b}) = (\sqrt{a})^2 - (\sqrt{b})^2 = a - b$$

That is, the product of two such conjugates is a rational number.

EXAMPLE 4 **Simplify:** **a.** $(2 + \sqrt{3})(2 - \sqrt{3})$ **b.** $(\sqrt{5} + \sqrt{2})(\sqrt{5} - \sqrt{2})$

a.
$$\begin{aligned} (2 + \sqrt{3})(2 - \sqrt{3}) &= 2^2 - (\sqrt{3})^2 && (a + b)(a - b) = a^2 - b^2 \\ &= 4 - 3 \\ &= 1 \end{aligned}$$

b.
$$\begin{aligned} (\sqrt{5} + \sqrt{2})(\sqrt{5} - \sqrt{2}) &= (\sqrt{5})^2 - (\sqrt{2})^2 && (a + b)(a - b) = a^2 - b^2 \\ &= 5 - 2 = 3 \end{aligned}$$

When an expression is written in simplest radical form, the denominator is rationalized. If the denominator is a binomial containing a radical with an index of 2, rationalize the denominator by multiplying the expression by 1 in the form $\frac{p}{p}$, where p is the conjugate of the denominator.

EXAMPLE 5 **Simplify:** $\frac{3+\sqrt{5}}{1-\sqrt{5}}$

$$\frac{3+\sqrt{5}}{1-\sqrt{5}} = \frac{3+\sqrt{5}}{1-\sqrt{5}} \cdot \frac{1+\sqrt{5}}{1+\sqrt{5}}$$

$1+\sqrt{5}$ is the conjugate of $1-\sqrt{5}$.

$$= \frac{(3+\sqrt{5})(1+\sqrt{5})}{(1-\sqrt{5})(1+\sqrt{5})}$$

$$= \frac{3+3\sqrt{5}+\sqrt{5}+(\sqrt{5})^2}{1^2-(\sqrt{5})^2}$$

$$= \frac{3+3\sqrt{5}+\sqrt{5}+5}{1-5} = \frac{8+4\sqrt{5}}{-4} = -2-\sqrt{5}$$

CLASS EXERCISES

Tell whether or not the radicals in each pair are like radicals. Give a reason for each answer.

1. $5\sqrt{6}, 5\sqrt{3}$ **2.** $6\sqrt{3}, 3\sqrt[3]{3}$ **3.** $4\sqrt[3]{5}, 5\sqrt[3]{5}$ **4.** $7\sqrt{x}, 8\sqrt{x^2}$

Predict the number of terms that will be in each product, after like terms have been combined.

5. $(3+\sqrt{5})(2-\sqrt{5})$

6. $(4+\sqrt{3})(6+\sqrt{3})$

7. $(\sqrt{3}+4\sqrt{2})(\sqrt{3}-4\sqrt{2})$

8. $(2\sqrt{5}-3\sqrt{2})^2$

9. $(\sqrt{3}+\sqrt{7})(\sqrt{3}-\sqrt{7})$

10. $(6\sqrt{2}-3\sqrt{5})(6\sqrt{2}+3\sqrt{5})$

Give the conjugate of each binomial.

11. $3-\sqrt{2}$ **12.** $\sqrt{3}+2$ **13.** $-1-\sqrt{5}$ **14.** $\sqrt{5}+\sqrt{3}$

PRACTICE EXERCISES

Use technology where appropriate.

Simplify.

1. $2\sqrt{7}+3\sqrt{7}$

2. $7\sqrt[4]{5}-2\sqrt[4]{5}$

3. $4\sqrt{2}+5\sqrt{2}$

4. $6\sqrt[3]{3}-2\sqrt[3]{3}$

5. $\sqrt{32}+\sqrt{8}$

6. $\sqrt{7x}-\sqrt{28x}$

7. $6\sqrt{18}+3\sqrt{50}$

8. $14\sqrt{20}-3\sqrt{125}$

9. $\sqrt{18}+\sqrt{32}$

10. $\sqrt{27}+\sqrt{48}$

11. $3\sqrt{18}+2\sqrt{72}$

12. $8\sqrt{45}-3\sqrt{80}$

13. $\sqrt[3]{54}+\sqrt[3]{16}$

14. $3\sqrt[3]{81}-2\sqrt[3]{54}$

15. $\sqrt[4]{32}+\sqrt[4]{48}$

16. $(3+\sqrt{5})(1+\sqrt{5})$

17. $(2+\sqrt{7})(1+3\sqrt{7})$

18. $(3-4\sqrt{2})(5-6\sqrt{2})$

19. $(\sqrt{5} - 1)(\sqrt{5} + 4)$

20. $(\sqrt{3} - \sqrt{7})(\sqrt{3} + 2\sqrt{7})$

21. $(2\sqrt{5} + 3\sqrt{2})(5\sqrt{5} - 7\sqrt{2})$

22. $(\sqrt{3} + \sqrt{5})^2$

23. $(\sqrt{8} - \sqrt{7})^2$

24. $(5 - \sqrt{11})^2$

25. $(\sqrt{13} + 6)^2$

26. $(5\sqrt{3} - 2)^2$

27. $(2\sqrt{5} + 3\sqrt{2})^2$

28. $(5 - \sqrt{11})(5 + \sqrt{11})$

29. $(4 - 2\sqrt{3})(4 + 2\sqrt{3})$

30. $(\sqrt{3} + \sqrt{5})(\sqrt{3} - \sqrt{5})$

31. $\dfrac{4}{1 + \sqrt{3}}$

32. $\dfrac{5}{1 - \sqrt{2}}$

33. $\dfrac{5 + \sqrt{3}}{2 - \sqrt{3}}$

34. $\dfrac{4 + \sqrt{5}}{2 + \sqrt{5}}$

35. $\sqrt{72} + \sqrt{32} + \sqrt{18}$

36. $\sqrt{75} + 2\sqrt{48} - 5\sqrt{3}$

37. $5\sqrt{32x} + 4\sqrt{98x}$

38. $4\sqrt{216y^2} + 3\sqrt{54y^2}$

39. $(3\sqrt{5} + 2\sqrt{10})(2\sqrt{5} + \sqrt{10})$

40. $(5\sqrt{6} - 3\sqrt{8})(3\sqrt{6} - 2\sqrt{8})$

41. $(1 + \sqrt{72})(5 + \sqrt{2})$

42. $(2 - \sqrt{98})(3 + \sqrt{18})$

43. $(\sqrt{x} + \sqrt{3})(\sqrt{x} + 2\sqrt{3})$

44. $(2\sqrt{y} - 3\sqrt{2})(4\sqrt{y} - 5\sqrt{2})$

45. $\dfrac{4}{3\sqrt{3} - 2}$

46. $\dfrac{5}{4\sqrt{7} + 5}$

47. $\dfrac{6 + \sqrt{15}}{4 - \sqrt{15}}$

48. $\dfrac{4\sqrt{18} - 2}{3 - \sqrt{18}}$

49. $\dfrac{3 + \sqrt{8}}{2 - 2\sqrt{8}}$

50. $\dfrac{4 + \sqrt{27}}{2 - 3\sqrt{27}}$

51. $\dfrac{4 + \sqrt{6}}{\sqrt{2} + \sqrt{3}}$

52. $\dfrac{5 - \sqrt{21}}{\sqrt{3} - \sqrt{7}}$

53. $\dfrac{2 + \sqrt{10}}{\sqrt{2} - 3\sqrt{5}}$

54. $\dfrac{3 + \sqrt{12}}{\sqrt{6} - 4\sqrt{2}}$

55. $\dfrac{-2 + \sqrt{8}}{-3 - \sqrt{2}}$

56. $\dfrac{-6 - \sqrt{27}}{-4 + \sqrt{3}}$

57. $\sqrt{\dfrac{x}{12}} + \sqrt{\dfrac{16x}{3}} - 3\sqrt{\dfrac{x}{27}}$

58. $\sqrt{12y} - \sqrt{\dfrac{y}{3}} + \dfrac{3\sqrt{y}}{\sqrt{27}}$

59. $\left(\dfrac{3 - 5\sqrt{2}}{3}\right)\left(\dfrac{3 + 5\sqrt{2}}{3}\right)$

60. $\left(\dfrac{7 + 8\sqrt{11}}{5}\right)\left(\dfrac{7 - 8\sqrt{11}}{5}\right)$

61. $\dfrac{a}{\sqrt{a - 1}} - \sqrt{a - 1}$

62. $\dfrac{3b}{\sqrt{b + 1}} + \sqrt{b + 1}$

63. $\dfrac{3 + \sqrt[3]{2}}{\sqrt[3]{2}}$

64. $\dfrac{5 + \sqrt[4]{x}}{\sqrt[4]{x}}$

65. $\dfrac{2}{(\sqrt{x} + \sqrt{y})^2}$

66. $\dfrac{3}{(\sqrt{a} - \sqrt{b})^2}$

67. $\dfrac{(\sqrt{a} + \sqrt{b})^3}{a + 2\sqrt{ab} + b}$

68. $\dfrac{\sqrt{a} - 2\sqrt{b}}{(\sqrt{a} + \sqrt{b})(\sqrt{a} + 2\sqrt{b})}$

69. $\sqrt{60 + 2\sqrt{3}} \cdot \sqrt{4 - 2\sqrt{3}}$

70. $\sqrt{18 - 2\sqrt{12}} \cdot \sqrt{18 + \sqrt{12}}$

Applications

71. Geometry A rectangular walk is $3\sqrt{7}$ m wide and $5\sqrt{7}$ m long. What is the perimeter of the walk?

72. Physics An object is moving at a speed of $(3 + \sqrt{2})$ ft/s. How long will it take the object to travel 20 ft?

73. Physics One resistance in a parallel circuit is $4\sqrt{6}$ ohms. A second resistance is $11\sqrt{6}$ ohms. Find the total resistance R_T in the parallel circuit. $\left(\frac{1}{R_T} = \frac{1}{R_1} + \frac{1}{R_2}\right)$

Developing Mathematical Power

Could you find the area of a triangle given only the lengths of its sides? Heron of Alexandria (about 60 A.D.) devised a formula involving a radical to do just that. For a triangle with sides a, b, c and $s = (a + b + c)/2$

$$\text{Area} = \sqrt{s(s - a)(s - b)(s - c)}$$

The computation in this formula could be long and tedious, so it is helpful to use a calculator. There are several possible approaches.

Method 1 Since $s = (a + b + c)/2$, Heron's formula can be rewritten with $(a + b + c)/2$ replacing s. If $a = 3$, $b = 4$, and $c = 5$, the calculator entry is $\sqrt{\ }(((3 + 4 + 5)/2)(((3 + 4 + 5)/2) - 3)(((3 + 4 + 5)/2) - 4)$ $(((3 + 4 + 5)/2 - 5))$. To find the result, it takes 59 keypresses.

Method 2 Find the value of s and use that value in your calculations. $(3 + 4 + 5)/2$ will yield 6. Then type in $\sqrt{\ }(6(6 - 3)(6 - 4)(6 - 5))$. With this method, it takes only 30 keypresses to find the result.

Method 3 If your calculator is capable of storing functions, store Heron's formula as a function (in the form shown in Method 1). After storing 3, 4, and 5 under the variable names a, b, and c, you can evaluate the function. Storing the function will take some time, but once it is stored, each result should take only 12 keypresses.

74. Find the area of a triangle with sides of 5, 12, and 13. Use any method.

75. Find the area of a triangle with sides of 4, 7, and 11. Explain.

76. Which method would be most practical if you needed to find the area of one triangle? five triangles?

77. Examine the expression $(a + b + c)/2$. Describe its meaning in geometric terms and create a descriptive name for the concept.

78. Write a step-by-step description of a fourth method, which involves storing $(a + b + c)/2$ and Heron's formula as separate functions.

8.4

Solving Equations with Radicals

Objective: To solve radical equations

A **radical equation** is an equation that has a variable *in a radicand*. The equations $\sqrt{x} = 5$ and $3 + x = \sqrt{x - 2}$ are radical equations. The equations $\sqrt{3} = x + 2$ and $3x + x\sqrt{3} = 5$ are *not* radical equations.

Many problems in physics require the solution of radical equations. In order to solve such an equation, you may have to solve a quadratic equation.

Capsule Review

To solve a quadratic equation of the form $ax^2 + bx + c = 0$ where $ax^2 + bx + c$ is factorable, factor the left side of the equation. Then set each factor equal to zero and solve the resulting linear equations.

EXAMPLE **Solve: $2x^2 = 7x + 15$**

$2x^2 = 7x + 15$

$2x^2 - 7x - 15 = 0$ *Bring all nonzero terms to one side.*

$(x - 5)(2x + 3) = 0$ *Factor.*

$x - 5 = 0$ or $2x + 3 = 0$ *Set each factor equal to zero.*

$x = 5$ or $x = -\frac{3}{2}$ Check to show that 5 and $-\frac{3}{2}$ are solutions.

Solve each quadratic equation by factoring.

1. $x^2 = -x + 6$ **2.** $x^2 = 5x + 14$ **3.** $2x^2 + x = 3$

4. $3x^2 - 2 = 5x$ **5.** $4x^2 = -8x + 5$ **6.** $6x^2 = 5x + 6$

In general, to solve a radical equation when just one term has a variable in the radicand, *isolate* that term on one side of the equation, if necessary. Then, if the index is 2, square both sides of the equation.

EXAMPLE 1 **Solve and check: $\sqrt{3x - 2} = 4$**

$\sqrt{3x - 2} = 4$

$(\sqrt{3x - 2})^2 = 4^2$ *Square both sides.*

$3x - 2 = 16$

$3x = 18$

$x = 6$ The solution is 6.

Check: $\sqrt{3x - 2} = 4$

$\sqrt{3(6) - 2} \stackrel{?}{=} 4$

$\sqrt{16} \stackrel{?}{=} 4$

$4 = 4$ ✓

In Example 1, the original equation and the equation obtained after squaring are equivalent, since they have the same solution. This is not always the case, as shown in the next example. It is very important to check solutions of radical equations, since solutions obtained after squaring both sides may not be solutions to the original equation.

EXAMPLE 2 **Solve and check:** $\sqrt{x-3}+5=x$

$$\begin{aligned} \sqrt{x-3}+5 &= x \\ \sqrt{x-3} &= x-5 && \textit{Isolate the radical.} \\ (\sqrt{x-3})^2 &= (x-5)^2 && \textit{Square both sides.} \\ x-3 &= x^2-10x+25 \\ 0 &= x^2-11x+28 && \textit{Bring all nonzero terms to one side.} \\ 0 &= (x-4)(x-7) && \textit{Factor.} \\ x-4=0 \quad &\text{or} \quad x-7=0 && \textit{Set each factor equal to zero.} \\ x=4 \quad &\text{or} \quad x=7 \end{aligned}$$

Check:

$$\begin{aligned} \sqrt{x-3}+5 &= x & \qquad \sqrt{x-3}+5 &= x \\ \sqrt{4-3}+5 &\stackrel{?}{=} 4 & \sqrt{7-3}+5 &\stackrel{?}{=} 7 \\ \sqrt{1}+5 &\stackrel{?}{=} 4 & \sqrt{4}+5 &\stackrel{?}{=} 7 \\ 6 &\neq 4 & 7 &= 7 \checkmark \end{aligned}$$

The solution is 7.

In Example 2, notice that 4 is a solution of $(\sqrt{x-3})^2=(x-5)^2$ but not of the original equation, $\sqrt{x-3}+5=x$. These equations are *not* equivalent. Since 4 is not a solution of the original equation, it is sometimes called an *extraneous solution*. Note also that $x=5$ fails to check in the original equation. Should 5 be considered an extraneous solution?

Sometimes a radical equation contains two radical expressions, each with a variable in a radicand and each with index 2. Rewrite such an equation with one of these terms isolated on one side of the equal sign. Then square both sides. If a variable remains in a radicand, repeat the process of isolating the term with the radical and squaring both sides of the equation.

EXAMPLE 3 **Solve and check:**

a. $\sqrt{3x+2}-\sqrt{2x+7}=0$ **b.** $\sqrt{x-5}-\sqrt{x}=2$

a.

$$\begin{aligned} \sqrt{3x+2}-\sqrt{2x+7} &= 0 && \textit{Isolate one radical.} \\ \sqrt{3x+2} &= \sqrt{2x+7} \\ (\sqrt{3x+2})^2 &= (\sqrt{2x+7})^2 && \textit{Square both sides.} \\ 3x+2 &= 2x+7 && \textit{Solve.} \\ x &= 5 \end{aligned}$$

Check:

$$\sqrt{3x+2} - \sqrt{2x+7} = 0$$
$$\sqrt{3(5)+2} - \sqrt{2(5)+7} \stackrel{?}{=} 0$$
$$\sqrt{15+2} - \sqrt{10+7} \stackrel{?}{=} 0$$
$$\sqrt{17} - \sqrt{17} \stackrel{?}{=} 0$$
$$0 = 0 \checkmark$$

The solution is 5.

b.

$$\sqrt{x-5} - \sqrt{x} = 2$$
$$\sqrt{x-5} = \sqrt{x} + 2$$
$$(\sqrt{x-5})^2 = (\sqrt{x}+2)^2$$
$$x - 5 = x + 4\sqrt{x} + 4$$
$$-9 = 4\sqrt{x}$$
$$(-9)^2 = (4\sqrt{x})^2$$
$$81 = 16x$$
$$\frac{81}{16} = x$$

Check:

$$\sqrt{x-5} - \sqrt{x} = 2$$
$$\sqrt{\frac{81}{16} - 5} - \sqrt{\frac{81}{16}} \stackrel{?}{=} 2$$
$$\sqrt{\frac{81-80}{16}} - \sqrt{\frac{81}{16}} \stackrel{?}{=} 2$$
$$\frac{1}{4} - \frac{9}{4} \stackrel{?}{=} 2$$
$$-2 \neq 2$$

There is no solution.

Similar techniques can be applied to radical equations involving radicals with indexes greater than 2. Isolate the term with the radical and then raise each side to a power equal to the index of the radical.

EXAMPLE 4 **Solve and check:** $\sqrt[3]{5y+2} - 3 = 0$

$$\sqrt[3]{5y+2} - 3 = 0$$
$$\sqrt[3]{5y+2} = 3 \qquad \textit{Isolate the radical.}$$
$$(\sqrt[3]{5y+2})^3 = 3^3 \qquad \textit{Cube both sides.}$$
$$5y + 2 = 27 \qquad \textit{Solve.}$$
$$5y = 25$$
$$y = 5$$

Check:

$$\sqrt[3]{5y+2} - 3 = 0$$
$$\sqrt[3]{5(5)+2} - 3 \stackrel{?}{=} 0$$
$$\sqrt[3]{27} - 3 \stackrel{?}{=} 0$$
$$0 = 0 \checkmark$$

The solution is 5.

CLASS EXERCISES

Identify each equation as a radical equation or a linear equation.

1. $3\sqrt{x} + 2x = 5$ **2.** $x\sqrt{8} - 5 = 3x$ **3.** $\sqrt[3]{3x} + 5 = 0$

Which of the following equations have no solutions? Why?

4. $\sqrt{x} = -2$ **5.** $\sqrt{x+3} + 1 = 0$ **6.** $\sqrt[3]{2x} = -4$

What would the first step be in solving each of the following equations?

7. $\sqrt{x} = 4$ **8.** $\sqrt{x} - 2 = 3$

9. $\sqrt{x-2} = 1$ **10.** $\sqrt{x+3} = 5$

11. $3\sqrt{x} + 3 = 15$

12. $4\sqrt{x} - 1 = 3$

13. $\sqrt[4]{x} = 2$

14. $2\sqrt[3]{x} = 8$

15. $\sqrt{7x + 1} - \sqrt{6x + 7} = 0$

16. $\sqrt{4x - 2} - \sqrt{3x + 9} = 0$

PRACTICE EXERCISES

Solve and check. Write *no solution*, where appropriate.

1. $\sqrt{x} = 6$
2. $\sqrt{x} = 7$
3. $\sqrt{x} = -4$
4. $\sqrt{2x} = 4$
5. $\sqrt{5x} = 10$
6. $\sqrt[3]{x} = 3$
7. $3\sqrt{2x} = 12$
8. $5\sqrt{3x} = 15$
9. $2\sqrt[3]{2x} = 4$
10. $\sqrt[3]{x} - 3 = 0$
11. $\sqrt{3x + 4} = 4$
12. $\sqrt{5x + 1} - 6 = 0$
13. $\sqrt{2x + 3} - 7 = 0$
14. $\sqrt{x^2 + 3} = x + 1$
15. $x + 8 = \sqrt{x^2 + 16}$
16. $\sqrt{x^2 + 9} = x + 1$
17. $\sqrt{x^2 - 9} - x = -3$
18. $\sqrt{x^2 + 12} - 2 = x$
19. $\sqrt{3x} = \sqrt{x + 6}$
20. $\sqrt{2x} = \sqrt{x + 5}$
21. $\sqrt{5x + 1} - \sqrt{4x + 3} = 0$
22. $\sqrt{7x + 6} - \sqrt{9 + 4x} = 0$
23. $5\sqrt{7x + 4} - 2 = 23$
24. $3\sqrt{7x + 2} - 3 = 9$
25. $\sqrt{4x - 10} - 3\sqrt{x - 5} = 0$
26. $\sqrt{10x} - 2\sqrt{5x - 25} = 0$
27. $\sqrt{3x + 2} = \dfrac{8}{\sqrt{3x + 2}}$
28. $\sqrt{x - 2} = \dfrac{-1}{\sqrt{x - 2}}$
29. $\sqrt{x + 7} - x = 1$
30. $\sqrt{3x + 13} - 5 = x$
31. $\sqrt{11x + 3} - 2x = 0$
32. $\sqrt{5x + 4} - 3x = 0$
33. $\sqrt[3]{2x} - \sqrt[3]{5x} = 0$
34. $\sqrt[3]{7x} - \sqrt[3]{5x + 2} = 0$
35. $1 = \sqrt{3 + x} - \sqrt{x}$
36. $\sqrt{x} = \sqrt{x - 8} + 2$
37. $\sqrt{x + 10} + \sqrt{3 - x} = 5$
38. $\sqrt{x - 2} - \sqrt{2x + 3} + 2 = 0$

Solve for x and check.

39. $\sqrt{2x + a} - \sqrt{x + b} = 0$
40. $\sqrt{x^2 + a} = x + 1$
41. $\sqrt{x + 1} + \sqrt{2x} = \sqrt{5x + 3}$
42. $2\sqrt{x + 1} = \sqrt{x - 2} + \sqrt{x + 6}$
43. $\sqrt[3]{x} = \sqrt{x}$
44. $\sqrt{x} - 2 = \sqrt[4]{x}$

Applications

45. Geometry The formula for the area A of a square whose side is s units long is $A = s^2$. Solve the formula for s.

46. Number Theory The square root of the sum of twice a number and 4 is equal to 6. Find the number.

47. Physics The velocity (v) of an object dropped from a tall building is given by the formula $v = \sqrt{64d}$, where d is the distance. Solve for d.

Developing Mathematical Power

Skylab, the space satellite launched in 1973, was powered almost entirely by solar-energy cells. Solar cells convert the energy of sunlight directly into electrical energy. For each square centimeter of a cell that is in direct sunlight, approximately 0.01 watt of electrical power is produced. Thus, if the amount of energy to be produced by a solar cell is known, the dimensions of the cell can be determined. Solar cells come in various shapes, including circles, squares, and hexagons.

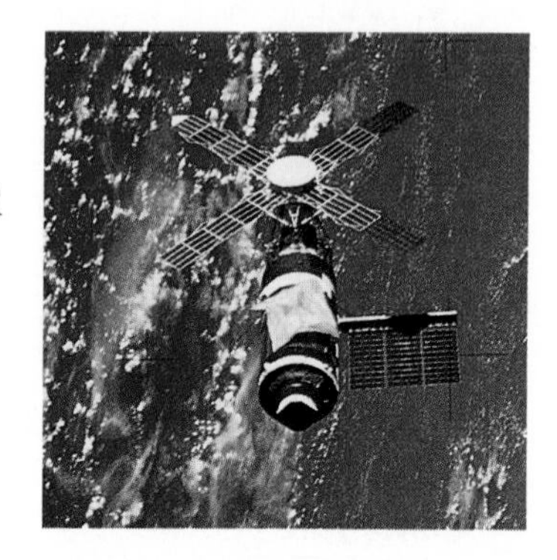

EXAMPLE A circular solar cell with radius r is to deliver 5 watts. Find the minimum radius of the circle, to the nearest 0.1 cm.

The total area A required, at 0.01 watt/cm^2, is $\frac{5}{0.01}$, or 500 cm^2.

$$A = \pi r^2$$
$$500 \approx \pi r^2$$
$$\sqrt{\frac{500}{\pi}} \approx r \qquad \textit{Calculation-ready form}$$
$$12.62 \approx r \qquad \text{The minimum radius is 12.6 cm.}$$

Solve. In Exercises 1–3, find the minimum radius r or side s of each solar cell. Round the answers to the nearest 0.1 cm, where necessary.

48. A circular cell that is to deliver 8 watts

49. A square cell that is to deliver 6 watts

50. A cell in the shape of a regular hexagon, built to deliver 18 watts *Hint:* The formula for the area of a regular hexagon is $A = \frac{3\sqrt{3}}{2}s^2$.

51. A cell in the shape of a regular hexagon has sides of length x. A square cell with the same output has sides of length y. Find x in terms of y.

8.5 Distance and Midpoint Formulas

Objectives: To find the distance between two points in a plane
To find the coordinates of the midpoint of a line segment

Surveyors, architects, mapmakers, and navigators need to find the distance between two places, or points, on a map or diagram. Sometimes right triangles are used to help determine distances.

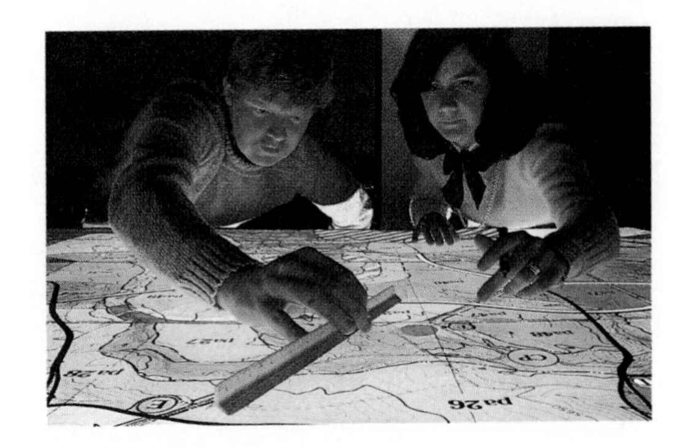

Capsule Review

Pythagorean Theorem: If a and b are the lengths of the legs, and c is the length of the hypotenuse of right triangle ABC, then

$$a^2 + b^2 = c^2$$

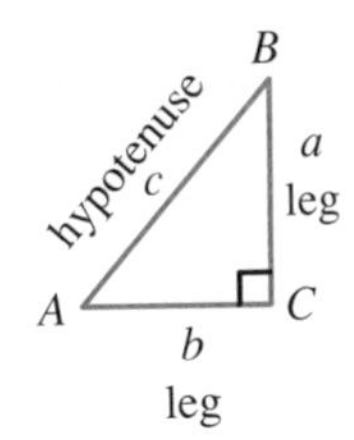

EXAMPLE **Find the length of the hypotenuse.**

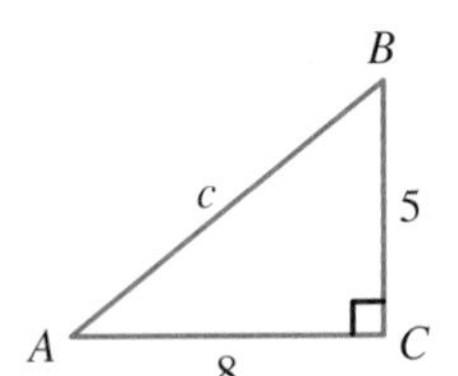

$c^2 = a^2 + b^2$ *Pythagorean Theorem*
$c^2 = 5^2 + 8^2$ *Substitute 5 for a and 8 for b.*
$c^2 = 25 + 64$
$c^2 = 89$
$c = \pm\sqrt{89}$ *Take the square root of each side.*

Reject the negative root, $-\sqrt{89}$, since the length of a side of a triangle cannot be a negative number. The length of the hypotenuse is $\sqrt{89}$ units.

Find the length of the unknown side.

1.

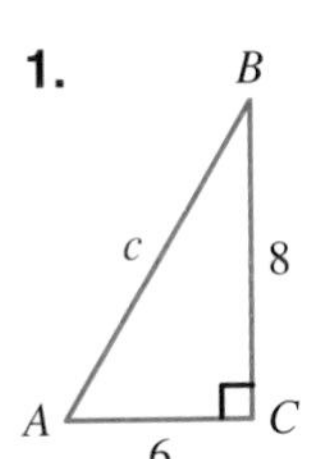

2.

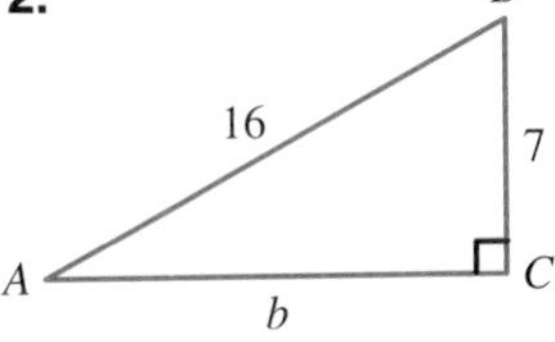

3.

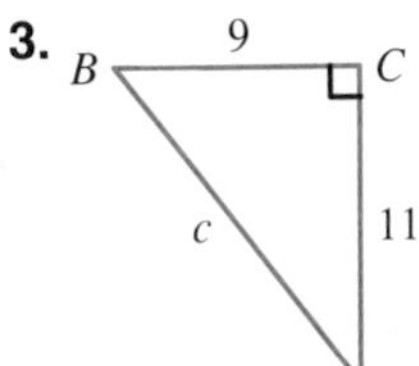

4.

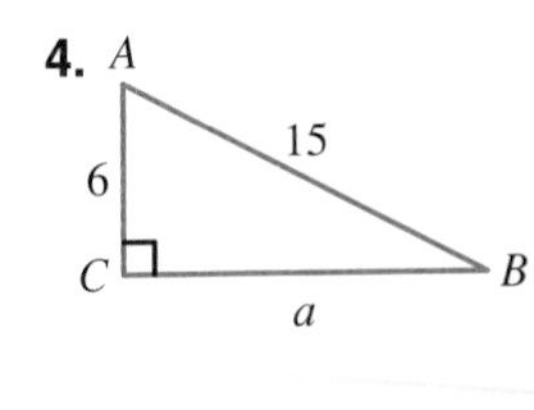

If you know the coordinates of two points A and B, then AB, the distance between them, can be found. Assume that A and B are points on a number line, with coordinates a and b, respectively. Then AB is the absolute value of the difference of the coordinates a and b.

$$AB = |b - a|$$

On the number line at the right,

$AB = |1 - (-6)| = 7$
$BC = |11 - 1| = 10$
$AC = |11 - (-6)| = 17$

Notice that distance is always a nonnegative number.

You can use this method to find the distance between two points in a coordinate plane that lie on the same horizontal or vertical line. Assume that the coordinates of A are (x_1, y_1) and the coordinates of B are (x_2, y_2).

If $\overleftrightarrow{AB}$ is horizontal, then $AB = |x_2 - x_1|$.
If $\overleftrightarrow{AB}$ is vertical, then $AB = |y_2 - y_1|$.

EXAMPLE 1 **Find the distance between the given points.**

a.

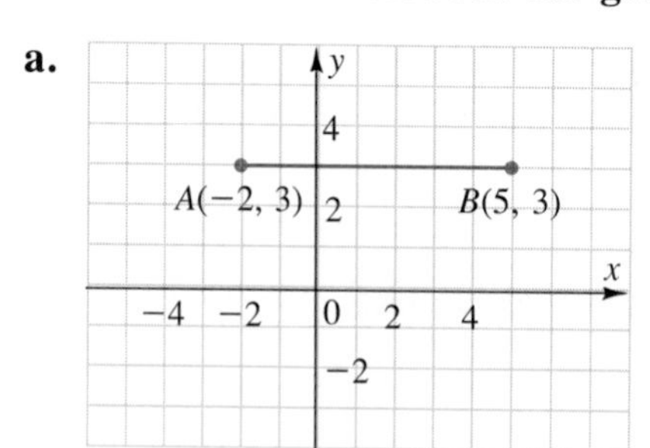

b.

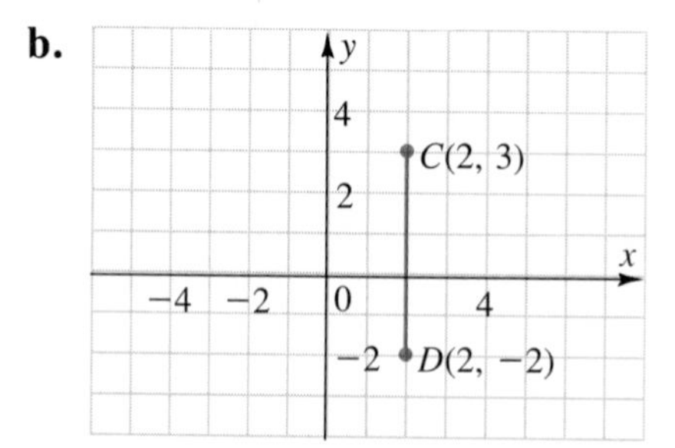

a. $AB = |x_2 - x_1|$
$AB = |5 - (-2)| = |5 + 2| = 7$
The distance from A to B is 7.

b. $CD = |y_2 - y_1|$
$CD = |-2 - 3| = |-5| = 5$
The distance from C to D is 5.

The Pythagorean theorem is used to derive a formula for the distance between points in the coordinate plane that are not on a horizontal or vertical line.

Let A and B be two points with coordinates (x_1, y_1) and (x_2, y_2), respectively.

Draw a line through A parallel to the x-axis. Draw a line through B parallel to the y-axis. The lines intersect at the point $C(x_2, y_1)$. Triangle ABC is a right triangle.

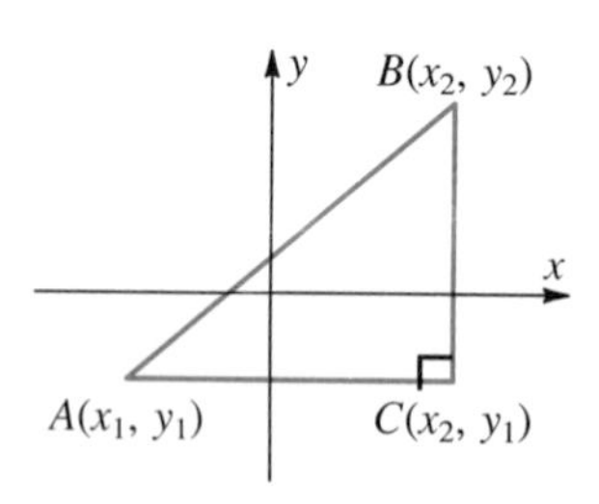

$$(AB)^2 = (AC)^2 + (BC)^2 \qquad \textit{Pythagorean Theorem}$$
$$(AB)^2 = |x_2 - x_1|^2 + |y_2 - y_1|^2$$
$$(AB)^2 = (x_2 - x_1)^2 + (y_2 - y_1)^2 \qquad \textit{For any real number } r, |r|^2 = r^2.$$
$$AB = \sqrt{(x_2 - x_1)^2 + (y_2 - y_1)^2}$$

Distance formula The distance d between two points $A(x_1, y_1)$ and $B(x_2, y_2)$ in a coordinate plane is given by the formula

$$d = \sqrt{(x_2 - x_1)^2 + (y_2 - y_1)^2}$$

EXAMPLE 2 **Find the distance between the points $P(-3, 2)$ and $Q(4, 6)$.**

$$d = \sqrt{(x_2 - x_1)^2 + (y_2 - y_1)^2} \qquad \textit{Let } (-3, 2) = (x_1, y_1) \textit{ and } (4, 6) = (x_2, y_2).$$
$$d = \sqrt{[4 - (-3)]^2 + (6 - 2)^2}$$
$$d = \sqrt{7^2 + 4^2}$$
$$d = \sqrt{65}$$

The distance between P and Q is $\sqrt{65}$.

A *midpoint* divides a line segment into two line segments of equal length. The definition of midpoint, along with the distance formula, can be used to verify a formula for the coordinates of the midpoint of a line segment. You will be asked to do this in Exercise 40 on page 382.

Midpoint formula The coordinates of the midpoint M of the line segment with endpoints $A(x_1, y_1)$ and $B(x_2, y_2)$ are

$$M\left(\frac{x_1 + x_2}{2}, \frac{y_1 + y_2}{2}\right)$$

Notice that the coordinates of the midpoint are the averages of the x- and y-coordinates, respectively, of the endpoints of the segment.

EXAMPLE 3 **Find the coordinates of the midpoint of the line segment with the endpoints $(5, -3)$ and $(-2, -7)$.**

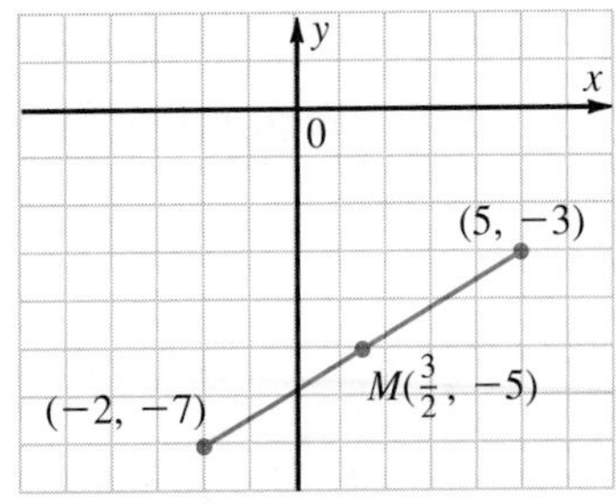

$$M\left(\frac{x_1 + x_2}{2}, \frac{y_1 + y_2}{2}\right) \qquad \textit{Let } (x_1, y_1) = (5, -3) \textit{ and } (x_2, y_2) = (-2, -7)$$
$$= \left(\frac{5 + (-2)}{2}, \frac{-3 + (-7)}{2}\right) = \left(\frac{3}{2}, -5\right)$$

The coordinates of the midpoint are $\left(\frac{3}{2}, -5\right)$.

The distance and midpoint formulas can be used to solve geometric problems.

EXAMPLE 4 The diagonals of a parallelogram *bisect* each other at the point (5, 4). One of the diagonals has an endpoint (1, 2). Find the coordinates of the other endpoint of that diagonal.

Understand the Problem Draw a parallelogram to help you understand the situation. You are asked to find the coordinates of the other endpoint of the diagonal. The diagonals *bisect* each other. That is, they intersect at their midpoints. You will need to use the midpoint formula.

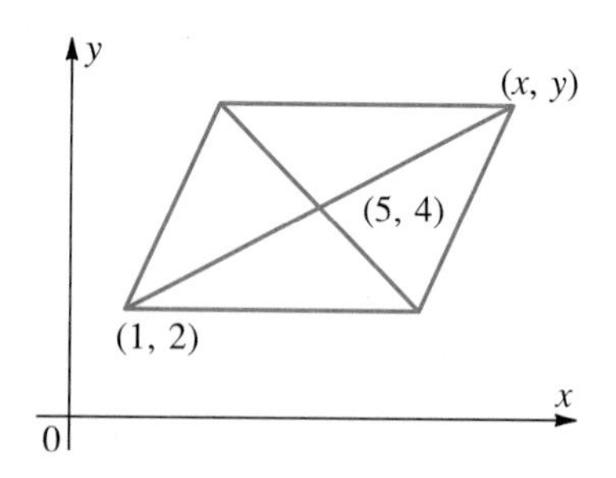

Plan Your Approach Let x be the x-coordinate of the other endpoint of the diagonal. Let y be the y-coordinate of the other endpoint. The coordinates of the midpoint M of the diagonal are (5, 4), so

$$\left(\frac{x_1 + x_2}{2}, \frac{y_1 + y_2}{2}\right) = (5, 4)$$

Complete the Work Therefore $\frac{1 + x}{2} = 5$ and $\frac{2 + y}{2} = 4$

$$1 + x = 10 \qquad 2 + y = 8$$
$$x = 9 \qquad y = 6$$

Interpret the Results The coordinates of the other endpoint are (9, 6).

Are (5, 4) the coordinates of the midpoint of the segment joining the points (1, 2) and (9, 6)?

$$\left(\frac{x_1 + x_2}{2}, \frac{y_1 + y_2}{2}\right) \longrightarrow \left(\frac{1 + 9}{2}, \frac{2 + 6}{2}\right) \stackrel{?}{=} (5, 4)$$

$$(5, 4) = (5, 4) \checkmark$$

CLASS EXERCISES

In the parallelogram at the right, $\overline{AC}$ and $\overline{BD}$ bisect each other at M. Complete each statement.

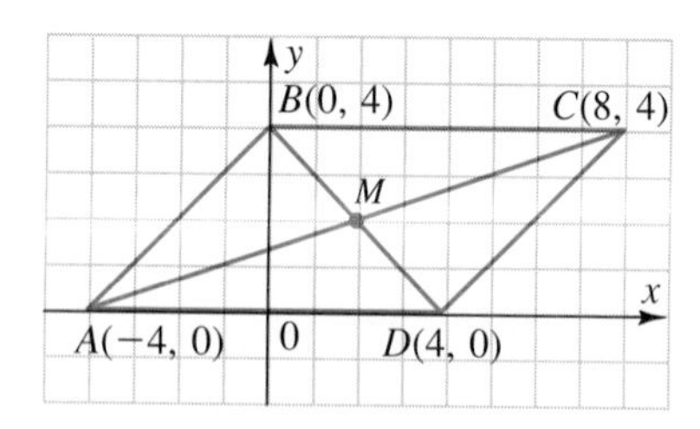

1. $AB =$ ___?___
2. $BC =$ ___?___
3. $CD =$ ___?___
4. $AD =$ ___?___
5. The coordinates of M are ___?___.
6. $AM = MC =$ ___?___
7. $DM = MB =$ ___?___

In the figure at the right, D is the midpoint of $\overline{BC}$. Complete each statement.

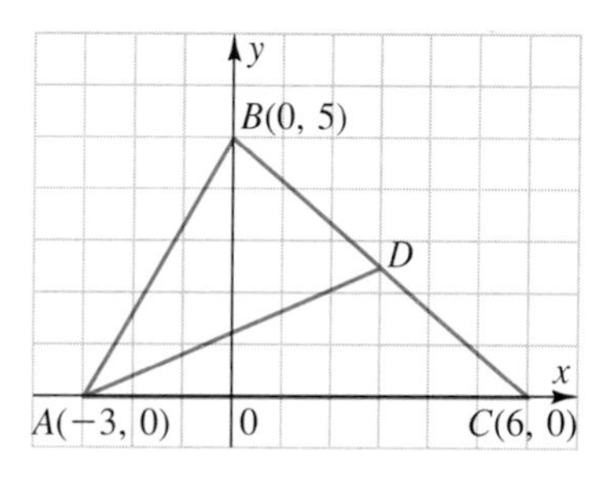

8. $AB =$ ___?___ **9.** $BC =$ ___?___

10. $CA =$ ___?___ **11.** $AD =$ ___?___

12. The coordinates of D are ___?___.

13. The perimeter of triangle ABC is ___?___.

PRACTICE EXERCISES

Find the distance between the points with the given coordinates.

1. (2, 7), (6, 7) **2.** (3, 5), (7, 5) **3.** (−3, −5), (2, −5)

4. (−4, −2), (−4, 3) **5.** (3, 5), (3, −7) **6.** (−6, −3), (−6, 0)

7. (2, −3), (5, −4) **8.** (−2, −3), (5, −1) **9.** (−1, 3), (5, 0)

Find the coordinates of the midpoint of the segment with the given endpoints.

10. (3, 1), (1, 5) **11.** (5, 2), (−1, −2) **12.** (11, 3), (1, 5)

13. (−5, 2), (−5, 8) **14.** (6, 4), (−4, 4) **15.** (0, 0), (−4, −6)

16. (8, 5), (12, 0) **17.** (0, 6), (−4, 13) **18.** (0, 0), (10, −9)

19. $\left(3, \frac{1}{2}\right), \left(5, \frac{1}{3}\right)$ **20.** $\left(\frac{1}{4}, -2\right), \left(\frac{1}{5}, 4\right)$ **21.** $\left(\frac{1}{2}, \frac{1}{3}\right), \left(\frac{2}{3}, \frac{3}{4}\right)$

Find the distance between the points with the given coordinates. Express all radicals in simplest form.

22. (15, −8), (17, 2) **23.** (25, −3), (16, −9)

24. $(6, -\sqrt{2}), (-2, -\sqrt{2})$ **25.** $(-3\sqrt{3}, \sqrt{5}), (4\sqrt{3}, \sqrt{5})$

26. $(\sqrt{6}, -1), (\sqrt{6}, 5)$ **27.** $(-3\sqrt{2}, 4), (-3\sqrt{2}, -5)$

If M is the midpoint of $\overline{AB}$, find the coordinates of B.

28. $A(-3, 5)$, $M(-6, -2)$ **29.** $A(-11, -9)$, $M(6, -7)$

30. $A(3, -7)$, $M\left(-\frac{2}{3}, 5\right)$ **31.** $A(-6, 9)$, $M\left(6, -\frac{3}{4}\right)$

32. $A(2\sqrt{3}, 5\sqrt{3})$, $M(6\sqrt{3}, 7\sqrt{3})$ **33.** $A(-3\sqrt{5}, \sqrt{5})$, $M(0, 7\sqrt{5})$

34. Find the length of a radius of a circle with its center at the origin if it passes through (5, 3).

35. The endpoints of a diameter of a circle are (2, 7) and (6, 5). Find the coordinates of the center of the circle.

36. A circle has a radius of length 8. One endpoint of a diameter is $(4, -2)$. If the x-coordinate of the other endpoint of that diameter is 1, find the two possible values of the y-coordinate of that point.

37. The altitude to the base of a triangle is 10 units long. The midpoint of the base is $(2, 5)$. Find the area of the triangle if one endpoint of the base is at $(3, 7)$.

38. The vertices of a triangle are $A(-4, -3)$, $B(3, 5)$, and $C(4, -2)$. If M, N, and P are the midpoints of $\overline{AB}$, $\overline{BC}$, and $\overline{AC}$, respectively, find the perimeter of triangle MNP.

39. The vertices of a triangle are $A(2, 3)$, $B(3, -3)$, and $C(-4, -1)$. D, E, and F are the midpoints of $\overline{AB}$, $\overline{BC}$, and $\overline{AC}$, respectively. Find the perimeter of quadrilateral $DECF$.

40. Verify that $\left(\frac{x_1 + x_2}{2}, \frac{y_1 + y_2}{2}\right)$ are the coordinates of the midpoint M of the segment joining points $A(x_1, y_1)$ and $B(x_2, y_2)$. Follow these steps:
   a. Show that M lies on $\overleftrightarrow{AB}$. *Hint:* Show that $\overleftrightarrow{AM}$ and $\overleftrightarrow{MB}$ have the same slope. Then M must lie on $\overleftrightarrow{AB}$, since there exists just one line in the plane that passes through a given point with a given slope.
   b. Show that $\overline{AM}$ and $\overline{MB}$ are equal in length.

Applications

Give answers to the nearest tenth.

41. **Sports** Find the distance from third base to first base on a baseball field if the distance between consecutive bases is 90 ft.

42. **Construction** A pool 18.25 m in length is 1 m deep at one end and 3.5 m deep at the other end. What is the length of the bottom of the pool, if the depth increases at a constant rate?

TEST YOURSELF

Simplify.

1. $\sqrt{32x^5y^9}$ 2. $\sqrt[3]{27x^6}$ 3. $\sqrt{2xy^5} \cdot \sqrt{8x^3y^2}$ **8.1–8.3**

4. $\frac{6\sqrt{2y}}{5\sqrt{7}}$ 5. $\frac{1 + \sqrt{6}}{2 - \sqrt{6}}$ 6. $2\sqrt{50} + 4\sqrt{98}$

Solve each equation.

7. $\sqrt{4x - 3} = 5$ 8. $\sqrt{3x + 7} = x + 3$ **8.4**

9. Find the distance between the points with coordinates $(-2, 7)$ and $(3, 5)$. **8.5**

10. The coordinates of the endpoints of a diameter of a circle are $(12, 8)$ and $(6, -10)$. Find the coordinates of the center of the circle.

8.6 Problem Solving Strategy: Use Coordinate Proofs

Many geometric theorems can be proved by using *analytic*, or *coordinate geometry*. In this type of geometry, a figure is drawn on a rectangular coordinate system and its vertices are assigned appropriate coordinates. Then the distance formula and the midpoint formula, together with other theorems from geometry, can be used to prove new theorems. It is often helpful to draw the figure so that one vertex is at the origin and, if possible, at least one side is on one of the coordinate axes.

EXAMPLE 1 To prepare for a coming storm, a hotel manager wishes to reinforce all the rectangular windows in his building with strips of tape. He plans to put a strip of tape along each diagonal of a window. He would like to cut all the tape strips to one length. Can he do so?

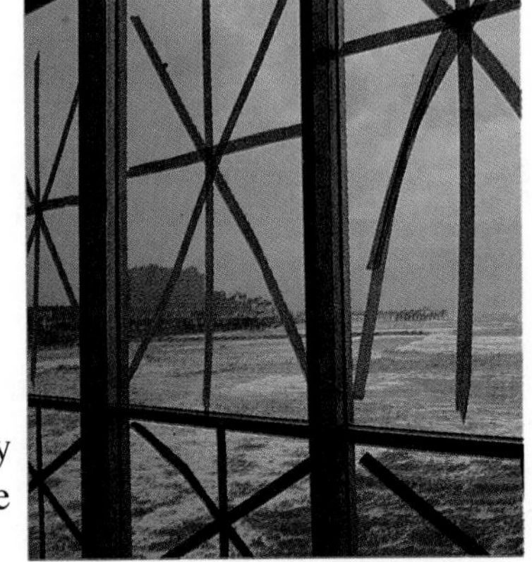

Understand the Problem The manager can cut all the strips the same length if the two diagonals of any rectangle are equal in length. So, prove that this is true.

Plan Your Approach A *rectangle* is a four-sided figure, or *quadrilateral,* with four right angles. Draw rectangle $ABCD$ so that A is at the origin and two adjacent sides are on the x- and y-axis, respectively.

Let w be the width of the rectangle and let l be the length. Then the coordinates of vertex B are $(l, 0)$ and the coordinates of vertex D are $(0, w)$. Since the other two sides are perpendicular to $\overline{AB}$ and to $\overline{AD}$, the coordinates of vertex C are (l, w).

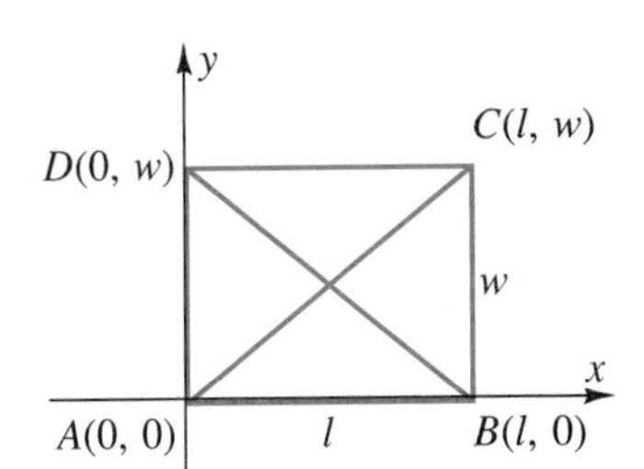

If the diagonals have equal length, then AC equals BD. Use the distance formula to calculate AC and BD.

$$d = \sqrt{(x_2 - x_1)^2 + (y_2 - y_1)^2}$$

Complete the Work

$$AC = \sqrt{(l - 0)^2 + (w - 0)^2} = \sqrt{l^2 + w^2} \qquad BD = \sqrt{(0 - l)^2 + (w - 0)^2} = \sqrt{l^2 + w^2}$$

Interpret the Results The two diagonals of any rectangle are equal in length, so the manager can cut all the tape strips the same length.

In Example 2, both the midpoint formula and the distance formula are used. When making a drawing of a geometric figure, be careful not to make any unwarranted assumptions about the figure.

EXAMPLE 2 Use coordinate geometry to prove that the line segment connecting the midpoints of two sides of any triangle is one-half the length of the third side of the triangle.

Understand the Problem In order to prove the statement for *any* triangle, you should draw a *scalene* triangle, which has three sides of unequal length.

Plan Your Approach Draw triangle ABC so that vertex A is at the origin and side $\overline{AB}$ is on the x-axis. Then draw $\overline{DE}$, with D and E the midpoints of $\overline{AC}$ and $\overline{BC}$, respectively.

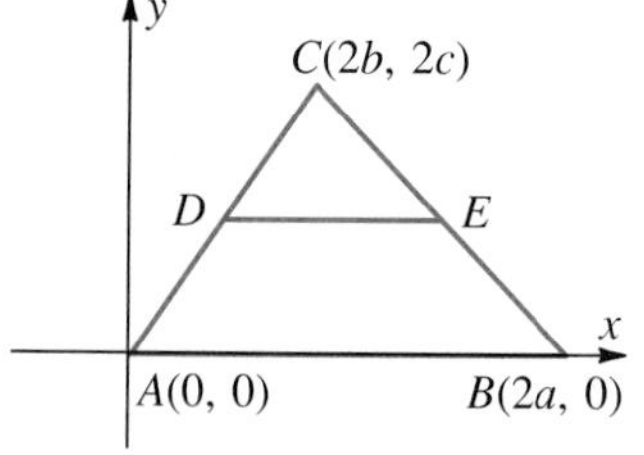

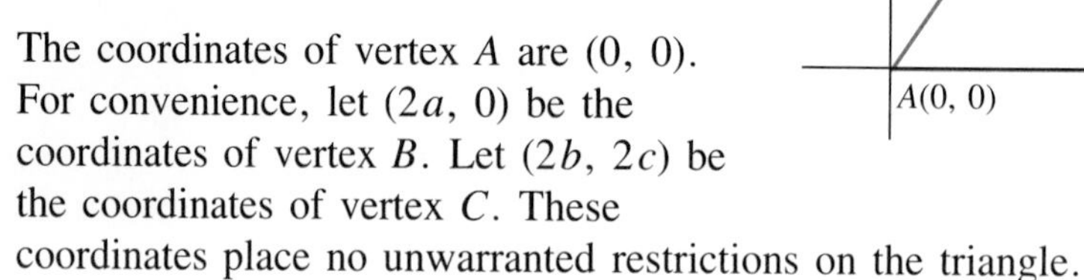

The coordinates of vertex A are $(0, 0)$. For convenience, let $(2a, 0)$ be the coordinates of vertex B. Let $(2b, 2c)$ be the coordinates of vertex C. These coordinates place no unwarranted restrictions on the triangle.

Use the midpoint formula to find the coordinates of D and E. Then find AB and DE to show that $DE = \frac{1}{2}AB$.

Complete the Work Use the midpoint formula to find the coordinates of D and E.

Coordinates of D: $\left(\frac{2b + 0}{2}, \frac{2c + 0}{2}\right) = \left(\frac{2b}{2}, \frac{2c}{2}\right) = (b, c)$

Coordinates of E: $\left(\frac{2a + 2b}{2}, \frac{0 + 2c}{2}\right) = \left(\frac{2(a + b)}{2}, \frac{2c}{2}\right) = (a + b, c)$

Since $\overleftrightarrow{AB}$ is a horizontal line, $AB = |2a - 0| = 2a$.

Use the distance formula to find DE.

$$DE = \sqrt{[(a + b) - b]^2 + (c - c)^2} = \sqrt{a^2 + 0} = a$$

Interpret the Results Since $AB = 2a$ and $DE = a$, the length of $\overline{DE}$ is one-half the length of $\overline{AB}$. Therefore, the length of the line segment connecting the midpoints of two sides of any triangle is one-half the length of the third side.

CLASS EXERCISES

The coordinates of the vertices of several figures are listed below. Match the most appropriate set of coordinates with each figure described in Exercises 1–6. *Hint:* Two lines are parallel if they have equal slopes.

a. $(0, 0), (a, 0), (a, a), (0, a)$

b. $(0, 0), (2a, 0), (2b, 2d), (2c, 2d)$

c. $(0, 0), (2a, 0), (0, 2b)$

d. $(0, 0), (a, 0), (a, b), (0, b)$

e. $(0, 0), (a, 0), (b, c)$

f. $(0, 0), (a, 0), (a + b, c), (b, c)$

1. Right triangle (a triangle with one right angle)

2. Parallelogram (a quadrilateral with two pairs of parallel sides)

3. Rectangle (a quadrilateral with four right angles)

4. Square (a rectangle with four sides of equal length)

5. Trapezoid (a quadrilateral with exactly one pair of parallel sides)

6. Scalene triangle (a triangle with no two sides of equal length)

PRACTICE EXERCISES

For some of these exercises, you may wish to use coordinates from the Class Exercises. Also, you may find the following definitions useful.

- A *median of a triangle* is a line segment from a vertex to the midpoint of the opposite side.
- The *legs* of an isosceles triangle are the two sides of equal length. The third side is the *base of the triangle*.
- The *median of a trapezoid* is the segment joining the midpoints of the nonparallel sides.
- The *bases of a trapezoid* are the two parallel sides.

Write a coordinate proof for each of the following.
For Exercises 1–6, use parallelogram *PQRS*.

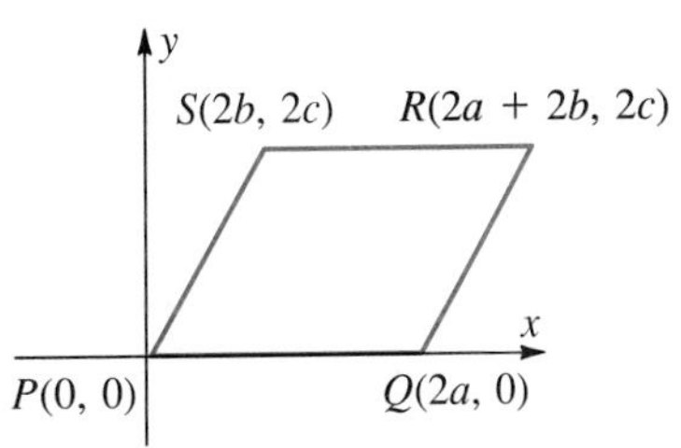

1. $PQ = SR$ **2.** $PS = QR$

3. A line segment connecting the midpoints of $\overline{PQ}$ and $\overline{SR}$ is the same length as $\overline{PS}$.

4. A line through the midpoints of $\overline{PQ}$ and $\overline{SR}$ is parallel to $\overline{QR}$.

5. The diagonals, $\overline{PR}$ and $\overline{QS}$, bisect each other.

6. The line segment joining the midpoints of $\overline{PQ}$ and $\overline{SR}$ and the line segment joining the midpoints of $\overline{PS}$ and $\overline{QR}$ bisect each other.

7. The line segment joining the midpoints of two sides of any triangle is parallel to the third side.

8. The diagonals of any square are perpendicular. *Hint:* If the slopes of two lines are negative reciprocals of each other, the lines are perpendicular.

9. The median to the hypotenuse of any right triangle is one-half the length of the hypotenuse.

10. The median of any trapezoid is parallel to the bases.

11. The length of the median of any trapezoid is one-half the sum of the lengths of the bases of the trapezoid.

12. An *isosceles triangle* is a triangle with two sides of equal length. Prove that the triangle with vertices $(-2a, 0)$, $(2a, 0)$, and $(0, 2b)$ is isosceles.

13. Use the isosceles triangle in Exercise 12 to prove that the medians to the legs of an isosceles triangle are the same length.

14. A *rhombus* is a quadrilateral with four sides of equal length. Prove that the quadrilateral with vertices $(0, 0)$, $(a, 0)$, $(a + b, \sqrt{a^2 - b^2})$, and $(b, \sqrt{a^2 - b^2})$ is a rhombus.

15. Use the rhombus in Exercise 14 to prove that the diagonals of any rhombus are perpendicular.

16. An *equilateral triangle* is a triangle with three sides of equal length. Prove that the triangle with vertices $(0, 0)$, $(4a, 0)$, $(2a, 2a\sqrt{3})$ is equilateral.

17. Use the equilateral triangle in Exercise 16 to prove that the line segments joining the midpoints of the sides of any equilateral triangle form an equilateral triangle.

18. An *isosceles trapezoid* is a trapezoid with legs of equal length. Prove that the diagonals of an isosceles trapezoid are equal in length. *Hint:* Let the endpoints of one base be $(-a, 0)$ and $(a, 0)$. Choose appropriate coordinates for the other two vertices.

19. The sum of the squares of the lengths of the four sides of any parallelogram is equal to the sum of the squares of the lengths of the two diagonals.

20. The sum of the squares of the lengths of the three medians of any triangle is equal to three-fourths the sum of the squares of the lengths of the three sides.

Mixed Problem Solving Review

1. Find three consecutive integers whose sum is equal to 324.
2. A market research firm found the demand for a new product was represented by a linear function. At a price of \$8, a total of 2500 items were sold. At \$6, 3000 items were sold. Find the potential sales at a price of \$9.
3. The sum of the digits of a three-digit number is 9. Four times the tens digit minus 3 times the units digit is 6. If 2 times the units digit is added to twice the hundreds digit, the result is 12. Find the number.
4. Mike can paint the front of his house in 10 hours and Tony can do it in 6 hours. How long will it take them to do the job together?

PROJECT

Construct an isosceles triangle with given base

If you are given the coordinates of the endpoints of a segment, you can construct an isosceles triangle with that segment as its base.

EXAMPLE **Points P and Q have coordinates (3, 0) and (−5, 2), respectively. Construct isosceles triangle PQR, with base $\overline{PQ}$.**

slope $\overline{PQ}$: $\frac{2-0}{-5-3} = -\frac{1}{4}$ midpoint M: $\left(\frac{3+(-5)}{2}, \frac{0+2}{2}\right) = (-1, 1)$

It can be shown that the median to the base of an isosceles triangle is perpendicular to the base. So, find an equation of the line that is perpendicular to $\overline{PQ}$ and passes through its midpoint, M.

$y - y_1 = m(x - x_1)$ *(x_1, y_1) are the coordinates of M.*

$y - 1 = 4[x - (-1)]$ *m is the negative reciprocal of $-\frac{1}{4}$, the slope of $\overline{PQ}$.*

$y = 4x + 5$

R is any point, other than M, on the line $y = 4x + 5$. If $x = -2$, then $y = 4(-2) + 5$, or $y = -3$. Let the coordinates of R be $(-2, -3)$.

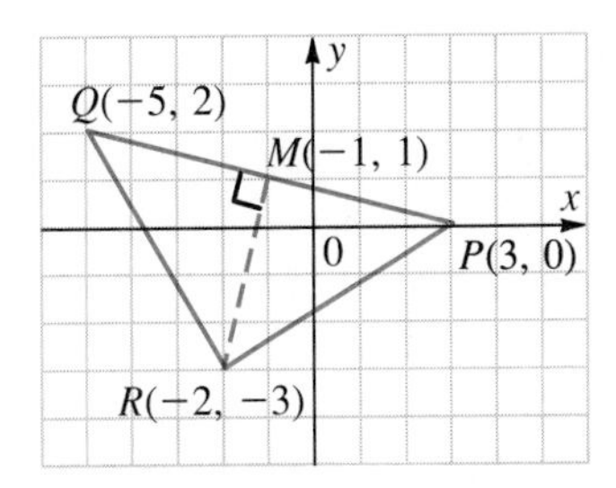

On a coordinate plane, plot the points P, Q, and R. Join them to construct triangle PQR. To check that the triangle is isosceles, use the distance formula to show that $PR = QR$.

For each, construct an isosceles triangle whose base has endpoints with the given coordinates. Check by using the distance formula.

1. (2, 5), (10, 9)
2. (1, −2), (11, −8)
3. (0, 0), (−8, −6)

8.7

Imaginary Numbers

Objective: To simplify square roots of negative numbers and powers of i

The simplest number system is the set of *natural numbers*. In order to have an identity element for addition, this set was expanded to the set of *whole numbers*. In order to be able to subtract any two whole numbers, negative numbers were introduced and the number system was expanded to the set of *integers*. In order to divide any two integers, *rational numbers* were needed. *Irrational numbers* were required for the solution of equations such as $x^2 = 2$. Therefore the number system was expanded to the set of *real numbers,* which includes the rationals and the irrationals.

Capsule Review

The set of real numbers is *closed* under addition and multiplication, since for all real numbers a and b, $a + b$ and $a \cdot b$ are unique real numbers. To show that a set of numbers is *not* closed under an operation, find two elements, a and b, of the set, such that the result of the operation is not an element of the set.

EXAMPLE **Show that the set of natural numbers is *not* closed under subtraction.**

Let $a = 1$ and $b = 5$, then $a - b = 1 - 5 = -4$. -4 is not a natural number, so the set of natural numbers is not closed under subtraction.

1. Show that the set of integers is *not* closed under division.
2. Show that the set of irrational numbers is *not* closed under multiplication.
3. Show that the set of odd integers is *not* closed under addition.

The equation $x^2 + 1 = 0$, or $x^2 = -1$, has no solution in the real number system because $x^2 \geq 0$ for any real number x. Therefore, there is no real number whose square is -1. To solve this equation, it is necessary to expand the number system to include the **imaginary number** i. The number i is defined in such a way that

$$i^2 = -1$$

Then the principal square root of -1 is i, or

$$i = \sqrt{-1}$$

Another square root of -1 is $-i$, or

$$-i = -\sqrt{-1}$$

since $(-i)^2 = (-1 \cdot i)^2 = (-1)^2 \cdot i^2 = 1(-1) = -1$.

Numbers such as $7i$, $-5i$, $i\sqrt{3}$, and $-i\sqrt{3}$ are called **pure imaginary numbers.** For all positive real numbers b, the principal square root of $-b$ is expressed as $i\sqrt{b}$, since

$$\sqrt{-b} = \sqrt{-1} \cdot \sqrt{b} = i\sqrt{b}$$

EXAMPLE 1 **Simplify:** **a.** $\sqrt{-1}$ **b.** $\sqrt{-7}$ **c.** $\sqrt{-16}$ **d.** $\sqrt{-24}$

a. $\sqrt{-1} = i$

b. $\sqrt{-7} = \sqrt{-1} \cdot \sqrt{7} = i\sqrt{7}$

c. $\sqrt{-16} = \sqrt{-1} \cdot \sqrt{16} = 4i$

d. $\sqrt{-24} = \sqrt{-1} \cdot \sqrt{24} = i \cdot \sqrt{4 \cdot 6} = 2i\sqrt{6}$

Notice that the product of i and $\sqrt{7}$ is written as $i\sqrt{7}$ not $\sqrt{7}i$. This is to avoid confusing the expressions $\sqrt{7}i$ and $\sqrt{7i}$.

You have just seen that $i^1 = i$ and $i^2 = -1$. What do i^3, i^4, i^5, i^6, i^7, and i^8 equal?

$$i^3 = i^2 \cdot i = -1 \cdot i = -i \qquad i^6 = i^4 \cdot i^2 = 1 \cdot -1 = -1$$
$$i^4 = i^2 \cdot i^2 = -1 \cdot -1 = 1 \qquad i^7 = i^4 \cdot i^3 = 1 \cdot -i = -i$$
$$i^5 = i^4 \cdot i = 1 \cdot i = i \qquad i^8 = i^4 \cdot i^4 = 1 \cdot 1 = 1$$

You can see that the numbers i, -1, $-i$, and 1 repeat in cycles of four.

EXAMPLE 2 **Simplify:** **a.** i^9 **b.** i^{10} **c.** i^{23}

a. $i^9 = (i^4)^2 \cdot i = 1 \cdot i = i$

b. $i^{10} = (i^4)^2 \cdot i^2 = 1 \cdot (-1) = -1$

c. $i^{23} = (i^4)^5 \cdot i^3 = 1 \cdot (-i) = -i$

To simplify expressions involving square roots of negative numbers, first write such numbers in terms of i. The commutative, associative, and distributive properties hold for pure imaginary numbers.

EXAMPLE 3 **Simplify:** **a.** $4i + 3i$ **b.** $\sqrt{-4} - 5i$ **c.** $\sqrt{-5} - 2\sqrt{-7}$

a. $4i + 3i = (4 + 3)i = 7i$

b. $\sqrt{-4} - 5i = 2i - 5i = (2 - 5)i = -3i$

c. $\sqrt{-5} - 2\sqrt{-7} = i\sqrt{5} - 2i\sqrt{7} = (\sqrt{5} - 2\sqrt{7})i$

If $x \geq 0$ and $y \geq 0$, then $\sqrt{x} \cdot \sqrt{y} = \sqrt{xy}$. However, $\sqrt{-x} \cdot \sqrt{-y} \neq \sqrt{(-x)(-y)}$. Therefore, always express square roots of negative numbers in terms of i before multiplying.

EXAMPLE 4 **Simplify:** $\sqrt{-4} \cdot \sqrt{-9}$

$$\begin{aligned} \sqrt{-4} \cdot \sqrt{-9} &= 2i \cdot 3i && \text{Express } \sqrt{-4} \text{ and } \sqrt{-9} \text{ in terms of } i. \\ &= 6i^2 \\ &= 6(-1) && i^2 = -1 \\ &= -6 \end{aligned}$$

When you divide with pure imaginary numbers, it may be necessary to rationalize the denominator.

EXAMPLE 5 **Divide:** **a.** $3i \div 5i$ **b.** $4 \div \sqrt{-3}$ **c.** $3\sqrt{5} \div \sqrt{-6}$

a. $\frac{3i}{5i} = \frac{3}{5}$ $\quad \frac{i}{i} = 1$

b. $\frac{4}{\sqrt{-3}} = \frac{4}{i\sqrt{3}} = \frac{4}{i\sqrt{3}} \cdot \frac{i\sqrt{3}}{i\sqrt{3}} = \frac{4i\sqrt{3}}{3i^2} = \frac{4i\sqrt{3}}{3(-1)} = \frac{-4i\sqrt{3}}{3}$

c. $\frac{3\sqrt{5}}{\sqrt{-6}} = \frac{3\sqrt{5}}{i\sqrt{6}} = \frac{3\sqrt{5}}{i\sqrt{6}} \cdot \frac{i\sqrt{6}}{i\sqrt{6}} = \frac{3i\sqrt{30}}{6i^2} = \frac{3i\sqrt{30}}{-6} = \frac{-i\sqrt{30}}{2}$

CLASS EXERCISES

Classify each number as one or more of the following: real, rational, irrational, or pure imaginary.

1. -36 **2.** $\sqrt{-36}$ **3.** i^7 **4.** $i\sqrt{-3}$ **5.** $8i$ **6.** $10i^4$

In Exercises 7–9, tell whether the solutions of each equation are real or imaginary.

7. $p^2 = -49$ **8.** $q^4 = 8$ **9.** $b^3 = b$

10. Show that i and $-i$ are reciprocals of each other; that is, show that $i(-i) = 1$.

PRACTICE EXERCISES

Simplify.

1. $\sqrt{-5}$ **2.** $\sqrt{-9}$ **3.** $\sqrt{-6}$ **4.** $\sqrt{-16}$

5. $\sqrt{-8}$ **6.** $\sqrt{-32}$ **7.** i^{11} **8.** i^{13}

9. i^{12}

10. i^{20}

11. i^{24}

12. i^{27}

13. $3i + 7i$

14. $5i + 6i$

15. $7i - 11i$

16. $\sqrt{-9} - 5i$

17. $\sqrt{-9} + \sqrt{-25}$

18. $\sqrt{-4} - \sqrt{-16}$

19. $\sqrt{-5} + 3\sqrt{-5}$

20. $3i\sqrt{-4}$

21. $\sqrt{-36} \cdot \sqrt{-25}$

22. $\sqrt{-16} \cdot \sqrt{-49}$

23. $\sqrt{-3} \cdot \sqrt{-5}$

24. $\sqrt{-7} \cdot \sqrt{-3}$

25. $6i \div 2i$

26. $12i \div 3i$

27. $7i \div 2i$

28. $15i \div 18i$

29. $\sqrt{-5} \div \sqrt{-2}$

30. $\sqrt{-7} \div \sqrt{-5}$

31. $2\sqrt{-5} \div \sqrt{-10}$

32. $\sqrt{-98}$

33. i^{55}

34. i^{82}

35. i^{132}

36. $(-i)^{16}$

37. $7\sqrt{-16} - 3\sqrt{-9}$

38. $5\sqrt{-8} + 2\sqrt{-50}$

39. $3\sqrt{-27} - 4\sqrt{-12}$

40. $-2\sqrt{-125} + 3\sqrt{-20}$

41. $-\sqrt{-28} - 2\sqrt{-63}$

42. $-\sqrt{-32} - \sqrt{-72}$

43. $-\sqrt{-300} + \sqrt{-243}$

44. $-\sqrt{-363} - \sqrt{-243}$

45. $2i\sqrt{7} - 3i\sqrt{7}$

46. $i\sqrt{5} + i\sqrt{5}$

47. $i\sqrt{-12} \cdot 2\sqrt{-10}$

48. $2i\sqrt{-3} \cdot \sqrt{-12}$

49. $4\sqrt{-8} \div 6\sqrt{-12}$

50. $8i \div \sqrt{-72}$

51. $6i \div \sqrt{-200}$

52. $6\sqrt{-50} \div 3\sqrt{-6}$

53. $4\sqrt{-45} \div 2\sqrt{-10}$

54. $\sqrt{-18} \div \sqrt{-9}$

55. i^{4n+1}

56. i^{4n+2}

57. i^{4n+3}

58. $(5i \cdot \sqrt{-3}) \div \sqrt{-6}$

59. $(\sqrt{-7} \cdot \sqrt{21}) \div 7\sqrt{-1}$

60. $(3i)^3 \div (2i)^4$

61. $i^{24} \div i^{18}$

62. $\sqrt{-x^2} + \sqrt{-y^2}$

63. $(\sqrt{-x^2})^5$

Applications

Electricity In an electrical circuit, reactance X is represented by a pure imaginary number. Reactance from an inductance X_L is represented by a positive, pure imaginary number and reactance from a capacitor X_C is represented as a negative, pure imaginary number. Therefore $X = X_L - X_C$.

64. In an electrical circuit, the reactance from capacitors is 8 ohms and the reactance from inductances is 12 ohms. Find the total reactance.

65. The reactance in an electrical circuit is 8 ohms. The reactance from capacitors is 9 ohms. Find the reactance from inductances.

Developing Mathematical Power

66. Thinking Critically Which, if any, of the following statements are true?

I. $(i^2 - i^3)^2 = (i^2 + i^3)^2$

II. $(i^3 + i^2 + i)^2 = (i^3 - i^2 + i)^2$

III. $(i + i^3)^2 = (i + i^3)^3$

a. III only

b. I and II only

c. II and III only

d. All are true.

e. None is true.

8.8

Addition and Subtraction of Complex Numbers

Objectives: To add and subtract complex numbers
To find the absolute value of a complex number

The sum of a real number and a pure imaginary number is a *complex number*. Before the addition of complex numbers is discussed, it will be useful to review the concepts of an additive identity and an additive inverse.

Capsule Review

The additive identity for a real number a is 0, since $a + 0 = a = 0 + a$. The additive inverse of a real number a is $-a$, since $a + (-a) = 0$.

Find the additive identity and the additive inverse of each real number.

1. 9 **2.** 12 **3.** -8 **4.** -20 **5.** $\sqrt{5}$ **6.** $-\sqrt{17}$

A **complex number** is a number that can be written in the form $a + bi$, where a and b are real numbers and $i = \sqrt{-1}$. The *real part* of a complex number $a + bi$ is a, and the *imaginary part* is bi.

A complex number $a + bi$ is *imaginary* if $b \neq 0$.
It is *pure imaginary* if $a = 0$ and $b \neq 0$.
It is a *real number* if $b = 0$.

Complex Numbers

Imaginary Numbers	Real Numbers
$5 + 6i$; $12 - i$; $3 - \sqrt{-4}$; Pure Imaginary Numbers: $3i$, $\sqrt{-1}$	17; -2; $4.\overline{3}$; 0; $\frac{7}{3}$; 9.6; Irrational Numbers: 0.191191119..., π, $\sqrt{5}$

EXAMPLE 1 **Write each complex number in the form $a + bi$.**

a. $3 - 4i$ **b.** $5i$ **c.** 4 **d.** $\sqrt{-9} + 6$

a. $3 - 4i = 3 + (-4)i$ **b.** $5i = 0 + 5i$

c. $4 = 4 + 0i$ **d.** $\sqrt{-9} + 6 = 3i + 6 = 6 + 3i$

The complex numbers $3 + 7i$ and $3 + \sqrt{-49}$ are equal because

$$3 + \sqrt{-49} = 3 + \sqrt{-1} \cdot \sqrt{49} = 3 + 7i$$

The complex numbers $3 + 7i$ and $3 - 7i$ are not equal, because $7 \neq -7$. Two complex numbers $a + bi$ and $c + di$ are *equal* if and only if $a = c$ and $b = d$.

The properties of addition for real numbers also hold for complex numbers.

Addition of complex numbers If a, b, c, and d are real numbers, then

$$(a + bi) + (c + di) = (a + c) + (b + d)i$$

EXAMPLE 2 **Simplify:** **a.** $(3 + 4i) + (2 + 5i)$ **b.** $(4 + 6i) + (1 - 3i)$ **c.** $(7 + \sqrt{-4}) + (5 + 3i)$

a. $(3 + 4i) + (2 + 5i) = (3 + 2) + (4 + 5)i$
$= 5 + 9i$

b. $(4 + 6i) + (1 - 3i) = (4 + 1) + (6 - 3)i$
$= 5 + 3i$

c. $(7 + \sqrt{-4}) + (5 + 3i) = (7 + 2i) + (5 + 3i)$ *Rewrite in terms of i.*
$= (7 + 5) + (2 + 3)i$
$= 12 + 5i$

As for the real numbers, the additive identity for the set of complex numbers is 0. In $a + bi$ form, it can be written $0 + 0i$.

$$(a + bi) + (0 + 0i) = (a + 0) + (b + 0)i = a + bi$$

Every complex number also has a unique additive inverse.

Additive inverse of complex numbers If a and b are real numbers, then the additive inverse of $a + bi$ is

$$-(a + bi), \text{ or } -a - bi$$

To subtract a complex number, add its additive inverse.

Subtraction of complex numbers If a, b, c, and d are real numbers, then

$$(a + bi) - (c + di) = (a + bi) + (-c - di)$$

EXAMPLE 3 **Simplify:** **a.** $(4 + 3i) - (1 + 2i)$ **b.** $(5 - 4i) - (6 - 2i)$

a. $(4 + 3i) - (1 + 2i) = (4 + 3i) + (-1 - 2i)$ *$-1 - 2i$ is the additive inverse of $1 + 2i$.*
$= (4 - 1) + (3 - 2)i$
$= 3 + i$

b. $(5 - 4i) - (6 - 2i) = (5 - 4i) + (-6 + 2i)$ *$-6 + 2i$ is the additive inverse of $6 - 2i$.*
$= (5 - 6) + (-4 + 2)i$
$= -1 - 2i$

Every complex number has a unique absolute value. The absolute value of a complex number is a real number and is defined as follows.

> **Absolute value of complex numbers** If a and b are real numbers, then
> $$|a + bi| = \sqrt{a^2 + b^2}$$

EXAMPLE 4 **Find the absolute value of each complex number.**
a. $|3 + 4i|$ **b.** $|5 - 2i|$

a. $|3 + 4i| = \sqrt{3^2 + 4^2} = \sqrt{9 + 16} = \sqrt{25} = 5$

b. $|5 - 2i| = \sqrt{5^2 + (-2)^2} = \sqrt{25 + 4} = \sqrt{29}$

CLASS EXERCISES

Tell in which of the sets of numbers listed at the right each of the following numbers belongs.

Complex numbers
Imaginary numbers
Pure imaginary numbers
Real numbers
Rational numbers
Irrational numbers

1. $3.121121112\ldots$
2. $\sqrt[3]{64}$
3. 5^{-3}
4. $\sqrt{32}$
5. $9 - 5i$
6. $-4.\overline{231}$
7. $-6i$
8. $8 + i\sqrt{7}$
9. $\sqrt[4]{32}$
10. $3 + \sqrt{-25}$

Find the additive inverse of each complex number.

11. $5 - 6i$
12. $-8 + 3i$
13. $-9 - 5i$
14. $8i$
15. $1 + 4i$
16. $0 + 7i$
17. $5 + 0i$
18. $10 - 10i$

Simplify.

19. $(5 - 2i) + (-7 + 5i)$
20. $(10 - 3i) - (8 + 3i)$
21. $|1 - i|$
22. $|-9 + 4i|$
23. $|-6 - 2i|$
24. $|3 - \sqrt{-16}|$

PRACTICE EXERCISES

Express each complex number in the form $a + bi$.

1. $4i + 6$ **2.** $-3i - 2$ **3.** -6 **4.** $5i$

5. $4i$ **6.** 0 **7.** $7 + \sqrt{-9}$ **8.** $-\sqrt{-4}$

Simplify.

9. $(9 + 2i) + (6 + i)$
10. $(4 + 2i) + (6 + 7i)$
11. $(-7 - 5i) + (-8 + 2i)$
12. $(-3 + 2i) + (5 + 6i)$
13. $(3 + \sqrt{-16}) + (2 + \sqrt{-4})$
14. $(3 + \sqrt{-9}) + (5 + \sqrt{-49})$
15. $(9 + \sqrt{-4}) + (16 + \sqrt{-25})$
16. $(9 + 5i) - (10 + 3i)$
17. $(-7i - 5) - (4 - i)$
18. $(8 + 6i) - (4i + 5)$
19. $(9 - 2i) - (-3i + 7)$
20. $(2i - 14) - (6 - 4i)$
21. $(7 - 2i) - (3 + i)$
22. $(5 - \sqrt{-9}) - (-2 - \sqrt{-36})$
23. $|6 + 8i|$ **24.** $|7 - 2i|$ **25.** $|-3 - 2i|$ **26.** $|-6 + i|$
27. $|5 + 12i|$ **28.** $|8 - 6i|$ **29.** $|-2 - 2i|$ **30.** $|5 + \sqrt{-4}|$
31. $|1 - \sqrt{-9}|$ **32.** $|-2 - \sqrt{-36}|$ **33.** $|\sqrt{-25}|$ **34.** $|-\sqrt{-25}|$
35. $(2 - 3\sqrt{-4}) + (4 + 2\sqrt{-16})$
36. $(7 + 2\sqrt{-9}) + (6 + 3\sqrt{-36})$
37. $7 + (2 + \sqrt{-81})$
38. $(6 - 2\sqrt{-49}) - 8$
39. $(4 + \sqrt{-8}) + (3 + \sqrt{-2})$
40. $(5 - \sqrt{-12}) + (3 + \sqrt{-27})$
41. $(3\sqrt{-50}) - (-2 - \sqrt{-32})$
42. $(6 + \sqrt{-20}) - (-7 - \sqrt{-45})$
43. $(2 - 3\sqrt{-98}) + (4\sqrt{-18})$
44. $(12 + 2\sqrt{-50}) - (-11 + 3\sqrt{-72})$
45. $(3\sqrt{-12}) + (-4\sqrt{12})$
46. $(5\sqrt{32}) - (3\sqrt{-32})$
47. $|2 + 5\sqrt{-9}|$ **48.** $|\sqrt{6} + 2\sqrt{-16}|$ **49.** $|7\sqrt{-4} + \sqrt{3}|$ **50.** $|\sqrt{98} + 2i|$
51. $|\sqrt{72} - 5i|$ **52.** $|\sqrt{100} - 3i|$ **53.** $|9 - 2\sqrt{-45}|$ **54.** $|6 - 4\sqrt{-50}|$

Find the values of m and n that will make each sentence true.

55. $5m + 3ni = 10 + 9i$
56. $6 - 4i = 4m + 4ni$
57. $8m + 6ni = 7 - 9i$
58. $9m - 15ni = -5 + 10i$

Simplify.

59. $[3(4 + 6\sqrt{-2})] - [2(5 + \sqrt{-8})]$

60. $[6(-2 - \sqrt{-12})] + [2(-3 + 5\sqrt{-27})]$

61. $[-4(\sqrt{7} + 5\sqrt{-2})] + [6(\sqrt{28} + 2\sqrt{-8})]$

62. $[\sqrt{3}(\sqrt{3} + 6\sqrt{-12})] - [\sqrt{2}(3\sqrt{18} - \sqrt{-8})]$

63. $(6 + i^7) + (4 - 2i^5)$

64. $(-3 - i^{11}) + (4i^4 - i^3)$

65. $(3 + i^{122}) - (-7 - i^{64})$

66. $(-1i^8 - 6i^{18}) - (-2i - 20i^{33})$

67. $|6i^{12} - i^7|$

68. $|7i^{16} + i^{11}|$

Applications

Electricity In an electrical circuit, current can be obstructed by resistors R, capacitors C, and inductances L. Impedance Z is the total effective resistance to the flow of current. Resistance R (in ohms) from resistors is represented by a real number. Reactance X (in ohms) from capacitors X_C and inductances X_L is represented by a pure imaginary number. Impedance Z is the sum of the resistances and reactors.

$$Z = R + Xi \quad \text{where} \quad X = X_L - X_C$$

The magnitude of the impedance is the $|Z|$ (in ohms) where

$$|Z| = \sqrt{R^2 + X^2}$$

69. A circuit has a resistance of 7 ohms and a reactance of 5 ohms. Find the magnitude of the impedance.

70. In a circuit, the resistance is 7.195 ohms, reactance from capacitors is 4.325 ohms and inductance is 9.362 ohms. Find the magnitude of the impedance.

Mixed Review

71. Graph $y = -\frac{2}{3}x + 4$.

72. Write an equation of the line that passes through $(4, -2)$ and has a slope of $\frac{2}{5}$.

73. Find the additive and multiplicative inverses of $\frac{1}{3}$.

74. Without solving, determine the replacement set for x:

$$\frac{x + 3}{x^2 - x - 6} = \frac{7}{x - 1} - x$$

75. Add $\frac{2x + 5}{xy} + \frac{x - 5}{xy}$.

76. Subtract $\frac{4x}{x - 7} - \frac{2x}{x + 1}$.

8.9

Multiplication and Division of Complex Numbers

Objective: To multiply and divide complex numbers

As with real numbers, the multiplication of complex numbers is commutative, associative, and distributive over addition. Therefore, complex-number denominators can be rationalized using a method similar to that used for irrational denominators.

Capsule Review

To rationalize a denominator that is an irrational number of the form $a + \sqrt{b}$, multiply by 1 in the form $\frac{p}{p}$, where p is the conjugate of the denominator. The conjugate of $a + \sqrt{b}$ is $a - \sqrt{b}$.

EXAMPLE **Rationalize the denominator:** $\frac{2 + \sqrt{2}}{4 + \sqrt{2}}$

$$\frac{2 + \sqrt{2}}{4 + \sqrt{2}} = \frac{2 + \sqrt{2}}{4 + \sqrt{2}} \cdot \frac{4 - \sqrt{2}}{4 - \sqrt{2}}$$

The conjugate of $4 + \sqrt{2}$ is $4 - \sqrt{2}$.

$$= \frac{8 - 2\sqrt{2} + 4\sqrt{2} - (\sqrt{2})^2}{4^2 - (\sqrt{2})^2}$$

$$= \frac{8 + 2\sqrt{2} - 2}{16 - 2}$$

$$= \frac{6 + 2\sqrt{2}}{14}$$

$$= \frac{3 + \sqrt{2}}{7}$$

Rationalize each denominator.

1. $\frac{2 + \sqrt{5}}{1 + \sqrt{5}}$

2. $\frac{3 - \sqrt{2}}{2 - \sqrt{2}}$

3. $\frac{4 + \sqrt{3}}{2 + \sqrt{3}}$

4. $\frac{5 + \sqrt{6}}{3 - \sqrt{6}}$

5. $\frac{8 - 3\sqrt{7}}{4 - 3\sqrt{7}}$

6. $\frac{3 - 2\sqrt{11}}{5 + 2\sqrt{11}}$

7. $\frac{1}{2 + \sqrt{2}}$

8. $\frac{1}{1 - \sqrt{3}}$

9. $\frac{1}{3 - \sqrt{2}}$

To multiply two complex numbers, use the FOIL method and the fact that $i^2 = -1$.

EXAMPLE 1 **Simplify:** **a.** $(4 + 2i)(3 + 5i)$ **b.** $(5 + 6i)(2 - 3i)$

a.
$$\begin{aligned}(4 + 2i)(3 + 5i) &= 4(3) + 4(5i) + 2i(3) + 2i(5i) && \textit{FOIL}\\ &= 12 + 20i + 6i + 10i^2\\ &= 12 + 20i + 6i + 10(-1) && i^2 = -1\\ &= 12 + 20i + 6i - 10\\ &= (12 - 10) + (20 + 6)i\\ &= 2 + 26i\end{aligned}$$

b.
$$\begin{aligned}(5 + 6i)(2 - 3i) &= 5(2) + 5(-3i) + 6i(2) + 6i(-3i) && \textit{FOIL}\\ &= 10 - 15i + 12i - 18i^2\\ &= 10 - 15i + 12i - 18(-1) && i^2 = -1\\ &= 10 - 15i + 12i + 18\\ &= (10 + 18) + (-15i + 12i)\\ &= 28 - 3i\end{aligned}$$

The numbers $2 + 3i$ and $2 - 3i$ are **complex conjugates.** They differ only in the sign of the imaginary part. The product of two complex conjugates is a real number, as shown in Example 2.

EXAMPLE 2 **Simplify:** $\mathbf{(2 + 3i)(2 - 3i)}$

$$\begin{aligned}(2 + 3i)(2 - 3i) &= 4 - 9i^2 && (a + b)(a - b) = a^2 - b^2\\ &= 4 - 9(-1) && i^2 = -1\\ &= 4 + 9\\ &= 13\end{aligned}$$

In general, if a and b are real numbers, $(a + bi)(a - bi) = a^2 + b^2$.

A complex number is squared in much the same way that a binomial is squared.

EXAMPLE 3 **Simplify:** **a.** $(3 + 5i)^2$ **b.** $(4 - 3i)^2$

a.
$$\begin{aligned}(3 + 5i)^2 &= 9 + 2(15i) + 25i^2 && (a + b)^2 = a^2 + 2ab + b^2\\ &= 9 + 30i - 25 && i^2 = -1\\ &= -16 + 30i\end{aligned}$$

b.
$$\begin{aligned}(4 - 3i)^2 &= 16 - 2(12i) + 9i^2 && (a - b)^2 = a^2 - 2ab + b^2\\ &= 16 - 24i - 9 && i^2 = -1\\ &= 7 - 24i\end{aligned}$$

To simplify the quotient of two complex numbers, rationalize the denominator. That is, multiply by 1 in the form $\frac{p}{p}$ where p is the conjugate of the denominator.

EXAMPLE 4 **Simplify:** **a.** $\frac{1+i}{2+3i}$ **b.** $\frac{4+3i}{1-2i}$

a. $\frac{1+i}{2+3i} = \frac{1+i}{2+3i} \cdot \frac{2-3i}{2-3i}$ *$2 - 3i$ is the conjugate of $2 + 3i$.*

$= \frac{2-3i+2i-3i^2}{4-9i^2}$ *$(a+b)(a-b) = a^2 - b^2$*

$= \frac{2-i-3(-1)}{4-9(-1)}$ *$i^2 = -1$*

$= \frac{5-i}{13}$, or $\frac{5}{13} - \frac{1}{13}i$

b. $\frac{4+3i}{1-2i} = \frac{4+3i}{1-2i} \cdot \frac{1+2i}{1+2i}$ *$1 + 2i$ is the conjugate of $1 - 2i$.*

$= \frac{4+8i+3i+6i^2}{1-4i^2}$

$= \frac{4+11i-6}{1+4}$

$= \frac{-2+11i}{5}$, or $-\frac{2}{5} + \frac{11}{5}i$

Every nonzero complex number has a multiplicative inverse, or reciprocal.

> **Multiplicative inverse, or reciprocal, of a complex number** If $a + bi$ is a nonzero complex number, then its multiplicative inverse is $\frac{1}{a+bi}$.

To express the reciprocal of a complex number in the form $a + bi$, follow the procedure for simplifying quotients.

EXAMPLE 5 **Express the reciprocal of $3 - 2i$ in the form $a + bi$.**

The reciprocal of $3 - 2i$ is $\frac{1}{3-2i}$.

$\frac{1}{3-2i} = \frac{1}{3-2i} \cdot \frac{3+2i}{3+2i}$ *$3 + 2i$ is the conjugate of $3 - 2i$.*

$= \frac{3+2i}{9+4}$ *$(a+bi)(a-bi) = a^2 + b^2$*

$= \frac{3+2i}{13}$, or $\frac{3}{13} + \frac{2}{13}i$

CLASS EXERCISES

Thinking Critically

1. Is the set of imaginary numbers closed under addition and multiplication? Give reasons for your answer.
2. Is the set of imaginary numbers closed under subtraction and division? Give reasons for your answer.

Give the conjugate of each complex number.

3. $3 + 7i$
4. $6 - 3i$
5. $4 - i$
6. $11 + 5i$

Express the reciprocal of each complex number in the form $a + bi$.

7. $2 + 4i$
8. $3 - 5i$
9. $-7 + 3i$
10. $0 - 4i$
11. $-3i$
12. $6 + 0i$
13. $\frac{1}{5 + i}$
14. $\frac{1}{3 - 2i}$

PRACTICE EXERCISES

Simplify.

1. $(3 + 5i)(1 + 2i)$
2. $(6 - i)(3 + 2i)$
3. $(7 + 3i)(3 + 4i)$
4. $(2 - 2i)(3 - 2i)$
5. $(5 - 3i)(6 + i)$
6. $(4 + 5i)(1 - 3i)$
7. $(7 - 6i)(3 + 2i)$
8. $(8 - i)(4 + 3i)$
9. $(1 - i)(1 + i)$
10. $(6 - 2i)(6 + 2i)$
11. $(3 + i)^2$
12. $(4 - 2i)^2$
13. $\frac{2 + i}{4 + i}$
14. $\frac{5 - 3i}{2 + 2i}$
15. $\frac{5 + 3i}{1 + 2i}$
16. $\frac{4 - 3i}{2 + 2i}$
17. $\frac{1}{4 + 2i}$
18. $\frac{1}{5 - 3i}$
19. $\frac{3 + 5i}{1 + 4i}$
20. $\frac{5 - 2i}{3 - 3i}$
21. $(3 + 2i) \div (3 - 2i)$
22. $(3 + 2i) \div (3 + 2i)$
23. $(3 + 4i) \div (3 + 2i)$
24. $(5 - 3i) \div (6 + 2i)$
25. $(\sqrt{3} + 2i)(\sqrt{3} - 4i)$
26. $(\sqrt{5} + 3i)(\sqrt{5} + 2i)$
27. $(\sqrt{8} + 2\sqrt{-1})(\sqrt{8} + 3\sqrt{-1})$
28. $(\sqrt{15} + \sqrt{-1})(2\sqrt{15} - \sqrt{-1})$
29. $(5 + 3i\sqrt{2})(3 + i\sqrt{2})$
30. $(6 - 2i\sqrt{2})(2 + i\sqrt{2})$
31. $(3\sqrt{2} + i) \div (4\sqrt{2} - i)$
32. $(12\sqrt{3} + 7i) \div (\sqrt{3} - 7i)$

33. $(0.2 + 6i) \div (0.5 - 2i)$

34. $(0.7 + 9i) \div (0.7 - 9i)$

35. $(4 + \sqrt{-4}) \div (3 + \sqrt{-9})$

36. $(3 - \sqrt{-16}) \div (2 + \sqrt{-25})$

37. $(1 + \sqrt{-18}) \div (2 - \sqrt{-8})$

38. $(3 + \sqrt{-27}) \div (1 + \sqrt{-75})$

39. Show that the reciprocal of a nonzero complex number $a + bi$ is $\frac{a - bi}{a^2 + b^2}$.

40. Show that the product of a complex number $a + bi$ and its conjugate is a real number.

41. $(2 + i)^3$

42. $(1 - 2i)^4$

43. $(2 + 3i)^2 \div (3 - 4i)$

44. $[(4 - i) \div (2 + i)]^2$

45. $(3 + \sqrt{-36}) \div (-\sqrt{-2})^5$

46. $(5 - \sqrt{-3}) \div (\sqrt{-1})^{15}$

47. Find real values for x and y such that $(x + yi)^2$ is a pure imaginary number.

48. If x is an integer and $(x + 3i)(x - 3i) = 34$, find x.

Applications

49. **Number Theory** The solutions of an equation are $3 + i$ and $3 - i$. Find the product of the solutions.

50. **Number Theory** One solution of an equation is $3 - \sqrt{-2}$. Find the cube of this number.

51. **Electricity** The total voltage of one circuit is $(9 + 3i)$ volts. The total voltage of a second circuit is $(5 - 2i)$ volts. Find the ratio of the total voltage of the first circuit to that of the second circuit.

TEST YOURSELF

1. Prove that $(4, 3)$, $(-6, 6)$, and $(-3, -4)$ are the vertices of an isosceles triangle. **8.6**

2. Prove that $(-a, 0)$, $(a, 0)$, $(a, 2a)$, and $(-a, 2a)$ are the vertices of a square.

Simplify. **8.7–8.9**

3. $\sqrt{-1} + \sqrt{-49}$

4. i^{28}

5. $|1 - 3i|$

6. $-5i - (3 - i)$

7. $\frac{5 + 3i}{2 - 2i}$

8. $\frac{1}{3 + i}$

9. $(5 + 3i)(2 - 2i)$

10. $(4 + i)(4 - i)$

11. $(3 - 4i)^2$

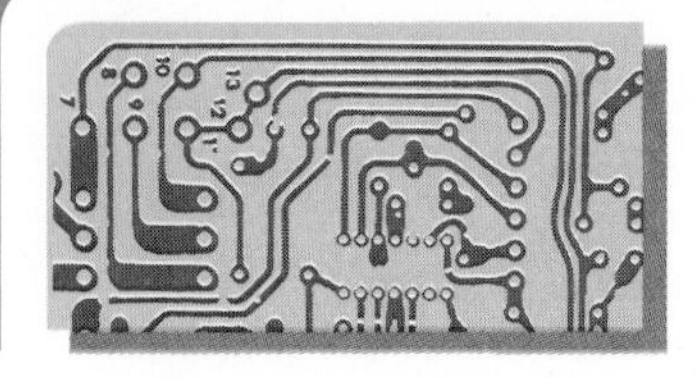

TECHNOLOGY: Fractals

We are surrounded by fractals! Fractal shapes occur in the curve of a mountain range, the jutting lines of a coastline or a landscape, the puffy shape of a cloud, and in many other natural objects. In the 1970s, Benoit B. Mandelbrot developed the concept of fractal to describe the shapes of natural objects. Fractals are geometric forms that are unbelievably complex yet can be created by the repetition of a simple geometric pattern or formula.

Fractal landscape

Real landscape

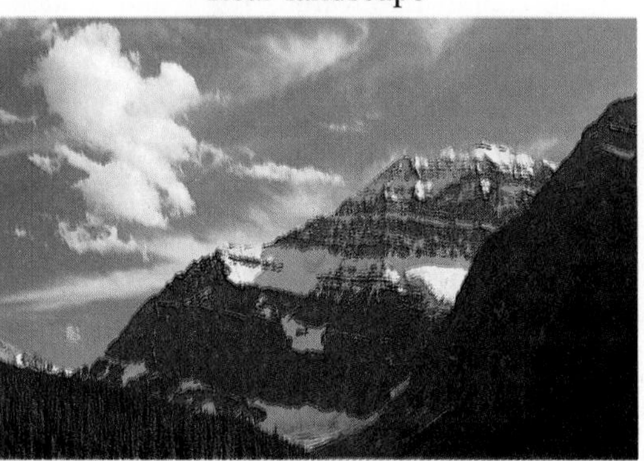

Mandelbrot defines a fractal curve as a curve that is not straight, yet has the property that its parts are small-scale replicas of the whole. A fractal curve has a "fractal dimension" greater than its topological dimension (the intuitive dimension you know from Euclidean geometry).

Topological Dimension	*Object*
0	point
1	line
2	square
3	cube

The topological dimension is always given by an integer. The fractal dimension, on the other hand, can be given by any real number. It is a precise gauge of how jagged a line is—how much it "wiggles about."

In 1904, H. von Koch created a fractal curve with a fractal dimension between 1 and 2. It is known as Koch's snowflake. For Koch's snowflake the topological dimension is 1 and the fractal dimension is about 1.26. Like any recursive fractal, it was constructed by a limit procedure in which a geometric shape is scaled and repeated. This produced the self-similarity property of fractal objects: the curve pattern repeats itself at any level of magnification.

The construction of Koch's snowflake illustrates the sheer simplicity of iteration. It is typical of fractals with dimension between 1 and 2.

Start with a line segment.

Trisect the line segment.

Construct $\overline{CE} \cong \overline{ED} \cong \overline{CD}$. Remove $\overline{CD}$.

Repeat these steps to the desired degree of complexity.

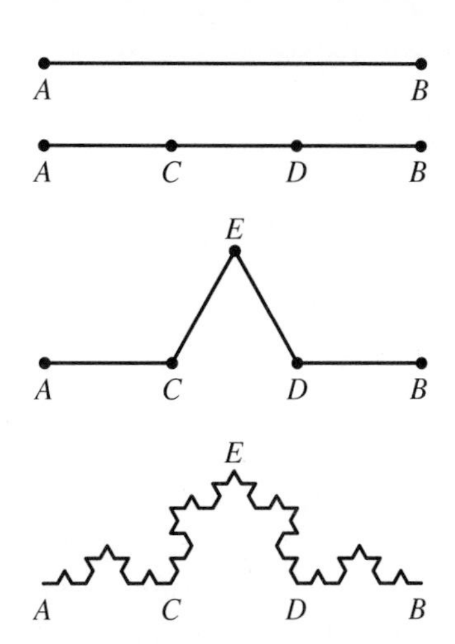

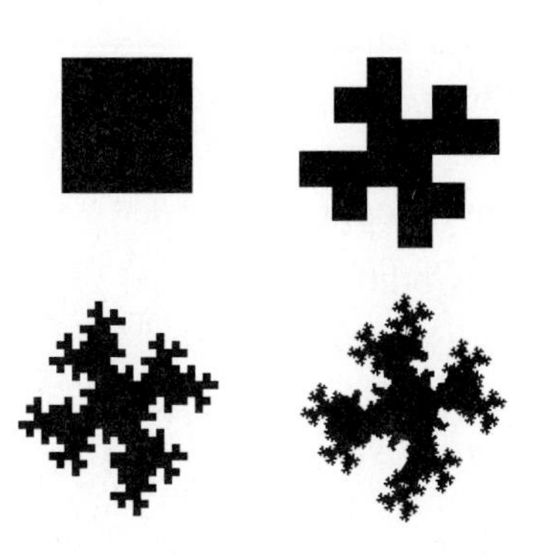

This Koch island is a fractal formed by replacing each segment of a figure with this pattern:

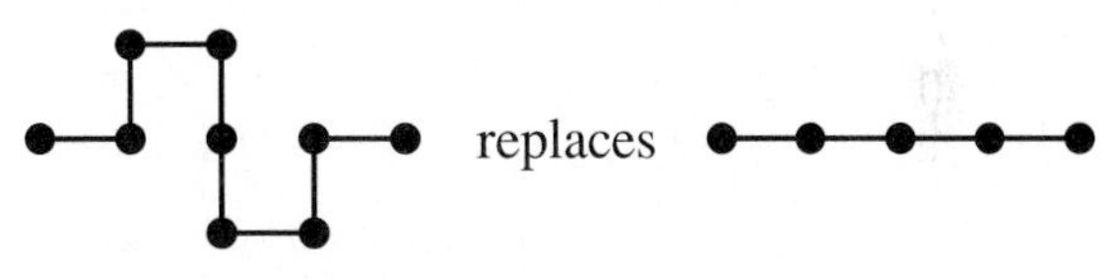

The original square and the first three applications of the process, or *iterations,* are shown here.

1. If one side of the original square has a length of 4, what is the area? the perimeter?
2. What are the area and perimeter of the first iteration? the second? the twentieth? Explain your reasoning.

Because fractals imitate nature so well, they have been applied to the study of chemistry, meteorology, physics, and statistics both to model and to predict other apparently random and chaotic events. Fractals help reveal order in chaos.

Investigate fractals and create your own fractal image from a geometric pattern.

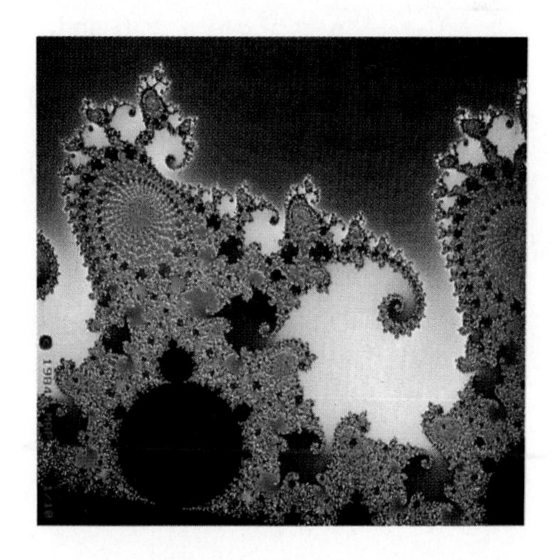

CHAPTER 8 SUMMARY AND REVIEW

Vocabulary

complex conjugates (398)
complex number (392)
conjugates (368)
distance formula (379)
imaginary number (388)
index (357)
like radicals (367)
midpoint formula (379)
*n*th root (357)
principal cube root (357)
principal *n*th root (357)
principal square root (357)
pure imaginary number (389)
radical (357)
radical equation (372)
radical expression (357)
radical sign (357)
radicand (357)
rationalize the denominator (364)
simplest radical form (364)

Principal *n*th Roots For any real number k and any integer n, $n > 1$ **8.1**

$\sqrt[n]{k} > 0$ if $k > 0$ $\qquad$ $\sqrt[n]{k^n} = k$ if n is odd
$\sqrt[n]{k} = 0$ if $k = 0$ $\qquad$ $\sqrt[n]{k^n} = |k|$ if n is even
$\sqrt[n]{k} < 0$ if $k < 0$ and n is odd
$\sqrt[n]{k}$ is not a real number if $k < 0$ and n is even

Simplify each radical expression if it is a real number.

1. $\sqrt{8}$ **2.** $\sqrt{0}$ **3.** $\sqrt[3]{-64}$ **4.** $\sqrt{-1}$

5. $\sqrt[3]{32}$ **6.** $\sqrt{4x^2y^2}$ **7.** $\sqrt[3]{27x^6y^9}$ **8.** $\sqrt[4]{-54x^5y^{18}}$

Operations with Radicals For any integer n, $n > 1$ **8.2–8.4**

$\sqrt[n]{a} \cdot \sqrt[n]{b} = \sqrt[n]{ab}$, $a \geq 0$ and $b \geq 0$

$\dfrac{\sqrt[n]{a}}{\sqrt[n]{b}} = \sqrt[n]{\dfrac{a}{b}}$, $a \geq 0$ and $b > 0$

$(\sqrt{a} + \sqrt{b})(\sqrt{a} - \sqrt{b}) = (\sqrt{a})^2 - (\sqrt{b})^2 = a - b$

Simplify.

9. $\sqrt{5x} \cdot \sqrt{5x}$ **10.** $\sqrt{3} \cdot \sqrt{12}$ **11.** $\dfrac{\sqrt{3}}{\sqrt{75}}$ **12.** $\dfrac{\sqrt{7x}}{\sqrt{98}}$

13. $3\sqrt{7} + 5\sqrt{7}$ **14.** $\sqrt{24} - \sqrt{54}$ **15.** $\sqrt{9y} - \sqrt{16y}$ **16.** $\sqrt{4x} + \sqrt{25x}$

17. $\dfrac{7}{1 + \sqrt{5}}$ **18.** $\dfrac{4 + \sqrt{3}}{2 + \sqrt{3}}$ **19.** $\dfrac{5 + \sqrt{8}}{3 - \sqrt{8}}$ **20.** $\dfrac{8 - \sqrt{7}}{5 + \sqrt{7}}$

Solve and check.

21. $5\sqrt{x} + 7 = 17$

22. $\sqrt{3x} = \sqrt{2x - 6}$

23. $-\sqrt{x^2 + 3} = x - 3$

24. $\sqrt{x + 1} = x + 1$

Distance and Midpoint Formulas For points $A(x_1, y_1)$ and $B(x_2, y_2)$, the distance AB and the midpoint of $\overline{AB}$ are given by these formulas: 8.5

Distance AB: $d = \sqrt{(x_2 - x_1)^2 + (y_2 - y_1)^2}$

Midpoint of $\overline{AB}$: $M\left(\frac{x_1 + x_2}{2}, \frac{y_1 + y_2}{2}\right)$

Find the distance between the points with the given coordinates. Then find the midpoint of the segment joining the two points.

25. $(3, 5), (4, 7)$

26. $(-2, 7), (3, -5)$

27. $(-1, -7), (0, 10)$

28. The endpoints of a diameter of a circle are (5, 3) and (9, 7). Find the coordinates of the center of the circle.

29. Find the perimeter of the triangle with vertices at $(3, 5)$, $(8, -7)$, and $(-1, 2)$.

30. Prove that $(c, -c)$, $(-c, c)$, $(c\sqrt{3}, c\sqrt{3})$, are vertices of an equilateral triangle. 8.6

31. Prove that the median to the hypotenuse of an isosceles right triangle forms two smaller isosceles right triangles with the sides of the original triangle.

Operations with Complex Numbers If $i = \sqrt{-1}$ and a, b, c, and d are real numbers, then 8.7–8.9

$$(a + bi) + (c + di) = (a + c) + (b + d)i$$
$$(a + bi) - (c + di) = (a + bi) + (-c - di) = (a - c) + (b - d)i$$
$$|a + bi| = \sqrt{a^2 + b^2}$$
$$(a + bi)(a - bi) = a^2 + b^2$$

Simplify.

32. i^3

33. i^4

34. i^{15}

35. i^{61}

36. $(2 + 3i) + (4 + 7i)$

37. $(5 + 6i) + (4 - 2i)$

38. $(6 + 5i) - (3 + 2i)$

39. $(2 - 7i) - (1 - 4i)$

40. $(3 + 2i)(5 - i)$

41. $(6 + 2i)(6 - 2i)$

42. $(2 + i) \div (3 - 4i)$

43. $(5 + 3i) \div (2 - 2i)$

CHAPTER 8 TEST

Simplify.

1. $\sqrt{81}$
2. $\sqrt[3]{-8x^3y^6}$
3. $\sqrt{24x^4y^{11}}$
4. $\sqrt{7x} \cdot \sqrt{14x^3}$
5. $\frac{\sqrt{15}}{\sqrt{3}}$
6. $\frac{\sqrt{2y}}{\sqrt{8}}$
7. $3\sqrt{28} + 2\sqrt{7}$
8. $\sqrt{54} - \sqrt{150}$
9. $\frac{4}{1 + \sqrt{3}}$
10. $(4 + 3i) \div (2 - 3i)$

Solve and check.

11. $2\sqrt{x} + 9 = 21$
12. $\sqrt{2x + 10} = x + 1$
13. Find the distance between the points (7, 2) and (10, 6).
14. Find the coordinates of the midpoint of the line segment with endpoints $(-5, 7)$ and $(5, 3)$.
15. Prove that $(0, 0)$, (a, b), $(a, b + c)$, and $(0, c)$ are the vertices of a parallelogram.

Simplify.

16. i^{33}
17. $(7 + 8i) + (1 + 2i)$
18. $(4 + 11i) - (9 - 2i)$
19. $(1 + 2i)(6 - 3i)$
20. $(1 + 6i) \div (3 + 4i)$
21. $|3 - 2i|$

Challenge

Use coordinate geometry to prove that in any triangle, the sum of the squares of the lengths of two sides is equal to twice the square of the length of the median to the third side, increased by twice the square of one-half the length of the third side. *Hint:* Let the coordinates of the vertices of the triangle be $(-a, 0)$, $(a, 0)$, and (b, c).

COLLEGE ENTRANCE EXAM REVIEW

Select the best choice for each question.

1. When $\frac{1 - 3i}{1 - i}$ is expressed in $a + bi$ form, then $a =$

A. 2 **B.** 1 **C.** $\frac{1}{2}$

D. -1 **E.** -2

2. Which of the following is NOT a factor of $x^{12} - 1$?

A. $x - 1$ **B.** $x + 1$ **C.** $x^2 + 1$

D. $x^4 + x^2 - 1$ **E.** $x^2 - x + 1$

3. If $\frac{x}{y} = \frac{2}{3}$ and $\frac{y}{z} = \frac{3}{4}$, then $\frac{x}{z} =$

A. $\frac{1}{2}$ **B.** $\frac{1}{2}y$ **C.** $\frac{8}{9}$

D. $\frac{8}{9}y$ **E.** 2

4. If M is the midpoint of $\overline{AB}$ and $M(-4.2, -1.9)$, $A(-3.4, 2.5)$ are given, then B is

A. $(-3.8, -0.6)$ **B.** $(-5, 6.9)$

C. $(-3.8, 0.3)$ **D.** $(-5, -6.3)$

E. $(1.3, -3.45)$

5. If the average of x, y, and 32 is 28, then the average of x and y would be

A. 24 **B.** 26 **C.** 28 **D.** 30

E. It cannot be determined from information given.

6. Express $(8 - 5i)^2$ in $a + bi$ form.

A. $39 + 80i$ **B.** $39 - 80i$

C. $69 + 80i$ **D.** $69 - 80i$

E. $-9 - 40i$

7. Simplify: $\sqrt{27} + \sqrt{48}$

A. $7\sqrt{3}$ **B.** $7\sqrt{6}$ **C.** $7\sqrt{2}$

D. $5\sqrt{3}$ **E.** $5\sqrt{6}$

8. If $z = a + bi$ and $\overline{z}$ is its conjugate, $a - bi$, where a and b are real numbers, which of the following is(are) true?

I. $z + \overline{z}$ is pure imaginary

II. $z \cdot \overline{z}$ is a real number

III. $(z)^2$ is never real

A. I only **B.** II only

C. III only **D.** I and II only

E. II and III only

9. The symbol $n!$ means the value of $n(n - 1)(n - 2) \ldots 3 \cdot 2 \cdot 1$.

For example,

$$5! = 5 \cdot 4 \cdot 3 \cdot 2 \cdot 1 = 120$$

Find the value of $6! - 4!$

A. 744 **B.** 720 **C.** 706

D. 696 **E.** 684

10. $\sqrt[4]{27x^2 \cdot \sqrt{9x^4}}$ simplifies to

A. $3\sqrt[4]{x}$ **B.** $3\sqrt{x}$ **C.** $\sqrt[4]{3x}$

D. $\sqrt{3x}$ **E.** $3x$

11. The fraction $\frac{12}{3 - \sqrt{5}}$ equals

A. $\frac{3 + \sqrt{5}}{4}$ **B.** $\frac{3 - \sqrt{5}}{4}$

C. $3 + \sqrt{5}$ **D.** $9 + 3\sqrt{5}$

E. $9 - 3\sqrt{5}$

CUMULATIVE REVIEW (CHAPTERS 1–8)

Solve and check.

1. $|2w - 3| = 5$

2. $|j| - 7 = 15$

3. $-3|8 - 5m| = 9$

Graph each equation in the coordinate plane.

4. $3y - 5x = 30$

5. $3 + x = 0$

6. $4x + y = 0$

7. $2x = 6y - 8$

8. $y - 2 = 0$

9. $2y - 3x = 0$

Multiply.

10. $-4y^2(5y^3 - 6y^2 + 8y - 1)$

11. $(3a - 5b)(3a + 5b)$

12. $(2x - 11y)^2$

13. $(2x - 3)(3x^2 + 4x - 9)$

Express in decimal form.

14. 7.2×10^{-2}

15. 1.003×10^{-6}

16. 6.250×10^3

$f(x) = -\frac{1}{x} + 1$, **and** $g(x) = \frac{1}{2}(x - 1)^2$. **Evaluate for the given value of** x.

17. $f(-2)$

18. $g(-4)$

19. $g\left(f\left(\frac{1}{3}\right)\right)$

Solve each system.

20. $\begin{cases} 3x + 7y = 4 \\ 5x - 2y = -7 \end{cases}$

21. $\begin{cases} x + y = 0 \\ 3x - 5y = 1 \end{cases}$

22. $\begin{cases} x = \frac{2}{5}y + 2 \\ 5x - 2y = 10 \end{cases}$

23. A 2-digit number is seven times the sum of its digits. If the digits are reversed, the new number is 36 less than the original. Find the original number.

Factor completely.

24. $8x^3 + 1$

25. $4a^2 - 16$

26. $-2x^2 + x + 3$

27. $4d^3 + 8d^2 + 4d$

28. $ax + a + bx + b$

29. $3c^2 + 2cd - d^2$

30. $6x^2 + 7x + 1$

31. $t^4 - v^4$

32. $15y^2 - 16y - 15$

Solve and check.

33. $\sqrt{y^2 - 12} = y - 6$

34. $\sqrt{x - 2} = x - 2$

Using the given matrices, perform the indicated operations, if possible.

$$C = \begin{bmatrix} 0 & 5 \\ -3 & 2 \\ 4 & 1 \end{bmatrix} \qquad D = \begin{bmatrix} -3 & 4 \\ 7 & 2 \end{bmatrix} \qquad E = \begin{bmatrix} 2 & 0 & -2 \\ -1 & 4 & -2 \\ 3 & 0 & 1 \end{bmatrix}$$

35. CD **36.** DC **37.** E^2 **38.** $-2C$

Solve each matrix equation.

39. $X - \begin{bmatrix} -3 & 1 \\ -2 & 4 \end{bmatrix} = \begin{bmatrix} 1 & 3 \\ 6 & 0 \end{bmatrix}$

40. $\begin{bmatrix} 3 & 0 \\ 7 & -2 \end{bmatrix} - 2X = \begin{bmatrix} -3 & -1 \\ 0 & 2 \end{bmatrix}$

Simplify, writing with positive exponents. Evaluate if possible.

41. $-4a^0 + (5b)^0$

42. $\left(\frac{2}{3}\right)^{-4}$

43. $(3c^{-2}b)(-3^2c^3b^{-4})$

44. $\frac{2x^{-1}y^3}{8x^{-2}y^4}$

45. $\frac{(2^{-4})^{-3}}{2^{-5}}$

46. $\left(\frac{2^{-3}x^2y^{-2}}{2^{-4}}\right)^{-2}$

47. Find a number if twice the number is 40 more than $\frac{2}{5}$ of the number.

Simplify.

48. $\frac{-3x^3y^2}{-27x^4y}$

49. $\frac{5y - 10}{2 - y}$

50. $\frac{y^2 - y - 2}{y^2 - 1}$

51. $\frac{6x^2}{6x + 4x^2 - x^3}$

52. $\frac{6x^2 + 7x - 3}{9x^2 - 6x + 1}$

53. $\frac{1 - x^3}{1 - x}$

Perform the indicated operations, and simplify.

54. $\frac{x^2 - 4x - 12}{3x^2} \div \frac{x^2 - x - 6}{12x}$

55. $\frac{y^2 + 6y + 9}{3y^2 - 27} \cdot \frac{4y^2 - 36}{5y + 15}$

56. $\frac{3}{2x} - \frac{4 - x}{2x}$

57. $\frac{2x}{3x - 6} + \frac{3x}{4x - 8}$

58. $\frac{4}{y^2 - 2y - 8} + \frac{2y}{3y + 6}$

59. $\frac{1}{3y^2 - 48} - \frac{5y}{3y + 12}$

Simplify.

60. $\sqrt{0.04x^7y^4z^3}$

61. $\sqrt[3]{-16b^7}$

62. $\sqrt[4]{32a^4b^{11}}$

63. $\sqrt[3]{2x^2}\,\sqrt[3]{4x}$

64. $\frac{\sqrt{6m}}{\sqrt{2mn}}$

65. $\frac{-\sqrt{2}}{1 - 2\sqrt{2}}$

66. $\sqrt{-8} \cdot \sqrt{-5}$

67. $-16i^7$

68. $|12 - 9i|$

Write, in standard form, the equation of each line described.

69. Has a slope of $-\frac{1}{3}$, and y-intercept of 11

70. Passes through (8, 3) and (10, −5)

71. Parallel to the graph of $5x - 3y = 7$ and with a y-intercept of 0

For each function tell whether y varies directly as x. If it does, find the constant of variation and write the equation.

72.

x	y
3	7.5
5	12.5
8	20

73.

x	y
2	2.6
11	14.3
15	19.5

74.

x	y
3	13
5	21
8	33

Solve each system.

75. $\begin{cases} 2x - 3y + 2z = 9 \\ -x + 4y + 3z = 1 \\ 3x - 2y - 3z = -1 \end{cases}$

76. $\begin{cases} 2x + 3y = -6 \\ 3x - 9z = 0 \\ 5y - 2z = 2 \end{cases}$

Solve and check.

77. $b^2 - 2b - 3 = 0$

78. $3w^2 = 147$

79. $8y^3 + 2y^2 - 15y = 0$

80. $t^4 - 20t^2 + 64 = 0$

81. $y^3 + 5y^2 - y - 5 = 0$

82. $(a + 2)^2 - (a + 2) = 0$

83. Find three consecutive integers if the product of the second and third is 2 more than 10 times the first.

Solve each inequality and graph its solution set.

84. $-3x + 4 < -8$

85. $5(4 - 3y) \geq 5y$

86. $-2(n - 11) \leq -2(n + 7)$

87. $-1 < 3 - 4x < 15$

88. $|r - 7| < 15$

89. $|2t + 1| \geq 5$

Solve and check.

90. $3 - \frac{2}{x} = \frac{4}{3x}$

91. $\frac{3}{y + 3} - \frac{5}{y^2 - 9} = 4$

92. If he increases his speed by 10 km/h, Pierre can cover a 42-km distance on his motor bike in $1\frac{1}{4}$ h less time. Find his two speeds.

Quadratic Functions

A jump start?

Developing Mathematical Power

Five hundred years ago, Leonardo da Vinci designed a parachute which he called a *tent roof,* but it was not until 1783 that the first successful parachute jump was made. Today the United States Parachute Association (USPA) licenses parachutists. USPA parachutists must follow certain regulations. The minimum altitude for opening a parachute is 3000 feet.

Project

Prepare a presentation about the USPA qualifications. For each of the four license classes, determine the minimum height at which a parachutist should jump from a plane, the distance a parachutist will fall during each free-fall period, and the parachutist's velocity at the time the parachute opens.

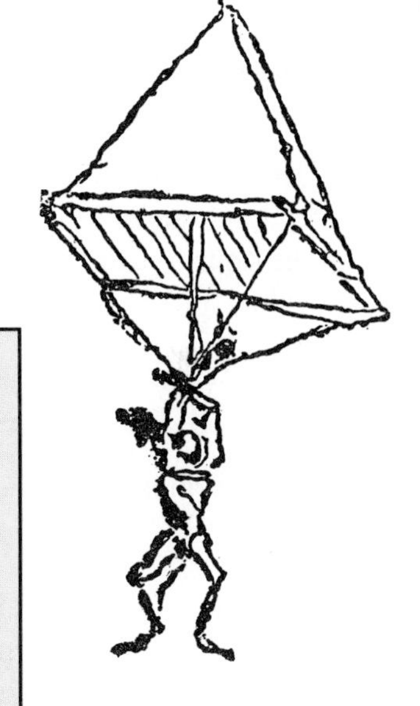

Leonardo da Vinci's sketch of a parachute, 1495.

USPA Qualifying Regulations

License Class	Number of Jumps	Number of Controlled Delays During Free Fall Before Opening Parachute
A (Basic)	20	3 of at least 40 s
B (Intermediate)	50	3 of at least 45 s
C (Advanced)	100	10 of at least 45 s
D (Master)	200	10 of at least 60 s

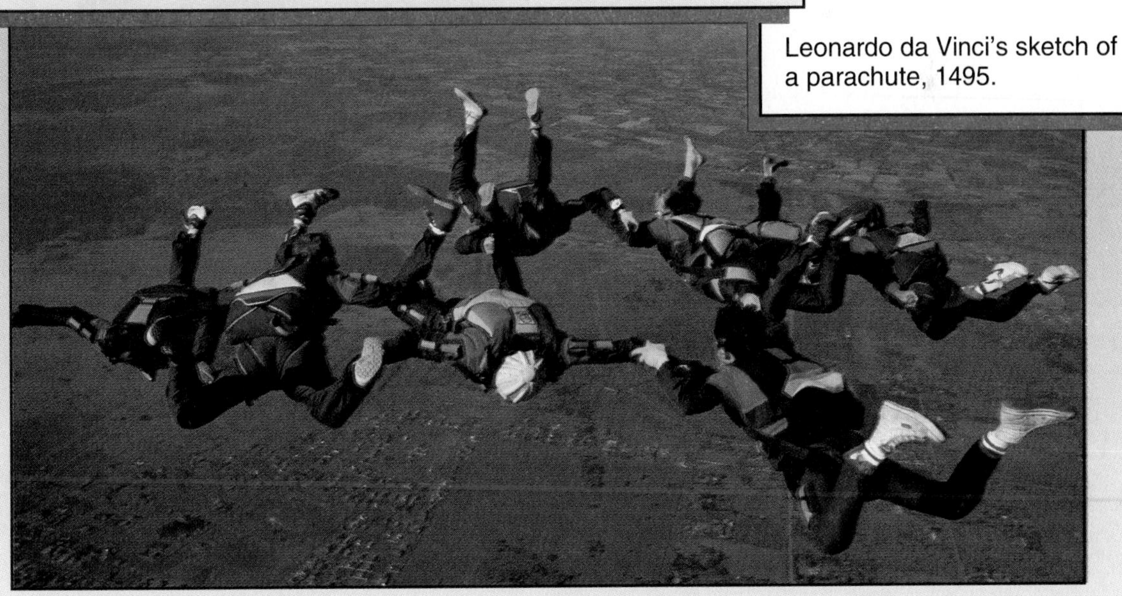

9.1

Graphing Quadratic Functions

Objectives: To graph quadratic functions of the form $y = ax^2 + c$
To find the axis of symmetry and the vertex of a parabola and tell whether the vertex is a maximum or a minimum point

The graph of a quadratic function is a **parabola** *(pa-rab'o-la)*. The shape of an antenna used for television reception from satellites is based on a parabola. Such an antenna is called a *parabolic reflector*.

To find points on the graph of a function, choose values of x and substitute them in the equation to find the corresponding values of y.

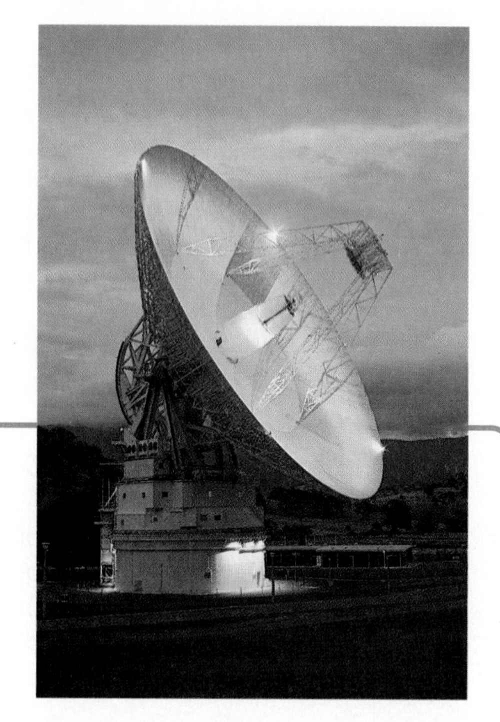

Capsule Review

For the linear function

$$y = 3x - 1$$

when $x = -3$, $y = -10$.

$$y = 3x - 1$$
$$y = 3(-3) - 1$$
$$y = -10$$

Therefore, the point with coordinates $(-3, -10)$ is on the graph of $y = 3x - 1$.

Find the y-coordinate of each point on the graph of the given equation.

1. $y = 5x - 4$; $(1, ?)$ **2.** $y = -2x - 1$; $(0, ?)$ **3.** $y = -3x + 9$; $(-2, ?)$

4. $y = \frac{1}{2}x - 3$; $(-6, ?)$ **5.** $y = \frac{2}{3}x + 5$; $(3, ?)$ **6.** $y = -\frac{3}{4}x - 3$; $(-4, ?)$

A **quadratic function** is defined by an equation of the form

$$y = ax^2 + bx + c$$

where a, b, and c are real numbers and $a \neq 0$. The graph of the function $y = x^2$ is the basic parabola.

EXAMPLE 1 **Graph the function:** **a.** $y = x^2$ **b.** $y = -x^2$

Make a table of ordered pairs. Then graph the points and connect them with a smooth curve.

a.

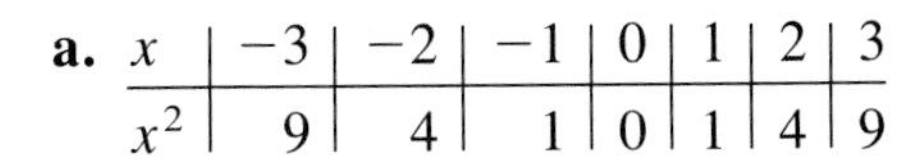

x	-3	-2	-1	0	1	2	3
x^2	9	4	1	0	1	4	9

b.

x	-3	-2	-1	0	1	2	3
$-x^2$	-9	-4	-1	0	-1	-4	-9

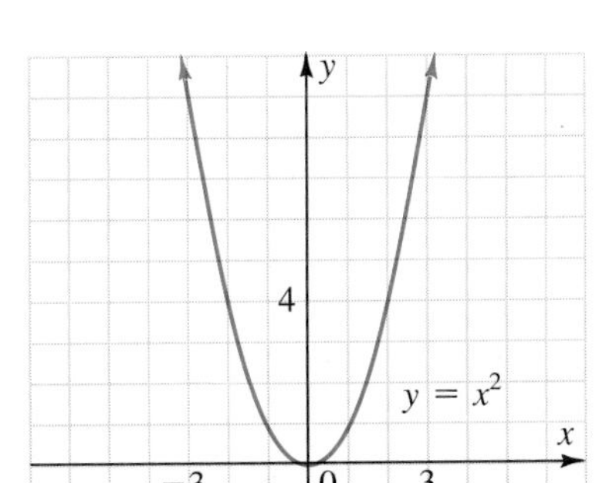

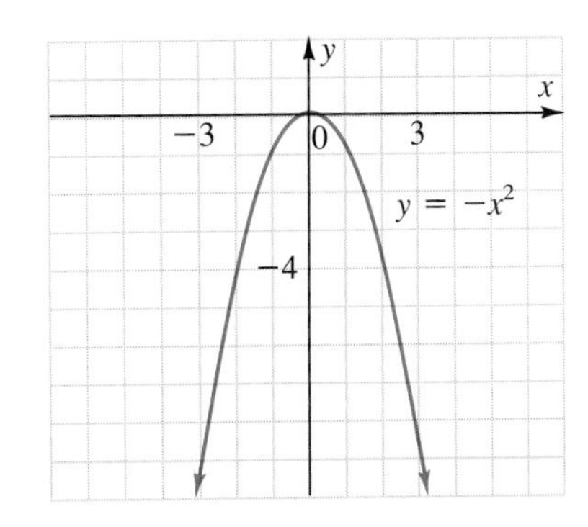

The two graphs in Example 1 are *congruent*. That is, they have the same size and shape. Each graph satisfies the vertical line test, so both $y = x^2$ and $y = -x^2$ are functions.

The y-axis is the **axis of symmetry,** or **axis,** of each parabola. This means that if each parabola were folded along the y-axis, its two parts would coincide. For each point on the graph, there is a corresponding point, or *mirror image,* on the other side of its axis. For example, on the graph of $y = x^2$, the point (2, 4) is the image of (−2, 4).

The **vertex** of a parabola is the point where the graph crosses its axis. Notice the following facts about each function.

$\mathbf{y = x^2}$	$\mathbf{y = -x^2}$
The graph opens upward. The axis of symmetry is $x = 0$. The vertex is at (0, 0). The vertex is the **minimum point,** or the point where y has the least value.	The graph opens downward. The axis of symmetry is $x = 0$. The vertex is at (0, 0). The vertex is the **maximum point,** or the point where y has the greatest value.

The graphs of linear functions are affected by changes in the coefficients in their equations. The graphs of quadratic functions are also affected by such changes. To determine the effect of the coefficient a on the graph of $y = ax^2$, consider different values of a.

EXAMPLE 2 **Graph $y = 2x^2$, $y = \frac{1}{2}x^2$, $y = -2x^2$, and $y = -\frac{1}{2}x^2$ on the same coordinate plane.**

x	-2	-1	0	1	2
$2x^2$	8	2	0	2	8
$\frac{1}{2}x^2$	2	$\frac{1}{2}$	0	$\frac{1}{2}$	2
$-2x^2$	-8	-2	0	-2	-8
$-\frac{1}{2}x^2$	-2	$-\frac{1}{2}$	0	$-\frac{1}{2}$	-2

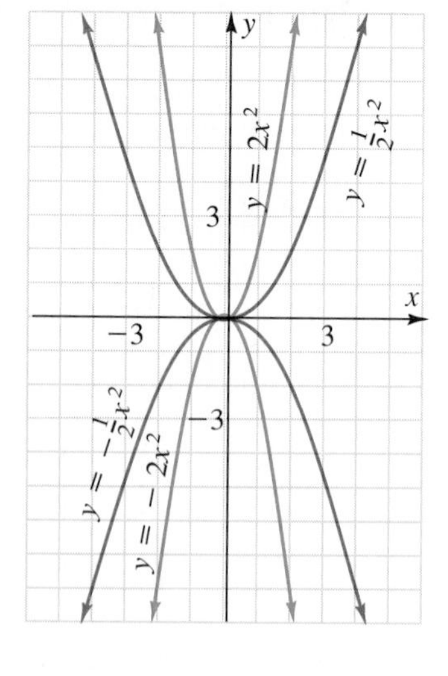

In Example 2, the axis of symmetry for each parabola is $x = 0$, and each vertex is at $(0, 0)$. Notice the effect of the coefficient a on the graph of $y = ax^2$.

For $a > 0$

The graph opens upward.
The vertex is a minimum point.

For $a < 0$

The graph opens downward.
The vertex is a maximum point.

The width of the graph is related to the absolute value of a.

As $|a|$ increases, the graph becomes narrower.
As $|a|$ decreases, the graph becomes wider.

To determine the effect of the constant c on the graph of $y = ax^2 + c$, consider different values of c. You can demonstrate the effect easily and clearly by using a graphing utility.

EXAMPLE 3 **Graph $y = x^2$, $y = x^2 + 3$ and $y = x^2 - 2$ on the same coordinate plane. Compare the graphs.**

All of the parabolas are congruent; they all open upward; and the axis for each is the line $x = 0$. However, the graphs have different positions on the coordinate plane. The vertex of the graph of $y = x^2 + 3$ is three units above that of $y = x^2$, and the vertex of the graph of $y = x^2 - 2$ is two units below that of $y = x^2$. The constant c has the effect of sliding the graph of $y = ax^2 + c$ up or down the y-axis.

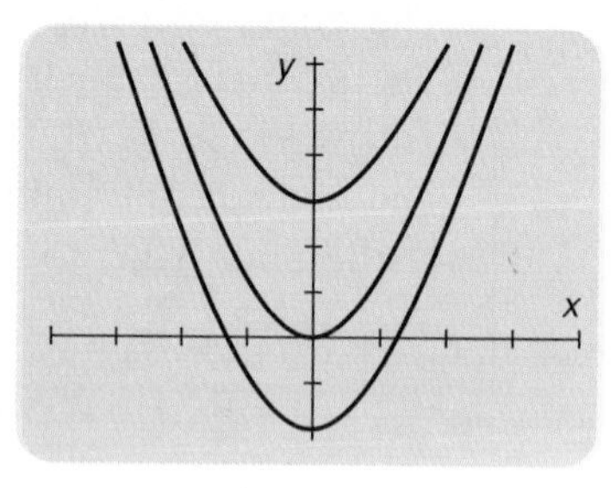

x scl: 1 y scl: 1

If c is positive, the vertex is c units above the point $(0, 0)$.
If c is negative, the vertex is c units below the point $(0, 0)$.

EXAMPLE 4 **Using a graphing utility, graph $y = x^2$, $y = (x - 2)^2$, and $y = (x + 4)^2$ on the same coordinate plane.**

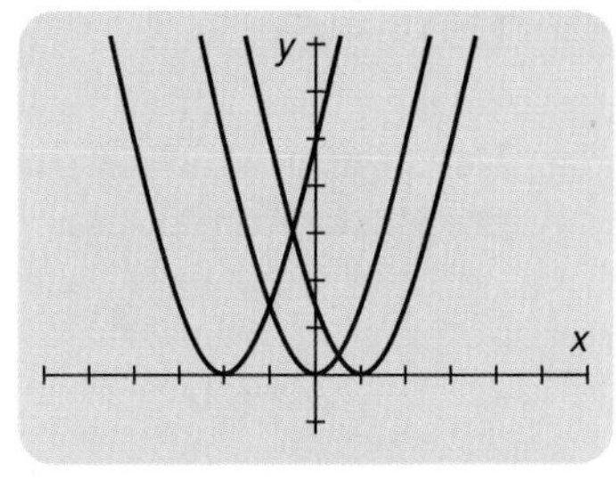

x scl: 2 y scl: 4

The second and third graphs are congruent to the basic parabola, $y = x^2$, but the vertex and axis of symmetry have shifted to the left or right along the x-axis.

It is evident from the graph that if a quadratic function is written in the form $y = (x - h)^2$, the constant h has the effect of sliding the graph of $y = x^2$ to the left or right, placing the axis of symmetry at $x = h$.

CLASS EXERCISES

Graph each quadratic function. Indicate whether the parabola opens up or down. Find the equation of the axis of symmetry and the coordinates of the vertex. Also, tell whether the vertex is a maximum or a minimum point.

1. $y = 3x^2$
2. $y = -\frac{1}{3}x^2$
3. $y = -x^2 + 1$
4. $y = \frac{3}{4}x^2 - 5$

For each quadratic function, indicate whether the parabola opens up or down; and find the equation of the axis of symmetry, and the coordinates of the vertex. Do not graph the function.

5. $y = -4x^2$
6. $y = \frac{1}{3}x^2$
7. $y = -\frac{1}{2}x^2 + 4$
8. $y = -\frac{2}{3}x^2 + 2$
9. $y = 6x^2$
10. $y = -5x^2$
11. $y = x^2 - 3$
12. $y = x^2 - 2$

PRACTICE EXERCISES

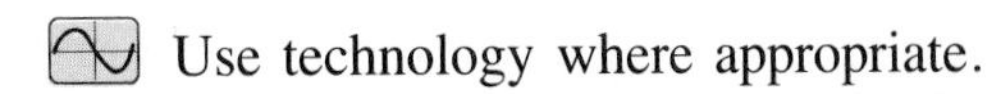
Use technology where appropriate.

Graph each quadratic function. Indicate whether the parabola opens up or down. Find the equation of the axis of symmetry and the coordinates of the vertex. Also, tell whether the vertex is a maximum or a minimum point.

1. $y = 4x^2$
2. $y = -4x^2$
3. $y = \frac{1}{4}x^2$
4. $y = -\frac{1}{4}x^2$
5. $y = x^2 + 2$
6. $y = -x^2 - 1$
7. $y = 4x^2 - 5$
8. $y = -3x^2 + 2$

For each quadratic function, indicate whether the parabola opens up or down; and find the equation of the axis of symmetry, and the coordinates of the vertex. Do not graph the function.

9. $y = 8x^2$
10. $y = -6x^2$
11. $y = \frac{3}{8}x^2$
12. $y = -\frac{1}{5}x^2$
13. $y = -4 - 4x^2$
14. $y = (x + 5)^2$
15. $y = (x - 3)^2$
16. $y = -4x^2 - 1$

17. $y = 5x^2 - \frac{1}{2}$ **18.** $y = \frac{1}{2}x^2 - 5$ **19.** $y = 5 + \frac{1}{2}x^2$ **20.** $y = \frac{1}{2} - 5x^2$

Determine which of the three functions has the narrowest graph. Do not graph the functions.

21. $y = x^2, y = 3x^2, y = \frac{1}{3}x^2$

22. $y = -x^2, y = -5x^2, y = -\frac{1}{5}x^2$

23. $y = x^2, y = 4x^2, y = -6x^2$

24. $y = -\frac{3}{4}x^2, y = \frac{1}{2}x^2, y = -\frac{3}{8}x^2$

Determine the value of a if the graph of $y = ax^2$ contains the given point.

25. $(-1, 8)$ **26.** $(-2, -2)$ **27.** $(3, 6)$ **28.** $\left(\frac{1}{2}, -4\right)$

Determine the value of c if the graph of the given equation contains the given point.

29. $y = 2x^2 + c; \left(-\frac{3}{4}, -\frac{1}{4}\right)$

30. $y = -\frac{3}{4}x^2 + c; \left(3, -\frac{1}{2}\right)$

31. Describe the difference between the graphs of $y = (x + 6)^2$ and $y = (x - 6)^2 - 7$.

32. Thinking Critically What values of h and k in the equation $y = (x - h)^2 + k$ would place the vertex at $(2, -3)$?

Applications

33. Business The revenue equation of a company, in terms of the price of their product, is $R = -15p^2 + 300p + 12000$. Graph the revenue equation and find the price that will yield the maximum profit.

34. Physics The equation for the motion of a projectile fired straight up at an initial velocity of 64 ft/s is $h = 64t - 16t^2$, where h is the height in feet and t is the time in seconds. Graph the equation and find when the projectile will reach its highest point, and how high it will be.

Mixed Review

Multiply.

35. $-2x^2(5x^3 - 2x^2 + x - 3)$

36. $(2x - 4y)(3x - 3y)$

37. $(x + 3)^2$

38. $(2x^3 + 4x - 1)^2$

39. $(x + 1)(x + 2)(x^2 - 3)$

40. $(x - 1)^3$

Simplify.

41. $\dfrac{5x^2 - 20}{3x^2 + 12x + 12}$ **42.** $\dfrac{(x^2 - 4)(x^2 - 4x - 5)}{(x^2 + 3x + 2)(x^2 - 7x + 10)}$ **43.** $\dfrac{2b - bx}{bx^2 - 2bx}$

44. Find the distance between $(-3, 5)$ and $(2, -7)$.

9.2

The Function $y = ax^2 + bx + c$

Objective: To sketch the graph of a quadratic function using the axis of symmetry, the vertex, and the intercepts

A baseball follows a parabolic path. It will approach the fielder at approximately the same angle at which it rose when it was hit.

To graph a line, it is often useful to locate the y-intercept. This is also true when a parabola is graphed.

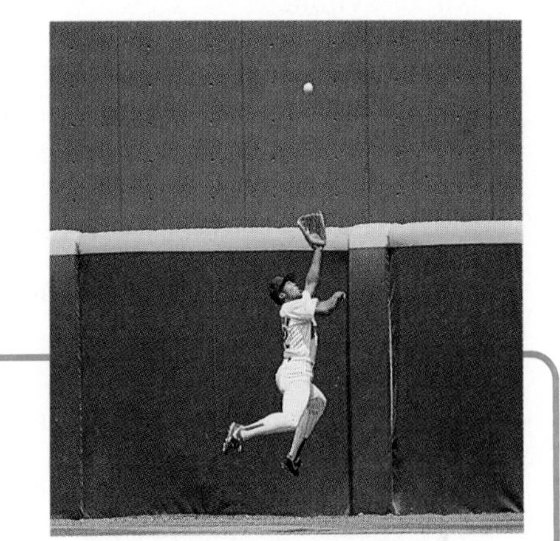

Capsule Review

To determine the y-intercept of the function $y = 2x - 3$, let $x = 0$ and solve for y.

$y = 2(0) - 3 = -3$ The y-intercept is -3.

Determine the y-intercept for each function.

1. $y = -x + 2$

2. $3y = 2x - 5$

3. $y + 7 = 4x$

4. $\frac{1}{2}y - 5x = 3$

5. $2y - 9 = -6x$

6. $\frac{2}{3}y - 3x + 8 = 0$

The graph of the quadratic function $y = ax^2 + bx + c$ is a parabola, and the sign of a determines the direction in which the parabola opens, up or down. It opens upward when a is positive and downward when a is negative.

Any x-intercept of the parabola occurs when $y = 0$. That is, an x-intercept is a solution to the equation $ax^2 + bx + c = 0$. An x-intercept is also called a *zero* of the function.

The y-intercept occurs when $x = 0$. Since $y = a(0)^2 + b(0) + c = c$, the y-intercept of a parabola is the constant c.

The equation of the axis of symmetry of a parabola can be expressed in terms of the two coefficients a and b. To illustrate, consider the function $y = ax^2 + bx$. Its graph crosses the x-axis when $y = 0$, or $ax^2 + bx = 0$.

$$ax^2 + bx = 0$$
$$x(ax + b) = 0$$
$$x = 0 \quad \text{or} \quad x = -\frac{b}{a}$$

The x-intercepts are 0 and $-\frac{b}{a}$. The axis of symmetry passes through the point halfway between (0, 0) and $\left(-\frac{b}{a}, 0\right)$. At this point, the x-coordinate is the average of 0 and $-\frac{b}{a}$, which is $-\frac{b}{2a}$, so the equation of the axis of symmetry is $x = -\frac{b}{2a}$. This is, in fact, the equation of the axis of any quadratic function $y = ax^2 + bx + c$. A more general case will be considered in Exercise 31 on page 440.

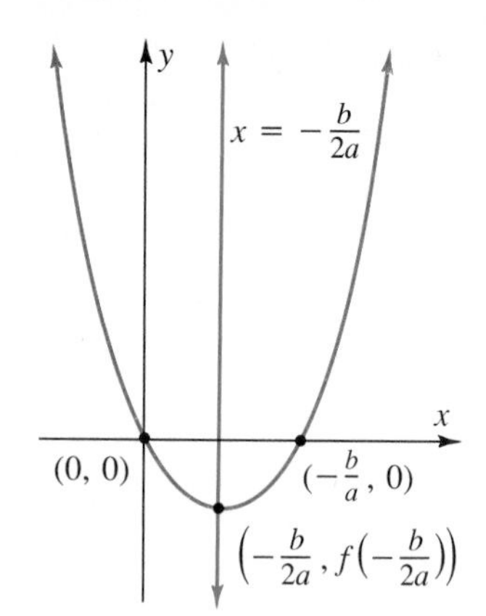

The vertex of a parabola lies on the axis of symmetry, so its x-coordinate is $-\frac{b}{2a}$ and its y-coordinate is the value of y when $x = -\frac{b}{2a}$. That is, the vertex is the point $\left(-\frac{b}{2a}, f\left(-\frac{b}{2a}\right)\right)$.

> To sketch the graph of a quadratic function $y = ax^2 + bx + c$
>
> **a.** Determine whether the parabola opens upward or downward. If $a > 0$, it opens upward. If $a < 0$, it opens downward.
>
> **b.** Graph the axis of symmetry, $x = -\frac{b}{2a}$.
>
> **c.** Plot the vertex, $\left(-\frac{b}{2a}, f\left(-\frac{b}{2a}\right)\right)$.
>
> **d.** Determine any x-intercepts and plot the corresponding points. An x-intercept is a solution to the equation $ax^2 + bx + c = 0$.
>
> **e.** Determine the y-intercept, c, and plot the corresponding point. Then use symmetry to plot the image of the point $(0, c)$.
>
> **f.** Connect the points with a smooth curve.

EXAMPLE 1 **Sketch the graph of $y = x^2 - 2x - 3$.**

a. $a > 0$ — *The graph opens upward.*

b. $-\frac{b}{2a} = \frac{-(-2)}{2(1)} = 1$ — *axis: $x = 1$*

c. $f(1) = 1^2 - 2(1) - 3 = -4$ — *vertex: $(1, -4)$*

d. $x^2 - 2x - 3 = 0$
$(x + 1)(x - 3) = 0$ — *x-intercepts: -1 and 3*

e. $c = -3$ — *y-intercept: -3*

$(0, -3)$ is 1 unit to the left of the axis of symmetry. Its image $(2, -3)$ is 1 unit to the right.

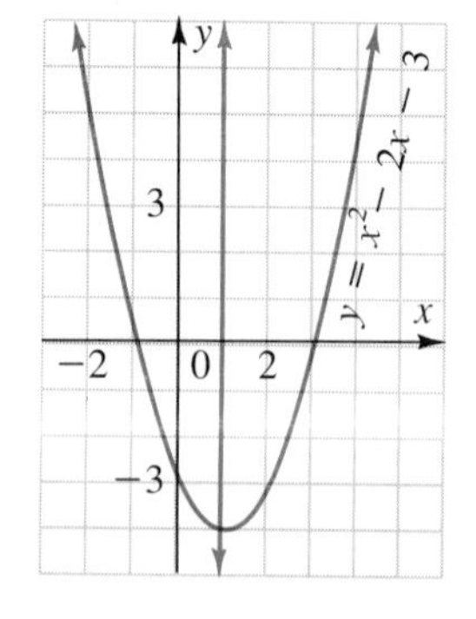

EXAMPLE 2 **Sketch the graph of the function $y = -2x^2 + 2x$. Check your work with a graphing utility.**

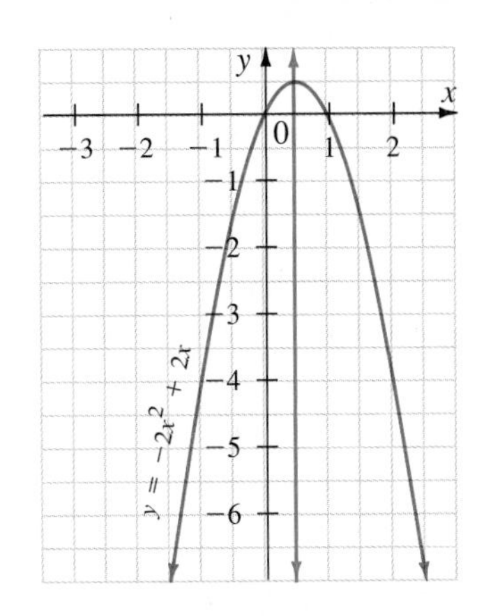

a. $a < 0$ *The graph opens downward.*

b. $-\frac{b}{2a} = -\frac{2}{2(-2)} = \frac{1}{2}$ *axis:* $x = \frac{1}{2}$

c. $f\left(\frac{1}{2}\right) = -2\left(\frac{1}{2}\right)^2 + 2\left(\frac{1}{2}\right) = \frac{1}{2}$ *vertex:* $\left(\frac{1}{2}, \frac{1}{2}\right)$

d. $-2x^2 + 2x = 0$
$-2x(x - 1) = 0$ *x-intercepts: 0 and 1*

e. $c = 0$ *y-intercept: 0*

Since (0, 0) and its image point (1, 0) have already been located, plot two other points.

For $x = -1$: $y = -2(-1)^2 + 2(-1) = -4$

The point $(-1, -4)$ is $1\frac{1}{2}$ units to the left of the axis. Its image point is $(2, -4)$, which is $1\frac{1}{2}$ units to the right of the axis.

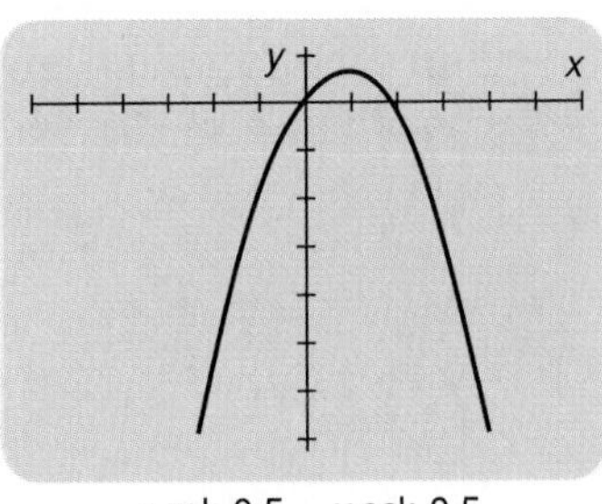

x scl: 0.5 y scl: 0.5

The calculator graph verifies the sketch.

For a parabola with the equation $y = ax^2 + bx + c$, the y-coordinate of the vertex is the maximum or minimum value of the function. You can solve many problems involving maximum or minimum values without graphing.

EXAMPLE 3 **Find two numbers whose sum is 18 and whose product is a maximum.**

Let x represent one number.
Then $18 - x$ represents the other number.
Let y represent the product of the two numbers.

$$y = x(18 - x) \qquad \textit{Write an equation.}$$
$$y = 18x - x^2$$
$$y = -x^2 + 18x$$

This is a quadratic function with $a = -1$ and $b = 18$. Since $a < 0$, the parabola opens downward and the vertex is the maximum point. This point occurs when

$$x = -\frac{b}{2a}$$
$$= -\frac{18}{2(-1)} = 9$$

When $x = 9$, then $18 - x$ is also equal to 9. The numbers are 9 and 9.

CLASS EXERCISES

For each function, give the equation of the axis of symmetry, the coordinates of the vertex, and the x- and y-intercepts.

1. $y = x^2 + 6x$
2. $y = -2x^2 - 8x$
3. $y = x^2 - 6x + 9$
4. $y = x^2 - 4x$
5. $y = -3x^2 + 10x$
6. $y = -4x^2 - 4x + 3$

Sketch the graph of each function.

7. $y = x^2 + 6x$
8. $y = -2x^2 - 8x$
9. $y = x^2 - 6x + 9$

PRACTICE EXERCISES

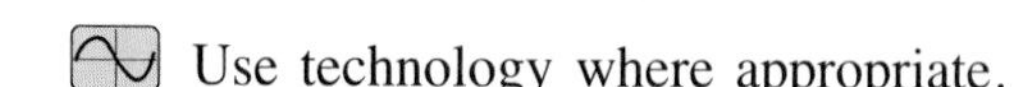

For each function, give the equation of the axis of symmetry, the coordinates of the vertex, and the x- and y-intercepts.

1. $y = -x^2 + 2x$
2. $y = x^2 - 3x$
3. $y = -2x^2 - 4x$
4. $y = 2x^2 + 4x$
5. $y = x^2 + 6x + 9$
6. $y = x^2 + 2x - 8$
7. $y = x^2 + 8x + 16$
8. $y = x^2 + 3x - 4$
9. $y = 2x^2 - 5x + 2$
10. $y = 3x^2 - 2x - 1$
11. $y = \frac{1}{2}x^2 + 4x$
12. $y = \frac{1}{2}x^2 - 6x$

Sketch the graph of each function.

13. $y = -x^2 + 2x$
14. $y = x^2 - 3x$
15. $y = -2x^2 - 4x$
16. $y = 2x^2 + 4x$
17. $y = x^2 + 6x + 9$
18. $y = x^2 + 2x - 8$
19. $y = 2x^2 - 7x + 6$
20. $y = -3x^2 + 4x - 1$
21. $y = 4x^2 - 12x + 9$
22. $y = 3x^2 - 2x - 1$
23. $y = \frac{1}{2}x^2 - \frac{3}{2}x + 1$
24. $y = x^2 - x - \frac{3}{4}$

Determine whether each function has a maximum or a minimum value. Then find that value.

25. $y = -x^2 + 2x + 5$
26. $y = 3x^2 - 4x - 2$
27. $y = -2x^2 - 3x + 4$
28. $y = 5 - x^2$
29. $y = 6x + 3x^2$
30. $y = 6 - x^2 - x$
31. Given the function $y = ax^2 + bx - 27$, for what values of a and b will the vertex be at $(2, -3)$?
32. Given the function $y = ax^2 + bx + 5$, for what values of a and b will the vertex be at $(-1, 4)$?
33. Given the function $y = ax^2 + bx$, for what values of a and b will the vertex be at $(-3, 2)$?
34. Given the function $y = 2x^2 + bx + c$, for what values of b and c will the vertex lie at the intersection of the x-axis and the line $x = 1$?

35. Given the function $y = ax^2 + bx + c$ where $a > 0$, does the function have a maximum value? Explain.

36. Does the relation $x = ay^2 + by + c$ express y as a function of x? Explain.

Applications

37. Construction Mary wants to build a fence around a rectangular area for a garden. She has 150 ft of fencing, and she wants to leave a 10-ft opening on one side for a gate. In order to make the area of the garden a maximum, what should the dimensions of the garden be?

38. Manufacturing The bottom of a box is to have a perimeter of 36 cm. If the box must be 4 cm high, what dimensions (length and width) would give the maximum volume?

Developing Mathematical Power

Thinking Critically Although the ancient Greeks did not have the advantage of our modern notation, they were able to solve quadratic equations geometrically. They were interested primarily in the application of mathematics, so they simply ignored negative solutions, which they thought had no practical value.

To illustrate, suppose a problem involved the quadratic equation $x^2 + 6x = 40$. The Greeks would start with a square x units on each side. Its area can be expressed as x^2. Then they would draw 6 rectangles each 1 unit by x units. The area of each rectangle can be expressed as $1 \cdot x$, or x. The area of the resulting figure is then $x^2 + 6x$, which is the left side of the original equation.

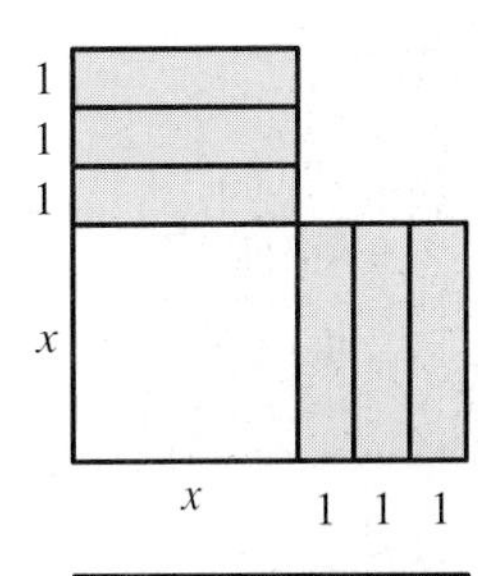

Finally, they would complete a new, larger square by adding 9 unit squares. The area of the larger square is $x^2 + 6x + 9$. This area can also be expressed as $(x + 3)^2$. Using modern algebraic notation

$$\begin{aligned} x^2 + 6x + 9 &= 40 + 9 \\ (x + 3)^2 &= 49 \\ x + 3 &= 7 \\ x &= 4 \end{aligned}$$

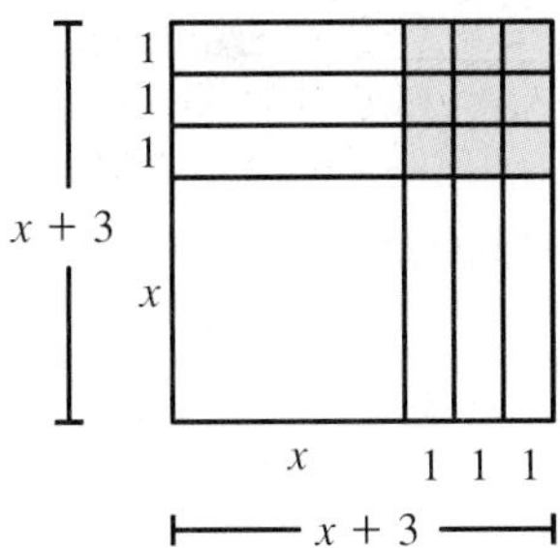

Find the positive solution for each equation by completing the square as the Greeks would have done. Sketch and label the related figure.

39. $x^2 + 2x = 8$ **40.** $x^2 + 4x = 5$ **41.** $x^2 + 8x = 9$

42. $x^2 + 10x = 11$ **43.** $x^2 + 3x = \frac{27}{4}$ **44.** $x^2 + 5x = 6$

9.3

Completing the Square

Objective: To solve quadratic equations by completing the square

Most quadratic equations cannot be easily solved by factoring. However, other methods can be used to solve them. Since many solutions to such equations involve irrational numbers, it will be useful to review the simplification of radical expressions.

Capsule Review

A radical expression is in simplest form when there are no powers in the radicand that are equal to or greater than the index, when no radicand contains a fraction or a negative sign, and when no denominator contains a radical.

EXAMPLES $\sqrt{32} = \sqrt{16} \cdot \sqrt{2} = 4\sqrt{2}$

$$\sqrt{\frac{27}{2}} = \frac{\sqrt{27}}{\sqrt{2}} \cdot \frac{\sqrt{2}}{\sqrt{2}} = \frac{\sqrt{54}}{2} = \frac{\sqrt{9} \cdot \sqrt{6}}{2} = \frac{3}{2}\sqrt{6}$$

$$\sqrt{-75} = \sqrt{-1} \cdot \sqrt{75} = i\sqrt{25} \cdot \sqrt{3} = 5i\sqrt{3}$$

Simplify.

1. $\sqrt{48}$ **2.** $2\sqrt{63}$ **3.** $\sqrt{\frac{24}{5}}$ **4.** $\frac{1}{2}\sqrt{\frac{98}{3}}$ **5.** $\sqrt{-27}$ **6.** $\sqrt{\frac{-32}{9}}$

One method of solving an equation in which one side is a perfect square trinomial is to take the square root of each side.

EXAMPLE 1 **Solve: $x^2 + 10x + 25 = 36$**

$x^2 + 10x + 25 = 36$

$(x + 5)^2 = 36$ *Factor the trinomial.*

$x + 5 = \pm 6$ *If $a^2 = k$, then $a = \pm\sqrt{k}$.*

$x = 6 - 5$ or $x = -6 - 5$

$x = 1$ or $x = -11$

Check:

$x^2 + 10x + 25 = 36$
$(1)^2 + 10(1) + 25 \stackrel{?}{=} 36$
$36 = 36$ ✓

$x^2 + 10x + 25 = 36$
$(-11)^2 + 10(-11) + 25 \stackrel{?}{=} 36$
$36 = 36$ ✓

The solutions are 1 and -11.

If one side of an equation is not a perfect square trinomial, it can be made into such a trinomial by **completing the square.** The following familiar relationship is used to do this

$$a^2 + 2ab + b^2 = (a + b)^2$$

Notice that *the constant term b^2 is the square of one-half the coefficient of a.* Example 2 shows how the method of completing the square can be used to solve a quadratic equation.

EXAMPLE 2 **Solve by completing the square: $x^2 - 12x + 7 = 0$**

$x^2 - 12x + 7 = 0$

$x^2 - 12x = -7$ — *Add -7 to both sides.*

$x^2 - 12x + 36 = -7 + 36$ — *Complete the square. Add $\left(\frac{1}{2} \cdot -12\right)^2$, or 36, to both sides. Then factor the left side.*

$(x - 6)^2 = 29$

$x - 6 = \pm\sqrt{29}$ — *If $a^2 = k$, then $a = \pm\sqrt{k}$.*

$x = 6 + \sqrt{29}$ or $x = 6 - \sqrt{29}$

Check:

$$x^2 - 12x + 7 = 0$$
$$(6 + \sqrt{29})^2 - 12(6 + \sqrt{29}) + 7 \stackrel{?}{=} 0$$
$$36 + 12\sqrt{29} + 29 - 72 - 12\sqrt{29} + 7 \stackrel{?}{=} 0$$
$$0 = 0 \checkmark$$

This check shows that $6 + \sqrt{29}$ is a solution. You should check to show that $6 - \sqrt{29}$ is also a solution.

This method is easier to use if the coefficient of x is 1. If it is not 1, then divide each side of the equation by the coefficient, and proceed as above.

EXAMPLE 3 **Solve by completing the square: $2x^2 - 3x - 11 = 0$**

$2x^2 - 3x - 11 = 0$

$x^2 - \frac{3}{2}x - \frac{11}{2} = 0$ — *Divide both sides by 2, the coefficient of x^2.*

$x^2 - \frac{3}{2}x = \frac{11}{2}$ — *Add $\frac{11}{2}$ to both sides.*

$x^2 - \frac{3}{2}x + \frac{9}{16} = \frac{11}{2} + \frac{9}{16}$ — *Add $\left(\frac{1}{2} \cdot \frac{-3}{2}\right)^2$, or $\frac{9}{16}$, to both sides.*

$\left(x - \frac{3}{4}\right)^2 = \frac{97}{16}$ — *Factor the trinomial.*

$x - \frac{3}{4} = \pm\sqrt{\frac{97}{16}}$ — *If $a^2 = k$, then $a = \pm\sqrt{k}$.*

$x = \frac{3}{4} \pm \sqrt{\frac{97}{16}} = \frac{3 \pm \sqrt{97}}{4}$

Check to show that the solutions are $\frac{3 + \sqrt{97}}{4}$ and $\frac{3 - \sqrt{97}}{4}$.

In the following example, the solutions are complex numbers.

EXAMPLE 4 **Solve by completing the square: $x^2 + 6x + 41 = 0$**

$$x^2 + 6x + 41 = 0 \qquad \textit{The coefficient of } x^2 \textit{ is 1.}$$
$$x^2 + 6x = -41$$
$$x^2 + 6x + 9 = -41 + 9 \qquad \textit{Add } \left(\tfrac{1}{2} \cdot 6\right)^2\textit{, or 9, to both sides.}$$
$$(x + 3)^2 = -32$$
$$x + 3 = \pm\sqrt{-32}$$
$$x = -3 + \sqrt{-32} \quad \text{or} \quad x = -3 - \sqrt{-32}$$
$$x = -3 + 4i\sqrt{2} \quad \text{or} \quad x = -3 - 4i\sqrt{2}$$

Check:

$$x^2 + 6x + 41 = 0$$
$$(-3 + 4i\sqrt{2})^2 + 6(-3 + 4i\sqrt{2}) + 41 \stackrel{?}{=} 0$$
$$9 + 2(-12i\sqrt{2}) + 16(i^2)(\sqrt{2})^2 - 18 + 24i\sqrt{2} + 41 \stackrel{?}{=} 0$$
$$9 - 24i\sqrt{2} - 32 - 18 + 24i\sqrt{2} + 41 \stackrel{?}{=} 0$$
$$0 = 0 \checkmark$$

The check above shows that $-3 + 4i\sqrt{2}$ is a solution. You should check to show that $-3 - 4i\sqrt{2}$ is also a solution.

CLASS EXERCISES

For each expression, find the number you would add to make it a perfect square trinomial.

1. $x^2 + 6x + \underline{?}$ **2.** $x^2 - 10x + \underline{?}$ **3.** $x^2 + 3x + \underline{?}$

For each equation, find the value of k that would make the left side a perfect square trinomial.

4. $x^2 + kx + 64 = 0$ **5.** $25x^2 - kx + 1 = 32$ **6.** $49 - kx + x^2 = 20$

Solve by completing the square.

7. $x^2 + 6x = 10$ **8.** $x^2 - 10x + 13 = 0$ **9.** $2x^2 + 6x - 8 = 0$

PRACTICE EXERCISES

Find the number you would add to make each expression a perfect square trinomial.

1. $x^2 - 14x + \underline{?}$ **2.** $x^2 + 12x + \underline{?}$ **3.** $y^2 - 9y + \underline{?}$

4. $z^2 + 7z + \underline{?}$ **5.** $x^2 + \frac{1}{2}x + \underline{?}$ **6.** $x^2 - \frac{2}{3}x + \underline{?}$

Find the value of k that would make the left side of each equation a perfect square trinomial.

7. $x^2 + kx + 16 = 0$
8. $x^2 - kx + 81 = 0$
9. $x^2 - kx + 25 = 7$
10. $x^2 + kx + 16 = -32$
11. $x^2 + kx + 36 = 10$
12. $x^2 - kx + 64 = 8$

Solve by taking the square root of each side.

13. $(x + 3)^2 = 9$
14. $(x - 6)^2 = 12$
15. $(y - 7)^2 = \frac{25}{4}$
16. $(y + 4)^2 = \frac{16}{9}$
17. $4x^2 + 4x + 1 = 4$
18. $4 - 12y + 9y^2 = 18$

Solve by completing the square. In Exercises 55–60, solve for x in terms of a.

19. $y^2 + 12y + 4 = 0$
20. $x^2 - 8x + 4 = 0$
21. $y^2 + 10y + 6 = 0$
22. $x^2 - 6x + 6 = 0$
23. $r^2 - 12r + 18 = 0$
24. $x^2 - 12x - 25 = 0$
25. $x^2 - 8x = 9$
26. $x^2 + 2x = 15$
27. $x^2 + 2x - 5 = 0$
28. $x^2 - 2x = -10$
29. $x^2 - 2x = -2$
30. $x^2 - 6x + 11 = 0$
31. $2x^2 - 8x - 14 = 0$
32. $2x^2 + 8x - 10 = 0$
33. $2y^2 - 20y + 24 = 0$
34. $2y^2 + 12y - 32 = 0$
35. $2z^2 + z - 28 = 0$
36. $3z^2 - 8z - 27 = 0$
37. $x^2 - 9x + 25 = 0$
38. $t^2 + 11t + 10 = 0$
39. $3s^2 + 2s + 18 = 0$
40. $2x^2 - 6x = 8$
41. $4x^2 + 4x = 3$
42. $2r^2 + 23r + 26 = 0$
43. $2y^2 - 5y - 3 = 0$
44. $3y^2 + 4y = -8$
45. $x^2 + 4 = 0$
46. $\frac{x^2}{3} + 8x - 3 = 0$
47. $\frac{x^2}{2} + 4x = 2$
48. $x^2 - \frac{x}{2} = \frac{1}{3}$
49. $x^2 + \frac{3}{4}x = \frac{1}{2}$
50. $9x^2 - 12x + 5 = 0$
51. $25x^2 - 20x = -9$
52. $x^2 + 2\sqrt{2}x = 35$
53. $x^2 = 4\sqrt{3}x - 12$
54. $3x^2 + x = \frac{2}{3}$
55. $2x^2 - ax = 6a^2$
56. $2a^2x^2 - 8ax = -6$
57. $3x^2 + ax = a^2$
58. $4a^2x^2 + 8ax + 3 = 0$
59. $3x^2 + ax^2 = 9x + 9a$
60. $6a^2x^2 - 11ax = 10$

Applications

61. Number Theory The square of a number increased by four times the number is equal to 96. What is the number?

62. Number Theory Twice a number increased by 24 is equal to the square of the number. Find the number.

Developing Mathematical Power

The method of finite differences, discussed in Chapter 6, is a way of analyzing number patterns. A sequence of numbers that is based on finite differences can be described by an ordered polynomial expression.

For example, the sequence 2, 5, 10, 17, 26, . . . can be described as $x^2 + 1$, where x is the number of the term. The sequence shows second-order difference; this indicates that its terms can be described by a quadratic expression.

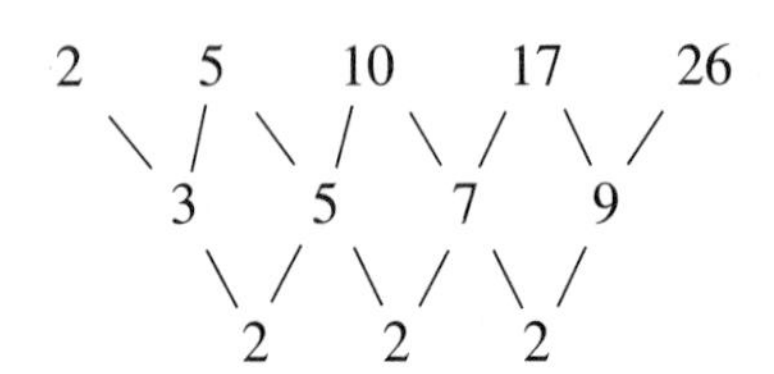

To find a quadratic expression that will describe a given sequence, it is helpful to determine first the general difference in a general sequence:

n (the number of the term)	1		2		3		4 . . .
$an^2 + bn + c$ (the term)	$a(1)^2 + b(1) + c$ or $a + b + c$		$a(2)^2 + b(2) + c$ or $4a + 2b + c$		$a(3)^2 + b(3) + c$ or $9a + 3b + c$		$a(4)^2 + b(4) + c$ . . . or $16a + 4b + c$
1st-order difference		$3a + b$		$5a + b$		$7a + b$ . . .	
2nd-order difference			$2a$		$2a$ . . .		

Once you determine the second-order difference for such a sequence, use it to find the value of a; then use the first-order difference and a to find the value of b; and finally, use any term of the sequence and the values of a and b to find the value of c. Working with the sequence above,

$$2a = 2;\ a = 1 \qquad \text{2nd-order difference}$$

$$3(1) + b = 3;\ b = 0 \qquad \text{1st-order difference}$$

$$1 + 0 + c = 2;\ c = 1 \qquad \text{first term of sequence}$$

The descriptive quadratic expression is $1x^2 + 0x + 1$, or $x^2 + 1$.

Find the quadratic expressions for each sequence.

63. 2, 9, 22, 41, 66, . . .

64. −6, −13, −18, −21, −22, . . .

65. 0, −1, −4, −9, −16, . . .

66. −2, 0, 6, 16, 30, . . .

9.4

The Quadratic Formula

Objective: To solve quadratic equations by using the quadratic formula

The method of completing the square can be used to derive a formula for finding the solutions to any quadratic equation. To use the formula, you must be able to evaluate certain types of expressions involving radicals.

Capsule Review

To simplify an expression such as $\frac{12 - \sqrt{-18}}{6}$, first simplify the radical.

$$\frac{12 - \sqrt{-18}}{6} = \frac{12 - \sqrt{-1} \cdot \sqrt{9} \cdot \sqrt{2}}{6} = \frac{12 - 3i\sqrt{2}}{6} = \frac{3(4 - i\sqrt{2})}{6} = \frac{4 - i\sqrt{2}}{2}$$

Simplify.

1. $\frac{-2 + \sqrt{64}}{6}$ **2.** $\frac{8 - \sqrt{12}}{4}$ **3.** $\frac{-6 + \sqrt{8}}{2}$ **4.** $\frac{9 - \sqrt{90}}{6}$

5. $\frac{4 + \sqrt{-4}}{2}$ **6.** $\frac{-4 - \sqrt{-8}}{4}$ **7.** $\frac{5 + \sqrt{-75}}{10}$ **8.** $\frac{-12 - \sqrt{-128}}{2}$

To derive the **quadratic formula,** start with the quadratic equation $ax^2 + bx + c = 0$, $a \neq 0$.

$$ax^2 + bx + c = 0$$

$$x^2 + \frac{b}{a}x + \frac{c}{a} = 0$$ *Divide both sides by a.*

$$x^2 + \frac{b}{a}x = -\frac{c}{a}$$ *Add $-\frac{c}{a}$ to both sides.*

$$x^2 + \frac{b}{a}x + \frac{b^2}{4a^2} = -\frac{c}{a} + \frac{b^2}{4a^2}$$ *Add $\left(\frac{1}{2} \cdot \frac{b}{a}\right)^2$, or $\frac{b^2}{4a^2}$, to both sides.*

$$\left(x + \frac{b}{2a}\right)^2 = \frac{b^2 - 4ac}{4a^2}$$ *Factor the left side. Rewrite the right side with a common denominator.*

$$x + \frac{b}{2a} = \pm\sqrt{\frac{b^2 - 4ac}{4a^2}}$$ *If $a^2 = k$, then $a = \pm\sqrt{k}$.*

$$x = -\frac{b}{2a} \pm \frac{\sqrt{b^2 - 4ac}}{2a}$$

$$x = \frac{-b + \sqrt{b^2 - 4ac}}{2a} \quad \text{or} \quad x = \frac{-b - \sqrt{b^2 - 4ac}}{2a}$$

The Quadratic Formula

The solutions of the quadratic equation $ax^2 + bx + c = 0$, $a \neq 0$ are given by the formula

$$x = \frac{-b \pm \sqrt{b^2 - 4ac}}{2a}$$

To solve a quadratic equation, determine the values of a, b, and c and then substitute these values in the formula.

EXAMPLE 1 **Solve using the quadratic formula: $2x^2 + 5x - 7 = 0$**

$2x^2 + 5x - 7 = 0$

$a = 2,\ b = 5,\ c = -7$ — *Determine the values of a, b, and c.*

$$x = \frac{-b \pm \sqrt{b^2 - 4ac}}{2a}$$ *Write the formula.*

$$x = \frac{-5 \pm \sqrt{5^2 - 4(2)(-7)}}{2(2)}$$ *Substitute.*

$$x = \frac{-5 \pm \sqrt{25 + 56}}{4}$$

$$x = \frac{-5 \pm \sqrt{81}}{4}$$

$$x = \frac{-5 + 9}{4} \quad \text{or} \quad x = \frac{-5 - 9}{4}$$

$$x = 1 \quad \text{or} \quad x = -\frac{7}{2}$$

Check:

$$\begin{aligned} 2x^2 + 5x - 7 &= 0 \\ 2(1)^2 + 5(1) - 7 &\stackrel{?}{=} 0 \\ 2 + 5 - 7 &\stackrel{?}{=} 0 \\ 0 &= 0 \ \checkmark \end{aligned}$$

$$\begin{aligned} 2x^2 + 5x - 7 &= 0 \\ 2\left(-\tfrac{7}{2}\right)^2 + 5\left(-\tfrac{7}{2}\right) - 7 &\stackrel{?}{=} 0 \\ 2\left(\tfrac{49}{4}\right) - \left(\tfrac{35}{2}\right) - 7 &\stackrel{?}{=} 0 \\ \tfrac{49}{2} - \tfrac{35}{2} - 7 &\stackrel{?}{=} 0 \\ 0 &= 0 \ \checkmark \end{aligned}$$

The solutions are 1 and $-\frac{7}{2}$.

When you use the quadratic formula keep in mind the fact that a negative radicand leads to solutions that are not real numbers.

EXAMPLE 2 **Solve using the quadratic formula: $4x^2 - 8x + 13 = 0$**

$$x = \frac{-b \pm \sqrt{b^2 - 4ac}}{2a}$$ *Write the quadratic formula.*

$$x = \frac{-(-8) \pm \sqrt{(-8)^2 - 4(4)(13)}}{2(4)}$$ *Substitute: $a = 4$, $b = -8$, $c = 13$*

$$x = \frac{8 \pm \sqrt{-144}}{8} = \frac{8 \pm 12i}{8} = \frac{4(2 \pm 3i)}{4(2)}$$ *Factor.*

$$x = \frac{2 + 3i}{2} \quad \text{or} \quad x = \frac{2 - 3i}{2}$$

Check:

$$4x^2 - 8x + 13 = 0$$
$$4\left(\frac{2 - 3i}{2}\right)^2 - 8\left(\frac{2 - 3i}{2}\right) + 13 \stackrel{?}{=} 0$$
$$4\left(\frac{4 - 12i + 9i^2}{4}\right) - 4(2 - 3i) + 13 \stackrel{?}{=} 0$$
$$4 - 12i - 9 - 8 + 12i + 13 \stackrel{?}{=} 0$$
$$0 = 0 \checkmark$$

This check shows that $\frac{2 - 3i}{2}$ is a solution. You should check to show that $\frac{2 + 3i}{2}$ is also a solution.

The solutions are $\frac{2 + 3i}{2}$ and $\frac{2 - 3i}{2}$.

A calculator can be used to find decimal approximations of solutions.

EXAMPLE 3 **Solve using the quadratic formula: $x^2 + 4x - 14 = 0$**

$$x = \frac{-b \pm \sqrt{b^2 - 4ac}}{2a}$$ *Write the quadratic formula.*

$$x = \frac{-4 \pm \sqrt{4^2 - 4(1)(-14)}}{2(1)}$$ *Substitute: $a = 1$, $b = 4$, $c = -14$*

Solutions in Radical Form

$$x = \frac{-4 \pm \sqrt{16 + 56}}{2}$$
$$x = \frac{-4 \pm \sqrt{72}}{2}$$
$$x = \frac{-4 \pm 6\sqrt{2}}{2}$$
$$x = -2 + 3\sqrt{2} \quad \text{or} \quad x = -2 - 3\sqrt{2}$$

Solutions Using a Calculator

Evaluate $\sqrt{4^2 - 4(1)(-14)}$ and store it in memory: 8.485281374

$$x = \frac{-4 + 8.485281374}{2(1)} = 2.242640687$$
$$x = \frac{-4 - 8.485281374}{2(1)} = -6.242640687$$

A calculator that can store formulas and variable values will save time when you need to solve more than one quadratic equation.

The solutions are 2.24 and −6.24, to the nearest hundredth.

CLASS EXERCISES

Determine the values of *a*, *b*, and *c*. Then solve each equation using the quadratic formula.

1. $2x^2 - 5x - 3 = 0$ **2.** $5x^2 - 7x = 6$ **3.** $x^2 + 6x = 10$

4. $x^2 - 10x = -13$ **5.** $2x^2 + 3x = -8$ **6.** $3x^2 - 2x = -1$

PRACTICE EXERCISES

 Use technology where appropriate.

Solve each equation using the quadratic formula. Write solutions in simplest form.

1. $2x^2 + 3x - 5 = 0$ **2.** $x^2 + 8x + 12 = 0$ **3.** $8x^2 - 2x - 3 = 0$

4. $2x^2 - 7x + 3 = 0$ **5.** $3x^2 - 4x - 2 = 0$ **6.** $4x^2 - 3x = 9$

7. $5x^2 + x = 3$ **8.** $3z^2 + 9z = 27$ **9.** $r^2 + 12r = 18$

10. $2r^2 + 13r + 16 = 0$ **11.** $2z^2 + z - 28 = 0$ **12.** $x^2 - 9x + 15 = 0$

13. $t^2 + 10t + 11 = 0$ **14.** $3s^2 + 4s + 10 = 0$ **15.** $x^2 - 12x + 25 = 0$

16. $8x^2 + 2x - 15 = 0$ **17.** $x^2 - 2x + 5 = 0$ **18.** $2x^2 + 4x + 15 = 0$

Solve each equation using the quadratic formula. Write the solutions in decimal form, to the nearest hundredth.

19. $2x^2 - 5x - 3 = 0$ **20.** $3x^2 - 10x + 5 = 0$ **21.** $3x^2 + 4x - 3 = 0$

22. $6x^2 - 5x - 1 = 0$ **23.** $7x^2 - x - 12 = 0$ **24.** $5x^2 + 8x - 11 = 0$

Solve each equation using the quadratic formula. Write solutions in simplest form. In Exercises 43–48, solve for *x* in terms of *a*.

25. $4x^2 = 4x + 3$ **26.** $2x^2 = 7x - 8$ **27.** $x^2 + 3x + 5 = 0$

28. $x^2 + 7x - 8 = 0$ **29.** $3x^2 + 5x = 7$ **30.** $4x^2 + 4x = 22$

31. $7x^2 - 2x = 25$ **32.** $5x^2 - 3x = 10$ **33.** $2x^2 - 1 = 5x$

34. $3x^2 + 2 = 8x$ **35.** $2x^2 + x = \frac{1}{2}$ **36.** $2x^2 - x = \frac{1}{8}$

37. $9x^2 + 3x + 4 = 0$ **38.** $15x^2 + 2x + 1 = 0$ **39.** $5x^2 = 2x - 8$

40. $6x^2 - x + 24 = 0$ **41.** $\frac{x + 2}{5} = \frac{3}{x + 1}$ **42.** $\frac{x - 3}{2} = \frac{6}{x - 2}$

43. $2a^2x^2 - 6ax = -5$ **44.** $3a^2x^2 + 8ax + 5 = 0$

45. $2x^2 + ax^2 = 4x + 4a$ **46.** $5a^2x^2 - 10ax = 12$

47. $x^2 + 2ax = 25a^2$ **48.** $ax^2 + 3a^2x - 10a^3 = 0$

Applications

Number Theory Solve each problem.

49. Find a number such that four more than twice its square is six times the original number.

50. Find a number such that its square increased by 34 is equal to ten times the original number.

51. A number decreased by its reciprocal equals $\frac{7}{12}$. Find the number.

52. Four more than three times the square of a certain number equals 15. Find the number.

53. Find the value of n if

$$\begin{vmatrix} 2 & n & 4 \\ n & 2 & -1 \\ 4 & n & 7 \end{vmatrix} = -37$$

Developing Mathematical Power

Thinking Critically Ancient Greek mathematicians, particularly Pythagoras and his students, believed that all geometrical concepts could be expressed using rational numbers. They were shocked to discover that some concepts could *not* be expressed in this way. For example, if a line segment, $\overline{AB}$, is separated into two smaller segments, $\overline{AC}$ and $\overline{CB}$, such that the ratio $\frac{AB}{AC}$ is equal to the ratio $\frac{AC}{CB}$, the measures of $\overline{AC}$ and $\overline{CB}$ are not rational numbers. The ratio $\frac{AB}{AC}\left(\text{or } \frac{AC}{CB}\right)$ is the same, regardless of the length of $\overline{AB}$. This ratio is called the *golden section,* and point C is sometimes called the *golden cut.*

A C B

$$\frac{AB}{AC} = \frac{AC}{CB}$$

54. If $\overline{AB}$ measures 1 unit, find the measures of $\overline{AC}$ and $\overline{CB}$.

55. If $\overline{AB}$ measures 2 units, find the measures of $\overline{AC}$ and $\overline{CB}$.

56. If $\overline{AB}$ measures 3 units, find the measures of $\overline{AC}$ and $\overline{CB}$.

57. Find the ratio $\frac{AB}{AC}\left(\text{or } \frac{AC}{CB}\right)$. Write it in simplest radical form.

9.5

Problem Solving Strategy: Use Quadratic Equations

Many problems can be solved by writing and solving quadratic equations. To solve a geometric problem, it is usually helpful to draw a diagram.

EXAMPLE 1 Tom increased the area of his garden by 165 ft^2. The rectangular garden was originally 16 ft by 12 ft, and he increased the length and the width by the same amount. Find the dimensions of the garden now.

Understand the Problem *What are the given facts?* Tom's garden was a 16 ft by 12 ft rectangle. He increased the area 165 ft^2 by adding an equal number of feet to its length and width.

What are you asked to find? The problem asks for the new dimensions of the garden.

Plan Your Approach *Choose a strategy.* Draw and label a diagram. Let x be the number of feet added to each side. Then $16 + x$ is the new length and $12 + x$ is the new width, both in feet. The area of a rectangle is length times width.

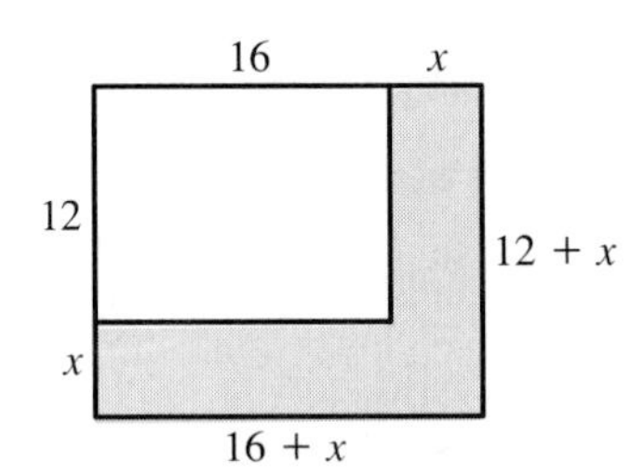

$$\text{new area} = \text{original area} + \text{added area}$$
$$(16 + x)(12 + x) = 16(12) + 165$$

Complete the Work

$$192 + 28x + x^2 = 192 + 165$$
$$x^2 + 28x - 165 = 0$$
$$(x + 33)(x - 5) = 0 \quad \textit{Factor.}$$
$$x = -33 \quad \text{or} \quad x = 5$$

Interpret the Results *State your answer.* Reject the negative solution. The amount added to each side was 5 ft.

Check your answer. Is the new area equal to the original area plus the added area? The new length is 16 + 5, or 21 ft, and the new width is 12 + 5, or 17 ft.

$$21(17) \stackrel{?}{=} 16(12) + 165$$
$$357 = 357 \checkmark$$

Quadratic equations can be solved by factoring only if they are factorable. The quadratic formula or completing the square, however, can be used to solve any quadratic equation.

EXAMPLE 2 Two chords intersect in a circle. One chord is 14 cm in length and the other is 16 cm. The ratio of the lengths of the two segments of the 14-cm chord is 3:4. Find the lengths of the two segments of the 16-cm chord.

Understand the Problem

What are the given facts? Recall from geometry that if two chords intersect inside a circle, then the product of the lengths of the segments of one chord is equal to the product of the lengths of the segments of the other chord.

Before applying this theorem, however, find the lengths of the segments of the 14-cm chord that are in the ratio of 3:4. Let x represent the length of one segment; the other must be $14 - x$.

$$\frac{x}{14 - x} = \frac{3}{4}$$
$$4x = 42 - 3x$$
$$7x = 42$$
$$x = 6$$

The segments of the 14-cm chord measure 6 cm and $14 - x$, or 8 cm. *What are you asked to find?* The problem asks for the lengths of the two segments of the 16-cm chord.

Plan Your Approach

Choose a strategy. Draw and label a diagram. Let x be the length of one segment of the 16-cm chord. Then $16 - x$ is the length of the other. Apply the geometric theorem.

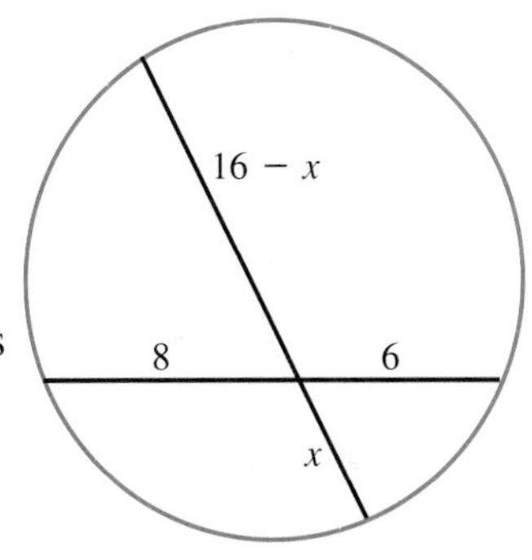

product of segments of 16-cm chord = product of segments of 14-cm chord.

$$x(16 - x) = 6(8)$$

Complete the Work

$$16x - x^2 = 48$$
$$0 = x^2 - 16x + 48$$
$$x = \frac{-(-16) \pm \sqrt{(-16)^2 - 4(1)(48)}}{2(1)} \qquad x = \frac{-b \pm \sqrt{b^2 - 4ac}}{2a}$$
$$x = 12 \quad \text{or} \quad x = 4$$

Interpret the Results

State your answer. If $x = 12$, then $16 - x = 4$. If $x = 4$, then $16 - x = 12$. The segments of the 16-cm chord are 12 cm and 4 cm long.

Check your answer. Are the products of the lengths of the segments of the two chords equal?

$$6(8) \stackrel{?}{=} 12(4)$$
$$48 = 48 \checkmark$$

CLASS EXERCISES

1. The lengths of two opposite sides of a square are doubled, and the lengths of the other two sides are each increased by 4 cm. The area of the resulting rectangle is 128 cm^2 greater than that of the original square. Find the length of a side of that square.
2. One leg of a right triangle is two-thirds the length of the other leg. The hypotenuse is 18 ft. Find the length of the longer leg. *Hint:* Use the Pythagorean theorem.
3. Find two consecutive even integers whose product is 168.
4. Find a number whose square increased by 32 is 12 times the number.

PRACTICE EXERCISES

1. The area of a square, expressed in square centimeters, is 165 cm greater than its perimeter, expressed in centimeters. What is the length of a side of the square?
2. The total area of two square fields is 18,000 ft^2. Each side of the larger field is 60 ft longer than a side of the smaller field. Find the dimensions of the two fields.
3. The base of a triangle is 6 cm shorter than the altitude to that base. Find the lengths of the base and the altitude if the area of the triangle is 56 cm^2.
4. Chords FG and HJ intersect at point I in a circle. If the lengths of the segments of $\overline{FG}$ are 4 cm and 5 cm, and the length of $\overline{HJ}$ is 12 cm, find the lengths of the two segments of $\overline{HJ}$.
5. The diagonal of a rectangle is 60 cm long. If the length of the rectangle is 12 cm greater than the width, find the dimensions.
6. A diagonal of a square is 3 in. longer than a side of the square. Find the length of a side and the length of a diagonal.
7. Find two consecutive odd integers whose product is 195.
8. Two numbers differ by 8, and the sum of their squares is 320. Find the two numbers.
9. If a number is added to its square, the sum is 156. Find the number.
10. Find a number whose square is 168 greater than 2 times the number.
11. One number is 3 less than twice another number. The product of the two numbers is 5. Find the numbers.

12. One number is equal to 6 times another number plus 7. The product of the two numbers is -2. Find the numbers.

13. The length of a rectangle is 1 cm more than twice its width and 2 cm less than the length of the diagonal of the rectangle. Find the dimensions of the rectangle.

14. The area of one square is one-half that of another, and the perimeter of the smaller one is 20 ft less than the perimeter of the other. Find the length of a side of each square.

15. Two chords, AB and CD, intersect in a circle at E. If $\overline{AE} = 6$ cm, $BE = 8$ cm, and $CE = 3(DE)$, what is the length of $\overline{CD}$?

16. A 21-ft chord and an 23-ft chord intersect within a circle. The lengths of the two segments of the 21-ft chord are in the ratio 2:5. Find the lengths of the segments of the 23-ft chord.

17. The product of two numbers is 2, and one number is 4 more than the other. Find the numbers.

18. The sum of a number and its reciprocal is $3\frac{11}{14}$. What is the number?

19. A sawhorse has 30-in. legs, with each pair arranged like an inverted V. The sawhorse is 21 in. high. If a carpenter wants to raise the height by 3 in., by how much must he decrease the distance between the bases of each pair of legs?

20. Mr. Garcia owns an irregular field that is in the shape of two squares, side by side. The area of the field is 17,396 yd^2, and the perimeter is 572 yd. What is the length of the longest side of the field?

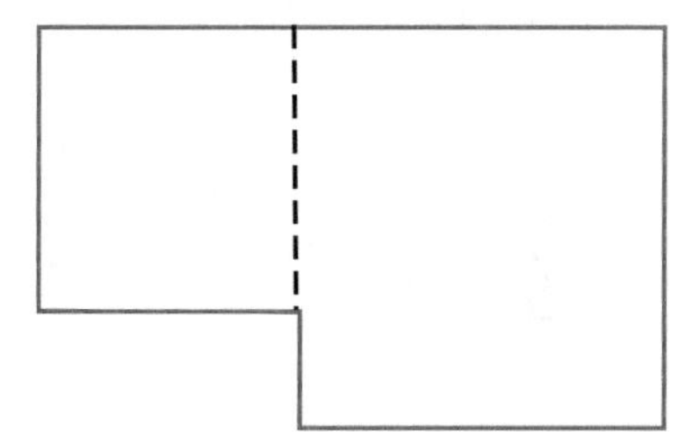

Mixed Problem Solving Review

1. Two trains start at the same time and travel in opposite directions. One train travels at 65 mi/h and the other travels at 75 mi/h. In how many hours will they be 490 mi apart?

2. When one-half of an integer is increased by 5, the result is greater than three-fourths of the integer decreased by 5. Find the positive integers that satisfy this condition.

3. A 75-ft flagpole breaks at a certain point, but remains attached. The top of the upper part of the pole hits the ground 40 ft from the base of the flagpole. How many feet above the ground did the flagpole break?

PROJECT

The engineering profession has an illustrious past and a very exciting future. Until the middle of the eighteenth century, engineers were primarily concerned with construction. Some of the greatest engineering feats took place thousands of years ago and include the construction of the Egyptian pyramids and the Great Wall of China. The start of the industrial revolution in 1750 sparked an explosion of scientific and technological advancements that dramatically increased the need for engineering professionals. Today, in an ever growing number of highly specialized fields, engineers apply current mathematical and scientific knowledge. The specialized branches of engineering include aerospace, biomedical, chemical, civil, electrical, environmental, industrial, mechanical, and nuclear, just to name a few.

Research three or four of the various fields of engineering and investigate the specialized activities of each branch.

TEST YOURSELF

For each quadratic function, give the direction of the opening of the parabola, the equation of the axis of symmetry, and the coordinates of the vertex.

1. $y = x^2 - 4$ **2.** $y = -8x^2$ **3.** $y = 3x^2 + 6x$ 9.1

Sketch the graph of each function.

4. $y = x^2 + 2x$ **5.** $y = x^2 - 4x - 5$ **6.** $y = 2x^2 + 7x - 4$ 9.2

Solve each equation by completing the square.

7. $x^2 + 8x = 5$ **8.** $8s^2 + 2s - 3 = 0$ **9.** $x^2 - 2x = 1$ 9.3

Solve each equation by using the quadratic formula.

10. $4z^2 + 4z - 15 = 0$ **11.** $3x^2 + 4x = -2$ **12.** $3t^2 - 5t + 27 = 0$ 9.4

Solve.

13. If each side of a square was increased by 3 cm, the area would be 4 times that of the original square. What is the length of a side of the original square? 9.5

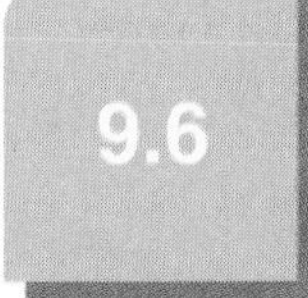

9.6 The Discriminant

Objectives: To use the discriminant to determine the nature of the solutions of a quadratic equation

To determine the relationship between the nature of the solutions and the graph of a quadratic function

An x-intercept of the graph of a linear or a quadratic function is a solution to the related equation.

Capsule Review

The x-intercept of the graph of $y = 2x - 6$ is the solution to $2x - 6 = 0$.

$$2x - 6 = 0$$
$$x = 3 \quad \text{The } x\text{-intercept is 3.}$$

Determine the x-intercept(s) of the graph of each equation.

1. $y = 5x - 4$ **2.** $3x = y - 9$ **3.** $2y + 3x = 6$ **4.** $y = 4x$

5. $y = x^2$ **6.** $y = x^2 - 4$ **7.** $y = x^2 - 4x$ **8.** $y = x^2 - 4x + 4$

$\frac{-b + \sqrt{b^2 - 4ac}}{2a}$ and $\frac{-b - \sqrt{b^2 - 4ac}}{2a}$ are solutions to the quadratic equation $ax^2 + bx + c = 0$. The expression $b^2 - 4ac$ is called the **discriminant.** Without actually solving the equation, you can obtain information about the solutions by examining the discriminant.

Nature of the Solutions of a Quadratic Equation

$ax^2 + bx + c = 0$, where a, b, and c are rational numbers

Discriminant, $b^2 - 4ac$	Nature of Solutions, $\frac{-b \pm \sqrt{b^2 - 4ac}}{2a}$
$b^2 - 4ac > 0$ and a perfect square	two real, rational, unequal numbers
$b^2 - 4ac > 0$ and not a perfect square	two real, irrational, unequal numbers
$b^2 - 4ac = 0$	one real, rational number (double)
$b^2 - 4ac < 0$	two complex conjugate numbers

EXAMPLE 1 **Use the discriminant to determine the nature of the solutions of each quadratic equation.**

a. $2x^2 + 7x + 6 = 0$ **b.** $3x^2 - 5x - 6 = 0$

c. $x^2 - 8x + 16 = 0$ **d.** $6x^2 - 2x + 5 = 0$

a. $2x^2 + 7x + 6 = 0$

$a = 2, b = 7, c = 6$

$b^2 - 4ac = 7^2 - 4(2)(6) = 1$

The solutions are two real, rational, unequal numbers.

b. $3x^2 - 5x - 6 = 0$

$a = 3, b = -5, c = -6$

$b^2 - 4ac = (-5)^2 - 4(3)(-6) = 97$

The solutions are two real, irrational, unequal numbers.

c. $x^2 - 8x + 16 = 0$

$a = 1, b = -8, c = 16$

$b^2 - 4ac = (-8)^2 - 4(1)(16) = 0$

The solution is one real, rational number (double).

d. $6x^2 - 2x + 5 = 0$

$a = 6, b = -2, c = 5$

$b^2 - 4ac = (-2)^2 - 4(6)(5) = -116$

The solutions are two complex conjugate numbers.

The discriminant can be used to find the required values of one of the coefficients of a quadratic equation if the nature of the solutions is known.

EXAMPLE 2 **Determine the values of k for which $3x^2 + kx + 12 = 0$ has exactly one solution.**

A quadratic equation has just one solution when $b^2 - 4ac = 0$.

$$b^2 - 4ac = 0$$
$$k^2 - 4(3)(12) = 0 \qquad a = 3, b = k, c = 12$$
$$k^2 - 144 = 0$$
$$k^2 = 144$$
$$k = \pm 12$$

There is one solution when $k = 12$ or $k = -12$.

The graph of a quadratic function is a parabola. If the parabola crosses the x-axis, it does so when $y = 0$, or $ax^2 + bx + c = 0$. The discriminant can be used to determine how the parabola is related to the x-axis.

If the discriminant is negative, there are *no* real solutions. The graph does *not* cross the x-axis.

If the discriminant is zero, there is *one* rational solution. The graph is tangent to the x-axis at the point that corresponds to the value of the solution.

If the discriminant is positive, there are *two* real solutions. The graph crosses the x-axis at the two points that correspond to the values of these solutions.

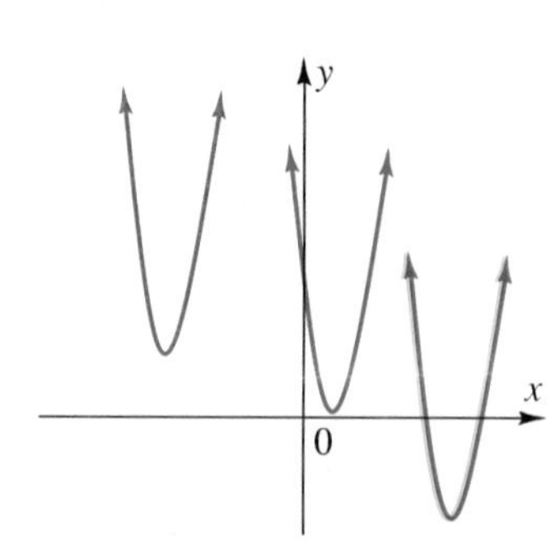

EXAMPLE 3 **Describe how the graph of each function is related to the x-axis.**

a. $y = 2x^2 - 3x - 7$ **b.** $y = 4x^2 + 5x + 2$

a. $y = 2x^2 - 3x - 7$

$a = 2, b = -3, c = -7$

$$b^2 - 4ac = (-3)^2 - 4(2)(-7)$$
$$= 65$$

The discriminant is positive. The graph crosses the x-axis at two points.

b. $y = 4x^2 + 5x + 2$

$a = 4, b = 5, c = 2$

$$b^2 - 4ac = 5^2 - 4(4)(2)$$
$$= -7$$

The discriminant is negative. The graph does not cross the x-axis.

CLASS EXERCISES

Use the discriminant to determine the nature of the solutions of each quadratic equation.

1. $x^2 - 7x + 6 = 0$ **2.** $x^2 - 8x + 12 = 0$ **3.** $x^2 - 8x + 32 = 0$

4. $x^2 + 4x + 24 = 0$ **5.** $2x^2 + 10x = -9$ **6.** $x^2 - 6x + 9 = 0$

Describe the relation of the graph of each function to the x-axis.

7. $y = x^2 + 15x + 36$ **8.** $y = x^2 - 8x + 16$ **9.** $y = 9x^2 + 3x + 9$

PRACTICE EXERCISES

Use the discriminant to determine the nature of the solutions of each quadratic equation.

1. $x^2 + 6x + 6 = 0$ **2.** $x^2 - 4x + 8 = 0$ **3.** $x^2 + 10x + 25 = 0$

4. $x^2 + 12x + 42 = 0$ **5.** $2x^2 - 8x = 14$ **6.** $2x^2 + 3x = 10$

7. $2x^2 - 20x + 24 = 0$ **8.** $2x^2 + 12x - 32 = 0$ **9.** $3x^2 - 9x = 27$

10. $2x^2 + x = 28$ **11.** $2x^2 + 23x + 26 = 0$ **12.** $x^2 - 12x + 36 = 0$

Describe the relation of the graph of each function to the x-axis.

13. $y = x^2 + 3x + 5$ **14.** $y = x^2 + 7x - 8$ **15.** $y = x^2 - 12x + 25$

16. $y = 2x^2 + x + 28$ **17.** $y = 4x^2 - 3x - 9$ **18.** $y = x^2 + 12x - 18$

Use the discriminant to determine the nature of the solutions of each quadratic equation.

19. $-x^2 + 2x + 5 = 4$ **20.** $-2x^2 - 7x - 5 = 0$ **21.** $2x(x + 1) + 8 = 0$

22. $(x - 3)(x + 6) + 4 = 0$ **23.** $2x^2 - 5x + 32 = 0$ **24.** $3x^2 - 5x + 5 = 0$

For each equation, determine the value(s) of k for which there will be only one solution.

25. $4x^2 + 8x + k = 0$ **26.** $kx^2 - 4x = 2$ **27.** $2x^2 + kx + 8 = 0$

28. $3x^2 + 2kx = -4$ **29.** $kx^2 - kx + 2 = 0$ **30.** $6x^2 - 2x + k + 1 = 0$

31. When the discriminant is positive, the graph of the quadratic function $y = ax^2 + bx + c$ crosses the x-axis in two points. Show that the equation of the axis of symmetry of the parabola is $x = -\frac{b}{2a}$.

32. Describe the relation of the graph of each function to the x-axis.

$y = x^2 - 4x + 3$ $y = x^2 - 4x + 4$ $y = x^2 - 4x + 5$

Then graph the three functions on the same coordinate plane.

Describe the relation of the graph of each function to the x-axis.

33. $y - 2 = 3x^2 - 4x$ **34.** $3x^2 = 5x + 7 + y$ **35.** $4x^2 = 4x - 22 - y$

Applications

36. Physics The equation for the motion of a projectile fired straight up at 80 ft/s is $h = 80t - 16t^2$, where h is the height and t is the time. Use the discriminant to determine if the projectile will reach a height of 100 ft.

37. Business The weekly revenue for a company is $R = -3p^2 + 60p + 1060$, where p is the price of the product. Use the discriminant to find if there is a price for which the weekly revenue would be \$1500.

Developing Mathematical Power

Thinking Critically The feature on page 431 introduced the *golden section*, by which any segment AB is separated into two smaller segments, AC and BC, such that

$$\frac{AB}{AC} = \frac{AC}{CB}$$

A C B

The golden section can be applied to a circle also. Consider a circle with a radius of 1 cm. Two points are chosen on the circle in such a way that the circumference and the lengths of the major (longer) and minor (shorter) arcs satisfy this proportion:

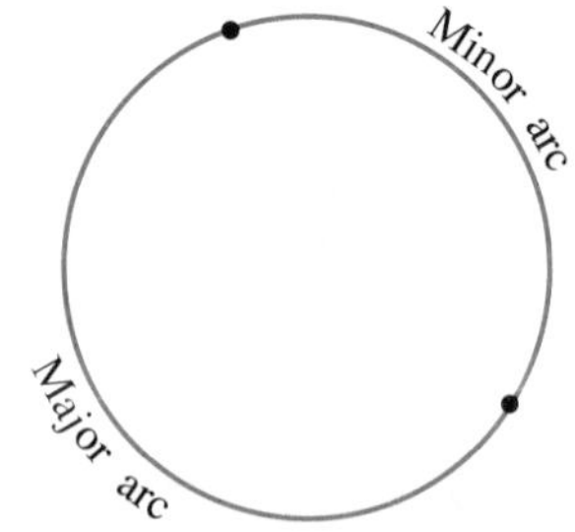

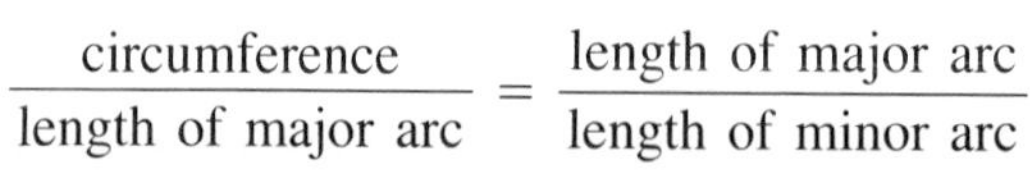

$$\frac{\text{circumference}}{\text{length of major arc}} = \frac{\text{length of major arc}}{\text{length of minor arc}}$$

38. Find the length of the major arc to the nearest 0.1 cm.

9.7 The Sum and Product of Solutions

Objectives: To find the sum and the product of the solutions of a quadratic equation without solving the equation
To write a quadratic equation that has given solutions

The sum and the product of the solutions of a quadratic equation are related to the coefficients of the equation. Before this relationship is considered, it will be helpful to review addition and multiplication with conjugate pairs of numbers.

Capsule Review

Add: $\frac{-3 + \sqrt{2}}{4} + \frac{-3 - \sqrt{2}}{4} = \frac{-3 + \sqrt{2} - 3 - \sqrt{2}}{4} = \frac{-6}{4} = -\frac{3}{2}$

Multiply: $\left(\frac{-3 + \sqrt{2}}{4}\right)\left(\frac{-3 - \sqrt{2}}{4}\right) = \frac{(-3)^2 - (\sqrt{2})^2}{16} = \frac{9 - 2}{16} = \frac{7}{16}$

Add or multiply, as indicated.

1. $(1 + \sqrt{3}) + (1 - \sqrt{3})$

2. $\frac{-8 + 2\sqrt{5}}{3} + \frac{-8 - 2\sqrt{5}}{3}$

3. $(9 + 2i) + (9 - 2i)$

4. $(1 + \sqrt{3})(1 - \sqrt{3})$

5. $\left(\frac{-8 + 2\sqrt{5}}{3}\right)\left(\frac{-8 - 2\sqrt{5}}{3}\right)$

6. $(9 + 2i)(9 - 2i)$

The solutions of the equation $2x^2 + 7x - 15 = 0$ can be found by factoring.

$$2x^2 + 7x - 15 = 0 \qquad a = 2, b = 7, c = -15$$
$$(2x - 3)(x + 5) = 0$$
$$2x - 3 = 0 \quad \text{or} \quad x + 5 = 0$$
$$x = \frac{3}{2} \quad \text{or} \quad x = -5$$

The sum of the solutions is $\frac{3}{2} + (-5)$, or $\frac{-7}{2}$.

The product of the solutions is $\frac{3}{2}(-5)$, or $\frac{-15}{2}$.

Notice that the sum of the solutions is $\frac{-b}{a}$ and the product is $\frac{c}{a}$. The general quadratic equation is used to show that this is always true.

The solutions to $ax^2 + bx + c = 0$, $a \neq 0$, are s_1 and s_2, where

$$s_1 = \frac{-b + \sqrt{b^2 - 4ac}}{2a} \text{ and } s_2 = \frac{-b - \sqrt{b^2 - 4ac}}{2a}$$

The sum of the solutions is $s_1 + s_2$.

$$\begin{aligned} s_1 + s_2 &= \frac{-b + \sqrt{b^2 - 4ac}}{2a} + \frac{-b - \sqrt{b^2 - 4ac}}{2a} \\ &= \frac{-b + \sqrt{b^2 - 4ac} - b - \sqrt{b^2 - 4ac}}{2a} \\ &= \frac{-2b}{2a} = \frac{-b}{a} \end{aligned}$$

The product of the solutions is $s_1 s_2$.

$$\begin{aligned} s_1 s_2 &= \left(\frac{-b + \sqrt{b^2 - 4ac}}{2a}\right)\left(\frac{-b - \sqrt{b^2 - 4ac}}{2a}\right) \\ &= \frac{(-b)^2 - (\sqrt{b^2 - 4ac})^2}{4a^2} \qquad (m + n)(m - n) = m^2 - n^2 \\ &= \frac{b^2 - (b^2 - 4ac)}{4a^2} \\ &= \frac{4ac}{4a^2} = \frac{c}{a} \end{aligned}$$

If s_1 and s_2 are solutions of the equation $ax^2 + bx + c = 0$, $a \neq 0$,

then $\quad s_1 + s_2 = \frac{-b}{a} \quad$ and $\quad s_1 s_2 = \frac{c}{a}$

EXAMPLE 1 **Find the sum and the product of the solutions of the equation $3x^2 - 4x + 5 = 0$.**

Sum of solutions: $-\frac{b}{a} = -\left(\frac{-4}{3}\right) = \frac{4}{3}$

Product of solutions: $\frac{c}{a} = \frac{5}{3}$

Since the sum and the product of the solutions of a quadratic equation are related to the coefficients a, b, and c, you can write a quadratic equation that has given solutions.

$$ax^2 + bx + c = 0$$

$$x^2 + \frac{b}{a}x + \frac{c}{a} = 0 \qquad \textit{Divide both sides by a.}$$

$$x^2 - (s_1 + s_2)x + s_1 s_2 = 0$$

Since $s_1 + s_2 = \frac{-b}{a}$, and $s_1 s_2 = \frac{c}{a}$, substitute $-(s_1 + s_2)$ for $\frac{b}{a}$ and $s_1 s_2$ for $\frac{c}{a}$.

Quadratic Equation With Solutions s_1 and s_2

$$x^2 - (s_1 + s_2)x + s_1 s_2 = 0$$

After you find a quadratic equation with given solutions, write the equation with integral coefficients.

EXAMPLE 2 **Write a quadratic equation with the given solutions.**

a. $\frac{-2}{3}$ and 6 **b.** $\frac{-3 + \sqrt{5}}{4}$ and $\frac{-3 - \sqrt{5}}{4}$ **c.** $-3 + 5i$ and $-3 - 5i$

a. Let $s_1 = \frac{-2}{3}$ and $s_2 = 6$.

$$s_1 + s_2 = \frac{-2}{3} + 6 = \frac{16}{3}$$

$$s_1 s_2 = \frac{-2}{3}(6) = -4$$

$$x^2 - (s_1 + s_2)x + s_1 s_2 = 0$$

$$x^2 - \frac{16}{3}x - 4 = 0 \qquad \textit{Substitute.}$$

$$3x^2 - 16x - 12 = 0 \qquad \textit{Multiply both sides by 3.}$$

The equation is $3x^2 - 16x - 12 = 0$.

b. Let $s_1 = \frac{-3 + \sqrt{5}}{4}$ and $s_2 = \frac{-3 - \sqrt{5}}{4}$.

$$s_1 + s_2 = \frac{-3 + \sqrt{5}}{4} + \frac{-3 - \sqrt{5}}{4} = \frac{-6}{4} = \frac{-3}{2}$$

$$s_1 s_2 = \left(\frac{-3 + \sqrt{5}}{4}\right)\left(\frac{-3 - \sqrt{5}}{4}\right) = \frac{(-3)^2 - (\sqrt{5})^2}{16} = \frac{1}{4}$$

$$x^2 - (s_1 + s_2)x + s_1 s_2 = 0$$

$$x^2 - \left(\frac{-3}{2}\right)x + \frac{1}{4} = 0 \qquad \textit{Substitute.}$$

$$4x^2 + 6x + 1 = 0 \qquad \textit{Multiply both sides by 4.}$$

The equation is $4x^2 + 6x + 1 = 0$.

c. Let $s_1 = -3 + 5i$ and $s_2 = -3 - 5i$.

$$s_1 + s_2 = (-3 + 5i) + (-3 - 5i) = -6$$
$$s_1 s_2 = (-3 + 5i)(-3 - 5i) = (-3)^2 - (5i)^2 = 9 + 25 = 34$$
$$x^2 - (s_1 + s_2)x + s_1 s_2 = 0$$
$$x^2 - (-6)x + 34 = 0$$
$$x^2 + 6x + 34 = 0 \quad \textit{Substitute.}$$

The equation is $x^2 + 6x + 34 = 0$.

CLASS EXERCISES

Find the sum and product of the solutions of each equation.

1. $2x^2 - 3x + 5 = 0$ **2.** $2x^2 + 4x - 7 = 0$ **3.** $x^2 + 5x - 10 = 0$

4. $5x^2 - 4x + 8 = 0$ **5.** $4x^2 + 2x = 0$ **6.** $x^2 - 2x = 0$

Write a quadratic equation with the given solutions.

7. 3 and -2 **8.** $2 + 3i$ and $2 - 3i$

9. $7 + i\sqrt{5}$ and $7 - i\sqrt{5}$ **10.** $\frac{-3 + \sqrt{7}}{2}$ and $\frac{-3 - \sqrt{7}}{2}$

PRACTICE EXERCISES

Find the sum and product of the solutions of each equation.

1. $x^2 + 7x - 4 = 0$ **2.** $x^2 - 8x + 7 = 0$ **3.** $2n^2 + 6n + 2 = 0$

4. $3t^2 + 5t + 2 = 0$ **5.** $4x^2 - 3x - 5 = 0$ **6.** $2x^2 - x - 15 = 0$

7. $2s^2 - 5s = -3$ **8.** $2m^2 + 3m = -7$ **9.** $3y^2 - 7y = 0$

10. $3x^2 - 2x = 5$ **11.** $x^2 + 12x - 6 = 0$ **12.** $4r^2 = 10 - r$

Write a quadratic equation with the given solutions.

13. 6 and 4 **14.** 3 and -7 **15.** $\frac{2}{3}$ and $\frac{1}{2}$

16. $\frac{3}{2}$ and $\frac{1}{3}$ **17.** $1 + \sqrt{2}$ and $1 - \sqrt{2}$ **18.** $2 + \sqrt{3}$ and $2 - \sqrt{3}$

19. $2 + 3i$ and $2 - 3i$ **20.** $1 + 4i$ and $1 - 4i$

21. $\frac{3 + \sqrt{5}}{2}$ and $\frac{3 - \sqrt{5}}{2}$ **22.** $\frac{2 + \sqrt{3}}{4}$ and $\frac{2 - \sqrt{3}}{4}$

23. $\frac{-3 + 2\sqrt{7}}{4}$ and $\frac{-3 - 2\sqrt{7}}{4}$ **24.** $\frac{6 + 3\sqrt{2}}{2}$ and $\frac{6 - 3\sqrt{2}}{2}$

25. $\frac{2 + 4\sqrt{2}}{3}$ and $\frac{2 - 4\sqrt{2}}{3}$

26. $\frac{-1 + 6\sqrt{5}}{6}$ and $\frac{-1 - 6\sqrt{5}}{6}$

27. $\frac{3 + i}{4}$ and $\frac{3 - i}{4}$

28. $\frac{2 + 4i}{3}$ and $\frac{2 - 4i}{3}$

29. $\frac{-2 + 2i}{3}$ and $\frac{-2 - 2i}{3}$

30. $\frac{-4 + 5i}{2}$ and $\frac{-4 - 5i}{2}$

31. $\frac{1 + i\sqrt{2}}{5}$ and $\frac{1 - i\sqrt{2}}{5}$

32. $\frac{2 + i\sqrt{5}}{3}$ and $\frac{2 - i\sqrt{5}}{3}$

33. $\frac{-3 + 2i\sqrt{3}}{3}$ and $\frac{-3 - 2i\sqrt{3}}{3}$

34. $\frac{-4 + 3i\sqrt{2}}{2}$ and $\frac{-4 - 3i\sqrt{2}}{2}$

Write a quadratic equation in x that has the given solution(s).

35. $2c + 3d$ and $2c - 3d$

36. $4a + 2a$ and $4a - 2a$

37. $\frac{-3a + 2\sqrt{c}}{4}$ and $\frac{-3a - 2\sqrt{c}}{4}$

38. $\frac{-2e + 3\sqrt{d}}{6}$ and $\frac{-2e - 3\sqrt{d}}{6}$

39. $-2 + 3i\sqrt{5}$ *Hint:* First find the other solution.

40. $5a - 2\sqrt{3}$

Applications

Number Theory For each problem, use the sum and the product of the numbers to write an equation that can be used to find those numbers. Then solve the problem.

41. The sum of two numbers is 24, and their product is 140.

42. The sum of two numbers is $\frac{13}{6}$, and their product is 1.

43. The sum of two numbers is -12, and their product is 34.

44. The sum of two numbers is 8, and their product is 25.

Mixed Review

Simplify.

45. $\sqrt{x^6y^4}$

46. $\sqrt{80x^3}$

47. $4x\sqrt{9x^2y}$

48. $\sqrt[3]{54x^6}$

49. $3x\sqrt{18} + 4\sqrt{2x^2}$

50. $(x + \sqrt{2})(x - \sqrt{2})$

51. Find the midpoint between (2, 5) and (7, -13).

52. Express 22,000 in scientific notation.

9.8

Solving Equations in Quadratic Form

Objective: To solve equations in quadratic form

Equations such as $x^4 - 2x^2 - 3 = 0$ can be written in quadratic form and solved by methods used to solve quadratic equations.

Capsule Review

The quadratic formula can be used to solve any equation in the form $ax^2 + bx + c = 0$, where a, b, and c are real numbers and $a \neq 0$.

EXAMPLE **Solve: $3h^2 - 3h - 2 = 0$**

$$h = \frac{-b \pm \sqrt{b^2 - 4ac}}{2a}$$

$$= \frac{-(-3) \pm \sqrt{(-3)^2 - 4(3)(-2)}}{2(3)} \qquad a = 3, b = -3, c = -2$$

$$= \frac{3 \pm \sqrt{9 + 24}}{6}$$

$$= \frac{3 \pm \sqrt{33}}{6}$$

Solve using any method.

1. $5s^2 - 7s - 6 = 0$ **2.** $2w^2 + 6w + 1 = 0$ **3.** $2x^2 - 7x = 15$

4. $7t^2 - 16t - 15 = 0$ **5.** $6x^2 - 11x = 10$ **6.** $4x^2 + 8x + 3 = 0$

Since $x^4 - 2x^2 - 3 = 0$ can be written as $(x^2)^2 - 2(x^2) - 3 = 0$, the equation is said to be in **quadratic form.** You can use factoring or the quadratic formula to solve for x^2, and then solve for x.

EXAMPLE 1 **Solve: $x^4 - 2x^2 - 3 = 0$**

$$x^4 - 2x^2 - 3 = 0$$

$$(x^2)^2 - 2(x^2) - 3 = 0 \qquad \textit{The equation is } \textbf{\textit{quadratic in } x^2}.$$

$$(x^2 - 3)(x^2 + 1) = 0$$

$$x^2 = 3 \quad \text{or} \quad x^2 = -1$$

$$x = \pm\sqrt{3} \quad \text{or} \quad x = \pm i$$

Check:

$$x^4 - 2x^2 - 3 = 0 \qquad\qquad x^4 - 2x^2 - 3 = 0$$
$$(\pm\sqrt{3})^4 - 2(\pm\sqrt{3})^2 - 3 \stackrel{?}{=} 0 \qquad (\pm i)^4 - 2(\pm i)^2 - 3 \stackrel{?}{=} 0$$
$$9 - 6 - 3 \stackrel{?}{=} 0 \qquad\qquad 1 + 2 - 3 \stackrel{?}{=} 0$$
$$0 = 0 \checkmark \qquad\qquad 0 = 0 \checkmark$$

The solutions are $\pm\sqrt{3}$ and $\pm i$.

One or more solutions that you find for an equation in quadratic form may not satisfy the original equation, as illustrated in the next example.

EXAMPLE 2 **Solve:** $x - 3\sqrt{x} - 4 = 0$

$$x - 3\sqrt{x} - 4 = 0$$
$$(\sqrt{x})^2 - 3(\sqrt{x}) - 4 = 0 \qquad \text{The equation is quadratic in } \sqrt{x}.$$
$$(\sqrt{x} - 4)(\sqrt{x} + 1) = 0 \qquad \text{Factor.}$$
$$\sqrt{x} = 4 \quad \text{or} \quad \sqrt{x} = -1$$
$$x = 16$$

Since the principal square root of a number is not negative, $\sqrt{x} = -1$ does not lead to a solution. Check the remaining solution, 16.

Check:

$$x - 3\sqrt{x} - 4 = 0$$
$$16 - 3\sqrt{16} - 4 \stackrel{?}{=} 0$$
$$16 - 3(4) - 4 \stackrel{?}{=} 0$$
$$0 = 0 \checkmark$$

The solution is 16.

Another technique is illustrated in the next example. The quadratic form of the equation is shown more clearly by substituting a single variable for a binomial.

EXAMPLE 3 **Solve:** $(z - 2)^2 - 6(z - 2) + 8 = 0$

The equation is quadratic in $(z - 2)$. To simplify the solution of the equation, start by substituting y for $z - 2$.

$$(z - 2)^2 - 6(z - 2) + 8 = 0$$
$$y^2 - 6y + 8 = 0 \qquad \text{Replace } z - 2 \text{ by } y.$$
$$(y - 2)(y - 4) = 0 \qquad \text{Factor.}$$
$$y - 2 = 0 \quad \text{or} \quad y - 4 = 0$$
$$y = 2 \quad \text{or} \quad y = 4$$
$$z - 2 = 2 \quad \text{or} \quad z - 2 = 4 \qquad \text{Replace } y \text{ by } z - 2.$$
$$z = 4 \quad \text{or} \quad z = 6$$

Check to show that the solutions are 4 and 6.

Some equations must be rewritten in order to show that they are quadratic.

EXAMPLE 4 **Solve.**

a. $(x + 1)(x - 3) = x(3x + 2) - 7$

b. $\frac{x - 2}{x} + \frac{1}{2} = \frac{x + 1}{x - 2}$

a. $(x + 1)(x - 3) = x(3x + 2) - 7$

$x^2 - 2x - 3 = 3x^2 + 2x - 7$ *Multiply.*

$0 = 2x^2 + 4x - 4$ *The equation is quadratic.*

$0 = x^2 + 2x - 2$ *Divide each side by 2.*

$x = \frac{-b \pm \sqrt{b^2 - 4ac}}{2a}$ *Use the quadratic formula.*

$x = \frac{-2 \pm \sqrt{2^2 - 4(1)(-2)}}{2(1)}$ *$a = 1, b = 2, c = -2$*

$x = \frac{-2 \pm \sqrt{12}}{2} = \frac{-2 \pm 2\sqrt{3}}{2}$

$x = -1 \pm \sqrt{3}$

Check to show that the solutions are $-1 + \sqrt{3}$ and $-1 - \sqrt{3}$.

b. $\frac{x - 2}{x} + \frac{1}{2} = \frac{x + 1}{x - 2}$ *Note that the solution set does not include 0 or 2.*

$2x(x - 2)\left(\frac{x - 2}{x} + \frac{1}{2}\right) = 2x(x - 2)\left(\frac{x + 1}{x - 2}\right)$ *Multiply both sides by the LCD, $2x(x - 2)$.*

$2(x - 2)(x - 2) + x(x - 2) = 2x(x + 1)$

$2x^2 - 8x + 8 + x^2 - 2x = 2x^2 + 2x$

$x^2 - 12x + 8 = 0$ *The equation is quadratic.*

$x = \frac{-b \pm \sqrt{b^2 - 4ac}}{2a}$ *Use the quadratic formula.*

$x = \frac{-(-12) \pm \sqrt{(-12)^2 - 4(1)(8)}}{2(1)}$ *$a = 1, b = -12, c = 8$*

$x = \frac{12 \pm \sqrt{112}}{2} = \frac{12 \pm 4\sqrt{7}}{2}$

$x = 6 \pm 2\sqrt{7}$

Check to show that the solutions are $6 + 2\sqrt{7}$ and $6 - 2\sqrt{7}$.

Solving an equation involving two radicals can lead to a quadratic equation. Remember to isolate one radical and square both sides. Then isolate the remaining radical and square again.

EXAMPLE 5 **Solve:** $\sqrt{y+1} - \sqrt{2y} + 1 = 0$

$$\begin{aligned}
\sqrt{y+1} - \sqrt{2y} + 1 &= 0 \\
\sqrt{y+1} &= \sqrt{2y} - 1 && \textit{Isolate one radical.} \\
y + 1 &= 2y - 2\sqrt{2y} + 1 && \textit{Square both sides.} \\
2\sqrt{2y} &= y && \textit{Isolate the remaining radical.} \\
8y &= y^2 && \textit{Square both sides again.} \\
y^2 - 8y &= 0 && \textit{The equation is quadratic.} \\
y(y-8) &= 0 \\
y = 0 \quad &\text{or} \quad y = 8
\end{aligned}$$

Check:

$$\begin{aligned}
\sqrt{y+1} - \sqrt{2y} + 1 &= 0 & \qquad \sqrt{y+1} - \sqrt{2y} + 1 &= 0 \\
\sqrt{0+1} - \sqrt{2(0)} + 1 &\stackrel{?}{=} 0 & \sqrt{8+1} - \sqrt{2(8)} + 1 &\stackrel{?}{=} 0 \\
1 - 0 + 1 &\stackrel{?}{=} 0 & 3 - 4 + 1 &\stackrel{?}{=} 0 \\
2 &\neq 0 & 0 &= 0 \checkmark
\end{aligned}$$

The solution is 8.

CLASS EXERCISES

Tell whether each equation can be written in quadratic form. Then solve.

1. $x + \sqrt{x} = 0$

2. $2x - 7\sqrt{x} + 6 = 0$

3. $x^4 + 2x^2 = 15$

4. $x^4 + 2x^2 = 3$

5. $\frac{1}{x+3} + \frac{1}{3-x} = 2$

6. $\frac{1}{x+4} - \frac{1}{4-x} = 4$

PRACTICE EXERCISES

Solve each equation.

1. $y^4 + 7y^2 + 6 = 0$

2. $x^4 + 5x^2 + 6 = 0$

3. $z^4 - 6z^2 = 27$

4. $z^4 - 8z^2 = 20$

5. $x + \sqrt{x} - 2 = 0$

6. $x - 2\sqrt{x} - 8 = 0$

7. $x + 5\sqrt{x} = -6$

8. $x + 7\sqrt{x} = -12$

9. $a^4 - 16 = 0$

10. $c^4 - 81 = 0$

11. $2x^4 = 3x^2 - 1$

12. $6x^4 + 3 = -11x^2$

13. $(x+1)^2 + 2(x+1) - 8 = 0$

14. $(x-1)^2 - 4(x-1) - 21 = 0$

15. $\frac{2}{x-1} + \frac{1}{x+1} = 3$

16. $\frac{3}{x-2} + \frac{1}{x+2} = 5$

17. $y^2 - \frac{1}{2}y = \frac{1}{12}$

18. $\frac{1}{2}x^2 - \frac{11}{2}x = 21$

19. $(x+7)(x-9) = 1 - 2x$

20. $(x+1)(x-2) = x + 6$

21. $\frac{2y+3}{4y-1} = \frac{3y-2}{3y+2}$

22. $\frac{9w+1}{w+9} = \frac{w+9}{9w+1}$

23. $5x^4 - 8x^2 - 48 = 0$

24. $3x^4 - 4x^2 = 15$

25. $\dfrac{5x}{5x-2} = \dfrac{10x}{2-5x} + \dfrac{3}{25x^2-4}$

26. $\dfrac{2x+10}{x+8} = \dfrac{x+5}{x+2} - \dfrac{9}{x^2+10x+16}$

27. $0.5x^2 - 0.2x - 0.1 = 0$

28. $0.3x^2 + 0.2x = 0.7$

29. $\sqrt{d-12} = \sqrt{d} - 2$

30. $\sqrt{y+4} = \sqrt{y} - 2$

31. $\dfrac{1}{x} + \dfrac{2}{\sqrt{x}} = 3$

32. $\dfrac{3}{y} - \dfrac{2}{\sqrt{y}} = 5$

33. $\sqrt{s+7} = 7 - \sqrt{3s}$

34. $\sqrt{5r+2} - 1 = \sqrt{r-3}$

35. $2 - \dfrac{1}{x^2} + \dfrac{1}{x} = 0$

36. $6 - \dfrac{4}{x^2} - \dfrac{23}{x} = 0$

37. $\sqrt{x} + 2x = 3$

38. $y - 3\sqrt{y} = 4$

39. $\sqrt{n+2} = 3 - \sqrt{3-n}$

40. $(x^2-1)^2 - 2(x^2-1) = 3$

Applications

41. **Electricity** In a certain electric circuit, the resistance R, in ohms, is found by solving the quadratic equation $(R-4)^2 = 5(14-R)$. Find R.

42. **Optics** One equation used in interference phenomena is $\sqrt{F} = \dfrac{2\sqrt{p}}{1-p}$. Find p if $F = 3$.

Developing Mathematical Power

Thinking Critically Sometimes conditions given in a problem can be correctly expressed in more than one way, resulting in different solutions. Consider the following:

> The sum of two integers is 8. The sum of the first integer and the quotient of the two integers is 5. Find the integers.

Let x represent the first integer. Then $8 - x$ represents the other integer.

43. The quotient of the two numbers could be represented by $\dfrac{8-x}{x}$. Based on this, write an equation to find the integers. Solve and check. How many pairs of integers satisfy the conditions in the problem? Name them.

44. Write a different equation that satisfies the conditions in the problem. Solve and check. Name any additional pairs of integers that satisfy the conditions in the problem.

45. Find all pairs of rational numbers that satisfy the following conditions: The sum of the two numbers is 5. If the first number is multiplied by the difference between the two numbers, the product is 3.

46. Write another number problem for which the given conditions can be expressed in more than one way.

9.9 Quadratic Inequalities

Objective: To solve and graph quadratic inequalities in one variable

Quadratic inequalities are sentences that can be written in one of the following forms, where $a \neq 0$:

$$ax^2 + bx + c < 0 \qquad ax^2 + bx + c > 0$$
$$ax^2 + bx + c \leq 0 \qquad ax^2 + bx + c \geq 0$$

There are several different methods for solving these inequalities.

Capsule Review

For the function $y = ax^2 + bx + c$, the constants a, b, and c have an effect on the shape and location of the parabola.

a. The graph opens upward when $a > 0$ and opens downward when $a < 0$. As $|a|$ increases or decreases, the graph becomes respectively narrower or wider.

b. The constant b, with a, determines the location of the axis of symmetry $\left(x = -\frac{b}{2a}\right)$.

c. The constant c represents the y-intercept.

For each function, locate the axis of symmetry, describe the shape, and name the y-intercept.

1. $y = x^2 + 5x + 6$ **2.** $y = 8x^2 - 4x - 20$ **3.** $y = -0.006x^2$

4. $y = 2x^2 - x + 5$ **5.** $y = x^2 + 10$ **6.** $y = 14x^2 + 7x$

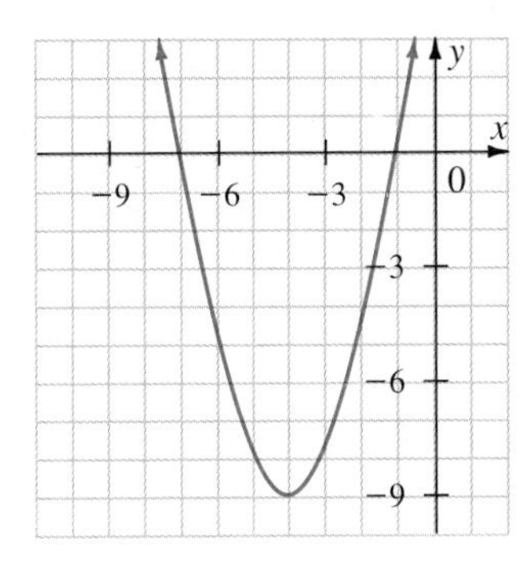

The graph of the function $y = x^2 + 8x + 7$ is shown. The part of the parabola that appears below the x-axis represents all negative values of the function: that is, all values of x such that $x^2 + 8x + 7 < 0$. Since those x-values lie between -7 and -1, the solution set of the inequality $x^2 + 8x + 7 < 0$ is $\{x: -7 < x < -1\}$.

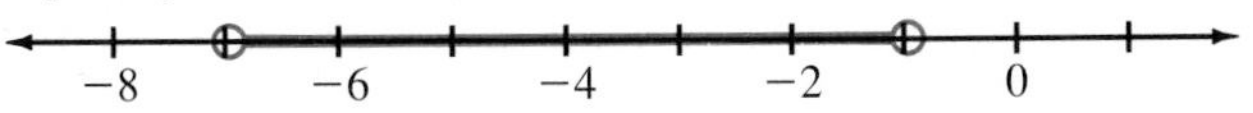

A similar examination of the same graph will show that the solution set of the inequality $x^2 + 8x + 7 > 0$ is $\{x: x < -7 \text{ or } x > -1\}$.

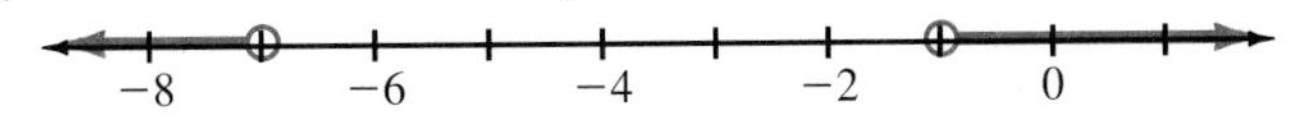

Every quadratic inequality has a quadratic function associated with it. The graph of this associated function is often helpful in graphing the inequality itself.

EXAMPLE 1 **Solve $4x^2 + 4x - 3 \geq 0$. Graph the solution set on a number line.**

First graph the associated function, $y = 4x^2 + 4x - 3$.

The axis of symmetry is $x = -\dfrac{b}{2a} = -\dfrac{1}{2}$;

the vertex is $\left(-\dfrac{b}{2a}, f\left(-\dfrac{b}{2a}\right)\right)$ or $\left(-\dfrac{1}{2}, -4\right)$.

The zeros, determined by factoring, are $-\dfrac{3}{2}$ and $\dfrac{1}{2}$.

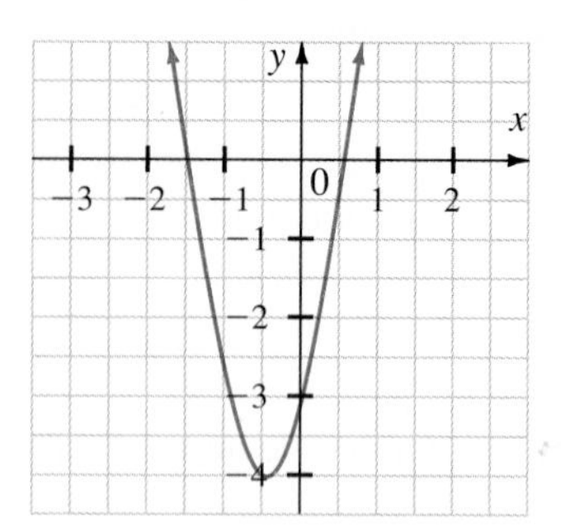

The positive values of the function, corresponding to the parts of the graph that lie above the x-axis, indicate a solution set of $\left\{x\colon x \leq -\dfrac{3}{2} \text{ or } x \geq \dfrac{1}{2}\right\}$. Use solid dots, since $-\dfrac{3}{2}$ and $\dfrac{1}{2}$ are also solutions of the inequality.

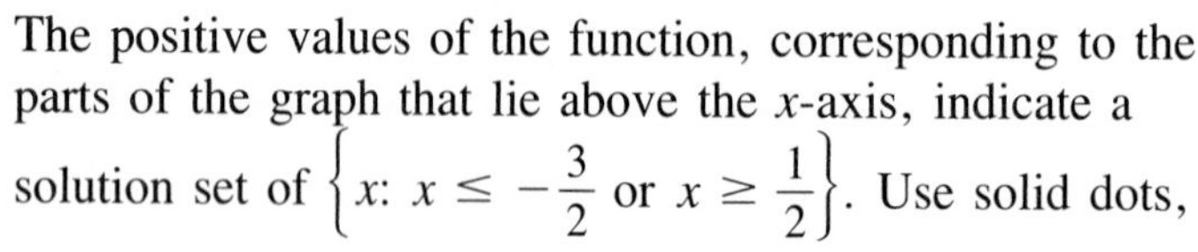

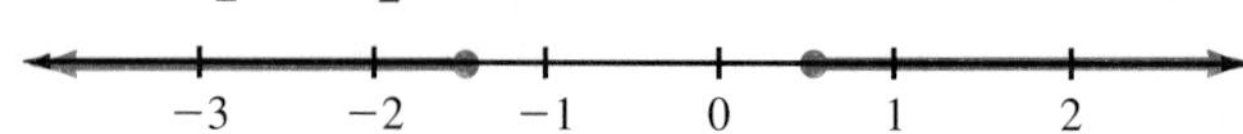

EXAMPLE 2 **Solve $-3x^2 + 4x + 6 > 0$. Graph the solution set on a number line.**

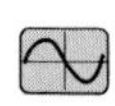

Graph with a graphing utility. Since the expression is not factorable, the zeros are apparently irrational.

Use the quadratic formula: $x = \dfrac{2 + \sqrt{22}}{3}$ and

$x = \dfrac{2 - \sqrt{22}}{3}$.

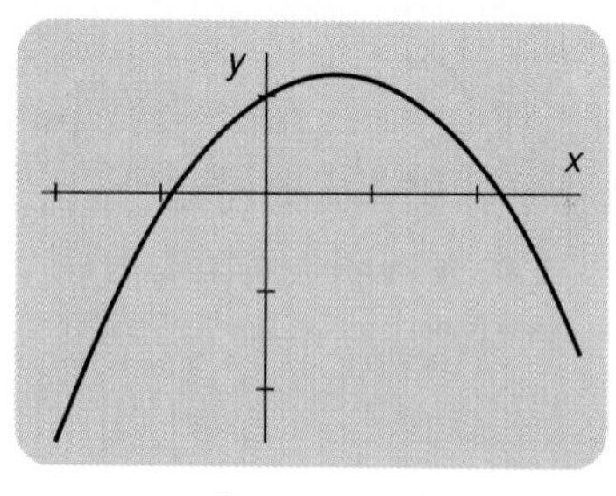

x scl: 1 y scl: 6

The positive values of the function, corresponding to the parts of the graph that lie above the x-axis, indicate a solution set of $\left\{x\colon \dfrac{2 - \sqrt{22}}{3} < x < \dfrac{2 + \sqrt{22}}{3}\right\}$.

Use "empty" dots, since $\dfrac{2 + \sqrt{22}}{3}$ and $\dfrac{2 - \sqrt{22}}{3}$ are not part of the solution set of the inequality.

$\dfrac{2 - \sqrt{22}}{3}$ $\dfrac{2 + \sqrt{22}}{3}$

−2 −1 0 1 2 3

The trace feature of your graphing utility will provide a decimal approximation of the zeros. You can verify your results by checking these numbers against your calculator's approximations of $\dfrac{2 \pm \sqrt{22}}{3}$.

The zero-product principle is used when a quadratic equation is solved by factoring. It can also be used to solve some quadratic inequalities. Consider the following statements:

If $ab = 0$, then $a = 0$ or $b = 0$.

If $ab < 0$, then $a < 0$ and $b > 0$ or $a > 0$ and $b < 0$.

If $ab > 0$, then $a > 0$ and $b > 0$ or $a < 0$ and $b < 0$.

To use these facts to solve a quadratic inequality, first solve the related equation and graph the solution points on a number line. Then use a chart to determine which parts of the number line represent solutions of the inequality.

If the inequality does not have 0 on one side, the first step is to rewrite it in that form.

EXAMPLE 3 **Solve $x^2 + 2x + 3 > 11$. Graph the solution set on a number line.**

$$x^2 + 2x + 3 > 11$$

$$x^2 + 2x - 8 = 0 \quad \textit{Rewrite inequality with 0 on one side.}$$

$$x^2 + 2x - 8 = 0 \quad \textit{Write and solve the related equation.}$$

$$(x + 4)(x - 2) = 0$$

$$x + 4 = 0 \quad \text{or} \quad x - 2 = 0$$

$$x = -4 \quad \text{or} \quad x = 2$$

Locate these points on a number line. Note that the two points separate the number line into three sections. Examine the values of the factors $x + 4$ and $x - 2$ in each section. Then determine the section, or sections, where the trinomial $x^2 + 2x - 8$ is positive. This chart is called a *sign graph*.

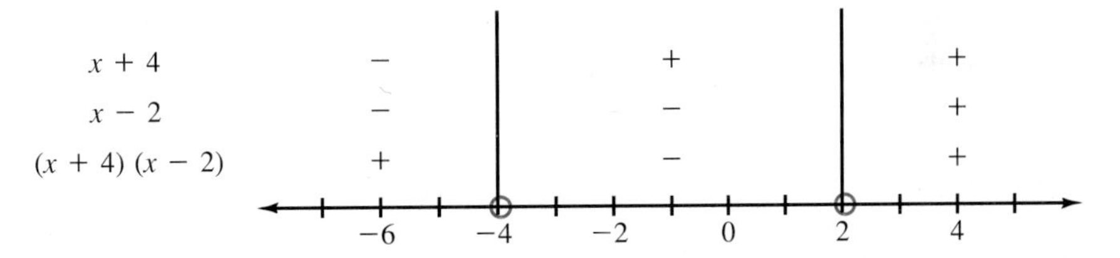

Since $x^2 + 2x - 8 > 0$, the graph of the solution set consists of the sections of the line where $(x + 4)(x - 2)$ is positive. The solution set is $\{x\text{: } x < -4 \text{ or } x > 2\}$.

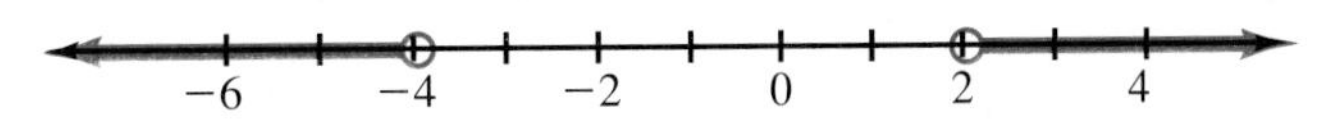

If the inequality symbol had been $<$ rather than $>$, the solution set would have been $\{x\text{: } -4 < x < 2\}$, which is the set of x-values resulting in negative values of the function.

CLASS EXERCISES

Solve each inequality. Graph the solution set on a number line.

1. $x^2 + 2x - 8 > 0$

2. $x^2 + 4x - 5 > 0$

3. $x^2 - 4 \geq 0$

4. $x^2 - 9 > 0$

5. $3x^2 + 8x - 3 < 0$

6. $x^2 + 2x - 3 \leq 0$

7. $x^2 + 3x - 10 > 0$

8. $x^2 - 5x + 6 > 0$

9. $3x^2 - 8x - 3 \leq 0$

PRACTICE EXERCISES

Use technology where appropriate.

Solve each inequality. Graph the solution set on a number line.

1. $x^2 + 3x + 2 > 0$

2. $x^2 - 4x - 5 > 0$

3. $x^2 - 16 > 0$

4. $x^2 - 25 > 0$

5. $x^2 - 6x + 5 \leq 0$

6. $x^2 - 8x - 20 \leq 0$

7. $3x^2 + 4x - 7 > 0$

8. $2x^2 - x - 6 > 0$

9. $-x^2 + 49 \leq 0$

10. $-x^2 + 36 \leq 0$

11. $x^2 + x \geq 0$

12. $x^2 + 2x \geq 0$

13. $-2x^2 + 3x < 0$

14. $-3x^2 + 4x < 0$

15. $x^2 + 3x > 4$

16. $x^2 < 10x - 24$

17. $-2x^2 - 9x + 5 \geq 0$

18. $-2x^2 + x + 10 \geq 0$

19. $2x^2 - 7x - 15 < 0$

20. $6x^2 - 11x - 10 < 0$

21. $3x^2 - 2x \geq 5$

22. $3x^2 + 4x \geq -1$

23. $4x^2 + 4x - 3 > 0$

24. $2x^2 - 7x + 5 > 0$

25. $5x^2 + 3x - 2 < 0$

26. $2x^2 + x - 10 \geq 0$

27. $6x^2 - x - 35 \leq 0$

Solve each inequality and graph the solution set on a number line. First solve the related equation using the quadratic formula. Use decimal approximations, to the nearest tenth, for irrational solutions of the equation, and graph these points.

28. $x^2 + 2x > 4$

29. $x^2 < 8x + 1$

30. $x^2 < 10x - 1$

31. $x^2 + x > 1$

32. $6x^2 + 3x \leq 1$

33. $5x^2 + 5x \leq 2$

34. $6x^2 + 5x - 5 < 0$

35. $2x^2 - 7x - 5 \geq 0$

36. $2x^2 - 6x + 3 > 0$

Applications

Number Theory Solve each problem.

37. The square of a number increased by 4 is not greater than 21. What is the greatest number that satisfies this condition?

38. Twice the square of a number decreased by the number is not greater than 6. What is the greatest number that satisfies this condition?

39. In a set of three numbers, the product of the second and third numbers is positive. The second number is 2 more than the first number, and the third number is 5 less than the second number. What are the possible values of the least number in the set? *Hint:* Let x represent the first number and express the other numbers in terms of x. Then write an inequality in which x is the only variable.

40. In a set of three numbers, the product of the second and third numbers is negative. The second number is 3 less than the first number, and the third number is 8 more than the second number. What are the possible values of the least number in the set?

TEST YOURSELF

Use the discriminant to determine the nature of the solutions of each quadratic equation.

1. $16x^2 - 40x = -25$ **2.** $3x^2 - 2x + 4 = 0$ **3.** $5x^2 = 8x - 2$ 9.6

Describe the relation of the graph of each function to the x-axis.

4. $y = 2x^2 - 11x - 6$ **5.** $y = 4x^2 - 32x + 9$ **6.** $y = 3x^2 + 2x + 1$

For each equation, determine the value(s) of k for which there is only one solution.

7. $8x^2 + kx + 2 = 0$ **8.** $2x^2 - 6x + k = 0$ **9.** $kx^2 - 6x = -3$

Find the sum and the product of the solutions of each equation.

10. $x^2 + 2x + 9 = 0$ **11.** $2x^2 - 6x = -5$ **12.** $9n^2 = 4n - 18$ 9.7

Write a quadratic equation with the given solution(s).

13. 3 and -8 **14.** $4 + \sqrt{10}$ and $4 - \sqrt{10}$ **15.** $\frac{1 + i\sqrt{5}}{4}$

Solve each equation.

16. $x^4 - x^2 - 12 = 0$ **17.** $2x - 5\sqrt{x} - 3 = 0$ 9.8

18. $6(n + 2)^2 + 5(n + 2) - 4 = 0$ **19.** $\frac{2n + 2}{n - 1} = \frac{n + 4}{n - 3}$

20. $\frac{1}{y + 1} + \frac{1}{3} = \frac{1}{y}$ **21.** $\sqrt{3d + 7} = \sqrt{6d + 1}$

Solve each inequality. Graph the solution set on a number line. 9.9

22. $x^2 + 3x - 10 < 0$ **23.** $4x^2 + 11x \geq 3$

INTEGRATING ALGEBRA
Transition Curves

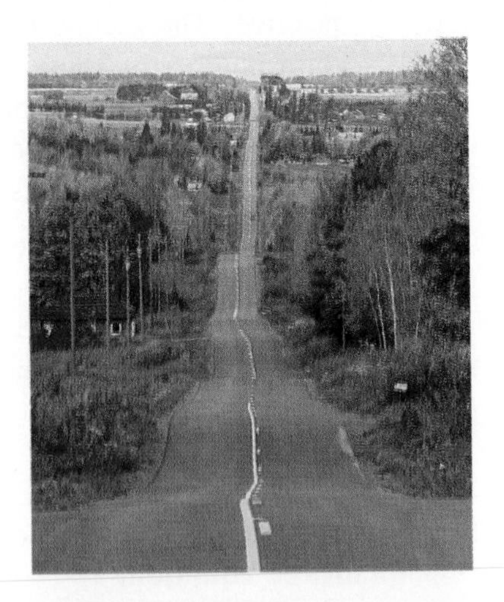

Did you know that the parabola is used by civil engineers in the design of highways? In most parts of the country, construction of level roads is not practical because the natural terrain has varying elevations over any given stretch of land. Before the development of modern engineering skills and powerful equipment, many roads simply followed the natural terrain. A ride on one of these older roads is much like a ride on a roller coaster.

The major task for a civil engineer involved in highway design today is to construct the road in a manner that is safe and comfortable for drivers and passengers riding in modern vehicles. This means that a portion of the road must consist of **transition curves** to smooth out peaks and valleys and eliminate the roller-coaster effect. The parabolic curve $y = ax^2 + bx + c$ has properties that provide a smooth transition for changing elevations.

Two terms used by highway-design engineers are station and gradient. Along a transition curve, points called **stations** are located every 100 ft. The **gradient** of a highway is its slope expressed as a percent, *positive* for uphill and *negative* for downhill. A 3% gradient, for example, means that between stations there is a rise of 3 ft in elevation. If the gradient were −3%, then there would be a decrease in elevation of 3 ft per station.

The problem is to find the curve that will serve as a smooth transition between two gradients in a valley or at a summit. In the drawing below, the segment of highway is assumed to lie in a vertical plane. If y is the elevation, then the transition curve is given by

$$y = ax^2 + bx + c$$

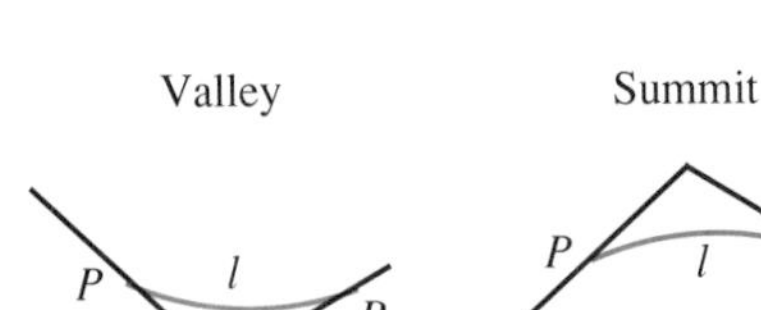

where $a = \dfrac{m - b}{2l}$, b is the gradient at point P, c is the elevation at point P, m is the gradient at point R, l is the length of the curve, and x is the 100 times the number of stations along the curve.

EXAMPLE A segment of highway under construction with a gradient of -4% at point P is to be joined smoothly to a segment with a gradient of 2% at point R. Find the equation for a parabolic curve 960 ft long where the elevation at point P is 1100 ft. Then use the equation to make a table for each station location and its corresponding elevation.

You are given $c = 1100$, $b = -4\% = -0.04$, $m = 2\% = 0.02$, and $l = 960$.

$$\text{Then } a = \frac{m - b}{2l} = \frac{0.02 - (-0.04)}{2(960)} = \frac{0.06}{1920} = 0.00003125.$$

Therefore, the transition curve is $y = 0.00003125x^2 - 0.04x + 1100$.

x	elevation y	x	elevation y
0	1100.0	600	1087.3
100	1096.3	700	1087.3
200	1093.3	800	1088.0
300	1090.8	900	1089.3
400	1089.0	960	1090.4
500	1087.8		

EXERCISES

1. A segment of highway under construction with a gradient of -3% at point P is to be joined smoothly to a segment with a gradient of 2% at point R. Find the equation for a parabolic curve 1000 ft long where the elevation at point P is 1200 ft. Make a chart to list the elevations of points on the curve for each station location beginning at point P.

2. A segment of highway with a gradient of 3% at point P is to be joined smoothly to a segment with a gradient of -5% at point R. Find the equation for a parabolic curve 1200 ft long where the elevation at point P is 1000 ft. Make a chart to list the elevations of points on the curve for each station location beginning at point P.

Tell whether each equation describes a valley or a summit.

3. $y = \frac{1}{25{,}000}x^2 - \frac{1}{20}x + 1100$

4. $y = \frac{1}{-25{,}000}x^2 + \frac{1}{20}x + 1000$

5. $y = \frac{1}{-75{,}000}x^2 + \frac{1}{25}x + 800$

6. $y = \frac{1}{90{,}000}x^2 - \frac{3}{100}x + 980$

CHAPTER 9 SUMMARY AND REVIEW

Vocabulary

axis of symmetry (413)
completing the square (423)
discriminant (437)
maximum point (413)
minimum point (413)
parabola (412)
quadratic form (446)
quadratic formula (427)
quadratic function (412)
quadratic inequality (451)
vertex of a parabola (413)

Graphs of Quadratic Functions A quadratic function is defined by an equation of the form $y = ax^2 + bx + c$, where $a \neq 0$. Its graph is a parabola. If $b = 0$, the axis of symmetry is the y-axis. If $a > 0$, the graph opens upward and the vertex is a minimum point. If $a < 0$, the graph opens downward and the vertex is a maximum point. 9.1

For each function, give the direction of the opening of the parabola, the equation of the axis of symmetry, and the coordinates of the vertex.

1. $y = 3x^2$ **2.** $y = -x^2 - 5$ **3.** $y = 4x^2 + 6$

The Parabola The equation of the axis of a function in the form $y = ax^2 + bx + c$ is $x = -\frac{b}{2a}$. The coordinates of the vertex are $\left(-\frac{b}{2a}, f\left(-\frac{b}{2a}\right)\right)$. The y-intercept is c. Any x-intercepts are solutions to the equation $ax^2 + bx + c = 0$. 9.2

For each function, give the equation of the axis of symmetry, the coordinates of the vertex, and the x- and y-intercepts of its graph. Then sketch the graph.

4. $y = 2x^2 - 4x$ **5.** $y = 4x^2 - 12x + 9$ **6.** $y = -x^2 + 3x + 10$

Completing the Square A quadratic equation can be solved by completing the square, using the fact that $x^2 + 2hx + h^2 = (x + h)^2$. 9.3

Solve by completing the square.

7. $x^2 - 8x - 20 = 0$ **8.** $2y^2 = 4y - 1$ **9.** $y^2 - 3y - 8 = 0$

The Quadratic Formula The solutions of a quadratic equation of the form $ax^2 + bx + c = 0$ are given by $x = \frac{-b \pm \sqrt{b^2 - 4ac}}{2a}$. 9.4

Solve by using the quadratic formula.

10. $3x^2 - 2x - 5 = 0$ **11.** $5w^2 + 5w + 1 = 0$ **12.** $2s - 7 = s^2$

Quadratic equations can be used to solve many types of problems. 9.5

13. The area of a square is equal to that of a rectangle. The length of the rectangle is twice the measure of a side of the square. The width of the rectangle is 5 ft less than the measure of a side of the square. Find the dimensions of the rectangle.

14. A 14-cm chord and a 19-cm chord intersect in a circle. The 14-cm chord is divided into a 6-cm segment and an 8-cm segment. Find the lengths of the two segments into which the other chord is divided.

The Discriminant The discriminant, $b^2 - 4ac$, of a quadratic equation is used to find the nature of the solutions. There can be one rational solution (double), two rational or two irrational solutions, or two complex conjugate solutions. 9.6

Use the discriminant to find the nature of the solutions of each equation.

15. $4x^2 - 17x - 15 = 0$ **16.** $x^2 - 6x + 9 = 0$ **17.** $4x^2 = 3x - 7$

Sum and Product If s_1 and s_2 are solutions of a quadratic equation $ax^2 + bx + c = 0$, then $s_1 + s_2 = -\frac{b}{a}$ and $s_1 s_2 = \frac{c}{a}$. A quadratic equation that has given solutions s_1 and s_2 can be written as $x^2 - (s_1 + s_2)x + s_1 s_2 = 0$. 9.7

Write a quadratic equation with the given solutions.

18. 3 and -4 **19.** $3 + \sqrt{2}$ and $3 - \sqrt{2}$ **20.** $2 + 6i$ and $2 - 6i$

Equations in Quadratic Form These can be solved using the same methods used to solve quadratic equations. 9.8

Solve each equation.

21. $z^4 - 7z^2 + 12 = 0$ **22.** $x + 2\sqrt{x} - 3 = 0$ **23.** $\frac{1}{x - 1} = \frac{2x}{x + 2}$

Quadratic Inequalities First solve the related equation. Graph the solutions on a number line and determine which of the three resulting sections of the line represent solutions to the inequality. Then complete the graph. 9.9

Solve each inequality. Graph the solution set on a number line.

24. $x^2 - 6x + 8 > 0$ **25.** $3x^2 - 4x - 7 \leq 0$

CHAPTER 9 TEST

Determine the equation of the axis of symmetry, the coordinates of the vertex, and the *x*- and *y*-intercepts of each parabola. Then sketch the graph of the parabola in Exercise 1 only.

1. $y = -2x^2 + 8$

2. $y = -3x^2 + 2x + 5$

3. $y = 2x^2 - 5x$

Solve by completing the square.

4. $x^2 - 3x = 4$

5. $y^2 - 2y - 10 = 0$

Solve by using the quadratic formula.

6. $5w^2 = 2w + 3$

7. $2x^2 - 3x + 5 = 0$

Solve.

8. Each side of a square is 3 ft less than twice the length of a side of a smaller square. The difference in the areas of the two squares is 45 ft^2. Find the length of a side of each square.

Use the discriminant to determine the nature of the solutions.

9. $2x^2 - 7x + 5 = 0$

10. $4x^2 = 6x - 3$

Write a quadratic equation with the given solutions.

11. -6 and 5

12. $2 + \sqrt{3}$ and $2 - \sqrt{3}$

Solve.

13. $z^4 - 4z^2 - 5 = 0$

14. $x + 3\sqrt{x} - 10 = 0$

Solve each inequality. Graph the solution set on a number line.

15. $2x^2 + 9x \geq 18$

16. $2x^2 - 3x - 5 < 0$

Challenge

If a, b, and c are real numbers and $a \neq 0$, then $ax^2 + bx + c = 0$ can be written as $\left(x - \frac{-b + \sqrt{b^2 - 4ac}}{2a}\right)\left(x - \frac{-b - \sqrt{b^2 - 4ac}}{2a}\right) = 0.$
Use this result to discuss the graphs of the possible solutions of the inequality $ax^2 + bx + c \leq 0$.

COLLEGE ENTRANCE EXAM REVIEW

In each item you are to compare a quantity in Column 1 with a quantity in Column 2. Write the letter of the correct answer from these choices:

A. The quantity in Column 1 is greater than the quantity in Column 2.
B. The quantity in Column 2 is greater than the quantity in Column 1.
C. The quantity in Column 1 is equal to the quantity in Column 2.
D. The relationship cannot be determined from the given information.

Notes: Information centered over both columns refers to one or both of the quantities being compared. A symbol that appears in both columns has the same meaning in each column. All variables represent real numbers. Most figures are not drawn to scale.

	Column 1	Column 2
1.	$\frac{1}{2-\sqrt{3}}$	$1+\sqrt{3}$

$$\begin{vmatrix} 3 & 0 & 0 \\ 1 & -2 & 0 \\ 0 & 4 & -1 \end{vmatrix} = D$$

	Column 1	Column 2
2.	D	6

$x > 0$

	Column 1	Column 2
3.	$\sqrt{9+x^2}$	$3+x$

Use this diagram for 4–6.

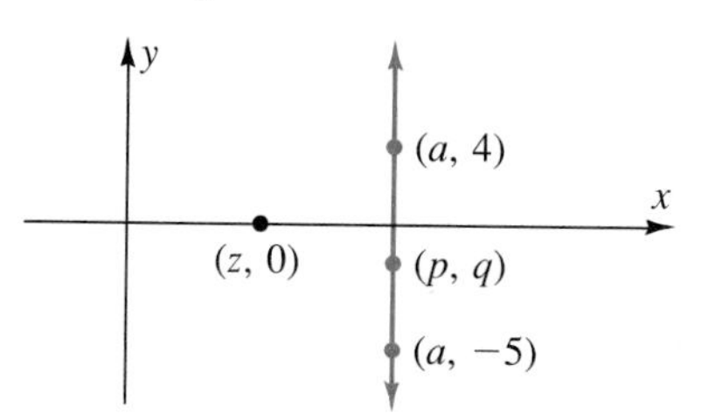

	Column 1	Column 2
4.	p	a
5.	z	p
6.	p	q

$c = 53$
$b = 47$

	Column 1	Column 2
7.	$\frac{126}{c}$	$\frac{126}{b}$

	Column 1	Column 2
8.	$(7-i)^2$	$(8+5i)(2-3i)$
9.	The maximum value of $-2x^2+8x-5$	The minimum value of $3x^2+6x+8$

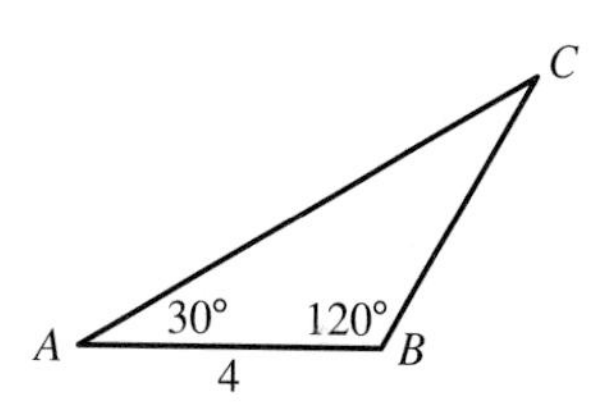

	Column 1	Column 2
10.	Area of $\triangle ABC$	4

Use this equation for 11–13.

$$3x^2 - 9x = 11$$

	Column 1	Column 2
11.	The sum of the roots	3
12.	The product of the roots	-11
13.	Discriminant	0

$3 < n < 7$

	Column 1	Column 2
14.	n	5

MIXED REVIEW

Compute.

Examples 20.04×7.5 $\qquad$ $4.2 \div 0.12$

$$\begin{array}{r} 20.04 \\ \times \quad 7.5 \\ \hline 10020 \\ 14028 \\ \hline 150.300 \end{array} \quad \begin{array}{l} \text{2 decimal places} \\ \text{1 decimal place} \\ \\ \\ \text{3 decimal places} \end{array}$$

$$\begin{array}{r} 35 \\ 0.12\overline{)4.20} \\ -36 \\ \hline 60 \\ -60 \\ \hline \end{array}$$

1. 9.34×0.7 **2.** 0.6×0.05 **3.** $6.3 \div 0.15$ **4.** $0.8 \div 25$

5. $45.6 - 36.48$ **6.** $14 - 8.9$

7. $7.8 + 21 + 18.39$ **8.** $20.25 + 9.7 + 15.05$

Factor.

Example $x^2 - 8x + 12 = (x - 2)(x - 6)$ *Find the factors of 12 whose sum is −8.*

(x^2, $+12$, $-8x$)

9. $x^2 - 5x + 6$ **10.** $2a^2 - 3a - 9$ **11.** $9m^2 - 24m + 16$

12. $8k^2 - 10k + 3$ **13.** $y^2 - 3x - 40$ **14.** $6t^3 + 11t^2 - 10t$

Solve.

Example
$$\begin{aligned} x^2 + 15 &= 6 \\ x^2 &= -9 \\ x &= \sqrt{-9} \\ x &= \pm 3i \end{aligned}$$
Remember: $i^2 = -1$

15. $x^2 + 9 = 5$ **16.** $5y - 8 = 12$ **17.** $a^2 - 4 = 21$

18. $d^2 + 18 = -18$ **19.** $3x^2 - 18 = 42$ **20.** $3m^2 + 5m = 2$

21. $y^2 = 5y$ **22.** $9c^2 + 4 = 12c$ **23.** $p^3 + p^2 = 9p + 9$

Solve.

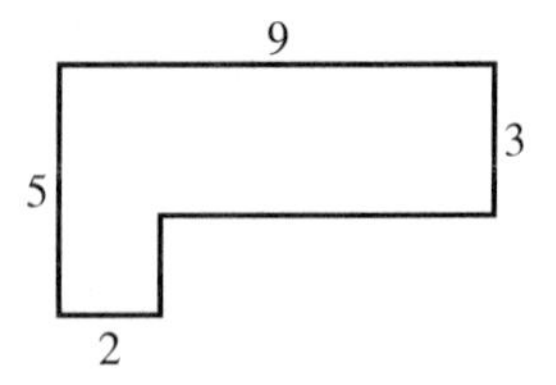

24. Find the perimeter of the figure at the right.

25. A rectangular swimming pool is 8 ft long, 5 ft wide and 4 ft deep. How much water is needed to fill the pool?

Polynomial Functions

Sizing up the situation!

Developing Mathematical Power

Manufacturers and advertisers consider many factors when designing the packaging for a product. Manufacturers have to consider the cost of the materials used and the amount of material needed. Their packaging must protect the product from damage during shipping, and its shape affects how much the package will hold.

Packaging is often designed to catch the shopper's eye. Does the way a product looks influence your decision to buy it?

Project

Suppose you are the designer for a series of new containers for popcorn. Each container will have twice the volume of the next smaller size. The containers should be made of as little material as possible. Prepare your presentation of the new containers, including models or diagrams.

Products are often packaged in two or more sizes.

10.1

Dividing Polynomials

Objective: To divide one polynomial by another

To solve polynomial equations of degree higher than 2, you will need to know how to divide polynomials. The properties of exponents are used when one monomial is divided by another. In this chapter, assume that no denominator is equal to zero.

Capsule Review

EXAMPLES

$$a^5 \div a^3 = a^{5-3} = a^2$$

$$8a^4 \div 2a^7 = 4a^{4-7} = 4a^{-3} = \frac{4}{a^3}$$

$$a^6 \div a^6 = a^{6-6} = a^0 = 1$$

Divide.

1. $49x^3 \div 7x^2$ **2.** $-36xy^3 \div 4x^2y^5$ **3.** $-27xz^2 \div -9xz$

4. $\frac{18wx^2y^3z^4}{-4w^2xy^4z^3}$ **5.** $\frac{81c^4d^3}{-3c^3d^4}$ **6.** $\frac{-25mr^2s}{-45mr^2st^2}$

To divide a polynomial with more than one term by a monomial, use the distributive property. That is, use the fact that if $c \neq 0$, then

$$\frac{a+b}{c} = \frac{1}{c}(a+b) = \frac{a}{c} + \frac{b}{c}$$

EXAMPLE 1 **Divide:** **a.** $\frac{28x^3y^5 - 21x^4y^3 + 49x^3y^6}{7x^2y^3}$ **b.** $\frac{y^3 - 3y^2 + 4y - 6}{y}$

a. $$\frac{28x^3y^5 - 21x^4y^3 + 49x^3y^6}{7x^2y^3} = \frac{28x^3y^5}{7x^2y^3} - \frac{21x^4y^3}{7x^2y^3} + \frac{49x^3y^6}{7x^2y^3}$$
$$= 4xy^2 - 3x^2 + 7xy^3$$

b. $$\frac{y^3 - 3y^2 + 4y - 6}{y} = \frac{y^3}{y} - \frac{3y^2}{y} + \frac{4y}{y} - \frac{6}{y} = y^2 - 3y + 4 - \frac{6}{y}$$

The algorithm, or process, for dividing a polynomial by a polynomial is similar to the long-division algorithm in arithmetic.

$$
\begin{array}{r l}
 & 33 \leftarrow \text{quotient} \\
\text{divisor} \nearrow 28\overline{)929} & \leftarrow \text{dividend} \\
 \underline{84} & \\
 89 & \\
 \underline{84} & \\
 5 & \leftarrow \text{remainder}
\end{array}
$$

Divide.
Multiply.
Subtract. Bring down the next digit.
Repeat the steps.
Stop when the remainder is 0 or when it is less than the divisor.

Thus, $\frac{929}{28} = 33\frac{5}{28} = 33 + \frac{5}{28}$ or $929 = 33 \times 28 + 5$

In general, $\dfrac{\text{dividend}}{\text{divisor}} = \text{quotient} + \dfrac{\text{remainder}}{\text{divisor}}$

or $\quad \text{dividend} = \text{quotient} \times \text{divisor} + \text{remainder}$

EXAMPLE 2 **Divide: $(6a^3 + 11a^2 - 4a - 9) \div (3a - 2)$**

$$
\begin{array}{r}
2a^2 \\
3a - 2\overline{)6a^3 + 11a^2 - 4a - 9} \\
\underline{6a^3 - 4a^2} \\
15a^2 - 4a
\end{array}
$$

Divide $6a^3$ by $3a$.
Multiply $3a - 2$ by $2a^2$.
Subtract and bring down the next term.

$$
\begin{array}{r}
2a^2 + 5a \\
3a - 2\overline{)6a^3 + 11a^2 - 4a - 9} \\
\underline{6a^3 - 4a^2} \\
15a^2 - 4a \\
\underline{15a^2 - 10a} \\
6a - 9
\end{array}
$$

Divide $15a^2$ by $3a$.
Multiply $3a - 2$ by $5a$.
Subtract and bring down the next term.

$$
\begin{array}{r}
2a^2 + 5a + 2 \\
3a - 2\overline{)6a^3 + 11a^2 - 4a - 9} \\
\underline{6a^3 - 4a^2} \\
15a^2 - 4a \\
\underline{15a^2 - 10a} \\
6a - 9 \\
\underline{6a - 4} \\
- 5
\end{array}
$$

Divide $6a$ by $3a$.
Multiply $3a - 2$ by 2. Subtract.
Stop when the remainder is 0 or its degree is lower than that of the divisor.

Check: dividend = quotient × divisor + remainder

$$
\begin{aligned}
6a^3 + 11a^2 - 4a - 9 &\stackrel{?}{=} (2a^2 + 5a + 2)(3a - 2) - 5 \\
6a^3 + 11a^2 - 4a - 9 &\stackrel{?}{=} (2a^2 + 5a + 2)3a + (2a^2 + 5a + 2)(-2) - 5 \\
6a^3 + 11a^2 - 4a - 9 &\stackrel{?}{=} 6a^3 + 15a^2 + 6a - 4a^2 - 10a - 4 - 5 \\
6a^3 + 11a^2 - 4a - 9 &= 6a^3 + 11a^2 - 4a - 9 \checkmark
\end{aligned}
$$

Then $\dfrac{6a^3 + 11a^2 - 4a - 9}{3a - 2} = 2a^2 + 5a + 2 - \dfrac{5}{3a - 2}$

Before dividing, arrange the terms of the polynomials in descending order of one variable.

EXAMPLE 3 **Divide:** $(15x^2 - 4x + 3x^4 + 3 - 5x^3) \div (4 - x + x^2)$

$$\begin{array}{r} 3x^2 - 2x + 1 \\ x^2 - x + 4\overline{)3x^4 - 5x^3 + 15x^2 - 4x + 3} \\ \underline{3x^4 - 3x^3 + 12x^2} \qquad\qquad \\ -2x^3 + 3x^2 - 4x \qquad \\ \underline{-2x^3 + 2x^2 - 8x} \qquad \\ x^2 + 4x + 3 \\ \underline{x^2 - x + 4} \\ 5x - 1 \end{array}$$

Arrange the terms in descending order.

Then $\dfrac{3x^4 - 5x^3 + 15x^2 - 4x + 3}{x^2 - x + 4} = 3x^2 - 2x + 1 + \dfrac{5x - 1}{x^2 - x + 4}$

If the dividend or the divisor has missing terms, insert these terms with zero coefficients.

EXAMPLE 4 **Divide:** $\dfrac{x^3 - 13x - 12}{x - 4}$

$$\begin{array}{r} x^2 + 4x + 3 \\ x - 4\overline{)x^3 + 0x^2 - 13x - 12} \\ \underline{x^3 - 4x^2} \qquad\qquad\quad \\ 4x^2 - 13x \qquad \\ \underline{4x^2 - 16x} \qquad \\ 3x - 12 \\ \underline{3x - 12} \\ 0 \end{array}$$

Insert $0x^2$.

The remainder is 0.

Then $\dfrac{x^3 - 13x - 12}{x - 4} = x^2 + 4x + 3$

CLASS EXERCISES

1. If $\dfrac{x^3 - 2x^2 + x - 1}{x} = x^2 - 2x + 1 - \dfrac{1}{x}$, identify the quotient.

2. If $\dfrac{x^2 - x - 1}{x - 3} = x + 2 + \dfrac{5}{x - 3}$, identify the remainder.

3. What would be your first step in dividing $16 - 5x^2 - x + x^3$ by $4 - x$?

4. Before dividing $x^3 - 8$ by $x - 2$, what should you do?

Divide.

5. $\dfrac{x^2 - 6x + 8}{x - 2}$

6. $\dfrac{x^2 - 7x + 10}{x + 3}$

7. $\dfrac{x^3 - 2x^2 + 7x - 8}{2x}$

8. $\dfrac{x^2 + 3}{x - 1}$

9. $\dfrac{6x^2 + 2x - 28}{3x + 7}$

10. $\dfrac{27a^4b^8 - 34a^3b^5 + 16a^2b^3}{2a^2b^2}$

PRACTICE EXERCISES

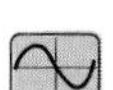 Use technology where appropriate.

Divide.

1. $(12x^5 - 15x^4 - 18x^3) \div 3x^2$

2. $(10s^4 - 12s^3 - 18s^2) \div 2s^2$

3. $\dfrac{4a^7b^6 - 6a^6b^8 + 12a^2b^{10}}{2a^2b^6}$

4. $\dfrac{24x^2y^3z + 12x^3y^6 - 15wx^4y^8}{-3x^2y^6}$

5. $\dfrac{9d^5 - 6d^4 + 12d^3 - 7}{d}$

6. $\dfrac{12x^3 + x^2 - 6x + 4}{x}$

7. $\dfrac{-39x^8 + 27x^6 + 12x^3 - 24}{-3x^3}$

8. $\dfrac{-24a^4 + 21a^3 + 18a^2 + 12}{3a^2}$

9. $(x^2 - 14x - 36) \div (x + 2)$

10. $(b^2 - 5b - 84) \div (b + 7)$

11. $(a^2 - 4a - 86) \div (a + 6)$

12. $(x^2 - 12x - 45) \div (x + 3)$

13. $\dfrac{5c^2 + 2c + 3}{c + 1}$

14. $\dfrac{6x^2 - 8x - 2}{x - 1}$

15. $\dfrac{7x - 5x^2 - 3 + 6x^3}{2x - 1}$

16. $\dfrac{4x^2 - 6 + 4x^3 - 7x}{2x + 3}$

17. $(x^2 - 6) \div (x - 2)$

18. $(x^2 - 12) \div (x - 3)$

19. $(4x^2 + 25) \div (2x - 5)$

20. $(9x^2 + 16) \div (3x - 4)$

21. $\dfrac{x^4 - x^3 - 9 + 21x - 15x^2}{3x - 4 + x^2}$

22. $\dfrac{-2x^3 - 4x^2 + x^4 - 10x + 3}{x^2 + 3 + 2x}$

23. $\dfrac{x^4 + 2x^3 - 2x^2 + 6x - 15}{x^2 + 3}$

24. $\dfrac{x^4 + 3x^3 - 9x^2 - 21x + 8}{x^2 - 7}$

25. $\dfrac{6n^3 + 5n^2 + 9}{2n + 3}$

26. $\dfrac{8n^3 + 5n^2 - 6}{2n - 1}$

27. $\dfrac{6x^3 - 3x^2 - 7}{3x - 2}$

28. $\dfrac{8x^3 - 4x^2 - 9}{2x - 3}$

29. $(x^3 - 1) \div (x - 1)$

30. $(y^3 + 27) \div (y + 3)$

31. $(64x^3 + 27) \div (4x + 3)$

32. $(27x^3 + 8) \div (3x + 2)$

33. $(y^4 - 1) \div (y + 1)$

34. $(x^4 - 1) \div (x - 1)$

In Exercises 35–40, assume that *a* and *b* are positive integers.

35. $\dfrac{x^{3a} - 4x^{2a} + 2x^a - 12}{x^a - 5}$

36. $\dfrac{y^{3b} - 2y^{2b} + 4y^b - 7}{y^b - 3}$

37. $\dfrac{x^{3a} - 5x^{2a} + 6x^a - 8x^{a-1}}{x^a}$

38. $\dfrac{y^{2b} - y^b + y^{b-1} - y^{b-2}}{y^b}$

39. $\dfrac{x^{4a} - 4x^{2a} - x^a + 2}{x^{2a} + x^a - 1}$

40. $\dfrac{x^{5a} - 2x^{2a} - x^a - 1}{x^{2a} + x^a + 1}$

Applications

41. **Geometry** A rectangular box has a volume of $(2x^3 + 3x^2 - 8x + 3)$ cm^3. If the length of the box is $(x + 3)$ cm and the width is $(x - 1)$ cm, find the height in terms of x.

42. **Engineering** When a section of a certain type of plastic is cut, the length of the section determines the strength of the plastic. An engineer has discovered that the function: $f(x) = x^3 - 14x^2 + 53x - 40$, describes the relative strength of a section of length x. By trial and error, he also found that 5-ft plastic sections are extremely weak. What are the other two lengths that result in extremely weak sections of plastic?

Developing Mathematical Power

Extension An example of a special function is the **greatest integer function.** The greatest integer function of a real number x, represented by $[x]$, is the greatest integer that is less than or equal to x. For example,

$$[4.25] = 4 \qquad [6] = 6 \qquad [-2.3] = -3$$

The greatest integer function is sometimes called a *step function,* because of the shape of its graph.

Determine whether each statement is true or false for real numbers *x* and *y*.

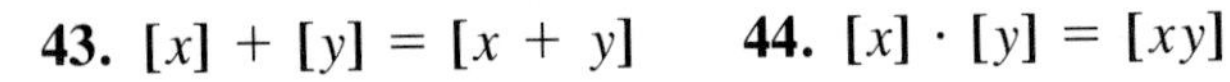

43. $[x] + [y] = [x + y]$

44. $[x] \cdot [y] = [xy]$

45. $[x] + [y] = [y] + [x]$

46. $-[x] = [-x]$

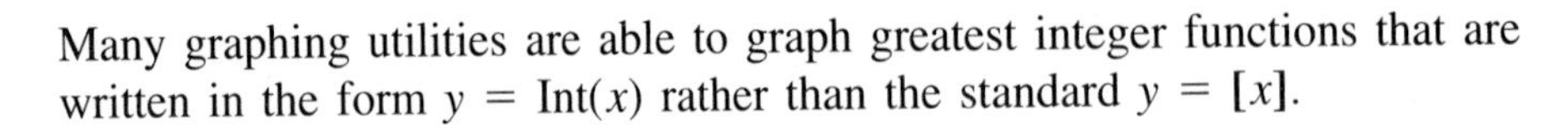

Many graphing utilities are able to graph greatest integer functions that are written in the form $y = \text{Int}(x)$ rather than the standard $y = [x]$.

Graph the following on paper or with a graphing utility.

47. $y = [2x]$

48. $y = 2[x]$

49. $y = [\sqrt{x}]$

50. $y = \sqrt{[x]}$

Synthetic Division

Objective: To use synthetic division to divide a polynomial by a first degree binomial

When a polynomial is divided by a polynomial of the form $x - a$, the terms are rewritten several times. To eliminate this repetition, *synthetic division* can be used. In synthetic division the terms of the dividend and the divisor are arranged in descending order of the variable.

Capsule Review

The terms of the polynomial $8n^3 - 22n^2 - 5n + 12$ are arranged in descending order of n. The degree of the polynomial is three, since the term $8n^3$ is the term of highest degree.

EXAMPLE **Write the polynomial in descending order of x and give its degree: $6x - 4x^2 - 9 + 7x^4$**

$7x^4 - 4x^2 + 6x - 9$ degree: 4

Write each polynomial in descending order of x and give its degree.

1. $5x - 2x^2 + 9 + 4x^3$

2. $-3x^2 + 8 + 2x^3 - 6x + x^4$

3. $7 - 9x - 3x^2$

4. $10 + 5x^3 - 9x^2$

5. $5x^2 - 2 + 3x^3 - 5x^4$

6. $7x^3 - 2x^2 + 8x^5$

Consider the division of $2x^3 + 3x^2 - 15x - 16$ by $x - 3$. There is no need to write the x's and their exponents, since the positions of the coefficients indicate what the exponents are. That is, the division on the left below can be written as shown on the right.

$$\begin{array}{rrrrr}
 & 2x^2 & +\ 9x & +\ 12 & \\
x - 3) & 2x^3 + 3x^2 & -\ 15x & -\ 16 & \\
 & 2x^3 - 6x^2 & & & \\
\hline
 & 9x^2 & -\ 15x & & \\
 & 9x^2 & -\ 27x & & \\
\hline
 & & 12x & -\ 16 & \\
 & & 12x & -\ 36 & \\
\hline
 & & & 20 &
\end{array}
\qquad
\begin{array}{rrrrr}
 & 2 & 9 & 12 & \\
-3) & 2 & 3 & -15 & -16 \\
 & 2 & -6 & & \\
\hline
 & & 9 & -15 & \\
 & & 9 & -27 & \\
\hline
 & & & 12 & -16 \\
 & & & 12 & -36 \\
\hline
 & & & & 20
\end{array}$$

At the bottom of page 469, the division form at the right repeats many of the coefficients more than once. Rewritten without such duplications, the division can be shown even more compactly. Also, if you divide by the opposite of -3, you can use addition instead of subtraction. This procedure for dividing a polynomial in one variable by a binomial of the form $x - a$ is called **synthetic division.**

The synthetic division process is illustrated below, using the example $(2x^3 + 3x^2 - 15x - 16) \div (x - 3)$. Think of the divisor in the form $x - a$, and write a at the left. In this case, $a = 3$. Then write the coefficients of the terms of the dividend, where those terms are in descending order.

Bring down the first coefficient. Multiply the divisor by this number. Write the product under the second coefficient. Then add.

$$\begin{array}{r|rrrr} 3 & 2 & 3 & -15 & -16 \\ & & 6 & & \\ \hline & 2 & 9 & & \end{array} \qquad \begin{array}{l} 2 \times 3 = 6 \\ 3 + 6 = 9 \end{array}$$

Repeat the process of multiplying and adding.

$$\begin{array}{r|rrrr} 3 & 2 & 3 & -15 & -16 \\ & & 6 & 27 & \\ \hline & 2 & 9 & 12 & \end{array} \qquad \begin{array}{l} 9 \times 3 = 27 \\ -15 + 27 = 12 \end{array}$$

Continue until there are no more coefficients in the dividend. The first three numbers in the bottom row are the coefficients of the terms of the quotient polynomial. The last number is the remainder.

$$\begin{array}{r|rrrr} 3 & 2 & 3 & -15 & -16 \\ & & 6 & 27 & 36 \\ \hline & 2 & 9 & 12 & 20 \end{array} \qquad \begin{array}{l} 12 \times 3 = 36 \\ -16 + 36 = 20 \end{array}$$

quotient: $2x^2 + 9x + 12$
remainder: 20

Then $$\frac{2x^3 + 3x^2 - 15x - 16}{x - 3} = 2x^2 + 9x + 12 + \frac{20}{x - 3}$$

or $$2x^3 + 3x^2 - 15x - 16 = (2x^2 + 9x + 12)(x - 3) + 20$$

Notice that the quotient polynomial is of degree one less than the original polynomial.

EXAMPLE 1 **Divide, using synthetic division:**

a. $\dfrac{x^3 + 4x^2 - 7x - 14}{x - 2}$ **b.** $\dfrac{2x^3 + 5x^2 - 7x - 12}{x + 3}$

a. $\dfrac{x^3 + 4x^2 - 7x - 14}{x - 2}$ *The divisor is $x - 2$, so $a = 2$.*

$$\begin{array}{r|rrrr} 2 & 1 & 4 & -7 & -14 \\ & & 2 & 12 & 10 \\ \hline & 1 & 6 & 5 & -4 \end{array}$$

Then $$\frac{x^3 + 4x^2 - 7x - 14}{x - 2} = x^2 + 6x + 5 - \frac{4}{x - 2}$$

b. $\dfrac{2x^3 + 5x^2 - 7x - 12}{x + 3}$ *The divisor is $x + 3$, or $x - (-3)$, so $a = -3$.*

$$\begin{array}{r|rrrr} -3 & 2 & 5 & -7 & -12 \\ & & -6 & 3 & 12 \\ \hline & 2 & -1 & -4 & 0 \end{array}$$

Zero remainder

Then $\dfrac{2x^3 + 5x^2 - 7x - 12}{x + 3} = 2x^2 - x - 4$

If a term is missing in the dividend, write a zero in that place.

EXAMPLE 2 **Divide, using synthetic division:** $\dfrac{x^4 - 5x^2 - 10x - 12}{x + 2}$

$$\begin{array}{r|rrrrr} -2 & 1 & 0 & -5 & -10 & -12 \\ & & -2 & 4 & 2 & 16 \\ \hline & 1 & -2 & -1 & -8 & 4 \end{array}$$

The x^3 term is missing, so insert 0 for $0x^3$.

Then $\dfrac{x^4 - 5x^2 - 10x - 12}{x + 2} = x^3 - 2x^2 - x - 8 + \dfrac{4}{x + 2}$

When the coefficient of x in the divisor is not 1, synthetic division can still be used. However, the example must be written in an equivalent form with the divisor in the form $x - a$. In Example 3 below, the divisor $2x - 3$ must be written as $2\left(x - \frac{3}{2}\right)$.

EXAMPLE 3 **Divide, using synthetic division:** $\dfrac{6x^3 + 3x^2 + 10x + 14}{2x - 3}$

$$\frac{6x^3 + 3x^2 + 10x + 14}{2x - 3} = \frac{1}{2}\left(\frac{6x^3 + 3x^2 + 10x + 14}{x - \frac{3}{2}}\right)$$

$$\begin{array}{r|rrrr} \frac{3}{2} & 6 & 3 & 10 & 14 \\ & & 9 & 18 & 42 \\ \hline & 6 & 12 & 28 & 56 \end{array}$$

Divide: $(6x^3 + 3x^2 + 10x + 14) \div \left(x - \frac{3}{2}\right)$

Then
$$\begin{aligned} \frac{6x^3 + 3x^2 + 10x + 14}{2x - 3} &= \frac{1}{2}\left(6x^2 + 12x + 28 + \frac{56}{x - \frac{3}{2}}\right) \\ &= \frac{1}{2}(6x^2) + \frac{1}{2}(12x) + \frac{1}{2}(28) + \frac{1}{2}\left(\frac{56}{x - \frac{3}{2}}\right) \\ &= 3x^2 + 6x + 14 + \frac{56}{2x - 3} \end{aligned}$$

CLASS EXERCISES

Using x as the variable, express each synthetic division exercise below in the form $\frac{\text{dividend}}{\text{divisor}} = \text{quotient} + \frac{\text{remainder}}{\text{divisor}}$.

1. $\begin{array}{r|rrrr} 3 & 1 & -6 & 3 & 5 \\ & & 3 & -9 & -18 \\ \hline & 1 & -3 & -6 & -13 \end{array}$

2. $\begin{array}{r|rrrr} 5 & 1 & -3 & 2 & 7 \\ & & 5 & 10 & 60 \\ \hline & 1 & 2 & 12 & 67 \end{array}$

3. $\begin{array}{r|rrrr} 2 & 3 & 0 & -1 & -2 \\ & & 6 & 12 & 22 \\ \hline & 3 & 6 & 11 & 20 \end{array}$

4. $\begin{array}{r|rrrrr} -1 & 5 & 0 & 2 & 0 & 7 \\ & & -5 & 5 & -7 & 7 \\ \hline & 5 & -5 & 7 & -7 & 14 \end{array}$

5. $\begin{array}{r|rrrr} -5 & 1 & 0 & 0 & 125 \\ & & -5 & 25 & -125 \\ \hline & 1 & -5 & 25 & 0 \end{array}$

6. $\begin{array}{r|rrrr} 3 & 1 & 0 & 0 & -27 \\ & & 3 & 9 & 27 \\ \hline & 1 & 3 & 9 & 0 \end{array}$

Divide, using synthetic division. Express the answer in the form $\frac{\text{dividend}}{\text{divisor}} = \text{quotient} + \frac{\text{remainder}}{\text{divisor}}$.

7. $(x^3 + 3x^2 + 5x + 7) \div (x - 1)$

8. $(x^3 + 2x^2 - 5x + 6) \div (x + 4)$

PRACTICE EXERCISES

Use technology where appropriate.

Divide, using synthetic division. Express the answer in the form $\frac{\text{dividend}}{\text{divisor}} = \text{quotient} + \frac{\text{remainder}}{\text{divisor}}$.

1. $(x^3 - 12x^2 - 5x + 8) \div (x - 2)$

2. $(x^3 - 4x^2 - 5x - 12) \div (x - 2)$

3. $(x^4 - 6x^2 + 7x - 12) \div (x + 3)$

4. $(x^3 - 6x^2 + 7x - 12) \div (x - 3)$

5. $(x^3 + 3x^2 - 8x - 10) \div (x - 2)$

6. $(x^3 - 7x^2 + 7x - 15) \div (x - 3)$

7. $\dfrac{2x^3 - 5x^2 - 7x - 3}{x + 1}$

8. $\dfrac{2x^3 - 7x^2 - 3x - 5}{x + 1}$

9. $\dfrac{2x^3 + 4x^2 - 10x - 9}{x - 3}$

10. $\dfrac{2x^3 - 12x^2 - 4x + 6}{x - 2}$

11. $\dfrac{2x^3 + 2x^2 - 8x + 6}{x - 2}$

12. $\dfrac{2x^3 + 3x^2 - 6x - 8}{x - 4}$

13. $\dfrac{3x^3 - 5x^2 - 17x - 12}{x - 4}$

14. $\dfrac{x^4 - 2x^2 - 6x + 15}{x + 5}$

15. $\dfrac{x^4 - 6x^3 + 3x - 9}{x + 3}$

16. $\dfrac{x^4 - 3x^2 - 2x + 8}{x + 2}$

17. $\dfrac{x^4 - 3x^2 - 5x + 20}{x + 5}$

18. $\dfrac{x^4 - 9x^3 - 7x^2 + 10}{x + 5}$

19. $\dfrac{2x^4 - 3x^3 - 4x + 10}{x - 2}$

20. $\dfrac{x^4 + 3x^2 - 5x + 10}{x + 1}$

21. $(y^3 - 1) \div (y - 1)$

22. $(x^3 - 8) \div (x - 2)$

23. $(x^3 + 64) \div (x + 4)$

24. $(x^3 + 27) \div (x + 3)$

25. $\dfrac{y^3 + 1}{y - 1}$

26. $\dfrac{y^3 + 8}{y - 2}$

27. $\dfrac{y^3 - 64}{y + 4}$

28. $\dfrac{y^3 - 27}{y + 3}$

29. $\dfrac{6a^3 - 13a^2 - 12a + 4}{2a + 1}$

30. $\dfrac{4c^3 + 4c^2 + 5c - 3}{2c + 1}$

31. $\dfrac{3x^4 + 4 - 10x + 7x^3}{3x - 2}$

32. $\dfrac{9x^2 - 21x + 4x^4 - 9}{2x - 3}$

33. $(8x^3 + 27) \div (2x - 3)$

34. $(27y^3 - 1) \div (3y + 1)$

Determine the value of k that will make the divisor a factor of the dividend.

35. $\dfrac{x^3 - 5x^2 - 2x + k}{x - 3}$

36. $\dfrac{x^3 + 2x^2 - 5x - k}{x - 2}$

37. $\dfrac{x^3 + 3x^2 - kx - 12}{x + 3}$

38. $\dfrac{x^3 + 2x^2 - kx - 2}{x + 2}$

39. $\dfrac{z^3 - kz^2 + 7z + 15}{z - 3}$

40. $\dfrac{x^3 - kx^2 - 21x - 18}{x - 6}$

Applications

Technology The repeated application of a single process, as in synthetic division, is called *iteration*. The iterative feature of a graphing utility can be used to solve *difference equations,* in which the value of one expression $(f(x_k))$ is used to find the value of the next expression $(f(x_{k+1}))$. In other words, each equation is defined in terms of the previous equation.

For example, the difference equation $T_{k+1} = \frac{8}{9}(T_k) + \frac{75}{9}$ is used to find the temperature of a can of club soda that comes out of the refrigerator at 36°F and gradually warms to room temperature. The chart on the next page gives the temperature (T_k) after k minutes.

To use the calculator for iteration, store the difference equation $Y = \frac{8}{9}X + \frac{75}{9}$ in the variable Y_1 or F_1. Then store the first value (36) in X. The iterative process will occur when you now store $Y_1 \rightarrow X$ (or $F_1 \rightarrow X$). As you continue to press ENTER or EXE, the temperature for successive minutes will appear. Check your calculator manual for instructions for storing.

$$T_0 = 36$$
$$T_1 = \frac{8}{9}(T_0) + \frac{75}{9} = 40.33333333$$
$$T_2 = \frac{8}{9}(T_1) + \frac{75}{9} = 44.18518519$$
$$T_3 = \frac{8}{9}(T_2) + \frac{75}{9} = 47.6090535$$
$$\vdots$$
$$T_{k+1} = \frac{8}{9}(T_k) + \frac{75}{9}$$

41. If you continue the process, the results will approach what number?

42. How would you define the difference equation in order to change the approached value to 82?

43. Compare the maximum values obtained using different initial temperature values for the soda can.

44. Describe the pattern for the maximum possible temperature when the "slope" of the difference equation is replaced with $\frac{1}{9}, \frac{2}{9}, \frac{3}{9}, \ldots, \frac{8}{9}$. Does a similar pattern exist when other fractions are used?

45. Health researchers have found that the difference equation $T_{k+1} = T_k(1 - T_k) + T_k$ models the normal spread of flu in a community, with T representing the percent of the community with the flu and k, the number of days. If 1% of a community has the flu now, about what percent of the community will have the flu after 1 day? 2 days? 10 days?

Mixed Review

Factor.

46. $18y^4 - 50x^6$

47. $128x^3y^7 + 54yf^3$

48. $6x^2 + x - 12$

49. $36x^3y^2 - 6x^2y^3 - 12xy^4$

Divide.

50. $\dfrac{3y^3 - 27y}{12y^2 + 30y - 18} \div \dfrac{y - 3}{2y - 1}$

51. $\dfrac{4y^4 + 2y^3 - 2y^2}{2x^3 + x^2} \div \dfrac{4y - 2}{2x^4 + x^3}$

Simplify.

52. i^6

53. $\sqrt{-32}$

54. $i\sqrt{-18} - 4\sqrt{2}$

55. $(9 + i)(3 - i)$

10.3 Remainder and Factor Theorems

Objectives: To evaluate a polynomial for a given value of the variable, using synthetic division
To show that a given binomial is a factor of a given polynomial

You know how to find factors of quadratic trinomials. In this lesson, you will learn how to use synthetic division and the factor theorem to find factors of polynomials of higher degree.

Capsule Review

To factor a quadratic expression, use the reverse of the FOIL method of multiplication. Try pairs of binomial factors until you find the pair that gives the correct middle term.

EXAMPLE **Factor:** $\overset{F}{3x^2} - \overset{O+I}{2x} - \overset{L}{5}$

Possible Factors	Middle Term (O + I)	
$(3x + 1)(x - 5)$	$-15x + x = -14x$	
$(3x - 1)(x + 5)$	$15x - x = 14x$	
$(3x + 5)(x - 1)$	$-3x + 5x = 2x$	
$(3x - 5)(x + 1)$	$3x - 5x = -2x$	*Correct middle term*

$3x^2 - 2x - 5 = (3x - 5)(x + 1)$

Factor.

1. $2x^2 - x - 3$ **2.** $6x^2 + 11x + 3$ **3.** $3x^2 - 7x + 2$

4. $5a^2 - 2a - 7$ **5.** $9x^2 + 24x + 16$ **6.** $6n^2 + n - 7$

The result of dividing a polynomial $P(x)$ by a binomial of the form $x - a$ is a quotient polynomial $Q(x)$ and a remainder. For example, if

$$P(x) = x^4 - 5x^2 - 17x - 12$$

is divided by $x + 3$, the remainder can be found using synthetic division.

$$\begin{array}{r|rrrrr} -3 & 1 & 0 & -5 & -17 & -12 \\ & & -3 & 9 & -12 & 87 \\ \hline & 1 & -3 & 4 & -29 & 75 \end{array}$$

The remainder is 75.

To find the value of $P(x)$ when $x = -3$, substitute -3 for x

$$\begin{aligned} P(x) &= x^4 - 5x^2 - 17x - 12 \\ P(-3) &= (-3)^4 - 5(-3)^2 - 17(-3) - 12 \\ &= 81 - 45 + 51 - 12 = 75 \end{aligned}$$

Notice that the remainder obtained by dividing $P(x)$ by $x + 3$ is equal to $P(-3)$, which is the value of the polynomial when $x = -3$.

> **Remainder Theorem** If $P(x)$ is a polynomial and a is a number, and if $P(x)$ is divided by $x - a$, then the remainder is $P(a)$.

To show that this theorem is true, let $Q(x)$ be the quotient and R be the remainder. Then

	dividend = quotient × divisor + remainder	
	$P(x) = Q(x) \cdot (x - a) + R$	
Now, let $x = a$. Then	$P(a) = Q(a) \cdot (a - a) + R$	
So	$P(a) = 0 + R$	$a - a = 0$
Therefore	$P(a) = R$	

The remainder theorem is true not only for polynomials with real coefficients, but also for polynomials with complex coefficients and for any complex number a.

Synthetic division is also called *synthetic substitution* because it can be used to evaluate polynomials for given values of the variable.

EXAMPLE 1 **Evaluate $P(x) = x^3 + 8x^2 - 5x - 84$ for the given values of x.**
a. $x = -5$ **b.** $x = 3$

a. $a = -5$

$$\begin{array}{r|rrrr} -5 & 1 & 8 & -5 & -84 \\ & & -5 & -15 & 100 \\ \hline & 1 & 3 & -20 & 16 \end{array}$$

$P(-5) = 16$

b. $a = 3$

$$\begin{array}{r|rrrr} 3 & 1 & 8 & -5 & -84 \\ & & 3 & 33 & 84 \\ \hline & 1 & 11 & 28 & 0 \end{array}$$

$P(3) = 0$

The following theorem is a corollary to the remainder theorem. You will be asked to prove this theorem in Exercises 43 and 44.

> **Factor Theorem** If $P(x)$ is a polynomial, then $x - a$ is a factor of $P(x)$ if and only if $P(a) = 0$.

EXAMPLE 2 **Determine whether the given binomial is a factor of $P(x)$.**

a. $P(x) = x^3 + 5x^2 - 17x - 21;\ x - 3$

b. $P(x) = x^3 - x^2 + 4x - 5;\ x + 2i$

a. $P(x) = x^3 + 5x^2 - 17x - 21$

$$\begin{array}{r|rrrr} 3 & 1 & 5 & -17 & -21 \\ & & 3 & 24 & 21 \\ \hline & 1 & 8 & 7 & 0 \end{array}$$

The remainder is 0, so $x - 3$ is a factor of $x^3 + 5x^2 - 17x - 21$.

b. $P(x) = x^3 - x^2 + 4x - 5$

$$\begin{array}{r|rrrr} -2i & 1 & -1 & 4 & -5 \\ & & -2i & -4 + 2i & 4 \\ \hline & 1 & -1 - 2i & 2i & -1 \end{array}$$

The remainder is not 0, so $x + 2i$ is not a factor of $x^3 - x^2 + 4x - 5$.

Notice that the quotient in Example 2a is $Q(x) = x^2 + 8x + 7$. The degree of the quotient is one less than the degree of the original polynomial, $P(x)$. For this reason, $Q(x)$ is called a **depressed polynomial.** To continue the factoring process, the polynomial $Q(x) = x^2 + 8x + 7$ can be factored by inspection.

$$x^2 + 8x + 7 = (x + 1)(x + 7)$$

Thus, the original polynomial can be shown in completely factored form

$$x^3 + 5x^2 - 17x - 21 = (x - 3)(x + 1)(x + 7)$$

EXAMPLE 3 **Show that $x + 2$ is a factor of $P(x) = 2x^3 + 3x^2 - 8x - 12$. Then factor $P(x)$ completely over the integers.**

$$\begin{array}{r|rrrr} -2 & 2 & 3 & -8 & -12 \\ & & -4 & 2 & 12 \\ \hline & 2 & -1 & -6 & 0 \end{array}$$

$a = -2$, since $x - a = x - (-2)$.

Since $P(-2) = 0$, $x + 2$ is a factor.

$$\begin{aligned} 2x^3 + 3x^2 - 8x - 12 &= (x + 2)(2x^2 - x - 6) \\ &= (x + 2)(2x + 3)(x - 2) \end{aligned}$$

Factor $2x^2 - x - 6$ by inspection.

If several factors are given, a polynomial with those factors can be written.

EXAMPLE 4 **Find a polynomial $P(x)$ with the given factors.**

a. $x - 5,\ x + 4,\ x - 1$ **b.** $x + 2i,\ x - 2i,\ 3x - 1$

a.
$$\begin{aligned} P(x) &= (x - 5)(x + 4)(x - 1) \\ &= (x - 5)(x^2 + 3x - 4) \\ &= x^3 - 2x^2 - 19x + 20 \end{aligned}$$

b. $P(x) = (x + 2i)(x - 2i)(3x - 1)$
$= (x^2 + 4)(3x - 1)$
$= 3x^3 - x^2 + 12x - 4$

An infinite number of polynomials can have the same binomial factors. Consider Example 4a. Any polynomial of the form $ax^3 - 2ax^2 - 19ax + 20a$, where $a \neq 0$, will also have $x - 5$, $x + 4$, and $x - 1$ as factors, since

$$ax^3 - 2ax^2 - 19ax + 20a = a(x^3 - 2x^2 - 19x + 20) = a(x - 5)(x + 4)(x - 1)$$

CLASS EXERCISES

Determine by synthetic division which of the given binomials are factors of $P(x)$.

1. $P(x) = x^3 - 4x^2 - 7x + 10$; $x - 1$, $x + 2$, $x - 5$
2. $P(x) = x^3 - x^2 - 18x - 18$; $x - 1$, $x - 3$, $x + 2$
3. $P(x) = 3x^3 - 4x^2 - 13x - 6$; $x - 3$, $x + 1$, $3x + 2$
4. $P(x) = 2x^3 - 3x^2 - 3x + 2$; $x - 1$, $x + 2$, $2x + 1$
5. $P(x) = x^4 + 3x^3 - x^2 + 11x - 4$; $x + 4$, $x - 4$, $x - 2$

PRACTICE EXERCISES

Evaluate $P(x)$ for each of the given values of x.

1. $P(x) = x^3 - 2x^2 - 5x + 6$; $x = -1$, $x = -2$
2. $P(x) = x^3 - 4x^2 + x + 6$; $x = -1$, $x = -3$
3. $P(x) = x^3 - 7x - 6$; $x = 1$, $x = 3$
4. $P(x) = x^3 - 7x + 6$; $x = 1$, $x = 3$
5. $P(x) = x^4 + 5x^3 + 5x^2 - 5x - 6$; $x = 1$, $x = -3$
6. $P(x) = x^4 - 5x^3 + 5x^2 + 5x - 6$; $x = 1$, $x = 2$

Determine whether the first polynomial is a factor of the second polynomial.

7. $x - 3$; $x^3 + x^2 - 9x - 9$
8. $x - 4$; $x^3 - 7x^2 + 8x + 16$
9. $x + 4$; $x^3 - 12x + 16$
10. $x + 6$; $x^3 - 6x^2 - 12x - 72$
11. $x + 1$; $x^3 + x^2 + x + 1$
12. $x - 1$; $x^3 - x^2 + x - 1$
13. $x - 1$; $x^3 + 5x^2 - 6x - 2$
14. $x - 3$; $x^3 + 3x^2 - 7x - 21$
15. $x - 5$; $2x^3 - 11x^2 + 6x - 5$
16. $x + 2$; $3x^3 - 6x^2 - 8x + 16$

Show that the first polynomial is a factor of the second polynomial. Then factor the second polynomial completely over the integers.

17. $t - 3;\ t^3 + t^2 - 8t - 12$

18. $y - 2;\ y^3 - 3y^2 - 10y + 24$

19. $x + 3;\ x^3 + 4x^2 + x - 6$

20. $s + 6;\ s^3 + 7s^2 - 36$

21. $x + 3;\ x^3 + 3x^2 - x - 3$

22. $y - 5;\ y^3 - 2y^2 - 13y - 10$

23. $t + 2;\ t^3 - 4t^2 - 7t + 10$

24. $y + 1;\ y^3 - 21y - 20$

25. $x + 2;\ 2x^3 - 3x^2 - 18x - 8$

26. $s + 3;\ 3s^3 + 5s^2 - 11s + 3$

27. $t + 1;\ t^3 + t^2 - 18t - 18$

28. $y + 3;\ 2y^3 + y^2 - 9y + 18$

29. $x - 2;\ 6x^3 - 7x^2 - 16x + 12$

30. $s - 2;\ 9s^3 - 18s^2 - s + 2$

31. $x - 1,\ x - 3;\ x^4 - 10x^3 + 35x^2 - 50x + 24$

32. $x + 2,\ x + 4;\ x^4 + 10x^3 + 35x^2 + 50x + 24$

33. $x - 1,\ x - 2;\ x^4 - 10x^3 + 35x^2 - 50x + 24$

34. $x + 3,\ x - 4;\ x^4 - 2x^3 - 13x^2 + 14x + 24$

Write a polynomial with the given factors.

35. $2x - 1,\ x + 2,\ x - 2$

36. $x - 3,\ x + 3,\ 3x + 1$

37. $x - \frac{2}{3},\ x + 1,\ x - 3$

38. $x - 1,\ x - 2,\ x - \frac{3}{2}$

39. $x + \sqrt{2},\ x - \sqrt{2},\ x + 3$

40. $x - \sqrt{3},\ x + \sqrt{3},\ x - 4$

41. $x + 3i,\ x - 3i,\ 2x + 1$

42. $x + 3,\ x - 2i,\ x + 2i$

Exercises 43 and 44 ask for the two parts of the proof of the factor theorem.

43. Prove that if $P(a) = 0$, then $x - a$ is a factor of the polynomial $P(x)$.

44. Prove that if $x - a$ is a factor of the polynomial $P(x)$, then $P(a) = 0$.

Use synthetic division to show that the first polynomial is a factor of the second polynomial.

45. $x - i;\ x^3 + 7x^2 + x + 7$

46. $x + i;\ 2x^3 - 3x^2 + 2x - 3$

47. $x - 3i;\ x^3 + 4x^2 + 9x + 36$

48. $x - 2i;\ 3x^3 - x^2 + 12x - 4$

49. $x - (1 - 2i);\ x^3 - 5x^2 + 11x - 15$

50. $x - (2 - i);\ x^3 - x^2 - 7x + 15$

51. $x - (4 + 2i);\ x^3 - 5x^2 - 4x + 60$

52. $x - (3 - 2i);\ x^3 - 9x^2 + 31x - 39$

Applications

53. Number Theory The product of three binomials is $2x^3 + 7x^2 - 7x - 12$. If one binomial is $x + 1$, what are the other two binomials?

54. Number Theory Martha represented three consecutive even integers in a unique way. If the product of the three integers is represented by $x^3 - 12x^2 + 44x - 48$, and one integer is represented by $x - 4$, what expressions did Martha use for the other integers? Could she have used a different set of expressions?

55. Number Theory Hector represented the product of four numbers as $x^4 - x^3 - 10x^2 - 8x$. One number was represented by x and another by $x + 1$. What expressions were used to represent the other two numbers?

56. Geometry If the volume of a box is represented by the expression $(x^3 - 3x^2 - 10x + 24)$ cm^3 and its width by $(x - 2)$ cm, what binomials can be used to represent the other two dimensions?

57. Geometry If the volume of a box is represented by the expression $(a^3x^3 - 6a^2x^2 + 11ax - 6)$ cm^3 and its length is $(ax - 1)$ cm, what binomials can be used to represent the other two dimensions?

TEST YOURSELF

Divide.

1. $(15z^6 - 12z^5 - 36z^3) \div 3z^2$ **2.** $(6x^3 + 5x^2 + 9) \div (2x + 3)$ **10.1**

Divide, using synthetic division.

3. $\dfrac{x^3 - 8x^2 - 12x + 15}{x - 3}$ **4.** $\dfrac{2x^4 + x^3 - 9x^2 - x + 6}{x + 2}$ **10.2**

Evaluate each polynomial for the given value of x.

5. $x^3 + 2x^2 - 7x + 6;\ x = -4$ **6.** $x^3 + 2x^2 - 5x - 6;\ x = 2$ **10.3**

Determine whether the given binomial is a factor of the other polynomial.

7. $x - 3;\ x^3 + x^2 - 9x - 9$ **8.** $x + 2;\ 3x^3 - 4x^2 - 12x + 8$

Show that the given binomial is a factor of the other polynomial. Then factor that polynomial completely over the integers.

9. $x^3 - 9x^2 + 23x - 15;\ x - 3$ **10.** $2x^3 - 10x^2 - 5x + 13;\ x - 1$

10.4

Solving Polynomial Equations

Objective: To solve polynomial equations of degree three or greater

Recall that to solve a quadratic equation in which the polynomial is factorable, you use the zero-product property: If $ab = 0$, then $a = 0$ or $b = 0$.

Capsule Review

EXAMPLE **Solve: $3x^2 - 7x = 6$**

$$3x^2 - 7x - 6 = 0$$
$$(3x + 2)(x - 3) = 0$$
$$3x + 2 = 0 \quad \text{or} \quad x - 3 = 0$$
$$x = -\frac{2}{3} \quad \text{or} \quad x = 3$$

Check to show that $-\frac{2}{3}$ and 3 are solutions to the equation.

Solve and check.

1. $2n^2 + 3n = 5$

2. $3y^2 + 7y = -4$

3. $4x^2 = 6x - 2$

4. $3 - 5y = 2y^2$

5. $x^2 = 18x$

6. $3n^2 = 10n - 3$

A **polynomial function** is a function that can be written in the form

$$P(x) = a_nx^n + a_{n-1}x^{n-1} + a_{n-2}x^{n-2} + \cdots + a_1x^1 + a_0$$

The **zeros** of a function $f(x)$ are the values of x for which $f(x) = 0$. The zeros of the polynomial function $P(x) = 8x^2 - 10x + 3$ can be found by solving the equation $8x^2 - 10x + 3 = 0$.

$$8x^2 - 10x + 3 = 0$$
$$(4x - 3)(2x - 1) = 0$$
$$4x - 3 = 0 \quad \text{or} \quad 2x - 1 = 0$$
$$x = \frac{3}{4} \quad \text{or} \quad x = \frac{1}{2}$$

The solutions of the equation, $\frac{3}{4}$ and $\frac{1}{2}$, are the zeros of the function. Notice that the numerators of these fractions are factors of 3, the **constant term** of the polynomial. The denominators are factors of 8, the **leading coefficient,** or the coefficient of the term of highest degree. This relationship holds for rational zeros of any polynomial function.

> **Rational Zeros Theorem** If a polynomial function has integral coefficients, and if it has a rational zero $\frac{p}{q}$, where p and q are relatively prime, then p is a factor of the constant term and q is a factor of the leading coefficient.

Recall that if $x - a$ is a factor of the polynomial $P(x)$, then $P(a) = 0$. That is, a is a zero of $P(x)$. You can use the rational zeros theorem and synthetic division to find zeros of polynomial functions.

EXAMPLE 1 **Find the rational zeros of $P(x) = 12x^3 - 8x^2 - 3x + 2$.**

The rational zeros theorem tells you that if $P(x)$ has rational zeros, then those zeros have numerators that are factors of 2, namely ± 1, ± 2, and denominators that are factors of 12, namely ± 1, ± 2, ± 3, ± 4, ± 6, ± 12.

Possible rational zeros: $\pm 1, \pm 2, \pm \frac{1}{2}, \pm \frac{1}{3}, \pm \frac{2}{3}, \pm \frac{1}{4}, \pm \frac{1}{6}, \pm \frac{1}{12}$ $\quad \frac{p}{q}$

When you have several possible numbers to test by synthetic division, it is convenient to write the coefficients of the polynomial just once and to perform the addition steps mentally or with the help of a calculator.

		12	−8	−3	2	← **Coefficients of $P(x)$**
Possible Rational Zeros	1	12	4	1	3	*Think:* $12 \times 1 = 12$; $-8 + 12 = 4$
	−1	12	−20	17	−15	
	2	12	16	29	60	
	−2	12	−32	61	−120	
	$\frac{1}{2}$	12	−2	−4	0	$\frac{1}{2}$ is a zero of the function.

Additional zeros of $P(x)$ are found by solving the depressed equation.

$$12x^2 - 2x - 4 = 0$$
$$6x^2 - x - 2 = 0$$
$$(3x - 2)(2x + 1) = 0$$
$$3x - 2 = 0 \quad \text{or} \quad 2x + 1 = 0$$
$$x = \frac{2}{3} \quad \text{or} \quad x = -\frac{1}{2}$$

Check to show that the zeros of the function are $\frac{1}{2}$, $\frac{2}{3}$, and $-\frac{1}{2}$.

If 1 is the leading coefficient of a polynomial function with integral coefficients, then the rational zeros theorem indicates that the denominator of any rational zero of the function must also be 1. This fact is stated below as a corollary to the rational zeros theorem.

Corollary of the Rational Zeros Theorem If the leading coefficient of a polynomial function with integral coefficients is 1, then any rational zeros of the function are integers.

EXAMPLE 2 **Find the rational zeros of $P(x) = x^3 + 3x^2 - 25x - 75$.**

Possible rational zeros: $\pm1, \pm3, \pm5, \pm15, \pm25, \pm75$ *All are integers.*

	1	3	−25	−75
1	1	4	−21	−96
−1	1	2	−27	−48
3	1	6	−7	−96
−3	1	0	−25	0

-3 is a zero of $P(x)$.

$$x^2 - 25 = 0$$ *Solve the depressed equation.*

$$(x + 5)(x - 5) = 0$$
$$x + 5 = 0 \quad \text{or} \quad x - 5 = 0$$
$$x = -5 \quad \text{or} \quad x = 5$$

Check to show that the zeros of the function are -3, -5, and 5.

Fundamental Theorem of Algebra Every polynomial function with complex coefficients has at least one zero in the set of complex numbers.

Corollary of the Fundamental Theorem of Algebra Every polynomial function containing a polynomial of degree n has exactly n complex zeros.

Keep in mind the fact that the zeros of a function $f(x)$ are the solutions to the equation $f(x) = 0$. Therefore, when you are solving a polynomial equation of degree n, you should look for n complex solutions. Sometimes the same solution must be counted more than once. For example

$$x^2 + 10x + 25 = 0$$
$$(x + 5)(x + 5) = 0$$
$$x = -5 \quad \text{or} \quad x = -5$$

The solution -5 is called a *multiple solution,* and -5 is said to have *multiplicity two*.

EXAMPLE 3 **Find the solutions: $x^4 - 11x^2 - 18x - 8 = 0$**

Possible rational solutions: $\pm1, \pm2, \pm4, \pm8$

	1	0	−11	−18	−8
1	1	1	−10	−28	−36
−1	1	−1	−10	−8	0

-1 is a solution.

Look for another rational solution. Use the coefficients of the first depressed equation, $x^3 - x^2 - 10x - 8 = 0$. Try -1 first to see if it is a multiple solution.

	1	−1	−10	−8
−1	1	−2	−8	0

-1 is a multiple solution.

Solve the second depressed equation, $x^2 - 2x - 8 = 0$.

$$\begin{aligned} x^2 - 2x - 8 &= 0 \\ (x - 4)(x + 2) &= 0 \\ x - 4 = 0 \quad &\text{or} \quad x + 2 = 0 \\ x = 4 \quad &\text{or} \quad x = -2 \end{aligned}$$

Check to show that the solutions are -1 (multiplicity two), 4, and -2.

Imaginary solutions of a quadratic equation with real coefficients occur in *conjugate pairs*. This is also true for all higher degree polynomial equations with real coefficients.

Complex Conjugate Theorem If a complex number, $a + bi$, is a solution of a polynomial equation with real coefficients, then the conjugate, $a - bi$, is also a solution of the equation.

EXAMPLE 4 Find the complex solutions: $x^4 - x^3 - x^2 - x - 2 = 0$

The highest power of x is 4, so there are 4 complex solutions. Possible rational solutions: ± 1, ± 2

	1	−1	−1	−1	−2
1	1	0	−1	−2	−4
−1	1	−2	1	−2	0

-1 is a solution.

Search for additional rational solutions. Use the coefficients of the first depressed equation, and use synthetic division again.

	1	−2	1	−2
−1	1	−3	4	−6
2	1	0	1	0

-1 is not a multiple solution.

2 is a solution.

$$\begin{aligned} x^2 + 1 &= 0 \\ x^2 &= -1 \\ x &= \pm i \end{aligned}$$

Solve the second depressed equation.

Check to show that the solutions are -1, 2, i, and $-i$.

The following rule may be used to limit the number of possible solutions that must be checked using synthetic division.

Descartes' Rule of Signs The number of *positive* real solutions of a polynomial equation $P(x) = 0$, with real coefficients, is equal to the number of sign changes between the coefficients of the terms of $P(x) = 0$ or is less than this number by a multiple of two.

The number of *negative* real solutions of such an equation is equal to the number of sign changes between the coefficients of the terms of $P(-x) = 0$ or is less than this number by a multiple of two.

The number of *sign changes* is determined by arranging the terms of the polynomial in descending order of the variable and counting the number of pairs of consecutive terms with different signs.

EXAMPLE 5 **Determine the possible number of positive and negative real solutions of the equation $x^4 - 4x^3 + 7x^2 - 6x - 18 = 0$. Then solve the equation.**

Possible rational roots: $\pm 1, \pm 2, \pm 3, \pm 6, \pm 9, \pm 18$

$$P(x) = \underbrace{x^4 - 4x^3}_{1}\underbrace{+ 7x^2}_{2}\underbrace{ - 6x}_{3} - 18 = 0$$

Three sign changes, so three or one positive real solutions

$$P(-x) = (-x)^4 - 4(-x)^3 + 7(-x)^2 - 6(-x) - 18 = 0$$
$$P(-x) = x^4 + 4x^3 + 7x^2 + \underbrace{6x - 18}_{1} = 0$$

One sign change, so one negative real solution

	1	−4	7	−6	−18
1	1	−3	4	−2	−20
−1	1	−5	12	−18	0

-1 is a solution.

Since there is only one negative real solution, it is -1. There is no need to try other negative possibilities. Now look for positive real solutions.

	1	−5	12	−18
2	1	−3	6	−6
3	1	−2	6	0

Coefficients of first depressed equation

3 is a solution.

The second depressed equation cannot be factored over the integers, so use the quadratic formula.

$$x^2 - 2x + 6 = 0$$
$$x = \frac{-b \pm \sqrt{b^2 - 4ac}}{2a}$$
$$= \frac{-(-2) \pm \sqrt{(-2)^2 - 4(1)(6)}}{2(1)} = \frac{2 \pm \sqrt{-20}}{2}$$
$$= 1 \pm i\sqrt{5}$$

Check to show that the solutions are -1, 3, $1 + i\sqrt{5}$, and $1 - i\sqrt{5}$.

CLASS EXERCISES

Thinking Critically

1. Explain why $P(x) = x^3 - x - 1$ has no rational zeros.

2. Explain why $P(x) = x^4 + x + 2$ has no rational zeros.

3. Explain why $P(x) = x^3 + k$ has at least one real zero.

For each equation, state the number of complex solutions, the *possible* rational solutions, the maximum number of positive real solutions, and the maximum number of negative real solutions.

4. $x^3 + 2x^2 - 4x - 6 = 0$

5. $x^4 - 3x^3 + x^2 - x + 3 = 0$

6. $3x^3 - x^2 + 5x - 26 = 0$

7. $2x^4 - x^3 + 2x^2 + 5x - 26 = 0$

A polynomial equation with real coefficients has four imaginary solutions, two of which are given. Identify the other two solutions.

8. $-3i,\ 1 + 2i$

9. $4 + 3i,\ -4 - 3i$

PRACTICE EXERCISES

Use technology where appropriate.

A polynomial function and one of its zeros are given. Find the remaining zeros.

1. $P(x) = x^3 - 7x^2 + 7x + 15;\ 5$

2. $P(x) = x^3 + 7x^2 + 7x - 15;\ -5$

3. $P(x) = x^3 - 3x^2 - 33x + 35;\ 1$

4. $P(x) = x^3 + 3x^2 - 4x - 12;\ 2$

5. $P(x) = x^3 - 3x^2 + 4x - 12;\ 3$

6. $P(x) = x^3 + 2x^2 + 9x + 18;\ -2$

Determine the maximum numbers of positive and negative real solutions of each equation.

7. $x^3 - 7x^2 + x + 5 = 0$

8. $2x^3 + 3x^2 - x - 7 = 0$

9. $2x^4 - x^3 - x + 3 = 0$

10. $3x^5 - 2x^4 - 3x^3 + x - 1 = 0$

Find the solutions of each equation.

11. $x^3 + 6x^2 - x - 6 = 0$

12. $x^3 + 4x^2 + x - 6 = 0$

13. $2d^3 + 3d^2 - 8d - 12 = 0$

14. $3c^3 - 5c^2 - 4c + 4 = 0$

15. $3t^3 + 10t^2 - t - 12 = 0$

16. $2a^3 - 11a^2 - a + 30 = 0$

17. $x^3 - 6x^2 + 11x - 6 = 0$

18. $x^3 + 3x^2 - 4x - 12 = 0$

19. $x^4 - 5x^2 + 4 = 0$

20. $x^4 - x^3 - 11x^2 + 9x + 18 = 0$

21. $z^3 - 12z + 16 = 0$

22. $y^3 - 27y - 54 = 0$

23. $r^3 - 12r - 16 = 0$

24. $a^3 - 3a - 2 = 0$

25. $4c^3 + 5c^2 - 2c - 3 = 0$

26. $6t^3 - 17t^2 - 4t + 3 = 0$

27. $x^3 - 5x^2 + 9x - 5 = 0$

28. $x^3 - 7x^2 + 16x - 10 = 0$

29. $z^3 - 4z^2 + 9z - 10 = 0$

30. $y^3 - 3y^2 + 12y - 10 = 0$

31. $z^3 - 5z^2 + 7z + 13 = 0$

32. $2a^3 + 3a^2 - 8a + 3 = 0$

33. $x^4 + x^3 + x^2 - 9x - 10 = 0$

34. $x^4 - 3x^2 - 4 = 0$

35. $2x^3 - 7x^2 + 10x - 6 = 0$

36. $y^3 + 4y^2 - 2y - 20 = 0$

37. $x^4 - 3x^3 + 7x^2 + 21x - 26 = 0$

38. $x^4 + 2x^3 + 5x^2 + 34x + 30 = 0$

39. $x^5 - 3x^4 + 4x^3 - 8x^2 + 16 = 0$

40. $x^5 - 2x^4 - 3x^3 + 6x^2 - 4x + 8 = 0$

Applications

Number Theory Find the following integers.

41. Three consecutive integers whose product is 60.

42. Three consecutive odd integers whose product is -105.

43. Four consecutive positive integers whose product is 120.

44. Four consecutive odd integers whose product is 9.

Mixed Review

Find real solutions if any exist.

45. $48 - 3x^4 = 0$

46. $2x^4 + 162 = 0$

47. $2x^3 + 3 = -51$

48. The measure of an angle is 44 less than three times its supplement. Find the measure of the angle.

49. Solve and graph on a number line: $2(4x + 3) < -17 + 10x$

50. Write an equation of the line passing through $(10, 1)$ and $(4, -5)$.

51. Evaluate $m(x) = x^3 + 3x^2 - 2x + \sqrt{-x}$ for $x = -4, -1, 0,$ and 1.

52. Find the values of x and y. $\begin{bmatrix} 2x & 7 \\ 3 & 1 \end{bmatrix} + \begin{bmatrix} 1 & -4 \\ -2y & 6 \end{bmatrix} = \begin{bmatrix} 2 & 3 \\ -1 & 7 \end{bmatrix}$

Use Cramer's rule to solve.

53. $\begin{aligned} 3x - 4y &= -9 \\ x + 8y &= 4 \end{aligned}$

54. $\begin{aligned} x + 2y - z &= 8 \\ 3x - y - 2z &= 7 \\ -x + y &= 1 \end{aligned}$

Solve.

55. $8x^2 + 10x = 3$

56. $x^2 + 4x + 4 = -5x^2 - 31x + 10$

10.5

Problem Solving Strategy: Use Deductive Reasoning

A mathematics statement that has to be *proved* requires the use of deductive reasoning. To reason deductively, start with a true statement. Then use other true statements to arrive at the desired conclusion. Here is a proof of the rational zeros theorem for a polynomial of degree 3.

EXAMPLE If a polynomial function of degree 3 has integral coefficients, and if it has a rational zero $\frac{p}{q}$ where p and q are relatively prime integers, then p is a factor of the constant term and q is a factor of the leading coefficient.

Understand the Problem *What are the given facts?* A polynomial function of degree 3 has integral coefficients and a rational zero $\frac{p}{q}$.

What are you asked to prove? Prove that p is a factor of the constant term and q is a factor of the leading coefficient.

Plan Your Approach *Choose a strategy*. Use deductive reasoning to write a proof.

Complete the Work

$ax^3 + bx^2 + cx + d = 0$	*a, b, c, and d are integers.*
$a\left(\frac{p}{q}\right)^3 + b\left(\frac{p}{q}\right)^2 + c\left(\frac{p}{q}\right) + d = 0$	*Replace x by $\frac{p}{q}$ since $\frac{p}{q}$ is a zero.*
$\frac{ap^3}{q^3} + \frac{bp^2}{q^2} + \frac{cp}{q} + d = 0$	*Eliminate the parentheses.*
$ap^3 + bp^2q + cpq^2 + dq^3 = 0$	*Multiply both sides by q^3.*
$dq^3 = -ap^3 - bp^2q - cpq^2$	*Subtract $ap^3 + bp^2q + cpq^2$.*
$\frac{dq^3}{p} = -ap^2 - bpq - cq^2$	*Divide both sides by p.*

Since a, p, b, q, and c are all integers, then $-ap^2 - bpq - cq^2$ is an integer. Therefore, $\frac{dq^3}{p}$ is an integer. Then all prime factors of p must cancel all prime factors of dq^3. But p and q are relatively prime and have no common prime factor. Therefore, p must be a factor of d. The proof that q is a factor of the leading coefficient a is Exercise 25.

Interpret the Results *Check the proof*. Review all the steps a few times to see that they are correct and no errors have been made.

CLASS EXERCISES

Answer each statement as *true* or *false*.

1. Mathematical proofs require the use of deductive reasoning.
2. Deductive reasoning uses a chain of true statements to reach a conclusion.
3. Deductive reasoning uses logical thinking and correct algebraic techniques to establish proofs of mathematical statements.
4. Five specific examples of a theorem can be used as a proof of the theorem.
5. A false statement in a proof does not invalidate the proof.
6. Writing proofs requires a good knowledge of mathematics.
7. Solving problems and writing proofs are not related activities.
8. Deductive reasoning is used to solve equations.

PRACTICE EXERCISES

For each exercise, select one of the given true statements on the right as a reason for justifying the exercise.

1. $\sqrt{2} + 7 = 7 + \sqrt{2}$
2. $\frac{-37}{-37} = 1$
3. $-3 + \left(12 + \frac{1}{2}\right) = (-3 + 12) + \frac{1}{2}$
4. $k = h$ and $h = n$, therefore $k = n$
5. $-3 < -1$ and $-1 < 2$, therefore $-3 < 2$
6. $x(yz) = (xy)z$
7. $-5(x + y) = -5x - 5y$
8. $(7q) \div 7 = q$
9. $cd = dc$
10. $-\sqrt{3} + \sqrt{3} = 0$

A. $a + (b + c) = (a + b) + c$
B. $a \cdot b = b \cdot a$
C. $a(bc) = (ab)c$
D. $a(b + c) = ab + ac$
E. $a + b = b + a$
F. $a + b = c$
G. If $a = b$ and $b = c$, then $a = c$.
H. $b \div b = 1,\ b \neq 0$
I. $(ab) \div b = a,\ b \neq 0$
J. If $a < b$ and $b < c$, then $a < c$.

Identify each statement as *true* or *false*.

11. When an even integer is divided by 2, the remainder is 0.
12. When an odd integer is divided by 2, the remainder is 1.
13. If n is an integer, then $2n$ is an even integer.
14. If n is an integer, then $2n + 1$ is an odd integer.
15. Zero is an odd integer.
16. Any integer is either even or odd.

Write a proof for each of the following statements.

17. If a number has two digits, then the sum of the number and its reversal is a multiple of 11.

18. If a number has six digits, then the sum of the number and its reversal is a multiple of 11.

19. The square of an even integer is even.

20. The square of an odd integer is odd. (*Hint:* Express the square in the *form* of an odd integer.)

21. The sum of an even integer and an odd integer is odd.

22. The product of two odd integers is odd.

Prove.

23. If the square of an integer is even, then the integer is even; if the square is odd, the integer is odd.

24. If the product of two integers is even, at least one of the integers is even.

25. In the Example, prove that q is a factor of the leading coefficient a for $ax^3 + bx^2 + cx + d = 0$.

Mixed Problem Solving Review

1. A shipping container must have a volume of at least 21 ft^3. Its width must exceed its height by 4 ft and its length must be 13 ft less than twice its width. Find the smallest possible dimensions of the container.

2. Louise is managing a pension fund of $100,000. She invests part of the fund in bonds paying 8%, another part in stocks paying 7% and the remainder in a mortgage fund paying 10%. The pension fund has an investment goal of $8000 in one year. Further, in order to minimize risk, Louise places three times as much money in stocks and bonds combined than in the mortgage fund. How much money should she invest in each instrument?

PROJECT

Deductive reasoning uses the idea of a *counterexample* to disprove statements. Look up the meaning of the term counterexample in a mathematics book. Then give three examples that illustrate the use of a counterexample.

How many examples are needed to prove that a statement is true? How many counterexamples are needed to prove that a statement is false?

10.6 Polynomial Functions

Objective: To graph a polynomial function and approximate its real zeros

The domain of any polynomial function is the set of all real numbers, and its graph is a continuous, unbroken curve that passes the vertical line test. You have already graphed two kinds of polynomial functions, linear and quadratic.

Capsule Review

EXAMPLE **Graph:** **a.** $f(x) = -5x + 3$ **b.** $f(x) = x^2 - 1$

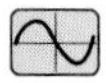

a.

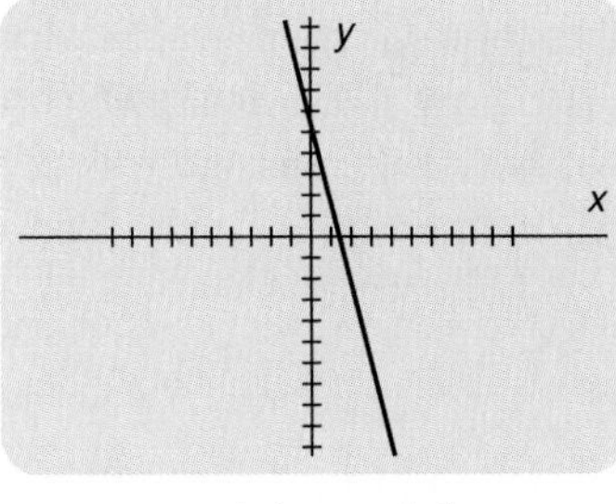

x scl: 1 y scl: 1

b.

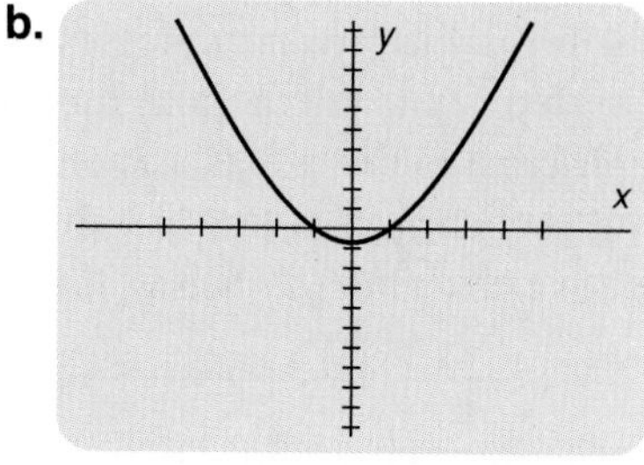

x scl: 1 y scl: 1

Use a graphing utility to graph each function.

1. $f(x) = 4x - 5$ **2.** $f(x) = -\frac{2}{3}x + 6$ **3.** $f(x) = \frac{5}{6}x - 4$

4. $f(x) = x^2 + 4$ **5.** $f(x) = -x^2 - 5$ **6.** $f(x) = 2x^2 + 5$

To graph a polynomial function of degree greater than two, evaluate the function to determine ordered pairs, then plot the points with these pairs as coordinates and connect the points with a smooth curve. The x-coordinates of the points where the graph meets the x-axis are the zeros of the function. A third-degree, or *cubic,* function has at most three real zeros, and its graph meets the x-axis no more than three times.

EXAMPLE 1 **Graph $f(x) = x^3 - x^2 - 9x + 9$ and find its real zeros.**

You can use synthetic division to determine some ordered pairs that satisfy the equation. Plot the points with these pairs as coordinates and join the points with a smooth curve.

Left column: values of x	1	−1	−9	9	Right column: values of f(x)
−4	1	−5	11	−35	
−3	1	−4	3	0	
−2	1	−3	−3	15	
−1	1	−2	−7	16	
0	1	−1	−9	9	
1	1	0	−9	0	
2	1	1	−7	−5	
3	1	2	−3	0	
4	1	3	3	21	

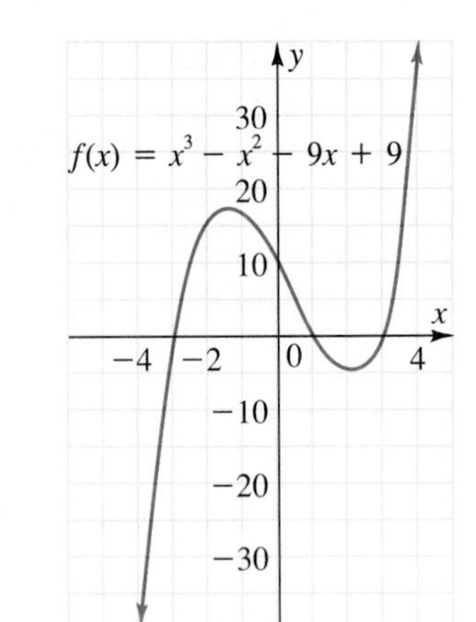

The graph crosses the x-axis three times, and the real zeros of $f(x) = x^3 - x^2 - 9x + 9$ are the integers -3, 1, and 3.

Real zeros that are not integers can be approximated graphically. When part of a graph of a polynomial function is below the x-axis and part is above, the graph crosses the x-axis somewhere in between, and there is a zero at that point. When $f(x)$ is positive the graph is above the x-axis, and when $f(x)$ is negative the graph is below the axis.

> **Location Theorem** If $f(x)$ is a polynomial function with real coefficients, and a and b are real numbers such that $f(a)$ is positive and $f(b)$ is negative, then at least one zero of $f(x)$ occurs between a and b.

EXAMPLE 2 **Graph $f(x) = x^3 - 3x^2 - 5x + 12$ using a graphing utility, and approximate its real zeros to the nearest half unit.**

	1	−3	−5	12	
−3	1	−6	13	−27	The sign changes. Real zero between −3 and −2
−2	1	−5	5	2	
−1	1	−4	−1	13	
0	1	−3	−5	12	
1	1	−2	−7	5	Real zero between 1 and 2
2	1	−1	−7	−2	
3	1	0	−5	−3	Real zero between 3 and 4
4	1	1	−1	8	

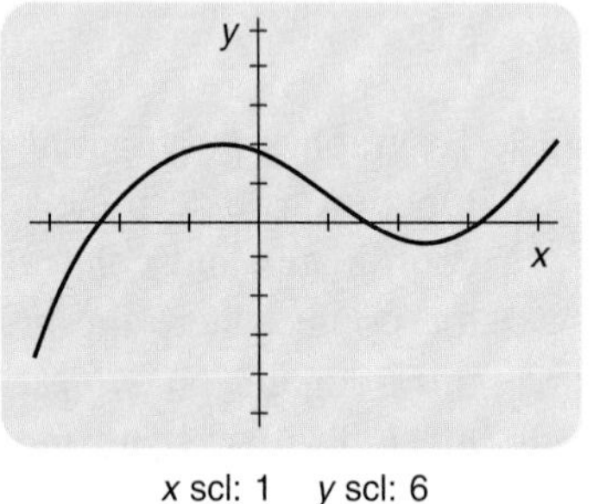

x scl: 1 y scl: 6

Note from the graph that the real zeros of $f(x) = x^3 - 3x^2 - 5x + 12$, to the nearest half unit, are -2.0, 1.5, and 3.5. The above table is consistent with the real zeros determined from the graph. Each tick mark on the y-axis represents an increment of six units.

The location theorem can be used repeatedly to determine real zeros to any desired accuracy. In Example 2, there is a real zero between 1 and 2. Since 0 is closer to $f(2) = -2$ than to $f(1) = 5$, select values closer to 2 than to 1. Use a calculator to find $f(x)$ for these values.

x	1.5	1.6	1.7	1.8
$f(x)$	1.125	0.416	−0.257	−0.888

Notice that a real zero occurs between 1.6 and 1.7 and that 0 is closer to $f(1.7)$ than to $f(1.6)$. You could obtain a closer approximation by finding $f(1.65)$, $f(1.66)$, $f(1.67)$, and so forth. A graphing utility will provide similar results more quickly if you use the zoom feature.

When you graph higher degree polynomial functions, the rational zero theorem and Descartes' rule can help you find any integral zeros. If there are many possibilities to try, it is also useful to know an *upper bound* and a *lower bound* for the zeros. An **upper bound** for the real zeros of a polynomial function is a number greater than or equal to the greatest real zero of the function. Similarly, a **lower bound** is a number less than or equal to the least real zero of the function.

Upper and Lower Bound Theorem Let a polynomial function be divided by $x - c$.

- If $c > 0$ and all the coefficients in the quotient and remainder are nonnegative, then c is an upper bound of the zeros.
- If $c < 0$ and the coefficients in the quotient and remainder alternate in sign, then c is a lower bound of the zeros.

EXAMPLE 3 **Graph the function and approximate its real zeros to the nearest tenth: $y = 2x^4 + 5x^3 - 3x^2 - 9x + 1$**

Synthetic division indicates that −3 is a lower bound and 2 is an upper bound. It is unnecessary to try divisors outside this range.

	2	5	−3	−9	1	
−3	2	−1	0	−9	28	← The signs alternate. −3 is a lower bound.
						← Real zero
−2	2	1	−5	1	−1	
						← Real zero
−1	2	3	−6	−3	4	
0	2	5	−3	−9	1	
						← Real zero
1	2	7	4	−5	−4	
						← Real zero
2	2	9	15	21	43	← All signs are positive. 2 is an upper bound.

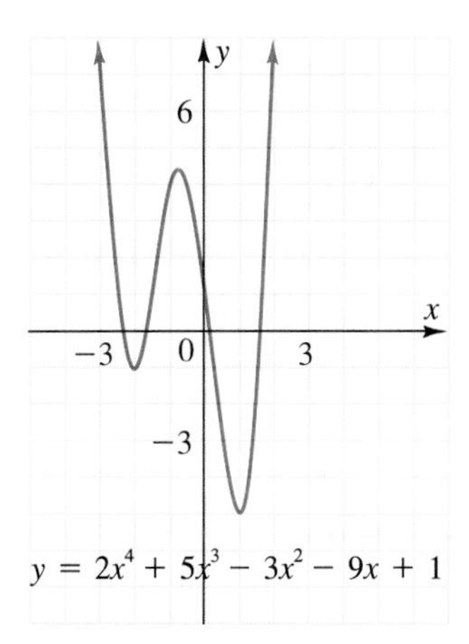

To find the real zeros to the nearest tenth, set the range on the graphing utility to reflect values of x from -3 to 2, the lower and upper bounds.

The trace feature of the graphing utility reveals that the real zeros of $y = 2x^4 + 5x^3 - 3x^2 - 9x + 1$ are -2.2, -1.7, 0.1, and 1.3.

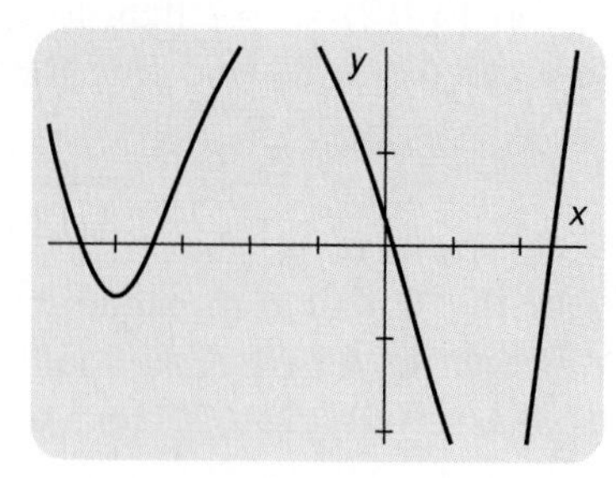

x scl: 0.5 y scl: 2

Manufacturers of precision tools find that rounding to the nearest tenth is much too inaccurate for most purposes. The zoom and trace features of the graphing utility can be used to find the zeros to the nearest hundredth: -2.21, -1.65, 0.11, and 1.26.

CLASS EXERCISES

For each function, determine if a real zero occurs between -1 and 0.

1. $f(x) = x^3 - x^2 - 1$

2. $f(x) = x^3 + x^2 - 1$

3. $f(x) = -x^3 + x^2 - 1$

4. $f(x) = x^3 - x^2 + 1$

5. $f(x) = x^4 - x^3 + x^2 - 1$

6. $f(x) = x^4 - 2x^3 + 6$

7. $f(x) = x^4 + x^2 + 2$

8. $f(x) = x^4 - 2x^2 - 3$

PRACTICE EXERCISES

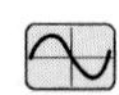 Use technology where appropriate.

Graph each function and find its real zeros.

1. $f(x) = x^3 - 2x^2 - 5x + 6$

2. $f(x) = x^3 - 4x^2 + x + 6$

3. $f(x) = x^3 + 3x^2 - 4x - 12$

4. $f(x) = x^3 + 3x^2 - 33x - 35$

Graph each function and approximate its real zeros to the nearest half unit.

5. $f(x) = x^3 - 2x^2 - 6x + 9$

6. $f(x) = x^3 - 3x^2 - 2x + 6$

Graph each function and find its real zeros. Approximate any nonintegral zeros to the nearest tenth.

7. $f(x) = x^3 - 3x^2 - 3x + 9$

8. $f(x) = x^3 + 4x^2 - 5x - 20$

Find a real zero, to the nearest tenth, between the given values.

9. $y = x^3 + 2x^2 - 5x + 1$; 1 and 2

10. $y = x^3 - x^2 - 5x - 4$; 3 and 4

11. $y = 2x^3 - x^2 - 3x - 1$; 1 and 2

12. $y = 2x^3 - 4x^2 - 13x - 5$; 0 and -1

Graph each function and find its real zeros. Approximate any nonintegral zeros to the nearest tenth.

13. $f(x) = x^3 - x^2 - 6x$

14. $f(x) = x^3 - x^2 - 12x$

15. $f(x) = x^4 - 5x^2 + 4$

16. $f(x) = x^4 - 10x^2 + 9$

17. $f(x) = x^3 - 4x^2 + 2x + 1$

18. $f(x) = x^4 - 3x^3 + x^2 + 6x - 5$

Find a real zero, to the nearest hundredth, of $y = x^4 - 3x^3 - 2x^2 + 3x + 8$:

19. between 1 and 2

20. between 3 and 4

Graph each function and find its real zeros. Approximate any nonintegral zeros to the nearest hundredth.

21. $y = 2x^3 - x^2 + 3x - 1$

22. $y = 2x^3 - 5x^2 - 2x - 7$

23. $y = 8x^3 - 22x^2 - 5x + 12$

24. $y = 8x^3 + 15x^2 - 3x - 15$

25. The graph of a polynomial function contains the points $(2, -6)$ and $(5, 7)$. Why must the function have a real zero between 2 and 5?

Applications

26. Geometry A 0.9-cm slice is cut from one end of a wood cube. The volume of the remaining piece of the cube is 10 cm^3. What was the length of each edge of the original cube?

27. Geometry A rectangular shed is 3 m long, 2 m wide, and 2 m high. If each dimension of the shed is increased by the same amount, the volume is doubled. Find, to the nearest 0.1 m, the amount by which each dimension must be increased.

Developing Mathematical Power

Thinking Critically You can write a cubic equation in x, with *complex* coefficients, that has $1 + i$ for a solution but not its conjugate, $1 - i$.

$[x - (1 - i)]x^2 = 0$ *If $1 + i$ is a solution, then $x - (1 + i)$ is a factor of the polynomial.*

$x^3 - (1 + i)x^2 = 0$ *An equation that meets the requirements*

	1	$-(1 + i)$	0	0	Check, using synthetic division.
$1 + i$	1	0	0	0	*$1 + i$ is a solution.*
$1 - i$	1	$-2i$	$-2 - 2i$	-4	*$1 - i$ is **not** a solution.*

Write a cubic equation that has the given number as a solution but does not have the conjugate of that number as a solution. Check your answer.

28. i **29.** $-2i$ **30.** $-1 + i$ **31.** $2 + i$ **32.** $2 + 3i$ **33.** $-3 + 2i$

10.7 Problem Solving: Using Polynomial Equations

Polynomial equations are used to solve many problems dealing with volume.

EXAMPLE 1 The width of a rectangular container is 2 m less than the length and the height is 1 m less than the length. Find the dimensions of the container if its volume is 60 m^3.

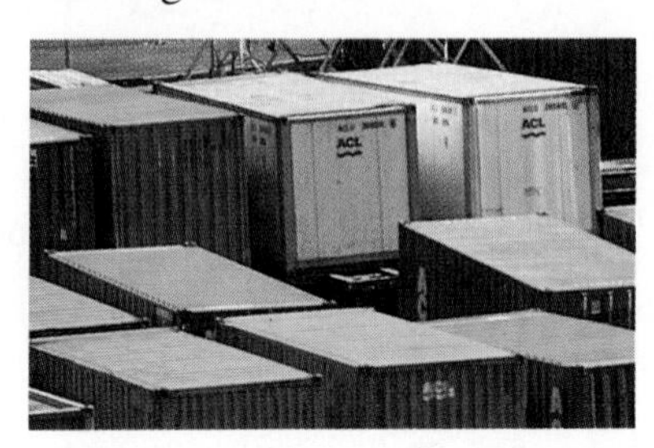

Understand the Problem Draw a diagram.

Plan Your Approach Let x represent the length. Then $x - 2$ represents the width and $x - 1$ the height. Label the diagram.

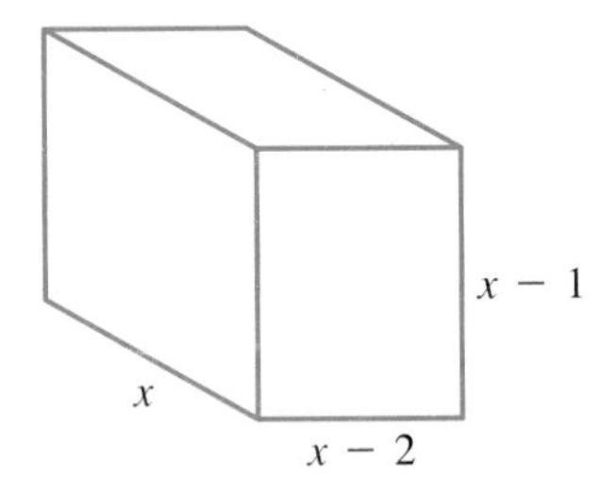

$$\text{length} \times \text{width} \times \text{height} = \text{volume}$$
$$x(x-2)(x-1) = 60$$
$$x^3 - 3x^2 + 2x = 60$$

Complete the Work

$$x^3 - 3x^2 + 2x = 60$$
$$x^3 - 3x^2 + 2x - 60 = 0$$

There are three sign changes in the polynomial, so there are at most three positive real solutions to the equation.

	1	−3	2	−60
1	1	−2	0	−60
2	1	−1	0	−60
3	1	0	2	−54
4	1	1	6	−36
5	1	2	12	0

Test factors of 60, using synthetic division.

5 is a solution.

$$x^2 + 2x + 12 = 0$$

Use the quadratic formula to solve the depressed equation.

$$x = \frac{-b \pm \sqrt{b^2 - 4ac}}{2a}$$
$$= \frac{-2 \pm \sqrt{2^2 - 4(1)(12)}}{2(1)}$$
$$= \frac{-2 \pm \sqrt{-44}}{2}$$

Reject these solutions, since they are not real numbers.

Therefore,
$$x = 5$$
$$x - 2 = 3$$
$$x - 1 = 4$$

Interpret the Results The container is 5 m long, 3 m wide, and 4 m high.

Is the volume of the container 60 m^3?

$$5 \cdot 3 \cdot 4 \stackrel{?}{=} 60$$
$$60 = 60 \checkmark$$

How could using the discriminant help you solve this problem?

Some problems have more than one solution. In such cases, your answer should include *all* the correct solutions.

EXAMPLE 2 A manufacturer has a piece of tin 12 ft wide and 16 ft long. An open box with a volume of 160 ft^3 is to be made by cutting a square of the same size from each corner and folding up the edges of the piece of tin. To the nearest tenth of a foot, find the length of a side of the square piece that is removed.

Understand the Problem Draw a diagram showing the square corners to be cut from the piece of tin.

Plan Your Approach Let x represent the length of a side of each square. Then $16 - 2x$ represents the length of the box and $12 - 2x$ the width.

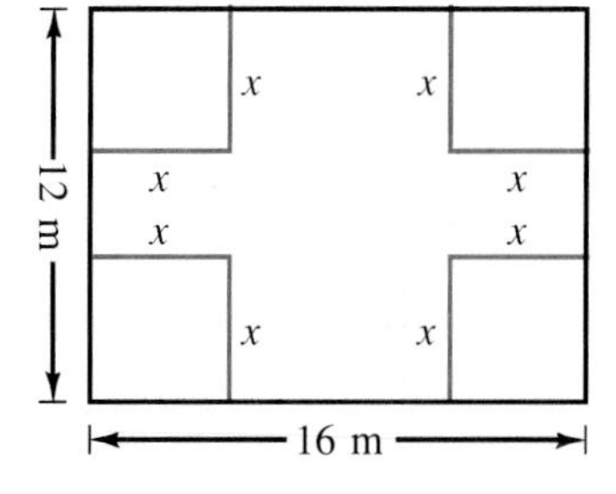

Label your diagram.

$$\text{Volume of box} = 160$$
$$x(16 - 2x)(12 - 2x) = 160$$

$$192x - 56x^2 + 4x^3 = 160$$
$$x^3 - 14x^2 + 48x - 40 = 0$$

The shortest side of the piece of tin measures 12 ft, so the value of x must be between 0 and 6. There are no rational solutions. However, there is a real solution between 1 and 2 and another between 3 and 4. Use a calculator to approximate these solutions.

	1	−14	48	−40	
1	1	−13	35	−5	←
2	1	−12	24	8	
3	1	−11	15	5	←
4	1	−10	8	−8	
5	1	−9	3	−25	

x	$f(x)$	
1.1	−2.809	
1.2	−0.832	0 is closer to −0.832 than to 0.937,
1.3	0.937	so one solution is approximately 1.2.
3.3	1.877	
3.4	0.664	0 is closer to −0.625 than to 0.664,
3.5	−0.625	so one solution is approximately 3.5.

Interpret the Results The length of a side of each square is either about 1.2 ft or about 3.5 ft.

Is the volume of the box 160 ft^3?

$$1.2[16 - 2(1.2)][12 - 2(1.2)] \stackrel{?}{=} 160$$
$$1.2(13.6)(9.6) \stackrel{?}{=} 160$$
$$156.672 \approx 160 \checkmark$$

$$3.5[16 - 2(3.5)][12 - 2(3.5)] \stackrel{?}{=} 160$$
$$3.5(9)(5) \stackrel{?}{=} 160$$
$$157.5 \approx 160 \checkmark$$

Some problems involving products of numbers can also be solved using polynomial equations.

EXAMPLE 3 Find four rational numbers such that the product of the first, third, and fourth numbers is 54. Also, the second number is 2 less than the first number, the third is 5 less than the second, and the fourth is 3 less than the third.

Understand the Problem You are asked to find four rational numbers that satisfy a given set of conditions.

Plan Your Approach Let n represent the first number.
Then $n - 2$ represents the second number.
$(n - 2) - 5$, or $n - 7$, represents the third number.
$(n - 7) - 3$, or $n - 10$, represents the fourth number.

The product of the first, third, and fourth numbers is 54.

$$n(n - 7)(n - 10) = 54$$
$$n^3 - 17n^2 + 70n - 54 = 0$$

Complete the Work

	1	−17	70	−54
1	1	−16	54	0

1 is a solution.

$$n = 1$$
$$n - 2 = -1$$
$$n - 7 = -6$$
$$n - 10 = -9$$

$$x^2 - 16x + 54 = 0$$

To see if there are other rational solutions, use the quadratic formula to solve the depressed equation.

$$x = \frac{-b \pm \sqrt{b^2 - 4ac}}{2a}$$

$$= \frac{-(-16) \pm \sqrt{(-16)^2 - 4(1)(54)}}{2(1)}$$

The discriminant is positive.

$$= \frac{16 \pm \sqrt{40}}{2}$$

Reject these solutions, since they are irrational.

Interpret the Results

The numbers are 1, -1, -6, and -9.

Is the product of the first, third, and fourth numbers 54?

$$1(-6)(-9) \stackrel{?}{=} 54$$
$$54 = 54 \checkmark$$

CLASS EXERCISES

Plan an approach for solving each problem, but do not solve the problem.

1. The volume of a cube is 7 cm^3 more than twice the volume of a rectangular box. The length of the box is 2 cm greater than the length of an edge of the cube, its width is 2 cm less than the length of an edge of the cube, and its height is 1 cm less than the length of an edge of the cube. Find the dimensions of the cube and of the box.
2. The length of a rectangular tank is 3 ft more than its width and the height is 5 ft more than its width. If the volume of the tank is 100 ft^3, what are its dimensions?
3. An open box with a volume of 64 cm^3 can be made by cutting a square of the same size from each corner of a square piece of metal 12 cm on a side and folding up the edges. What is the length of a side of the square that is cut from each corner?
4. Find four numbers such that the product of the first, third, and fourth numbers is 35. The first number is 2 less than the second number, the third is 4 more than the second, and the fourth is 3 more than twice the first.

PRACTICE EXERCISES

Use technology where appropriate.

1. The volume of a rectangular tank is 14 ft^3 less than twice the volume of a cube. The length of the tank is 1 ft more than the length of an edge of the cube, its width is 1 ft less than the length of an edge of the cube, and its height is 2 ft greater than the length of an edge of the cube. What are dimensions of the cube and of the tank?

2. The volume of a rectangular box is 6 cm^3 more than twice the volume of a cube. The length of the box is 3 cm more than the length of an edge of the cube, its width is 2 cm more than the length of an edge of the cube, and its height is 1 cm less than the length of an edge of the cube. Find the dimensions of the cube and of the box.

3. The length of a rectangular tank is 6 ft more than its height and the width is 3 ft more than its height. If the volume of the tank is 80 ft^3, what are the dimensions of the tank?

4. The length of a rectangular bin is 2 ft more than its width and the height is 3 ft less than its width. Find the dimensions of the bin if its volume is 70 ft^3.

5. An open box with a volume of 144 cm^3 can be made by cutting a square of the same size from each corner of a square piece of tin 14 cm on a side and folding up the edges of the tin. What is the length of a side of the square that is cut from each corner?

6. A manufacturer has a square piece of cardboard 12 ft on a side. An open box with a volume of 12 ft^3 is to be made by cutting out a square of the same size from each corner and folding up the edges of the piece of cardboard. What is the length of a side of the square that is cut from each corner?

7. Find four rational numbers such that the product of the first, third, and fourth numbers is -6. The second number is 3 less than the first number, the third is 2 less than the second, and the fourth is 2 less than the third.

8. Find four rational numbers such that the product of the first, third, and fourth numbers is 54. The second number is 3 less than the first number, the third number is 3 less than the second, and the fourth is 3 less than the third.

9. Find four consecutive multiples of 3 such that the product of the first, third, and fourth is 1080.

10. Find four consecutive even numbers such that the product of the first, third, and fourth is 2240.

11. Find four rational numbers such that the product of the second, third, and fourth numbers is -60. The second number is 1 more than twice the first number, the third number is 3 less than the first, and the fourth is 4 more than the third.

12. Find four rational numbers such that the product of the second, third, and fourth numbers is -63. The second number is 2 more than the first number, the third is 5 less than twice the second, and the fourth is 6 less than the first.

13. Nancy has a piece of tin 14 cm wide and 18 cm long. She wants to make an open box with a volume of 180 cm^3 by cutting out a square of the same size from each corner and folding up the edges of the piece of tin. To the nearest tenth of a centimeter, find the length of a side of the square that is cut from each corner.

14. Peter has a piece of metal 18 in. wide and 20 in. long. He wants to make an open box with a volume of 200 in.^3 by cutting out a square of the same size from each corner and folding up the edges of the piece of metal. To the nearest tenth of an inch, what is the length of a side of the square cut from each corner?

15. If a 5-in. slice is cut from one face of a cube, the remaining solid has a volume of 93 in.^3 To the nearest tenth of an inch, find the length of an edge of the original cube.

16. If a 6-in. slice is cut from one face of a cube, the remaining solid has a volume of 258 in.^3 To the nearest tenth of an inch, find the length of an edge of the original cube.

17. The volume of a cube is 170 ft^3 less than the volume of a rectangular box. The length of the box is 1 ft less than twice the length of an edge of the cube, its width is 2 ft less than the length of an edge of the cube, and its height is 3 ft more than the length of an edge of the cube. To the nearest tenth of a foot, find the dimensions of the cube and the box.

18. A rectangular box is 6 ft long, 5 ft wide, and 4 ft high. If each of the three dimensions is increased by the same amount, the volume of the box is doubled. To the nearest 0.1 ft, find the amount by which each dimension must be increased.

HISTORICAL NOTE

The Renaissance was a period of tremendous advances in mathematics as well as in the arts and sciences. The year 1545, with the publication of the *Ars magna* by Geronimo Cardan [1501–1576], is often considered to mark the beginning of the modern period in mathematics. The *Ars magna* contains the solution to cubic polynomial equations, which had stumped mathematicians since the time of the Babylonians. It also contains the solution to quartic equations. In his work, Cardan acknowledges that he did not discover these solutions and that Niccolo Tartaglia [ca. 1500–1557] provided him with an important hint to the solution of cubics. Interestingly enough, Tartaglia himself may have received a hint from an earlier mathematician, Scipione del Ferro [ca. 1465–1526], a professor of mathematics at Bologna whose name is scarcely remembered today.

Investigate the lives of Cardan, Tartaglia and del Ferro.

10.8

Graphing Rational Algebraic Functions

Objective: To sketch the graphs of rational algebraic functions

Graphing an equation that involves the quotient of expressions containing the variable is very tedious if certain characteristics of the graph are not used to simplify the process. The fact that a graph does or does not have x- or y-intercepts can be particularly important.

Capsule Review

An x-intercept of a graph is the x-coordinate of a point whose y-coordinate is zero. A y-intercept of a graph is the y-coordinate of a point whose x-coordinate is zero.

Determine the intercepts of the graph of each equation, if they exist.

1. $y = 2x + 3$
2. $y = -3x - 8$
3. $2x + 3y = 12$
4. $x - 3y = 10$
5. $y = x^2 - 3x$
6. $y = x^3 - x$
7. $y = x^2 - 3x - 10$
8. $x^2y - xy = 2$
9. $x^2 + 25y^2 = 25$

A **rational algebraic function** is a function that can be represented as the quotient of two polynomials. The degree of the denominator must be greater than or equal to 1. Consider the rational algebraic function

$$y = \frac{1}{x}$$

The graph of this function has no x- or y-intercepts, since neither x nor y can equal zero. However, as the absolute value of x increases, y gets closer and closer to 0.

x	-1	-5	-10	-100	-1000
y	-1	-0.2	-0.1	-0.01	-0.001

x	1	5	10	100	1000
y	1	0.2	0.1	0.01	0.001

Also, when the absolute value of x approaches 0, from the right or from the left, the absolute value of y increases without bound.

x	-1	-0.2	-0.1	-0.01	-0.001
y	-1	-5	-10	-100	-1000

x	1	0.2	0.1	0.01	0.001
y	1	5	10	100	1000

The graph of $y = \frac{1}{x}$ is shown at the right. The y-axis, with the equation $x = 0$, is called the **vertical asymptote** of the graph. The x-axis, with the equation $y = 0$, is called the **horizontal asymptote.** The distance between the graph and its asymptotes approaches zero; that is, the graph gets closer and closer to its asymptotes without ever reaching them.

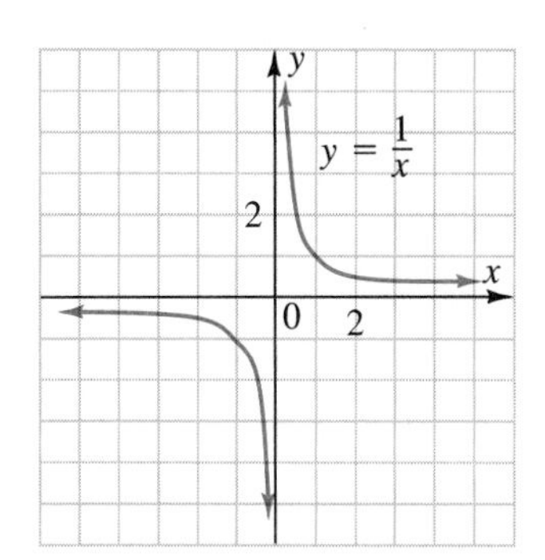

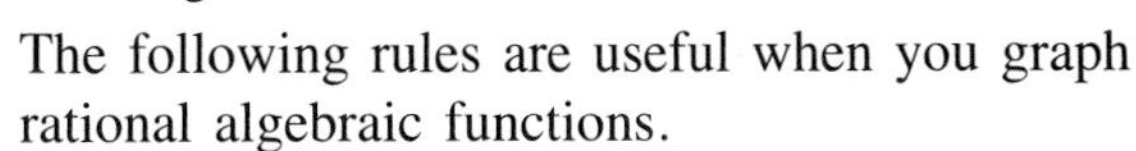
The following rules are useful when you graph rational algebraic functions.

Asymptote Rules

Vertical Asymptotes If a is a value for which the function is not defined, then $x = a$ is a vertical asymptote.

Horizontal Asymptotes Consider the function

$$y = \frac{bx^n + \text{(a polynomial of lower degree)}}{cx^m + \text{(a polynomial of lower degree)}}$$

Three possible conditions determine a horizontal asymptote:

- If $n = m$, then $y = \frac{b}{c}$ is a horizontal asymptote.
- If $n < m$, then $y = 0$ is a horizontal asymptote.
- If $n > m$, then there is no horizontal asymptote.

The function $y = \frac{1}{x}$ is undefined when $x = 0$. Therefore, the rule shows that the vertical asymptote is $x = 0$. Also, since $n = 0$ and $m = 1$, the horizontal asymptote is $y = 0$. You can use a graphing calculator or a computer to show the graph of a hyperbola.

EXAMPLE 1 **Determine the intercepts and the asymptotes of the graph of $y = \frac{3x}{2x - 1}$. Then sketch the graph.**

x-intercept: 0 y-intercept: 0

Vertical asymptote: $x = \frac{1}{2}$ *When $x = \frac{1}{2}$, the function is not defined.*

Horizontal asymptotes: $y = \frac{3}{2}$ *$m = n$*

It is helpful to find some other points on the graph, especially a few points near the asymptotes. Plot the intercept and any other points and sketch the asymptotes. Then sketch the graph of the function.

x	y
-2	1.2
-1	1
0.2	-1
0.8	4
1	3
2	2
3	1.8

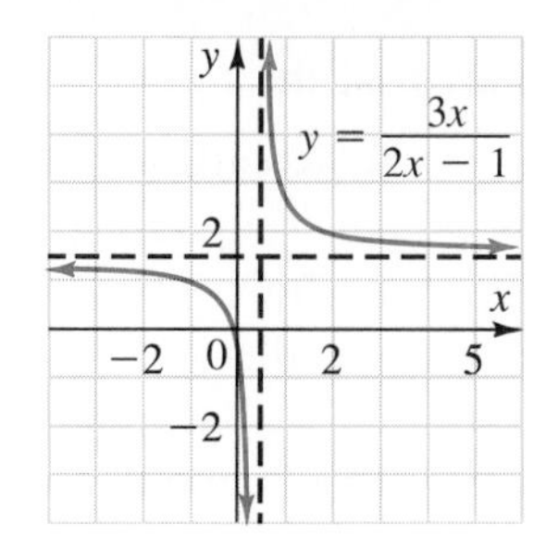

The graph in Example 1 has two separate *branches*. Example 2 illustrates a graph with three branches.

EXAMPLE 2 **Graph $y = \frac{x}{x^2 - 1}$. Determine the intercepts and the asymptotes of the function.**

The intercepts can be found algebraically or with the trace feature of your graphing utility.

x-intercept: 0
y-intercept: 0

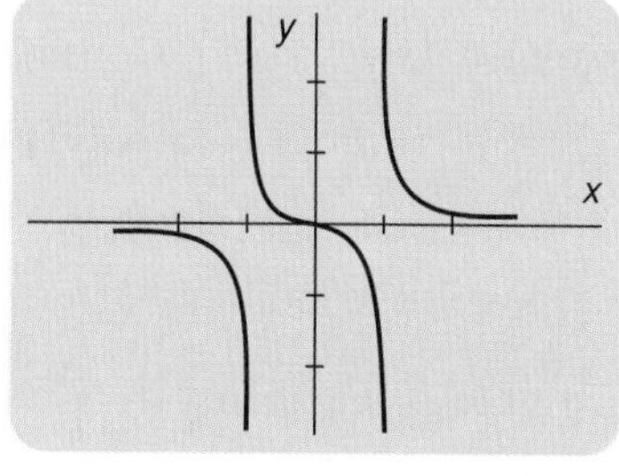

x scl: 1 y scl: 3

The function is not defined for $x^2 - 1 = 0$ or $x = \pm 1$. Therefore, the vertical asymptotes are $x = 1$ and $x = -1$.

Since $n < m$, $(1 < 2)$, the line $y = 0$ is a horizontal asymptote.

CLASS EXERCISES

Thinking Critically

Consider the graphs of the following functions:

a. $y = \frac{3x}{5x - 4}$ **b.** $y = \frac{x^2 + 1}{x + 2}$ **c.** $y = \frac{8}{x^2}$

1. Which graph has an x-intercept?
2. Which graph has a y-intercept but no x-intercept?
3. Which graph has a vertical asymptote but no horizontal asymptote?
4. For which graph is the y-axis a vertical asymptote?
5. For which graph is the x-axis a horizontal asymptote?

PRACTICE EXERCISES

 Use technology where appropriate.

For the graph of each function, determine the intercepts and the asymptotes, if they exist.

1. $y = \frac{4}{x}$

2. $y = \frac{x + 1}{x^2}$

3. $y = \frac{x^2}{x - 1}$

4. $y = \frac{10}{x^2}$

5. $y = \frac{x^2 + 2}{x^2 - 4}$

6. $y = \frac{x + 6}{x^2}$

7. $y = \frac{x^2 - 9}{x^2 - 4}$

8. $y = \frac{2x}{3x + 2}$

9. $y = \frac{25}{x}$

10. $y = \frac{x}{x^2 - 4}$

11. $y = \frac{x + 3}{(x + 1)(x + 2)}$

12. $y = \frac{2x - 3}{(x - 1)^2}$

Graph each function. Use the asymptotes, the intercepts, and other points, as necessary, or use a graphing utility.

13. $y = \frac{4}{x^2}$

14. $y = \frac{3}{x^2}$

15. $y = \frac{x - 4}{x^2}$

16. $y = \frac{x + 3}{x^2}$

17. $y = \frac{2x}{3x - 1}$

18. $y = \frac{3x}{2x - 3}$

If $-x$ is substituted for x in an equation, and the resulting equation is equivalent to the original equation, then the graph is symmetric with respect to the y-axis.

19. In Exercises 1–18, the graphs of which equations are symmetric with respect to the y-axis?

20. If the graph of a rational algebraic function has symmetry with respect to the y-axis, how can the graphing process be simplified?

Graph each function. Use the asymptotes, the intercepts, and other points, as necessary, or use a graphing utility.

21. $y = \frac{3x - 2}{x^2 - 4}$

22. $y = \frac{2x - 1}{x^2 - 1}$

23. $y = \frac{4}{x^2 - 1}$

24. $y = \frac{2}{x^2 - 4}$

Applications

25. Number Theory Show that the horizontal asymptote rule is equivalent to dividing each term by the highest power of the variable and considering what happens as x becomes increasingly great.

26. **Physics** If an object is above the earth's surface, its weight w is given by the formula $w = \frac{K}{d^2}$, where K is a constant and d is the distance from the earth's center. If a person weighs 100 lb at the earth's surface, at what distance from the surface would the person weigh 25 lb? ($d = 4000$ mi at the earth's surface)

27. **Physics** A person's weight w_h at a specific height h above the earth's surface can be found using the formula

$$w_h = \frac{r}{r + h} w_0$$

where r is the radius of the earth (about 6400 km) and w_0 is the weight at sea level. If a person's weight at sea level is 50 kg, graph the person's weight at 0, 1, 2, 3, and so on, earth radii above sea level. Explain the meaning of the asymptotes.

TEST YOURSELF

Find the rational solutions of each equation.

1. $2x^3 - x^2 - 13x - 6 = 0$
2. $27x^3 - 18x^2 - 3x + 2 = 0$ **10.4**

Find the complex solutions of each equation.

3. $x^3 - 6x^2 + x - 6 = 0$
4. $x^4 + 2x^3 + x^2 + 8x - 12 = 0$

Graph each function and find its real zeros. Approximate any nonintegral zeros to the nearest tenth.

5. $f(x) = x^3 - 6x^2 - x + 30$ **10.6**
6. $f(x) = 18x^3 + 21x^2 - 37x - 14$

Solve.

7. The length of a 400-ft^3 rectangular storage bin is 2 ft greater than its width and 5 ft greater than its height. Find its dimensions. **10.7**
8. The product of three rational numbers is 384. The second number is 2 more than the first number, and the third number is the square of the first number. Find the numbers.

Determine the intercepts and the asymptotes, if any, for the graph of each function. Then sketch the graph.

9. $y = \frac{2}{x}$
10. $y = \frac{x}{3x - 2}$ **10.8**

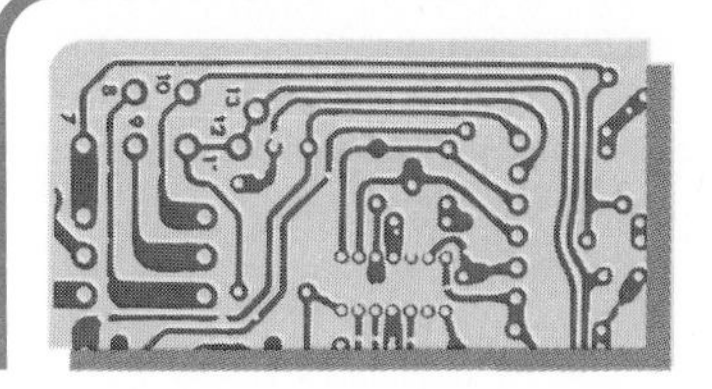

TECHNOLOGY: Correcting Codes

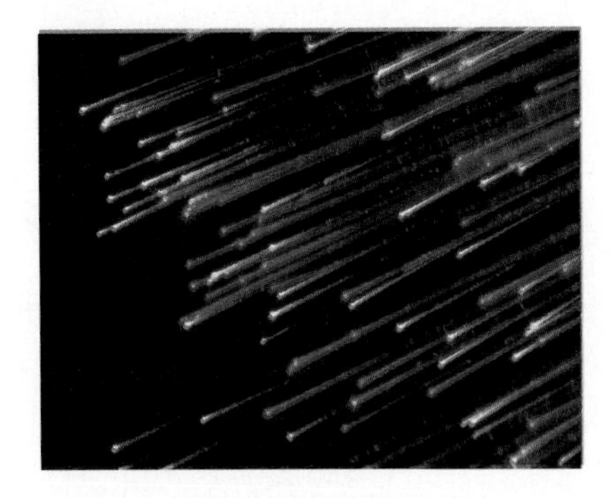

Did you know that just two hundred years ago, when Benjamin Franklin operated the first U.S. postal service, it could take weeks for information to be sent from one place to another? Today vast amounts of information can be sent in a matter of seconds, through telephone wire or via radio signals, not only between states, but between planets. In spite of the sophisticated equipment used to transmit data today, such transmissions are subject to error. Therefore, error codes have been developed to detect such errors. Data is transformed into binary signals, strings of 1's and 0's, and sent by a device called an encoder. It is received by another device called a decoder that, by use of the code, detects errors as they occur.

In error-detection schemes using polynomials, the encoder multiplies an input polynomial by a fixed polynomial for each line of data sent. When the data is received the decoder divides and checks the quotient against an acceptable serial pattern. The division, however, uses base two modular arithmetic. For instance, when a coefficient of 1 is subtracted from 0, the result is 1 instead of -1.

EXAMPLE Find the acceptable serial pattern for the error code of the input polynomial $x^{13} + x^{12} + 0 + x^{10} + x^9 + 0 + 0 + x^6 + 0 + x^4 + 0 + 0 + x + 1$ divided by the fixed polynomial $x^6 + 0 + x^4 + x^3 + 0 + 0 + 1$.

The quotient is $x^7 + x^6 + x^5 + x^4 + x^3 + 0 + x + 1$.

The serial pattern of the input polynomial consists of its coefficients, 1 1 0 1 1 0 0 1 0 1 0 0 1 1. The quotient, with x^7 as its first nonzero term, has the serial pattern of 0 0 0 0 0 0 1 1 1 1 1 0 1 1. If the decoder does not get this pattern with a remainder of 0, then it "knows" a transmission error has occurred.

EXERCISES

Find the serial pattern for the error codes of the following input polynomials. Use the fixed polynomial in the example problem.

1. $x^{13} + 0 + x^{11} + x^{10} + 0 + x^8 + x^7 + x^6 + x^5 + 0 + 0 + x^2 + 0 + 0$

2. $x^{13} + 0 + 0 + x^{10} + 0 + 0 + x^7 + x^6 + x^5 + 0 + 0 + x^2 + x + 1$

CHAPTER 10 SUMMARY AND REVIEW

Vocabulary

complex conjugates theorem (484)
constant term (481)
depressed polynomial (477)
Descartes' rule of signs (485)
factor theorem (476)
fundamental theorem of algebra (483)
horizontal asymptote (503)
leading coefficient (481)
location theorem (492)
lower bound (493)
polynomial function (481)
rational algebraic function (502)
rational zeros theorem (482)
remainder theorem (476)
synthetic division (470)
upper and lower bound theorem (493)
upper bound (493)
vertical asymptote (503)
zeros of a function (481)

Dividing Polynomials To divide a polynomial by a monomial, divide each term of the polynomial by the monomial. To divide by a polynomial, arrange the terms of each polynomial in descending order of one variable. **10.1**

Divide.

1. $\dfrac{x^3y^5 - x^4y^6 + x^4y^3}{x^2y^3}$

2. $\dfrac{9s^3t^6 - 15s^5t^4 + 6s^6t^4}{3s^3t^2}$

3. $(y^3 - 81) \div (y + 1)$

4. $(8 - w + w^3) \div (w + 2)$

Synthetic Division Synthetic division is a shortcut method for dividing a polynomial in one variable by a binomial of the form $x - a$, where a can be any real or imaginary number. **10.2**

Divide, using synthetic division.

5. $(x^3 - 14x^2 + 6x + 4) \div (x - 2)$

6. $(2x^3 - 6x^2 + 12x - 8) \div (x + 4)$

Remainder and Factor Theorems *Remainder Theorem:* If a polynomial $P(x)$ is divided by $x - a$, the remainder is $P(a)$. *Factor Theorem:* $x - a$ is a factor of a polynomial $P(x)$ if and only if $P(a) = 0$. **10.3**

7. Evaluate $P(x) = x^3 + 2x^2 - 9x - 18$ for $x = 2$.

8. Determine whether $x + 2$ is a factor of $6x^3 + 4x^2 - 14x + 4$.

Show that the given binomial is a factor of the other polynomial. Then factor that polynomial completely over the integers.

9. $x - 1$; $x^3 - 6x^2 + 11x - 6$

10. $t + 2$; $t^3 + 3t^2 - 4t - 12$

Solving Polynomial Equations The rational zeros theorem, the corollary to the fundamental theorem of algebra, the complex conjugate theorem, and Descartes' rule of signs are helpful when you must find zeros of polynomial functions or solutions to polynomial equations. 10.4

Find the rational solutions of each equation.

11. $2x^3 - 11x^2 + 18x - 9 = 0$

12. $8x^3 - 4x^2 - 2x + 1 = 0$

Find the complex solutions of each equation.

13. $x^3 + x^2 + 4x + 4 = 0$

14. $x^4 + 15x^2 - 16 = 0$

Graphing Polynomial Functions To graph a polynomial function, plot ordered pairs of the form $(x, f(x))$ and join the points with a smooth curve. The location theorem can be used to approximate real zeros. 10.6

Graph each function and find its real zeros.

15. $f(x) = x^3 + 2x^2 - 5x - 6$

16. $f(x) = x^4 - 4x^3 - 4x^2 + 16x$

Find a real zero, to the nearest tenth, between the given values.

17. $f(x) = x^3 - 4x^2 + 4$; 1 and 2

18. $f(x) = 2x^3 + 7x^2 + 7x + 5$; -2 and -3

Solve.

19. The volume of a cube is 48 ft^3 less than twice the volume of a rectangular box. An edge of the cube is 2 ft shorter than the length of the box, 4 ft longer than the width of the box, and 1 ft longer than the height of the box. Find the dimensions of the cube and the box. 10.7

20. Find four rational numbers such that the product of the first, third, and fourth numbers is -70. The second number is 4 more than the first number, the third is 3 less than twice the first, and the fourth is 7 less than the first.

Graphing Rational Algebraic Functions To graph a rational algebraic function, first determine and graph any intercepts and asymptotes. 10.8

Graph each function.

21. $y = \frac{1}{x^2}$

22. $y = \frac{1 - x}{2x + 1}$

CHAPTER 10 TEST

Divide.

1. $\dfrac{2x^3 - 3x^2 - 4x - 6}{x - 2}$

2. $\dfrac{4n^3 + 6n^2 + 3}{2n + 1}$

Divide, using synthetic division.

3. $\dfrac{x^3 - 5x^2 - 3x + 25}{x - 3}$

4. $\dfrac{2a^3 - 11a^2 - a + 30}{a + 3}$

Show that the given binomial is a factor of the other polynomial. Then factor that polynomial completely over the integers.

5. $x - 4$; $x^3 - 8x^2 + 11x + 20$

6. $x + 2$; $3x^3 + 17x^2 + 18x - 8$

Find all the solutions, real and imaginary, of each equation.

7. $2x^3 + 3x^2 - 8x - 12 = 0$

8. $x^3 + x^2 + 25x + 25 = 0$

9. Graph the function $f(x) = x^3 + 3x^2 - 2x - 6$ and find its real zeros. Approximate any nonintegral zeros to the nearest tenth.

10. For $f(x) = x^3 - 4x^2 + x + 3$, find a real zero, to the nearest tenth, between 1 and 2.

Solve.

11. The length of a rectangular storage bin is 1 m less than twice its width, and the height is 3 m more than its width. The volume of the bin is 90 m^3. Find the dimensions of the bin.

12. Find four rational numbers such that the product of the first, third, and fourth is 126. Also, the second number is two more than the first, the third is one more than the second, and the fourth is one more than the third.

13. Find the intercept(s) and the asymptote(s) of the rational algebraic function $y = \dfrac{3}{x + 4}$. Then sketch its graph.

Challenge

Find all the complex solutions of the equation $x^5 - 2x^2 = 9x + 6$.

COLLEGE ENTRANCE EXAM REVIEW

Solve. Grid in your response on the answer sheet.

1. What is the value of x when $p \| g$?

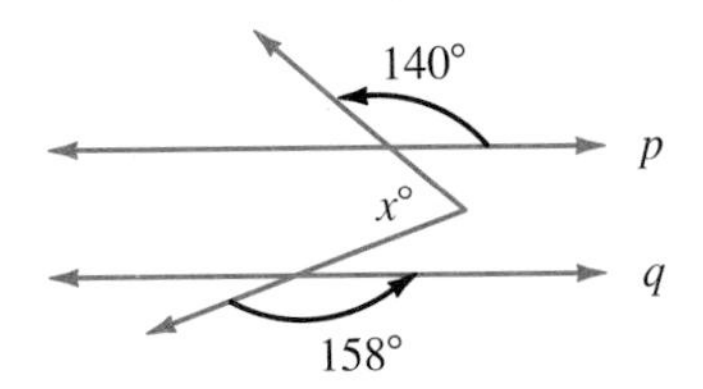

2. $f(x) = 5x^4 - 3x^3 - x^2 + 2x + 5$. Find $f(-1) - f(1)$.

3. What is the remainder when $2x^4 - x^3 + 3x - 1$ is divided by $x + 2$?

4. An isosceles triangle with base 10 cm has an area of 60 cm^2. Find its perimeter.

5. x will be less than or equal to what value when $3(2x - 1) + 8 \leq 4x + 11$?

6. In the diagram below, $\overline{PA}$ is a tangent to circle O at A and secant $\overline{PC}$ passes through O. If $PA = 15$ and $PB = 9$, find the radius of circle O.

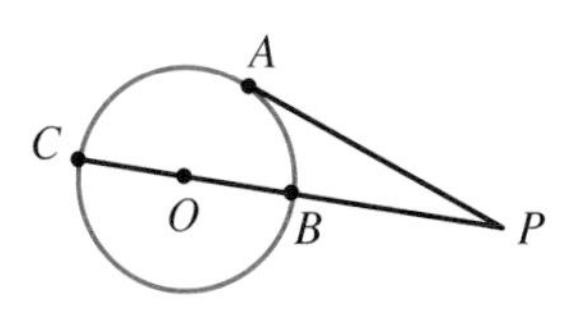

7. A biker took 2 h to ride 24 mi and then returned by the same route. If he rode at an average speed of 8 mi/h on the return trip, what was his average speed in mi/h for the round trip?

8. 5% of 6% is what percent?

9. If $\frac{a}{3} = \frac{12}{b}$ and $b - 2 = 2$, then a is equal to what value?

10. Find the least of four rational numbers such that the product of the first, third, and fourth numbers is -6. The second number is 3 less than the first number, the third is 2 less than the second, and the fourth is 2 less than the third.

11. If $100 \leq m \leq 400$, and m is a multiple of 5, 6, 7, and 10, then what is the value of m?

12. $f(x) = x^3 + 2x^2 - 5x + 1$. Find a real zero to the nearest tenth between 1 and 2.

Use this diagram for 13.

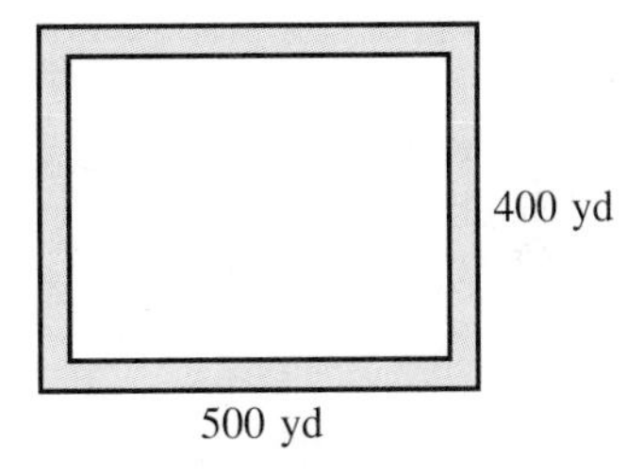

13. A farmer has a rectangular field 400 yd by 500 yd. On this field a uniform strip on the outside will be left unplanted so that half of the total area will be planted. To the nearest tenth of a yard, what is the width of the strip?

14. If $\sqrt{R} = \frac{2\sqrt{t}}{1 - t}$, find the least value of t when $R = 3$.

CUMULATIVE REVIEW

Solve and check.

1. $4y + 4 = 19 - y$

2. $|4x + 2| = 6$

3. $y^2 - 7y + 6 = 0$

4. Solve using an augmented matrix: $\begin{cases} 2x - y - z = 1 \\ x - 2z = -7 \\ -2y - z = 2 \end{cases}$

5. Divide, using synthetic division: $\dfrac{x^3 - 3x^2 + 4x - 12}{x - 1}$

6. Solve using Cramer's Rule: $\begin{cases} 2x - y = 8 \\ 3x + 2y = -2 \end{cases}$

7. Given $f(x) = -3x$, $g(x) = x - 1$, evaluate $g[f(-2)]$ and $f[g(1)]$.

8. If a number is added to its square, the sum is 182. Find the number.

9. Evaluate $3x - 5x^2 + 4x^3$ when $x = -3$.

10. Given $R = \begin{bmatrix} 5 & -1 & 0 \\ 2 & 3 & -2 \\ 0 & -3 & 1 \end{bmatrix}$, find $3R$.

11. Find the equation of the line having slope 2 and passing through $(-2, 3)$.

12. Graph the solution set: $x < -2$ or $x > 3$

13. Simplify: $\dfrac{x^2 + x - 12}{x^2 - 3x - 10} \cdot \dfrac{x^2 + x - 2}{x^2 - 4x + 3}$

14. Solve the linear system by substitution. Determine whether the system is consistent and independent, inconsistent, or consistent and dependent.
$\begin{cases} 3x - 2y = -6 \\ y = x - 5 \end{cases}$

15. Solve and check: $\sqrt{x + 2} = x + 2$

16. Three times a number increased by eight is 29. Find the number.

17. y varies directly as x. If $y = 4$ when $x = 3$, find y when $x = 9$.

18. Simplify: $\dfrac{2 + \sqrt{3}}{\sqrt{2} + \sqrt{3}}$

Factor.

19. $25y^2 - 4$

20. $c^2 + 5c - 14$

21. $y^3 - 64$

22. $5a^2 - 80$

Conic Sections

A suspenseful equation!

Developing Mathematical Power

The Golden Gate Bridge is an example of a parabola formed by tying weights to a catenary at equal spaces. In this case, the weight comes from the bridge itself. An equation that approximates the parabola formed is

$$y = 700 + 5\left(\frac{x}{200}\right)^2 - \frac{x}{2}.$$

Project

Graph the equation

$$y = 700 + 5\left(\frac{x}{200}\right)^2 - \frac{x}{2}.$$

Construct a scale model of the catenary between the towers of the Golden Gate Bridge and then suspend weights to approximate the parabola formed. Compare your model to the graph.

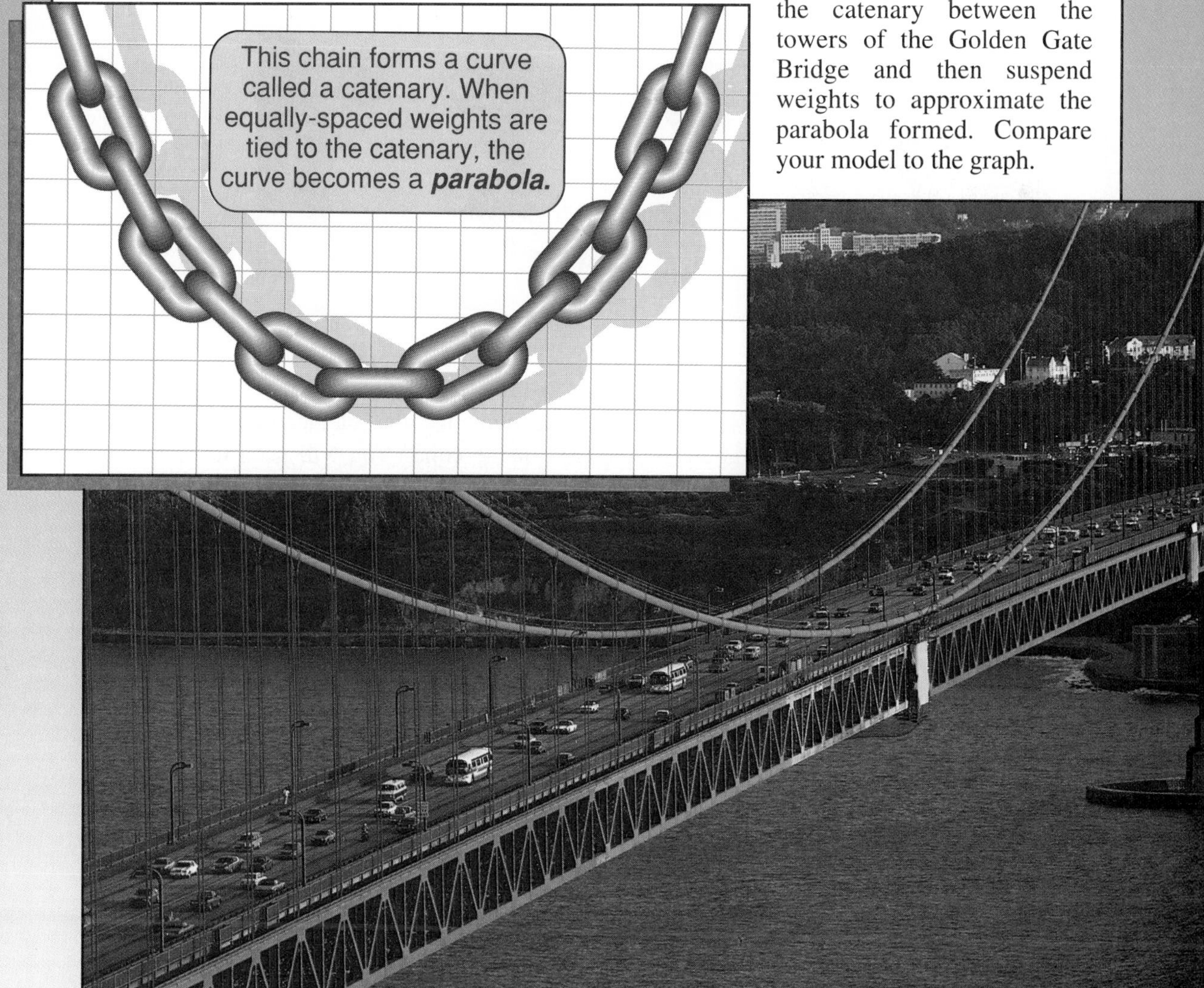

11.1

Circles

Objective: To determine the relationships among the center, the radius, and the equation of a circle

Four *double right circular cones* are shown below. Each cone consists of all lines, or *elements,* through a point V that form a given angle with line a. Point V is called the *vertex* of the cone, and line a is the *axis*. Curves formed by the intersection of a plane and a cone of this type are called **conic sections,** or **conics.**

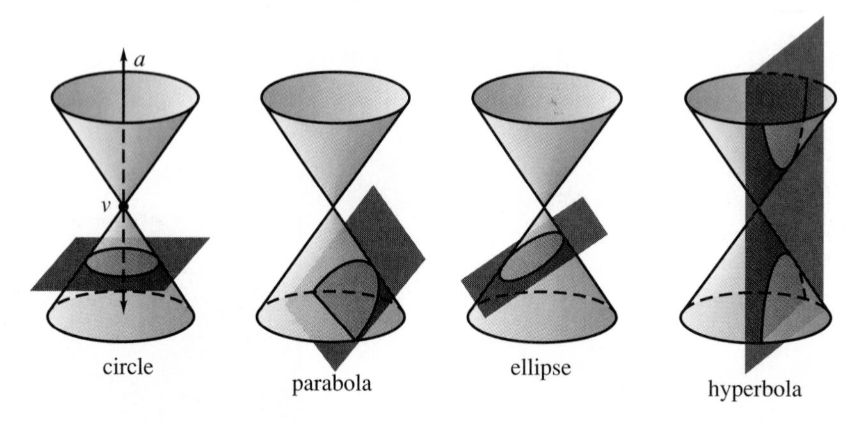

The conic considered in this lesson is the circle, which is the intersection of the cone and a plane perpendicular to the axis of the cone. You will need to use the distance formula to find an equation for a circle.

Capsule Review

EXAMPLE **Find the distance between (3, −5) and (−2, 4).**

$d = \sqrt{(-2 - 3)^2 + [4 - (-5)]^2}$ $\quad d = \sqrt{(x_2 - x_1)^2 + (y_2 - y_1)^2}$

$d = \sqrt{25 + 81}$

$d = \sqrt{106}$

Find the distance between the two points.

1. (3, 7), (5, 10)

2. (−1, 4), (−3, 9)

3. (4, 5), (−3, −4)

4. (6, −3), (−2, 8)

5. (−3, −6), (−5, −2)

6. (7, −3), (5, −2)

7. $(2a, 4b)$, $(3a, 2b)$

8. $(3a, 4)$, $(5a, 6)$

9. $(10c, 13d)$, $(5c, 3d)$

A **circle** is the set of all points P in a plane that are the same distance from a given point. The given distance is the **radius** of the circle, and the given point is the **center** of the circle. To derive the equation of a circle with center $C(h, k)$ and radius r, use the distance formula. Let $P(x, y)$ represent an arbitrary point on the circle. The distance from P to C is equal to the radius, r.

$$d = \sqrt{(x_2 - x_1)^2 + (y_2 - y_1)^2}$$
$$r = \sqrt{(x - h)^2 + (y - k)^2}$$
$$r^2 = (x - h)^2 + (y - k)^2 \qquad \textit{Square both sides.}$$

> The standard form of the equation of a circle with center $C(h, k)$ and radius r is
>
> $$(x - h)^2 + (y - k)^2 = r^2$$

EXAMPLE 1 **Express in standard form the equation of the circle shown below.**

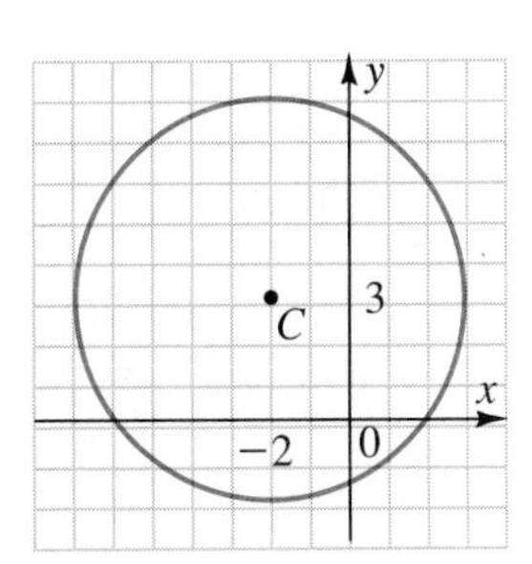

The center is $(-2, 3)$.

The circle passes through the point $(3, 3)$, so the radius is $3 - (-2)$, or 5.

Substitute in the equation.

$$(x - h)^2 + (y - k)^2 = r^2$$
$$[x - (-2)]^2 + (y - 3)^2 = 5^2$$
$$(x + 2)^2 + (y - 3)^2 = 25 \qquad \textit{Standard form}$$

To find an equation of the circle with center at the origin, $(0, 0)$, replace h with 0 and k with 0.

$$(x - 0)^2 + (y - 0)^2 = r^2 \qquad (x - h)^2 + (y - k)^2 = r^2$$
$$x^2 + y^2 = r^2$$

> The standard form of the equation of a circle with center at the origin $(0, 0)$ and radius r is
>
> $$x^2 + y^2 = r^2$$

EXAMPLE 2 **Express in standard form the equation of the circle with center at the origin and radius 4. Draw the graph.**

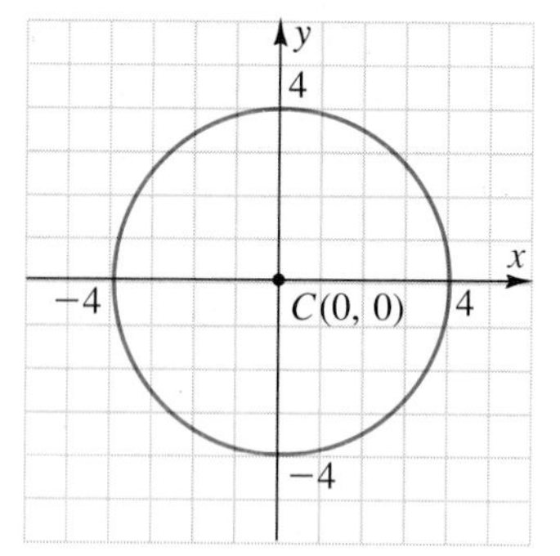

$x^2 + y^2 = 4^2$ *Substitute 4 for r in $x^2 + y^2 = r^2$.*

$x^2 + y^2 = 16$ *Standard form of the equation*

The center of the circle is the origin and the radius is 4, so the circle intersects the coordinate axes at the points (4, 0), (0, 4), (−4, 0), and (0, −4). To make a rough graph of the circle, plot these four points and sketch the circle that contains them. To draw the graph more accurately, use a compass.

A conic section can be graphed with a graphing utility. To do this, you usually need to express the equation in *function form*. That is, solve for y before entering the equation.

$$x^2 + y^2 = 16$$
$$y^2 = 16 - x^2$$
$$y = \pm\sqrt{16 - x^2} \quad \textit{Function form}$$

Then you would input the two functions $y = \sqrt{16 - x^2}$ and $y = -\sqrt{16 - x^2}$.

Given the standard form of the equation of a circle, you can find the center and the radius of the circle easily.

EXAMPLE 3 **Find the center and radius of the circle with the equation $(x - 5)^2 + (y + 1)^2 = 16$. Graph the circle.**

$(x - h)^2 + (y - k)^2 = r^2$

$(x - 5)^2 + (y + 1)^2 = 16$ *$h = 5, k = -1, r = 4$*

The center is (5, −1) and the radius is 4.

Solve the equation for y and graph both results on the same screen.

$y = -1 \pm \sqrt{16 - (x - 5)^2}$

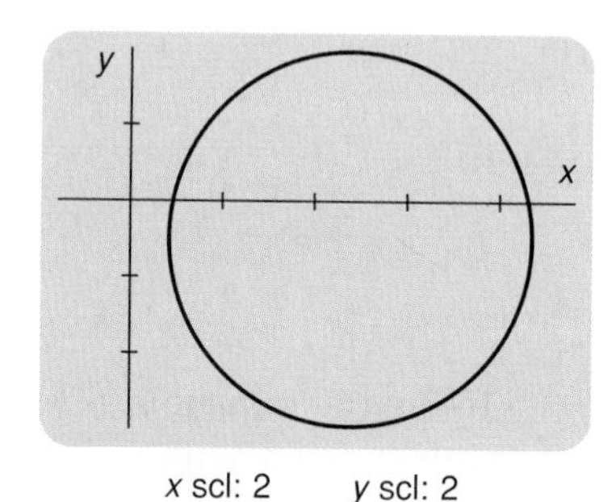

x scl: 2 y scl: 2

EXAMPLE 4 **Find the center and radius of the circle with the equation $x^2 + y^2 + 4x - 6y + 5 = 0$.**

$$x^2 + y^2 + 4x - 6y + 5 = 0$$
$$(x^2 + 4x + \underline{?}) + (y^2 - 6y + \underline{?}) = -5 \quad \textit{Group the terms. Complete the squares.}$$
$$(x^2 + 4x + 4) + (y^2 - 6y + 9) = -5 + 4 + 9 \quad \left[\tfrac{1}{2}(4)\right]^2 = 4; \left[\tfrac{1}{2}(-6)\right]^2 = 9$$
$$(x + 2)^2 + (y - 3)^2 = 8$$

The center is (−2, 3) and the radius is $\sqrt{8}$, or $2\sqrt{2}$.

The graph of the equation $x^2 + y^2 = r^2$ is a circle with center at the origin and radius r. If x is replaced with $x - h$ and y is replaced with $y - k$, the equation $(x - h)^2 + (y - k)^2 = r^2$ is obtained. As noted on page 515, the graph of this equation is a circle with center (h, k) and radius r. The substitution of $x - h$ for x and $y - k$ for y has the effect of *translating* the graph of $x^2 + y^2 = r^2$ a distance of h units horizontally and k units vertically. Why? When h is positive, the shift is to the right, and when h is negative, the shift is to the left. When k is positive, the shift is upwards, and when k is negative, the shift is downwards.

EXAMPLE 5 **The graph of $x^2 + y^2 = 16$ is translated 3 units to the right and 2 units down. Find the equation of the new circle and sketch its graph.**

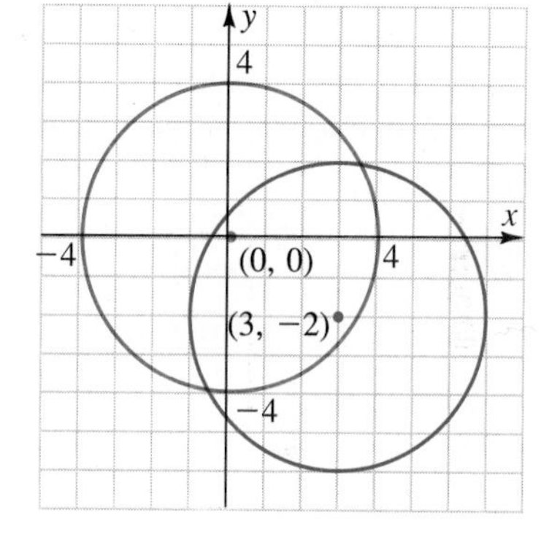

$$(x - h)^2 + (y - k)^2 = r^2$$

$$(x - 3)^2 + [y - (-2)]^2 = 16 \quad \text{Substitute 3 for } h,$$

$$(x - 3)^2 + (y + 2)^2 = 16 \quad -2 \text{ for } k, \text{ 16 for } r^2.$$

To graph the circle in its new position, you can translate each of the four points (4, 0), (0, 4), (−4, 0), and (0, −4) 3 units to the right and 2 units down, and then sketch the circle that contains them. Alternately, you can translate the center from (0, 0) to (3, −2) and use a compass.

CLASS EXERCISES

Express in standard form the equation of the circle with the given center and radius.

1. (3, 4); 8 **2.** (−2, 5); 7 **3.** (0, 0); 9 **4.** (0, −1); 2

Find the center and radius of the circle with the given equation.

5. $x^2 + y^2 = 81$

6. $x^2 + y^2 = 49$

7. $x^2 + (y + 3)^2 = 25$

8. $(x - 6)^2 + y^2 = 64$

9. $(x - 3)^2 + (y - 1)^2 = 36$

10. $(x + 2)^2 + (y - 4)^2 = 16$

11. $x^2 + y^2 + 6x - 10y - 4 = 0$

12. $x^2 + y^2 - 2x + 6y - 3 = 0$

13. The graph of $x^2 + y^2 = 4$ is translated 3 units to the right and 4 units down. Find the equation of the new circle.

14. The graph of $x^2 + y^2 = 9$ is translated 2 units to the left and 5 units up. Find the equation of the new circle.

PRACTICE EXERCISES 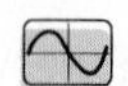 Use technology where appropriate.

Express the equation of each circle in standard form.

1.

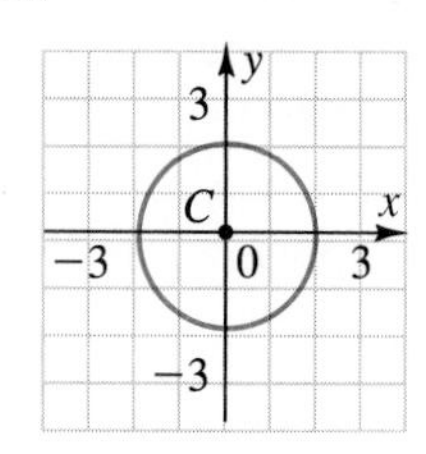

2.

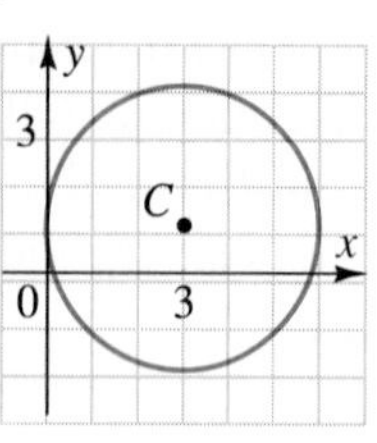

3.

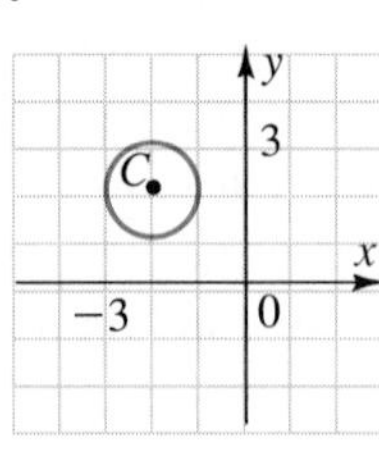

4.

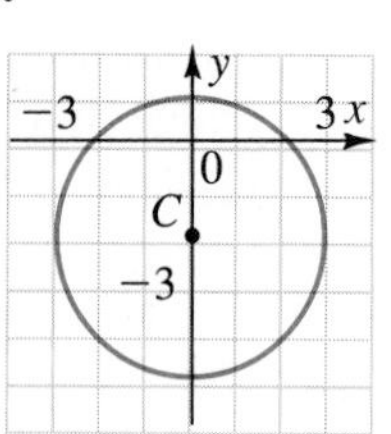

Express in standard form the equation of the circle with the given center and radius. Sketch the graph.

5. (0, 0); 7 **6.** (0, 0); 4 **7.** (5, −1); 6 **8.** (−4, 2); 3

Find the center and radius of the circle with the given equation.

9. $x^2 + y^2 = 9$

10. $x^2 + y^2 = 16$

11. $x^2 + y^2 = 25$

12. $x^2 + y^2 = 36$

13. $(x - 1)^2 + y^2 = 49$

14. $x^2 + (y - 2)^2 = 64$

15. $x^2 + (y + 7)^2 = 36$

16. $(x + 6)^2 + y^2 = 81$

17. $(x + 3)^2 + (y - 9)^2 = 49$

18. $(x + 8)^2 + (y + 3)^2 = 16$

19. $x^2 + y^2 - 6x - 2y + 5 = 0$

20. $x^2 + y^2 - 8x - 12y - 3 = 0$

21. $x^2 + y^2 - 4x + 4y - 6 = 0$

22. $x^2 + y^2 + 8x - 3y + 8 = 0$

23. $x^2 + y^2 - 5x - 6y + 4 = 0$

24. $x^2 + y^2 + 8x - 12y - 7 = 0$

25. $x^2 + y^2 + 6x + 3y - 6 = 0$

26. $x^2 + y^2 + 7x - 8y + 5 = 0$

27. $x^2 + y^2 - 5x - 3y + 8 = 0$

28. $x^2 + y^2 + 8x - 10y - 12 = 0$

29. $x^2 + y^2 - 8y + 5 = 0$

30. $x^2 + y^2 + 10x + 7 = 0$

31. The graph of $x^2 + y^2 = 25$ is translated 5 units to the left. Find the equation of the new circle. Graph the circle.

32. The graph of $(x + 4)^2 + (y + 5)^2 = 9$ is translated 4 units to the right and 5 units up. Find the equation of the new circle. Graph the circle.

Find the center and radius of the circle with the given equation.

33. $4x^2 + 4y^2 - 16x + 8y - 8 = 0$

34. $3x^2 + 3y^2 + 24x - 6y + 9 = 0$

35. $x^2 + y^2 - 4ax - 6ay + 4a = 0$

36. $x^2 + y^2 + 8bx - 12by - 7b = 0$

37. Find the equation of the circle with center (3, 4) and containing the point (7, 6).

38. A circle with center (2, 1) is tangent to the line $y = x + 2$. Find the equation of the circle.

Applications

39. Drafting A mechanical drawing is needed of a disk that is represented by a circle. Find the equation of the circle if the radius is 9 cm and the center is at the origin.

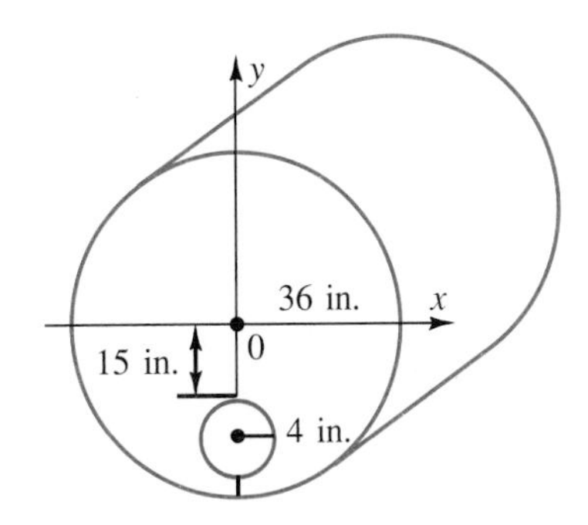

40. Storage A cylindrical oil tank has a radius of 36 in. A circular hole with radius 4 in. has been drilled 15 in. below the center of one base of the tank. Find an equation of the circular hole, using the center of the base of the tank as the origin and the axes as shown in the figure.

Developing Mathematical Power

Extension Under certain conditions, the intersection of a plane and a cone may be a point or one or two lines. These are called *degenerate conics.*

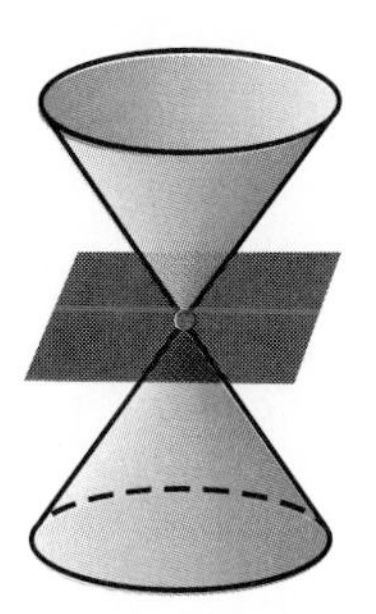

Degenerate circle or ellipse

Degenerate parabola

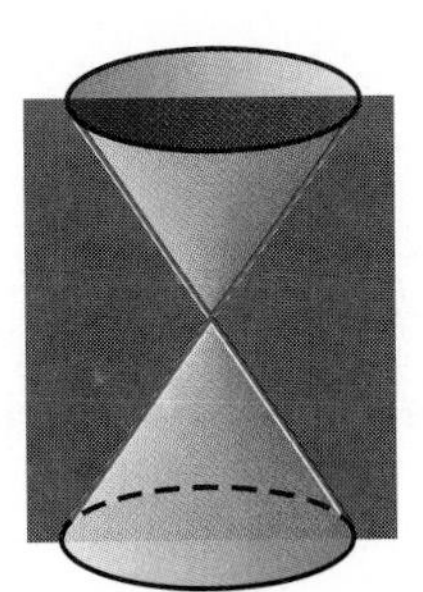

Degenerate hyperbola

A degenerate circle, which is a single point, may be thought of as a circle with zero radius. $(x - 3)^2 + (y + 1)^2 = 0$ is the equation of a degenerate circle. Only one ordered pair, $(3, -1)$, is a solution to this equation, so its graph is a single point.

Determine which of the following are equations of degenerate circles. For each such equation, give the coordinates of the point that is its graph.

41. $x^2 + y^2 = 0$ **42.** $x^2 + y^2 = 4x + 4$ **43.** $(x + 1)^2 = 1 - y^2$

44. $x^2 = 3 - y^2$ **45.** $x^2 + y^2 - 2y = -1$ **46.** $x^2 + y^2 + 4x - 4y = -8$

11.2

Parabolas

Objective: To determine the relationships among the focus, directrix, vertex, axis of symmetry, and the equation of a parabola

Telescopic mirrors, automobile headlights, disk antennas for receiving satellite television signals, and radar antennas are all applications of the *parabola*.

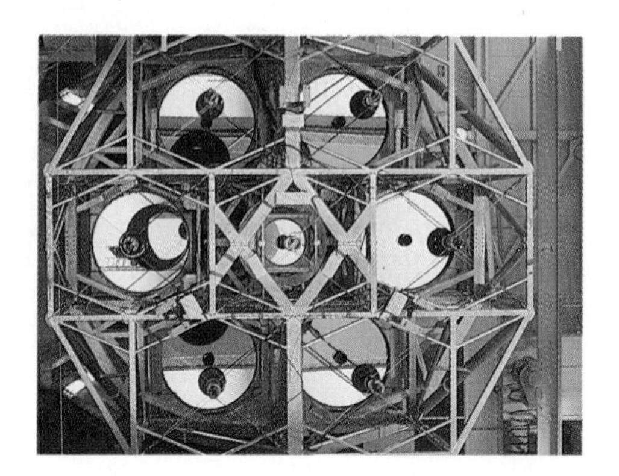

The intersection of a right circular cone and a plane parallel to an element of the cone is a parabola. (See page 514.) In order to find the coordinates of the vertex of a parabola from its equation, the method of completing the square is often used.

Capsule Review

EXAMPLE **Write the equation $y = 3x^2 - 12x + 16$ in the form $y = a(x - h)^2 + k$.**

$y = 3x^2 - 12x + 16$

$y = 3(x^2 - 4x) + 16$ *Factor $3x^2 - 12x$.*

$y = 3(x^2 - 4x + 4) + 16 - 12$ *Complete the square.*

$y = 3(x - 2)^2 + 4$

Write each equation in the form $y = a(x - h)^2 + k$.

1. $y = 5x^2 - 10x + 8$ **2.** $y = -3x^2 - 24x - 49$ **3.** $y = 2x^2 + 8x + 2$

4. $y = -x^2 + 10x - 15$ **5.** $2y = x^2 - 8x + 20$ **6.** $4y = -x^2 - 2x$

The **parabola,** which was introduced in Chapter 9, can be defined as the set of all points P in a plane that are the same distance from a given line and a fixed point not on the line. The fixed point is called the **focus,** and the given line is called the **directrix** of the parabola. Recall that every parabola has an *axis of symmetry* and that the *vertex* of a parabola is the point at which it crosses this axis.

In communications and astronomy, one property of the focus of a parabola is very important. When light rays or radio waves parallel to the axis of symmetry of a parabolic antenna reflect off the interior of the antenna, they meet at one point, the focus.

In the figure at the right, PF is the distance from an arbitrary point $P(x, y)$ on a parabola to the focus $F(0, c)$. PF is equal to the distance PA from point P to the directrix, line ℓ, which has equation $y = -c$. The vertex is halfway between the focus and the directrix. Thus, the distance between the vertex and focus is $|c|$ and between the vertex and directrix is $|c|$. To find an equation of this parabola with vertex at the origin and focus on the y-axis, use the distance formula.

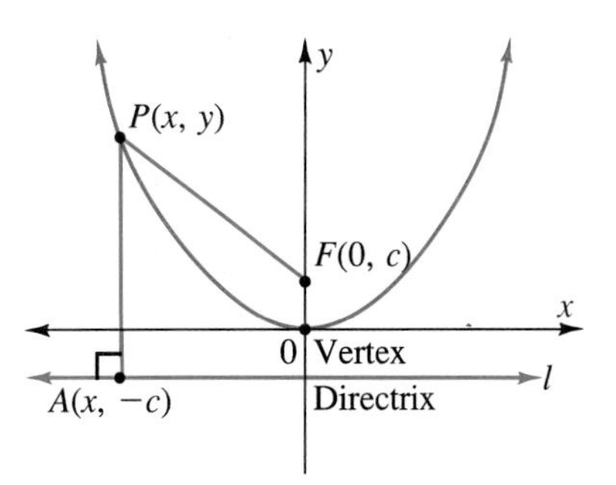

$$PF = PA$$
$$\sqrt{(x-0)^2 + (y-c)^2} = \sqrt{(x-x)^2 + (y+c)^2}$$
$$x^2 + (y-c)^2 = (y+c)^2 \qquad \textit{Square both sides.}$$
$$x^2 + y^2 - 2cy + c^2 = y^2 + 2cy + c^2 \qquad \textit{Simplify.}$$
$$x^2 = 4cy$$

Note that the equation $x^2 = 4cy$, or $y = \frac{1}{4c}x^2$, defines y as a function of x. Recall that if the coefficient of x^2 is positive ($c > 0$), the parabola opens upward; if it is negative ($c < 0$), the parabola opens downward.

The equation of a parabola with vertex $(0, 0)$, focus $(0, c)$, directrix $y = -c$, and axis of symmetry the y-axis can be written

$$x^2 = 4cy \quad \text{or} \quad y = \frac{1}{4c}x^2$$

EXAMPLE 1 **Find the vertex, the axis, the focus, and the directrix of a parabola with the equation $x^2 = 8y$. Sketch the graph.**

Compare $x^2 = 8y$ with $x^2 = 4cy$, or $y = \frac{1}{4c}x^2$.

Vertex: $(0, 0)$
Axis: $x = 0$
Focus: $(0, 2)$ *$4c = 8$, so $c = 2$.*
Directrix: $y = -2$

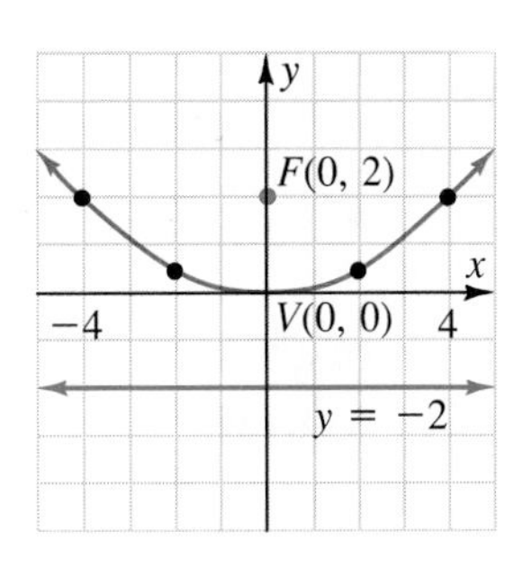

Since $c > 0$, the coefficient of x^2 is positive and the parabola opens upward. Plot the vertex and a few extra points, keeping in mind that the parabola is symmetric with respect to its axis. Join the points with a smooth curve.

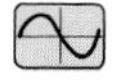 The graph can be checked with a graphing utility.

x	± 2	± 4
y	0.5	2

Note that the parabola with vertex (0, 0) and vertical axis was introduced in Lesson 9.1, where the general equation for such a parabola was given as $y = ax^2$. You can see now that the value of a, the coefficient of x^2, depends on the location of the focus. That is, the equation of such a parabola can be written $y = \frac{1}{4c}x^2$, where c is the y-coordinate of the focus.

The equation for a parabola with its focus on the x-axis is derived in a similar way. You will be asked to show the derivation in Exercise 41.

> The equation of a parabola with vertex (0, 0), focus $(c, 0)$, directrix $x = -c$, and axis of symmetry the x-axis, can be written
>
> $$y^2 = 4cx \quad \text{or} \quad x = \frac{1}{4c}y^2$$

EXAMPLE 2 **Write an equation for the parabola shown below.**

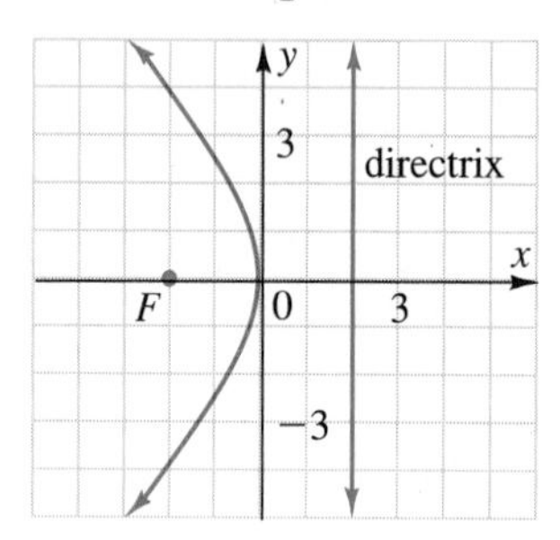

The focus is $(-2, 0)$, so $c = -2$. Since the focus is on the x-axis, substitute in the equation $y^2 = 4cx$.

$$y^2 = 4cx$$
$$y^2 = 4(-2)x$$
$$y^2 = -8x, \text{ or } x = -\tfrac{1}{8}y^2$$

A parabola with an equation of the form $y^2 = 4cx$ does not define y as a function of x, since it fails the vertical line test. Note also that if $c < 0$ the parabola opens to the left; if $c > 0$ it opens to the right.

The definition of a parabola and the distance formula are used to derive an equation for a parabola that does not have its vertex at the origin. To illustrate, for a parabola with focus $F(2, 3)$ and directrix $y = -1$, let $P(x, y)$ represent an arbitrary point on the parabola. Then

$$PF = PA$$
$$\sqrt{(x-2)^2 + (y-3)^2} = \sqrt{(x-x)^2 + (y+1)^2}$$
$$(x-2)^2 + (y-3)^2 = (x-x)^2 + (y+1)^2$$
$$x^2 - 4x + 4 + y^2 - 6y + 9 = 0^2 + y^2 + 2y + 1$$
$$x^2 - 4x + 4 = 8y - 8$$
$$(x-2)^2 = 8y - 8$$
$$8y = (x-2)^2 + 8$$
$$y = \tfrac{1}{8}(x-2)^2 + 1$$

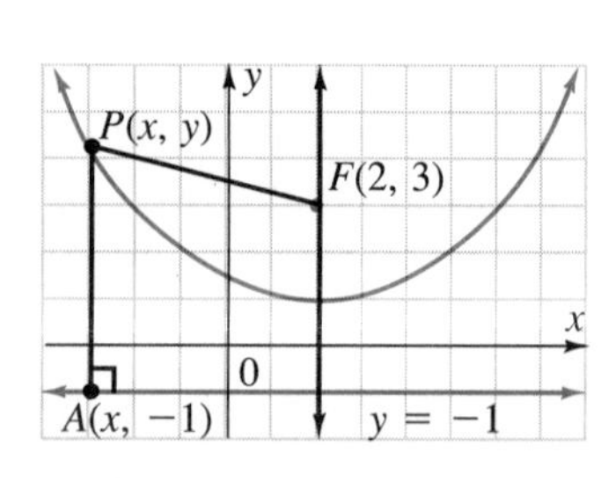

This equation of the parabola is of the form $y = a(x - h)^2 + k$. The vertex of the parabola is (h, k), or $(2, 1)$, and the axis of symmetry is the line $x = h$, or $x = 2$. Recall that the vertex of the basic parabola $y = ax^2$ is $(0, 0)$. With the vertex at (h, k), the basic graph is shifted h units horizontally and k units vertically. The standard form of the equation of a parabola with vertex (h, k) and axis parallel to a coordinate axis is

$$y = a(x - h)^2 + k \quad \text{or} \quad x = a(y - k)^2 + h$$

Information about such parabolas is summarized in the table below.

Parabolas Centered at (h, k)

Standard Form of Equation	$y = a(x - h)^2 + k$	$x = a(y - k)^2 + h$
Vertex	(h, k)	(h, k)
Axis of Symmetry	$x = h$	$y = k$
Focus	$\left(h, k + \frac{1}{4a}\right)$	$\left(h + \frac{1}{4a}, k\right)$
Directrix	$y = k - \frac{1}{4a}$	$x = h - \frac{1}{4a}$
Opening	Upward if $a > 0$ Downward if $a < 0$	Right if $a > 0$ Left if $a < 0$

EXAMPLE 3 **For the parabola $y^2 - 4x - 4y + 16 = 0$, determine the vertex, the axis of symmetry, the focus, and the directrix. Sketch the graph.**

$y^2 - 4x - 4y + 16 = 0$

$x = \frac{1}{4}y^2 - y + 4$ *Solve for x in terms of y.*

$x = \frac{1}{4}(y^2 - 4y + 4) + 4 - 1$ *Complete the square in y.*

$x = \frac{1}{4}(y - 2)^2 + 3$ *Standard form*

Vertex: $(3, 2)$ (h, k)

Axis: $y = 2$ $y = k$

Focus: $(3 + 1, 2)$, or $(4, 2)$ $\left(h + \frac{1}{4a}, k\right)$

Directrix: $x = 3 - 1$, or $x = 2$ $x = h - \frac{1}{4a}$

Since $a > 0$, the graph opens to the right.

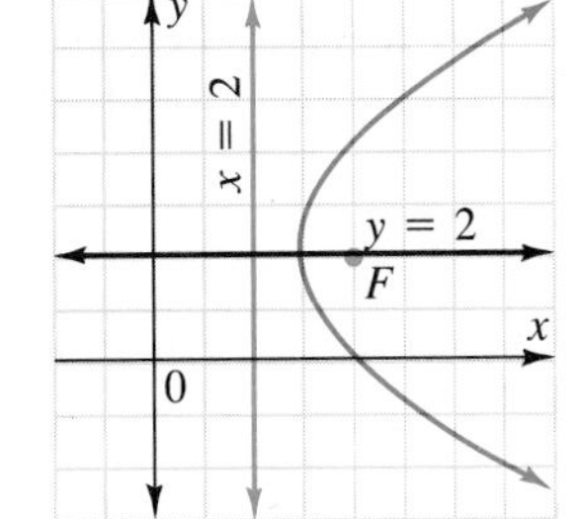

A parabola with a horizontal axis is not a function; to graph such a parabola on a utility, solve for y and graph both results together.

CLASS EXERCISES

Express in standard form the equation of each parabola.

1. Focus (0, 3); directrix $y = -3$

2. Focus (4, 0); directrix $x = -4$

3. Focus (3, 4); directrix $y = -2$

4. Focus (1, 2); directrix $x = -3$

Find the vertex, the axis of symmetry, the focus, and the directrix of each parabola. Sketch the graph.

5. $x^2 = -8y$

6. $y^2 = 4x$

7. $y^2 = 4(x - 1)$

8. $y = 4(x - 2)^2 - 3$

PRACTICE EXERCISES

 Use technology where appropriate.

Write in standard form the equation of each parabola. Point F is the focus and line ℓ is the directrix.

1.

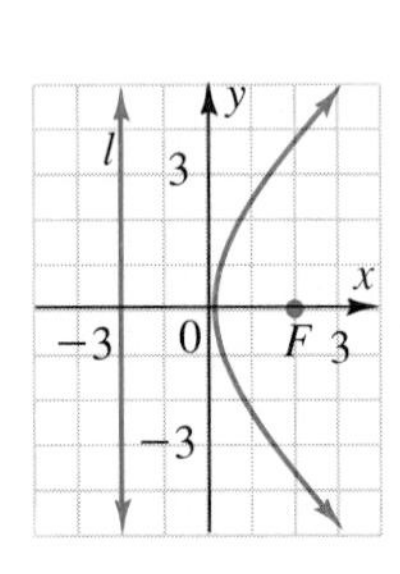

2.

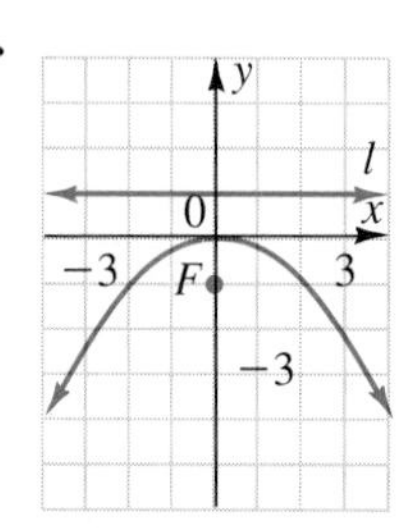

3.

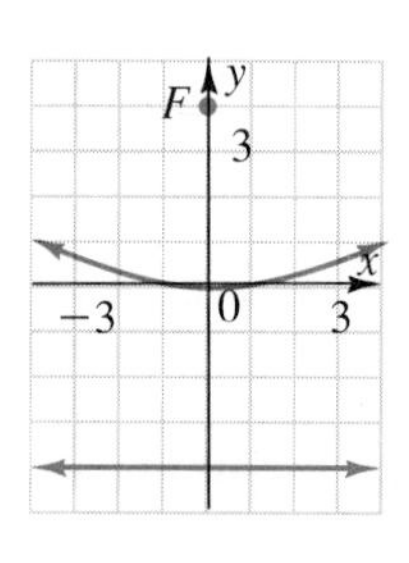

4.

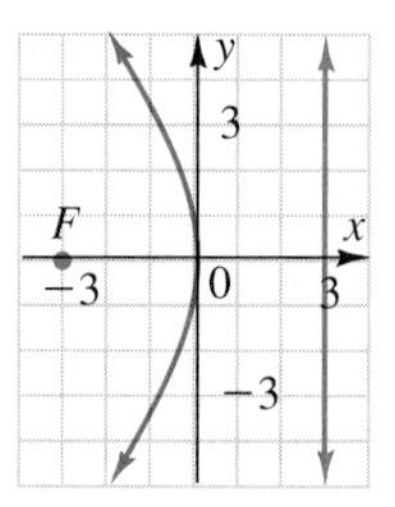

Write in standard form the equation of each parabola.

5. Focus (0, 2); directrix $y = -2$

6. Focus (3, 0); directrix $x = -3$

7. Focus (4, 3); directrix $y = -3$

8. Focus (2, 1); directrix $x = -2$

9. Focus (−1, 0); directrix $y = 1$

10. Focus (0, −2); directrix $y = -4$

11. Focus (−2, 0); directrix $x = 4$

12. Focus (2, 0); directrix $x = -2$

Find the vertex, the axis of symmetry, the focus, and the directrix of each parabola. Sketch the graph.

13. $x^2 = 12y$

14. $y^2 = -4x$

15. $y^2 = 12x$

16. $x^2 = -4y$

17. $x^2 = -20y$

18. $y^2 = -16x$

19. $y = 4(x - 2)^2$

20. $y = 4(x - 3)^2 - 2$

21. $(x + 2)^2 = y - 3$

22. $(x - 3)^2 = y - 2$

23. $y = x^2 + 4x + 1$

24. $x = y^2 - 4y + 1$

25. $(y + 3)^2 = 4(x - 2)$

26. $(y - 2)^2 = 4(x + 3)$

27. $(x - 5)^2 = 12(y - 6)$

28. $(x + 4)^2 = 8(y + 2)$

29. $x^2 - 6x - y + 11 = 0$

30. $x^2 - 8x - y + 19 = 0$

31. $y^2 - 6y - x + 4 = 0$

32. $y^2 + 2y - x - 1 = 0$

Write in standard form the equation of each parabola.

33. Focus $(0, -4)$; vertex $(0, 0)$

34. Focus $(0, 4)$; vertex $(0, 0)$

35. Focus $(2, 0)$; vertex $(0, 0)$

36. Focus $(-2, 0)$; vertex $(0, 0)$

37. Focus $(3, 2)$; vertex $(3, 0)$

38. Focus $(4, 3)$; vertex $(4, 0)$

39. Write in standard form the equation of the parabola with vertex $(3, 0)$ and directrix $y = -2$.

40. Write in standard form the equation of the parabola with vertex $(0, 5)$ and directrix $x = -1$.

41. Use the distance formula and the definition of a parabola to derive the standard form of the equation of a parabola with focus $F(c, 0)$ and directrix $x = -c$.

Find the vertex, axis of symmetry, focus, and directrix of each parabola. Indicate whether the vertex is a maximum or a minimum point.

42. $y = 2x^2 + 4x - 3$

43. $2x^2 + x + y - 2 = 0$

44. $4y^2 - 12y + 2x + 3 = 0$

45. $x = 4y^2 - 6y + 15$

46. Derive an equation for the parabola with focus (a, b) and directrix $x = c$.

47. Use the definition of parabola to show that the parabola with vertex (h, k) and focus $(h, k + c)$ has the equation $(x - h)^2 = 4c(y - k)$.

48. Use the definition of parabola to show that the parabola with vertex (h, k) and focus $(h + c, k)$ has the equation $(y - k)^2 = 4c(x - h)$.

49. Find an equation of the parabola with vertex $(2, 1)$ that passes through the point $(4, 5)$ and that has an axis parallel to the y-axis.

Applications

50. **Physics** A ball is tossed upward with an initial velocity of 112 ft/s from a platform that is 700 ft above the surface of the earth. After t seconds, the height h of the ball above the ground is given by the equation $h = -16t^2 + 112t + 700$. What is the maximum height reached by the ball?

51. **Civil Engineering** The base of a parabolic arch measures 120 ft, and the vertex of the arch is 90 ft above the ground. Write an equation for this parabola. *Hint:* Use coordinate axes as shown on the figure. Find the coordinates of point P and substitute values of x and y in the equation $x^2 = 4cy$ to find the value of c.

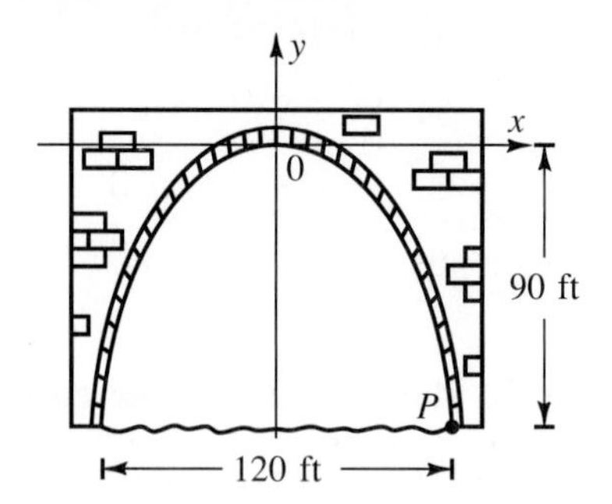

Mixed Review

52. Solve $\frac{8}{3}|2x - 5| = 4x + 32$.

53. If two vertices of an equilateral triangle have the coordinates $(-2, -2)$ and $(8, -2)$, name one possible set of coordinates for the third vertex.

54. Determine whether $(4, 28)$ lies on the parabola $y = 4\left(x - \frac{3}{2}\right)^2 + 4$.

55. Write an equation of a line passing through $(2, 3)$ with the same y-intercept as the graph of $y = 3x^4 - 2x^3 + 3x^2 - 4x + 2$.

56. Solve by addition: $\begin{cases} 2x - y = 14 \\ 3x + 2y = -7 \end{cases}$

57. Express 3.25×10^{-6} in decimal notation.

58. Simplify $(3 + 4i) + (5 - 3i)$.

59. Simplify $\frac{2 + 4i}{4 - 3i}$.

60. Solve $4\sqrt{2x} - x = 8$.

61. Solve and graph the solution set on a number line: $x^2 - 4 < 0$.

62. For what values of the variables is $\frac{2ab}{3c(x^2 - 8)}$ undefined?

63. Add $\frac{2x}{x + 1} + \frac{3x}{x + 2}$.

64. Write an algebraic expression for the sum of three consecutive even integers.

Developing Mathematical Power

65. **Thinking Critically** Is it possible for two different parabolas to have the same focus and vertex? Explain.

11.3 Ellipses

Objective: To determine the relationships among the foci, the intercepts, and the equation of an ellipse

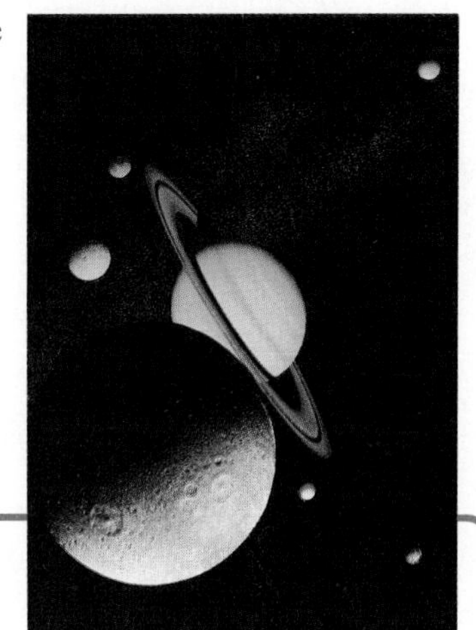

The orbits of the planets in our solar system are in the shape of *ellipses,* and many objects in space travel in elliptical paths.

If a plane is tilted in such a way that its intersection with a right circular cone is a closed curve, then that intersection is an ellipse. (See page 514.) You will need to solve an equation involving radicals in order to find the standard equation of an ellipse.

Capsule Review

To solve equations containing one or more radical expressions, isolate one radical expression and square both sides of the equation. Repeat this process as necessary and check each solution.

EXAMPLE **Solve:** $\sqrt{x+4} + \sqrt{x-3} = 7$

$$\sqrt{x+4} + \sqrt{x-3} = 7$$
$$\sqrt{x+4} = 7 - \sqrt{x-3}$$
$$x + 4 = 49 - 14\sqrt{x-3} + x - 3$$
$$14\sqrt{x-3} = 42$$
$$\sqrt{x-3} = 3$$
$$x - 3 = 9$$
$$x = 12$$

Check:

$$\sqrt{x+4} + \sqrt{x-3} = 7$$
$$\sqrt{12+4} + \sqrt{12-3} \stackrel{?}{=} 7$$
$$\sqrt{16} + \sqrt{9} \stackrel{?}{=} 7$$
$$4 + 3 \stackrel{?}{=} 7$$
$$7 = 7 \checkmark$$

Solve and check.

1. $3\sqrt{n} = 5\sqrt{2}$

2. $\sqrt{x-4} + \sqrt{x} = 2$

3. $\sqrt{x+2} = 4 - x$

4. $2\sqrt{w} = 3 - 2w$

5. $\sqrt{2x-1} - 2 = \sqrt{x+4}$

6. $\sqrt{2n+7} = 2\sqrt{n-1}$

An **ellipse** is the set of all points P in a plane such that the sum of the distances from P to two fixed points is a constant. Each of the fixed points is a **focus** (plural: **foci**).

Let $P(x, y)$ be any point on an ellipse with foci F_1 $(-4, 0)$ and F_2 $(4, 0)$ and with the sum of the distances PF_1 and PF_2 equal to 10. The definition of an ellipse and the distance formula can be used to derive an equation of the ellipse.

$$\begin{aligned} PF_1 \quad + \quad PF_2 &= 10 \\ \sqrt{(x+4)^2 + y^2} + \sqrt{(x-4)^2 + y^2} &= 10 \end{aligned}$$

Isolate one radical, square both sides, and simplify.

$$\begin{aligned} \sqrt{(x+4)^2 + y^2} &= 10 - \sqrt{(x-4)^2 + y^2} \\ (x+4)^2 + y^2 &= 10^2 - 20\sqrt{(x-4)^2 + y^2} + (x-4)^2 + y^2 \\ x^2 + 8x + 16 + y^2 &= 100 - 20\sqrt{(x-4)^2 + y^2} + x^2 - 8x + 16 + y^2 \\ 20\sqrt{(x-4)^2 + y^2} &= 100 - 16x \\ 5\sqrt{(x-4)^2 + y^2} &= 25 - 4x \end{aligned}$$

Square both sides again and simplify.

$$\begin{aligned} 25[(x-4)^2 + y^2] &= 625 - 200x + 16x^2 \\ 9x^2 + 25y^2 &= 225 \\ \frac{x^2}{25} + \frac{y^2}{9} &= 1 \qquad \textit{Divide by 225.} \end{aligned}$$

The x- and y-intercepts can be determined from the equation.

If $y = 0$, then $\frac{x^2}{25} + 0 = 1$

$x^2 = 25$, so $x = \pm 5$ *x-intercepts*

If $x = 0$, then $0 + \frac{y^2}{9} = 1$

$y^2 = 9$, so $y = \pm 3$ *y-intercepts*

Line segment $\overline{A_1A_2}$, which contains the foci, is the **major axis** of the ellipse. The shorter segment $\overline{B_1B_2}$ is the **minor axis.** The midpoint of $\overline{F_1F_2}$ is the **center** of the ellipse, and for this ellipse the center is the origin. Note the relationship among the squares of the intercepts and the square of the distance from a focus to the center.

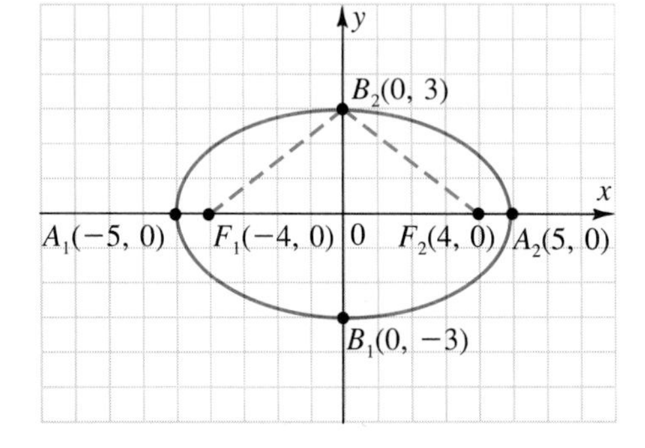

$$4^2 = 5^2 - 3^2$$

The standard form of the equation of an ellipse centered at the origin is

$$\frac{x^2}{a^2} + \frac{y^2}{b^2} = 1 \text{ or } \frac{x^2}{b^2} + \frac{y^2}{a^2} = 1, \text{ where } a > b$$

The major axis is horizontal if the greater denominator is in the x^2-term, and it is vertical if it is in the y^2-term.

Ellipses Centered at the Origin

Equation in Standard Form	$\frac{x^2}{a^2} + \frac{y^2}{b^2} = 1 \quad (a > b)$	$\frac{x^2}{b^2} + \frac{y^2}{a^2} = 1 \quad (a > b)$
x-intercepts	$\pm a$	$\pm b$
y-intercepts	$\pm b$	$\pm a$
Major axis	on x-axis; length $2a$	on y-axis; length $2a$
Minor axis	on y-axis; length $2b$	on x-axis; length $2b$
Foci	$(\pm c, 0)$, where $c^2 = a^2 - b^2$	$(0, \pm c)$, where $c^2 = a^2 - b^2$

EXAMPLE 1 **Find the x- and y-intercepts, the foci, and the length of the major axis of the ellipse with the equation $\frac{x^2}{100} + \frac{y^2}{36} = 1$. Sketch the graph.**

$$\frac{x^2}{100} + \frac{y^2}{36} = 1 \qquad a^2 = 100;\ b^2 = 36$$

Major axis horizontal

x-intercepts: 10 and -10
y-intercepts: 6 and -6

$$c^2 = a^2 - b^2$$
$$c^2 = 100 - 36$$
$$c^2 = 64$$
$$c = \pm 8$$

Foci: (8, 0) and (−8, 0)
Length of major axis: 20 $\quad 2a = 2(10)$

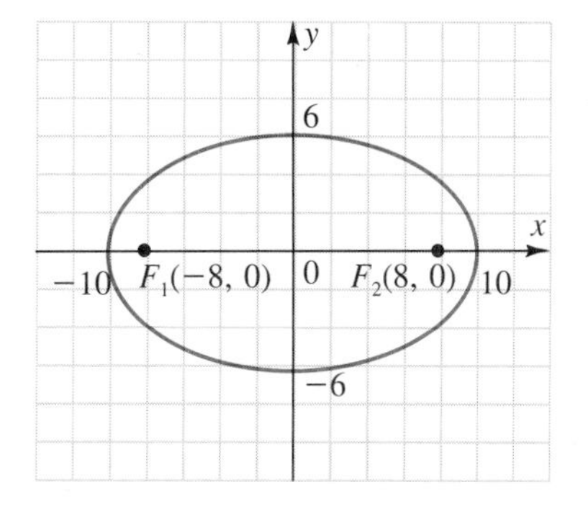

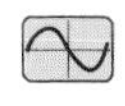

Ellipses are not functions. To graph an ellipse on a utility, therefore, you must first solve for y and graph both results.

$$36x^2 + 100y^2 = 3600 \qquad \textit{Multiply by 3600.}$$
$$y = \pm\sqrt{[(3600 - 36x^2)/100]}$$

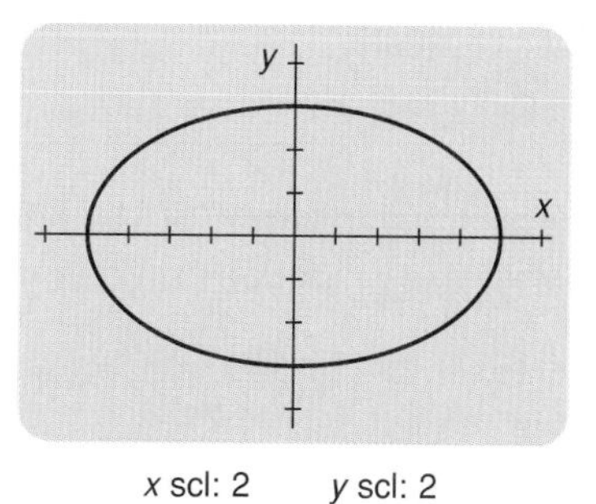

x scl: 2 y scl: 2

EXAMPLE 2 **Find the x- and y-intercepts, the foci, and the length of the major axis of the ellipse with the equation $16x^2 + 9y^2 = 144$. Sketch the graph.**

$$16x^2 + 9y^2 = 144$$

Divide both sides by 144 to write the equation in standard form.

$$\frac{16x^2}{144} + \frac{9y^2}{144} = \frac{144}{144}$$

$$\frac{x^2}{9} + \frac{y^2}{16} = 1 \qquad a^2 = 16;\ b^2 = 9;\ \textit{major axis vertical}$$

x-intercepts: 3 and -3
y-intercepts: 4 and -4

$$c^2 = a^2 - b^2$$
$$c^2 = 16 - 9$$
$$c^2 = 7$$
$$c = \pm\sqrt{7}$$

Foci: $(0, \sqrt{7})$ and $(0, -\sqrt{7})$
Length of major axis: 8 $\quad$ *$2a = 2(4)$*

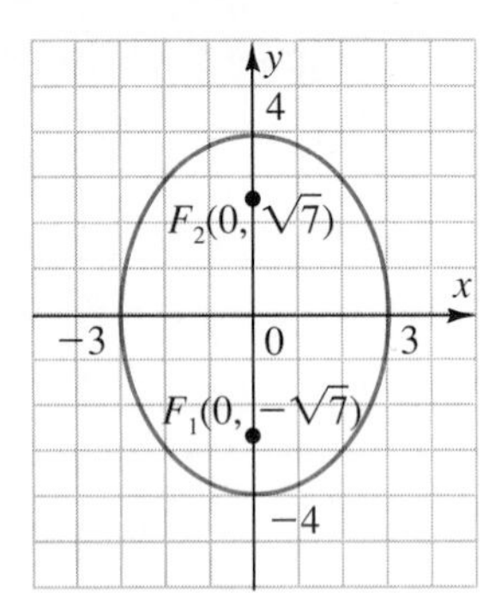

You can find the equation of an ellipse from its graph.

EXAMPLE 3 **Express in standard form the equation of the ellipse shown below.**

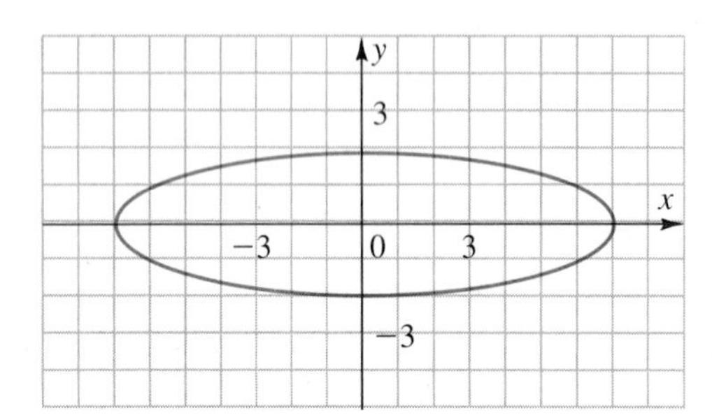

The major axis is on the x-axis.

$$\frac{x^2}{a^2} + \frac{y^2}{b^2} = 1 \qquad \textit{Use the appropriate equation.}$$

$$\frac{x^2}{7^2} + \frac{y^2}{2^2} = 1 \qquad \textit{From the graph: } a = 7,\ b = 2$$

$$\frac{x^2}{49} + \frac{y^2}{4} = 1 \qquad \textit{Standard form}$$

As shown on page 517, if x is replaced by $x - h$ and y by $y - k$ in an equation, the graph of the equation is translated h units horizontally and k units vertically.

> The standard form of the equation of an ellipse with center (h, k) and with axes parallel to the coordinate axes is
>
> $$\frac{(x-h)^2}{a^2} + \frac{(y-k)^2}{b^2} = 1 \quad \text{or} \quad \frac{(x-h)^2}{b^2} + \frac{(y-k)^2}{a^2} = 1 \quad (a > b)$$

EXAMPLE 4 **Express in standard form the equation of the ellipse with center $(-2, 3)$, if the major axis is horizontal and $a = 5$, $b = 4$. Sketch the graph.**

$$\frac{(x-h)^2}{a^2} + \frac{(y-k)^2}{b^2} = 1 \qquad \textit{The major axis is horizontal.}$$

$$\frac{(x+2)^2}{25} + \frac{(y-3)^2}{16} = 1 \qquad \textit{Substitute in the appropriate equation. } h = -2,\ k = 3,\ a = 5,\ b = 4$$

First sketch the basic ellipse with the equation $\frac{x^2}{25} + \frac{y^2}{16} = 1$. The x-intercepts are 5 and -5; the y-intercepts are 4 and -4. Then sketch the graph of $\frac{(x + 2)^2}{25} + \frac{(y - 3)^2}{16} = 1$ by translating the basic graph 2 units to the left and 3 units up.

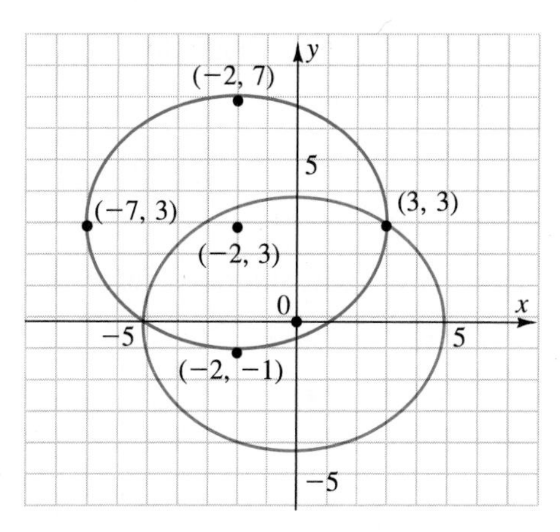

To find the center and foci of the ellipse with the equation $4x^2 + 9y^2 + 16x - 54y + 61 = 0$, first write the equation in standard form.

EXAMPLE 5 **Find the center and foci of the ellipse with the equation $4x^2 + 9y^2 + 16x - 54y + 61 = 0$.**

$$4x^2 + 9y^2 + 16x - 54y + 61 = 0$$
$$(4x^2 + 16x + \underline{?}) + (9y^2 - 54y + \underline{?}) = -61 \quad \text{Complete the square in } x \text{ and in } y.$$
$$4(x^2 + 4x + 4) + 9(y^2 - 6y + 9) = -61 + 16 + 81$$
$$4(x + 2)^2 + 9(y - 3)^2 = 36$$
$$\frac{(x + 2)^2}{9} + \frac{(y - 3)^2}{4} = 1 \quad \text{Divide both sides by 36.}$$

The coordinates of the center are $(-2, 3)$. *(h, k)*

$$c^2 = a^2 - b^2$$
$$c^2 = 9 - 4 \quad a^2 = 9,\ b^2 = 4$$
$$c = \pm\sqrt{5}$$

Since the major axis is horizontal, the foci are to the right and to the left of the center. The coordinates of the foci are $(-2 + \sqrt{5}, 3)$ and $(-2 - \sqrt{5}, 3)$.

CLASS EXERCISES

Find the intercepts, the foci, and the length of the major axis of each ellipse. Sketch the graph.

1. $\frac{x^2}{81} + \frac{y^2}{16} = 1$
2. $\frac{x^2}{49} + \frac{y^2}{25} = 1$
3. $\frac{x^2}{25} + \frac{y^2}{144} = 1$
4. $\frac{x^2}{64} + \frac{y^2}{289} = 1$
5. $4x^2 + 9y^2 = 36$
6. $16x^2 + 4y^2 = 64$

7. Thinking Critically Discuss the graph of $9x^2 + 4y^2 + 18x = 8y - 49$.

PRACTICE EXERCISES

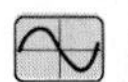 Use technology where appropriate.

Express in standard form the equation of each ellipse.

1.

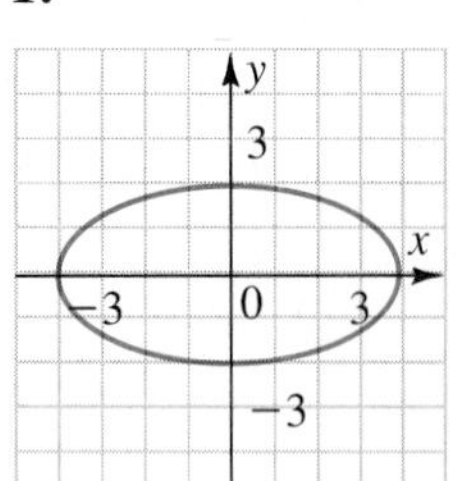

2.

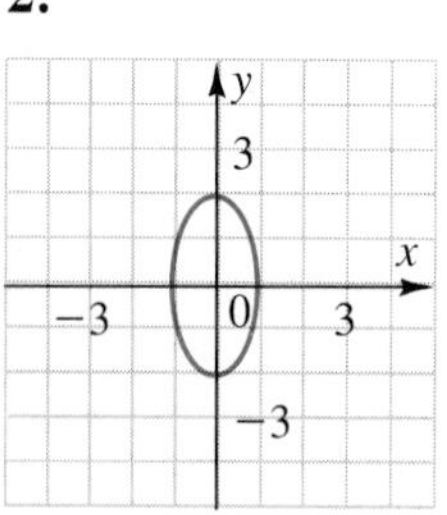

3.

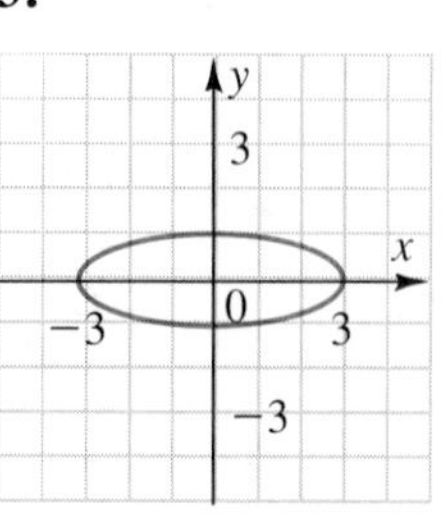

4.

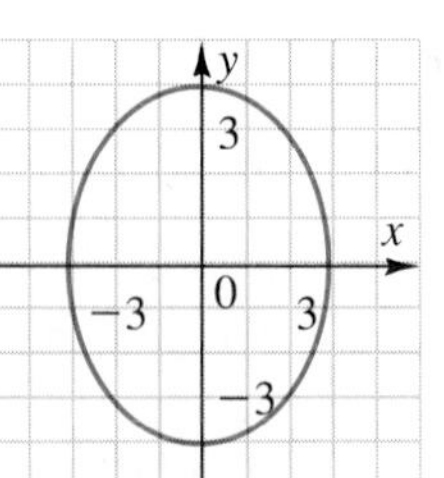

Find the intercepts, the foci, and the length of the major axis of each ellipse. Sketch the graph.

5. $\dfrac{x^2}{25} + \dfrac{y^2}{16} = 1$

6. $\dfrac{x^2}{100} + \dfrac{y^2}{64} = 1$

7. $\dfrac{x^2}{9} + \dfrac{y^2}{36} = 1$

8. $\dfrac{x^2}{16} + \dfrac{y^2}{64} = 1$

9. $\dfrac{x^2}{121} + \dfrac{y^2}{81} = 1$

10. $\dfrac{x^2}{144} + \dfrac{y^2}{49} = 1$

11. $4x^2 + 25y^2 = 100$

12. $4x^2 + 36y^2 = 144$

13. $25x^2 + y^2 = 25$

14. $16x^2 + y^2 = 64$

15. $25x^2 + 9y^2 = 225$

16. $25x^2 + 4y^2 = 100$

17. $x^2 + 4y^2 = 100$

18. $4x^2 + y^2 = 144$

19. $2x^2 + 3y^2 = 12$

20. $2x^2 + 6y^2 = 18$

Find the center and foci for each ellipse. Sketch the graph.

21. $\dfrac{(x-3)^2}{4} + \dfrac{(y-4)^2}{9} = 1$

22. $\dfrac{(x+2)^2}{16} + \dfrac{(y-1)^2}{36} = 1$

23. $\dfrac{(x+3)^2}{64} + \dfrac{(y+2)^2}{49} = 1$

24. $\dfrac{(x-4)^2}{81} + \dfrac{(y+1)^2}{25} = 1$

25. $9x^2 + 16y^2 + 18x = 64y + 71$

26. $x^2 + 4y^2 + 6x - 7 = 0$

27. $25x^2 + 16y^2 + 150x = 160y - 225$

28. $2x^2 + 8x + y^2 + 4 = 0$

Express in standard form the equation of each ellipse.

29. Foci $(\pm 2, 0)$; x-intercepts ± 4

30. Foci $(\pm 3, 0)$; x-intercepts ± 6

31. Foci $(0, \pm 3)$; y-intercepts ± 5

32. Foci $(0, \pm 5)$; y-intercepts ± 8

33. Center $(-5, 7)$, $a = 5$, $b = 2$, major axis vertical

34. Center $(-1, -1)$, $a = 3$, $b = 2$, major axis horizontal

35. Center $(3, 4)$, $a = 2\sqrt{5}$, $b = 3\sqrt{2}$, major axis vertical

36. Center $(-c, d)$, $a = 2c$, $b = 4d$, major axis horizontal

37. Foci $(\pm 1, 0)$, length of major axis 4

38. Foci $(\pm 3, 0)$; length of major axis 8

39. Foci $(\pm 2, 0)$; length of major axis 6

40. Foci $(0, \pm 3)$; length of major axis 10

41. Foci $(0, \pm 1)$; length of major axis 6

42. Center $(0, 0)$; length of horizontal major axis 8; length of minor axis 6

43. Center $(0, 0)$; length of horizontal major axis 10; length of minor axis 8

Applications

44. Sports Tutor College is building a new track for cycling teams. The track is to be elliptical, with the ratio between the lengths of the major and minor axes to be 4:3. If the available land is 200 yd long and 100 yd wide, what are the maximum lengths of the axes? Sketch a graph of the track and indicate the location of the foci.

45. Space Science A satellite has an elliptical orbit about the earth, with the earth located at one of the foci. Sketch a graph of the orbit, with center at the origin, foci on the x-axis, and the position of the earth twice as far from the center of the ellipse as from the closest point at which the ellipse intersects the x-axis.

Developing Mathematical Power

46. Extension The area of an ellipse with the equation $\frac{x^2}{a^2} + \frac{y^2}{b^2} = 1$ can be shown to be πab. A circle can be considered a special case of the ellipse in which $a = b$. Find the radius of the circle with area equal to that of the ellipse with the equation $\frac{x^2}{81} + \frac{y^2}{16} = 1$.

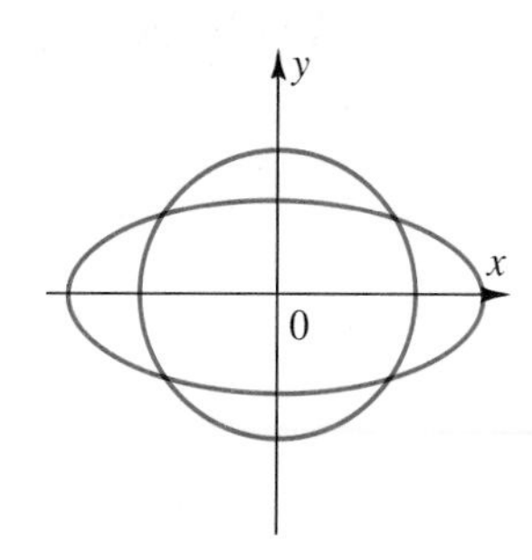

Hyperbolas

Objective: To determine the relationships among the foci, the intercepts, the asymptotes, and the equation of a hyperbola

If the intersection of a plane and a double right cylindrical cone contains points in both the top and the bottom of the cone (excluding the vertex), that intersection is called a *hyperbola*. Before hyperbolas are introduced, it will be helpful to review the concept of the intercepts of a graph.

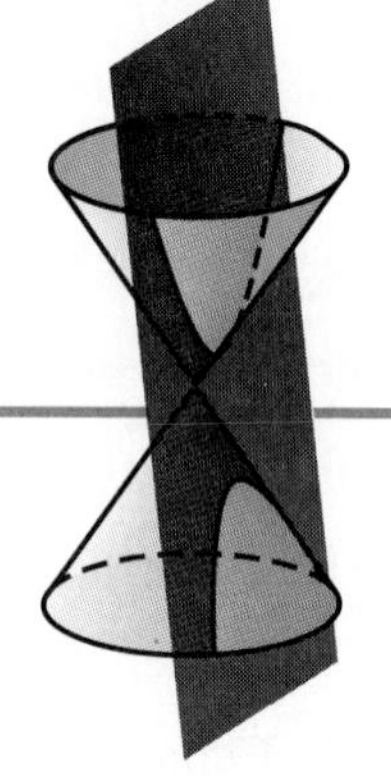

Capsule Review

An *intercept* is a point where a graph crosses an axis.

EXAMPLE **How many intercepts does the graph of $y^2 = 4x$ have?**

Let $x = 0$. Then $y^2 = 4(0)$, so $y = 0$.

Let $y = 0$. Then $0 = 4x$, so $x = 0$.

There is one intercept, the point (0, 0).

How many intercepts does the graph have?

1. $y = 4x$ **2.** $x + y = 4$ **3.** $xy = 4$ **4.** $x - y = 4$

5. $x^2 = 4y$ **6.** $x^2 + y^2 = 4$ **7.** $x^2 - y^2 = 4$ **8.** $x^2 = y + 4$

A **hyperbola** is the set of all points P in a plane such that the difference of the distances from P to two fixed points is a constant. As in the case of the ellipse, the two fixed points are called **foci**.

Let $P(x, y)$ be an arbitrary point on a hyperbola with foci F_1 $(-5, 0)$ and F_2 $(5, 0)$ and with the difference between the distances PF_1 and PF_2 equal to 8.

The definition of a hyperbola and the distance formula can be used to derive an equation of the hyperbola.

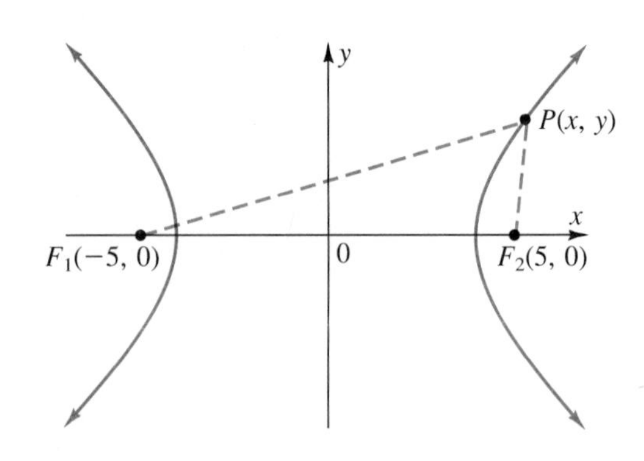

$$\begin{aligned} PF_1 \quad - \quad PF_2 &= 8 \\ \sqrt{(x+5)^2+y^2} - \sqrt{(x-5)^2+y^2} &= 8 \end{aligned}$$

Isolate one radical, square both sides, and simplify.

$$\begin{aligned} \sqrt{(x+5)^2+y^2} &= 8 + \sqrt{(x-5)^2+y^2} \\ (x+5)^2+y^2 &= 64 + 16\sqrt{(x-5)^2+y^2} + (x-5)^2 + y^2 \\ x^2+10x+25+y^2 &= 64 + 16\sqrt{(x-5)^2+y^2} + x^2 - 10x + 25 + y^2 \\ 20x - 64 &= 16\sqrt{(x-5)^2+y^2} \\ 5x - 16 &= 4\sqrt{(x-5)^2+y^2} \end{aligned}$$

Square both sides again and simplify.

$$\begin{aligned} 25x^2 - 160x + 256 &= 16[(x-5)^2+y^2] \\ 9x^2 - 16y^2 &= 144 \\ \frac{x^2}{16} - \frac{y^2}{9} &= 1 \qquad \text{Divide by 144.} \end{aligned}$$

The intercepts can be determined from the equation.

If $y = 0$, then $\frac{x^2}{16} - 0 = 1$

$$x^2 = 16$$
$$x = \pm 4$$

The x-intercepts are 4 and -4.

If $x = 0$, then $0 - \frac{y^2}{9} = 1$

$$y^2 = -9$$

The square of a real number must be nonnegative. There are no y-intercepts.

The hyperbola has two parts, or *branches*. $\overleftrightarrow{A_1A_2}$ contains the foci, and $\overline{A_1A_2}$ is called the **transverse axis** of the hyperbola. $\overline{B_1B_2}$ is the **conjugate axis.** The **center** of the hyperbola is the midpoint of $\overline{F_1F_2}$, in this case the origin. Note the relationship among the coordinates of the endpoints of the transverse and conjugate axes and the coordinates of the foci: $5^2 = 4^2 + 3^2$

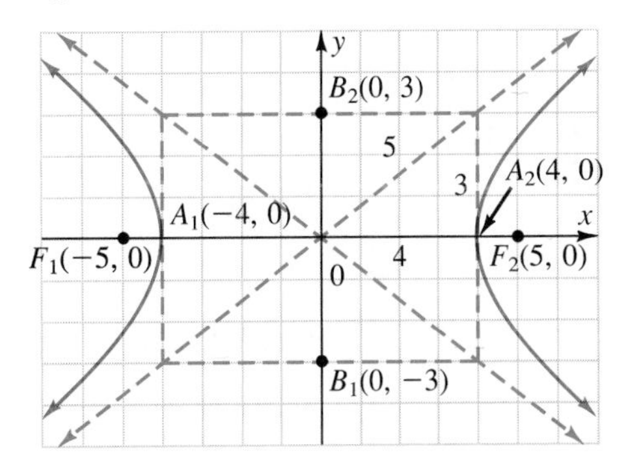

Lines perpendicular to the coordinate axes at (4, 0), (−4, 0), (0, 3), and (0, −3) form a rectangle whose extended diagonals are the **asymptotes** of the hyperbola. The hyperbola never meets its asymptotes, but approaches them more and more closely as $|x|$ and $|y|$ increase. The slope of one diagonal is $\frac{3}{4}$ and the slope of the other is $-\frac{3}{4}$, so the equations of the asymptotes can be written $y = \frac{3}{4}x$ and $y = -\frac{3}{4}x$.

A hyperbola may have its foci on the y-axis, in which case the transverse axis is vertical. The equations for hyperbolas centered at the origin, and with foci on a coordinate axis, are given in the table below.

Hyperbolas Centered at the Origin

Equation in Standard Form	$\frac{x^2}{a^2} - \frac{y^2}{b^2} = 1$	$\frac{y^2}{a^2} - \frac{x^2}{b^2} = 1$
x-intercepts	$\pm a$	none
y-intercepts	none	$\pm a$
Transverse axis	on x-axis; length $2a$	on y-axis; length $2a$
Conjugate axis	on y-axis; length $2b$	on x-axis; length $2b$
Foci	$(\pm c, 0)$, where $c^2 = a^2 + b^2$	$(0, \pm c)$, where $c^2 = a^2 + b^2$
Asymptotes	$y = \pm \frac{b}{a}x$	$y = \pm \frac{a}{b}x$

EXAMPLE 1 **Express in standard form the equation of the hyperbola shown below.**

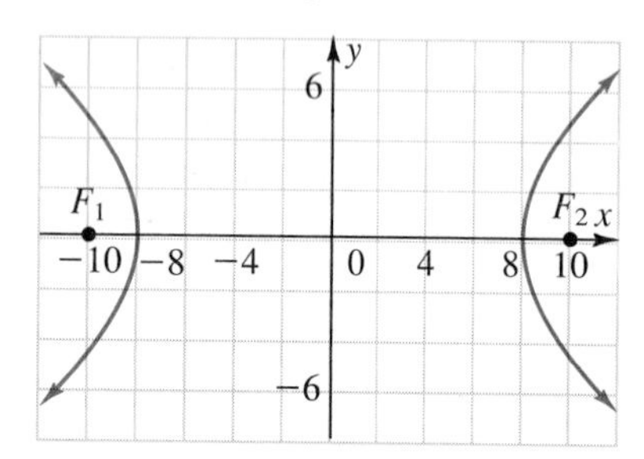

$$c^2 = a^2 + b^2$$ *Find b^2.*

$$100 = 64 + b^2$$ *$a = \pm 8$, $c = \pm 10$*

$$36 = b^2$$

$$\frac{x^2}{a^2} - \frac{y^2}{b^2} = 1$$ *The transverse axis is horizontal.*

$$\frac{x^2}{64} - \frac{y^2}{36} = 1$$ *Substitute 64 for a^2 and 36 for b^2.*

To draw the graph of a hyperbola, use a and b to create a rectangle and draw the asymptotes. Draw the intercepts of the hyperbola and extend each branch to approach, but not meet, the asymptotes.

EXAMPLE 2 **Find the intercepts, the foci, and the asymptotes of the hyperbola with the equation $144y^2 - 25x^2 = 3600$. Sketch the graph.**

To write the equation in standard form, divide both sides by 3600.

$$144y^2 - 25x^2 = 3600$$

$$\frac{y^2}{25} - \frac{x^2}{144} = 1$$ *$a^2 = 25$; $b^2 = 144$*

x-intercepts: none $\quad$ y-intercepts: 5 and -5

$$c^2 = a^2 + b^2$$
$$c^2 = 25 + 144 = 169$$
$$c = \pm 13$$

Foci: $(0, 13)$ and $(0, -13)$

Asymptotes: $y = \frac{5}{12}x$ and $y = -\frac{5}{12}x$

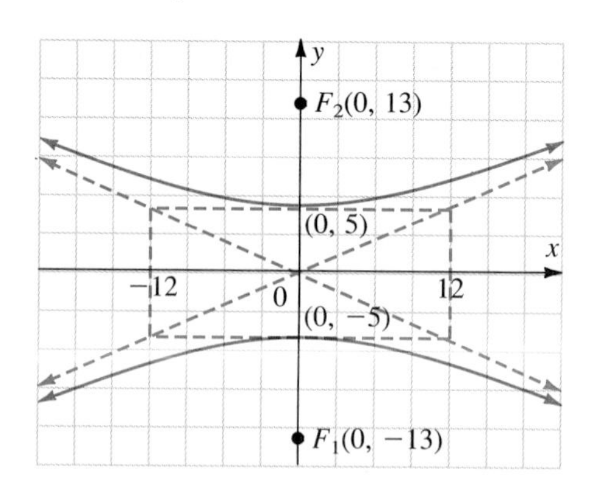

Hyperbolas may be centered at points other than the origin.

The standard form of the equation of a hyperbola with center (h, k) and with axes parallel to the coordinate axes is

$$\frac{(x-h)^2}{a^2} - \frac{(y-k)^2}{b^2} = 1 \quad \text{or} \quad \frac{(y-k)^2}{a^2} - \frac{(x-h)^2}{b^2} = 1$$

EXAMPLE 3 **Sketch the graph of the hyperbola:** $\frac{(x-3)^2}{36} - \frac{(y+2)^2}{16} = 1$

The graph is congruent to that of the hyperbola $\frac{x^2}{36} - \frac{y^2}{16} = 1$, but the center is $(3, -2)$. Sketch the hyperbola with center at the origin and x-intercepts 6 and -6. Then, sketch the graph of $\frac{(x-3)^2}{36} - \frac{(y+2)^2}{16} = 1$ by translating the basic graph 3 units to the right and 2 units down.

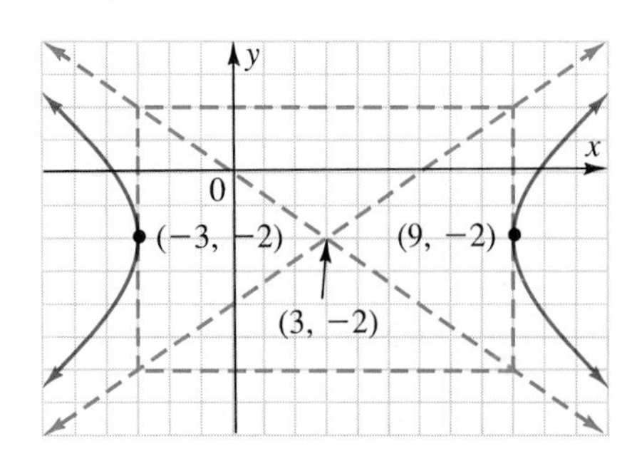

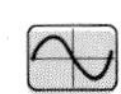

The graph of a hyperbola may be verified with a graphing utility. As with most conic sections, you must first express the equation of a hyperbola as two functions in terms of y, and then graph both.

$$16(x-3)^2 - 36(y+2)^2 = 576$$
$$y = -2 \pm \sqrt{[-576 + 16(x-3)^2]/36}$$

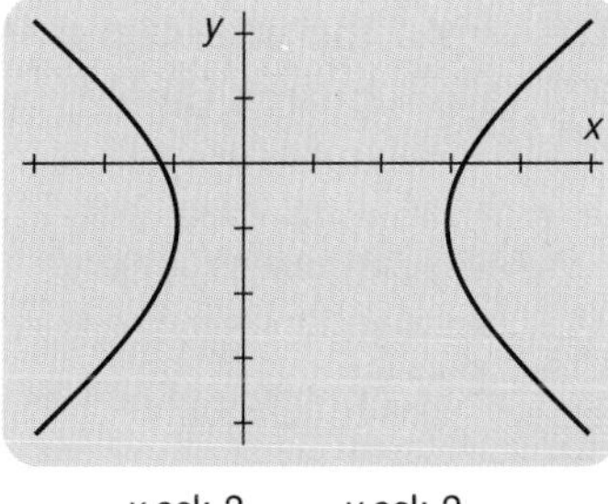

x scl: 3   y scl: 2

In the next example, the technique of completing the square is used in order to express the equation of a hyperbola in standard form.

EXAMPLE 4 **Find the coordinates of the center of the hyperbola with the equation $x^2 - 4y^2 - 2x - 8y - 7 = 0$.**

$$x^2 - 4y^2 - 2x - 8y - 7 = 0$$
$$(x^2 - 2x + \underline{?}) - (4y^2 + 8y + \underline{?}) = 7 \qquad \textit{Complete the square in x and in y.}$$
$$(x^2 - 2x + 1) - 4(y^2 + 2y + 1) = 7 + 1 - 4$$
$$(x-1)^2 - 4(y+1)^2 = 4$$
$$\frac{(x-1)^2}{4} - \frac{(y+1)^2}{1} = 1$$

The coordinates of the center are $(1, -1)$. *(h, k)*

The graphs of equations of the form $xy = k$, where $k \neq 0$, were considered in Chapter 10. The graph of such an equation is a hyperbola. It is called a **rectangular hyperbola,** and its asymptotes are the coordinate axes.

EXAMPLE 5 **Graph the rectangular hyperbola $xy = 6$.**

$$xy = 6$$

$$y = \frac{6}{x} \qquad \textit{Make a table of values.}$$

x	-3	-2	-1	1	2	3
y	-2	-3	-6	6	3	2

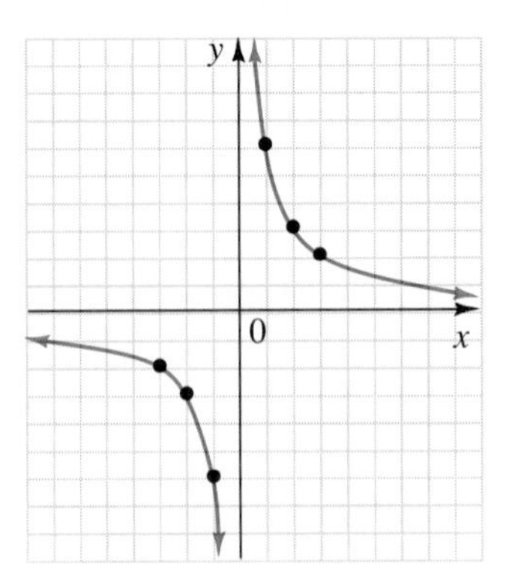

The graph of this function is entirely in the first and third quadrants. A graphing calculator or a computer can be used to verify the graph.

The graph of a rectangular hyperbola of the form $xy = k$ lies in Quadrants I and III if $k > 0$, and in Quadrants II and IV if $k < 0$.

CLASS EXERCISES

Find the intercepts, the foci, and the asymptotes of each hyperbola. Sketch the graph.

1. $\frac{x^2}{49} - \frac{y^2}{16} = 1$

2. $\frac{y^2}{81} - \frac{x^2}{25} = 1$

3. $\frac{x^2}{25} - \frac{y^2}{144} = 1$

4. $\frac{y^2}{64} - \frac{x^2}{289} = 1$

5. $4x^2 - 16y^2 = 64$

6. $16x^2 - 9y^2 = 144$

Sketch the graph of each rectangular hyperbola.

7. $xy = 3$

8. $xy = -4$

PRACTICE EXERCISES

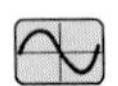 Use technology where appropriate.

Write in standard form the equation of each hyperbola.

1.

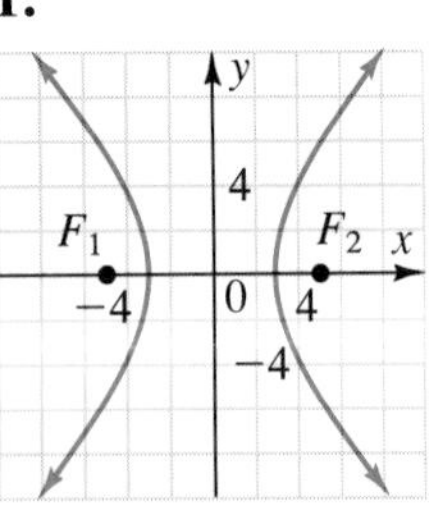

2.

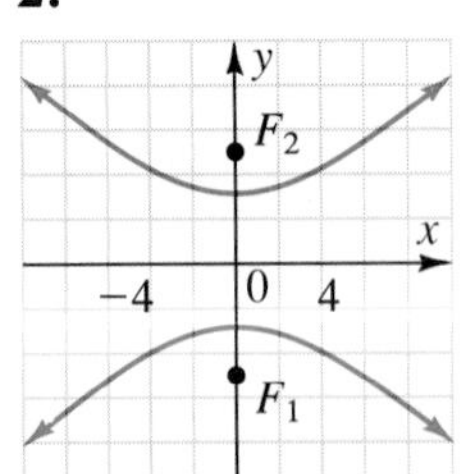

3.

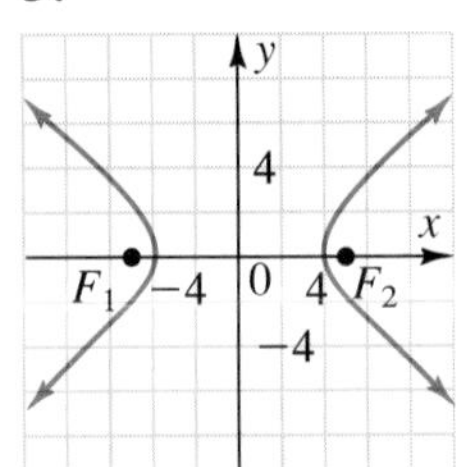

4.

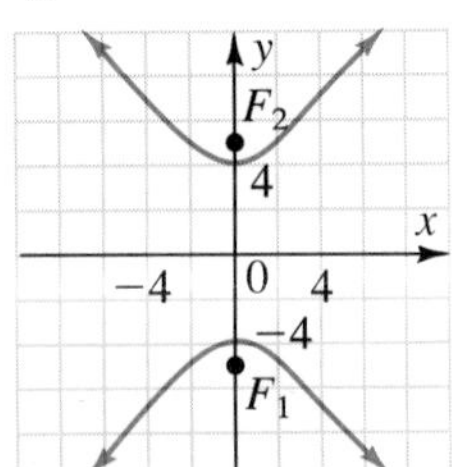

Find the intercepts, the foci, and the asymptotes for each hyperbola.

5. $\frac{x^2}{25} - \frac{y^2}{16} = 1$

6. $\frac{x^2}{49} - \frac{y^2}{25} = 1$

7. $\frac{x^2}{9} - \frac{y^2}{36} = 1$

8. $\frac{x^2}{16} - \frac{y^2}{64} = 1$

9. $\frac{x^2}{81} - \frac{y^2}{49} = 1$

10. $\frac{x^2}{100} - \frac{y^2}{81} = 1$

11. $\frac{y^2}{25} - \frac{x^2}{16} = 1$

12. $\frac{y^2}{64} - \frac{x^2}{36} = 1$

13. $\frac{y^2}{4} - \frac{x^2}{9} = 1$

14. $\frac{y^2}{16} - \frac{x^2}{36} = 1$

15. $\frac{y^2}{4} - \frac{x^2}{25} = 1$

16. $\frac{y^2}{9} - \frac{x^2}{4} = 1$

Sketch the graph of each hyperbola.

17. $\frac{x^2}{16} - \frac{y^2}{4} = 1$

18. $\frac{y^2}{100} - \frac{x^2}{4} = 1$

19. $x^2 - 9y^2 = 81$

20. $y^2 - 4x^2 = 64$

21. $xy = 9$

22. $xy = 16$

23. $xy = -8$

24. $xy = -14$

25. $x^2 - y^2 = 1$

26. $y^2 - x^2 = 1$

27. $\frac{(x-2)^2}{36} - \frac{(y-3)^2}{25} = 1$

28. $\frac{(x+4)^2}{49} - \frac{(y+3)^2}{16} = 1$

29. $\frac{(y-5)^2}{81} - \frac{(x+4)^2}{64} = 1$

30. $\frac{(y+6)^2}{64} - \frac{(x-3)^2}{36} = 1$

31. $x^2 - 16y^2 - 2x + 128y = 271$

32. $y^2 - 9x^2 - 8y + 36x - 29 = 0$

33. $x^2 - y^2 + 6x + 10y - 17 = 0$

34. $y^2 - x^2 + 6x - 4y - 6 = 0$

Classify each equation as that of a circle, a parabola, an ellipse, a hyperbola, a degenerate conic, or none of these.

35. $x^2 + 4y^2 - 2x - 15 = 0$

36. $-3xy = 0$

37. $y = 2x^2 - 4x + 3$

38. $x^2 + 2x + y^2 + 2y + 2 = 0$

39. $x^2 + y^2 + 2x - 4y + 8 = 0$

40. $9x^2 - 4y^2 - 24y = 72$

Write in standard form the equation of the hyperbola.

41. x-intercepts: 3 and -3
foci: $(4, 0)$ and $(-4, 0)$

42. x-intercepts: 5 and -5
foci: $(6, 0)$ and $(-6, 0)$

43. y-intercepts: 3 and -3
foci: $(0, 5)$ and $(0, -5)$

44. y-intercepts: 4 and -4
foci: $(0, 6)$ and $(0, -6)$

45. center: $(0, 0)$; $a = 4$, $b = 2$, horizontal transverse axis

46. center: $(3, 2)$; $a = 6$, $b = 1$, horizontal transverse axis

47. center: $(-2, 1)$; $a = 5$, $c = 8$, vertical transverse axis

48. center: $(4, -3)$; $a = 8$, $c = 12$, vertical transverse axis

Applications

49. **Forestry** An explosion is heard at a forest ranger station. Seven seconds later it is heard at another station. If the speed of sound is approximately 350 m/s, the site of the explosion is 2450 m closer to one station than to the other. It is not known which station heard the sound first, and the ranger stations are approximately 4 km apart. Make a diagram to show all possible locations of the explosion.

50. **Analytic Geometry** Use the distance formula and the definition of the hyperbola to write the equation in standard form of a hyperbola with foci F_1 $(0, -10)$ and F_2 $(0, 10)$. Assume that $P(x, y)$ is an arbitrary point on the hyperbola and that the difference between the distances PF_1 and PF_2 is 12.

TEST YOURSELF

Write the equation of each circle in standard form. **11.1**

1. Center $(4, 3)$; radius 5

2. Center $(0, 0)$; radius 4

Find the center and radius of each circle.

3. $(x + 1)^2 + (y - 3)^2 = 36$

4. $x^2 + y^2 + 8x - 16y + 12 = 0$

11.2

5. Write in standard form the equation of the parabola with focus $(5, 0)$ and directrix $y = -2$.

6. Find the vertex, the axis of symmetry, the focus, and the directrix of the parabola $x^2 - 4x - 2y + 13 = 0$.

Find the intercepts, the foci, and the length of the major axis of each ellipse. Sketch the graph. **11.3**

7. $\dfrac{x^2}{81} + \dfrac{y^2}{25} = 1$

8. $\dfrac{x^2}{100} + \dfrac{y^2}{36} = 1$

Sketch the graph of each hyperbola. **11.4**

9. $\dfrac{x^2}{64} - \dfrac{y^2}{49} = 1$

10. $xy = 16$

11.5 Graphing Quadratic Systems

Objectives: To determine graphically the number of real solutions of a quadratic system
To find or to approximate solutions graphically

In Chapter 4, you studied systems of linear equations. Systems can also include quadratic equations. Such a system is called a **quadratic system.** The system may consist of only quadratic equations, or it may contain both quadratic and linear equations. To determine the solution of a system of equations graphically, graph each equation in the same coordinate plane.

Capsule Review

EXAMPLE **Identify and sketch the graph of each equation:**

a. $y = (x + 1)^2 - 2$ **b.** $\frac{x^2}{25} + \frac{y^2}{9} = 1$

a. $y = (x + 1)^2 - 2$ is a parabola.
$a = 1,\ h = -1,\ k = -2$

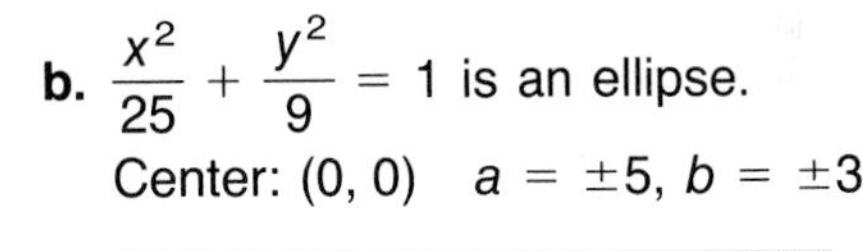
b. $\frac{x^2}{25} + \frac{y^2}{9} = 1$ is an ellipse.
Center: (0, 0) $a = \pm 5,\ b = \pm 3$

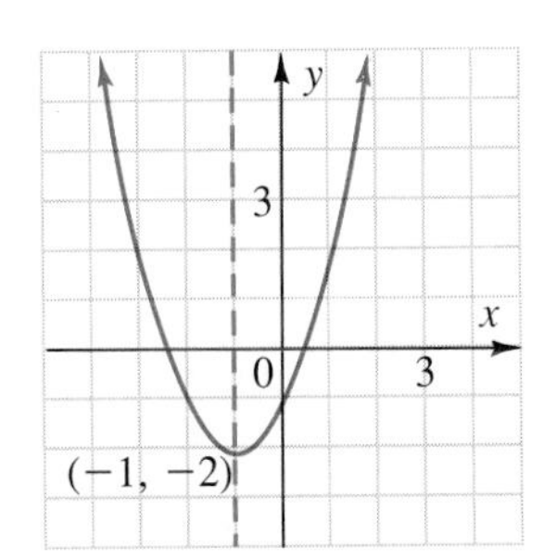

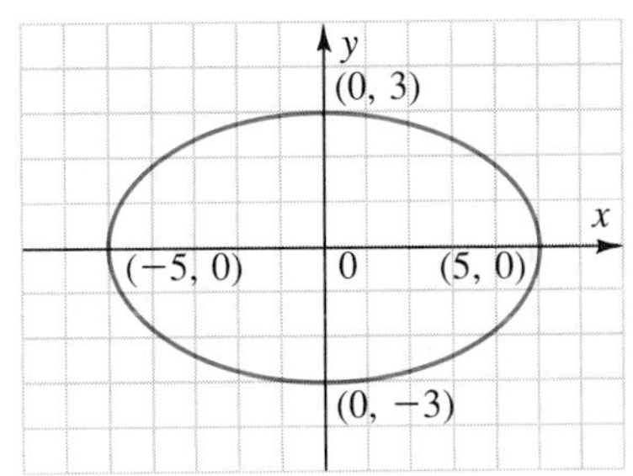

Identify and sketch the graph of each equation.

1. $x^2 + y^2 = 16$ **2.** $y = (x - 3)^2 - 1$ **3.** $y = -2x + 3$

4. $\frac{y^2}{16} - \frac{x^2}{9} = 1$ **5.** $\frac{x^2}{49} + \frac{y^2}{16} = 1$ **6.** $xy = 2$

Recall from the discussion of linear systems in Chapter 4 that two lines can intersect in one point, no points or infinitely many points. In how many points can a line and a circle intersect?

The graphs of two quadratic equations may intersect in four points, three points, two points, one point, or no points. To illustrate, consider the possible points of intersection of a hyperbola and a circle.

4 points

3 points

2 points

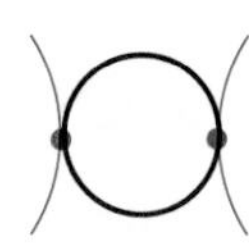
2 points

1 point

0 points

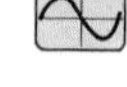 Each point of intersection of the graphs of the equations in a system represents a real solution of the system. A graphing utility can be used to show the points of intersection.

EXAMPLE 1 **Identify and sketch the graph of the system of equations. Then, determine the number of solutions.**

a. $\begin{cases} x^2 + y^2 = 36 \\ y = (x - 2)^2 - 3 \end{cases}$ **b.** $\begin{cases} 9x^2 + 25y^2 = 225 \\ y = -x^2 + 5 \end{cases}$

a. $\begin{cases} x^2 + y^2 = 36 \\ y = (x - 2)^2 - 3 \end{cases}$

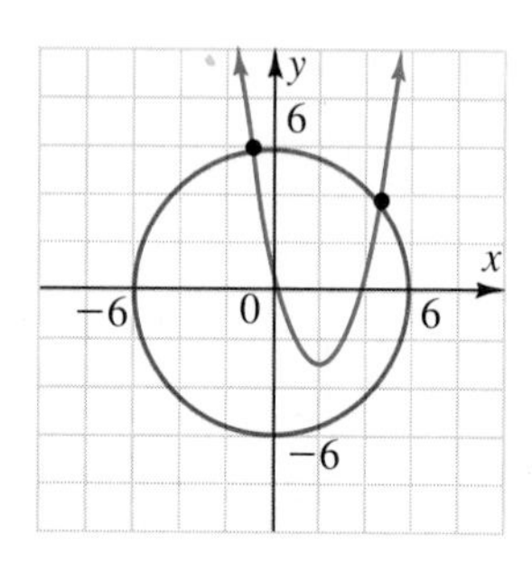

The graph of $x^2 + y^2 = 36$ is a circle with center at the origin and radius 6. The graph of $y = (x - 2)^2 - 3$ is a parabola that opens upward. Its vertex is $(2, -3)$ and its axis of symmetry is $x = 2$. Since the vertex of the parabola is in the interior of the circle, the parabola intersects the circle in two points. This system has two real solutions. It is convenient to check graphs using a computer or a graphing calculator.

b. $\begin{cases} 9x^2 + 25y^2 = 225 \\ y = -x^2 + 5 \end{cases}$

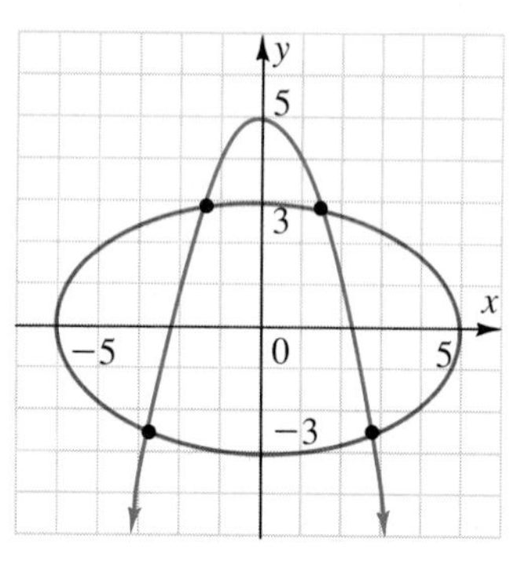

Write the first equation in standard form.

$$\frac{x^2}{25} + \frac{y^2}{9} = 1$$

The graph of this equation is an ellipse with center $(0, 0)$, x-intercepts ± 5, and y-intercepts ± 3. The graph of $y = -x^2 + 5$ is a parabola that opens downward. Its vertex is $(0, 5)$, its axis is the y-axis, and its x-intercepts are $\pm\sqrt{5}$, or approximately ± 2.2. The parabola intersects the ellipse in four points, so the system has four real solutions.

As a rule, it is difficult to find exact solutions of a quadratic system using hand-drawn graphs. Even a graphing utility can provide only a close decimal approximation. However, it is possible by either method to find solutions that are pairs of integers or to approximate solutions that are not pairs of integers. The next example illustrates the graphical solution of a system containing a linear equation and a quadratic equation. There can be at most two points of intersection of a line and a conic, so the system can have at most two real solutions.

EXAMPLE 2 **Use a utility to graph the given system of equations and to determine the coordinates of all solutions.**

$$\begin{cases} x^2 + y^2 = 25 \\ 3x - 2y = 6 \end{cases}$$

The form of the first equation indicates that its graph is a circle; the graph of the second is a line. The line is a function and can easily be solved for y; the equation of the circle can be expressed as two functions in terms of y and graphed on the same screen with the line.

$$y = \frac{3}{2}x - 3 \qquad y = \pm\sqrt{25 - x^2}$$

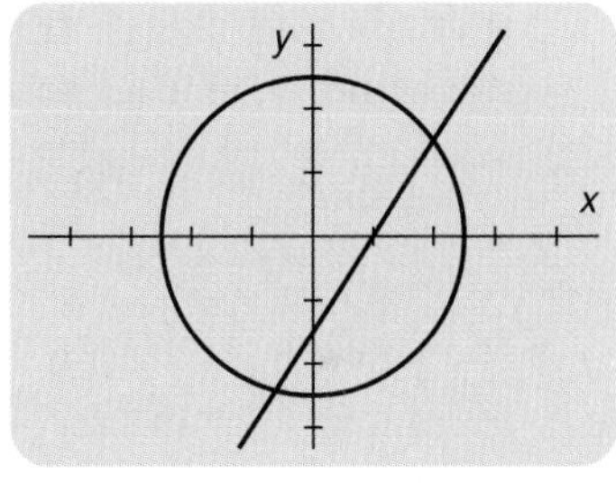

x scl: 2 y scl: 2

Trace to the first point; the coordinates appear to be (4, 3). Show that (4, 3) *is,* in fact, a solution.

$$\begin{aligned} x^2 + y^2 &= 25 \\ 4^2 + 3^2 &\stackrel{?}{=} 25 \\ 25 &= 25 \end{aligned} \qquad \begin{aligned} 3x - 2y &= 6 \\ 3(4) - 2(3) &\stackrel{?}{=} 6 \\ 6 &= 6 \end{aligned}$$

Trace to the second point. The coordinates appear to be $(-1.2, -4.8)$. Substitution will show that these are only approximations.

Systems of inequalities may contain quadratic inequalities as well as linear inequalities.

EXAMPLE 3 **Identify and sketch the graph of the system of inequalities:**

$$\begin{cases} 16x^2 + 9y^2 \le 144 \\ 25x^2 - 4y^2 \ge 100 \end{cases}$$

The graph of $16x^2 + 9y^2 = 144$, or $\frac{x^2}{9} + \frac{y^2}{16} = 1$, is an ellipse with its center at the origin, x-intercepts ± 3, and y-intercepts ± 4. The graph of the inequality $16x^2 + 9y^2 \le 144$ is this ellipse and its interior.

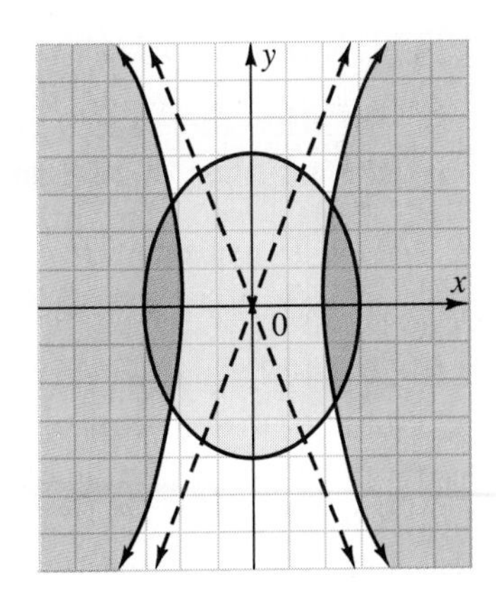

The graph of $25x^2 - 4y^2 = 100$, or $\frac{x^2}{4} - \frac{y^2}{25} = 1$, is a hyperbola with its center at the origin, horizontal transverse axis, and x-intercepts ± 2. The graph of the inequality $25x^2 - 4y^2 \ge 100$ is this hyperbola and the two regions containing its foci.

The ellipse and hyperbola overlap, so the intersection of the two solution sets is the set of points within the two darkly shaded regions. The coordinates of all points in and on the boundaries of this region are solutions of the system.

CLASS EXERCISES

1. Make sketches to illustrate the different numbers of points in which a line and an ellipse can intersect.

2. Make sketches to illustrate the different numbers of points in which a circle and a parabola can intersect.

Identify and sketch the graph of each system of equations, and determine the number of real solutions.

3. $\begin{cases} x^2 + y^2 = 100 \\ y = 2x - 1 \end{cases}$

4. $\begin{cases} x = y^2 - 1 \\ x + y = 2 \end{cases}$

5. $\begin{cases} x^2 - y^2 = 16 \\ x^2 + 4y^2 = 36 \end{cases}$

6. $\begin{cases} 9x^2 + 4y^2 = 36 \\ y = x^2 + 2 \end{cases}$

7. $\begin{cases} x^2 + y^2 = 16 \\ 2x - y = 4 \end{cases}$

8. $\begin{cases} x = y^2 - 1 \\ 9x^2 + 4y^2 = 36 \end{cases}$

Solve each system. Round nonintegral solutions to the nearest tenth.

9. $\begin{cases} x^2 + y^2 = 4 \\ 2x + y = 2 \end{cases}$

10. $\begin{cases} x^2 + y^2 = 4 \\ (x - 3)^2 + y^2 = 1 \end{cases}$

Sketch the graph of each system of inequalities.

11. $\begin{cases} x^2 + y^2 \leq 16 \\ x - y \leq 2 \end{cases}$

12. $\begin{cases} 16x^2 + 4y^2 \geq 64 \\ 3y + x \geq 6 \end{cases}$

PRACTICE EXERCISES

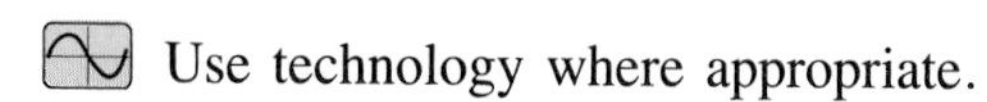

Use technology where appropriate.

Identify and sketch the graph of each system of equations, and determine the number of real solutions.

1. $\begin{cases} x^2 + y^2 = 36 \\ y = 5 \end{cases}$

2. $\begin{cases} y = x^2 - 5 \\ y - x = 1 \end{cases}$

3. $\begin{cases} x^2 - y^2 = 1 \\ 2x - y = 4 \end{cases}$

4. $\begin{cases} 25x^2 - 16y^2 = 400 \\ y = x \end{cases}$

5. $\begin{cases} x^2 - y^2 = 16 \\ x = 2 \end{cases}$

6. $\begin{cases} 9x^2 + 4y^2 = 36 \\ y = x^2 \end{cases}$

7. $\begin{cases} xy = 1 \\ 3x = 6 - 2y \end{cases}$

8. $\begin{cases} 4x^2 + 4y^2 = 100 \\ xy = 5 \end{cases}$

Graph each system and find the real solutions, if any.

9. $\begin{cases} x^2 + y^2 = 25 \\ x - y = -5 \end{cases}$

10. $\begin{cases} x^2 + y^2 = 25 \\ x + y = 1 \end{cases}$

11. $\begin{cases} x^2 + y^2 = 4 \\ x + y = 2 \end{cases}$

12. $\begin{cases} x^2 + y^2 = 25 \\ 3x - 4y = -25 \end{cases}$

13. $\begin{cases} x^2 + y^2 = 9 \\ x^2 + y^2 = 4 \end{cases}$

14. $\begin{cases} x^2 + y^2 = 16 \\ x^2 + 4y^2 = 16 \end{cases}$

Sketch the graph of each system of inequalities.

15. $\begin{cases} x^2 + y^2 \leq 9 \\ xy \leq 18 \end{cases}$

16. $\begin{cases} x^2 - y \leq 4 \\ x^2 + y^2 \leq 25 \end{cases}$

Identify and sketch the graph of each system. Find the real solutions that are pairs of integers, and approximate any other real solutions in tenths.

17. $\begin{cases} x^2 + y^2 = 4 \\ (x - 5)^2 + y^2 = 9 \end{cases}$

18. $\begin{cases} 9x^2 + 4y^2 = 36 \\ y = -x^2 + 4 \end{cases}$

19. $\begin{cases} 4x^2 + y^2 = 16 \\ 9x^2 - 4y^2 = 36 \end{cases}$

20. $\begin{cases} (x - 4)^2 + y^2 = 25 \\ (x + 4)^2 + y^2 = 25 \end{cases}$

21. $\begin{cases} x^2 + y^2 = 16 \\ xy = 8 \end{cases}$

22. $\begin{cases} x^2 + y^2 = 4 \\ 4x - y = -8 \end{cases}$

23. $\begin{cases} x^2 + y^2 = 9 \\ x + y = 2 \end{cases}$

24. $\begin{cases} x^2 + y^2 = 16 \\ x - y = 2 \end{cases}$

Sketch the graph of each system of inequalities.

25. $\begin{cases} y \geq x^2 \\ x^2 + 2y^2 \geq 4 \end{cases}$

26. $\begin{cases} 25x^2 - 4y^2 \leq 100 \\ y \leq (x - 1)^2 \end{cases}$

27. $\begin{cases} y \leq 2 - x^2 \\ y \geq x^2 - 1 \end{cases}$

28. $\begin{cases} x^2 + y^2 \geq 9 \\ x^2 + y^2 \leq 4 \end{cases}$

Sketch the graph of each system. Approximate the real solutions in tenths.

29. $\begin{cases} (x - 3)^2 + (y + 4)^2 = 36 \\ x^2 + (y - 3)^2 = 9 \end{cases}$

30. $\begin{cases} (x - 1)^2 + (y + 2)^2 = 4 \\ (x + 1)^2 + y^2 = 9 \end{cases}$

Sketch the graph of each system of inequalities.

31. $\begin{cases} xy \geq 12 \\ 4x^2 + 9y^2 \leq 36 \end{cases}$

32. $\begin{cases} 4x^2 + y^2 \geq 4 \\ 4x^2 - y^2 \leq 4 \end{cases}$

Applications

33. Astronomy In a certain solar system the orbit of a planet is described by the equation $4x^2 + 9y^2 - 16x - 18y = 11$. The path of a comet approaching the solar system is described by $y = x^2 + 2x$. Is a collision possible? If so, how many opportunities exist?

34. Astronautics A rocket may be launched at any time between 2 AM and 10 AM, provided the speed of the wind at launch time is below 30 mi/h. If the speed of the wind between 2 AM and 10 AM can be determined by the equation $s = t^2 - 14t + 74$ where t is time in hours, can the rocket be launched? If so, during which hours?

Developing Mathematical Power

Extension Thus far, the graphing of mathematical relations in this book has been limited to the two-dimensional, coordinate plane. Most objects, however, occur in three-dimensional space. One way to generate three-dimensional graphs is to rotate the graph of a region in the coordinate plane about one of the axes. The graph of $x^2 + y^2 \leq 9$ when rotated about the y-axis, for example, yields a solid sphere of radius 3.

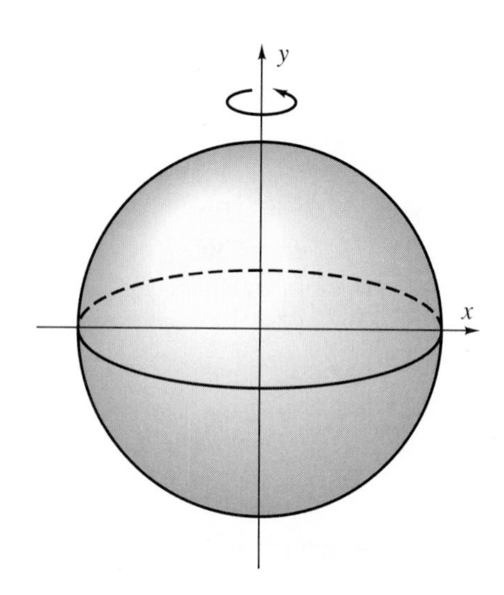

For each of the following, graph the region given by the equation or system on a set of coordinate axes. Then rotate the region about the given axis and sketch the resulting three-dimensional solid.

35. $\begin{cases} y \geq x^2 \\ y \leq 9 \end{cases}$, y-axis

36. $\begin{cases} 25x^2 - 9y^2 \geq 225 \\ |x| \leq 8 \end{cases}$, x-axis

37. $4x^2 + 9(y - 3)^2 \leq 36$, x-axis

38. $(x - 5)^2 + y^2 \leq 4$, y-axis

Note: the doughnut-shaped solid in exercise 38 is called a *torus*.

11.6

Solving Quadratic Systems

Objective: To solve a quadratic system algebraically

Many quadratic systems can be solved by the same methods that are used to solve linear systems. The methods of substitution and addition as they apply to the solution of quadratic systems are shown in this lesson.

Capsule Review

EXAMPLE **Solve using substitution:** $\begin{cases} 2x + 7y = 21 \\ x - 2y = 5 \end{cases}$

$x - 2y = 5$
$x = 2y + 5$ *Solve the second equation for x in terms of y.*

$2x + 7y = 21$ *Substitute 2y + 5 for x in the first equation.*
$2(2y + 5) + 7y = 21$
$4y + 10 + 7y = 21$
$11y = 11$
$y = 1$

$x - 2y = 5$ *Substitute 1 for y in one equation to find the value of x.*
$x - 2(1) = 5$
$x = 7$

Check in both equations to show that the solution is (7, 1).

Solve each system of equations.

1. $\begin{cases} 2x - y = 4 \\ 5x - y = 13 \end{cases}$

2. $\begin{cases} 3x + 2y = 7 \\ x - 3y = -5 \end{cases}$

3. $\begin{cases} 9a = 2b + 3 \\ 7a = 8b + 11 \end{cases}$

4. $\begin{cases} 3n - 5m = -13 \\ 4n + 3m = 2 \end{cases}$

5. $\begin{cases} 7s + 4t = 11 \\ 5s - 2t = 20 \end{cases}$

6. $\begin{cases} 2x + 3y = 12 \\ 3x - 2y = 5 \end{cases}$

When solving a quadratic system, it is usually more convenient to use one method, substitution or addition, then it is to use the other. If one of the equations is linear, the substitution method is often used. Before finding the solution to a system, think about the graphs of the equations in order to determine the maximum number of solutions.

EXAMPLE 1 **Identify the graph and solve the system:** $\begin{cases} 3x + y = 6 \\ x^2 = 4x + y \end{cases}$

The system contains the equations of a line and a parabola. The maximum number of intersections is two, so there are at most two solutions.

$3x + y = 6$

$y = -3x + 6$ *Solve the linear equation for y.*

$x^2 = 4x + y$

$x^2 = 4x + (-3x + 6)$ *Substitute $-3x + 6$ for y in the quadratic equation.*

$x^2 - x - 6 = 0$ *Solve by factoring.*

$x = -2$ or $x = 3$

Substitute into the linear equation to find the corresponding values of y.

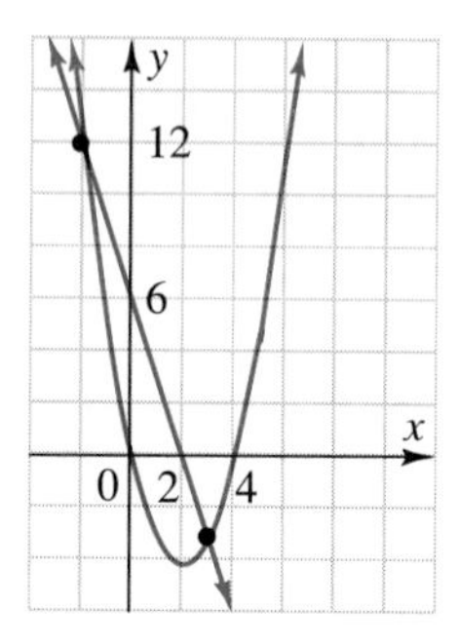

$y = -3x + 6$	$y = -3x + 6$
$y = -3(-2) + 6$	$y = -3(3) + 6$
$y = 12$	$y = -3$

Check to show that $(-2, 12)$ and $(3, -3)$ are solutions. You can use a graphing utility to show the graph of this system and to verify the solution.

The addition method is illustrated in the next example, which involves a system of two quadratic equations.

EXAMPLE 2 **Identify the graph and solve the system:** $\begin{cases} x^2 - y^2 = 9 \\ x^2 + 9y^2 = 169 \end{cases}$

The system contains the equations of a hyperbola and an ellipse. The maximum number of intersections, and therefore solutions, is four.

$-x^2 + y^2 = -9$ *Multiply both sides of the first equation by -1.*

$x^2 + 9y^2 = 169$

$10y^2 = 160$ *Add the equations.*

$y^2 = 16$

$y = 4$ or $y = -4$

Substitute 4 and -4 into one of the original equations to find the corresponding values of x.

$x^2 - y^2 = 9$	$x^2 - y^2 = 9$
$x^2 - 4^2 = 9$	$x^2 - (-4)^2 = 9$
$x^2 = 25$	$x^2 = 25$
$x = 5$ or $x = -5$	$x = 5$ or $x = -5$

Check $(5, 4)$, $(-5, 4)$, $(5, -4)$, and $(-5, -4)$ in the original equations. Use a computer or graphing calculator to verify the solution.

The system in the next example has both real and imaginary solutions. However, you are only asked to find the real solutions.

EXAMPLE 3 **Identify the graphs and find the real solutions of the system:**

$$\begin{cases} 4x^2 - y^2 = 20 \\ xy = 12 \end{cases}$$

The system contains the equations of a hyperbola and a rectangular hyperbola. There are at most four intersections and solutions.

$xy = 12$ *Solve the second equation for y in terms of x.*

$y = \frac{12}{x}$

$4x^2 - \left(\frac{12}{x}\right)^2 = 20$ *Substitute $\frac{12}{x}$ for y in the first equation.*

$4x^2 - \frac{144}{x^2} = 20$

$4x^4 - 144 = 20x^2$ *Multiply both sides by x^2.*

$4x^4 - 20x^2 - 144 = 0$

$x^4 - 5x^2 - 36 = 0$

$(x^2 - 9)(x^2 + 4) = 0$ *Factor.*

$x^2 - 9 = 0$ or $x^2 + 4 = 0$

The equation $x^2 + 4 = 0$ has no real solutions, so solve $x^2 - 9 = 0$.

$x^2 - 9 = 0.$

$x = 3$ or $x = -3$

Substitute 3 and -3 for x in $y = \frac{12}{x}$.

When $x = 3$, $y = \frac{12}{3}$ When $x = -3$, $y = \frac{12}{-3}$

$y = 4$ $y = -4$

Check (3, 4) and $(-3, -4)$ in the original system to show that they are solutions.

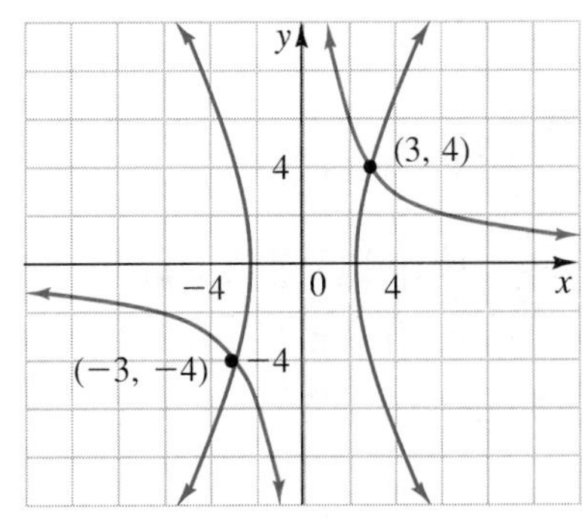

You may wish to use a computer or a graphing calculator to show the graph of this system and to verify your results in the exercises.

CLASS EXERCISES

Identify the graph and find the real solutions of each system.

1. $\begin{cases} x^2 + y^2 = 10 \\ y = 2x - 5 \end{cases}$

2. $\begin{cases} x^2 + y^2 = 13 \\ y = x + 1 \end{cases}$

3. $\begin{cases} 5x^2 + y^2 = 36 \\ x^2 + 2y^2 = 36 \end{cases}$

4. $\begin{cases} 2x^2 + 3y^2 = 23 \\ 3x^2 - y^2 = 7 \end{cases}$

5. $\begin{cases} xy = 8 \\ x^2 + y^2 = 16 \end{cases}$

6. $\begin{cases} y = x - 4 \\ y = x^2 - 4x + 2 \end{cases}$

PRACTICE EXERCISES

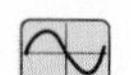 Use technology where appropriate.

Find the real solutions, if any, of each system.

1. $\begin{cases} x^2 + y^2 = 25 \\ y = x - 1 \end{cases}$

2. $\begin{cases} x^2 + y^2 = 13 \\ y = 2x - 4 \end{cases}$

3. $\begin{cases} 4x^2 + 9y^2 = 36 \\ y = x + 3 \end{cases}$

4. $\begin{cases} 4x^2 + 25y^2 = 100 \\ y = x + 2 \end{cases}$

5. $\begin{cases} 2a^2 - b^2 = 2 \\ a^2 + b^2 = 25 \end{cases}$

6. $\begin{cases} 4c^2 + 3d^2 = 43 \\ 3c^2 - d^2 = 3 \end{cases}$

7. $\begin{cases} 3s^2 - 5t^2 = 30 \\ s^2 + 5t^2 = 70 \end{cases}$

8. $\begin{cases} w^2 - 25z^2 = 20 \\ 2w^2 + 25z^2 = 88 \end{cases}$

9. $\begin{cases} 3m^2 - 2n^2 = 9 \\ m^2 - n^2 = -1 \end{cases}$

10. $\begin{cases} 9t^2 + 4v^2 = 36 \\ t^2 - v^2 = 4 \end{cases}$

11. $\begin{cases} x^2 + 64y^2 = 64 \\ x^2 + y^2 = 64 \end{cases}$

12. $\begin{cases} 3x^2 - y^2 = 9 \\ x^2 + 2y^2 = 38 \end{cases}$

13. $\begin{cases} xy = 2 \\ x^2 + y^2 = 4 \end{cases}$

14. $\begin{cases} xy = 3 \\ x^2 - 4y^2 = 5 \end{cases}$

15. $\begin{cases} a^2 + 2b^2 = 10 \\ 3a^2 - b^2 = 9 \end{cases}$

16. $\begin{cases} r + 4 = s^2 - 2s \\ r - 2 = s^2 + 4s \end{cases}$

17. $\begin{cases} t + 4 = (v - 1)^2 \\ t + 1 = -v \end{cases}$

18. $\begin{cases} 2m^2 - n^2 = 11 \\ m^2 + 3n^2 = 9 \end{cases}$

19. $\begin{cases} 2w^2 = 10 - z^2 \\ 3w^2 - z^2 = 9 \end{cases}$

20. $\begin{cases} x^2 + y^2 = 64 \\ x^2 + y^2 = 16 \end{cases}$

21. $\begin{cases} 4x^2 + 4y^2 = 100 \\ 3x^2 + 3y^2 = 27 \end{cases}$

22. $\begin{cases} \dfrac{2}{x} + \dfrac{2}{y} = 1 \\ y = 2x \end{cases}$

23. $\begin{cases} \dfrac{2}{r} - \dfrac{4}{t} = 2 \\ t = -2r \end{cases}$

24. $\begin{cases} \dfrac{1}{x} + \dfrac{1}{y} = \dfrac{5}{6} \\ xy = 6 \end{cases}$

25. $\begin{cases} \dfrac{1}{x} - \dfrac{1}{y} = \dfrac{1}{4} \\ xy = 8 \end{cases}$

26. $\begin{cases} (x - 10)(y - 11) = 8 \\ y = x + 1 \end{cases}$

27. $\begin{cases} (x - 2)(y - 3) = 14 \\ x = 2y - 1 \end{cases}$

28. $\begin{cases} (x - 4)^2 + (y - 6)^2 = 100 \\ y = x \end{cases}$

29. $\begin{cases} 4(x + 2)^2 - 9(y - 3)^2 = 28 \\ y - 3 = x \end{cases}$

Applications

30. Geometry Find the dimensions of a rectangle if the area is 12 cm^2 and its length is 2 cm less than twice its width.

31. Number Theory The sum of the squares of two numbers is 25. The difference of twice the first and three times the second is -1. Find the numbers.

32. Number Theory Find two numbers such that the sum of their squares is 25 and the difference of their squares is 7.

33. Number Theory Find two numbers such that 4 times the square of the first plus 7 times the square of the second is 211 and the difference of their squares is -16.

34. Physical Fitness In a 16-block neighborhood of Bennington the blocks are of uniform rectangular shape. The area of one block is 147,000 ft^2. The jogging distance from the school to the football field is 1470 ft. In order to get to the football field from the school, one must jog down the lengths of three blocks, turn right, and jog the width of one block. What are the dimensions of each block in the neighborhood?

Developing Mathematical Power

Extension The equations in this chapter are special cases of the general second-degree equation $Ax^2 + Bxy + Cy^2 + Dx + Ey + F = 0$. Now, consider only the case where $B = 0$:

$$Ax^2 + Cy^2 + Dx + Ey + F = 0$$

This equation represents (with some exceptions) a conic section with axes on or parallel to the coordinate axes. In order to identify the conic, it is not necessary to rewrite the equation in standard form. The relationship between the coefficients A and C is of particular importance.

- If $A = C \neq 0$, the equation is that of a circle.
- If $A \neq C$ and $AC > 0$, the equation is that of an ellipse.
- If $A = 0$ or $C = 0$, but not both, the equation is that of a parabola.
- If $AC < 0$, the equation is that of a hyperbola.

The exceptions to these rules occur when, under certain conditions, the equation has no graph or it represents a point or one or two lines, which are known as the *degenerate conics*. (See page 519.)

Each of the following equations represents a conic section. Without rewriting the equation in standard form, identify the curve represented.

35. $5x^2 - y = 0$

36. $x^2 + 4y^2 - 4 = 0$

37. $x^2 + 4x + y^2 + 4y - 8 = 0$

38. $x^2 - y^2 - 1 = 0$

39. $25x^2 + 9y^2 - 225 = 0$

40. $2x^2 - 12x - y + 32 = 0$

41. $x^2 - 9y^2 - 36 = 0$

42. $x^2 - 4x + y^2 = 0$

43. $x^2 + y^2 - 2x - 24 = 0$

44. $2x^2 - 4x + y + 1 = 0$

45. $x^2 + 4y^2 + 24y + 32 = 0$

46. $x^2 - 16y^2 - 2x - 32y - 31 = 0$

47. $3x^2 - 5x + 3y^2 = 8$

48. $5y^2 + 6 = 3x^2 - 4x$

11.7

Inverse and Joint Variation

Objective: To solve problems involving inverse and joint variation

In Chapter 4, you learned that the linear function $y = kx$, or $\frac{y}{x} = k$, expresses direct variation. In this function, a change in one variable produces a proportional change in the other variable.

Capsule Review

The equation $y = 3x$ expresses direct variation, since it is of the form $y = kx$. The constant of variation is k, or 3.

Determine if each function is a direct variation. If so, give the constant of variation.

1. $y = -5x$ **2.** $y = 3x - 1$ **3.** $y + x = 0$ **4.** $y - 2x = 4$

5. If y varies directly as x, and $y = 6$ when $x = 4$, find y when $x = 8$.

6. If y varies directly as x, and $y = 12$ when $x = -6$, find x when $y = 22$.

7. If y varies directly as x, and $y = 1$ when $x = 2$, find y when $x = 1$.

For each set of ordered pairs (x, y), write a direct variation equation.

8. {. . . (2, 4), (3, 6), (4, 8) . . .} **9.** {. . . (100, 25), (96, 24), (92, 23) . . .}

In a different type of function, a change in one variable causes an *inverse* change in another variable. Recall that the graph of the equation $xy = 1$, or $y = \frac{1}{x}$, is a rectangular hyperbola. One branch of this hyperbola is shown at the right. Notice that, as the value of x increases, the value of y decreases; as the value of x decreases, the value of y increases. An equation of the form

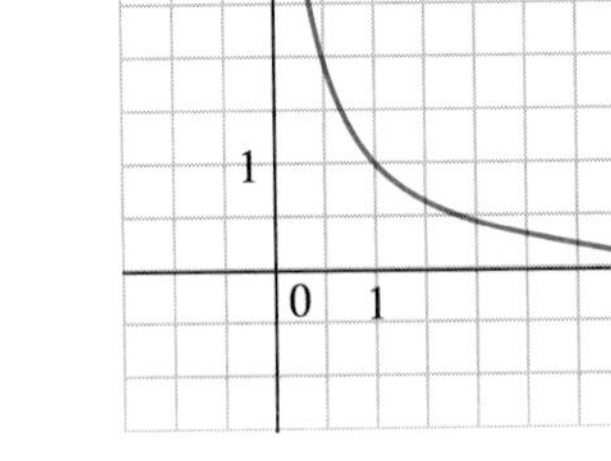

$$xy = k, \quad \text{or} \quad y = \frac{k}{x} \qquad k \neq 0, x \neq 0$$

expresses **inverse variation,** where k is the **constant of variation.** The variable y *varies inversely as* x, or y *is inversely proportional to* x.

EXAMPLE 1 A laboratory technician recorded in a table the volume of a gas at constant temperature and under several different pressures. Do the results suggest that the volume varies inversely as the pressure?

Volume, V (cm^3)	210	252	315	420
Pressure, P (in. of mercury)	30	25	20	15

If the volume varies inversely with the pressure, then the product VP will be equal to a constant. See if the given data satisfy this condition.

$$\begin{aligned} &\mathbf{V \times P} \\ 210 \times 30 &= 6300 \\ 252 \times 25 &= 6300 \\ 315 \times 20 &= 6300 \\ 420 \times 15 &= 6300 \end{aligned}$$

In each case, $VP = 6300$, so it appears that the volume varies inversely as the pressure.

Some equations show that y varies inversely *as a function of* x. In the next example, one variable varies inversely as the square of the other variable.

EXAMPLE 2 **The illumination, I, from a light source varies inversely as the square of the distance, d. If the illumination is 10 foot-candles when the distance is 2 m, find the illumination when the distance is 4 m.**

$$I = \frac{k}{d^2}$$

$$Id^2 = k \qquad \textit{Write an equation and solve for } k.$$

$$10(2)^2 = k \qquad \textit{Substitute: } I = 10,\ d = 2$$

$$40 = k$$

$$Id^2 = k$$

$$I(4)^2 = 40 \qquad \textit{Substitute: } d = 4,\ k = 40$$

$$I = \frac{40}{16}, \text{ or } 2.5$$

The illumination is 2.5 foot-candles.

Proportions can also be used to solve inverse variation problems. If the ordered pairs (x_1, y_1) and (x_2, y_2) satisfy the equation $xy = k$, then $x_1y_1 = k$ and $x_2y_2 = k$. Therefore,

$$x_1y_1 = x_2y_2$$

$$\frac{x_1}{x_2} = \frac{y_2}{y_1} \qquad \textit{Divide both sides by } x_2y_1.$$

EXAMPLE 3 **If y varies inversely as x, and $y = 3$ when $x = 8$, what is the value of y when $x = 15$?**

$$\frac{8}{15} = \frac{y}{3} \qquad \frac{x_1}{x_2} = \frac{y_2}{y_1}$$

$$y = 3\left(\frac{8}{15}\right) \qquad \text{Calculator-ready form}$$

$$y = 1.6$$

The variable y is said to **vary jointly** as x and z if $y = kxz$. In the equation $y = kxz$, k is the constant of variation. An equation can also show that y varies jointly as *functions* of other variables.

EXAMPLE 4 **If y varies jointly as x and z, and $y = 54$ when $x = 2$ and $z = 9$, find y when $x = 7$ and $z = 10$.**

$$y = kxz \qquad \text{Write an equation and solve for } k.$$

$$54 = k(2)(9) \qquad \text{Substitute: } y = 54,\ x = 2,\ z = 9$$

$$3 = k$$

$$y = kxz$$

$$y = 3(7)(10) \qquad \text{Substitute: } k = 3,\ x = 7,\ z = 10$$

$$y = 210$$

In **combined variation** the relation involves both direct and inverse variation. The equation $y = k\frac{rt}{v^2}$ expresses combined variation. The variation may be read: *y varies directly as r and t and inversely as the square of v*. Or, it may be read: *y varies jointly as r and t and inversely as the square of v*.

EXAMPLE 5 **If y varies directly as x and inversely as z, and $y = 30$ when $x = 20$ and $z = 50$, find y when $x = 40$ and $z = 25$.**

$$y = k\frac{x}{z} \qquad \text{Write an equation and solve for } k.$$

$$30 = k\left(\frac{20}{50}\right) \qquad \text{Substitute: } y = 30,\ x = 20,\ z = 50$$

$$30\left(\frac{50}{20}\right) = k$$

$$75 = k$$

$$y = k\frac{x}{z}$$

$$y = 75\left(\frac{40}{25}\right) \qquad \text{Substitute: } k = 75,\ x = 40,\ z = 25$$

$$y = 120$$

CLASS EXERCISES

State whether each equation represents inverse variation, joint variation, or combined variation.

1. $xy = 2$ **2.** $z = kxy$ **3.** $y = \frac{x}{z}$

4. $t = k\frac{r^2}{s}$ **5.** $V = \frac{1}{3}\pi r^2 h$ **6.** $F = kdg^2$

7. If z varies inversely as t, and $t = 10$ when $z = 4$, find the constant of variation and write an equation for the variation.

8. If r varies inversely as s, and $s = 4$ when $r = 12$, find s when $r = 6$.

9. If x varies jointly as y and the square of z, and $x = 15$ when $y = 5$ and $z = 1$, find x when $y = 1$ and $z = 2$.

PRACTICE EXERCISES 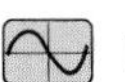 Use technology where appropriate.

Does the data in the table suggest that y varies inversely as x? If so, find the constant of variation and the equation.

1.

x	1	2	4	8	12
y	4	2	1	$\frac{1}{2}$	$\frac{1}{3}$

2.

x	50	25	20	10
y	2	4	5	10

State whether the formula expresses inverse variation, joint variation, or neither. State the constant of variation.

3. $V = e^3$ **4.** $24 = lw$ **5.** $xy = 18$

6. $p = 3s$ **7.** $V = \pi r^2 h$ **8.** $A = \frac{1}{2}bh$

In Exercises 9–12, y varies inversely as x.

9. If $y = 4$ when $x = 2$, find y when $x = 6$.

10. If $y = 8$ when $x = 3$, find y when $x = 12$.

11. If $y = 20.4$ when $x = -6.8$, find x when $y = 47.6$.

12. If $y = 82.0$ when $x = 32.8$, find x when $y = 180.4$.

In Exercises 13–16, x varies inversely as y^2.

13. If $x = 4$ when $y = 2$, find x when $y = 3$.

14. If $x = 6$ when $y = 4$, find x when $y = 8$.

15. If $x = 42$ when $y = 21$, find y when $x = 378$.

16. If $x = 72$ when $y = 27$, find y when $x = 54$.

In Exercises 17–22, z varies jointly as x and y.

17. If $z = 12$ when $x = 3$ and $y = 4$, find z when $x = 5$ and $y = 6$.

18. If $z = 15$ when $x = 5$ and $y = 2$, find z when $x = 7$ and $y = 4$.

19. If $z = 198$ when $x = 33$ and $y = 9$, find z when $x = 24$ and $y = 36$.

20. If $z = 1088$ when $x = 34$ and $y = 4$, find z when $x = 28$ and $y = 18$.

21. If $z = 1.1$ when $x = 55$ and $y = 2$, find z when $x = 75$ and $y = 3$.

22. If $z = 742.5$ when $x = 33$ and $y = 15$, find z when $x = 12$ and $y = 15$.

In Exercises 23–30, use k as the constant of variation and write an equation for the variation.

23. t varies directly as q and inversely as s.

24. w varies jointly as x and y.

25. r varies directly as the square of t and inversely as the cube of v.

26. g varies jointly as a and b and inversely as the square of c.

27. The weight of an object varies inversely as the square of the distance of the object from the center of the earth.

28. The load that a beam of constant length can support varies jointly as its width and the square of its height.

29. The resistance of a wire to the passage of an electric current varies directly as the length of the wire and inversely as the square of its diameter.

30. The volume of a right circular cylinder varies jointly as its height and the square of its radius.

31. If y varies inversely as x, and $y = 42$ when $x = 3\frac{1}{2}$, find y when $x = 1\frac{2}{5}$.

32. If y varies inversely as the square of x, and $y = 50$ when $x = 4$, find y when $x = 5$.

33. If w varies jointly as x and y, and $w = 28$ when $x = 4$ and $y = 21$, find w when $x = 12$ and $y = 17$.

34. If c varies jointly as d and the square of e, and $c = 30$ when $d = 15$ and $e = 2$, find d when $c = 6$ and $e = 8$.

35. If y varies directly as x and inversely as z, and $y = 49$ when $x = 14$ and $z = 4$, find y when $x = 16$ and $z = 7$.

36. If a varies directly as b and inversely as the square of c, and $a = 46$ when $b = 12$ and $c = 6$, find b when $a = 23$ and $c = 6$.

37. If x varies directly as the square of y, and $x = 4$ when $y = \frac{1}{2}$, find x when $y = \frac{1}{4}$.

38. If a varies jointly as b and c, and $a = 100$ when $b = 10$ and $c = 5$, find c when $a = 150$ and $b = 15$.

39. If d varies jointly as r and t, and $d = 110$ when $r = 55$ and $t = 2$, find r when $d = 40$ and $t = 3$.

40. If y varies directly as x and inversely as z^2, and $x = 48$ when $y = 8$ and $z = 3$, find x when $y = 12$ and $z = 2$.

41. It t varies directly as s and inversely as the square of r, how is the value of t changed when the value of s is doubled?

42. If x varies directly as the square of y and inversely as z, what happens to the value of x if the value of y is halved?

Applications

43. Geometry The area of a rectangle varies jointly as its length and width. If the area of a rectangle is 72 cm^2 when the length is 12 cm and the width is 6 cm, find the length when the area is 62.5 cm^2 and the width is 2.5 cm.

44. Geometry The volume of a cube varies directly as the cube of the length of an edge. If the length of the edge is tripled, what happens to the volume of the cube?

45. Physics The distance traveled varies jointly as the rate and the time. If John travels 135 mi in 3 h at a rate of 45 mi/h, find the rate at which he travels to cover a distance of 189 mi in 3.5 h.

Mixed Review

46. How can you determine from the equation of a parabola whether it opens upward or downward?

47. Add $\begin{bmatrix} 1 & -3 & 5 \\ 8 & 3 & -6 \end{bmatrix} + \begin{bmatrix} 2 & 2 & -4 \\ 4 & -3 & 7 \end{bmatrix}$

48. Factor $8x^3y^6 + 1$.

49. Multiply $(3x^4 - 1)(2x - 4)$.

11.8

Problem Solving Strategy: Select a Function

The concept of a function is one of the most important ideas in mathematics. A linear function such as $x + y = 12$ relates the two variables x and y. Other functions relate other variables to one another. Functional relationships are important in many fields of study and can be used to solve an endless variety of problems. Many problems in chemistry and physics involve the functions of direct, inverse, or joint variation.

EXAMPLE 1 How far from the fulcrum of a lever must a 7.2-kg object be placed in order to balance a 6.4-kg object that is 2.7 m from the fulcrum?

Understand the Problem *What are the given facts?* A 6.4-kg object is 2.7 m from the fulcrum of a lever. A 7.2-kg object is being placed on the lever.
What are you asked to find? The problem is to find how far from the fulcrum the 7.2-kg object should be placed in order to balance the lever.

Plan Your Approach *Choose a strategy.* If the 7.2-kg object were placed at the same distance from the fulcrum as the 6.4-kg object, the lever would not balance—it would go down on the side of the 7.2-kg object.

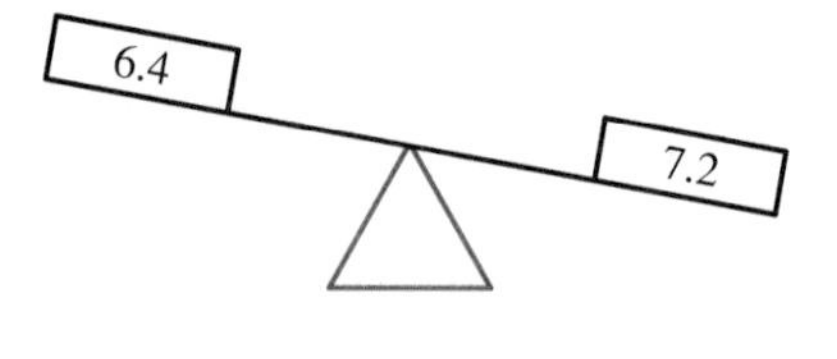

In order to balance the fulcrum, the 7.2-kg object needs to be placed closer to the fulcrum than the 6.4-kg object.

The force exerted on the lever at a certain distance from the fulcrum is called *torque*. The torque depends on the mass of the object chosen and the distance it is placed from the fulcrum. In order to reduce the torque on a lever, either reduce the distance between the object and the fulcrum or choose an object with less mass. Therefore, the torque varies jointly with mass and distance. Note that the torque on both sides of the fulcrum must be equal in order for the lever to balance.

Let m_1, d_1 and T_1 represent the mass, distance and torque of the 6.4-kg object and let m_2, d_2 and T_2 represent the mass, distance and torque of the 7.2-kg object. Also, let k represent the constant of variation. Then

$$T_1 = T_2$$

$$k(m_1)(d_1) = k(m_2)(d_2) \quad T_n \text{ varies jointly with } m_n \text{ and } d_n.$$

$$\frac{(m_1)(d_1)}{m_2} = d_2 \quad \text{Solve for } d_2.$$

Complete the Work

$$d_2 = \frac{(6.4)(2.7)}{7.2} \quad m_1 = 6.4,\ d_1 = 2.7,\ m_2 = 7.2.$$

$$d_2 = 2.4$$

Interpret the Results *State your answer.* The 7.2-kg object should be placed 2.4 m from the fulcrum.

Check your answer. Is the torque on both sides of the fulcrum equal?

$$k(6.4)(2.7) \stackrel{?}{=} k(7.2)(2.4)$$

$$17.28 = 17.28 \checkmark$$

Many relationships in science are governed by functions of variation. The following examples illustrate two of these scientific relationships.

The volume of a gas varies inversely with the pressure and directly with the absolute temperature. Therefore, this relationship is an example of combined variation.

EXAMPLE 2 A gas with a volume of 500 cm^3 is under a pressure of 30 in. of mercury at an absolute temperature of 300°. Find the volume of the gas if the pressure is decreased to 25 in. of mercury and the absolute temperature is decreased to 285°.

Understand the Problem The volume, V, of a gas varies inversely with the pressure, P, and directly with the absolute temperature, t.

Plan Your Approach Write an equation, substitute one set of given values, and find the constant of variation. Then solve the equation for V, using the second set of values for P and t.

$$V = k\left(\frac{t}{p}\right)$$

Complete the Work

$$500 = k\left(\frac{300}{30}\right) \quad V = 500,\ t = 300,\ P = 30$$

$$50 = k$$

$$V = k\left(\frac{t}{p}\right)$$

$$V = 50\left(\frac{285}{25}\right) \quad k = 50,\ t = 285,\ P = 20$$

$$V = 570$$

Interpret the Results The new volume is 570 cm^3.

The next example illustrates a problem involving joint variation. The kinetic energy of an object varies jointly with its mass and the square of its velocity.

EXAMPLE 3 An object with a mass of 10 kg is moving at a rate of 5 m/s and has a kinetic energy of 125 joules. What is the kinetic energy of an object with a mass of 18 kg, moving at a rate of 25 m/s?

Understand the Problem The kinetic energy, K_e, varies jointly with the mass, s, and the square of the velocity, v, of an object in motion.

Plan Your Approach Write an equation and find the constant of variation. Then use the equation again to find the kinetic energy.

$$K_e = ksv^2$$

Complete the Work

$$125 = k(10)(5^2) \qquad K_e = 125,\ s = 10,\ v = 5$$

$$0.5 = k$$

$$K_e = ksv^2$$

$$K_e = 0.5(18)(25^2) \qquad K_e = 0.5,\ s = 18,\ v = 25$$

$$K_e = 5625$$

Interpret the Results The kinetic energy is 5625 joules.

CLASS EXERCISES

Solve. Round your answers to the nearest tenth, when necessary.

1. A 3.2-kg object, placed 24.6 cm from the fulcrum of a lever, balances a rock 32.4 cm from the fulcrum. What is the mass of the rock?
2. A gas with a volume of 427 cm^3 is under pressure of 25 in. of mercury at an absolute temperature of 305°. If the volume of the gas is to be decreased to 390 cm^3 while the absolute temperature is increased to 312°, what pressure is required?
3. An object with a mass of 12.4 kg moves at a rate of 8.0 m/s and has a kinetic energy of 396.8 joules. If another object with kinetic energy of 295.5 joules moves at a rate of 6.4 m/s, what is its mass?

PRACTICE EXERCISES

Use technology where appropriate.

Solve. Round your answers to the nearest tenth, when necessary.

1. How far from the fulcrum of a lever must a 5.7-kg object be placed in order to balance a 4.6-kg object that is 1.4 m from the fulcrum?

2. How far from the fulcrum of a lever must a 3.3-kg object be placed in order to balance a 1.5-kg object that is 0.8 m from the fulcrum?

3. A 7.1-kg object, placed 35.0 cm from the fulcrum of a lever, balances a rock 40.5 cm from the fulcrum. What is the mass of the rock?

4. A 6.8-kg object, placed 52.9 cm from the fulcrum of a lever, balances a steel cube 29.8 cm from the fulcrum. What is the mass of the cube?

5. A gas with a volume of 400 cm^3 is under a pressure of 31 in. of mercury at an absolute temperature of 310°. Find the volume of the gas at a pressure of 30 in. of mercury and absolute temperature 300°.

6. A gas with a volume of 840 cm^3 is under a pressure of 32 in. of mercury at an absolute temperature of 280°. Find the volume of the gas at a pressure of 30 in. of mercury and absolute temperature 300°.

7. A gas with a volume of 600 cm^3 is under a pressure of 33 in. of mercury at an absolute temperature of 275°. At what pressure is the volume 700 cm^3 and the absolute temperature 350°?

8. A gas with a volume of 500 cm^3 is under a pressure of 28 in. of mercury at an absolute temperature of 280°. At what pressure is the volume 450 cm^3 and the absolute temperature 288°?

9. An 8-kg object is moving at 4 m/s and has a kinetic energy of 64 joules. What is the kinetic energy of a 30-kg object moving at 5 m/s?

10. A 10-kg object is moving at 6 m/s and has a kinetic energy of 180 joules. What is the kinetic energy of a 20-kg object moving at 15 m/s?

In a mechanical device involving meshed gears, the rotational speed of each gear is inversely proportional to the number of its teeth.

11. Two meshed gears have 40 teeth and 48 teeth, respectively. At what speed will the second gear be driven if the first gear runs at 1200 revolutions per minute (rpm)?

12. Two meshed gears have 45 teeth and 50 teeth, respectively. At what speed will the second gear be driven if the first gear runs at 1500 rpm?

The load that a beam of constant length can support varies jointly as its width and the square of its height.

13. If a beam 4 ft wide and 2 ft high can support a load of 1760 lb, find the height of a beam 1 ft wide that can support a load of 990 lb.

14. If a beam 3 ft wide and 4 ft high can support a load of 1500 lb, find the height of a beam 1 ft wide that can support a load of 900 lb.

15. The intensity of illumination on a surface varies inversely as the square of the distance of the source of illumination from the surface. The

surface of a book is 70 cm from a lamp. How far from the lamp should the book be placed so that the illumination is doubled?

16. When a given object moves on a circular path, the centripetal force varies directly as the square of the velocity and inversely as the radius of the curve. If the force is 640 lb for a velocity of 20 mi/h and radius 5 m, find the force for a velocity of 30 mi/h and radius 4 m.

Mixed Problem Solving Review

1. The sum of three numbers is 85. When the third is divided by the first, the quotient is 2 with a remainder of 2. When the third is divided by the second, the quotient is 1 with a remainder of 19. Find the numbers.

2. Barry makes boats with two different types of wood, teak and oak. When he uses 7 parts of teak and 5 parts of oak, the cost of wood is $2.50 per foot. If he uses 5 parts of teak and 7 parts of oak, the cost is $2.30 per foot. Find the cost per foot of each type of wood.

3. Telegraph poles along a railway line are placed at intervals of 150 yd. A passenger in a train observes that 7 poles pass by in 42 seconds. Calculate the speed of the train in miles per hour.

PROJECT

Look up the definition of *function* in this textbook. Write the definition in your notebook and then give one example of each function you have studied in Chapters 1–11. Give both general information about the function and a specific example.

TEST YOURSELF

Graph each system. For Exercise 1, determine the number of real solutions.

1. $\begin{cases} 4x^2 + 9y^2 = 36 \\ y = x^2 \end{cases}$ **2.** $\begin{cases} x^2 + y^2 \leq 16 \\ xy \leq 4 \end{cases}$ 11.5

Find the real solutions of each system.

3. $\begin{cases} x^2 + y^2 = 49 \\ y = x - 2 \end{cases}$ **4.** $\begin{cases} x^2 - 3y^2 = 2 \\ 2x^2 + y^2 = 4 \end{cases}$ 11.6

5. z varies jointly as x and y, and $z = 2.4$ when $x = 0.3$ and $y = 0.4$. Find z when $x = 0.5$ and $y = 0.8$. 11.7

6. How far from the fulcrum of a lever must a 0.8-kg object be placed in order to balance a 1.2-kg object 4.0 m from the fulcrum? 11.8

INTEGRATING ALGEBRA
Newton's Law of Universal Gravitation

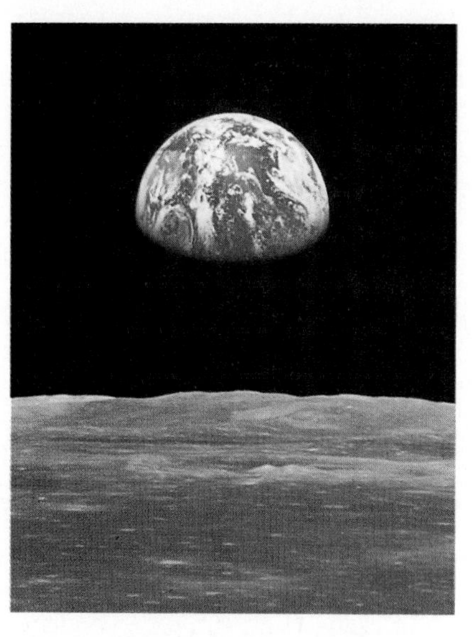

Did you know that the same force that pulls an object to earth keeps the planets in orbit about the sun? Without gravitational force, the earth would fly off in a direction tangent to its orbit about the sun and a ball thrown upward from the surface of the earth would sail out of the earth's atmosphere and continue to move through the universe.

Sir Isaac Newton [1642–1727] discovered that a force of gravitation causes particles in the universe to attract each other. Newton's discovery states that each pair of objects attracts one another with a force that varies directly as the product of their masses and inversely as the square of the distance between their centers. Let F represent the force of attraction in units of newtons (N), m_1 and m_2 the masses of the two bodies in kilograms (kg), and d the distance between their centers in meters (m). Further, let G represent the constant of proportionality. Therefore,

$$F = G\frac{m_1 \cdot m_2}{d^2} \qquad \text{where } G = 6.67 \times 10^{-11}\,\frac{\text{N} \cdot \text{m}^2}{\text{kg}^2}$$

Use the data listed below to work the exercises that follow.

Mass	Distance
earth: 5.98×10^{24} kg	earth to sun: 1.50×10^{8} km
sun: 1.99×10^{30} kg	earth to moon: 3.85×10^{5} km
moon: 7.36×10^{22} kg	radius of earth: 1.27×10^{4} km

1. Calculate the force of attraction between the moon and the earth.

2. Find the force of attraction between the earth and a 50-kg rock whose center is 50 cm above the earth's surface.

3. Find the force of attraction between two 50-kg rocks whose centers are 10 m apart.

4. The rocks in Exercise 3, in spite of the force of attraction between them, do not move toward one another. Why not? *Hint:* See Exercise 2.

CHAPTER 11 SUMMARY AND REVIEW

Vocabulary

asymptotes of a hyperbola (535)
center of a circle (515)
center of a hyperbola (535)
center of an ellipse (528)
circle (515)
combined variation (554)
conic section, or conic (514)
conjugate axis of a hyperbola (535)
constant of variation (552)
directrix of a parabola (520)
ellipse (527)
foci of a hyperbola (534)
foci of an ellipse (527)
focus of a parabola (520)
hyperbola (534)
inverse variation (552)
joint variation (554)
major axis of an ellipse (528)
minor axis of an ellipse (528)
parabola (520)
quadratic system (541)
radius of a circle (515)
rectangular hyperbola (537)
transverse axis of a hyperbola (535)

Circle Standard forms of the equation: **11.1**

$x^2 + y^2 = r^2$ center $(0, 0)$; radius r

$(x - h)^2 + (y - k)^2 = r^2$ center (h, k); radius r

1. Write in standard form the equation of the circle with center (2, 3) and radius 4.

2. Find the center and radius of the circle $x^2 + y^2 + 4x - 12y - 2 = 0$.

Parabola Equations (vertex (0, 0)): **11.2**

$x^2 = 4cy$ focus $(0, c)$ $y^2 = 4cx$ focus $(c, 0)$

Standard forms of the equation (vertex (h, k)):

$y = a(x - h)^2 + k$ $x = a(y - k)^2 + h$

3. Write in standard form the equation of the parabola with focus (0, 2) and directrix $y = -4$.

4. Find the vertex, the axis of symmetry, the focus, and the directrix of the parabola $y = 8(x - 1)^2 - 4$. Sketch the graph.

Ellipse Standard forms of the equation ($a > b$, $c^2 = a^2 - b^2$): **11.3**

$\frac{x^2}{a^2} + \frac{y^2}{b^2} = 1$ foci $(\pm c, 0)$ $\frac{x^2}{b^2} + \frac{y^2}{a^2} = 1$ foci $(0, \pm c)$ center $(0, 0)$

$\frac{(x - h)^2}{a^2} + \frac{(y - k)^2}{b^2} = 1$ $\frac{(x - h)^2}{b^2} + \frac{(y - k)^2}{a^2} = 1$ center (h, k)

5. Find the intercepts, the foci, and the length of the major axis of the ellipse $\frac{x^2}{36} + \frac{y^2}{100} = 1$. Sketch the graph.

6. Write in standard form the equation of the ellipse with center (0, 0), x-intercepts ± 4, and y-intercepts ± 6.

Hyperbola Standard forms of the equation ($c^2 = a^2 + b^2$): 11.4

$\frac{x^2}{a^2} - \frac{y^2}{b^2} = 1$ foci $(\pm c, 0)$ $\frac{y^2}{a^2} - \frac{x^2}{b^2} = 1$ foci $(0, \pm c)$ center (0, 0)

$\frac{(x-h)^2}{a^2} - \frac{(y-k)^2}{b^2} = 1$ $\frac{(y-k)^2}{a^2} - \frac{(x-h)^2}{b^2} = 1$ center (h, k)

7. Find the intercepts, the foci, and the asymptotes of the hyperbola $\frac{x^2}{169} - \frac{y^2}{25} = 1$. Sketch the graph.

8. Write in standard form the equation of the hyperbola with center (0, 0), x-intercepts ± 3, and asymptotes $y = \pm \frac{4}{3}x$.

9. Classify the equation $3y^2 + 2y - 1 + 3x^2 = 0$ as that of a circle, a parabola, an ellipse, a hyperbola, or none of these.

Graphing Quadratic Systems The graphs of two quadratic equations may intersect in at most four points. The graphs of a linear equation and a quadratic equation may intersect in at most two points. 11.5

10. Graph the system at the right. Then approximate the real solutions in tenths.
$\begin{cases} x^2 + y^2 = 9 \\ x + 2y = 1 \end{cases}$

Solving Quadratic Systems Quadratic systems are solved by the substitution and the addition methods. 11.6

11. Find the real solutions of the system:
$\begin{cases} 9m^2 - 4n^2 = 36 \\ m^2 - n^2 = 1 \end{cases}$

Inverse and Joint Variation If y varies inversely as x, then $xy = k$, or $y = \frac{k}{x}$, where k is the constant of variation. If y varies jointly as x and z, then $y = kxz$. 11.7

12. If z varies inversely as t, and $t = 10$ when $z = 4$, find z when $t = 8$.

13. If x varies jointly as y and z, and $x = 36$ when $y = 36$ and $z = 4$, find x when $y = 12$ and $z = 8$.

14. A 6.7-kg object, placed 0.8 m from the fulcrum of a lever, balances an object 1.4 m from the fulcrum. What is the mass of this object? 11.8

CHAPTER 11 TEST

1. Write in standard form the equation of the circle with center (5, 3) and radius 2.

2. Find the center and radius of the circle $x^2 + y^2 = 81$.

3. Write in standard form the equation of the parabola with focus (0, 3) and directrix $y = -3$.

4. Find the vertex, the focus, the axis of symmetry, and the directrix of the parabola $y = -8x^2$.

5. Find the intercepts, the foci, and the length of the major axis of the ellipse $\frac{x^2}{100} + \frac{y^2}{36} = 1$.

6. Find the intercepts, the foci, and the asymptotes of the hyperbola $\frac{x^2}{625} - \frac{y^2}{225} = 1$.

7. Identify $3x^2 - x = 4 - 2y^2$ as the equation of a circle, a parabola, an ellipse, a hyperbola, or none of these.

8. Sketch the graph of the system $\begin{cases} x^2 + y^2 = 16 \\ y = 2x + 3 \end{cases}$

Determine the number of real solutions it has.

Find the real solutions of each system.

9. $\begin{cases} x^2 + 4y^2 = 16 \\ x - 2y = -4 \end{cases}$

10. $\begin{cases} x^2 + y^2 = 20 \\ 2x^2 - y^2 = 16 \end{cases}$

Solve.

11. If y varies inversely as the square of x, and $y = 4$ when $x = 3$, find y when $x = 6$.

12. How far from the fulcrum of a lever must a 2.5-kg object be placed to balance a 3.0-kg object 0.5 m from the fulcrum?

Challenge

The segments perpendicular to the major axis of an ellipse at the foci are called *focal chords*. Find the coordinates of the endpoints of the focal chords of the ellipse with the equation

$$\frac{x^2}{25} + \frac{y^2}{16} = 1$$

Find the length of the focal chords.

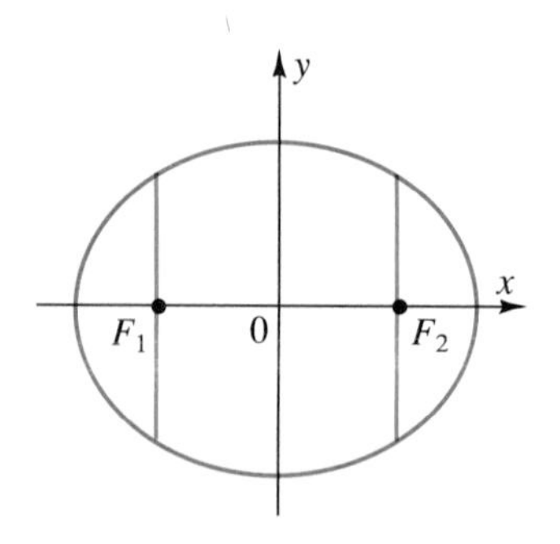

COLLEGE ENTRANCE EXAM REVIEW

Select the best choice for each question.

1. $\dfrac{a}{a-b} + \dfrac{b}{b-a}$ equals
   A. 1 **B.** 0 **C.** -1
   D. $\dfrac{a+b}{a-b}$ **E.** $\dfrac{b+a}{b-a}$

2. If $\sqrt{a-b} = 2$ when a and b are positive integers, then which of the following cannot equal ab?
   A. 21 **B.** 32 **C.** 48
   D. 60 **E.** 77

3. If the graph of $y = 2(x-1)^2 + k$ passes through the point $(3, -7)$, then k equals
   A. 15 **B.** 1 **C.** -1
   D. -11 **E.** -15

4. In the formula $A = \frac{4}{3}\pi r^3$, if r is doubled, then A will be
   A. doubled
   B. multiplied by 4
   C. increased by 4
   D. multiplied by 8
   E. increased by 8

5. If $x^2 + y^2 - 8x + 6y = 0$ is the equation of a circle with center at C and radius r, then
   A. $C(4, -3)$ and $r = 25$
   B. $C(4, -3)$ and $r = 5$
   C. $C(-4, 3)$ and $r = 25$
   D. $C(-4, 3)$ and $r = 5$
   E. it's just the point, $(4, -3)$

6. Find the length of the altitude to the hypotenuse of a right triangle with legs of length 15 and 20.
   A. 9.5 **B.** 12 **C.** 12.5
   D. 18 **E.** 24

7. If the graph of $y = -x^2$ is translated 4 units to the left and 1 unit up, the resulting graph has the equation
   A. $y = -(x+1)^2 + 4$
   B. $y = -(x+4)^2 + 1$
   C. $y = -(x+4)^2 - 1$
   D. $y = (x+1)^2 + 4$
   E. $y = (x+4)^2 + 1$

8. When $f(x) = 3x^3 - x^2 - 12x + 4$, then $f(1) + f(-1) =$
   A. 18 **B.** 6 **C.** 0
   D. -6 **E.** -18

9. The graph of $16x^2 + 9y^2 = 144$ is an ellipse having one focus at
   A. $(5, 0)$ **B.** $(0, 5)$ **C.** $(\sqrt{7}, 0)$
   D. $(0, \sqrt{7})$ **E.** $(12, 0)$

10. When the denominator of $\dfrac{4-\sqrt{5}}{2+\sqrt{5}}$ is rationalized, the result is
    A. $11 - 6\sqrt{5}$ **B.** $6\sqrt{5} - 13$
    C. $13 - 6\sqrt{5}$ **D.** $\dfrac{6\sqrt{5} - 13}{3}$
    E. $\dfrac{13 - 6\sqrt{5}}{3}$

11. A committee is made up of people 18 years old or older, no two the same age. If the sum of their ages is 145, what is the maximum possible number of people on the committee? (Assume ages are whole years)
    A. 6 **B.** 7 **C.** 8 **D.** 9
    E. It cannot be determined from the information given.

MIXED REVIEW

Express each fraction as a percent and each percent as a decimal.

Examples $\frac{17}{200}$

$$\begin{array}{r} 0.085 \\ 200\overline{)17.000} \\ -1600 \\ \hline 1000 \\ -1000 \\ \hline \end{array}$$

$0.085 = 8.5\%$

$6\frac{1}{4}\%$ *Think:* $\frac{1}{4} = 0.25$

$6\frac{1}{4}\% = 6.25\%$

$= 0.0625$

$6\frac{1}{4}\% = 0.0625$

1. $\frac{3}{8}$ **2.** $\frac{12}{25}$ **3.** $\frac{19}{200}$ **4.** $\frac{7}{20}$ **5.** $\frac{3}{5}$ **6.** $\frac{7}{25}$

7. 12.5% **8.** 3.25% **9.** $10\frac{1}{2}\%$ **10.** $8\frac{3}{4}\%$ **11.** 55.5% **12.** $9\frac{1}{4}\%$

Express each number in scientific notation.

Examples 4,870,000 0.00034

4,870,000 0.00034

4.87×10^6 *scientific notation* 3.4×10^{-4}

13. 0.0000289 **14.** 76,000,000 **15.** 0.00005632

16. $\frac{9}{1,000,000}$ **17.** 800,000 **18.** $\frac{15}{100}$

Evaluate.

Example $(\sqrt[3]{-8})^2$

$\sqrt[3]{-8} = -2$ *since* $(-2)(-2)(-2) = -8$

$(\sqrt[3]{-8})^2 = (-2)^2 = 4$

19. $(\sqrt{9})^3$ **20.** $(\sqrt[3]{64})^2$ **21.** $(\sqrt{81})^2$

22. $(\sqrt[4]{10,000})^5$ **23.** $(\sqrt[4]{81})^2$ **24.** $(\sqrt[3]{-27})^2$

Solve.

25. Orangeburg High won 24 games this season. If this was 75% of the games played, how many games did Orangeburg play?

26. 65% of the students at Hillside High go on to college. If the total enrollment at Hillside High is 800, how many students go on to college?

12 Exponential and Logarithmic Functions

On your own terms?

Developing Mathematical Power

When buying a house, a down payment is usually required. The rest of the cost is often financed by a bank loan, called a mortgage, which is paid back monthly over a number of years.

How much money will a bank lend for a home mortgage? A rule of thumb is that the monthly payment should be no more than 28% of the borrower's gross monthly income.

Project

Research the average income in your community. Check local newspaper ads for house prices and current mortgage rates. Prepare a report that would help buyers determine their price range and the term of their mortgage.

Monthly payments (*MP*) on a home mortgage:

$$MP = P\left[\frac{\frac{r}{12}\left(1+\frac{r}{12}\right)^{12T}}{\left(1+\frac{r}{12}\right)^{12T-1}}\right]$$

P = amount of mortgage

r = interest rate

T = term of mortgage in years

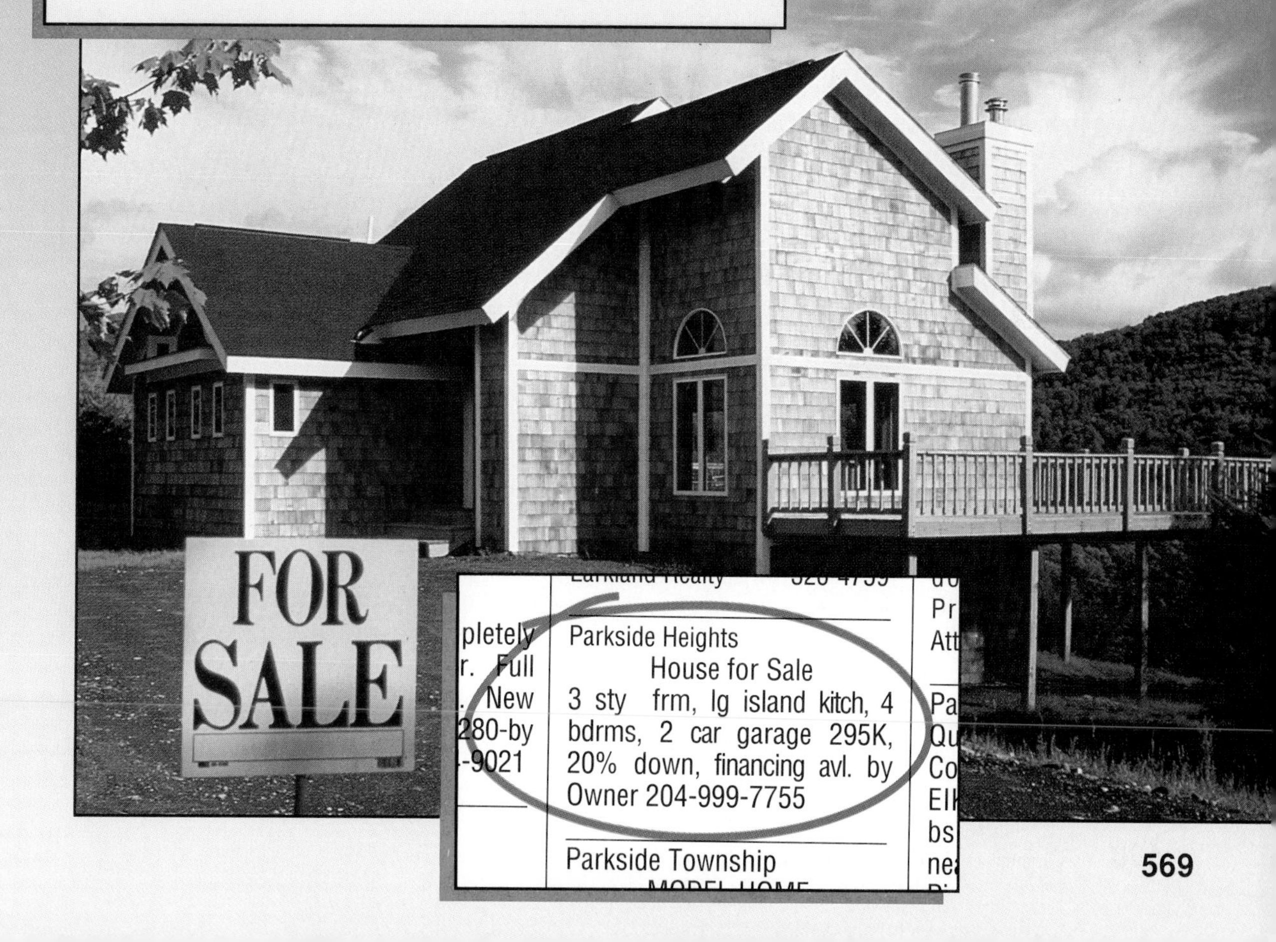

12.1 Rational Exponents

Objectives: To simplify expressions with rational exponents
To evaluate expressions with rational exponents

Before nonintegral exponents are introduced, it will be helpful to review expressions containing exponents that are negative integers.

Capsule Review

By definition, if $a \neq 0$ and n is an integer, then $a^{-n} = \frac{1}{a^n}$.

EXAMPLES **Simplify:** **a.** $2^{-2} = \frac{1}{2^2} = \frac{1}{4}$ **b.** $(5x)^{-3} = \frac{1}{(5x)^3} = \frac{1}{125x^3}$

Simplify.

1. 2^{-4}
2. $(3x)^{-2}$
3. $(5x^2y)^{-3}$
4. $2^{-2} + 4^{-1}$
5. $(2x)^{-2} + (5x)^{-1}$
6. $3x^{-2} - (2x)^{-2}$

Assume that the properties of integral exponents also hold for rational-number exponents. Then, since $(a^m)^n = a^{mn}$,

$$(7^{\frac{1}{2}})^2 = 7^{\frac{1}{2}\cdot 2} = 7^1 = 7$$

Since both equal 7, $(7^{\frac{1}{2}})^2 = (\sqrt{7})^2$. Although numbers whose squares are equal may not always be equal themselves, $7^{\frac{1}{2}}$ is defined as follows:

$$7^{\frac{1}{2}} = \sqrt{7}$$

Similarly, $7^{\frac{1}{3}} = \sqrt[3]{7}$, $7^{\frac{1}{4}} = \sqrt[4]{7}$, and so forth.

The expression $27^{\frac{2}{3}}$ may be written as $(27^2)^{\frac{1}{3}}$ or as $(27^{\frac{1}{3}})^2$, since $\frac{2}{3} = 2\left(\frac{1}{3}\right) = \frac{1}{3}(2)$. Therefore, $27^{\frac{2}{3}}$ may be written in radical form as $\sqrt[3]{27^2}$ or as $(\sqrt[3]{27})^2$. Check to show that $\sqrt[3]{27^2} = 9$ and $(\sqrt[3]{27})^2 = 9$.

Rational Exponent

If m and n are positive integers, and $\sqrt[n]{a}$ is a real number, then

$$a^{\frac{1}{n}} = \sqrt[n]{a} \quad \text{and} \quad a^{\frac{m}{n}} = \sqrt[n]{a^m}, \text{ or } a^{\frac{m}{n}} = (\sqrt[n]{a})^m$$

When rational exponents are defined in this way, they have all the properties of integral exponents. In the examples and exercises in this lesson, assume that all variables are positive real numbers.

EXAMPLE 1 **Write in radical form:** **a.** $x^{\frac{1}{2}}$ **b.** $x^{\frac{3}{5}}$ **c.** $y^{2.5}$ **d.** $y^{-\frac{4}{5}}$

a. $x^{\frac{1}{2}} = \sqrt{x}$

b. $x^{\frac{3}{5}} = \sqrt[5]{x^3}$, or $(\sqrt[5]{x})^3$

c. $y^{2.5} = y^{\frac{5}{2}}$
$= \sqrt{y^5}$, or $(\sqrt{y})^5$

d. $y^{-\frac{4}{5}} = \dfrac{1}{y^{\frac{4}{5}}}$ $\qquad a^{-n} = \dfrac{1}{a^n}$
$= \dfrac{1}{\sqrt[5]{y^4}}$, or $\dfrac{1}{(\sqrt[5]{y})^4}$

In Example 2, the definitions are used to write radical expressions in exponential form.

EXAMPLE 2 **Write in exponential form:**
a. $\sqrt[3]{x}$ **b.** $\sqrt[3]{x^2}$ **c.** $\sqrt[5]{8y}$ **d.** $8(\sqrt{y})^3$

a. $\sqrt[3]{x} = x^{\frac{1}{3}}$

b. $\sqrt[3]{x^2} = x^{\frac{2}{3}}$

c. $\sqrt[5]{8y} = (8y)^{\frac{1}{5}}$

d. $8(\sqrt{y})^3 = 8y^{\frac{3}{2}}$

The definitions of $a^{\frac{1}{n}}$ and $a^{\frac{m}{n}}$ are also used to evaluate numbers or expressions with rational exponents. When you evaluate an expression of the form $a^{\frac{m}{n}}$, it is usually easier to find the *n*th root first.

EXAMPLE 3 **Evaluate:** **a.** $27^{\frac{1}{3}}$ **b.** $(-32)^{\frac{3}{5}}$ **c.** $25^{-\frac{1}{2}}$ **d.** $4^{-3.5}$

a. $27^{\frac{1}{3}} = \sqrt[3]{27}$
$= 3$

b. $(-32)^{\frac{3}{5}} = (\sqrt[5]{-32})^3$
$= (-2)^3$
$= -8$

c. $25^{-\frac{1}{2}} = \dfrac{1}{25^{\frac{1}{2}}}$
$= \dfrac{1}{\sqrt{25}}$
$= \dfrac{1}{5}$

d. $4^{-3.5} = 4^{-\frac{7}{2}}$
$= \dfrac{1}{4^{\frac{7}{2}}}$
$= \dfrac{1}{(\sqrt{4})^7}$
$= \dfrac{1}{2^7} = \dfrac{1}{128}$

For an expression with rational exponents to be in simplest form, every exponent must be positive.

EXAMPLE 4 **Simplify:** **a.** $3x^{\frac{1}{5}} \cdot 8^{-\frac{2}{3}} \cdot x^{\frac{2}{5}}$ **b.** $\dfrac{x^{\frac{3}{4}}y^{-\frac{2}{3}}}{x^{\frac{1}{2}}y^{-\frac{1}{6}}}$

c. $\left(\dfrac{64x^{15}}{125y^{12}}\right)^{\frac{2}{3}}$ **d.** $(16x^{16}y^{-8})^{-\frac{3}{4}}$

a.
$$3x^{\frac{1}{5}} \cdot 8^{-\frac{2}{3}} \cdot x^{\frac{2}{5}} = 3 \cdot 8^{-\frac{2}{3}} \cdot x^{\frac{1}{5}} \cdot x^{\frac{2}{5}}$$
$$= \frac{3x^{\frac{1}{5}+\frac{2}{5}}}{8^{\frac{2}{3}}} \qquad a^{-n} = \frac{1}{a^n};\ a^m \cdot a^n = a^{m+n}$$
$$= \frac{3x^{\frac{3}{5}}}{(\sqrt[3]{8})^2}$$
$$= \frac{3x^{\frac{3}{5}}}{4}$$

b.
$$\frac{x^{\frac{3}{4}}y^{-\frac{2}{3}}}{x^{\frac{1}{2}}y^{-\frac{1}{6}}} = x^{\frac{3}{4}-\frac{1}{2}} \cdot y^{-\frac{2}{3}-\left(-\frac{1}{6}\right)} \qquad \frac{a^m}{a^n} = a^{m-n}$$
$$= x^{\frac{3}{4}-\frac{2}{4}} \cdot y^{-\frac{4}{6}+\frac{1}{6}}$$
$$= x^{\frac{1}{4}}y^{-\frac{3}{6}}$$
$$= x^{\frac{1}{4}}y^{-\frac{1}{2}}$$
$$= \frac{x^{\frac{1}{4}}}{y^{\frac{1}{2}}}$$

c.
$$\left(\frac{64x^{15}}{125y^{12}}\right)^{\frac{2}{3}} = \frac{(64x^{15})^{\frac{2}{3}}}{(125y^{12})^{\frac{2}{3}}} \qquad \left(\frac{a}{b}\right)^m = \frac{a^m}{b^m}$$
$$= \frac{64^{\frac{2}{3}}x^{15\cdot\frac{2}{3}}}{125^{\frac{2}{3}}y^{12\cdot\frac{2}{3}}} \qquad (a^mb^n)^p = a^{mp}b^{np}$$
$$= \frac{(\sqrt[3]{64})^2x^{10}}{(\sqrt[3]{125})^2y^8} = \frac{16x^{10}}{25y^8}$$

d.
$$(16x^{16}y^{-8})^{-\frac{3}{4}} = 16^{-\frac{3}{4}} \cdot x^{16\left(-\frac{3}{4}\right)} \cdot y^{-8\left(-\frac{3}{4}\right)}$$
$$= \frac{1}{16^{\frac{3}{4}}} \cdot x^{-12} \cdot y^6$$
$$= \frac{1}{(\sqrt[4]{16})^3} \cdot \frac{1}{x^{12}} \cdot y^6 = \frac{y^6}{8x^{12}}$$

You can use a scientific calculator to evaluate an expression with a rational exponent. If you know the decimal equivalent for a fractional exponent, use that decimal. If not, use the calculator keys for parentheses.

EXAMPLE 5 **Evaluate to the nearest thousandth.** **a.** $6^{\frac{3}{4}}$ **b.** $18^{\frac{3}{7}}$ **c.** $(-12)^{\frac{3}{5}}$

a. $6^{\frac{3}{4}} \approx 3.834$ — *Think of $6^{\frac{3}{4}}$ as $6^{0.75}$. Use the exponent key (usually ^ or y^x) and the exponent 0.75. Round the result.*

b. $18^{\frac{3}{7}} \approx 3.451$ — *Use the exponent key. Enter the exponent in parentheses: (3 ÷ 7). Round the result.*

c. $(-12)^{\frac{3}{5}} \approx -4.441$ — *When the base is negative the calculator may require this format: (− 12)^3^(1/5)*

CLASS EXERCISES

Tell whether each statement is true or false. If it is false, tell why it is false.

1. $8^{\frac{1}{3}} \cdot 27^{\frac{1}{3}} = (8 \cdot 27)^{\frac{1}{3}}$

2. $27^{\frac{1}{3}} - 8^{\frac{1}{3}} = (27 - 8)^{\frac{1}{3}}$

3. $\sqrt{\sqrt[3]{3}} = 3^{\frac{1}{6}}$

4. $(x^{-4})^{-\frac{1}{4}} = x$

5. $(2 + 2)^{-2} = 2^{-2} + 2^{-2}$

6. $(27^{\frac{1}{3}} + 8^{\frac{1}{3}})^3 = (3 + 2)^3$

7. $-16x^{-\frac{1}{4}} = \dfrac{-2}{x^{\frac{1}{4}}}$

8. $\left(\dfrac{c^2}{27d^{-3}}\right)^{\frac{1}{3}} = \dfrac{c^{\frac{2}{3}}d}{3}$

9. $2 \cdot \left(\dfrac{b^2}{a^3}\right)^{-2} = \dfrac{4a^6}{b^4}$

10. $\left(\dfrac{1}{2}\right)^{-2} \cdot \left(\dfrac{m}{n^{-3}}\right)^2 = 4m^2n^6$

Identify the exponent.

11. $\sqrt{\sqrt{2}} = 2^?$

12. $3\sqrt[3]{9} = 3^?$

13. $\sqrt[3]{\sqrt{11}} = 11^?$

14. $(0.01)^{\frac{1}{2}} = 10^?$

PRACTICE EXERCISES

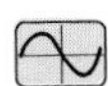

Use technology where appropriate.

Write in radical form.

1. $x^{\frac{1}{6}}$

2. $x^{\frac{1}{5}}$

3. $x^{\frac{2}{7}}$

4. $y^{\frac{3}{5}}$

5. $y^{-\frac{3}{8}}$

6. $t^{-\frac{3}{4}}$

7. $x^{1.5}$

8. $y^{1.2}$

Write in exponential form.

9. $\sqrt{y}$

10. $\sqrt[4]{x}$

11. $\sqrt[5]{x^2}$

12. $(\sqrt[5]{y})^4$

13. $\sqrt[6]{3t}$

14. $\sqrt[5]{8y^{-1}}$

15. $7\sqrt[3]{x}$

16. $-3\sqrt[5]{p}$

Evaluate.

17. $36^{\frac{1}{2}}$ **18.** $27^{\frac{1}{3}}$ **19.** $49^{\frac{1}{2}}$ **20.** $8^{\frac{2}{3}}$

21. $64^{\frac{2}{3}}$ **22.** $32^{\frac{3}{5}}$ **23.** $(-8)^{\frac{2}{3}}$ **24.** $(-32)^{\frac{4}{5}}$

25. $(81)^{-\frac{1}{4}}$ **26.** $(125)^{-\frac{1}{3}}$ **27.** $4^{1.5}$ **28.** $16^{1.5}$

Simplify.

29. $x^{\frac{2}{7}} \cdot x^{\frac{3}{14}}$ **30.** $y^{\frac{1}{2}} \cdot y^{\frac{3}{10}}$ **31.** $x^{\frac{3}{5}} \div x^{\frac{1}{10}}$ **32.** $y^{\frac{5}{7}} \div y^{\frac{3}{14}}$

33. $\dfrac{x^{\frac{2}{3}}y^{-\frac{1}{4}}}{x^{\frac{1}{2}}y^{-\frac{1}{2}}}$ **34.** $\dfrac{x^{\frac{1}{2}}y^{-\frac{1}{3}}}{x^{\frac{3}{4}}y^{\frac{1}{2}}}$ **35.** $\left(\dfrac{16x^{14}}{81y^{18}}\right)^{\frac{1}{2}}$ **36.** $\left(\dfrac{81y^{16}}{16x^{12}}\right)^{\frac{1}{4}}$

37. $(8x^{15}y^{-9})^{-\frac{1}{3}}$ **38.** $(-27x^{-9}y^{6})^{\frac{1}{3}}$ **39.** $(-32x^{-10}y^{15})^{\frac{1}{5}}$ **40.** $(32x^{20}y^{-10})^{-\frac{1}{5}}$

Evaluate each expression to the nearest thousandth.

41. $7^{\frac{2}{5}}$ **42.** $12^{\frac{3}{2}}$ **43.** $8^{\frac{7}{10}}$ **44.** $17^{\frac{5}{6}}$

45. $5^{\frac{4}{7}}$ **46.** $2.93^{\frac{4}{11}}$ **47.** $0.13^{\frac{16}{9}}$ **48.** $0.57^{\frac{15}{4}}$

Evaluate.

49. $(-343)^{\frac{1}{3}}$ **50.** $(-243)^{\frac{1}{5}}$ **51.** $32^{1.2}$ **52.** $243^{1.2}$

53. $64^{3.5}$ **54.** $100^{4.5}$ **55.** $32^{-0.4}$ **56.** $64^{-0.5}$

57. $(4 \cdot 64)^{\frac{3}{4}}$ **58.** $(-6 \cdot 36)^{-\frac{2}{3}}$ **59.** $(-216)^{-\frac{2}{3}}$ **60.** $2(16)^{-\frac{3}{4}}$

61. $3(-343)^{-\frac{2}{3}}$ **62.** $-(-27)^{-\frac{4}{3}}$ **63.** $-(-32)^{-\frac{6}{5}}$ **64.** $(10{,}000)^{-\frac{3}{4}}$

Simplify.

65. $x^{\frac{2}{3}} \cdot x^{\frac{4}{9}}$ **66.** $x^{\frac{5}{8}} \cdot x^{\frac{13}{16}}$ **67.** $(x^{\frac{2}{3}})^{2}$ **68.** $(x^{\frac{4}{7}})^{7}$

69. $(3x^{\frac{2}{3}})^{-1}$ **70.** $5(x^{\frac{2}{3}})^{-1}$ **71.** $5x^{\frac{1}{4}} \cdot 7^{-\frac{2}{3}} \cdot x^{\frac{1}{3}}$ **72.** $8x^{\frac{2}{3}} \cdot 5^{-\frac{1}{2}}x^{\frac{1}{6}}$

73. $(x^{\frac{1}{2}} \cdot y^{\frac{2}{3}})^{-2}$ **74.** $(x^{\frac{2}{3}} \cdot y^{\frac{1}{6}})^{-3}$ **75.** $\left(\dfrac{x^{-\frac{1}{4}}}{y^{-\frac{3}{4}}}\right)^{12}$ **76.** $\left(\dfrac{x^{-\frac{2}{3}}}{y^{-\frac{1}{3}}}\right)^{15}$

Evaluate.

77. $(-243^{\frac{2}{5}}) - (\sqrt{9})^{3}$ **78.** $\left(\dfrac{125}{125^{-\frac{1}{3}}}\right)^{-\frac{1}{4}}$ **79.** $\left(\dfrac{32}{256^{-\frac{1}{2}}}\right)^{\frac{1}{3}}$

80. $(64^{-\frac{1}{6}}) + (\sqrt[3]{8})^{-2}$ **81.** $\sqrt[5]{\sqrt{1024}}$ **82.** $\sqrt{\sqrt[4]{256}}$

Simplify.

83. $(x^{\frac{1}{2}} \cdot x^{\frac{5}{12}}) \div x^{\frac{2}{3}}$

84. $(x^{\frac{3}{4}} \div x^{\frac{7}{8}}) \cdot x^{-\frac{1}{6}}$

85. $[(x^{-\frac{1}{2}})^2]^{\frac{1}{3}}$

86. $[(\sqrt{x^3y^3})^{\frac{1}{3}}]^{-1}$

87. $\sqrt[3]{x \cdot \sqrt{x}}$

88. $\sqrt[4]{\sqrt[5]{\sqrt{x^3}}}$

Applications

89. **Biology** The expression $0.036\ m^{\frac{3}{4}}$ is used in the study of biological fluids. Evaluate the expression for $m = 46 \times 10^4$.

90. **Physics** The expression $PV^{\frac{7}{5}}$ is used to measure gases in thermodynamics. Evaluate the expression for $P = 6$ and $V = 32$.

91. **Archaeology** The number of milligrams of carbon 14 left in a fossil A can be determined by $A = A_0(2)^{-\frac{3}{4}}$. Find A when $A_0 = 120$.

CAREER

Dr. Carolyn Meyers had no idea that engineering careers existed until she was a senior in high school. It was then that she participated in a summer science institute sponsored by the National Science Foundation and the National Aeronautics and Space Administration at what was then the manned spaceflight center in Langley, Virginia. It was there that she had a chance to watch the original seven astronauts in training and received her first exposure to engineering. She remembers, "In school, I loved the problem solving part of mathematics but not the theory part. I discovered that I could make a career out of the fun part by becoming an engineer."

Carolyn attended the School of Engineering and Architecture at Howard University in Washington, D.C., and graduated with a Bachelor of Science degree in Mechanical Engineering. Shortly thereafter, she was employed as a steam generator analyst in Schenectady, NY, and then as a systems engineer in Bethesda, MD.

She briefly left the engineering profession to devote time to her husband and growing family. When all of her children were of school age, she pursued graduate studies at the Georgia Institute of Technology and earned a Master of Science degree in Mechanical Engineering and then a Doctor of Philosophy degree from the School of Chemical Engineering, Metallurgy Program. She now holds the position of tenured Associate Professor of Mechanical Engineering at the George W. Woodruff School of Mechanical Engineering at the Georgia Institute of Technology.

12.2

Real Exponents and Exponential Functions

Objectives: To graph exponential functions
To solve exponential equations

Polynomial functions are widely used as models in problem solving. Another type of function, an *exponential function,* is often particularly useful in the solution of problems in physics, biology, and other sciences.

Capsule Review

To evaluate $f(x)$ for given values of the independent variable x, use substitution.

EXAMPLE **If $f(x) = x^2 + 3x$, find $f(2)$ and $f(-1)$.**

$$f(x) = x^2 + 3x$$
$$f(2) = 2^2 + 3(2) = 4 + 6 = 10$$
$$f(-1) = (-1)^2 + 3(-1) = 1 - 3 = -2$$

Find $f(3)$, $f(-1)$, and $f(k)$ for each function.

1. $f(x) = 2x$
2. $f(x) = 3x + 2$
3. $f(x) = 4x - 3$
4. $f(x) = x^2$
5. $f(x) = -x^2 + 4$
6. $f(x) = 2x^2 - 3x$
7. $f(x) = -5x^3$
8. $f(x) = 6x^3 - 12$
9. $f(x) = 8x^3 - 5x^2$

An equation in which the variable is in the exponent is called an **exponential equation.** Examples of exponential equations are

$$4^x = 16 \qquad 2^x = \frac{1}{8} \qquad y = 3^{2x}$$

If x and b are real numbers with $b > 0$ and $b \neq 1$, then an equation of the form

$$y = b^x$$

defines the **exponential function** $f(x) = b^x$ with base b.

So far, b^x has been defined for rational values of x. It can also be defined for irrational values so that the graph of $y = b^x$ is a continuous curve, as shown in Example 1.

EXAMPLE 1 **Graph: $y = 2^x$**

Make a table of values. Write nonintegral values of y in decimal form. Plot the points and join them with a smooth curve.

x	y
-3	0.125
-2	0.25
-1	0.5
0	1
1	2
2	4
3	8

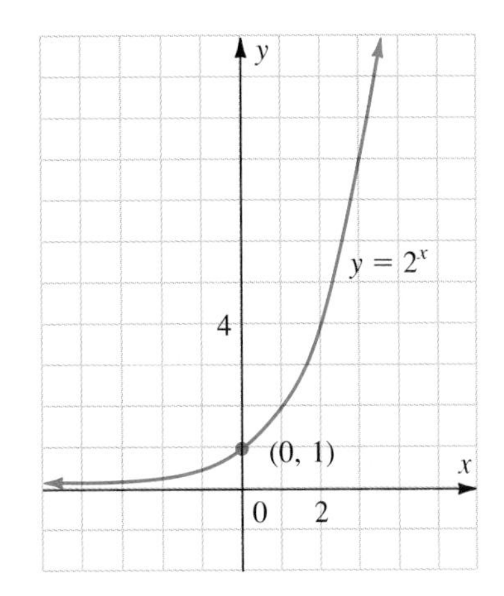

The vertical line test shows that $y = 2^x$ is a function. The domain is $\{x\text{: } x \text{ is a real number}\}$ and the range is $\{y\text{: } y \text{ is a positive real number}\}$. Furthermore, as x increases, y increases. As x decreases, y decreases and the curve gets closer and closer to the negative part of the x-axis but never crosses it. The x-axis is an asymptote of the curve. Also notice that any horizontal line intersects the graph in at most one point, so $y = 2^x$ is a *one-to-one function*.

Since irrational numbers are included in the domain and range of the function $y = 2^x$, many points on the graph have irrational coordinates. For example, there is a point on the graph where $x = \sqrt{2}$ and $y = 2^{\sqrt{2}}$. When x is

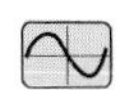

irrational, values of b^x can be approximated with a calculator.

EXAMPLE 2 **Given that $f(x) = 2^x$, find $f(\sqrt{2})$ to the nearest thousandth.**

$f(\sqrt{2}) = 2^{\sqrt{2}}$, so use the exponent key on a scientific calculator to find $2^{\sqrt{2}}$. Many calculators have a square root key that can be used to enter the exponent, $\sqrt{2}$. On other calculators, you must use the inverse key and the x^2 key to enter $\sqrt{2}$.

$$2^{\sqrt{2}} \approx 2.665144143$$

$f(\sqrt{2}) = 2.665$, to the nearest thousandth. The approximate location of the point $(\sqrt{2}, 2^{\sqrt{2}})$ on the graph of $y = 2^x$ is shown at the right.

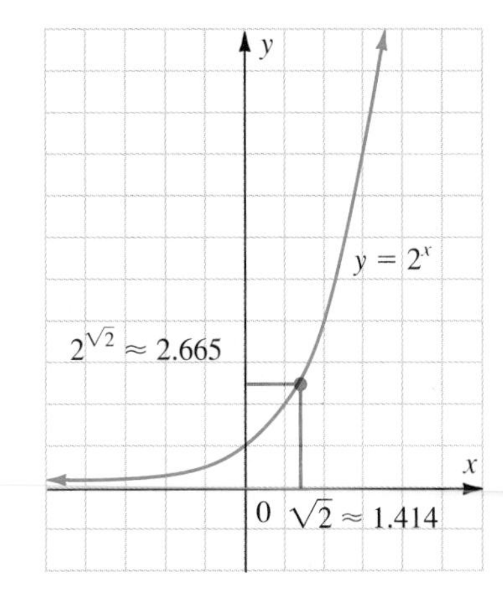

Example 3 shows the graph of an exponential function in which the base b is a positive number less than 1. Explain why $y = 2^{-x}$ and $y = \left(\frac{1}{2}\right)^x$ are equivalent.

EXAMPLE 3 **Graph $y = \left(\frac{1}{2}\right)^x$ using a graphing calculator.**

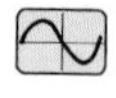

x	y
−3	8
−2	4
−1	2
0	1
1	0.5
2	0.25
3	0.125

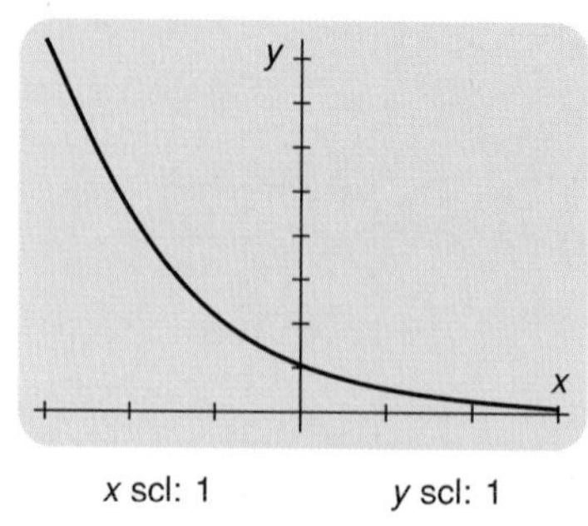

Note that the graph intersects the y-axis at the point (0, 1). This is consistent with the table of values. The function $y = \left(\frac{1}{2}\right)^x$ is a one-to-one function. The domain is $\{x\colon x \text{ is a real number}\}$ and the range is $\{y\colon y \text{ is a positive real number}\}$. Note that as x decreases, y increases. As x increases, y decreases and the curve gets closer and closer to the positive part of the x-axis. The x-axis is an asymptote. In general, the graph of $y = b^x$ rises to the right for $b > 1$ and to the left for $0 < b < 1$.

Properties of the Exponential Function $f(x) = b^x$

- The domain is the set of all real numbers.
- The range is the set of positive real numbers.
- The y-intercept of the graph is 1.
- The x-axis is an asymptote of the graph.
- The function is one-to-one.

All exponential functions are one-to-one, so $b^m = b^p$ if and only if $m = p$. You can use this fact to solve an exponential equation if both sides of the equation can be expressed in terms of the same base.

EXAMPLE 4 **Solve and check:** **a.** $16 = 8^x$ **b.** $6^x = \frac{1}{36}$ **c.** $3^{4x-3} = 27^{2x-4}$

a.
$$16 = 8^x$$
$$2^4 = (2^3)^x$$
$$2^4 = 2^{3x}$$
$$4 = 3x \qquad \textit{If } b^m = b^p, \textit{ then } m = p.$$
$$\frac{4}{3} = x \qquad \text{The solution is } \frac{4}{3}.$$

Check:
$$16 = 8^x$$
$$16 \stackrel{?}{=} 8^{\frac{4}{3}}$$
$$16 \stackrel{?}{=} (\sqrt[3]{8})^4$$
$$16 \stackrel{?}{=} 2^4$$
$$16 = 16 \checkmark$$

b. $6^x = \frac{1}{36}$

$6^x = \frac{1}{6^2}$

$6^x = 6^{-2}$

$x = -2$

The solution is -2.

Check: $6^x = \frac{1}{36}$

$6^{-2} \stackrel{?}{=} \frac{1}{36}$

$\frac{1}{6^2} \stackrel{?}{=} \frac{1}{36}$

$\frac{1}{36} = \frac{1}{36}$ ✓

c. $3^{4x-3} = 27^{2x-4}$

$3^{4x-3} = (3^3)^{2x-4}$

$3^{4x-3} = 3^{6x-12}$

$4x - 3 = 6x - 12$

$-2x = -9$

$x = \frac{9}{2}$

The solution is $\frac{9}{2}$.

Check: $3^{4x-3} = 27^{2x-4}$

$3^{4\cdot\frac{9}{2}-3} \stackrel{?}{=} 27^{2\cdot\frac{9}{2}-4}$

$3^{15} \stackrel{?}{=} 27^5$

$3^{15} \stackrel{?}{=} (3^3)^5$

$3^{15} = 3^{15}$ ✓

CLASS EXERCISES

1. **Thinking Critically** Which of the five properties of an exponential function do *not* hold for the function $y = b^x$ when $b = 1$?

Simplify. Give each answer in simplest exponential form.

2. $2^{3\sqrt{2}} \cdot 2^{-3\sqrt{2}}$
3. $8^{-\sqrt{3}} \cdot 8^{\sqrt{3}}$
4. $5^{1+\sqrt{5}} \cdot 5^{1-\sqrt{5}}$
5. $9^{2-\sqrt{3}} \cdot 9^{2+\sqrt{3}}$
6. $\frac{6^{4\sqrt{3}}}{6^{3\sqrt{3}}}$
7. $\frac{10^{5\pi}}{10^{4\pi}}$
8. $\frac{27^{\sqrt{5}}}{3^{\sqrt{5}}}$
9. $\frac{64^{\pi}}{8^{\pi}}$

Find the value of each function for the given value of *x*. Give the answers to the nearest thousandth.

10. If $f(x) = 3^x$, find $f(\sqrt{2})$.
11. If $f(x) = \left(\frac{1}{2}\right)^x$, find $f(\sqrt{2})$.

PRACTICE EXERCISES

Use technology where appropriate.

Find the value of each function for the given value of *x*. Give the answers to the nearest thousandth.

1. If $f(x) = 5^x$, find $f(\sqrt{2})$.
2. If $f(x) = 5^x$, find $f(\sqrt{5})$.
3. If $f(x) = \left(\frac{1}{10}\right)^x$, find $f(\sqrt{3})$.
4. If $f(x) = \left(\frac{1}{10}\right)^x$, find $f(\sqrt{2})$.

Graph.

5. $y = 5^x$
6. $y = 10^x$
7. $y = \left(\frac{1}{5}\right)^x$
8. $y = \left(\frac{1}{10}\right)^x$

Solve and check.

9. $2^x = 8$
10. $3^x = 9$
11. $5^x = 125$
12. $4^x = 64$
13. $4^x = \frac{1}{16}$
14. $3^x = \frac{1}{27}$
15. $2^x = \frac{1}{2}$
16. $3^x = \frac{1}{9}$
17. $32 = 4^x$
18. $125 = 25^x$
19. $128 = 4^x$
20. $27 = 9^x$
21. $3^{3x-1} = 9^{x-3}$
22. $4^{3x+1} = 16^{x-1}$
23. $5^{2x-2} = 25^{2x+5}$
24. $3^{3x-2} = 27^{3x+2}$

Graph. Use a graphing calculator or computer.

25. $y = 2^{2x}$
26. $y = 3^{2x}$
27. $y = 2^{x+1}$
28. $y = 3^{x-1}$
29. $y = 2^{-x}$
30. $y = 3^{-x}$
31. $y = \left(\frac{1}{2}\right)^{2x}$
32. $y = \left(\frac{1}{3}\right)^{2x}$

Solve and check.

33. $2^{3x+1} = 4^x$
34. $3^{4x-2} = 27^x$
35. $9^{2x+1} = 27^{x+2}$
36. $4^{3x-1} = 8^{x+2}$
37. $100^{5x+2} = 1000^{4x+6}$
38. $1000^{4x} = 10{,}000^{6x+4}$
39. $100^x = 0.001$
40. $1000^x = 0.000001$

Sketch each exponential function.

41. $y = a^x,\ 0 < a < 1$
42. $y = a^x,\ a > 1$

Solve and check.

43. $\left(\frac{1}{3}\right)^{x+6} = 81^{-x}$
44. $9^{2-x} = 27^{\frac{x}{3}}$
45. $a^x = a^3 \cdot a^{2x+1}$
46. $a^{x-4} = a^6 \cdot a^{3x-6}$
47. $(2a)^x = (4a^2)^{x+2}$
48. $(3a)^{x+4} = (27a^3)^{x+6}$

Applications

49. **Sets** The function for the number of subsets y of a set containing x elements is $y = 2^x$. If there are 64 subsets, how many elements are there in the original set?

50. **Number Theory** A number raised to twice a certain power is equal to the same number raised to 3 more than that power. Find the power.

51. **Number Theory** If a number is multiplied by itself and the product is raised to a certain power, the result is equal to the same number raised to 5 less than that power. Find the power.

52. **Biology** The number N of bacteria present in a certain culture after t hours is given by the function $N = 10(2^t)$. Graph this function and determine N when $t = 2$.

53. **Optics** One function used in the study of the optical intensity of objects is $I = 10^{-d}$. Graph this function and find the value of I, to the nearest thousandth, when $d = 1.5$.

54. **Archaeology** Archaeologists can use the formula $P = a^k$ to estimate the population P that once occupied a given site. In this formula, a represents the area of the site, in acres, and k is a constant determined by the geography of the region. Graph the function for $a = 20$, and determine the population for $k = 0.7$.

Developing Mathematical Power

Extension If you wanted to determine the number of direct ancestors you have in a given generation, you might use the exponential function $y = 2^x$. In such a function, the only meaningful domain elements are integers.

Those who measure *continuous* growth or decay often use an exponential function with a base of e, an irrational number approximately equal to 2.718. For example, psychologists have developed a mathematical model to help explain the learning process. The learning curve function is $f(t) = c(1 - e^{-kt})$,

where c is the total number of tasks to be learned
k is the rate of learning
t is the amount of time
and $f(t)$ is the number of tasks learned.

Use the learning curve function to answer the following.

55. Suppose you move to a new school, and you want to learn the names of the 30 classmates in your homeroom. If your learning rate for new tasks is 0.2, how many complete names will you know after 2 days? 8 days? $10\frac{1}{2}$ days?

56. How many days will it take to learn everyone's name? Explain.

57. Does this function seem to describe your own learning rate? If not, how could you adapt it to reflect your learning rate?

12.3 Logarithmic Functions

Objectives: To define and graph logarithmic functions
To evaluate logarithms

The *pH* of a chemical solution is a measure of the concentration of hydrogen ions; it indicates whether the solution is acidic or basic. The *pH* is defined by a special type of function, a *logarithmic function.* A logarithmic function is the inverse of an exponential function.

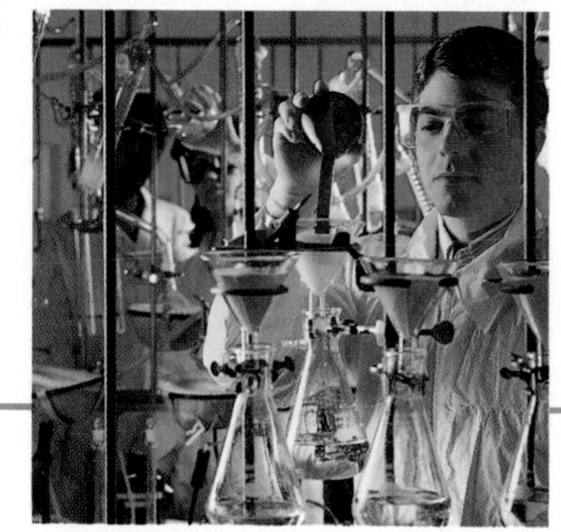

Capsule Review

To find the inverse of a function $y = f(x)$, interchange x and y and solve for y.

EXAMPLE **Find the inverse and state whether the inverse is also a function:**
a. $y = 2x + 1$ **b.** $y = x^2$

a.
$$y = 2x + 1$$
$$x = 2y + 1 \quad \textit{Interchange x and y.}$$
$$x - 1 = 2y \quad \textit{Solve for y.}$$
$$\frac{x-1}{2} = y$$

Since for each value of x there is only one value of y, $y = \dfrac{x-1}{2}$ is a function.

b.
$$y = x^2$$
$$x = y^2$$
$$\pm\sqrt{x} = y$$

Since for each positive value of x there is more than one value of y, $y = \pm\sqrt{x}$ is *not* a function.

Find the inverse of each function. State whether the inverse is a function.

1. $y = 3x$ **2.** $y = 5x - 1$ **3.** $y = 2x^2$

4. $y = x^3$ **5.** $y = x^2 + 1$ **6.** $y = -8x^3$

Since the exponential function $y = b^x$ is a one-to-one function, it has an inverse function. In order to write the inverse function, $x = b^y$, in the form $y = f(x)$, the *logarithmic function* is defined.

For all positive real numbers x and b, $b \neq 1$, the inverse of the exponential function $y = b^x$ is the **logarithmic function,**

$y = \log_b x$ *Read: "y equals the logarithm to the base b of x" or "y equals log base b of x."*

$y = \log_b x$ if and only if $x = b^y$

The graphs of the exponential function $y = b^x$ for $b > 1$ and for $0 < b < 1$ are shown below. Recall that the graph of a function is symmetric to the graph of its inverse about the line $y = x$. Therefore the graph of $y = \log_b x$ is symmetric to the graph of $y = b^x$ about that line. You can use a graphing calculator or a computer to draw graphs of exponential and logarithmic functions.

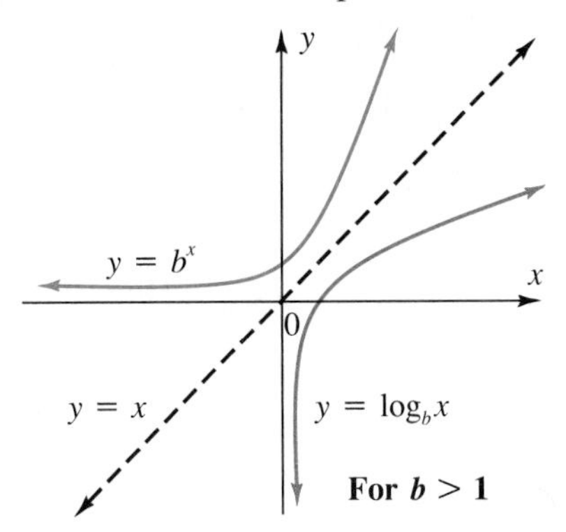

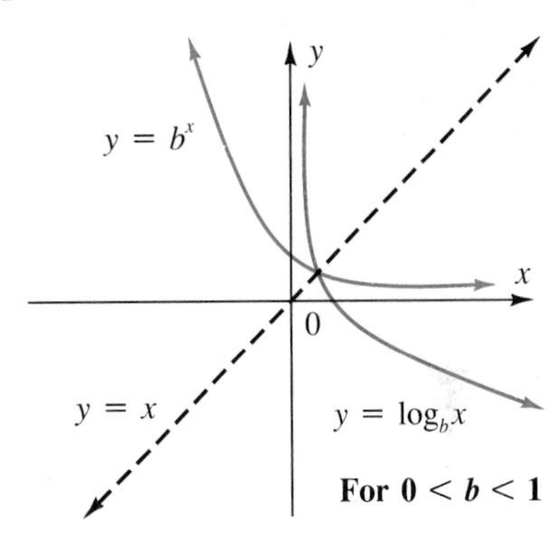

For the logarithmic function with $b > 1$, notice that as x increases, y increases. As x decreases, y decreases and the curve approaches the y-axis, its asymptote. For the logarithmic function with $0 < b < 1$, as x increases, y decreases. As x decreases, y increases and the curve approaches its asymptote, the y-axis. Here is a list of properties of any logarithmic function.

Properties of the Logarithmic Function $f(x) = \log_b x$

- The domain is the set of positive real numbers.
- The range is the set of all real numbers.
- The x-intercept of the graph is 1.
- The y-axis is an asymptote of the graph.
- The function is one-to-one.

Since $y = \log_b x$ is equivalent to $x = b^y$, you can express in exponential form an equation given in logarithmic form.

EXAMPLE 1 **Express in exponential form:**

a. $\log_2 8 = 3$ **b.** $\log_{10} 100 = 2$ **c.** $\log_{10} 0.001 = -3$

a. $\log_2 8 = 3$
$2^3 = 8$

b. $\log_{10} 100 = 2$
$10^2 = 100$

c. $\log_{10} 0.001 = -3$
$10^{-3} = 0.001$

You can also express in logarithmic form an equation given in exponential form.

EXAMPLE 2 **Express in logarithmic form:**

a. $2^4 = 16$ **b.** $16^{\frac{3}{4}} = 8$ **c.** $27^{-\frac{2}{3}} = \frac{1}{9}$ **d.** $10^{\frac{3}{2}} = 10\sqrt{10}$

a. $2^4 = 16$ **b.** $16^{\frac{3}{4}} = 8$ **c.** $27^{-\frac{2}{3}} = \frac{1}{9}$ **d.** $10^{\frac{3}{2}} = 10\sqrt{10}$

$\log_2 16 = 4$ $\log_{16} 8 = \frac{3}{4}$ $\log_{27} \frac{1}{9} = -\frac{2}{3}$ $\log_{10} 10\sqrt{10} = \frac{3}{2}$

To evaluate a logarithm, set it equal to a variable and rewrite the equation in exponential form.

EXAMPLE 3 **Evaluate:** **a.** $\log_6 36$ **b.** $\log_2 \frac{1}{16}$ **c.** $\log_3 3\sqrt{3}$

a. $\log_6 36$

Let $x = \log_6 36$

$$6^x = 36$$
$$6^x = 6^2$$
$$x = 2$$

b. $\log_2 \frac{1}{16}$

Let $y = \log_2 \frac{1}{16}$

$$2^y = \frac{1}{16}$$
$$2^y = \frac{1}{2^4}$$
$$2^y = 2^{-4}$$
$$y = -4$$

c. $\log_3 3\sqrt{3}$

Let $p = \log_3 3\sqrt{3}$

$$3^p = 3\sqrt{3}$$
$$3^p = 3 \cdot 3^{\frac{1}{2}}$$
$$3^p = 3^{\frac{3}{2}}$$
$$p = \frac{3}{2}$$

Some equations involving logarithms can be solved readily if they are first rewritten in exponential form.

EXAMPLE 4 **Solve:** **a.** $\log_x 125 = 3$ **b.** $\log_2 x = 4$ **c.** $\log_3 81 = x$

a. $\log_x 125 = 3$

$$x^3 = 125$$
$$x^3 = 5^3$$
$$x = 5$$

b. $\log_2 x = 4$

$$2^4 = x$$
$$16 = x$$

c. $\log_3 81 = x$

$$3^x = 81$$
$$3^x = 3^4$$
$$x = 4$$

Check to show that the solutions to the equations $\log_x 125 = 3$, $\log_2 x = 4$, and $\log_3 81 = x$ are 5, 16, and 4, respectively.

To make a table of values for the graph of a logarithmic function, use the exponential form. If $y = \log_{10} x$, then $x = 10^y$. Choose values for y and find the corresponding x-values.

$x = 10^y$	y
0.01	−2
0.1	−1
1	0
10	1
100	2

EXAMPLE 5 **Compare the graphs of $y = \log_{10} x$, $y = 5 + \log_{10} x$, and $y = \log_{10}(x + 5)$ using a graphing calculator.**

The graph of $y = 5 + \log_{10} x$ is five units above the graph of $y = \log_{10} x$, and the graph of $y = \log_{10}(x + 5)$ is five units to the left of $y = \log_{10} x$.

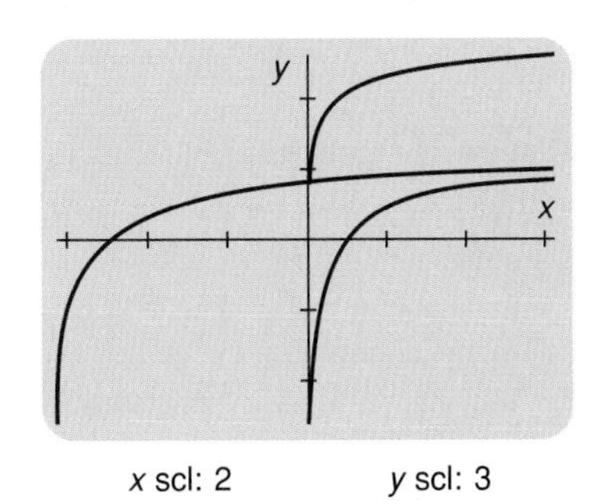

x scl: 2 y scl: 3

CLASS EXERCISES

Thinking Critically

Tell whether each statement is true or false. If false, tell why it is false.

1. The logarithm of a positive number is a real number.
2. Logarithms of negative numbers are defined.
3. If $x > 1$ and $b > 1$, then $\log_b x$ is a positive number.
4. If $0 < x < 1$ and $b > 1$, then $\log_b x$ is a negative number.
5. If $x = b$, then $\log_b x = 0$.
6. If $x = 1$, then the logarithm of x is 0 regardless of the base.
7. The domain of $y = \log_b x$ is the set of real numbers.
8. The range of $y = \log_b x$ is the set of real numbers.

PRACTICE EXERCISES

Use technology where appropriate.

Express in exponential form.

1. $\log_3 27 = 3$ **2.** $\log_6 36 = 2$ **3.** $\log_3 81 = 4$ **4.** $\log_5 125 = 3$

5. $\log_3 \frac{1}{9} = -2$ **6.** $\log_2 \frac{1}{8} = -3$ **7.** $\log_b m = n$ **8.** $\log_a k = 1$

Express in logarithmic form.

9. $3^4 = 81$ **10.** $2^7 = 128$ **11.** $8^{-\frac{2}{3}} = \frac{1}{4}$ **12.** $16^{-\frac{3}{4}} = \frac{1}{8}$

13. $10^{-1} = 0.1$ **14.** $10^{-2} = 0.01$ **15.** $6^{\frac{3}{2}} = 6\sqrt{6}$ **16.** $15^{\frac{3}{2}} = 15\sqrt{15}$

Evaluate.

17. $\log_4 64$ **18.** $\log_3 81$ **19.** $\log_5 125$ **20.** $\log_3 \frac{1}{27}$

21. $\log_2 \frac{1}{32}$ **22.** $\log_{10} \frac{1}{100}$ **23.** $\log_5 5\sqrt{5}$ **24.** $\log_{15} 15\sqrt{15}$

Graph.

25. $y = \log_2 x$

26. $y = \log_{\frac{1}{2}} x$

Solve for x.

27. $\log_{10} 0.001 = x$

28. $\log_{10} 0.1 = x$

29. $\log_5 125 = x + 4$

30. $\log_3 27 = 3x + 6$

31. $\log_4 16 = 2x - 2$

32. $\log_6 36 = 4x - 2$

33. $\log_3 \frac{1}{9} = \frac{x}{3}$

34. $\log_4 \frac{1}{4} = \frac{x}{8}$

35. $\log_7 \frac{1}{49} = -x - 2$

36. $\log_{11} \frac{1}{121} = -x - 4$

37. $\log_{0.1} x = 3$

38. $\log_{0.1} x = -2$

Graph and find the domain and range.

39. $y = \log_5 x$

40. $y = \log_6 x$

41. $y = 1 + \log_{10} x$

42. $y = 3 + \log_{10} x$

43. $y = \log_{10} (x + 1)$

44. $y = \log_{10} (x + 3)$

Solve for x.

45. $\log_{\frac{3}{4}} \frac{9}{16} = 2x$

46. $\log_{\frac{6}{7}} \frac{36}{49} = 2x$

47. $\log_{\frac{3}{5}} \frac{25}{9} = x + 4$

48. $\log_{\frac{2}{3}} \frac{9}{4} = x + 3$

49. $\log_{\frac{3}{2}} \frac{8}{27} = x - 3$

50. $\log_{\frac{2}{3}} \frac{27}{8} = x - 2$

51. $\log_{\frac{2}{3}} \frac{8}{27} = 3x + 6$

52. $\log_{\frac{5}{7}} \frac{125}{343} = 2x + 5$

53. $\log_{\frac{4}{3}} \frac{64}{27} = 2x - 1$

Applications

54. Chemistry The formula $\text{pH} = \log_{10} \frac{1}{\text{H}^+}$ relates the pH of a solution to the concentration of hydrogen ions, H^+, in gram atoms per liter. If the pH of a solution is 5, what is the concentration of hydrogen ions?

55. Physics The electric current i in a circuit is given by the formula $\log_2 i = -t$. Find the current when t is 3 s.

56. Biology A single-celled bacterium doubles every hour. The number N of bacteria after t hours is given by the formula $\log_2 N = t$. After how many hours will there be 32 bacteria?

Mixed Review

Graph each quadratic function. Find the equation of the axis of symmetry and the coordinates of the vertex, and decide whether the vertex is a maximum or minimum point.

57. $y = 2x^2$

58. $y = -x^2 - 2$

59. $y = -\frac{1}{2}x^2 + 1$

60. $y = 3x^2 + 2$

61. Find the sum and product of the solutions of $x^2 - 2x - 12 = 0$.

12.4

Properties of Logarithms

Objectives: To express logarithms in expanded form
To solve equations involving logarithms

Logarithmic functions and exponential functions are inverses, so the properties of logarithms follow directly from the properties of exponents. Keep in mind that logarithms are defined only for positive real numbers and that the base of a logarithm can be any positive real number except 1.

Capsule Review

To simplify expressions involving exponents, use the following properties.

$$a^m \cdot a^n = a^{m+n} \qquad (a^m)^n = a^{mn} \qquad \left(\frac{a}{b}\right)^m = \frac{a^m}{b^m}$$

$$a^m \div a^n = a^{m-n} \qquad (ab)^m = a^m b^m$$

Simplify.

1. $x^2 \cdot x^7$ **2.** $10^6 \cdot 10^5$ **3.** $2^7 \div 2^3$ **4.** $(3^4)^5$ **5.** $(10^3)^2$

6. $\left(\frac{2}{3}\right)^3$ **7.** 10^{-3} **8.** $16^{\frac{1}{8}}$ **9.** $(5y)^4$ **10.** $32^{\frac{3}{4}}$

The following properties can be used to express in different forms the logarithms of products, quotients, and powers of positive real numbers.

Properties of Logarithms

If M, N, and b $(b \neq 1)$ are positive real numbers and r is any real number, then

$$\log_b MN = \log_b M + \log_b N$$

$$\log_b \frac{M}{N} = \log_b M - \log_b N$$

$$\log_b N^r = r \log_b N$$

The expression on the right side of each of the above equations is called the *expanded form* of the logarithm on the left side.

Each of these properties can be derived from the properties of exponents. To derive the first property, $\log_b MN = \log_b M + \log_b N$

Let	$\log_b M = u$	and	$\log_b N = v$
Then	$b^u = M$	and	$b^v = N$
So	$MN = b^u b^v = b^{u+v}$		
Thus	$\log_b MN = u + v$		
Therefore	$\log_b MN = \log_b M + \log_b N$	*Substitute $\log_b M$ for u and $\log_b N$ for v.*	

Derivations of the other two properties are considered in exercises 75 and 76.

EXAMPLE 1 **Express in expanded form:**

a. $\log_2 xy$ **b.** $\log_3 \frac{y}{8}$ **c.** $\log_{10} x^5$ **d.** $\log_a \sqrt[3]{x}$

a. $\log_2 xy = \log_2 x + \log_2 y$

b. $\log_3 \frac{y}{8} = \log_3 y - \log_3 8$

c. $\log_{10} x^5 = 5 \log_{10} x$

d. $\log_a \sqrt[3]{x} = \log_a x^{\frac{1}{3}} = \frac{1}{3} \log_a x$

Sometimes more than one property of logarithms must be used to express a logarithm in expanded form.

EXAMPLE 2 **Express in expanded form:** **a.** $\log_2 x^2 y$ **b.** $\log_{10} \sqrt{\frac{y}{x^3}}$

a. $\log_2 x^2 y = \log_2 x^2 + \log_2 y$ *$\log_b MN = \log_b M + \log_b N$*

$= 2 \log_2 x + \log_2 y$ *$\log_b N^r = r \log_b N$*

b. $\log_{10} \sqrt{\frac{y}{x^3}} = \log_{10} \left(\frac{y}{x^3}\right)^{\frac{1}{2}}$

$= \frac{1}{2} \log_{10} \frac{y}{x^3}$ *$\log_b N^r = r \log_b N$*

$= \frac{1}{2}(\log_{10} y - \log_{10} x^3)$ *$\log_b \frac{M}{N} = \log_b M - \log_b N$*

$= \frac{1}{2}(\log_{10} y - 3 \log_{10} x)$ *$\log_b N^r = r \log_b N$*

If the terms of a logarithmic expression in expanded form have the same base, you can rewrite the expression as a single logarithm.

EXAMPLE 3 **Express as a single logarithm:**

a. $\log_2 x + 5 \log_2 y$ **b.** $\log_4 y - \frac{1}{4} \log_4 x$

a. $\log_2 x + 5 \log_2 y = \log_2 x + \log_2 y^5$ *$r \log_b N = \log_b N^r$*

$= \log_2 xy^5$ *$\log_b M + \log_b N = \log_b MN$*

b. $\log_4 y - \frac{1}{4} \log_4 x = \log_4 y - \log_4 x^{\frac{1}{4}}$ *$r \log_b N = \log_b N^r$*

$= \log_4 \frac{y}{x^{\frac{1}{4}}}$ *$\log_b M - \log_b N = \log_b \frac{M}{N}$*

$= \log_4 \frac{y}{\sqrt[4]{x}}$

For inverse functions recall that $f(f^{-1}(x)) = x$ and $f^{-1}(f(x)) = x$. Since logarithmic functions and exponential functions are inverses, the following properties are true:

$$\log_b b^x = x \quad \text{and} \quad b^{\log_b x} = x$$

Two other important properties of logarithms are

$\log_b b = 1$ *Recall that $b^1 = b$.*

and $\log_b 1 = 0$ *Recall that $b^0 = 1$.*

You can use these properties to evaluate logarithmic expressions.

EXAMPLE 4 **Evaluate:** **a.** $4 \log_2 2 - \log_2 4$ **b.** $\log_{10} 25 + \log_{10} 40$

a. $4 \log_2 2 - \log_2 4 = 4(1) - \log_2 4$ *$\log_b b = 1$*

$= 4 - \log_2 2^2$

$= 4 - 2$ *$\log_b b^x = x$*

$= 2$

b. $\log_{10} 25 + \log_{10} 40 = \log_{10} 25 \cdot 40$

$= \log_{10} 1000$

$= \log_{10} 10^3$

$= 3$ *$\log_b b^x = x$*

Since the logarithmic function is continuous and one-to-one, every positive real number has a unique logarithm to the base b. Therefore

$$\log_b N = \log_b M \quad \text{if and only if} \quad N = M$$

This property can be used to solve *logarithmic equations*. It is important to check all possible solutions, since the logarithm of a variable expression is defined only when the variable expression is positive.

EXAMPLE 5 **Solve:** **a.** $\log_a x = 2\log_a 4 - \log_a 2$ **b.** $\log_2 x + \log_2 (x - 2) = 3$

a.

$$\begin{aligned} \log_a x &= 2\log_a 4 - \log_a 2 \\ \log_a x &= \log_a 4^2 - \log_a 2 \\ \log_a x &= \log_a 16 - \log_a 2 \\ \log_a x &= \log_a \frac{16}{2} \\ \log_a x &= \log_a 8 \\ x &= 8 \end{aligned}$$

The solution is 8.

Check:

$$\begin{aligned} \log_a x &= 2\log_a 4 - \log_a 2 \\ \log_a 8 &\stackrel{?}{=} \log_a 4^2 - \log_a 2 \\ \log_a 8 &\stackrel{?}{=} \log_a 16 - \log_a 2 \\ \log_a 8 &\stackrel{?}{=} \log_a \frac{16}{2} \\ \log_a 8 &= \log_a 8 \; \checkmark \end{aligned}$$

b.

$$\begin{aligned} \log_2 x + \log_2 (x - 2) &= 3 \\ \log_2 x(x - 2) &= 3 \\ \log_2 (x^2 - 2x) &= 3 \\ x^2 - 2x &= 2^3 \\ x^2 - 2x - 8 &= 0 \\ (x + 2)(x - 4) &= 0 \\ x = -2 \quad \text{or} \quad x &= 4 \end{aligned}$$

The solution is 4.

Check: When $x = -2$, $\log_2 x$ is undefined, so -2 is not a solution.

$$\begin{aligned} \log_2 x + \log_2 (x - 2) &= 3 \\ \log_2 4 + \log_2 (4 - 2) &\stackrel{?}{=} 3 \\ \log_2 4 \cdot 2 &\stackrel{?}{=} 3 \\ \log_2 8 &\stackrel{?}{=} 3 \\ 3 &= 3 \; \checkmark \end{aligned}$$

CLASS EXERCISES

Thinking Critically

Tell whether each equation is true or false. If an equation is false, explain why it is false.

1. $\log_2 4 + \log_2 8 = \log_2 2 \cdot 3$
2. $\log_b x^3 y = 3\log_b x + \log_b y$
3. $\log_{10} \sqrt[3]{3x} = \frac{1}{3}\log_{10} 3x$
4. $\log_6 \sqrt[4]{5x} = \frac{1}{4}\log_6 5 + \frac{1}{4}\log_6 x$
5. $\log_3 \sqrt[5]{xy} = \frac{1}{5}\log_3 x + \log_3 y$
6. $\log_2 \frac{x}{y} = \frac{\log_2 x}{\log_2 y}$
7. $\frac{\log_b x}{\log_b y} = \log_b x - \log_b y$
8. $\log_5 \frac{x}{y} = \frac{1}{y}\log_5 x$
9. $\log_3 \sqrt[n]{x^p} = \frac{p}{n}\log_3 x$
10. $\log_2 \sqrt{5x^3} = 3(\log_2 5 + \log_2 x)$

PRACTICE EXERCISES

Express in expanded form.

1. $\log_a xz$
2. $\log_b yz$
3. $\log_4 3y$
4. $\log_{10} 7x$
5. $\log_a \frac{y}{c}$
6. $\log_b \frac{x}{z}$
7. $\log_b \frac{x}{4}$
8. $\log_b \frac{y}{6}$

9. $\log_6 x^3$
10. $\log_3 r^8$
11. $\log_{10} 17^5$
12. $\log_8 19^3$
13. $\log_2 x^3$
14. $\log_3 y^5$
15. $\log_3 \sqrt{5}$
16. $\log_{10} \sqrt[4]{8}$
17. $\log_3 y^3 z$
18. $\log_b p^4 x$
19. $\log_3 (2x)^2$
20. $\log_4 2x^2$
21. $\log_{10} \sqrt{\dfrac{m^3}{n}}$
22. $\log_{10} \sqrt{\dfrac{2x}{y}}$
23. $\log_a x^2 y^3 z$
24. $\log_a 4x^2 y$

Express as a single logarithm.

25. $\log_3 y + 4 \log_3 t$
26. $7 \log_{10} p + \log_{10} q$
27. $5 \log_2 m - \log_2 p$
28. $\log_{10} x - 4 \log_{10} y$
29. $\log_5 x - \frac{1}{5} \log_5 y$
30. $\frac{1}{2} \log_4 p - \log_4 q$

Evaluate.

31. $3 \log_2 2 - \log_2 4$
32. $5 \log_3 3 - \log_3 9$
33. $\log_{10} 1 + \log_{10} 100$
34. $\log_{10} 250 + \log_{10} 40$
35. $\log_2 4 - \log_2 16$
36. $\log_5 5 - \log_5 125$

Solve.

37. $\log_b x = 4 \log_b 2 - \log_b 2$
38. $\log_a x = 2 \log_a 9 - \log_a 3$

Express in expanded form.

39. $\log_b 2x^2 y^3$
40. $\log_b 3m^3 p^2$
41. $\log_b (2xy)^4$
42. $\log_b (4mn)^5$
43. $\log_b \dfrac{x^2}{2y}$
44. $\log_b \dfrac{3x^3}{y^2}$
45. $\log_b \dfrac{(xy)^4}{2}$
46. $\log_b \left(\dfrac{x}{4}\right)^{2y}$
47. $\log_b \sqrt[5]{x^3}$
48. $\log_b \sqrt[7]{y^2}$
49. $\log_b \sqrt[3]{x^4 y^2}$
50. $\log_b \sqrt{8x^3}$
51. $\log_b \dfrac{4\sqrt{x}}{y^2}$
52. $\log_b \dfrac{(\sqrt[5]{x})(y^2)}{z^3}$
53. $\log_b \dfrac{7x^3 y^2}{\sqrt{z}}$
54. $\log_b \dfrac{(9x)^2}{\sqrt[5]{y^3}}$

Express as a single logarithm.

55. $(\log_2 x + \log_2 y) - \log_2 z$
56. $(\log_3 x - \log_3 y) + \log_3 z$
57. $\dfrac{3 \log_2 x}{2}$
58. $\dfrac{5 \log_3 x}{3}$

Solve.

59. $\log_3 x + \log_3 (x - 2) = 1$
60. $\log_4 (x + 6) + \log_4 x = 2$
61. $\log_4 (x + 1) + \log_4 5 = 2$
62. $\log_2 (x - 3) + \log_2 3 = 3$
63. $\log_3 (x^2 - 22) = 3$
64. $\log_4 (x^2 - 17) = 3$

Express in expanded form.

65. $\log_b \frac{(\sqrt{x})(\sqrt[3]{y^2})}{\sqrt[5]{z^2}}$

66. $\log_a \frac{\sqrt{x^5 y^7}}{zw^4}$

67. $\log_3 [(xy)^{\frac{1}{3}} \div z^{\frac{1}{2}}]^3$

68. $\log_2 [(4x)^5 \div \sqrt{5z}]^{\frac{1}{2}}$

Express as a single logarithm, if possible.

69. $\left(\frac{2 \log_a x}{3} + \frac{3 \log_a y}{4}\right) - 5 \log_a z$

70. $\frac{\log_a z - \log_a 3}{4} - 5\left(\frac{\log_a x}{2} + \frac{3 \log_a y}{7}\right)$

71. $\frac{4 \log_a x}{5 \log_a y}$

72. $\frac{\log_a x + \log_a y}{\log_a 2}$

Solve.

73. $4 \log_2 x = 8$

74. $\log_a (x + 4) + \log_a 3 = 0$

75. Using properties of exponents, show that $\log_b \frac{M}{N} = \log_b M - \log_b N$.

76. Using properties of exponents, show that $\log_b N^r = r \log_b N$.

Applications

77. Physics One formula for the decibel level D of sound is $D = \log_{10} I - \log_{10} 10^{-16}$. In this formula, I represents the intensity level in watts per square meter. Find D if $I = 10^{-13}$.

78. Biology The number N of bacteria present in a culture after t hours is given by the formula $\log_{10} N = t(\log_{10} 2 + \log_{10} 4)$. Find N if $t = 1$.

79. Chemistry The pH of a solution is given by the formula $pH = \log_{10} \frac{1}{H^+}$, where H^+ is the concentration of hydrogen ions in gram atoms per liter. Find the pH, if $H^+ = 10^{-7}$.

TEST YOURSELF

Evaluate.

1. $16^{\frac{3}{4}}$ **2.** $64^{-\frac{4}{3}}$ **3.** $x^{\frac{1}{6}} \div x^{\frac{2}{3}}$ **4.** $(x^{\frac{2}{5}})^{-3}$ **12.1**

5. Graph: $y = 2^x$ **6.** Solve: $9^{x+1} = 81$ **12.2**

7. Express in logarithmic form: $3^6 = 729$ **8.** Evaluate: $\log_2 128$ **12.3**

9. Express in expanded form: $\log_2 xy^2$ **10.** Solve: $\log_b \frac{3x - 4}{x + 2} = \log_b 4$ **12.4**

12.5

Evaluating Logarithms

Objectives: To evaluate common logarithms
To evaluate natural logarithms

Before electronic calculators became widely available, logarithms were used to simplify lengthy or difficult calculations. Since our number system is based on powers of 10, base-10 logarithms, or **common logarithms,** were found to be particularly useful, and tables of common logarithms were developed. For a discussion of the use of such tables, see pages 858–859 in the Appendix at the back of this book.

When using a table of common logarithms, you must express numbers in scientific notation. When a calculator is used to find common logarithms, it is not generally necessary to express numbers in this way. It should be kept in mind, however, that a calculator uses scientific notation to display very large and very small numbers.

Capsule Review

A positive number of the form $a \times 10^b$, where $1 \le a < 10$ and b is an integer, is in scientific notation.

Decimal form	Scientific notation
720,000	7.2×10^5
0.00532	5.32×10^{-3}

Express each number in scientific notation.

1. 5820 **2.** 23,000,000 **3.** 0.00462 **4.** 0.0000000007

Express each number in decimal form.

5. 1.24×10^3 **6.** 3.6×10^{-3} **7.** 5.24×10^{13} **8.** 6.27×10^{-1}

When a common logarithm is written, it is customary to omit the base.

$$\log N \text{ means } \log_{10} N$$

You can find the common logarithm of a number by entering the number and using the log key on a scientific calculator. Most logarithms displayed by a calculator are approximations. However, the = symbol, rather than the ≈ symbol, is customarily used.

The approximate value of a base-10 logarithm can be determined by considering the relationship between common logarithms and scientific notation.

$\log 10 = \log (1 \times 10^1) = 1$	$\log 1 = 0$
$\log 100 = \log (1 \times 10^2) = 2$	$\log 0.1 = \log (1 \times 10^{-1}) = -1$
$\log 1000 = \log (1 \times 10^3) = 3$	$\log 0.01 = \log (1 \times 10^{-2}) = -2$
$\log 10{,}000 = \log (1 \times 10^4) = 4$	$\log 0.001 = \log (1 \times 10^{-3}) = -3$

Therefore log 27 is between 1 and 2; log 270 is between 2 and 3; and log 0.0270 is between −2 and −1. Thus the common logarithm of a number can be expressed as the sum of an integer, called the *characteristic,* and a nonnegative number less than 1, called the *mantissa*. So, log 27 = 1.4314.

EXAMPLE 1 **Find each logarithm to four decimal places.**

a. log 6230 **b.** log 0.2

a. log 6230 = 3.794488047 *10-digit display*

To four decimal places, log 6230 = 3.7945.

b. log 0.2 = −0.698970004

To four decimal places, log 0.2 = −0.6990

If log x is known, you can find x. The number x is called the *antilogarithm,* or *antilog,* of log x. 10^x is the inverse of $\log_{10} x$, or log x. If your calculator has a 10^x key, use it to find the antilog of a common logarithm. On other calculators, the inverse or second-function key is used along with the log key.

EXAMPLE 2 **Find x to four significant digits.**

a. $\log x = -3.864$ **b.** $\log x = 12.86$

a. $\log x = -3.864$

$x = 0.000136773$

To four significant digits, $x = 0.0001368$.

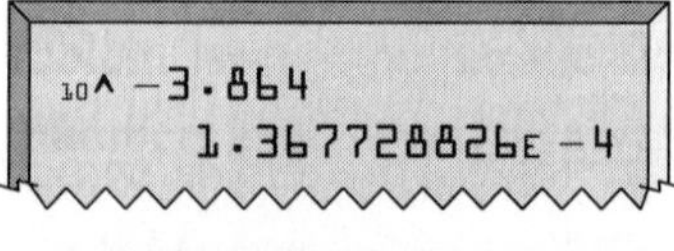

b. $\log x = 12.86$

$x = 7.244359601 \times 10^{12}$

To four significant digits, $x = 7.244 \times 10^{12}$, or $x = 7{,}244{,}000{,}000{,}000$.

Logarithms to another base, e, are used in calculus and are important in mathematical theory and applications. e is an irrational number, and its value can be approximated by evaluating the expression $\left(1 + \frac{1}{n}\right)^n$. As n becomes greater, the value of this expression becomes closer to the value of e.

n	100	1000	100,000	1,000,000
$\left(1 + \frac{1}{n}\right)^n$	2.70481	2.71692	2.71827	2.71828

Correct to 9 significant digits, e is equal to 2.71828183. Logarithms to the base e are called **natural logarithms.** Just as the symbol *log x* is used for common logarithms, the symbol *ln x* is used for natural logarithms.

$$\ln x = \log_e x$$ *ln is read "el en."*

Since natural logarithms are widely used in applications, most scientific calculators have an ln key. To find a natural logarithm, enter the number and use that key.

EXAMPLE 3 **Find each logarithm to four decimal places.**
a. ln 365,200 **b.** ln 0.4209

a. ln 365,200 = 12.80820043

To four decimal places, ln 365,200 = 12.8082.

b. ln 0.4209 = −0.865360003

To four decimal places, ln 0.4209 = −0.8654.

Since e is the base of the natural logarithm, the inverse of $\log_e x$, or $\ln x$, is e^x. To find a number when its natural logarithm is known, enter the given logarithm and use the e^x key or use the inverse and ln keys.

EXAMPLE 4 **Find x to four significant digits.**
a. $\ln x = 2.4802$ **b.** $\ln x = -30.8316$

a. $\ln x = 2.4802$
$x = 11.94365291$

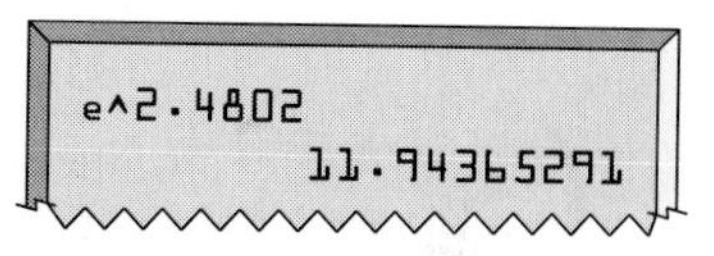

To four significant digits, $x = 11.94$.

b. $\ln x = -30.8316$
$x = 4.073861422 \times 10^{-14}$

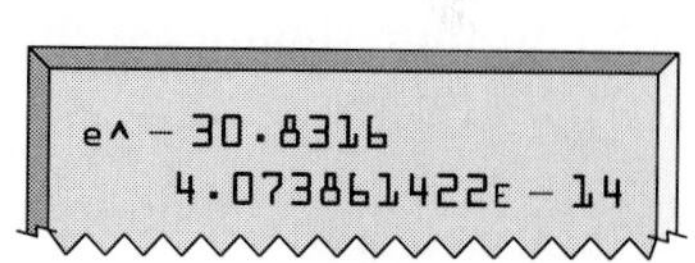

To four significant digits, $x = 4.074 \times 10^{-14}$, or $x = 0.00000000000004074$.

A scientific calculator makes it easy to evaluate more complicated expressions involving logarithms. There are usually several ways you can make the entries. Sometimes you must use the parentheses keys (as suggested in the next example) or the memory feature. If you are unsure about the operation of your calculator, experiment with it to find the sequence of keys that gives the correct results in Example 5. The reference guide or manual that came with your calculator should also be helpful.

EXAMPLE 5 **Evaluate. Give each answer to four decimal places.**

a. $3(\log 15 - \log 11)$ **b.** $\dfrac{\ln 23}{\ln 8 + \ln 5}$

a. $3(\log 15 - \log 11) = 0.404095721$ *Use parentheses keys:* $3 \times (\log 15 - \log 11) = \underline{?}$

To four decimal places, the answer is 0.4041.

b. $\dfrac{\ln 23}{\ln 8 + \ln 5} = 0.849985545$ *Use parentheses. Think:* $\ln 23 \div (\ln 8 + \ln 5) = \underline{?}$

To four decimal places, the answer is 0.8500.

CLASS EXERCISES

Find each logarithm to four decimal places.

1. $\log 34.95$ **2.** $\log 7892$ **3.** $\log 0.002487$ **4.** $\log 0.00002$

5. $\ln 63$ **6.** $\ln 348$ **7.** $\ln 0.085$ **8.** $\ln 0.000088$

Find x to four significant digits.

9. $\log x = 0.8654$ **10.** $\log x = 6.8769$ **11.** $\log x = -10.8794$

12. $\ln x = 0.4526$ **13.** $\ln x = 29.2554$ **14.** $\ln x = -1.7395$

PRACTICE EXERCISES

Use technology where appropriate.

Find each logarithm to four decimal places.

1. $\log 5.92$ **2.** $\log 599{,}000$ **3.** $\log 0.135$ **4.** $\log 0.000093$

5. $\ln 256$ **6.** $\ln 1{,}250{,}000$ **7.** $\ln 0.000093$ **8.** $\ln 0.0777$

Find x to four significant digits.

9. $\log x = 3.9263$ **10.** $\log x = 0.0149$ **11.** $\log x = -0.2236$

12. $\ln x = 1.2004$ **13.** $\ln x = 9.0663$ **14.** $\ln x = -1.3252$

Evaluate. Give each answer to four decimal places.

15. $\log 2 + \log 9$ **16.** $8(\log 8 - \log 5)$ **17.** $\dfrac{\log 4}{\log 6}$

18. $\dfrac{5(\log 12)}{\log 9}$ **19.** $\dfrac{\ln 5 + \ln 19}{\ln 13}$ **20.** $\dfrac{\ln 11}{\ln 32 - \ln 27}$

Find x to four significant digits.

21. $\log x = 13.0125$ **22.** $\log x = 22.3306$ **23.** $\log x = -14.1216$

24. $\ln x = 28.3187$ **25.** $\ln x = -33.1264$ **26.** $\ln x = -40.2055$

Evaluate. Give each answer to four decimal places.

27. $\dfrac{2 \log 16}{\log 4 + \log 14}$

28. $\dfrac{6 \log 20}{\log 36 - \log 34}$

29. $\dfrac{\log 8.4 + \log 2.5}{\log 3.6 - \log 1.5}$

30. $\dfrac{\log 1.4 - \log 0.2}{\log 2.7 + \log 3.8}$

31. $\dfrac{13(\ln 8 + \ln 9)}{2(\ln 10 - \ln 24)}$

32. $\dfrac{3(\ln 5 - \ln 7)}{14(\ln 18 + \ln 45)}$

33. If a calculator has no keys labeled e or e^x, how could you use the inverse and ln keys to find the approximate value of e?

34. For what value of x is the equation $\log x = \ln x$ true?

Evaluate. Give each answer to four decimal places.

35. $\dfrac{(\log 42)^2}{1 + \log 13}$

36. $\dfrac{(3 + \log 2)^2}{\log 5}$

37. $\sqrt{\dfrac{3 + \log 4}{\log 9}}$

38. $\sqrt{\dfrac{\log 7}{2 + \log 16}}$

39. $\dfrac{5 + 5 \ln 5}{\sqrt{\ln 5} - 5}$

40. $\dfrac{2 - 2 \ln 2}{2 + \sqrt{\ln 2}}$

Applications

41. Astronomy The limiting magnitude L of an optical telescope with a lens diameter d inches is $L = 8.8 + 5.1 \log d$. Find the limiting magnitude for an 8-in. reflecting telescope.

42. Business The amount of money at the end of 5 yr in an account where the interest is compounded continuously at a rate r can be determined by $\ln A = \ln 50 + 5r$. Find the amount of money in this account for an interest rate of 8%.

43. Optics The optical intensity I of an object can be determined by $\log I = \log 2.3 - d$ where d is the distance in inches. Find the optical intensity for a distance of 4 in.

Mixed Review

Solve by completing the square.

44. $x^2 + 5x - 1 = 0$

45. $x^2 + 12x - 6 = 0$

Solve using the quadratic formula.

46. $8x^2 - 6x + 1 = 0$

47. $5x^2 - 11x - 4 = 0$

48. Use the discriminant to find the nature of the solutions of $y = -4y^2 - 3y - 4$.

49. Determine the value of a if $y = ax^2$ contains $\left(-2, \frac{3}{2}\right)$.

50. Solve $y^4 + 3y^2 - 28 = 0$.

12.6

Exponential Equations

Objectives: To use logarithms to solve exponential equations
To express the logarithm of a number to one base as the logarithm of that number to another base

In Lesson 12.2, some exponential equations were solved by writing both sides of the equation in terms of the same base.

Capsule Review

EXAMPLE **Solve for x:** **a.** $5^x = 125$ **b.** $2^x = 16$

a. $5^x = 125$
$5^x = 5^3$
$x = 3$

b. $2^x = 16$
$2^x = 2^4$
$x = 4$

Solve for x.

1. $3^x = 27$ **2.** $2^x = 32$ **3.** $5^{x+1} = 625$

4. $7^{x-1} = 343$ **5.** $10^{2x-1} = 1000$ **6.** $4^{x-3} = 64$

Sometimes you cannot write both sides of an exponential equation in terms of the same base. In such cases the solution may be found using logarithms, since

$$\log_b M = \log_b N \text{ if and only if } M = N$$

The first step is to take the logarithm of each side of the equation. Then write the equation in *calculation-ready* form. That is, before the logarithms are evaluated, write the equation with the variable alone on one side.

EXAMPLE 1 **Solve for x to three significant digits.**
a. $2^x = 7$ **b.** $3^{x+1} = 84.2$

a.
$2^x = 7$
$\log 2^x = \log 7$ *Take the common log of each side.*
$x \log 2 = \log 7$ *$\log N^r = r \log N$*
$x = \dfrac{\log 7}{\log 2}$ *Divide both sides by log 2. The equation is now in calculation-ready form.*
$x = 2.807354922$

The solution is 2.81, to three significant digits.

Check: Use your calculator to show that $2^{2.81}$ is approximately equal to 7.

b.

$$3^{x+1} = 84.2$$
$$\log 3^{x+1} = \log 84.2$$
$$(x + 1) \log 3 = \log 84.2$$
$$x \log 3 + \log 3 = \log 84.2 \quad \textit{Distributive property}$$
$$x \log 3 = \log 84.2 - \log 3$$
$$x = \frac{\log 84.2 - \log 3}{\log 3} \quad \textit{Calculation-ready form}$$
$$x = 3.035267917$$

The solution is 3.04, to three significant digits.

Check: If $x = 3.04$, then $3^{x+1} = 3^{4.04}$. Use the y^x key on your calculator to show that $3^{4.04}$ is close to 84. (Small differences result from rounding.)

If you know the logarithm of a number to one base (the *reference base*), you can use the *change-of-base formula* to find the logarithm of that number to some other base.

Change-of-Base Formula

$$\log_b N = \frac{\log_a N}{\log_a b} \quad \textit{The reference base is a.}$$

This formula is derived as follows.

Let $\log_b N = x$

Then $b^x = N$ *Definition of a log function*

$$\log_a b^x = \log_a N \quad \textit{Take the log to base a of each side.}$$
$$x \log_a b = \log_a N$$
$$x = \frac{\log_a N}{\log_a b}$$
$$\log_b N = \frac{\log_a N}{\log_a b} \quad \textit{Replace x with } \log_b N.$$

Since a scientific calculator can be used to find the approximate values of common logarithms, it is convenient to use 10 as the reference base.

EXAMPLE 2 **Find the logarithm to three significant digits: $\log_3 6$**

Let $N = 6$, $b = 3$, and $a = 10$.

$$\log_3 6 = \frac{\log_{10} 6}{\log_{10} 3} = 1.630929754 \qquad \log_3 6 = 1.63, \text{ to three significant digits.}$$

CLASS EXERCISES

Thinking Critically

1. Suppose that a scientific calculator and natural logarithms, instead of common logarithms, were used in Examples 1 and 2. Would the results have been the same? Could logarithms to some base other than 10 or e have been used? Explain.

Express each exponential equation in calculation-ready form. Do *not* solve.

2. $3^x = 5$	**3.** $7^x = 15$	**4.** $2^{x+1} = 22$
5. $3^{x-1} = 34$	**6.** $5^x = 62.5$	**7.** $6^{x-1} = 51.3$
8. $7^{2x+1} = 436$	**9.** $11^{3x-2} = 7982$	**10.** $13^{4x-3} = 8793$

Express each logarithm as the quotient of two common logarithms. Do *not* compute.

11. $\log_2 7$	**12.** $\log_3 8$	**13.** $\log_5 14$
14. $\log_4 32$	**15.** $\log_3 123$	**16.** $\log_9 456$

PRACTICE EXERCISES

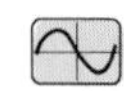 Use technology where appropriate.

For the exercises in this lesson, give each answer to three significant digits.

Solve and check.

1. $2^x = 5$	**2.** $3^x = 4$	**3.** $4^x = 19$
4. $4^x = 47$	**5.** $5^x = 81.2$	**6.** $3^x = 27.3$
7. $4^{2x} = 17$	**8.** $6^{2x} = 21$	**9.** $5^{x+1} = 24$
10. $7^{x+2} = 54$	**11.** $3^{x+4} = 101$	**12.** $4^{x-2} = 89$
13. $5^{x-3} = 75$	**14.** $6^{3x+1} = 243$	**15.** $7^{2x-1} = 316$

Use the change-of-base formula to find each logarithm.

16. $\log_2 9$	**17.** $\log_4 8$	**18.** $\log_3 54$
19. $\log_5 62$	**20.** $\log_4 72$	**21.** $\log_6 120$
22. $\log_3 29$	**23.** $\log_5 63$	**24.** $\log_6 84$

Solve and check.

25. $1.3^x = 7$	**26.** $2.1^x = 9$	**27.** $4.2^x = 93.1$
28. $6.3^x = 69.5$	**29.** $4^{3x} = 77.2$	**30.** $5^{3x} = 61.4$

31. $6^{x-1} = 71.3$

32. $5^{2x+1} = 79.6$

33. $3^{3x+1} = 41.7$

34. $2.3^{x} + 0.7 = 4.9$

35. $3.1^{x} - 0.2 = 6.7$

36. $1.4^{2x} - 15 = 72$

Use the change-of-base formula to solve for x.

37. $x = \log_3 72.4$

38. $x = \log_2 834$

39. $x = \log_2 6 + \log_3 5$

40. $x = \log_4 7 - \log_3 8$

Solve and check.

41. $(29.3)^{4x-1} = (17.3)^{5x+2}$

42. $(3^{x})(5^{2x}) = (271)^{x+1}$

43. $7^{3x} \div 2^{7x-1} = (43.2)^{x+2}$

44. $(5.3)^{x^2} = (22.1)^{x}$

Use the change-of-base formula to find each logarithm.

45. $\log_{3.14} 0.0072$

46. $\log_{6.51} (8 \times 10^6)$

47. $\log_{\sqrt{7}} \sqrt{5}$

48. Prove the change-of-base formula in another way. *Hint:* Start with the true equation $b^{\log_b N} = N$.

Applications

49. **Biology** A biologist wants to determine the number of hours t needed for a given culture to grow to 452 bacteria. If the number N of bacteria in the culture is given by the formula $N = 8(2)^t$, find t.

50. **Electricity** The current i in a certain circuit is given by the formula $i = 2^{-t}$, where t is time measured in seconds. Find t for a current of 0.3 amperes.

51. **Physics** At a constant temperature, the atmospheric pressure p, in pascals, is given by the formula $p = 101.3e^{-0.001h}$, where h is the altitude in meters. Find h when p is 74.3 pascals. *Hint:* Start by taking the natural logarithms of the expressions on each side of the equation.

Developing Mathematical Power

Extension Consider the following functions that are neither polynomial nor exponential. Make a table of values for each function and sketch its graph, if possible. Check your graph on a computer or graphing calculator.

52. $y = x^x$

53. $y = \sqrt[x]{x}$

54. $y = (-x)^x$

12.7

Problem Solving Strategy: Use Exponential and Logarithmic Equations

When money is deposited in a bank account, the bank pays the depositor interest. Interest may be compounded periodically, such as semiannually, quarterly, or daily. The following formula gives the total amount of money in the account after a given period of time.

Compound Interest Formula

$$A = P\left(1 + \frac{r}{n}\right)^{nt}$$

A = total amount (interest + principal), in dollars
P = principal invested, in dollars
r = rate of annual interest, expressed as a decimal
n = number of times a year interest is compounded
t = number of years principal is invested

EXAMPLE 1 *Banking* Rita puts \$3000 in a bank account at 8% annual interest, compounded semiannually. How long will it take to earn \$1800 interest?

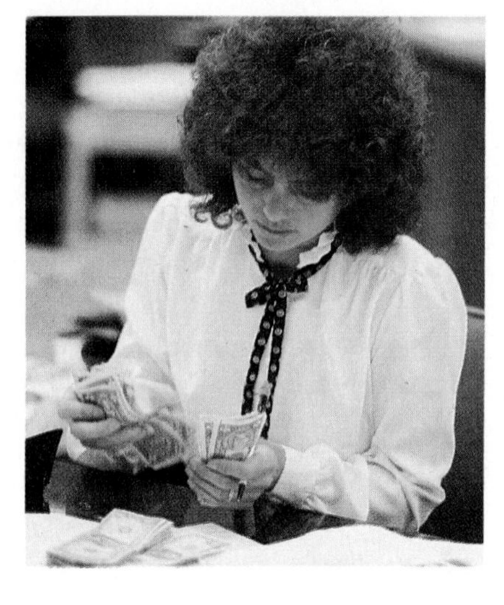

Understand the Problem *What are the given facts?* The principal is \$3000, the interest rate is 8%, or 0.08, and the interest is compounded semiannually, or twice a year. *What are you asked to find?* The number of years, t, it will take to earn \$1800. The amount in the account after t years is \$3000 + \$1800, or \$4800.

Plan Your Approach *Choose a strategy.* Use the compound interest formula.

$$A = P\left(1 + \frac{r}{n}\right)^{nt}$$ Assign known values to the variables. $A = 4800$, $P = 3000$, $r = 0.08$, $n = 2$

$$4800 = 3000\left(1 + \frac{0.08}{2}\right)^{2t}$$ *Substitute in the formula.*

Complete the Work

$$1.6 = (1 + 0.04)^{2t}$$ *Divide both sides by 3000.*

$$1.6 = 1.04^{2t}$$

$$\log 1.6 = \log 1.04^{2t}$$

$$\log 1.6 = 2t \log 1.04$$

$$t = \frac{\log 1.6}{2 \log 1.04} = 5.991778214$$ *Calculation-ready form*

Interpret the Results *State your answer.* It will take almost 6 years for the account to earn \$1800.

Check your answer. To check that the answer is reasonable, note that the *simple interest* for 1 year on \$3000 invested at 8% would be $0.08 \times \$3000$, or \$240. Therefore, it would take $1800 \div 240$, or 7.5 years to earn \$1800 *simple interest* at the given rate. Since compounding makes the interest grow faster, 6 years is reasonable.

In 1947, the chemist Willard Libby made an important discovery that led to the use of radiocarbon dating to help determine the age of organic matter such as dinosaur bones. By measuring the amount of the radioactive isotope carbon-14 that is left in an organism, and applying a formula called the *exponential decay formula,* it is possible to approximate how many years earlier an organism died.

Exponential Decay Formula

$$A = A_0 2^{-\frac{t}{k}}$$

A = present amount of the radioactive isotope

A_0 = original amount of the radioactive isotope, measured in the same units as A

t = time it takes to reduce original amount of the isotope to present amount

k = half-life of the isotope, measured in the same units as t

The half-life is the average time required for half the atoms of a sample of a radioactive substance to decay.

EXAMPLE 2 *Archaeology* A bone that originally contained 150 mg of carbon-14 now contains 85 mg of that isotope. Determine the age of the bone, to the nearest 100 years, if the half-life of carbon-14 is 5570 years.

Understand the Problem You are asked to find the approximate age of the bone. Use the exponential decay formula.

Plan Your Approach

$A = 85 \qquad A_0 = 150 \qquad k = 5570$ *Assign known values to the variables.*

$$A = A_0 2^{-\frac{t}{k}}$$

$$85 = 150(2)^{-\frac{t}{5570}}$$ *Substitute in the formula.*

Complete the Work

$$\log 85 = \log 150(2)^{-\frac{t}{5570}}$$

$$\log 85 = \log 150 - \frac{t}{5570}\log 2$$

$$\frac{t}{5570}\log 2 = \log 150 - \log 85$$

$$t = \frac{5570(\log 150 - \log 85)}{\log 2}$$ *Calculation-ready form*

$$t = 4564.212592$$

Interpret the Results Rounded to the nearest hundred, the bone is approximately 4600 years old. Since there are now 85 mg of carbon-14 in the bone, it has lost somewhat less than one-half of its carbon-14. The half-life of carbon 14 is 5570 years; therefore, 4600 years is a reasonable answer.

Experimentation has shown that the decibel voltage gain (or loss) of an amplifier or a length of transmission line can be calculated using a formula involving logarithms.

Decibel voltage gain = 10 log $\frac{E_O}{E_I}$ E_O = the output voltage
E_I = the input voltage

EXAMPLE 3 If the input to an amplifier is 0.6 volts and the output is 40 volts, find the decibel voltage gain.

Understand the Problem You are asked to find the decibel voltage gain. Use the formula.

$$\text{Decibel voltage gain} = 10 \log \frac{E_O}{E_I}$$

Plan Your Approach $E_O = 40$ $E_I = 0.6$ *Assign known values to the variables.*

$$\text{Decibel voltage gain} = 10 \log \frac{40}{0.6}$$ *Substitute in the formula.*

Complete the Work

$$= 18.23908741$$

Interpret the Results The amplifier provides an 18.2 decibel voltage gain, to the nearest tenth. One way to check the calculations is to use a different calculation-ready form and see if the results are the same.

$$10 \log \frac{40}{0.6} = 10(\log 40 - \log 0.6)$$

Use your calculator to show that the expression on the right has the same value as the one on the left.

CLASS EXERCISES

For each problem, write an equation that can be used to solve it. Do not solve the problem.

1. Bill invested $10,000 at 6.75% annual interest, compounded quarterly. How long will it take the investment to double in value?
2. At 6% annual interest, compounded daily, how long will it take for a $100 investment to be worth $300?

3. A radioactive isotope has a half-life of 4 days. How many days will it take for a 10-g sample of the isotope to decay to 3 g?
4. About how old is a bone that originally contained 200 g of carbon-14 and now contains 80 g of that isotope?
5. What is the decibel voltage gain of an amplifier with an output voltage of 48 volts and an input voltage of 0.8 volts?

PRACTICE EXERCISES

Use technology where appropriate.

Solve. Round answers to the nearest tenth, unless otherwise specified.

1. Kevin deposits $1000 at 9% annual interest, compounded quarterly. How long will it take for the account to earn $1000 interest?
2. Diane invests $4000 at 10% annual interest, compounded monthly. How long will it take for the account to earn $1000 interest?
3. Jon invests $2300 at 7.5% annual interest, compounded semiannually. How long will it take this money to double in value?
4. The Lichtfoots put $8500 into an investment plan paying 8.5% annual interest, compounded quarterly. How long will it take this money to double in value?
5. A sample of a bone found at an archaeological site contains 50 mg of carbon-14. The sample originally contained 175 mg of that isotope, and the half-life of carbon-14 is 5570 years. Find the approximate age of the bone, to the nearest 100 years.
6. A fossil sample now contains 60 mg of carbon-14. There were originally 133 mg of carbon-14 in the sample. Find the approximate age of the fossil, to the nearest 100 years.
7. An isotope has a half-life of 80 days. How many days will it take for a 7-mg sample of this isotope to decay to 1 mg? Round the answer to the nearest whole number.
8. An isotope has a half-life of 90 days. How many days will it take for a 5-g sample of this isotope to decay to 1 g? Round the answer to the nearest whole number.
9. The voltage input of an amplifier is 0.5 volts and its voltage output is 52 volts. Find the decibel voltage gain.
10. If the voltage output of an amplifier is 46 volts and its voltage input is 0.8 volts, what is the decibel voltage gain?

11. A transmission line has an input voltage of 0.6 volts and an output voltage of 54 volts. Find the decibel voltage gain.

12. A transmission line has an input voltage of 0.4 volts and an output voltage of 29 volts. Find the decibel voltage gain.

13. A technician found 11 mg of a radioactive isotope in a soil sample. After 4 hours another measurement was made, and there was just 7.3 mg of the isotope. Find the approximate half-life of the isotope.

14. If 6.00 g of radon decays to 2.28 g in 5.3 days, what is the approximate half-life of radon?

15. An amplifier delivers a 33-decibel voltage gain. If the input voltage is 0.7 volts, find the output voltage.

16. If an amplifier delivers a 29-decibel voltage gain and has an output voltage of 240 volts, what is the input voltage?

17. The following formula is used when interest is compounded *continuously:*

$$A = Pe^{rt}$$

P is the principal, A is the total amount (principal + interest), r is the annual rate (expressed as a decimal), and t is the number of years the principal is invested.

If $1000 is deposited at an annual rate of 8.25% compounded continuously, how long will it take for the account to double in value? *Hint:* Substitute in the formula and then take the natural logarithms of the expressions on the two sides of the equation.

18. Marla deposited $8000 in an account for which interest is compounded continuously. At the end of 1 year, interest in the amount of $410.17 had been credited to the account. Use the formula in Exercise 17 to find the annual interest rate as a percent.

19. The Richter Scale gives the magnitude of an earthquake as the common logarithm of the ratio of the intensity I of its shock waves to a minimum intensity I_0 that is a constant used for comparison.

$$R = \log \frac{I}{I_0}$$

How many times more intense was the 1971 Los Angeles earthquake ($R = 6.7$) than the 1987 Los Angeles earthquake ($R = 6.0$)?

20. A scientist is given a piece of fossilized tree trunk that is thought to be over 5000 years old. The scientist determines that the sample contains 65% of the original amount of carbon-14. Is the reputed age of the tree correct? Justify your answer.

Mixed Problem Solving Review

1. The department of parks and recreation in a town doubles the area of a rectangular park by adding strips of equal width to one end and one side. Find the original dimensions of the park if it now measures 160 m by 120 m.

2. The circumference of a bicycle tire is twice the circumference of an automobile tire. When the car and bicycle travel 100 ft, the car tire makes 15 more revolutions than the bicycle tire. Find the radius of each tire in inches.

3. A worker laying telephone lines throws a stone out of a ditch, releasing it at ground level. The stone achieves a maximum height of 3 ft and lands 6 ft away from the ditch. Assume the stone travels in a parabolic path. Place the origin of a coordinate system at the point of release of the stone and find an equation that describes its path.

PROJECT

An interesting field of study is the mathematics of finance and investments. See if you can find any books in your school library or a public library on finance or investments. Write a brief essay on the meaning of the future value and present value of money. Include in your essay a discussion of *annuity*. Include any mathematical formulas that are pertinent to your discussion of the subject.

TEST YOURSELF

Find each logarithm to four decimal places.

1. log 6270 **2.** log 0.000025 **3.** ln 1.08 **4.** ln 0.00417 — 12.5

Find x to four significant digits.

5. $\log x = 1.2671$ **6.** $\log x = -16.0304$ **7.** $\ln x = 27.3363$

Solve for x to three significant digits.

8. $4^x = 28.2$ **9.** $3^{2x} = 473$ **10.** $5^{2x+3} = 9.74$ — 12.6

Use the change-of-base formula to find each logarithm to three significant digits.

11. $\log_3 7$ **12.** $\log_9 46$ **13.** $\log_4 348$

14. Ted Yokamati invested \$5000 at 7.5% interest, compounded quarterly. To the nearest tenth, how many years will it take for the principal to earn \$2000 interest? — 12.7

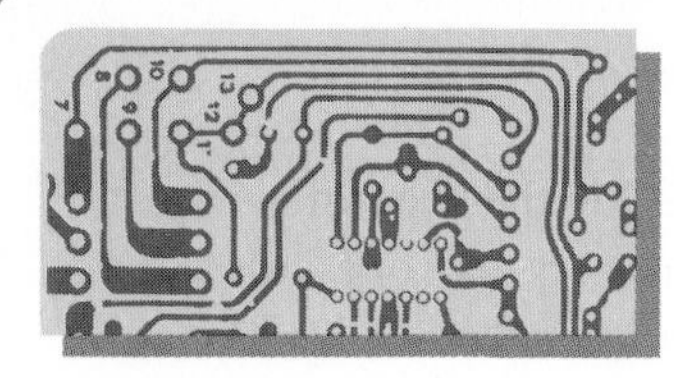

TECHNOLOGY: Exponential Decay Formula

Did you know that some nuclear wastes take 24,000 years to be reduced to half their original volume?

Archaeologists use their knowledge of radioactive half-life to determine the approximate ages of fossils. The same concepts can be applied to the safe management of nuclear material.

Nuclear engineers use the exponential decay formula to determine how long nuclear wastes must be stored before the amount of radioactivity is negligible.

The exponential decay formula is

$$A = A_0 2^{-\frac{t}{k}}$$

where A_0 is the original amount of a given substance, A is the amount that remains after decay, k is the half-life of the substance, and t is the time in years.

EXAMPLE A typical nuclear power plant produces about 10 lb of Krypton-85 per year. The half-life of Krypton-85 is 11 yr. How long must the Krypton be contained so that only 0.1 lb will remain?

$$A = A_0 2^{-\frac{t}{k}}$$ *Use the exponential decay formula.*

where $A = 0.1$, $A_0 = 10$, and $k = 11$

$$0.1 = 10(2)^{-\frac{t}{11}}$$ *Substitute in the formula.*

$$\log 0.1 = \log 10(2)^{-\frac{t}{11}}$$

$$\log 0.1 = \log 10 - \frac{t}{11} \log 2$$

$$\frac{t}{11} \log 2 = \log 10 - \log 0.1$$

$$t = \frac{11(\log 10 - \log 0.1)}{\log 2}$$ *Calculation-ready form*

$$t = 73.082417$$ The Krypton must be contained about 73 yr.

The exponential decay formula lends itself well to the iterative features of a calculator. By entering the exponential decay formula as a function, using a given half-life and the period of a year, you can see how quickly a particular isotope decays.

In the previous example, $k = 11$. If you wish to iterate once for each year, use $t = 1$. Store the function $y = x(2^{-\frac{1}{11}})$, or Y1=X*2^(−1/11).

After storing the original weight in X as 10, store the new value of Y_1 in X. Repeatedly pressing the ENTER or EXE key will result in a display of the amounts of Krypton remaining, year by year.

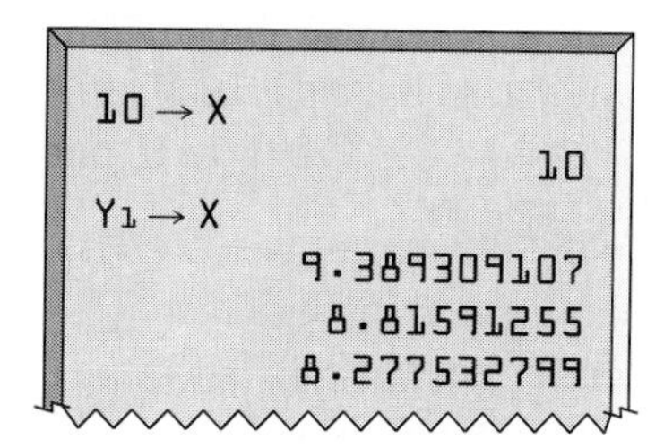

EXERCISES

1. Approximately 4,000,000 curies (a measure of the quantity of a radioactive gas) of Hydrogen-3 were released by nuclear power plants in a recent year. How long will it take for this quantity of Hydrogen-3 gas (half-life = 12 yr) to be reduced to 31,250 curies by decay?

2. Another nuclear waste, Plutonium-239, has a half-life of 24,000 yr. A rule of thumb is that radioactive wastes are virtually harmless after 10 half-lives. How long must 1 gram of Plutonium be securely stored before it is virtually harmless? How much of the Plutonium will remain at that time?

3. Some radioactive waste products of nuclear plants have half-lives of about 30 yr. If a stockpile of 120 m^3 of these nuclear wastes has accumulated at a given time, how much will be present 30 yr later? How much will be present 300 yr later?

4. A 50-gram sample of radium decays to 5 grams in approximately 5615 yr. What is the half-life of this substance?

5. After 100 yr of storage, the nuclear wastes of a nuclear plant have diminished to 247 m^3. The half-life of the waste is 40 yr. Find the original amount of waste that was stored.

6. Another example of exponential decay is the decrease in atmospheric pressure with increasing height above sea level. In the formula $P = P_0 2^{-\frac{h}{4795}}$, where P is the atmospheric pressure (millimeters of mercury) at height h above sea level (meters). If the atmospheric pressure is 42 mm at a height of 20 km, what is the pressure P_0 at sea level?

7. Use the result of Exercise 6 to find the atmospheric pressure at 50 km above sea level.

8. Use the result of Exercise 6 to find the atmospheric pressure at 10 km above sea level.

9. Use the iterative capability of a calculator to find the remaining amounts of a 25-lb quantity of Krypton-85 after 10, 20, 30, 40, and 50 years.

CHAPTER 12 SUMMARY AND REVIEW

Vocabulary

change-of-base formula (599)
common logarithm (593)
exponential equation (576)
exponential function (576)
logarithmic function (583)
natural logarithm (595)
rational exponent (570)

Rational Exponents If m and n are positive integers, and $\sqrt[n]{a}$ is a real number, then **12.1**

$$a^{\frac{1}{n}} = \sqrt[n]{a} \quad \text{and} \quad a^{\frac{m}{n}} = \sqrt[n]{a^m} = (\sqrt[n]{a})^m$$

Evaluate.

1. $(-8)^{\frac{1}{3}}$ **2.** $(16)^{\frac{1}{2}}$ **3.** $(81)^{-\frac{3}{4}}$ **4.** $\left(\frac{4}{9}\right)^{-\frac{1}{2}}$

Simplify.

5. $x^{\frac{1}{5}} \cdot x^{\frac{3}{5}}$ **6.** $x^{-\frac{1}{4}} \cdot x^{\frac{3}{8}}$ **7.** $x^{\frac{2}{7}} \div x^{\frac{9}{14}}$ **8.** $(x^4)^{-\frac{1}{2}}$

Exponential Function If x and b are real numbers with $b > 0$ and $b \neq 1$, then an equation of the form $y = b^x$ defines the exponential function $f(x) = b^x$ with base b. The graph of $y = b^x$ rises to the right for $b > 1$ and to the left for $0 < b < 1$. The x-axis is an asymptote. **12.2**

Graph.

9. $y = 4^x$ **10.** $y = \left(\frac{1}{3}\right)^x$

Solve and check.

11. $2^x = 32$ **12.** $9^x = 27$ **13.** $2^{2x+1} = 8$ **14.** $4^x = \frac{1}{16}$

Logarithmic Function For all positive real numbers x and b, $b \neq 1$, the inverse of the exponential function $y = b^x$ is the logarithmic function $y = \log_b x$. $y = \log_b x$ if and only if $x = b^y$. **12.3**

Express in exponential form.

15. $\log_2 16 = 4$ **16.** $\log_6 36 = 2$

Express in logarithmic form.

17. $5^3 = 125$

18. $3^{-3} = \frac{1}{27}$

Evaluate.

19. $\log_4 64$

20. $\log_2 64$

21. $\log_3 \frac{1}{81}$

22. $\log_4 \frac{1}{16}$

Graph.

23. $y = \log_2 x$

24. $y = \log_5 x$

Properties of Logarithms — 12.4

$$\log_b MN = \log_b M + \log_b N$$

$$\log_b \frac{M}{N} = \log_b M - \log_b N$$

$$\log_b N^r = r \log_b N$$

Express in expanded form.

25. $\log_b x^2y$

26. $\log_b \frac{3x}{y}$

27. $\log_b \sqrt[4]{xy}$

Find each logarithm to four decimal places. — 12.5

28. $\log 0.043$

29. $\log 782{,}000$

30. $\ln 1.04$

Find x to four significant digits.

31. $\log x = 8.2237$

32. $\log x = -11.1064$

33. $\ln x = 33.2691$

Solve for x to three significant digits. — 12.6

34. $3^x = 54.2$

35. $4^{2x} = 671$

36. $7^{2x+1} = 8.93$

Change-of-Base Formula

$$\log_b x = \frac{\log_a x}{\log_a b}$$

Use the change-of-base formula to find each logarithm to three significant digits.

37. $\log_2 9$

38. $\log_7 74$

39. $\log_5 743$

Solve. — 12.7

40. Approximate the age of a bone that now contains 84 g of carbon-14 if it originally contained 192 g of that isotope. The half-life of carbon-14 is 5570 yr. Round the answer to the nearest 100 years.

CHAPTER 12 TEST

Evaluate.

1. $(27)^{\frac{1}{3}}$

2. $(-32)^{\frac{1}{5}}$

3. $(16)^{-\frac{3}{4}}$

Simplify.

4. $x^{\frac{1}{7}} \cdot x^{-\frac{3}{7}}$

5. $x^{\frac{2}{5}} \div x^{\frac{7}{10}}$

6. $(x^3)^{-\frac{1}{2}}$

7. Graph: $y = 3^x$

8. Solve: $4^{3x} = 16^{x+2}$

9. Write in logarithmic form: $2^{-4} = \frac{1}{16}$

10. Evaluate: $\log_3 27$

Express in expanded form.

11. $\log 2xy$

12. $\log \frac{x^2}{y}$

13. $\log \sqrt[6]{3x}$

Solve for *x* to four significant digits.

14. $\log x = 1.6675$

15. $\ln x = 2.1281$

16. $\log x = -12.8852$

Solve for *x* to three significant digits.

17. $4^x = 275$

18. $2^{2x-1} = 3.2$

19. Use the change-of-base formula to find $\log_4 245$ to three significant digits.

Solve.

20. A company invests \$10,000 at 8% annual interest, compounded semiannually. How long will it take for the principal to earn \$5000 interest? Use the compound interest formula, $A = P\left(1 + \frac{r}{n}\right)^{nt}$, and round the answer to the nearest tenth of a year.

Challenge

Ann invests some money at 10% annual interest, compounded *annually*. Bob invests the same amount of money at 9.75% annual interest, compounded *daily*. Which person's investment will take less time to double? Assume that there are 365 days in a year.

COLLEGE ENTRANCE EXAM REVIEW

In each item you are to compare a quantity in Column 1 with a quantity in Column 2. Write the letter of the correct answer from these choices:

A. The quantity in Column 1 is greater than the quantity in Column 2.
B. The quantity in Column 2 is greater than the quantity in Column 1.
C. The quantity in Column 1 is equal to the quantity in Column 2.
D. The relationship cannot be determined from the information given.

Notes: Information centered over both columns refers to one or both of the quantities being compared. A symbol that appears in both columns has the same meaning in each column. All variables represent real numbers. Most figures are not drawn to scale.

	Column 1	Column 2
1.	$\log_3 27$	$\log_{27} 3$

$$\begin{cases} 2x + 3y = 12 \\ 3x - 2y = 5 \end{cases}$$

	Column 1	Column 2
2.	x	y
3.	$\begin{vmatrix} 3 & 1 \\ 4 & 2 \end{vmatrix}$	$\begin{vmatrix} -4 & 2 \\ 5 & -3 \end{vmatrix}$

$$f(x) = \lvert 2x - 3 \rvert$$

	Column 1	Column 2
4.	$f(2)$	$f(0)$
5.	$\sqrt[6]{16\sqrt[3]{64}}$	$81^{\frac{1}{4}}$

$$-3k + 5 > 20$$

	Column 1	Column 2
6.	k	-4
7.	$\log_3 5$	$\log_2 (7 + 13)^0$

$$n < -1$$

	Column 1	Column 2
8.	$\frac{1}{n^2 - 1}$	0

	Column 1	Column 2

Use this information for 9–10.

Coplanar lines a, b, c

$a \parallel b$

$a \perp c$

	Column 1	Column 2
9.	slope of a	slope of b
10.	slope of b	slope of c
11.	$\log 2 + \log 3$	$\log 5$
12.	$125^{-\frac{2}{3}}$	40×10^{-3}

Use this diagram for 13–15.

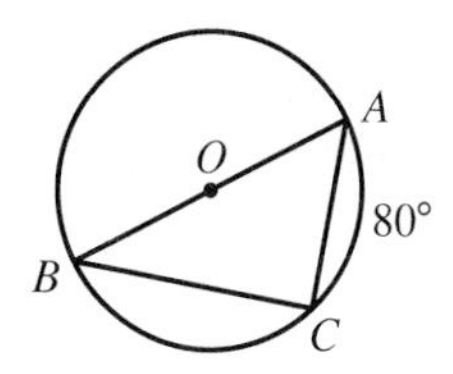

	Column 1	Column 2
13.	$m\angle C$	90°
14.	$m\angle A$	40°
15.	degrees in $\widehat{BC}$	120°

CUMULATIVE REVIEW (CHAPTERS 1–12)

Solve each system by an appropriate method.

1. $\begin{cases} x + 3y = 7 \\ y + \frac{1}{3}x = 0 \end{cases}$

2. $\begin{cases} 3x - 5y = 2 \\ 5x + 4y = -9 \end{cases}$

3. $\begin{cases} 3x + 4y = -4 \\ y = -\frac{3}{4}x - 1 \end{cases}$

Simplify, and write with positive exponents.

4. $(5a^{-1}b)(2a^2b^{-1})$
5. $3x^0 - y^0$
6. $(5^{-2})\left(\frac{1}{2}\right)^{-3}$
7. $(4^{-2})^3$

Factor completely.

8. $b^2 - 2b - 15$
9. $5r^4 - 625rs^3$
10. $c^2 - 6cd + 9d^2$
11. $3a^4 + 3a^2$
12. $-2x^2 + x + 3$
13. $pq - rt + pt - rq$

Express in scientific notation. If an integer ends in zeros, assume the zeros are not significant.

14. 718500
15. 0.0135
16. 5000
17. 0.0000230

$f(x) = 1 - 2x$, and $g(x) = 3x^2$. Find the values of the functions.

18. $f(-3)$
19. $g\left(\frac{1}{2}\right)$
20. $f(g(-1))$
21. $g\left(f\left(\frac{1}{3}\right)\right)$

For matrices A, B, C and D below, perform the indicated operations, if they are defined.

$$A = \begin{bmatrix} -3 & 2 \\ 0 & 4 \\ 5 & -1 \end{bmatrix} \quad B = \begin{bmatrix} -1 & 3 & 0 \\ -2 & 1 & 4 \end{bmatrix} \quad C = \begin{bmatrix} -4 & -2 & 1 \\ 3 & 3 & -1 \\ 0 & 4 & 5 \end{bmatrix} \quad D = \begin{bmatrix} 5 & 3 & -2 \\ 0 & -3 & 4 \\ -5 & 1 & 0 \end{bmatrix}$$

22. $C + D$
23. AB
24. CB
25. D^2
26. A^2B
27. $B + C$
28. $D - C$
29. $-3A$

Simplify each radical expression.

30. $\sqrt[3]{-27a^6}$
31. $\sqrt[3]{3x^2}\sqrt[3]{9x^5}$
32. $\dfrac{\sqrt{2} - 2\sqrt{3}}{\sqrt{3}}$
33. $\dfrac{\sqrt{5}}{\sqrt{10x}}$
34. $\sqrt{8d} + \sqrt{2d} - \sqrt{50d}$
35. $3\sqrt[3]{8a^4} - 5a\sqrt[3]{a}$

Solve. Graph the solution set if it is not the empty set.

36. $-5r + 2 < -8$ or $7 - 2r \geq 5r$
37. $-5 \leq 4 - 3q \leq 16$
38. $|x + 11| < 6$
39. $|3x - 8| \geq -1$
40. $-4|5t - 10| \leq -20$

Perform the indicated operations, and simplify.

41. $\dfrac{5x^2}{x^2 - 3x + 2} \cdot \dfrac{x^2 + x - 6}{10x}$

42. $\dfrac{4y + 10}{4y^3} \div \dfrac{4y^2 + 20y + 25}{2y^3 + 5y^2}$

43. $\dfrac{y}{x - y} - \dfrac{x}{y - x}$

44. $\dfrac{3x}{x^2 - 6x + 9} + \dfrac{6}{x - 3}$

Identify the graphs of the equations of each system, and find the real solutions.

45. $\begin{cases} x^2 + y^2 = 25 \\ xy = -12 \end{cases}$

46. $\begin{cases} x^2 + 4y^2 = 36 \\ x - 2y = 6 \end{cases}$

47. $\begin{cases} x^2 - y^2 = 1 \\ x^2 + y = 3 \end{cases}$

Solve each system of inequalities graphically.

48. $\begin{cases} y + 3x \geq 2 \\ y - x < 1 \end{cases}$

49. $\begin{cases} 3y - 2x < 6 \\ x \geq -1 \end{cases}$

50. $\begin{cases} y - 2 > 1 \\ 5x + 2y < -4 \end{cases}$

Solve and check.

51. $3t^2 = 48t$

52. $6x^2 - 5x = 4$

53. $(x - 1)(x + 7) = -16$

54. $25x^2 + 10x + 1 = 0$

Determine the intercepts and asymptotes of the graph of each equation. Then sketch the graph.

55. $y = \dfrac{x}{x + 1}$

56. $y = \dfrac{1}{4 - x^2}$

57. $y = \dfrac{1}{2x - 1}$

Solve each equation.

58. $2x^2 + 7x + 3 = 0$

59. $-9x^2 + 24x - 28 = 0$

60. $m^4 + 3m^2 - 28 = 0$

61. $0.25x^2 - x + 0.25 = 0$

62. Solve: $x^4 + 6x^3 + 9x^2 - 6x - 10 = 0$

63. A 12-mi canoe trip upstream took 6 h while the return trip downstream took 4 h. Find the rate of the canoe in still water and the rate of the current.

Simplify.

64. $(-2 + \sqrt{-9}) - (3 - \sqrt{-4})$

65. $|-5 + 12i|$

66. $(-2 + 3i)(5 - i^{17})$

67. $\dfrac{i}{3 - i\sqrt{5}}$

Solve each system of equations.

68. $\begin{cases} 3x - y + 2z = -1 \\ -2x + 3y - z = 4 \\ 4x + 5y + 2z = 3 \end{cases}$

69. $\begin{cases} 2x + 3y = -6 \\ 3x - 9z = 0 \\ 5y - 2z = 2 \end{cases}$

Solve and check.

70. $\sqrt{x - 1} = x - 7$

71. $\sqrt[4]{2x + 2} = 2$

72. $\frac{5}{x + 1} = \frac{8}{3} - \frac{1}{x - 1}$

73. $\frac{3}{y^2 - 9} = \frac{4}{y + 3}$

74. The width of a rectangular box is 1 in. longer than its height, and its length is three times its width. If its volume is 54 in.^{3}, find its dimensions.

For each function, find the coordinates of the vertex. Then graph.

75. $y = -\frac{1}{2}x^2 + 2$

76. $y = 2x^2 - 8x + 10$

77. $y = -x^2 - 8x - 10$

Write, in standard form, the equation of the line described.

78. Passes through $(-1, 1)$ and $(2, -4)$

79. Perpendicular to the graph of $3y - x = 1$ and passes through $(0, 0)$

Write a quadratic equation that has the given solutions.

80. $-2, \frac{3}{5}$

81. $2 \pm 4\sqrt{3}$

82. $\frac{3 + i\sqrt{2}}{3}$

Find the inverse of each function, and determine whether the inverse is also a function.

83. $y = 3x - 4$

84. $2y + 5x = 0$

85. $y + 2x^2 = -4$

86. A walk of uniform width surrounds a rectangular garden 12 ft by 16 ft. Find the width of the walk if its area is 78 ft^2.

Solve for *y*, and indicate any restrictions on the variables.

87. $A = 2x + 2y$

88. $2ay - 3by = a$

89. $\frac{3}{y} - 2a = b$

Find the distance between the points with the given coordinates. Then find the midpoint of the segment joining the points.

90. $(5, -3), (-3, 12)$

91. $(-3\sqrt{6}, \sqrt{5}), (-3\sqrt{6}, 3\sqrt{5})$

Solve and check.

92. $\log_3 81 = x$

93. $\log_4 \frac{1}{2} = x$

94. $5^{-2k} = \frac{1}{125}$

13 Sequences and Series

Pyramid Power $

Developing Mathematical Power

In network marketing, distributors make a profit by buying a product at a discount and selling it at a markup. A distributor can become a supervisor by recruiting and training others. Supervisors make a profit on their own sales and by selling the product to the distributors they supervise. Supervisors also earn a royalty on the sales of their distributors.

Project

You have been asked to become a Sweet Teeth Toothpaste distributor. Make a prediction for your first six months in business: How many people could you recruit? What level could you attain? How much money could you earn each month? How many hours would you work per month? What financial problems might occur?

Sweet Teeth Toothpaste

Retail Price $3.29/tube; 48 tubes/carton

Wholesale Discount per Carton

Number of cartons	1–5	6–10	11–15	Above 15
Discount	25%	30%	35%	40%

Royalty: 5% of all sales made by distributors you supervise

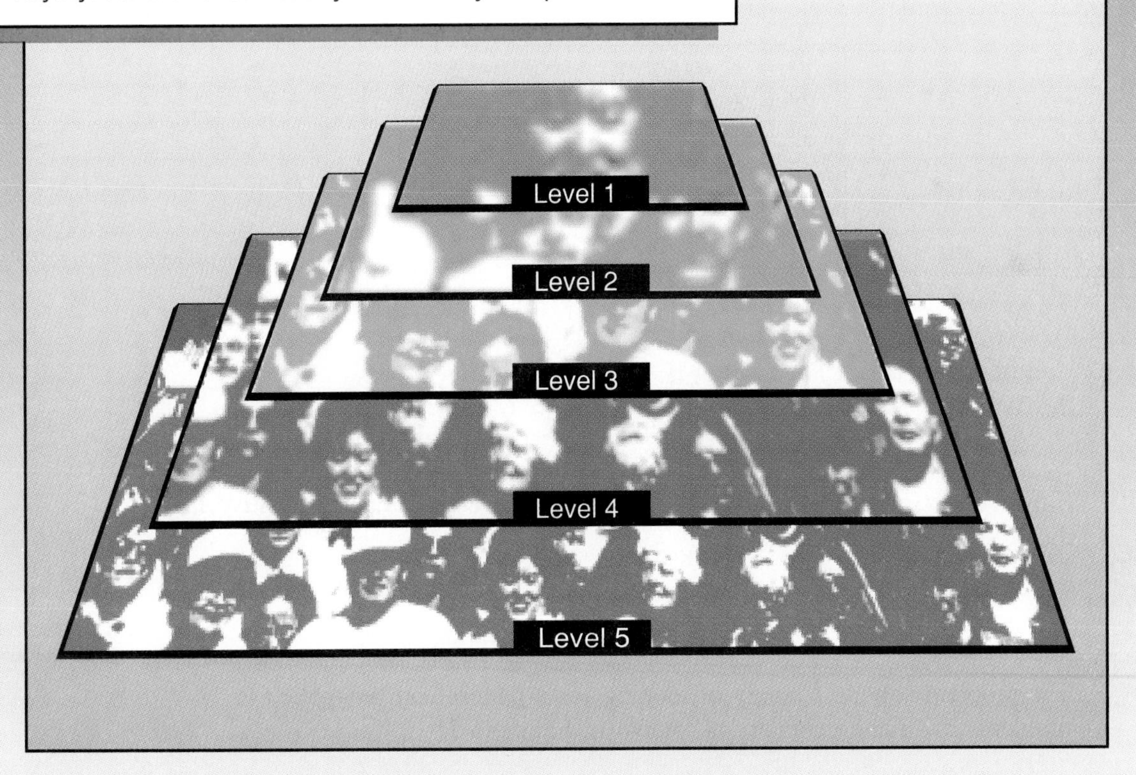

13.1

Problem Solving Strategy: Look for a Pattern

The numbers 2, 3, 5, 8, 12, ... are related by a pattern: the differences between consecutive numbers are 1, 2, 3, and 4. If the pattern continues, the next group of three numbers following 12 is 17, 23, and 30. Looking for patterns is a frequently used problem solving strategy. If a pattern exists and is identified in a problem situation, then the pattern can often be used to solve the problem.

EXAMPLE 1 **Suppose x is any positive integer. Which is greater, x^{x+1} or $(x + 1)^x$?**

Understand the Problem *What are you given?* It is given that x is any positive integer. *What are you asked to find?* The problem is to determine which is greater, x^{x+1} or $(x + 1)^x$.

Plan Your Approach *Choose a strategy.* Substitute some specific values for x to see if there is a pattern to the values obtained.

x	x^{x+1}	$(x + 1)^x$
1	$1^2 = 1$	$2^1 = 2$
2	$2^3 = 8$	$3^2 = 9$
3	$3^4 = 81$	$4^3 = 64$
4	$4^5 = 1024$	$5^4 = 625$
5	$5^6 = 15,625$	$6^5 = 7776$
6	$6^7 = 279,936$	$7^6 = 117,649$
7	$7^8 = 5,764,801$	$8^7 = 2,097,152$
8	$8^9 = 134,217,728$	$9^8 = 43,046,721$

Complete the Work *Look for a pattern.*

$1^2 < 2^1$ $\quad 3^4 > 4^3$ $\quad 5^6 > 6^5$ $\quad 7^8 > 8^7$

$2^3 < 3^2$ $\quad 4^5 > 5^4$ $\quad 6^7 > 7^6$ $\quad 8^9 > 9^8$

Interpret the Results *State your answer.* A pattern does exist. For any positive integer greater than 2,

$$x^{x+1} \text{ is greater than } (x + 1)^x$$

That is, $x^{x+1} > (x + 1)^x$ if $x > 2$.

The problem solving strategy of looking for a pattern can be applied to problems involving geometry as shown in Example 2.

EXAMPLE 2 **What is the greatest number of regions into which 7 line segments can divide a circle?**

Understand the Problem *What are the given facts?* Seven line segments divide a circle into regions.

What are you asked to find? The problem is to find the greatest number of regions into which the circle is divided.

Plan Your Approach *Choose a strategy*. Dividing a circle into regions by drawing seven line segments would produce a complicated drawing and counting the regions would be very difficult. However, the strategy of looking for a pattern can be used.

One line segment gives 2 regions.

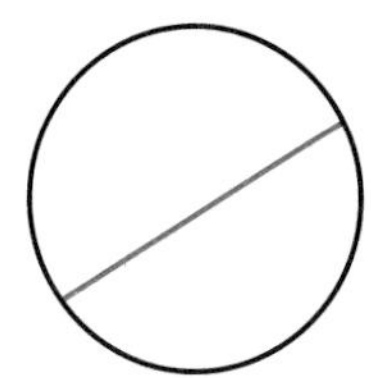

Two line segments give 4 regions.

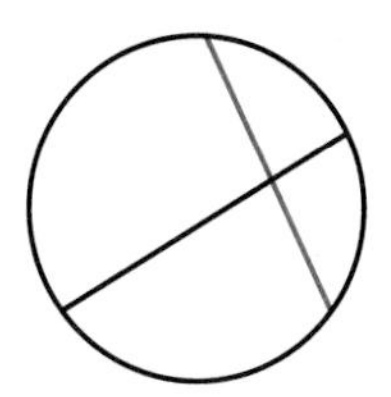

Three line segments give 7 regions.

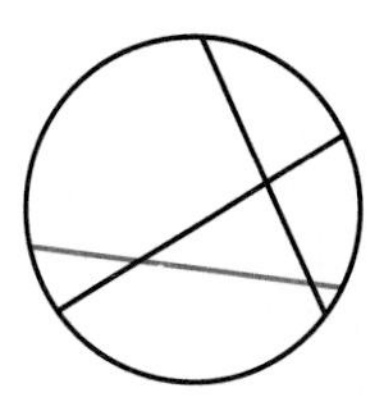

Four line segments give 11 regions.

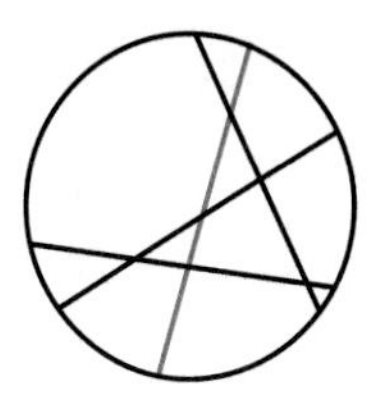

Complete the Work

segments		regions	
1	⟶	2	
			+ 2
2	⟶	4	
			+ 3
3	⟶	7	
			+ 4
4	⟶	11	
			+ 5
5	⟶	16	
			+ 6
6	⟶	22	
			+ 7
7	⟶	29	

Look for a pattern.

The pattern is to add n more regions for each n segments.

Interpret the Results *State your answer*. A pattern does exist and can be used to determine that 7 line segments divide a circle into 29 regions.

CLASS EXERCISES

Classify each statement as *true* or *false*.

1. It is always possible to find a pattern in a set of numbers.
2. Patterns can be used to solve some mathematical problems.
3. $9^{10} > 10^9$
4. Eight line segments divide a circle into at most 37 regions.
5. Identification of a pattern will never lead to a solution.

Identify the pattern and use it to find the next three numbers.

6. $-1, 0, -2, 1, -3, \ldots$

7. $\frac{3}{4}, \frac{5}{6}, \frac{7}{8}, \frac{9}{10}, \ldots$

8. $1, 8, 27, 64, \ldots$

9. $1, 1, 2, 3, 5, 8, 13, \ldots$

10. $\frac{1}{2}, \frac{1}{4}, \frac{1}{6}, \frac{1}{8}, \ldots$

11. $\frac{1}{100}, \frac{1}{81}, \frac{1}{64}, \frac{1}{49}$

PRACTICE EXERCISES

1. $9, 6, 3, 0, \ldots$

2. $0, 3, 8, 15, 24, \ldots$

3. $10, 100, 1000, \ldots$

4. $1, 7, 9, 3, 1, 7, 9, 3, \ldots$

5. $2, 5, 7, 12, 19, \ldots$

6. $30, 60, 90, 21, 51, 81, \ldots$

7. $5, 2, 6, 3, 9, 6, \ldots$

8. $5, 10, 9, 18, 17, 34, 33, 66, \ldots$

9. $0, 4, 1, 5, 2, 6, 3, \ldots$

10. $40, 35, 42, 37, 44, \ldots$

Solve each problem using a pattern.

11. The symbol used for a dollar sign is a capital S with one stroke through it. The stroke separates the symbol into 4 parts. If 100 strokes were drawn through the S, how many parts would be formed?

4 parts

7 parts

12. Frank received a very unusual job offer. He is told that his pay on the first day will be 10¢. However, his pay will be twice as much the next day and will double each day thereafter. How much will Frank's pay be on the 21st day?

13. The sum of the numbers 1 to 10 can be found by grouping the numbers to form five sums of 11. For example,

$1 + 10 = 11 \quad 2 + 9 = 11 \quad 3 + 8 = 11 \quad 4 + 7 = 11 \quad 5 + 6 = 11$

Therefore, $1 + 2 + 3 + 4 + 5 + 6 + 7 + 8 + 9 + 10 = 5 \times 11 = 55$. Use this pattern to find the sum of the numbers from 1 to 100.

14. The numbers 1, 1, 2, 3, 5, 8, 13, ... form a pattern. Use the pattern to find the fifteenth number. Then calculate the sum of the first thirteen numbers. What relationship does the sum have to the fifteenth and second numbers?

15. Use these figures to find the first 10 *rectangular* numbers.

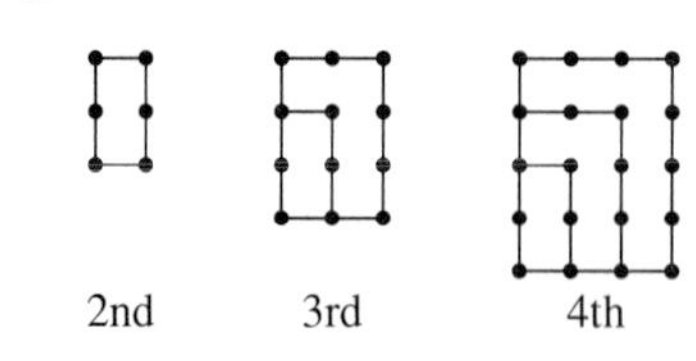

16. Use these figures to find the first 10 *pentagonal* numbers.

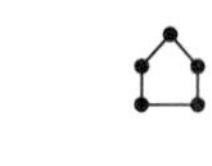
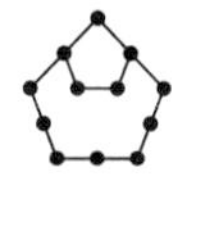
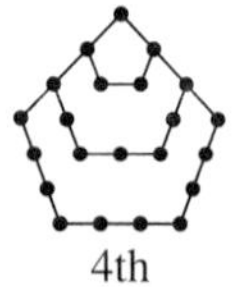
1st 2nd 3rd 4th

17. Use a pattern to find the sum of the first n odd positive integers.

18. The figure at the right is a decagon. How many diagonals can be drawn in it?

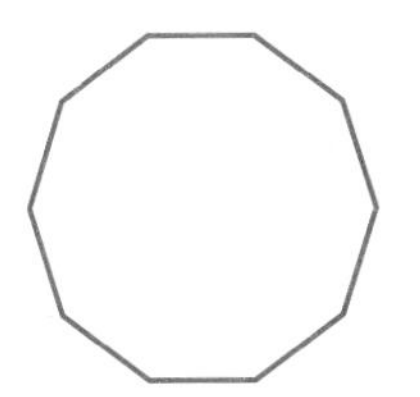

Mixed Problem Solving Review

1. With musical strings of the same length, material, and diameter, the number of vibrations per second varies as the square root of the tension in the string. If a tension of 25 units in a certain string yields 60 vibrations per second, find the tension that will result in 36 vibrations per second.

2. The half-life of radium is approximately 1760 years. In how many years will 10 g of radium weigh 7.2 g?

3. A rectangular box is constructed such that the length is twice the width and the height is five times the width. The total surface area of the six sides of the box is 136 m^2. Find the dimensions of the box.

4. In 10 hours, an oil tanker travels 110 km with the current and 70 km against the current. On another trip requiring the same amount of time and with a similar current, the tanker travels 88 km with the current and 84 km against. Find the rate of the current and that of the tanker in still water.

5. An apple farmer in Michigan calculates that he can produce 52 boxes of apples per tree if he plants 40 trees per acre. If he plants one more tree per acre the yield drops one box per tree due to congestion. Find the number of trees per acre he should plant in order to maximize the total production.

6. A florist maintains three square flower beds. Two of the beds are the same size and the third bed is larger. The sum of the areas of the three beds is 216 ft^2 and the sum of the perimeters is 96 ft. Find the dimensions of the flower beds.

PROJECT

The number pattern 1, 1, 2, 3, 5, 8, 13, ... is called a **Fibonacci sequence** in honor of the mathematician who discovered it. This sequence is particularly interesting because it is a model for many physical things in nature. Look up the name Fibonacci in a book on the history of mathematics. Write a brief essay about his life. Try to find some things in nature that are described by Fibonacci pattern.

13.2

Sequences

Objective: To write the rule of a sequence as a recursive or explicit formula and use it to find specified terms

The honeycombs in a beehive and the petals of a flower form patterns. Many patterns are found in nature, and such patterns can often be expressed as mathematical relations.

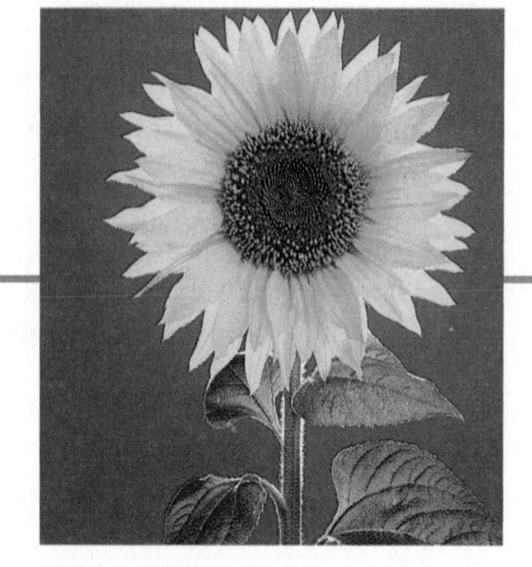

Capsule Review

A *relation* is a set of ordered pairs. For example, consider the relation R, where

$$R = \{(1, 5), (2, 7), (3, 9), (4, 11)\}$$

The *domain* of relation R is $\{1, 2, 3, 4\}$ and the *range* of R is $\{5, 7, 9, 11\}$.

A relation is a *function* if for every element in the domain there is one and only one corresponding element in the range. Therefore, the relation R is a function. R can also be written as a rule, or equation.

$$R = f(n) = 2n + 3 \qquad \text{for } n \in \{1, 2, 3, 4\}$$

Determine which relations, if any, are functions. Assume that n is a real number.

1. $\{(1, 1), (2, 8), (3, 27), (4, 64)\}$ **2.** $\{(1, -1), (1, 3), (2, 4), (2, 0)\}$

3. $f(n) = 3n$ **4.** $f(n) = n^2$ **5.** $f(n) = \pm n$ **6.** $f(n) = 3n^3$

Yuan has been elected President of the Chess Club. Her first job is to schedule the tournaments for the month of March. The tournaments are played every Tuesday night. If the first Tuesday in March is March 4, find the dates of all the tournaments.

Tuesdays are 7 days apart. *Look for a pattern.*

1st Tuesday	2nd Tuesday	3rd Tuesday	4th Tuesday	
4	$4 + 7 = 11$	$11 + 7 = 18$	$18 + 7 = 25$	*Apply the pattern.*

The dates of the tournaments are March 4, 11, 18, and 25.

The ordered set of numbers {4, 11, 18, 25} is called a *sequence*. It can be thought of as the set of ordered pairs {(1st Tuesday, 4), (2nd Tuesday, 11), (3rd Tuesday, 18), (4th Tuesday, 25)}, or {(1, 4), (2, 11), (3, 18), (4, 25)}. A **sequence** is a function defined on the set of consecutive positive integers or on a subset of consecutive positive integers. Each value in a sequence is called a **term.** The first term is a_1, the second is a_2, and the *n*th term is a_n. The *n*th term is also called the **general term.** The subscripts of the terms $a_1, a_2, \ldots, a_n, \ldots$ are positive integers.

The sequence {4, 11, 18, 25} is a **finite sequence** because it has a last term. The sequence consisting of the set of positive odd integers, {1, 3, 5, 7, ...}, is an **infinite sequence** because it does *not* have a last term. The three dots are used to show that the sequence goes on and on indefinitely.

A formula for finding the *n*th term of a sequence is a **rule** of the sequence. Sometimes a sequence can be defined by stating the first term, a_1, and a rule that shows how to obtain the $(n + 1)$th term, a_{n+1}, from the preceding term, a_n. This type of rule is called a **recursive formula.**

EXAMPLE 1 **Write the recursive formula for the sequence of the tournament dates shown on page 622.**

Dates: 4, 11, 18, 25 *Look at the pattern. You get from each term to the next by adding 7.*

$$\begin{aligned} a_1 &= 4 \\ a_2 &= a_1 + 7 = 4 + 7 = 11 \\ a_3 &= a_2 + 7 = 11 + 7 = 18 \\ a_4 &= a_3 + 7 = 18 + 7 = 25 \\ a_{n+1} &= a_n + 7 \end{aligned}$$

Recursive formula: $a_1 = 4$; $a_{n+1} = a_n + 7$ for $n \in \{1, 2, 3, 4\}$

After you find the rule for a sequence or write the rule as a formula, you can use it to find additional terms.

EXAMPLE 2 **Write a recursive formula for the sequence 1, 2, 4, 8, ... and find the next three terms.**

Look for a pattern. Note that each term is twice the preceding term.

$$\begin{aligned} a_1 &= 1 \\ a_2 &= 2a_1 = 2(1) = 2 \\ a_3 &= 2a_2 = 2(2) = 4 \\ a_4 &= 2a_3 = 2(4) = 8 \\ &\vdots \\ a_{n+1} &= 2a_n \end{aligned}$$

Recursive formula: $a_1 = 1;\ a_{n+1} = 2a_n$

$a_5 = 2a_4 = 2(8) = 16$ — *Substitute in the formula to find a_5, a_6, and a_7.*
$a_6 = 2a_5 = 2(16) = 32$
$a_7 = 2a_6 = 2(32) = 64$

To give the rule of a sequence as an **explicit formula,** express the *n*th term, a_n, as a function of n, where n is a positive integer. To illustrate, an explicit formula for the sequence in Example 2 can be written by noting that each of the given terms is a power of 2. To find a formula, write each value in terms of n for $n = 1$ through $n = 4$.

$a_1 = 1 = 2^0 = 2^{1-1}$ — *$n = 1$*
$a_2 = 2 = 2^1 = 2^{2-1}$ — *$n = 2$*
$a_3 = 4 = 2^2 = 2^{3-1}$ — *$n = 3$*
$a_4 = 8 = 2^3 = 2^{4-1}$ — *$n = 4$*
$\vdots$
$a_n = 2^{n-1}$ — *Explicit formula*

It should be noted that if nothing is known about a sequence except the values of a finite number of terms, then there may be more than one formula that expresses the rule of the sequence. For example, the explicit formula $a_n = \frac{n(n^2 - 3n + 8)}{6}$ gives the first four terms of the sequence in Example 2, as you can verify. If this formula were used, however, the fifth term would be 15, not 16, and corresponding subsequent terms would also differ from those found in Example 2.

EXAMPLE 3 Joe is training for a marathon. The first week he runs 3 mi, the second week 6 mi, the third week 9 mi, and the fourth week 12 mi. He maintains this pattern for 6 weeks. Write a rule for the sequence of the number of miles he runs as an explicit formula and use it to find the number of miles he runs the sixth week.

3, 6, 9, 12 — *Write the sequence.*
$a_1 = 3 \cdot 1 \quad a_2 = 3 \cdot 2 \quad a_3 = 3 \cdot 3 \quad a_4 = 3 \cdot 4$ — *Look for a pattern.*
$a_n = 3 \cdot n = 3n$ — *Write the explicit formula (a_n as a function of n).*
$a_6 = 3 \cdot 6 = 18$ — *Substitute in the formula.*

Joe runs 18 mi the sixth week.

CLASS EXERCISES

1. **Thinking Critically** Is it possible to write a rule of the sequence 3, 6, 9, 12, . . . , 18 both as a recursive formula and an explicit formula? Justify your answer.

Classify each formula as recursive or explicit.

2. $a_1 = 4; a_{n+1} = a_n + 2$
3. $a_n = 2n + 1$
4. $a_n = 2n^2$
5. $a_1 = -5; a_{n+1} = 2a_n + 1$

Write a recursive formula for each sequence. Then use the formula to write the next three terms.

6. 5, 6, 7, 8, ...
7. 2, 5, 8, 11, ...
8. −4, 8, −16, 32, ...
9. 1, −3, 9, −27, ...

Use the given formula to find the indicated term for each sequence.

10. $a_n = \frac{1}{n + 1}$, 8th term
11. $a_n = \frac{n}{n + 1}$, 4th term

PRACTICE EXERCISES

Write a rule for each sequence as a recursive formula and use it to find the next three terms of the sequence.

1. 2, 6, 10, 14, ...
2. 5, 8, 11, 14, ...
3. 1, 0, −1, −2, ...
4. −3, −7, −11, −15, ...
5. −4, 8, 20, 32, ...
6. 0, −8, −16, −24, ...
7. $1, 1\frac{1}{2}, 2, 2\frac{1}{2}, \ldots$
8. $1, \frac{1}{2}, 0, -\frac{1}{2}, \ldots$

Write a rule for each sequence as an explicit formula and use it to find the indicated term.

9. 2, 4, 6, 8, ...10th term
10. 3, 6, 9, 12, ...7th term
11. 2, 3, 4, 5, ...11th term
12. 0, 1, 2, 3, ...15th term
13. 3, 5, 7, 9, ...14th term
14. 1, 4, 7, 10, ...12th term

Use the rule to find the first three terms of each sequence.

15. $a_1 = 4; a_{n+1} = a_n + 1$
16. $a_1 = -2; a_{n+1} = -a_n$
17. $a_n = n^3$
18. $a_n = 3n - 1$
19. $a_1 = 0; a_{n+1} = a_n - 1$
20. $a_1 = -2; a_{n-1} = a_n + 5$

Write a recursive or an explicit rule for each sequence and use it to find the next three terms.

21. 0, 0.5, 1, 1.5, ...
22. 3, 3.25, 3.5, 3.75, ...

23. $1, \frac{1}{2}, \frac{1}{4}, \frac{1}{8}, \ldots$

24. $3, 1, \frac{1}{3}, \frac{1}{9}, \ldots$

25. $5, -5, 5, -5, \ldots$

26. $-2, -4, -6, -8, \ldots$

Use the rule to write the first four terms of each sequence.

27. $a_1 = -1;\ a_{n+1} = (a_n)^2 + 1$

28. $a_1 = -2;\ a_{n+1} = 3(a_n + 2)$

29. $a_n = (n + 1)^2$

30. $a_n = 2(n - 1)^3$

31. $a_n = \dfrac{n^2}{n + 1}$

32. $a_n = \dfrac{n + 1}{n + 2}$

33. The first term in a sequence is 20. Each term after the first term is 4 more than the term before it. Write a recursive formula for the sequence and find the first three terms.

34. The first term in a sequence is 28. Each term after the first term is 6 less than the term before it. Write a recursive formula for the sequence and find the first three terms.

The recursive formula for the rule of a sequence is given. Write an explicit formula for the rule of each sequence.

35. $a_1 = 10;\ a_{n+1} = 2a_n$

36. $a_1 = -2;\ a_{n+1} = \frac{1}{2}a_n$

37. $a_1 = 1;\ a_{n+1} = a_n + 4$

38. $a_1 = -5;\ a_{n+1} = a_n - 1$

39. Given the sequence 9, 11, 13, 15, . . . , write a recursive formula for a_{n-2} in terms of a_{n-3}. Assume that n is an integer greater than 3.

40. Given the sequence $0, -3, -6, -9, \ldots$, write a recursive formula for a_{n-5} in terms of a_{n-6}. Assume that n is an integer greater than 6.

Applications

41. Business The starting salary at a supermarket is $4 per hour. Every six months, an employee gets a raise of 50¢ per hour. Write a recursive formula for the rule for the salary sequence and use it to find the salary after the third 6-month period.

42. Horticulture The number of flowers a certain rose bush produces doubles every year for the first 7 years. If the bush produces 8 flowers the first year, write an explicit formula for the rule of the sequence and use it to find the number of flowers the bush will produce the 4th year.

43. **Travel** A travel club has a vacation plan. After an initial deposit of \$25, each member deposits \$10 per week. Write a recursive formula for the rule of the sequence.

44. **Coin collecting** A coin dealer has a collection of 50 old silver dollars. Suppose he were able to sell 2 of the coins each day. Write a recursive formula for the rule of the sequence that would represent the number of coins in the collection on successive days. For what values of n would the rule be valid? Write an explicit formula for the rule of the same sequence.

Developing Mathematical Power

45. **Investigation** The Tower of Hanoi is a famous problem invented in 1883 and apparently sold as a children's toy. The task is to move the eight disks to one of the vacant pegs in the fewest possible moves. Move one disk at a time and never place a disk on top of a smaller disk.

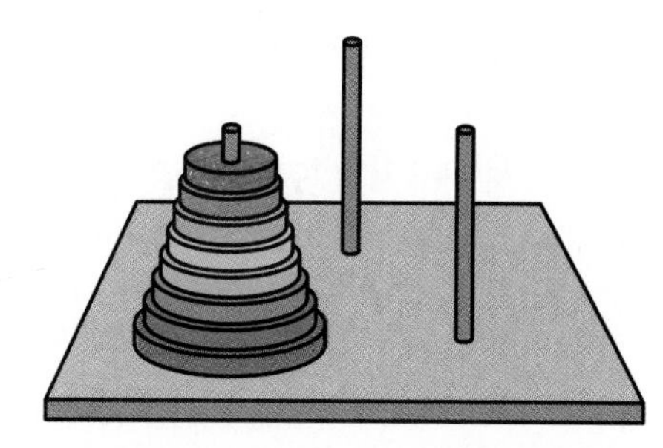

The inventor, who used the alias "Professor Claus," was Edouard Lucas, a professor of mathematics at St. Louis College in France. He based it on the mythical Tower of Brahma of a temple in Benares, India. The Tower was supposedly made of 64 golden disks, which the temple priests were transferring. The task would take so long, according to the myth, that the temple would crumble to dust and the world vanish in a clap of thunder before it was done. Not a bad estimate—if the priests worked unceasingly, moving one disk per second, the task would require about 585 billion years!

You do it—solve the Tower of Hanoi, 8 disks and 3 pegs as shown in the picture.

You may want to solve simpler problems first and look for a pattern. Using only one disk, the solution is trivial. Solve for two disks, then for three disks, and so on. You will find it helpful to maintain a table of the moves in each subproblem as well as a table summarizing the solutions of the subproblems.

While some patterns are obvious, many are not. You will want to look across the columns as well as down the columns of the tables to search for a relation. What could be done to the first member of the pair to produce the second? Think about each of the four basic operations, exponents, and roots, and then combinations of them.

13.3

Arithmetic Sequences

Objectives: To find specified terms of an arithmetic sequence
To find arithmetic means

Sequences can be classified according to the way in which the terms are formed. You can find any specified term in a sequence by evaluating the formula for the *n*th term.

Capsule Review

To evaluate a formula for one variable, substitute known values and solve.

EXAMPLE **Given the formula $p = (k + 1)m$, find p if $k = 3$ and $m = 7$.**

$p = (k + 1)m$ *Write the formula.*
$p = (3 + 1)7$ *Substitute: $k = 3$, $m = 7$*
$p = 28$ *Solve for p.*

Evaluate each formula for the indicated variable.

1. $d = rt$ for d, if $r = 25$ and $t = 4$
2. $y = mx + b$ for y, if $m = 1$, $x = 2$, and $b = 5$
3. $r - t = (k + 3)n$ for r, if $t = 2$, $k = 1$, and $n = 1$
4. $m(r - 2) = 3n$ for m, if $r = -4$ and $n = 6$
5. $ks^2 + 9 = t$ for k, if $s = 3$ and $t = 27$
6. $pq^2 = r + 4$ for q, if $p = 2$ and $r = 14$

In the sequence $1, 1\frac{1}{2}, 2, 2\frac{1}{2}, \ldots$, each term after the first is found by adding $\frac{1}{2}$ to the preceding term. The difference between any two successive terms is $\frac{1}{2}$. This sequence is an example of an *arithmetic sequence,* or *arithmetic progression.* An **arithmetic sequence** is a sequence in which the difference between successive terms is a constant. This constant is called the **common difference.**

Since a sequence is a function whose domain is the set of positive integers, it is possible to show a sequence graphically.

The sequence $1, 1\frac{1}{2}, 2, 2\frac{1}{2}, \ldots$ can be represented by the explicit formula $a_n = \frac{1}{2}(n + 1)$. The graph is comprised of isolated points, since the terms of the sequence are defined only for positive integral values of n. The graph shows that as n increases without bound, the terms of the sequence also increase without bound.

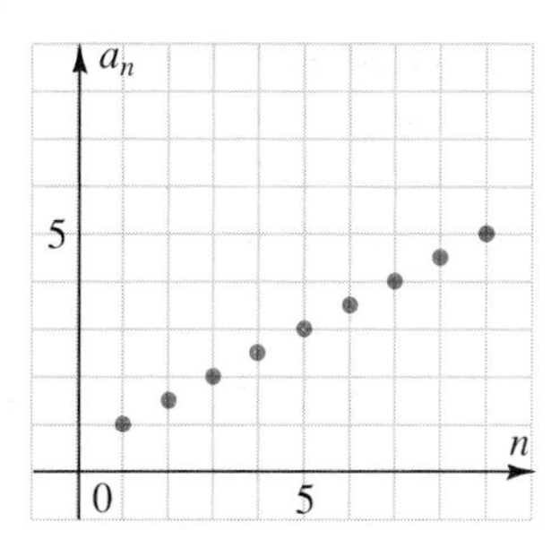

You can find the common difference, d, by subtracting any two successive terms of an arithmetic sequence. That is

$$d = a_{n+1} - a_n$$

EXAMPLE 1 **Find the common difference, d, for the arithmetic sequence 6, 4, 2, 0, −2, ... Then graph the sequence.**

Subtract any two successive terms to find d.

$$d = a_2 - a_1 = 4 - 6 = -2$$

The graph of the sequence is shown at the right. Since $d < 0$, the terms of this sequence decrease without bound as n increases without bound.

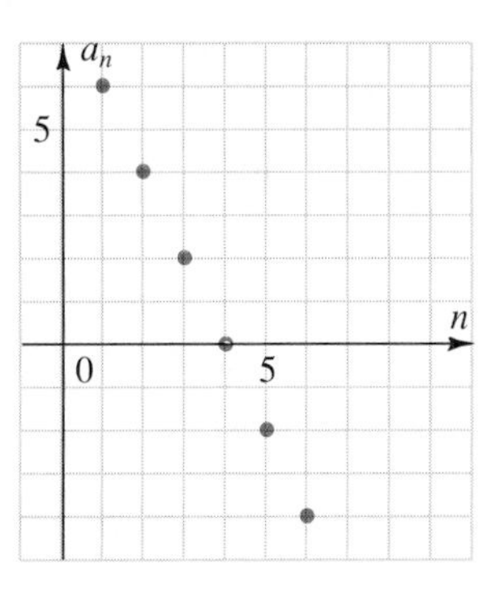

Any term after the first term in an arithmetic sequence is the sum of the preceding term and the common difference. You can use this fact to develop a formula for the nth or general term of the sequence.

First term	a_1	$= a_1$
Second term	$a_2 = a_1 + d$	$= a_1 + d$
Third term	$a_3 = a_2 + d = (a_1 + d) + d$	$= a_1 + 2d$
Fourth term	$a_4 = a_3 + d = (a_1 + 2d) + d$	$= a_1 + 3d$
Fifth term	$a_5 = a_4 + d = (a_1 + 3d) + d$	$= a_1 + 4d$

The formula for the nth, or general term of an arithmetic sequence is

$$a_n = a_1 + (n - 1)d$$

where a_1 is the first term and d is the common difference.

You can use this formula to find any specified term of a given sequence, if you know the first term and the common difference.

EXAMPLE 2 **The first term, a_1, of an arithmetic sequence is 4 and the common difference, d, is 2. Find the ninth term.**

$a_n = a_1 + (n - 1)d$ — Use the formula for the nth term.
$a_9 = 4 + (9 - 1)2$ — Substitute: $n = 9$, $a_1 = 4$, $d = 2$
$a_9 = 4 + 16 = 20$ — The nth term is 20.

If the common difference is not known, it must be found before the formula is applied.

EXAMPLE 3 **Find the fifteenth term of the arithmetic sequence 1.00, 1.25, 1.50, 1.75, ...**

$d = 1.25 - 1.00 = 0.25$ — Find the common difference.

$a_n = a_1 + (n - 1)d$ — Use the formula for the nth term.
$a_{15} = 1.00 + (15 - 1)0.25$ — Substitute: $n = 15$, $a_1 = 1.00$, $d = 0.25$
$a_{15} = 1.00 + 3.50 = 4.50$ — The fifteenth term is 4.50.

If two terms of an arithmetic sequence are known, you can find a_1 and d by using the formula for the nth term twice and solving the resulting system of equations.

EXAMPLE 4 **In an arithmetic sequence, $a_3 = 10$ and $a_6 = 19$. Find a_1 and d.**

$a_n = a_1 + (n - 1)d$ — Use the formula for the nth term.
$10 = a_1 + (3 - 1)d$ — $a_3 = 10, n = 3$
$19 = a_1 + (6 - 1)d$ — $a_6 = 19, n = 6$

$$\begin{cases} 10 = a_1 + 2d \\ 19 = a_1 + 5d \end{cases}$$ Solve the system of equations.

$$\begin{array}{r} 10 = a_1 + 2d \\ -19 = -a_1 - 5d \\ \hline -9 = -3d \\ 3 = d \end{array}$$

Multiply the second equation by -1.
Add the equations
d is 3.

$10 = a_1 + 2d$ — Then use one of the original equations.
$10 = a_1 + 2(3)$ — Substitute 3 for d.
$4 = a_1$ — a_1 is 4.

In the arithmetic sequence 3, 10, 17, 24, 31, . . . , the terms 10, 17, and 24 are *arithmetic means* between 3 and 31. **Arithmetic means** are the terms between any two given terms of an arithmetic sequence.

EXAMPLE 5 **Insert the three arithmetic means between 6 and 26.**

Let $a_1 = 6$ and $a_5 = 26$. Then the sequence can be represented by

$6, a_2, a_3, a_4, 26$ *There are three missing terms.*

Use the formula for the nth term to find the common difference, d.

$$a_n = a_1 + (n - 1)d$$
$$a_5 = a_1 + (5 - 1)d \quad \text{Substitute: } n = 5$$
$$26 = 6 + (5 - 1)d \quad \text{Substitute: } a_5 = 26, a_1 = 6$$
$$5 = d$$

Therefore, $a_2 = 6 + 5 = 11$ $\quad a_{n+1} = a_n + d$

$a_3 = 11 + 5 = 16$

$a_4 = 16 + 5 = 21$

The three arithmetic means are 11, 16, and 21.

Any number of arithmetic means can be inserted between two given numbers. A single arithmetic mean between any two given numbers is called **the arithmetic mean** of the numbers.

The arithmetic mean of two real numbers, a and b, is $\frac{a + b}{2}$, since a, $\frac{a + b}{2}$, b is an arithmetic sequence with constant difference $\frac{-a + b}{2}$, as you can easily verify.

CLASS EXERCISES

1. **Thinking Critically** To determine whether a sequence is arithmetic, is it sufficient to find the difference between only one pair of successive terms? Explain, and illustrate your answer.

Determine whether or not each sequence is arithmetic. Find the common difference, d, for each arithmetic sequence.

2. 4, 7, 10, 13, ...
3. 5, 7, 9, 11, ...
4. 6, 16, 26, 36, ...
5. 4, 12, 36, 108, ...
6. 7.5, 5.5, 3.5, 1.5, ...
7. 7, −14, 28, −56, ...

Find the eighth term of each arithmetic sequence.

8. $a_1 = 3, d = 2$
9. $a_1 = 9, d = 9$
10. 5, 7, 9, 11, ...

Given two terms of each arithmetic sequence, find a_1 and d.

11. $a_3 = 5$ and $a_5 = 11$
12. $a_4 = 8$ and $a_7 = 20$
13. Insert the two arithmetic means between 25 and 70.

PRACTICE EXERCISES

Graph each arithmetic sequence.

1. $-4, -3, -2, -1, \ldots$

2. $1, \frac{1}{2}, 0, -\frac{1}{2}, \ldots$

Find the specified term of each arithmetic sequence.

3. $a_1 = 2, d = 5; a_7$

4. $a_1 = 4, d = 6; a_8$

5. $a_1 = 1, d = -3; a_8$

6. $a_1 = -2, d = -3; a_{15}$

7. $4, 6, 8, 10, \ldots; a_{12}$

8. $15, 18, 21, 24, \ldots; a_{19}$

Find a_1 and d for an arithmetic sequence with the given terms.

9. $a_4 = 9$ and $a_7 = 18$

10. $a_3 = 14$ and $a_6 = 29$

11. $a_5 = 19$ and $a_8 = 40$

12. $a_7 = -5$ and $a_{11} = 11$

13. $a_7 = 6$ and $a_{12} = 21$

14. $a_3 = -8$ and $a_7 = 32$

15. Find the arithmetic mean between 18 and 28.

16. Find the arithmetic mean between 16 and 26.

17. Insert the two arithmetic means between 50 and 62.

18. Insert the three arithmetic means between 20 and 60.

Find the indicated term of each arithmetic sequence.

19. $a_1 = 3, d = 0.5; a_7$

20. $a_1 = -2.75, d = 0.25; a_9$

21. $a_1 = x, d = 3x; a_{11}$

22. $a_1 = 2x, d = -x; a_{15}$

23. $\frac{1}{2}, 1, 1\frac{1}{2}, 2, \ldots; a_{18}$

24. $1, 1\frac{1}{4}, 1\frac{1}{2}, 1\frac{3}{4}, \ldots; a_{22}$

25. $2, 2, 2, 2, \ldots; a_{60}$

26. $-3, -3, -3, -3, \ldots; a_{40}$

27. The third term of an arithmetic sequence is 10 and the ninth term is 46. Find the fifth term.

28. The eighth term of an arithmetic sequence is 55 and the tenth term is 77. Find the third term.

29. The arithmetic mean between two terms in an arithmetic sequence is 42. If one term is 30, find the other term.

30. The arithmetic mean between two terms of an arithmetic sequence is -6. One term is -20. Find the other term.

31. In an arithmetic sequence with $a_1 = 4$ and $d = 9$, which term is 184?

32. In an arithmetic sequence with $a_1 = 2$ and $d = -2$, which term is -82?

Find the indicated term of each arithmetic sequence.

33. $a_1 = k$, $d = k + 4$; a_9

34. $a_1 = k + 7$, $d = 2k - 5$; a_{11}

Find a_1 and d in terms of r and s for an arithmetic sequence with the given terms.

35. $a_3 = r$ and $a_8 = s$

36. $a_4 = r + s$ and $a_9 = 2s$

Applications

Technology A spreadsheet program provides a very quick and efficient way to find specified terms of an arithmetic sequence. Enter the first term in C1, the common difference in C3, and the number of the term in C5. You can then enter a formula in E5 to find the nth term and another formula in E7 to find the sum of the terms.

```
========= A ========= B========= C=========D========= E
1: FIRST TERM:                    9
2:
3:   THE COMMON DIFFERENCE IS:    4
4:
5:                                            THE 50th TERM IS 205.
6:
7:                                   THE SUM OF THE TERMS IS 5350.
```

What formula or value would go in each of the following?

37. C3

38. E5

39. E7

Use the spreadsheet to find the value of the given term.

40. 15th term of 57, 59, 61, ...

41. 21st term of 9, 13, 17, ...

Mixed Review

42. Divide $\dfrac{3m^4 + 2m^3 + 5m^2 - 2}{m + 3}$

43. Divide using synthetic division: $\dfrac{4x^3 + 3x^2 + x - 3}{x + 2}$

44. Determine whether $x + 3$ is a factor of $2x^3 + 3x^2 - 4x + 15$.

45. Determine whether $2x + 1$ is a factor of $3x^3 - 2x^2 + 5x + 8$.

46. Solve $2x^3 + x^2 - 25x + 12 = 0$.

47. To the nearest tenth, find a zero of $y = x^3 - 2x^2 - 5x - 1$ between -2 and -1.

13.4 Geometric Sequences

Objectives: To find specified terms of a geometric sequence
To find geometric means

One new model of a sports car sells for \$19,200. It is said that the car will depreciate by $\frac{1}{4}$ of its value per year.

$$\$19{,}200, \$14{,}400, \$10{,}800, \ldots$$

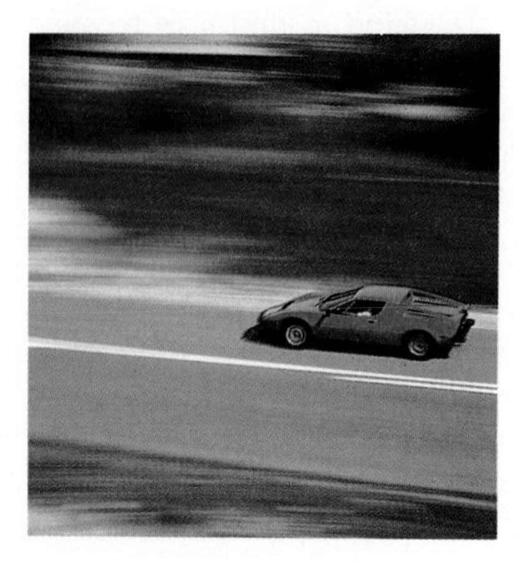

In this case, the sequence is *not* arithmetic because the difference between successive terms is not constant. The depreciating values of the sports car is an example of a *geometric sequence,* or *geometric progression*. In a geometric sequence, consecutive terms are compared by dividing to find the common ratio of the numbers.

Capsule Review

To find the ratio of two nonzero numbers or expressions a and b, write them as a quotient $\frac{a}{b}$ and express the quotient in simplest form.

The ratio of 10 to 12 is $\frac{10}{12} = \frac{5}{6}$.

The ratio of x^3 to x is $\frac{x^3}{x} = \frac{x^2}{1} = x^2$.

Express each ratio in simplest form. Assume that all variables represent nonzero numbers.

1. 125 to 25 **2.** -3 to 27 **3.** 1 to 5 **4.** 1.5 to 2.5

5. x^4 to x^2 **6.** y^2 to y^5 **7.** $2x$ to $4x^2$ **8.** $-8n^3$ to $6n^2$

A **geometric sequence** is a sequence in which the ratio of consecutive terms is a constant. This constant is called the **common ratio.**

In the geometric sequence 1, 2, 4, 8, . . . , each term after the first term is found by multiplying the preceding term by 2. That is, the common ratio is $\frac{2}{1}$, or 2.

The geometric sequence

$$1, 2, 4, 8, \ldots$$

has the explicit formula $a_n = 2^{n-1}$.

The graph of the sequence shows that as n increases without bound, the terms of the sequence also increase without bound.

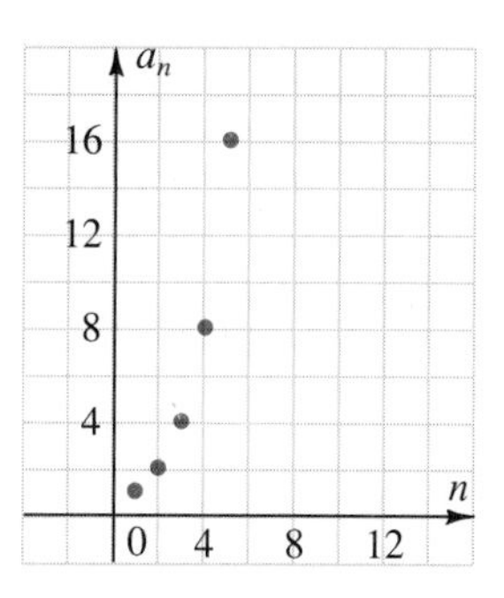

You can find the common ratio, r, by dividing any term in a geometric sequence by the preceding term. That is,

$$r = \frac{a_{n+1}}{a_n}$$

EXAMPLE 1 **Find the common ratio for the geometric sequence 8, 4, 2, 1, $\frac{1}{2}$, ... Then graph the sequence.**

Find the common ratio by dividing any term by the preceding term.

$$r = \frac{a_2}{a_1}$$

$$r = \frac{4}{8} = \frac{1}{2}$$

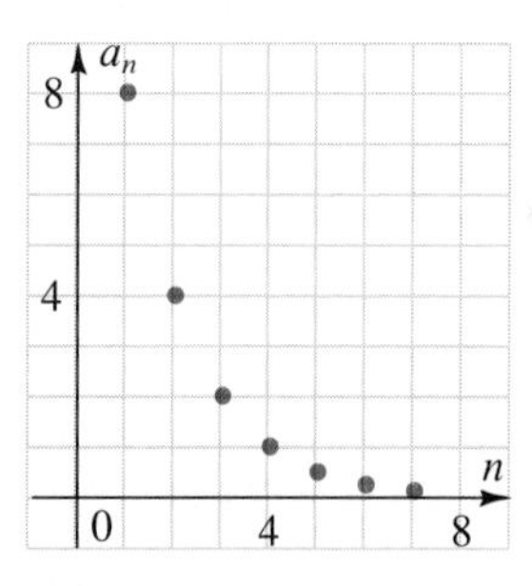

The graph of this sequence shows that as n increases without bound, the terms of the sequence get smaller and smaller. However, the terms do *not* decrease without bound. As n becomes greater, the terms get closer and closer to 0, but never equal 0. That is, a_n approaches 0 as a *limit*.

Any term after the first term in a geometric sequence is the product of the preceding term and the common ratio. You can use this fact to find a formula for the nth, or general, term of a geometric sequence.

First term	a_1	$= a_1$
Second term	$a_2 = a_1 \cdot r$	$= a_1 r$
Third term	$a_3 = a_2 \cdot r = a_1 r \cdot r$	$= a_1 r^2$
Fourth term	$a_4 = a_3 \cdot r = a_1 r^2 \cdot r$	$= a_1 r^3$
Fifth term	$a_5 = a_4 \cdot r = a_1 r^3 \cdot r$	$= a_1 r^4$

A formula for the nth, or general term of a geometric sequence is

$$a_n = a_1 r^{n-1}$$

where a_1 is the first term and r is the common ratio.

EXAMPLE 2 **Consider the sequence formed by the depreciating values of the car described on page 634. Write the first term of this sequence and the common ratio. Then find the value of the car during the fifth year.**

The first term a_1 is the purchase price of the car. The car depreciates by $\frac{1}{4}$ of its value each year, so its value each year is $1 - \frac{1}{4}$, or $\frac{3}{4}$ of its value the preceding year. The common ratio is $\frac{3}{4}$. That is,

$a_1 = 19{,}200$
$a_2 = 14{,}400$
$a_3 = 10{,}800$

$$\frac{14{,}400}{19{,}200} = \frac{3}{4} = r$$

$a_n = a_1 r^{n-1}$ *Use the formula for the nth term.*

$a_5 = 19{,}200\left(\frac{3}{4}\right)^{5-1}$ *Substitute: $n = 5$, $a_1 = 19{,}200$, $r = \frac{3}{4}$*

$a_5 = 19{,}200(0.75)^4$ *Calculation-ready form*

$a_5 = 6075$

During the fifth year, the car will be worth about \$6000.

The calculation of higher powers of numbers can be tedious, so it is helpful to use a calculator when you work with geometric sequences.

EXAMPLE 3 **Find the eighth term of the geometric sequence 2, −6, 18, −54, ...**

$r = \frac{a_{n+1}}{a_n}$ *Find the common ratio.*

$r = \frac{-6}{2} = -3$

$a_n = a_1 r^{n-1}$ *Use the formula for the nth term.*

$a_8 = 2(-3)^{8-1}$ *Substitute: $n = 8$, $a_1 = 2$, $r = -3$*

$a_8 = 2(-3)^7$ *Calculation-ready form*

$a_8 = -4374$ The eighth term is −4374.

Geometric means are the terms between any two given terms of a geometric sequence. Any number of geometric means can be inserted between two given numbers.

EXAMPLE 4 **Insert the three geometric means between 3 and 768.**

Let $a_1 = 3$ and $a_5 = 768$. Then the sequence can be represented by

$3, a_2, a_3, a_4, 768$ *There are three missing terms.*

Use the formula for the nth term to find the common ratio, r.

$$a_n = a_1 r^{n-1}$$
$$a_5 = a_1 r^{5-1} \quad \text{Substitute: } n = 5$$
$$768 = 3r^4 \quad \text{Substitute: } a_5 = 768,\ a_1 = 3$$
$$256 = r^4$$
$$\pm\sqrt[4]{256} = r$$
$$\pm 4 = r \quad \text{Use the positive and negative fourth roots.}$$

When $r = 4$, $a_1 = 3$
$a_2 = 3 \cdot 4 = 12$
$a_3 = 12 \cdot 4 = 48$
$a_4 = 48 \cdot 4 = 192$
$a_5 = 192 \cdot 4 = 768$

When $r = -4$, $a_1 = 3$
$a_2 = 3 \cdot -4 = -12$
$a_3 = -12 \cdot -4 = 48$
$a_4 = 48 \cdot -4 = -192$
$a_5 = -192 \cdot -4 = 768$

The three geometric means are 12, 48, and 192 or -12, 48, and -192.

A single geometric mean between two terms of a geometric sequence is called **the geometric mean** or **mean proportional** between the two terms. m is the mean proportional between the two real numbers a and b, $b \neq 0$, if

$$\frac{a}{m} = \frac{m}{b}$$

Thus, $m = \sqrt{ab}$ or $m = -\sqrt{ab}$. In order to make the geometric mean unique, it is customary to let $m = \sqrt{ab}$ when a and b are positive and $m = -\sqrt{ab}$ when a and b are negative. In this book, only *real* geometric means (geometric means of pairs of real numbers with the same sign) will be considered.

EXAMPLE 5 Find the mean proportional, m, between 50 and 98.

$$m = \sqrt{ab} \quad a \text{ and } b \text{ are positive.}$$
$$m = \sqrt{50 \cdot 98}$$
$$m = \sqrt{25 \cdot 2 \cdot 2 \cdot 49}$$
$$m = \sqrt{25 \cdot 4 \cdot 49} = 5 \cdot 2 \cdot 7 = 70$$

Check: $\frac{50}{70} \stackrel{?}{=} \frac{70}{98}$

$\frac{5}{7} = \frac{5}{7}$ ✓

The mean proportional is 70.

CLASS EXERCISES

1. If an odd number of real geometric means are inserted between two numbers, there are two possible values of the common ratio, r. How many values of r are there if an even number of real geometric means are inserted between two numbers? Give a reason for your answer.

Determine whether or not each sequence is geometric. Find r for each geometric sequence.

2. 2, 10, 50, 250

3. 1, 6, 36, 216

4. 2, 4, 6, 8

Find the specified term of each geometric sequence.

5. $a_1 = 1, r = 3; a_5$

6. $a_1 = -1, r = 2; a_7$

7. $a_1 = 2, r = \frac{1}{2}; a_9$

8. $a_1 = 4, r = -\frac{1}{2}; a_{10}$

9. $3, 15, 75, 375, \ldots; a_{11}$

10. $512, 256, 128, 64, \ldots; a_{20}$

11. Insert the three geometric means between 4 and 324.

12. Find the mean proportional of 17 and 612. Check your answer.

PRACTICE EXERCISES

Use technology where appropriate.

Graph each geometric sequence.

1. $\frac{1}{4}, \frac{1}{2}, 1, 2, \ldots$

2. $16, -8, 4, -2, \ldots$

Find the specified term of each geometric sequence.

3. $a_1 = 7, r = 2; a_4$

4. $a_1 = -3, r = -2; a_7$

5. $a_1 = 9, r = \frac{1}{3}; a_5$

6. $a_1 = 64, r = -\frac{1}{2}; a_{11}$

7. $a_1 = 7, r = 3; a_8$

8. $a_1 = -4, r = -3; a_{10}$

9. $2, 8, 32, 128, \ldots; a_9$

10. $1, \frac{1}{4}, \frac{1}{16}, \frac{1}{64}, \ldots; a_7$

11. $-16, -8, -4, -2, \ldots; a_{15}$

12. $-3, 3, -3, 3, \ldots; a_{30}$

13. Insert the three geometric means between 1 and 256.

14. Insert the three geometric means between 2 and 162.

15. Insert the four geometric means between 3 and 96.

16. Insert the two geometric means between 10 and 10,000.

17. Find the mean proportional between 6 and 216. Check your answer.

18. Find the mean proportional between -10 and -1000. Check your answer.

Find the specified term of each geometric sequence.

19. $a_1 = 0.3, r = 0.2; a_5$

20. $a_1 = 0.5, r = -0.4; a_7$

21. $a_1 = 10^5$, $r = 10^3$; a_{11}

22. $a_1 = 10^{-3}$, $r = 10^{-5}$; a_{15}

23. $a_1 = 10$, $r = 1$; a_{25}

24. $a_1 = 25$, $r = -1$; a_{37}

25. $3x, 6x, 12x, 24x, \ldots; a_{11}$

26. $2z, 6z^2, 18z^3, 54z^4, \ldots; a_9$

Find a_1 for a geometric sequence with the given terms.

27. $a_5 = 112$ and $a_7 = 448$

28. $a_9 = \frac{1}{2}$ and $a_{12} = \frac{1}{16}$

29. $a_4 = -375$ and $a_7 = 46{,}875$

30. $a_{15} = 25$ and $a_{20} = -25$

31. If $a_1 = -10$ and $r = 0.1$, which term in the geometric sequence is -10^{-16}?

32. If $a_1 = x$ and $r = x^3$, which term in the geometric sequence is x^{22}?

33. Prove that if $a_1, a_2, a_3, a_4, \ldots$ is a geometric sequence, then $ka_1, ka_2, ka_3, ka_4, \ldots$ is also a geometric sequence.

34. Prove that if $a_1, a_2, a_3, a_4, \ldots$ is a geometric sequence, then $(a_1)^2, (a_2)^2, (a_3)^2, (a_4)^2, \ldots$ is also a geometric sequence.

Applications

35. Investment Karen invests \$100 in a mutual fund which claims her money will double by the end of each year. If this is what happens, how much money will she have at the beginning of the 8th year?

36. Geometry Concentric circles are drawn so that each circle has a radius 3 times as long as a radius of the circle preceding it. The smallest circle has a 2-in. radius. Find the length of a diameter of the eleventh circle.

37. Consumerism Victor bought a new car for \$12,000. If the car's value depreciates 20% per year, what will it be worth during the fourth year?

Developing Mathematical Power

38. Thinking Critically Prove that if a and b are positive real numbers, then their arithmetic mean is greater than or equal to their geometric mean. That is, show that

$$\frac{a + b}{2} \geq \sqrt{ab}$$

Also, give the circumstances under which the arithmetic mean is equal to the geometric mean.

Hint: Start with the true statement $(a - b)^2 \geq 0$.

13.5

Arithmetic Series

Objectives: To use sigma notation to represent the sum of a series
To find a partial sum of an arithmetic series

The general terms of many sequences are described by formulas. You can generate any specified number of terms of a sequence by evaluating its formula for the required values of n.

Capsule Review

EXAMPLE **Find the first four terms of the sequence with the formula:** $a_n = \frac{1}{n^2 - 2}$

$a_1 = \frac{1}{1^2 - 2} = \frac{1}{-1} = -1$ $\quad n = 1$

$a_2 = \frac{1}{2^2 - 2} = \frac{1}{4 - 2} = \frac{1}{2}$ $\quad n = 2$

$a_3 = \frac{1}{3^2 - 2} = \frac{1}{9 - 2} = \frac{1}{7}$ $\quad n = 3$

$a_4 = \frac{1}{4^2 - 2} = \frac{1}{16 - 2} = \frac{1}{14}$ $\quad n = 4$

Find the first four terms of the sequence with the given formula.

1. $a_n = 7n$ **2.** $a_n = 6n + 3$ **3.** $a_n = 8n - 2$ **4.** $a_n = 4n - 1$

5. $a_n = n^3 - 2$ **6.** $a_n = \frac{n}{3} + 1$ **7.** $a_n = \frac{3}{n}$ **8.** $a_n = 2(3)^n$

A **series** is the indicated sum of the terms of a sequence. For example

Finite sequence: 3, 7, 11, 15
Finite series: $3 + 7 + 11 + 15$

Infinite sequence: 6, 8, 10, 12, . . .
Infinite series: $6 + 8 + 10 + 12 + \cdots$

An **arithmetic series** is the indicated sum of an arithmetic sequence. The arithmetic series $1 + 3 + 5 + 7 + \cdots$ (indicated sum of the positive odd integers) is an infinite arithmetic series. The sum of the first n terms of an arithmetic series, denoted S_n, is the nth **partial sum** of the series. The partial sum of the first four terms, S_4, of the series $1 + 3 + 5 + 7 + \cdots$ is 16, since $1 + 3 + 5 + 7 = 16$.

EXAMPLE 1 **Find the partial sum, S_7, for the arithmetic series $1 + 3 + 5 + 7 + \cdots$.**

$d = a_{n+1} - a_n = 3 - 1 = 2$ *Find d, the common difference.*

$a_5 = 7 + 2 = 9$ *Use d to find a_5, a_6, and a_7.*

$a_6 = 9 + 2 = 11$

$a_7 = 11 + 2 = 13$

$S_7 = 1 + 3 + 5 + 7 + 9 + 11 + 13 = 49$ The partial sum is 49.

The sum of a series may be written using a shorthand method called **sigma (Σ) notation,** or **summation notation.** For example, to write S_7 in Example 1 using sigma notation, first observe that the general term of the series is the same as the general term of the corresponding sequence, $2n - 1$. Then S_7 may be written as

$$S_7 = \sum_{n=1}^{7} (2n - 1)$$

Upper limit of summation (7); Sum of first seven terms (S_7); General, or nth term ($2n - 1$); Index of summation (n); Lower limit of summation ($n = 1$)

This is read "the sum from $n = 1$ to $n = 7$ of $2n - 1$." Note that any letter may be used as the index of summation.

EXAMPLE 2 **Use sigma notation to write the arithmetic series $3 + 6 + \cdots$ for 33 terms.**

$$3 + 6 + 9 + 12 + \cdots + 99 = \sum_{n=1}^{33} 3n$$ *$3 \cdot 1 = 3$; $3 \cdot 2 = 6$; $3 \cdot 3 = 9$*

To find the partial sum of a series, first write the series in *expanded form* by replacing the index successively with the integers, starting with the lower limit of summation and ending with the upper limit. Then add the resulting terms.

EXAMPLE 3 **Write $\sum_{k=1}^{5} (3k + 1)$ in expanded form and find S_5.**

$$S_5 = \sum_{k=1}^{5} (3k + 1)$$ *The limits of summation are 1 and 5, so write the indicated sum of the first five terms. Let $k = 1, 2, 3, 4$, and 5.*

$= [3(1) + 1] + [3(2) + 1] + [3(3) + 1] + [3(4) + 1] + [3(5) + 1]$

$= 4 + 7 + 10 + 13 + 16 = 50$ The partial sum, S_5, is 50.

To find a formula for the partial sum, S_n, of an arithmetic series, write S_n in two ways, as shown below, and add.

$$\begin{aligned}
S_n &= a_1 + (a_1 + d) + (a_1 + 2d) + \cdots + (a_n - d) + a_n \\
+\ S_n &= a_n + (a_n - d) + (a_n - 2d) + \cdots + (a_n + d) + a_1 \\
\hline
2S_n &= (a_1 + a_n) + (a_1 + a_n) + (a_1 + a_n) + \cdots + (a_1 + a_n) + (a_1 + a_n)
\end{aligned}$$

Since there are n terms in the series S_n, there are n terms of the form $(a_1 + a_n)$ in $2S_n$. That is,

$$2S_n = n(a_1 + a_n)$$

$$S_n = \frac{n(a_1 + a_n)}{2} \qquad \textit{Sum of the first n terms of an arithmetic series}$$

EXAMPLE 4 **Find the sum of the first fifteen terms, S_{15}, of an arithmetic series if $a_1 = -5$ and $a_{15} = 23$.**

$$S_n = \frac{n(a_1 + a_n)}{2} \qquad \textit{Use the formula.}$$

$$S_{15} = \frac{15(-5 + 23)}{2} \qquad n = 15,\ a_1 = -5,\ a_{15} = 23$$

$$S_{15} = \frac{15(18)}{2} = 135 \qquad \text{The sum of the first fifteen terms is 135.}$$

A second formula for the sum of the first n terms of an arithmetic series can be derived from the first sum formula and the formula for the nth term of an arithmetic sequence. This formula is useful when the nth term is not given.

$$S_n = \frac{n(a_1 + a_n)}{2}$$

$$S_n = \frac{n[a_1 + a_1 + (n - 1)d]}{2} \qquad a_n = a_1 + (n - 1)d$$

$$S_n = \frac{n[2a_1 + (n - 1)d]}{2}$$

EXAMPLE 5 **Find the sum of the first ten terms, S_{10}, of the arithmetic series $2 + 5 + 8 + 11 + \cdots$.**

$$d = a_{n+1} - a_n = 8 - 5 = 3 \qquad \textit{Find d.}$$

$$S_n = \frac{n[2a_1 + (n - 1)d]}{2} \qquad \textit{Use the formula.}$$

$$S_{10} = \frac{10[2(2) + (10 - 1)3]}{2} \qquad n = 10,\ a_1 = 2,\ d = 3$$

$$S_{10} = 5(4 + 27) = 155 \qquad \text{The sum of the first ten terms is 155.}$$

The formula $S_n = \frac{n(a_1 + a_n)}{2}$ can also be used to find the partial sum of a series expressed in sigma notation without writing the expanded form.

EXAMPLE 6 **Find the sum of the arithmetic series $\sum_{n=1}^{20} (2n - 1)$.**

$a_1 = 2(1) - 1 = 1; a_{20} = 2(20) - 1 = 39$ *Find a_1 and a_{20}, $a_n = 2n - 1$*

$S_n = \frac{n(a_1 + a_n)}{2}$ *Use the formula.*

$S_{20} = \frac{20(1 + 39)}{2} = 400$ The sum of the arithmetic series is 400.

> The formulas for finding the partial sum, S_n, of an arithmetic series are
>
> $$S_n = \frac{n(a_1 + a_n)}{2} \quad \text{and} \quad S_n = \frac{n[2a_1 + (n - 1)d]}{2}$$
>
> where n is the number of terms, a_1 is the first term, a_n is the nth term, and d is the common difference.

CLASS EXERCISES

Complete each statement.

1. In the expression $\sum_{p=4}^{9} (3p^2 - 2)$, the general term is __?__.

2. In the expression $\sum_{a=5}^{10} (2a^3 + 6)$, the limits of summation are __?__.

3. In the expression $\sum_{n=1}^{6} n^2$, n is the __?__.

4. In the expression $\sum_{x=2}^{15} (x^2 - 1)$, the index of summation is __?__

Express the sum of each series using sigma notation.

5. $a_n = n + 7$, from $n = 4$ to $n = 15$

6. $a_n = -2n + 1$, from $n = 1$ to $n = 23$

Write each arithmetic series in expanded form.

7. $\sum_{n=1}^{4} 3n$

8. $\sum_{k=1}^{7} (2k + 1)$

9. $\sum_{k=1}^{5} (5k + 3)$

10. $\sum_{n=1}^{6} (4n - 2)$

PRACTICE EXERCISES

Use technology where appropriate.

Write each arithmetic series in expanded form and find the indicated partial sum.

1. $\sum_{n=1}^{3} 4n$ **2.** $\sum_{k=1}^{7} 5k$ **3.** $\sum_{k=1}^{6} (3k + 2)$ **4.** $\sum_{k=1}^{5} (4k - 3)$

Find the specified partial sum for each arithmetic series.

5. $5 + 10 + 15 + 20 + \cdots; S_8$

6. $15 + 17 + 19 + 21 + \cdots; S_{11}$

7. $20 + 30 + 40 + 50 + \cdots; S_{20}$

8. $1 + 7 + 13 + 19 + \cdots; S_{16}$

9. $-6 - 9 - 12 - 15 + \cdots; S_{12}$

10. $1 + 1 + 1 + 1 + \cdots; S_{100}$

11. $a_1 = 6, a_6 = 30; S_6$

12. $a_1 = 4, a_{11} = 64; S_{11}$

13. $a_1 = -1, a_{14} = -55; S_{14}$

14. $a_1 = -9, a_{10} = -40; S_{10}$

15. $a_1 = 9, a_4 = -27; S_4$

16. $a_1 = -16, a_5 = -84; S_5$

17. $\sum_{n=1}^{8} 5n$ **18.** $\sum_{n=1}^{8} (2n + 3)$ **19.** $\sum_{k=1}^{10} (k - 4)$ **20.** $\sum_{k=1}^{15} (20 - k)$

Write each arithmetic series for the specified number of terms. Use sigma $\left(\sum\right)$ notation.

21. $1 + 2 + 3 + 4 + \cdots; n = 6$

22. $2 + 4 + 6 + 8 + \cdots; n = 4$

23. $8 + 9 + 10 + 11 + \cdots; n = 8$

24. $5 + 6 + 7 + 8 + \cdots; n = 7$

25. $1 + 4 + 7 + 10 + \cdots; n = 11$

26. $3 + 8 + 13 + 18 + \cdots; n = 9$

Find the specified term for each arithmetic series.

27. $a_1 = -6$ and $S_{20} = -790; a_{20}$

28. $a_1 = -432$ and $S_{30} = -660; a_{30}$

29. $a_1 = 4$ and $S_{40} = 6080; a_{40}$

30. $a_1 = -6$ and $S_{50} = -5150; a_{50}$

Find a_1 for each arithmetic series.

31. $S_{20} = 80$ and $d = 2$

32. $S_{30} = 240$ and $d = -2$

33. $S_8 = 440$ and $d = 6$

34. $S_{100} = 5050$ and $d = 1$

Find the specified partial sum for each arithmetic series.

35. $x + (x + y) + (x + 2y) + \cdots; S_{10}$

36. $3x + (3x - 2y) + (3x - 4y) + (3x - 6y) + \cdots; S_{15}$

37. If $\sum_{n=5}^{8} (2nx - 6) = 28$, find the value of x.

38. Prove that the sum of the first n positive integers is $\frac{n(n + 1)}{2}$.

39. Prove that the sum of the first n positive odd integers is n^2.

40. Prove that the sum of the first n positive even integers is $n(n + 1)$.

Applications

41. **Geometry** A triangular garden has 20 plants in the first row. Each succeeding row has 12 more plants than the one before it. How many plants are there if the garden consists of 8 rows?

42. **Sales** Bill was earning \$80 per month delivering newspapers. He planned to add \$5 worth of new customers per month for one year. How much will he have earned at the end of one year if his plan works out? *Hint:* The first month he earns \$85.

43. **Geometry** Equilateral triangle ABC is drawn with each side 50 cm long. Equilateral triangle DEF is drawn with each side 5 cm shorter than a side of triangle ABC. Equilateral triangle GHI is drawn with each side 5 cm shorter than a side of triangle DEF. If this pattern is continued, find the sum of the perimeters of the first six triangles.

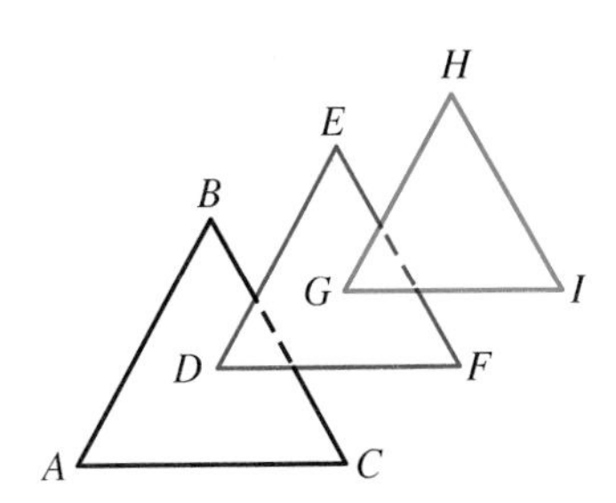

TEST YOURSELF

Use the given formula to find the first three terms of each sequence.

1. $a_n = n^2 - 1$

2. $a_1 = 0;\ a_{n+1} = a_n + 3$ **13.1–13.2**

3. Find the fifteenth term of the arithmetic sequence 3, 8, 13, 18, ... **13.3**

4. Find the two arithmetic means between 15 and 30.

5. Find the eighth term of the geometric sequence −24, −12, −6, −3, ... **13.4**

6. Find the three geometric means between 3 and 48.

Find the indicated sum of each arithmetic series.

7. $1 + 8 + 15 + 22 + \cdots;\ S_{12}$

8. $\sum_{n=1}^{15} (3n - 2)$ **13.5**

13.6 Geometric Series

Objective: To find a partial sum of a geometric series

The indicated sum of the terms of an arithmetic sequence is an arithmetic series. In a similar way, a *geometric series* can be formed from a geometric sequence.

Capsule Review

The formula for the nth, or general, term of a geometric sequence is $a_n = a_1 r^{n-1}$, where n is the number of the term, a_1 is the first term, and r is the common ratio. Any term after the first term in a geometric sequence can be found by multiplying the preceding term by r. That is, $a_{n+1} = ra_n$.

EXAMPLE **Find the specified term of the geometric sequence:**

a. $a_1 = 10, r = \frac{1}{2}; a_5$ **b.** $a_{10} = -8, r = -3; a_{11}$

a. $a_n = a_1 r^{n-1}$

$a_5 = 10\left(\frac{1}{2}\right)^{5-1}$

$a_5 = \frac{10}{16} = \frac{5}{8}$

b. $a_{n+1} = ra_n$

$a_{11} = (-3)a_{10}$

$a_{11} = (-3)(-8) = 24$

Find the specified term of each geometric sequence.

1. $a_1 = 5, r = 4; a_7$ **2.** $a_1 = -3, r = -2; a_8$ **3.** $a_{10} = 9, r = -2; a_{11}$

4. $a_8 = 1, r = -1; a_9$ **5.** $a_4 = -1, r = 3; a_6$ **6.** $a_n = a_n, r = r; a_{n-1}$

A **geometric series** is the indicated sum of the terms of a geometric sequence. The series $2 + 6 + 18 + 54$ is a finite geometric series, since there is a limited, or finite, number of terms. The series $1 + 5 + 25 + 125 + \cdots$ is an infinite geometric series, since there is an unlimited, or infinite, number of terms.

The partial sum, S_n, of a geometric series is the sum of the first n terms of the series. Therefore,

$$S_n = a_1 + a_1 r + a_1 r^2 + \cdots + a_1 r^{n-1} \quad \text{Sum of a geometric series}$$

EXAMPLE 1 **Find S_6 for the geometric series $1 + 2 + 4 + 8 + \cdots$.**

$$r = \frac{a_{n+1}}{a_n} = \frac{2}{1} = 2 \qquad \textit{Find the common ratio.}$$

$$a_5 = 2(8) = 16 \qquad a_6 = 2(16) = 32 \qquad \textit{Find } a_5 \textit{ and } a_6.\ a_{n+1} = ra_n$$

$$S_6 = 1 + 2 + 4 + 8 + 16 + 32 = 63 \qquad \text{The partial sum is 63.}$$

To find a formula for the partial sum, S_n, of a geometric series, multiply S_n by r and subtract the product from S_n.

$$\begin{aligned} S_n &= a_1 + a_1r + a_1r^2 + a_1r^3 + \cdots + a_1r^{n-1} \\ rS_n &= \phantom{a_1 + {}} a_1r + a_1r^2 + a_1r^3 + \cdots + a_1r^{n-1} + a_1r^n \\ \hline S_n - rS_n &= a_1 - a_1r^n \\ S_n(1 - r) &= a_1(1 - r^n) \qquad \textit{Factor.} \\ S_n &= \frac{a_1(1 - r^n)}{1 - r} \quad r \neq 1 \qquad \textit{Divide each side by } (1 - r). \end{aligned}$$

Note that if S_n were subtracted from rS_n, the formula would have the form

$$S_n = \frac{a_1(r^n - 1)}{r - 1} \quad r \neq 1$$

EXAMPLE 2 **Find the sum of the first five terms, S_5, of the geometric series $2 + 6 + 18 + \cdots$.**

$$r = \frac{a_{n+1}}{a_n} = \frac{6}{2} = 3 \qquad \textit{Find the common ratio, r.}$$

$$S_n = \frac{a_1(1 - r^n)}{1 - r} \qquad \textit{Use the formula.}$$

$$S_5 = \frac{2(1 - 3^5)}{1 - 3} \qquad n = 5,\ a_1 = 2,\ r = 3$$

$$S_5 = \frac{2(1 - 243)}{-2} = 242 \qquad \text{The sum of the first five terms is 242.}$$

A second formula for finding the partial sum, S_n, of a geometric series can be derived from the first formula. This formula is useful when the first and nth terms are known. The nth term of a geometric series is the same as the nth term of the corresponding sequence.

Since $a_n = a_1r^{n-1}$, then $ra_n = r(a_1r^{n-1}) = a_1r^n$. Substitute in the formula.

$$S_n = \frac{a_1(1 - r^n)}{1 - r} = \frac{a_1 - a_1r^n}{1 - r} = \frac{a_1 - ra_n}{1 - r},\ r \neq 1$$

EXAMPLE 3 **In a geometric series, $a_1 = 9$, $r = -2$, and $a_8 = -1152$. Find S_8.**

$$S_n = \frac{a_1 - ra_n}{1 - r} \quad \text{Use the formula.}$$

$$S_8 = \frac{9 - (-2)(-1152)}{1 - (-2)} = -765 \quad \text{The partial sum, } S_8\text{, is } -765.$$

> The formulas for the partial sum, S_n, of a geometric series are
>
> $$S_n = \frac{a_1(1 - r^n)}{1 - r},\ r \neq 1 \quad \text{and} \quad S_n = \frac{a_1 - ra_n}{1 - r},\ r \neq 1$$
>
> where n is the number of terms, a_1 is the first term, r is the common ratio, and a_n is the last term.

If a geometric series has exactly n terms, then S_n is *the sum of the series.* A geometric series can also be written using sigma $\left(\sum\right)$ notation.

EXAMPLE 4 **Find the sum of the geometric series $\sum_{n=1}^{9} 7(2)^{n-1}$.**

$r = 2$

$a_1 = 7(2)^{1-1} = 7(1) = 7$

$$S_n = \frac{a_1(1 - r^n)}{1 - r} \quad \text{Use the formula.}$$

$$S_9 = \frac{7(1 - 2^9)}{1 - 2} \quad n = 9$$

$$S_9 = \frac{7(-511)}{-1} = 3577 \quad \text{The sum of the series is 3577.}$$

CLASS EXERCISES

1. **Thinking Critically** Why is the restriction $r \neq 1$ needed for the formulas for S_n?

Find S_6 for each geometric series.

2. $5 + 15 + 45 + \cdots$
3. $8 + 24 + 72 + \cdots$
4. $9 - 36 + 144 - \cdots$
5. $2 - 16 + 128 - \cdots$
6. $1 + \frac{1}{2} + \frac{1}{4} + \cdots$
7. $1 - \frac{1}{2} + \frac{1}{4} - \cdots$

Identify a_1, r, and n for each geometric series.

8. $\sum_{n=1}^{6} (2)^{n-1}$ **9.** $\sum_{n=1}^{7} 2(3)^{n-1}$ **10.** $\sum_{n=1}^{5} 6^n$ **11.** $\sum_{n=1}^{15} \left(\frac{1}{2}\right)^n$

PRACTICE EXERCISES

Use technology where appropriate.

Find the indicated sum for each geometric series.

1. $1 + 2 + 4 + \cdots; S_8$

2. $3 + 6 + 12 + \cdots; S_7$

3. $4 + 12 + 36 + \cdots; S_6$

4. $7 - 35 + 175 - \cdots; S_5$

5. $7 + 70 + 700 + \cdots; S_7$

6. $11 + 33 + 99 + \cdots; S_8$

7. $-2 - 6 - 18 - \cdots; S_9$

8. $-5 - 10 - 20 - \cdots; S_{11}$

9. $4 + 2 + 1 + \cdots; S_7$

10. $36 - 6 + 1 - \cdots; S_5$

11. $a_1 = 2, a_7 = 1458, r = 3; S_7$

12. $a_1 = -1, a_{10} = 512, r = -2; S_{10}$

13. $a_1 = 1, a_{11} = \frac{1}{1024}, r = \frac{1}{2}; S_{11}$

14. $a_1 = 1024, a_{11} = 1, r = \frac{1}{2}; S_{11}$

15. $\sum_{n=1}^{6} 3(4)^{n-1}$ **16.** $\sum_{k=1}^{5} 5(3)^{k-1}$ **17.** $\sum_{k=1}^{7} \left(-\frac{1}{2}\right)^{k-1}$ **18.** $\sum_{n=1}^{8} 2\left(\frac{1}{3}\right)^{n-1}$

Classify the series as arithmetic or geometric and find S_n.

19. $\sum_{n=1}^{7} (2n - 1)$ **20.** $\sum_{k=1}^{4} 3^k$ **21.** $\sum_{n=1}^{6} (3n - 1)$ **22.** $\sum_{n=1}^{7} 5(2)^{n+1}$

23. $\sum_{n=1}^{7} 3(3)^{n-2}$ **24.** $\sum_{n=1}^{5} 4^{n-2}$ **25.** $\sum_{n=1}^{4} x^{n-1}$ **26.** $\sum_{k=1}^{8} 2(x)^{k+1}$

Find the specified value, using the given information about the geometric series.

27. $r = 2, S_8 = 1020; a_1 = \underline{\ ?\ }$

28. $r = 3, S_6 = 364; a_1 = \underline{\ ?\ }$

29. $r = -2, S_7 = 86; a_1 = \underline{\ ?\ }$

30. $r = -1, S_{11} = 7; a_1 = \underline{\ ?\ }$

Find the indicated sum for each geometric series.

31. $x + 2x^2 + 4x^3 + \cdots; S_8$

32. $xy + xy^3 + xy^5 + \cdots; S_{11}$

33. $\sum_{n=1}^{5} y(x^3)^{n-1}$

34. $\sum_{n=1}^{4} y^3(x^4)^n$

35. $a_1 = xy^4, r = x^2y^{-2}; S_6$

36. $a_1 = x^3y^{-2}, r = x^{-2}y^5; S_7$

Applications

37. Art Eileen wants to make a collage using squares. She cuts paper squares in such a way that the area of each square is 4 times the area of the preceding square. She makes five squares, and the area of the first square is 2 in.^{2}. What is the total area of the squares?

38. Physics On the first swing, a pendulum makes an arc 8 in. long. On each successive swing, the arc is $\frac{1}{4}$ the length of the preceding arc. Find the total distance the pendulum has traveled at the end of the fourth swing.

39. Finance Jon is promised a 10% raise after each of the first five years he works for a company. If his initial yearly salary if $15,500, what is the total amount of money the company will have paid him by the end of the fifth year?

40. Finance Jennifer has received a 5% salary increase after each year that she has worked at the state automobile license bureau. If she earned $68,962 during the four years she has worked there, how much did she earn during her first year on the job?

Mixed Review

Simplify.

41. $\dfrac{3x^2 - 6x - 24}{2x^2 - 2x - 4} \cdot \dfrac{4x^2 - 8x - 12}{6x^2 - 42x + 72}$

42. $\dfrac{5\sqrt{2x^3}}{2x\sqrt{5xy^5}}$

43. $\sqrt[3]{16x^2} \cdot \sqrt[3]{6x^2y}$

44. $\sqrt{24x^3y^{-1}}$

45. $\sqrt{27} + 3\sqrt[3]{54x} - \sqrt[3]{x} + 3\sqrt{12}$

46. $\dfrac{3}{2 + \sqrt{2}}$

47. $\sqrt[3]{-16x} + \sqrt[3]{-54x} + \sqrt[3]{2x}$

48. $\sqrt{-24y}$

49. $i\sqrt{-72} - 2\sqrt{50}$

50. i^{23}

51. Find the number you would add to make $x^2 - \frac{3}{4}x + \underline{\ ?\ }$ a perfect square trinomial.

52. Determine whether $y = -2x^2 + 5$ has a maximum or a minimum. Then find that value.

53. Determine the value of k for which $kx^2 - 2x = -4$ has exactly one solution.

54. Write a quadratic equation for which -2 and $\frac{4}{5}$ are solutions.

55. Solve $c^4 - 3c^2 - 10 = 0$.

13.7

Infinite Geometric Series

Objective: To find the sum of an infinite geometric series

When you deposit money in a bank, the bank may use part of that money for loans to businesses, which in turn may loan part of the money to their suppliers, and so forth. The total effect this has on the economy is called the *multiplier effect*. The multiplier effect is the sum of a certain type of infinite geometric series.

When you evaluate the partial sum of an infinite geometric series, it is sometimes necessary to simplify a complex fraction.

Capsule Review

EXAMPLE **Simplify:** $\dfrac{1-\frac{1}{5}}{\frac{1}{3}}$

$$\frac{1-\frac{1}{5}}{\frac{1}{3}} = \frac{15\cdot\left(1-\frac{1}{5}\right)}{15\cdot\frac{1}{3}} = \frac{15-3}{5} = \frac{12}{5}$$

Multiply the numerator and the denominator by 15, the LCD of $\frac{1}{5}$ and $\frac{1}{3}$.

Simplify.

1. $\dfrac{1}{1-\frac{1}{4}}$

2. $\dfrac{2+\frac{1}{16}}{\frac{1}{3}}$

3. $\dfrac{4+\frac{3}{8}}{\frac{5}{6}}$

4. $\dfrac{\frac{1}{2}-\frac{1}{3}}{\frac{1}{4}}$

5. $\dfrac{x+\frac{1}{x}}{\frac{1}{y}}$

6. $\dfrac{\frac{1}{x}+\frac{1}{y}}{\frac{1}{xy^2}}$

Consider the geometric series $1 + \frac{1}{2} + \frac{1}{4} + \cdots$. Since this series is infinite, it is not possible to find the sum by adding all the terms. Instead, look at the partial sums of the series and try to find a pattern. Note that the partial sums of a geometric series, $S_1, S_2, S_3, \ldots, S_n, \ldots$, form a sequence.

EXAMPLE 1 **Find the partial sums S_1, S_2, S_3, S_{12}, S_{20}, and S_n of the infinite geometric series $1 + \frac{1}{2} + \frac{1}{4} + \cdots$.**

$$a_1 = 1$$

$$r = \frac{a_{n+1}}{a_n} = \frac{\frac{1}{2}}{1} = \frac{1}{2} \qquad \textit{Find r.}$$

$$S_1 = 1$$

$$S_2 = 1 + \frac{1}{2} = 1\frac{1}{2}$$

$$S_3 = 1 + \frac{1}{2} + \frac{1}{4} = 1\frac{3}{4}$$

$$S_{12} = 1 + \frac{1}{2} + \frac{1}{4} + \cdots + \frac{1}{2048} = 1\frac{2047}{2048} \approx 2$$

$$S_{20} = 1 + \frac{1}{2} + \frac{1}{4} + \cdots + \frac{1}{524{,}288} = 1\frac{524{,}287}{524{,}288} \approx 2$$

$$S_n = 1 + \frac{1}{2} + \frac{1}{4} + \cdots + \left(\frac{1}{2}\right)^{n-1} = \frac{1\left[1 - \left(\frac{1}{2}\right)^n\right]}{1 - \frac{1}{2}} = 2 - \left(\frac{1}{2}\right)^{n-1}$$

$$a_n = a_1 r^{n-1}$$

$$S_n = a_1 \frac{(1 - r^n)}{1 - r}$$

The graph for the sequence of partial sums in Example 1 shows that as the number of terms, n, gets very large, the terms of the sequence of partial sums get closer and closer to 2 but never equal 2. In particular, by choosing an appropriate value of n, you can make S_n as close in value to 2 as you wish. That is, as n increases without bound, S_n approaches 2 as a *limit*. Since the series in Example 1 has a limit, it is said to *converge,* or to be *convergent*.

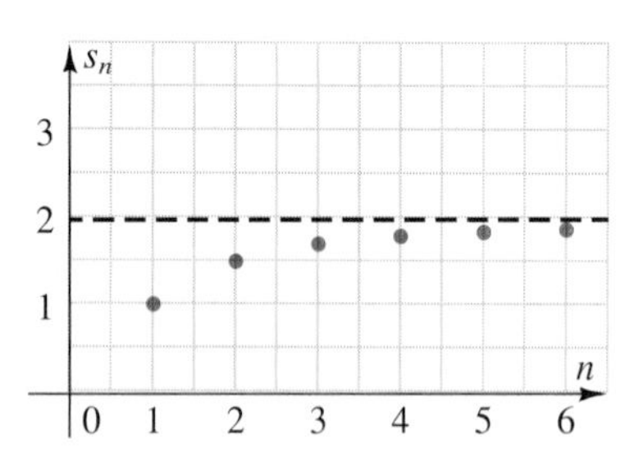

An infinite geometric series **converges** if the sequence of partial sums approaches a limit as n increases without bound. It can be shown that an infinite geometric series *converges if* $|r| < 1$, where r is the common ratio. As illustrated in Example 1, where $r = \frac{1}{2}$, as n increases without bound, $\left(\frac{1}{2}\right)^{n-1}$ gets increasingly small and approaches the limit 0. Then the sequence of partial sums also approaches a limit, since $S_n = 2 - \left(\frac{1}{2}\right)^{n-1}$ approaches the limit $2 - 0$, or 2.

EXAMPLE 2 **Find the partial sums S_1, S_2, S_3, S_4, S_5, and S_n of the infinite geometric series $3 + 6 + 12 + \cdots$.**

$$a_1 = 3 \qquad r = \frac{a_{n+1}}{a_n} = \frac{6}{3} = 2 \qquad \textit{Find r.}$$

$$
\begin{aligned}
S_1 &= 3 \\
S_2 &= 3 + 6 = 9 \\
S_3 &= 3 + 6 + 12 = 21 \\
S_4 &= 3 + 6 + 12 + 24 = 45 \\
S_5 &= 3 + 6 + 12 + 24 + 48 = 93 \\
S_n &= 3 + 6 + 12 + \cdots + 3(2)^{n-1} \\
&= \frac{3(1 - 2^n)}{1 - 2} = 3(2^n - 1) \qquad S_n = \frac{a_1(1 - r^n)}{1 - r}
\end{aligned}
$$

The graph of the sequence of partial sums in Example 2 shows that as the number of terms, n, gets very large, the terms of the sequence of partial sums also get very large. That is, as n increases without bound, S_n also increases without bound. The series in this example has no finite sum and is said to *diverge,* or to be *divergent.*

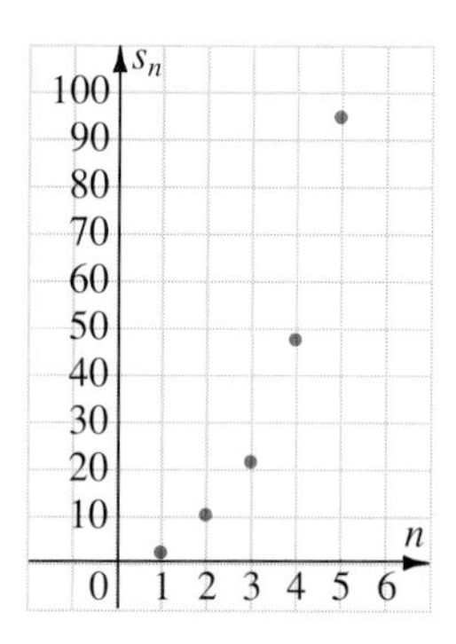

A series **diverges** if the sequence of partial sums increases or decreases without limit as the number of terms, n, gets very large.

It can be shown that an infinite geometric series *diverges if* $|r| \geq 1$, where r is the common ratio. To illustrate, in Example 2 above, r has the value 2. As n increases without bound, 2^n also increases without bound, and so does $S_n = 3(2^n - 1)$.

The sum of an infinite geometric series in which the sequence of partial sums converges to a limit L is defined as L. To find a formula for the sum, S, of a convergent infinite geometric series, consider the formula for the partial sum

$$S_n = \frac{a_1(1 - r^n)}{1 - r}, \; r \neq 1$$

You know that $|r| < 1$ for a convergent geometric series. For $|r| < 1$ the value of r^n approaches 0 as n increases without bound. Furthermore, the value of S_n approaches the limit $\frac{a_1(1 - 0)}{1 - r}$, or $\frac{a_1}{1 - r}$. This limit is the sum of a convergent geometric series.

The sum, S, of a convergent geometric series is given by the formula

$$S = \frac{a_1}{1 - r}, \; |r| < 1,$$

where a_1 is the first term and r is the common ratio.

When an infinite series is written using sigma notation, the symbol ∞ is used for the upper limit, to show that the summation does not end. This is illustrated in Example 3b.

EXAMPLE 3 **Find the sum of each infinite geometric series, if it exists:**

a. $1 - \frac{1}{3} + \frac{1}{9} - \cdots$ **b.** $\sum_{n=1}^{\infty} 5(2)^{n-1}$

a. $1 - \frac{1}{3} + \frac{1}{9} - \cdots$ *$a_1 = 1$ $r = -\frac{1}{3}$*

$S = \dfrac{1}{1 - \left(-\frac{1}{3}\right)} = \dfrac{3}{4}$ *$|r| < 1$, so the series converges and the sum exists. Use the formula $S = \dfrac{a_1}{1 - r}$.*

b. $\sum_{n=1}^{\infty} 5(2)^{n-1}$ *$a_1 = 5$ $r = 2$*

Since $|r| \geq 1$, the series diverges and the sum does not exist.

You can use the formula for the sum of a convergent geometric series to write repeating decimals as rational numbers.

EXAMPLE 4 **Write $0.\overline{3}$ (or $0.333\cdots$) as a rational number.**

$0.\overline{3} = 0.3 + 0.03 + 0.003 + \cdots$ *Write $0.\overline{3}$ as an infinite geometric series.*

$a_1 = 0.3$ $r = 0.1$ *$|r| < 1$ The series converges. Find the sum.*

$S = \dfrac{a_1}{1 - r} = \dfrac{0.3}{1 - 0.1} = \dfrac{0.3}{0.9} = \dfrac{1}{3}$ $S = \frac{1}{3}$, so $0.\overline{3} = \frac{1}{3}$.

CLASS EXERCISES

Predict whether each infinite geometric series has a sum. Verify your answer.

1. $1 + \frac{1}{4} + \frac{1}{16} + \cdots$

2. $1 - \frac{1}{2} + \frac{1}{4} - \cdots$

3. $4 + 2 + 1 + \cdots$

4. $1 + 2 + 4 + \cdots$

5. $6 + 18 + 54 + \cdots$

6. $-54 - 18 - 6 - \cdots$

7. $1 + 1 + 1 + \cdots$

8. $1 - 1 + 1 - \cdots$

Express each repeating decimal as an infinite geometric series.

9. $0.\overline{6}$ **10.** $0.\overline{18}$ **11.** $0.\overline{9}$ **12.** $0.\overline{14}$

PRACTICE EXERCISES

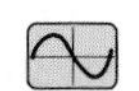 Use technology where appropriate.

Find the sum, if it exists, of each infinite geometric series. If the sum does not exist, so state.

1. $1 + \frac{1}{5} + \frac{1}{25} + \cdots$

2. $2 + 1 + \frac{1}{2} + \cdots$

3. $4 + 8 + 16 + \cdots$

4. $16 + 8 + 4 + \cdots$

5. $1 + \frac{1}{10} + \frac{1}{100} + \cdots$

6. $25 - 5 + 1 - \cdots$

7. $5 + 5 + 5 + \cdots$

8. $4 - 4 + 4 - \cdots$

9. $\frac{1}{7} + \frac{1}{49} + \frac{1}{343} + \cdots$

10. $7 + 49 + 343 + \cdots$

11. $3 - 1 + \frac{1}{3} - \cdots$

12. $-8 + 1 - \frac{1}{8} + \cdots$

13. $3 + 0.3 + 0.03 + \cdots$

14. $20 - 2 + 0.2 - \cdots$

Write each repeating decimal as a rational number.

15. $0.\overline{2}$ **16.** $0.\overline{7}$ **17.** $0.\overline{57}$ **18.** $0.\overline{43}$ **19.** $0.\overline{075}$ **20.** $0.\overline{142}$

Find the sum, if it exists, of each infinite series.

21. $\sum_{n=1}^{\infty} \left(\frac{1}{5}\right)^{n-1}$

22. $\sum_{n=1}^{\infty} 3\left(\frac{1}{4}\right)^{n-1}$

23. $\sum_{n=1}^{\infty} \left(-\frac{1}{3}\right)^{n-1}$

24. $\sum_{n=1}^{\infty} 7(2)^{n-1}$

25. $\sum_{n=1}^{\infty} (-0.2)^{n-1}$

26. $\sum_{n=1}^{\infty} 2(1.2)^{n-1}$

Find the specified value for each infinite geometric series.

27. $a_1 = 12$, $S = 96$, find r

28. $a_1 = -50$, $S = -45$, find r

29. $r = \frac{1}{6}$, $S = 12$, find a_1

30. $r = -0.3$, $S = 3$, find a_1

Write each repeating decimal as a rational number.

31. $0.\overline{83}$ **32.** $7.\overline{47}$ **33.** $0.174\overline{9}$ **34.** $0.4\overline{53}$

35. The sum of an infinite geometric series is twice its first term. Find the common ratio of the series.

36. If the common ratio, r, of a convergent series is multiplied by $\frac{4}{3}$, the new series diverges. If the common ratio is positive, find the range of the values of r.

Find the sum of each geometric series if x is a positive integer.

37. $(x + 1)^3 + (x^2 + 2x + 1) + (x + 1) + \cdots$

38. $8(x^2 + 4x + 4)^2 + (x^3 + 6x^2 + 12x + 8) + \frac{1}{8}(x^2 + 4x + 4) + \cdots$

Find the sum, if it exists, of each infinite geometric series.

39. $\sum_{n=1}^{\infty} \left(\frac{3^n}{4^{n-1}}\right)$

40. $\sum_{n=1}^{\infty} \left(\frac{5^{n+1}}{2^{n-1}}\right)$

Applications

41. Economics \$1000 is deposited in a bank and 80% is then loaned to a corporation, which then loans 80% to a supplier, who then loans 80% to a builder, and so forth. Find the *multiplier effect* of the original \$1000. *Hint:* Find the sum of a convergent geometric series, where $a_1 = 1000$ and $r = 0.8$.

42. Physics A ball bounces from the ground 50 cm into the air. Each time it bounces it rises $\frac{1}{3}$ the height of the previous bounce. If it continues in this manner, how far will the ball travel in all? (Consider only the *vertical* distance.)

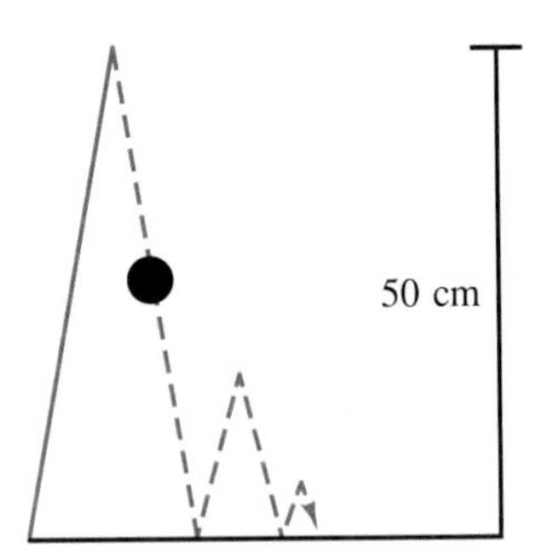

43. Physics The length of the arc of the swing of a pendulum is 120 mm. If the length of the arc of each succeeding swing is decreased by 10%, find the total distance the pendulum travels.

Developing Mathematical Power

Writing in Mathematics Understanding algebraic symbols and terms is important to the study of sequences and series. These questions will test your understanding of some of the symbols and terms used so far in this chapter.

44. Explain in your own words the difference between a sequence and a series.

45. Explain in your own words the difference between an arithmetic and a geometric series.

46. Give an example of a recursive formula for an arithmetic sequence.

47. Give an example of an explicit formula for a geometric sequence.

48. What is the meaning of a_{n+2}?

49. Use sigma notation to write the series $-2 + 4 - 8 + 16 - 32 + 64$.

13.8

Problem Solving: Using Sequences and Series

To solve problems involving sequences and series, it is important that you recognize the type of sequence or series the information fits. Once you know the pattern you can use the appropriate formulas that will lead to a solution.

EXAMPLE 1 A live filming of a TV show is to take place in the park. Due to the shape of the open area, seats are set up so that the first row has 16 seats and each successive row has 3 more seats than the row before it. The park's open area can accommodate 20 rows of seats in this pattern. What is the maximum number of tickets that can be distributed if everyone who has a ticket is to be guaranteed a seat?

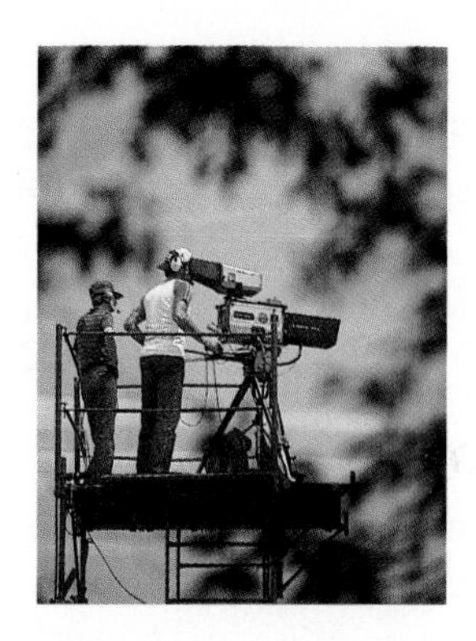

Understand the Problem The maximum number of tickets must be the same as the total number of seats. Since there is a constant difference between the number of seats in each succeeding row, the total number of seats is the sum of an arithmetic series. The first three terms are

$$16 + 19 + 22$$

Plan Your Approach

$$S_n = \frac{n[2a_1 + (n - 1)d]}{2}$$ *Use a formula for the sum of an arithmetic series.*

Complete the Work

$$S_{20} = \frac{20[2(16) + (20 - 1)3]}{2}$$ *$n = 20, a_1 = 16, d = 3$*

$$S_{20} = 10[32 + (19)3] = 890$$

Interpret the Results A maximum of 890 tickets can be given out if everyone who has a ticket is to be guaranteed a seat.

One way to check is to find the number of seats in the 20th row and use the formula $S_n = \frac{n(a_1 + a_n)}{2}$.

$$a_{20} = 16 + (20 - 1)3 = 73$$ *$a_n = a_1 + (n - 1)d$*

$$S_{20} = \frac{20(16 + 73)}{2} = 890 \checkmark$$

EXAMPLE 2 As part of a design of an amusement park, a level monorail is to be built on a hill. The supports for the rail are to be in the shape of isosceles triangles. The height of the tallest triangle is 22 ft and the height of each succeeding triangle is 90% of the height of the previous triangle. Twenty-five supports are to be built following this pattern. Find the height of the shortest support to the nearest tenth of a foot.

Understand the Problem Since the height of each succeeding support structure is 90% of the height of the previous support structure, the heights form a geometric sequence.

$$22, 19.8, 17.82, \ldots$$

Plan Your Approach The height of the shortest (25th) triangular support will be the 25th term of the geometric sequence.

Complete the Work

$$a_{25} = a_1 r^{n-1}$$

$$a_{25} = 22(0.9)^{25-1} \qquad n = 25, a_1 = 22, r = 90\% = 0.9$$

$$\approx 1.8$$

Interpret the Results The height of the shortest support, to the nearest tenth of a foot, is 1.8 ft.

To check, work backwards. The shortest support is about 1.8 ft. The ratio of the sequence going from shorter to taller support structures is $\frac{22}{19.8} = \frac{10}{9}$, or about 1.11. Therefore, the height of the tallest support structure is

$$a_{25} = a_1 r^{n-1} = 1.8(1.11)^{24} \approx 22 \checkmark$$

EXAMPLE 3 Halcombes, a manufacturer of handballs, added a chemical to its product to give the ball a better bounce. To test the bounce, a ball is dropped from a height of 64 ft. The height of its first bounce is decreased by $\frac{1}{8}$ of 64 ft, and the height of each successive bounce is decreased by $\frac{1}{8}$ the height of the previous bounce. If this pattern continues, how far will the ball travel in a vertical direction after it hits the ground the first time?

Understand the Problem The height of each bounce is $1 - \frac{1}{8}$, or $\frac{7}{8}$ the height of the previous bounce. After the 64 ft drop, the ball travels $\frac{7}{8} \cdot 64$, or 56 ft up and 56 ft down again, then $\frac{7}{8} \cdot 56$, or 49 ft up and 49 ft down again, and so forth. The total vertical distance traveled by the ball after it first hits the ground is the sum of the infinite geometric series

$$2(56) + 2(56)\left(\frac{7}{8}\right) + 2(56)\left(\frac{7}{8}\right)^2 + \cdots$$

$$\text{or} \quad 112 + 112\left(\frac{7}{8}\right) + 112\left(\frac{7}{8}\right)^2 + \cdots$$

Plan Your Approach

$$S = \frac{a_1}{1 - r}$$ *Use the formula for the sum of an infinite geometric series.*

Complete the Work

$$S = \frac{112}{1 - \frac{7}{8}}$$ *$a_1 = 112$, $r = \frac{7}{8}$*

$$S = \frac{8 \cdot 112}{8 \cdot \left(1 - \frac{7}{8}\right)} = 896$$

Interpret the Results

The total vertical distance the ball will travel after it first hits the ground is 896 ft.

To check, use a calculator to approximate the answer by finding S_n for a very large value of n, such as 100.

$$S_n = \frac{a_1(1 - r^n)}{1 - r}$$

$$S_{100} = \frac{112\left[1 - \left(\frac{7}{8}\right)^{100}\right]}{1 - \frac{7}{8}}$$ *Calculation-ready form*

$$S_{100} \approx 896$$ The answer is reasonable.

CLASS EXERCISES

What formula(s) would you use to solve each problem? Why? Do not solve the problems.

1. To replace the trees destroyed in a fire, the forestry department has developed a ten-year plan. The first year they will plant 100 trees. Each succeeding year they will plant 50 more trees than they planted the year before.
   a. How many trees will they plant during the fifth year?
   b. How many trees will they have planted by the end of the tenth year?

2. A television commercial claims its product is so good people have told their friends about it. Two people heard about the product the first day. On the second day, these two each told two friends, and the pattern continued.
   a. How many people were told about the product on the sixth day?
   b. How many people in all would have heard about the product at the end of the twelfth day?

PRACTICE EXERCISES

Use technology where appropriate.

1. Silva runs 2 mi farther each week than she ran the week before. The first week she ran 3 mi. If she keeps up this pattern, how many miles will she be able to run at the end of the fifth week?

2. For each hour that Sam practices his typing his speed increases by 2 words per minute. If his speed during the first hour was 40 words per minute, what will his speed be during the seventh hour?

3. A clock chimes as many times as the hour, every hour on the hour. How many times has it chimed by 9:15 PM if it is programmed to start at 1 PM that same afternoon?

4. Suppose the clock in Exercise 3 is programmed to start chiming at 2 PM. How many times will it have chimed by 12:15 AM the next morning?

5. There are 12 bacteria in a culture at the end of one hour, and that number triples every hour. Find the total number of bacteria in the culture at the end of the fifth hour.

6. The population of a town doubled each year for 7 years. If the population at the end of the first year was 500, find the population at the end of the seventh year.

7. Every year the cost of manufacturing one case of soup has increased by \$0.75. During the first year, the cost for one case was \$8.75. Find the cost of manufacturing a case during the seventh year.

8. Diana's starting salary was \$12,000 per year. She received an 8% raise at the beginning of each year. Find her total salary for the first 5 years.

9. Rudy saved 5 quarters one day. Each day thereafter, he saved 3 more quarters than on the previous day. How many dollars will Rudy have saved at the end of 8 days?

10. Anita saved 1 penny the first day, 2 pennies the second day, 3 pennies the third day, and so on for a period of 200 days. What was the dollar value of the pennies at the end of 200 days?

11. After each week that a certain car is driven without changing the air filter, the gas mileage decreases by 0.5 miles per gallon. If the gas mileage was 24 miles per gallon after one week, what would it be after 24 weeks of driving without changing the air filter?

12. The first prize in a contest is \$100,000. Each succeeding prize is 75% of the value of the preceding prize. If the number of prizes could be unlimited, find the total amount of money that would be given away.

13. A wheel turns 64 revolutions in the first minute. Each succeeding minute the number of revolutions per minute is decreased by 20%. To the nearest whole number, how many revolutions will it make in the sixth minute and how many revolutions will it make, in all, before it stops?

14. When it is full, a storage tank holds 963 gal of gasoline. If $\frac{1}{3}$ of the gasoline is released each time the valve is operated, how much gasoline will remain in the tank after the fourth time the valve is opened? Give your answer to the nearest tenth of a gallon.

15. A ball is dropped from a height of 35 ft. It bounces $\frac{5}{7}$ of that height, and on each successive bounce it bounces $\frac{5}{7}$ of the height of the previous bounce. Find the total vertical distance the ball will travel after first striking the ground.

Developing Mathematical Power

Thinking Critically In 1772, the German astronomer Johann Titius announced a startling discovery: he had related planetary distances to the simple numerical sequence 4, 7, 10, 16, 28, 52, 100, ...

He found that when the distance from Earth to the center of the sun was given as 10, the distances for the other known planets were all elements of the same sequence. The distances used by Titius are not the figures astronomers accept today; yet his theory, known as the Bode-Titius law or simply Bode's law, led to several surprising discoveries.

Planet	Bode-Titius Number
Mercury	4
Venus	7
Earth	10
Mars	16
Jupiter	52
Saturn	100

The obvious question about the series is the whereabouts of the "missing" planet that theoretically ought to lie at 28, between Mars and Jupiter. Astronomers searching for this hypothetical planet were rewarded in 1801 with the discovery of Ceres, the largest asteroid, orbiting the sun at the predicted distance of 28.

16. Sir William Herschel, a British astronomer, discovered Uranus in 1781, exactly where Bode's law predicted it should be. Using the same scale as given above, how far is Uranus from the sun?

17. Subsequent astronomers searching for the next planets found Neptune and Pluto at distances of 305 and 388. Were these planets orbiting at distances predicted by Bode's law?

18. The first element of the sequence demonstrated by Titius is troublesome: most rules for the sequence do not hold for the element 4. Write a rule that will generate each element except 4.

13.9

Binomial Expansion

Objective: To expand powers of binomials using Pascal's triangle

By studying the patterns of sequences and series, formulas were developed to find the *n*th, or general, terms. When binomials are raised to positive integral powers, the expanded forms also show patterns. These patterns can be used to simplify the expansion of binomials.

Capsule Review

You can *expand* $(x + y)^3$ by multiplying.

$$\begin{aligned}(x + y)^3 &= (x + y)^2(x + y)\\ &= (x^2 + 2xy + y^2)(x + y)\\ &= x^3 + 2x^2y + xy^2 + x^2y + 2xy^2 + y^3\\ &= x^3 + 3x^2y + 3xy^2 + y^3\end{aligned}$$

Then you can use the expansion of $(x + y)^3$ to expand $(x + y)^4$.

$$\begin{aligned}(x + y)^4 &= (x + y)^3(x + y)\\ &= (x^3 + 3x^2y + 3xy^2 + y^3)(x + y)\\ &= x^4 + 3x^3y + 3x^2y^2 + xy^3 + x^3y + 3x^2y^2 + 3xy^3 + y^4\\ &= x^4 + 4x^3y + 6x^2y^2 + 4xy^3 + y^4\end{aligned}$$

Expand each binomial.

1. $(x + y)^5$ **2.** $(x - y)^3$ **3.** $(x - y)^4$

4. $(x - y)^5$ **5.** $(2x + y)^2$ **6.** $(2x + y)^3$

The expansion of a binomial raised to a positive integral power is a finite series.

EXAMPLE 1 **Expand $(a + b)^n$ for $n = 0, 1, 2, 3, 4,$ and 5.**

$(a + b)^0 = 1$ *$x^0 = 1$*

$(a + b)^1 = a + b$

$(a + b)^2 = a^2 + 2ab + b^2$

$(a + b)^3 = a^3 + 3a^2b + 3ab^2 + b^3$

$(a + b)^4 = a^4 + 4a^3b + 6a^2b^2 + 4ab^3 + b^4$

$(a + b)^5 = a^5 + 5a^4b + 10a^3b^2 + 10a^2b^3 + 5ab^4 + b^5$

In Example 1, notice that each expression has $n + 1$ terms. The exponents of the variables for each term of the expansion form a pattern.

The first term is $a^n b^0$, written a^n.
The last term is $a^0 b^n$, written b^n.
The exponent of each a decreases by 1 and the exponent of each b increases by 1 for each succeeding term in the series.
The sum of the exponents of each term is n.

The above observations are true for every positive integral exponent n.

EXAMPLE 2 **Write the variable parts of the terms of the expansion: $(a + b)^7$**

First term: a^7 Last term: b^7 Number of terms: $7 + 1 = 8$

$a^7, a^6b, a^5b^2, a^4b^3, a^3b^4, a^2b^5, ab^6, b^7$ *Follow the pattern.*

A pattern for the coefficients of the terms can be found by writing the coefficients in triangular form. This pattern is called **Pascal's triangle.**

Coefficients of Expansion

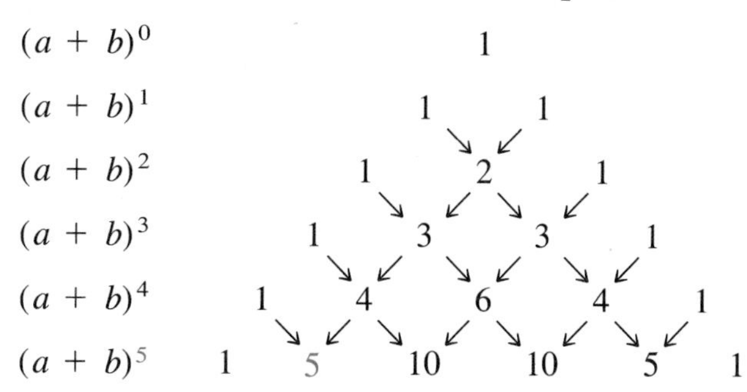

Notice that each row is symmetrical. That is, the coefficients that are "equidistant" from the ends of the row are equal. The coefficients of the terms have the following pattern.

The coefficients of the first and last terms are both 1.
The other elements in the row are the sum of the two closest numbers in the row directly above it.
The row of coefficients for $(a + b)^n$ is the row whose second entry is n.

EXAMPLE 3 **Find the row of coefficients for $(a + b)^6$ from Pascal's triangle.**

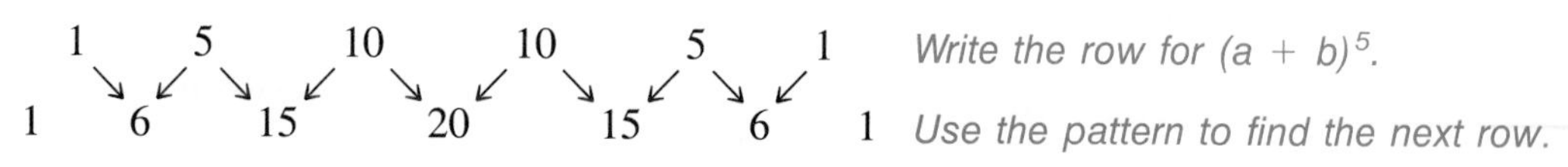

Write the row for $(a + b)^5$.
Use the pattern to find the next row.

The row of coefficients for $(a + b)^6$ is: 1, 6, 15, 20, 15, 6, 1.

By combining the pattern for the variable parts of the terms and the values of the coefficients of each term taken from Pascal's triangle, an expansion of a binomial can be written without repeated multiplication.

EXAMPLE 4 **Expand: $(x + y)^7$**

$x^7 \quad x^6y \quad x^5y^2 \quad x^4y^3 \quad x^3y^4 \quad x^2y^5 \quad xy^6 \quad y^7$ — *Write the variable parts.*

$$\begin{array}{ccccccccccccccc} & 1 & & 6 & & 15 & & 20 & & 15 & & 6 & & 1 & \\ & \searrow & \swarrow & \searrow & \swarrow & \searrow & \swarrow & \searrow & \swarrow & \searrow & \swarrow & \searrow & \swarrow & & \\ 1 & & 7 & & 21 & & 35 & & 35 & & 21 & & 7 & & 1 \end{array}$$

To find the coefficients, write the row for $(a + b)^6$. Then use the pattern.

Write the expansion by multiplying the variable terms by their corresponding coefficients and summing them.

$$(x + y)^7 = x^7 + 7x^6y + 21x^5y^2 + 35x^4y^3 + 35x^3y^4 + 21x^2y^5 + 7xy^6 + y^7$$

Example 5 shows how to use the same method to expand a binomial in which the coefficients of a and b are not 1.

EXAMPLE 5 **Expand: $(2x + 3y)^3$**

$(2x + 3y)^3 = [(2x) + (3y)]^3$ — *Write in $(a + b)^n$ form.*

$(2x)^3 \quad (2x)^2(3y) \quad (2x)(3y)^2 \quad (3y)^3$ — *Write the variable parts.*

$8x^3 \quad 12x^2y \quad 18xy^2 \quad 27y^3$ — *Simplify.*

$1 \quad 3 \quad 3 \quad 1$ — *Find the coefficients from Pascal's triangle.*

To write the expansion, multiply the variable part of each term by its corresponding coefficient.

$$(2x + 3y)^3 = 1(8x^3) + 3(12x^2y) + 3(18xy^2) + 1(27y^3)$$

$$(2x + 3y)^3 = 8x^3 + 36x^2y + 54xy^2 + 27y^3$$ *Simplify.*

In Example 6, notice that the signs of the terms of the expansion of a binomial power of the form $(a - b)^n$ alternate.

EXAMPLE 6 **Expand: $(x - 2y)^4$**

$(x - 2y)^4 = [x + (-2y)]^4$ — *Write in $(a + b)^n$ form.*

$x^4 \quad x^3(-2y) \quad x^2(-2y)^2 \quad x(-2y)^3 \quad (-2y)^4$ — *Write the variable parts.*

$x^4 \quad -2x^3y \quad 4x^2y^2 \quad -8xy^3 \quad 16y^4$ — *Simplify.*

$1 \quad 4 \quad 6 \quad 4 \quad 1$ — *Find the coefficients.*

Multiply the variable part of each term by its corresponding coefficient.

$$(x - 2y)^4 = 1(x^4) + 4(-2x^3y) + 6(4x^2y^2) + 4(-8xy^3) + 1(16y^4)$$

$$(x - 2y)^4 = x^4 - 8x^3y + 24x^2y^2 - 32xy^3 + 16y^4$$ *Simplify.*

CLASS EXERCISES

State the number of terms in each expansion and give the first and last terms.

1. $(x + y)^3$

2. $(x + y)^7$

3. $(x - y)^5$

Give the variable parts of the terms in each expansion.

4. $(x + y)^5$

5. $(x - y)^5$

6. $(2x - 3y)^4$

7. $(-x + 5y)^3$

8. $(x^2 + y)^3$

9. $(x^2 - y^2)^4$

Use the information in Exercises 4–9 to write each expansion.

10. $(x + y)^5$ (Exercise 4)

11. $(x - y)^5$ (Exercise 5)

12. $(2x - 3y)^4$ (Exercise 6)

13. $(-x + 5y)^3$ (Exercise 7)

14. $(x^2 + y)^3$ (Exercise 8)

15. $(x^2 - y^2)^4$ (Exercise 9)

PRACTICE EXERCISES

1. Write rows 1–11 of Pascal's triangle.

2. Which row would be used for the coefficients of $(a + b)^6$?

Expand each binomial.

3. $(x + y)^4$

4. $(x - y)^4$

5. $(x - y)^5$

6. $(x + y)^5$

7. $(x + 1)^6$

8. $(x - 1)^6$

9. $(x + 2)^5$

10. $(x - 2)^5$

11. $(2x + 3y)^4$

12. $(3x + 5y)^3$

13. $(2x + 2y)^6$

14. $(3x + 2y)^4$

15. $(2x + y)^5$

16. $(3x + y)^7$

17. $(x + 3y)^6$

18. $(x + 5y)^3$

Expand each binomial and simplify if possible.

19. $(x - y)^{10}$

20. $(-x + y)^{10}$

21. $(x^2 + 3)^4$

22. $(x^3 - 2)^5$

23. $(x^2 - y^3)^6$

24. $(2x^2 + 2y)^4$

25. $(x^{-1} + y^{-1})^5$

26. $(x^{-2} - y^{-3})^4$

27. $\left(\frac{1}{x} - \frac{1}{2}\right)^3$

28. $\left(1 - \frac{1}{z}\right)^3$

29. $(x^{\frac{1}{2}} + y^{\frac{1}{2}})^4$

30. $(x^{\frac{3}{2}} + y^{\frac{1}{2}})^5$

31. $(3 + i)^5$

32. $(1 - i)^{10}$

33. $(i - 2)^4$

34. $(1 + i)^8$

Find the sum of the expansions.

35. $(x + \sqrt{2})^4 + (x - \sqrt{2})^4$

36. $(\sqrt{x} + \sqrt{y})^6 + (\sqrt{x} - \sqrt{y})^6$

37. $(3 + 2i)^5 + (3 - 2i)^5$

38. $(a + bi)^6 + (a - bi)^6$

Applications

Combinations Pascal's triangle will help you count the number of ways you can choose 3 goldfish out of a bowl that contains 5 of them. Counting the 1 at the top of the triangle as row 0, move down to row 5 (for the 5 possibilities) and over to term 3 + 1, or 4 (for the 3 fish chosen). There are 10 ways to choose 3 goldfish out of a possible 5, although you probably could not tell one choice from another.

39. A restaurant offers a choice of 2 vegetables with any entree out of 10 vegetables altogether. How many combinations are possible?

40. There are 11 players on the basketball team, but only 5 can be starters. How many starting combinations are possible?

41. Todd is offered his choice of two days off per week on his summer job. How many different combinations of days off are possible? How else might this have been calculated?

Developing Mathematical Power

Extension Examine the sequence 1, 1, 2, 3, 5, 8, 13, 21, 34, ... Each term beyond the second is equal to the sum of the two preceding terms. This is called a **Fibonacci sequence,** and its elements are called Fibonacci numbers. These numbers occur in unexpected places, not just in mathematics, but also in nature.

As a leaf grows from a stem, it creates an angle with the leaf below. On a given plant, the angle between consecutive leaves remains the same; the most common angles are shown here, in ratio with 360°.

$$\frac{180^\circ}{360^\circ} = \frac{1}{2} \qquad \frac{120^\circ}{360^\circ} = \frac{1}{3} \qquad \frac{144^\circ}{360^\circ} = \frac{2}{5} \qquad \frac{135^\circ}{360^\circ} = \frac{3}{8} \qquad \frac{138\frac{6}{13}^\circ}{360^\circ} = \frac{5}{13}$$

42. Examine the numbers in the reduced ratios. Predict the next ratio. The next most frequent leaf pattern has an angle of $137\frac{1}{7}^\circ$. Does this fit the predicted ratio?

43. Examine the circled diagonals of Pascal's triangle in this diagram. Describe the pattern.

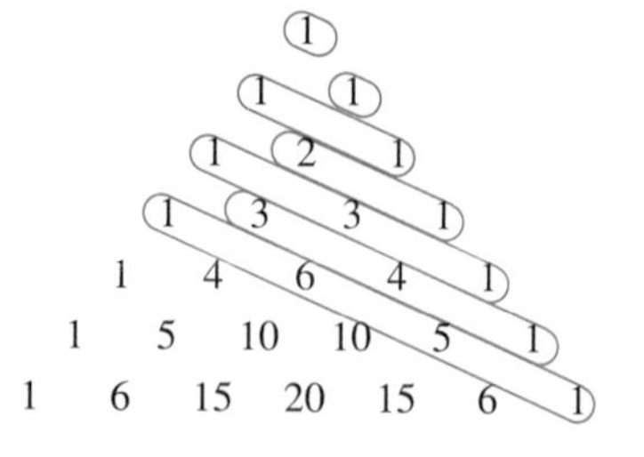

Binomial Theorem

Objective: To use the binomial theorem to write a binomial expansion and to find a specified term of a binomial expansion

The entries in each row of Pascal's triangle form a pattern that is related to a formula for writing the coefficients of the terms of a binomial expansion or for finding any term of a binomial expansion. After the formula is applied, the coefficients of the terms must be simplified.

Capsule Review

To simplify the expression $\frac{8 \cdot 7 \cdot 6 \cdot 5 \cdot 4 \cdot 3 \cdot 2 \cdot 1}{(5 \cdot 4 \cdot 3 \cdot 2 \cdot 1)(3 \cdot 2 \cdot 1)}$, divide out factors common to the numerator and the denominator.

$$\frac{8 \cdot 7 \cdot \cancel{6} \cdot \cancel{5} \cdot \cancel{4} \cdot \cancel{3} \cdot \cancel{2} \cdot \cancel{1}}{(\cancel{5} \cdot \cancel{4} \cdot \cancel{3} \cdot \cancel{2} \cdot \cancel{1})(\cancel{3} \cdot \cancel{2} \cdot 1)} = \frac{8 \cdot 7}{1} = 56$$

Simplify.

1. $\frac{5 \cdot 4 \cdot 3 \cdot 2 \cdot 1}{(3 \cdot 2 \cdot 1)(2 \cdot 1)}$

2. $\frac{6 \cdot 5 \cdot 4 \cdot 3 \cdot 2 \cdot 1}{(3 \cdot 2 \cdot 1)(3 \cdot 2 \cdot 1)}$

3. $\frac{8 \cdot 7 \cdot 6 \cdot 5 \cdot 4 \cdot 3 \cdot 2 \cdot 1}{(4 \cdot 3 \cdot 2 \cdot 1)(4 \cdot 3 \cdot 2 \cdot 1)}$

4. $\frac{7 \cdot 6 \cdot 5 \cdot 4 \cdot 3 \cdot 2 \cdot 1}{(6 \cdot 5 \cdot 4 \cdot 3 \cdot 2 \cdot 1)(1)}$

5. $\frac{9 \cdot 8 \cdot 7 \cdot 6 \cdot 5 \cdot 4 \cdot 3 \cdot 2 \cdot 1}{(3 \cdot 2 \cdot 1)(6 \cdot 5 \cdot 4 \cdot 3 \cdot 2 \cdot 1)}$

6. $\frac{10 \cdot 9 \cdot 8 \cdot 7 \cdot 6 \cdot 5 \cdot 4 \cdot 3 \cdot 2 \cdot 1}{(6 \cdot 5 \cdot 4 \cdot 3 \cdot 2 \cdot 1)(4 \cdot 3 \cdot 2 \cdot 1)}$

To find a pattern for the entries in the rows of Pascal's triangle, it is necessary to work with products of the form $7 \cdot 6 \cdot 5 \cdot 4 \cdot 3 \cdot 2 \cdot 1$. **Factorial notation** provides a shorthand method for writing such products.

$7! = 7 \cdot 6 \cdot 5 \cdot 4 \cdot 3 \cdot 2 \cdot 1$ *Read: seven factorial*

By definition:

$n! = n(n - 1)(n - 2) \cdot \cdots \cdot 3 \cdot 2 \cdot 1$, where n is a positive integer
$0! = 1$

EXAMPLE 1 **Write each factorial in expanded form.** **a.** $6!$ **b.** $9!$ **c.** $1!$

a. $6! = 6 \cdot 5 \cdot 4 \cdot 3 \cdot 2 \cdot 1$ *$n! = n(n - 1)(n - 2) \cdot \cdots \cdot 3 \cdot 2 \cdot 1$*

b. $9! = 9 \cdot 8 \cdot 7 \cdot 6 \cdot 5 \cdot 4 \cdot 3 \cdot 2 \cdot 1$

c. $1! = 1$

Factorial notation can be used to simplify the writing of products and quotients.

EXAMPLE 2 **Find the value of each expression.** **a.** $3!4!$ **b.** $\frac{10!}{5!}$

a. $3! = 3 \cdot 2 \cdot 1 = 6$ and $4! = 4 \cdot 3 \cdot 2 \cdot 1 = 24$.
Therefore, $3!\,4! = 6 \cdot 24 = 144$

Note that $3!\,4! \neq 12!$.

b. $\frac{10!}{5!} = \frac{10 \cdot 9 \cdot 8 \cdot 7 \cdot 6 \cdot \cancel{5} \cdot \cancel{4} \cdot \cancel{3} \cdot \cancel{2} \cdot \cancel{1}}{\cancel{5} \cdot \cancel{4} \cdot \cancel{3} \cdot \cancel{2} \cdot \cancel{1}} = 30{,}240$

Note that $\frac{10!}{5!} \neq 2!$.

In the lesson before this, you learned to use Pascal's triangle to find the coefficients of the terms of a binomial expansion. A general formula for finding these coefficients can be developed. First, consider the expansion of $(a + b)^4$, which was developed on page 662 by the use of repeated multiplication. The coefficient of each term can be expressed in terms of the exponents of a and b and the power of the binomial.

$$(a + b)^4 = 1a^4b^0 + 4a^3b^1 + 6a^2b^2 + 4a^1b^3 + 1a^0b^4$$

$$(a + b)^4 = \frac{4!}{4!\,0!}a^4b^0 + \frac{4!}{3!\,1!}a^3b^1 + \frac{4!}{2!\,2!}a^2b^2 + \frac{4!}{1!\,3!}a^1b^3 + \frac{4!}{0!\,4!}a^0b^4$$

You should verify that the coefficients of the terms in the two forms of the expansion are equivalent.

The pattern shown in the second form of the expansion of $(a + b)^4$ holds for the expansion of any binomial $(a + b)^n$, where n is a positive integer. The coefficient of any term in such an expansion can be given by the expression

$$\frac{n!}{(\exp a)!\,(\exp b)!}$$

where n is the power of the binomial, exp a is the exponent of a for that term, and exp b is the exponent of b. This pattern is reflected in the general formula for expanding a binomial. This formula, called the **binomial theorem,** is proved in more advanced mathematics courses.

Binomial Theorem

For every positive integer n,

$$(a + b)^n = a^n + \frac{n!}{(n-1)!\,1!}a^{n-1}b^1 + \frac{n!}{(n-2)!\,2!}a^{n-2}b^2 + \cdots + \underbrace{\frac{n!}{(n-r+1)!\,(r-1)!}a^{n-r+1}b^{r-1}}_{r\text{th term}} + \cdots + b^n$$

EXAMPLE 3 **Use the binomial theorem to write the expansion of $(2x + y)^3$.**

$$(2x + y)^3 = [(2x) + y]^3 \quad \textit{Write in } (a + b)^n \textit{ form.}$$

$$= (2x)^3 + \frac{3!}{2!\,1!}(2x)^2y + \frac{3!}{1!\,2!}(2x)y^2 + y^3 \quad \textit{Use the binomial theorem.}$$

$$= 8x^3 + 3(4x^2)y + 3(2xy^2) + y^3$$

$$= 8x^3 + 12x^2y + 6xy^2 + y^3$$

Example 4 shows how to use the formula for the rth term

$$r\text{th term} = \frac{n!}{(n - r + 1)!(r - 1)!}a^{n-r+1}b^{r-1}$$

to find a term of a binomial expansion without actually expanding the binomial. A calculator with a factorial function will simplify calculations.

EXAMPLE 4 **Find the fifth term of the binomial expansion $(x - 2y)^9$.**

$$(x - 2y)^9 = [x + (-2y)]^9 \quad \textit{Write in } (a + b)^n \textit{ form.}$$

Use the formula for the rth term. Substitute 9 for n and 5 for r.

$$\frac{n!}{(n - r + 1)!\,(r - 1)!}x^{n-r+1}y^{r-1} = \frac{9!}{(9 - 5 + 1)!\,(5 - 1)!}x^{9-5+1}(-2y)^{5-1}$$

$$= \frac{9!}{5!\,4!}x^5(-2y)^4$$

$$= 126x^5(16y^4)$$

$$= 2016x^5y^4$$

Since $0! = 1$ by definition, the formula for the rth term holds for the first and last terms of the expansion of $(a + b)^n$ as well as for the other terms.

Sigma notation can be used to write the binomial theorem in more compact form, as shown below.

$$(a + b)^n = \sum_{r=0}^{n} \frac{n!}{(n - r)!\,r!}a^{n-r}b^r, \text{ for every positive integer } n$$

CLASS EXERCISES

Evaluate.

1. $6!$ **2.** $9!$ **3.** $2!\,5!$ **4.** $4!\,3!$ **5.** $\frac{10!}{4!}$

6. $\frac{12!}{8!}$ **7.** $\frac{10!}{2!\,4!}$ **8.** $\frac{13!}{4!\,3!}$ **9.** $\frac{7!\,3!}{4!}$ **10.** $\frac{9!\,4!}{8!}$

11. Give the third term of $(x + y)^5$?

12. Give the fifth term of $(x + y)^7$?

PRACTICE EXERCISES

Use technology where appropriate.

Use the binomial theorem to expand each binomial.

1. $(x + y)^5$
2. $(x - y)^5$
3. $(2x + y)^3$
4. $(x + 3y)^4$
5. $(x - 2y)^5$
6. $(2x - y)^5$
7. $(x - 3y)^4$
8. $(4x - y)^5$
9. $(x - 1)^7$
10. $(1 - x)^6$
11. $(x^2 + 1)^5$
12. $(y^2 + a)^4$

Find the specified term of each binomial expansion.

13. Third term of $(x + 3)^{12}$
14. Fourth term of $(x + 2)^5$
15. Second term of $(x + 3)^9$
16. Third term of $(x - 2)^{12}$
17. Twelfth term of $(2 + x)^{11}$
18. Seventh term of $(x - 2y)^6$
19. Eighth term of $(x - 2y)^{15}$
20. Third term of $(3x - 2)^9$

Find the first three terms of each binomial expansion.

21. $(2x + 3)^{18}$
22. $(3x - 1)^{20}$
23. $(x^2 + y)^{12}$
24. $(x^2 - y^2)^8$
25. $(2x^2 + y^4)^{11}$
26. $(y^4 - x^3)^{11}$
27. $(2 + xy)^{15}$
28. $(3 - x^2y)^7$

Find the specified term of each binomial expansion.

29. Seventh term of $(x^2 - 2y)^{11}$
30. Eighth term of $(x^2 + y^2)^{13}$
31. Sixth term of $\left(x - \frac{1}{2}\right)^9$
32. Seventh term of $\left(\frac{x}{3} - y\right)^{11}$
33. Fifteenth term of $\left(\frac{x}{9} + y^2\right)^{14}$
34. Eighth term of $\left(\frac{2x}{7} - \frac{y}{2}\right)^7$

The values of powers of some numbers can be approximated by using the first two terms of binomial expansion.

Example **Approximate the value of $(1.015)^{20}$ by using the first two terms of a binomial expansion.**

$$(1.015)^{20} = (1 + 0.015)^{20} = 1^{20} + 20(1)^{19}(0.015)$$
$$\approx 1 + 0.3 \approx 1.3$$

Approximate each value to the nearest tenth using the first two terms of a binomial expansion.

35. 1.04^5
36. 0.98^{10}
37. 1.02^{20}
38. 1.01^{50}
39. 0.99^{30}
40. 1.001^{200}

Applications

41. Advertising The projections of advertising dollars to be spent over the next five years are given by the five terms of the expansion of $(x + 2)^4$, where x is the number of types of media used. Find the projections for the first three years.

42. Botany The number of seeds produced by a certain species of plant is approximated by the expression $(x + 4)^3$, where x is the number of buds on the plant. The first term of the expansion is the number of seeds produced the first year, the second term is the number produced the second year, and so forth. Find the number of seeds produced in the second year if there are 14 buds.

43. Geometry Each edge of a cube is x cm in length. If each edge is increased by 3 cm, find the binomial expansion that represents the volume of the new cube.

TEST YOURSELF

Find the indicated sum of each geometric series.

1. $4 - 12 + 36 - \cdots; S_6$

2. $\sum_{k=1}^{8} 3(2)^{k-1}$

13.6

Find the sum of each infinite geometric series, if it exists. If the sum does not exist, so state.

3. $1 + \frac{1}{8} + \frac{1}{64} + \cdots$

4. $6 - 6 + 6 - \cdots$

13.7

5. $3 - 9 + 27 - \cdots$

6. $12 - 6 + 3 - \cdots$

7. Dawn Field planted a tree in her yard. At the end of the first year, the tree was 5 ft high. If the height of the tree increases by 70% each year thereafter, how tall will it be at the end of the sixth year? Give your answer to the nearest foot.

13.8

8. A ball is dropped from a height of 24 ft. It bounces up $\frac{3}{4}$ of that distance, and every subsequent bounce is $\frac{3}{4}$ as high as the preceding bounce. Find the vertical distance traveled by the ball after it first hits the ground.

Use Pascal's triangle to expand each binomial.

9. $(x - y)^5$

10. $(2x + 3y)^4$

13.9

11. Use the binomial theorem to expand $(3x + y)^5$.

13.10

12. Find the fourth term in the expansion of $(x - 1)^{10}$.

INTEGRATING ALGEBRA
Musical Tuning Systems

Did you know that the evolution of tuning systems in western music has taken place over thousands of years and is continuing to evolve? The most current trends propose an entire reorganization of the octave into many more pitches than currently exist.

Sound is the result of the vibration of materials such as wood or metal. The human ear senses the wavelike increases and decreases of air pressure caused by such vibrations. The frequency of these vibrations determines whether a tone will be perceived as having a high or low pitch. For example, the note A above "middle C" has a frequency of 440 cycles per second or hertz, named after the German physicist Heinrich R. Hertz.

The most fundamental interval in music is that of the octave. Doubling the frequency of a pitch produces a tone one octave above the original pitch. For example, the note A vibrating at 880 hertz is one octave above the note A vibrating at 440 hertz.

In the fifth century BC the Greek philosopher and mathematician Pythagoras investigated harmony. He proposed that tones of a scale be determined by multiplying the starting pitch by ratios of the form $\frac{3^m}{2^n}$, for integer values of m and n. Another tuning system, which is based on the acoustic concept of the overtone series, is called the method of *Just Intonation*. This method generates scalar tones by multiplying the starting pitch by simple ratios of whole numbers. Although these two tuning systems produce pure and natural-sounding intervals, they have shortcomings. For example, an instrument tuned by either the Pythagorean or Just-Intonation method in the key of C sounds exceptionally in tune when played in the key of C or in keys very close to C, such as G and F. However, due to unequal frequency distances between notes, the same instrument sounds extremely out of tune in keys far away from C.

After the passage of centuries during which these "natural" scales were used, musicians became increasingly reluctant to accept the limited choice of keys that these scales imposed. Finally, in the early eighteenth century a new tuning system, the *method of equal temperament,* replaced the older systems. This method divides the octave into 12 intervals, such that the ratio of the frequencies of any two successive intervals is a constant. Thus, if A is chosen to be 440 hertz, then A♯ is 466.2 hertz and B is 493.9.

$$\frac{466.2}{440} = \frac{493.9}{466.2} \approx 1.059$$

The method of equal temperament is particularly important for keyboard instruments such as the piano. However, a minor disadvantage of the equally tempered (or "well-tempered") scale is that, in natural terms, intervals are slightly out of tune. For example, A♯ and B♭, which are not exactly the same note, are represented by a single black key on the piano.

EXERCISES

1. The note names of the 12-tone scale, beginning with A, are: A, A♯, B, C, C♯, D, D♯, E, F, F♯, G, G♯, and then back to A. Find, in simplified radical form, the common ratio of the sequence of pitch frequencies in the equal temperament tuning system, given A equals 440 hertz and A one octave above equals 880 hertz.

2. The note E, often referred to as the fifth in a diatonic scale that begins with A, is determined by multiplying the starting pitch by $\frac{3}{2}$ in both the Pythagorean and Just-Intonation methods. Find its frequency in all three systems. Is the equal-temperament fifth higher or lower than a "pure" fifth?

3. The note C♯, often referred to as the major third in a diatonic scale that begins with A, is determined by multiplying the starting pitch by $\frac{81}{64}$ in the Pythagorean system and multiplying by $\frac{5}{4}$ in the Just-Intonation system. Find its frequency in all three systems. Which major third has the highest frequency? Which has the lowest?

4. Some music theorists are proposing two new divisions of the octave as a way to increase the means of musical expression. Find the common ratios for the proposed 19-tone and 31-tone scales.

5. Find the term number of the frequencies closest to E in both of the 19-tone and 31-tone scales.

6. Find the frequency closest to C♯ in the 19-tone scale. Which major third of the three tuning systems is this C♯ closest to?

7. Find the frequency closest to C♯ in the 31-tone scale. Which major third of the three tuning systems is this C♯ closest to?

CHAPTER 13 SUMMARY AND REVIEW

Vocabulary

arithmetic mean (630)
arithmetic sequence (628)
arithmetic series (640)
binomial theorem (668)
common difference (628)
common ratio (634)
converges (652)
diverges (653)
explicit formula (624)
factorial notation (667)
Fibonacci sequence (621)
finite sequence (623)
general term (623)
geometric mean (637)
geometric sequence (634)
geometric series (646)
infinite sequence (623)
mean proportional (637)
partial sum (640)
Pascal's triangle (663)
recursive formula (623)
rule of a sequence (623)
sequence (623)
series (640)
sigma $\left(\sum\right)$ notation (641)
summation notation (641)
term of a sequence (623)

Sequences To give the rule of a sequence as a recursive formula, state the first term and an expression for finding the $(n + 1)$th term, a_{n+1}, from the preceding term, a_n. To give the rule as an explicit formula, express the nth term as a function of n, where n is a positive integer. **13.1–13.2**

1. Write a rule of the sequence 2, 7, 12, 17, ... as a recursive formula and use the rule to find the next three terms.
2. The explicit formula for a sequence is $a_n = 3n + 2$. Find the twelfth term of the sequence.

Arithmetic Sequences If a_1 is the first term, n is the number of terms, d is the common difference, and a_n is the nth term of an arithmetic sequence, then $a_n = a_1 + (n - 1)d$. **13.3**

Find the specified term of each arithmetic sequence.

3. $a_1 = 2$, $d = 4$; a_6
4. $a_1 = 10$, $d = 5$; a_7
5. 10, 5, 0, ...; a_{15}
6. Insert the three arithmetic means between -6 and 18.

Geometric Sequences If a_1 is the first term, n is the number of terms, r is the common ratio, and a_n is the nth term of a geometric sequence, then $a_n = a_1 r^{n-1}$. **13.4**

Find the specified term of each geometric sequence.

7. $a_1 = 2$, $r = 3$; a_5
8. 10, 20, 40, ...; a_8
9. 8, −4, 2, ...; a_7
10. Insert the two geometric means between 3 and 192.
11. Find the mean proportional of −8 and −200.

13.5

Arithmetic Series If a_1 is the first term, n is the number of terms, d is the common difference, a_n is the nth term, and S_n is the partial sum of an arithmetic series, then $S_n = \frac{n(a_1 + a_n)}{2}$ or $S_n = \frac{n[2a_1 + (n-1)d]}{2}$.

Find the specified partial sum for each arithmetic series

12. $7 + 9 + 11 + 13 + \cdots$; s_8
13. $\sum_{n=1}^{7} (4n + 1)$
14. $\sum_{n=1}^{9} (10 - 5n)$
15. $a_1 = 4$ and $S_6 = 144$. Find a_6.
16. $a_7 = 28$ and $S_7 = 91$. Find a_1.

13.6–13.7

Geometric Series If S_n is the sum of the first n terms of a geometric series and S is the sum of a convergent infinite geometric series, then $S_n = \frac{a_1(1 - r^n)}{1 - r} = \frac{a_1 - ra_n}{1 - r}$, $r \neq 1$ and $S = \frac{a_1}{1 - r}$, $r \neq 1$. An infinite geometric series converges if $|r| < 1$.

Find S_n for each geometric series.

17. $4 - 4 + 4 - 4 + \cdots$; S_{51}
18. $\sum_{n=1}^{6} 5(2)^{n-1}$
19. $\sum_{k=2}^{5} -3\left(\frac{2}{3}\right)^{k-1}$

Find the sum of the infinite geometric series, if it exists.

20. $1 + 2 + 4 + \cdots$
21. $4 + 2 + 1 + \cdots$

13.8

22. The cost of a can of premium paint has increased $2.50 per year for the last 6 years. If the cost of one can was $4.50 at the end of the first year, find the cost at the end of the sixth year.

13.9

23. Use Pascal's triangle to expand $(x + 2)^4$.

13.10

Binomial Theorem For all positive integers n,

$$(a + b)^n = a^n + \frac{n!}{(n-1)!\,1!}a^{n-1}b^1 + \frac{n!}{(n-2)!\,2!}a^{n-2}b^2 + \cdots + \frac{n!}{(n-r+1)!\,(r-1)!}a^{n-r+1}b^{r-1} + \cdots + b^n$$

24. Use the binomial theorem to expand $(-2x + 1)^6$.
25. Find the ninth term of $(x - y)^{12}$.

CHAPTER 13 TEST

1. Write the rule of the sequence 3, 8, 13, 18, ... as a recursive formula and find the next three terms.

2. Write the first three terms of the sequence for which the explicit formula is $a_n = 2n - 3$.

3. Find a_{10} for the arithmetic sequence $-2, 2, 6, 10, \ldots$

4. Find the two arithmetic means between 20 and 41.

5. For a geometric sequence, $a_1 = 32$ and $r = -\frac{1}{4}$. Find a_6.

6. Find the three geometric means between 5 and 3125.

7. For an arithmetic series, $a_1 = 12$ and $a_6 = -3$. Find S_6.

8. For an arithmetic series, $a_1 = 18$ and $S_5 = 20$. Find a_5.

9. Find the partial sum S_8 for the geometric series $-3 + 6 - 12 + \cdots$.

10. Find the sum: $\sum_{k=1}^{5} 4\left(\frac{1}{5}\right)^{k-1}$

Find the sum of the infinite geometric series, if it exists. If the sum does not exist, so state.

11. $\frac{1}{3} + \frac{1}{9} + \frac{1}{27} + \cdots$

12. $2 - 2 + 2 - 2 + \cdots$

13. The first row of seats in a theater has 25 seats. Each successive row has 2 more seats than the previous row. If there are 15 rows, how many seats does the theater have?

14. The value of an oil painting has doubled during each of the past 6 years. If it was valued at $200 during the first year, what was its value during the sixth year?

15. Use Pascal's triangle to expand $(3x + 1)^5$.

16. Use the binomial theorem to expand $(2x - 1)^6$.

17. Find the fifth term of $(x + b)^8$.

18. Find the fifteenth term of $(2x - y)^{14}$.

Challenge

At the end of the year 1800, a $5 deposit was made in a bank account. No further deposits or withdrawals were made. If the money doubles every 15 years, what will the account balance be at the end of the year 1995?

COLLEGE ENTRANCE EXAM REVIEW

Solve. Grid in your response on the answer sheet.

1. Find the sum of 14 terms of the series $-8, -3, 2, 7, \ldots$

2. Solve for x:

$$6x - 7(2x - 3) = 3(4 - 3x) + 7$$

3. Find $f(g(-2))$ when $f(x) = \dfrac{1}{3x + 4}$ and $g(x) = \sqrt{5 - x^2}$.

4. What is the value of the discriminant in the equation $2x^2 - 7x = 10$?

5. Solve for x:

$$\frac{3}{x+2} - \frac{5}{x^2 - 4} = \frac{2}{x - 2}$$

6. If $4x + 3yi = (6 - 4i)^2$, then what is the value of $x + y$?

7. If $2x^4 - 3x^3 + x - 5$ is divided by $x - 2$, what is the remainder?

8. Find the minimum value of the function f when $f(x) = 3x^2 - 6x + 1$.

9. Solve for x if $25^{2x+7} = 125^{x+6}$.

10. What is the coefficient of the term containing a^2 when $\left(\frac{1}{a} - a^2\right)^7$ is expanded and simplified?

11. If $f(x + 1) = f(x - 1) + 2$ and $f(5) = 6$, find $f(1)$.

12. In the figure below, $m\angle DOB = 70$, $m\angle COA = 80$, and $m\angle DOA = 110$. What is the measure in degrees of $\angle COB$?

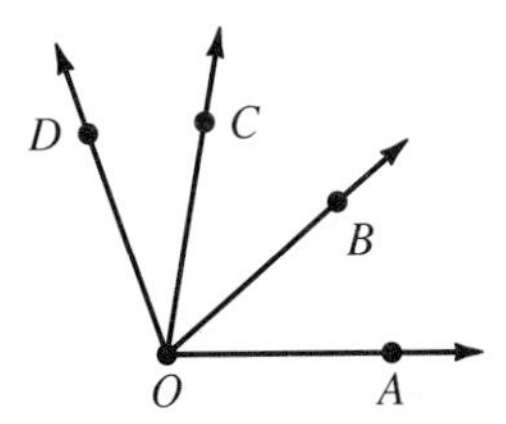

13. The intensity of illumination I on a surface varies inversely as the square of the distance d from the source. If $I = 12$ footcandles when $d = 12.5$ ft, find I (in footcandles) when $d = 10$ ft.

14. A small company makes a total of 96 rugs a day in sizes small, medium, and large. If the number of rugs in the small and large sizes combined is equal to the number of rugs in medium size, what is the daily production of rugs in the medium size?

15. A woman is paid time and a half for all hours worked in excess of 7.5 h per day. If she works 10 h in one day, by what percent are her regular wages for the day increased?

16. One leg of a right triangle is 4 ft longer than the other. The hypotenuse is 20 ft. What is the length (in feet) of the shorter leg?

MIXED REVIEW

Name all of the distinct line segments in the figure. How many line segments are there?

Example A B C D E

$\overline{AB}$ is the same as $\overline{BA}$.

$\overline{AB}$, $\overline{AC}$, $\overline{AD}$, $\overline{AE}$, $\overline{BC}$, $\overline{BD}$, $\overline{BE}$, $\overline{CD}$, $\overline{CE}$, $\overline{DE}$; 10 line segments

1. A B C D E F

A marble is picked at random from a bag containing 3 blue marbles, 4 yellow marbles, and 5 red marbles. Find the probability of each event.

Example $P(Y)$ means the probability of picking a yellow marble

$$P(Y) = \frac{4 \text{ yellow}}{12 \text{ total}} = \frac{4}{12} \text{ or } \frac{1}{3}$$

2. $P(B)$ **3.** $P(R)$ **4.** $P(B \text{ or } Y)$ **5.** $P(B \text{ or } R)$ **6.** $P(Y \text{ or } R)$

Find the mean (average) for each set of data.

Example 126 cm; 249 cm; 184 cm; 157 cm

$$\text{mean} = \frac{126 + 249 + 184 + 157}{4} = 179 \text{ cm}$$

Find the sum of the values. Divide by the number of values.

7. 13.5 kg; 25.8 kg; 17.6 kg; 14.1 kg; 9 kg

8. 1354; 998; 1005; 1203

9. 562 lb; 539 lb; 525 lb; 547 lb; 559 lb

10. 0.8 l; 0.75 l; 1 l; 3.2 l; 0.65 l

Use the bar graph to solve each word problem.

11. How many people rode the train in February?

12. How many more people rode the train in February than in May?

13. During which month did 25,000 people ride the train?

14. What was the average number of people riding the train each month?

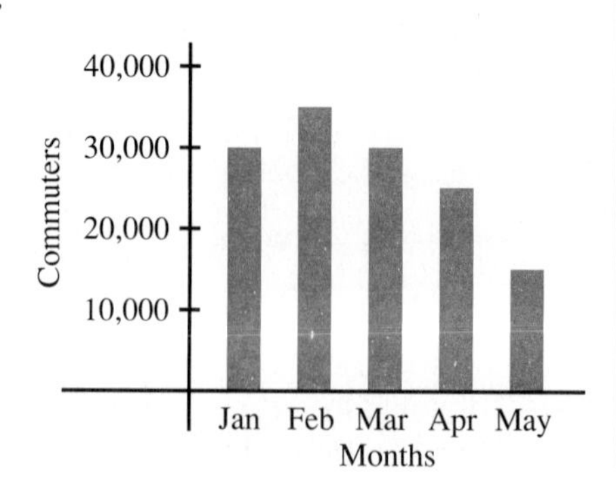

Probability and Statistics

How commercial can it get?

Developing Mathematical Power

Until July 1, 1951, television was commercial-free. Since then commercials have become big business. Until 1984, the Federal Communications Commission (FCC) limited the amount of commercial time for children's programming to 9.5 minutes per hour on weekends and 12 minutes per hour on weekdays. Today, time limits are set by each station.

1984 Limits of TV commercial time	
Weekends	9.5 min/h
Weekdays	12 min/h

Project

You are attending an FCC hearing that will determine whether television stations allow too much commercial time during children's programming. What is your point of view? What statistics will you present to defend your point of view? Put together a convincing argument.

14.1

The Fundamental Counting Principle and Permutations

Objectives: To use the fundamental counting principle
To find the number of permutations of n elements

The *fundamental counting principle* can make it possible to count elements of a set without actually listing them. Sometimes the application of this principle involves the evaluation of factorial expressions.

Capsule Review

In *factorial notation:*

$$n! = n \cdot (n-1) \cdot (n-2) \cdot \cdots \cdot 3 \cdot 2 \cdot 1$$
$$0! = 1$$

These definitions can be used to evaluate expressions such as $\frac{8!}{5!}$.

$$\frac{8!}{5!} = \frac{8 \cdot 7 \cdot 6 \cdot \cancel{5} \cdot \cancel{4} \cdot \cancel{3} \cdot \cancel{2} \cdot \cancel{1}}{\cancel{5} \cdot \cancel{4} \cdot \cancel{3} \cdot \cancel{2} \cdot \cancel{1}} = 8 \cdot 7 \cdot 6 = 336$$

or $$\frac{8!}{5!} = \frac{8 \cdot 7 \cdot 6 \cdot \cancel{5!}}{\cancel{5!}} = 8 \cdot 7 \cdot 6 = 336$$

Evaluate.

1. $6!$ **2.** $6 \cdot 5 \cdot 4!$ **3.** $\frac{6!}{4!}$ **4.** $\frac{7!}{4!}$ **5.** $\frac{10!}{9!}$ **6.** $\frac{0!}{3!}$

A printer offers the following options for the cover of a yearbook.

Composition of Cover: smooth vinyl or crinkled vinyl
Lettering Style: Roman, block, script, or *italic*

To determine how many different choices there are, make a *tree diagram*.

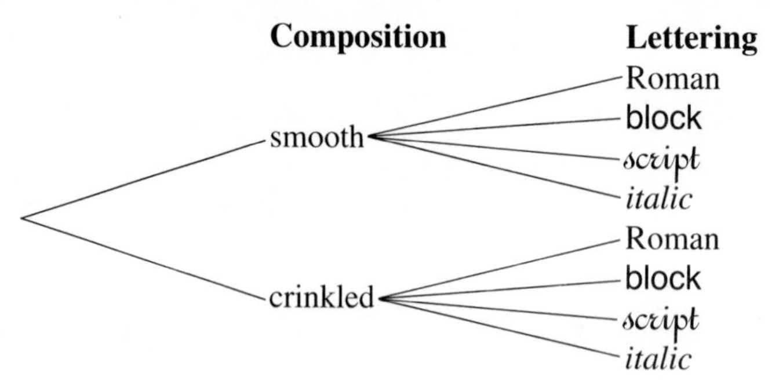

The diagram shows 8 "branches," such as *crinkled—script,* so there are 8 different choices. The number of choices could have been determined without making a diagram. That is, for *each* of the 2 choices of cover composition there are 4 choices of lettering style. Thus, there are $2 \cdot 4$, or 8 choices in all. This illustrates the *fundamental counting principle*.

Fundamental Counting Principle

If one event can occur in m different ways and a second event can occur in n different ways, then together the events can occur in $m \cdot n$ different ways, assuming that the second event is not influenced by the first event.

The fundamental counting principle can be extended to include more than two events, as in the following example.

EXAMPLE 1 A new economy car has just come on the market. To keep the price low, the manufacturer offers only the following options:

Color	Engine Size	Transmission	Radio
white	4 cylinders	manual	stereo, 4 speakers
red	6 cylinders	automatic	stereo, 8 speakers
blue			stereo/cassette, 4 speakers
			stereo/cassette, 8 speakers

How many different choices are there?

Count the number of possibilities for each option. Then multiply.

$3 \cdot 2 \cdot 2 \cdot 4 = 48$ *3 colors, 2 engines, 2 transmissions, 4 radios*

There are 48 different choices.

A **permutation**, P, of some or all of the elements of a set is any arrangement of the elements in a definite order. To illustrate, there are 6 permutations of all of the elements of the set $\{a, b, c\}$.

$$abc \quad acb \quad bac \quad bca \quad cab \quad cba$$

To find the number of permutations without listing them, the fundamental counting principle can be used. This is shown below for the set $\{a, b, c\}$.

There are 3 ways to fill the first position.	For *each* way to fill the first position there are 2 ways to fill the second position.	For *each* way to fill the first *and* second positions there is only 1 way to fill the third position.
3 ·	**2** ·	**1**

Thus, the number of permutations of the elements of $\{a, b, c\}$ is

$$3 \cdot 2 \cdot 1 = 3! = 6$$

In general, the number of permutations of n distinct elements of a set, denoted ${}_nP_n$, is given by the formula

$${}_nP_n = n!$$

EXAMPLE 2 **How many permutations of the letters a, b, c, d, e, and f are there?**

$${}_6P_6 = 6! = 6 \cdot 5 \cdot 4 \cdot 3 \cdot 2 \cdot 1 = 720$$

There are 720 permutations of the letters a, b, c, d, e, and f.

Sometimes it is necessary to find the number of permutations of n elements taken r at a time, ${}_nP_r$, when $r < n$.

EXAMPLE 3 A museum has 7 paintings to hang and 3 vacant locations, each of which will hold one painting. In how many different ways can these 3 locations be filled by the paintings?

Use the fundamental counting principle. The number of permutations of 7 paintings taken 3 at a time is

$${}_7P_3 = \underbrace{7}_{\substack{\text{choices for} \\ \text{1st location}}} \cdot \underbrace{6}_{\substack{\text{choices for} \\ \text{2nd location}}} \cdot \underbrace{5}_{\substack{\text{choices for} \\ \text{3rd location}}} = 210$$

There are 210 different ways to fill the vacant locations.

Note that ${}_7P_3$ can be written as a factorial expression:

$$\begin{aligned} {}_7P_3 = 7 \cdot 6 \cdot 5 &= \frac{7 \cdot 6 \cdot 5 \cdot 4 \cdot 3 \cdot 2 \cdot 1}{4 \cdot 3 \cdot 2 \cdot 1} \\ &= \frac{7!}{4!} = \frac{7!}{(7-3)!} \end{aligned}$$

In general, the number of permutations of n distinct elements taken r at a time, denoted ${}_nP_r$, is given by the formula

$${}_nP_r = \frac{n!}{(n-r)!} \qquad \text{for } 0 \leq r \leq n$$

EXAMPLE 4 How many different ways can 8 people be seated in a row of 5 chairs?

Find the number of permutations of 8 things taken 5 at a time.

$${}_8P_5 = \frac{8!}{(8-5)!} \qquad {}_nP_r = \frac{n!}{(n-r)!}$$

$$= 6720$$ There are 6720 different ways to seat them.

When the objects to be arranged are not all distinct, the formula for the number of permutations has to be adjusted. Listed below are all the arrangements, or permutations, of the letters of the words *READ* and *REED*.

READ						**REED**		
READ	*REDA*	*RDEA*	*RDAE*	*RADE*	*RAED*	*REED*	*REDE*	*RDEE*
ERAD	*ERDA*	*EADR*	*EARD*	*EDRA*	*EDAR*	*ERED*	*ERDE*	*EDER*
ARED	*ARDE*	*AEDR*	*AERD*	*ADRE*	*ADER*	*EDRE*	*EEDR*	*EERD*
DEAR	*DERA*	*DARE*	*DAER*	*DRAE*	*DREA*	*DEER*	*DERE*	*DREE*

There are 24 permutations of the letters in the word *READ*, which has 4 distinguishable letters, and only 12 permutations of the letters in the word *REED*, in which the letter *E* occurs twice. The two *E*'s can be permuted in 2 ways. The number of permutations of the letters of the word *READ* is ${}_4P_4 = 4!$, whereas the number of permutations of the letters of the word *REED* is $\frac{4!}{2!} = \frac{4 \cdot 3 \cdot 2 \cdot 1}{2 \cdot 1} = 12$.

In general, the number of distinguishable permutations P of n elements taken n at a time, with r_1 like elements, r_2 like elements of another kind, and so on, is given by the formula

$$P = \frac{n!}{r_1!\, r_2!\, r_3! \cdots}$$

EXAMPLE 5 Find the number of permutations of the letters in the word *MISSISSIPPI*.

$$P = \frac{n!}{r_1!\, r_2!\, r_3! \cdots}$$

$$= \frac{11!}{4!\, 4!\, 2!}$$ *There are 11 letters in MISSISSIPPI. There are 4 I's, 4 S's, and 2 P's.*

$$= 34{,}650$$

There are 34,650 permutations of the letters in *MISSISSIPPI*.

CLASS EXERCISES

1. A school has received a grant to buy one piece of new equipment for each of its three shops. Each shop teacher has submitted a list of items, as shown below. Make a tree diagram to show the different ways the grant can be used.

   Woodworking: circular saw or sander
   Car Repair: tachometer, emissions tester, metric tool set, or manuals
   Electrical: circuit tester or oscilloscope

Use the fundamental counting principle to solve each problem.

2. A company manufactures sneakers in 3 colors, 2 styles, and 8 sizes. How many different sneakers are made?
3. A quiz consists of 10 true/false questions. If a student guesses all of the answers, in how many different ways can she complete the quiz?

Evaluate.

4. ${}_5P_5$
5. ${}_7P_4$
6. ${}_9P_8$
7. ${}_6P_2$
8. ${}_8P_6$
9. ${}_{10}P_7$

PRACTICE EXERCISES 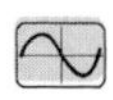 Use technology where appropriate.

Use the fundamental counting principle to solve each problem.

1. Harry is selecting new uniforms for his team. Pants come in 3 styles, shirts in 2 styles, and hats in 4 styles. In how many different ways can a 3-piece uniform be selected?
2. A store makes custom paints using a base, a texture, and a pigment. If the store has 3 different bases, 2 textures, and 50 pigments, how many different custom paints can be mixed?
3. There are 12 questions on a true/false test. If all questions are answered, in how many different ways can the test be completed?
4. Each of the 8 questions on a multiple-choice test has 3 possible answers. If all questions are answered, in how many ways can the test be completed?

Find the number of permutations of the elements in each set.

5. $\{C, A, R, B, O, N\}$
6. $\{1, 2, 3, 4, 5\}$
7. In how many different ways can 9 baseball players' names be listed in a column on the roster?
8. Six people volunteer to help put out a fire. In how many different ways can they be lined up in a row to hold the hose?

9. How many ways can 8 different books be arranged on a shelf?

10. How many ways can 10 students be seated in a row?

11. How many 3-letter code words can be made from the letters b, c, d, e, and f, if repetition of a letter is allowed?

12. How many 2-letter code words can be made from the letters b, c, d, e, and f, if repetition of a letter is not allowed?

13. How many 4-digit permutations of the 10 digits are there, if no digit may be repeated?

14. How many 3-digit permutations of the 10 digits are there, if any digit may be repeated?

15. How many permutations of the letters of the word *BABBLING* are there?

16. How many permutations of the letters of the word *CARICATURE* are there?

Use the fundamental counting principle to solve each problem.

17. A company is setting up phone numbers for a new town. Each number must have 7 digits and must not start with 0. How many different numbers are possible? (Repetition of a digit is allowed.)

18. A 7-digit phone number is to be chosen so that the first and last digits are not 0 and the last digit is even. How many different numbers can be chosen? (Repetition of a digit is allowed.)

19. On a 15-item test, the first five items have 4 choices each, the next five have 3 choices each, and the last five are true or false. If Joe answers items 2, 7, and 10 correctly and guesses all the others, how many different ways can he complete the test?

20. A code is constructed so that each word has exactly 5 letters. A word cannot start with a vowel or end with a consonant. If the letters q, x, and z are excluded, how many different code words can be formed?

21. How many different 6-digit license plates can be made if the first digit must not be 0 and no digits may be repeated?

22. How many different 6-digit license plates can be made if the first digit must not be 0, but digits may be repeated?

23. How many permutations of the letters in the word *DAZZLED* are there?

24. How many permutations of the letters in the word *REARRANGEMENT* are there?

25. The manager of a baseball team wants the best hitter up fifth. If the lineup consists of 9 players, how many different lineups are possible?

26. An insurance company is giving each customer a 9-digit code number. Each number is odd and does not start with 0, and repetition of digits is not allowed. How many code numbers are possible?

Circular permutations occur when objects are arranged in a circular pattern. The figures at the right show all of the possibilities for 3 objects so arranged. Since the *relative* positions of the objects are the same in the 3 figures in the first column, these arrangements are considered to be the same. Similarly, the 3 arrangements in the second column are considered the same. Thus, there are just 2 distinct circular permutations of 3 objects. In general, the number of distinct circular permutations of n different objects is

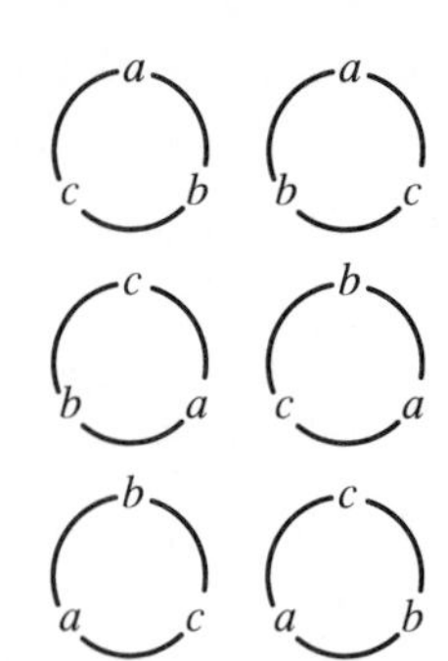

$$(n - 1)!$$

27. In how many different ways can 6 knights be seated in 6 chairs around a round table?

28. If there are 11! different seating arrangements around a round table, how many people are there?

29. In how many ways can 9 people be lined up in a row if 2 of them must stand next to each other?

30. In how many ways can 9 people be seated at a round table if 2 of them must sit next to each other?

Prove each of the following statements:

31. ${}_nP_r = \dfrac{n!}{(n - r)!}$

32. ${}_nP_{n-r} = \dfrac{n!}{r!}$

Applications

33. Banking A bank plans to assign an identification number to each account. Each number will have 5 digits, and no digit will be repeated. How many different account numbers can be formed?

34. Government A state's license plate has 3 letters, followed by 3 digits. If neither letters nor digits can be repeated, and the first digit cannot be a zero, how many different license plates can the state issue?

35. Sports Only 8 members of a softball team showed up for a practice game. How many different batting lineups were possible, if the best hitter batted fifth and the next best hitter batted eighth?

Developing Mathematical Power

36. Thinking Critically There are just 17 coins in a bank. Eight are 1980 quarters, five are 1984 quarters, and four are 1988 quarters. You want to shake out three coins minted in the same year. What is the maximum number of coins you must shake out of the bank?

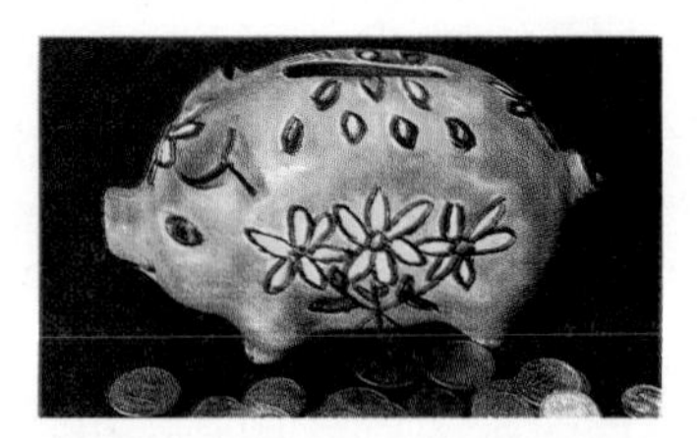

37. Project How many different license plates can your state issue?

14.2

Combinations

Objective: To find the number of combinations of n elements taken r at a time

A is a *subset* of B if every element of A is also an element of B. For example, the subsets of $\{a, b, c\}$ are

$\{a, b, c\}$	*Every set is a subset of itself.*
$\{a, b\}, \{a, c\}, \{b, c\}$	*Two-element subsets*
$\{a\}, \{b\}, \{c\}$	*One-element subsets*
$\emptyset$	*The empty set is considered a subset of every set.*

Permutations of the elements of a set are subsets of that set. However, when you count permutations, the order in which the elements are listed is important. When you count *combinations* of elements of a set, the order in which they are listed is disregarded. A **combination** of n elements of a set taken r at a time, denoted ${}_nC_r$, is any r-element subset of the given set.

When you calculate the number of combinations of a given number of elements of a set it is helpful to use factorial notation.

Capsule Review

To evaluate the expression $\frac{8!}{3!\,5!}$, use the fact that $8! = 8 \cdot 7 \cdot 6 \cdot 5!$.

$$\frac{8!}{3!\,5!} = \frac{8 \cdot 7 \cdot \cancel{6} \cdot \cancel{5!}}{\cancel{3} \cdot \cancel{2} \cdot 1 \cdot \cancel{5!}}$$
$$= 8 \cdot 7 = 56$$

Evaluate.

1. $\frac{7!}{4!\,3!}$ **2.** $\frac{8!}{4!\,4!}$ **3.** $\frac{9!}{8!\,1!}$ **4.** $\frac{6!}{6!\,0!}$ **5.** $\frac{10!}{8!\,2!}$ **6.** $\frac{12!}{5!\,7!}$

7. $\frac{6!}{4!\,2!} + \frac{3!}{2!\,1!} + \frac{15!}{8!\,7!}$ **8.** $\frac{11!}{6!\,5!} - \frac{12!}{11!\,1!} - \frac{8!}{5!\,3!}$

EXAMPLE 1 **List all the combinations of the elements of the set $\{a, b, c\}$ if the elements are taken 2 at a time.**

The number of combinations is the number of 2-element subsets, $\{a, b\}$, $\{a, c\}$, and $\{b, c\}$. Therefore, ${}_3C_2 = 3$.

The number of permutations of the elements of the set $\{a, b, c\}$ taken 2 at a time is

$$_3P_2 = \frac{3!}{(3-2)!} = \frac{3!}{1!}$$
$$= 3 \cdot 2 = 6$$

The number of combinations is

$$_3C_2 = 3$$

Since there are $_2P_2$ permutations of each subset of 2 elements,

$$_2P_2 \cdot {_3C_2} = {_3P_2}$$

or $$_3C_2 = \frac{_3P_2}{_2P_2} = \frac{_3P_2}{2!}$$ $_nP_n = n!$

In general, the number of combinations of n elements taken r at a time, denoted $_nC_r$, is given by the formula

$$_nC_r = \frac{_nP_r}{r!} = \frac{n!}{r!\,(n-r)!} \quad \text{for } 0 \leq r \leq n$$

Since combinations are subsets in which order is not considered, you count combinations when you wish to find the number of committees, teams, or groups of objects or persons that can be chosen from a given set.

EXAMPLE 2 Fifteen people entered a talent contest. The top 3 contestants will each win \$50, and everyone else will get an honorable mention.

a. In how many different ways can 3 winners be chosen?

b. In how many different ways can 12 people be chosen for honorable mention?

a. $_{15}C_3 = \frac{15!}{3!\,12!} = 455$ $\quad {_nC_r} = \frac{n!}{r!\,(n-r)!};\ n = 15,\ r = 3$

b. $_{15}C_{12} = \frac{15!}{12!\,3!} = 455$ $\quad n = 15,\ r = 12$

Notice that $_{15}C_3 = {_{15}C_{12}} = {_{15}C_{15-3}}$. In general, $_nC_r = {_nC_{n-r}}$.

The following notation is often used for combinations of n elements taken r at a time:

$$\binom{n}{r} = {}_nC_r = \frac{n!}{r!\,(n-r)!}$$

EXAMPLE 3 **How many lines are determined by 5 points, no 3 of which are collinear?**

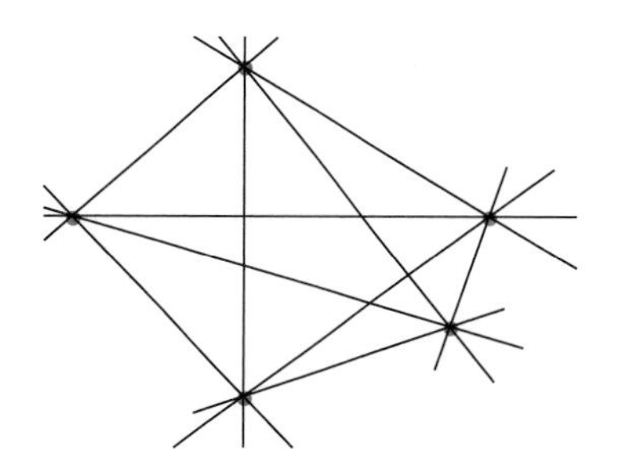

Two points determine a line.
Therefore; $n = 5$ and $r = 2$.

$\binom{5}{2} = \frac{5!}{2!\,3!} = 10$ $\quad \binom{n}{r} = \frac{n!}{r!\,(n-r)!}$

Ten lines are determined.

CLASS EXERCISES

Evaluate.

1. ${}_6C_2$ **2.** ${}_6C_4$ **3.** ${}_5C_4$ **4.** $\binom{8}{5}$ **5.** $\binom{10}{10}$ **6.** $\binom{20}{0}$

PRACTICE EXERCISES  Use technology where appropriate.

1. How many 4-element subsets can be formed from the set $\{a, b, c, d, e, f, g\}$?

2. How many 7-element subsets can be formed from the set $\{1, 2, 3, 4, 5, 6, 7, 8, 9\}$?

3. How many different committees of 3 can be chosen from 12 people?

4. How many different teams of 9 can be chosen from 12 students?

5. How many different 3-card hands can be chosen from a 52-card deck?

6. How many different sets of 4 books can be chosen from a shelf on which there are 24 books?

7. There are 14 different pens in a carton. How many different sets of 11 pens can be chosen?

8. A box contains 10 different fishing hooks. How many different sets of 6 hooks can be chosen?

9. How many lines are determined by 8 points, no 3 of which are collinear?

10. How many different diagonals can be drawn in a seven-sided polygon?

11. Nine students are eligible to play doubles tennis. How many different 2-person teams can be chosen?

12. A box contains 7 flashlights, all of different colors. How many different sets of 3 flashlights can be chosen?

13. In a lottery, 4 winners will get equal prizes. If 20 people enter the lottery, how many different groups of 4 winners can be chosen?

14. There are 7 men and 6 women in a club. How many different 3-member committees can be chosen so that all are women? so that all are men?

15. There are 10 soccer players and 8 volleyball players in a room. How many different groups of 2 players can be chosen so that there are no soccer players in the group? so that there are no volleyball players in the group?

16. A club has 8 executive board members and 15 general members. How many committees of 5 can be chosen so that only executive board members are included? so that executive board members are excluded?

17. How many 5-card hands that contain exactly 2 aces and 3 kings can be chosen from a 52-card deck? (There are 4 aces and 4 kings in the deck.)

18. How many 5-card hands that contain exactly 3 red and 2 black cards can be chosen from a 52-card deck? (Half the cards in the deck are red, half are black.)

19. A wallet contains a nickel, a dime, a penny, and a quarter. How many different sums of money can be made from the change in the wallet?

20. A wallet contains a $1 bill, a $5 bill, and two $10 bills. How many different sums of money can be made if each sum can contain only one of each type of bill?

21. How many different sets of 6 balls can you choose from a box containing only 7 gloves and 3 bats?

22. A plane contains 5 collinear points. How many triangles can be drawn, using any three of these points as vertices?

Prove.

23. ${}_nC_r = {}_nC_{n-r}$

24. ${}_nC_n = 1 = {}_nC_0$

25. If $n = r$, then $n!\,({}_nC_r) = {}_nP_r$

26. ${}_nC_r + {}_nC_{r+1} = {}_{n+1}C_{r+1}$

27. There are 6 points in a plane, no 3 of which are collinear. How many different polygons can be drawn using only these points as vertices?

28. How many different 5-card hands can be chosen from a 52-card deck so that each hand has at least one ace? (There are 4 aces in a deck.)

Applications

29. **Sports** A team manager has 11 students who are qualified to play basketball. How many different 5-person teams can be chosen?

30. **Parades** Six horses are needed to pull a float in a parade. If there are 10 horses in the stable, how many different teams of 6 can be selected?

31. **Construction** A construction supervisor needs a crew of 4 people to do a job, and 15 people volunteer. How many different 4-person crews can be chosen?

32. **Manufacturing** A computer manufacturer makes 3 types of monitors and 4 types of printers. How many different packages containing one monitor and one printer can be made?

33. **Sports** The summer Olympics games of 1988 had 16 countries qualified to compete in soccer. How many different ways can teams of 2 be selected?

Technology Most graphing calculators offer a probability menu that includes functions for permutations and combinations. To calculate ${}_nP_r$ or ${}_nC_r$, simply enter n, followed by the function, followed by r, and press ENTER or EXE.

Use a utility to calculate each of the following.

34. ${}_6P_4$

35. ${}_{10}P_2$

36. ${}_{52}C_4$

37. ${}_{13}C_4$

Mixed Review

State whether each compound sentence represents an intersection or a union of sets. Given a universal set of $-10 \le x \le 10$, find each intersection or union.

38. x is an odd integer or x is a negative integer.

39. x is a whole number and x is an even integer.

40. x is divisible by 3 and x is divisible by 2.

Write a rule for each relation and state its domain and range.

41. $\{(1, 6), (2, 8), (3, 10), (4, 12)\}$

42. $\{(2, 2), (3, 4), (4, 8), (5, 14)\}$

43. Write an equation of the line that passes through the y-axis at 7 with a slope of zero.

44. Find the distance between $(2\sqrt{3}, 5)$ and $(9\sqrt{3}, -2)$.

45. Find the midpoint of the segment with endpoints $(5, 6\sqrt{2})$ and $(-2, 0)$.

Developing Mathematical Power

Investigation The graphs of combination functions can reveal surprising patterns. Since the domain consists of integers, it is important to choose screen dimensions that will display integral domain values.

Use a utility to graph each function over $\{x\text{: } 0 \le x \le 19\}$.

46. $y = {}_xC_{x-2}$

47. $y = \sqrt{{}_xC_2 + {}_{20}C_x}$

48. $y = \sqrt[12]{{}_{18}P_x}$

Probability

Objectives: To specify the sample space for a random experiment
To calculate the probability that a given event will occur

If you toss a coin, there are 2 possible outcomes: heads or tails. If you roll a die, there are 6 possible outcomes: 1, 2, 3, 4, 5, or 6. Experiments in which the outcome is not necessarily the same when you repeat the experiment, such as tossing a coin or rolling a die, are called *random experiments*. You can use the fundamental counting principle to count all of the possible outcomes for some random experiments.

Capsule Review

If there are 2 roads from A to B, and 3 roads from B to C, you can use the fundamental counting principle to find the number of different ways you can drive from A to C.

$$\underbrace{\text{No. of ways } A \text{ to } B}_{2} \cdot \underbrace{\text{No. of ways } B \text{ to } C}_{3} = 6$$

There are 6 different ways to drive from A to C.

Solve.

1. A bicycle can be ordered in one of 4 colors and with one of 2 gear mechanisms. How many different bicycles are available?
2. A menu offers 5 appetizers, 4 main courses, and 3 desserts. How many different meals, (appetizer, main course, and dessert) are available?
3. A party invitation is offered in 5 colors, 6 type styles, and 4 kinds of paper. How many choices are there?
4. In how many different ways can a sandwich be made, if there is a choice of 4 kinds of filling and 4 kinds of bread?
5. How many positive 3-digit numbers can be written, if each digit is greater than 5 and repetition is allowed?
6. How many positive 4-digit numbers can be written if the first digit is not 0 and repetition is not allowed?

The set of all possible outcomes of a random experiment is called a **sample space.** Any subset of the sample space is called an **event.**

EXAMPLE 1 **A die is rolled. Specify the following:**

a. the sample space, S

b. event A, which is the event of rolling an even number

c. event B, which is the event of rolling a number greater than 4

a. Each outcome is the number of dots on the top face, so $S = \{1, 2, 3, 4, 5, 6\}$

b. $A = \{2, 4, 6\}$

c. $B = \{5, 6\}$

Note that it is assumed that a "fair die" is used in the random experiment. When such a die is rolled, the six possible outcomes are *equally likely*. If the die were weighted or damaged, one or more of the outcomes would be more likely to occur than the others.

Assuming each outcome is equally likely, if you compare the number of outcomes in an event to the number of outcomes in the sample space, you are expressing the likelihood, or the *probability*, that the event will occur.

> If a sample space consists of equally likely outcomes which can be counted, then the **probability** that an event E will occur, $P(E)$, is the ratio of the number of outcomes in the event to the number of outcomes in the sample space.
>
> $$P(E) = \frac{\text{number of outcomes in the event}}{\text{number of outcomes in the sample space}}$$

EXAMPLE 2 **A die is rolled. Find the probability of each event.**

a. getting a 4

b. getting an odd number

c. getting 1, 2, 3, 4, 5, or 6

d. getting a 7

a. $P(\text{getting } 4) = \frac{1}{6}$ — *There is only one way to get a 4.*

b. $P(\text{getting an odd number}) = \frac{3}{6} = \frac{1}{2}$ — *There are three odd numbers: 1, 3, and 5.*

c. $P(\text{getting } 1, 2, 3, 4, 5, \text{ or } 6) = \frac{6}{6} = 1$ — *There are six ways to roll one of these numbers.*

d. $P(\text{getting } 7) = \frac{0}{6} = 0$ — *There is no way to get a 7.*

The probability of an event that is certain to occur is 1. The probability of an event that cannot occur is 0. Thus, the probability of an event E, $P(E)$, is a number such that $0 \leq P(E) \leq 1$.

Getting an outcome in a specified event is called a *success*. Getting an outcome that is not in a specified event is called a *failure*. The **odds** that an event will occur, or the odds in favor of an event, are expressed as the ratio of the number of successes to the number of failures.

Odds in favor of an event = (number of successes):(number of failures)

EXAMPLE 3 **Two coins are tossed. Specify the following:**

a. the sample space, S
b. the probability of getting exactly 2 heads
c. the odds in favor of getting exactly 2 heads
d. the probability of getting exactly 1 head
e. the odds in favor of getting exactly 1 head

a. $S = \{(H, H), (H, T), (T, H), (T, T)\}$

b. $P(\text{getting exactly 2 heads}) = \frac{1}{4}$

There is only one way to get 2 heads: (H, H).

c. The odds in favor of getting exactly 2 heads are 1:3.

Just 1 outcome is a success, while 3 are failures.

d. $P(\text{getting exactly 1 head}) = \frac{2}{4} = \frac{1}{2}$

There are two ways to get exactly 1 head: (H, T), (T, H).

e. The odds in favor of getting exactly 1 head are 2:2 or 1:1.

Two outcomes are successes and 2 are failures.

You can sometimes use the fundamental counting principle, permutations, or combinations to find the probability of an event.

EXAMPLE 4 A bridge deck consists of 52 cards, with 13 in each of four suits: spades, hearts, diamonds, and clubs. Each suit consists of an ace, nine cards numbered 2 through 10, and three face cards (jack, queen, and king). If 3 cards are picked from a bridge deck, what is the probability that they will all be diamonds?

The sample space, S, is the set of all possible 3-card hands in a 52-card deck. Therefore, the number of outcomes in S is ${}_{52}C_3$.

$$\text{Number of outcomes in } S = {}_{52}C_3 = \frac{52!}{3!\,49!} = 22{,}100$$

Let E be the event of picking 3 diamonds from the 13 in the deck. The number of outcomes in E is ${}_{13}C_3$.

Number of outcomes in $E = {}_{13}C_3 = \frac{13!}{3!\ 10!} = 286$

Therefore, $P(E) = \frac{286}{22{,}100} = \frac{11}{850}$

The probability, expressed as a decimal, is about 0.013.

CLASS EXERCISES

A red die and a blue die are rolled.

1. Specify the sample space. Write each outcome as an ordered pair, (r, b).

Write the ordered pairs for each event.

2. The red die shows 3.
3. The sum is 8.

Find the probability of each event and the odds in favor of the event.

4. The red die shows 3.
5. The sum is 8.
6. The outcome is (3, 4).
7. The sum is less than 8.
8. Both numbers are even.
9. The product is 12.

PRACTICE EXERCISES

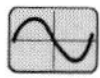 Use technology where appropriate.

Each of the numbers from 1 through 10 is written on a slip of paper and the slips are placed in a covered bowl. The bowl is shaken and one slip of paper is drawn at random. Specify:

1. the sample space
2. the event of drawing an even number
3. the event of drawing a number less than 5
4. the probability of getting an even number and the odds in favor of getting an even number
5. the probability of getting a number less than 5 and the odds in favor of getting a number less than 5

Each of the 10 letters *A* through *J* is written on a slip of paper and the slips are placed in a bag. The bag is shaken and one slip is drawn at random. Specify:

6. the sample space
7. the event of drawing a vowel
8. the event of *not* drawing a vowel

9. the probability of drawing a vowel and the odds in favor of drawing a vowel

10. the probability of *not* drawing a vowel and the odds in favor of *not* drawing a vowel

11. Six books labeled *A* through *F* are arranged at random on a shelf. What is the probability that they are arranged in alphabetical order?

12. If a card is picked at random from a bridge deck, what is the probability of getting a 7?

13. Ten people, including Alice, Bob, and Lisa, are in a room. If a committee of 3 people is selected at random, what is the probability that Alice, Bob, and Lisa will all be on the committee?

14. A car dealer can order a car in 3 colors and 2 engine sizes. If one of the colors is red and one engine is a V-6, what is the probability a car shipped at random will be red, with a V-6 engine?

15. If two dice are tossed, what is the probability that the sum of the dots showing will be 2? that the sum will be 5?

16. A coin is tossed and a die is rolled. What is the probability of getting heads and 6? heads and an odd number?

17. A 4-card hand is to be picked at random from a bridge deck. What is the probability that the cards will all be face cards?

18. A 5-card hand is to be picked at random from a bridge deck. What is the probability that the cards will all be hearts?

19. Two dice are rolled. What is the probability that the sum of the dots showing will be an odd number? will be less than 9?

20. Three dimes are tossed. What is the probability that all the faces will be the same? that there will be exactly 2 heads?

21. Ten chemists and 15 biologists apply for positions in a company. If 5 people are hired at random, what is the probability that 2 are chemists and 3 are biologists?

22. Twelve pumpkins and 8 squash are left in a bin at a farm stand at the end of the day. If 6 vegetables are picked at random from this bin, what is the probability that 4 are pumpkins and 2 are squash?

23. Irene, Cathy, Charlene, and Kisha are to be seated in a row. If they are seated at random, what is the probability that Irene and Cathy will sit next to each other? that Charlene will sit in the last seat on the left?

24. Three coins are tossed. What is the probability of getting at least one head? at most one head?

25. Two dice are rolled. What is the probability of rolling at least one 6?

26. Two cards have been taken from a bridge deck, an 8 and a jack. What is the probability that the next card chosen will be a 3?

27. If the letters E, A, and R are permuted, what is the probability that the letters will spell an English word?

28. In a random arrangement of the letters T, E, E, and M, what is the probability that the word *TEEM* *or* the word *MEET* will be spelled?

29. If the odds in favor of an event are 7:12, what is the probability of that event occurring?

30. If the odds in favor of an event are $a:b$, what is the probability of that event occurring?

Five tags lettered *A*, *B*, *C*, *D*, and *E* are placed in a bowl. A tag is drawn at random and then a second tag is drawn.

31. What is the probability that the tags will be drawn in alphabetical order?

32. If the tags do not have to be drawn in alphabetical order, what is the probability that C will be drawn?

Applications

Computer The computer is an excellent tool to use in simulating a method of approximating pi. Because this method plays a game of chance (as in Monte Carlo) to solve a mathematical problem, it is called the Monte Carlo method.

Let a square that is superimposed on a unit circle whose center is at the origin represent a dartboard.

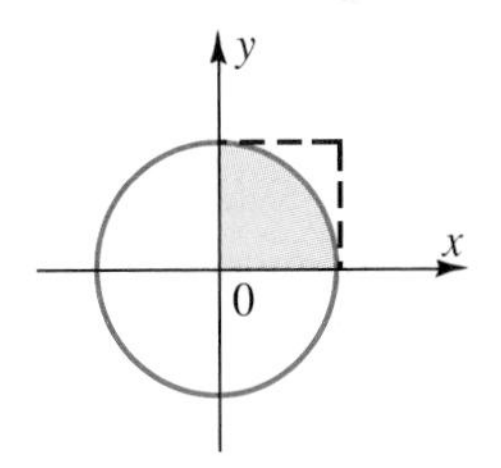

$$\begin{aligned}\text{Inside circle:} &\quad x^2 + y^2 < 1\\ \text{On circle:} &\quad x^2 + y^2 = 1\\ \text{Outside circle:} &\quad x^2 + y^2 > 1\\ A = \pi r^2 &= \pi(1)^2 = \pi\end{aligned}$$

The program simulates throws of a dart inside the circle in quadrant one (shaded in the drawing.) Notice in the program that random numbers are positive, so that values of x and y are positive.

```
10 LET H = 0
20 INPUT "ENTER NUMBER OF THROWS:
   ";N: PRINT
30 FOR I = 1 TO N
40 X = RND (1) * 2 - 1:Y = RND (1) * 2 - 1
50 IF X ^ 2 + Y ^ 2 < = 1 THEN H = H + 1
60 NEXT I
70 P = 4 * (H / N)
80 PRINT "IN A SAMPLE SPACE OF ";N;" THROWS,
   PI IS APPROXIMATELY ";P
90 END
```

For each throw, the computer will randomly generate the numbers that represent the coordinates of the point where the dart lands. The throw is a "hit" if the dart lands inside or on the circle. When the throws are completed, the computer will calculate

$$\pi = 4\left(\frac{\text{number of hits}}{\text{number of throws}}\right)$$

33. Does a greater number of throws necessarily improve the approximation of pi? Is there a number of throws that seems to consistently give the closest approximation of pi?

34. Do you get more accurate results if your sample space is a number that ends in a zero? Why or why not?

Developing Mathematical Power

Writing in Mathematics A meter attendant kept track of parked cars by noting the positions of the air valves on the front and rear tires on one side of each car. She recorded her observations using a system favored by pilots. To illustrate, the position of the valve on the front tire of the car in the picture would be considered *2 o'clock;* the position of the rear tire would be considered *9 o'clock.*

A driver who received a ticket for parking overtime went to court and testified that she had left the parking space on time but returned to the same space later. She maintained that it was just by chance that the tires came to rest in the positions they were in when she left. A court-appointed expert testified that the probability of such a coincidence is $\frac{1}{12} \cdot \frac{1}{12}$, or $\frac{1}{144}$, since there are 12 hour-positions on each tire. The judge acquitted the driver, saying that the probability of a coincidence was great enough to avoid conviction.

For each question, write a short paragraph defending your position.

35. If you had been the judge, would you have acquitted the defendant if the meter attendant had noted that all four tire valves were in the same positions that they had been in earlier?

36. The court-appointed expert assumed that the tires on a car rotate at the same rate. Is this a valid assumption? Is it true when the car travels in a straight path? Is it true when the car travels around a curve or turns a corner?

37. Do you agree with the judge's decision? Why or why not?

Mutually Exclusive and Independent Events

Objective: To calculate the probability of mutually exclusive events and of independent events

A sample space and events that are subsets of that sample space can be represented by a *Venn diagram*. The sample space is represented by a rectangular region and the events are represented by circular regions. If the events have outcomes in common, the circular regions overlap. If the events do not have outcomes in common, the sets are **disjoint** and the circular regions do not overlap.

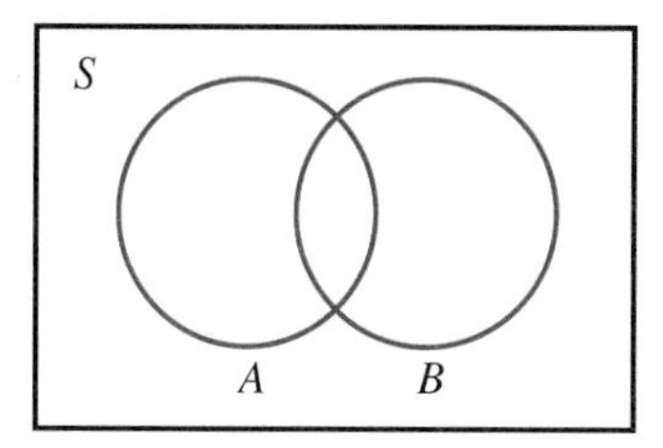

Sets A and B are *not* disjoint. Events A and B have outcomes in common.

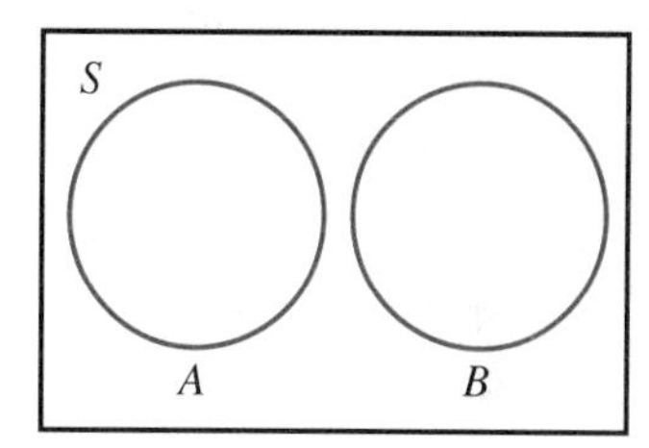

Sets A and B are disjoint. Events A and B have *no* outcomes in common.

Capsule Review

Let the sample space be the set of positive integers, and let

$$C = \{2, 4, 6, 8, 10\} \quad \text{and} \quad D = \{4, 8, 12\}$$

Events C and D have two elements in common, so they are not disjoint. The Venn diagram is shown at the right.

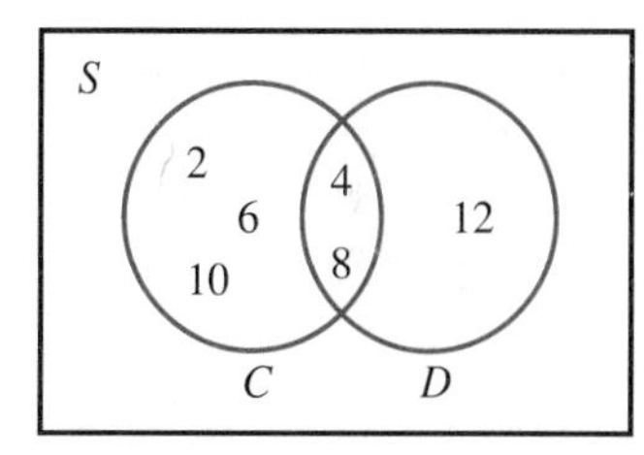

Draw a Venn diagram for each pair of events. The sample space is the set of positive integers.

1. $E = \{2, 4, 6, 8\}$, $F = \{1, 3, 5, 7\}$

2. $G = \{3, 6, 9, 12\}$, $H = \{2, 4, 6, 8, 10\}$

3. $I = \{10, 20, 30, 40\}$, $J = \{50\}$

4. $K = \{5, 10, 15\}$, $L = \{15, 20, 25\}$

5. $M = \{36, 49, 64\}$, $N = \{1, 8, 27, 64\}$

6. $P = \{1, 2, 3, 4\}$, $Q = \{2, 3, 4, 5\}$

7. $R = \{8, 16, 24\}$, $S = \{12, 24\}$

8. $T = \{4, 16, 64\}$, $V = \{9, 25, 49\}$

The *union* of sets A and B, written $A \cup B$, is the set whose elements belong to A *or* B or both A and B. The *intersection* of sets A and B, written $A \cap B$, is the set whose elements belong to A *and* B.

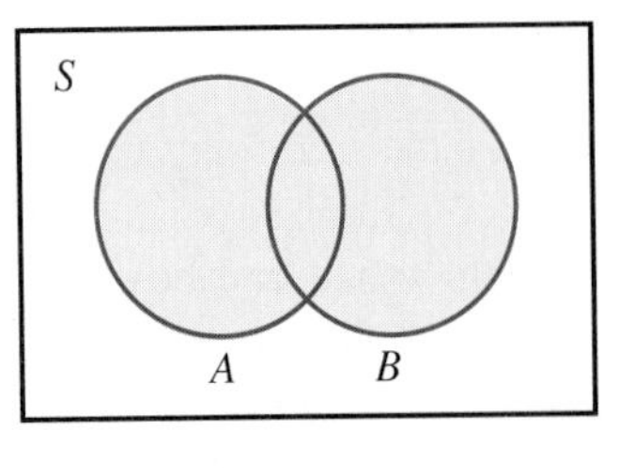

$A \cup B$

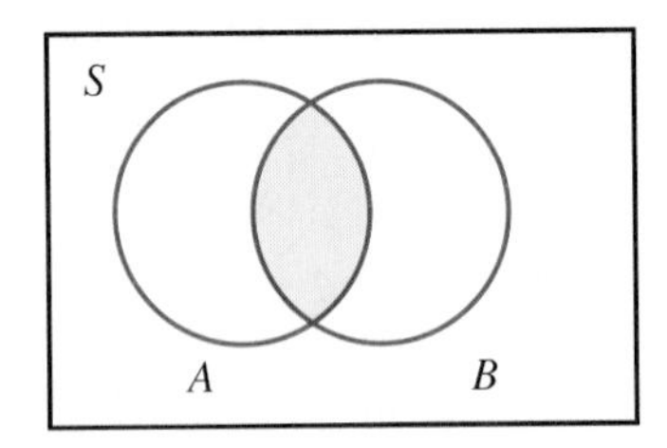

$A \cap B$

You can often use the union and intersection of sets to represent events and to find their probabilities.

EXAMPLE 1 **Roll a die once. Let A be the event of getting 1, 2, 3, or 4. Let B be the event of getting 2, 4, or 6. Find**

a. $P(A)$ **b.** $P(B)$ **c.** $P(A \text{ or } B)$ **d.** $P(A \text{ and } B)$

Draw a Venn diagram.

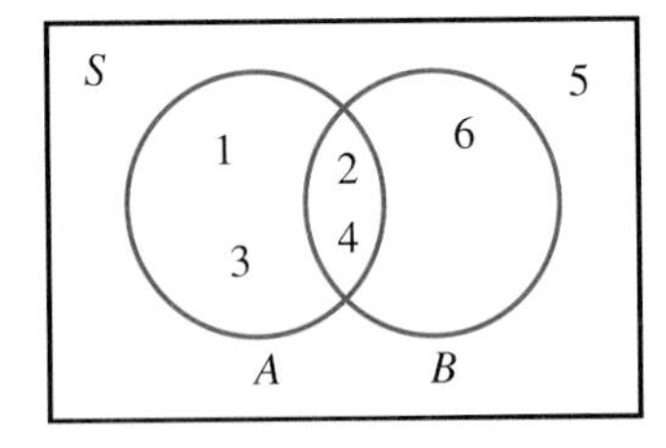

a. $P(A) = \frac{4}{6} = \frac{2}{3}$ **b.** $P(B) = \frac{3}{6} = \frac{1}{2}$

c. $P(A \text{ or } B) = P(A \cup B)$ $A \cup B = \{1, 2, 3, 4, 6\}$

$$= \frac{\text{outcomes in } A \cup B}{\text{outcomes in sample space}} = \frac{5}{6}$$

d. $P(A \text{ and } B) = P(A \cap B)$ $A \cap B = \{2, 4\}$

$$= \frac{\text{outcomes in } A \cap B}{\text{outcomes in sample space}} = \frac{2}{6} = \frac{1}{3}$$

In Example 1, notice that $P(A) + P(B) = \frac{4}{6} + \frac{3}{6} = \frac{7}{6}$; that is,

$$P(A) + P(B) > P(A \cup B)$$

Events A and B both contain 2 and 4. Therefore 2 and 4 were counted twice in the computation of $P(A) + P(B)$. If the probability of the outcomes in the intersection of A and B, $\frac{2}{6}$, is subtracted from the sum of the probabilities of events A and B, then

$$P(A \cup B) = \frac{5}{6} = \frac{4}{6} + \frac{3}{6} - \frac{2}{6}$$

where $\frac{4}{6} = P(A)$, $\frac{3}{6} = P(B)$, and $\frac{2}{6} = P(A \cap B)$.

In general, for two events A and B in a sample space,

$$P(A \cup B) = P(A) + P(B) - P(A \cap B)$$

Two events that have no elements in common are called **mutually exclusive** events. If A and B are mutually exclusive, then their intersection is the empty set. That is, two events A and B are mutually exclusive if and only if $A \cap B = \emptyset$.

If two events, A and B, are mutually exclusive, then

$$P(A \cup B) = P(A) + P(B)$$

If E is an event in a sample space S, then any outcome in S that is not in E is said to be in ***E*-complement**, written $\overline{E}$. E and $\overline{E}$ are mutually exclusive events.

$$P(E \cup \overline{E}) = P(S) \quad \text{since} \quad E \cup \overline{E} = S$$

Therefore, $P(E) + P(\overline{E}) = 1$ and $P(\overline{E}) = 1 - P(E)$.

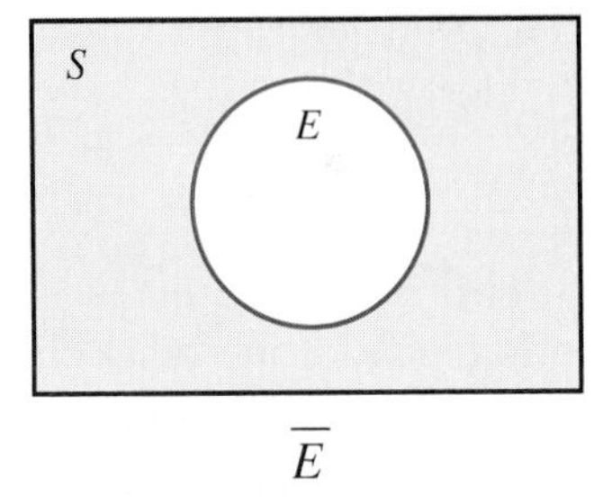

The probability of an event is sometimes considered to be the ratio of the number of times the event occurred in the past to the number of times it *could* have occurred in the past. To illustrate, suppose that in the past, 13 out of 1000 computers shipped by a certain manufacturer were found to be defective. On the basis of this information, it may be reasonable to infer that the probability of getting a defective computer from the manufacturer is about $\frac{13}{1000}$, or 0.013. Notice that this type of probability is based on a statistical approach and that past experience is used to find the probability of an event.

EXAMPLE 2 **Past records have shown that the probability a car manufactured at a particular factory will be defective is 0.14. Find the probability that a car from that factory will not be defective.**

Let event E be the car is defective.
Let event $\overline{E}$ be the car is not defective.

$$P(\overline{E}) = 1 - P(E) = 1 - 0.14 = 0.86 \qquad P(E) = 0.14$$

Another type of diagram of a sample space may be useful in finding probabilities: you can circle outcomes that belong to a particular event.

EXAMPLE 3 **A jar contains just 2 red marbles and 3 green marbles. One marble is selected at random and replaced. Then a second marble is selected. Find the probability that**

a. the first marble is green

b. both marbles are green

Use a diagram to determine the sample space. There are $5 \cdot 5 = 25$ outcomes in the sample space. Let A be the event that the first marble is green and B be the event that the second marble is green.

a. Using the diagram, count the outcomes in event A, shown circled in red.

$$P(A) = \frac{15}{25} = \frac{3}{5}$$

b. The outcomes in event B are shown circled in blue. The intersection of A and B contains 9 outcomes.

$$P(A \text{ and } B) = P(A \cap B) = \frac{9}{25}$$

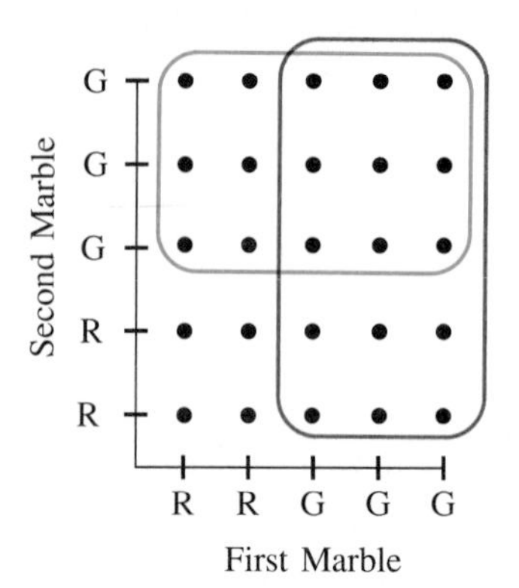

Two events are *independent* if the outcome of one has no effect on the outcome of the other. For example, the toss of a coin and the roll of a die are independent events. However, if a particular card is drawn from a bridge deck and then, *without* replacement of that card, another card is drawn, the two draws are not independent events. Once a card is removed from the deck, there is a different sample space for the second draw. In Example 3, the probability that the first marble is green and the second marble is green is $P(A) \cdot P(B)$, since A and B are independent events. Notice that $P(A \cap B) = \frac{9}{25} = \frac{3}{5} \cdot \frac{3}{5} = P(A) \cdot P(B)$.

> A and B are **independent events** if and only if
>
> $$P(A \cap B) = P(A) \cdot P(B)$$

Now, consider the case where A and B are not independent events, that is, A and B are *dependent events*. For the jar with 2 red marbles and 3 green marbles, consider the case where the first marble is *not* replaced. Then the probability that the first marble is green is $\frac{3}{5}$. If the first marble is green, then only 2 green marbles remain, and only 4 marbles remain in all. Then the probability that the second marble is green is $\frac{2}{4}$. The probability that both marbles are green is $\frac{3}{5} \cdot \frac{2}{4} = \frac{3}{10}$.

If two events A and B are **dependent events,** then the probability of both occurring is

$$P(A \cap B) = P(A) \cdot P(B \textit{ after } A)$$

CLASS EXERCISES

One card is drawn at random from a bridge deck. State whether or not the events are mutually exclusive.

1. The card is a diamond. The card is a face card.
2. The card is the ace of spades. The card is a heart.

State whether the events are independent or dependent.

3. A green die and a red die are rolled. The outcomes are the same.
4. A card is drawn from a bridge deck. The card is replaced and another card is drawn. Both cards are aces.
5. A card is drawn from a bridge deck. The card is not replaced, and another card is drawn. Both cards are aces.

PRACTICE EXERCISES

The wheel at the right is spun once. Find the probability that the point stops at

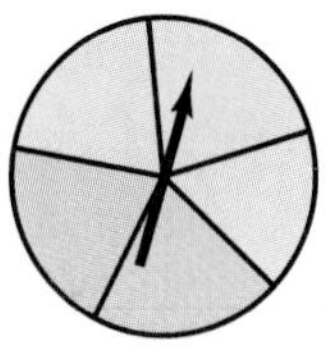

1. red *or* blue
2. green *or* red
3. not green *and* not red
4. red *or* not blue

A die is rolled and a coin is tossed. Find the probability of getting

5. a 6 *and* heads
6. an even number *and* heads
7. an odd number *and* tails
8. a number less than 3 *and* tails
9. If $P(A) = 0.32$, find $P(\overline{A})$.
10. If $P(\overline{B}) = 0.002$, find $P(B)$.

A jar contains just 4 blue marbles and 2 red marbles. A marble is chosen at random and replaced before a second marble is chosen. Find each probability.

11. P(both marbles are red)
12. P(both are blue)
13. P(one marble is red *and* one is blue)
14. P(the second marble is red, given that the first marble is blue)

A jar contains just 4 blue marbles and 2 red marbles. A marble is chosen at random and is *not* replaced. A second marble is chosen. Find each probability.

15. P(both marbles are red)

16. P(both are blue)

17. P(one marble is red *and* one is blue)

18. P(the second marble is blue, given that the first marble is red)

Five tags numbered 1 through 5 are in a bowl. A tag is drawn at random and is *not* replaced. Then a second tag is drawn. Find each probability.

19. P(the first number is odd *and* the second is even)

20. P(the first number is 4 *and* the second is less than 4)

21. P(both numbers are odd)

22. P(both numbers are less than 4)

23. P(the second number is odd, given that the first is even)

24. P(the second number is less than 4, given that the first is less than 4)

Determine whether the two events are independent or not.

25. $P(A) = 0.2$, $P(B) = 0.45$, $P(A \cap B) = 0.056$

26. $P(C) = 0.4$, $P(D) = 0.31$, $P(C \cap D) = 0.124$

Eight balls numbered 1 to 8 are in an automatic random selector. A ball is selected at random and is not replaced before a second ball is selected. Find the probability that

27. the sum of the numbers is greater than 5;

28. the sum of the numbers is greater than 6;

29. the sum of the numbers is greater than 5, given that the first number is 1;

30. the sum of the numbers is greater than 6, given that the first number is 3.

A card is chosen at random from a bridge deck and is *not* replaced before a second card is chosen. Find the probability that

31. both cards are face cards;

32. neither card is a face card;

33. exactly one card is a face card;

34. neither card is an ace, given that the first card is a 7.

35. If S is a sample space and events A, B, and C are not mutually exclusive, draw a Venn diagram and write a formula for $P(A \cup B \cup C)$.

36. A bowl contains tags numbered 1 to 10 inclusive. One tag is chosen at random from the bowl. Let A be the event the number is less than 7; B be the event the number is 2, 5, 6, or 9; and C the event that the number is greater than 2 and less than 9. Find $P(A \cup B \cup C)$.

Applications

37. **Quality Control** The probability that a can of beets is underweight is 0.3. Find the probability that a can is not underweight.

38. **Education** In order to pass a quiz a student must answer 3 questions out of 5 correctly. The probability of answering any one question correctly by guessing is 0.5. If a student has answered 2 questions correctly, what is the probability that he or she will answer exactly 3 correctly if that student guesses on all the other questions?

39. **Quality Control** A box contains 75 batteries. Twenty of them are dead and another 15 are weak. Three batteries are chosen at random from the box. What is the probability that the third battery is weak, given that the first battery is dead and the second is weak?

TEST YOURSELF

1. You can take 5 different roads from City A to City B, 4 different roads from City B to City C, and 2 different roads from City C to City D. How many different routes can you take from City A to City D? 14.1

2. How many different 4-letter code words can you make from the letters of the word *DESIGN*? Repetition of a letter is not allowed.

3. How many different ways can 10 people be seated in a row of 10 chairs?

4. How many different lines are determined by 9 points, no 3 of which are collinear? 14.2

5. How many 7-card hands that contain exactly 3 diamonds and 4 clubs can be chosen from a 52-card deck?

Two coins are tossed.

6. Find the probability of tossing exactly one head. 14.3

7. Find the odds in favor of tossing exactly one head.

8. Two dice are rolled. Find the probability that the sum of the two numbers is 8.

9. The probability that an airplane will land on time or ahead of time is 0.68. What is the probability that it will be late? 14.4

A jar contains just 4 red tags and 1 green tag. A tag is drawn at random and is not replaced before a second tag is drawn. Find the probability that

10. both are red

11. both are green

14.5 Central Tendencies

Objectives: To find the mean, median, and mode of a set of data
To construct a histogram from a frequency distribution

The collection, summarization, and characterization of data is a field of mathematics called *descriptive statistics*. A **statistic** is a single number that characterizes a set of data. To find certain statistics it is necessary to find the summation of data.

Capsule Review

Sigma, or summation, notation indicates a sum of numbers.

$$\sum_{i=1}^{n} x_i = x_1 + x_2 + x_3 + \cdots + x_n$$

The index of summation is i; the limits of summation are 1 and n.

EXAMPLE **Use the table below to find the sum:** $\sum_{i=1}^{4} x_i$

Number	x_1	x_2	x_3	x_4	x_5	x_6	x_7
Value	3	4	7	2	5	6	8

$$\sum_{i=1}^{4} x_i = 3 + 4 + 7 + 2 = 16$$

Use the table above to find each sum.

1. $\sum_{i=1}^{5} x_i$

2. $\sum_{i=1}^{6} x_i$

3. $\sum_{i=1}^{7} x_i$

4. $\sum_{i=1}^{3} (x_i - 1)$

5. $\sum_{i=1}^{4} (x_i + 1)$

6. $\sum_{i=1}^{5} (x_i)^2$

Twenty-four students tried out for the basketball team. Each student was asked to shoot for the basket 5 times. The coach collected data, shown at the right, on the number of baskets scored by each student.

5	2	1	5	5	4
1	0	2	5	3	3
3	4	5	5	2	3
5	3	3	2	3	4

A table called a **frequency distribution**, shown at the left below, can be used to organize the data. The first column shows each score and the second column shows how many students got that score, or the **frequency** of that score. The same information is shown in the *histogram* at the right below. A **histogram** is a vertical bar graph with no spaces between the bars. The horizontal axis is used to represent the scores, or values, and the vertical axis is used to represent the frequencies. The scales for the two axes do not have to be the same.

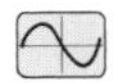 A graphing utility may be used to generate a histogram.

Frequency Distribution

Score, x	Frequency, f
0	1
1	2
2	4
3	7
4	3
5	7
	$24 = n$

Histogram

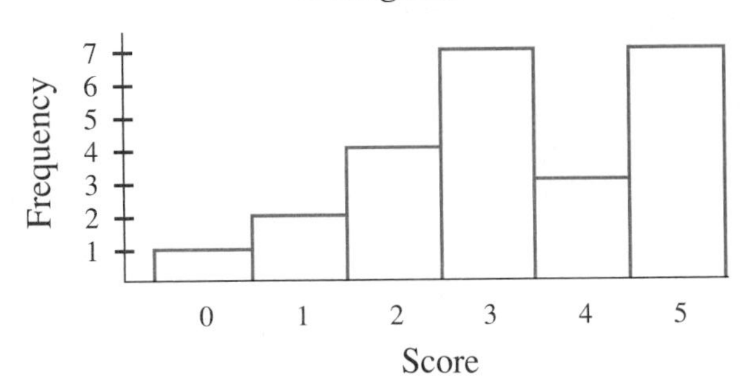

The sum of the frequencies is equal to the total number of scores, n.

Three types of "averages" or *measures of central tendency*—the *mean,* the *median,* and the *mode*—are often used to describe data.

> The **mean**, $\bar{x}$ (read *x-bar*), is the arithmetic average of the data.
>
> $\bar{x} = \frac{1}{n} \sum_{i=1}^{n} x_i$, where n is the number of values, each x_i is an individual value, and $\sum_{i=1}^{n} x_i$ is the sum of the n values.

When the same values occur more than once, each value can be multiplied by its frequency to obtain the required sum. Then, in the case of the basketball team, $\bar{x} = \frac{1}{n} \sum_{i=1}^{6} s_i f_i$, where s_i represents each of the six possible scores and f_i represents the frequency of each score.

The **median** is the middle value if there is an odd number of values and they are listed in ascending or descending order. It is the mean of the two middle values if there is an even number of values.

The **mode** is the value with the greatest frequency. If more than one value has this frequency, then there is more than one mode. If each value occurs only once, there is no mode.

Of the three measures of central tendency, the mean is usually the most stable, or reliable. Also, the mean has certain characteristics which make it more useful than the median or the mode in more advanced statistical analysis. However, when just a few values in a set of values are much higher or much lower than the rest, the median may be more representative of a typical, or ''average'' member of the set. The mode is particularly useful when one is interested only in the most typical case, or cases. For example, if the owner of a shop is ordering T-shirts, he is probably more interested in finding the mode of the sizes sold than the mean or the median.

EXAMPLE **Find the mean, median, and mode for the data the coach collected.**

To find the mean, use the frequency distribution and add a third column, *sf*.

Score, s	Frequency, f	Score · frequency, sf
0	1	0
1	2	2
2	4	8
3	7	21
4	3	12
5	7	35
	$n = 24$	$\sum_{i=1}^{6} s_i f_i = 78$

$$\bar{x} = \frac{1}{n}\sum_{i=1}^{6} s_i f_i = \frac{78}{24} = 3.25$$

To find the median, write the scores in ascending order. Since there is an even number of scores, find the mean of the two middle scores.

middle scores

0, 1, 1, 2, 2, 2, 2, 3, 3, 3, 3, 3, 3, 3, 4, 4, 4, 5, 5, 5, 5, 5, 5, 5

$$\text{median} = \frac{3 + 3}{2} = 3$$

To find the mode, look at the frequency distribution to see which score, or scores, have the greatest frequency.

3 occurs seven times
5 occurs seven times

This distribution has two modes, 3 and 5, and is said to be **bi-modal**.

CLASS EXERCISES

Thinking Critically

1. The salaries of the 10 employees of a company are as follows: president, \$1,000,000 per year; all other employees, \$20,000 per year. Which measure, or measures, of central tendency would most accurately represent the salary of a typical employee? Why?

Construct a frequency distribution for each set of data.

2. Grades on an exam: 90, 85, 67, 92, 85, 63, 94, 64, 90, 80
3. Number of field goals for the year: 5, 6, 11, 5, 5, 5, 6, 5
4. Number of books read by seniors: 15, 16, 14, 20, 10, 15, 14, 14, 15, 15, 18, 20, 22, 13, 14, 15, 16, 18, 18, 19, 20, 22, 15, 11, 10

PRACTICE EXERCISES

Use technology where appropriate.

Round answers to the nearest tenth, when necessary.

For Exercises 1–4, use the following quiz scores:

5, 7, 10, 10, 7, 9, 5, 6, 7, 7, 4

1. Construct a frequency distribution and a histogram for the scores.
2. Find the mean.
3. Find the median.
4. Find the mode, or modes.

For Exercises 5–8, use the following numbers of hits in eight games for members of a softball team: 8, 10, 7, 12, 8, 8, 5, 13, 12, 10, 10, 5.

5. Construct a frequency distribution and a histogram for the scores.
6. Find the mean.
7. Find the median.
8. Find the mode, or modes.

Make a frequency distribution for each set of data. Find the mean, the median, and the mode(s). Then answer the given question.

9. Sneaker sizes sold:

   5, 6, 2, 13, 11, 2, 3, 3, 5, 11, 6, 3, 3, 5, 6, 7, 3, 3

   Which measure of central tendency would be most helpful to the store manager who orders sneakers from the manufacturer?

10. A store manager coded the colors of shirts: 1 = red, 2 = blue, 3 = green, 4 = pink, 5 = white, and 6 = plaid. These were the color codes of the shirts sold one week:

    3, 2, 6, 4, 3, 2, 5, 5, 5, 6, 2, 2, 1, 5, 5, 6, 6, 6, 2, 2

    Which measure of central tendency would tell the manager the most popular color or colors?

11. Numbers of strokes needed to get a golf ball in the third hole at a miniature golf course:

7, 2, 9, 3, 1, 4, 6, 8, 6, 5, 4, 3, 2, 4, 4, 5, 6, 7, 8, 6

What would be a reasonable number of strokes to consider *par* (average)?

12. Prices, in cents, of a quart of milk in area stores:

60, 59, 47, 62, 54, 54, 59, 60, 59, 50,
53, 54, 55, 67, 60, 59, 59, 59, 60, 50

If you were to open a new store in the same area, what would be a reasonable amount for you to charge for a quart of milk?

The scores shown below were received by three math teams on a standardized examination. Each coach reported a team "average" of 93.

Team 1: 90, 91, 92, 93, 99
Team 2: 93, 93, 80, 93, 81
Team 3: 60, 80, 93, 95, 100

13. State which measure of central tendency was used by each coach.

14. Which of the three math teams do you think did consistently better on the exam?

Fertilizers from two different companies were tested on apple trees. The numbers of edible apples per tree are shown below. Each company reported an "average" of 82 edible apples per tree.

Using fertilizer from Company *A*: 81, 82, 83, 80, 84
Using fertilizer from Company *B*: 70, 90, 85, 71, 94

15. State which measure of central tendency was used by each company.

16. Which company's product showed more consistent results?

17. Three groups of students were surveyed to determine how many hours they studied each week. Find the mean number of hours studied by all students in the groups.

Group 1 (25 students)	Mean: 10.8 hours
Group 2 (32 students)	Mean: 8.7 hours
Group 3 (29 students)	Mean: 9.1 hours

18. Four groups participated in a study of the numbers of cavities members had at age 16. Find the mean number of cavities for all the people in the groups.

Group *A* (20 people)	Mean: 5.0	Group *B* (30 people)	Mean: 6.3
Group *C* (25 people)	Mean: 2.0	Group *D* (33 people)	Mean: 4.7

19. List seven numbers, not all alike, for which the mean, the median, and the mode are equal.

20. List seven numbers for which the mean is 2 more than the median and the median is 2 more than the mode.

Applications

21. **Marketing** A market survey produced the following data on the ages of people who watched a particular television show: 21, 34, 25, 26, 22, 33, 34, 24, 26, 18. What is the mean age of these people?

22. **Quality Control** A quality control team opened cans of peas at random. The team found the following numbers of peas in the cans: 65, 75, 65, 76, 65, 68, 67, 68, 69, 72, 56, 72, 72, 73, 67, 76, 82. Find the median number of peas in a can.

23. **Sales** The manager of a men's store listed the following sleeve lengths of shirts sold one week: 32, 33, 34, 35, 33, 33, 34, 35, 36, 33, 31, 33, 33, 31. Which sleeve length appears to be the most popular?

24. **Consumerism** A survey of local stores showed that a pound of butter sold at these prices: \$1.09, \$0.99, \$1.29, \$1.59, \$0.99, \$1.42, \$1.33, \$1.25, \$1.10, \$1.19, \$1.12. Find the mean, median, and mode for the data.

CAREER

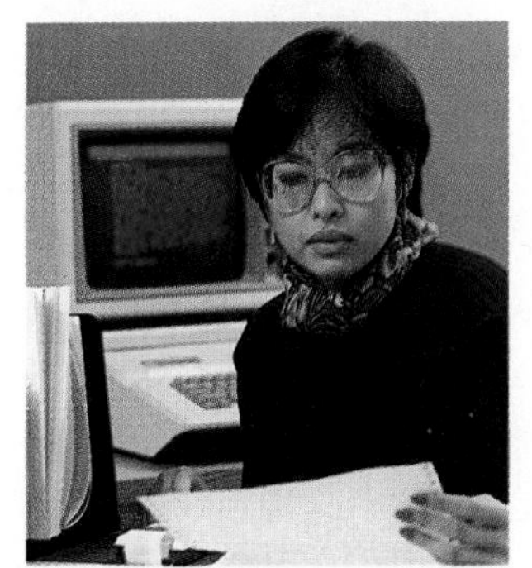

The **actuary** is at the center of the insurance industry. Some of an actuary's duties are to set premium and dividend rates, develop new forms of insurance, and determine how much money a company should set aside to assure payment of future claims. In particular, an actuary in a life insurance company is concerned with the interpretation of data on life expectancy for different segments of the population and with solving problems on the basis of such data.

Typically, an actuarial trainee earns a college degree in mathematics, statistics, or actuarial science. As a rule, a trainee learns by working in the different departments of a company under the supervision of fully-qualified actuaries. He or she must pass a series of ten examinations over a period of eight to ten years.

Out of 250 jobs listed in the *Jobs Rated Almanac,* published by World Almanac in 1988, the position of actuary is rated *number one* in terms of work environment, salary, outlook, stress, and physical demands.

14.6

Standard Deviation

Objective: To find the standard deviation for a set of data

The mean, median, and mode are not sufficient to give a true picture of the distribution of the values in a set of data. Statistics that describe the variability or dispersion of data are needed. Again, it is convenient to use sigma notation with such statistics.

Capsule Review

Find each sum, using the following values of x:

$$x_1 = 2,\ x_2 = 6,\ x_3 = 7,\ x_4 = 1,\ x_5 = 4,\ x_6 = 10,\ x_7 = 12$$

EXAMPLE $\frac{1}{4}\sum_{i=1}^{4}(x_i - 4)^2 = \frac{1}{4}[(2-4)^2 + (6-4)^2 + (7-4)^2 + (1-4)^2]$

$$= \frac{1}{4}(4 + 4 + 9 + 9) = 6.5$$

1. $\frac{1}{3}\sum_{i=1}^{3}(x_i - 5)^2$ **2.** $\frac{1}{5}\sum_{i=1}^{5}(x_i - 4)^2$ **3.** $\frac{1}{6}\sum_{i=1}^{6}(x_i - 5)^2$ **4.** $\frac{1}{7}\sum_{i=1}^{7}(x_i - 6)^2$

The mean of each set of data shown at the right is 5, but the distribution of the values within each set is very different. One measure of variance is the *range*. The **range** of a set of data is the difference between the highest value and the lowest value.

Data Set 1	Data Set 2	Data Set 3
5	10	7
5	2	5
5	3	6
5	9	3
5	1	4
$\overline{x} = 5$	$\overline{x} = 5$	$\overline{x} = 5$

The range for Data Set 1 is $5 - 5 = 0$, for Set 2 it is $10 - 1 = 9$, and for Set 3 it is $7 - 3 = 4$. Thus, the values in Set 2 are the most variable and those in Set 1 are the least variable.

The range is not always useful as a measure of dispersion because it does not measure the variability of all the data from the mean. A statistic used as a measure of such variability is the *variance,* which is denoted σ^2, where σ is the lowercase Greek letter *sigma*.

The **variance,** σ^2, of a set of n values, $x_1, x_2, x_3, \ldots, x_n$, for which the mean is $\overline{x}$, is given by the formula

$$\sigma^2 = \frac{1}{n} \sum_{i=1}^{n} (x_i - \overline{x})^2$$

Each value $x_i - \overline{x}$ is called a **deviation from the mean.**

EXAMPLE 1 **Find the variance for Data Sets 1, 2, and 3 on page 712.**

Make tables. The mean, $\overline{x}$, of each set of data is 5. For each table, the sum of the squares of the deviations from the mean is found by adding the values in the last column.

Data Set 1

x	$x - \overline{x}$	$(x - \overline{x})^2$
5	0	0
5	0	0
5	0	0
5	0	0
5	0	0

$$\sum_{i=1}^{n} (x_i - \overline{x})^2 = 0$$

$$\sigma^2 = \frac{1}{n} \sum_{i=1}^{n} (x_i - \overline{x})^2$$

$$\sigma^2 = \frac{1}{5}(0) = 0$$

Data Set 2

x	$x - \overline{x}$	$(x - \overline{x})^2$
10	5	25
2	−3	9
3	−2	4
9	4	16
1	−4	16

$$\sum_{i=1}^{n} (x_i - \overline{x})^2 = 70$$

$$\sigma^2 = \frac{1}{n} \sum_{i=1}^{n} (x_i - \overline{x})^2$$

$$\sigma^2 = \frac{1}{5}(70) = 14$$

Data Set 3

x	$x - \overline{x}$	$(x - \overline{x})^2$
7	2	4
5	0	0
6	1	1
3	−2	4
4	−1	1

$$\sum_{i=1}^{n} (x_i - \overline{x})^2 = 10$$

$$\sigma^2 = \frac{1}{n} \sum_{i=1}^{n} (x_i - \overline{x})^2$$

$$\sigma^2 = \frac{1}{5}(10) = 2$$

The more the scores vary from the mean, the greater the variance. Notice that in each set of data, the sum of the deviations from the mean (the values in the center column of each table) is zero. That is why the square of the deviations is used in the formula for the variance.

Since the data are in units of measure, and the deviations are squared when computing the variance, then the variance is a quantity in square units of measure. For example, if each value of x represents a number of inches, then each value of $(x_i - \overline{x})^2$ represents a number of square inches. For this reason, the positive square root of the variance, called the *standard deviation,* is more commonly used as a measure of dispersion.

The **standard deviation,** σ, of a set of n values, $x_1, x_2, x_3, \ldots, x_n$, for which the mean is $\overline{x}$, is given by the formula

$$\sigma = \sqrt{\frac{1}{n} \sum_{i=1}^{n} (x_i - \overline{x})^2}$$

Most scientific calculators have a *statistics mode* that makes it possible to find the mean and the standard deviation of a set of data directly.

EXAMPLE 2 A birdwatcher counted the number of species of birds she saw during a four-week period. She spotted 25 the first week, 47 the second week, 13 the third week, and 19 the fourth week. Find the mean and the standard deviation for this data.

Solution

$$\overline{x} = \frac{1}{n} \sum_{i=1}^{n} x_i = \frac{1}{4}(25 + 47 + 13 + 19) = 26$$

$$\sigma = \sqrt{\frac{1}{n} \sum_{i=1}^{n} (x_i - \overline{x})^2}$$

$$= \sqrt{\frac{1}{4}[(25 - 26)^2 + (47 - 26)^2 + (13 - 26)^2 + (19 - 26)^2]}$$

$$= \sqrt{\frac{1}{4}(1 + 441 + 169 + 49)} = \sqrt{165} \approx 12.8$$

Solution, Using Statistics Mode of a Scientific Calculator

Put the calculator into the statistics mode and enter the data (the four numbers of weekly sightings). Consult the instruction manual for your calculator, if necessary. Then use the keys for $\overline{x}$ and σ.

$$\overline{x} = 26 \qquad \sigma \approx 12.8$$

CLASS EXERCISES

Thinking Critically

1. Show that the sum of the deviations from the mean is always zero.

Find the range, the mean, the variance, and the standard deviation for each set of data. Round answers to the nearest tenth, if necessary.

2. Number of completed passes each quarter: 6, 8, 15, 7

3. Scores on an exam: 90, 92, 93, 80, 100

PRACTICE EXERCISES

Use technology where appropriate.

Find the range for each set of data.

1. Scores on a quiz: 10, 9, 6, 6, 7, 8, 8, 8, 9, 8
2. Number of points per game: 16, 18, 10, 20, 15, 7, 16, 24
3. Temperatures, in degrees Fahrenheit: 42, 50, 66, 58, 63, 49, 55, 68
4. Distances traveled, in miles: 100, 85, 160, 320, 92, 68, 283, 57

Find the mean, the variance, and the standard deviation for each set of data. Round answers to the nearest tenth, if necessary.

5. Scores on an exam: 65, 62, 83, 80, 78
6. Number of baskets scored: 12, 18, 15, 10, 14
7. Number of homeruns scored: 5, 3, 2, 1, 8, 4, 6
8. Number of hours of study per day: 5, 3, 4, 4, 6, 2
9. Number of cars per hour to pass an intersection: 12, 15, 14, 20, 10, 15
10. Temperatures, in degrees Fahrenheit: 75, 78, 83, 92, 85, 75, 90
11. Number of movies rented per week: 54, 54, 54, 55, 55, 56, 56, 56
12. Number of VCRs sold per week: 8, 10, 12, 13, 15, 7, 6, 14, 18, 20
13. Number of minutes a commuter waited for a train during two weeks: 10, 6, 2, 15, 9, 7, 3, 1, 11, 9
14. Number of games played in the World Series from 1905 through 1917: 5, 6, 4, 5, 7, 5, 6, 7, 5, 4, 5, 5, 6
15. Number of words spelled incorrectly on a test: 6, 8, 19, 3, 4, 5, 9, 10, 2, 7, 1, 0, 13, 11, 14
16. Ages of members: 20, 21, 24, 35, 43, 27, 26, 41, 18, 14, 19, 22, 33, 37
17. The five highest waterfalls in the world measure 3212, 3110, 2625, 2425, and 1904 ft.
18. The minimum distance a batter has to hit the ball down the center of the field to get a home run in 8 different stadiums is 410, 420, 406, 400, 440, 421, 402, and 425 ft.

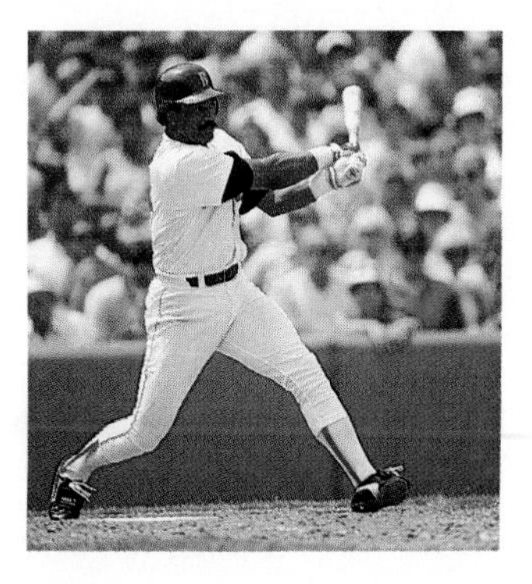

19. The mean number of games played in a World Series for the years 1914, 1922, 1927, 1928, 1932, and 1938 is 4, with a standard deviation of 0. How many games did the losing team win in each of these years?

20. If $b = a + 3$ and $c = a + 6$, what is the standard deviation of the set of numbers a, b, and c?

Use the following rules for Exercises 21 and 22; c represents a constant.

$$\sum_{i=1}^{n} c = nc \qquad \sum_{i=1}^{n} cx_i = c\sum_{i=1}^{n} x_i \qquad \sum_{i=1}^{n} (x_i + c) = \sum_{i=1}^{n} x_i + \sum_{i=1}^{n} c$$

21. Show that if each value in a set of data is decreased by a constant, c, then the mean of the original set of data is equal to the mean of the new set, plus c.

22. Show that if each value in a set of data is decreased by a constant, c, then the variance of the set of data is unchanged.

Applications

23. Quality control To assure a uniform product, a company measures each extension wire as it comes off the production line. The lengths in inches of the first batch of ten wires were: 10, 15, 14, 11, 13, 10, 10, 11, 12, and 13. Find the mean, the variance, and the standard deviation for this data.

24. Sports In eight games, a hockey player scored the following numbers of goals: 2, 4, 8, 6, 2, 1, 7, 3. Find the mean, the variance, and the standard deviation for the data.

25. Merchandizing The manager of a new store kept the following records of sales, in dollars, for the first seven days the store was open: 150, 125, 135, 160, 155, 166, and 170. Find the mean, the variance, and the standard deviation for the sales during the first week.

Mixed Review

Simplify.

26. $(3x^3y^0z)^2$ **27.** $5(3x^4y^2)^0$ **28.** $9x^2y^{-4}z^{-1}$

29. Find the inverse of $y = 2x^2 + 1$. Is it a function?

30. If $f(x) = 2x^2$ and $g(x) = 2x + 4$, find $f[g(x)]$.

31. Without actually graphing the equations, describe the graphs of $y = 2x - 5$ and $x = 2y - 5$ relative to each other.

32. It is given that y varies directly as the cube of x. If $y = 32$ when $x = -2$, find y when $x = 5$.

33. Solve for x and for y: $\begin{bmatrix} 3 & 2 \\ 4 & 0 \end{bmatrix} \begin{bmatrix} 2x & -2 \\ 3 & -y \end{bmatrix} = \begin{bmatrix} 18 & 6 \\ 16 & -8 \end{bmatrix}$

14.7 Normal Distribution

Objective: To use the normal distribution to find probabilities

Using statistical methods and computers, the winner in an election can be predicted with a high rate of accuracy from a small, but carefully selected, sample of the voters. *Inferential statistics* is a branch of mathematics that uses statistics from a sample of the population to infer information about the population. An important tool of inferential statistics is a theoretical distribution called a *normal distribution,* which can be related to certain histograms.

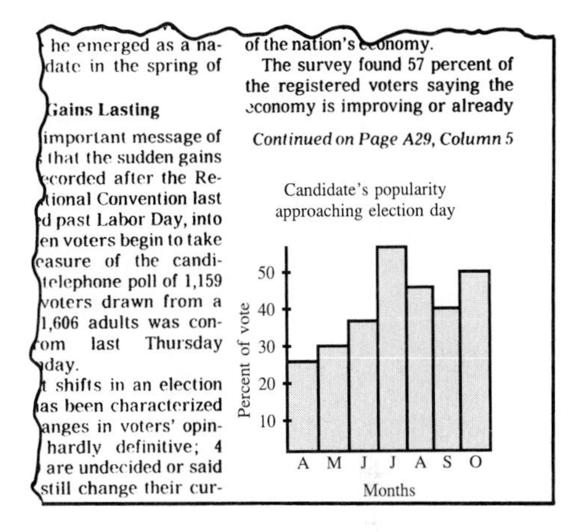
he emerged as a na-
date in the spring of

Gains Lasting

important message of
that the sudden gains
corded after the Re-
tional Convention last
d past Labor Day, into
en voters begin to take
easure of the candi-
telephone poll of 1,159
voters drawn from a
1,606 adults was con-
om last Thursday
day.
shifts in an election
as been characterized
anges in voters' opin-
hardly definitive; 4
are undecided or said
still change their cur-

of the nation's economy.
The survey found 57 percent of the registered voters saying the economy is improving or already

Continued on Page A29, Column 5

Capsule Review

A histogram is a graph that displays information in a frequency distribution.

1. Copy and complete this sample space for an experiment in which 4 coins are tossed once.

HHHH	*HHHT*	*HHTH*	*HTHH*
THHH	?	?	?
?	?	?	?
?	?	?	?

2. Copy and complete this frequency distribution. Use information from the sample space.

Event	Frequency
0 heads	1
1 head	4
2 heads	?
3 heads	?
4 heads	?

3. Copy and complete the histogram at the right to display the information in the frequency distribution.

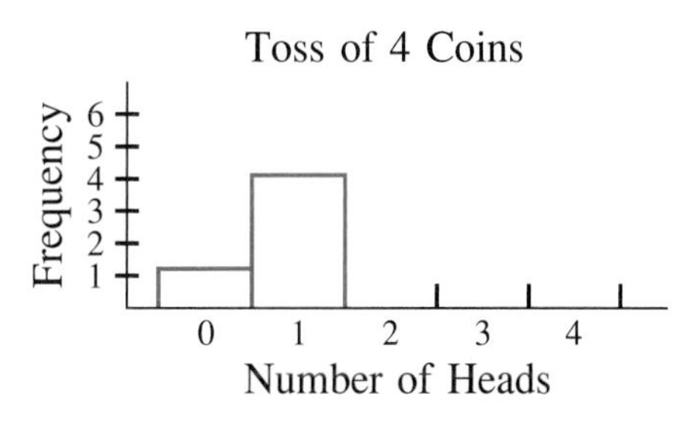

If you toss a greater number of coins, the sample space will be larger and there will be more bars in the related histogram. For example, there are 256 possible outcomes in the sample space for the toss of 8 coins. The related histogram is shown at the right. The graph formed by joining the midpoints of the tops of the bars forms what is called a *bell-shaped curve*. If you actually toss 8 coins *many* times and show the results in a similar way, the histogram and curve will be similar to those shown here for the theoretical results. Try it!

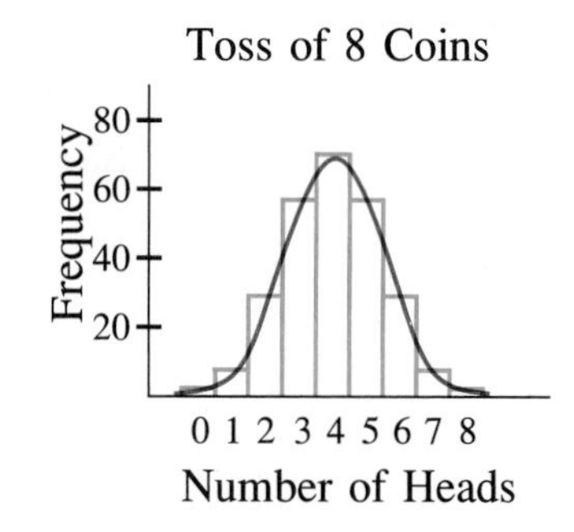

A distribution such as the one shown above is called a **normal distribution**, and the curve is called a **normal curve**. The highest point of the normal curve is directly over the mean. The curve is symmetric with respect to a vertical line through the mean, and it approaches the x-axis asymptotically as $|x|$ increases. Although all normal curves have these characteristics, they vary in shape, some being tall and narrow and others short and spread out. The mean and the standard deviation determine the shape of the curve.

A *standard normal distribution* is a normal distribution with a mean of 0 and a standard deviation of 1. The *standard normal curve* for such a distribution is shown below. The area between the curve and the x-axis is equal to 1, so the areas of regions under the curve can be used to approximate the *probability* that an *observation*, or a value chosen at random, will occur within a certain interval of values.

It can be shown that in any normal distribution, approximately 68% of the values are within 1 standard deviation of the mean. Furthermore, about 95% of the values are within 2 standard deviations of the mean, and about 99.8% are within 3 standard deviations of the mean. The table below gives the areas under the standard normal curve for values of observations from $x = 0$ through $x = 3.0$. Since the curve is symmetric, the area between 0 and a positive number x is equal to the area between 0 and $-x$.

Standard Normal Curve

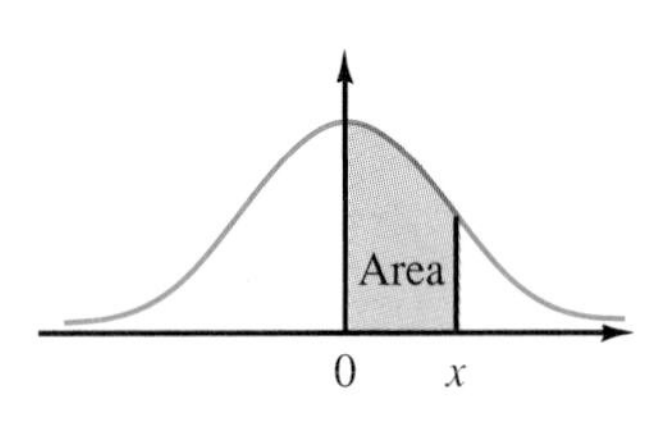

Area Under Standard Normal Curve

Observation, x	Area	Observation, x	Area
0.0	0.0000	1.6	0.4452
0.2	0.0793	1.8	0.4641
0.4	0.1554	2.0	0.4772
0.6	0.2257	2.2	0.4861
0.8	0.2881	2.4	0.4918
1.0	0.3413	2.6	0.4953
1.2	0.3849	2.8	0.4974
1.4	0.4192	3.0	0.4987

EXAMPLE 1 **Use the table to find each value for a standard normal curve.**

a. the area under the curve between $x = 0$ and $x = 1.2$
b. $P(0 < x < 1.2)$
c. $P(-1.2 < x < 0)$
d. the percentage of observations between 0 and 1.2
e. $P(x > 1.2)$

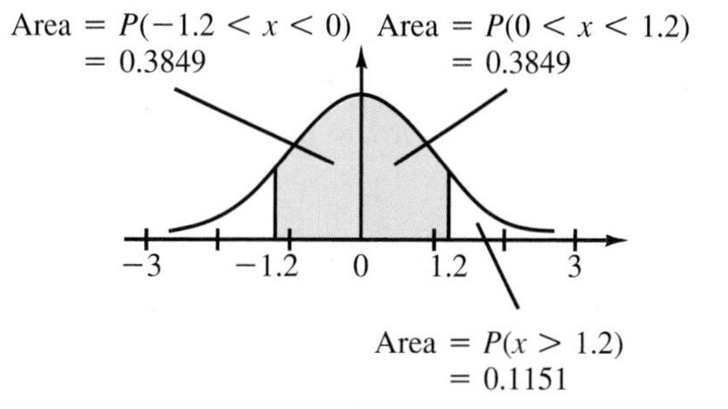

a. The area under the curve between $x = 0$ and $x = 1.2$ is 0.3849.

b. $P(0 < x < 1.2) = 0.3849$

c. Use symmetry.

$$P(-1.2 < x < 0) = 0.3849$$

d. The percentage of observations between 0 and 1.2 is 38.49%.

e. The area under the curve to the right of $x = 0$ is 0.5000. The area under the curve between 0 and 1.2 is 0.3849.

$$P(x > 1.2) = 0.5000 - 0.3849 = 0.1151$$

The standard normal curve can be used to find probabilities relating to normal distributions that are *not* standard. Keep in mind the fact that the standard deviation for a standard normal distribution is 1, and the mean is 0. If a normal distribution has mean $\overline{x}$ and standard deviation σ, then the probability that an observation falls between $\overline{x}$ and $\overline{x} + \sigma$ is the same as the probability that an observation falls between 0 and 1 in a *standard* normal distribution. Similarly, the probability that an observation falls between $\overline{x} + 2\sigma$ and $\overline{x} + 3\sigma$ is the same as the probability that an observation falls between 2 and 3 in a *standard* normal distribution.

EXAMPLE 2 The scores on a standardized exam are normally distributed, with a mean of 100 and a standard deviation of 10. Find the probability that a score will be between 90 and 120.

A score of 90 is 1 standard deviation below the mean, and a score of 120 is 2 standard deviations above the mean. The area under the curve for $90 < x < 100$ is 0.3413, and the area for $100 < x < 120$ is 0.4772. The probability that a score will be between 90 and 120 is

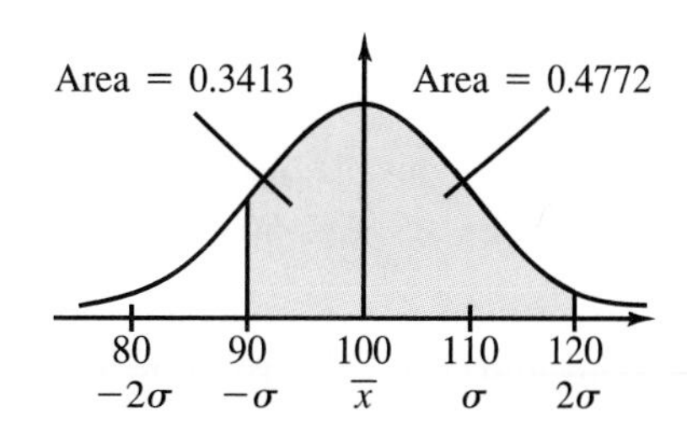

$$0.3413 + 0.4772 = 0.8185$$

CLASS EXERCISES

Use the table on page 718 to find each value for a standard normal curve.

1. $P(0 < x < 1)$
2. $P(0 < x < 2)$
3. $P(-1.4 < x < 0)$
4. $P(-2 < x < 1)$
5. $P(-1 < x < 1.8)$
6. $P(-1.6 < x < 3)$
7. The number of paper clips in a box is normally distributed, with a mean of 100 and a standard deviation of 5. Find the probability that a box will contain
   a. between 100 and 105 clips;
   b. between 95 and 100 clips;
   c. more than 110 clips;
   d. less than 90 clips.

PRACTICE EXERCISES

 Use technology where appropriate.

Use the table on page 718 to find each value for a standard normal curve.

1. $P(0 < x < 3)$
2. $P(x > -1)$
3. $P(x < 2.2)$
4. $P(-0.6 < x < 0.4)$
5. $P(-2.8 < x < 1.8)$
6. $P(-1.8 < x < 0.2)$
7. $P(-0.4 < x < 1.6)$
8. $P(x > 1.6)$
9. $P(x > 2.8)$
10. $P(x < 0.8)$

Groups of scores are normally distributed, with the given means and standard deviations. For each group, find the given probability, *P*.

11. Score Group *A*: mean 40; standard deviation 2; *P*(greater than 42)
12. Score Group *B*: mean 90; standard deviation 5; *P*(between 95 and 100)
13. Score Group *C*: mean 75; standard deviation 3; *P*(less than 69)
14. Score Group *D*: mean 78; standard deviation 6; *P*(between 66 and 84)

A battery manufacturer claims the life span of his batteries is normally distributed, with a mean of 45 months and a standard deviation of 5 months.

15. What is the probability that a particular battery will have a life span between 45 and 52 months?
16. What is the probability that a particular battery will have a life span between 48 and 50 months?
17. To qualify as a contestant in a race, a runner has to be in the top 16% of all entrants. The running times are normally distributed, with a mean of 63 min and a standard deviation of 4 min. To the nearest minute, what is the qualifying time for the race?

18. Every year the top 8% of the graduating seniors get awards. The grades of the graduates are normally distributed, with a mean of 85 and a standard deviation of 5. To the nearest whole number, what is the lowest score that qualifies for an award?

19. Gail and George each took the same standardized exam, but with two different groups of students. Each scored 83. In Gail's group, the mean was 78 and the standard deviation was 6. In George's group, the mean was 74 and the standard deviation was 4. Determine whether Gail or George, if either, is in the top 10% of his or her group.

20. Two tomato plants were grown in different soils. Each produced 23 tomatoes. The mean number of tomatoes for the first soil was 17, with a standard deviation of 5. The mean number of tomatoes for the second soil was 18, with a standard deviation of 6. Determine which plant, if either, is in the top 20% of its group.

Applications

21. **Education** The scores on an exam are normally distributed, with a mean of 85 and a standard deviation of 5. What is the probability that a student will score between 85 and 95?

22. **Agriculture** The weights of sacks of potatoes are normally distributed, with a mean of 20 lb and a standard deviation of 1 lb. What is the probability that a sack of potatoes will weigh less than 19 lb?

23. **Agriculture** To win a prize, a tomato must be greater than 4 in. in diameter. If the sizes of a crop of tomatoes grown in a special soil are normally distributed, with a mean of 3.2 in. and a standard deviation of 0.4 in., determine the probability that the soil will produce a winning tomato.

BIOGRAPHY: Ronald A. Fisher

Ronald A. Fisher (1890–1962), who was among the key scientists of this century, is known for his work in genetics and for the development of a field of knowledge called experimental design. He refined some of the statistical functions used by earlier mathematicians and developed the famous *F* function.

Learn as much as you can about the *F* function and its frequency distribution. Prepare a report that compares the frequency distribution of *F* with the normal distribution.

14.8

Problem Solving Strategy: Use a Random Sample

Problems that involve large sets of data can often be solved by studying a **random sample** of the data. Then the results of studying the random sample can be analyzed to infer information about the original set of data. A sample is random if each member of the sample has an equal chance of being included and if each sample of the same size is equally likely to be selected.

A random sample is often determined by using a table of random numbers. Such tables are found in books of tables or are generated by a computer or a programmable calculator. Example 1 illustrates the use of random numbers.

EXAMPLE 1 A science class wants to track the typical random movements of a particle of rice floating on the surface of water in a dish. The school is not equipped with the sophisticated microscopes and slow-motion cameras required for the task. Use random numbers to simulate the first five directions of the motion of the particle.

Understand the Problem *What are the given facts?* A particle of rice is floating on a surface of water in a dish.
What are you asked to find? The problem is to simulate the first five directions a particle of rice will move.

Plan Your Approach *Choose a strategy*. The strategy is to use random numbers to simulate the movement of the particle. Draw a circle and one diameter. Think of the particle as a movable point at the center of the circle. Use a table of random numbers to determine the first direction in which the particle moves (relative to the given diameter) and the next four changes of direction (each relative to the previous path of the particle). Express the directions as whole-number angle measures from 1° through 359°.

Complete the Work *Apply the strategy*. Start at a random location in the table—say the second row, third column. Take the numbers in sequence and use the first three digits of each number, since the greatest possible value, 359, has three digits. Skip the three-digit numbers that are greater than 359.

Random Numbers

9048	6239	5973	3555	4236
9431	4670	0624 →	1149 →	0712
7005 →	8589 →	2070 →	8313 →	8808
4020 →	1886	1391	4144	4360
0153	9392	5066	7937	4901

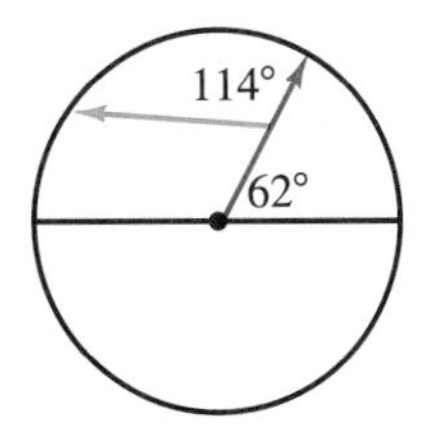

Interpret the Results *State your answer*. In the simulation, the particle's first direction is 62°. The subsequent directions are 114°, 71°, 207°, and 188°.

A statement of what is expected to happen is called a *hypothesis*. A **null hypothesis**, $\boldsymbol{H_0}$, states that a population mean is equal to some given standard. The **alternate hypothesis**, $\boldsymbol{H_a}$, states that the population mean is not equal to that standard. In *hypothesis testing,* various statistics are calculated in order to determine whether to accept or reject a particular null hypothesis.

If many random samples are taken, the means of the samples have an approximately normal distribution, and the mean of the sample means is approximately equal to the population mean. The measure of variability of the means in the distribution of sample means is called the **standard error**, σ_e, and

$$\sigma_e = \frac{\sigma}{\sqrt{n}}$$

where σ is the standard deviation of the population and n is the number of elements in the sample.

The population mean is designated by μ, the Greek letter *mu*. In a group of sample means that is normally distributed, 95% of the sample means fall between $\mu - 1.96\sigma_e$ and $\mu + 1.96\sigma_e$. This interval is called the *95% confidence interval,* and it represents the *0.05 level of significance* for hypothesis testing.

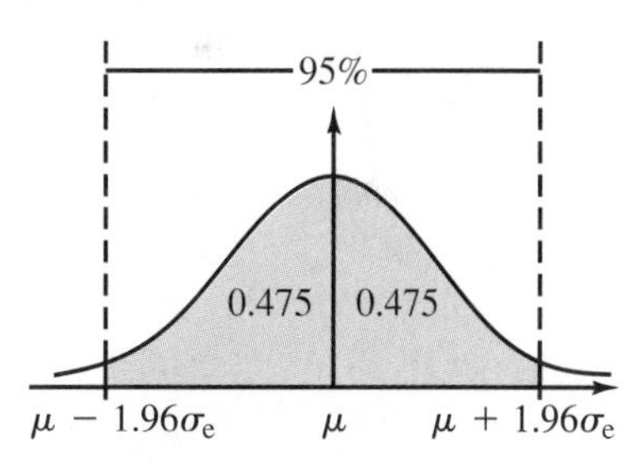

To test whether a population mean meets a specific standard at the 0.05 level of significance, it is necessary to find the theoretical interval where 95% of the sample means are likely to fall. Then relate the sample mean actually found to this 95% confidence interval. If the sample mean lies inside the interval you can say with 95% confidence that the population mean is equal to the standard. That is, the null hypothesis is accepted. If the sample mean lies outside the interval, there is a significant difference between the population mean and the standard at the 0.05 level of significance, and the null hypothesis is rejected.

The following steps illustrate one way to determine if the mean of a random sample meets a certain standard at the 0.05 level of significance.

1. State the null hypothesis, H_0, and the alternative hypothesis, H_a.
2. Find the standard error, σ_e; $\sigma_e = \frac{\sigma}{\sqrt{n}}$, where σ is the standard deviation and n is the number of elements in the sample.
3. Use the standard error to determine the 95% confidence interval. This interval lies between $\mu - 1.96\sigma_e$ and $\mu + 1.96\sigma_e$, where μ is the population mean.
4. Determine whether the sample mean lies inside the confidence interval. If it does, then the null hypothesis is accepted. If not, the null hypothesis is rejected.

EXAMPLE 2 An apple wholesaler received a shipment of 2000 bags of apples. The grower said that the mean weight of the bags was 50.0 lb, with a standard deviation of 1.8 lb. The wholesaler took a random sample of 40 bags and found that the mean weight of the sample bags was 49.7 lb. Determine at the 0.05 level of significance if the shipment meets the grower's standard.

Understand the Problem A wholesaler receives a shipment of apples. He wants to test a sample to see if it meets the standard of a mean weight of 50.0 lb per bag.

Plan Your Approach

1. State the null hypothesis and the alternative hypothesis.
   H_0: mean weight of the bags = 50.0 lb *Null hypothesis*
   H_a: mean weight of the bags $\neq$ 50.0 lb *Alternative hypothesis*

Complete the Work

2. Find the standard error.

$$\sigma_e = \frac{\sigma}{\sqrt{n}} = \frac{1.8}{\sqrt{40}} \approx 0.28$$ *Calculation-ready form*

3. Use the standard error to determine the 95% confidence interval
   $\mu - 1.96\sigma_e = 50.0 - 1.96(0.28) \approx 49.45$
   $\mu + 1.96\sigma_e = 50.0 + 1.96(0.28) \approx 50.55$
   The interval is between 49.45 and 50.55.

4. Determine whether the sample mean, 49.7, lies within the 95% confidence interval.
   $49.45 < 49.7 < 50.55$

Interpret the Results Since the sample mean falls within the 95% confidence interval, the null hypothesis is accepted. It can be said with 95% confidence that the mean weight of the bags in the shipment is not significantly different from 50.0 lb.

CLASS EXERCISES

Tell whether each statement is true or false and state the reason.

To find the average score of California students on a statewide exam, the scores of all students in San Diego were studied.

1. The group of students from San Diego represents the sample for this study.
2. The group of students from San Diego represents a random sample of the population for this study.

The food preferences of children in the *ABC* School are to be studied. All students in the *ABC* School are interviewed.

3. The group interviewed represents the population for this study.
4. The group interviewed represents a random sample of the population for this study.

State the null hypothesis and the alternative hypothesis for each situation.

5. The norm for the mean number of hours of homework done by a high school student per day is 2 h. A random sample of students in a high school is taken to determine the mean number of hours they do homework. Do the students in this high school differ significantly from the norm?
6. The mean life span of a particular breed of dog is 9 years. A random sample of the case histories of such dogs who were fed *Long Life Dog Food* is taken to determine their mean life span. Did the dogs who were fed *Long Life* differ significantly from the norm?

PRACTICE EXERCISES

Use technology where appropriate.

1. Construct a simulation of the first 10 random movements of a speck of dust floating on the surface of a fluid.
2. Construct a simulation of the first 15 random movements of a grain of wheat floating in water.

The mean quantity of juice in a bottle is supposed to be 64.0 oz, with a standard deviation of 2.0 oz. A random sample of 80 bottles from a certain shipment has a mean of 64.5 oz. Determine if the shipment is significantly different from the standard at the 0.05 level of significance.

3. State the null hypothesis and the alternative hypothesis.
4. Find the standard error.
5. Find the 95% confidence interval.
6. Determine whether the sample mean lies inside the confidence interval and state whether to accept or reject the null hypothesis.

The mean number of sheets in a ream of paper is supposed to be 500, with a standard deviation of 15 sheets. A random sample of 40 reams from a shipment has a mean of 497 sheets. Determine if the shipment is significantly different from the standard at the 0.05 level of significance.

7. State the null hypothesis and the alternative hypothesis.

8. Find the standard error.

9. Find the 95% confidence interval.

10. Determine whether the sample mean lies inside the confidence interval and state whether to accept or reject the null hypothesis.

11. A bottle of vitamin tablets should contain 100 tablets, with a standard deviation of 2 tablets. A random sample of 60 bottles from a case has a mean of 100.3 tablets. Is this case significantly different from the standard at the 0.05 level of significance?

12. A shirt manufacturer makes shirts with a standard sleeve length of 33.0 in. and a standard deviation of 0.5 in. In one shipment, a random sample of 45 shirts marked "33-in. sleeve" has a mean sleeve length of 32.7 in. Is this shipment significantly different from the standard at the 0.05 level of significance?

13. The diameter of a hose is supposed to be 1.25 in., with a standard deviation of 0.06 in. A batch of hoses were shipped, and a random sample of 45 had a mean diameter of 1.21 in. Is this batch significantly different from the standard at the 0.05 level of significance?

14. The gear speed for a piece of machinery should be 1500 revolutions per minute (rpm), with a standard deviation of 26 rpm. A random sample of 60 machines had a mean gear speed of 1494 rpm. Is this sample significantly different from the standard at the 0.05 level of significance?

The standard height of a computer table should be 73.0 cm, with a standard deviation of 2.3 cm. A random sample of 32 tables was taken. 15 tables measured 73.1 cm, 8 tables measured 72.9 cm, 5 tables measured 73.0 cm, and 4 tables measured 73.2 cm. Determine if this sample is significantly different from the standard at the 0.05 level of significance.

15. State the null hypothesis and the alternative hypothesis.

16. Approximate the standard error using the formula $\sigma_e \approx \frac{\sigma}{\sqrt{n}}$, where σ is the standard deviation of the population.

17. Find the 95% confidence interval.

18. Test at the 0.05 level of confidence and state whether or not to accept the null hypothesis.

19. The 95% confidence interval for the mean weight of a bag of onions is between 2.93 and 3.07 lb. The standard deviation is 0.25 lb. Find the size of the random sample that was used to determine this interval.

20. The 95% confidence interval for the mean score on a standardized exam is between 62.04 and 62.56. If a random sample of 38 scores was used to determine this interval, what was the standard deviation?

Mixed Problem Solving Review

1. Mr. North invests \$5000 at $9\frac{1}{2}$% interest compounded quarterly. How long will it take before he earns a 50% profit?

2. Lydia's starting salary was \$18,000 per year. She received a 7% raise each year. Find her yearly salary after 8 years.

3. A 26-g object placed 40 cm from the fulcrum of a lever balances an object 55 cm from the fulcrum. What is the mass of the second object?

PROJECT

An important idea associated with random samples is that of a statistical inference. Use a book on elementary statistics to look up the concept of statistical inference. Write a brief statement that describes the meaning of the concept. Include three examples that illustrate the method of making a statistical inference.

TEST YOURSELF

1. Construct a frequency distribution and find the mean, median, and mode for this set of scores on an algebra test: — 14.5

6, 9, 7, 6, 9, 10, 10, 13, 15, 14, 12, 8, 8, 8, 15

2. Find the variance and the standard deviation for the data in Exercise 1. — 14.6

Use the table on page 718 to find each probability.

3. $P(-1.8 < x < 0)$ **4.** $P(x < 0.4)$ — 14.7

5. A jar is supposed to contain 8.0 oz of garlic powder, with a standard deviation of 0.21 oz. A random sample of 45 jars from the production line is found to have a mean weight of 7.96 oz. Does this production lot differ significantly from the standard at the 0.05 level of significance? — 14.8

INTEGRATING ALGEBRA
Binomial Experiments

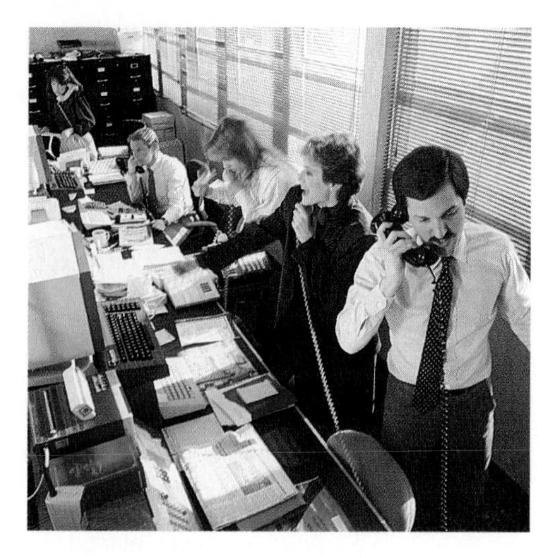

Did you know that in spite of all the thought, research, and analysis into the behavior of the financial markets, an investment analyst must at some point make a yes-or-no decision; recommend that a client buy or sell, invest more or pull out, move now or stand pat. This type of binary (two-valued) decision will have a binary outcome. That is, an investment will make money or lose money. It will be a success or a failure.

Suppose that an analyst is researching three independent investment instruments, such as a stock or bond. Each instrument has a 0.60 probability of growth. Therefore, the probability of decline is 0.40. What is the probability that at least two of the investments will grow? *Note:* An instrument that neither rises nor falls in value may be regarded as a declining investment because of commissions and other costs.

In the situation described above, a trial is assumed to have only two possible outcomes, that is, the value of each instrument can go either up or down. Such a situation is referred to as a *binomial probability experiment* or a *Bernoulli experiment,* named for the Swiss mathematician Jacques Bernoulli [1654–1705]. In binomial experiments, the outcome of the trials must be independent. Furthermore, since each trial has one of only two possible outcomes, success or failure, the probability of success plus the probability of failure must equal 1. That is, if $p = P(\text{success})$ and $q = P(\text{failure})$, then $p + q = 1$.

For the financial analyst above,

$$P(\text{success}) = P(\text{growth}) = p = 0.60 \quad \text{and} \quad P(\text{failure}) = P(\text{decline}) = q = 0.40$$

Therefore, $p + q = 0.60 + 0.40 = 1$.

In order to determine the probability that at least two investments will go up, examine all possible ways in which this could occur.

1. All three go up. This can occur in only one way: (up, up, up).
2. Two go up and one goes down. This can occur in three ways: (up, up, down), (up, down, up), (down, up, up).

Therefore,

$$\begin{array}{c}\text{the probability} \\ \text{of at least 2 up}\end{array} = \begin{array}{c}\text{the probability} \\ \text{of all 3 up}\end{array} + \begin{array}{c}\text{the probability of 2 up and} \\ \text{1 down in any of three ways}\end{array}$$

$$\begin{aligned} P(\text{at least 2 up}) &= P(3 \text{ up}) + 3P(2 \text{ up, 1 down}) \\ &= p^3 + 3p^2q \\ &= (0.60)^3 + 3(0.60)^2(0.40) \\ &= 0.216 + 0.432 = 0.648 \end{aligned}$$

Thus the probability that at least two investments grow is 0.648.

Binomial experiments have what is known as a *probability distribution* that describes in terms of probabilities all possible outcomes of the experiment. In order to determine the probability distribution for an experiment, first consider all possible results and the number of ways in which each result can be obtained. In the example, four possible outcomes exist

1. All three investments go up.	one way:	(up, up, up)
2. Two go up and one down.	3 ways:	(up, up, down)
		(up, down, up)
		(down, up, up)
3. One goes up and two down.	3 ways:	(up, down, down)
		(down, up, down)
		(down, down, up)
4. All three go down.	one way:	(down, down, down)

Since these four outcomes represent every possible result of the experiment, the sum of the probabilities must equal 1. Therefore,

$$p^3 + 3p^2q + 3pq^2 + q^3 = 1$$ *Notice that the left side equals $(p + q)^3$*

EXERCISES

1. Show that the probability distribution of the example is equal to 1.

2. In the example, what is the probability that at least one investment will grow?

3. Another analyst is researching four different possible investments, each of which has a 0.58 probability of increasing in value. Find the probability distribution for this Bernoulli experiment.

4. What is the probability that at least half of the investments in Exercise 3 will grow?

5. If seven independent investment instruments have a 0.78 probability of growth, what is the probability that exactly five will grow? What is the probability that more than five will grow?

CHAPTER 14 SUMMARY AND REVIEW

Vocabulary

alternate hypothesis (723)
bi-modal (708)
combination (687)
complement (701)
dependent events (702)
deviation from the mean (713)
disjoint (699)
event (692)
frequency (707)
frequency distribution (707)
fundamental counting principle (681)
histogram (707)
independent events (702)
mean (707)
median (708)
mode (708)
mutually exclusive (701)
normal curve (718)
normal distribution (718)
null hypothesis (723)
odds (694)
permutation (681)
probability (693)
random sample (722)
range (712)
sample space (692)
standard deviation (714)
standard error (723)
statistic (706)
variance (713)

Fundamental Counting Principle and Permutations The number of permutations of n distinct elements taken n at a time is given by the formula ${}_nP_n = n!$. If the elements are taken r at a time, where $0 \leq r \leq n$, the formula is ${}_nP_r = \dfrac{n!}{(n - r)!}$. The number of distinguishable permutations of n elements taken n at a time, with r_1 like elements, r_2 like elements of another kind, and so on, is given by the formula $P = \dfrac{n!}{r_1!\, r_2!\, r_3! \cdots}$. **14.1**

1. Evaluate ${}_5P_2$, ${}_9P_9$, ${}_{10}P_4$, and ${}_{12}P_3$.
2. Find the number of permutations of the letters of the word *COMPUTER*.
3. Find the number of permutations of the letters of the word *REWRITE*.

Combinations The number of combinations of n objects taken r at a time is given by the formula ${}_nC_r = \binom{n}{r} = \dfrac{n!}{r!\,(n - r)!}$, for $0 \leq r \leq n$. **14.2**

4. Evaluate ${}_8C_1$, ${}_9C_4$, $\binom{6}{4}$, and $\binom{10}{10}$.
5. There are 10 men and 8 women in a cooking class. How many different committees of 3 can be chosen if all members must be women? if at least 2 members must be men?

Probability If the event E is a subset of a sample space S, and all outcomes are equally likely, then $P(E)$ is given by the formula 14.3

$$P(E) = \frac{\text{number of outcomes in the event}}{\text{number of outcomes in the sample space}}, \text{ where } 0 \leq P(E) \leq 1$$

6. Six tags numbered 1 to 6 are placed in a bowl. If one tag is drawn from the bowl, find the probability that it is a 2; that it is an odd number.

Mutually Exclusive and Independent Events If A and B are events, then: $P(A \cup B) = P(A) + P(B) - P(A \cap B)$ 14.4

$P(A \cup B) = P(A) + P(B)$, if A and B are mutually exclusive events.

If $\overline{A}$ is the complement of A then $P(A) + P(\overline{A}) = 1$.

A and B are independent events if and only if $P(A \cap B) = P(A) \cdot P(B)$.

$P(A \cap B) = P(A) \cdot P(B \textit{ after } A)$, if A and B are dependent events.

7. Four red and five blue marbles are in a jar. One marble is drawn at random and then another. If the first marble is replaced before the second draw, what is the probability that both marbles are red? If the first marble is *not* replaced, what is the probability both marbles are red?

Central Tendencies The mean of n values is $\overline{x} = \frac{1}{n}\sum_{i=1}^{n} x_i$. The median is the middle value of an odd number of values, or the mean of the two middle values of an even number of values. The mode is the value or values with the greatest frequency. If each value occurs only once, there is no mode. 14.5

8. Make a frequency distribution and a histogram for the following values: 14, 13, 14, 13, 15, 16, 12, 13, 14, 15, 16, 15, 14, 13, 13, 13, 14, 12, 12, 13. Then find the mean, the median, and the mode.

Standard Deviation Range = highest value − lowest value 14.6

$$\text{Variance} = \sigma^2 = \frac{1}{n}\sum_{i=1}^{n}(x_i - \overline{x})^2 \qquad \text{Standard deviation} = \sigma = \sqrt{\frac{1}{n}\sum_{i=1}^{n}(x_i - \overline{x})^2}$$

9. Find the range, variance, and standard deviation for Exercise 8.

Normal Distribution For a standard normal distribution, the mean is 0, the standard deviation is 1, and the area between the curve and the x-axis is 1. 14.7

10. The scores on an exam are normally distributed, with a mean of 50 and a standard deviation of 3. Find the probability that a student chosen at random will score between 47 and 53; will score less than 44.

11. A tub of butter is supposed to weigh 1.0 lb, with a standard deviation of 0.06 lb. A random sample of 40 tubs of butter taken from a shipment was found to have a mean of 1.01 lb. Does this shipment differ significantly from the standard at the 0.05 level of significance? 14.8

CHAPTER 14 TEST

1. Evaluate ${}_6P_3$.
2. Evaluate ${}_7P_7$.
3. Find the number of permutations of the letters in the word *VOTER*.
4. Evaluate ${}_{10}C_4$.
5. Evaluate $\binom{8}{5}$.
6. There are 6 angelfish and 3 tetras in a tank. How many different selections of 5 fish, including 3 angelfish and 2 tetras, can be made?
7. A die is rolled. Find the probability of getting a number less than 5.
8. A coin is tossed and a die is rolled. Write a sample space and find the probability of getting heads and an even number.
9. A card is chosen at random from a bridge deck. Then a second card is chosen. If the first card is *not* replaced before the second draw, what is the probability that both cards are face cards?

Some tomato plants were found to have the following heights, in inches:

12, 13, 14, 9, 7, 8, 15, 10, 11, 6

10. Find the mean height of the plants.
11. Find the median height.
12. Find the variance of the data, to the nearest tenth.
13. Find the standard deviation, to the nearest tenth.
14. The weights of seven-year-old girls are normally distributed, with a mean of 35 lb and a standard deviation of 2 lb. Find the probability that a girl, chosen at random from a group of girls who are all seven years old, will weigh between 31 and 39 lb.
15. A tube of toothpaste is supposed to contain 5.0 oz, with a standard deviation of 0.14 oz. A random sample of 50 tubes taken from one shipment has a mean of 4.97 oz. Does this shipment differ significantly from the standard at the 0.05 level of significance?

Challenge

A restaurant serves mostly chicken and veal. From past experience, the probability that a customer will order chicken is 0.2 and the probability that a customer will order veal is 0.7. What is the probability that everyone in a party of 6 will order chicken? will *not* order veal?

COLLEGE ENTRANCE EXAM REVIEW

In each item you are to compare a quantity in Column 1 with a quantity in Column 2. Write the letter of the correct answer from these choices:

A. The quantity in Column 1 is greater than the quantity in Column 2.
B. The quantity in Column 2 is greater than the quantity in Column 1.
C. The quantity in Column 1 is equal to the quantity in Column 2.
D. The relationship cannot be determined from the given information.

Notes: Information centered over both columns refers to one or both of the quantities being compared. A symbol that appears in both columns has the same meaning in each column. All variables represent real numbers. Most figures are not drawn to scale.

	Column 1	**Column 2**

$$f(x) = 3x^3 - 7x^2 - 5x + 1$$

1. $f(1)$ — $f(-1)$

$$\begin{cases} 4x + 7y = 12 \\ 2x - y = 15 \end{cases}$$

2. x — y

3. Probability of exactly 2 heads on toss of 2 dimes — Probability of exactly 2 heads on toss of 3 dimes

Use this information for 4–6.

History Test Scores

Score	*Frequency*
100	1
95	2
90	7
85	6
80	9
75	4
70	1

4. mean — mode

5. median — 80

6. range — 30

	Column 1	**Column 2**

7. ${}_6C_2$ — ${}_6P_n$

8. Area of this regular hexagon — $6\sqrt{3}$

9. $\sum_{k=1}^{5} (2k - 1)$ — $\sum_{t=1}^{\infty} \left(\frac{50}{3t}\right)$

10. $(5 - 4i)^2$ — $(20 - i)^2$

$$f(x) = -x^2 - 4x + 3$$
$$g(x) = -2x^2 + 6x + 3$$

11. maximum $f(x)$ — maximum $g(x)$

Use this figure for 12–13.

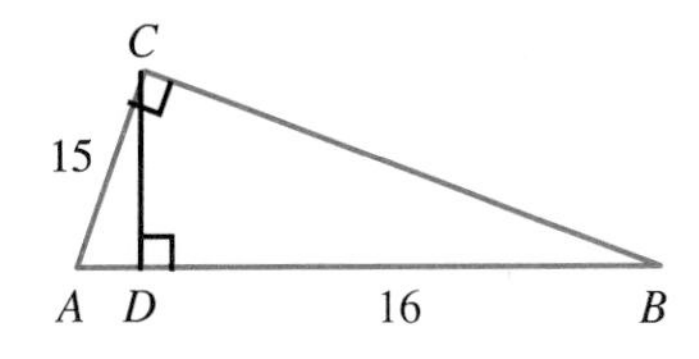

12. CD — AD

13. BC — 20

CUMULATIVE REVIEW (CHAPTERS 1–14)

1. If y varies inversely as the cube of x and $x = 2.5$ when $y = 0.015$, find y when $x = 5$.
2. In standard form, write the equation of a line parallel to the graph of $3x + y = 5$ and containing the point $(-6, 4)$.

Determine whether the first polynomial is a factor of the second.

3. $x + 3$; $2x^3 + 5x^2 - 8x - 15$
4. $x - 2$; $3x^5 - 6x^4 - 4x + 8$
5. 4 lb of peanuts and 3 lb of cashews cost \$16.50, while 5 lb of peanuts and 7 lb of cashews cost \$32.00. Find the cost per lb of each.

For a club of 20 members, determine:

6. In how many ways a president and a secretary can be chosen
7. In how many ways three member committees can be chosen
8. Find a number which is 3 less than its reciprocal.

Express in logarithmic form.

9. $6^{-2} = \frac{1}{36}$
10. $8^{\frac{4}{3}} = 16$
11. $b^e = c, b > 0$

Find the specified partial sum for each series.

12. $3 + 3.25 + 3.50 + \cdots; S_{34}$
13. $\sum_{n=1}^{8} 2^{-n}$
14. Graph this system of inequalities: $\begin{cases} y - 2 < 2x - x^2 \\ y \geq \frac{1}{2}x \end{cases}$

Find the center and radius of a circle with the given equation.

15. $(x + 4)^2 + y^2 = 72$
16. $x^2 + y^2 - 6y + 12x = 0$
17. Mario invests \$8750 at 7.20% compounded monthly. How long will it take for his account to earn \$1500 interest?

Expand and simplify each binomial.

18. $(a + 3)^5$
19. $(2x - y)^4$
20. $(x^2 + 2y^2)^3$
21. The population of an area is now 850 people and is increasing each year by 250 people. What is the population after 12 yr?

15 Trigonometric Functions and Graphs

All you need is an angle!

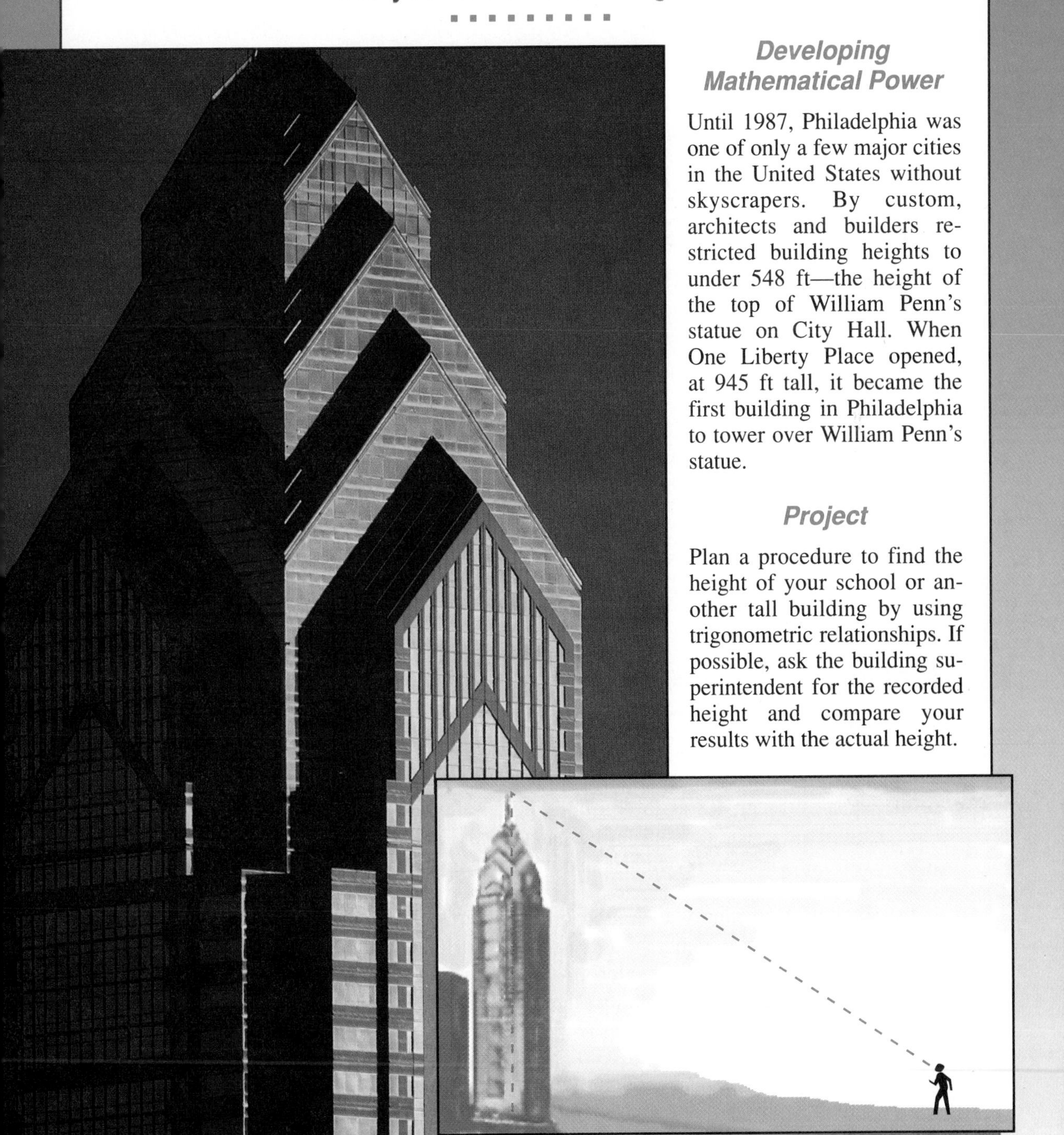

Developing Mathematical Power

Until 1987, Philadelphia was one of only a few major cities in the United States without skyscrapers. By custom, architects and builders restricted building heights to under 548 ft—the height of the top of William Penn's statue on City Hall. When One Liberty Place opened, at 945 ft tall, it became the first building in Philadelphia to tower over William Penn's statue.

Project

Plan a procedure to find the height of your school or another tall building by using trigonometric relationships. If possible, ask the building superintendent for the recorded height and compare your results with the actual height.

Angle Measures

Objective: To measure angles in degrees and in radians

In earlier mathematics courses you learned that an *angle* is the union of two rays with a common endpoint and that a protractor shows measures of angles from 0 to 180 degrees.

Capsule Review

You can use a protractor to find the degree measure of an angle. For example, the measure of angle AOB is 30 degrees ($\angle AOB = 30°$).

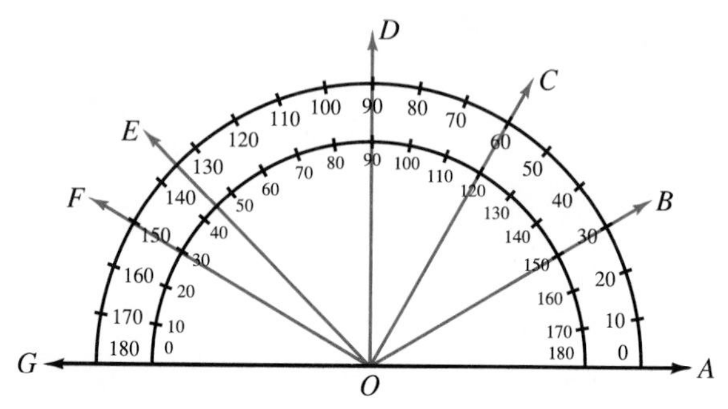

Find the measure of each angle.

1. $\angle AOC$
2. $\angle AOD$
3. $\angle AOE$
4. $\angle AOF$
5. $\angle BOC$
6. $\angle COD$
7. $\angle BOE$
8. $\angle BOF$
9. $\angle COG$
10. $\angle DOG$
11. $\angle DOE$
12. $\angle EOF$

In this course, the rays of an angle may be collinear or noncollinear. One ray of the angle is designated as the **initial side** and the other ray is designated as the **terminal side.** The measure of the angle is the amount of opening between the rays as one ray rotates about the other.

If the amount of opening is measured in a counterclockwise direction from the initial side to the terminal side, the measure of the angle is *positive*. If it is measured in a clockwise direction, the measure is *negative*. Thus, an angle can have any real number for its measure.

A Greek letter, such as θ (*theta*), is often used to represent an angle and also to represent its measure. It is generally clear from the context whether the angle or its measure is being considered.

An angle is in **standard position** on a Cartesian coordinate plane if its vertex is at the origin and its initial side is on the positive x-axis. The terminal side of the angle will fall in one of the four quadrants or on one of the axes. An angle whose terminal side falls on an axis is called a **quadrantal angle.**

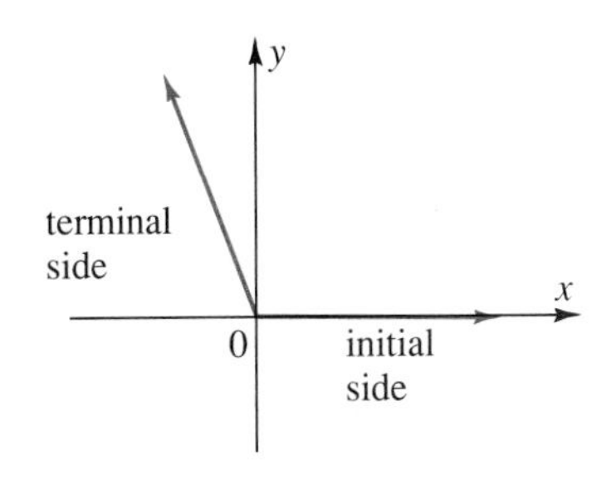

There are 360° in one complete revolution, and the degree measures of the quadrantal angles from 0° through 360° are

$$0°, 90°, 180°, 270°, \text{ and } 360°$$

To determine in which quadrant the terminal side of a nonquadrantal angle between 0° and 360° lies, simply determine between which of these values the measure of the angle lies.

If θ represents the measure of an angle in degrees, then the terminal side of the angle is in

quadrant I if $0° < \theta < 90°$
quadrant II if $90° < \theta < 180°$
quadrant III if $180° < \theta < 270°$
quadrant IV if $270° < \theta < 360°$

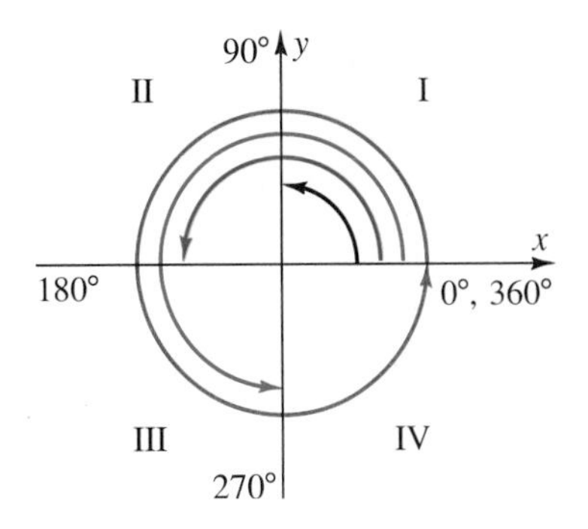

EXAMPLE 1 **In which quadrant, or on which axis, does the terminal side of each angle lie?** **a.** 33° **b.** 270° **c.** 335°

a. $0° < 33° < 90°$, so the terminal side of a 33° angle lies in quadrant I.

b. A 270° angle is quadrantal. Its terminal side lies on the negative y-axis.

c. $270° < 335° < 360°$, so the terminal side of a 335° angle lies in quadrant IV.

If the terminal sides of two angles in standard position coincide, the angles are said to be **coterminal.** For example, angles of 420° and −300° are both coterminal with a 60° angle. The measures of all angles coterminal with an angle θ, measured in degrees, are of the form

$$\theta + n \cdot 360°$$

where n is an integer.

EXAMPLE 2 **In which quadrant, or on which axis, does the terminal side of each angle lie?** **a.** $-122°$ **b.** $470°$ **c.** $720°$

a. $-122° + 360° = 238°$, so a $-122°$ angle is coterminal with a $238°$ angle and its terminal side lies in quadrant III.

b. $470° - 360° = 110°$, so a $470°$ angle is coterminal with a $110°$ angle and its terminal side lies in quadrant II.

c. $720° - 2(360°) = 720° - 720° = 0°$, and a $0°$ angle is quadrantal. The terminal side of a $720°$ angle lies on the positive x-axis.

A *central angle* of a circle is an angle whose vertex is the center of the circle and whose sides are radii of the circle. Therefore, an angle in standard position is a central angle of any circle whose center is at the origin of a coordinate plane. When the central angle intercepts an arc that has the same length as a radius of the circle, the measure of this angle is defined to be one **radian.**

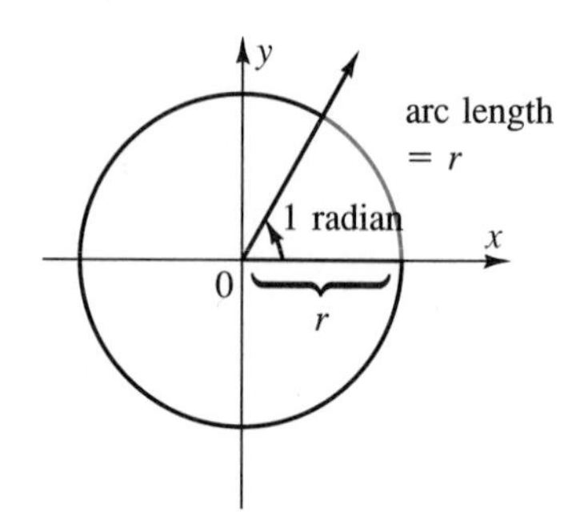

The circumference of a circle is $2\pi r$, where r is the length of a radius. Thus, there are 2π radians in one complete revolution about a point. Since one complete revolution is equal to $360°$,

$$2\pi \text{ radians} = 360° \qquad \text{and} \qquad \pi \text{ radians} = 180°$$

Therefore,

$$1° = \frac{\pi}{180} \text{ radians}$$

$$1 \text{ radian} = \left(\frac{180}{\pi}\right)^{\circ} \qquad \text{(approximately } 57.3°\text{)}$$

You can use these facts to convert from degrees to radians and vice versa. To convert degrees to radians, multiply the number of degrees by $\frac{\pi}{180}$.

EXAMPLE 3 **Convert each degree measure to radian measure.**
a. $120°$ **b.** $-245°$

a. $120° = \left(120 \cdot \frac{\pi}{180}\right)\text{radians}$
$= \frac{2\pi}{3}$ radians

b. $-245° = \left(-245 \cdot \frac{\pi}{180}\right)\text{radians}$
$= -\frac{49\pi}{36}$ radians

To convert radians to degrees, multiply the number of radians by $\frac{180}{\pi}$.

EXAMPLE 4 **Convert each radian measure to degree measure:**

a. $\frac{\pi}{3}$ radians **b.** $-\frac{3\pi}{4}$ radians

a. $\frac{\pi}{3}$ radians $= \left(\frac{\pi}{3} \cdot \frac{180}{\pi}\right)^\circ$
$= 60^\circ$

b. $-\frac{3\pi}{4}$ radians $= \left(-\frac{3\pi}{4} \cdot \frac{180}{\pi}\right)^\circ$
$= -135^\circ$

If an angle measure is simply given as a number without a degree symbol, it is assumed that the units are radians. For example, $\theta = 3\pi$ means that the measure of the angle is 3π radians.

If θ represents the measure of an angle, then the terminal side of the angle is in

quadrant I if $0 < \theta < \frac{\pi}{2}$

quadrant II if $\frac{\pi}{2} < \theta < \pi$

quadrant III if $\pi < \theta < \frac{3\pi}{2}$

quadrant IV if $\frac{3\pi}{2} < \theta < 2\pi$

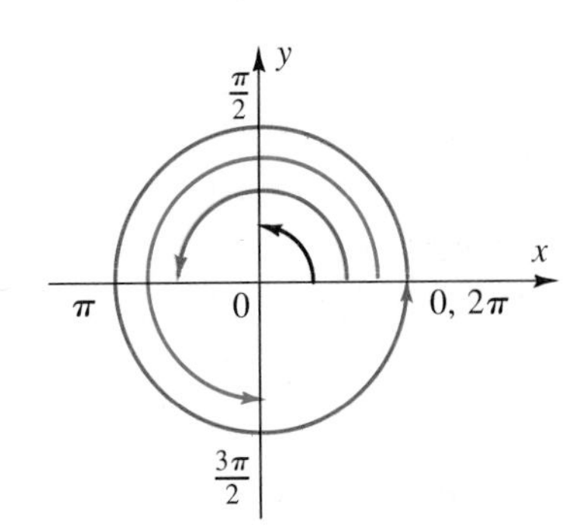

The measures of all angles coterminal with an angle θ, measured in radians, are of the form $\theta + n \cdot 2\pi$ radians, where n is an integer. If the measure of an angle is not between 0 and 2π, you determine the quadrant in which its terminal side lies by first finding a coterminal angle with a measure between 0 and 2π.

EXAMPLE 5 **In which quadrant, or on which axis, does the terminal side of each angle lie?** **a.** $\frac{4\pi}{3}$ **b.** $-\frac{5\pi}{4}$ **c.** $\frac{9\pi}{2}$

a. $\pi < \frac{4\pi}{3} < \frac{3\pi}{2}$, so the terminal side of an angle of $\frac{4\pi}{3}$ radians is in quadrant III.

b. $-\frac{5\pi}{4} + 2\pi = -\frac{5\pi}{4} + \frac{8\pi}{4} = \frac{3\pi}{4}$, so an angle of $-\frac{5\pi}{4}$ radians is coterminal with an angle of $\frac{3\pi}{4}$ radians, and its terminal side is in quadrant II.

c. $\frac{9\pi}{2} - 2(2\pi) = \frac{9\pi}{2} - \frac{8\pi}{2} = \frac{\pi}{2}$, so an angle of $\frac{9\pi}{2}$ radians is coterminal with an angle of $\frac{\pi}{2}$ radians, and its terminal side is on the positive y-axis.

It is sometimes necessary to use greater precision than the nearest degree when measuring an angle. You can do this by expressing the angle measure in decimal degrees or in minutes and seconds, where

$$1 \text{ minute } (1') = \left(\frac{1}{60}\right)^{\circ} \text{ and } 1 \text{ second } (1'') = \left(\frac{1}{60}\right)' \text{ or } \left(\frac{1}{3600}\right)^{\circ}$$

EXAMPLE 6 **Convert each angle measure as indicated.**

a. 12.464° to degrees, minutes, and seconds, to the nearest second
b. 23°42′45″ to decimal degrees, to the nearest tenth

a.
$$\begin{aligned} 12.464^{\circ} &= 12^{\circ} + (0.464 \cdot 60)' \\ &= 12^{\circ} + 27.84' \\ &= 12^{\circ} + 27' + (0.84 \cdot 60)'' \\ &= 12^{\circ} + 27' + 50.4'' \\ &= 12^{\circ}27'50'', \text{ to the nearest second} \end{aligned}$$

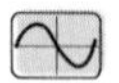

To do this on a calculator:

- Be sure the calculator is in *degree mode.* Write the number of degrees. **12°**
- Multiply 0.464 · 60 to find the number of minutes, 27.84. Write 27′. **12°27′**
- Subtract 27 from the 27.84 shown in the display, and multiply 0.84 · 60 to find the number of seconds, 50.4. Round to 50 seconds. **12°27′50″**

b.
$$\begin{aligned} 23^{\circ}42'45'' &= 23^{\circ} + \left(\frac{42}{60}\right)^{\circ} + \left(\frac{45}{3600}\right)^{\circ} \\ &= 23.7^{\circ} \text{ to the nearest tenth} \end{aligned}$$

If a calculator is used, be sure it is in degree mode.

Conversions such as those in Example 6 can be done directly on most scientific calculators if special keys are employed. Check your calculator manual to see if your calculator has this feature.

CLASS EXERCISES

In which quadrant does the terminal side of each angle lie?

1. 240° **2.** 340° **3.** 415° **4.** −215° **5.** $\frac{5\pi}{6}$ **6.** $-\pi$

Sketch each angle in standard position.

7. 150° **8.** −300° **9.** 720° **10.** 665° **11.** $\frac{6\pi}{5}$ **12.** $-\frac{5\pi}{2}$

13. Convert 200° to radian measure.

14. Convert $\frac{4\pi}{3}$ to degree measure.

PRACTICE EXERCISES 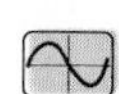 Use technology where appropriate.

In which quadrant, or on which axis, does the terminal side of each angle lie?

1. $150°$ **2.** $210°$ **3.** $-60°$ **4.** $180°$ **5.** $-240°$ **6.** $540°$

7. 2π **8.** $\frac{\pi}{3}$ **9.** $\frac{3\pi}{4}$ **10.** $\frac{7\pi}{3}$ **11.** $\frac{5\pi}{4}$ **12.** $\frac{10\pi}{3}$

Convert each degree measure to radian measure.

13. $150°$ **14.** $210°$ **15.** $45°$ **16.** $240°$

Convert each radian measure to degree measure.

17. $\frac{\pi}{6}$ **18.** $\frac{\pi}{4}$ **19.** $\frac{5\pi}{6}$ **20.** $\frac{7\pi}{6}$

Convert to degrees, minutes, and seconds, to the nearest second.

21. $23.42°$ **22.** $15.27°$ **23.** $48.35°$ **24.** $62.73°$

Convert to decimal degrees, to the nearest tenth of a degree.

25. $14°33'45''$ **26.** $38°24'36''$ **27.** $35°45'10''$ **28.** $28°32'20''$

Sketch the angle with the given revolution in standard position, and give its degree measure.

29. $\frac{1}{4}$ clockwise revolution

30. $\frac{2}{3}$ clockwise revolution

31. $\frac{3}{4}$ counterclockwise revolution

32. $\frac{5}{6}$ counterclockwise revolution

33. $1\frac{1}{2}$ clockwise revolutions

34. 3 counterclockwise revolutions

Sketch the angle with the given revolution in standard position, and give its radian measure.

35. $1\frac{1}{8}$ clockwise revolutions

36. $1\frac{2}{3}$ clockwise revolutions

37. 3 clockwise revolutions

38. 4 counterclockwise revolutions

39. $3\frac{1}{2}$ counterclockwise revolutions

40. $2\frac{3}{4}$ clockwise revolutions

Convert each degree measure to radian measure.

41. $-600°$ **42.** $-880°$ **43.** $28°30'$ **44.** $8°15'$

Convert each radian measure to the nearest tenth of a degree.

45. 3 **46.** -10 **47.** 0.8 **48.** 1.5

Start at the terminal side of the given angle θ in standard position. Find the radian measure of the resulting angle after the given number of revolutions.

49. $\theta = \frac{\pi}{2}$; 1 clockwise revolution

50. $\theta = \frac{\pi}{3}$; 2 clockwise revolutions

51. $\theta = -\frac{2\pi}{3}$; $1\frac{1}{2}$ clockwise revolutions

52. $\theta = -\frac{3\pi}{4}$; $1\frac{2}{3}$ clockwise revolutions

53. $\theta = \frac{\pi}{4}$; 1 counterclockwise revolution

54. $\theta = \frac{\pi}{6}$; 2 counterclockwise revolutions

55. $\theta = -\frac{3\pi}{2}$; $2\frac{1}{2}$ counterclockwise revolutions

56. $\theta = \frac{5\pi}{6}$; $2\frac{1}{3}$ counterclockwise revolutions

Applications

57. Physics A point on the outside edge of a wheel moves through a 43° angle. What is the radian measure of the angle?

Construction Mioka made a wooden wheel to spin at the cake-walk booth at the school carnival. She divided the wheel into 3 congruent sectors and then divided each part into 5 congruent sectors.

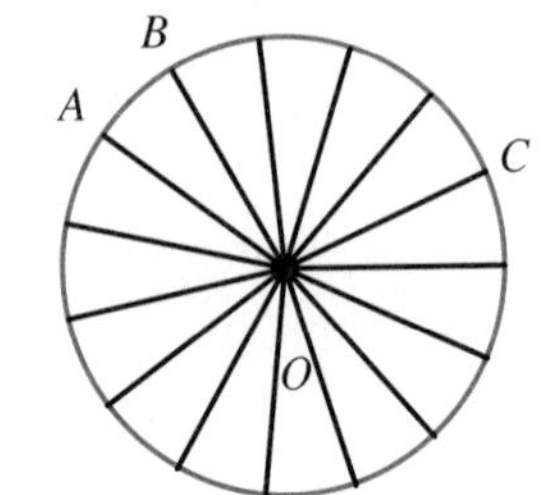

58. Find the measure of central angle AOC in degrees.

59. Find the measure of central angle AOB in degrees.

HISTORICAL NOTE

James Muir, a mathematician, and James T. Thomson, a physicist, were working independently during the late nineteenth century to develop a new unit of angle measure. They met and agreed on the name *radian,* a shortened form of the phrase *radial angle*. Different names were used for the new unit until about 1900. Today the term *radian* is in common usage. This unit of angle measure has been found to simplify many formulas, such as those used in optics and in the design of gears. Find out more about angle measure.

15.2

Definition of Trigonometric Functions

Objective: To define the six trigonometric functions

The trigonometric functions are defined in terms of the ratios of the lengths of the sides of a **reference triangle.** This triangle is formed by drawing a perpendicular to the x-axis from any point P on the terminal side of an angle in standard position. The position of P may change, but the ratios remain the same for a given angle in standard position, since the reference triangles formed are similar triangles.

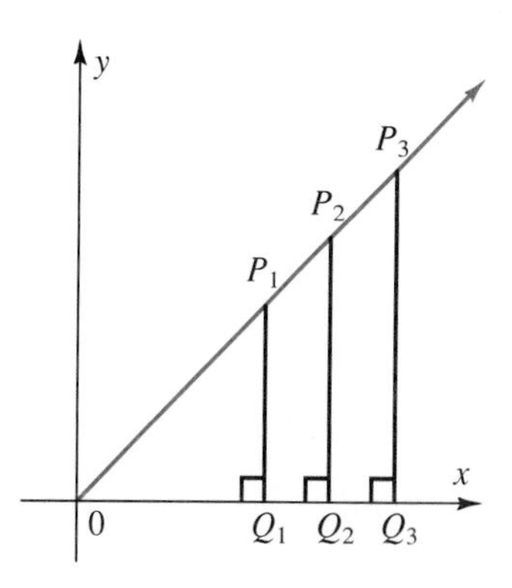

Capsule Review

Recall that two triangles are similar if two angles of one are congruent to two angles of the other. If two triangles are similar, the lengths of their corresponding sides are proportional. For example, triangles AED and ABC are similar, so

$$\frac{AE}{AB} = \frac{AD}{AC} = \frac{ED}{BC}$$

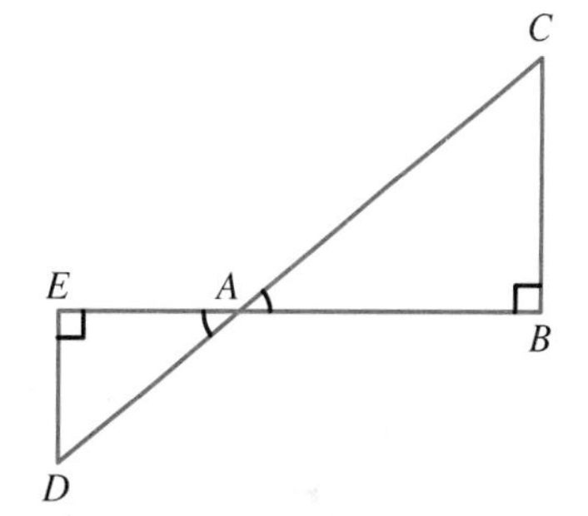

Write the proportion relating the lengths of the corresponding sides of each pair of similar triangles.

1.

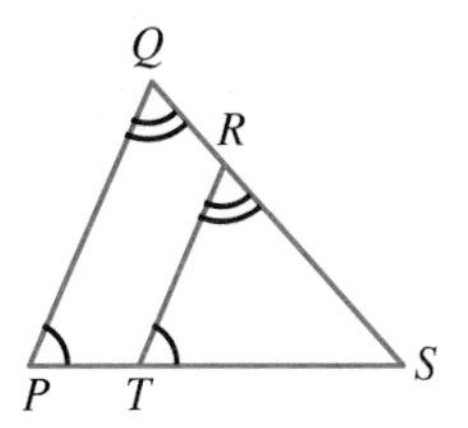

2.

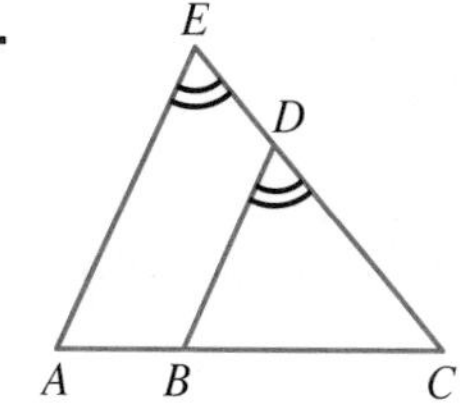

3.

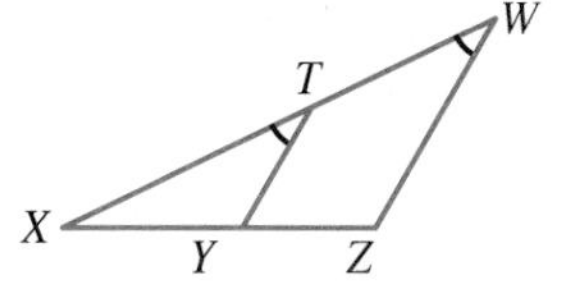

4.

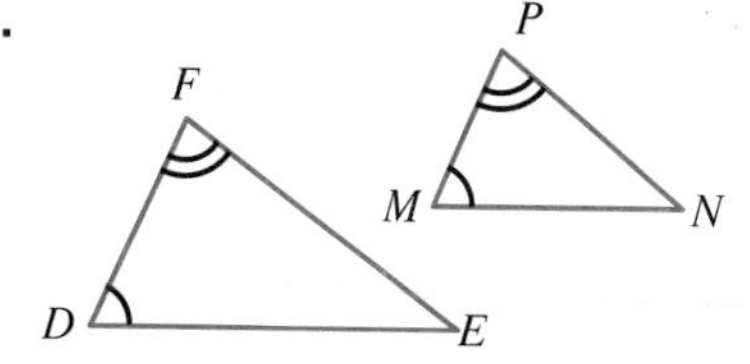

Let θ be the measure of an angle in standard position with $P(x, y)$ a point on its terminal side, and let r be the length of the line segment from the origin to P. Then r is the length of the hypotenuse of the reference triangle for angle θ and, by the Pythagorean theorem, $r^2 = x^2 + y^2$, or $r = \sqrt{x^2 + y^2}$.

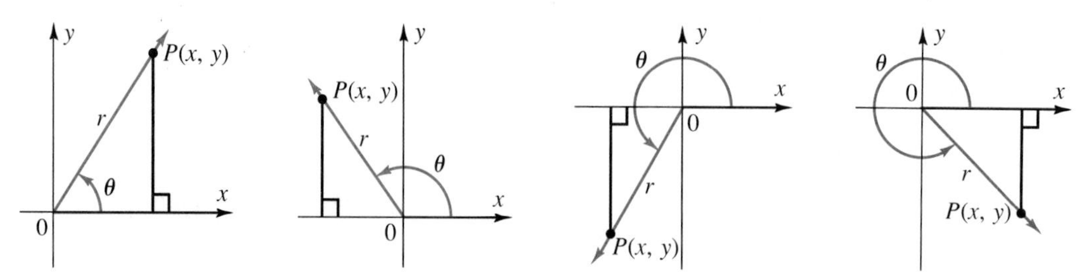

Six ratios can be written using r and the coordinates of P: $\frac{y}{r}$, $\frac{x}{r}$, $\frac{y}{x}$, $\frac{r}{y}$, $\frac{r}{x}$, and $\frac{x}{y}$. These ratios are used so often in mathematics that they are given special names. Each ratio is defined only if the denominator is not zero.

$\sin \theta = \frac{y}{r}$	*sine*	$\csc \theta = \frac{r}{y}$	*cosecant*
$\cos \theta = \frac{x}{r}$	*cosine*	$\sec \theta = \frac{r}{x}$	*secant*
$\tan \theta = \frac{y}{x}$	*tangent*	$\cot \theta = \frac{x}{y}$	*cotangent*

The six ratios above are called **trigonometric functions** because each ratio depends only on the measure of the angle, and for any given angle of measure θ, each ordered pair $(\theta, \sin \theta)$, $(\theta, \cos \theta)$, $(\theta, \tan \theta)$, and so on, is unique. Since r is always positive, the sign of each ratio is determined by the signs of the coordinates of P.

EXAMPLE 1 **Find the values of the six trigonometric functions for an angle θ in standard position if point $P(-5, 12)$ is on its terminal side.**

The coordinates of P are $(-5, 12)$, so P is in quadrant II.

$r = \sqrt{(-5)^2 + (12)^2} = 13$

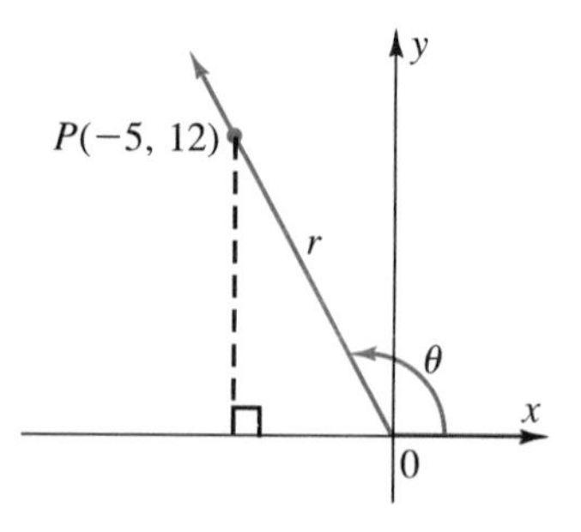

$\sin \theta = \frac{y}{r} = \frac{12}{13}$ $\qquad$ $\csc \theta = \frac{r}{y} = \frac{13}{12}$

$\cos \theta = \frac{x}{r} = \frac{-5}{13} = -\frac{5}{13}$ $\qquad$ $\sec \theta = \frac{r}{x} = \frac{13}{-5} = -\frac{13}{5}$

$\tan \theta = \frac{y}{x} = \frac{12}{-5} = -\frac{12}{5}$ $\qquad$ $\cot \theta = \frac{x}{y} = \frac{-5}{12} = -\frac{5}{12}$

Note that $\csc \theta = \frac{1}{\sin \theta}$, $\sec \theta = \frac{1}{\cos \theta}$, and $\cot \theta = \frac{1}{\tan \theta}$, where no denominator is 0. These three pairs—the sine and cosecant, the cosine and secant, and the tangent and cotangent—are called **reciprocal functions.**

If angle θ is in standard position, with point $P(x, y)$ on its terminal side, then the signs of x and y determine the signs of the trigonometric functions of the angle. All of the functions are positive if P is in quadrant I, only the sine and cosecant are positive for quadrant II, only the tangent and cotangent for quadrant III, and only the cosine and secant for quadrant IV.

sin and csc positive	all positive
tan and cot positive	cos and sec positive

EXAMPLE 2 **Find the values of the other five trigonometric functions for $\angle A$ if $\tan A = \frac{3}{4}$ and the terminal side of $\angle A$ is in quadrant III.**

Since $\angle A$ is in quadrant III, both x and y are negative.

$\tan A = \frac{y}{x} = \frac{-3}{-4} = \frac{3}{4}$ If $x = -4$ and $y = -3$, then $r = \sqrt{(-4)^2 + (-3)^2} = 5$.

$\cot A = \frac{-4}{-3} = \frac{4}{3}$ $\sin A = -\frac{3}{5}$ $\cos A = -\frac{4}{5}$ $\csc A = -\frac{5}{3}$ $\sec A = -\frac{5}{4}$

You can use the measures of quadrantal angles to determine the sign of the value of a trigonometric function for a specific angle measure.

EXAMPLE 3 **State whether each value is positive or negative.**

a. $\cos 340°$ **b.** $\tan \frac{2\pi}{3}$

a. $270° < 340° < 360°$; a $340°$ angle is in quadrant IV and $\cos 340°$ is positive.

b. $\frac{\pi}{2} < \frac{2\pi}{3} < \pi$; an angle of $\frac{2\pi}{3}$ radians is in quadrant II and $\tan \frac{2\pi}{3}$ is negative.

CLASS EXERCISES

1. Find the values of the six trigonometric functions for an angle in standard position with point $Q(6, -8)$ on its terminal side.

2. The terminal side of $\angle B$ is in quadrant II and $\cos B = -\frac{8}{17}$. Find the values of the other five trigonometric functions for $\angle B$.

State whether each value is positive or negative.

3. $\cos 130°$ **4.** $\sec 25°$ **5.** $\sin \frac{7\pi}{6}$ **6.** $\tan \frac{4\pi}{3}$

PRACTICE EXERCISES

Find the values of the six trigonometric functions for $\angle R$ if $\angle R$ is in standard position and point Q is on its terminal side.

1. $Q(8, 6)$ **2.** $Q(-8, 15)$ **3.** $Q(-3, -4)$ **4.** $Q(12, -5)$

Find the values of the other five trigonometric functions for $\angle S$.

5. $\cos S = \frac{3}{5}$ and the terminal side of $\angle S$ is in quadrant I.

6. $\tan S = \frac{24}{7}$ and the terminal side of $\angle S$ is in quadrant III.

7. $\sin S = -\frac{5}{13}$ and the terminal side of $\angle S$ is in quadrant IV.

8. $\tan S = -\frac{15}{8}$ and the terminal side of $\angle S$ is in quadrant II.

State whether each value is positive or negative.

9. $\sin 140°$ **10.** $\cos 240°$ **11.** $\tan 120°$ **12.** $\sec 113°$

13. $\cot 208°$ **14.** $\csc 215°$ **15.** $\sin \frac{5\pi}{3}$ **16.** $\tan \frac{7\pi}{4}$

17. $\cos \frac{5\pi}{6}$ **18.** $\sec \frac{4\pi}{3}$ **19.** $\cot \frac{5\pi}{4}$ **20.** $\csc \frac{5\pi}{6}$

Find the values of the six trigonometric functions for $\angle R$ if $\angle R$ is in standard position and point Q is on its terminal side. Express answers involving radicals in simplest radical form.

21. $Q(1, 1)$ **22.** $Q(-3, 3)$ **23.** $Q(-1, \sqrt{3})$ **24.** $Q(-2\sqrt{3}, -2)$

Find the values of the other five trigonometric functions for $\angle S$.

25. $\cos S = \frac{1}{3}$ and the terminal side of $\angle S$ is in quadrant I.

26. $\tan S = -\frac{10}{7}$ and the terminal side of $\angle S$ is in quadrant II.

27. $\sin S = -\frac{5}{14}$ and the terminal side of $\angle S$ is in quadrant III.

28. $\sec S = \frac{13}{11}$ and the terminal side of $\angle S$ is in quadrant IV.

State whether each value is positive or negative.

29. $\cos 640°$ **30.** $\sin 820°$ **31.** $\tan 520°$

32. $\sec 714°$ **33.** $\cot 910°$ **34.** $\csc 1089°$

Find the value of each expression if $\cos C = \frac{15}{17}$ and the terminal side of $\angle C$ is in quadrant IV. ***Note:*** **$\sin^2 C$ means $(\sin C)^2$.**

35. $\sin^2 C + \cos^2 C$

36. $\csc^2 C - \cot^2 C$

Find the value of each expression if $\cos A = \frac{3}{5}$ and $\sin B = \frac{5}{13}$.

37. $\dfrac{\cos A \tan A}{\csc A}$

38. $\dfrac{\sin B \sec B}{\cot B}$

39. If $\tan C = -\frac{8}{15}$ and $\sin D = -\frac{4}{5}$, in which quadrant(s) might the terminal side of an angle whose measure is $(C + D)$ lie?

40. Find the values of the other five trigonometric functions for $\angle A$ if $\tan A = \dfrac{\sqrt{1 - p^2}}{p}$ and the terminal side of $\angle A$ is in quadrant I.

Applications

41. Boating A sail is in the shape of a right triangle. If the cosine of the smallest angle of the triangle is $\frac{24}{25}$, what is the sine of that angle?

42. Construction A 20-ft ladder forms an angle θ with the side of a building, such that $\sin \theta = \frac{3}{5}$. How far is the foot of the ladder from the base of the building?

Mixed Review

Write in radical form.

43. $m^{\frac{1}{2}}$

44. $z^{\frac{3}{4}}$

45. $x^{-\frac{2}{3}}$

Write in exponential form.

46. $\sqrt{3}$

47. $\sqrt[3]{w}$

48. $\sqrt[z]{m^2}$

Solve for x.

49. $3^{x+2} = 27^{2x}$

50. $16^{2x-1} = 32^{x+4}$

51. $\left(\frac{1}{2}\right)^{4x+2} = 4^{-3x}$

Evaluate.

52. $\log_3 81$

53. $\log_2 \frac{1}{2}$

54. $\log_7 1$

Express in expanded form.

55. $\log_3 2x$

56. $\log_m \frac{3}{c}$

57. $\log_2 m^3$

15.3 Trigonometric Functions of Special Angles

Objective: To find the values of the trigonometric functions for 30°, 45°, 60°, and quadrantal angles

Angles that measure 30°, 45°, and 60° appear frequently in mathematics. The values of the trigonometric functions for these angles can be derived easily using information about the lengths of the sides of the right triangles that have angles with these measures.

Capsule Review

Recall that an altitude of an equilateral triangle bisects both an angle and a side of the triangle, thus forming two 30°-60°-90° triangles. If d represents the length of the hypotenuse of a 30°-60°-90° triangle, then the length of the side opposite the 30° angle is $\frac{d}{2}$. By the Pythagorean theorem, the length of the side opposite the 60° angle is $\sqrt{d^2 - \left(\frac{d}{2}\right)^2}$, or $\frac{d\sqrt{3}}{2}$.

Recall also that if two angles of a triangle are equal in measure, then the sides opposite these angles are equal in length. If s represents the length of each of the congruent sides of a 45°-45°-90° triangle, then the length of the hypotenuse is $\sqrt{s^2 + s^2}$, or $s\sqrt{2}$.

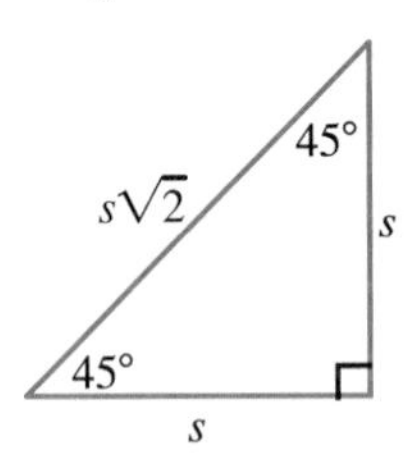

For a triangle ABC, find the lengths of sides a and b.

1. $\angle B = 45°, \angle C = 90°, c = 16$
2. $\angle B = 45°, \angle C = 90°, c = 10$
3. $\angle B = 30°, \angle C = 90°, c = 12$
4. $\angle B = 30°, \angle C = 90°, c = 18$
5. $\angle B = 45°, \angle C = 90°, c = \sqrt{2}$
6. $\angle B = 60°, \angle C = 90°, c = 2$

The relationships between the lengths of each pair of sides of a 30°-60°-90° triangle and between the lengths of each pair of sides of a 45°-45°-90° triangle can be used to find the values of the six trigonometric functions of 30°, 45°, and 60° angles in standard position. Since all reference triangles for a given angle are similar, the size of the reference triangle does not affect the values of the trigonometric ratios for a given angle. Therefore, convenient measures are chosen, as shown in the figures on page 749.

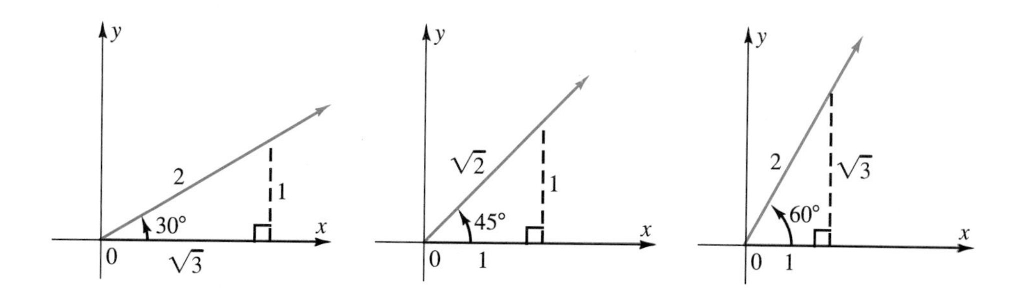

	$\sin\theta$	$\cos\theta$	$\tan\theta$	$\csc\theta$	$\sec\theta$	$\cot\theta$
$\theta = 30°$	$\frac{1}{2}$	$\frac{\sqrt{3}}{2}$	$\frac{\sqrt{3}}{3}$	2	$\frac{2\sqrt{3}}{3}$	$\sqrt{3}$
$\theta = 45°$	$\frac{\sqrt{2}}{2}$	$\frac{\sqrt{2}}{2}$	1	$\sqrt{2}$	$\sqrt{2}$	1
$\theta = 60°$	$\frac{\sqrt{3}}{2}$	$\frac{1}{2}$	$\sqrt{3}$	$\frac{2\sqrt{3}}{3}$	2	$\frac{\sqrt{3}}{3}$

To find the values of the special angles in the other three quadrants, make use of reference triangles and keep in mind the signs associated with the values of the functions in the different quadrants.

EXAMPLE 1 **Find the values of the six trigonometric functions for 225°.**

$\sin 225° = -\frac{\sqrt{2}}{2}$ $\qquad \csc 225° = -\sqrt{2}$

$\cos 225° = -\frac{\sqrt{2}}{2}$ $\qquad \sec 225° = -\sqrt{2}$

$\tan 225° = 1$ $\qquad \cot 225° = 1$

Since the terminal side of a quadrantal angle lies on an axis, no reference triangle can be formed. The values of the trigonometric functions of quadrantal angles are given by the ratios $\frac{y}{r}, \frac{x}{r}, \frac{y}{x}, \frac{r}{y}, \frac{r}{x}$, and $\frac{x}{y}$, when these ratios are defined. Clearly, certain functions are not defined when x or y is 0. For example, any point on the terminal side of a 90° angle has 0 as its x-coordinate. Therefore, tan 90° and sec 90° $\left(\frac{y}{x} \text{ and } \frac{r}{x}, \text{ respectively}\right)$ are not defined.

EXAMPLE 2 **Find the values of the six trigonometric functions for π.**

Choose a point, such as $(-1, 0)$, on the terminal side of the angle in standard position. Then $x = -1$, $y = 0$, and $r = 1$.

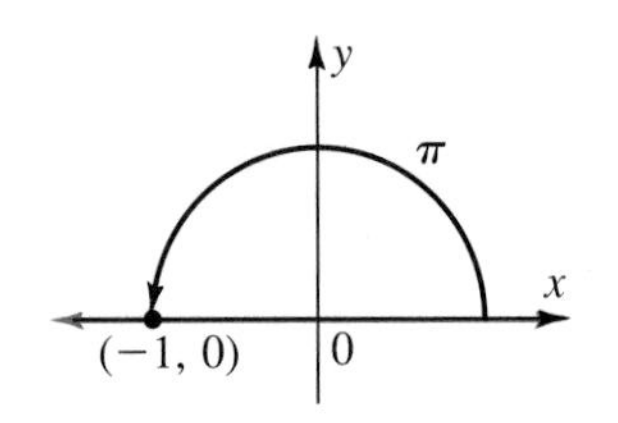

$\sin \pi = \frac{0}{1} = 0$ $\csc \pi = \frac{1}{0}$, so $\csc \pi$ is undefined.

$\cos \pi = \frac{-1}{1} = -1$ $\sec \pi = \frac{1}{-1} = -1$

$\tan \pi = \frac{0}{-1} = 0$ $\cot \pi = \frac{-1}{0}$, so $\cot \pi$ is undefined.

To find the values of trigonometric functions for negative angle measures and for measures greater than 360°, first determine where the terminal side of the angle lies.

EXAMPLE 3 **Find the values of the six trigonometric functions for each angle.**

a. $-\frac{\pi}{6}$ **b.** 495°

a. The terminal side of an angle measuring $-\frac{\pi}{6}$ radians (−30°) lies in quadrant IV, where only the cosine and the secant are positive.

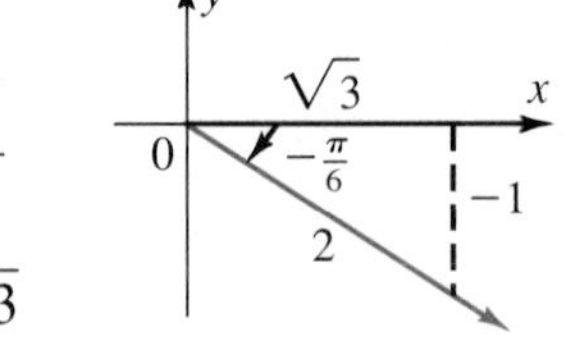

$\sin\left(-\frac{\pi}{6}\right) = -\frac{1}{2}$ $\csc\left(-\frac{\pi}{6}\right) = -2$

$\cos\left(-\frac{\pi}{6}\right) = \frac{\sqrt{3}}{2}$ $\sec\left(-\frac{\pi}{6}\right) = \frac{2\sqrt{3}}{3}$

$\tan\left(-\frac{\pi}{6}\right) = -\frac{\sqrt{3}}{3}$ $\cot\left(-\frac{\pi}{6}\right) = -\sqrt{3}$

b. 495° − 360° = 135°, so a 495° angle is coterminal with a 135° angle, and its terminal side is in quadrant II. Only the sine and the cosecant are positive.

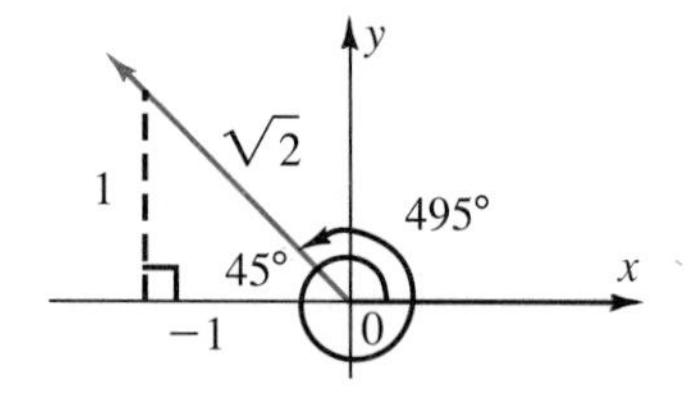

$\sin 495° = \frac{\sqrt{2}}{2}$ $\csc 495° = \sqrt{2}$

$\cos 495° = -\frac{\sqrt{2}}{2}$ $\sec 495° = -\sqrt{2}$

$\tan 495° = -1$ $\cot 495° = -1$

Given that $\tan \theta = 1$, the measure of θ cannot be uniquely determined without additional information. There are an infinite number of values for which $\tan \theta = 1$; for example, θ could equal 45°, 225°, 405°, or −135°. However, if you know that θ is between 0° and 360°, then the value of θ must be either 45° or 225°.

EXAMPLE 4 **If $0° \leq \theta \leq 360°$, find all values of θ for which $\sin \theta = \frac{\sqrt{3}}{2}$.**

The sine is positive only in quadrants I and II, so draw the reference triangles in these quadrants.

Since $\sin \theta = \frac{y}{r}$, let $y = \sqrt{3}$ and $r = 2$.

The values of θ for which $\sin \theta = \frac{\sqrt{3}}{2}$ are 60° and 120°.

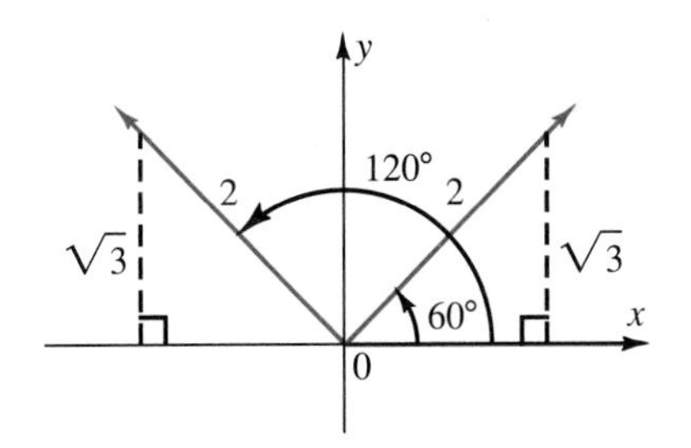

CLASS EXERCISES

Complete each statement.

1. If the length of the hypotenuse of a 30°-60°-90° triangle is 30 cm, then the length of the leg opposite the 60° angle is __?__ cm.
2. If the length of the leg opposite the 30° angle in a 30°-60°-90° triangle is 5 m, then the length of the side opposite the 60° angle is __?__ m.
3. If the length of a leg of a 45°-45°-90° triangle is 6 in., then the length of the hypotenuse is __?__ in.
4. If the length of the hypotenuse of a 45°-45°-90° triangle is 12 ft, then the length of each leg is __?__ ft.
5. If $\tan \theta = \frac{\sqrt{3}}{3}$, then $\cot \theta =$ __?__.
6. If $\sec \theta = 2$, then $\cos \theta =$ __?__.

PRACTICE EXERCISES

Find the values of the six trigonometric functions for each angle. When radicals occur, express the answer in simplest radical form.

1. 135° **2.** 150° **3.** 240° **4.** 210° **5.** 315° **6.** 300°

7. 390° **8.** 405° **9.** 360° **10.** 0° **11.** −45° **12.** −60°

13. $\frac{\pi}{6}$ **14.** $\frac{\pi}{3}$ **15.** $\frac{\pi}{4}$ **16.** $\frac{4\pi}{3}$ **17.** $\frac{\pi}{2}$ **18.** $\frac{3\pi}{2}$

19. 420° **20.** 450° **21.** −540° **22.** $-\frac{2\pi}{3}$ **23.** $-\frac{5\pi}{6}$ **24.** $-\frac{4\pi}{3}$

If $0° \leq \theta \leq 360°$, find all values of θ for which each of the following is true.

25. $\sin \theta = -\frac{1}{2}$ **26.** $\cos \theta = -\frac{1}{2}$ **27.** $\tan \theta = -1$

28. $\cot \theta = -\sqrt{3}$

29. $\sec \theta = \sqrt{2}$

30. $\sin \theta = 0$

If $0° < \theta < 720°$, find all values of θ for which each of the following is true.

31. $\csc \theta = 2$ and $\sin \theta > 0$

32. $\cos \theta = -\frac{\sqrt{2}}{2}$ and $\sin \theta < 0$

33. $\sin \theta = \frac{\sqrt{3}}{2}$ and $\cos \theta < 0$

34. $\tan \theta = \frac{\sqrt{3}}{3}$ and $\sin \theta > 0$

35. $\cot \theta = -\frac{\sqrt{3}}{3}$ and $\cos \theta < 0$

36. $\sin \theta = -\frac{\sqrt{2}}{2}$ and $\cos \theta < 0$

Applications

37. Geometry Which trigonometric function(s) could you use to find the height h of the parallelogram shown below?

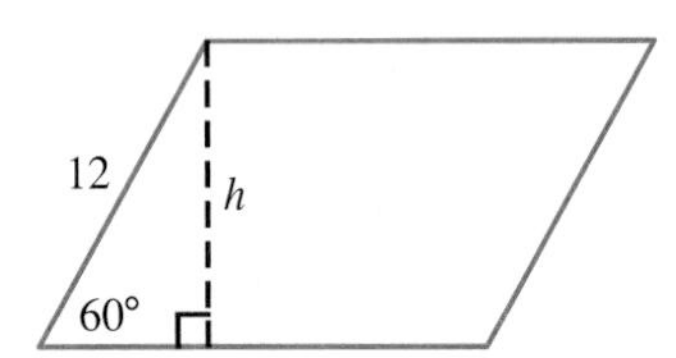

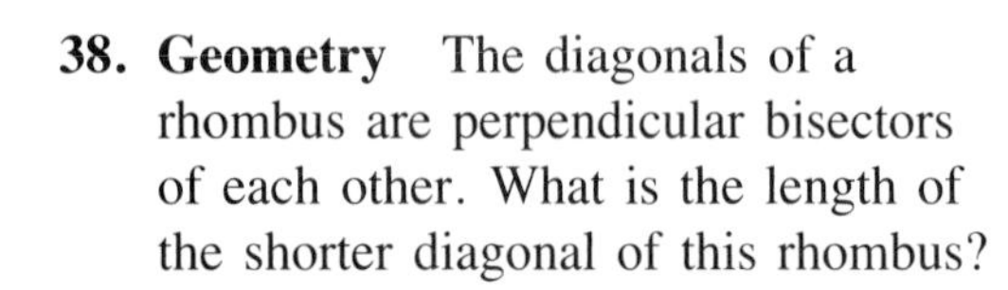

38. Geometry The diagonals of a rhombus are perpendicular bisectors of each other. What is the length of the shorter diagonal of this rhombus?

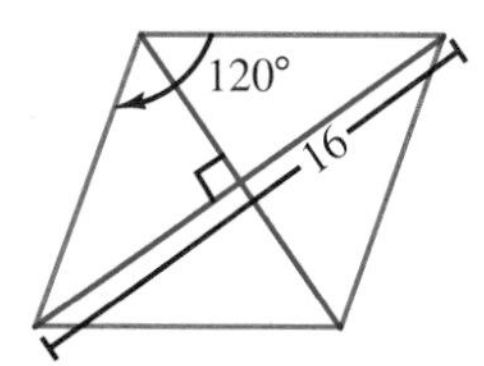

39. Construction A roofer leaned a 20-ft ladder against the wall of a house in such a way that it made a 60° angle with the ground. How far up the wall did the ladder reach?

BIOGRAPHY: Mary Fairfax Somerville

Mary Fairfax Somerville was fascinated by mathematics all her life, although she lived in an era when women were not encouraged to study technical subjects. Her brother's tutor provided her with a copy of Euclid's *Elements* and taught her a little algebra and geometry, but she was mainly self-taught. She became famous for her exceptional ability to write popularized accounts of current developments in mathematics and science in language that could be understood by almost anyone.

Mary Somerville described in detail the curve called a *cycloid,* which is generated by a point on a rolling circle. Find out more about this curve.

15.4

Values of Trigonometric Functions

Objectives: To find decimal approximations for the values of the trigonometric functions of angles
To find the measure of an acute angle, given the value of one of its trigonometric functions

Trigonometry is an invaluable tool in such fields as navigation and surveying. If the measures of certain angles are known, the values of their trigonometric functions can be used to calculate distances that cannot be measured directly.

You can use a scientific calculator to find the values of the trigonometric functions of angles. A trigonometric table, such as the one in the Appendix of this book, can also be employed. Refer to pages 870–874 for the table and pages 864–865 for a discussion of its use.

Capsule Review

To express in decimal degrees an angle measure given in degrees and minutes, use the fact that $60' = 1°$.

EXAMPLE **Express 125°37′ in decimal degrees. Round to four decimal places.**

$125°37' = \left(125 + \frac{37}{60}\right)^{\circ}$ *Calculation-ready form*

$= 125.6166667°$

$= 125.6167°$, to four decimal places

Express each angle measure in decimal degrees. Round to four decimal places.

1. 87°30′ **2.** 65°45′ **3.** 108°12′ **4.** 82°13′ **5.** 236°8′ **6.** 110°59′

When using a calculator to find the values of the trigonometric functions of an angle whose measure is given in degrees, be sure that the calculator is in the *degree mode,* not the *radian mode.*

EXAMPLE 1 **Find each value. Round to four decimal places.**

a. sin 28.5° **b.** tan 49°17′

a. Be sure your calculator is in the *degree mode*. The angle measure is given in decimal degrees, so simply enter 28.5, using the *sin* key.

$$\sin 28.5° = 0.47715876$$
$$\sin 28.5° = 0.4772, \text{ to four decimal places}$$

b. Be sure your calculator is in the *degree mode*. Unless your calculator accepts angle measures given in degrees and minutes, it is necessary to convert to decimal degrees. Use the *tan* key.

$$\begin{aligned}\tan 49°17' &= \tan\left(49 + \frac{17}{60}\right)^\circ \\ &= \tan 49.28333333° \\ &= 1.161923419 \\ &= 1.1619, \text{ to four decimal places}\end{aligned}$$

In the example above, 0.47715876 is not exactly equal to sin 28.5°, and 1.161923419 is not exactly equal to tan 49°17′. The equals sign is customarily used in examples of this type, but you should keep in mind the fact that most of the values of the trigonometric functions are irrational numbers and cannot be written exactly as decimals, no matter how many decimal places are shown.

When using a calculator to find the value of a trigonometric function of an angle whose measure is given in radians, your calculator must be set in the *radian mode*.

EXAMPLE 2 **Find $\cos \frac{\pi}{8}$. Round to four decimal places.**

Be sure the calculator is in the *radian mode*. Use the *cos* key.

$$\begin{aligned}\cos \frac{\pi}{8} &= \cos 0.392699081 \qquad \textit{Use the key for } \pi \textit{ and the } \div \textit{ key.} \\ &= 0.9239, \text{ to four decimal places}\end{aligned}$$

When using a calculator to find the values of the cosecant, secant, and cotangent, use the fact that csc θ, sec θ, and cot θ are the reciprocal functions of sin θ, cos θ, and tan θ, respectively. Therefore, you use the reciprocal key, which is usually labeled $\frac{1}{x}$ or x^{-1}.

EXAMPLE 3 **Find csc 42.25°. Round to four decimal places.**

Be sure your calculator is in the *degree mode*.
Enter 42.25 and use the *sin* key and the reciprocal key.
csc 42.25° = 1.4873, to four decimal places

The angles in the first three examples are all *acute;* that is, their measures are between $0°$ and $90°$. Therefore, the values of the trigonometric functions for these angles are all positive. If the entries are made correctly, a calculator will give the correct sign for the value of a trigonometric function of an angle with measure greater than $90°$ or less than $0°$. However, it is wise to check the results, using your knowledge of the sign of the given function value for the quadrant in which the terminal side of the angle is known to lie.

EXAMPLE 4 **Find each value. Round to four decimal places.**
a. $\cos 226°30'$ **b.** $\sin(-200°21')$

a. Be sure the calculator is in the *degree mode*. Use the *cos* key.

$$\cos 226°30' = \cos 226.5°$$
$$= -0.6884, \text{ to four decimal places}$$

Check that the sign is correct: $180° < 226°30' < 270°$, so the terminal side of a $226°30'$ angle is in the third quadrant and its cosine is negative.

b. Be sure the calculator is in the *degree mode*. Use the *sin* key.

$$\sin(-200°21') = \sin\left[-\left(200 + \tfrac{21}{60}\right)\right]^\circ$$ *Use the calculator to add $\frac{21}{60}$ to 200.*

$$= \sin -200.35°$$ *Use the +/− key to show the result as a negative number.*

$$= 0.3478, \text{ to four decimal places}$$

To check, note that $-200°21' + 360° = -200°21' + 359°60' = 159°39'$. Since a $-200°21'$ angle is coterminal with a $159°39'$ angle, its terminal side is in the second quadrant and the sine is positive.

You can use a calculator to find the measure of an acute angle if the value of one of its trigonometric functions is given.

EXAMPLE 5 **Find the measure of each acute angle θ to the nearest tenth of one degree.** **a.** $\sin\theta = 0.4305$ **b.** $\cot\theta = 1.508$

a. The sine of an acute angle θ is 0.4305. You want to find the measure of the angle. Finding the measure of an angle with a known sine and finding the sine of an angle of known measure are *inverse* operations.

Check to see that the calculator is in the *degree mode*.
Enter 0.4305 and use the inverse key and the *sin* key.

$\theta = 25.5°$, to the nearest tenth of one degree

b. With the calculator in the *degree mode,* enter 1.508. Use the reciprocal key, the inverse key, and the *tan* key.

$\theta = 33.5°$, to the nearest tenth of one degree

CLASS EXERCISES

Find each value. Round to four decimal places.

1. $\sin 24.25°$ **2.** $\cos 42.33°$ **3.** $\tan 54.5°$ **4.** $\sec 63°10'$

5. $\cot 72°50'$ **6.** $\cos 128°20'$ **7.** $\sin 212°30'$ **8.** $\csc 352°50'$

Find the measure of the acute angle to the nearest tenth of one degree.

9. $\sin x = 0.5422$ **10.** $\tan y = 1.0416$ **11.** $\cos z = 0.6799$

12. $\sec p = 1.184$ **13.** $\cot q = 4.1653$ **14.** $\csc x = 3.742$

PRACTICE EXERCISES  Use technology where appropriate.

Find each value. Round to four decimal places.

1. $\sin 35.75°$ **2.** $\cos 33.1°$ **3.** $\tan 45.5°$ **4.** $\tan 15.38°$

5. $\sin 72°30'$ **6.** $\cos 73°40'$ **7.** $\sin 5°30'$ **8.** $\cos 68°20'$

9. $\tan \frac{\pi}{6}$ **10.** $\sin \frac{2\pi}{3}$ **11.** $\cos \frac{3\pi}{4}$ **12.** $\tan \frac{4\pi}{5}$

13. $\cos 67°38'$ **14.** $\tan 59°43'$ **15.** $\cos 273°10'$ **16.** $\sin 155°40'$

17. $\csc 45°30'$ **18.** $\sec 24°40'$ **19.** $\cot 26°50'$ **20.** $\csc 57°10'$

Find the measure of the acute angle to the nearest tenth of one degree.

21. $\cos x = 0.7528$ **22.** $\sin y = 0.8192$ **23.** $\tan z = 0.8342$

24. $\cos x = 0.8984$ **25.** $\sin y = 0.3052$ **26.** $\tan z = 0.9145$

Find each value. Round to four decimal places.

27. $\cos 258°32'$ **28.** $\tan 239°23'$ **29.** $\sin 335°17'$ **30.** $\cos 146°24'$

31. $\cos 127°38'$ **32.** $\tan 259°43'$ **33.** $\sec 325°17'$ **34.** $\cot 286°23'$

35. $\cos(-123°40')$ **36.** $\tan(-62°20')$ **37.** $\sin(-59°10')$ **38.** $\csc(-125°20')$

Find the measure of the acute angle to the nearest tenth of one degree.

39. $\sec x = 1.335$ **40.** $\cot y = 1.611$ **41.** $\csc z = 2.475$

42. $\sec x = 1.234$ **43.** $\cot y = 1.323$ **44.** $\csc z = 1.8255$

Find the measures of angle x, to the nearest ten minutes, such that $0° < x < 360°$.

45. $\sec x = -1.468$ **46.** $\csc x = -1.193$ **47.** $\cot x = -1.7315$

48. $\sec x = -5.565$ **49.** $\csc x = -3.158$ **50.** $\cot x = -0.2578$

51. $\csc x = -6.162$ **52.** $\cot x = -4.3312$ **53.** $\sec x = -14.87$

Applications

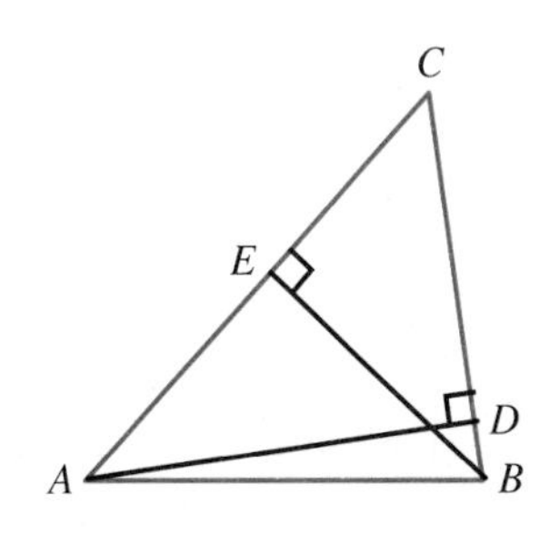

54. Geometry In triangle ABC, $AC = 34$ and $\angle C = 53°15'$. To find the length of altitude AD, Samantha wrote the equation $\sin 53°15' = \frac{AD}{34}$. What value should she substitute for $\sin 53°15'$?

55. Geometry In triangle ABC, $AB = 27$ and $\angle ABE = 46°30'$. To find the length of altitude BE, Jonathan wrote the equation $\cos 46°30' = \frac{BE}{27}$. What value should he substitute for $\cos 46°30'$?

Developing Mathematical Power

Extension If a regular polygon with n sides is inscribed in a circle with a radius of 1, the sine function would help you find the length of a side. Since each θ is equal to $\frac{360°}{n}$,

$$\beta = \frac{360°}{2n} = \frac{180°}{n} \text{ and } \sin\left(\frac{180°}{n}\right) = \frac{x}{1} = x.$$

Therefore, one side $= 2x = 2 \sin\left(\frac{180°}{n}\right)$.

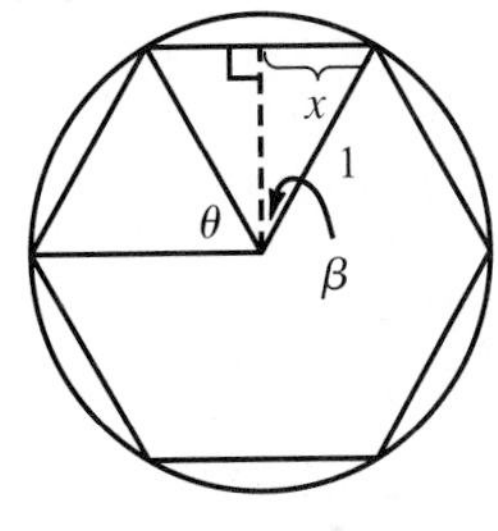

The more sides a regular polygon has, the more it begins to resemble a circle. Likewise, the perimeter of a many-sided regular polygon approaches the circumference of a circle of equal radius.

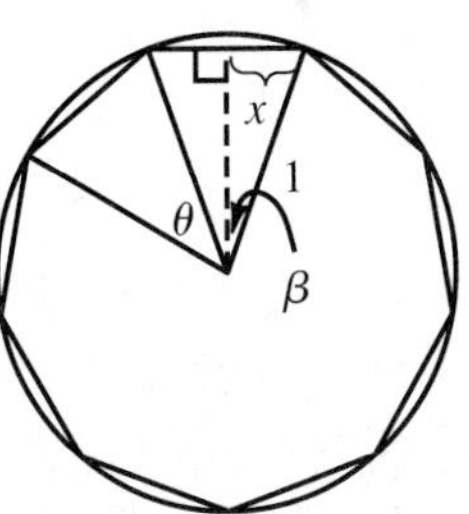

56. Complete the chart, rounding to three decimal places.

Number of sides	3	6	12	20	50	300
Perimeter of inscribed polygon	?	?	?	?	?	?

57. Remembering that the circumference of a circle is $2\pi r$, how many sides must the inscribed polygon have to approximate the circumference of the circle to two decimal places? three? six?

15.5 Graphing the Sine and Cosine Functions

Objective: To graph the sine and cosine and related functions

An electrocardiogram (EKG) is a graph of a person's heartbeat. The graph shows the repetitive, or *periodic,* nature of the electrical impulses from the heart. The sine and the cosine functions are also periodic, and their graphs exhibit certain similarities to the EKG. Before you study the graphs of the sine and cosine functions, it will be helpful to review the meaning of the terms *relation* and *function*.

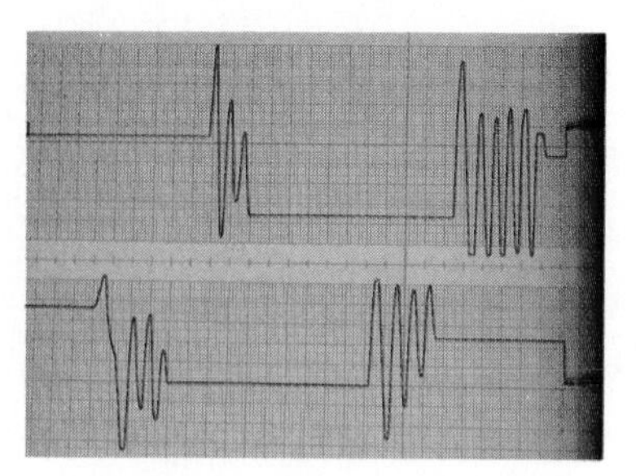

Capsule Review

A set of ordered pairs (a, b) is a relation. The relation is also a function if for each first coordinate a there is one and only one second coordinate b. Relations and functions can also be defined by rules.

Tell whether or not each relation is also a function.

1. $\{(2, 4), (1, 3), (-3, -1), (4, 6)\}$

2. $\{(2, 6), (-3, 1), (-2, 2)\}$

3. $\{(x, y): x = 3\}$

4. $\{(x, y): y = 8\}$

5. $\{(x, y): x = y^2\}$

6. $\{(x, y): x^2 + y^2 = 36\}$

7. $\{(a, b): a = b^3\}$

8. $\{(w, z): w = z - 36\}$

To graph the function $y = \sin x$, make a table of values for ordered pairs of the form $(x, \sin x)$. Plot the points and connect them with a smooth curve.

EXAMPLE 1 **Graph $y = \sin x$ for $-2\pi \le x \le 2\pi$.**

x	-2π	$-\frac{7\pi}{4}$	$-\frac{3\pi}{2}$	$-\frac{5\pi}{4}$	$-\pi$	$-\frac{3\pi}{4}$	$-\frac{\pi}{2}$	$-\frac{\pi}{4}$	0
$\sin x$	0	0.71	1	0.71	0	−0.71	−1	−0.71	0

x	$\frac{\pi}{4}$	$\frac{\pi}{2}$	$\frac{3\pi}{4}$	π	$\frac{5\pi}{4}$	$\frac{3\pi}{2}$	$\frac{7\pi}{4}$	2π
$\sin x$	0.71	1	0.71	0	−0.71	−1	−0.71	0

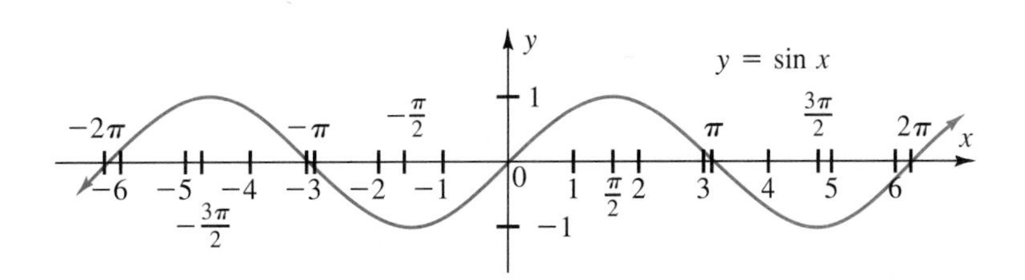

The portion of the graph of $y = \sin x$ from $x = 0$ to $x = 2\pi$ is repeated over and over again, to the left and to the right. Such a function is called a **periodic function,** and the horizontal distance from a point on the curve to the point at which the curve repeats is called the **period** of the function. The period of $y = \sin x$ is 2π, or 360°. The maximum value of $y = \sin x$ is 1, and the minimum value is -1. The *amplitude* of the function is one half the difference between the maximum and minimum values. Therefore, the amplitude of $y = \sin x$ is 1. A graphing utility can be used to draw the graph of the sine and cosine functions. While such utilities usually cannot provide exact values of irrational numbers, they are useful indicators of maximum and minimum values and of repeating cycles.

> The **amplitude** of a periodic function with maximum value M and minimum value m is $\frac{M - m}{2}$.

EXAMPLE 2 **Graph $y = \cos x$ for $-2\pi \leq x \leq 2\pi$, and give its period and amplitude.**

Some graphing utilities have preset ranges for trigonometric graphs. If your utility does not, and you are working in radians, you may have to use decimal approximations of π and 2π when you set screen dimensions.

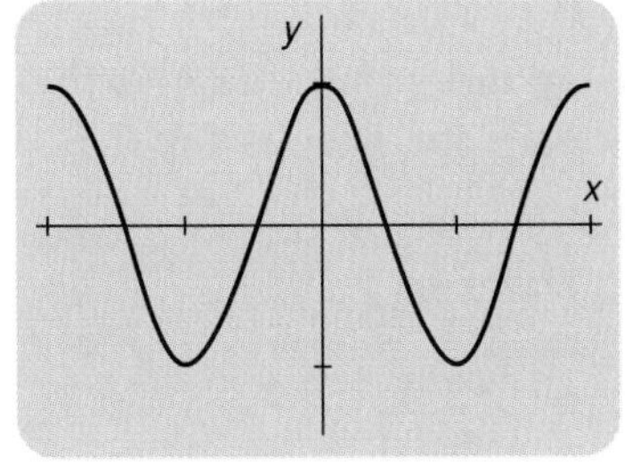

x scl: π y scl: 1

Though the trace feature of your utility may not provide exact values (depending on the range you have chosen), the graph provides a strong indication that the period of $y = \cos x$ is 2π and the amplitude is 1.

If the graph of $y = \cos x$ were shifted $\frac{\pi}{2}$ units to the right, it would coincide with that of $y = \sin x$. That is, the two graphs are congruent.

EXAMPLE 3 **Graph each function for $0 \le x \le 2\pi$, and give its period and amplitude.** **a.** $y = 2 \sin x$ **b.** $y = \sin 2x$

a. For $y = 2 \sin x$, each y-value is twice the corresponding y-value for $y = \sin x$, so double each y-value for $x = 0$ to $x = 2\pi$ in the table in Example 1 and plot the resulting points. The period is 2π, since the curve between 0 and 2π repeats endlessly to the left and to the right. The amplitude is 2, since $\frac{2 - (-2)}{2} = 2$.

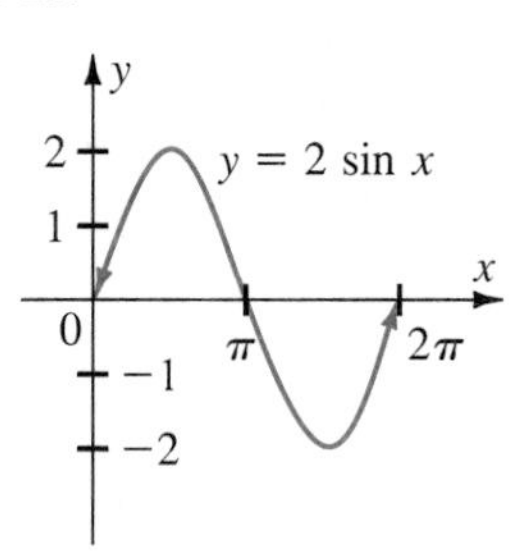

b. To see how the coefficient of x in $y = \sin 2x$ affects the graph of the function, make a table of values and plot the points.

x	0	$\frac{\pi}{4}$	$\frac{\pi}{2}$	$\frac{3\pi}{4}$	π
$2x$	0	$\frac{\pi}{2}$	π	$\frac{3\pi}{2}$	2π
$\sin 2x$	0	1	0	-1	0

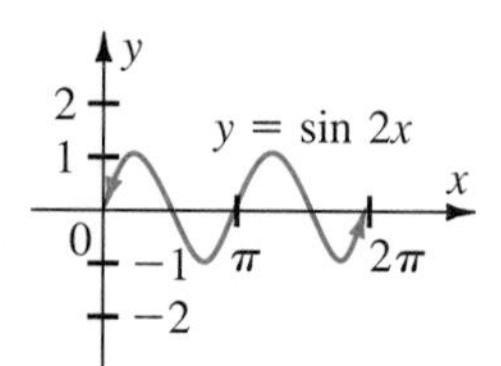

Notice that the period of $y = \sin 2x$ is π, which is one-half the period of $y = \sin x$. The amplitude is $\frac{1 - (-1)}{2} = 1$.

> For the functions $y = a \sin bx$ and $y = a \cos bx$, where a is a real number and b is a positive real number, the amplitude is $|a|$ and the period is $\frac{2\pi}{b}$, or $\left(\frac{360}{b}\right)^{\circ}$.

EXAMPLE 4 **Give the amplitude and the period of $y = 2 \cos 4x$ and $y = -2 \cos 4x$. Sketch the graphs for $0 \le x \le \pi$.**

$y = 2 \cos 4x$	$y = -2 \cos 4x$	
$\lvert 2 \rvert = 2$	$\lvert -2 \rvert = 2$	*Amplitude, $\lvert a \rvert$*
$\frac{2\pi}{4} = \frac{\pi}{2}$	$\frac{2\pi}{4} = \frac{\pi}{2}$	*Period, $\frac{2\pi}{b}$*

Use the amplitude and period, along with your knowledge of the general shape of the cosine curve, to sketch the graph of $y = 2 \cos 4x$. Then note that a is negative in $y = -2 \cos 4x$, so points on this graph are on the opposite side of the x-axis from the corresponding points of $y = 2 \cos 4x$.

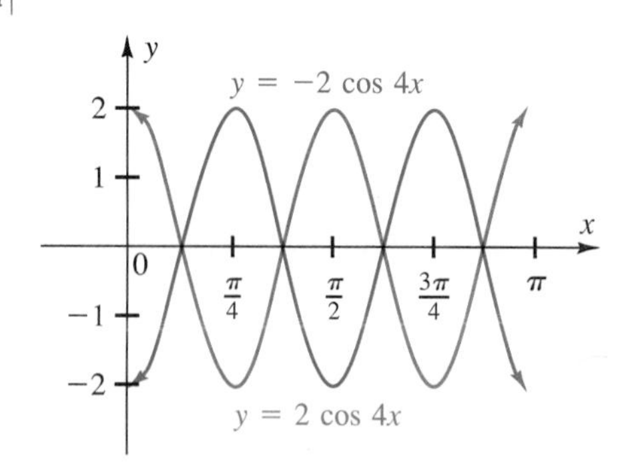

CLASS EXERCISES

Give the amplitude and period for each function. Then sketch the graph for $0 \leq x \leq 2\pi$.

1. $y = \cos 2x$

2. $y = -\sin x$

3. $y = 2 \sin x$

4. $y = 2 \cos x$

5. $y = 3 \sin 2x$

6. $y = -2 \sin 4x$

7. $y = 3 \cos \frac{1}{2}x$

8. $y = \frac{3}{4} \cos 8x$

9. $y = -\frac{1}{4} \cos x$

PRACTICE EXERCISES

Use technology where appropriate.

Give the amplitude and period for each function. Then graph the function.

1. $y = \sin x$ for $0° \leq x \leq 360°$

2. $y = \cos x$ for $0° \leq x \leq 360°$

3. $y = -\cos x$ for $0 \leq x \leq 2\pi$

4. $y = -\sin x$ for $0 \leq x \leq 2\pi$

5. $y = 3 \sin x$ for $-2\pi \leq x \leq 2\pi$

6. $y = 3 \cos x$ for $-2\pi \leq x \leq 2\pi$

7. $y = -2 \cos x$ for $-360° \leq x \leq 360°$

8. $y = -2 \sin x$ for $-360° \leq x \leq 360°$

9. $y = \frac{1}{2} \sin x$ for $0 \leq x \leq 2\pi$

10. $y = \frac{1}{2} \cos x$ for $0 \leq x \leq 2\pi$

11. $y = \sin 3x$ for $0 \leq x \leq 2\pi$

12. $y = \cos 3x$ for $0 \leq x \leq 2\pi$

13. $y = \sin \frac{1}{2}x$ for $-2\pi \leq x \leq 2\pi$

14. $y = \cos \frac{1}{2}x$ for $-2\pi \leq x \leq 2\pi$

15. $y = 3 \sin 2x$ for $-2\pi \leq x \leq 2\pi$

16. $y = 3 \cos 2x$ for $-2\pi \leq x \leq 2\pi$

17. $y = 2 \cos 3x$ for $-2\pi \leq x \leq 2\pi$

18. $y = 2 \sin 3x$ for $-2\pi \leq x \leq 2\pi$

19. $y = 3 \sin \frac{1}{2}x$ for $0 \leq x \leq 2\pi$

20. $y = 3 \cos \frac{1}{2}x$ for $0 \leq x \leq 2\pi$

21. $y = \frac{2}{3} \sin 2x$ for $-2\pi \leq x \leq 2\pi$

22. $y = \frac{2}{3} \cos 3x$ for $-2\pi \leq x \leq 2\pi$

23. $y = 2 \sin \frac{1}{2}x$ for $-360° \leq x \leq 360°$

24. $y = 4 \cos \frac{1}{2}x$ for $-360° \leq x \leq 360°$

25. $y = \frac{4}{3} \sin 3x - 2$ for $-2\pi \leq x \leq 2\pi$

26. $y = \frac{2}{5} \cos 2x - 1$ for $-2\pi \leq x \leq 2\pi$

27. $y = 2 \sin \frac{3}{4}x + 4$ for $-2\pi \leq x \leq 2\pi$

28. $y = 3 \cos \frac{3}{4}x - 5$ for $-2\pi \leq x \leq 2\pi$

Applications

Acoustics On an oscilloscope, an electronic instrument that converts sound waves into electrical pulses and displays them on a screen, pure musical tones have sine- and cosine-shaped graphs. The equation for the graph of the sound generated by a tuning fork, for example, is $y = \pm r \sin(2\pi f t)$ where r is amplitude, t is time in seconds, and f is the frequency or number of periods completed in one second. Note that the frequency and period are reciprocals of one another.

For each of the following sound wave equations find its amplitude, frequency, and period. Then make a table of values and sketch its graph for $0 \le t \le 0.01$. Increment t in steps of 0.0005 second. Be sure to use radians.

29. $y = 0.02 \sin(500\pi t)$

30. $y = 0.06 \cos(300\pi t)$

Find an equation of the sound waves for the following frequencies.

31. frequency: 440; amplitude: 0.027

32. frequency: 4138.4; amplitude: 0.001

TEST YOURSELF

In which quadrant does the terminal side of each angle lie?

1. 160° **2.** 205° **3.** 340° **4.** −120° **5.** $\frac{2\pi}{3}$ **6.** $\frac{5\pi}{4}$ 15.1

Convert to radian measure.

7. 150° **8.** 210° **9.** 130°

Convert to degrees.

10. $\frac{2\pi}{3}$ **11.** $\frac{5\pi}{4}$ **12.** $\frac{4\pi}{3}$

13. Find the values of the other five trigonometric functions for $\angle S$, if $\cos S = \frac{6}{10}$ and the terminal side of $\angle S$ is in quadrant IV. 15.2

Find each value. Express answers in simplest form.

14. $\sin 150°$ **15.** $\tan(-60°)$ **16.** $\cos \frac{\pi}{2}$ **17.** $\sin \frac{3\pi}{4}$ 15.3

Find each value. Round to four decimal places.

18. $\sin 45°20'$

19. $\tan 125°10'$ 15.4

20. $\cos 230°50'$

21. $\csc 325°40'$

Give the amplitude and period of each function and sketch the graph.

22. $y = 3 \cos x$ for $0 \le x \le 2\pi$

23. $y = \sin 3x$ for $0 \le x \le 2\pi$ 15.5

15.6

Graphing Other Trigonometric Functions

Objective: To graph the tangent, cotangent, secant, and cosecant functions

The graphs of the sine and cosine functions were introduced in the previous lesson. The other four trigonometric relations are also functions, and their graphs are discussed in this lesson.

Capsule Review

For an angle θ in standard position with point $P(x, y)$ on the terminal side, the length OP and the coordinates of P are used to define six trigonometric ratios for θ.

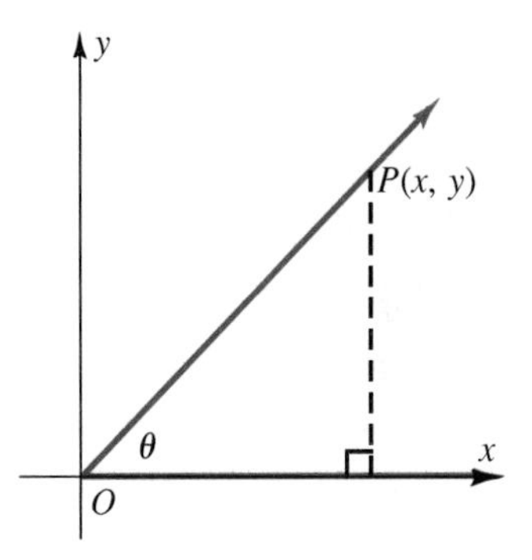

$$\sin\theta = \frac{y}{OP} \qquad \csc\theta = \frac{OP}{y}$$

$$\cos\theta = \frac{x}{OP} \qquad \sec\theta = \frac{OP}{x}$$

$$\tan\theta = \frac{y}{x} \qquad \cot\theta = \frac{x}{y}$$

If the given point $P(x, y)$ is on the terminal side of angle θ, write the six trigonometric ratios associated with θ.

1. $P(5, -12)$ **2.** $P(-3, -4)$ **3.** $P(-7, 7)$

4. $P(-8, 15)$ **5.** $P\left(\frac{1}{2}, \frac{3}{2}\right)$ **6.** $P(-5, 8)$

The graphs of the tangent, cotangent, secant, and cosecant function—unlike those of the sine and cosine functions—are *discontinuous* at certain points. That is, these graphs have breaks where the functions are not defined. To illustrate, the value of tan x is not defined when $x = \frac{\pi}{2}$ or $x = \frac{3\pi}{2}$. When the function $y = \tan x$ is graphed, it is helpful to draw dashed vertical lines at the points where tan x is not defined. These lines are *asymptotes;* that is, the graph of $y = \tan x$ gets closer and closer to the lines, but never meets them.

When these functions are graphed with a utility, lines that resemble asymptotes often appear with the graph. Remember that the asymptotes are not part of the function.

EXAMPLE 1 **Graph $y = \tan x$ for $-\pi \le x \le 2\pi$.**

x	$\tan x$
$-\pi$	0
$-\frac{3\pi}{4}$	1
$-\frac{\pi}{2}$	undefined
$-\frac{\pi}{4}$	-1
0	0
$\frac{\pi}{4}$	1
$\frac{\pi}{2}$	undefined

x	$\tan x$
$\frac{3\pi}{4}$	-1
π	0
$\frac{5\pi}{4}$	1
$\frac{3\pi}{2}$	undefined
$\frac{7\pi}{4}$	-1
2π	0

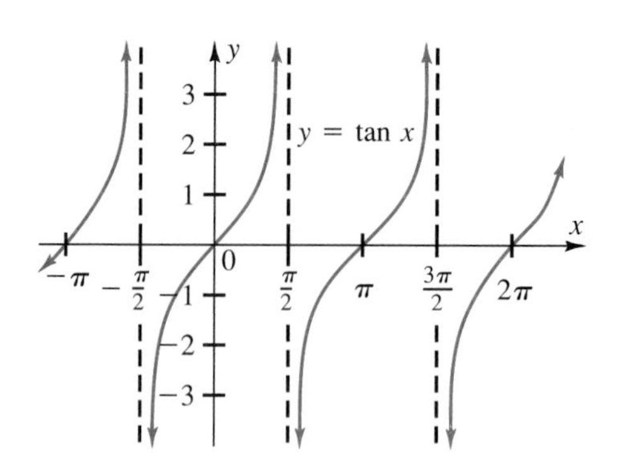

Notice that the tangent function repeats at intervals of π units, or 180°. Therefore, the period of $y = \tan x$ is π. The tangent function increases without bound over each interval, so the amplitude is not defined.

The graphs of the cotangent, secant, and cosecant functions are drawn in a similar manner. These graphs are shown below, along with the graphs of their reciprocal functions. The amplitude is undefined for $y = \cot x$, $y = \sec x$, and $y = \csc x$. The period of $y = \cot x$ is π. The period of

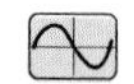

$y = \sec x$ and $y = \csc x$ is 2π. You can use a graphing calculator or a computer to draw the graphs of the trigonometric functions.

For the functions $y = a \tan bx$ and $y = a \cot bx$, where a is a real number and b is a positive real number, the period is $\frac{\pi}{b}$, or $\left(\frac{180}{b}\right)^\circ$.

For the functions $y = a \sec bx$ and $y = a \csc bx$, where a is a real number and b is a positive real number, the period is $\frac{2\pi}{b}$, or $\left(\frac{360}{b}\right)^\circ$.

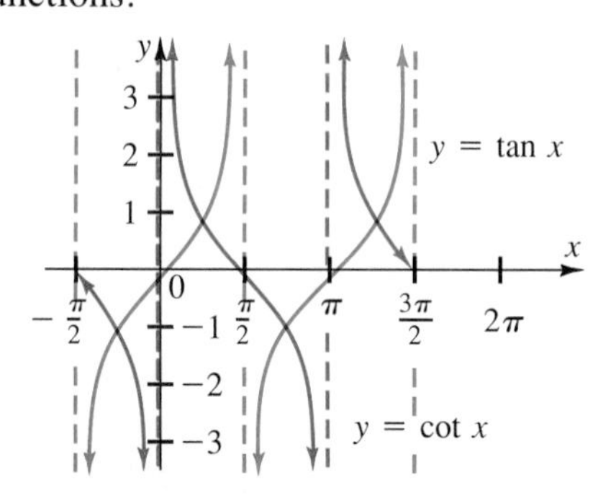

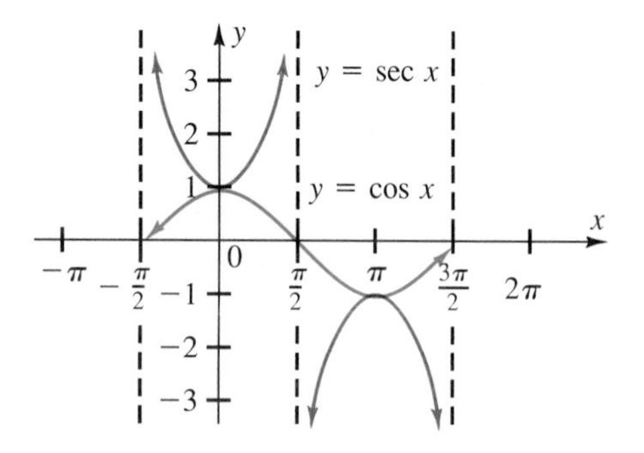

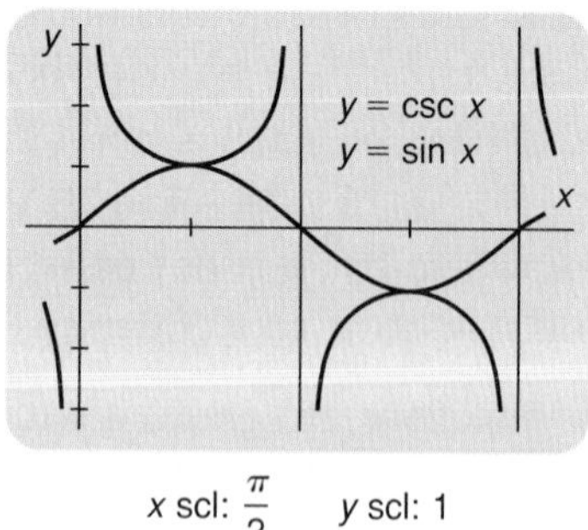

x scl: $\frac{\pi}{2}$ y scl: 1

The six trigonometric functions illustrate the concept of even and odd functions. A function f is an **even function** if $f(-x) = f(x)$ for all x in the domain of f; it is symmetric with respect to the x-axis. A function f is an **odd function** if $f(-x) = -f(x)$ for all x in the domain of f; it is symmetric with respect to the origin.

EXAMPLE 2 **Determine whether $y = \tan x$ is an odd or an even function.**

If x is the measure of an angle with terminal side in quadrant I, then $-x$ is the measure of an angle with terminal side in quadrant IV. Similarly, if x is the measure of a quadrant-IV angle, then $-x$ is the measure of a quadrant-I angle. In either case, since $\tan x$ is negative in quadrant IV and positive in quadrant I, $\tan(-x) = -\tan x$.

If x is the measure of an angle with terminal side in quadrant II, then $-x$ is the measure of an angle with terminal side in quadrant III. Also, if x is the measure of a quadrant-III angle, then $-x$ is the measure of a quadrant-II angle. In either case, $\tan(-x) = -\tan x$.

Since $\tan(-x) = -\tan x$ for all values of x in the domain of the function, $y = \tan x$ is an odd function.

CLASS EXERCISES

Thinking Critically

1. How can the graph of the secant function be shifted to produce the graph of the cosecant function?
2. Will a shift to the left or right of the graph of the tangent function produce the graph of the cotangent function?

Give the period of each function.

3. $y = \tan 3x$
4. $y = \cot 2x$
5. $y = \csc \frac{x}{2}$
6. $y = \frac{3}{2} \sec 2x$

For what values of x, $-2\pi \le x \le 2\pi$, is the function undefined?

7. $y = \sec x$
8. $y = \csc x$
9. $y = \tan 2x$
10. $y = \cot \frac{x}{2}$

PRACTICE EXERCISES

Give the period for each. Then sketch the graph for $-2\pi \le x \le 2\pi$.

1. $y = \tan x$
2. $y = \cot x$
3. $y = \sec x$
4. $y = \csc x$
5. $y = 2 \sec x$
6. $y = 3 \cot x$

7. $y = 2 \csc x$

8. $y = 2 \cot x$

9. $y = -\csc x$

10. $y = -\tan x$

11. $y = \frac{1}{2} \sec x$

12. $y = \frac{1}{2} \cot x$

Determine whether each function is odd or even.

13. $y = \cot x$

14. $y = \sin x$

15. $y = \cos x$

16. $y = \sec x$

Give the period for each. Then sketch the graph for $-2\pi \leq x \leq 2\pi$.

17. $y = \cot 2x$

18. $y = \sec 2x$

19. $y = \csc 3x$

20. $y = \tan 3x$

21. $y = \cot \frac{1}{2}x$

22. $y = \tan \frac{1}{2}x$

23. $y = 2 \csc 2x$

24. $y = -\tan 3x$

25. $y = 3 \cot \frac{1}{2}x$

26. $y = -\sec x$

27. $y = -2 \tan x$

28. $y = 2 \tan 2x$

29. $y = \csc x + 2$

30. $y = \sec x + 1$

31. $y = \sec 2x - 3$

Applications

32. Space Engineering A rocket launched into the atmosphere forms an angle x with the ground. An equation for this angle is $y = -\frac{1}{2} \tan x + \frac{3}{2}$, where y is the horizontal path of the rocket. Sketch the graph of the function for $0 \leq x \leq 2\pi$.

Mixed Review

Solve, rounding to three decimal places.

33. $3^x = 7$

34. $4^{x+1} = 12$

35. $5.3^x = 17$

36. Write a recursive formula for this sequence: 3, 10, 17, 24 . . .

37. Write an explicit formula for this sequence: $-7, -4, -1, 2$ . . .

38. The third term of a geometric sequence is 72 and the sixth term is 243. Find the first term.

39. The sum of the digits of a two-digit number is 11. If the digits are reversed, the new number is 45 more than the old number. Find the new number.

40. Simplify: $\dfrac{2x^3 - 6x^2 - 8x + 24}{4x^3 - 32x^2 + 84x - 72}$

41. Multiply: $(2x^3 + 4x - 3)^2$

42. For what value(s) of x is $\dfrac{2x - 4}{x^2 - 4x + 3}$ undefined?

43. Simplify: $\dfrac{\frac{2}{x} - \frac{4}{y}}{\frac{4}{x} - \frac{3}{y}}$

Solving Right Triangles

Objective: To find the measures of the sides and angles of a right triangle

If you know the lengths of two sides of a right triangle or the length of one side and the measure of an acute angle, then you can *solve the triangle*. That is, you can find the measures of the remaining side(s) and angle(s).

Capsule Review

The sum of the measures of the three angles of a triangle is 180°, so the sum of the measures of the two acute angles of a right triangle is 180° − 90°, or 90°. Therefore, if one angle of a right triangle measures 17°46′, then the measure of the other acute angle is 72°14′, since

$$90° - 17°46' = 89°60' - 17°46' = 72°14'$$

The measure of one acute angle of a right triangle is given. Find the measure of the other acute angle of the triangle.

1. 32° **2.** 83° **3.** 9° **4.** 61°

5. 43°15′ **6.** 25°23′ **7.** 78°37′ **8.** 54°38′

Every angle θ in standard position, except a quadrantal angle, has a right triangle as its reference triangle. In this triangle, the length of the hypotenuse is r, the length of the side opposite θ is $|y|$, and the length of the side adjacent to θ is $|x|$. Therefore, the trigonometric functions of an acute angle can be defined in terms of the lengths of the sides of a right triangle.

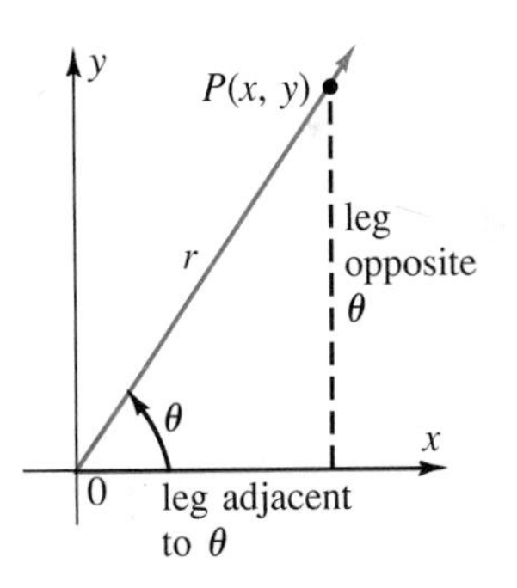

$$\sin\theta = \frac{\text{length of leg opposite } \theta}{\text{length of hypotenuse}} \qquad \cos\theta = \frac{\text{length of leg adjacent to } \theta}{\text{length of hypotenuse}}$$

$$\tan\theta = \frac{\text{length of leg opposite } \theta}{\text{length of leg adjacent to } \theta} \qquad \cot\theta = \frac{\text{length of leg adjacent to } \theta}{\text{length of leg opposite } \theta}$$

$$\csc\theta = \frac{\text{length of hypotenuse}}{\text{length of leg opposite } \theta} \qquad \sec\theta = \frac{\text{length of hypotenuse}}{\text{length of leg adjacent to } \theta}$$

EXAMPLE 1 **Solve right triangle RST if $\angle R = 90°$, $\angle S = 42°$, and $r = 15$. Give the measures of the sides to two significant digits.**

The sum of the measures of the acute angles of a right triangle is 90°, so $\angle T = 90° - 42° = 48°$.

$\sin 42° = \frac{s}{15}$ *$\sin S = \frac{s}{r}$*

$s = 15 \sin 42°$

$s = 10$, to two significant digits

$\cos 42° = \frac{t}{15}$ *$\cos S = \frac{t}{r}$*

$t = 15 \cos 42°$

$t = 11$, to two significant digits

Therefore, $\angle T = 48°$, $s = 10$, and $t = 11$.

When the lengths of the two legs of a right triangle are known, the tangent or the cotangent can be used to find the measure of an acute angle.

EXAMPLE 2 **Solve right triangle BCD if $\angle C = 90°$, $b = 12$, and $d = 16$. Give c to two significant digits and angle measures to the nearest degree.**

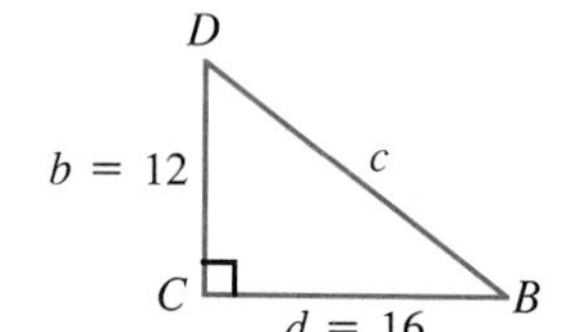

$\tan B = \frac{12}{16}$ *$\tan B = \frac{b}{d}$*

$\tan B = 0.75$

$\angle B = 36.86989765°$ *Find the measure of the acute angle B for which $\tan B = 0.75$.*

$\angle B = 37°$, to the nearest degree

$\angle D = 90° - 37° = 53°$

$\sin 37° = \frac{12}{c}$ *Use the equation $\sin B = \frac{b}{c}$ to find c.*

$c = \frac{12}{\sin 37°} = 20$, to two significant digits

Therefore, $\angle B = 37°$, $\angle D = 53°$, and $c = 20$.

You can solve an isosceles triangle if the measures of the sides or the measures of one side and an angle are known.

EXAMPLE 3 **Solve isosceles triangle STV if $s = 12$, $v = 12$, and $t = 14$. Give angle measures to the nearest degree.**

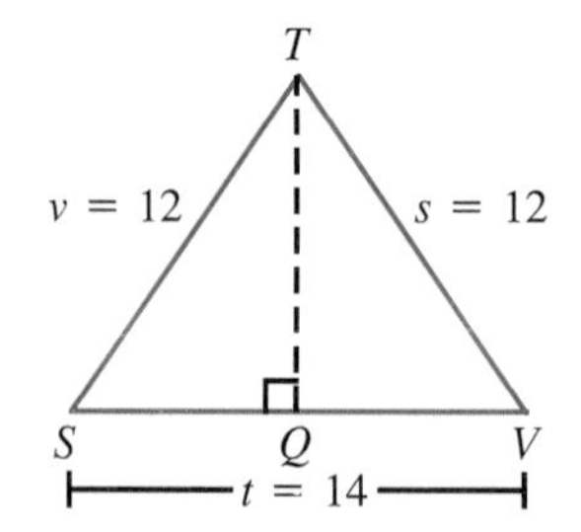

The altitude from T to $\overline{SV}$ divides the triangle into two congruent right triangles.

In right triangle SQT: $SQ = \frac{1}{2}t = \frac{1}{2}(14) = 7$

$\cos S = \dfrac{7}{12}$ $\quad$ *$\cos S = \dfrac{SQ}{ST}$*

$$\angle S = 54°, \text{ to the nearest degree}$$
$$\angle V = 54°, \text{ to the nearest degree}$$
$$\angle STV = 180° - (54° + 54°) = 72°$$

The sum of the angle measures of any triangle is 180°.

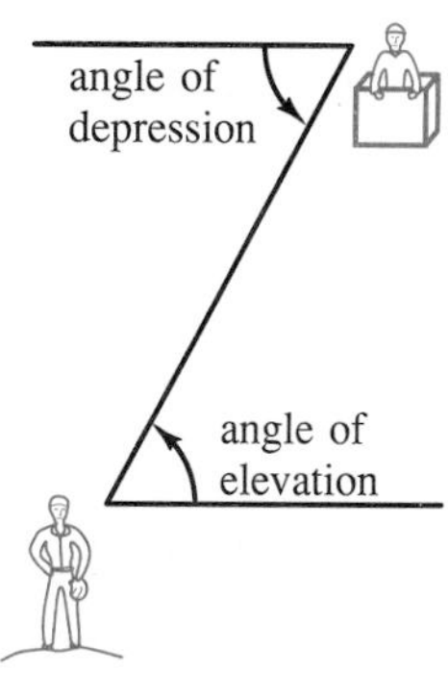

A pitcher standing on the pitcher's mound on a baseball field sights the announcer in the press booth at the top of the stadium. The angle formed by the pitcher's line of sight to the booth and the horizontal is called the **angle of elevation.** The angle formed by the announcer's line of sight to the pitcher's mound and the horizontal is called the **angle of depression.** Since the horizontal lines are parallel, the angles of elevation and depression are equal in measure.

EXAMPLE 4 **Ricky's kite is flying above a field at the end of 65 m of string. If the angle of elevation to the kite measures 70°, how high is the kite above Ricky's waist? Give the answer to two significant digits.**

65m x 70°

The length of the hypotenuse is given and you want to find the length of a leg. Use the sine or the cosecant.

$$\sin 70° = \frac{x}{65}$$

$$x = 65 \sin 70° = 61, \text{ to two significant digits}$$

The kite is 61 m above Ricky's waist.

CLASS EXERCISES

Give an equation that could be used to find the required part of triangle *DEF*.

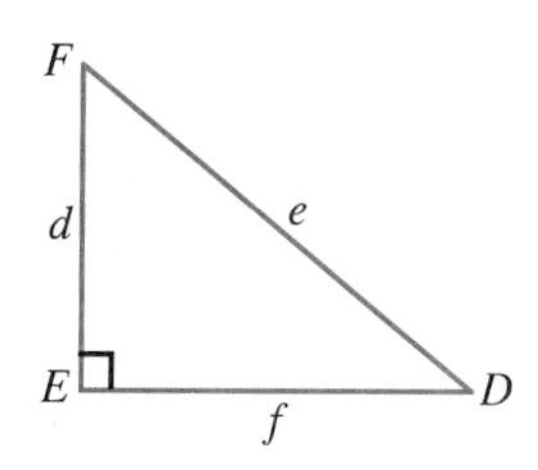

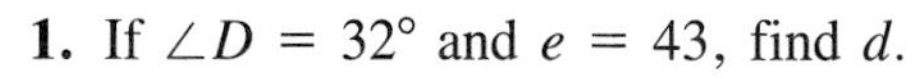

1. If $\angle D = 32°$ and $e = 43$, find d.
2. If $\angle D = 70°$ and $f = 25$, find d.
3. If $\angle F = 60°$ and $d = 20$, find e.
4. If $\angle F = 25°$ and $e = 16$, find f.
5. If $e = 12$ and $f = 8$, find $\angle F$.
6. If $f = 15$ and $d = 9$, find $\angle D$.
7. If $\angle F = 43°$ and $e = 25$, find d.
8. If $\angle D = 68°$ and $f = 13$, find e.
9. If $\angle F = 32°$ and $d = 20$, find e.
10. If $d = 14$ and $f = 18$, find $\angle F$.

PRACTICE EXERCISES

 Use technology where appropriate.

For Exercises 1–14, give angle measures to the nearest degree and lengths to two significant digits.

Solve right triangle RST, using the given information.

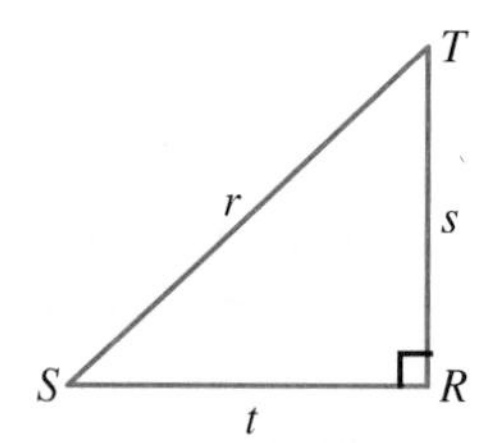

1. $\angle S = 32°$, $r = 43$
2. $\angle S = 70°$, $t = 25$
3. $\angle T = 60°$, $s = 20$
4. $\angle T = 25°$, $r = 16$
5. $r = 12$, $t = 8$
6. $t = 15$, $s = 9$
7. $s = 14$, $t = 18$
8. $s = 12$, $t = 20$

Solve isosceles triangle PQR, using the given information.

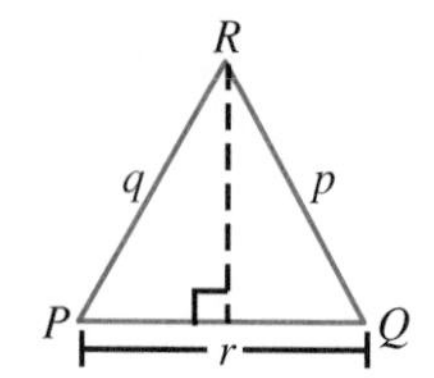

9. $p = 15$, $r = 18$
10. $q = 16$, $r = 14$
11. $q = 18$, $\angle P = 40°$
12. $p = 14$, $\angle Q = 80°$

13. From a point on the ground 12 ft from the base of a flagpole, the angle of elevation of the top of the pole measures 53°. How tall is the flagpole?

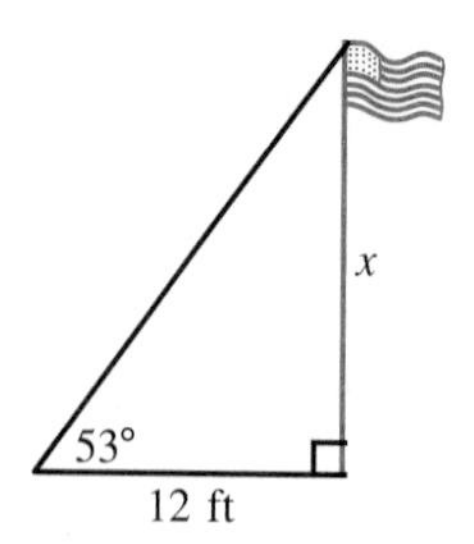

14. From an airplane at an altitude of 1200 m, the angle of depression of a rock on the ground measures 28°. Find the distance from the plane to the rock.

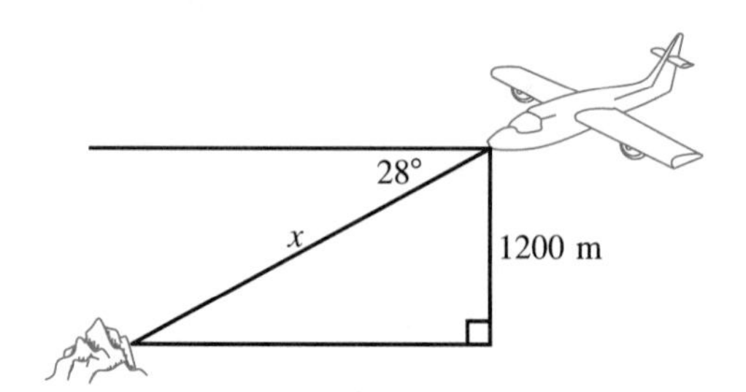

For Exercises 15–28, give angle measures to the nearest ten minutes and lengths to three significant digits.

Solve right triangle ABC, using the given information.

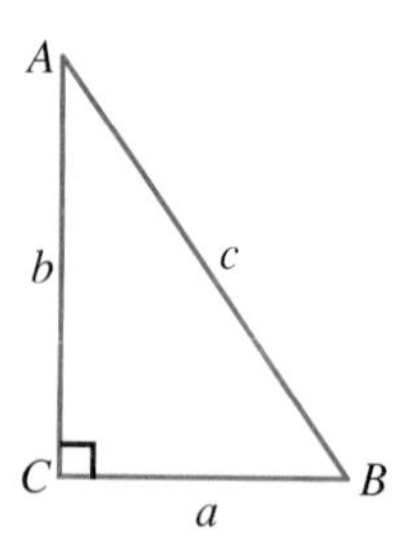

15. $\angle A = 43°20'$, $c = 25.1$
16. $\angle B = 68°40'$, $a = 13.3$
17. $\angle A = 32°50'$, $b = 19.9$
18. $\angle B = 52°10'$, $c = 54.8$
19. $b = 30.5$, $\angle A = 30°30'$
20. $a = 22.2$, $\angle A = 22°20'$
21. $\angle B = 18°50'$; $c = 42.5$
22. $\angle A = 62°40'$, $c = 12.6$

23. If a 150-ft pole casts a shadow 210 ft long, find the measure of the angle of elevation of the sun.

24. From the top of a utility pole 250 m tall, a rock is sighted on the ground below. If the rock is 170 m from the base of the pole, find the angle of depression of the rock from the top of the pole.

Solve right triangle PQR, using the given information.

25. $\angle P = 40°10'$, $\angle QSR = 70°30'$, $RS = 14.2$

26. $\angle P = 23°10'$, $\angle QSR = 46°20'$, $RS = 12.4$

27. From an observation post 330 ft from the base of a building, the angles of elevation of the top and bottom of a flagpole situated on top of the building are 63°20′ and 53°40′. Find the height of the flagpole.

28. From a plane flying due east at 265 m above sea level, the angles of depression of two ships sailing due east measure 35°20′ and 25°30′. How far apart are the ships?

Applications

29. **Forestry** From a ranger's station 1.3 mi above sea level, a forest fire is sighted 0.2 mi above sea level. The angle of depression to the fire is 18°15′. To the nearest tenth of a mile, how far is the fire from the ranger's station?

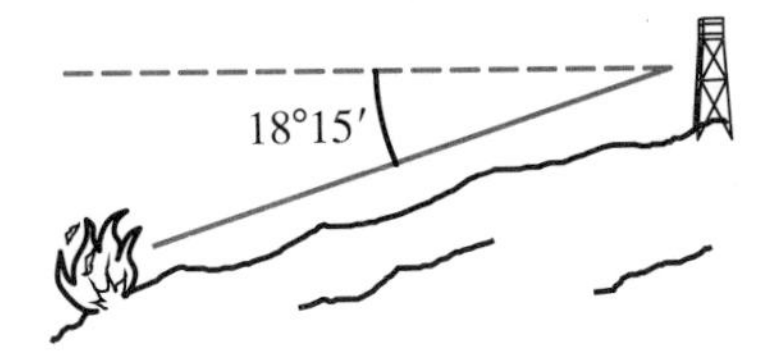

30. **Rescue** A lifeboat is sighted from a rescue helicopter directly overhead at a range of 75 m. After rescuing two survivors, the helicopter returns to the same coordinates and sights the lifeboat at an angle of depression of 26°25′. How far has the lifeboat drifted?

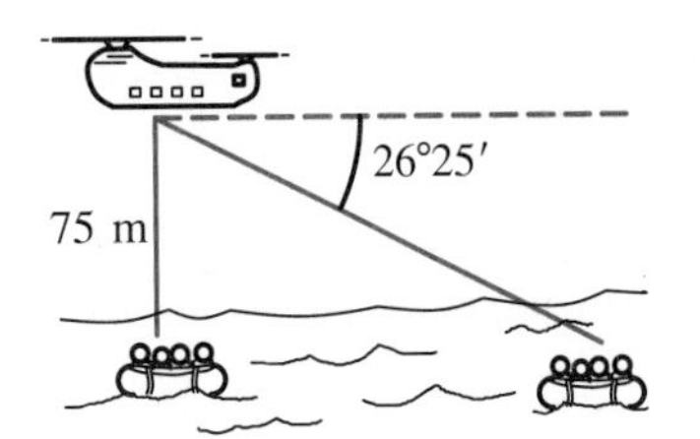

Developing Mathematical Power

31. **Thinking Critically** The edges of a cube have the same length. In triangle ABC, shown in red, $\angle ABC$ is a right angle. Find the measures of angles ACB and BAC, to the nearest degree.

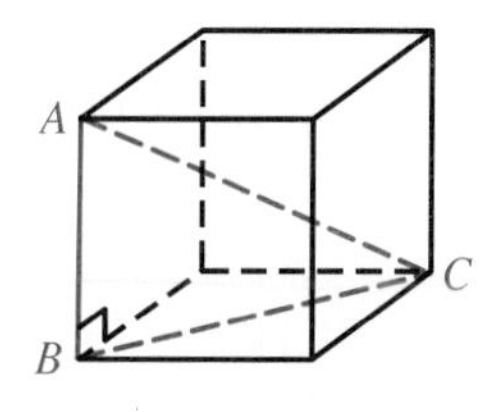

15.8

Law of Sines

Objectives: To find the area of a triangle, given the measure of two sides and the included angle
To use the law of sines to solve a triangle

The trigonometric ratios can be used to solve triangles that are not right triangles. One method involves the law of sines, which is derived by using expressions for the area of a triangle.

Capsule Review

Any triangle has three altitudes, and the area of a triangle is equal to one-half the product of the length of any one of these altitudes and the length of the base to which that altitude is drawn.

Use each of the given altitudes to write an expression for area *K* of triangle *ABC*.

1. $\overline{AD}$
2. $\overline{BE}$
3. $\overline{CF}$

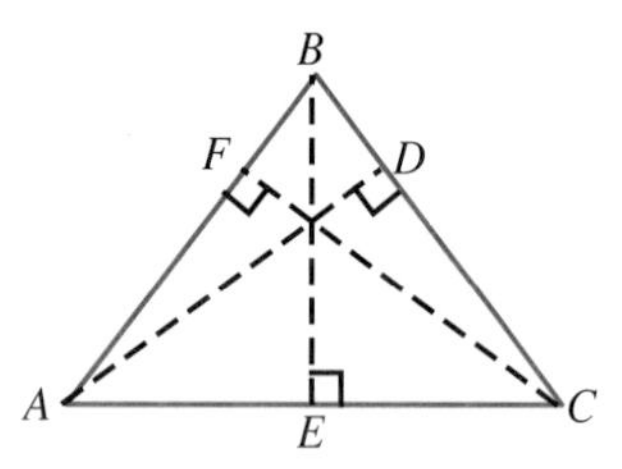

Use each of the given altitudes to write an expression for area *K* of triangle *PQR*.

4. $\overline{PX}$
5. $\overline{QY}$
6. $\overline{RZ}$

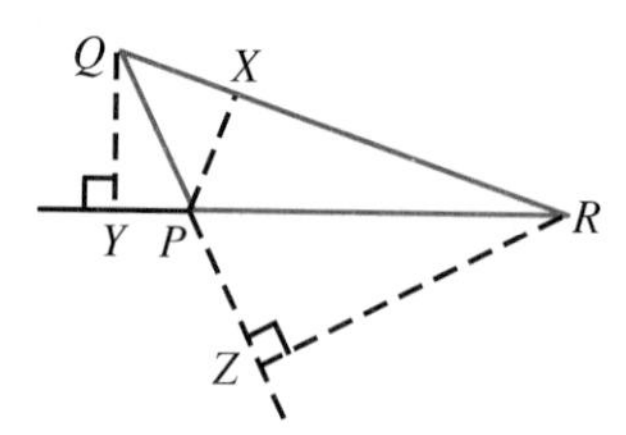

In triangle ABC, an altitude of measure h is drawn to base $\overline{AC}$, with measure b. If the area of the triangle is K, then

$$K = \frac{1}{2}bh$$

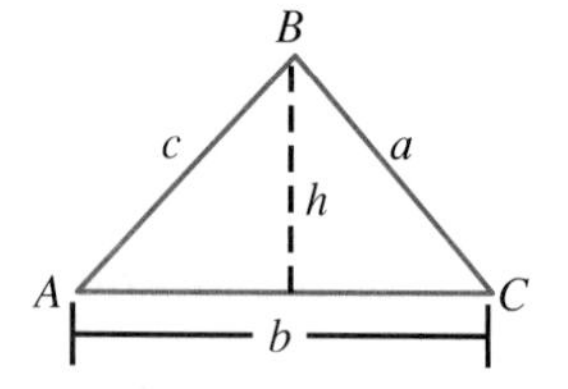

But $\sin A = \frac{h}{c}$, or $h = c \sin A$.

Therefore, $K = \frac{1}{2}bc \sin A$ *Substitute c sin A for h.*

Using the same method with each of the other two expressions for the area, you can write two other equations involving trigonometric ratios. The results would be the same if an angle of the triangle is obtuse.

The area K of any triangle ABC is given by any one of these formulas:

$$K = \frac{1}{2}bc \sin A \qquad K = \frac{1}{2}ac \sin B \qquad K = \frac{1}{2}ab \sin C$$

EXAMPLE 1 **Find the area of triangle ABC.**

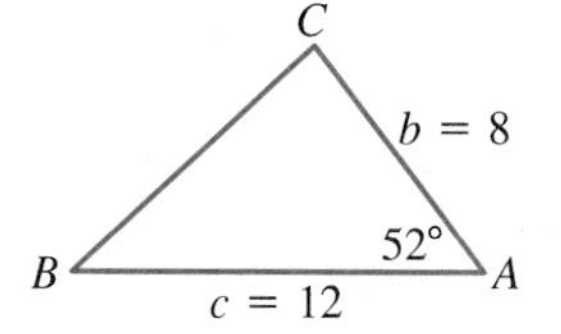

$K = \frac{1}{2}(8)(12) \sin 52°$ *$K = \frac{1}{2}bc \sin A$*

$K = 38$, to two significant digits

For a specific triangle ABC, $\frac{1}{2}bc \sin A$, $\frac{1}{2}ac \sin B$, and $\frac{1}{2}ab \sin C$ are all equal to the area K so $\frac{1}{2}bc \sin A = \frac{1}{2}ac \sin B = \frac{1}{2}ab \sin C$. Each part of the equation can be divided by $\frac{1}{2}abc$ to get $\frac{\sin A}{a} = \frac{\sin B}{b} = \frac{\sin C}{c}$, which is called the *law of sines*.

Law of Sines

For any triangle ABC, where a, b, and c are the lengths of the sides opposite the angles with measures A, B, and C, respectively,

$$\frac{\sin A}{a} = \frac{\sin B}{b} = \frac{\sin C}{c}$$

EXAMPLE 2 **Solve triangle ABC.**

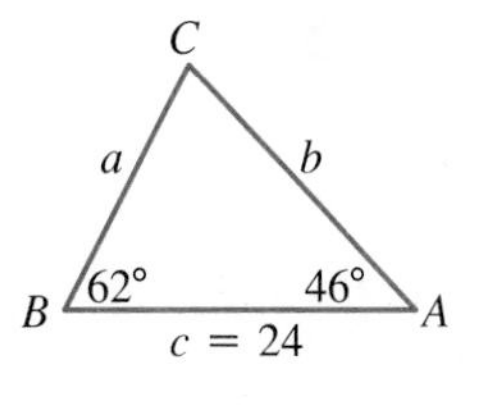

$\angle C = 180° - (62° + 46°) = 72°$

$\frac{\sin 46°}{a} = \frac{\sin 72°}{24}$ *Use the law of sines to find a.*

$a = \frac{24 \sin 46°}{\sin 72°} = 18$, to two significant digits

$\frac{\sin 62°}{b} = \frac{\sin 72°}{24}$ *Use the law of sines to find b.*

$b = \frac{24 \sin 62°}{\sin 72°} = 22$, to two significant digits

Therefore, $\angle C = 72°$, $a = 18$, and $b = 22$.

When the measures of two sides and the angle opposite one of them are given, you may be able to use the law of sines to find the measure of another angle of the triangle. Two angle measures, one, or none may satisfy the given value of the sine ratio. This is called the *ambiguous case,* since it is not always possible to solve a triangle under these conditions.

EXAMPLE 3 **Solve the given triangle, if possible.**

a. Triangle RST, where $s = 10$, $t = 15$, and $\angle T = 66°$

b. Triangle DEF, where $d = 24$, $e = 36$, and $\angle D = 25°$

c. Triangle ABC, where $a = 17$, $b = 21$, and $\angle A = 64°$

a. $s = 10$, $t = 15$, and $\angle T = 66°$, so use the law of sines to find $\angle S$.

$$\frac{\sin 66°}{15} = \frac{\sin S}{10} \qquad \frac{\sin T}{t} = \frac{\sin S}{s}$$

$$\sin S = \frac{10 \sin 66°}{15}$$

$$\sin S = 0.609030305$$

$$\angle S = 38° \text{ or } \angle S = 142°$$

An angle of a triangle may have any measure between 0° and 180°, so two measures seem to be possible for $\angle S$. However, the measure of 142° is not possible in this particular triangle, since $66° + 142° = 208°$, and 208° is greater than 180°. Therefore, $\angle S = 38°$, and

$$\angle R = 180° - (66° + 38°) = 76°$$

$$\frac{\sin 66°}{15} = \frac{\sin 76°}{r} \qquad \text{Find } r.$$

$$r = \frac{15 \sin 76°}{\sin 66°}$$

$$r = 16, \text{ to two significant digits}$$

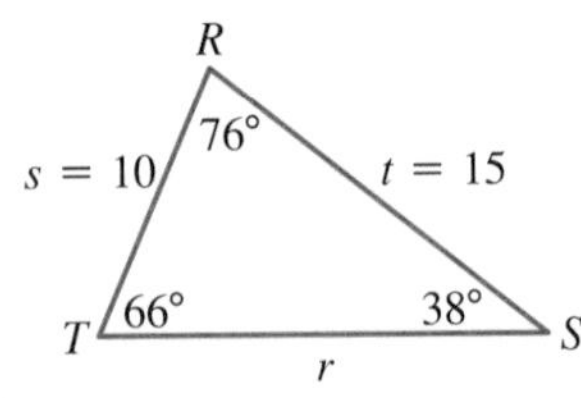

Therefore, $\angle S = 38°$, $\angle R = 76°$, and $r = 16$.

b. $d = 24$, $e = 36$, and $\angle D = 25°$, so use the law of sines to find $\angle E$.

$$\frac{\sin 25°}{24} = \frac{\sin E}{36} \qquad \frac{\sin D}{d} = \frac{\sin E}{e}$$

$$\sin E = \frac{36 \sin 25°}{24}$$

$$\sin E = 0.633927392$$

$$\angle E = 39° \text{ or } \angle E = 141°$$

If $\angle E = 39°$, then $\angle F = 180° - (25° + 39°) = 116°$.
If $\angle E = 141°$, then $\angle F = 180° - (25° + 141°) = 14°$.

Therefore, both measures are possible for $\angle E$, and there are two triangles.

For $\angle E = 39°$ and $\angle F = 116°$

$$\frac{\sin 25°}{24} = \frac{\sin 116°}{f}$$

$$f = \frac{24 \sin 116°}{\sin 25°} = 51$$

For $\angle E = 141°$ and $\angle F = 14°$

$$\frac{\sin 25°}{24} = \frac{\sin 14°}{f}$$

$$f = \frac{24 \sin 14°}{\sin 25°} = 14$$

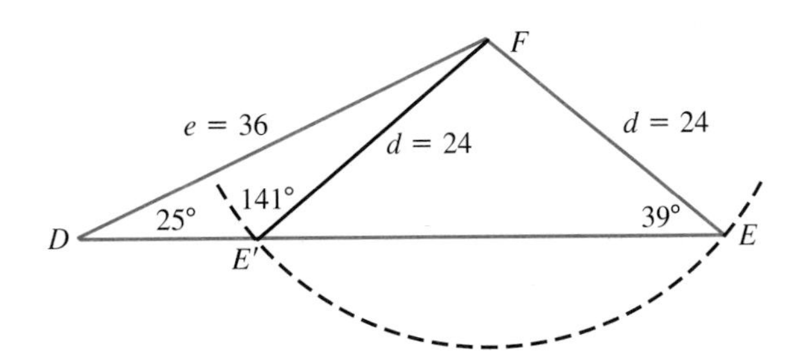

Triangles DEF and $DE'F$ are both possible, with the given conditions. Thus $\angle E = 39°$, $\angle F = 116°$, and $f = 51$, or $\angle E' = 141°$, $\angle F = 14°$, and $f = 14$.

c. $a = 17$, $b = 21$, and $\angle A = 64°$, so use the law of sines to find $\angle B$.

$$\frac{\sin 64°}{17} = \frac{\sin B}{21}$$

$$\sin B = \frac{21 \sin 64°}{17} = 1.110274998$$

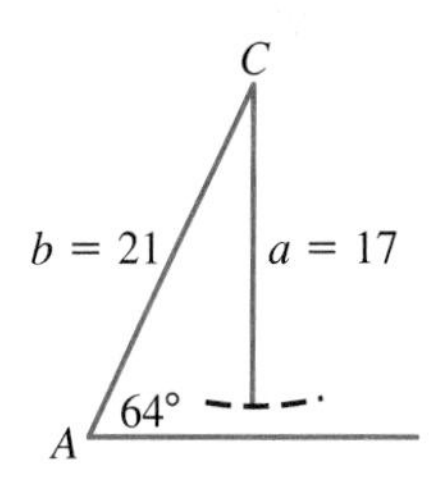

This is not possible, since the sine of an angle can be no greater than 1. Therefore, there is no triangle.

CLASS EXERCISES

Give the equation you would use to find the area of triangle DEF, using the given information.

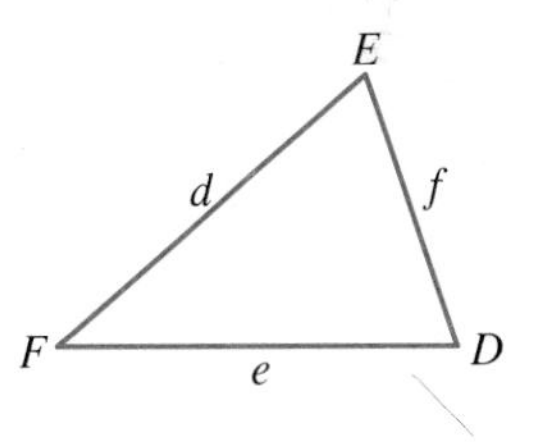

1. $\angle F = 54°$, $d = 14$, $e = 20$

2. $\angle D = 32°$, $f = 10$, $e = 17$

Give the equation you would use to find the indicated part of triangle DEF, using the given information.

3. $\angle F = 43°$, $d = 16$, and $f = 24$. Find $\angle D$.

4. $\angle D = 52°$, $f = 10$, and $d = 15$. Find $\angle F$.

5. $\angle F = 64°$, $d = 18$, and $\angle E = 36°$. Find e.

6. $\angle E = 58°$, $e = 20$, and $\angle D = 42°$. Find d.

PRACTICE EXERCISES

Use technology where appropriate.

For Exercises 1–12, give angle measures to the nearest degree, and give areas and lengths to two significant digits.

Find the area of triangle ABC, using the given information.

1. $\angle A = 48°$, $b = 20$, and $c = 16$
2. $\angle B = 52°$, $a = 15$, and $c = 10$
3. $\angle B = 56°$, $a = 12$, and $c = 9$
4. $\angle C = 75°$, $a = 14$, and $b = 16$

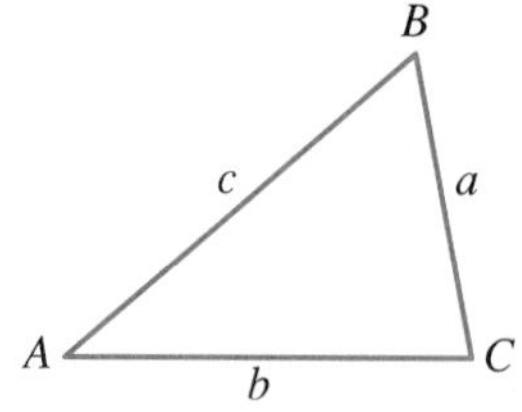

Solve each triangle.

5. In triangle DEF, $\angle D = 54°$, $\angle E = 54°$, and $d = 20$.
6. In triangle ABC, $\angle A = 62°$, $\angle B = 43°$, and $b = 15$.
7. In triangle DEF, $\angle F = 32°$, $\angle E = 68°$, and $f = 12$.
8. In triangle ABC, $\angle A = 58°$, $\angle B = 46°$, and $a = 14$.
9. In triangle DEF, $\angle D = 54°$, $e = 8$, and $d = 10$.
10. In triangle ABC, $\angle A = 62°$, $a = 15$, and $b = 12$.
11. In triangle DEF, $\angle F = 32°$, $d = 16$, and $f = 20$.
12. In triangle ABC, $\angle A = 40°$, $a = 20$, and $b = 16$.

For Exercises 13–20, give angle measures to the nearest ten minutes or to the nearest tenth of one degree, and give areas and lengths to three significant digits.

Find the area of triangle ABC, using the given information.

13. $\angle C = 68.1°$, $b = 12.9$, and $c = 15.2$
14. $\angle A = 52.8°$, $a = 9.71$, and $c = 9.33$

Solve each triangle, if possible.

15. In triangle DEF, $\angle F = 16°30'$, $\angle D = 24°50'$, and $e = 25.0$.
16. In triangle ABC, $\angle A = 72°20'$, $a = 22.9$, and $\angle C = 24°40'$.
17. In triangle DEF, $\angle D = 54.1°$, $e = 11.9$, and $d = 10.2$.
18. In triangle ABC, $\angle A = 39.9°$, $a = 15.9$, and $b = 20.2$.

19. In triangle DEF, $\angle F = 58.1°$, $f = 10.4$, and $d = 15.6$.

20. In triangle ABC, $\angle A = 50.2°$, $a = 2.06$, and $b = 10.2$.

For given measures of $\angle A$, a, and c, either one triangle ABC can be drawn, two can be drawn, or no triangle can be drawn. Explain why each of the following statements is true.

21. Just one triangle can be drawn if $0° < \angle A < 90°$ and $a = c \sin A$, or if $0° < \angle A < 90°$ and $a \geq c$.

22. No triangle can be drawn if $0° < \angle A < 90°$ and $a < c \sin A$.

23. Two different triangles can be drawn if $0° < \angle A < 90°$, $a > c \sin A$, and $a < c$.

24. If $90° \leq \angle A < 180°$, then only one triangle can be drawn if $a > c$, and no triangle can be drawn if $a \leq c$.

Applications

25. **Geometry** Find the perimeter of an isosceles triangle if one of the congruent sides is 10 cm long and one of the congruent angles is 54°. Give the answer to two significant digits.

26. **Geometry** The sides of a triangle are 15 cm, 17 cm, and 16 cm long. If the smallest angle has a measure of 53°, find the measure of the largest angle to the nearest degree.

27. **Surveying** A vacant lot shaped like an isosceles triangle is between two streets that intersect at an angle that measures 85.9°. If each of the sides of the lot that face these streets is 150 ft long, find the length of the third side to three significant digits.

Developing Mathematical Power

28. **Thinking Critically** Buoys are located in the sea at points A, B, and C, such that $\angle ACB$ is a right angle and $AC = 3.0$ mi, $BC = 4.0$ mi, and $AB = 5.0$ mi. A ship is located at point D on $\overline{AB}$ in such a position that $\angle ACD = 30°$. How far is the ship from the buoy at point C? Give your answer to two significant digits.

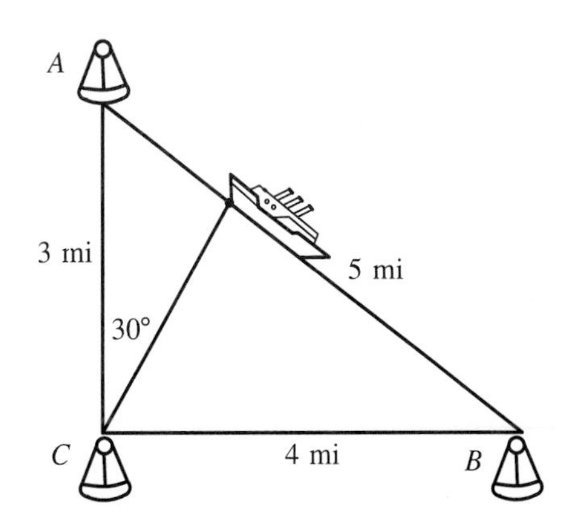

15.9

Law of Cosines

Objective: To use the law of cosines to solve a triangle

The law of sines cannot be used to solve every triangle. If the measures of two sides and the included angle or the measures of all three sides are given, another formula called the *law of cosines* is used. The derivation of this law is based on the Pythagorean theorem.

Capsule Review

The lengths of the sides of a right triangle are related by the Pythagorean theorem

$$c^2 = a^2 + b^2$$

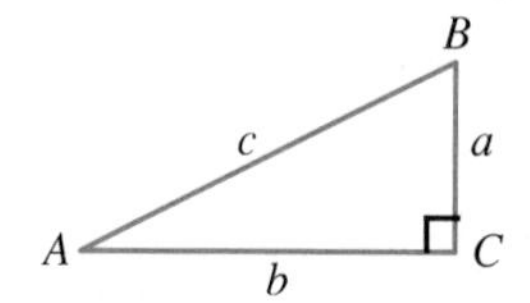

a and b are the lengths of the legs of a right triangle, and c is the length of the hypotenuse. Find the length of the indicated side. Round to the nearest whole number, when necessary.

1. If $a = 6$ and $b = 8$, find c.

2. If $a = 10$ and $b = 24$, find c.

3. If $c = 15$ and $a = 12$, find b.

4. If $c = 20$ and $a = 12$, find b.

5. If $a = 5$ and $b = 5$, find c.

6. If $c = 8$ and $b = 4$, find a.

Altitude $\overline{AD}$ of length h separates triangle ABC into two right triangles with the common side $\overline{AD}$. Using the Pythagorean theorem and the cosine ratio, you can derive a relationship among a, b, c and the measure of $\angle C$. Note that x is the measure of $\overline{DC}$, so $a - x$ is the measure of $\overline{BD}$.

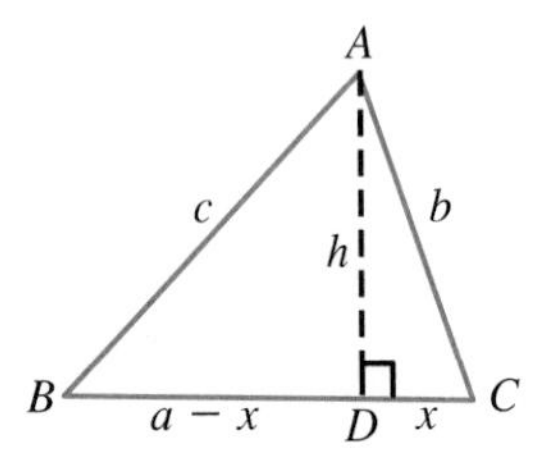

$c^2 = (a - x)^2 + h^2$ *Apply the Pythagorean theorem to triangle ABD.*

$= a^2 - 2ax + x^2 + h^2$ *For triangle ADC,* $b^2 = x^2 + h^2$.

$= a^2 - 2ax + b^2$ *So, substitute* b^2 *for* $x^2 + h^2$.

$= a^2 - 2a(b \cos C) + b^2$ $\cos C = \frac{x}{b}$, *so* $x = b \cos C$.

$= a^2 + b^2 - 2ab \cos C$

In a similar manner, you can derive expressions for a^2 and for b^2.

Law of Cosines

For any triangle ABC, where a, b, and c are the lengths of the sides opposite the angles with measures A, B, and C, respectively,

$$a^2 = b^2 + c^2 - 2bc \cos A$$
$$b^2 = a^2 + c^2 - 2ac \cos B$$
$$c^2 = a^2 + b^2 - 2ab \cos C$$

EXAMPLE 1 **In triangle ABC, $a = 15$, $b = 12$, and $\angle C = 56°$. Find c.**

$$c^2 = a^2 + b^2 - 2ab \cos C$$
$$c^2 = 15^2 + 12^2 - 2(15)(12) \cos 56°$$
$$c = 13, \text{ to two significant digits}$$

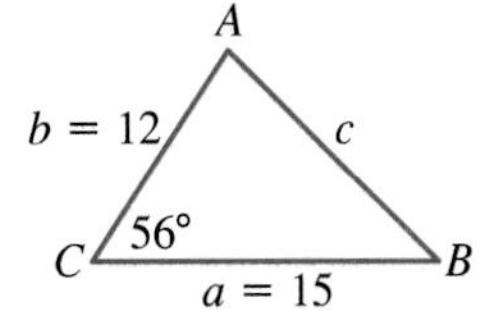

When the lengths of the three sides of a triangle are given, use the law of cosines to find the measure of one angle. Then solve the triangle by using the law of sines or the law of cosines to find the measure of a second angle.

EXAMPLE 2 **Solve triangle DEF, where $d = 10$, $e = 16$, and $f = 14$.**

Use the law of cosines to find the measure of one angle.

$$d^2 = e^2 + f^2 - 2ef \cos D$$
$$\cos D = \frac{e^2 + f^2 - d^2}{2ef}$$
$$\cos D = \frac{16^2 + 14^2 - 10^2}{2(16)(14)}$$
$$\angle D = 38°, \text{ to the nearest degree}$$

Now, use the law of cosines or the law of sines to find the measure of $\angle E$ or $\angle F$. The law of cosines is used below to find the measure of $\angle E$.

$$\cos E = \frac{10^2 + 14^2 - 16^2}{2(10)(14)} \qquad \cos E = \frac{d^2 + f^2 - e^2}{2df}$$
$$\angle E = 82°, \text{ to the nearest degree}$$
$$\angle F = 180° - (38° + 82°) = 60°$$

Therefore, $\angle D = 38°$, $\angle E = 82°$, and $\angle F = 60°$.

If the measures of two sides and the included angle are given, use the law of cosines to find the length of the third side. Use the law of cosines or the law of sines to find the measure of a second angle.

EXAMPLE 3 **Solve triangle ABC, when $b = 6.2$, $c = 7.8$, and $\angle A = 45°$.**

Use the law of cosines to find a.

$$a^2 = 6.2^2 + 7.8^2 - 2(6.2)(7.8)\cos 45°$$
$$a = 5.6, \text{ to two significant digits}$$

Then find the measure of $\angle B$ or $\angle C$, using either the law of cosines or the law of sines. The law of sines is used below to find the measure of $\angle B$.

$$\frac{\sin 45°}{5.6} = \frac{\sin B}{6.2} \qquad \frac{\sin A}{a} = \frac{\sin B}{b}$$

$$\sin B = \frac{6.2 \sin 45°}{5.6}$$

$$\angle B = 52°$$

$$\angle C = 180° - (45° + 52°) = 83°$$

Therefore, $a = 5.6$, $\angle B = 52°$, and $\angle C = 83°$.

CLASS EXERCISES

Write the equation you would use to find the indicated part of triangle *DEF*, using the given information.

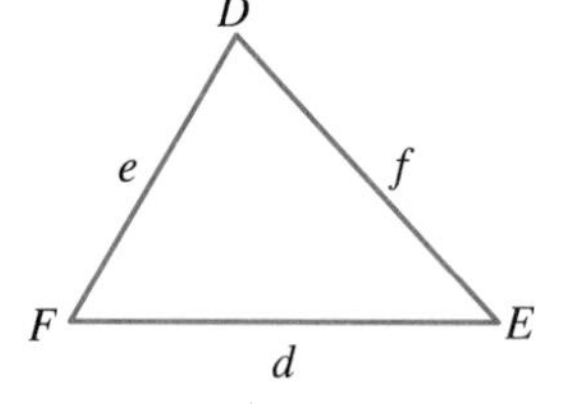

1. $\angle E = 54°$, $d = 14$, and $f = 20$. Find e.
2. $\angle F = 32°$, $d = 10$, and $e = 17$. Find f.
3. $\angle F = 43°$, $d = 16$, and $e = 24$. Find f.
4. $e = 18$, $f = 10$, and $d = 15$. Find $\angle F$.

PRACTICE EXERCISES

Use technology where appropriate.

For Exercises 1–10, give angle measures to the nearest degree, and give lengths to two significant digits.

Use the given information to find the indicated part of triangle *ABC*.

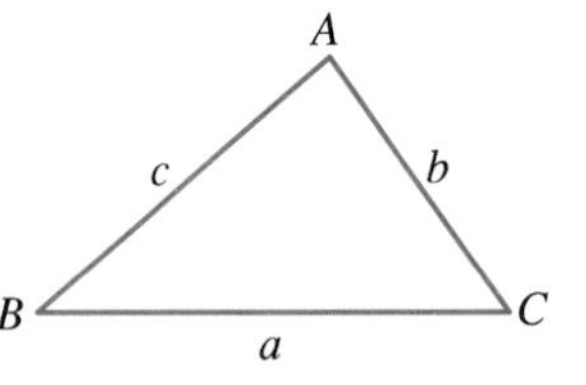

1. $\angle A = 48°$, $b = 20$, and $c = 16$. Find a.
2. $\angle B = 52°$, $a = 15$, and $c = 10$. Find b.
3. $a = 20$, $b = 14$, and $c = 16$. Find $\angle A$.
4. $a = 12$, $b = 10$, and $c = 9$. Find $\angle C$.

Solve each triangle.

5. In triangle DEF, $\angle D = 54°$, $e = 18$, and $f = 20$.
6. In triangle ABC, $\angle A = 62°$, $c = 12$, and $b = 15$.

7. In triangle DEF, $\angle F = 32°$, $d = 16$, and $e = 12$.

8. In triangle ABC, $\angle A = 58°$, $b = 14$, and $c = 16$.

9. In triangle DEF, $f = 24$, $d = 16$, and $e = 20$.

10. In triangle ABC, $a = 10$, $b = 14$, and $c = 12$.

Solve each triangle. Give angle measures to the nearest ten minutes or to the nearest tenth of one degree, and give lengths to three significant digits.

11. In triangle DEF, $\angle E = 64°40'$, $d = 15.1$, and $f = 17.2$.

12. In triangle ABC, $\angle C = 72°20'$, $a = 22.7$, and $b = 17.8$.

13. In triangle ABC, $\angle A = 119.8°$, $c = 8.05$, and $b = 8.05$.

14. In triangle DEF, $\angle E = 104.8°$, $f = 16.1$, and $d = 16.1$.

15. In triangle ABC, $a = 26.3$, $b = 28.2$, and $c = 14.1$.

16. In triangle ABC, $a = 36.4$, $b = 16.8$, and $c = 24.6$.

17. In triangle DEF, $\angle E = 48°20'$, $d = 24.8$, and $f = 24.8$.

18. In triangle DEF, $\angle D = 68°20'$, $e = 42.6$, and $f = 42.6$.

19. Find x in triangle ABC.

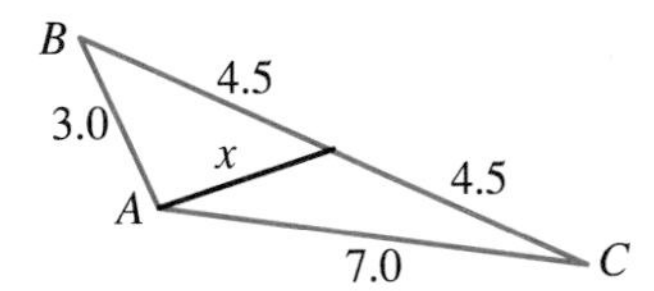

20. For any parallelogram, prove that the sum of the squares of the lengths of the diagonals equals twice the sum of the squares of the lengths of two adjacent sides.

Applications

21. Geometry Two adjacent sides of a parallelogram are 21 cm and 14 cm long. Find the length of the shorter diagonal, to two significant digits, if the smaller angle formed by the sides measures 58°.

22. Navigation A ship in the gulf is 16 mi from one lighthouse and 28 mi from another. Find the distance between the lighthouses, to two significant digits, if the measure of the angle between them is 120°.

Developing Mathematical Power

23. Extension In a regular polygon, each side is the same length. Each vertex of regular pentagon $ABCDE$ is on circle O, which has a radius of 17 cm. Use the law of cosines to find the perimeter of the pentagon.

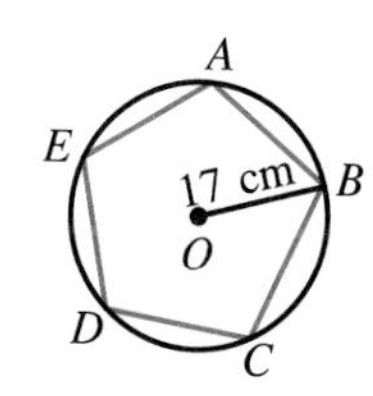

15.10 Problem Solving Strategy: Use Trigonometry

Many practical problems involving distances or measures of angles can be solved by using trigonometric functions. One of the first problems solved by using trigonometry was to compare the distance between Earth and the moon with the distance between Earth and the sun. To solve problems involving trigonometry it is often necessary to draw diagrams that illustrate the conditions of the problem.

EXAMPLE 1 Two guy wires bracing a television antenna each make an angle measuring $62°30'$ with the ground. If the wires are anchored to the ground and are 80.0 ft apart, how long is each wire?

Understand the Problem The wires and the ground between them form a triangle. You are asked to find the lengths of two sides of the triangle.

Plan Your Approach Draw and label a diagram. The sides of a triangle opposite angles of equal measure are equal in length, so let x represent the length of each wire. The measure of angle θ is $180° - 2(62°30') = 55°$. Since you know the measures of the angles and the side opposite one of them, use the law of sines.

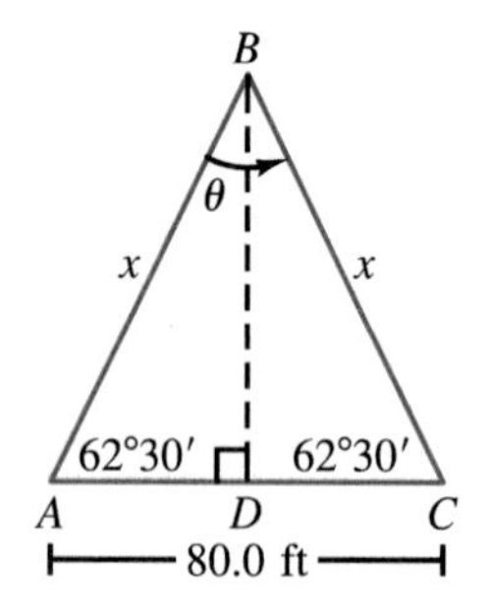

Complete the Work

$$\frac{\sin 55°}{80.0} = \frac{\sin 62°30'}{x} \qquad \frac{\sin \theta}{AC} = \frac{\sin 62°30'}{x}$$

$$x = \frac{80.0 \sin 62°30'}{\sin 55°}$$

$x = 86.6$, to three significant digits

Interpret the Results Each wire is 86.6 ft long.

One way to check this answer is to find the ratio $\frac{AD}{x}$, which should be equal to the cosine of $62°30'$.

$$\frac{AD}{x} = \frac{40.0}{86.6} = 0.462 = \cos 62°30'$$

$$0.462 = 0.462 \checkmark$$

There is often more than one way to solve a problem. To illustrate, the problem in Example 1 could have been solved using the equation $\frac{40.0}{x} = \cos 62°30'$ or the equation $\frac{40.0}{x} = \sin 27°30'$.

EXAMPLE 2 From a point on top of a 48-ft cliff a camper sights two bears in a field, one directly beyond the other. If the angle of depression of one bear measures 38° and that of the other bear measures 27°, how far apart are the bears?

Understand the Problem Triangles are formed by a vertical line, the ground, and the lines of sight. You are to find the distance between the bears, which can be represented by the length of one line segment.

Plan Your Approach Draw and label a diagram. If D is the point directly under the camper, let x represent the distance from D to the bear at B, and let y represent the distance from D to the bear at C. Then the distance between the bears is $x - y$.

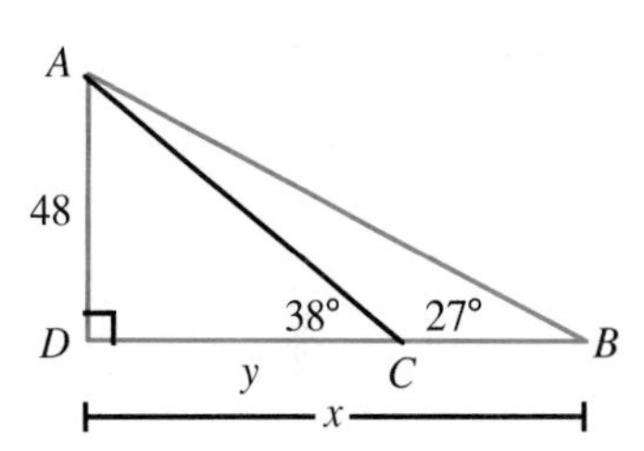

Complete the Work In right triangle ADB

$$\tan 27° = \frac{48}{x}$$

$$x = \frac{48}{\tan 27°} = 94$$

In right triangle ADC

$$\tan 38° = \frac{48}{y}$$

$$y = \frac{48}{\tan 38°} = 61$$

$$x - y = 94 - 61 = 33$$

Interpret the Results The bears are 33 ft apart.

One way to check is to find the length of the hypotenuse of each right triangle and use the Pythagorean theorem.

In triangle ADB, $AB = \frac{48}{\sin 27°} \approx 106$. $\quad 48^2 + 94^2 \overset{?}{\approx} 106^2$

$$11{,}140 \approx 11{,}236 \checkmark$$

In triangle ADC, $AC = \frac{48}{\sin 38°} \approx 78$. $\quad 48^2 + 61^2 \overset{?}{\approx} 78^2$

$$6025 \approx 6084 \checkmark$$

CLASS EXERCISES

Draw and label a diagram for each problem. Then write an equation that can be used to solve the problem, but do not solve it.

1. A ladder 20 ft long leans against the wall of a building. The foot of the ladder rests 10 ft from the base of the wall. What is the measure of the angle that the ladder makes with the ground?
2. The base of an isosceles triangle is 25 cm long and the vertex angle measures 40°. Find the length of each leg.
3. The lengths of the sides of a triangle are 7.6 cm, 8.2 cm, and 5.2 cm. Find the measure of the largest angle.
4. The lengths of the adjacent sides of a parallelogram are 54 cm and 78 cm. Find the length of the longest diagonal if the larger angle of the parallelogram measures 110°.
5. A building is 100 ft tall. Find the measure of the angle of elevation from a point 80 ft from the base of the building to the top of the building.
6. A cliff rises 55 m above the bay. The angle of depression from a point on top of the cliff to a boat in the bay is 35°. How far from the base of the cliff is the ship?

PRACTICE EXERCISES 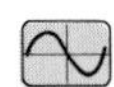 Use technology where appropriate.

For Exercises 1–16, give angle measures to the nearest degree, and give lengths and areas to two significant digits.

Use the diagrams and equations from the Class Exercises for Exercises 1–6.

1. Solve Class Exercise 1.
2. Solve Class Exercise 2.
3. Solve Class Exercise 3.
4. Solve Class Exercise 4.
5. Solve Class Exercise 5.
6. Solve Class Exercise 6.
7. Find the area of this regular octagon with 36-mm sides.

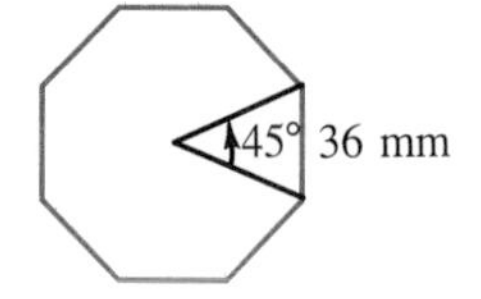

8. In the figure below, find ST.

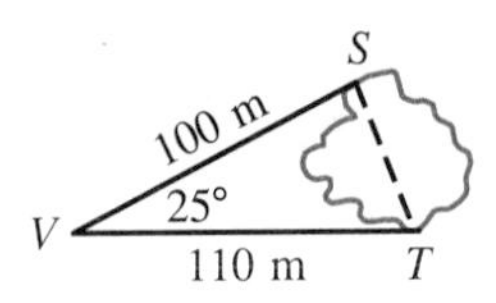

9. From a point on the ground 150 m from the base of a tower, the angle of elevation of the top of the tower is 75°. How tall is the tower?

10. Main and Commerce streets meet to form a triangular region with the river bank, as shown. Find x, the distance between the streets along the river front.

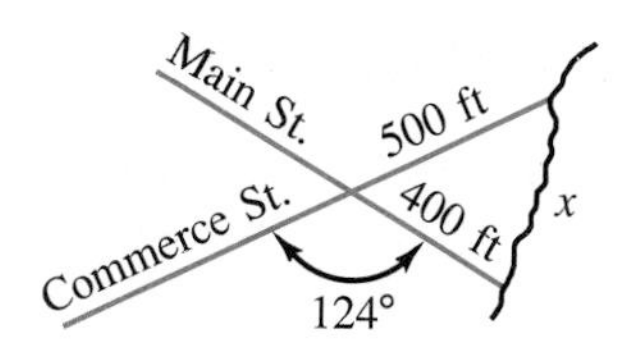

11. What is the area of a parallogram if two adjacent sides are 46 cm and 72 cm long, and the angle between them measures 62°?

12. To find the *ceiling* (the height of the bottom of the lowest cloud), a spotlight is aimed at the clouds directly overhead, and the angle of elevation of the spot of light on the cloud is measured. If the spotlight is 250 ft from the observer, what is the angle of elevation when the ceiling is 1000 ft?

13. Find the perimeter of the quadrilateral at the right.

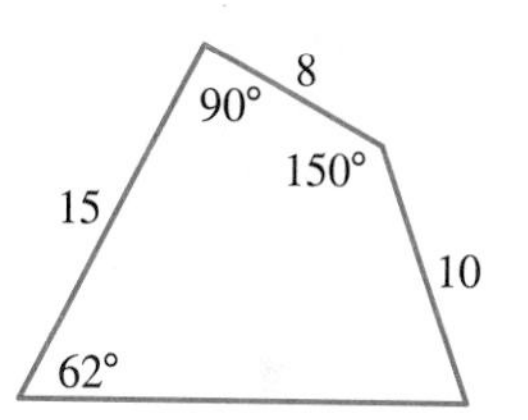

14. A tree was broken in a hurricane and the broken part made a 35° angle with the ground. The section of the tree that remained upright was 6 ft high. How tall was the tree before it was broken?

15. A baseball diamond is in the shape of a square 90 ft on a side. The pitcher's mound is on the diagonal from home plate to second base. If the mound is 60 ft from home plate, how far is it from each base?

16. Two rangers on towers 12 mi apart see the same fire. The angle of depression of the fire is 35° from one tower and 50° from the other. What is the shortest distance each ranger could be from the fire?

17. A pilot returning to an aircraft carrier is 800 ft above the water when he finds that the angle of depression of the carrier is 54°50′. To three significant digits, how far away is the carrier?

18. A pendulum 36 in. long swings 30° from the vertical. How high above the horizontal is the pendulum at the end of its swing? Give the answer to two significant digits.

Mixed Problem Solving Review

1. Fred, Leslie, and Howard are colleagues. When Fred's age triples, he will be 5 years older than the sum of the present ages of Leslie and Howard. If Leslie is 3 years older than Howard and 14 years older than Fred, find the ages of the three colleagues.

2. Angus makes Lou the following offer: "Give me as much money as I have and I will spend $20 with you." Lou accepts. Angus then makes the same arrangement with Patty and then with Neil. How much money did Angus start with if he has $4 left?

3. When rolling three dice, what is the probability that at least two show the same number?

4. Cary doubled the number of pumpkin plants in his garden of pumpkin and squash. He then removed the four weakest pumpkin plants and added 5 squash plants. If the total number of plants is now 10, how many plants did his garden originally contain?

5. Two ferries leave from opposite sides of a bay at the same time. In 15 min the two boats pass each other. If the trip across the bay is 4 mi and one ferry takes 6 min longer to complete the trip, find the rates of the two ferries.

6. A deep sea fishing boat operator offers expeditions for $35 per person with a 4 person minimum. For each additional person, the price per person is reduced by $2.50. The fishing boat can have as many as 12 people on board. Find the number of people for which the operator will obtain maximum revenue. Find the maximum revenue.

PROJECT

In a book on the history of mathematics, look up the name Hipparchus of Nicaea. Write a short essay that tells who this person was and what he did. List two problems that were first solved by using trigonometry at about the time that Hipparchus lived.

TEST YOURSELF

Give angle measures to the nearest degree, and give lengths and areas to two significant digits.

1. Give the period for the function $y = 2 \csc x$. Then sketch the graph for $-\pi \leq x \leq \pi$. 15.6

2. Solve triangle RST, if $\angle R = 24°$, $\angle S = 90°$, and $r = 47$. 15.7

3. Find the area of triangle ABC, if $\angle A = 56°$, $b = 30$, and $c = 18$. 15.8

4. Solve triangle ABC, if $\angle A = 48°$, $a = 16$, and $c = 14$.

5. In triangle ABC, $\angle A = 32°$, $b = 10$, and $c = 17$. Find a. 15.9

6. From two positions, A and B, that are 150 ft apart and are in line with the base of a mountain, the angles of elevation of the peak measure 25° and 55°, respectively. What is the distance from A to the peak? 15.10

INTEGRATING ALGEBRA
Navigation

Did you know that one of the most vital positions on board any craft, whether in space, in the air, or on the sea, is that of navigator?

Given two sides and the included angle of a triangle, the law of cosines can be used to find the missing side. Both missing angles can be found using the *law of tangents*.

Law of Tangents: For any triangle ABC, $\dfrac{a+b}{a-b} = \dfrac{\tan \frac{1}{2}(A+B)}{\tan \frac{1}{2}(A-B)}$

EXAMPLE The cutter Chinook (C) sends out distress signals, to trawlers Albacore (A) and Bonita (B). Return signals indicate that the Bonita is at an angle of 50.0° SW with respect to the position of the Albacore. The Albacore is 5.0 mi away, and the Bonita is 9.0 mi away, south of the Albacore. In what directions should the trawlers head for rescue?

A
C
B
5.0 mi
50.0°
9.0 mi

Since $C = 50°$, $A + B = 130°$. Thus, $\frac{1}{2}(A + B) = 65°$.

$$\frac{9+5}{9-5} = \frac{\tan 65°}{\tan \frac{1}{2}(A-B)} \qquad \textit{Use the law of tangents.}$$

$$\tan \tfrac{1}{2}(A-B) = 0.6127 \qquad \textit{Solve for } \tan \tfrac{1}{2}(A-B).$$

$$\tfrac{1}{2}(A-B) = 31.5°, \quad \text{so } A - B = 63.0° \qquad \textit{Solve for } A - B.$$

Solve the system: $\begin{cases} A + B = 130.0° \\ A - B = 63.0° \end{cases}$ $\quad A = 96.5°$, $B = 33.5°$

Thus the Albacore should head 96.5° counterclockwise from line AB and the Bonita should head 33.5° clockwise from line AB.

EXERCISES

1. Find the distance between the trawlers in the example.
2. In what direction should the trawlers head if their distances from the Chinook are 10 mi and 2.0 mi, respectively, and if the distress signals create a 40° angle?
3. Find the distance between the trawlers in Exercise 2.

CHAPTER 15 SUMMARY AND REVIEW

Vocabulary

amplitude (759)
angle of depression (769)
angle of elevation (769)
cosecant function (744)
cosine function (744)
cotangent function (744)
coterminal angles (737)
even function (765)
initial side of an angle (736)
law of cosines (779)
law of sines (773)
odd function (765)
period (759)
periodic function (759)
quadrantal angle (737)
radian (738)
reciprocal functions (745)
reference triangle (743)
secant function (744)
sine function (744)
standard position of an angle (737)
tangent function (744)
terminal side of an angle (736)
trigonometric functions (744)

Angle Measures An angle is in standard position if its vertex is the origin and its initial side is the positive x-axis. Degree and radian measure are related by the equations $1° = \frac{\pi}{180}$ radians and 1 radian $= \left(\frac{180}{\pi}\right)^{\circ}$.

Convert to radian measure.

1. 150° **2.** 310° **3.** 135°

Convert to degree measure.

4. $\frac{5\pi}{6}$ **5.** $\frac{5\pi}{3}$ **6.** $\frac{2\pi}{3}$ 15.1

Definitions of Trigonometric Functions If $P(x, y)$ is on the terminal side of an angle θ and r is the distance from the origin to P, then $\sin\theta = \frac{y}{r}$, $\cos\theta = \frac{x}{r}$, $\tan\theta = \frac{y}{x}$, $\csc\theta = \frac{r}{y}$, $\sec\theta = \frac{r}{x}$ and $\cot\theta = \frac{x}{y}$ $(x, y, r \neq 0)$.

7. Find the values of the other five trigonometric functions for $\angle K$ if $\tan K = -\frac{4}{3}$ and the terminal side of $\angle K$ is in quadrant II. 15.2

Trigonometric Functions of Special Angles Special angles measure $30°\left(\frac{\pi}{6}\right)$, $45°\left(\frac{\pi}{4}\right)$, $60°\left(\frac{\pi}{3}\right)$. Quadrantal angles measure 0°, $90°\left(\frac{\pi}{2}\right)$, $180°(\pi)$, $270°\left(\frac{3\pi}{2}\right)$, and $360°(2\pi)$.

Find the sine, cosine, and tangent of each angle measure. When radicals are involved, give answers in simplest radical form.

8. 150° **9.** 315° **10.** $\frac{\pi}{2}$ **11.** $\frac{5\pi}{4}$ 15.3

Find each value. Round to four decimal places.

12. $\cos 33°10'$ **13.** $\sin 65°20'$ **14.** $\cos 48°24'$ 15.4

Graphing the Sine and Cosine Functions For $y = a \sin bx$ and $y = a \cos bx$, where a is a real number and b is a positive real number, the amplitude is $|a|$ and the period is $\frac{2\pi}{b}$.

Give the amplitude and period. Then sketch the graph for $0 \leq x \leq 2\pi$.

15. $y = \sin 2x$ **16.** $y = 3 \cos x$ **17.** $y = 2 \sin x$ 15.5

Graphing Other Trigonometric Functions If a is a real number and b is a positive real number, the period for $y = a \tan bx$ and $y = a \cot bx$ is $\frac{\pi}{b}$, and the period for $y = a \sec bx$ and $y = a \csc bx$ is $\frac{2\pi}{b}$.

Give the period for each function. Then sketch the graph for $-2\pi \leq x \leq 2\pi$.

18. $y = \tan 2x$ **19.** $y = 3 \sec x$ **20.** $y = \csc x$ 15.6

The values of the six trigonometric functions can be defined in terms of the lengths of the sides of right triangle ABC, with right angle at C: $\sin A = \frac{a}{c}$, $\cos A = \frac{b}{c}$, $\tan A = \frac{a}{b}$, $\csc A = \frac{c}{a}$, $\sec A = \frac{c}{b}$, and $\cot A = \frac{b}{a}$.

For Exercises 21–27, give angle measures to the nearest degree, and give lengths to two significant digits.

Solve right triangle ABC, using the given information.

21. $c = 25$, $\angle A = 35°$, $\angle C = 90°$ **22.** $a = 8$, $b = 6$, $\angle C = 90°$ 15.7

Law of Sines For any triangle ABC, $\frac{\sin A}{a} = \frac{\sin B}{b} = \frac{\sin C}{c}$.

23. Solve triangle ABC, if possible, if $a = 10$, $b = 12$, and $\angle A = 56°$. 15.8

24. Solve triangle DEF, if $d = 15$, $\angle D = 42°$, and $\angle E = 52°$.

Law of Cosines For any triangle ABC, $a^2 = b^2 + c^2 - 2bc \cos A$, $b^2 = a^2 + c^2 - 2ac \cos B$, and $c^2 = a^2 + b^2 - 2ab \cos C$.

25. Solve triangle ABC if $a = 10$, $b = 12$, and $\angle C = 52°$. 15.9

26. Solve triangle DEF if $d = 15$, $e = 12$, and $f = 14$.

27. Points X and Y lie on opposite sides of a swampy piece of land, and a surveyor stands at point P, which is 48 m from X and 35 m from Y. If $\angle XPY = 115°$, find the distance from X to Y. 15.10

CHAPTER 15 TEST

Convert to radian measure.

1. $90°$ **2.** $320°$

Convert to degree measure.

3. $\frac{4\pi}{3}$ **4.** $\frac{3\pi}{4}$

5. Find the values of the other five trigonometric functions for $\angle K$ if $\cot K = -\frac{5}{12}$ and the terminal side of $\angle K$ is in quadrant IV.

Find each value. When radicals are involved, give answers in simplest radical form.

6. $\sin 60°$ **7.** $\cos 315°$ **8.** $\tan(-30°)$

9. $\cos \frac{\pi}{2}$ **10.** $\sin \frac{5\pi}{4}$ **11.** $\tan \frac{2\pi}{3}$

Find each value. Round to four decimal places.

12. $\cos 53°20'$ **13.** $\tan 42°50'$ **14.** $\sin 63°24'$

Give the period and amplitude (if defined) for each function. Then sketch the graph for $0 \leq x \leq 2\pi$.

15. $y = \cos 2x$ **16.** $y = \tan x$

For Exercises 17–20, give angle measures to the nearest degree, and give lengths to two significant digits.

17. In triangle ABC, $\angle C = 90°$, $c = 20$, and $\angle A = 35°$. Find a.

18. In triangle ABC, $a = 12$, $b = 14$, and $\angle A = 52°$. Find $\angle B$, if possible.

19. In triangle DEF, $d = 10$, $e = 12$, and $f = 8$. Find $\angle E$.

20. In a triangle, two 12-cm sides meet to form a 70° angle. What is the length of the third side of the triangle?

Challenge

In triangle ABC, $\angle C = 90°$, $BC = a$, $AC = b$, and $AB = c$. If $\cos^2 A$ means the same as $(\cos A)^2$ and $\sin^2 A$ means the same as $(\sin A)^2$, show that $\sin^2 A + \cos^2 A = 1$.

COLLEGE ENTRANCE EXAM REVIEW

In each item you are to compare a quantity in Column 1 with a quantity in Column 2. Write the letter of the correct answer from these choices:

A. The quantity in Column 1 is greater than the quantity in Column 2.
B. The quantity in Column 2 is greater than the quantity in Column 1.
C. The quantity in Column 1 is equal to the quantity in column 2.
D. The relationship cannot be determined from the given information.

Notes: Information centered over both columns refers to one or both of the quantities being compared. A symbol that appears in both columns has the same meaning in each column. All variables represent real numbers. Most figures are not drawn to scale.

	Column 1	Column 2
1.	$\sin 30°$	$\cos 60°$
2.	$\tan 60°$	$\sin 60°$
3.	$\cos \frac{6\pi}{5}$	$\cos \frac{9\pi}{5}$

$$x^2 - 7x > 30 \text{ and } x > 0$$

	Column 1	Column 2
4.	x	9
5.	$\left(\frac{4}{9}\right)^{-\frac{3}{2}}$	$\left(\frac{27}{8}\right)^{\frac{2}{3}}$

Use this information for 6–8.

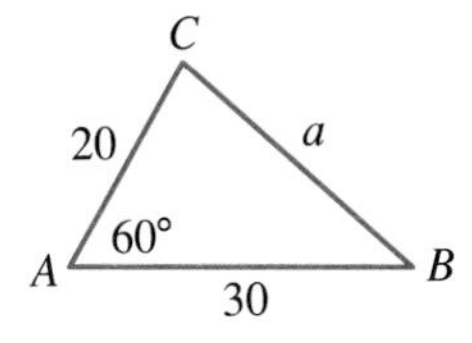

	Column 1	Column 2
6.	a	25
7.	$\sin B$	$\frac{1}{2}$
8.	area $\triangle ABC$	$150\sqrt{3}$

$$f(x) = \frac{x - 1}{x + 1}$$
$$g(x) = \sqrt{x^2 + 1}$$

	Column 1	Column 2
9.	$f(g(0))$	$g(f(0))$

$$2x^2 - 8x + 7 = 0$$

	Column 1	Column 2
10.	sum of the roots	product of the roots

Use this figure for 11–12.

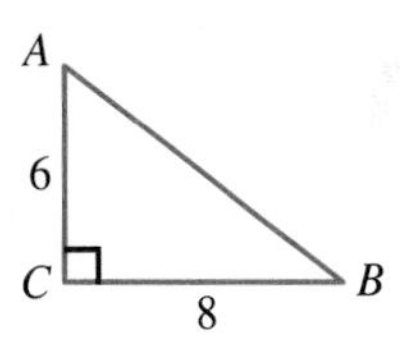

	Column 1	Column 2
11.	$\cos B$	$\frac{4}{5}$
12.	$\tan A$	$\frac{3}{4}$

Use this sequence for 13–14.
$-21, -18, -15, -12, \ldots$

	Column 1	Column 2
13.	12th term	12
14.	nth term	n

MIXED REVIEW

Simplify.

Example $\dfrac{1+\left(-\frac{3}{4}\right)}{1-\left(-\frac{1}{2}\right)} = \dfrac{4\left[1+\left(-\frac{3}{4}\right)\right]}{4\left[1-\left(-\frac{1}{2}\right)\right]}$ *Multiply numerator and denominator by the LCD of all the denominators.*

$$= \frac{4+(-3)}{4-(-2)} = \frac{1}{6}$$

1. $\left(\frac{1}{3}\right)^2 - \left(\frac{\sqrt{2}}{3}\right)^2$ **2.** $2\left(-\frac{2}{3}\right)^2 - 1$ **3.** $\dfrac{\frac{3}{4}}{1+\left(-\frac{7}{8}\right)}$ **4.** $\dfrac{1-\frac{2}{3}}{\frac{5}{12}}$

5. $\dfrac{1+\left(-\frac{5}{6}\right)}{1-\left(-\frac{3}{8}\right)}$ **6.** $\dfrac{2+\left(-\frac{1}{2}\right)^2}{1+\left(-\frac{3}{4}\right)}$ **7.** $\dfrac{1-\left(\frac{\sqrt{2}}{5}\right)^2}{2+\left(-\frac{3}{5}\right)}$ **8.** $\dfrac{1+\left(-\frac{\sqrt{2}}{3}\right)^2}{1-\left(\frac{2}{\sqrt{3}}\right)^2}$

Example $\dfrac{3+\sqrt{5}}{3-\sqrt{5}} = \dfrac{3+\sqrt{5}}{3-\sqrt{5}} \cdot \dfrac{3+\sqrt{5}}{3+\sqrt{5}}$ *Multiply by the conjugate.*

$$= \frac{9+6\sqrt{5}+5}{3-5}$$

$$= \frac{14+6\sqrt{2}}{-2} = -7-3\sqrt{5}$$

9. $\dfrac{5\sqrt{18}-3\sqrt{2}}{\sqrt{6}}$ **10.** $\dfrac{3\sqrt{27}-5\sqrt{3}}{\sqrt{24}}$ **11.** $\dfrac{2+\sqrt{3}}{2-\sqrt{3}}$ **12.** $\dfrac{3-\sqrt{2}}{1+\sqrt{2}}$

13. $\sqrt{\dfrac{1-\frac{1}{3}}{2}}$ **14.** $\sqrt{\dfrac{2-\frac{1}{4}}{3}}$ **15.** $\dfrac{\sqrt{1-\left(\frac{2}{3}\right)^2}}{\frac{3}{4}}$ **16.** $\dfrac{\sqrt{1+\left(\frac{1}{2}\right)^2}}{1-\frac{1}{4}}$

17. A recipe that feeds 4 people calls for 12 oz of meat and 6 oz of macaroni. Kelly is planning a dinner for 12 people. How many pounds of meat will she need?

18. Ben cut a 9-in. shelf from a piece of wood 1 ft 3 in. long. How long was the piece that was left over?

16 Trigonometric Identities and Equations

Time for a change?

Developing Mathematical Power

The High Noon Clock Company is creating a new line of clocks. Each clock will have a triangular face framed in silver. In order to keep within the budget, each clock's perimeter will be no more than 3 ft.

Project

Create a list of triangles that meet the company's restrictions. Diagram each triangle, using the relationships below to find the angles. Choose the triangle of your preference and use it to create a clock face.

The following relationships are true for any triangle *ABC* with sides opposite *a, b,* and *c* where

$$s = \frac{1}{2}(a + b + c)$$

and *h* is the perpendicular from angle *A* to opposite side *a*.

$$\sin\frac{A}{2} = \sqrt{\frac{(s-b)(s-c)}{bc}} \qquad h = a\,\frac{\sin B \sin C}{\sin(B+C)}$$

16.1

Fundamental Identities

Objective: To use the fundamental identities to simplify trigonometric expressions

The six trigonometric functions of an angle in standard position can be expressed in terms of the coordinates of a point on its terminal side and the radius vector.

Capsule Review

If $\angle A$ is in standard position with point $P(x, y)$ on its terminal side such that $PA = r$, then

$\sin A = \frac{y}{r}$ $\cos A = \frac{x}{r}$ $\tan A = \frac{y}{x}$

$\csc A = \frac{r}{y}$ $\sec A = \frac{r}{x}$ $\cot A = \frac{x}{y}$

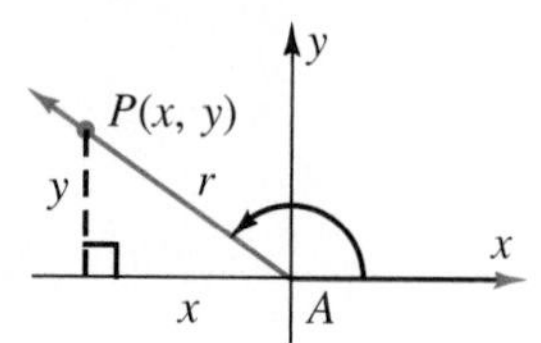

Write the six trigonometric functions for $\angle A$ in standard position if the given point is on its terminal side.

1. $P(6, 8)$ **2.** $P(-5, 12)$ **3.** $P(-8, -15)$

4. $P(4, -5)$ **5.** $P(-3, -3)$ **6.** $P(6, -6)$

Any equation that is true for all values of the variable for which both sides are defined is an **identity.** Some of the trigonometric functions are reciprocals of each other. Using this fact and the *property of reciprocals* gives the *reciprocal identities*.

Reciprocal Identities

$\sin A = \frac{1}{\csc A}$	$\cos A = \frac{1}{\sec A}$	$\tan A = \frac{1}{\cot A}$
$\csc A = \frac{1}{\sin A}$	$\sec A = \frac{1}{\cos A}$	$\cot A = \frac{1}{\tan A}$
$\sin A \csc A = 1$	$\cos A \sec A = 1$	$\tan A \cot A = 1$

The definitions of the trigonometric functions can be used to prove the reciprocal identities.

EXAMPLE 1 **Prove:** $\sin A \csc A = 1$

$$\sin A \csc A = \frac{y}{r} \cdot \frac{r}{y} = 1$$ *Substitute using the definition of sine and cosecant.*

The reciprocal identities are called *fundamental identities*. Two other fundamental identities, called the *quotient identities*, are listed below.

Quotient Identities

$$\tan A = \frac{\sin A}{\cos A} \qquad \cot A = \frac{\cos A}{\sin A}$$

EXAMPLE 2 **Prove:** $\cot A = \frac{\cos A}{\sin A}$

$$\frac{\cos A}{\sin A} = \frac{\frac{x}{r}}{\frac{y}{r}}$$ *Substitute using the definition of sine and cosine.*

$$= \frac{x}{r} \cdot \frac{r}{y}$$ *To divide, multiply by the reciprocal of the divisor.*

$$= \frac{x}{y} = \cot A$$ *This is the definition of cotangent.*

The Pythagorean theorem helps to prove three fundamental identities called the *Pythagorean identities*. Each proof is accomplished by dividing both sides of the Pythagorean formula, $x^2 + y^2 = r^2$, by x^2 or y^2 or r^2.

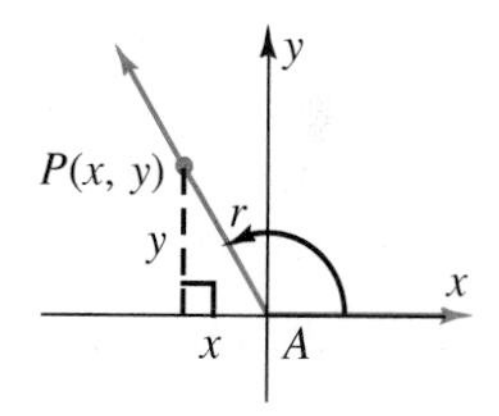

$$x^2 + y^2 = r^2 \qquad x^2 + y^2 = r^2 \qquad x^2 + y^2 = r^2$$

$$\frac{x^2}{r^2} + \frac{y^2}{r^2} = \frac{r^2}{r^2} \qquad \frac{x^2}{x^2} + \frac{y^2}{x^2} = \frac{r^2}{x^2} \qquad \frac{x^2}{y^2} + \frac{y^2}{y^2} = \frac{r^2}{y^2}$$

$$\sin^2 A + \cos^2 A = 1 \qquad 1 + \tan^2 A = \sec^2 A \qquad \cot^2 A + 1 = \csc^2 A$$

Pythagorean Identities

$$\sin^2 A + \cos^2 A = 1 \qquad 1 + \tan^2 A = \sec^2 A \qquad 1 + \cot^2 A = \csc^2 A$$

The cofunctions of complementary angle measures are equal.

Cofunction Identities

$\sin A = \cos (90° - A)$	$\csc A = \sec (90° - A)$
$\cos A = \sin (90° - A)$	$\sec A = \csc (90° - A)$
$\tan A = \cot (90° - A)$	$\cot A = \tan (90° - A)$

EXAMPLE 3 **Express $\cos C$ in terms of $\sin C$.**

$\sin^2 C + \cos^2 C = 1$ — *Select an identity that contains sin C and cos C.*

$\cos^2 C = 1 - \sin^2 C$

$\cos C = \pm\sqrt{1 - \sin^2 C}$ — *Solve for cos C.*

The use of + or − depends on the quadrant in which $\angle C$ is located.

To simplify a trigonometric expression, substitute by using the basic identities. Then, simplify the result.

EXAMPLE 4 **Simplify: $\cos A + \sin A \tan A$**

$\cos A + \sin A \tan A = \cos A + \sin A \dfrac{\sin A}{\cos A}$ — *Substitute $\frac{\sin A}{\cos A}$ for tan A.*

$= \cos A + \dfrac{\sin^2 A}{\cos A}$ — *Multiply.*

$= \dfrac{\cos^2 A + \sin^2 A}{\cos A}$ — *Add using cos A as a common denominator.*

$= \dfrac{1}{\cos A} = \sec A$

Note that $\sec A$ is not defined for $\cos A = 0$. Therefore, $A \neq \frac{\pi}{2} + k\pi$, where k is an integer. Such restrictions must always be considered.

CLASS EXERCISES

1. **Thinking Critically** Explain how the last three cofunction identities follow from the first three.

Simplify.

2. $\csc^2 A - 1$
3. $1 - \cos^2 A$
4. $\cos^2 A - 1$
5. $\csc A \cos A \tan A$
6. $\tan A(\cot A + \tan A)$
7. $\sin^2 A + \cos^2 A + \tan^2 A$

PRACTICE EXERCISES

1. Use the definitions for cosine and secant to prove $\cos A \sec A = 1$.
2. Use the definitions for tangent and cotangent to prove $\tan A \cot A = 1$.

Express the first trigonometric function in terms of the second.

3. $\sin A$; $\cos A$
4. $\tan A$; $\cos A$
5. $\cot A$; $\sin A$
6. $\csc A$; $\cot A$
7. $\cot A$; $\csc A$
8. $\sec A$; $\tan A$

Simplify.

9. $\cos A \tan A$
10. $\sec A \cot A$
11. $\cos^2 A \sec A \csc A$
12. Prove: $\tan A = \dfrac{\sin A}{\cos A}$
13. Prove: $\sin B \cot B = \cos B$
14. Express in terms of $\sin A$: $\cos A \csc A \cot A$

Simplify.

15. $\sin A(1 + \cot^2 A)$
16. $\cot A \tan A + \sec^2 A$
17. $\sec A \cos A - \cos^2 A$
18. $\sin^2 A \csc A \sec A$
19. $\cos B(1 + \tan^2 B)$
20. $\sin C(1 + \cot^2 C)$
21. Prove: $\dfrac{1 + \cos C}{\cos C} = 1 + \sec C$
22. Express in terms of $\cot A$: $\dfrac{1 - \tan A}{1 + \tan A}$
23. Express in terms of $\sin A$: $\dfrac{\cos A}{\sec A + \tan A}$

Applications

24. **Geometry** In the figure at the right, the circle has a radius of 1 and its center is at the origin. $\overrightarrow{DE}$ and $\overrightarrow{HF}$ are tangents. Use similar triangles to show why the following are true:
$BC = \sin \theta$ $AB = \cos \theta$ $DE = \tan \theta$
$AE = \sec \theta$ $FH = \cot \theta$ $AF = \csc \theta$

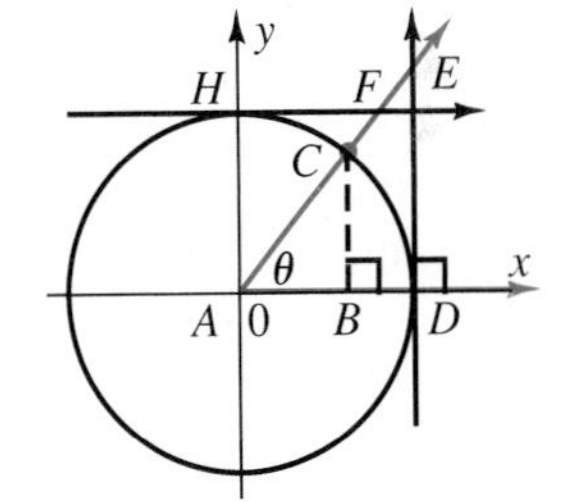

25. **Coordinate Geometry** Show that the slope of a line on the Cartesian coordinate plane is the same as the tangent of the angle that the line makes with the x-axis.

Developing Mathematical Power

26. **Extension** What is the value of:
$(\tan 45°)(\cos 30°)(\sec 120°)(\sin 150°)(\cot 90°)(\csc 135°)$?

16.2 Proving Trigonometric Identities

Objective: To prove trigonometric identities by using the fundamental identities

The reciprocal identities, the Pythagorean identities, and the quotient identities are used to prove that certain equations are identities. Factoring, simplifying expressions, and working with fractions are among the algebraic procedures used in proving identities.

Capsule Review

EXAMPLE **Perform the indicated procedure.**

a. Factor and simplify: $\cos^3 A + \cos A \sin^2 A$

b. Combine: $1 - \dfrac{\cos A}{\sin A}$

a. $\cos^3 A + \cos A \sin^2 A$
$= \cos A(\cos^2 A + \sin^2 A)$
$= \cos A(1) = \cos A$

b. $1 - \dfrac{\cos A}{\sin A}$
$= \dfrac{\sin A}{\sin A} - \dfrac{\cos A}{\sin A} = \dfrac{\sin A - \cos A}{\sin A}$

Factor.

1. $\sin^2 A - \sin^2 A \cos^2 A$ **2.** $\tan^4 A + \tan^2 A$ **3.** $\cos^3 A - \cos A$

Perform the indicated operation.

4. $\cos B + \dfrac{\sin^2 B}{\cos B}$ **5.** $\dfrac{\cos A}{1 + \sin A} - \cos A$ **6.** $1 + \dfrac{\cos A}{\sin A} - \dfrac{2}{\sin^2 A}$

One way to prove that an equation is an identity involves transforming the more complicated side into the simpler side. Work on one side only.

EXAMPLE 1 **Prove: $\cos B + \tan B \sin B = \sec B$**

$\cos B + \tan B \sin B$	$\sec B$	
$\cos B + \dfrac{\sin B}{\cos B} \sin B$		*Substitute using $\dfrac{\sin B}{\cos B} = \tan B$.*
$\dfrac{\cos^2 B + \sin^2 B}{\cos B}$		*Multiply by $\dfrac{\cos B}{\cos B}$ to add fractions.*
$\dfrac{1}{\cos B}$		*Substitute using $\sin^2 B + \cos^2 B = 1$.*
$\sec B$		*Substitute using $\dfrac{1}{\cos B} = \sec B$.*

Both sides of an equation may be equally complicated. Select either side and transform it into the other.

EXAMPLE 2 **Prove:** $\sec^4 A - \sec^2 A = \tan^2 A + \tan^4 A$

$\sec^4 A - \sec^2 A$	$\tan^2 A + \tan^4 A$	*Work on the left side.*
$\sec^2 A(\sec^2 A - 1)$		*Factor out* $\sec^2 A$.
$\sec^2 A \tan^2 A$		*Substitute using* $\tan^2 A = \sec^2 A - 1$.
$(1 + \tan^2 A)\tan^2 A$		*Substitute using* $1 + \tan^2 A = \sec^2 A$.
$\tan^2 A + \tan^4 A$		*Multiply.*

Transforming the right side could also be done to prove that the equation in Example 2 is an identity. Another method to prove that an equation is an identity is to transform both sides of the equation into the same expression.

EXAMPLE 3 **Prove:** $\dfrac{1 + \cos A}{\sin A} = \dfrac{\sin A}{1 - \cos A}$

Work on the left side.

$\dfrac{1 + \cos A}{\sin A}$	$\dfrac{\sin A}{1 - \cos A}$
$\dfrac{\sin A(1 + \cos A)}{\sin^2 A}$	
$\dfrac{\sin A(1 + \cos A)}{1 - \cos^2 A}$	

Work on the right side.

$\dfrac{\sin A(1 + \cos A)}{1 - \cos^2 A}$	$\dfrac{\sin A}{1 - \cos A}$
	$\dfrac{\sin A(1 + \cos A)}{(1 - \cos A)(1 + \cos A)}$
	$\dfrac{\sin A(1 + \cos A)}{1 - \cos^2 A}$

If you think an equation may be an identity, use a utility to graph both sides of the equation on the same screen. If only one curve appears, this is a strong indication (but not proof) that the equation is an identity.

Strategies for Proving Identities

- Transform the complicated side into the simpler side of the equation.
- Transform either side of the equation.
- Transform both sides into the same expression.

CLASS EXERCISES

Prove.

1. $\sin x \sec x \cot x = 1$

2. $\csc^2 A(1 - \cos^2 A) = 1$

3. $\tan Q \sin Q + \cos Q = \sec Q$

4. $\dfrac{\cos A \csc A}{\cot A} = 1$

PRACTICE EXERCISES

 Use technology where appropriate.

Prove.

1. $\sin A \sec A = \tan A$
2. $\cot B \sec B = \csc B$
3. $1 - 2\sin^2 C = 2\cos^2 C - 1$
4. $\dfrac{\sin A - 1}{\cos A} = \tan A - \sec A$
5. $\dfrac{\sin B + \tan B}{1 + \cos B} = \tan B$
6. $\dfrac{\sin^2 A + \cos^2 A}{\cos A} = \sec A$
7. $\cos Z(\csc Z - \sec Z) = \cot Z - 1$
8. $(1 + \csc A)(1 - \sin A) = \cot A \cos A$
9. $\dfrac{\sec A}{\cot A + \tan A} = \sin A$
10. $(\cot A + 1)^2 = \csc^2 A + 2\cot A$
11. $\sec^2 Q(1 - \cos^2 Q) = \tan^2 Q$
12. $(\cot B + 1)^2 - 2\cot B = \csc^2 B$
13. $\dfrac{\csc A \cos A}{\sec A \sin A} = (\csc^2 A - 1)$
14. $\dfrac{\cos^2 A}{(1 - \sin A)^2} = (\sec A + \tan A)^2$
15. $\dfrac{\csc A \sin B}{\sec A \cos B} = \dfrac{\tan B}{\tan A}$
16. $\dfrac{\tan A - 1}{\tan A + 1} = \dfrac{1 - \cot A}{1 + \cot A}$
17. $\tan A \sin A + \cos A = \sec A$
18. $\sin A \cos A(\tan A + \cot A) = 1$
19. $\csc^2 A(1 - \cos^2 A) = 1$
20. $(1 - \cos B)(1 + \sec B)\cot B = \sin B$
21. $\dfrac{1 - \sin B}{\cos B} = \dfrac{\cos B}{1 + \sin B}$
22. $\dfrac{\tan A - \sin A}{\sin^2 A} = \dfrac{\tan A}{1 + \cos A}$
23. $\dfrac{\tan x + \sec x - 1}{\tan x - \sec x + 1} = \tan x + \sec x$
24. $\dfrac{\sec B + \csc B}{\tan B + \cot B} = \sin B + \cos B$
25. $\dfrac{\sin^3 A + \cos^3 A}{\sin A + \cos A} = 1 - \sin A \cos A$
26. $\dfrac{\sin B \cos B}{\cos^2 B - \sin^2 B} = \dfrac{\tan B}{1 - \tan^2 B}$

Applications

27. **Physics** When a ray of light passes from one medium into a second, the angle of incidence and refraction (θ_1 and θ_2) are related by Snell's Law, $n_1 \sin \theta_1 = n_2 \sin \theta_2$, where n_1 is the index of refraction of the first medium and n_2 is the index of refraction of the second medium. How are θ_1 and θ_2 related if $n_2 > n_1$? $n_2 < n_1$? $n_2 = n_1$?

Mixed Review

28. Write $\sum_{m=1}^{4} (3m + 4)$ in expanded form.
29. Find the sum of $\sum_{x=1}^{11} 3(2)^{x+1}$.
30. Express $0.\overline{3}$ as an infinite geometric series.

16.3

Sum and Difference Identities

Objective: To use the sum and difference identities to find exact values of the trigonometric functions for certain angles

Many objects in everyday life are in periodic motion. Examples include the vibration of a guitar string, the motion of a pendulum in a grandfather clock, and the motion of a piston in an engine. A kind of periodic motion is called *simple harmonic motion*. The position x of a mass moving in simple harmonic motion at time t can be found using the formula

$$x = A \cos (\omega t + \gamma)$$

where A, ω (omega), and γ (gamma) are constants. To evaluate the cosine of a sum such as $\omega t + \gamma$, you may apply your knowledge of special angles and the sum and difference identities in this section.

Capsule Review

Use the same procedure to square a trigonometric expression that is used to square an algebraic expression.

EXAMPLE $(\cos A - \cos B)^2 = (\cos A - \cos B)(\cos A - \cos B)$
$= \cos^2 A - 2 \cos A \cos B + \cos^2 B$

Multiply.

1. $(\sin A - \sin B)^2$ **2.** $(\cos A + \cos B)^2$ **3.** $(\sin A + \sin B)^2$

4. $[\cos (A - B)]^2$ **5.** $[\cos (A - B) - 1]^2$ **6.** $[\sin (A - B)]^2$

The circle at the right is a unit circle with its center at the origin. Central angle PRQ intersects an arc with chord $\overline{PQ}$. In the unit circle $r = 1$, so the coordinates of P and Q are $P(\cos A, \sin A)$ and $Q(\cos B, \sin B)$. Draw chord $\overline{P'Q'}$ congruent to $\overline{PQ}$, where Q' is $(1, 0)$ and P' is as shown. Since $\triangle RPQ \cong \triangle RP'Q'$ (by SSS), $m\angle P'RQ' = A - B$, and the coordinates of P' are $(\cos (A - B), \sin (A - B))$.

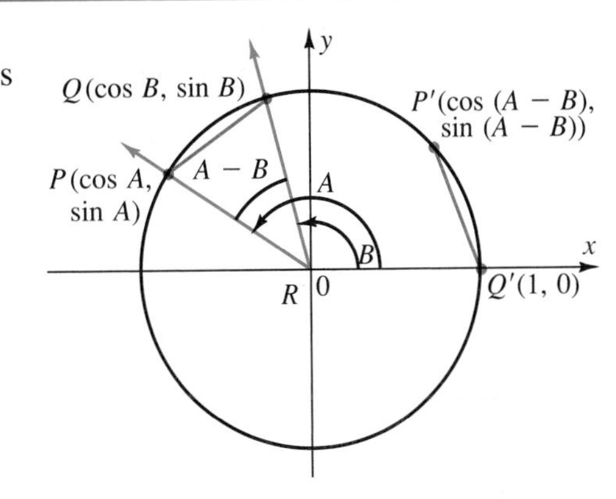

Using the distance formula, you can find the lengths of PQ and $P'Q'$.

$$\begin{aligned}(PQ)^2 &= (\cos A - \cos B)^2 + (\sin A - \sin B)^2 \\ &= \cos^2 A - 2 \cos A \cos B + \cos^2 B + \sin^2 A - 2 \sin A \sin B + \sin^2 B \\ &= \cos^2 A + \sin^2 A + \cos^2 B + \sin^2 B - 2 \cos A \cos B - 2 \sin A \sin B \\ &= 2 - 2 \cos A \cos B - 2 \sin A \sin B\end{aligned}$$

Similar calculations will show that $(P'Q')^2 = 2 - 2 \cos (A - B)$.

$$PQ = P'Q' \text{ so } (PQ)^2 = (P'Q')^2$$
$$2 - 2 \cos (A - B) = 2 - 2 \cos A \cos B - 2 \sin A \sin B$$
$$\cos (A - B) = \cos A \cos B + \sin A \sin B$$

The formula for $\cos (A + B)$ can be derived by using the formula for $\cos (A - B)$.

$$\begin{aligned}\cos (A + B) &= \cos [A - (-B)] \\ &= \cos A \cos (-B) + \sin A \sin (-B) \\ &= \cos A \cos B - \sin A \sin B \qquad \textit{\cos (-B) = \cos B; \sin (-B) = -\sin B}\end{aligned}$$

Use the cofunctions identities to find $\sin (A - B)$.

$$\begin{aligned}\sin (A - B) &= \cos [90 - (A - B)] && \text{Cofunction identity} \\ &= \cos [(90 - A) + B] \\ &= \cos (90 - A) \cos B - \sin (90 - A) \sin B && \text{Cosine sum} \\ &= \sin A \cos B - \cos A \sin B && \text{Cofunction identity}\end{aligned}$$

Substitute $A + (-B)$ for $A - B$ in the above formula to find $\sin (A + B)$.

$$\sin (A + B) = \sin A \cos B + \cos A \sin B$$

The sum and difference identities for sine and cosine can be used to find the exact values of the sine and cosine of angles that are not special angles.

EXAMPLE 1 **Find the exact value of:** **a.** $\sin 15°$ **b.** $\cos \frac{5\pi}{12}$

a.
$$\begin{aligned}\sin 15° &= \sin (45° - 30°) \\ &= \sin 45° \cos 30° - \cos 45° \sin 30° \\ &= \left(\frac{\sqrt{2}}{2}\right)\left(\frac{\sqrt{3}}{2}\right) - \left(\frac{\sqrt{2}}{2}\right)\left(\frac{1}{2}\right) = \frac{\sqrt{6}}{4} - \frac{\sqrt{2}}{4} = \frac{\sqrt{6} - \sqrt{2}}{4}\end{aligned}$$

b.
$$\begin{aligned}\cos \frac{5\pi}{12} &= \cos \left(\frac{\pi}{4} + \frac{\pi}{6}\right) \\ &= \cos \frac{\pi}{4} \cos \frac{\pi}{6} - \sin \frac{\pi}{4} \sin \frac{\pi}{6} \\ &= \left(\frac{\sqrt{2}}{2}\right)\left(\frac{\sqrt{3}}{2}\right) - \left(\frac{\sqrt{2}}{2}\right)\left(\frac{1}{2}\right) = \frac{\sqrt{6}}{4} - \frac{\sqrt{2}}{4} = \frac{\sqrt{6} - \sqrt{2}}{4}\end{aligned}$$

The formulas for $\tan(A - B)$ and $\tan(A + B)$ are derived by using the formulas for sine and cosine.

$$\tan(A - B) = \frac{\tan A - \tan B}{1 + \tan A \tan B} \qquad \tan(A + B) = \frac{\tan A + \tan B}{1 - \tan A \tan B}$$

EXAMPLE 2 **Find the exact value of tan 105°.**

$$\tan 105° = \tan(45° + 60°) = \frac{\tan 45° + \tan 60°}{1 - \tan 45° \tan 60°} = \frac{1 + \sqrt{3}}{1 - \sqrt{3}}$$

$$= \frac{1 + \sqrt{3}}{1 - \sqrt{3}} \cdot \frac{1 + \sqrt{3}}{1 + \sqrt{3}} = \frac{1 + 2\sqrt{3} + 3}{1 - 3} = \frac{4 + 2\sqrt{3}}{-2} = -2 - \sqrt{3}$$

Suppose the value of one trigonometric function is known for the measure of $\angle A$ and another is known for the measure of $\angle B$; then the value of a trigonometric function of the sum or difference can be found.

EXAMPLE 3 **If $\sin A = -\frac{1}{3}$, $180° < \angle A < 270°$ and $\cos B = \frac{1}{5}$, $270° < \angle B < 360°$. Find $\cos(A - B)$.**

If $\sin A = -\frac{1}{3}$, and $180° < \angle A < 270°$, then $\cos A = \frac{-2\sqrt{2}}{3}$.

First find cos A and sin B.

If $\cos B = \frac{1}{5}$, and $270° < \angle B < 360°$, then $\sin B = \frac{-2\sqrt{6}}{5}$.

$$\cos(A - B) = \cos A \cos B + \sin A \sin B$$

$$= \left(\frac{-2\sqrt{2}}{3}\right)\left(\frac{1}{5}\right) + \left(-\frac{1}{3}\right)\left(\frac{-2\sqrt{6}}{5}\right) = \frac{-2\sqrt{2}}{15} + \frac{2\sqrt{6}}{15} = \frac{-2\sqrt{2} + 2\sqrt{6}}{15}$$

The Sum and Difference Identities

$$\cos(A - B) = \cos A \cos B + \sin A \sin B$$
$$\cos(A + B) = \cos A \cos B - \sin A \sin B$$
$$\sin(A - B) = \sin A \cos B - \cos A \sin B$$
$$\sin(A + B) = \sin A \cos B + \cos A \sin B$$
$$\tan(A - B) = \frac{\tan A - \tan B}{1 + \tan A \tan B}$$
$$\tan(A + B) = \frac{\tan A + \tan B}{1 - \tan A \tan B}$$

CLASS EXERCISES

1. **Thinking Critically** Discuss the method of derivation of the formula for $\sin(A + B)$.

Express each measure as the sum or difference of two special angle measures.

2. 75° **3.** 150° **4.** 135° **5.** 60° **6.** 30°

7. $\frac{\pi}{12}$ **8.** $\frac{\pi}{6}$ **9.** $\frac{3\pi}{4}$ **10.** $\frac{7\pi}{12}$ **11.** $\frac{4\pi}{3}$

Evaluate.

12. $\cos 50^\circ \cos 40^\circ - \sin 50^\circ \sin 40^\circ$

13. $\sin 80^\circ \cos 35^\circ - \cos 80^\circ \sin 35^\circ$

14. $\dfrac{\tan 40^\circ + \tan 20^\circ}{1 - \tan 40^\circ \tan 20^\circ}$

15. $\dfrac{\tan 80^\circ - \tan 20^\circ}{1 + \tan 80^\circ \tan 20^\circ}$

PRACTICE EXERCISES

Evaluate by using special angle measures and a sum or difference identity.

1. $\sin 105^\circ$ **2.** $\sin 75^\circ$ **3.** $\sin 195^\circ$ **4.** $\sin 135^\circ$

5. $\cos 15^\circ$ **6.** $\cos 255^\circ$ **7.** $\cos 285^\circ$ **8.** $\cos 315^\circ$

9. $\sin \frac{5\pi}{12}$ **10.** $\sin \frac{8\pi}{3}$ **11.** $\cos \frac{11\pi}{6}$ **12.** $\cos \frac{13\pi}{12}$

13. $\tan 75^\circ$ **14.** $\tan 195^\circ$ **15.** $\tan 15^\circ$ **16.** $\tan 120^\circ$

17. $\sin A = \frac{4}{5}$, $0^\circ < \angle A < 90^\circ$ and $\cos B = \frac{12}{13}$, $0^\circ < \angle B < 90^\circ$. Find $\sin (A - B)$.

18. $\tan A = \frac{12}{5}$, $0^\circ < \angle A < 90^\circ$, and $\tan B = \frac{3}{4}$, $0^\circ < \angle B < 90^\circ$. Find $\tan (A + B)$.

19. $\sin A = \frac{3}{5}$, $90^\circ < \angle A < 180^\circ$, and $\cos B = -\frac{12}{13}$, $180^\circ < \angle B < 270^\circ$. Find $\cos (A + B)$.

20. $\tan A = \frac{5}{12}$, $0^\circ < \angle A < 90^\circ$, and $\tan B = -\frac{3}{4}$, $90^\circ < \angle B < 180^\circ$. Find $\tan (A - B)$.

21. $\sin A = -\frac{5}{13}$, $180^\circ < \angle A < 270^\circ$, and $\cos B = \frac{4}{5}$, $270^\circ < \angle B < 360^\circ$. Find $\sin (A + B)$.

22. $\sin A = -\frac{4}{5}$, $180^\circ < \angle A < 270^\circ$, and $\cos B = -\frac{12}{13}$, $180^\circ < \angle B < 270^\circ$. Find $\cos (A - B)$.

Derive a formula for each sum or difference by using $\sin (A + B)$, $\sin (A - B)$, $\cos (A + B)$, or $\cos (A - B)$.

23. $\tan (A + B)$ **24.** $\tan (A - B)$ **25.** $\cot (A + B)$ **26.** $\cot (A - B)$

Evaluate by using special angle measures and a sum or difference identity.

27. $\cot 15^\circ$ **28.** $\cot 75^\circ$ **29.** $\cot \frac{5\pi}{12}$ **30.** $\cot \frac{8\pi}{3}$

Express each as a trigonometric function of a single angle measure.

31. $\sin 2\theta \cos \theta + \cos 2\theta \sin \theta$

32. $\sin 3\theta \cos 2\theta + \cos 3\theta \sin 2\theta$

33. $\cos 3A \cos 4A - \sin 3A \sin 4A$

34. $\cos 2B \cos 3B - \sin 2B \sin 3B$

35. $\dfrac{\tan 5C + \tan 6C}{1 - \tan 5C \tan 6C}$

36. $\dfrac{\tan 3A - \tan A}{1 + \tan 3A \tan A}$

Use the sum and difference formulas to show that each of the following is true.

37. $\cos(\pi - \theta) = -\cos \theta$

38. $\sin(\pi - \theta) = \sin \theta$

39. $\sin(\pi + \theta) = -\sin \theta$

40. $\cos(\pi + \theta) = -\cos \theta$

Applications

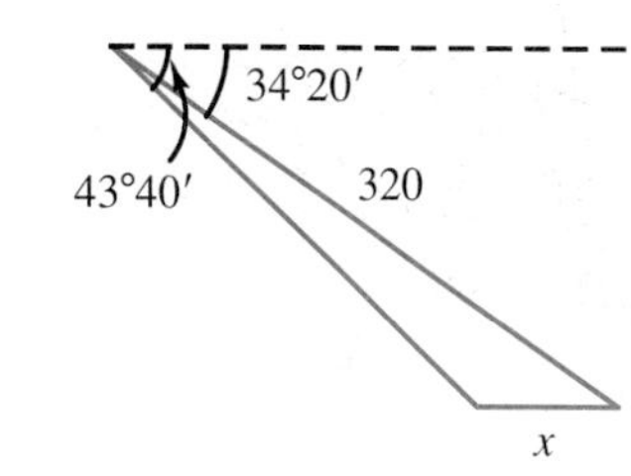

41. Navigation When a plane is 320 mi from the runway, the angle of depression from the cockpit to the airport is 34°20′. At the same time, the angle of depression from the cockpit to the center of a lake located between the plane and the airport is 43°40′. How far is the center of the lake from the airport?

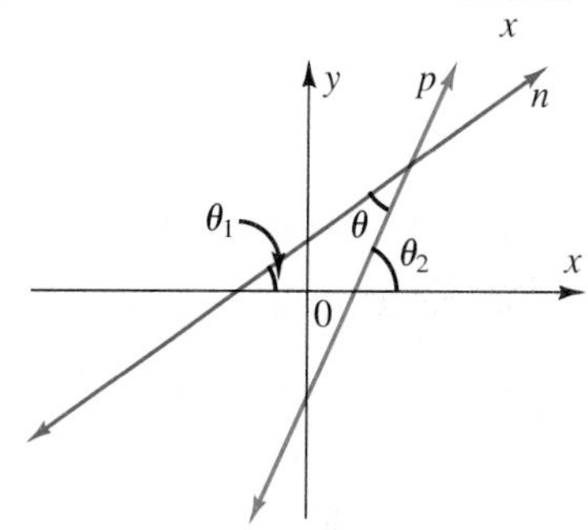

42. Coordinate Geometry The slope m of a line on a coordinate plane is equal to the tangent of the angle that it makes with the x-axis. Let θ be the smallest angle formed by the intersection of the lines n and p that make angles of θ_1 and θ_2, respectively, with the x-axis. These lines have slopes m_1 and m_2, respectively. Show that $\tan \theta = \dfrac{m_2 - m_1}{1 + m_2 m_1}$.

CAREER

Electrical Engineer

Engineers apply scientific principles and use mathematics as a tool to design goods, services, and methods of production. Electrical engineers are concerned with electrical and electronic equipment such as motors, generators, medical monitoring equipment, medical aids (including artificial hearts and kidney dialysis machines), pollution monitoring equipment, computers, lasers, radars, missile guidance systems, and satellite power systems.

Engineers need at least a four-year college education, with courses in physical sciences, technology, mathematics (including trigonometry and calculus), humanities, and computer science.

16.4

Double-Angles and Half-Angles

Objective: To use the double- and half-angle formulas to find the exact value of certain angles

Double-angle formulas can be derived from sum identities and half-angle formulas can be derived from double-angle formulas.

Capsule Review

The values of the trigonometric functions associated with certain angles can be found by using the relationships among the sides of a 30°-60°-90° or a 45°-45°-90° triangle.

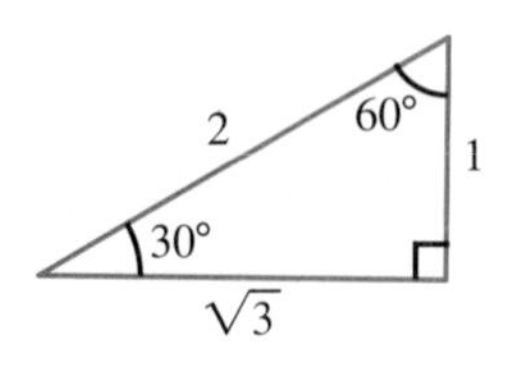

$\sin 30° = \frac{1}{2}$ $\quad \cos 45° = \frac{\sqrt{2}}{2}$ $\quad \tan 60° = \sqrt{3}$

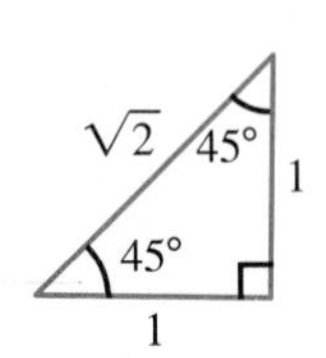

Find the value of each trigonometric function.

1. sin 60° **2.** cos 30° **3.** sin 135° **4.** tan 30°

5. tan 120° **6.** cos 150° **7.** sin 225° **8.** tan 330°

To derive formulas for the sine, cosine, and tangent of twice the measure of an angle, substitute A for B in the formulas $\sin(A + B)$, $\cos(A + B)$, and $\tan(A + B)$.

$$\begin{aligned}\sin 2A &= \sin(A + A)\\ &= \sin A \cos A + \sin A \cos A\\ &= 2 \sin A \cos A\end{aligned}$$

$$\begin{aligned}\cos 2A &= \cos(A + A)\\ &= \cos A \cos A - \sin A \sin A\\ &= \cos^2 A - \sin^2 A\end{aligned}$$

$$\tan 2A = \tan(A + A) = \frac{\tan A + \tan A}{1 - \tan A \tan A} = \frac{2 \tan A}{1 - \tan^2 A}$$

Two useful equivalent forms of the double-angle formula for cosine are

$$\cos 2A = 2\cos^2 A - 1 \qquad \cos 2A = 1 - 2\sin^2 A$$

Double-angle Formulas

$$\cos 2A = \cos^2 A - \sin^2 A$$
$$\cos 2A = 2\cos^2 A - 1$$
$$\cos 2A = 1 - 2\sin^2 A$$
$$\sin 2A = 2 \sin A \cos A$$
$$\tan 2A = \frac{2 \tan A}{1 - \tan^2 A}$$

EXAMPLE 1 **Use the double-angle formulas to find the exact value of:**
a. $\sin 120^\circ$ **b.** $\tan 120^\circ$

a. $\sin 120^\circ = \sin 2(60^\circ)$
$= 2 \sin 60^\circ \cos 60^\circ$
$= 2\left(\frac{\sqrt{3}}{2}\right)\left(\frac{1}{2}\right) = \frac{\sqrt{3}}{2}$

b. $\tan 120^\circ = \tan 2(60^\circ)$
$= \frac{2 \tan 60^\circ}{1 - \tan^2 60^\circ} = \frac{2\sqrt{3}}{1 - (\sqrt{3})^2}$
$= \frac{2\sqrt{3}}{-2} = -\sqrt{3}$

The double-angle formulas are applied in Example 2.

EXAMPLE 2 **Sin** $B = \frac{3}{5}$ **and** $\angle B$ **is in the second quadrant. Find the exact value of:** **a.** $\sin 2B$ **b.** $\cos 2B$

If $\sin B = \frac{3}{5}$, $\cos B = -\frac{4}{5}$. *Cosine is negative in quadrant II.*

a. $\sin 2B = 2 \sin B \cos B$
$= 2\left(\frac{3}{5}\right)\left(-\frac{4}{5}\right) = -\frac{24}{25}$

b. $\cos 2B = 2 \cos^2 B - 1$
$= 2\left(-\frac{4}{5}\right)^2 - 1 = \frac{7}{25}$

Replacing A by $\frac{x}{2}$ in two formulas for $\cos 2A$ gives half-angle formulas for sine and cosine.

$$\cos 2A = 1 - 2 \sin^2 A$$
$$\cos 2\left(\frac{x}{2}\right) = 1 - 2 \sin^2 \frac{x}{2}$$
$$\cos x = 1 - 2 \sin^2 \frac{x}{2}$$
$$2 \sin^2 \frac{x}{2} = 1 - \cos x$$
$$\sin^2 \frac{x}{2} = \frac{1 - \cos x}{2}$$
$$\sin \frac{x}{2} = \pm\sqrt{\frac{1 - \cos x}{2}}$$

$$\cos 2A = 2 \cos^2 A - 1$$
$$\cos 2\left(\frac{x}{2}\right) = 2 \cos^2 \frac{x}{2} - 1$$
$$\cos x = 2 \cos^2 \frac{x}{2} - 1$$
$$2 \cos^2 \frac{x}{2} = 1 + \cos x$$
$$\cos^2 \frac{x}{2} = \frac{1 + \cos x}{2}$$
$$\cos \frac{x}{2} = \pm\sqrt{\frac{1 + \cos x}{2}}$$

The use of a + or − depends on the quadrant in which $\frac{x}{2}$ is located.

The half-angle formulas for sine and cosine can be used to derive the half-angle formula for tangent. There are three equivalent forms for this formula.

$$\tan \frac{x}{2} = \pm\sqrt{\frac{1 - \cos x}{1 + \cos x}} \qquad \tan \frac{x}{2} = \frac{1 - \cos x}{\sin x} \qquad \tan \frac{x}{2} = \frac{\sin x}{1 + \cos x}$$

Half-angle Formulas

$$\sin\frac{x}{2} = \pm\sqrt{\frac{1-\cos x}{2}} \qquad \tan\frac{x}{2} = \pm\sqrt{\frac{1-\cos x}{1+\cos x}}$$

$$\cos\frac{x}{2} = \pm\sqrt{\frac{1+\cos x}{2}} \qquad \tan\frac{x}{2} = \frac{1-\cos x}{\sin x}$$

$$\tan\frac{x}{2} = \frac{\sin x}{1+\cos x}$$

EXAMPLE 3 **Use the half-angle formulas to find the exact value of each.**

a. $\sin 15°$ **b.** $\cos 150°$

a. $\sin 15° = +\sqrt{\frac{1-\cos 30°}{2}}$ *$\sin 15°$ is positive. $15° = \frac{1}{2}\cdot 30°$*

$$= \sqrt{\frac{1-\frac{\sqrt{3}}{2}}{2}} = \sqrt{\frac{2-\sqrt{3}}{4}} = \frac{\sqrt{2-\sqrt{3}}}{2}$$ *$\cos 30° = \frac{\sqrt{3}}{2}$*

b. $\cos 150° = -\sqrt{\frac{1+\cos 300°}{2}}$ *$\cos 150°$ is negative. $150° = \frac{1}{2}\cdot 300°$*

$$= -\sqrt{\frac{1+\left(\frac{1}{2}\right)}{2}} = -\sqrt{\frac{2+1}{4}} = -\frac{\sqrt{3}}{2}$$ *$\cos 300° = \frac{1}{2}$*

The next example illustrates the use of half-angle formulas for an angle located in the third quadrant.

EXAMPLE 4 **$\sin Y = -\frac{24}{25}$ and $180° < \angle Y < 270°$. Find the exact value of:**

a. $\sin\frac{Y}{2}$ **b.** $\cos\frac{Y}{2}$

Since $\sin Y = -\frac{24}{25}$ and $180° < \angle Y < 270°$, then $\cos Y = -\frac{7}{25}$ and $90° < \angle\frac{Y}{2} < 135°$. Therefore, $\angle\frac{Y}{2}$ is a second-quadrant angle.

a. $\sin\frac{Y}{2} = \pm\sqrt{\frac{1-\cos Y}{2}}$

$$\sin\frac{Y}{2} = +\sqrt{\frac{1-\left(-\frac{7}{25}\right)}{2}}$$

$$= \sqrt{\frac{32}{50}} = \frac{4}{5}$$

b. $\cos\frac{Y}{2} = \pm\sqrt{\frac{1+\cos Y}{2}}$

$$\cos\frac{Y}{2} = -\sqrt{\frac{1+\left(-\frac{7}{25}\right)}{2}}$$

$$= -\sqrt{\frac{18}{50}} = -\frac{3}{5}$$

CLASS EXERCISES

1. **Thinking Critically** Discuss how $\cos 2A = 2\cos^2 A - 1$ and $\cos 2A = 1 - 2\sin^2 A$ can both be derived from $\cos 2A = \cos^2 A - \sin^2 A$.

If $90° < \angle Q < 180°$ and $\sin Q = \frac{12}{13}$ find the exact value:

2. $\sin \frac{Q}{2}$
3. $\cos \frac{Q}{2}$
4. $\tan \frac{Q}{2}$

PRACTICE EXERCISES

Use a special angle and a double-angle formula to find the exact value.

1. $\sin 90°$
2. $\cos 90°$
3. $\tan 90°$
4. $\sin 300°$
5. $\cos 300°$
6. $\tan 300°$

If $\angle B$ is in the first quadrant and $\sin B = \frac{12}{13}$, find the exact value.

7. $\sin 2B$
8. $\cos 2B$
9. $\tan 2B$

If $\angle A$ is in the second quadrant and $\cos A = -\frac{4}{5}$, find the exact value.

10. $\cos 2A$
11. $\sin 2A$
12. $\cot 2A$

Use a special angle and a half-angle formula to find the exact value.

13. $\cos 15°$
14. $\sin 75°$
15. $\tan 30°$

If $0° < \angle B < 90°$ and $\sin B = \frac{3}{5}$, find the exact value.

16. $\cos \frac{B}{2}$
17. $\sin \frac{B}{2}$
18. $\tan \frac{B}{2}$

If $90° < \angle A < 180°$ and $\cos A = -\frac{4}{5}$, find the exact value.

19. $\sin \frac{A}{2}$
20. $\tan \frac{A}{2}$
21. $\cot \frac{A}{2}$

Use a special angle and a half-angle formula to find the exact value.

22. $\sin 22.5°$
23. $\tan 67.5°$
24. $\cos 202.5°$

If $180° < \angle C < 270°$ and $\cos C = -\frac{15}{17}$, find the exact value.

25. $\sin 2C$
26. $\cos 2C$
27. $\tan 2C$

28. $\cos \frac{C}{2}$

29. $\sin \frac{C}{2}$

30. $\tan \frac{C}{2}$

If $270° < \angle B < 360°$ and $\cos B = \frac{3}{5}$, find the exact value.

31. $\sin 2B$

32. $\cos 2B$

33. $\tan 2B$

34. $\tan \frac{B}{2}$

35. $\sin \frac{B}{2}$

36. $\cos \frac{B}{2}$

37. Prove: $\tan \frac{x}{2} = \pm\sqrt{\frac{1 - \cos x}{1 + \cos x}}$

38. Prove: $\tan \frac{A}{2} = \frac{\sin A}{1 + \cos A}$

Use the double-angle formulas to derive a formula for each.

39. $\sin 4A$

40. $\cos 4A$

41. $\tan 4A$

Use the half-angle formulas to derive a formula for each.

42. $\sin \frac{A}{4}$

43. $\cos \frac{A}{4}$

44. $\tan \frac{A}{4}$

Applications

Geometry In right $\triangle RST$ with $\angle T = 90°$, show that:

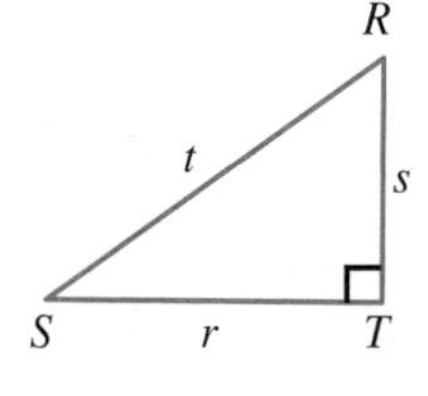

45. $\sin 2R = \frac{2rs}{t^2}$

46. $\cos 2R = \frac{s^2 - r^2}{t^2}$

47. $\sin 2S = \sin 2R$

48. $\sin^2 \frac{S}{2} = \frac{t - r}{2t}$

49. $\tan \frac{R}{2} = \frac{r}{t + s}$

Developing Mathematical Power

Extension There are many ways to define the trigonometric functions. When the angle x is in radian measure,

$$\sin x = x - \frac{x^3}{3!} + \frac{x^5}{5!} - \cdots$$

$$\cos x = 1 - \frac{x^2}{2!} + \frac{x^4}{4!} - \cdots$$

Definitions in terms of infinite series can be very useful for computing, hence for building tables of values for the trigonometric functions.

50. What is $\sin x$ when $x = \frac{\pi}{2}$ radians?

51. Use the infinite series to show $\sin 2x = 2 \sin x \cos x$.

16.5 Inverse Trigonometric Functions

Objective: To define and use inverse trigonometric functions

To define the inverse of a trigonometric function, it is helpful to recall how to define the inverse of other functions.

Capsule Review

EXAMPLE **Find f^{-1} for $f(x) = 3x - 6$.**

$$y = 3x - 6$$
$$x = 3y - 6 \qquad \textit{Interchange } x \textit{ and } y.$$
$$x + 6 = 3y \qquad \textit{Solve for } y.$$
$$\frac{x}{3} + 2 = y \qquad \text{So, } f^{-1} \text{ is } y = \frac{x}{3} + 2.$$

Find the inverse of each function.

1. $y = -5x + 3$ **2.** $y = 3x^2$ **3.** $f(x) = -2x + 8$ **4.** $f(x) = \frac{x}{2} + 1$

5. The graphs of f and f^{-1} are symmetric with respect to the line _?_.

The inverse of a function is formed by interchanging x and y and then solving for y.

$y = \cos x$ means: x is an angle whose cosine is y.

inverse: $x = \cos y$ — y is an angle whose cosine is x.

The *inverse cosine* function, that is, the inverse of the function $y = \cos x$, is written as $y =$ *arc cos x* or $y = \cos^{-1} x$. Note that $\cos^{-1} x$ *does not mean* the reciprocal of $\cos x$, $\frac{1}{\cos x}$. The inverse of each of the other trigonometric functions can be formed in the same manner.

EXAMPLE 1 **Evaluate:** **a.** $\text{arc cos } \frac{\sqrt{2}}{2}$ **b.** $\sin^{-1} 1$

a. Let $A = \text{arc cos } \frac{\sqrt{2}}{2}$. Then $\cos A = \frac{\sqrt{2}}{2}$.

Recall that $\cos\left(\pm\frac{\pi}{4}\right) = \frac{\sqrt{2}}{2}$ and $\cos\left(\pm\frac{7\pi}{4}\right) = \frac{\sqrt{2}}{2}$.

Therefore $A = \frac{\pi}{4} + 2\pi k$ or $-\frac{\pi}{4} + 2\pi k$, where k represents an integer.

So, $\text{arc cos } \frac{\sqrt{2}}{2} = \frac{\pi}{4} + 2\pi k$ or $-\frac{\pi}{4} + 2\pi k$, where k is an integer.

b. Let $A = \sin^{-1} 1$. Then $\sin A = 1$.

Recall that $\sin \frac{\pi}{2} = 1$, $\sin \frac{5\pi}{2} = 1$, and $\sin\left(-\frac{3\pi}{2}\right) = 1$.

Therefore $A = \frac{\pi}{2} + 2\pi k$, where k represents an integer.

So, $\sin^{-1} 1 = \frac{\pi}{2} + 2\pi k$, where k is an integer.

The graphs of $y = \cos x$ and $y = \cos^{-1} x$ are shown below. Notice that $y = \cos^{-1} x$ is *not* a function. Remember that $y = \cos x$ is a many-to-one function. That is, for any given value of y there is more than one value of x. The domain of $y = \cos x$ can be restricted so that it is a *one-to-one* function and then its inverse will define a function.

If the domain of $y = \cos x$ is restricted so that $0 \leq x \leq \pi$, $y = \cos x$ will have an inverse that is a function. Note that other restrictions can also be made. When the restriction is made, the notation of the function is changed to $y = \text{Cos } x$ and the notation of the inverse cosine is changed to

$$y = \text{Cos}^{-1}\, x \quad \text{or} \quad y = \text{Arccos } x$$

The graphs of these functions are shown with a solid line. The values of $\text{Cos}^{-1}\, x$ are called the **principal values** of $\cos^{-1} x$.

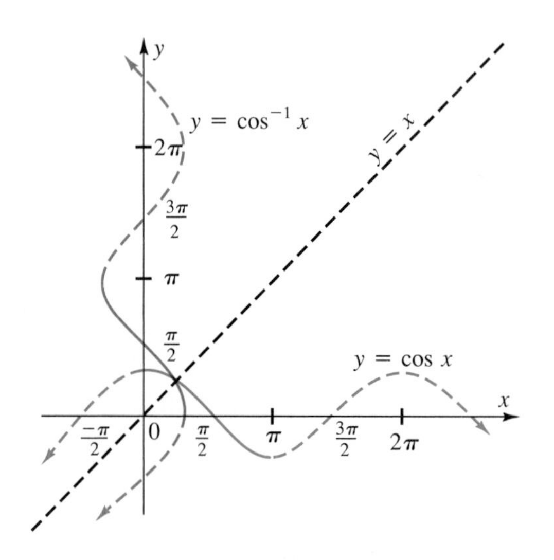

The domains of the other trigonometric functions can be restricted so that their inverses are functions. The graphs of $y = \text{Sin}^{-1}\, x$ (or $y = \text{Arcsin } x$) and $y = \text{Tan}^{-1}\, x$ (or $y = \text{Arctan } x$) are shown below.

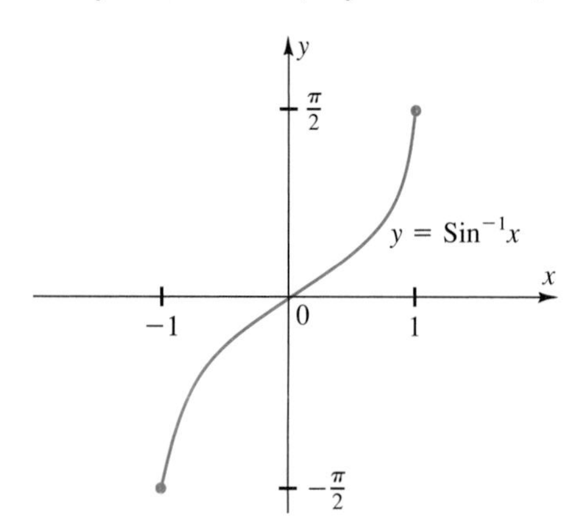

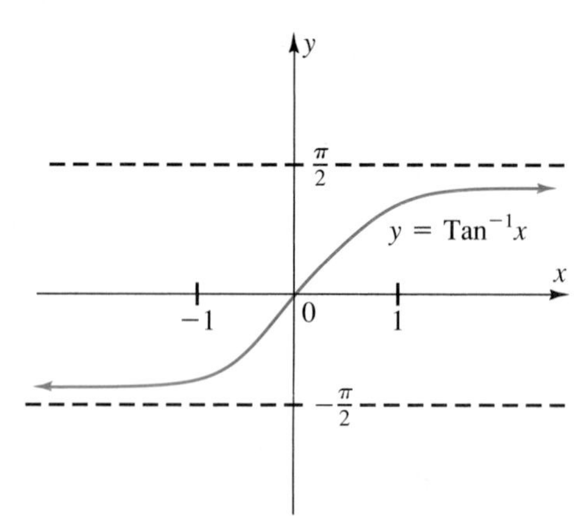

Inverse Trigonometric Functions

Function	Domain	Range
$y = \text{Sin } x$	$-\frac{\pi}{2} \leq x \leq \frac{\pi}{2}$	$-1 \leq y \leq 1$
$y = \text{Sin}^{-1} x$	$-1 \leq x \leq 1$	$-\frac{\pi}{2} \leq y \leq \frac{\pi}{2}$
$y = \text{Cos } x$	$0 \leq x \leq \pi$	$-1 \leq y \leq 1$
$y = \text{Cos}^{-1} x$	$-1 \leq x \leq 1$	$0 \leq y \leq \pi$
$y = \text{Tan } x$	$-\frac{\pi}{2} < x < \frac{\pi}{2}$	all real numbers
$y = \text{Tan}^{-1} x$	all real numbers	$-\frac{\pi}{2} < y < \frac{\pi}{2}$
$y = \text{Cot } x$	$0 < x < \pi$	all real numbers
$y = \text{Cot}^{-1} x$	all real numbers	$0 < y < \pi$
$y = \text{Sec } x$	$0 \leq x \leq \pi,\ x \neq \frac{\pi}{2}$	$\lvert y \rvert \geq 1$
$y = \text{Sec}^{-1} x$	$\lvert x \rvert \geq 1$	$0 \leq y \leq \pi,\ y \neq \frac{\pi}{2}$
$y = \text{Csc } x$	$-\frac{\pi}{2} \leq x \leq \frac{\pi}{2}$	$\lvert y \rvert \geq 1$
$y = \text{Csc}^{-1} x$	$\lvert x \rvert \geq 1$	$-\frac{\pi}{2} \leq y \leq \frac{\pi}{2}$

EXAMPLE 2 **Evaluate:** **a.** $\text{Tan}^{-1} 1$ **b.** $\text{Csc}^{-1} 2$

a. Let $A = \text{Tan}^{-1} 1$. Then $\text{Tan } A = 1$.
Recall $\text{Tan } \frac{\pi}{4} = 1$. $\text{Tan}^{-1} 1 = \frac{\pi}{4}$.

b. Let $A = \text{Csc}^{-1} 2$. Then $\text{Csc } A = 2$.
Recall $\text{Csc } \frac{\pi}{6} = 2$. $\text{Csc}^{-1} 2 = \frac{\pi}{6}$.

When using a calculator to find values of inverse trigonometric functions, the calculator displays only principal values. For example, consider part (a) of Example 2. Enter 1 using the inv and tan keys. The display will show 45 (calculator in degree mode) or 0.78539816 (calculator in radian mode).

EXAMPLE 3 **Evaluate:** $\textbf{Sin}^{-1} \textbf{0.7314}$

Let $A = \text{Sin}^{-1} 0.7314$
Then $\text{Sin } A = 0.7314$
$A = 47°$ *Use a calculator in degree mode.*
$\text{Sin}^{-1} 0.7314 = 47°$

When an expression contains parentheses, first evaluate the expression inside the parentheses.

EXAMPLE 4 **Evaluate:** $\text{Cos}^{-1}\left[\cos\left(-\frac{\pi}{6}\right)\right]$

$$\text{Cos}^{-1}\left[\cos\left(-\frac{\pi}{6}\right)\right] = \text{Cos}^{-1}\frac{\sqrt{3}}{2} = \frac{\pi}{6}$$

CLASS EXERCISES

Thinking Critically

1. Explain the difference between "$y = \text{arc} \cos x$" and "$y = \text{Arc} \cos x$."
2. Explain how to restrict the domains of $y = \csc x$, $y = \sec x$, and $y = \cot x$ so that the inverses are functions.
3. Find $\text{Sin}^{-1}\, 0.5$
4. Find $\text{Sec}^{-1}\, 2$
5. Find $\tan\left(\text{Arc} \cos \frac{3}{5}\right)$

PRACTICE EXERCISES

Use technology where appropriate.

Evaluate.

1. $\text{arc} \cos 0.5$
2. $\text{arc} \cos 0$
3. $\cos^{-1}\frac{\sqrt{3}}{2}$
4. $\cos^{-1} 1$
5. $\text{Arc} \cos 0.5$
6. $\text{Arc} \cos (-1)$
7. $\text{Cos}^{-1}\, 1$
8. $\text{Cos}^{-1}\, 0$
9. $\text{arc} \sin \frac{1}{2}$
10. $\text{arc} \sin\left(-\frac{\sqrt{2}}{2}\right)$
11. $\text{Tan}^{-1}\, 1$
12. $\text{Csc}^{-1}\, 2$
13. $\text{Arc} \cot (-\sqrt{3})$
14. $\text{Arc} \sec 2$
15. $\text{arc} \sin 0.6157$
16. $\text{arc} \cos 0.7547$
17. $\text{arc} \tan 0.3839$
18. $\text{arc} \sec 1.079$
19. $\text{Arc} \sin 0.3584$
20. $\text{Arc} \sin 0.2250$
21. $\text{Sec}^{-1}\, 1.103$
22. $\text{Tan}^{-1}\, 0.2493$
23. $\text{Sin}^{-1}\left[\sin\left(-\frac{\pi}{6}\right)\right]$
24. $\text{Cos}^{-1}\left[\cos\left(-\frac{\pi}{3}\right)\right]$
25. $\sin\left(\text{Sin}^{-1}\frac{3}{5}\right)$
26. $\cos\left(\text{Cos}^{-1}\frac{2}{5}\right)$
27. $\tan\left(\text{Arc} \sin \frac{3}{5}\right)$
28. $\csc\left(\text{Arc} \cos \frac{5}{13}\right)$
29. $\sin (\text{Arc} \sin 0.6)$
30. $\cos (\text{Arc} \cos 0.8)$
31. $\tan\left(\text{Sin}^{-1}\frac{12}{13}\right)$
32. $\cos\left(\text{Tan}^{-1}\frac{8}{17}\right)$
33. $\sin\left(\text{Sec}^{-1}\frac{17}{15}\right)$
34. $\tan\left(\text{Csc}^{-1}\frac{13}{5}\right)$
35. $\sec\left(\text{Cos}^{-1}\frac{4}{5}\right)$
36. $\cot\left(\text{Tan}^{-1}\frac{8}{15}\right)$
37. $\cot\left[\text{Csc}^{-1}\left(-\frac{13}{5}\right)\right]$
38. $\sec\left[\text{Cos}^{-1}\left(-\frac{3}{5}\right)\right]$
39. $\csc\left[\text{Tan}^{-1}\left(-\frac{8}{15}\right)\right]$
40. $\text{Sin}^{-1}\left(\cos\frac{\pi}{4}\right)$
41. $\text{Tan}^{-1}\left(\cot\frac{\pi}{4}\right)$
42. $\text{Cos}^{-1}\left(\sin\frac{\pi}{6}\right)$

43. $\cos\left(\frac{\pi}{4} + \text{Cos}^{-1} 0.5\right)$

44. $\sin\left(\frac{\pi}{4} + \text{Sin}^{-1} 0.5\right)$

45. $\cos(\text{Tan}^{-1} 1 + \text{Cot}^{-1} 1)$

46. $\tan(\text{Sin}^{-1} 0.5 - \text{Cos}^{-1} 0.5)$

47. $\sin\left(\text{Sin}^{-1} \frac{3}{5} + \text{Cos}^{-1} \frac{5}{13}\right)$

48. $\tan\left(\text{Tan}^{-1} \frac{3}{2} + \text{Tan}^{-1} \frac{4}{3}\right)$

Show that each of the following is true.

49. $\text{Cos}^{-1} x = \frac{\pi}{2} - \text{Sin}^{-1} x$

50. $\text{Arc sin } \frac{8}{17} + \text{Arc tan } \frac{15}{8} = \frac{\pi}{2}$

Applications

Physics The surface of the pavement of an auto raceway is tilted, or banked, so that the outer part is higher than the inner part. This allows cars to go around a curve at a relatively high speed. The following formula gives the correct banking angle for a curve of radius r and vehicle speed v with g representing the force of gravity: $\theta = \tan^{-1} \frac{v^2}{rg}$.

51. If r and g remain the same, how does θ change as v changes?

52. Other than those in the formula, what other factors would influence the banking of a highway instead of a raceway?

TEST YOURSELF

Prove each identity.

16.1

1. $\sec^2 A - \tan^2 A = 1$

2. $\tan S = \frac{\sin S}{\cos S}$

16.2

3. $\cos B + \tan B \sin B = \sec B$

4. $\frac{\sec S}{\sin S} - \frac{\sin S}{\cos S} = \cot S$

16.3

5. $\angle A$ is a second-quadrant angle, $\sin A = \frac{4}{5}$, $\angle B$ is a third-quadrant angle, and $\cos B = -\frac{12}{13}$. Find the exact value of $\cos(A - B)$.

6. $\angle A$ is a fourth-quadrant angle, $\sin A = -\frac{8}{17}$, $\angle B$ is a second-quadrant angle, and $\cos B = -\frac{5}{13}$. Find the exact value of $\tan(A + B)$.

16.4

7. If $\angle A$ is a fourth-quadrant angle and $\sin A = -\frac{4}{5}$, find the exact value of $\sin 2A$.

8. If $90° < \angle C < 180°$ and $\cos C = -\frac{3}{5}$, find the exact value of $\cos \frac{C}{2}$.

Evaluate.

16.5

9. $\cos\left(\text{Arc sin } \frac{12}{13}\right)$

10. $\text{Cos}^{-1}\left(\sin \frac{\pi}{2}\right)$

16.6

Solving Trigonometric Equations

Objective: To solve trigonometric equations

Equations involving trigonometric functions can be solved by using methods from algebra and trigonometric identities.

Capsule Review

EXAMPLE **Solve:** **a.** $x^2 - 7x + 12 = 0$ **b.** $6x^2 - 5x = -1$

a.
$$x^2 - 7x + 12 = 0$$
$$(x - 4)(x - 3) = 0$$
$$x - 4 = 0 \quad \text{or} \quad x - 3 = 0$$
$$x = 4 \quad \text{or} \quad x = 3$$

b.
$$6x^2 - 5x = -1$$
$$6x^2 - 5x + 1 = 0$$
$$(3x - 1)(2x - 1) = 0$$
$$3x - 1 = 0 \quad \text{or} \quad 2x - 1 = 0$$
$$x = \frac{1}{3} \quad \text{or} \quad x = \frac{1}{2}$$

Solve.

1. $x^2 + 8x + 15 = 0$ **2.** $x^2 - 4x = 21$ **3.** $15m^2 = 8m - 1$

4. $4x^2 = 16x$ **5.** $(t + 3)(t^2 - 16) = 0$ **6.** $(y - 6)^2 = y$

When a trigonometric equation contains only one function, one strategy for solving the equation is to use algebraic methods. The solution is expressed in radians or degrees depending on the specified interval.

EXAMPLE 1 **Solve:** $\sin^2 x + 3 \sin x + 2 = 0,\ 0 \le x < 2\pi$

$$\sin^2 x + 3 \sin x + 2 = 0$$
$$(\sin x + 2)(\sin x + 1) = 0 \qquad \textit{Factor.}$$
$$\sin x + 2 = 0 \quad \text{or} \quad \sin x + 1 = 0 \qquad \textit{Use the zero-product property.}$$
$$\sin x = -2 \quad \text{or} \quad \sin x = -1 \qquad \textit{Reject } -2 \textit{ since } -1 \le \sin x \le 1.$$
$$x = \frac{3\pi}{2}$$

The solution within the specified interval is $\frac{3\pi}{2}$.

In Example 1, $\frac{3\pi}{2}$ is called a **primary solution** because it is a solution within the given interval. All angles that are coterminal with the angle that is a primary solution would also be a solution. These solutions differ from the primary solution by integral multiples of the period of the function and are called *general solutions* of the equation. The general solution of the equation in Example 1 is

$$x = \frac{3\pi}{2} + 2k\pi, \text{ where } k \text{ is an integer.}$$

When a trigonometric equation contains more than one function, transform it into an equation containing only one trigonometric function. Use the identities and substitute.

EXAMPLE 2 **Solve: $5 \sec^2 x + 2 \tan x = 8$, $0° \leq x < 360°$**

$$\begin{aligned} 5 \sec^2 x + 2 \tan x &= 8 \\ 5(\tan^2 x + 1) + 2 \tan x - 8 &= 0 \quad \textit{Substitute } \tan^2 x + 1 \textit{ for } \sec^2 x. \\ 5 \tan^2 x + 5 + 2 \tan x - 8 &= 0 \\ 5 \tan^2 x + 2 \tan x - 3 &= 0 \\ (5 \tan x - 3)(\tan x + 1) &= 0 \end{aligned}$$

$$\tan x = \frac{3}{5} \quad \text{or} \quad \tan x = -1$$

$$\tan x = 0.6 \quad \text{or} \quad \tan x = -1$$

$$x = 31° \qquad x = 135°$$

$$x = 211° \qquad x = 315°$$

Since the tangent function has a period of 180°, the solutions within the specified interval are 31°, 135°, 211°, and 315°.

CLASS EXERCISES

If $0° \leq x < 360°$, how many solutions does the equation have?

1. $\sin x \cos x = 0$

2. $\cos x - \sec x = 0$

3. $5 \sin x - 2 = 3$

4. $\cot^2 x = 4$

If $0 \leq x < 2\pi$, how many solutions does the equation have?

5. $4 \cos x = 3$

6. $3 \tan^2 x = 1$

7. $2 \sin x \cos x - \sin x = 0$

8. $2 \sec^2 x - 3 \sec x - 2 = 0$

Solve for x, $0 \leq x \leq 2\pi$.

9. $2 \cos x = 1$

10. $\sin^2 x - 1 = 0$

PRACTICE EXERCISES

Use technology where appropriate.

Solve for x, $0° \leq x < 360°$.

1. $2 \sin x = 1$

2. $2 \csc x = -4$

3. $\tan^2 x - 4 = 0$

4. $\cot^2 x - 3 = 0$

5. $2 \sin x \cos x = 1$

6. $2 \cot^2 x - 1 = 0$

7. $3 \sin^2 x + 5 \sin x - 2 = 0$

8. $\cos^2 x - 4 \sin^2 x = 0$

Solve for x, $0 \leq x < 2\pi$.

9. $\tan x = 2 \sin x$

10. $2 \sec x + 4 = \sec x + 6$

11. $\tan x - \cot x = 0$

12. $2 \cos x \cot x - 3 = 0$

13. $2 \cos^2 x - 5 \cos x + 2 = 0$

14. $\sec x = \csc x$

15. $1 + \sin x + \cos^2 x = 0$

16. $5 \sin x = 1 + 3 \sin x$

17. $4 \sin^2 x + 1 = 4 \sin x$

18. $2 \cos^2 x + 9 \cos x - 5 = 0$

19. $\sin x \csc x - \csc x + 2 \sin x = 0$

20. $\tan x \cot x - \tan x + 2 \cot x = 0$

21. $\sin 2x = 2 \sin x$

22. $\tan 2x = \cot x$

23. $2 \cos^2 \frac{x}{2} = 2 \cos x + 2$

24. $\cos x = \sin \frac{x}{2}$

25. $\sec^2 x = 2 \tan x$

26. $\sin x \cot^2 x - 3 \sin x = 0$

27. $\cos 2x + \sin x = 1$

28. $\tan 2x = 2 \sin x$

29. $\sec x + 2 \cos x = 3$

30. $\sin 2x + \sin x = 0$

31. $2 \sin x + 1 = \csc x$

32. $2 \cos^2 x + \sin x = 1$

33. $\sin 3x \cos x - \cos 3x \sin x = 0.5$

34. $\sin 2x \cos x + \cos 2x \sin x = 1$

35. $\tan^2 x - \sin^2 x = 0$

36. $\sin 2x = 2 \cos x$

37. $\sin x = 1 - \cos x$

38. $\sin^2 x + \cos 2x = \cos x$

Applications

39. Music A tuning fork generates a wave represented by $\sin x + \cos x = y$. If $0 \leq x < 2\pi$, what values of x satisfy the equation $\sin x + \cos x = \sqrt{2}$?

40. Sound Consider the sound waves described by $f(x) = \sin 3x$ and $g(x) = \cos 3x$. Research how to describe the sound waves when they are played together. Then, graph this behavior over the interval $0 \leq x \leq 2\pi$.

Mixed Review

41. In how many ways can 10 symbols be used to represent the 7 days of the week?

42. A box of 40 granola bars has been invaded by 30 cockroaches. What is the probability that a bar chosen at random has a cockroach on it?

43. Expand $(x^3 + 4)^4$.

44. Find the sixth term of $(x + 3)^{14}$.

16.7 Vector Operations

Objective: To find the sum and difference of two vectors, multiply a vector by a scalar, and find the norm and bearing of a vector

Vectors can be used to describe quantities that have both magnitude and direction. Examples of vectors are force, velocity, and acceleration. The laws of sines and cosines can be used to solve some vector problems.

Capsule Review

Use the law of sines to find the indicated measure.

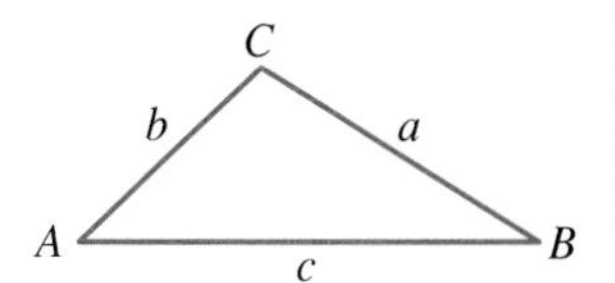

1. In $\triangle ABC$, $a = 20$, $b = 13$, and $\angle A = 26°$, find $\angle B$.

2. In $\triangle DEF$, $d = 32$, $\angle D = 50°$, and $\angle E = 64°$, find e.

Use the law of cosines to find the indicated measure.

3. In $\triangle ABC$, $a = 16$, $b = 12$, and $\angle C = 26°$, find c.

4. In $\triangle DEF$, $d = 15$, $e = 18$, and $f = 10$, find $\angle F$.

Any quantity that has both magnitude and direction is called a **vector quantity.** A vector quantity is represented by a **directed line segment,** or arrow, called a **vector.** The length of the vector is proportional to the magnitude of the vector quantity. The direction of the arrow gives the direction of the vector quantity.

Vector AB, denoted $\overrightarrow{AB}$, is named by first naming its initial point A (the tail) and then naming its terminal point B (the head). The length of $\overrightarrow{AB}$ can be found by using the distance formula. $\overrightarrow{AB}$ and $\overrightarrow{CD}$ are **equivalent** because they have the same length and the same direction.

$$\overrightarrow{AB} = \overrightarrow{CD}$$

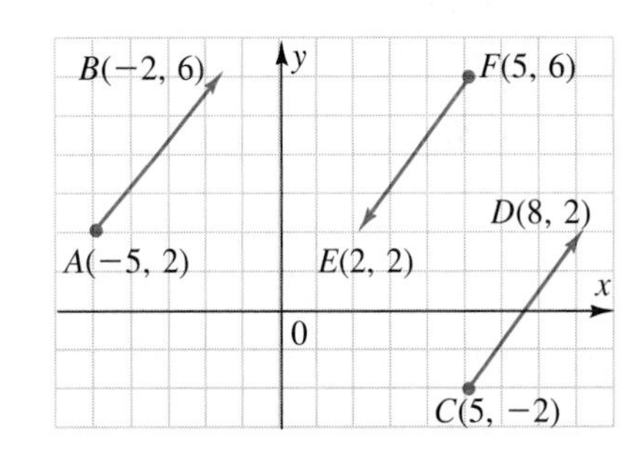

$\overrightarrow{AB}$ is the **opposite** of $\overrightarrow{FE}$ because they have the same length but their directions are opposite. This is expressed as $\overrightarrow{AB} = -\overrightarrow{FE}$.

A single vector can be used to represent the **sum** or **resultant** of two vectors. If the vectors have the same initial point, use equivalent vectors and form a parallelogram by the parallelogram rule at the top of page 820. If the vectors do not have the same initial point, move the vectors, and complete a triangle. When moving a vector, keep it parallel to its original direction.

Parallelogram Rule Use the given vectors as adjacent sides and complete a parallelogram. The diagonal is the sum or resultant vector.

A vector can also be named with a single letter.

EXAMPLE 1 **Find:** **a.** $\overrightarrow{AB} + \overrightarrow{AC}$ **b.** $\mathbf{t} + \mathbf{v}$

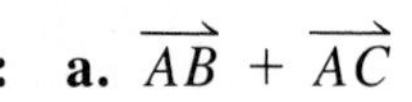

a. $\overrightarrow{AB} + \overrightarrow{AC}$

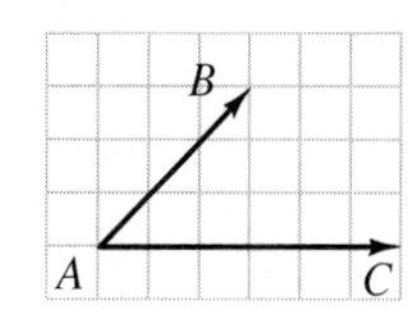

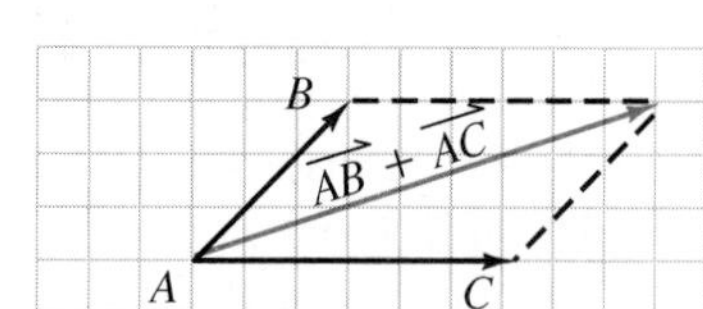

b. $\mathbf{t} + \mathbf{v}$

To get the resultant vector, connect the tail of one vector to the head of the other.

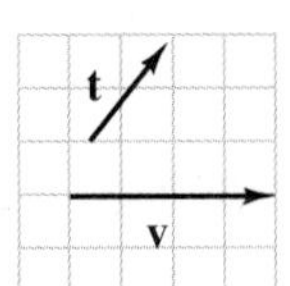

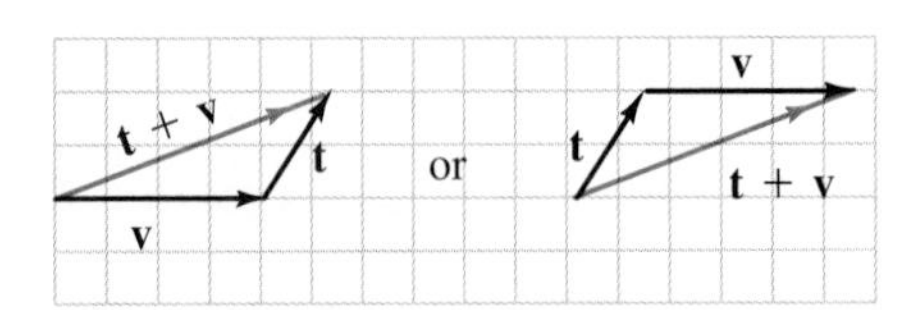

Subtracting vectors is similar to subtracting real numbers. Just as $a - b = a + (-b)$ for all real numbers a and b, $\mathbf{a} - \mathbf{b} = \mathbf{a} + (-\mathbf{b})$ for all vectors **a** and **b.** Also, a vector can be multiplied by a **scalar.** A scalar is a real number. To multiply a vector by a scalar, multiply the magnitude of the vector by the scalar. If the scalar is negative, the direction of the vector representing the product is reversed.

EXAMPLE 2 **Use the vectors at the right to draw:**
a. $\mathbf{c} - \mathbf{d}$ **b.** $2 \cdot \mathbf{c}$ **c.** $-3 \cdot \mathbf{d}$

a. $\mathbf{c} - \mathbf{d}$

Draw $-\mathbf{d}$.
Then draw $\mathbf{c} + (-\mathbf{d})$.

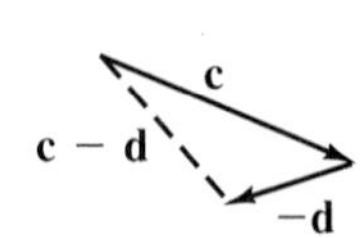

b. $2 \cdot \mathbf{c}$

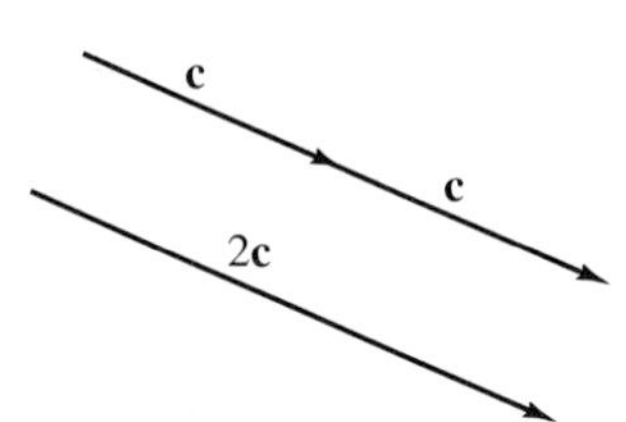

c. $-3 \cdot \mathbf{d}$

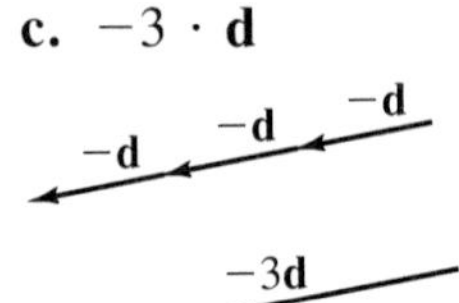

The symbol $\|\mathbf{a}\|$ indicates the **magnitude,** or length, of **a.**

Any vector, $\overrightarrow{OA}$, can be resolved into its *components* by placing the vector on a coordinate axis with its initial point at the origin. The coordinates of its terminal point are (x, y). In the figure at the right, $\overrightarrow{OA}$ has a magnitude of r and a direction of $\theta°$ and $\overrightarrow{OA} = \overrightarrow{OC} + \overrightarrow{OB}$. The ***x*-component** is $\overrightarrow{OB}$ and the ***y*-component** is $\overrightarrow{OC}$. Since $\sin\theta = \frac{y}{r}$ and $\cos\theta = \frac{x}{r}$, $y = r\sin\theta$ and $x = r\cos\theta$. The scalars x and y may also be referred to as the x- and y-components of $\overrightarrow{OA}$.

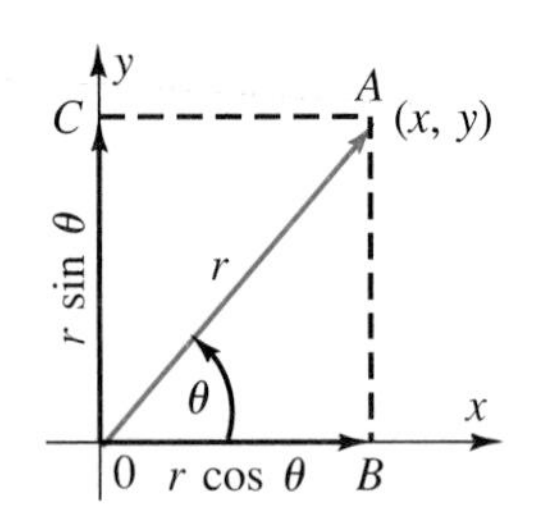

EXAMPLE 3 **If $\overrightarrow{OC}$ has a magnitude of 5 and a direction of 125°, find the x- and y-components of $\overrightarrow{OC}$ to the nearest whole number.**

x-component $= r \cos \theta$ $\qquad$ y-component $= r \sin \theta$

$x = 5 \cos 125° = -2.8679$ $\qquad$ $y = 5 \sin 125° = 4.0958$

The x-component is -3 and the y-component is 4.

The **bearing** of a vector is the angle measured clockwise from the north around to the vector. In the diagram, the bearing of $\overrightarrow{OA}$ is 60°. The bearing of an airplane's path in still air is called its *heading*. The bearing of a plane's actual path under the influence of wind is called the plane's course angle, or *course*.

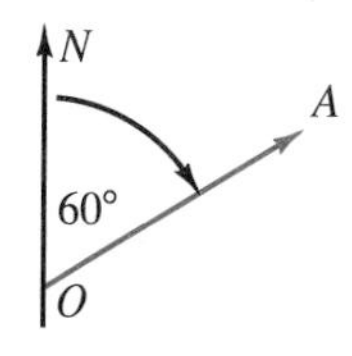

EXAMPLE 4 **An airplane with an air speed of 200 mi/h is flying on a heading of 58°. The wind is blowing from due north at 26 mi/h. What is the ground speed of the plane and its course?**

$$a^2 = b^2 + c^2 - 2bc \cos A$$

$$\|\overrightarrow{OB}\|^2 = 200^2 + 26^2 - 2(200)(26) \cos 58°$$

$$= 35{,}164.84$$

$$\|\overrightarrow{OB}\| = 187.5$$

$$\frac{\sin 58°}{187.5} = \frac{\sin \theta}{26} \qquad \frac{\sin A}{a} = \frac{\sin B}{b}$$

$$\sin \theta = \frac{26 \sin 58°}{187.5} \qquad \textit{Calculation-ready form}$$

$\theta = 6°45'$ $\qquad$ The course is approximately $58° + 6°45' = 64°45'$.

The ground speed of the plane is about 188 mi/h and its course is about 65°.

CLASS EXERCISES

In Exercises 1–8, use parallelogram $ACEG$. $\overrightarrow{HD} \| \overrightarrow{GE}$ and $\overrightarrow{FB} \| \overrightarrow{EC}$

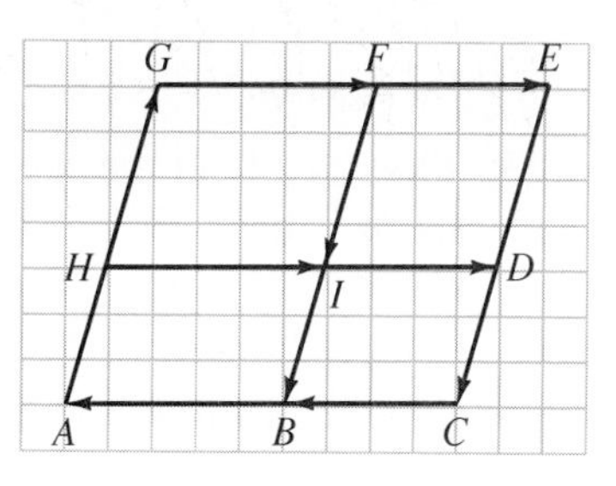

1. Name two vectors that are equivalent to $\overrightarrow{AB}$.
2. Name three vectors that are opposites of $\overrightarrow{HI}$.
3. Name a vector that is collinear with $\overrightarrow{AH}$ and has the same direction as $\overrightarrow{AH}$.
4. Name two vectors that are collinear with $\overrightarrow{ED}$ and have the opposite direction from $\overrightarrow{ED}$.

Find the following sums.

5. $\overrightarrow{FI} + \overrightarrow{IB}$ $\qquad$ 6. $\overrightarrow{CB} + \overrightarrow{BA}$ $\qquad$ 7. $\overrightarrow{AF} + \overrightarrow{FD}$ $\qquad$ 8. $\overrightarrow{GI} + \overrightarrow{IA}$

PRACTICE EXERCISES

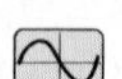 Use technology where appropriate.

Use the vectors at the right to draw each of the following.

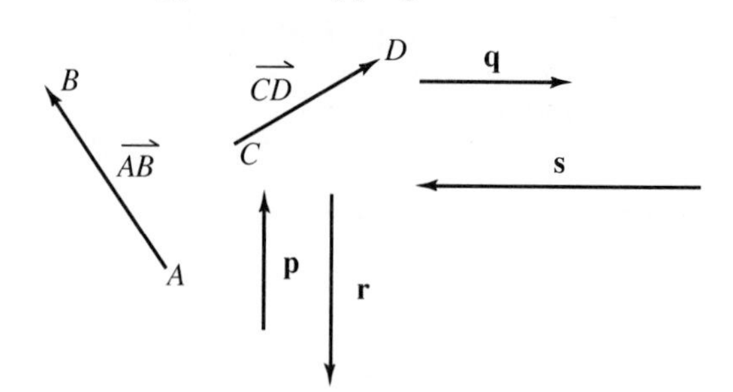

1. $\overrightarrow{AB} + \overrightarrow{CD}$
2. $\mathbf{p} + \mathbf{q}$
3. $\mathbf{r} + \mathbf{s}$
4. $\mathbf{q} + \overrightarrow{CD}$
5. $\mathbf{p} - \mathbf{q}$
6. $\mathbf{s} - \mathbf{r}$
7. $3 \cdot \mathbf{s}$
8. $2 \cdot \mathbf{p}$

Find the *x*- and *y*-components, to the nearest integer, of each vector.

9. $\overrightarrow{AB}$: magnitude 3, direction 40°
10. $\overrightarrow{CD}$: magnitude 8, direction 35°
11. **v**: magnitude 18, direction 130°
12. **w**: magnitude 24, direction 145°

In Exercises 13–18, draw each using the vectors given above.

13. $\mathbf{p} + \mathbf{q} + \mathbf{r}$
14. $\mathbf{p} + \mathbf{q} + \mathbf{s}$
15. $\mathbf{p} + \mathbf{q} - \mathbf{r}$
16. $\mathbf{p} + \mathbf{s} - \mathbf{q}$
17. $3(\mathbf{p} + \mathbf{r})$
18. $2(\mathbf{q} + \mathbf{s})$
19. A plane with an air speed of 250 mi/h is flying on a heading of 130°. The wind is blowing from due south at 50 mi/h. What is the ground speed of the plane and its course?
20. Two planes left an airport at the same time. One flew due east at 225 mi/h, and the other flew at 190 mi/h. On what heading did the second plane fly if the two planes were 400 mi apart after 1 h?
21. The air speed of a plane is 245 mi/h and its heading is 188°. If the wind is blowing at 20 mi/h and the ground speed of the plane is 250 mi/h, what is the course of the plane?

Applications

22. **Physics** Forces of 16 lb and 30 lb act on a body at right angles. Find the resultant force.
23. **Physics** Find the measure of the angle between forces of 78 lb and 67 lb if the resultant force is 94 lb.

Developing Mathematical Power

Writing in Mathematics Write a short paragraph about each word or phrase explaining what purpose each serves. Then add a specific example.

24. resultant vectors
25. double-angle formula
26. trigonometric identity

16.8

Problem Solving Strategy: Use Trigonometry

You can use trigonometry to solve problems involving geometric concepts from the real world such as surveying, navigation, and engineering. Such problems can be solved more easily with the aid of a diagram.

EXAMPLE 1 Find the length and direction of the third side of a pasture with boundaries defined as follows: Beginning at the brass stake by the old well walk, proceed 334 ft at a bearing of 110° from north, then proceed 917 ft at a bearing of 200° from north. This point is due south of the brass stake.

Understand the Problem *What are the given facts?* You are given a pasture whose sides are a 334-ft vector at a bearing of 110° from north and a 917-ft vector at a bearing of 200° from north. *What are you asked to find?* You are asked to find the length and direction of the third side of a pasture.

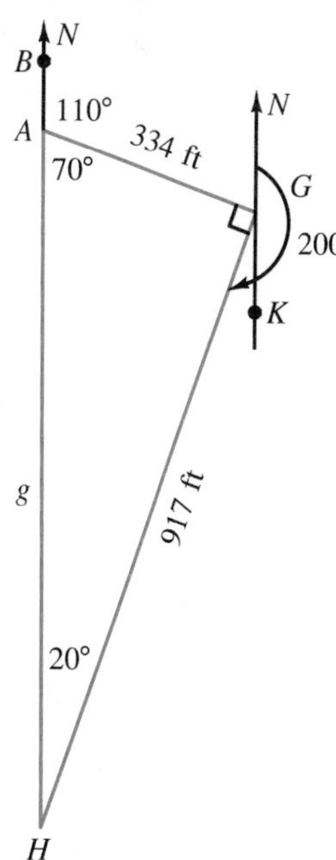

Plan Your Approach *Choose a strategy.* Draw and label a diagram to show the given boundaries. Since lines BAH and NGK are parallel, alternate interior angles BAG and KGA are equal; each is 110°. Also, $\angle HGK$ is $200° - 180°$, or 20°. Thus,

$$\angle AGH = 110° - 20° = 90°$$

Therefore, the triangle is a right triangle.

$$\sin 70° = \frac{917}{g}$$

Complete the Work

$$g = \frac{917}{\sin 70°} = 975.8$$ *Calculation-ready form*

Interpret the Results *State your answer.* The length of the third side of the pasture is about 976 ft at a bearing due north.

Trigonometry can also be used in problems in physics that involve forces.

EXAMPLE 2 Two forces, one of 150 lb and the other of 200 lb act on a body and make an angle measuring 56°20′ with each other. What is the magnitude to the nearest pound of the resultant of the forces and what is the measure to the nearest 10 min of the angle that it makes with the 200-lb force?

Understand the Problem You are asked to find the resultant of the forces and its direction.

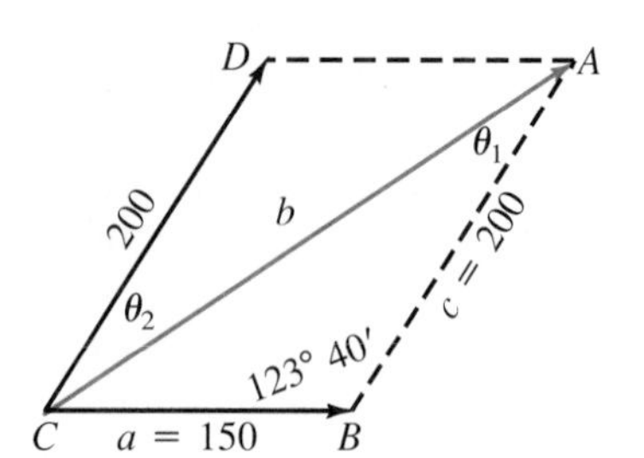

Plan Your Approach Draw a diagram to show the data. Let b represent the length of the resultant of the two forces. Label the drawing. In $\square ABCD$

$$\angle B = 180° - 56°20' = 123°40'$$

Complete the Work

$$b^2 = a^2 + c^2 - 2ac \cos B \qquad \text{Use the law of cosines.}$$
$$b^2 = (150)^2 + (200)^2 - 2(150)(200) \cos 123°40' \qquad \text{Calculation-ready form}$$
$$b^2 = 95{,}761.6$$
$$b = 309$$

$$\frac{\sin \theta_1}{150} = \frac{\sin 123°40'}{309} \qquad \frac{\sin A}{a} = \frac{\sin B}{b}$$

$$\sin \theta_1 = \frac{150 \sin 123°40'}{309} \qquad \text{Calculation-ready form}$$

$$\sin \theta_1 = 0.4040$$
$$\theta_1 = 23°50'$$
$$\theta_2 = 23°50' \qquad \theta_1 = \theta_2$$

Interpret the Results The resultant of the forces is 309 lb and it makes an angle of 23°50′ with the 200-lb force.

CLASS EXERCISES

Draw a diagram to represent the data in each problem and explain how to solve it. Do not solve the problem.

1. Each of the congruent sides of an isosceles trapezoid is 62.3 cm long and the parallel bases are 25.6 cm long and 50.2 cm long. What is the measure of each angle of the trapezoid?

2. The angle of elevation of the top of a tree is 35° at one point and 53° at a point 30 ft closer to the tree. How high is the tree if both points are on a horizontal plane with the base of the tree?

3. Two forces, one of 75 lb and one of 100 lb, act on a body at an angle 45° with each other. Find the magnitude of the resultant of the forces and the measure of the angle it makes with each of the component forces.

4. A switch engine pulls a freight car which is on a track parallel to the track that the engine is on. If a force of 2500 lb is needed to start the freight car when the cable makes a 40° angle with the tracks, what force is needed for the engine to pull the car when it is hitched to the car on the same track?

PRACTICE EXERCISES

Use technology where appropriate.

1. A force of 200 lb is resolved into components of 150 lb and 100 lb. What is the measure of the angle that the components make with each other?

2. An airplane with a speed of 300 mi/h is on a heading of 56° and the wind is blowing from the north at a speed of 50 mi/h. What is the plane's ground speed and in what direction is it actually flying?

3. Two forces, one of 250 lb and the other of 100 lb, act on a body and make an angle of 62°10′ with each other. What is the magnitude of the resultant of the forces and what is the measure to the nearest 10 min that it makes with the 100-lb force?

4. A plane with an air speed of 300 mi/h is flying on a heading of 110°. The wind is blowing at 60 mi/h from due south. What is the ground speed of the plane and its course?

5. Two planes left an airport at the same time. One flew due east at 200 mi/h, and the other flew at 230 mi/h. On what heading did the second plane fly if the two were 400 mi apart after 1 h?

6. A school marquee is supported by two steel cables, of equal length, that make an angle of 35° with the roof of the building. If the marquee weighs 500 lb, what is the magnitude of the force on each cable?

7. How long is each side of a regular pentagon that is inscribed in a circle with a radius 320 cm?

8. Find the length of each side of a regular octagon that is inscribed in a circle with a radius 325 cm.

9. The air speed of a plane is 300 mi/h and its heading is 150°. If the wind is blowing from due south at 30 mi/h, what is the ground speed of the plane and its course?

10. A plane with an air speed of 400 mi/h is flying on a heading of 110°. The wind is blowing from due north at 25 mi/h. What is the ground speed of the plane and its course?

11. The radii of two circles O and P are 16.8 cm and 12.6 cm, respectively. If $\angle AOP$ measures 35°, what is the length of $\overline{OP}$?

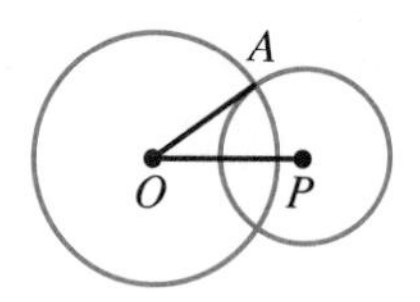

12. When two spheres intersect, the intersection is a circle with its center on the line that passes through the centers of the two spheres. If this line is 78 cm and the radii are 41 cm and 53 cm, what is the length of the radius of the circle?

Mixed Problem Solving Review

1. A ball is dropped from a height of 100 cm and rebounds to three quarters of its previous height each time. Find the total distance the ball travels before it comes to rest.

2. After 3.5 yr 68.8 g of a poisonous material have decayed to 43.7 grams. In how many years will half the poisonous material remain?

3. A survey of 135 high school students indicated that 83 like rock music, 27 like jazz and 12 like both. How many students like rock only? How many students do not like either rock or jazz?

4. Two roads intersect at right angles. A car is traveling toward the intersection on one road at a rate of 40 mi/h and a truck is approaching the intersection on the other road at a rate of 25 mi/h. At noon both vehicles are 4 mi from the intersection. At what time will the distance between them be the least? How far apart will they be at that time?

5. An apple farmer estimates he will have 155 bushels worth two dollars a bushel if he picks now. The crop will increase by 25 bushels a week if he waits. At the same time the selling price will drop twenty cents a bushel and the labor cost of picking will increase 5 cents per bushel per week. When should the farmer harvest to realize maximum income?

PROJECT

In the second century B.C., Hipparchus, the greatest astronomer of the ancient world, developed the branch of mathematics known as trigonometry from a single geometric theorem. His use of the similar-triangles theorem enabled him to calculate the heights of mountains, the radius of the earth and even the distance between the earth and the moon. Research the techniques of Hipparchus in the library. Use his methods to calculate the heights of three buildings without measuring their heights directly.

16.9

Polar Coordinates and Complex Numbers

Objectives: To define and use polar coordinates
To express complex numbers in polar form

A number of methods can be used to describe the location of a point in a plane. The rectangular coordinate system has already been discussed. Another method is the system of *polar coordinates*.

Capsule Review

A point located on a rectangular coordinate plane is defined by an ordered pair of numbers. The ordered pair (x, y) defines the point P, written $P(x, y)$, that is x units from the y-axis and y units from the x-axis.

EXAMPLE **Describe the location of the point $P(-3, 5)$.**

$P(-3, 5)$ is 3 units to the left of the y-axis and 5 units above the x-axis.

Describe the location of each point.

1. $P(3, 4)$ **2.** $P(5, 0)$ **3.** $P(-2, 8)$ **4.** $P(-6, -9)$

5. $P(4, -3)$ **6.** $P(-7, 0)$ **7.** $P(0, 0)$ **8.** $P(-4, 0)$

In the rectangular coordinate system, the reference system is a pair of perpendicular number lines called axes. In the polar coordinate system, the reference system is a point O, called the **pole,** and a horizontal ray with the pole as its left endpoint. This ray is called the **polar axis.** The **polar coordinates** of a point in the polar coordinate system is an ordered pair (r, θ), where r is the distance from the pole to point P and θ is a measure of the angle formed by the polar axis and $\overrightarrow{OP}$.

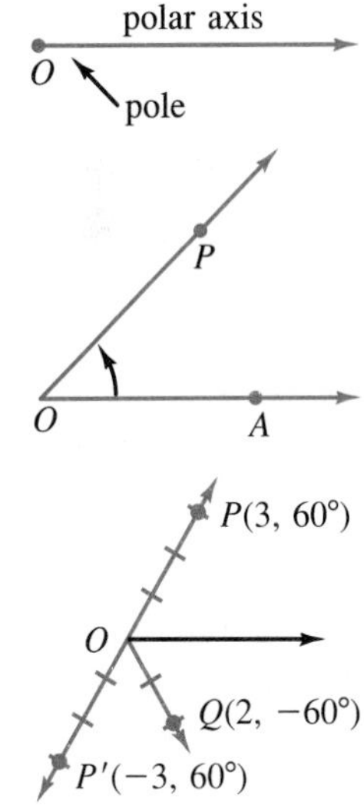

Any given point, $P(r, \theta)$, has infinitely many pairs of coordinates. For example, $(4, 240°)$, $(4, -120°)$, and $(4, 600°)$ are all polar coordinates for the same point.

As just noted, an angle is *positive,* if it is measured in a counterclockwise direction and *negative,* if measured in a clockwise direction. Similarly, for $P(r, \theta)$, the polar distance r is *positive* if measured along $\overrightarrow{OP}$ and *negative* if measured along $\overrightarrow{OP}'$ (the ray opposite $\overrightarrow{OP}$). Note that in the figure, $P(3, 60°)$ could be represented also as $P(-3, 240°)$.

EXAMPLE 1 **Locate the points in the same polar coordinate system: $P(3, 50°)$, $Q(4, 130°)$, $R(-3, 50°)$, $S(4, -30°)$**

First sketch the angle using the polar axis as the initial ray. Then mark off the distance on the terminal side of the angle.

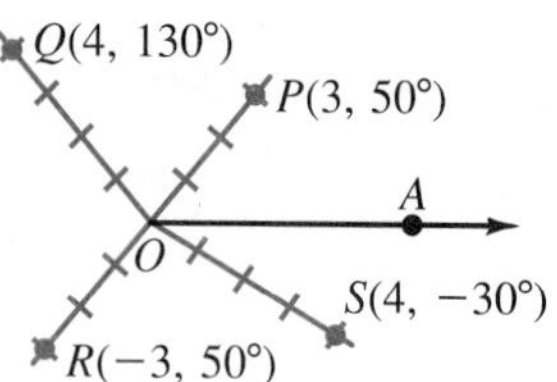

The polar and rectangular coordinate systems are related. The coordinates in one system can be expressed in terms of the other. Let the polar axis coincide with the positive half of the x-axis in the rectangular coordinate system. Then to change from rectangular to polar coordinates, use

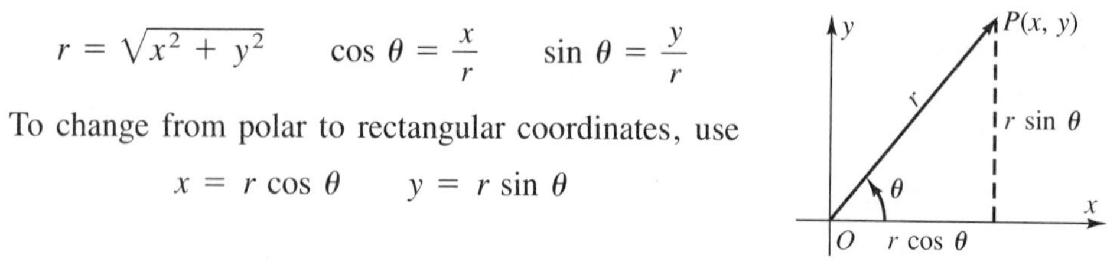

$$r = \sqrt{x^2 + y^2} \qquad \cos\theta = \frac{x}{r} \qquad \sin\theta = \frac{y}{r}$$

To change from polar to rectangular coordinates, use

$$x = r\cos\theta \qquad y = r\sin\theta$$

EXAMPLE 2 **Change from rectangular coordinates to polar coordinates: $P(-3, 4)$**

$r = \sqrt{x^2 + y^2}$

$r = \sqrt{(-3)^2 + (4)^2} = 5$

Since $r > 0$, $\sin\theta = \frac{4}{5} = 0.8$, and $\theta = 180° - 53° = 127°$.

The polar coordinates are $P(5, 127°)$, $P(5, 487°)$, $P(5, -233°)$, etc.

EXAMPLE 3 **Change from polar coordinates to rectangular coordinates: $P(12, 120°)$**

$x = r\cos\theta \qquad y = r\sin\theta$

$x = r\cos 120° \qquad y = r\sin 120°$

$x = 12\left(-\frac{1}{2}\right) \qquad y = 12\left(\frac{\sqrt{3}}{2}\right)$

$x = -6 \qquad y = 6\sqrt{3}$

The rectangular coordinates are $P(-6, 6\sqrt{3})$.

Any complex number $a + bi$, where a and b are real numbers, can be expressed as an ordered pair (a, b). You can represent complex numbers in a coordinate plane called a **complex number plane.** In the complex plane, the horizontal axis is called the **real axis** and the vertical axis is called the **imaginary axis.**

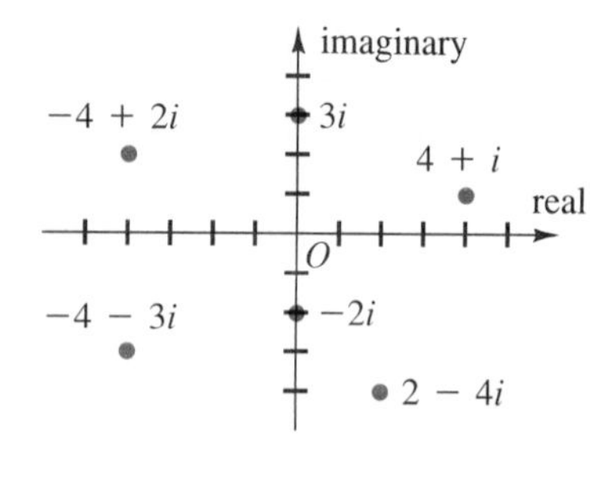

If $z = a + bi$ is any complex number, then

$$|z| = \sqrt{a^2 + b^2}$$

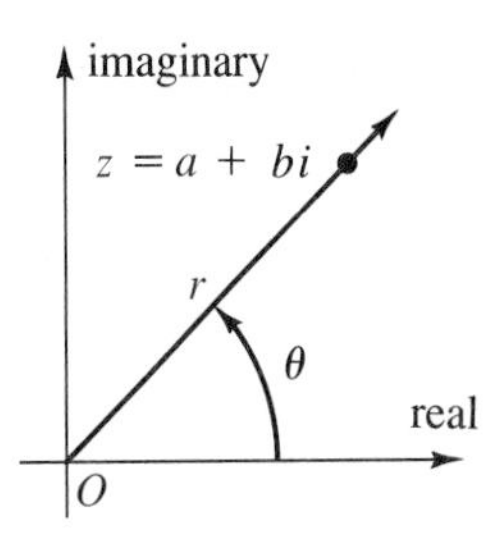

Note that $|z|$ is the distance from the origin to point z. Since $\overline{Oz}$ has length r, $r = |z| = \sqrt{a^2 + b^2}$. The absolute value of z is called the **modulus** of z. $\overrightarrow{Oz}$ makes an angle θ, called the **argument** with the positive real axis so $a = r \cos \theta$ and $b = r \sin \theta$. Thus $z = a + bi$ can be expressed as

$$z = r \cos \theta + (r \sin \theta)i = r(\cos \theta + i \sin \theta)$$

When a nonzero complex number z is expressed as $z = r(\cos \theta + i \sin \theta)$, it is said to be expressed in **polar,** or **trigonometric, form.**

EXAMPLE 4 **Express in polar form: $-3 + 5i$**

$$r = \sqrt{(-3)^2 + (5)^2} = \sqrt{34} \qquad r = \sqrt{a^2 + b^2}$$

$$\sin \theta = \frac{5}{\sqrt{34}} = 0.8575$$

$$\theta = 180° - 59° = 121°$$

The reference angle is 59° and θ is a second-quadrant angle.

The polar form is $\sqrt{34}(\cos 121° + i \sin 121°)$.

A complex number in polar form can be expressed in standard form.

EXAMPLE 5 **Express in standard form: $3(\cos 60° + i \sin 60°)$**

$$3(\cos 60° + i \sin 60°) = 3\left(\frac{1}{2} + \frac{\sqrt{3}}{2}i\right) = \frac{3}{2} + \frac{3\sqrt{3}}{2}i \qquad \cos 60° = \frac{1}{2};\ \sin 60° = \frac{\sqrt{3}}{2}$$

The standard form is $\frac{3}{2} + \frac{3\sqrt{3}}{2}i$.

CLASS EXERCISES

1. Thinking Critically How can the integer -5 be expressed in polar form?

Give three equivalent pairs of polar coordinates for each point, including one with $-r$ and one with $-\theta$.

2. $P(3, 30°)$ **3.** $Q(-2, 60°)$ **4.** $R(5, 300°)$ **5.** $S(2, 180°)$

Change from rectangular coordinates to polar coordinates.

6. $A(5, 12)$ **7.** $B(-2, 0)$ **8.** $C(-\sqrt{2}, -\sqrt{2})$ **9.** $D(4, -4)$

Change from polar coordinates to rectangular coordinates.

10. $P(8, 150°)$ **11.** $Q(-3, 135°)$ **12.** $R(-6, 120°)$ **13.** $S(-2, -60°)$

Find the modulus and argument of each complex number.

14. $1 + i$ **15.** $1 - i$ **16.** $-1 - i$ **17.** $3i$

PRACTICE EXERCISES

Locate the points in the same polar coordinate system.

1. $P(5, 60°)$ **2.** $P(2, 120°)$ **3.** $P(4, 225°)$ **4.** $P(5, 180°)$

5. $P(-3, 30°)$ **6.** $P(-2, 300°)$ **7.** $P(3, -30°)$ **8.** $P(4, -45°)$

Change from rectangular coordinates to polar coordinates.

9. $P(3, 3)$ **10.** $P(1, \sqrt{3})$ **11.** $P(-\sqrt{3}, -\sqrt{3})$ **12.** $P(-5, 5\sqrt{3})$

13. $P(2\sqrt{3}, -2)$ **14.** $P(8, -8)$ **15.** $P(0, 4)$ **16.** $P(-5, 0)$

Change from polar coordinates to rectangular coordinates.

17. $P(3, 30°)$ **18.** $P(-2, 60°)$ **19.** $P(-4, 120°)$ **20.** $P(-6, 225°)$

21. $P(8, 300°)$ **22.** $P(-6, 180°)$ **23.** $P(-3, -120°)$ **24.** $P(8, -300°)$

Express in polar form.

25. $-2 + 3i$ **26.** $3 - 8i$ **27.** -6 **28.** 4

Express in standard form.

29. $2(\cos 30° + i \sin 30°)$ **30.** $3(\cos 45° + i \sin 45°)$

31. $5(\cos 225° + i \sin 225°)$ **32.** $4(\cos 0° + i \sin 0°)$

Express in polar form.

33. $1 + \sqrt{2}i$ **34.** $-1 + \sqrt{2}i$ **35.** $-1 + 2\sqrt{3}i$ **36.** $1 - 3\sqrt{3}i$

Express in standard form.

37. $\cos 6° + i \sin 6°$ **38.** $\cos 5° + i \sin 5°$

39. $\cos 12° + i \sin 12°$ **40.** $\cos 32° + i \sin 32°$

41. $3(\cos 40° + i \sin 40°)$ **42.** $3(\cos 50° + i \sin 50°)$

43. $0.4(\cos 140° + i \sin 140°)$ **44.** $0.4(\cos 230° + i \sin 230°)$

45. $\sqrt{2}[\cos(-30°) + i \sin(-30°)]$ **46.** $\sqrt{3}[\cos(-45°) + i \sin(-45°)]$

47. $3[\cos(\text{Arc tan } \sqrt{3}) + i \sin(\text{Arc tan } \sqrt{3})]$

48. $5[\cos(\text{Arc tan } 1) + i \sin(\text{Arc tan } 1)]$

49. $5[\cos(\text{Tan}^{-1} 0.6814) + i \sin(\text{Tan}^{-1} 0.6814)]$

50. $6[\cos(\text{Tan}^{-1} 0) + i \sin(\text{Tan}^{-1} 0)]$

51. $3[\cos(\text{Tan}^{-1} 0) + i \sin(\text{Tan}^{-1} 0)]$

52. $P_2(3, 225°)$, $P_1(-4, 150°)$

53. $P_2(-5, 120°)$, $P_1(12, 150°)$

54. $P_2(-3, 150°)$, $P_1(4, 200°)$

55. $P_2(6, 220°)$, $P_1(-8, 300°)$

56. $P_2(-3, -80°)$, $P_1(2, -120°)$

57. $P_2(4, -225°)$, $P_1(-4, 200°)$

Graph using polar coordinates.

58. $r = 2 + 2 \sin \theta$

59. $r = 1 + 2 \cos \theta$

60. $r = 3 + \cos \theta$

Applications

61. **Design** Graph the equation $r = 1 - \cos \theta$ on a polar coordinate system. Use the following values for θ: 0°, 45°, 60°, 90°, 120°, 180°, 240°, 270°, 300°, 315°, 360°. Connect the points with a smooth curve. The curve that results is called a **cardioid.**

62. **Design** Graph the equation $r = 4 \sin 3\theta$. Choose convenient values for θ. Connect the points with a smooth curve. Describe the result.

TEST YOURSELF

Solve each equation.

16.6

1. $\sin x - \csc x = 0$, $0° \le x < 360°$

2. $\tan x \cot x - \tan x = 0$, $0 \le x < 2\pi$

For each vector, find the x- and y-components to the nearest whole number.

16.7

3. $\overrightarrow{AB}$: magnitude 4, direction 35°

4. $\overrightarrow{CD}$: magnitude 9, direction 50°

16.8

5. Forces of 18 lb and 24 lb act on a body at right angles. Find the magnitude of the resultant force and the measure to the nearest 10 min of the angle that the resultant makes with the 24-lb force.

6. The magnitude of the resultant of two forces is 20 lb. Find the measure of the angle between the two forces if they are 15 lb and 10 lb.

Express in polar form.

16.9

7. $P(4, 4\sqrt{3})$

8. $-2 + 5i$

9. Change $P(4, 60°)$ to rectangular coordinates.

10. Express $4(\cos 30° + i \sin 30°)$ in standard form.

INTEGRATING ALGEBRA
De Moivre's Theorem

De Moivre's theorem can be used to raise a complex number in polar form to a power.

De Moivre's Theorem If n is a positive integer and $z = r(\cos \theta + i \sin \theta)$ then

$z^n = r^n(\cos n\theta + i \sin n\theta)$

EXAMPLE 1 **Find: $(1 + i)^9$**

Let $z = 1 + i$

Then $|z| = \sqrt{2}$ and $z = \sqrt{2}\left(\frac{\sqrt{2}}{2} + i\frac{\sqrt{2}}{2}\right)$

So $z = \sqrt{2}(\cos 45° + i \sin 45°)$ *Express z in polar form.*

Then, by De Moivre's theorem,

$$\begin{aligned} z^9 &= (\sqrt{2})^9[\cos (9 \cdot 45°) + i \sin (9 \cdot 45°)] \\ &= 16\sqrt{2}(\cos 405° + i \sin 405°) \qquad 405° = 360° + 45° \\ &= 16\sqrt{2}\left(\frac{\sqrt{2}}{2} + i\frac{\sqrt{2}}{2}\right) \qquad \text{Change back to standard form.} \\ &= 16 + 16i \end{aligned}$$

So, $(1 + i)^9 = 16 + 16i$.

De Moivre's theorem can also be used to find the roots of a complex number.

EXAMPLE 2 **Find the fifth roots of $32(\cos 120° + i \sin 120°)$.**

$$\begin{aligned} \text{Let } 32(\cos 120° + i \sin 120°) &= [r(\cos \theta + i \sin \theta)]^5 \\ &= r^5(\cos \theta + i \sin \theta)^5 \end{aligned}$$

Then, by De Moivre's theorem,

$$32(\cos 120° + i \sin 120°) = r^5(\cos 5\theta + i \sin 5\theta)$$

Therefore, $r^5 = 32$, so $r = 2$ and

$$\begin{aligned} 5\theta &= 120° + 360° \cdot k, \text{ where } k \text{ is a positive integer} \\ \theta &= 24° + 72° \cdot k \end{aligned}$$

For $k = 0$, $\theta = 24°$; for $k = 1$, $\theta = 96°$; for $k = 2$, $\theta = 168°$; for $k = 3$, $\theta = 240°$; for $k = 4$, $\theta = 312°$

The five roots of $32(\cos 120° + i \sin 120°)$ are shown as graphed points below.

The modulus of each of the roots is 2. The figure shows that these roots are in a circle of radius 2 with the center at the origin. The roots are equally spaced on the circle. Notice that for $k = 5, 6, \ldots$ the roots are coterminal with one of the original five roots.

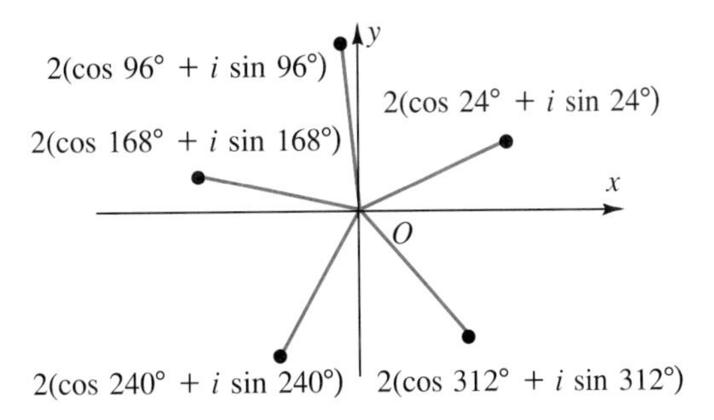

Finding the roots of the number one is called finding the **roots of unity.** De Moivre's theorem can be used to show that the expression

$$\cos \frac{k \cdot 360°}{n} + i \sin \frac{k \cdot 360°}{n} \qquad \text{for } k = 0, 1, 2, \ldots, n - 1$$

gives the n nth roots of unity.

EXAMPLE 3 Find the three cube roots of 1.

3 cube roots ($n = 3$): $\cos (k \cdot 120°) + i \sin (k \cdot 120°) \qquad$ for $k = 0, 1, 2$

$$\text{For } k = 0: \cos 0° + i \sin 0° = 1$$

$$k = 1: \cos 120° + i \sin 120° = \frac{-1}{2} + i\frac{\sqrt{3}}{2}$$

$$k = 2: \cos 240° + i \sin 240° = \frac{-1}{2} - i\frac{\sqrt{3}}{2}$$

EXERCISES

1. Verify De Moivre's theorem for $n = 2$ and $n = 3$.

Find each of the following using De Moivre's theorem.

2. $(-1 + i\sqrt{3})^6$ **3.** $(2 + i)^3$ **4.** $[3(\cos 20° + i \sin 20°)]^4$

5. Find the fourth roots of $16(\cos 150° + i \sin 150°)$

6. Find the cube roots of $64(\cos 324° + i \sin 324°)$

7. Find the fourth roots of $8 + 8i\sqrt{3}$ **8.** Find the cube roots of i.

9. Find the fourth roots of unity.

10. Show that $\cos 60° + i \sin 60°$ is a sixth root of unity.

CHAPTER 16 SUMMARY AND REVIEW

Vocabulary

argument (829)
bearing (821)
cofunction identity (796)
complex plane (828)
De Moivre's theorem (832)
difference formulas (803)
directed line segment (819)
double-angle formulas (806)
equivalent vectors (819)
half-angle formulas (808)
identity (794)
imaginary axis (828)
inverse trigonometric functions (813)
magnitude (820)
modulus (829)
opposite (819)
parallelogram rule (820)
polar axis (827)
polar coordinates (827)
polar form (829)
pole (827)
primary solution (816)
principal value (812)
Pythagorean identities (795)
quotient identities (795)
real axis (828)
reciprocal identities (794)
resultant vector (819)
scalar (820)
sum of two vectors (819)
vector (819)
vector quantity (819)
x- and y-components (820)

Fundamental Identities The fundamental identities are the reciprocal identities, the quotient identities, and the Pythagorean identities. 16.1

Simplify.

1. $\sin^2 B \csc B \sec B \tan B + \cos^2 B$
2. $(1 - \tan B)^2$

Proving Trigonometric Identities One strategy for proving an identity is to try to transform the more complicated side into the simpler expression. 16.2

Prove.

3. $\dfrac{\sin A \csc A}{\cos A} = \sec A$
4. $\dfrac{\tan x + \sin x}{1 + \cos x} = \tan x$

Evaluate by using special angle measures and a sum or difference identity. When radicals occur, express the answer in simplest radical form. 16.3

5. $\sin 75°$
6. $\cos 15°$
7. If $\sin A = \frac{4}{5}$, $0° < A < 90°$, and $\cos B = \frac{15}{17}$, $0° < B < 90°$, find $\sin (A - B)$.

Use a special angle measure and a double or half-angle identity to find each of the following. Express the answer in simplest radical form. 16.4

8. $\sin 300°$
9. $\cos \frac{\pi}{8}$
10. $\tan 75°$

If Q is in the second quadrant and $\sin Q = \frac{5}{13}$, find

11. $\sin 2Q$ **12.** $\cos 2Q$ **13.** $\tan \frac{Q}{2}$

Inverse Trigonometric Functions The inverse trigonometric relations provide a method for determining the angle measure when the value of the function is given. The principal values of the inverse trigonometric functions are indicated by the use of a capital letter. 16.5

Evaluate.

14. Arc sec 1.188 **15.** $\sec\left(\text{Cos}^{-1} \frac{5}{13}\right)$

Solving Trigonometric Equations To solve a trigonometric equation write the equation in terms of one trigonometric function and then use the same methods for solving other equations. Use the concept of inverse relations to find the measure of the angle. 16.6

Solve for x, $0° \leq x < 360°$.

16. $2 \sin^2 x + \cos x = 1$ **17.** $\sin 2x + \sin x = 0$

Vector Operations A vector is a directed line segment, since it has both magnitude and direction. 16.7

For each vector, find the x- and y-components to the nearest whole number.

18. $\overrightarrow{OA}$: magnitude 5, direction 43° **19.** $\overrightarrow{OP}$: magnitude 10, direction 125°

20. A plane with an air speed of 420 mi/h is flying on a heading of 120°. The wind is blowing from due east at 30 mi/h. What is the ground speed of the plane and in what direction is it actually flying?

21. Two forces, one of 48 lb and the other of 80 lb, act on a body and make an angle measuring 54° with each other. Find the magnitude of the resultant of the forces and the measures of the angles that the resultant makes with each of the component forces. 16.8

Complex Numbers and Polar Coordinates In the polar coordinate system, a point P is represented by an ordered pair (r, θ), where r is the distance from the pole to point P and θ is the measure of the angle from the polar axis to $\overrightarrow{OP}$. The polar, or trigonometric form of a nonzero complex number z, where $z = x + yi$, is $z = r(\cos \theta + i \sin \theta)$. 16.9

22. Express $-3 + 2i$ in polar form.

23. Express $2(\cos 60° + i \sin 60°)$ in standard form.

CHAPTER 16 TEST

Simplify.

1. $\cos^2 A \sec A \csc A$
2. $\cot B \cos B + \sin B$
3. $\dfrac{\sin A}{\sec A} + \dfrac{\cos A}{\csc A}$

Evaluate by using special angle measures and a sum or difference identity. When radicals occur, express the answer in simplest radical form.

4. $\cos 105°$
5. $\sin 75°$
6. $\tan 15°$
7. If $\sin A = -\frac{3}{5}$, $180° < A < 270°$, and $\cos B = -\frac{5}{13}$, $180° < B < 270°$, find $\cos (A + B)$.
8. If $\angle A$ is in the third quadrant and $\sin A = -\frac{4}{5}$, find $\sin 2A$.

Prove.

9. $\dfrac{1 - \tan A}{\sec A} + \dfrac{\sec A}{\tan A} = \dfrac{1 + \tan A}{\sec A \tan A}$
10. $\tan x + \cot x = \sec x \csc x$

Evaluate.

11. Arc tan 0.6318
12. $\text{Cos}^{-1} \frac{3}{5}$
13. Solve for x. $\sin 2x = \sin x$, $0 \leq x < 2\pi$

Find the x- and y-components.

14. OA: magnitude 9, direction 52°
15. A plane with an air speed of 325 mi/h is flying on a heading of 110°. The wind is blowing from due west at 40 mi/h. What is the ground speed of the plane and in what direction is it actually flying?
16. Two forces act on a body at an angle of 34°20′. One force is 32 lb and the other is 14 lb. Find the magnitude of the resultant of the forces.
17. Express in polar form: $2 - 3i$.
18. Change $P(4, -3)$ to polar coordinates.
19. Change $P(3, 120°)$ to rectangular coordinates.

Challenge

Prove: $\sin x \cos^3 x - \cos x \sin^3 x = \frac{1}{4} \sin 4x$

COLLEGE ENTRANCE EXAM REVIEW

Select the best choice for each question.

1. Which of the following does *not* equal 1 for all A in each domain?
A. $\sin^2 A + \cos^2 A$
B. $\sec^2 A - \tan^2 A$
C. $\csc^2 A - \cot^2 A$
D. $\sin A \cdot \sec A$
E. $\tan A \cdot \cot A$

2. In radians, Arc sin $\frac{\sqrt{3}}{2}$ equals:

A. $-\frac{\pi}{3}$ **B.** $-\frac{\pi}{6}$ **C.** 0
D. $\frac{\pi}{6}$ **E.** $\frac{\pi}{3}$

3. What is the probability of having a sum of 8 on a toss of two dice?

A. $\frac{1}{9}$ **B.** $\frac{5}{36}$ **C.** $\frac{1}{6}$ **D.** $\frac{7}{36}$ **E.** $\frac{2}{9}$

4. In how many different ways can a committee of 6 be seated at a round table with six chairs?
A. 60 **B.** 120 **C.** 180
D. 360 **E.** 720

5. Find all solutions for $0° \leq A < 180°$ when $\sin^2 A - \sin A \cdot \cos A = 0$.
A. 0° **B.** 45° **C.** 135°
D. 0°, 45° **E.** 0°, 45°, 135°

6. In this vector diagram $\vec{a} + \vec{b}$ is equal to:
A. $-\vec{c}$ **B.** $\vec{d} - \vec{e}$
C. $\vec{e} - \vec{d}$ **D.** $\vec{d} + \vec{e}$
E. $-(\vec{d} + \vec{e})$

7. Which of the following cannot possibly be a rational solution of $3x^5 - kx^3 - 2x^2 + x - 4 = 0$?

A. $\frac{2}{3}$ **B.** $\frac{3}{4}$ **C.** 1 **D.** 2 **E.** 4

8. Find x when $\log(\log x) = 0$.
A. 10^{-2} **B.** 10^{-1} **C.** 10^0
D. 10^1 **E.** 10^{10}

9. The solutions of a quadratic equation are 5 and -2. The equation could be:
A. $x^2 - 3x - 10 = 0$
B. $x^2 + 10x - 3 = 0$
C. $x^2 + 3x - 10 = 0$
D. $x^2 - 3x + 10 = 0$
E. $x^2 + 10x + 3 = 0$

10. If $x = 2 + 3i$, then find $\frac{x - 1}{x + 1}$.

A. $\frac{-2 - i}{3}$ **B.** $\frac{2 - i}{6}$ **C.** $\frac{2 + i}{3}$
D. $\frac{-2 + i}{6}$ **E.** $\frac{-2 + i}{3}$

11. Which of the following is *not* a factor of $x^6 - 64$?
A. $x + 2$ **B.** $x - 2$
C. $x^2 - 4x + 4$ **D.** $x^2 - 2x + 4$
E. $x^2 - 4$

12. Find the 10th term of the sequence $-6, 2, -\frac{2}{3}, \ldots$

A. $\frac{-512}{6561}$ **B.** $\frac{512}{6561}$ **C.** $\frac{6}{6561}$
D. $\frac{-2}{6561}$ **E.** $\frac{2}{6561}$

13. Find the domain for the system of inequalities $\begin{cases} x - y \geq 0 \\ y \geq x^2 - 2x \end{cases}$.
A. $0 \leq x \leq 3$
B. $-3 \leq x \leq 0$
C. $x \geq 3$ or $x \leq 0$
D. $x \leq 3$ or $x \geq 0$
E. $x \leq -3$ and $x \geq 0$

CUMULATIVE REVIEW (CHAPTERS 1–16)

Simplify, writing without negative exponents.

1. $y^{\frac{2}{3}} \cdot y^{-\frac{3}{4}}$

2. $\left(\dfrac{4x^2y^{\frac{3}{2}}}{16x^{-\frac{1}{2}}y}\right)^2$

3. $(25x^{\frac{4}{3}}y^{-2})^{-\frac{3}{2}}$

Simplify each radical expression.

4. $\sqrt{-81} \cdot \sqrt{-36}$

5. $(2\sqrt{3} - 7\sqrt{2})^2$

6. $\sqrt{5} - 2\sqrt{5} + \dfrac{3}{\sqrt{20}}$

7. $\dfrac{6 + i\sqrt{3}}{1 - 2i\sqrt{3}}$

A card is chosen at random from a bridge deck. Find the probability of choosing:

8. a red five

9. a face card

10. an ace

Give the period of each function. Sketch the graph.

11. $y = -2 \cos \frac{1}{2}x$

12. $y = \frac{1}{2} \sin 2x$

13. $y = \cot x$

Evaluate.

14. Arc csc 2

15. $\sin\left(\text{Cos}^{-1}\left(-\frac{3}{5}\right)\right)$

16. $\sec(\text{Tan}^{-1}\sqrt{3})$

Write each repeating decimal as a rational number.

17. $0.\overline{7}$

18. $0.\overline{204}$

19. $3.\overline{5}$

20. $5.0\overline{2}$

21. Find the mean proportional between 6 and 108.

Using the conditions given, write, in standard form, the equation of each line described.

22. has a slope of -1 and contains the point $(-4, 5)$

23. has a y-intercept of 3 and contains the point $(2, -1)$

Solve each equation.

24. $2x^2 - 5x = 12$

25. $3x^2 + 8 = 0$

26. $4x^2 - 20x = 23$

27. $\dfrac{2x}{x - 4} - \dfrac{3}{x - 2} = 9$

Simplify.

28. $\dfrac{2 - \dfrac{2}{a^2}}{-\dfrac{1}{2a^2} + \dfrac{1}{2a}}$

29. $\dfrac{\dfrac{2y}{y^2 - 4} + \dfrac{4}{y - 2}}{\dfrac{1}{2y - 4} - \dfrac{2}{2 + y}}$

30. If $\tan P = \frac{12}{5}$ and $\angle P > 90°$, find the five other trigonometric functions of $\angle P$.

Simplify.

31. $\sin B \tan B + \cos B$

32. $(\csc A - \cot A)(\sec A + 1)$

Solve for x, $0° \leq x \leq 360°$.

33. $\tan 3x = 1$

34. $2 \cos 2x + 1 = 0$

35. $\sin x = \csc x$

Find the indicated term of each arithmetic sequence.

36. $a_1 = -\frac{3}{4}$, $d = \frac{3}{8}$; a_6

37. $a_4 = 11$; $a_{12} = 51$; a_1

For each infinite geometric series, find the sum, if it exists.

38. $\sum_{n=1}^{\infty} 5\left(\frac{3}{4}\right)^n$

39. $\sum_{k=1}^{\infty} (-2.1)^{k+2}$

40. $-32 + 16 - 8 + \cdots$

Expand and simplify each binomial.

41. $\left(\frac{1}{2} - b\right)^4$

42. $(3a^3 + a)^3$

43. $\left(\sqrt{2} - \frac{\sqrt{2}}{2}\right)^5$

How many distinct four-letter arrangements can be made from each word?

44. HEAT

45. COOL

46. BOTHER

47. From 15 different baseball cards, how many 5-card sets can be assembled?

The ages of students in a gymnastics class are as follows:

12, 13, 11, 12, 14, 12, 10, 9, 12, 13, 11, 11

For this age distribution, find:

48. the mean

49. the standard deviation

50. A pharmacist has 30 mL of a 25% solution of antibiotic in alcohol. In order to dilute the solution to 20%, how much alcohol should she add?

Without solving, tell the nature of the solutions of each equation.

51. $x^2 - 6x = -9$

52. $2x^2 - 3x - 4 = 0$

53. $6x^2 + 2x = -3$

Solve, and check.

54. $\sqrt{5 - 2x} = 5$

55. $\sqrt{x + 1} - \sqrt{2x + 1} = 0$

56. $y + 1 = \sqrt{y^2 + 9}$

57. $\sqrt{x} - \sqrt{x - 7} = 1$

58. Find the length of a radius of a circle with center at the origin if it passes through the point $(5, -12)$.

59. If the lengths of one pair of opposite sides of a square are each increased by 4 m, the rectangle formed will have an area of 320 m^2. What are the dimensions of the original square?

Find the foci for each ellipse whose equation is given. Sketch the graph.

60. $\frac{y^2}{169} + \frac{x^2}{25} = 1$

61. $16x^2 + 36y^2 = 576$

Write the equation of each parabola, given its focus and directrix.

62. $F(0, 2)$; $y = -2$

63. $F(-1, 0)$; $x = 1$

64. $F(3, -3)$; $y = 1$

Graph each function, and find its real zeros to the nearest tenth.

65. $f(x) = 9x^3 + 9x^2 - 4x - 4$

66. $f(x) = x^3 + 2x^2 - 8x - 19$

Express as a single logarithm. Evaluate if possible.

67. $\log_{10} x - \log_{10} 5$

68. $2 \log_b x - 3 \log_b y$

69. $\frac{1}{2} \log_5 a + \frac{1}{2} \log_5 1$

70. $3 \log_4 2 + \frac{1}{2} \log_4 16$

In $\triangle ABC$, $\angle B = 120°$, $a = 6$, and $c = 4$. Find:

71. the area of $\triangle ABC$

72. the length of side b

Express in polar form.

73. $1 + i\sqrt{3}$

74. $3 - 4i$

75. $-\sqrt{2} + i\sqrt{2}$

Solve to the nearest thousandth. Check.

76. $3^x = 25$

77. $(0.52)^x = 38$

78. $\log_5 100 = x$

Chapter 1 Real Numbers and Equations

Show that each decimal can be written as the quotient of two integers.

1. 3.375 **2.** 0.12 **3.** 0.252525... **4.** 0.24

Convert each fraction to decimal form.

5. $\frac{5}{8}$ **6.** $\frac{3}{11}$ **7.** $\frac{1}{7}$ **8.** $\frac{7}{25}$

9. Write the additive inverse and the multiplicative inverse of $-\frac{5}{7}$.

Simplify.

10. $-212 + (-15)$

11. $\left(-\frac{7}{15}\right)\left(\frac{5}{21}\right)$

12. $|-5| - |8|$

13. $\frac{|15 - 4|}{10} \div \frac{|5 - 7|}{5}$

Evaluate each expression using the value given for each variable.

14. $3a^2 - 5a + 2$ for $a = -3$

15. $|3m + 2| + |5 - 2m|$ for $m = -2$

16. Name the property of real numbers illustrated by $(8)(-7) = (-7)(8)$.

Solve and check.

17. $4a + 11 = 39$

18. $3y - 2(4 - y) = 17$

19. $3y - \frac{1}{5} = 2y + \frac{3}{10}$

20. $6(c - 2) - 2c = 4 + 2(c - 1)$

21. Using n for the variable, write an algebraic expression for *six less than twice a number.*

22. Write an algebraic sentence for *five more than three times a number is twenty-three.*

23. The measures of an angle and its complement differ by eighteen degrees. Find the angle and its complement.

24. Find four consecutive even integers such that the sum of the second and the fourth is 52.

25. Two times a number, increased by eight, is equal to four times the number, decreased by six. Find the number.

Chapter 2 Equations and Inequalities

Solve for x and indicate any restrictions on the values of the variables.

1. $r - x = s$ **2.** $ax + b = c$ **3.** $cx - dx = 0$ **4.** $\frac{x}{m} - n = p$

Solve and write the solution set using set-builder notation. Graph the solution set, if it is not the empty set.

5. $4x - 2 > 18$

6. $3(y - 2) + 7 < 11$

7. $7c + 4c - 13 \leq 240$

8. $4m + 12 \geq 2m - 6$

Classify each sentence as a conjunction or a disjunction and state whether it is true, false, or open.

9. $x + 5 > 7$ *and* $x - 2 < 3$

10. $2(-1 + 2) < 0$ *or* $3(-3 + 1) > 0$

11. $x - 2 < 6$ *or* $x - 6 > 4$

12. $-(10 - 3) < -(-7)$ *and* $-(-4 + 1) > 2$

13. Given $A = \{1, 3, 4, 7, 8, 9\}$ and $B = \{2, 4, 5, 6, 8, 10\}$, find $A \cap B$ and $A \cup B$.

Solve and graph the solution set, if it is not empty.

14. $x > 3$ *and* $x < 9$

15. $3x > -15$ *and* $-3x > 3$

16. $7x > -14$ *and* $3x < -12$

17. $2x < 8$ *or* $x + 4 > 14$

18. The perimeter of Brian's new rectangular deck is 36 yd. If the length is 3 yd more than twice the width, find the dimensions of the deck.

Solve and check. If an equation has no solution, so state.

19. $|2x| = 10$ **20.** $|x - 3| = 12$ **21.** $|4n - 3| = 9$ **22.** $|5x - 4| = -8$

Solve and graph the solution set.

23. $|x + 2| < 7$ **24.** $|2x - 5| \geq 3$ **25.** $3|w - 2| > 12$ **26.** $|4x| - 5 > 3$

27. Find all sets of three consecutive even integers whose sum is at least 25 less than 5 times the third integer.

28. At 10:30 AM two cars pass each other as they travel in opposite directions along a straight highway. If one car averages 48 mi/h and the other 52 mi/h, at what time will they be 25 mi apart?

29. Janice has $74 in one-, five-, and ten-dollar bills. She has 3 times as many one-dollar bills as ten-dollar bills, and she has 2 fewer five-dollar bills than one-dollar bills. How many five-dollar bills does she have?

30. The average of three fractions is $\frac{9}{20}$. If two of the fractions are $\frac{1}{2}$ and $\frac{1}{4}$, what is the third fraction?

Chapter 3 Functions and Graphs

Graph each ordered pair and name the quadrant or axis where each point lies.

1. $A(5, 4)$ **2.** $B(-3, 1)$ **3.** $C(-5, -3)$ **4.** $D(2, -4)$ **5.** $E(-2, 0)$

6. Graph the relation and state its domain and range.
$\{(-4, 2), (-2, 0), (0, -2), (2, -4), (4, -6)\}$

7. Determine whether the relation is a function.
$\{(0, 3), (1, -2), (2, 4), (1, -1), (3, 5), (-1, -2), (-2, 3)\}$

8. Write a rule for the relation and state its domain and range.
$\{(1, 3), (2, 6), (3, 9), (4, 12), (5, 15)\}$

Graph each relation.

9. $y = -3x$ **10.** $x - 3y = -12$ **11.** $x = 3$ **12.** $-2x + 4y = -8$

Evaluate $f(x)$ and $g(x)$ for $x = 0, 1, 2, c$.

13. $f(x) = -2x$, $g(x) = x + 5$ **14.** $f(x) = 3x^2 + 1$, $g(x) = x^2 - 2x + 3$

15. Evaluate $g[f(2)]$, $g[f(-1)]$ and $g[f(0)]$ for $f(x) = -3x^2$ and $g(x) = 2x - 1$.

16. Find the inverse of $\{(1, -3), (2, -6), (3, -9), (4, -12), (5, -15)\}$.

17. Find the inverse of the function and state its domain and range. Use a mapping diagram to determine if the inverse is also a function.
$\{(1, -3), (2, -3), (3, -5), (6, -5)\}$

18. Find the inverse of $y = 3x + 1$. Graph f, f^{-1} and $y = x$ on the same coordinate plane.

19. Find the slope of the line passing through $(-1, -5)$ and $(2, 3)$.

20. Find the slope and y-intercept of the line $2y + 5x = -8$.

21. Write in standard form the equation of the line with slope $m = \frac{3}{2}$ and y-intercept $b = -2$.

22. Sketch the graph of $y = -2x + 3$.

23. Determine whether the lines passing through the pairs of points $(-1, 5)$, $(3, 3)$ and $(2, 4)$, $(1, 2)$ are parallel, perpendicular or neither.

24. Write in standard form the equation of the line parallel to $y = -3x - 2$ with y-intercept $b = 3$.

25. Write in slope-intercept form the equation having slope $m = 3$ that passes through $(4, -1)$.

26. Write in standard form the equation of the line passing through $(2, -3)$ and $(-4, 1)$.

Chapter 4 Systems of Equations and Inequalities

For each function, determine whether y varies directly as x. If so, find the constant of variation and write the equation.

1.

x	y
3	12
4	16
5	20

2.

x	y
2	2.4
3	3.6
4	4.8

3.

x	y
1	4
2	6
3	8

4. Given: y varies directly as x. If $y = 9$ when $x = 2$, find y when $x = 6$.

Solve each linear system graphically. Determine whether the system is consistent and independent, inconsistent, or consistent and dependent.

5. $\begin{cases} 4x - 2y = 10 \\ 8y = 3x - 14 \end{cases}$

6. $\begin{cases} y = 3 - x \\ x + y = 9 \end{cases}$

Solve each linear system by addition. If the system is inconsistent, write *no solution*.

7. $\begin{cases} 5x - 2y = 7 \\ 3x - 2y = 3 \end{cases}$

8. $\begin{cases} x + 3y = 6 \\ 2x + 6y = 9 \end{cases}$

Solve each system by substitution. If the system is inconsistent, write *no solution*.

9. $\begin{cases} y = x + 4 \\ 3x - y = 8 \end{cases}$

10. $\begin{cases} 3x + y = 12 \\ x - 2y = 4 \end{cases}$

11. A teacher orders pens and pencils to be used as class prizes. An order of 20 pencils and 10 pens costs $7.00. An order of 5 pencils and 12 pens from the same company costs $6.50. How much is each pencil and each pen?

12. The sum of the digits of a two-digit number is 15. If the digits of the number are reversed, the new number is 9 less than the original number. What is the original number?

13. Graph $4y \geq 8x - 12$ on the coordinate plane.

14. Solve the system of inequalities by graphing on the coordinate plane.

$\begin{cases} -4x + 2y > 0 \\ y - x < 3 \end{cases}$

15. Linear programming: If profit is represented by $P = 2x + y$, find the maximum profit under the constraints shown at the right.

$\begin{cases} y \leq -x + 4 \\ y \geq x - 2 \\ x \geq 0 \\ y \geq 0 \end{cases}$

Chapter 5 Matrices and Determinants

Perform the indicated operations, if they are defined, for matrices *A* and *B*.

$$A = \begin{bmatrix} 2 & -3 & 1 \\ 0 & 4 & -1 \end{bmatrix} \qquad B = \begin{bmatrix} 1 & 3 & -2 \\ 4 & -3 & 0 \end{bmatrix}$$

1. $A + B$ **2.** $B - A$ **3.** $2A$ **4.** $5A - 3B$

5. Solve: $X + \begin{bmatrix} 4 & -1 \\ 3 & -2 \end{bmatrix} = \begin{bmatrix} -1 & 1 \\ 0 & 3 \end{bmatrix}$

Perform the indicated operations, if they are defined, for matrices *R* and *S*.

$$R = \begin{bmatrix} 1 & 3 & -2 \\ 0 & 1 & 4 \\ -2 & 3 & -1 \end{bmatrix} \qquad S = \begin{bmatrix} 2 & -3 & 6 \\ 5 & -1 & 3 \\ 0 & 2 & -2 \end{bmatrix}$$

6. RS **7.** R^2

8. The direct communications lines for executives r, s, t, and u from Firm Z is given below. Write a matrix for Firm Z.

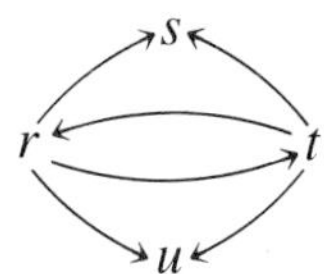

Solve each system of equations and state whether the system is consistent and independent, consistent and dependent, or inconsistent.

9. $\begin{cases} x + y - 2z = 3 \\ 3x - 2y + z = 8 \\ x - 3y = -5 \end{cases}$

10. $\begin{cases} x + y = 7 \\ y - z = 4 \\ x + z = -2 \end{cases}$

Solve each system of equations using the augmented matrix method. Determine whether the system is consistent and independent, consistent and dependent, or inconsistent.

11. $\begin{cases} 3x + y - 2z = -3 \\ x - 3y - z = -2 \\ 2x + 2y + 3z = 11 \end{cases}$

12. $\begin{cases} x - 2y + z = 5 \\ 2x - 4y + 2z = 3 \\ 3x - y - z = 2 \end{cases}$

13. Find the multiplicative inverse of $\begin{bmatrix} 3 & 2 \\ -8 & 4 \end{bmatrix}$, if it exists.

Solve each system of equations using the multiplicative inverse of a matrix.

14. $\begin{cases} 3x - y = 7 \\ 2x + 2y = 10 \end{cases}$

15. $\begin{cases} 2x + y = 11 \\ -3x - 2y = -17 \end{cases}$

16. Evaluate the determinant using minors.

$$\begin{vmatrix} 3 & -1 & 0 \\ 4 & 3 & -2 \\ -1 & 5 & 2 \end{vmatrix}$$

17. Solve using Cramer's rule:

$$\begin{cases} 4x + 3y + z = 7 \\ x - 2y + 3z = 10 \\ 5x + 4y - 2z = 2 \end{cases}$$

Chapter 6 Polynomials

Simplify.

1. $a^5 \cdot a^7$ **2.** $\frac{r^8}{r^2}$ **3.** $(c^6)^3$ **4.** $\left(\frac{3y}{4}\right)^3$

5. $(-a^2b^3)^2$ **6.** $(7a^3b^2)(4ab^4)$ **7.** $(-ab^3)(-3a^3b^2)^3$

8. Give the degree and coefficient of $-4r^2st$.

9. Classify $5x^2 - \sqrt{7}x + 2$ as a monomial, a binomial, or a trinomial. Give the degree of the polynomial.

10. Simplify: $r^2s + 3rs^2 - 6r^2s - rs^2$

11. Add. $\begin{array}{r} 9x^2 - 6x - 13 \\ \underline{5x^2 + 8x - \ \ 7} \end{array}$

12. Subtract. $\begin{array}{r} 11y^2 - 4y - 3 \\ \underline{7y^2 - 3y + 7} \end{array}$

Perform the indicated operations.

13. $(5x^2 - 7x - 3) + (-3x^2 + 6x - 4)$

14. $(-a^3 - 3a^2 + a) - (-4a^2 - 3a + 6)$

15. $y(8 - 5y)$

16. $(x + 6)(2x - 3)$

17. $(r + 3s)^2$

18. $(5a - 2)(5a + 2)$

19. $(r - s)(3c + d)$

20. $(3 - 2c)(9 + 6c + 4c^2)$

Factor.

21. $32a^2 + 40a$

22. $9t^2 - 4$

23. $c^3 + 27$

24. $y^2 - 8y + 16$

25. $a^2 + 4a - 21$

26. $y^2 - 5y - 36$

27. $18a^2 - 3a - 10$

28. $12y^2 + 5y - 2$

29. $r^2 - rt + rs - st$

30. $11y^2 - 44$

31. $-y^3 - 64$

32. $32a^2 - 18b^2$

Solve and check.

33. $y^2 + y - 6 = 0$

34. $r^2 - 5r = 0$

35. $m^2 - 6m = -9$

36. $x^3 - 2x^2 - x + 2 = 0$

37. Find three consecutive odd integers if the product of the first and third integers is 88 more than 19 times the second.

38. A walk of uniform width surrounds a school garden 20 ft wide by 40 ft long. Find the width of the walk if its area is 396 ft^2.

Chapter 7 Rational Expressions

Simplify, leaving no negative exponents.

1. $(2x)^{-3}$ **2.** $(2^{-3}x^{-1})^2$ **3.** $-7x^0$ **4.** $\dfrac{3^{-4}xy^2}{3x^2y^{-2}}$

Express in scientific notation. If an integer ends in zeros, assume that they are not significant.

5. 0.584 **6.** 6280 **7.** 0.000035 **8.** 7,783,000,000

Express in decimal form.

9. 3.8×10^2 **10.** 5.13×10^{-1} **11.** 6×10^4 **12.** 4.9120×10^{-6}

Perform the indicated operations, and express in scientific notation.

13. $(2.1 \times 10^3)(5.91 \times 10^{-1})$ **14.** $\dfrac{1.985 \times 10^{-3}}{3.602 \times 10^{-1}}$

State values of the variable(s) for which each expression is undefined. Then simplify.

15. $\dfrac{-14x^4yz^2}{21x^3y^2z^2}$ **16.** $\dfrac{4xy - 12}{18 - 6xy}$ **17.** $\dfrac{5y^3 - 5y}{10y^2 + 20y + 10}$

Perform the indicated operations, and simplify.

18. $\dfrac{(2y-3)^2}{5y-30} \div \dfrac{2y-3}{y^2 - 8y + 12}$

19. $\dfrac{3x^2 + 5x - 2}{4x - x^3} \cdot \dfrac{2x^3 + 4x^2}{9x^2 - 6x + 1}$

20. $\dfrac{x}{6y} - \dfrac{2y^2}{9x^2} + \dfrac{5}{3xy} - 3$

21. $\dfrac{3}{x-2} - \dfrac{x}{x-2} - \dfrac{1}{x-2}$

22. $\dfrac{6y+9}{16y^2 - 36} + \dfrac{3}{2y+3}$

23. $\dfrac{3}{3x+1} \cdot \dfrac{3x^2 - x - 10}{9x} \div \dfrac{6 - 3x}{x^2}$

24. $\dfrac{\dfrac{3}{x^2y} - \dfrac{9}{2xy}}{\dfrac{6y}{x^2} + 12}$

25. $\dfrac{x - 3 + \dfrac{3x}{9 - x^2}}{\dfrac{x^2}{x+3} - \dfrac{x-2}{3x}}$

State the replacement set for each equation. Then solve.

26. $\dfrac{3}{15}x - \dfrac{1}{3}x = \dfrac{3}{10}x - \dfrac{13}{2}$

27. $\dfrac{x+1}{x-1} = \dfrac{x}{3} + \dfrac{2}{x-1}$

28. $\dfrac{10}{2y} - 17 = \dfrac{8}{5y}$

29. $\dfrac{-7x}{4x^2 - 4x - 15} + \dfrac{7}{2x-5} = \dfrac{2x}{3+2x}$

Classify this problem by naming the type of mathematical model you would use to solve it.

30. Moira needs a formula for computing her weekly earnings e if she works for h hours. Last week she earned $69.00 for working $5\frac{3}{4}$ h.

Chapter 8 Irrational and Complex Numbers

Solve over the real numbers.

1. $9y^2 = 49$ **2.** $x^6 + 128 = 0$ **3.** $3y^3 + 81 = 0$ **4.** $7x^4 + 3 = 3$

Simplify if possible.

5. $\sqrt[3]{-216}$ **6.** $\sqrt{25x^2y^4z^6}$ **7.** $\sqrt{0.0225}$ **8.** $\sqrt[5]{32x^{10}y^{15}}$

9. $5\sqrt{6} \cdot 11\sqrt{30}$ **10.** $\frac{1}{\sqrt{5x}}$ **11.** $(2\sqrt{5} - 4)\sqrt{10}$

12. $\frac{-3\sqrt{27a^2b}}{\sqrt{3b^3}}$ **13.** $\frac{6 - 7\sqrt{2}}{3\sqrt{2}}$ **14.** $\frac{3y}{\sqrt{3y^3}}$

15. $2\sqrt{3} - 4\sqrt{3}$ **16.** $7\sqrt{27x^3} + 2\sqrt{3x}$ **17.** $(2\sqrt{5} - 6\sqrt{10})^2$

18. $\sqrt[3]{40x^4y^2} + \sqrt[3]{135x^4y^2}$ **19.** $\frac{2\sqrt{5}}{3 + 2\sqrt{5}}$ **20.** $(3\sqrt{2y} - 4\sqrt{6y})(\sqrt{6y})$

Solve each equation if possible.

21. $\sqrt[3]{2y} = -2$ **22.** $\sqrt{1 - 3x} = \sqrt{x - 1}$ **23.** $2\sqrt{3x + 6} = \sqrt{20x}$

24. $\sqrt{3x + 4} + 8 = 3x$ **25.** $\sqrt{5 - 2y} = \frac{9}{\sqrt{5 - 2y}}$ **26.** $\sqrt[3]{9x} - \sqrt[3]{x - 24} = 0$

For each pair of points representing the endpoints of a segment, find: (a) the midpoint of the segment (b) the length of the segment

27. $A(-3, 2), B(2, -10)$ **28.** $C(5, -1), D(-2, 2)$ **29.** $E(2\sqrt{3}, 0), F(\sqrt{3}, 1)$

30. The endpoints of a diameter of a circle are $(-6, 5)$ and $(-1, -7)$. Find its center and the length of a radius.

31. The endpoints of a diameter of a circle are $A(0, 0)$ and $B(2a, 0)$. Draw a sketch showing the coordinates of the center and the endpoint coordinates of the diameter perpendicular to $\overline{AB}$. Show that $F\left(a + \frac{a\sqrt{2}}{2}, \frac{a\sqrt{2}}{2}\right)$ is a point on this circle.

Simplify.

32. $\sqrt{-12} \cdot \sqrt{-3}$ **33.** $-3\sqrt{-18} + \sqrt{-98}$ **34.** $\frac{4i^{29}}{i^3}$

35. $i\sqrt{-20} \cdot i\sqrt{-45}$ **36.** $i\sqrt{3} - 2i\sqrt{3}$ **37.** $8i^3 - 6i$

Simplify, and write in $a + bi$ form.

38. $(3 + 7i) + (-4 + i)$ **39.** $5i\sqrt{2} - (3\sqrt{2} + 2i\sqrt{2})$ **40.** $3\sqrt{-18} + 7\sqrt{-8}$

41. $|10 - 24i|$ **42.** $|-4 + 2i\sqrt{6}|$ **43.** $|3i\sqrt{17}|$

44. $(2 - 3i)(5 + 7i)$ **45.** $(3 + 8i)^2$ **46.** $(5\sqrt{2} - 2i\sqrt{3})^2$

47. $(6 + 3i) \div (2 - i)$ **48.** $(3 - 4i)(3 + 4i)$ **49.** $(1 - i)^3$

Chapter 9 Quadratic Functions

Determine the value of a if the graph of $y = ax^2$ contains the given point.

1. $(-2, 4)$ **2.** $(5, -10)$ **3.** $(-2, -8)$

4. Sketch the graph of the function $y = -x^2$.

Sketch the graph of each quadratic function and determine the y-intercept.

5. $y = -x^2 + 2x$ **6.** $y = 2x^2 - 4x$

7. $y = x^2 + 6x + 9$ **8.** $y = -x^2 + 2x + 8$

Solve by completing the square.

9. $2x^2 - 12x - 7 = 0$ **10.** $2x^2 + 5x - 3 = 0$ **11.** $x^2 + 20x = -91$

Solve each equation using the quadratic formula.

12. $2x^2 - 7x + 3 = 0$ **13.** $2x^2 - 9x + 7 = 0$ **14.** $3x^2 + 7x + 3 = 0$

15. Find two consecutive positive integers such that the sum of their squares is 145.

16. The square of a number exceeds 16 times the number by 225. Find the number.

17. Two chords intersect in a circle. One chord is 7 cm in length and the other is 8 cm. The ratio of the lengths of the two segments of the 7 cm chord is 3:4. Find the lengths of the segments of the 8 cm chord.

Use the discriminant to determine the nature of the solutions of each.

18. $2x^2 + 6x + 6 = 0$ **19.** $-x^2 + 10x - 20 = 0$ **20.** $4x^2 + 24x - 64 = 0$

For each equation, determine the value(s) of k for which there will be only one solution.

21. $6x^2 + 4kx = -8$ **22.** $4x^2 + kx + 16 = 0$

23. Find the sum and the product of the solutions of the equation $6x^2 - 4x - 10 = 0$.

24. Solve $(x - 2)^2 - 6(x - 2) + 8 = 0$ for x.

Solve and check.

25. $x^4 - 2x^2 - 24 = 0$ **26.** $\frac{n + 2}{n - 1} = \frac{n + 4}{n + 2}$

Solve each inequality.

27. $2x^2 + 4x - 16 > 0$ **28.** $x^2 - 9 \geq 0$ **29.** $4x^2 - 14x - 30 < 0$

Chapter 10 Polynomial Functions

Divide.

1. $(2x^3 - 4x^2y + 8xy^2) \div 4xy$
2. $(9x^2 - 3x - 1) \div (3x - 2)$
3. $\dfrac{2x^3 - 9x^2 + 9x + 3}{2x^2 - 3}$
4. $\dfrac{8a^3 - 1}{2a - 1}$

Divide, using synthetic division.

5. $\dfrac{3x^3 - 4x^2 - 16x + 7}{x - 3}$
6. $\dfrac{2y^4 - 3y^3 + y^2 + 5y + 2}{y - 1}$
7. $(y^3 + 125) \div (y + 5)$
8. $(x^4 + 1) \div (x + 1)$
9. $P(x) = -2x^4 + 3x^2 - 5$. Use the remainder theorem to find $P(-2)$.
10. Show that $y + 5$ is a factor of $4y^3 + 20y^2 - 9y - 45$, and factor the polynomial over the integers.
11. Write a polynomial with factors $3z + 1$, $2z - 3$, and $2z + 3$.
12. Find the maximum number of positive and negative real roots of the equation $-3x^4 + 2x^3 - 5x^2 - x + 9 = 0$.

Find all solutions of the following equations.

13. $y^3 + y^2 - y + 15 = 0$
14. $2x^4 - 9x^3 - 2x^2 + 39x - 18 = 0$
15. The four-digit number 3737 is divisible by 101. Prove that any four-digit number of this form is divisible by 101.
16. For $y = x^3 + 3x^2 + 4x + 3$, show that a real zero exists between $x = -2$ and $x = -1$ and find this zero to the nearest tenth.

Graph each function and find its zeros to the nearest tenth.

17. $f(x) = 2x^3 + 7x^2 - 2x - 1$
18. $f(x) = 2x^4 - 5x^3 - 5x^2 + 5x + 3$
19. The volume of a rectangular box is 225 cm^3. If its height and width are the same and its length is 1 cm less than twice its height, find the box's dimensions.

Find all intercepts and asymptotes of each function.

20. $y = \dfrac{5x + 4}{x - 2}$
21. $y = \dfrac{2x - 3}{x^2 - 4}$
22. Use asymptotes, intercepts, and any necessary points to sketch the graph of $y = \dfrac{3 - x}{x^2}$.

Chapter 11 Conic Sections

Write the equation of the circle with the given center and radius.

1. $C(-2, 0)$, 3

2. $C(5, -6)$, $\frac{1}{2}$

3. $C(7, 0)$, $\sqrt{11}$

Find the center and the radius of the circle with the given equation.

4. $(x + 1)^2 + (y - 2)^2 = 50$

5. $x^2 + y^2 + 4x - 10y = 20$

Write in standard form the equation of each parabola from its given focus and directrix or from its given focus and vertex.

6. $F(-1, 0)$, $x = 1$

7. $F(2, 2)$, $y = -6$

8. $F(0, 0)$, $V(-3, 0)$

Find the vertex, the axis of symmetry, the focus, and the directrix of each parabola, and sketch its graph.

9. $y^2 = -2x$

10. $(x - 3)^2 = y$

11. $x - 1 = 3y^2 + 6y$

Find the center, the intercepts, and the foci of each ellipse, and sketch its graph.

12. $16x^2 + y^2 = 16$

13. $3x^2 + 8y^2 + 32y = 160$

Find the intercepts, the foci, and the asymptotes of each hyperbola. Sketch each graph.

14. $x^2 - 4y^2 = 4$

15. $(y + 2)^2 - (x - 1)^2 = 1$

Graph each system and find any real solutions to the nearest tenth.

16. $\begin{cases} x^2 + y^2 = 5 \\ y - 2x = 5 \end{cases}$

17. $\begin{cases} \dfrac{x^2}{16} + \dfrac{y^2}{9} = 1 \\ x^2 + y^2 = 9 \end{cases}$

18. $\begin{cases} x^2 - \dfrac{y^2}{4} = 1 \\ y = 2x \end{cases}$

19. Graph this system of inequalities: $\begin{cases} 2y \geq x^2 - 2 \\ xy \leq -1 \end{cases}$

Identify the graphs of each system and find any real solutions.

20. $\begin{cases} x^2 - y^2 = 4 \\ x^2 + y^2 = 4 \end{cases}$

21. $\begin{cases} y^2 - x^2 = 6 \\ xy = 4 \end{cases}$

22. $\begin{cases} y + x^2 = 4 \\ y - 6x = 13 \end{cases}$

23. b varies jointly as a and the square of c. If $b = \frac{2}{25}$ when $a = \frac{3}{8}$ and $c = \frac{4}{5}$, find b when $a = 8$ and $c = \frac{3}{4}$.

24. z varies directly as the square of y and inversely as x. If $z = 9$ when $y = \frac{1}{2}$ and $x = \frac{1}{2}$, what is x when $z = 16$ and $y = \frac{1}{3}$?

Chapter 12 Exponential and Logarithmic Functions

Simplify.

1. $3x^{-\frac{2}{3}} \cdot x^{\frac{4}{3}} \cdot x^{-\frac{1}{6}}$

2. $27x^3y^{-\frac{3}{2}} \div \frac{1}{3}x^{\frac{1}{4}}y^{-2}$

3. $\left(\frac{1}{8}x^{\frac{5}{6}}y^{-\frac{1}{4}}\right)^{-\frac{2}{3}}$

Evaluate. Give decimal answers to the nearest thousandth.

4. $\left(\frac{9}{25}\right)^{\frac{1}{2}}$

5. $0.001^{-\frac{2}{3}}$

6. $11^{0.65}$

7. $0.35^{\frac{3}{4}}$

8. Graph the function $f(x) = \left(\frac{1}{2}\right)^x$.

9. Use a calculator to find $g(3.4)$, $g(\sqrt{5})$, and $g(\pi)$ if $g(x) = 3^{2x}$.

Solve and check.

10. $\left(\frac{1}{9}\right)^{2x} = 27$

11. $32^{3-x} = \left(\frac{1}{4}\right)^{2x}$

12. $0.01^{2x+1} = 10{,}000^{4x}$

Express in exponential form.

13. $\log_5 25 = 2$

14. $\log_3 \frac{1}{27} = -3$

15. $\log_{100} 10 = \frac{1}{2}$

Express in logarithmic form.

16. $625^{\frac{1}{4}} = 5$

17. $\left(\frac{1}{4}\right)^{-2} = 16$

18. $b^m = c,\ b > 0$

Evaluate.

19. $\log_{10} 1000$

20. $\log_2 \frac{1}{32}$

21. $\log_7 \sqrt{7}$

22. $\log_8 2\sqrt[3]{4}$

Express in expanded form.

23. $\log_b xyz$

24. $\log_6 \frac{x^3}{y^2}$

25. $\log_3 5\sqrt{x^3}$

26. $\log_b \frac{\sqrt{x}}{y^3 z}$

Express as a single logarithm. Evaluate if possible.

27. $\frac{1}{2}\log_b x + \frac{1}{2}\log_b y - \frac{1}{2}\log_b z$

28. $\frac{1}{3}\log_2 8 - 3\log_2 \frac{1}{4} + \log_2 16$

Solve and check.

29. $\log_6 x = 3\log_6 2 - \log_6 5$

30. $\log_4 x + \log_4 (x + 6) = 2$

Solve to the nearest thousandth and check.

31. $11^x = 258$

32. $72^{x-1} = 28$

33. $0.19^{x+1} = 5.72$

34. An isotope has a half-life of 94 s. How long will it take for 50 mg of this isotope to decay to 5 mg?

Chapter 13 Sequences and Series

1. Write the first four terms of a sequence if $a_1 = -2$ and $a_n = 3 - (a_{n-1})^2$.

Find the specified term of each sequence.

2. $-5, 1, 7, \ldots, a_{29}$

3. $\frac{2}{27}, \frac{2}{9}, \frac{2}{3}, \ldots, a_7$

4. $\frac{1}{2}, \frac{2}{3}, \frac{3}{4}, \ldots, a_{15}$

5. Find the mean proportional between 6 and 27.

6. Find a_2 of an arithmetic sequence if $a_4 = -4$ and $a_{10} = -14\frac{1}{2}$.

Find the specified partial sum of each series shown.

7. $-8 - 1 + 6 + 13 + \cdots; S_{21}$

8. $-8 + 4 - 2 + 1 - \cdots; S_8$

9. $\sum_{k=1}^{7} \left(\frac{1}{2}k + 2\right)$

10. $\sum_{n=1}^{6} 2 \cdot 3^{6-n}$

11. $\sum_{j=1}^{11} (-1)^j$

For the given arithmetic series, find the specified variable.

12. $a_1 = -1, S_{25} = 875, a_{25} = \underline{?}$

13. $d = -1, S_{12} = -18, a_1 = \underline{?}$

For the given geometric series, find the specified variable.

14. $r = \frac{1}{2}, S_7 = 127, a_1 = \underline{?}$

15. $a_1 = 3, a_5 = 48, S_5 = 93, r = \underline{?}$

For each infinite geometric series, find the sum if it exists.

16. $\sum_{n=1}^{\infty} \frac{1}{4}(2^{n-1})$

17. $\sum_{k=1}^{\infty} \left(\frac{3}{5}\right)^{k+1}$

18. $\sum_{j=1}^{\infty} 4 \cdot 3^{-j}$

Write each repeating decimal as a common fraction in lowest terms.

19. $0.\overline{17}$

20. $0.5\overline{2}$

21. $-2.\overline{106}$

22. $11.2\overline{03}$

23. The midpoints of the sides of a square are connected by segments to form another square, and the midpoints of the new square are connected in the same way, and so on. The side of the original square is 16 cm. Find the perimeter of the fourth square and the sum of the perimeters of the first five squares.

Expand and simplify each binomial expansion.

24. $(y + 3)^3$

25. $(2x^2 - 3y)^4$

26. $(i - 1)^5, i = \sqrt{-1}$

Write and simplify the specified term of each binomial expansion.

27. 4th term of $(x + y)^7$

28. middle term of $\left(\frac{1}{3}a - b^2\right)^6$

Chapter 14 Probability and Statistics

1. Find the number of permutations of the letters in the word LESSON.

2. Find the number of permutations of the elements in $\{1, 2, 3, 6, 7, 8, 9\}$.

3. Eight people stand in line at a theater. In how many different ways can they be arranged?

4. In a class election, a president, vice-president, and secretary are chosen from among 8 people. Find the number of possible election results.

5. In how many different orders can 3 books be read out of a total of 6?

6. How many different 4-card hands can be chosen from a 52-card deck?

7. A record club lists 11 records from which you may choose any 5. In how many ways can your selection be made?

8. Three coins are tossed. Specify:
   a. The sample space
   b. The probability of getting at least 2 heads
   c. The probability of getting at least 1 tail

9. Each of the numbers 1 through 15 is written on a slip of paper and all of the slips are placed in a covered bowl. The bowl is shaken and one slip of paper is drawn at random. Specify the probability of drawing:
   a. An even number **b.** A number greater than 7
   c. A prime number

10. A jar contains just 6 red marbles and 2 blue marbles. A marble is chosen at random and replaced before a second marble is chosen. Find each probability:
    a. P(both marbles are red) **b.** P(both are blue)
    c. P(the first is blue and the second is red)

11. Determine the mean, median and mode of the following sample:
    5, 5, 6, 7, 8, 9, 11, 9, 9, 7, 6, 5, 3, 9, 9.

12. Determine the mean, median and mode of this frequency distribution. Also, draw a histogram.

Score	Frequency
0	2
1	5
2	6
3	7
4	8
5	9

13. Determine the range, variance, and standard deviation of the following:
    65, 62, 88, 82, 72.

14. Given $\overline{x} = 88$ and $\sigma = 5$, find the area under the curve between $x = 90$ and $x = 95$.

Chapter 15 Trigonometric Functions and Graphs

Convert each degree measure to radian measure.

1. 240° **2.** -270° **3.** 315°

Convert each radian measure to degree measure.

4. $\frac{\pi}{3}$ **5.** $\frac{5\pi}{3}$ **6.** $-\frac{3\pi}{4}$

Give the sign of each of the following and sketch each angle.

7. $\sin 120^\circ$ **8.** $\cos 580^\circ$ **9.** $\tan 410^\circ$

Find the values of the six trigonometric functions of each angle. When radicals occur, express the answer in simplest radical form.

10. 30° **11.** 300°

Find each value.

12. $\sin 58^\circ 40'$ **13.** $\cos 296^\circ 30'$ **14.** $\tan 860^\circ 20'$

Find the measure of the acute angle to the nearest degree.

15. $\sin x = 0.6996$ **16.** $\tan x = 1.244$

17. Graph the function $y = \sin 3x$ for $-2\pi \leq x \leq 2\pi$.

Give the period and amplitude of each function. Then sketch the curve.

18. $y = 5 \sin x$ **19.** $y = \cos 2x$ **20.** $y = 4 \cos \frac{1}{2}x$

21. Solve right $\triangle ABC$ with right angle at B, $\angle C = 23^\circ$, and $c = 12$.

Solve each triangle.

22. $A = 120^\circ$, $B = 40^\circ$, $c = 35$

23. $C = 35^\circ$, $a = 11$, $b = 10.5$

Solve.

24. A ladder 17 ft in length leans against the wall of a building. The foot of the ladder rests 12 ft from the base of the wall. To the nearest degree, what is the measure of the angle that the ladder makes with the ground?

25. What is the area of a parallelogram if two adjacent sides measure 12 cm and 18 cm and the angle between them measures 30°?

26. A tower is 200 ft tall. To the nearest degree, find the measure of the angle of elevation from a point 50 ft from the base of the tower to the top of the tower.

Chapter 16 Trigonometric Identities and Equations

Prove each identity.

1. $\cos A \sec A = 1$

2. $1 + \tan^2 A = \sec^2 A$

Simplify.

3. $\sin x \sec x$

4. $\sec x - \sin x \cdot \tan x$

5. $\dfrac{1}{1 + \tan^2 x} + \dfrac{1}{1 + \cot^2 x}$

6. Evaluate $\tan 120°$. If radicals occur, give answer in simplest radical form.

7. If $\tan A = \frac{1}{4}$ and $\tan B = \frac{1}{2}$, find $\tan (A + B)$.

Evaluate the following using special angle measures and a sum or difference identity. When radicals occur, express the answer in simplest radical form.

8. $\sin 105°$

9. $\cos 75°$

10. $\tan \dfrac{\pi}{12}$

Use a double-angle or half-angle formula to find the values of the following. If radicals occur, give answers in simplest radical form.

11. $\sin 450°$

12. $\cos 15°$

Prove.

13. $\dfrac{\sin 2A}{1 + \cos 2A} = \tan A$

14. $\sec^4 x - \sec^2 x = \tan^4 x + \tan^2 x$

Evaluate.

15. $\text{Arccos}\ \dfrac{5}{13}$

16. $\text{Arccsc}\ -2$

17. $\text{Arcsec}\ 1.0324$

18. An airplane with an air speed of 350 mi/h is flying on a heading of 53°. The wind is blowing from due north at 28 mi/h. What is the ground speed of the plane and its course?

19. Two forces, one of 120 lb and the other of 200 lb, act on a body and make an angle measuring 52° with each other. What is the magnitude of the resultant of the forces and what is the measure, to the nearest degree, of the angle that it makes with the 200-lb force?

20. Change the polar coordinates P(4, 120°) to rectangular coordinates.

21. Express in standard form: $3(\cos 300° + i \sin 300°)$

22. Express $2 + 3i$ in polar form.

Table of Squares, Cubes, Square and Cube Roots

N	N^2	N^3	$\sqrt{N}$	$\sqrt[3]{N}$	N	N^2	N^3	$\sqrt{N}$	$\sqrt[3]{N}$
1	1	1	1.000	1.000	**51**	2,601	132,651	7.141	3.708
2	4	8	1.414	1.260	**52**	2,704	140,608	7.211	3.733
3	9	27	1.732	1.442	**53**	2,809	148,877	7.280	3.756
4	16	64	2.000	1.587	**54**	2,916	157,464	7.348	3.780
5	25	125	2.236	1.710	**55**	3,025	166,375	7.416	3.803
6	36	216	2.449	1.817	**56**	3,136	175,616	7.483	3.826
7	49	343	2.646	1.913	**57**	3,249	185,193	7.550	3.849
8	64	512	2.828	2.000	**58**	3,364	195,112	7.616	3.871
9	81	729	3.000	2.080	**59**	3,481	205,379	7.681	3.893
10	100	1,000	3.162	2.154	**60**	3,600	216,000	7.746	3.915
11	121	1,331	3.317	2.224	**61**	3,721	226,981	7.810	3.936
12	144	1,728	3.464	2.289	**62**	3,844	238,328	7.874	3.958
13	169	2,197	3.606	2.351	**63**	3,969	250,047	7.937	3.979
14	196	2,744	3.742	2.410	**64**	4,096	262,144	8.000	4.000
15	225	3,375	3.873	2.466	**65**	4,225	274,625	8.062	4.021
16	256	4,096	4.000	2.520	**66**	4,356	287,496	8.124	4.041
17	289	4,913	4.123	2.571	**67**	4,489	300,763	8.185	4.062
18	324	5,832	4.243	2.621	**68**	4,624	314,432	8.246	4.082
19	361	6,859	4.359	2.668	**69**	4,761	328,509	8.307	4.102
20	400	8,000	4.472	2.714	**70**	4,900	343,000	8.367	4.121
21	441	9,261	4.583	2.759	**71**	5,041	357,911	8.426	4.141
22	484	10,648	4.690	2.802	**72**	5,184	373,248	8.485	4.160
23	529	12,167	4.796	2.844	**73**	5,329	389,017	8.544	4.179
24	576	13,824	4.899	2.884	**74**	5,476	405,224	8.602	4.198
25	625	15,625	5.000	2.924	**75**	5,625	421,875	8.660	4.217
26	676	17,576	5.099	2.962	**76**	5,776	438,976	8.718	4.236
27	729	19,683	5.196	3.000	**77**	5,929	456,533	8.775	4.254
28	784	21,952	5.292	3.037	**78**	6,084	474,552	8.832	4.273
29	841	24,389	5.385	3.072	**79**	6,241	493,039	8.888	4.291
30	900	27,000	5.477	3.107	**80**	6,400	512,000	8.944	4.309
31	961	29,791	5.568	3.141	**81**	6,561	531,441	9.000	4.327
32	1,024	32,768	5.657	3.175	**82**	6,724	551,368	9.055	4.344
33	1,089	35,937	5.745	3.208	**83**	6,889	571,787	9.110	4.362
34	1,156	39,304	5.831	3.240	**84**	7,056	592,704	9.165	4.380
35	1,225	42,875	5.916	3.271	**85**	7,225	614,125	9.220	4.397
36	1,296	46,656	6.000	3.302	**86**	7,396	636,056	9.274	4.414
37	1,369	50,653	6.083	3.332	**87**	7,569	658,503	9.327	4.431
38	1,444	54,872	6.164	3.362	**88**	7,744	681,472	9.381	4.448
39	1,521	59,319	6.245	3.391	**89**	7,921	704,969	9.434	4.465
40	1,600	64,000	6.325	3.420	**90**	8,100	729,000	9.487	4.481
41	1,681	68,921	6.403	3.448	**91**	8,281	753,571	9.539	4.498
42	1,764	74,088	6.481	3.476	**92**	8,464	778,688	9.592	4.514
43	1,849	79,507	6.557	3.503	**93**	8,649	804,357	9.644	4.531
44	1,936	85,184	6.633	3.530	**94**	8,836	830,584	9.695	4.547
45	2,025	91,125	6.708	3.557	**95**	9,025	857,375	9.747	4.563
46	2,116	97,336	6.782	3.583	**96**	9,216	884,736	9.798	4.579
47	2,209	103,823	6.856	3.609	**97**	9,409	912,673	9.849	4.595
48	2,304	110,592	6.928	3.634	**98**	9,604	941,192	9.899	4.610
49	2,401	117,649	7.000	3.659	**99**	9,801	970,299	9.950	4.626
50	2,500	125,000	7.071	3.684	**100**	10,000	1,000,000	10.000	4.642

Using a Table of Common Logarithms

The table on pages 866 and 867 contains the common logarithms of numbers from 1 to 9.99. A portion of the table is shown below. The logarithm values are rounded to four decimal places and the decimal points are omitted.

N	0	1	2	3	4	5	6	7	8	9
4.1	6128	6138	6149	6160	6170	6180	6191	6201	6212	6222

To find the logarithm of a number N between 1 and 10, find the first two digits of N in the column on the left and the third digit in the top row. The logarithm of N is shown at the intersection of that row and column. To illustrate, the logarithm of 4.17 is found at the intersection of the row for 41 and the column for 7. To four decimal places, $\log 4.17 = 0.6201$.

To find the logarithm of a number greater than 10 or between 0 and 1, write the number in scientific notation and then use the properties of logarithms.

EXAMPLE 1 **Use a table to find each logarithm:**
a. $\log 4170$ **b.** $\log 0.0417$

a. $\log 4170 = \log(4.17 \times 10^3)$
$= \log 4.17 + \log 10^3$ *$\log_b MN = \log_b M + \log_b N$*
$= 0.6201 + 3$ *$\log_b b^x = x$*
$= 3.6201$ *To four decimal places*

b. $\log 0.0417 = \log(4.17 \times 10^{-2})$
$= \log 4.17 + \log 10^{-2}$
$= 0.6201 + (-2)$
$= 0.6201 - 2$, or -1.3799 *To four decimal places*

A common logarithm can be written as the sum of a *nonnegative* number less than 1, called the **mantissa,** and an integer, called the **characteristic.**

	mantissa		*characteristic*
$\log 4170 =$	0.6201	+	3
$\log 0.0417 =$	0.6201	+	(−2)

The logarithm of a number such as 0.0417 can be written in many ways. Sometimes it is more convenient to use one form than another.

$\log 0.0417 = -1.3799$	*Not* in mantissa/characteristic form	
$\log 0.0417 = 0.6201 - 2$	Mantissa: 0.6201	Characteristic: -2
$\log 0.0417 = 2.6201 - 4$	Mantissa: 0.6201	Characteristic: $2 - 4 = -2$
$\log 0.0417 = 8.6201 - 10$	Mantissa: 0.6201	Characteristic: $8 - 10 = -2$

The table of common logarithms can also be used to find a number N if $\log N$ is known. If $\log N$ is between 0 and 1, then N is between 1 and 10 and its value can be read directly from the table. For instance, if $\log N = 0.7932$, the table on page 867 shows that the logarithm closest to this value is found at the intersection of the row for 62 and the column for 1. That is, N is approximately 6.21, to three significant digits.

EXAMPLE 2 **Use a table to find N:**
a. $\log N = 5.1904$ **b.** $\log N = -3.2165$.

First write each number in mantissa/characteristic form.

a.
$$\begin{aligned} \log N &= 5.1905 \\ &= 0.1905 + 5 && \textit{Mantissa/characteristic form} \\ &= \log 1.55 + 5 && \textit{Use the table.} \\ &= \log 1.55 + \log 10^5 \\ &= \log (1.55 \times 10^5) \\ &= \log 155{,}000 \end{aligned}$$
So, $N = 155{,}000$ *Three significant digits*

b.
$$\begin{aligned} \log N &= -3.2165 \\ &= 4 + -3.2165 - 4 \\ &= 0.7835 - 4 \\ &= \log 6.07 - 4 \\ &= \log 6.07 + \log 10^{-4} \\ &= \log (6.07 \times 10^{-4}) \\ &= \log 0.000607 \end{aligned}$$
So, $N = 0.000607$

Tables

EXERCISES

Use a table to find each common logarithm to four decimal places.

1. log 2.58 **2.** log 9.04 **3.** log 16.8 **4.** log 340

5. log 72,500 **6.** log 0.143 **7.** log 0.00329 **8.** log 0.0605

Use a table to find N to three significant digits.

9. $\log N = 0.8982$ **10.** $\log N = 0.5528$ **11.** $\log N = 1.9805$

12. $\log N = 3.6944$ **13.** $\log N = 0.9842 - 2$ **14.** $\log N = 0.1732 - 1$

Use a table to find each common logarithm to four decimal places.

15. log 300,000 **16.** log 75,000,000 **17.** log 0.000086 **18.** 0.0000006

Use a table to find N to three significant digits.

19. $\log N = 8.8131$ **20.** $\log N = 9.5265$ **21.** $\log N = 4.7378 - 10$

22. $\log N = -1.7714$ **23.** $\log N = -2.3100$ **24.** $\log N = 6.9272 - 10$

25. $\log N = 11.9559$ **26.** $\log N = 13.0042$

27. $\log N = -12.5406$ **28.** $\log N = -10.7393$

Tables

Computation Using Common Logarithms

Common logarithms can be used to simplify computations. In this process, addition and subtraction replace the operations of multiplication and division, while multiplication replaces the operation of exponentiation.

EXAMPLE 1 **Use common logarithms from a table to evaluate:**

a. $0.00496 \times 23{,}600$ **b.** $\dfrac{0.000408}{7.73}$ **c.** $\sqrt[3]{0.425}$

a. Let $N = 0.00496 \times 23{,}600$

Then $\log N = \log (0.00496 \times 23{,}600)$

$= \log 0.00496 + \log 23{,}600$ *$\log MN = \log M + \log N$*

$= \log (4.96 \times 10^{-3}) + \log (2.36 \times 10^{4})$

$= (0.6955 - 3) + (4.3729)$ *Use the table.*

$= 2.0684$

$= 0.0684 + 2$ *Mantissa/characteristic form*

So, $N = 1.17 \times 10^{2}$ *Use the table again. Find the closest entry.*

$N = 117$ *Three significant digits*

b. Let $N = \dfrac{0.000408}{7.73}$

Then $\log N = \log \left(\dfrac{0.000408}{7.73}\right)$

$= \log 0.000408 - \log 7.73$ *$\log \dfrac{M}{N} = \log M - \log N$*

$= \log (4.08 \times 10^{-4}) - \log 7.73$

$= (0.6107 - 4) - 0.8882$ *Use the table.*

$= (1.6107 - 5) - 0.8882$ *To obtain a positive mantissa, write $0.6107 - 4$ as $1.6107 - 5$.*

$= 0.7225 - 5$ *Mantissa/characteristic form*

So, $N = 5.28 \times 10^{-5}$ *Use the table again. Find the closest entry.*

$N = 0.0000528$ *Three significant digits*

c. Let $N = \sqrt[3]{0.425} = 0.425^{\frac{1}{3}}$

Then $\log N = \log 0.425^{\frac{1}{3}}$

$= \frac{1}{3} \log 0.425$ *$\log N^{r} = r \log N$*

$= \frac{1}{3} (0.6284 - 1)$ *Use the table.*

$= \frac{1}{3} (2.6284 - 3)$ *To get an integer for the negative part of the characteristic, write $0.6284 - 1$ as $2.6284 - 3$.*

$= 0.8761 - 1$ *Mantissa/characteristic form*

So, $N = 7.52 \times 10^{-1}$ *Use the table again. Find the closest entry.*

$N = 0.752$ *Three significant digits*

In the next example, several operations are performed using common logarithms.

EXAMPLE 2 **Use common logarithms from a table to evaluate:** $\dfrac{0.0493(3.71)^4}{52.6}$

$$\text{Let} \qquad N = \frac{0.0493(3.71)^4}{52.6}$$

$$\begin{aligned} \text{Then} \quad \log N &= \log\left(\frac{0.0493(3.71)^4}{52.6}\right) \\ &= \log 0.0493 + \log 3.71^4 - \log 52.6 \\ &= \log 0.0493 + 4 \log 3.71 - \log 52.6 \\ &= (0.6928 - 2) + 4(0.5694) - 1.7210 \\ &= 0.6928 - 2 + 2.2776 - 1.7210 \\ &= -0.7506, \quad \text{or } 0.2494 - 1 \end{aligned}$$

$$\text{So,} \qquad N = 0.178 \qquad \textit{Three significant digits}$$

EXERCISES

Use common logarithms from a table to evaluate to three significant digits.

1. 2.96×9.74

2. 7.45×8.93

3. 0.0309×768

4. 0.00265×37.6

5. 3.49^3

6. 6.27^4

7. $\sqrt[5]{925{,}000}$

8. $\sqrt[4]{34{,}800}$

9. $256(4.95)^2$

10. $0.747(8.71)^2$

11. $3760\sqrt[4]{46{,}800}$

12. $0.0129\sqrt[3]{5080}$

13. $\dfrac{5620}{3.85}$

14. $\dfrac{8.38}{0.125}$

15. $\dfrac{0.714(7.36)}{247}$

16. $\dfrac{0.00368(2.54)}{1.13}$

17. $\dfrac{56.5(5.66)^4}{363{,}000}$

18. $\dfrac{0.296(4.87)^3}{1370}$

19. $0.0165\sqrt[6]{7280}$

20. $45{,}200\sqrt[8]{0.968}$

21. $\sqrt{26.8(127)^3}$

22. $\sqrt[3]{0.0561}(4.99)^2$

23. $56.3(4.26)^{-6}$

24. $7250(36.2)^{-5}$

25. $\dfrac{4.72}{0.928(42.6)^2}$

26. $\dfrac{38.3}{2.09(0.0256)^3}$

27. $\dfrac{7.66(0.0478)}{0.395(26.7)}$

28. $\dfrac{686(0.00261)}{0.985(49.7)}$

29. $\dfrac{\sqrt{21.3}}{\sqrt[7]{26{,}700}}$

30. $\dfrac{\sqrt[5]{0.796}}{\sqrt[3]{545}}$

31. $\left(\dfrac{\sqrt[3]{751}}{275}\right)^4$

32. $\left(\dfrac{368}{\sqrt[4]{911}}\right)^3$

33. $\sqrt{\dfrac{92.6(1.49)^3}{0.368}}$

34. $\sqrt[3]{\dfrac{476(0.155)^2}{76.7}}$

35. $\sqrt{\dfrac{529(1.48)^5}{1.37\sqrt[3]{0.0245}}}$

36. $\sqrt[4]{\dfrac{29.4\sqrt[5]{0.766}}{301(0.135)^3}}$

Linear Interpolation

The table on pages 866 and 867 contains logarithms of numbers with three significant digits. To approximate the logarithms of numbers with more than three significant digits, you can use a method called **linear interpolation.**

Over very small intervals, the graph of a logarithmic function is nearly a straight line. Consider the graph of $y = \log x$ between the points $A(3.25, 0.5119)$ and $C(3.26, 0.5132)$. In order to estimate the value of $\log x$ at a point between A and C where $x = 3.257$, assume that the graph between points A and C is a straight line. Then, since triangles ABD and ACE are similar, corresponding sides are proportional. That is,

$$\frac{AD}{AE} = \frac{BD}{CE}$$

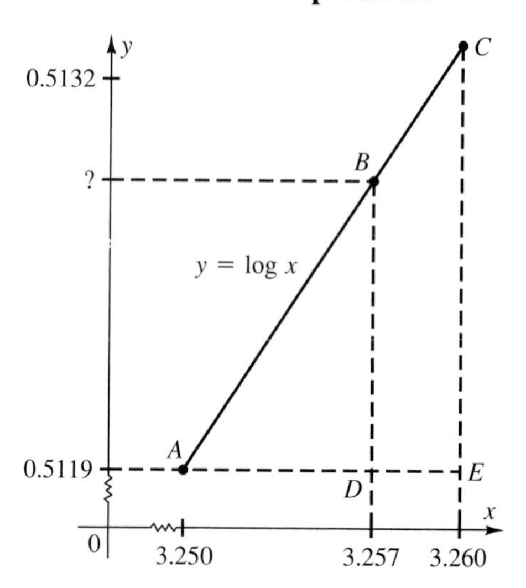

EXAMPLE 1 **Use linear interpolation to approximate log 3.257.**

First set up a table.

		x	$\log x$		
0.010	0.007	3.250	0.5119	d	0.0013
		3.257	?		
		3.260	0.5132		

Then write and solve a proportion.

$$\frac{0.007}{0.010} = \frac{d}{0.0013} \qquad \frac{AD}{AE} = \frac{BD}{CE}$$

$$d = \frac{7}{10}(0.0013) \approx 0.0009$$ *Round to four decimal places, as in the log table.*

So, $\log 3.257 = 0.5119 + d = 0.5119 + 0.0009 = 0.5128$

Linear interpolation can also be used to approximate a number when its mantissa falls between entries in the table of common logarithms.

EXAMPLE 2 **Use linear interpolation to approximate x if $\log x = 3.1714$.**

Use a table of common logarithms to find the logarithms just less than and just greater than 3.1714.

		x	$\log x$		
10	d	1480	3.1703	0.0011	0.0029
		?	3.1714		
		1490	3.1732		

Write and solve a proportion.

$$\frac{d}{10} = \frac{0.0011}{0.0029}$$

$$d = \frac{11}{29}(10) \approx 4$$ *Round so that the value of x will have four significant digits.*

So, $x = 1480 + d = 1480 + 4 = 1484$

EXERCISES

Use linear interpolation to find each logarithm to four decimal places.

1. $\log 1.675$ **2.** $\log 3.724$ **3.** $\log 8.508$ **4.** $\log 7.736$

5. $\log 2.444$ **6.** $\log 9.065$ **7.** $\log 6.333$ **8.** $\log 5.247$

Use linear interpolation to find the value of x to four significant digits.

9. $\log x = 0.1265$ **10.** $\log x = 0.9346$ **11.** $\log x = 0.5162$

12. $\log x = 0.8114$ **13.** $\log x = 0.6601$ **14.** $\log x = 0.1029$

Use linear interpolation to find each logarithm to four decimal places.

15. $\log 12.66$ **16.** $\log 485.3$ **17.** $\log 0.03257$ **18.** $\log 0.1056$

Use linear interpolation to find the value of x to four significant digits.

19. $\log x = 2.1269$ **20.** $\log x = 4.3070$ **21.** $\log x = 0.6747 - 3$

22. $\log x = 0.1065 - 2$ **23.** $\log x = -1.2643$ **24.** $\log x = -2.4692$

Use logarithms and linear interpolation to find each value to four significant digits.

25. $\sqrt[3]{2.376}$ **26.** $0.004762(32.45)^3$

27. $\sqrt{\frac{1.246(2377)}{42{,}880}}$ **28.** $\sqrt{\frac{4.262}{(0.1204)^3}}$

Using a Table of Trigonometric Values

The table on pages 870 to 874 contains the values of the six trigonometric functions from 0°00′ to 90°00′ in 10-minute increments. The radian measure of each angle value is also listed. A portion of the table is shown below.

θ Deg.	θ Rad.	Sin θ	Cos θ	Tan θ	Cot θ	Sec θ	Csc θ		
19°00′	0.3316	0.3256	0.9455	0.3443	2.9042	1.0576	3.0716	1.2392	**71°00′**
10′	0.3345	0.3283	0.9446	0.3476	2.8770	1.0587	3.0458	1.2363	**50′**
20′	0.3374	0.3311	0.9436	0.3508	2.8502	1.0598	3.0206	1.2334	**40′**
30′	0.3403	0.3338	0.9426	0.3541	2.8239	1.0608	2.9957	1.2305	**30′**
40′	0.3432	0.3365	0.9417	0.3574	2.7980	1.0619	2.9713	1.2275	**20′**
50′	0.3462	0.3393	0.9407	0.3607	2.7725	1.0631	2.9474	1.2246	**10′**
20°00′	0.3491	0.3420	0.9397	0.3640	2.7475	1.0642	2.9238	1.2217	**70°00′**
10′	0.3520	0.3448	0.9387	0.3673	2.7228	1.0653	2.9006	1.2188	**50′**
		Cos θ	**Sin θ**	**Cot θ**	**Tan θ**	**Csc θ**	**Sec θ**	**θ Rad.**	**θ Deg.**

To find the value of a function for an angle between 0°00′ and 45°00′, read *down* the first column to locate the angle and across the *top* row to locate the function. The value of the function is at the intersection of that row and column. To illustrate, sin 19°30′ = 0.3338. To find the value of a function for an angle between 45°00′ and 90°00′, read *up* the last column to locate the angle and across the *bottom* row to locate the function. For instance, tan 70°50′ = 2.8770.

To find the value for a trigonometric function of an angle expressed in tenths of degrees, convert the angle measure to degrees and minutes first. To find the value for a trigonometric function of an angle expressed in radians, convert to decimal radian measure and then use the second column of the table. Use linear interpolation, if necessary.

EXAMPLE 1 **Use a table to find the following.**
a. cot 54.5° **b.** cos 32.2°

a. cot 54.5° = cot 54°30′ *Convert to degrees and minutes.*
cot 54.5° = 0.7133 *Use the table.*

b. cos 32.2° = cos 32°12′ *Convert to degrees and minutes.*

cos 32°10′ = 0.8465
cos 32°20′ = 0.8450 *Use the table for values of cosine around 32°12′.*

cos 32.2° = 0.8462 *Interpolate.*

To find the value for a trigonometric function of an angle greater than 90°, use a reference triangle in the appropriate quadrant. Find the value of the *reference angle,* which is the acute angle of the reference triangle that the terminal side makes with the x-axis.

EXAMPLE 2 **Use a table to find the following.** **a.** $\sin 245°40'$ **b.** $\sec \frac{23\pi}{12}$

a. $\sin 245°40' = -\sin 65°40'$ *Reference triangle is in quadrant III.*

$= -0.9112$ *Use the table.*

b. $\sec \frac{23\pi}{12} = \sec \frac{\pi}{12}$ *Reference triangle is in quadrant IV.*

$= \sec 0.2618$ *Convert to decimal radian measure.*

$= 1.0353$ *Use second column of table for radians.*

The table of trigonometric function values can also be used to solve trigonometric equations. The acute reference angle θ of a reference triangle can be found directly if the value of a function is known.

EXAMPLE 3 **Solve for θ, $0° \leq \theta < 360°$, to the nearest minute:**
a. $\tan \theta = 0.8642$ **b.** $\sin \theta = -0.5628$

a. $\tan \theta = 0.8642$

reference angle $= 40°50'$ *Use the table.*

$\theta = 40°50', 220°50'$ *Tangent is positive in quadrants I and III.*

b. $\sin \theta = -0.5628$ *Ignore negative sign to find reference angle.*

reference angle $= 34°15'$ *Use the table and interpolate.*

$\theta = 214°15', 325°45'$ *Sine is negative in quadrants III and IV.*

EXERCISES

Use a table to find the following.

1. $\cos 9°20'$ **2.** $\sin 74°10'$ **3.** $\tan 43°50'$ **4.** $\csc 88°40'$

5. $\sec \frac{\pi}{18}$ **6.** $\cot \frac{5\pi}{24}$ **7.** $\sin \frac{17\pi}{9}$ **8.** $\csc \frac{23\pi}{30}$

9. $\tan 223°10'$ **10.** $\cos 147°20'$ **11.** $\csc 289°50'$ **12.** $\sin 315°30'$

13. $\sec 12.8°$ **14.** $\cot 53.4°$ **15.** $\sin 237.6°$ **16.** $\tan 302.1°$

Solve for θ, $0° \leq \theta < 360°$, to the nearest minute.

17. $\sin \theta = 0.9013$ **18.** $\cot \theta = 14.301$ **19.** $\sec \theta = -5.3813$

20. $\tan \theta = -2.5487$ **21.** $\csc \theta = -1.4579$ **22.** $\cos \theta = 0.8853$

Table of Common Logarithms

N	0	1	2	3	4	5	6	7	8	9
1.0	0000	0043	0086	0128	0170	0212	0253	0294	0334	0374
1.1	0414	0453	0492	0531	0569	0607	0645	0682	0719	0755
1.2	0792	0828	0864	0899	0934	0969	1004	1038	1072	1106
1.3	1139	1173	1206	1239	1271	1303	1335	1367	1399	1430
1.4	1461	1492	1523	1553	1584	1614	1644	1673	1703	1732
1.5	1761	1790	1818	1847	1875	1903	1931	1959	1987	2014
1.6	2041	2068	2095	2122	2148	2175	2201	2227	2253	2279
1.7	2304	2330	2355	2380	2405	2430	2455	2480	2504	2529
1.8	2553	2577	2601	2625	2648	2672	2695	2718	2742	2765
1.9	2788	2810	2833	2856	2878	2900	2923	2945	2967	2989
2.0	3010	3032	3054	3075	3096	3118	3139	3160	3181	3201
2.1	3222	3243	3263	3284	3304	3324	3345	3365	3385	3404
2.2	3424	3444	3464	3483	3502	3522	3541	3560	3579	3598
2.3	3617	3636	3655	3674	3692	3711	3729	3747	3766	3784
2.4	3802	3820	3838	3856	3874	3892	3909	3927	3945	3962
2.5	3979	3997	4014	4031	4048	4065	4082	4099	4116	4133
2.6	4150	4166	4183	4200	4216	4232	4249	4265	4281	4298
2.7	4314	4330	4346	4362	4378	4393	4409	4425	4440	4456
2.8	4472	4487	4502	4518	4533	4548	4564	4579	4594	4609
2.9	4624	4639	4654	4669	4683	4698	4713	4728	4742	4757
3.0	4771	4786	4800	4814	4829	4843	4857	4871	4886	4900
3.1	4914	4928	4942	4955	4969	4983	4997	5011	5024	5038
3.2	5051	5065	5079	5092	5105	5119	5132	5145	5159	5172
3.3	5185	5198	5211	5224	5237	5250	5263	5276	5289	5302
3.4	5315	5328	5340	5353	5366	5378	5391	5403	5416	5428
3.5	5441	5453	5465	5478	5490	5502	5514	5527	5539	5551
3.6	5563	5575	5587	5599	5611	5623	5635	5647	5658	5670
3.7	5682	5694	5705	5717	5729	5740	5752	5763	5775	5786
3.8	5798	5809	5821	5832	5843	5855	5866	5877	5888	5899
3.9	5911	5922	5933	5944	5955	5966	5977	5988	5999	6010
4.0	6021	6031	6042	6053	6064	6075	6085	6096	6107	6117
4.1	6128	6138	6149	6160	6170	6180	6191	6201	6212	6222
4.2	6232	6243	6253	6263	6274	6284	6294	6304	6314	6325
4.3	6335	6345	6355	6365	6375	6385	6395	6405	6415	6425
4.4	6435	6444	6454	6464	6474	6484	6493	6503	6513	6522
4.5	6532	6542	6551	6561	6571	6580	6590	6599	6609	6618
4.6	6628	6637	6646	6656	6665	6675	6684	6693	6702	6712
4.7	6721	6730	6739	6749	6758	6767	6776	6785	6794	6803
4.8	6812	6821	6830	6839	6848	6857	6866	6875	6884	6893
4.9	6902	6911	6920	6928	6937	6946	6955	6964	6972	6981
5.0	6990	6998	7007	7016	7024	7033	7042	7050	7059	7067
5.1	7076	7084	7093	7101	7110	7118	7126	7135	7143	7152
5.2	7160	7168	7177	7185	7193	7202	7210	7218	7226	7235
5.3	7243	7251	7259	7267	7275	7284	7292	7300	7308	7316
5.4	7324	7332	7340	7348	7356	7364	7372	7380	7388	7396

Table of Common Logarithms

N	0	1	2	3	4	5	6	7	8	9
5.5	7404	7412	7419	7427	7435	7443	7451	7459	7466	7474
5.6	7482	7490	7497	7505	7513	7520	7528	7536	7543	7551
5.7	7559	7566	7574	7582	7589	7597	7604	7612	7619	7627
5.8	7634	7642	7649	7657	7664	7672	7679	7686	7694	7701
5.9	7709	7716	7723	7731	7738	7745	7752	7760	7767	7774
6.0	7782	7789	7796	7803	7810	7818	7825	7832	7839	7846
6.1	7853	7860	7868	7875	7882	7889	7896	7903	7910	7917
6.2	7924	7931	7938	7945	7952	7959	7966	7973	7980	7987
6.3	7993	8000	8007	8014	8021	8028	8035	8041	8048	8055
6.4	8062	8069	8075	8082	8089	8096	8102	8109	8116	8122
6.5	8129	8136	8142	8149	8156	8162	8169	8176	8182	8189
6.6	8195	8202	8209	8215	8222	8228	8235	8241	8248	8254
6.7	8261	8267	8274	8280	8287	8293	8299	8306	8312	8319
6.8	8325	8331	8338	8344	8351	8357	8363	8370	8376	8382
6.9	8388	8395	8401	8407	8414	8420	8426	8432	8439	8445
7.0	8451	8457	8463	8470	8476	8482	8488	8494	8500	8506
7.1	8513	8519	8525	8531	8537	8543	8549	8555	8561	8567
7.2	8573	8579	8585	8591	8597	8603	8609	8615	8621	8627
7.3	8633	8639	8645	8651	8657	8663	8669	8675	8681	8686
7.4	8692	8698	8704	8710	8716	8722	8727	8733	8739	8745
7.5	8751	8756	8762	8768	8774	8779	8785	8791	8797	8802
7.6	8808	8814	8820	8825	8831	8837	8842	8848	8854	8859
7.7	8865	8871	8876	8882	8887	8893	8899	8904	8910	8915
7.8	8921	8927	8932	8938	8943	8949	8954	8960	8965	8971
7.9	8976	8982	8987	8993	8998	9004	9009	9015	9020	9025
8.0	9031	9036	9042	9047	9053	9058	9063	9069	9074	9079
8.1	9085	9090	9096	9101	9106	9112	9117	9122	9128	9133
8.2	9138	9143	9149	9154	9159	9165	9170	9175	9180	9186
8.3	9191	9196	9201	9206	9212	9217	9222	9227	9232	9238
8.4	9243	9248	9253	9258	9263	9269	9274	9279	9284	9289
8.5	9294	9299	9304	9309	9315	9320	9325	9330	9335	9340
8.6	9345	9350	9355	9360	9365	9370	9375	9380	9385	9390
8.7	9395	9400	9405	9410	9415	9420	9425	9430	9435	9440
8.8	9445	9450	9455	9460	9465	9469	9474	9479	9484	9489
8.9	9494	9499	9504	9509	9513	9518	9523	9528	9533	9538
9.0	9542	9547	9552	9557	9562	9566	9571	9576	9581	9586
9.1	9590	9595	9600	9605	9609	9614	9619	9624	9628	9633
9.2	9638	9643	9647	9652	9657	9661	9666	9671	9675	9680
9.3	9685	9689	9694	9699	9703	9708	9713	9717	9722	9727
9.4	9731	9736	9741	9745	9750	9754	9759	9763	9768	9773
9.5	9777	9782	9786	9791	9795	9800	9805	9809	9814	9818
9.6	9823	9827	9832	9836	9841	9845	9850	9854	9859	9863
9.7	9868	9872	9877	9881	9886	9890	9894	9899	9903	9908
9.8	9912	9917	9921	9926	9930	9934	9939	9943	9948	9952
9.9	9956	9961	9965	9969	9974	9978	9983	9987	9991	9996

Table of Natural Logarithms (ln x)

x	0.00	0.01	0.02	0.03	0.04	0.05	0.06	0.07	0.08	0.09
1.0	0.0000	0.0100	0.0198	0.0296	0.0392	0.0488	0.0583	0.0677	0.0770	0.0862
1.1	0.0953	0.1044	0.1133	0.1222	0.1310	0.1398	0.1484	0.1570	0.1655	0.1740
1.2	0.1823	0.1906	0.1989	0.2070	0.2151	0.2231	0.2311	0.2390	0.2469	0.2546
1.3	0.2624	0.2700	0.2776	0.2852	0.2927	0.3001	0.3075	0.3148	0.3221	0.3293
1.4	0.3365	0.3436	0.3507	0.3577	0.3646	0.3716	0.3784	0.3853	0.3920	0.3988
1.5	0.4055	0.4121	0.4187	0.4253	0.4318	0.4383	0.4447	0.4511	0.4574	0.4637
1.6	0.4700	0.4762	0.4824	0.4886	0.4947	0.5008	0.5068	0.5128	0.5188	0.5247
1.7	0.5306	0.5365	0.5423	0.5481	0.5539	0.5596	0.5653	0.5710	0.5766	0.5822
1.8	0.5878	0.5933	0.5988	0.6043	0.6098	0.6152	0.6206	0.6259	0.6313	0.6166
1.9	0.6419	0.6471	0.6523	0.6575	0.6627	0.6678	0.6729	0.6780	0.6831	0.6881
2.0	0.6931	0.6981	0.7031	0.7080	0.7130	0.7178	0.7227	0.7275	0.7324	0.7372
2.1	0.7419	0.7467	0.7514	0.7561	0.7608	0.7655	0.7701	0.7747	0.7793	0.7839
2.2	0.7885	0.7930	0.7975	0.8020	0.8065	0.8109	0.8154	0.8198	0.8242	0.8286
2.3	0.8329	0.8372	0.8416	0.8459	0.8502	0.8544	0.8587	0.8629	0.8671	0.8713
2.4	0.8755	0.8796	0.8838	0.8879	0.8920	0.8961	0.9002	0.9042	0.9083	0.9123
2.5	0.9163	0.9203	0.9243	0.9282	0.9322	0.9361	0.9400	0.9439	0.9478	0.9517
2.6	0.9555	0.9594	0.9632	0.9670	0.9708	0.9746	0.9783	0.9821	0.9858	0.9895
2.7	0.9933	0.9969	1.0006	1.0043	1.0080	1.0116	1.0152	0.0188	1.0225	1.0260
2.8	1.0296	1.0332	1.0367	1.0403	1.0438	1.0473	1.0508	1.0543	1.0578	1.0613
2.9	1.0647	1.0682	1.0716	1.0750	1.0784	1.0818	1.0852	1.0886	1.0919	1.0953
3.0	1.0986	1.1019	1.1053	1.1086	1.1119	1.1151	1.1184	1.1217	1.1249	1.1282
3.1	1.1314	1.1346	1.1378	1.1410	1.1442	1.1474	1.1506	1.1537	1.1569	1.1600
3.2	1.1632	1.1663	1.1694	1.1725	1.1756	1.1787	1.1817	1.1848	1.1878	1.1909
3.3	1.1939	1.1970	1.2000	1.2030	1.2060	1.2090	1.2119	1.2149	1.2179	1.2208
3.4	1.2238	1.2267	1.2296	1.2326	1.2355	1.2384	1.2413	1.2442	1.2470	1.2499
3.5	1.2528	1.2556	1.2585	1.2613	1.2641	1.2669	1.2698	1.2726	1.2754	1.2782
3.6	1.2809	1.2837	1.2865	1.2892	1.2920	1.2947	1.2975	1.3002	1.3029	1.3056
3.7	1.3083	1.3110	1.3137	1.3164	1.3191	1.3218	1.3244	1.3271	1.3297	1.3324
3.8	1.3350	1.3376	1.3403	1.3429	1.3455	1.3481	1.3507	1.3533	1.3558	1.3584
3.9	1.3610	1.3635	1.3661	1.3686	1.3712	1.3737	1.3762	1.3788	1.3813	1.3838
4.0	1.3863	1.3888	1.3913	1.3938	1.3962	1.3987	1.4012	1.4036	1.4061	1.4085
4.1	1.4110	1.4134	1.4159	1.4183	1.4207	1.4231	1.4255	1.4279	1.4303	1.4327
4.2	1.4351	1.4375	1.4398	1.4422	1.4446	1.4469	1.4493	1.4516	1.4540	1.4563
4.3	1.4586	1.4609	1.4633	1.4656	1.4679	1.4702	1.4725	1.4748	1.4770	1.4793
4.4	1.4816	1.4839	1.4861	1.4884	1.4907	1.4929	1.4952	1.4974	1.4996	1.5019
4.5	1.5041	1.5063	1.5085	1.5107	1.5129	1.5151	1.5173	1.5195	1.5217	1.5239
4.6	1.5261	1.5282	1.5304	1.5326	1.5347	1.5369	1.5390	1.5412	1.5433	1.5454
4.7	1.5476	1.5497	1.5518	1.5539	1.5560	1.5581	1.5602	1.5623	1.5644	1.5665
4.8	1.5686	1.5707	1.5728	1.5748	1.5769	1.5790	1.5810	1.5831	1.5851	1.5872
4.9	1.5892	1.5913	1.5933	1.5953	1.5974	1.5994	1.6014	1.6034	1.6054	1.6074
5.0	1.6094	1.6114	1.6134	1.6154	1.6174	1.6194	1.6214	1.6233	1.6253	1.6273
5.1	1.6292	1.6312	1.6332	1.6351	1.6371	1.6390	1.6409	1.6429	1.6448	1.6467
5.2	1.6487	1.6506	1.6525	1.6544	1.6563	1.6582	1.6601	1.6620	1.6639	1.6658
5.3	1.6677	1.6696	1.6715	1.6734	1.6752	1.6771	1.6790	1.6808	1.6827	1.6845
5.4	1.6864	1.6882	1.6901	1.6919	1.6938	1.6956	1.6974	1.6993	1.7001	1.7029

Table of Natural Logarithms (ln *x*)

x	0.00	0.01	0.02	0.03	0.04	0.05	0.06	0.07	0.08	0.09
5.5	1.7047	1.7066	1.7084	1.7102	1.7120	1.7138	1.7156	1.7174	1.7192	1.7210
5.6	1.7228	1.7246	1.7263	1.7281	1.7299	1.7317	1.7334	1.7352	1.7370	1.7387
5.7	1.7405	1.7422	1.7440	1.7457	1.7475	1.7492	1.7509	1.7527	1.7544	1.7561
5.8	1.7579	1.7596	1.7613	1.7630	1.7647	1.7664	1.7682	1.7699	1.7716	1.7733
5.9	1.7750	1.7766	1.7783	1.7800	1.7817	1.7834	1.7851	1.7867	1.7884	1.7901
6.0	1.7918	1.7934	1.7951	1.7967	1.7984	1.8001	1.8017	1.8034	1.8050	1.8066
6.1	1.8083	1.8099	1.8116	1.8132	1.8148	1.8165	1.8181	1.8197	1.8213	1.8229
6.2	1.8245	1.8262	1.8278	1.8294	1.8310	1.8326	1.8342	1.8358	1.8374	1.8390
6.3	1.8406	1.8421	1.8437	1.8453	1.8469	1.8485	1.8500	1.8516	1.8532	1.8547
6.4	1.8563	1.8579	1.8594	1.8610	1.8625	1.8641	1.8656	1.8672	1.8687	1.8703
6.5	1.8718	1.8733	1.8749	1.8764	1.8779	1.8795	1.8810	1.8825	1.8840	1.8856
6.6	1.8871	1.8886	1.8901	1.8916	1.8931	1.8946	1.8961	1.8976	1.8991	1.9006
6.7	1.9021	1.9036	1.9051	1.9066	1.9081	1.9095	1.9110	1.9125	1.9140	1.9155
6.8	1.9169	1.9184	1.9199	1.9213	1.9228	1.9242	1.9257	1.9272	1.9286	1.9301
6.9	1.9315	1.9330	1.9344	1.9359	1.9373	1.9387	1.9402	1.9416	1.9430	1.9445
7.0	1.9459	1.9473	1.9488	1.9502	1.9516	1.9530	1.9544	1.9559	1.9573	1.9587
7.1	1.9601	1.9615	1.9629	1.9643	1.9657	1.9671	1.9685	1.9699	1.9713	1.9727
7.2	1.9741	1.9755	1.9769	1.9782	1.9796	1.9810	1.9824	1.9838	1.9851	1.9865
7.3	1.9879	1.9892	1.9906	1.9920	1.9933	1.9947	1.9961	1.9974	1.9988	2.0001
7.4	2.0015	2.0028	2.0042	2.0055	2.0069	2.0082	2.0096	2.0109	2.0122	2.0136
7.5	2.0149	2.0162	2.0176	2.0189	2.0202	2.0215	2.0229	2.0242	2.0255	2.0268
7.6	2.0282	2.0295	2.0308	2.0321	2.0334	2.0347	2.0360	2.0373	2.0386	2.0399
7.7	2.0412	2.0425	2.0438	2.0451	2.0464	2.0477	2.0490	2.0503	2.0516	2.0528
7.8	2.0541	2.0554	2.0567	2.0580	2.0592	2.0605	2.0618	2.0631	2.0643	2.0665
7.9	2.0669	2.0681	2.0694	2.0707	2.0719	2.0732	2.0744	2.0757	2.0769	2.0782
8.0	2.0794	2.0807	2.0819	2.0832	2.0844	2.0857	2.0869	2.0882	2.0894	2.0906
8.1	2.0919	2.0931	2.0943	2.0956	2.0968	2.0980	2.0992	2.1005	2.1017	2.1029
8.2	2.1041	2.1054	2.1066	2.1078	2.1090	2.1102	2.1114	2.1126	2.1138	2.1150
8.3	2.1163	2.1175	2.1187	2.1199	2.1211	2.1223	2.1235	2.1247	2.1258	2.1270
8.4	2.1282	2.1294	2.1306	2.1318	2.1330	2.1342	2.1353	2.1365	2.1377	2.1389
8.5	2.1401	2.1412	2.1424	2.1436	2.1448	2.1459	2.1471	2.1483	2.1494	2.1506
8.6	2.1518	2.1529	2.1541	2.1552	2.1564	2.1576	2.1587	2.1599	2.1610	2.1622
8.7	2.1633	2.1645	2.1656	2.1668	2.1679	2.1691	2.1702	2.1713	2.1725	2.1736
8.8	2.1748	2.1759	2.1770	2.1782	2.1793	2.1804	2.1815	2.1827	2.1838	2.1849
8.9	2.1861	2.1872	2.1883	2.1894	2.1905	2.1917	2.1928	2.1939	2.1950	2.1961
9.0	2.1972	2.1983	2.1994	2.2006	2.2017	2.2028	2.2039	2.2050	2.2061	2.2072
9.1	2.2083	2.2094	2.2105	2.2116	2.2127	2.2138	2.2148	2.2159	2.2170	2.2181
9.2	2.2192	2.2203	2.2214	2.2225	2.2235	2.2246	2.2257	2.2268	2.2279	2.2289
9.3	2.2300	2.2311	2.2322	2.2332	2.2343	2.2354	2.2364	2.2375	2.2386	2.2396
9.4	2.2407	2.2418	2.2428	2.2439	2.2450	2.2460	2.2471	2.2481	2.2492	2.2502
9.5	2.2513	2.2523	2.2534	2.2544	2.2555	2.2565	2.2576	2.2586	2.2597	2.2607
9.6	2.2618	2.2628	2.2638	2.2649	2.2659	2.2670	2.2680	2.2690	2.2701	2.2711
9.7	2.2721	2.2732	2.2742	2.2752	2.2762	2.2773	2.2783	2.2793	2.2803	2.2814
9.8	2.2824	2.2834	2.2844	2.2854	2.2865	2.2875	2.2885	2.2895	2.2905	2.2915
9.9	2.2925	2.2935	2.2946	2.2956	2.2966	2.2976	2.2986	2.2996	2.3006	2.3016

Table of Values of the Trigonometric Functions

θ Deg.	θ Rad.	Sin θ	Cos θ	Tan θ	Cot θ	Sec θ	Csc θ		
0°00′	0.0000	0.0000	1.0000	0.0000		1.000		1.5708	**90°00′**
10′	0.0029	0.0029	1.0000	0.0029	343.77	1.000	343.8	1.5679	**50′**
20′	0.0058	0.0058	1.0000	0.0058	171.89	1.000	171.9	1.5650	**40′**
30′	0.0087	0.0087	1.0000	0.0087	114.59	1.000	114.6	1.5621	**30′**
40′	0.0116	0.0116	0.9999	0.0116	85.940	1.000	85.95	1.5592	**20′**
50′	0.0145	0.0145	0.9999	0.0145	68.750	1.000	68.76	1.5563	**10′**
1°00′	0.0175	0.0175	0.9998	0.0175	57.290	1.000	57.30	1.5533	**89°00′**
10′	0.0204	0.0204	0.9998	0.0204	49.104	1.000	49.11	1.5504	**50′**
20′	0.0233	0.0233	0.9997	0.0233	42.964	1.000	42.98	1.5475	**40′**
30′	0.0262	0.0262	0.9997	0.0262	38.188	1.000	38.20	1.5446	**30′**
40′	0.0291	0.0291	0.9996	0.0291	34.368	1.000	34.38	1.5417	**20′**
50′	0.0320	0.0320	0.9995	0.0320	31.242	1.001	31.26	1.5388	**10′**
2°00′	0.0349	0.0349	0.9994	0.0349	28.636	1.001	28.65	1.5359	**88°00′**
10′	0.0378	0.0378	0.9993	0.0378	26.432	1.001	26.45	1.5330	**50′**
20′	0.0407	0.0407	0.9992	0.0407	24.542	1.001	24.56	1.5301	**40′**
30′	0.0436	0.0436	0.9990	0.0437	22.904	1.001	22.93	1.5272	**30′**
40′	0.0465	0.0465	0.9989	0.0466	21.470	1.001	21.49	1.5243	**20′**
50′	0.0495	0.0494	0.9988	0.0495	20.206	1.001	20.23	1.5213	**10′**
3°00′	0.0524	0.0523	0.9986	0.0524	19.081	1.001	19.11	1.5184	**87°00′**
10′	0.0553	0.0552	0.9985	0.0553	18.075	1.002	18.10	1.5155	**50′**
20′	0.0582	0.0581	0.9983	0.0582	17.169	1.002	17.20	1.5126	**40′**
30′	0.0611	0.0610	0.9981	0.0612	16.350	1.002	16.38	1.5097	**30′**
40′	0.0640	0.0640	0.9980	0.0641	15.605	1.002	15.64	1.5068	**20′**
50′	0.0669	0.0669	0.9978	0.0670	14.924	1.002	14.96	1.5039	**10′**
4°00′	0.0698	0.0698	0.9976	0.0699	14.301	1.002	14.34	1.5010	**86°00′**
10′	0.0727	0.0727	0.9974	0.0729	13.727	1.003	13.76	1.4981	**50′**
20′	0.0756	0.0756	0.9971	0.0758	13.197	1.003	13.23	1.4952	**40′**
30′	0.0785	0.0785	0.9969	0.0787	12.706	1.003	12.75	1.4923	**30′**
40′	0.0814	0.0814	0.9967	0.0816	12.251	1.003	12.29	1.4893	**20′**
50′	0.0844	0.0843	0.9964	0.0846	11.826	1.004	11.87	1.4864	**10′**
5°00′	0.0873	0.0872	0.9962	0.0875	11.430	1.004	11.47	1.4835	**85°00′**
10′	0.0902	0.0901	0.9959	0.0904	11.059	1.004	11.10	1.4806	**50′**
20′	0.0931	0.0929	0.9957	0.0934	10.712	1.004	10.76	1.4777	**40′**
30′	0.0960	0.0958	0.9954	0.0963	10.385	1.005	10.43	1.4748	**30′**
40′	0.0989	0.0987	0.9951	0.0992	10.078	1.005	10.13	1.4719	**20′**
50′	0.1018	0.1016	0.9948	0.1022	9.7882	1.005	9.839	1.4690	**10′**
6°00′	0.1047	0.1045	0.9945	0.1051	9.5144	1.006	9.567	1.4661	**84°00′**
10′	0.1076	0.1074	0.9942	0.1080	9.2553	1.006	9.309	1.4632	**50′**
20′	0.1105	0.1103	0.9939	0.1110	9.0098	1.006	9.065	1.4603	**40′**
30′	0.1134	0.1132	0.9936	0.1139	8.7769	1.006	8.834	1.4573	**30′**
40′	0.1164	0.1161	0.9932	0.1169	8.5555	1.007	8.614	1.4544	**20′**
50′	0.1193	0.1190	0.9929	0.1198	8.3450	1.007	8.405	1.4515	**10′**
7°00′	0.1222	0.1219	0.9925	0.1228	8.1443	1.008	8.206	1.4486	**83°00′**
10′	0.1251	0.1248	0.9922	0.1257	7.9530	1.008	8.016	1.4457	**50′**
20′	0.1280	0.1276	0.9918	0.1287	7.7704	1.008	7.834	1.4428	**40′**
30′	0.1309	0.1305	0.9914	0.1317	7.5958	1.009	7.661	1.4399	**30′**
40′	0.1338	0.1334	0.9911	0.1346	7.4287	1.009	7.496	1.4370	**20′**
50′	0.1367	0.1363	0.9907	0.1376	7.2687	1.009	7.337	1.4341	**10′**
8°00′	0.1396	0.1392	0.9903	0.1405	7.1154	1.010	7.185	1.4312	**82°00′**
10′	0.1425	0.1421	0.9899	0.1435	6.9682	1.010	7.040	1.4283	**50′**
20′	0.1454	0.1449	0.9894	0.1465	6.8269	1.011	6.900	1.4254	**40′**
30′	0.1484	0.1478	0.9890	0.1495	6.6912	1.011	6.765	1.4224	**30′**
40′	0.1513	0.1507	0.9886	0.1524	6.6506	1.012	6.636	1.4195	**20′**
50′	0.1542	0.1536	0.9881	0.1554	6.4348	1.012	6.512	1.4166	**10′**
9°00′	0.1571	0.1564	0.9877	0.1584	6.3138	1.012	6.392	1.4137	**81°00′**
		Cos θ	**Sin θ**	**Cot θ**	**Tan θ**	**Csc θ**	**Sec θ**	**θ Rad.**	**θ Deg.**

Table of Values of the Trigonometric Functions

θ Deg.	θ Rad.	Sin θ	Cos θ	Tan θ	Cot θ	Sec θ	Csc θ		
9°00′	0.1571	0.1564	0.9877	0.1584	6.3138	1.012	6.392	1.4137	**81°00′**
10′	0.1600	0.1593	0.9872	0.1614	6.1970	1.013	6.277	1.4108	**50′**
20′	0.1629	0.1622	0.9868	0.1644	6.0844	1.013	6.166	1.4079	**40′**
30′	0.1658	0.1650	0.9863	0.1673	5.9758	1.014	6.059	1.4050	**30′**
40′	0.1687	0.1679	0.9858	0.1703	5.8708	1.014	5.955	1.4021	**20′**
50′	0.1716	0.1708	0.9853	0.1733	5.7694	1.015	5.855	1.3992	**10′**
10°00′	0.1745	0.1736	0.9848	0.1763	5.6713	1.015	5.759	1.3963	**80°00′**
10′	0.1774	0.1765	0.9843	0.1793	5.5764	1.016	5.665	1.3934	**50′**
20′	0.1804	0.1794	0.9838	0.1823	5.4845	1.016	5.575	1.3904	**40′**
30′	0.1833	0.1822	0.9833	0.1853	5.3955	1.017	5.487	1.3875	**30′**
40′	0.1862	0.1851	0.9827	0.1883	5.3093	1.018	5.403	1.3846	**20′**
50′	0.1891	0.1880	0.9822	0.1914	5.2257	1.018	5.320	1.3817	**10′**
11°00′	0.1920	0.1908	0.9816	0.1944	5.1446	1.019	5.241	1.3788	**79°00′**
10′	0.1949	0.1937	0.9811	0.1974	5.0658	1.019	5.164	1.3759	**50′**
20′	0.1978	0.1965	0.9805	0.2004	4.9894	1.020	5.089	1.3730	**40′**
30′	0.2007	0.1994	0.9799	0.2035	4.9152	1.020	5.016	1.3701	**30′**
40′	0.2036	0.2022	0.9793	0.2065	4.8430	1.021	4.945	1.3672	**20′**
50′	0.2065	0.2051	0.9787	0.2095	4.7729	1.022	4.876	1.3643	**10′**
12°00′	0.2094	0.2079	0.9781	0.2126	4.7046	1.022	4.810	1.3614	**78°00′**
10′	0.2123	0.2108	0.9775	0.2156	4.6382	1.023	4.745	1.3584	**50′**
20′	0.2153	0.2136	0.9769	0.2186	4.5736	1.024	4.682	1.3555	**40′**
30′	0.2182	0.2164	0.9763	0.2217	4.5107	1.024	4.620	1.3526	**30′**
40′	0.2211	0.2193	0.9757	0.2247	4.4494	1.025	4.560	1.3497	**20′**
50′	0.2240	0.2221	0.9750	0.2278	4.3897	1.026	4.502	1.3468	**10′**
13°00′	0.2269	0.2250	0.9744	0.2309	4.3315	1.026	4.445	1.3439	**77°00′**
10′	0.2298	0.2278	0.9737	0.2339	4.2747	1.027	4.390	1.3410	**50′**
20′	0.2327	0.2306	0.9730	0.2370	4.2193	1.028	4.336	1.3381	**40′**
30′	0.2356	0.2334	0.9724	0.2401	4.1653	1.028	4.284	1.3352	**30′**
40′	0.2385	0.2363	0.9717	0.2432	4.1126	1.029	4.232	1.3323	**20′**
50′	0.2414	0.2391	0.9710	0.2462	4.0611	1.030	4.182	1.3294	**10′**
14°00′	0.2443	0.2419	0.9703	0.2493	4.0108	1.031	4.134	1.3265	**76°00′**
10′	0.2473	0.2447	0.9696	0.2524	3.9617	1.031	4.086	1.3235	**50′**
20′	0.2502	0.2476	0.9689	0.2555	3.9136	1.032	4.039	1.3206	**40′**
30′	0.2531	0.2504	0.9681	0.2586	3.8667	1.033	3.994	1.3177	**30′**
40′	0.2560	0.2532	0.9674	0.2617	3.8208	1.034	3.950	1.3148	**20′**
50′	0.2589	0.2560	0.9667	0.2648	3.7760	1.034	3.906	1.3119	**10′**
15°00′	0.2618	0.2588	0.9659	0.2679	3.7321	1.035	3.864	1.3090	**75°00′**
10′	0.2647	0.2616	0.9652	0.2711	3.6891	1.036	3.822	1.3061	**50′**
20′	0.2676	0.2644	0.9644	0.2742	3.6470	1.037	3.782	1.3032	**40′**
30′	0.2705	0.2672	0.9636	0.2773	3.6059	1.038	3,742	1.3003	**30′**
40′	0.2734	0.2700	0.9628	0.2805	3.5656	1.039	3.703	1.2974	**20′**
50′	0.2763	0.2728	0.9621	0.2836	3.5261	1.039	3.665	1.2945	**10′**
16°00′	0.2793	0.2756	0.9613	0.2867	3.4874	1.040	3.628	1.2915	**74°00′**
10′	0.2822	0.2784	0.9605	0.2899	3.4495	1.041	3.592	1.2886	**50′**
20′	0.2851	0.2812	0.9596	0.2931	3.4124	1.042	3.556	1.2857	**40′**
30′	0.2880	0.2840	0.9588	0.2962	3.3759	1.043	3.521	1.2828	**30′**
40′	0.2909	0.2868	0.9580	0.2994	3.3402	1.044	3.487	1.2799	**20′**
50′	0.2938	0.2896	0.9572	0.3026	3.3052	1.045	3.453	1.2770	**10′**
17°00′	0.2967	0.2924	0.9563	0.3057	3.2709	1.046	3.420	1.2741	**73°00′**
10′	0.2996	0.2952	0.9555	0.3089	3.2371	1.047	3.388	1.2712	**50′**
20′	0.3025	0.2979	0.9546	0.3121	3.2041	1.048	3.356	1.2683	**40′**
30′	0.3054	0.3007	0.9537	0.3153	3.1716	1.049	3.326	1.2654	**30′**
40′	0.3083	0.3035	0.9528	0.3185	3.1397	1.049	3.295	1.2625	**20′**
50′	0.3113	0.3062	0.9520	0.3217	3.1084	1.050	3.265	1.2595	**10′**
18°00′	0.3142	0.3090	0.9511	0.3249	3.0777	1.051	3.236	1.2566	**72°00′**
		Cos θ	**Sin θ**	**Cot θ**	**Tan θ**	**Csc θ**	**Sec θ**	**θ Rad.**	**θ Deg.**

Table of Values of the Trigonometric Functions

θ Deg.	θ Rad.	Sin θ	Cos θ	Tan θ	Cot θ	Sec θ	Csc θ		
18°00′	0.3142	0.3090	0.9511	0.3249	3.0777	1.051	3.236	1.2566	**72°00′**
10′	0.3171	0.3118	0.9502	0.3281	3.0475	1.052	3.207	1.2537	**50′**
20′	0.3200	0.3145	0.9492	0.3314	3.0178	1.053	3.179	1.2508	**40′**
30′	0.3229	0.3173	0.9483	0.3346	2.9887	1.054	3.152	1.2479	**30′**
40′	0.3258	0.3201	0.9474	0.3378	2.9600	1.056	3.124	1.2450	**20′**
50′	0.3287	0.3228	0.9465	0.3411	2.9319	1.057	3.098	1.2421	**10′**
19°00′	0.3316	0.3256	0.9455	0.3443	2.9042	1.058	3.072	1.2392	**71°00′**
10′	0.3345	0.3283	0.9446	0.3476	2.8770	1.059	3.046	1.2363	**50′**
20′	0.3374	0.3311	0.9436	0.3508	2.8502	1.060	3.021	1.2334	**40′**
30′	0.3403	0.3338	0.9426	0.3541	2.8239	1.061	2.996	1.2305	**30′**
40′	0.3432	0.3365	0.9417	0.3574	2.7980	1.062	2.971	1.2275	**20′**
50′	0.3462	0.3393	0.9407	0.3607	2.7725	1.063	2.947	1.2246	**10′**
20°00′	0.3491	0.3420	0.9397	0.3640	2.7475	1.064	2.924	1.2217	**70°00′**
10′	0.3520	0.3448	0.9387	0.3673	2.7228	1.065	2.901	1.2188	**50′**
20′	0.3549	0.3475	0.9377	0.3706	2.6985	1.066	2.878	1.2159	**40′**
30′	0.3578	0.3502	0.9367	0.3739	2.6746	1.068	2.855	1.2130	**30′**
40′	0.3607	0.3529	0.9356	0.3772	2.6511	1.069	2.833	1.2101	**20′**
50′	0.3636	0.3557	0.9346	0.3805	2.6279	1.070	2.812	1.2072	**10′**
21°00′	0.3665	0.3584	0.9336	0.3839	2.6051	1.071	2.790	1.2043	**69°00′**
10′	0.3694	0.3611	0.9325	0.3872	2.5826	1.072	2.769	1.2014	**50′**
20′	0.3723	0.3638	0.9315	0.3906	2.5605	1.074	2.749	1.1985	**40′**
30′	0.3752	0.3665	0.9304	0.3939	2.5386	1.075	2.729	1.1956	**30′**
40′	0.3782	0.3692	0.9293	0.3973	2.5172	1.076	2.709	1.1926	**20′**
50′	0.3811	0.3719	0.9283	0.4006	2.4960	1.077	2.689	1.1897	**10′**
22°00′	0.3840	0.3746	0.9272	0.4040	2.4751	1.079	2.669	1.1868	**68°00′**
10′	0.3869	0.3773	0.9261	0.4074	2.4545	1.080	2.650	1.1839	**50′**
20′	0.3898	0.3800	0.9250	0.4108	2.4342	1.081	2.632	1.1810	**40′**
30′	0.3927	0.3827	0.9239	0.4142	2.4142	1.082	2.613	1.1781	**30′**
40′	0.3956	0.3854	0.9228	0.4176	2.3945	1.084	2.595	1.1752	**20′**
50′	0.3985	0.3881	0.9216	0.4210	2.3750	1.085	2.577	1.1723	**10′**
23°00′	0.4014	0.3907	0.9205	0.4245	2.3559	1.086	2.559	1.1694	**67°00′**
10′	0.4043	0.3934	0.9194	0.4279	2.3369	1.088	2.542	1.1665	**50′**
20′	0.4072	0.3961	0.9182	0.4314	2.3183	1.089	2.525	1.1636	**40′**
30′	0.4102	0.3987	0.9171	0.4348	2.2998	1.090	2.508	1.1606	**30′**
40′	0.4131	0.4014	0.9159	0.4383	2.2817	1.092	2.491	1.1577	**20′**
50′	0.4160	0.4041	0.9147	0.4417	2.2637	1.093	2.475	1.1548	**10′**
24°00′	0.4189	0.4067	0.9135	0.4452	2.2460	1.095	2.459	1.1519	**66°00′**
10′	0.4218	0.4094	0.9124	0.4487	2.2286	1.096	2.443	1.1490	**50′**
20′	0.4247	0.4120	0.9112	0.4522	2.2113	1.097	2.427	1.1461	**40′**
30′	0.4276	0.4147	0.9100	0.4557	2.1943	1.099	2.411	1.1432	**30′**
40′	0.4305	0.4173	0.9088	0.4592	2.1775	1.100	2.396	1.1403	**20′**
50′	0.4334	0.4200	0.9075	0.4628	2.1609	1.102	2.381	1.1374	**10′**
25°00′	0.4363	0.4226	0.9063	0.4663	2.1445	1.103	2.366	1.1345	**65°00′**
10′	0.4392	0.4253	0.9051	0.4699	2.1283	1.105	2.352	1.1316	**50′**
20′	0.4422	0.4279	0.9038	0.4734	2.1123	1.106	2.337	1.1286	**40′**
30′	0.4451	0.4305	0.9026	0.4770	2.0965	1.108	2.323	1.1257	**30′**
40′	0.4480	0.4331	0.9013	0.4806	2.0809	1.109	2.309	1.1228	**20′**
50′	0.4509	0.4358	0.9001	0.4841	2.0655	1.111	2.295	1.1199	**10′**
26°00′	0.4538	0.4384	0.8988	0.4877	2.0503	1.113	2.281	1.1170	**64°00′**
10′	0.4567	0.4410	0.8975	0.4913	2.0353	1.114	2.268	1.1141	**50′**
20′	0.4596	0.4436	0.8962	0.4950	2.0204	1.116	2.254	1.1112	**40′**
30′	0.4625	0.4462	0.8949	0.4986	2.0057	1.117	2.241	1.1083	**30′**
40′	0.4654	0.4488	0.8936	0.5022	1.9912	1.119	2.228	1.1054	**20′**
50′	0.4683	0.4514	0.8923	0.5059	1.9768	1.121	2.215	1.1025	**10′**
27°00′	0.4712	0.4540	0.8910	0.5095	1.9626	1.122	2.203	1.0996	**63°00′**
		Cos θ	**Sin θ**	**Cot θ**	**Tan θ**	**Csc θ**	**Sec θ**	**θ Rad.**	**θ Deg.**

Tables

Table of Values of the Trigonometric Functions

θ Deg.	θ Rad.	Sin θ	Cos θ	Tan θ	Cot θ	Sec θ	Csc θ		
27°00′	0.4712	0.4540	0.8910	0.5095	1.9626	1.122	2.203	1.0996	63°00′
10′	0.4741	0.4566	0.8897	0.5132	1.9486	1.124	2.190	1.0966	50′
20′	0.4771	0.4592	0.8884	0.5169	1.9347	1.126	2.178	1.0937	40′
30′	0.4800	0.4617	0.8870	0.5206	1.9210	1.127	2.166	1.0908	30′
40′	0.4829	0.4643	0.8857	0.5243	1.9074	1.129	2.154	1.0879	20′
50′	0.4858	0.4669	0.8843	0.5280	1.8940	1.131	2.142	1.0850	10′
28°00′	0.4887	0.4695	0.8829	0.5317	1.8807	1.133	2.130	1.0821	62°00′
10′	0.4916	0.4720	0.8816	0.5354	1.8676	1.134	2.118	1.0792	50′
20′	0.4945	0.4746	0.8802	0.5392	1.8546	1.136	2.107	1.0763	40′
30′	0.4974	0.4772	0.8788	0.5430	1.8418	1.138	2.096	1.0734	30′
40′	0.5003	0.4797	0.8774	0.5467	1.8291	1.140	2.085	1.0705	20′
50′	0.5032	0.4823	0.8760	0.5505	1.8165	1.142	2.074	1.0676	10′
29°00′	0.5061	0.4848	0.8746	0.5543	1.8040	1.143	2.063	1.0647	61°00′
10′	0.5091	0.4874	0.8732	0.5581	1.7917	1.145	2.052	1.0617	50′
20′	0.5120	0.4899	0.8718	0.5619	1.7796	1.147	2.041	1.0588	40′
30′	0.5149	0.4924	0.8704	0.5658	1.7675	1.149	2.031	1.0559	30′
40′	0.5178	0.4950	0.8689	0.5696	1.7556	1.151	2.020	1.0530	20′
50′	0.5207	0.4975	0.8675	0.5735	1.7437	1.153	2.010	1.0501	10′
30°00′	0.5236	0.5000	0.8660	0.5774	1.7321	1.155	2.000	1.0472	60°00′
10′	0.5265	0.5025	0.8646	0.5812	1.7205	1.157	1.990	1.0443	50′
20′	0.5294	0.5050	0.8631	0.5851	1.7090	1.159	1.980	1.0414	40′
30′	0.5323	0.5075	0.8616	0.5890	1.6977	1.161	1.970	1.0385	30′
40′	0.5352	0.5100	0.8601	0.5930	1.6864	1.163	1.961	1.0356	20′
50′	0.5381	0.5125	0.8587	0.5969	1.6753	1.165	1.951	1.0327	10′
31°00′	0.5411	0.5150	0.8572	0.6009	1.6643	1.167	1.942	1.0297	59°00′
10′	0.5440	0.5175	0.8557	0.6048	1.6534	1.169	1.932	1.0268	50′
20′	0.5469	0.5200	0.8542	0.6088	1.6426	1.171	1.923	1.0239	40′
30′	0.5498	0.5225	0.8526	0.6128	1.6319	1.173	1.914	1.0210	30′
40′	0.5527	0.5250	0.8511	0.6168	1.6212	1.175	1.905	1.0181	20′
50′	0.5556	0.5275	0.8496	0.6208	1.6107	1.177	1.896	1.0152	10′
32°00′	0.5585	0.5299	0.8480	0.6249	1.6003	1.179	1.887	1.0123	58°00′
10′	0.5614	0.5324	0.8465	0.6289	1.5900	1.181	1.878	1.0094	50′
20′	0.5643	0.5348	0.8450	0.6330	1.5798	1.184	1.870	1.0065	40′
30′	0.5672	0.5373	0.8434	0.6371	1.5697	1.186	1.861	1.0036	30′
40′	0.5701	0.5398	0.8418	0.6412	1.5597	1.188	1.853	1.0007	20′
50′	0.5730	0.5422	0.8403	0.6453	1.5497	1.190	1.844	0.9977	10′
33°00′	0.5760	0.5446	0.8387	0.6494	1.5399	1.192	1.836	0.9948	57°00′
10′	0.5789	0.5471	0.8371	0.6536	1.5301	1.195	1.828	0.9919	50′
20′	0.5818	0.5495	0.8355	0.6577	1.5204	1.197	1.820	0.9890	40′
30′	0.5847	0.5519	0.8339	0.6619	1.5108	1.199	1.812	0.9861	30′
40′	0.5876	0.5544	0.8323	0.6661	1.5013	1.202	1.804	0.9832	20′
50′	0.5905	0.5568	0.8307	0.6703	1.4919	1.204	1.796	0.9803	10′
34°00′	0.5934	0.5592	0.8290	0.6745	1.4826	1.206	1.788	0.9774	56°00′
10′	0.5963	0.5616	0.8274	0.6787	1.4733	1.209	1.781	0.9745	50′
20′	0.5992	0.5640	0.8258	0.6830	1.4641	1.211	1.773	0.9716	40′
30′	0.6021	0.5664	0.8241	0.6873	1.4550	1.213	1.766	0.9687	30′
40′	0.6050	0.5688	0.8225	0.6916	1.4460	1.216	1.758	0.9657	20′
50′	0.6080	0.5712	0.8208	0.6959	1.4370	1.218	1.751	0.9628	10′
35°00′	0.6109	0.5736	0.8192	0.7002	1.4281	1.221	1.743	0.9599	55°00′
10′	0.6138	0.5760	0.8175	0.7046	1.4193	1.223	1.736	0.9570	50′
20′	0.6167	0.5783	0.8158	0.7089	1.4106	1.226	1.729	0.9541	40′
30′	0.6196	0.5807	0.8141	0.7133	1.4019	1.228	1.722	0.9512	30′
40′	0.6225	0.5831	0.8124	0.7177	1.3934	1.231	1.715	0.9483	20′
50′	0.6254	0.5854	0.8107	0.7221	1.3848	1.233	1.708	0.9454	10′
36°00′	0.6283	0.5878	0.8090	0.7265	1.3764	1.236	1.701	0.9425	54°00′
		Cos θ	Sin θ	Cot θ	Tan θ	Csc θ	Sec θ	θ Rad.	θ Deg.

Tables

Table of Values of the Trigonometric Functions

θ Deg.	θ Rad.	Sin θ	Cos θ	Tan θ	Cot θ	Sec θ	Csc θ		
36°00′	0.6283	0.5878	0.8090	0.7265	1.3764	1.236	1.701	0.9425	**54°00′**
10′	0.6312	0.5901	0.8073	0.7310	1.3680	1.239	1.695	0.9396	**50′**
20′	0.6341	0.5925	0.8056	0.7355	1.3597	1.241	1.688	0.9367	**40′**
30′	0.6370	0.5948	0.8039	0.7400	1.3514	1.244	1.681	0.9338	**30′**
40′	0.6400	0.5972	0.8021	0.7445	1.3432	1.247	1.675	0.9308	**20′**
50′	0.6429	0.5995	0.8004	0.7490	1.3351	1.249	1.668	0.9279	**10′**
37°00′	0.6458	0.6018	0.7986	0.7536	1.3270	1.252	1.662	0.9250	**53°00′**
10′	0.6487	0.6041	0.7969	0.7581	1.3190	1.255	1.655	0.9221	**50′**
20′	0.6516	0.6065	0.7951	0.7627	1.3111	1.258	1.649	0.9192	**40′**
30′	0.6545	0.6088	0.7934	0.7673	1.3032	1.260	1.643	0.9163	**30′**
40′	0.6574	0.6111	0.7916	0.7720	1.2954	1.263	1.636	0.9134	**20′**
50′	0.6603	0.6134	0.7898	0.7766	1.2876	1.266	1.630	0.9105	**10′**
38°00′	0.6632	0.6157	0.7880	0.7813	1.2799	1.269	1.624	0.9076	**52°00′**
10′	0.6661	0.6180	0.7862	0.7860	1.2723	1.272	1.618	0.9047	**50′**
20′	0.6690	0.6202	0.7844	0.7907	1.2647	1.275	1.612	0.9018	**40′**
30′	0.6720	0.6225	0.7826	0.7954	1.2572	1.278	1.606	0.8988	**30′**
40′	0.6749	0.6248	0.7808	0.8002	1.2497	1.281	1.601	0.8959	**20′**
50′	0.6778	0.6271	0.7790	0.8050	1.2423	1.284	1.595	0.8930	**10′**
39°00′	0.6807	0.6293	0.7771	0.8098	1.2349	1.287	1.589	0.8901	**51°00′**
10′	0.6836	0.6316	0.7753	0.8146	1.2276	1.290	1.583	0.8872	**50′**
20′	0.6865	0.6338	0.7735	0.8195	1.2203	1.293	1.578	0.8843	**40′**
30′	0.6894	0.6361	0.7716	0.8243	1.2131	1.296	1.572	0.8814	**30′**
40′	0.6923	0.6383	0.7698	0.8292	1.2059	1.299	1.567	0.8785	**20′**
50′	0.6952	0.6406	0.7679	0.8342	1.1988	1.302	1.561	0.8756	**10′**
40°00′	0.6981	0.6428	0.7660	0.8391	1.1918	1.305	1.556	0.8727	**50°00′**
10′	0.7010	0.6450	0.7642	0.8441	1.1847	1.309	1.550	0.8698	**50′**
20′	0.7039	0.6472	0.7623	0.8491	1.1778	1.312	1.545	0.8668	**40′**
30′	0.7069	0.6494	0.7604	0.8541	1.1708	1.315	1.540	0.8639	**30′**
40′	0.7098	0.6517	0.7585	0.8591	1.1640	1.318	1.535	0.8610	**20′**
50′	0.7127	0.6539	0.7566	0.8642	1.1571	1.322	1.529	0.8581	**10′**
41°00′	0.7156	0.6561	0.7547	0.8693	1.1504	1.325	1.524	0.8552	**49°00′**
10′	0.7185	0.6583	0.7528	0.8744	1.1436	1.328	1.519	0.8523	**50′**
20′	0.7214	0.6604	0.7509	0.8796	1.1369	1.332	1.514	0.8494	**40′**
30′	0.7243	0.6626	0.7490	0.8847	1.1303	1.335	1.509	0.8465	**30′**
40′	0.7272	0.6648	0.7470	0.8899	1.1237	1.339	1.504	0.8436	**20′**
50′	0.7301	0.6670	0.7451	0.8952	1.1171	1.342	1.499	0.8407	**10′**
42°00′	0.7330	0.6691	0.7431	0.9004	1.1106	1.346	1.494	0.8378	**48°00′**
10′	0.7359	0.6713	0.7412	0.9057	1.1041	1.349	1.490	0.8348	**50′**
20′	0.7389	0.6734	0.7392	0.9110	1.0977	1.353	1.485	0.8319	**40′**
30′	0.7418	0.6756	0.7373	0.9163	1.0913	1.356	1.480	0.8290	**30′**
40′	0.7447	0.6777	0.7353	0.9217	1.0850	1.360	1.476	0.8261	**20′**
50′	0.7476	0.6799	0.7333	0.9271	1.0786	1.364	1.471	0.8232	**10′**
43°00′	0.7505	0.6820	0.7314	0.9325	1.0724	1.367	1.466	0.8203	**47°00′**
10′	0.7534	0.6841	0.7294	0.9380	1.0661	1.371	1.462	0.8174	**50′**
20′	0.7563	0.6862	0.7274	0.9435	1.0599	1.375	1.457	0.8145	**40′**
30′	0.7592	0.6884	0.7254	0.9490	1.0538	1.379	1.453	0.8116	**30′**
40′	0.7621	0.6905	0.7234	0.9545	1.0477	1.382	1.448	0.8087	**20′**
50′	0.7650	0.6926	0.7214	0.9601	1.0416	1.386	1.444	0.8058	**10′**
44°00′	0.7679	0.6947	0.7193	0.9657	1.0355	1.390	1.440	0.8029	**46°00′**
10′	0.7709	0.6967	0.7173	0.9713	1.0295	1.394	1.435	0.7999	**50′**
20′	0.7738	0.6988	0.7153	0.9770	1.0235	1.398	1.431	0.7970	**40′**
30′	0.7767	0.7009	0.7133	0.9827	1.0176	1.402	1.427	0.7941	**30′**
40′	0.7796	0.7030	0.7112	0.9884	1.0117	1.406	1.423	0.7912	**20′**
50′	0.7825	0.7050	0.7092	0.9942	1.0058	1.410	1.418	0.7883	**10′**
45°00′	0.7854	0.7071	0.7071	1.0000	1.0000	1.414	1.414	0.7854	**45°00′**
		Cos θ	Sin θ	Cot θ	Tan θ	Csc θ	Sec θ	θ Rad.	θ Deg.

SYMBOLS

Symbol	Meaning	Page
$\{\ \}$	set	2
$\in$	is an element of	2
$=$	equals or is equal to	4
$<$	is less than	4
$>$	is greater than	4
$\lvert a \rvert$	absolute value of *a*	8
$-a$	additive inverse of *a* or the opposite of *a*	9
$\neq$	is not equal to	11
a^n	the *n*th power of *a*	11
$\emptyset$	empty set or null set	25
$\leq$	is less than or equal to	54
$\geq$	is greater than or equal to	54
$\cap$	intersection	58
$\cup$	union	59
$f(x)$	*f* of *x* or the value of *f* at *x*	109
$f \circ g$	composite of *f* and *g*	110
f^{-1}	inverse function of *f*	114
$\pm$	plus or minus sign	115
A^{-1}	inverse of matrix *A*	227
D	determinant	234
$\approx$	is approximately equal to	309
$\sqrt[n]{k}$	the *n*th root of *k*	357
i	imaginary unit ($i^2 = -1$)	388
$a^{\frac{m}{n}}$	*n*th root of *m*th power of *a*	570
$\log_b x$	logarithm of *x* to the base *b*	583
Σ	summation sign	641
!	factorial	667
${}_nP_r$	number of permutations of *n* elements taken *r* at a time	682
${}_nC_r$	number of combinations of *n* elements taken *r* at a time	688
$P(E)$	probability of event *E*	693
$\overline{E}$	complement of event *E*	701
σ	standard deviation	714
$^\circ$	degree	736
$'$	minute	740
$''$	second	740
$\overrightarrow{AB}$	vector *AB*	819
$\lVert A \rVert$	norm of vector *A*	820
$\overline{z}$	conjugate of the complex number *z*	829

Greek letters: α, β, γ, ϵ, θ, π, σ, ϕ, ω, μ
alpha, beta, gamma, epsilon, theta, pi, sigma, phi, omega, mu

ANSWERS TO SELECTED EXERCISES

Chapter 1 Real Numbers

Practice Exercises, pages 5–7 **1.** nat., whole, int., rat., real **3.** irrat., real **5.** whole, int., rat., real **7.** rat., real **9.** rat., real **17.** −5 **19.** −1.5 **23.** 4 **25.** > **27.** > **29.** > **31.** $\frac{8}{25}$ **33.** $\frac{11}{50}$ **35.** $\frac{2}{3}$ **37.** $\frac{5}{11}$ **39.** 0.375 **41.** $0.\overline{54}$ **43.** T **45.** T **47.** F **49.** F **51.** T **53.** T **55.** T **57.** F **59.** −20, −17, −9.5, −8 **61.** −3, −0.3, −0.03, −0.003 **63.** $\frac{9}{16}, \frac{15}{24}, \frac{3}{4}, \frac{7}{8}$ **65.** $\frac{157}{495}$ **67.** F **69.** F **71.** F **73.** F **75.** T **79.** 9 **83.** 0.350

Practice Exercises, pages 11–12 **1.** $-23; \frac{1}{23}$ **3.** $5; -\frac{1}{5}$ **5.** $-3\frac{2}{5}; \frac{5}{17}$ **7.** $-\frac{6}{7}; \frac{7}{6}$ **9.** 58 **11.** 101 **13.** −11.9 **15.** −61 **17.** 0 **19.** −35 **21.** $-\frac{1}{8}$ **23.** 1 **25.** $\frac{5}{24}$ **27.** 64 **29.** 91 **31.** 12 **33.** $\frac{23}{6}$ **35.** −3 **37.** 100 **39.** 144 **41.** 9 **43.** 16 **45.** true **47.** false; $a = 5$, $b = -3$ **49.** false; $a = -5$ **51.** 108°F **53.** $\frac{12}{25}$ **54.** $\frac{53}{10}$ **55.** $\frac{1}{200}$ **56.** $\frac{1}{6}$

Practice Exercises, pages 15–16 **1.** 26 **3.** 1.72 **5.** 64 **7.** −40 **9.** 317 **11.** 5 **13.** 41 **15.** −1 **17.** $4\sqrt{3} - 9$ **19.** 1 **21.** $\frac{15}{64}$ **23.** $3 \times (5 + 6) \times (3 + 1) = 132$ **25.** $(1 + 8) \times (2 - 3) \times 4 = -36$ **29.** \$4685 **31.** 59 **32.** −19 **33.** −306 **34.** −0.5 **35.** = **36.** < **37.** > **38.** int., rat., real **39.** $-\frac{1}{7}; 7$

Practice Exercises, pages 21–22 **1.** mult. ident. **3.** comm. **5.** comm. **7.** add. inv. **9.** assoc. **11.** comm. **13.** no; $5 + 7 = 12$; $5 \div 3 = 1\frac{2}{3}$ **15.** yes **17.** add. inv., mult. inv. **19.** no; all except closure for add. and inv. for add. **21. a.** dist.; **b.** assoc.; **c.** comm.; **d.** assoc.; **e.** subst. **27.** no **29.** yes

Test Yourself, page 22 **1.** irrat., real **2.** rat., real **3.** nat., whole, int., rat., real **4.** int., rat., real **5.** $\frac{7}{40}$ **6.** $\frac{8}{9}$ **7.** $\frac{13}{33}$ **8.** −1 **9.** 5 **10.** 8 **11.** 3 **12.** −5 **13.** −12 **14.** mult. inv. **15.** add. ident. **16.** closure **17.** dist.

Practice Exercises, pages 25–26 **1.** 8 **3.** 35 **5.** 2 **7.** 3 **9.** ∅ **11.** $-\frac{11}{18}$ **13.** $\frac{27}{5}$ **15.** 7; natural, whole, integers, rational, real **17.** −1; integers, rational, real **19.** 1; natural, whole, integers, rational, real **21.** ∅ **23.** 4, −12 **25.** 5, 2 **27.** $v = \frac{h + 16t^2}{t}$ **31.** $\frac{6333}{1000}$ **32.** $-\frac{71}{99}$ **33.** $\frac{55}{9}$ **34.** 36 **35.** $-\frac{4}{15}$ **36.** add. inv. **37.** comm. add. **38.** closure add. **39.** distributive **40.** closure mult. **41.** comm. mult.

Practice Exercises, pages 29–30 **1.** $2n + 8$ **3.** $5n - 6$ **5.** $4(x + 8) = x + 20$ **7.** $6y - 10$ **9.** $y^2 - 2y = 24$ **11.** $\sqrt{y} = y^2 - 4$ **13.** $(y + 7)^2 = 81$ **15.** nine times a number decreased by three **17.** the sum of a number and ten times the number **21.** eight times a number increased by 3 is equal to 12 **23.** a number increased by 2 is equal to 3 times that number decreased by 1 **27.** $\sqrt{4n - 3}$ **29.** $5n - 7 = 2n + 32$ **31.** $x + y > x - y$ **33.** $12x$ **35.** $\frac{z}{36}$ **37.** $x + 1$ **41.** $A = \frac{1}{2}h(b_1 + b_2)$ **43.** $\text{ERA} = \frac{9r}{i}$ **44.** −1 0 **45.** −4 −3 **46.** 6 7 **47.** 221 **48.** 10

Practice Exercises, pages 35–36 **1.** 12.75 **3.** 40° and 140° **5.** 43, 45, 47, 49 **7.** 11.5 **9.** 4 h **11.** $36\frac{1}{3}$; rat., real **13.** 31, 65, 146 **15.** 18, 18, 29 **17.** 52, 54, 56 **19.** 2, 4, 6, 8 **21.** 20 mi

Test Yourself, page 37 **1.** 8 **2.** −18 **3.** −1 **4.** 16 **5.** $5x - 12$ **6.** $3 + 6x$ **7.** $2x^2 - 7 = 15$ **8.** $4(x + 9) = 16 + x$ **9.** 17, 44 **10.** 50, 52, 54

Summary and Review, pages 40–41 **1.** nat., rat., whole, int., real **3.** irrat., real **5.** F **7.** T **9.** 48 **11.** 2 **13.** $-\frac{5}{3}; \frac{3}{5}$ **15.** 2 **17.** −1

19. closure **21.** add. ident. **23.** $\emptyset$ **25.** $x^2 + 20$ **27.** 14, 15, 16

Mixed Review, page 44 **1.** −29, −18, 25, 31 **3.** −143, −132, −128, −116 **5.** −3.5, −2.1, −1.4, −0.8 **7.** −9 **9.** 14 **11.** 0 **13.** 6 **15.** 15 **17.** $\frac{1}{5}$ **19.** 15 ft **21.** \$400

Chapter 2

Practice Exercises, pages 49–51 **1.** $x = \frac{c}{a}$; $a \neq 0$ **3.** $x = e - d$ **5.** $x = m - n$ **7.** $x = bk$ **9.** $x = \frac{r - p}{n}$; $n \neq 0$ **11.** $x = \frac{e + d}{c}$; $c \neq 0$ **13.** $x = ac - ab$ **15.** $x = ab + 5a$ **19.** $h = \frac{2A}{b}$; $b \neq 0$ **23.** $h = \frac{V}{\pi r^2}$; $r \neq 0$ **27.** $x = \frac{c}{9b}$; $b \neq 0$ **29.** $x = \frac{3a - b - 8}{a - b}$; $a \neq b$ **33.** $x = \frac{b}{bc + 1}$; $bc \neq -1$ **35.** $r_2 = \frac{Rr_1}{r_1 - R}$; $r_1 \neq R$ **39.** $b_2 = \frac{2A - b_1 h}{h}$; $h \neq 0$ **41.** $x = \frac{12ad + bc}{2ad}$; $ad \neq 0$ **43.** $x = \frac{y - c + ma}{m}$; $m \neq 0$ **45.** $h = \frac{S - 2\pi r^2}{2\pi r}$; $r \neq 0$ **47.** $r = \sqrt[3]{\frac{3V}{4\pi}}$ **49.** $r = \frac{A - p}{pt}$; $p \neq 0$; $t \neq 0$ **53.** 41°F

Practice Exercises, pages 55–56 **1.** $\{x: x > 11\}$ **3.** $\{z: z > 4\}$ **5.** $\{y: y \leq -14\}$ **7.** $\left\{s: s > -\frac{7}{5}\right\}$ **9.** $\{n: n > -2\}$ **11.** $\{y: y < 10\}$ **13.** $\{z: z \geq 11\}$ **15.** $\{x: x > 2\}$ **17.** $\{r: r < 16\}$ **19.** $\{x: x \geq 2\}$ **21.** $\{n: n \in$ real numbers$\}$ **23.** $\{y: y > -10\}$ **25.** $\{y: y > -35\}$ **27.** $\{z: z \geq 6\}$ **29.** $\{x: x \geq -48\}$ **31.** $\emptyset$ **33.** $\left\{x: x > -\frac{10b}{a}\right\}$ **37.** 8 pens **39.** $AB > 2$ and $AB < 6$ **41.** 30; 18

Practice Exercises, pages 60–61 **1.** conj.; F **3.** conj.; T **5.** disj.; F **7.** disj.; T **9.** {4, 6, 8}; {2, 3, 4, 5, 6, 7, 8, 10} **11.** {3, 4, 5, 6}; {1, 2, 3, 4, 5, 6, 7, 8} **23.** union; {−9, −7, −5, −3, −1, 0, 1, 3, 5, 7, 9} **27.** union; {−9, −8, −7, −6, −5, −4, −3, −2, −1, 0, 2, 4, 6, 8} **29.** union; {−9, −8, −7, −6, −5, −4, −3, −2, −1, 0, 1, 2, 3, 9} **31.** intersection; {−9, −8, −7, −6, −5, −4, −3, −2, −1} **33.** intersection; $\emptyset$ **43.** $4|2x|$ **44.** $12x - 6$ **45.** $x + (x + 2) + (x + 4)$ **46.** $90 - x$ **47.** $\frac{8 - 5a}{3 - d}$, $d \neq 3$ **48.** $\frac{24t - c}{b}$, $b \neq 0$ **49.** $\frac{3f - 5}{35c}$, $c \neq 0$ **50.** $\frac{3c^3 - 19}{2}$

Practice Exercises, pages 65–66 **1.** $\{x: -5 < x < 2\}$ **5.** $\{x: -1 < x < 8\}$ **9.** $\{y: y < 4$ or $y > 12\}$ **11.** $\{x: x \in$ real numbers$\}$ **13.** $\{y: y \in$ real numbers$\}$ **15.** $\{x: x \in$ real numbers$\}$ **17.** $\emptyset$ **19.** $\{z: z > -1\}$ **21.** $\{w: w < -3$ or $w > 1\}$ **23.** $\emptyset$ **29.** $\{w: w \in$ real numbers$\}$ **31.** $\{z: -4 < z < 2\}$ **33.** $\{x: 3 < x < 4\}$ **37.** $\{y: y \in$ real numbers$\}$ **39.** $\{w: w \in$ real numbers, $w \neq 7\}$ **41.** $\{w: w < 6\}$ **43.** $\{t: b < t < 2b\}$ **45.** {0, 1, or 2}

Test Yourself, page 66 **1.** $t = \frac{b - c}{a}$; $a \neq 0$ **2.** $x = \frac{bc + cs}{a}$; $a \neq 0$ **3.** $h = \frac{A}{\pi r^2}$; $r \neq 0$ **4.** $\{x: x > 7\}$ **5.** $\{x: x > -6\}$ **6.** conj.; F **7.** disj.; T **8.** inter.; {4, 5, 6} **9.** union; {−5, −4, −3, −2, −1, 0, 1, 2, 3, 4, 5, 6, 7} **10.** $\{x: x \geq 1\}$ **11.** $\{y: -1 < y < 3\}$

Practice Exercises, pages 70–72 **1.** length: 236 yd; width: 76 yd **3.** 6 h **5.** Joe: 45 km/h; Jim: 60 km/h **7.** 9 ft **9.** 6 **11.** \$5000 **13.** 150 mi **15.** 84 mi **17.** \$4000 **19.** $AB = 6$ cm; $BC = 12$ cm; $AC = 10$ cm

Practice Exercises, pages 75–76 **1.** −6, 6 **3.** −8, 8 **5.** −6, 12 **7.** −2 **9.** $-\frac{4}{5}$, 4 **11.** $-3, \frac{5}{3}$ **13.** −11, −1 **15.** −3, 4 **17.** no solution **19.** $\frac{3}{2}$ **21.** −18, 10 **23.** $-\frac{13}{3}$, 5 **25.** no solution **27.** $-\frac{5}{2}, \frac{11}{2}$ **29.** $-\frac{3}{2}$, −1 **31.** no solution **33.** $-\frac{1}{3}$ **35.** $\frac{-c - b}{a}, \frac{c + b}{a}$ **37.** $\frac{ac - d}{ab}, \frac{ac + d}{ab}$ **39.** $\frac{5 - 3a}{3}$ **43.** a square **44.** 118 **45.** 2 **46.** 4 **47.** $\{x: x < -2\}$ **48.** $\{x: x > -2\frac{1}{3}\}$

−2 0

−3 −2 −1

49. F **50.** T **51.** Open **52.** $|(-3)^2| = |-3|^2$; yes **53.** $|2(-1)| = |2|\,|-1|$; yes

Practice Exercises, pages 79–80 **1.** $\{x: -12 < x < 6\}$ **3.** $\{x: -4 < x < 8\}$ **5.** $\{x: -2 < x < 5\}$ **7.** $\{y: y < -5$ or $y > 15\}$

9. $\{y: -14 < y < 24\}$ 11. $\left\{z: -\frac{23}{2} < z < \frac{17}{2}\right\}$
13. $\left\{x: -6 \le x \le \frac{26}{3}\right\}$ 15. $\{x: -15 \le x \le 7\}$
17. $\{w: w < 0 \text{ or } w > 18\}$ 19. $\{z: -4 \le z \le 4\}$
21. $\left\{y: -\frac{8}{3} < y < \frac{10}{3}\right\}$ 23. $\left\{w: -\frac{7}{2} < w < \frac{1}{2}\right\}$
25. $\emptyset$ 27. $\left\{x: x < -\frac{1}{2} \text{ or } x > \frac{3}{2}\right\}$
29. $\{x: -2 < x < 14\}$ 31. $\left\{x: x \le -\frac{17}{2} \text{ or } x \ge \frac{19}{2}\right\}$ 33. $\{x: -5 < x < 11\}$
35. $\{x: -7 \le x \le -4 \text{ or } 4 \le x \le 7\}$
37. $\{x: -7 < x < 9\}$ 39. $\{x: x \in \text{real numbers}\}$
43. 36.85 mm 45. $4\frac{2}{3}, -3$ 46. 4, 5
47. 0 2 48. −2 −1 0 49. −3 0 3 6
50. −10 3 7 51. assoc. 52. add. ident.
53. subst. 54. add. 55. 2 in./s; 4 in./s

Practice Exercises; pages 83–85 1. 52 in.
3. 41, 42, 43 5. 51, 53, 55, 57 7. 12 9. 1936
11. 130 13. 85, 86, 87, 88 15. $\{w: w < 8 \text{ ft}\}$
17. R 114: 69; R 214: 54; R 314: 27 19. 98
21. 7, 8, 9, 10

Test Yourself, page 85 1. 5 h 2. $-\frac{3}{2}, \frac{9}{2}$
3. $\frac{11}{8}$ 4. $\left\{x: -\frac{5}{3} < x < 5\right\}$ 5. $\{x: x \le -3 \text{ or } x \ge 2\}$ 6. $\{x: 2 < x < 3\}$ 7. 51, 52, 53
8. {37, 39, 41}, {35, 37, 39}. . . .

Summary and Review, pages 88–89 1. $x = \frac{c-b}{a}; a \ne 0$ 3. $x = \frac{3c}{a}; a \ne 0$
5. $t > 7$ 7. $t < -2$ 9. conjunction; true
11. conjunction; open 13. $\{y: 5 < y \le 8\}$
15. $\{y: -9 < y < -5\}$ 17. 20 nickels 19. 3, −7 21. $\{z: -10 < z < 15\}$ 23. $\{z: z \in \text{real numbers}\}$ 25. $\{x: 130 < x < 160\}$

Cumulative Review (Ch. 1–2), page 92 1. 3
3. $\frac{1}{4}$ 5. $3x - 5 > 10$ 7. $\frac{4}{5}$ 9. 700 mi
11. $-\frac{121}{7}$ 13. 23 15. −261 17. $-\frac{5}{2}$
21. −2 23. $x > -4$ or $x < -7$

Chapter 3 Functions and Graphs

Practice Exercises, pages 96–97 1. quadrant I
3. quadrant III 5. y-axis 7. x-axis 9. quadrant III
11. quadrant I 13. (4, 4) 15. (−4, −3)
17. (0, −5) 19. (0, 0) 21. quadrant IV
23. quadrant I 25. quadrant III 31. (3, 4)
33. (6, 3) 35. (−2, −3); (2, 3); (−14, −3)
37. x-axis: 4; y-axis: 2 39. x-axis: 4; y-axis: 0
Answers may vary: 44. $\frac{23}{10{,}000}$ 45. $\frac{5}{1}$
46. $\frac{4100}{333}$ 47. $\left\{x: x < \frac{15}{19}\right\}$ 48. $\left\{x: x \le -\frac{7}{2}\right\}$
49. $\left\{x: 4 > x > \frac{39}{25}\right\}$ 50. $\{x: -2 \le x \le 7\}$

Practice Exercises, pages 101–103 1. $D = \{-1, -2, -3, -4\}$; $R = \{1, 2, 3, 4\}$ 3. $D = \{0, 2, 3, 5\}$; $R = \{-2, 0, 1, 3\}$ 5. $S = \{(x, y): y = x + 4\}$; $D = \{0, 1, 2, 3\}$; $R = \{4, 5, 6, 7\}$ 7. $S = \{(x, y): y = x - 3\}$; $D = \{1, 2, 4, 5\}$; $R = \{-2, -1, 1, 2\}$ 9. $\{(-3, -3), (-2, 1), (1, 0), (1, 1), (3, 3), (3, -3)\}$; $D = \{-3, -2, 1, 3\}$; $R = \{-3, 0, 1, 3\}$ 11. function 13. function 15. function
17. not a function 19. function 21. $S = \{(x, y): y = 3x\}$; $D = \{-2, -4, 6\}$; $R = \{-6, -12, 18\}$
23. $S = \{(x, y): y = 2x + 1\}$; $D = \{-1, 0, 2, 3\}$; $R = \{-1, 1, 5, 7\}$
25. $3 \to 5$; function
$-7 \to 1$
$2 \to 6$
$4 \to 2$
27. $0 \to 1$; function
1, −2, 3 → −3
29. function 31. function 33. function
35. $\{(x, y): y = |x|\}$ 37. $\{(x, y): y = x^3\}$
39. yes 41. function 43. $\{(x, y): y = 2x, x = 18, 20, 22\}$; function

Practice Exercises, pages 107–108 1. solution
3. solution 5. solution
7. 9.

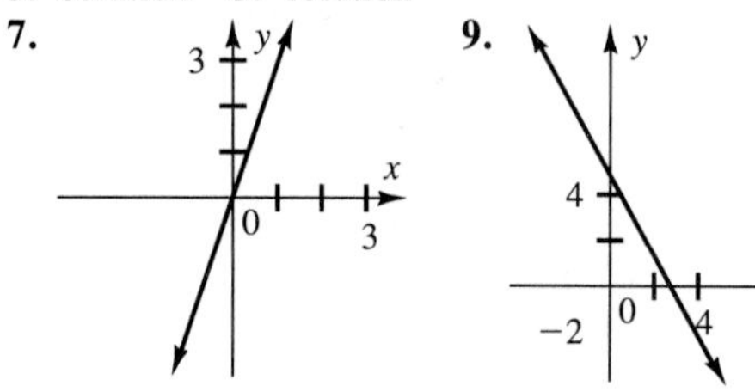

11. 13.

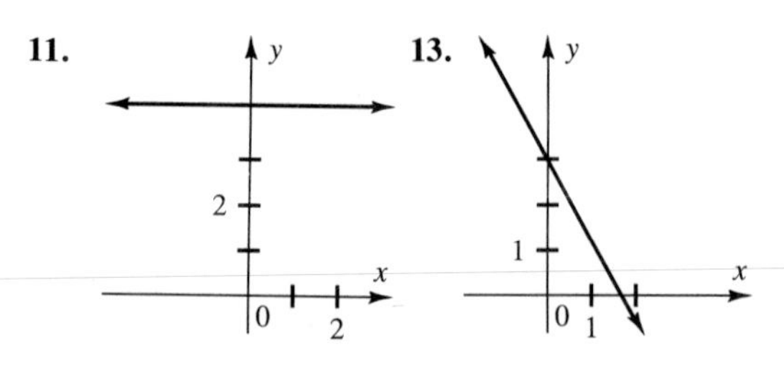

15.

17.

19. function **21.** function **23.** solution **25.** solution

27.

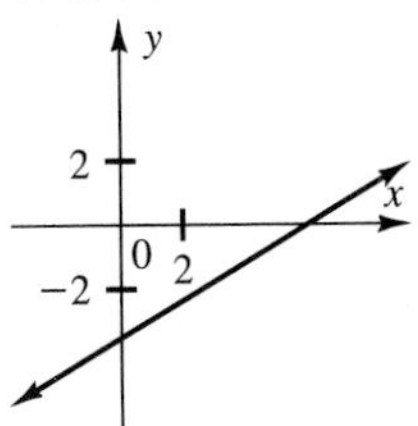

29.

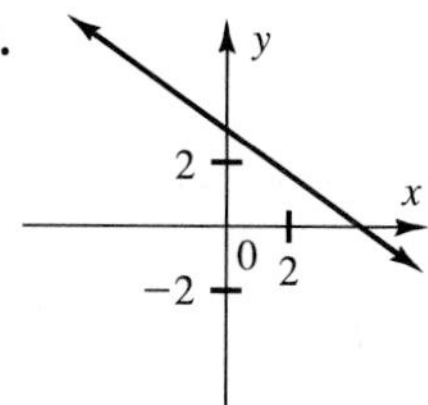

31.

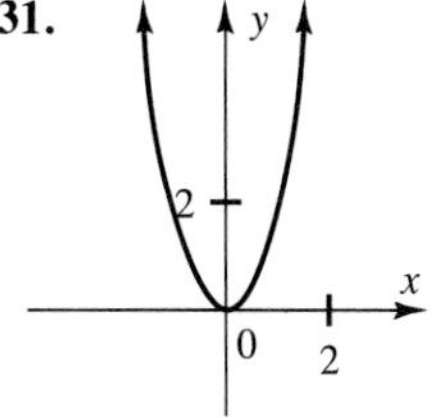

33. linear **35.** not linear **37.** linear

39.

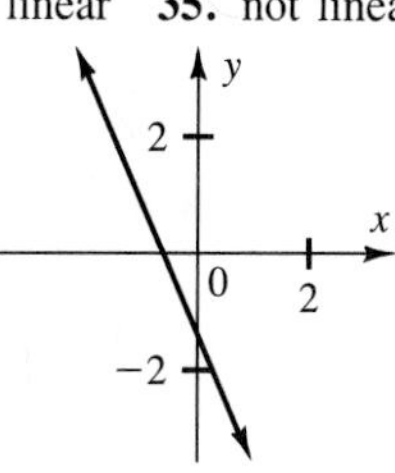

41.

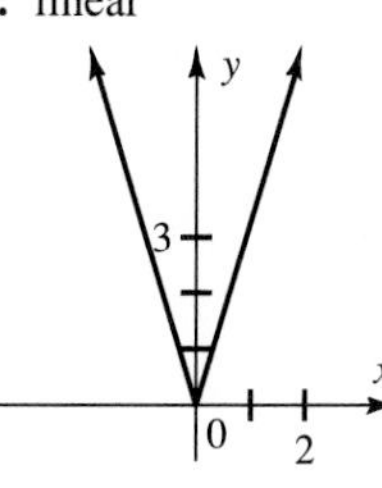

43.

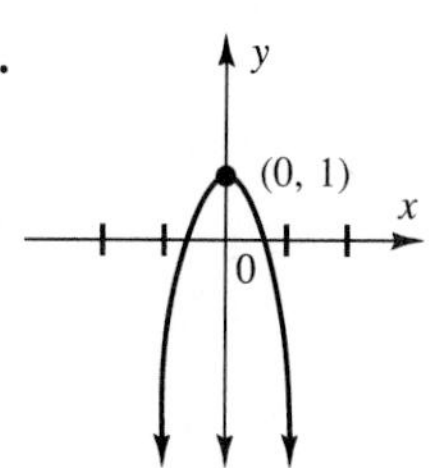

45.

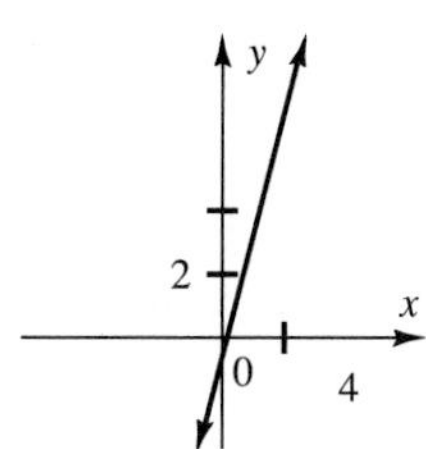

47.

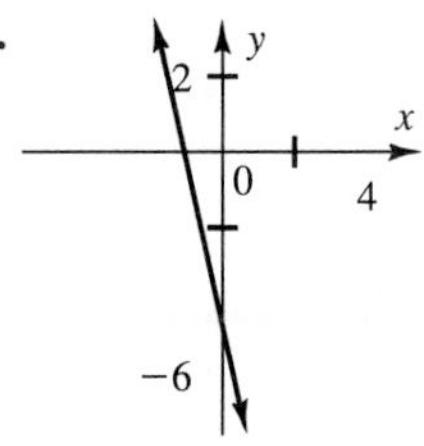

Practice Exercises, pages 111–112 **1.** 3, 0, −6, $-3a$; 1, 2, 4, $a + 2$ **3.** 1, 0, 4, a^2; 4, 5, 7, $a + 5$ **5.** −2, 1, 25, $3a^3 + 1$; −2, 0, 4, $2a$ **7.** −3, −1, 15, $2a^2 + 4a - 1$; 3, 0, 12, $3a^2$ **9.** −1, 8, 2 **11.** 10, −5, 5 **13.** 4, 16, 0 **15.** 8, 26, 2 **17.** −6, 12, 0 **19.** 14, −10, 6 **21.** −3, −21, −9 **23.** 2, −4, 0 **25.** 1, 1, 4, $k^2 + 2k + 1$; 5, 1, 2, $k^2 + 1$ **27.** 2, 18, 50, $8k^2 + 24k + 18$; 19, 3, 7, $4k^2 + 3$ **31.** 80, 8, −1, $9k^2 - 18k + 8$; 2, 2, −7, $-3k^2 - 6k + 2$ **33.** $2x^2$; $4x^2$ **35.** $-12x^2$; $48x^2$

37.

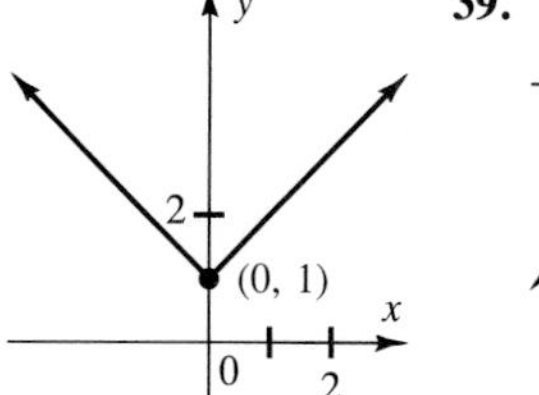

39.

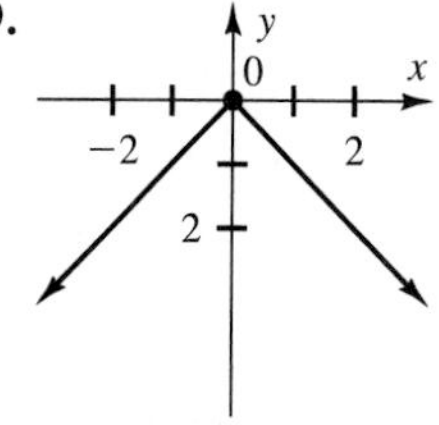

41. $2x^4 + 3$ **43.** x^4 **45.** $2x^2 + 4x + 5$ **47.** 2 **49.** $10 **51.** 0.66 lb **52.** −16, 8 **53.** $7\frac{2}{3}$, −3 **54.** $\frac{1}{15}$ **55.** $-\frac{7}{12}$ **56.** no solution **57.** 10, −6 **58.** 17 boys **59.** about 1454 ft

Practice Exercises, pages 116–117 **1.** {(4, 2), (8, 4), (16, 8)} **3.** $y = \frac{x}{4} + \frac{1}{4}$ **5.** $y = \frac{x}{5} + \frac{2}{5}$ **7.** $y = -\frac{x}{3} + 1$ **9.** {(3, 3), (4, 4), (5, 5), (6, 6)}; $D = \{3, 4, 5, 6\}$; $R = \{3, 4, 5, 6\}$; function **11.** {(3, 1), (3, 2), (5, 4), (5, 9)}; $D = \{3, 5\}$; $R = \{1, 2, 4, 9\}$; not a function **13.** {(3, d), (3, e), (3, f), (4, g)}; $D = \{3, 4\}$; $R = \{d, e, f, g\}$; not a function **15.** $y = \frac{x}{2} + \frac{3}{2}$ **17.** $y = \frac{x}{9} + \frac{4}{9}$ **19.** $y = -\frac{x}{4} + \frac{13}{4}$ **21.** $y = \frac{x}{2} - \frac{3}{2}$; $x = 5, 7, 9$ **25.** $y = \frac{x}{3} - \frac{4}{3}$; function **27.** $y = 7 - x$; function **29.** $y = \frac{3}{2}x + \frac{9}{2}$; function **31.** (x, y): $y = \frac{1}{2}x$, $D = \{2, 4\}$; $R = \{1, 2\}$; function **33.** (x, y): $y = \frac{1}{3}x + \frac{4}{3}$, $D = \{1, -2, -5, -8, -11\}$; $R = \{1, 2, 3, 4, 5\}$; function **37.** 2, 5 **39.** $D = \{x: x \geq 0\}$; $R = \{y: y \geq 0\}$ **41.** $y = \pm\sqrt{x}$; not a function **43.** $y = \frac{5}{9}(x - 32)$; function

Test Yourself, page 117 **1.** II **2.** I **3.** III **4.** yes **5.** no **6.** yes **7.** yes **8.** no **9.** yes **10.** 28 **11.** −12 **12.** −21 **13.** {(5, 4), (2, 3), (0, 1)} **14.** $y = \frac{x}{2} - 2$

Practice Exercises, pages 121–122 **1.** 1 **3.** 4 **5.** 3 **7.** 0 **9.** undefined **11.** $y = x + 5$ **13.** $y = x - 4$ **15.** $m = 3, b = -4$
19.

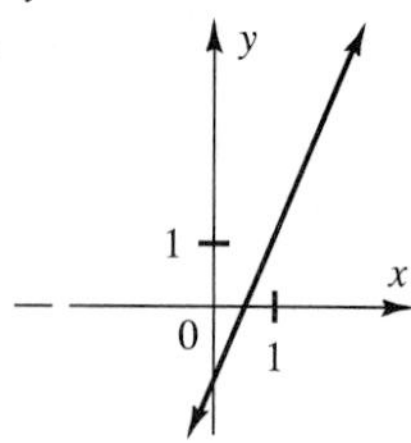

21. $\frac{1}{2}$ **23.** -1 **25.** undef. **27.** $y = -\frac{3}{2}x - 1$ **29.** $m = -\frac{2}{3}, b = 2$ **31.** $m = \frac{5}{2}, b = 2$ **33.** $m = 1, b = 0$ **35.** undef. **37.** undef. **39.** 0 **41.** $\frac{3}{50}$ **43.** $2xy^2$ **44.** $x + x + 2 + x + 4 + x + 6$ **45.** $\frac{x + 2x + 4x + 8x}{4}$ **46.** $\frac{3x + 10}{2}$ **47.** $-\frac{1}{2}x^2$ **48.** 2 **49.** $|x + (-x^2)|$

Practice Exercises, pages 125–126 **1.** parallel **3.** perp. **5.** neither **7.** $y = 3x - 3$ **11.** $y = -x + 1$ **13.** $y = \frac{1}{2}x + 2$ **17.** $y = -\frac{3}{2}x - 1$ **19.** perp. **21.** par. **23.** $y = \frac{2}{7}x + 3$ **25.** $y = \frac{3}{2}x - 4$ **27.** $y = -\frac{5}{3}x + 4$ **31.** $y = -\frac{2}{7}x + 3$ **33.** $y = -\frac{1}{5}x$ **35.** yes **37.** Slopes are 1 and -1. **39.** Slopes are $\frac{3}{2}, \frac{3}{2}$ and $-\frac{2}{3}, -\frac{2}{3}$. **41.** -2

Practice Exercises, pages 129–130 **1.** $y = -x - 5$ **3.** $y = 3x$ **5.** $y = -4x + 3$ **7.** $y = -3x - 7$ **9.** $y = 4x + 2$ **11.** $x - y = -4$ **13.** $2x + y = -1$ **15.** $3x + y = -2$ **17.** $3x - y = -3$ **19.** $2x + y = -1$ **21.** $3x - 5y = -5$ **23.** $y = -5$ **25.** $y = \frac{1}{2}x + 4$ **27.** $y = -\frac{3}{4}x - 3$ **29.** $y = -\frac{3}{5}x + 7$ **31.** $2x - 9y = -18$ **33.** $x - 3y = -15$ **35.** $5x + 2y = 2$ **37.** $2x - 9y = 81$ **39.** $5x + 12y = 17$ **41.** $8x + 3y = 0$ **43.** $4x - 5y = 25$ **45.** $2x + y = 4$ **47.** $9x - 2y = 0$ **49.** \$140; \$250

Practice Exercises, pages 133–135 **1.** $f(x) = -10x + 50$; 40 **3.** $m = 2, 2$; $m = \frac{1}{2}, \frac{1}{2}$ **5.** 150 **7.** $y = 4x - 7$ **9.** $y = \frac{1}{3}x - 6$ **11.** \$550 **13.** $y = 46x + 40$ **15.** \$3950; \$3250 **17.** $y = \frac{1}{2}x + 7$ **19.** $y = 4x - 20$; $x = 5.75$ **21.** 324

Test Yourself, page 136 **1.** 1 **2.** -2 **3.** 0 **4.** $y = -3x + 2$ **5.** $y = \frac{2}{3}x - 1$ **6.** $y = -4$ **7.** $y = 2x + 5$ **8.** $y = \frac{1}{2}x - 3$ **9.** $y = \frac{4}{3}x - 5$ **10.** $x + y = 3$ **11.** $x + 2y = -2$ **12.** $x + 5y = 45$

Summary and Review, pages 138–139 **1.** quadrant I **3.** quadrant III **5.** $D = \{-1, 0, 1\}$; $R = \{2, 3, 4\}$ **7.** function
9. **11.**

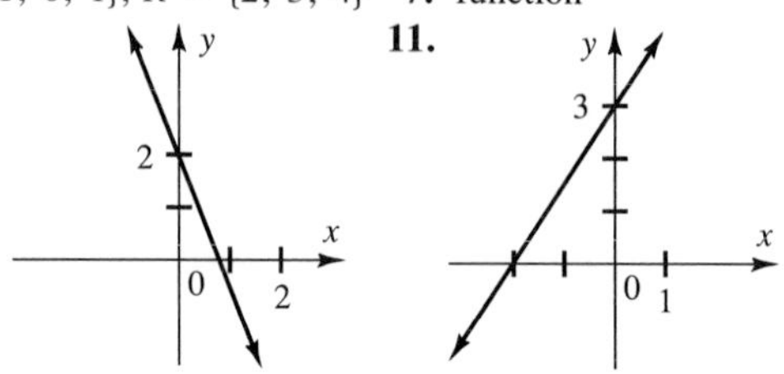

13. not linear **15.** $-5, 4, 13$ **17.** 1, 4, 25 **19.** $-5, 12$ **21.** $y = -\frac{x}{2} + 2$ **23.** 3 **25.** 7 **27.** $y = -4x + 5$ **29.** $y = 2x - 2$ **31.** $y = \frac{1}{3}x + 4$ **33.** $y = 2x - 2$ **35.** $y = -4x - 11$

Mixed Review, page 142 **1.** 0.5 **3.** 0.4 **5.** 0.625 **7.** 0.45 **9.** 48 **11.** 5 **13.** $x = -12$ **15.** $x = 2$ **17.** $x = -2$ **19.** \$147 **21.** 16 oz

Chapter 4 Systems of Equations and Inequalities

Practice Exercises, pages 147–149 **1.** direct; 7; $y = 7x$ **3.** direct; -2; $y = -2x$ **5.** direct; $\frac{2}{3}$; $y = \frac{2}{3}$ **7.** not direct **9.** direct; 12 **11.** direct; -2 **13.** not direct **15.** direct; 1 **17.** direct; $\frac{4}{3}$ **19.** direct; 6 **21.** 8 **23.** -3 **25.** $\frac{21}{2}$ **27.** not direct **29.** not direct **31.** no, varies as x^3 **33.** 33 **35.** 9 **37.** 90 **39.** 1.2 **41.** 9 **43.** ± 5 **45.** y is doubled **47.** y is divided by 7 **49.** y is multiplied by 4 **51.** y is multiplied by 9 **53.** 350 mi **55.** \$10,400

Practice Exercises, page 152 **1.** 3; 7 **3.** $\frac{1}{2}$; -3 **5.** 3; $\frac{9}{2}$ **7.** $-\frac{1}{2}$; 1 **9.** $\frac{4}{3}$; -8 **11.** intersect **13.** intersect, perpendicular **15.** none **17.** yes; one **19.** yes; infinite number **21.** 0, 1, 2

Practice Exercises, pages 157–158 **1.** (1, 1); consistent; independent **3.** (3, -4); consistent; independent **5.** (-2, 4); consistent; independent **7.** (2, 1); consistent; independent **9.** $\{(x, y)$: $x + 3y = 6\}$; consistent; dependent **11.** (-3, 7); consistent; independent **13.** (3, -1); consistent; independent **15.** inconsistent; no solution **17.** (-2, 2); consistent; independent **19.** (2, 3); cons.; ind. **21.** incons.; no solution **23.** (0, 0); cons.; ind. **25.** (2, 3); cons.; ind. **27.** (0, 0); cons.; ind. **29.** (-2, 1); cons.; ind. **31.** cons.; ind. **33.** incons. **35.** cons.; ind. **39.** For $x < 10$, demand > supply; for $x > 10$, supply > demand **42.** $y = \frac{2}{3}x + 5$; $y = -2x + 6$; $y = x + 7$

Practice Exercises, pages 162–163 **1.** (7, 5) **3.** (5, 2) **5.** (2, 4) **7.** (2, -2) **9.** (2, 3) **11.** (1, 0) **13.** (-2, 3) **15.** $\{(x, y)$: $-2x + 3y = 13\}$ **17.** no solution **19.** (4, -2) **21.** (-8, -6) **23.** (-10, 12) **25.** no solution **27.** $\{(x, y)$: $15x - 23y = 34\}$ **29.** (0, 4) **31.** $\left(\frac{1}{3}, 1\right)$ **33.** (5, -7) **35.** (2, 1) **37.** $x - y = 8$; $3y = 2x - 1$; 23 quarters; 15 dimes

Practice Exercises, pages 167–169 **1.** (3, 5) **3.** (16, 13) **5.** (-1, 1) **7.** (5, 0) **9.** (2, 6) **11.** (1, 5) **13.** (-2, -8) **15.** $\{(x, y)$: $x - 3y = 1\}$ **17.** no solution **19.** (2, 0) **21.** (0, -4) **23.** (-9, 2) **25.** (3, 2) **27.** $\left(-1, -\frac{1}{2}\right)$ **29.** (3, 7) **31.** no solution **33.** (2, 2) **35.** (-5, -1) **37.** (20, -18) **39.** $\left(\frac{1}{2}, 1\right)$ **41.** $\left(\frac{1}{2}, \frac{1}{4}\right)$ **43.** -2 **45.** 8 **47.** $\frac{5}{2}$ **49.** Noreen; 12 mi; Heather: 8 mi **51.** 2875

Test Yourself, page 169 **1.** direct; 5 **2.** not direct **3.** -8 **4.** 10 **5.** (-2, 2); cons. ind. **6.** no sol.; incons. **7.** (2, 1) **8.** (-1, -3) **9.** (2, 5) **10.** (2, 8)

Practice Exercises, pages 173–176 **1.** recorder: \$6; harmonica; \$8 **3.** poodle: \$300; cat: \$150 **5.** girls: 7; boys: 14 **7.** chairs: 5; tables: 4 **9.** basic: \$25; movie channel: \$10 **11.** air speed: 130 mi/h; wind speed: 30 mi/h **15.** 36 **17.** 27 **19.** quarters: 18; dimes: 8 **21.** one-bedroom: 28; two-bedroom: 22 **23.** Mindy's rowing rate: 6 mi/h; rate of current: 2 mi/h **25.** 48 **27.** 8%: \$12,000; 6%: \$8000 **28. a.** 14 **b.** 94 **c.** $3t^2 + 4t - 1$ **29.** $y = \frac{x - 3}{2}$; yes **30.** AI; BII; CIV; DIII

31. 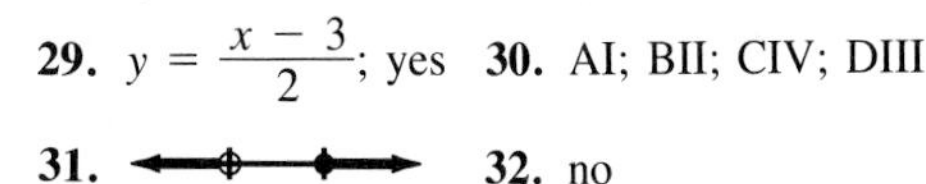

32. no

33. 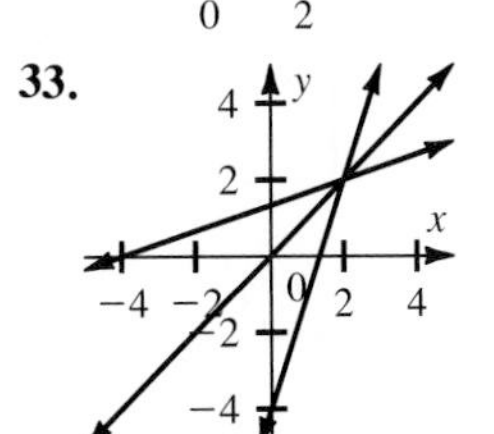

34. 3 **35.** $\frac{1}{2}$ **36.** (0, 0) or (6, 8) or (20, -8) **37.** add. ident.

Practice Exercises, pages 180–181

1. 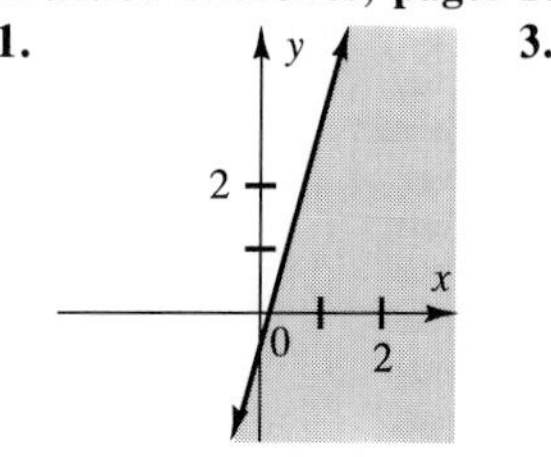

3. 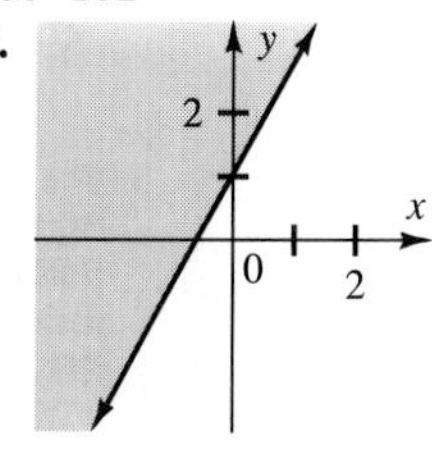

5. 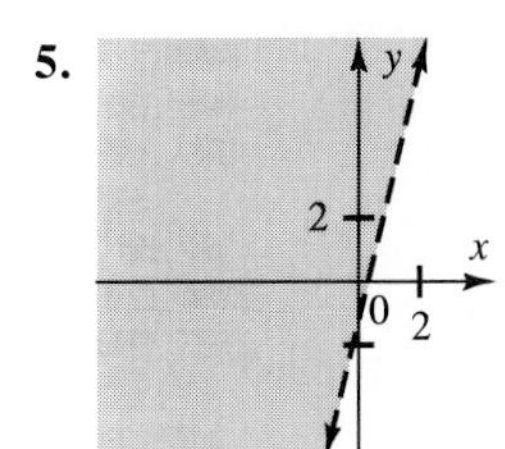

7. 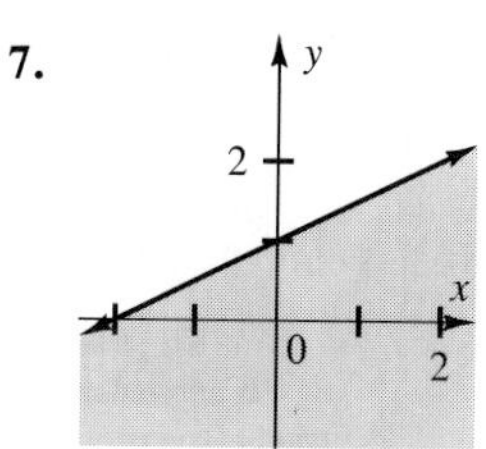

9. 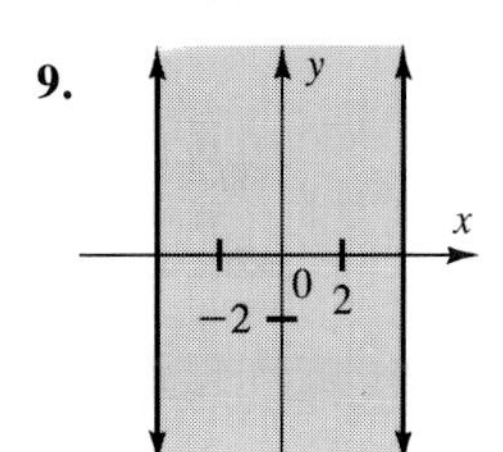

11. 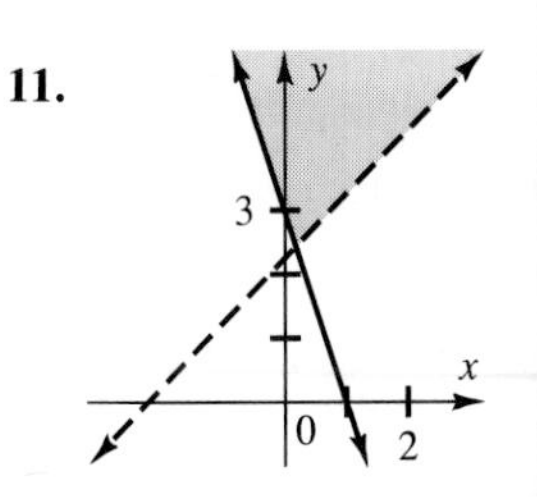

13.

19.

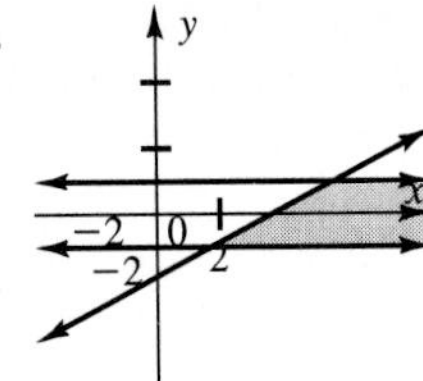

21.

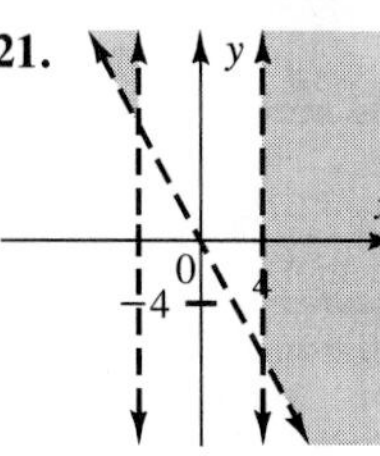

23.

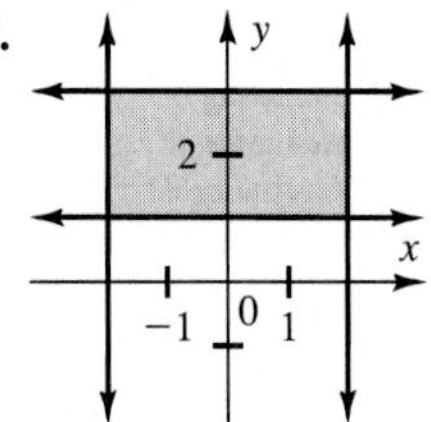

25.

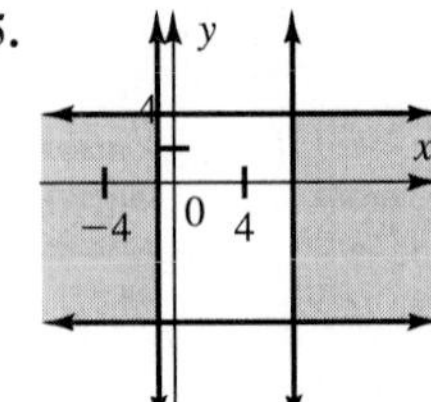

27.

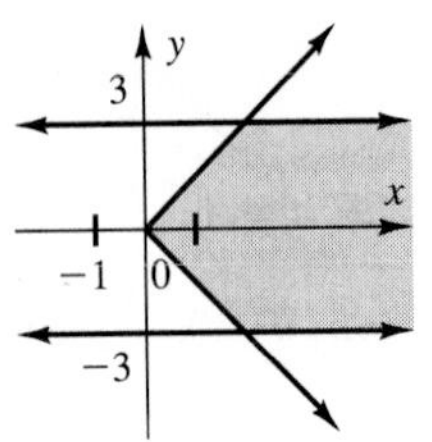

29.

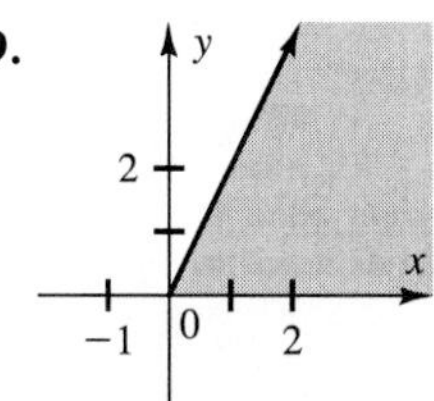

31. admitted: no scholarship **33.** full

Practice Exercises, pages 184–186 **1.** max. (0, 4); $P = 12$ **3.** min. (0, 2); $C = 4$ **5.** max. (1, 2); $P = 11$ **7.** banana bread: 3; nut bread: 2 **9.** pine: 8; oak: 2 **11.** perch: 10; bass: 30 **13.** whole wheat: 18 boxes; sesame: 20 boxes

Test Yourself, page 186 **1.** plane's speed: 145 mi/h; wind speed: 5 mi/h **2.** 14

3.

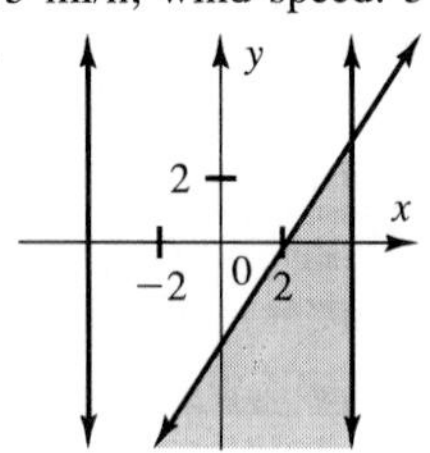

4. max. (7, 0); $P = 21$

Summary and Review, pages 188–189 **1.** direct; 9 **3.** not direct **5.** 36 **7.** (3, −2); consistent; independent **9.** (3, −5) **11.** (−2, −4) **13.** milk: $1.29; juice: $2.36

15.

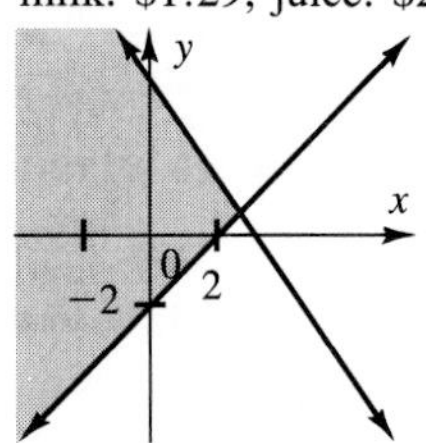

17. max. (1, 2); $P = 10$

Cumulative Review (Ch. 1–4), pages 192–194

1.

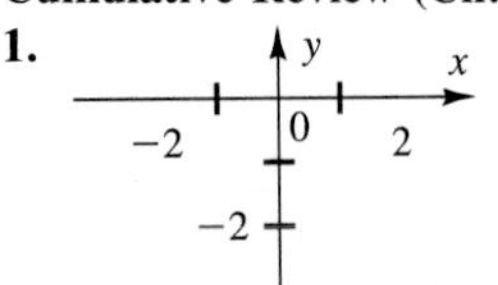

3.

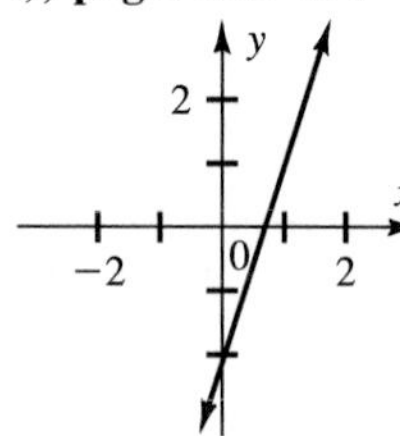

5. $7x - y = 11$ **7.** $\emptyset$ **9.** 33, 35, 37 **11.** quadrant III **13.** quadrant II **15.** x-axis **17.** $4y + 6 = 32$ **19.** 10 oz first type, 8 oz second type **21.** 2, $-\frac{1}{2}$

23.

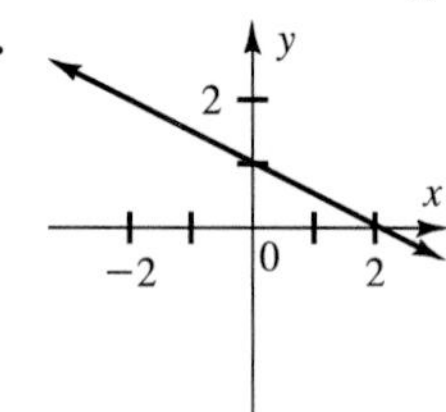

25. 68° and 22° **27.** $2x - 1 = y$; $D = \{0, 1, 2, 3\}$; $R = \{-1, 1, 3, 5\}$

29.

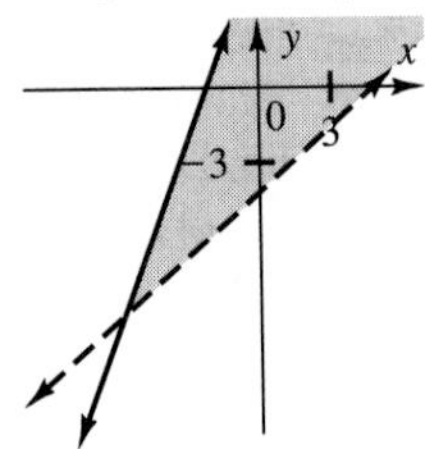

31. 4, 2, 0, −2, $-2c$; 9, 3, 1, 3, $2c^2 + 1$

33. $m = -\frac{3}{2}$; $b = 2$ **35.** yes; −3 **37.** 0

39. $m = -1$; $b = -3$; **41.** $y = -\frac{1}{2}x + 2$

43. disk: $0.25; ribbon: $4.00 **45.** parallel **47.** function **49.** $2x + 3y = -3$ **51.** (1, 2)

53. $x = \frac{r}{a+b}$; $a \neq -b$ **55.** $x = c(h + g)$

57. distrib. **59.** mult. inv. **61.** $y = \frac{x + 10}{2}$

63. Slopes of consecutive sides are $\frac{2}{3}$ and $-\frac{3}{2}$.

Chapter 5 Matrices and Determinants

Practice Exercises, pages 201–203

1. $\begin{bmatrix} 0 & 5 \\ 8 & -6 \\ 0 & 5 \end{bmatrix}$ **3.** undefined **5.** $\begin{bmatrix} 6 & 3 \\ -3 & 3 \end{bmatrix}$

7. $\begin{bmatrix} -6 & -3 \\ -4 & -2 \\ -2 & 5 \end{bmatrix}$ **9.** $\begin{bmatrix} -4 & 1 \\ -3 & -1 \end{bmatrix}$ **11.** $\begin{bmatrix} 9 & 12 \\ 18 & -6 \\ 3 & 0 \end{bmatrix}$

13. $\begin{bmatrix} 3 & 14 \\ 22 & -14 \\ 1 & 10 \end{bmatrix}$ **15.** $\begin{bmatrix} 21 & 3 \\ 2 & 16 \\ 7 & -25 \end{bmatrix}$ **17.** $\begin{bmatrix} 6 & 2 \\ -1 & 3 \end{bmatrix}$

19. $\begin{bmatrix} 10 & -2 & 3 \\ 5 & 6 & -1 \end{bmatrix}$ **21.** $x = 2$; $y = 4$

23. $x = 5$; $y = 1$ **25.** $\begin{bmatrix} 1 \\ 4 \\ 12 \end{bmatrix}$ **27.** $[-1 \quad 2 \quad 3]$

29. $[5 \quad -10 \quad -15]$ **31.** $x = 2$; $y = -1$

41. $\begin{bmatrix} 1100 \\ 1300 \\ 1600 \end{bmatrix}$; $\begin{bmatrix} 1500 \\ 2000 \\ 1900 \end{bmatrix}$ **43.** 2nd row; 2nd column **45.** b_{13} **46.** 21, 23, 25, 27 **47.** 13.5 **48.** 40.5

49. nat., whole, int., rat., real; irrat., real

Practice Exercises, pages 208–209

1. $\begin{bmatrix} 1 & -6 & -5 \\ 6 & 1 & -5 \\ -3 & -12 & 0 \end{bmatrix}$ **3.** $\begin{bmatrix} -15 & 0 \\ -13 & 7 \\ -5 & -5 \end{bmatrix}$ **5.** undefined

7. $\begin{bmatrix} 17 & -24 \\ -33 & -7 \\ 69 & -18 \end{bmatrix}$ **9.** $\begin{bmatrix} 1 & 2 & -1 \\ 0 & 3 & 1 \\ 2 & -1 & -2 \end{bmatrix}$

11. $\begin{bmatrix} 4 & -1 \\ -20 & 1 \\ -17 & 6 \end{bmatrix}$ **13.** $\begin{bmatrix} -1 & 9 & 3 \\ 2 & 8 & 1 \\ -2 & 3 & 1 \end{bmatrix}$

15. $\begin{bmatrix} 16 & 8 & -15 \\ 15 & -9 & -15 \\ 2 & 11 & -5 \end{bmatrix}$ **17.** equal **19.** equal

21. equal **23.** equal **25.** $x = 4$; $y = 1$ **27.** $x = 2$; $y = 1$ **35.** \$115

Practice Exercises, pages 213–215

1. $\begin{array}{cccc} w & c & a.r. & c.r. \\ \$5 & \$6 & \$10 & \$7 \end{array}$ **3.** \$851

5. $\begin{array}{c|ccc} & 10 & 11 & 12 \\ \hline w & 18 & 33 & 27 \\ c & 24 & 45 & 30 \\ a.r. & 30 & 57 & 9 \\ c.r. & 42 & 18 & 27 \end{array}$ **7.** $\begin{array}{c|cccc} & a & b & c & d \\ \hline a & 0 & 1 & 0 & 1 \\ b & 1 & 0 & 1 & 0 \\ c & 0 & 0 & 0 & 1 \\ d & 1 & 0 & 1 & 0 \end{array}$

9. $\begin{array}{c|cccc} & a & b & c & d \\ \hline a & 2 & 0 & 2 & 0 \\ b & 0 & 1 & 0 & 2 \\ c & 1 & 0 & 1 & 0 \\ d & 0 & 1 & 0 & 2 \end{array}$; 2; 0 **11.** \$103; \$137

13. $\begin{array}{c|ccc} & p & s & h \\ \hline g & 3 & 2 & 0 \\ b & 4 & 5 & 9 \\ b & 7 & 1 & 6 \end{array}$ **15.** $\begin{array}{c|ccc} & a & b & c \\ \hline a & 0 & 2 & 0 \\ b & 2 & 0 & 2 \\ c & 0 & 2 & 0 \end{array}$; $\begin{array}{c|ccc} & a & b & c \\ \hline a & 0 & 0 & 0 \\ b & 1 & 0 & 1 \\ c & 0 & 1 & 0 \end{array}$

Practice Exercises, pages 219–220 **1.** (5, −2, 0); cons.; ind. **3.** (−2, 1, 3); cons.; ind. **5.** (4, 0, −2); cons.; ind. **7.** no solution; incons. **9.** (2, 2, 5); cons.; ind. **11.** (−1, 2, 0); cons.; ind. **13.** $\left(\frac{35}{17}, \frac{24}{17}, \frac{29}{17}\right)$; cons. ind. **15.** $y = 0$; $x + 2z = 5$; cons.; dep. **17.** (2, 4, 6); cons.; ind. **19.** (1, 0, 2, −2); cons.; ind. **21.** nut: 10¢; bolt: 5¢; washer: 5¢ **23.** x: triangle; y: pentagon; z: hexagon

Test Yourself, page 220 **1.** $\begin{bmatrix} 6 & 2 & 5 \\ 3 & 2 & 0 \\ -1 & 1 & 2 \end{bmatrix}$

2. $\begin{bmatrix} 4 & 6 & 0 \\ 6 & 8 & 2 \end{bmatrix}$ **3.** $\begin{bmatrix} 8 & 3 & 8 \\ 11 & 7 & 14 \end{bmatrix}$ **4.** $\begin{bmatrix} 1 & -6 & 1 \\ 11 & 5 & 2 \\ 1 & -1 & 0 \end{bmatrix}$

5. undefined **6.** $\begin{bmatrix} 21 & 10 & 10 \\ 29 & 15 & 17 \end{bmatrix}$

7. $[205 \quad 73 \quad 140 \quad 58]$ **8.** (1, −1, 2) **9.** (0, 1, −2)

Practice Exercises, pages 224–226 **1.** (1, 1, 0); cons.; ind. **3.** no solution; incons. **5.** z = any real number; $x + y = 1$; cons.; dep. **7.** (1, 1, −1); cons.; ind. **9.** (4, 0, 1); cons.; ind. **11.** (1, 3, 2); cons.; ind. **13.** $3x + 6y = 2$; $2x - y = 3$ **15.** (2, 0, 3) **17.** no solution **19.** $z = 0$; $x + y = 5$ **21.** (1, 4, 5) **23.** (1, 1, 1, 0) **25.** $10x + 4y + 12z = 120$; $11x + 77y = 220$; $8x + 2y + 32z = 160$ **27.** 20 **29.** 72

Practice Exercises, pages 230–232

1. $\begin{bmatrix} 0 & 1 \\ -1 & 2 \end{bmatrix}$ **3.** $\begin{bmatrix} -3 & 4 \\ 1 & -1 \end{bmatrix}$ **5.** $\begin{bmatrix} -3 & -2 \\ 8 & 5 \end{bmatrix}$

7. $\begin{bmatrix} \frac{1}{2} & 0 \\ 0 & \frac{1}{2} \end{bmatrix}$ **9.** (2, 1) **11.** (3, 2) **13.** (2, 4)

17. $\begin{bmatrix} 7 & 6 & -3 \\ 2 & 2 & -1 \\ -6 & -5 & 3 \end{bmatrix}$ **19.** $\begin{bmatrix} -\frac{5}{2} & \frac{15}{2} & -\frac{1}{2} \\ \frac{3}{2} & -\frac{9}{2} & \frac{1}{2} \\ \frac{5}{2} & -\frac{13}{2} & \frac{1}{2} \end{bmatrix}$

21. (3, 0, 1) **23.** (1, 1, 1) **25.** (4, 1, 3) **27.** (1, 1, 1, 1) **29.** (2, 3) **31.** large: $0.18; small: $0.14

Practice Exercises, pages 237–238 **1.** 13 **3.** −6 **5.** 0 **7.** 25 **9.** −30 **11.** (3, 1) **13.** (4, −1) **15.** (5, 3) **17.** (3, −2) **19.** (10, −5) **21.** (6, 1, 1) **23.** (4, 4, 1) **25.** (−4, 5, 1) **29.** (5, −10) **33.** almonds: $2; pecans: $4; pistachios: $6 **34.** $x = -2$; $y = 5$ **35.** $x = 3$; $y = -1$ **36.** $x = 2$; $y = 2$ **37.** $x = 8$; $y = -3$

Practice Exercises, pages 242–245 **1.** 5%: $4000; 6%: $1000; 7%: $5000 **3.** triangles: 40; squares: 30; pentagons: 10 **5.** nickels: 50; dimes: 10; quarters: 15 **7.** 792 **9.** length: 20 in.; width: 10 in.; height: 5 in. **11.** wheat: $1; corn: $1.50; oat: $2 **13.** *A*: $10; *B*: $20; *C*: $18 **15.** 6.25%: $2000; 8.5%: $2000; 9.5%: $2000

Test Yourself, page 245 **1.** (3, −2, 3) **2.** (5, −1, 1) **3.** (3, 4) **4.** (−2, 3) **5.** 26 **6.** 10 **7.** 242 **8.** (3, −5) **9.** 372 **10.** pennies: 78; nickels: 33; dimes: 45

Summary and Review, pages 248–249

1. $\begin{bmatrix} -1 & 5 \\ 6 & -2 \\ -1 & 6 \end{bmatrix}$ **3.** undefined **5.** $\begin{bmatrix} -10 & 5 \\ 15 & -10 \\ 5 & 0 \end{bmatrix}$

7. $\begin{bmatrix} 37 & -27 \\ 26 & -19 \end{bmatrix}$ **9.** $\begin{bmatrix} 13 & 6 \\ 11 & 5 \end{bmatrix}$ **11.** undefined

13. $\begin{array}{c} \\ C \\ J \\ P \end{array}\begin{array}{c} \begin{array}{ccc} C & J & P \end{array} \\ \begin{bmatrix} 0 & 1 & 1 \\ 1 & 0 & 0 \\ 1 & 0 & 0 \end{bmatrix} \end{array}$; $\begin{array}{c} \\ C \\ J \\ P \end{array}\begin{array}{c} \begin{array}{ccc} C & J & P \end{array} \\ \begin{bmatrix} 2 & 0 & 0 \\ 0 & 1 & 1 \\ 0 & 1 & 1 \end{bmatrix} \end{array}$ **15.** (−1, 2, 0) **17.** (1, 6, 2) **19.** (1, −2, 2) **21.** (3, 1, −2)

Mixed Review, page 252 **1.** 16 **3.** 49 **5.** 10,000 **7.** $-2a + b + 7$ **9.** $2s - t + 3$ **11.** $4c - 3d - 8$ **13.** $3(5a - 3b + 7)$ **15.** $7(4r - 5s + 6)$ **17.** $5p + 6(4q - 1)$ **19.** 7 ft **21.** 19, 21, 23

Chapter 6 Polynomials

Practice Exercises, Pages 257–258 **1.** x^{13} **3.** y^{11} **5.** x **7.** t^6 **9.** m^{35} **11.** x^8 **13.** $81x^4$ **15.** $9a^2b^2$ **17.** $\frac{x^2}{16}$ **19.** $\frac{4t^2}{9}$ **21.** $-x^8y^6$ **23.** $-18m^3n^3$ **25.** 2; 3 **27.** 3; −3 **31.** $a^{17}b^{13}$ **33.** $0.0128x^5y^7z$ **35.** $-a^{10}b^{11}$ **37.** $-864x^5y^2z^3$ **39.** $\frac{1}{2}a^6b^4$ **41.** $\frac{27}{64}r^6s^3$ **47.** no; y^3 is in the denominator. **49.** x^{24} **51.** a^{2k} **53.** $s^{an}t^{bn}m^{cn}$

Practice Exercises, pages 262–264 **1.** monomial; 15 **3.** monomial; 0 **5.** $-4a^2b + 3ab^2$ **7.** $2x^2$ **9.** $-2n^3 + 3n^2 - n - 3$ **11.** $14x^2 - x - 39$ **13.** $13x + 1$ **15.** $8x^2 - 3x - 14$ **17.** $-23x - 27$ **19.** $a + 4b$ **21.** $-c^2 + 16$ **23.** $8x^2 - 6y$ **25.** $-13x^2y - 4xy^2 - 4xy + 7$ **29.** $-4x^4 - 3x^3 + 5x - 54$ **33.** $10a^2 - 3ab + 10$ **35.** $-5x^3 + 7x^2 - 5x - 6$ **37.** $y^2 - 2y - 20$ **39.** $-4b + 14c + 8d$ **41.** $P + Q$ is a polynomial; $P + Q = 3x^2 + 4x + 10$ **45.** $x^3 - 3x^2 - 7x + 9$ **47.** No **49.** $15{,}000x + 15{,}000$

53. **55.**

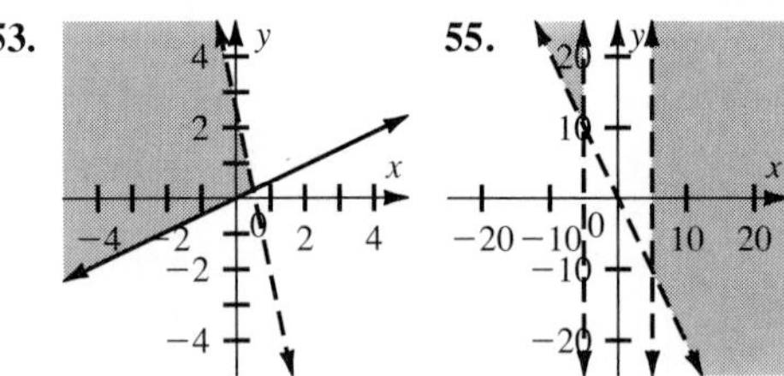

56. $\begin{bmatrix} 5 & 1 \\ 4 & 4 \\ 0 & -11 \end{bmatrix}$ **57.** $\begin{bmatrix} 4 & 4 & 0 \\ 12 & 0 & -8 \\ 16 & 12 & -20 \end{bmatrix}$ **58.** $\begin{bmatrix} 0 & 0.5 \\ 1 & 0 \\ 1.5 & -6 \end{bmatrix}$

59. $\begin{bmatrix} -5 & 5 \\ 5 & -4 \\ 12 & -37 \end{bmatrix}$ **60.** $\begin{bmatrix} \frac{48}{13} & \frac{40}{13} & -\frac{16}{13} \\ \frac{56}{13} & -\frac{40}{13} & \frac{16}{13} \\ \frac{72}{13} & \frac{8}{13} & -\frac{24}{13} \end{bmatrix}$ **61.** $\begin{bmatrix} 2 & 1 \\ -6 & 27 \\ -9 & 64 \end{bmatrix}$

Practice Exercises, pages 268–269 **1.** $4x - 3x^2$ **5.** $2x^2 + 2x - 12$ **7.** $33 + 8x - x^2$ **11.** $5x^2 - 7x - 34$ **15.** $2a^2 - 11ab - 21b^2$ **19.** $x^3 - x^2y - xy^2 + y^3$ **23.** $x^2 - 4x + 4$ **27.** $c^2 - 81$ **31.** $8a^3 - 6a^2 - 41a + 15$ **33.** $8x^3 + 27$ **35.** $a^3 + b^3$ **37.** $x^3 - y^3$ **41.** $36 + 21y - 20y^2 - 12y^3$ **45.** $2x^5 + 11x^4 - 4x^3 - 88x^2 - 96x$ **47.** $9a^4 - 60a^3 + 52a^2 + 160a + 64$ **49.** $16a^4 - 32a^3 + 24a^2 - 8a + 1$ **51.** $y^{2a} - 49$

53. $16x^{4a} + 24x^{2a} - 7$ **57.** $a^2 - b^2 - 2bc - c^2$ **59.** $w^{3c} + z^{3e}$ **61.** $4x + 12$ **63.** $2x^3 - x^2 - x$ **65.** $16x^2 + 96x + 144$

Practice Exercises, pages 273–274 **1.** $3a(a + 1)$ **5.** $6w(5w - 2h + h^2w)$ **9.** $(x + 2)^2$ **11.** $(x - 7)^2$ **15.** $(5x - 4y)^2$ **17.** $(2a + 1)(2a - 1)$ **19.** $(c + 2d)(c - 2d)$ **23.** $(c + 2)(c^2 - 2c + 4)$ **27.** $(m - 5)(m^2 + 5m + 25)$ **29.** $(3x - y)(9x^2 + 3xy + y^2)$ **33.** $(a^2 - 2)(a^2 + 2)$ **35.** $(2f^2 - 5e^2)^2$ **39.** $(y^3 - 8z^4)(y^3 + 8z^4)$ **41.** $(4x^2y - 3z)(16x^4y^2 + 12x^2yz + 9z^2)$ **45.** $\frac{1}{4}(2x + 1)^2$ **49.** $0.01(4y + 1)(4y - 1)$ **53.** $x^n(2x^3 + 3)$ **55.** $(x^a + y^b)(x^a - y^b)$ **57.** $(2x^ay^{4b} - 5z^c)^2$ **59.** $(4x^{2m} - 11y^{m+3})^2$ **61.** $5x - 1$ cm **65.** $x(x + 2y)$

Test Yourself, page 274 **1.** 6 **2.** 5 **3.** $35a^3b^4c^5$ **4.** $108m^7n^{12}$ **5.** $-4a^2 + 6b$ **6.** $3t^2 - 8v^3 + 8w^4$ **7.** $6x^2 + 11x - 10$ **8.** $4x^2 + 12xy + 9y^2$ **9.** $9a^2 - 24ab - 9b^2$ **10.** $w^3 + 8a^3$ **11.** $5(x^2 - 3x + 9)$ **12.** $(3a + b)(3a - b)$ **13.** $(3x - 4y)^2$ **14.** $(3x + 2)(9x^2 - 6x + 4)$

Practice Exercises, pages 278–279 **1.** $(x + 3)(x + 2)$ **3.** $(t - 9)(t - 3)$ **7.** $(x - 7)(x + 2)$ **11.** $(d + 15)(d - 5)$ **13.** $(3y + 4)(y + 9)$ **15.** $(5r + 13)(r + 2)$ **17.** $(5t + 8)(t + 4)$ **21.** $(7z + 6)(z - 2)$ **23.** $(3c + 4)(2c + 1)$ **25.** not factorable **29.** $(x^2 - 5)(x^2 - 3)$ **31.** $(6y + 7)(2y - 5)$ **33.** $(5c + 2)(3c - 8)$ **35.** $(8m - 3p)(3m + 5p)$ **37.** $(7xy - 8)(3xy - 5)$ **39.** $(4x^2 - 5y^2)(3x^2 + 4y^2)$ **41.** not factorable **43.** $(x^4 - 5y^4)(x^4 + 3y^4)$ **45.** $(12x^2 + y)(x^2 - y)$ **47.** $(4xy + 3z)(3xy - z)$ **49.** $(4abc - 1)(abc - 6)$ **51.** $(x^n - 3)(x^n - 5)$ **53.** $(2x^{3a} - 7)(x^{3a} - 2)$ **55.** $(3x^{2m} - 10)(2x^{2m} + 5)$ **57.** $(8x^{n+2} - 5)(3x^{n+2} + 4)$ **59.** not factorable **61.** $x - 15$ m **63.** $3x - 17$ cm

Practice Exercises, pages 282–284 **1.** $(x + b)(x - a)$ **3.** $(m - 2n)(2s + 3t)$ **5.** $(f + g)(f + g + m)$ **7.** $(a - 2b)(a - 2b - 1)$ **11.** $(p - 2q)(p^2 + 2pq + 4q^2 + p + 2q)$ **13.** $(5x - 1 + 2y)(5x - 1 - 2y)$ **15.** $(a + 2b + 5c)(a + 2b - 5c)$ **17.** $9(x + 2)(x - 2)$ **21.** $3(2x + 3)^2$ **23.** $2b(a - 4)^2$ **25.** $-(x - 3)(x^2 + 3x + 9)$ **27.** $3(x - 9)(x + 1)$ **29.** $4(f - 3)(f - 2)$ **33.** $-(x - 1)(x - 4)$ **35.** $(5x + 6y)(x + y)(x - y)$ **37.** $(y + 4)(y - 4)(y + 2)(y^2 - 2y + 4)$ **39.** $(a - 2b + c + d)(a - 2b - c - d)$ **43.** $(x - 4y)(x - 9y)$ **45.** $-(3c + 4)(3c - 4)(2c + 5)(2c - 5)$ **47.** $-6q^2(q + 10)(q^2 - 10q + 100)$ **49.** $(9y^4 - 8)(y^4 - 2)$ **53.** $(2x + 5)^2(2x - 5)^2$ **55.** $\frac{1}{2}(x + 1)(x - 1)$ **57.** $0.3(3x - 7)(2x + 5)$ **59.** $6(x^{2a} + 3y^{2b})(x^{2a} - 3y^{2b})$ **63.** $(3x + 3y - z)(2x + 2y - z)$ **65.** $(a - 2b - 4)(a + 2b + 2)$ **67.** $4(y - 1)^2$ **69.** $\pi(M + m)(M - m)$ **71.** 172 **72.** $h^{-1}(x) = \frac{1}{4}x + \frac{1}{4}$ **73.** $3x + 4y = 6$ **74.** $x = \frac{4 - 5c - 2a^2}{a^2 - c}, c \neq a^2$ **75.** $x = \frac{3 + 3t}{5y - 6}, y \neq \frac{6}{5}$

76. −6 0 $4\frac{1}{2}$ **77.** $\begin{bmatrix} 2 & 1 \\ 5 & 9 \\ 3 & 0 \end{bmatrix}$

78. $\begin{bmatrix} 0 & 0 & 0 \\ 0 & 0 & 0 \end{bmatrix}$ **79.** −2 **80.** −42 **81.** 56

Practice Exercises, pages 288–289 **1.** 3, 4 **3.** 3, 5 **5.** 1, $\frac{1}{2}$ **7.** 0, 6 **9.** 0, 2 **11.** 0, −5 **13.** −5; double **15.** $\frac{3}{2}$; double **17.** 5, −5 **19.** 3, 1, −1 **21.** $-\frac{5}{2}, \frac{2}{3}$ **23.** 2, $\frac{2}{3}, -\frac{2}{3}$ **25.** 1, double; −1, double **27.** 3, double; 4 **29.** $\frac{1}{2}$, −1 **31.** 0, $\frac{14}{3}$ **33.** $\frac{\sqrt{6}}{2}, -\frac{\sqrt{6}}{2}$, 1, −1 **35.** $x = 2 - b$; $x = 2 + b$ **37.** $x = -3$; $a = b$ **39.** $x = 3 - a$; $x = 5 - a$ **41.** length: 18 cm; width: 6 cm **43.** 72 in., 21 in.

Practice Exercises, pages 291–292 **1.** *T* **3.** *T* **5.** *F* **7.** 30 **9.** 56 **11.** −5, 7 **13.** 2, double **15.** −6, 2 **17.** 29 men; 38 women **19.** $\frac{n(n + 1)}{2}$

Practice Exercises, pages 296–297 **1.** 11, 13, 15 or −3, −1, 1 **3.** 20, 25 or −25, −20 **5.** height: 6 ft; base: 16 ft **7.** 1 ft **9.** 8 cm × 16 cm **11.** 3 s; 11 s **13.** *A*'s: 7; *B*'s: 8 **15.** 6, 8, 10; reject −2, 0, 2 **17.** 4 s; 59 s **19.** 144 sq ft

Test Yourself, page 297 **1.** $(3x + 2)(x - 4)$ **2.** $3(2x + y)(2x - y)$ **3.** $(x + 1)(x - 1)(x + y)$ **4.** $x(x + y)(x - y)$ **5.** −6, 3 **6.** $\frac{1}{2}$, double **7.** 0, 2, −2 **8.** 17 cm by 9 cm **9.** 6, 7, 8

Summary and Review, pages 300–301 **1.** $-15a^3b^6c^2$ **3.** $-8m^3n^6$ **5.** 10 **7.** $a + 6b + c$ **9.** $10a$ **11.** $15x^2 + 7xy - 2y^2$ **13.** $x^2 - 16y^2$ **15.** $27y^3 - z^3$ **17.** $(ab + 5c)(ab - 5c)$ **19.** $(6p + 5q)^2$ **21.** $(2a - 9)(a - 1)$ **25.** $3(2x^2 - 5)(x + 2)(x - 2)$ **27.** $\frac{1}{2}, -\frac{1}{3}$ **29.** 12, 14, 16 or −16, −14, −12

Cumulative Review (Ch. 1–6), page 304

1.

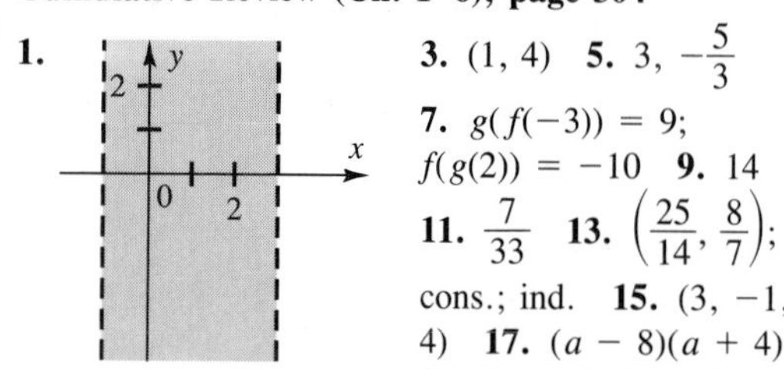

3. (1, 4) **5.** $3, -\frac{5}{3}$ **7.** $g(f(-3)) = 9$; $f(g(2)) = -10$ **9.** 14 **11.** $\frac{7}{33}$ **13.** $\left(\frac{25}{14}, \frac{8}{7}\right)$; cons.; ind. **15.** (3, −1, 4) **17.** $(a - 8)(a + 4)$ **19.** $3(c + 3)(c - 3)$ **21.** (0, 1, −2)

Chapter 7 Rational Expressions

Practice Exercises, pages 310–311 **1.** 1 **3.** 1 **5.** 5 **7.** 9 **9.** $\frac{1}{x^2}$ **11.** $\frac{1}{x}$ **13.** $\frac{x}{y}$ **15.** $\frac{1}{x^8}$ **17.** $\frac{1}{x^6}$ **19.** $\frac{1}{x}$ **21.** 3.205×10^4 **25.** 250 **27.** 0.000541 **29.** $\frac{9y^2}{x^4}$ **31.** 1 **33.** $\frac{1}{x^2y^2}$ **35.** $\frac{1}{2x^6y}$ **37.** $\frac{-72}{x^7}$ **39.** $\frac{8}{x^3y^9}$ **41.** $\frac{1}{4}$ **43.** 729 **45.** 0.0070 **49.** 920,000 **51.** $\frac{1}{x^{17}y^6}$ **53.** $\frac{1}{x^6} + \frac{1}{y^2}$ **57.** 0.015625 **59.** 0.00024414063

Practice Exercises, pages 314–316 **1.** $3xy^3$ **3.** $\frac{x}{2y}$ **5.** $\frac{3}{2xy}$ **7.** $\frac{5}{2}$ **9.** $\frac{7}{3}$ **11.** −2 **13.** $\frac{2}{y - 2}$ **15.** $\frac{2}{x + 5}$ **17.** $\frac{2(x - 6)}{x - 3}$ **19.** $\frac{5(x - 2)}{x - 7}$ **21.** $\frac{y + 5}{y + 4}$ **23.** $\frac{x + 2}{x - 2}$ **25.** $\frac{y + 7}{y - 7}$ **27.** $\frac{x + 4}{x - 6}$ **29.** $x = 4$; $x = 2$ **31.** $\frac{2x - 1}{2x + 3}$ **33.** $\frac{x + 5y}{x - y}$ **35.** $\frac{2(x + 2y)}{3(x + y)}$ **37.** $\frac{-2(x - 7)}{x + 3}$ **39.** $\frac{-5y(y - 3)}{y - 2}$ **41.** $\frac{y(y + 3)}{12(y + 4)}$ **43.** $\frac{x + 4}{x - 4}$ **45.** $\frac{xy + 5}{1 - xy}$ **47.** $\frac{x^{3a}y^{2b}}{9}$ **49.** $x^{2n} - y^{3n}$ **51.** $\frac{(x^n - 2y^{2n})^2}{x^n + 2y^{2n}}$ **53.** $\frac{x^n + 4}{x^n - 5}$ **55.** $\frac{2}{ax}\%$ **57.** $\frac{2d}{a + b}$ **58.** (−5, 0) **59.** (−2, −12) **60.** (2, 3)

Practice Exercises, pages 319–321 **1.** $\frac{7}{15x^2}$ **3.** $\frac{2}{3x^2y^2}$ **5.** $\frac{4}{3}$ **7.** $\frac{5(x + y)}{3}$ **9.** $\frac{x(x - 1)}{3(x + 1)}$ **11.** $\frac{x - 2}{x(x - 1)}$ **13.** $\frac{3(x + 7)}{x - 7}$ **15.** 1 **17.** 1 **19.** 1 **21.** 2 **23.** $\frac{x}{2(x - 5)}$ **25.** $\frac{x + 4}{2(x - 1)}$ **27.** $\frac{18x}{(x + 9)(x + 3)}$ **29.** $\frac{2x + 3}{x - 6}$ **31.** $\frac{3x - 2}{3x + 2}$ **33.** $\frac{2y + 1}{2y - 3}$ **35.** 1 **37.** $\frac{2}{1 - y}$ **39.** $\frac{2(x + 1)(x - 1)}{(x - 2)(x + 2)}$ **41.** $\frac{x^4y^4(x^a + 2)}{x^a + 1}$ **43.** $\frac{(x + y)^2}{(x - y)^2}$ **45.** $\frac{(x + 2y)(x + 3y)}{(x + y)(x + 5y)}$ **47.** yes

Practice Exercises, pages 325–326 **1.** $18y$ **3.** $(x - 1)(x - 2)$ **5.** $(x + 3)^2$ **7.** $\frac{3x}{5}$ **9.** $\frac{3x + 11}{4y}$ **11.** $\frac{3}{x^2 + 2}$ **13.** $\frac{24y^2 + 4x^2}{9x^3y^3}$ **15.** $\frac{18y - 3x}{10x^2y^2}$ **17.** $\frac{y + 6}{2(y + 2)}$ **19.** $\frac{-y - 13}{4(y + 1)}$ **21.** $\frac{8}{3(x + 3)}$ **23.** $\frac{y^2 + 10}{2y(y + 2)}$ **25.** $\frac{5xy - 12}{2y(y + 2)}$ **27.** $\frac{7x - 17}{x^2 - 9}$ **29.** $\frac{-3x^2 - 7x}{(x - 2)(x^2 - 1)}$ **31.** $\frac{-2x^2 + 4x}{(x - 3)(x^2 - 1)}$ **33.** $\frac{2y + 14}{3y(y - 2)}$ **35.** $\frac{3x^3 + x^2 - x}{x^2 - 2}$ **37.** $\frac{14x + 20}{3(x^2 - 4)}$ **39.** $\frac{8x^3 - 32x^2 + 23x + 6}{(x - 1)(x + 2)(x - 2)^2}$ **41.** $\frac{1}{2y - x}$ **45.** $\frac{-2x^4 + 12x^3 + 83x + 441}{x(x - 3)(x^2 - 49)}$ **47.** $\frac{-2x^2 + 6x + 1}{(x + 1)(x^2 - x + 1)}$ **49.** $\frac{-3x - 27}{x(x + 1)(3x + 4)}$ **51.** $\frac{-1}{x + 6}$ **53.** $\frac{3x^2 - 6x + 30}{x(x - 5)^2}$ ft

Test Yourself, page 326 **1.** $\frac{1}{x^3}$ **2.** x^{20} **3.** $\frac{1}{x^3}$ **4.** $\frac{8}{x^9y^3}$ **5.** 4.56×10^{-5} **6.** 2.79×10^8 **7.** 378,000 **8.** 0.0000941 **9.** $\frac{x + 4}{2(x + 5)}$ **10.** $\frac{3(2x + 5)}{2x - 5}$ **11.** $\frac{2}{x + 7}$ **12.** $\frac{3x - 11}{x^2 - 49}$

Practice Exercises, pages 330–331 **1.** $\frac{1}{2}$ **3.** $\frac{3y}{2x}$ **5.** $\frac{4y}{2y + x}$ **7.** $\frac{2y + 3x}{-5y + 7x}$ **9.** $\frac{2x + 4}{4x + 3}$ **11.** $\frac{2x + 10}{x + 7}$ **13.** $\frac{5 + 6x^2}{-8x^2 + x}$ **15.** $\frac{x}{3x + 9}$ **17.** $\frac{2x + 10}{7x}$ **19.** $\frac{-y - 10}{5y - 2}$ **21.** $\frac{2x}{3x + 7}$ **23.** $\frac{7y}{5y + 21}$ **25.** $\frac{x^2 - 2x + 2}{x^2 + 6x + 6}$ **27.** $\frac{2x^3 + 24x^2 + 102x + 160}{3x^3 + 29x^2 + 65x - 15}$ **29.** $\frac{4x^3 + x^2 - 12x - 12}{-x^4 + 4x^3 - 11x^2 + 38x - 28}$ **31.** $\frac{28x + 20}{69x + 345}$ **33.** $\frac{6x - 17}{-8x - 80}$ **35.** $\frac{y - 10}{16y - 96}$ **37.** $\frac{-216y^2 + 428y}{27y - 54}$ **39.** $\frac{x^4 + 3x^2 + 1}{x^3 + 2x}$

Practice Exercises, pages 335–337 **1.** $x \neq 0$; $x = 2$ **3.** all reals; $x = 20$ **5.** all reals; $x = 1$ **7.** all reals; $x = 6$ **9.** all reals; $x = 8$ **11.** all reals; $y = 11$ **13.** $y \neq \pm 3$; $y = 15$ **15.** all reals; $x = 2$ **17.** $x \neq \frac{5}{3}$; $x \neq 15$; $x = -1$ **19.** $y \neq \pm 2$; $y = -1$ **21.** all reals; $x = 38$ **23.** $x \neq -1$; no solution **25.** $x \neq 4$; $x \neq 3$; no solution **27.** $x \neq 0$; $x \neq 1$; no solution **29.** $x \neq 3$; $x \neq 5$; $x = 1$; $x = -\frac{2}{3}$ **31.** $x \neq -1$; $x \neq 2$; $x \neq 5$; $x = 3$ **33.** $x \neq -3$; $x \neq 4$; $x = -1$ **35.** $x \neq -5$; $x \neq 2$; $x = 3$ **37.** $y \neq 0$; $y = -\frac{2}{3}$ **39.** $x =$ all reals; $b \neq 0$; $x = -31b$ **41.** Scott: \$30,000; Alan: \$70,000 **43.** $9\frac{1}{2}$ ft **45.** -8; $\frac{1}{8}$ **46.** $\frac{4}{3}$; $-\frac{3}{4}$ **47.** $-\frac{c^2}{x}$; $\frac{x}{c^2}$ **48.** $\begin{bmatrix} -1 & 0 \\ 2 & 1 \end{bmatrix}$; $\begin{bmatrix} 1 & 0 \\ -2 & -1 \end{bmatrix}$ **49.** $2\sqrt{x} = (90 - y) + 3$ **50.** $-x^2 = y + \frac{1}{y}$ **51.** 4 **52.** no **53.** yes **54.** yes **55.** $-\frac{2}{7}$ **56.** yes; $\frac{3\pi}{4}$ **57.** no **58.** yes; $\frac{2}{9}$

Practice Exercises, pages 340–341 **11.** linear equation **13.** calculations **15.** rational equation **17.** polynomial equation **19.** 20 ft by 45 ft

Practice Exercises, pages 345–347 **1.** walk: 6 mi/h; car: 9 mi/h **3.** uphill: 6 mi/h; downhill: 10 mi/h **5.** 12 min **7.** yes, $8\frac{4}{7}$ min **9.** 750 L **11.** 12 mL **13.** $4\frac{8}{13}$ h **15.** $8\frac{4}{7}$ h **17.** 8%: \$2200; 7%: \$1400 **19.** 4 h **21.** 5 L **23.** 3 mi **25.** carpenter: 8 h; apprentice: 24 h

Test Yourself, page 347 **1.** $\frac{10}{3}$ **2.** $\frac{3x + 9}{6x + 2}$ **3.** $\frac{x - 4}{7x - 2}$ **4.** $x \neq 0$; $x = 1$ **5.** $x \neq 0$, $x \neq \pm 1$; no solution **6.** $y \neq \pm 3$; $y = 57$ **7.** 18 min **8.** 600 mL of 5%; 100 mL of 12%

Summary and Review, pages 350–351 **1.** $\frac{1}{x^7}$ **3.** 1 **5.** 6.5×10^{10} **7.** 5.709×10^6 **9.** 34,000,000 **11.** 0.0090 **13.** $\frac{-3y}{x}$ **15.** $\frac{2(x + 3)}{x + 6}$ **17.** $\frac{x + 3}{4x}$ **19.** $\frac{-x - 8}{x - 3}$ **21.** $\frac{-5x^2 - 32x - 3}{(x + 4)(x - 5)}$ **23.** $\frac{x + 7}{7x + 3}$ **25.** $x \neq 0$; $x = 10$ **27.** $x \neq 7$; no solution **29.** 8

Mixed Review, page 354 **1.** $4\frac{1}{3}$ **3.** $\frac{3}{8}$ **5.** $10\frac{17}{24}$ **7.** $4\frac{1}{6}$ **9.** 7 **11.** 0.4 **13.** $\frac{2}{3}$ **15.** $|x|$ **17.** $-7m^6$ **19.** $12x^3y^2$ **21.** $3ab^2$ **23.** $-2x^5y^6$ **25.** 7.5 cm

Chapter 8

Practice Exercises, pages 359–360 **1.** ± 10 **3.** ± 6 **5.** none **7.** 2 **9.** 0 **11.** -6 **13.** 6 **15.** ± 6 **17.** not real **19.** -11 **21.** 4 **23.** $4x$ **25.** x^4y^9 **27.** $x^{40}y^{25}$ **29.** $11a^{45}$ **31.** $8x^{18}y^{48}$ **33.** $2y^5$ **35.** $0.08x^{20}$ **37.** $x + 3$ **39.** $(x - 5)^2$ **41.** $(x - 4)^6$ **43.** $\pm m^{3a}$ **45.** $2q^{2a+9}$ **47.** s^n **49.** $p + q$ **51.** a^2 **53.** 0.7 m/s

Practice Exercises, pages 365–366 **1.** 3 **3.** 11 **5.** $10\sqrt{2}$ **7.** $5y^2\sqrt[3]{2}$ **9.** $7xy$ **11.** $42x^3y\sqrt{3}$ **13.** $40y^2\sqrt[3]{2y}$ **15.** 10 **17.** $2\sqrt{3}$ **19.** $2x^2y^2\sqrt{2}$ **21.** $\frac{\sqrt{2x}}{2}$ **23.** $\frac{\sqrt{10}}{5}$ **25.** $3\sqrt{2} - 9\sqrt{5}$ **27.** $\frac{\sqrt{2} + 2}{2}$ **29.** $\frac{2\sqrt{7} - 7}{7}$ **31.** $8y^3\sqrt{5y}$ **33.** $2\sqrt[3]{12}$ **35.** $3x^6y^5\sqrt{2y}$ **37.** $10 + 7\sqrt{2}$ **39.** $5 + 5\sqrt{3}$ **41.** $2x\sqrt[3]{2}$ **43.** $5x^2\sqrt{5}$ **45.** $\frac{x\sqrt{10}}{2y}$ **47.** $\frac{2\sqrt{21y}}{5y^2}$ **49.** $\frac{\sqrt[3]{3x^2}}{3x}$ **51.** $\frac{\sqrt[3]{2xy^2}}{xy}$ **53.** $\frac{4\sqrt{2} + \sqrt{10}}{12}$ **55.** $\frac{12x^2\sqrt{5}}{5}$ **57.** $2xy$

59. $2x^2y^5$ **61.** $\frac{\sqrt[3]{x^2y}}{xy}$ **63.** $\frac{\sqrt[6]{x^4y^3}}{y}$ **65.** $\frac{3a^2b}{2}$ **67.** yes **68.** yes **69.** no

Practice Exercises, pages 369–371 **1.** $5\sqrt{7}$ **3.** $9\sqrt{2}$ **5.** $6\sqrt{2}$ **7.** $33\sqrt{2}$ **9.** $7\sqrt{2}$ **11.** $21\sqrt{2}$ **13.** $5\sqrt[3]{2}$ **15.** $2\sqrt[4]{2} + 2\sqrt[4]{3}$ **17.** $23 + 7\sqrt{7}$ **19.** $1 + 3\sqrt{5}$ **21.** $8 + \sqrt{10}$ **23.** $15 - 4\sqrt{14}$ **25.** $49 + 12\sqrt{13}$ **27.** $38 + 12\sqrt{10}$ **29.** 4 **31.** $-2 + 2\sqrt{3}$ **33.** $13 + 7\sqrt{3}$ **35.** $13\sqrt{2}$ **37.** $48\sqrt{2x}$ **39.** $50 + 35\sqrt{2}$ **41.** $17 + 31\sqrt{2}$ **43.** $x + 6 + 3\sqrt{3x}$ **45.** $\frac{12\sqrt{3} + 8}{23}$ **47.** $39 + 10\sqrt{15}$ **49.** $\frac{-11 - 8\sqrt{2}}{14}$ **51.** $-\sqrt{2} + 2\sqrt{3}$ **53.** $\frac{-17\sqrt{2} - 8\sqrt{5}}{43}$ **55.** $\frac{10 - 8\sqrt{2}}{7}$ **57.** $\frac{7\sqrt{3x}}{6}$ **59.** $-\frac{41}{9}$ **61.** $\frac{\sqrt{a} - 1}{a - 1}$ **63.** $\frac{3\sqrt[3]{4} + 2}{2}$ **65.** $\frac{2x - 4\sqrt{xy} + 2y}{x^2 - 2xy + y^2}$ **67.** $\sqrt{a} + \sqrt{b}$ **69.** $2\sqrt{57 - 28\sqrt{3}}$ **71.** $16\sqrt{7}$ m **73.** $\frac{44\sqrt{6}}{15}$ ohms

Practice Exercises, pages 375–376 **1.** 36 **3.** no solution **5.** 20 **7.** 8 **9.** 4 **11.** 4 **13.** 23 **15.** −3 **17.** 3 **19.** 3 **21.** 2 **23.** 3 **25.** 7 **27.** 2 **29.** 2 **31.** 3 **33.** 0 **35.** 1 **37.** −1, −6 **39.** $b - a$ **41.** 1 **43.** 0, 1 **45.** $s = \sqrt{A}$ **47.** $d = \frac{v^2}{64}$ **49.** 24.5 cm **51.** $x = 0.62y$

Practice Exercises, pages 381–382 **1.** 4 **3.** 5 **5.** 12 **7.** $\sqrt{10}$ **9.** $3\sqrt{5}$ **11.** (2, 0) **13.** (−5, 5) **15.** (−2, −3) **19.** $\left(4, \frac{5}{12}\right)$ **23.** $3\sqrt{13}$ **27.** 9 **31.** $\left(18, -\frac{21}{2}\right)$ **33.** $(3\sqrt{5}, 13\sqrt{5})$ **35.** (4, 6) **37.** $10\sqrt{5}$ **39.** $2\sqrt{13} + \sqrt{53}$ **41.** 127.3 ft

Test Yourself, page 382 **1.** $4x^2y^4\sqrt{2xy}$ **2.** $3x^2$ **3.** $4x^2y^3\sqrt{y}$ **4.** $\frac{6\sqrt{14y}}{35}$ **5.** $\frac{-8 - 3\sqrt{6}}{2}$ **6.** $38\sqrt{2}$ **7.** 7 **8.** −2, −1 **9.** $\sqrt{29}$ **10.** (9, −1)

Practice Exercises, pages 385–387 **1.** $\overline{PQ} = \sqrt{(2a - 0)^2 + (0 - 0)^2} = \sqrt{(2a)^2} = 2a$; $\overline{SR} = \sqrt{(2a + 2b - 2b)^2 + (2c - 2c)^2} = \sqrt{(2a)^2} = 2a$ **3.** T: midpt. $\overline{SR} = \left(\frac{2b + 2a + 2b}{2}, \frac{2c + 2c}{2}\right) = (a + 2b, 2c)$; W: midpt. $\overline{PQ} = \left(\frac{0 + 2a}{2}, \frac{0 + 0}{2}\right) = (a, 0)$; $\overline{TW} = \sqrt{(a + 2b - a)^2 + (2c - 0)^2} = \sqrt{4b^2 + 4c^2} = 2\sqrt{b^2 + c^2}$; $\overline{PS} = \sqrt{(2b - 0)^2 + (2c - 0)^2} = \sqrt{4b^2 + 4c^2} = 2\sqrt{b^2 + c^2}$ **5.** midpt. $\overline{PR} = \left(\frac{2a + 2b + 0}{2}, \frac{2c + 0}{2}\right) = (a + b, c)$; midpt. $\overline{QS} = \left(\frac{2a + 2b}{2}, \frac{2c + 0}{2}\right) = (a + b, c)$

Practice Exercises, pages 390–391 **1.** $i\sqrt{5}$ **3.** $i\sqrt{6}$ **5.** $2i\sqrt{2}$ **7.** $-i$ **9.** 1 **11.** 1 **13.** $10i$ **15.** $-4i$ **17.** $8i$ **19.** $4i\sqrt{5}$ **21.** −30 **23.** $-\sqrt{15}$ **25.** 3 **27.** $\frac{7}{2}$ **29.** $\frac{\sqrt{10}}{2}$ **31.** $\sqrt{2}$ **33.** $-i$ **35.** 1 **37.** $19i$ **39.** $i\sqrt{3}$ **41.** $-8i\sqrt{7}$ **43.** $-i\sqrt{3}$ **45.** $-i\sqrt{7}$ **47.** $-4i\sqrt{30}$ **49.** $\frac{2\sqrt{6}}{9}$ **51.** $\frac{3\sqrt{2}}{10}$ **53.** $3\sqrt{2}$ **55.** i **57.** $-i$ **59.** $\sqrt{3}$ **61.** −1 **63.** ix^5 **65.** 17 ohms

Practice Exercises, pages 395–396 **1.** $6 + 4i$ **7.** $7 + 3i$ **9.** $15 + 3i$ **13.** $5 + 6i$ **17.** $-9 - 6i$ **21.** $4 - 3i$ **23.** 10 **27.** 13 **31.** $\sqrt{10}$ **33.** 5 **35.** $6 + 2i$ **39.** $7 + 3i\sqrt{2}$ **43.** $2 - 9i\sqrt{2}$ **47.** $\sqrt{229}$ **49.** $\sqrt{199}$ **51.** $\sqrt{97}$ **53.** $3\sqrt{29}$ **55.** 2; 3 **57.** $\frac{7}{8}; -\frac{3}{2}$ **59.** $2 + 14i\sqrt{2}$ **61.** $8\sqrt{7} + 4i\sqrt{2}$ **63.** $10 - 3i$ **65.** 10 **67.** $\sqrt{37}$ **69.** $Z = 7 + 5i$; $|Z| = 8.6$ ohms

71. [graph]

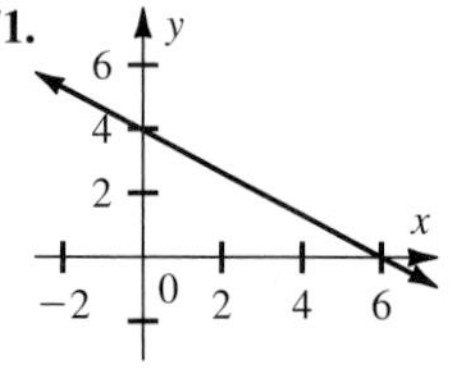

72. $2x - 5y = 18$ **73.** $-\frac{1}{3}; 3$ **74.** $\{x: x \in \text{real numbers}; x \neq -2, 1, 3\}$ **75.** $\frac{3}{y}$, $x \neq 0$, $y \neq 0$ **76.** $\frac{2x(x + 9)}{(x - 7)(x + 1)}$, $x \neq 7$, $x \neq -1$

Practice Exercises, pages 400–401 **1.** $-7 + 11i$ **5.** $33 - 13i$ **9.** 2 **11.** $8 + 6i$ **13.** $\frac{9}{17} + \frac{2i}{17}$ **15.** $\frac{11}{5} - \frac{7i}{5}$ **17.** $\frac{1}{5} - \frac{i}{10}$ **21.** $\frac{5}{13} + \frac{12i}{13}$ **23.** $\frac{17}{13} + \frac{6i}{13}$ **27.** $2 + 10i\sqrt{2}$ **31.** $\frac{23}{33} + \frac{7i\sqrt{2}}{33}$

33. $-2.8 + 0.8i$ **35.** $1 - \frac{i}{3}$ **37.** $-\frac{5}{6} + \frac{2i\sqrt{2}}{3}$ **41.** $2 + 11i$ **43.** $-\frac{63}{25} + \frac{16i}{25}$ **45.** $-\frac{3\sqrt{2}}{4} + \frac{3i\sqrt{2}}{8}$ **47.** any real values such that $x = y$ **49.** 10 **51.** $\frac{39}{29} + \frac{33i}{29}$

Test Yourself, page 401 **3.** $8i$ **4.** 1 **5.** $\sqrt{10}$ **6.** $-3 - 4i$ **7.** $\frac{1}{2} + 2i$ **8.** $\frac{3}{10} - \frac{i}{10}$ **9.** $16 - 4i$ **10.** 17 **11.** $-7 - 24i$

Summary and Review, pages 404–405 **1.** $2\sqrt{2}$ **3.** -4 **5.** $2\sqrt[3]{4}$ **7.** $3x^2y^3$ **9.** $5x$ **11.** $\frac{1}{5}$ **13.** $8\sqrt{7}$ **15.** $-\sqrt{y}$ **17.** $\frac{-7 + 7\sqrt{5}}{4}$ **19.** $23 + 16\sqrt{2}$ **21.** 4 **23.** 1 **25.** $\sqrt{5}; \left(\frac{7}{2}, 6\right)$ **27.** $\sqrt{290}; \left(-\frac{1}{2}, \frac{3}{2}\right)$ **29.** $18 + 9\sqrt{2}$ **33.** 1 **35.** i **37.** $9 + 4i$ **39.** $1 - 3i$ **41.** 40 **43.** $\frac{1}{2} + 2i$

Cumulative Review (Ch. 1–8), pages 408–410
1. $-1, 4$ **3.** no solution
5.

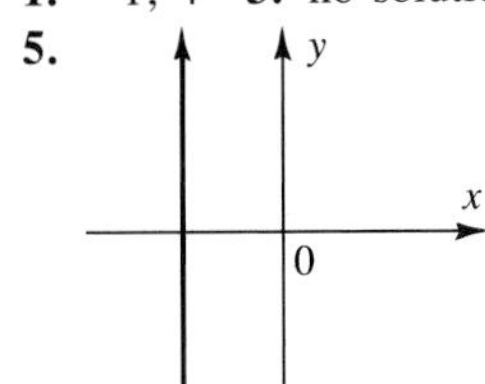

7.

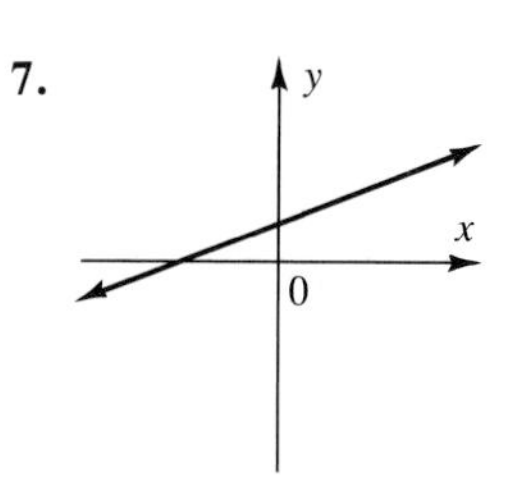

9.

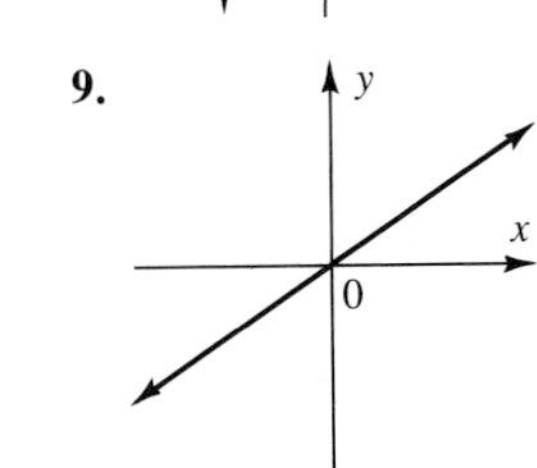

11. $9a^2 - 25b^2$ **13.** $6x^3 - x^2 - 30x + 27$ **15.** 0.000001003 **17.** $\frac{3}{2}$ **19.** $\frac{9}{2}$ **21.** $\left(\frac{1}{8}, -\frac{1}{8}\right)$ **23.** 84 **25.** $4(a + 2)(a - 2)$ **27.** $4d(d + 1)^2$ **29.** $(3c - d)(c + d)$ **31.** $(t - v)(t + v)(t^2 + v^2)$ **33.** no solution **35.** $\begin{bmatrix} 35 & 10 \\ 23 & -8 \\ -5 & 18 \end{bmatrix}$ **37.** $\begin{bmatrix} -2 & 0 & -6 \\ -12 & 16 & -8 \\ 9 & 0 & -5 \end{bmatrix}$ **39.** $\begin{bmatrix} -2 & 4 \\ 4 & 4 \end{bmatrix}$ **41.** -3 **43.** $\frac{-27c}{b^3}$ **45.** 131,072 **47.** 25 **49.** -5 **51.** $\frac{6x}{6 + 4x - x^2}$ **53.** $1 + x + x^2$ **55.** $\frac{4}{15}(y + 3)$ **57.** $\frac{17x}{12(x - 2)}$ **59.** $\frac{-5y^2 + 20y + 1}{3(y - 4)(y + 4)}$ **61.** $-2b^2\sqrt[3]{2b}$ **63.** $2x$ **65.** $\frac{4 + \sqrt{2}}{7}$ **67.** $16i$ **69.** $x + 3y = 33$ **71.** $5x - 3y = 0$ **73.** yes; $y = 1.3x$ **75.** $(1, -1, 2)$ **77.** $-1, 3$ **79.** $-\frac{3}{2}, 0, \frac{5}{4}$ **81.** $-5, -1, 1$ **83.** 7, 8, 9 or 0, 1, 2 **85.** $y \le 1$ **87.** $-3 < x < 1$ **89.** $t \ge 2$ or $t \le -3$

Chapter 9 Quadratic Functions

Practice Exercises, pages 415–416 **1.** up; $x = 0$; $v(0, 0)$; min. **3.** up; $x = 0$; $v(0, 0)$; min. **5.** up; $x = 0$; $v(0, 2)$; min. **7.** up; $x = 0$; $v(0, -5)$; min. **9.** up; $x = 0$; $v(0, 0)$ **11.** up; $x = 0$; $v(0, 0)$ **13.** down; $x = 0$; $v(0, -4)$ **15.** up; $x = 3$; $v(3, 0)$ **17.** up; $x = 0$; $v\left(0, -\frac{1}{2}\right)$ **19.** up; $x = 0$; $v(0, 5)$ **21.** $y = 3x^2$ **25.** 8 **27.** $\frac{2}{3}$ **29.** $-\frac{11}{8}$ **35.** $-10x^5 + 4x^4 - 2x^3 + 6x^2$ **36.** $6x^2 - 18xy + 12y^2$ **37.** $x^2 + 6x + 9$ **38.** $4x^6 + 16x^4 - 4x^3 + 16x^2 - 8x + 1$ **39.** $x^4 + 3x^3 - x^2 - 9x - 6$ **40.** $x^3 - 3x^2 + 3x - 1$ **41.** $\frac{5(x - 2)}{3(x + 2)}$ **42.** 1 **43.** $\frac{-1}{x}$ **44.** 13

Practice Exercises, pages 420–421 **1.** $x = 1$; $v(1, 1)$; 0, 2; 0 **3.** $x = -1$; $v(-1, 2)$; $0, -2$; 0 **5.** $x = -3$; $v(-3, 0)$; -3; 9 **7.** $x = -4$; $v(-4, 0)$; -4; 16 **9.** $x = \frac{5}{4}$; $v\left(\frac{5}{4}, -\frac{9}{8}\right)$; $\frac{1}{2}$, 2; 2 **11.** $x = -4$; $v(-4, -8)$; -8, 0; 0

13.

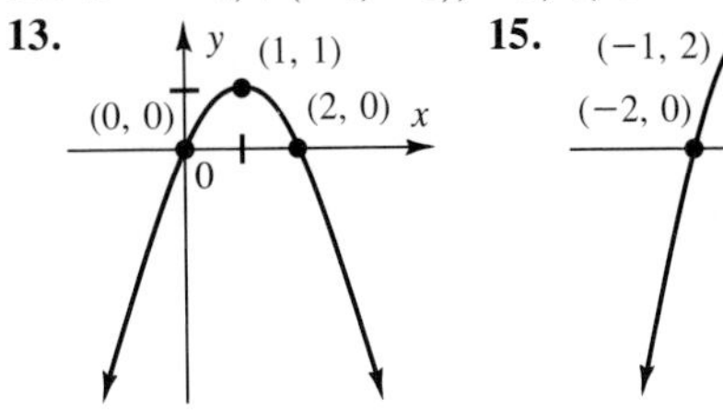

15.

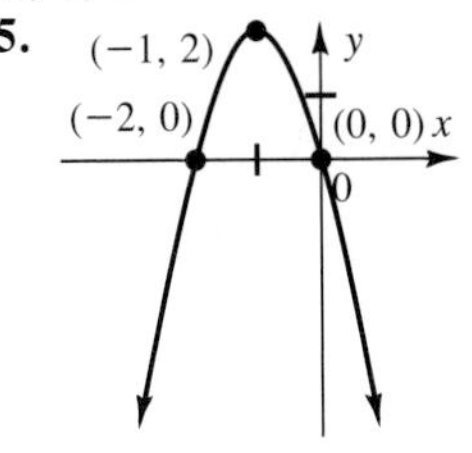

17.

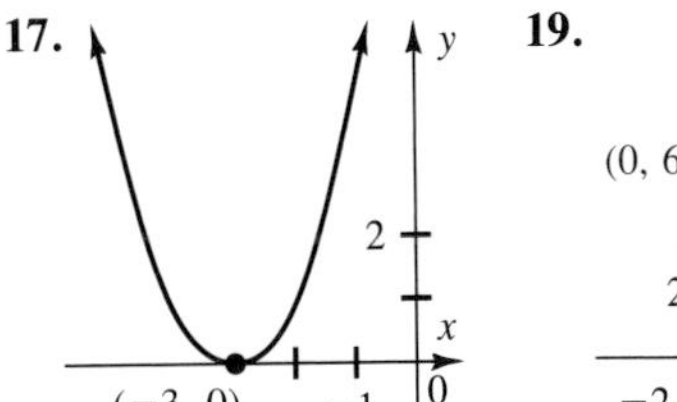

19.

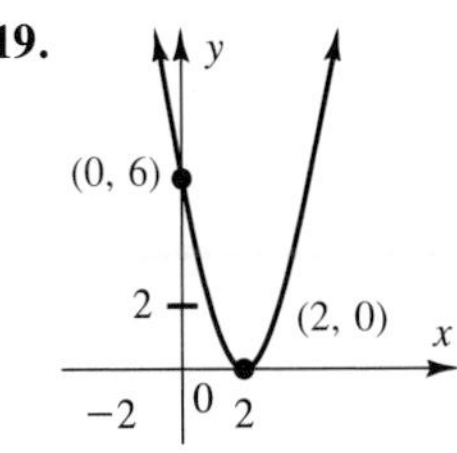

Answers

21. 23.

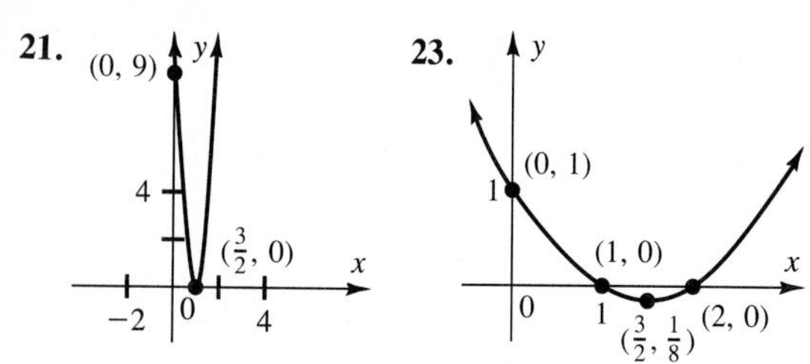

25. max.; 6 **27.** max.; $\frac{41}{8}$ **29.** min.; −3

31. $a = -6$; $b = 24$ **33.** $a = -\frac{2}{9}$; $b = -\frac{4}{3}$

35. No; since it is an upward parabola, there is no upper bound. **37.** l: 40 ft; w: 40 ft

Practice Exercises, pages 424–426 **1.** 49 **3.** $\frac{81}{4}$

5. $\frac{1}{16}$ **7.** ± 8 **9.** ± 10 **11.** ± 12 **13.** 0, −6

15. $\frac{19}{2}, \frac{9}{2}$ **17.** $\frac{1}{2}, -\frac{3}{2}$ **19.** $-6 \pm 4\sqrt{2}$

21. $-5 \pm \sqrt{19}$ **25.** 9, −1 **27.** $-1 \pm \sqrt{6}$

29. $1 \pm i$ **31.** $2 \pm \sqrt{11}$ **35.** $\frac{7}{2}, -4$

37. $\frac{9 \pm i\sqrt{19}}{2}$ **39.** $\frac{-1 \pm i\sqrt{53}}{3}$ **41.** $\frac{1}{2}, -\frac{3}{2}$

45. $\pm 2i$ **47.** $-4 \pm 2\sqrt{5}$ **49.** $\frac{-3 \pm \sqrt{41}}{8}$

51. $\frac{2 \pm i\sqrt{5}}{5}$ **53.** $2\sqrt{3}$ **55.** $2a, -\frac{3a}{2}$

57. $\frac{-a \pm a\sqrt{13}}{6}$ **59.** $3, -\frac{3a}{a + 3}$ **61.** −12 or 8

63. $3x^2 - 2x + 1$ **65.** $-x^2 + 2x - 1$

Practice Exercises, pages 430–431 **1.** $-\frac{5}{2}, 1$

3. $-\frac{1}{2}, \frac{3}{4}$ **5.** $\frac{2 \pm \sqrt{10}}{3}$ **7.** $\frac{-1 \pm \sqrt{61}}{10}$

11. $-4, \frac{7}{2}$ **15.** $6 \pm \sqrt{11}$ **17.** $1 \pm 2i$ **19.** 3, −0.5 **21.** 0.54, −1.87 **23.** 1.38, −1.24

25. $-\frac{1}{2}, \frac{3}{2}$ **27.** $\frac{-3 \pm i\sqrt{11}}{2}$

31. $\frac{1 \pm 4\sqrt{11}}{7}$ **33.** $\frac{5 \pm \sqrt{33}}{4}$ **35.** $\frac{-1 \pm \sqrt{5}}{4}$

37. $\frac{-1 \pm i\sqrt{15}}{6}$ **39.** $\frac{1 \pm i\sqrt{39}}{5}$ **41.** $\frac{-3 \pm \sqrt{61}}{2}$

43. $\frac{3 \pm i}{2a}$ **45.** $2, \frac{-2a}{a + 2}$ **47.** $-a \pm a\sqrt{26}$

49. 2 or 1 **51.** $\frac{4}{3}$ or $-\frac{3}{4}$ **53.** 3 or $-\frac{11}{3}$

Practice Exercises, pages 434–436 **1.** 15 cm **3.** base: 8 cm; alt.: 14 cm **5.** l: 48 cm; w: 36 cm **7.** 13, 15 or −15, −13 **9.** −13 or 12 **11.** −5, −1 or 2, $\frac{5}{2}$ **13.** w: $4 + 2\sqrt{6}$ cm; l: $9 + 4\sqrt{6}$ cm **15.** 16 cm **17.** $-2 + \sqrt{6}, 2 + \sqrt{6}$ or $-2 - \sqrt{6}, 2 - \sqrt{6}$ **19.** 6.8 in.

Test Yourself, page 436 **1.** up; $x = 0$; (0, −4) **2.** down; $x = 0$; (0, 0) **3.** up; $x = -1$; (−1, −3)

4.

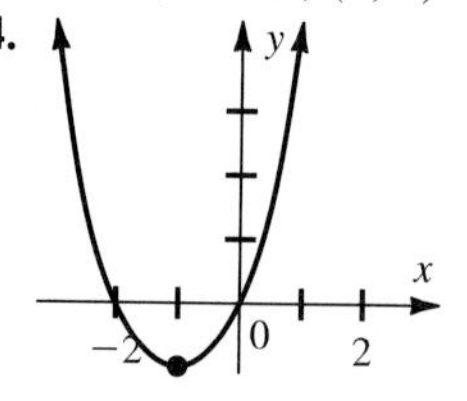

5.

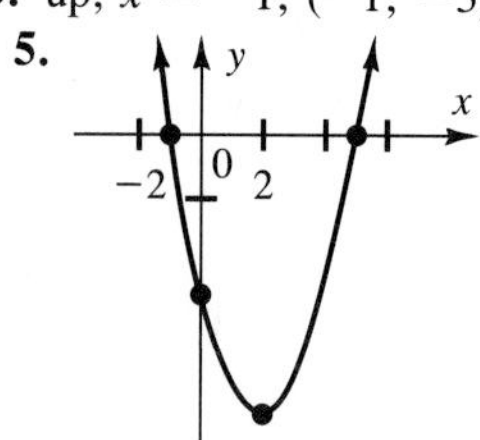

6.

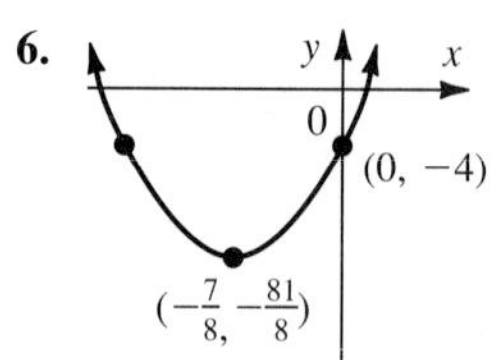

7. $-4 \pm \sqrt{21}$

8. $-\frac{3}{4}, \frac{1}{2}$ **9.** $1 \pm \sqrt{2}$

10. $-\frac{5}{2}, \frac{3}{2}$

11. $\frac{-2 \pm i\sqrt{2}}{3}$

12. $\frac{5 \pm i\sqrt{299}}{6}$ **13.** 3 cm

Practice Exercises, pages 439–440 **1.** 2 irrat. unequal **3.** 1 real rat. **5.** 2 real irrat. unequal **7.** 2 real irrat. unequal **9.** 2 real irrat. unequal **11.** 2 real irrat. unequal **13.** does not cross x-axis **15.** crosses x-axis twice **17.** crosses x-axis twice **19.** 2 real irrat. unequal **21.** 2 complex conj. **23.** 2 complex conj. **25.** 4 **27.** ± 8 **29.** 8 **33.** does not cross x-axis **35.** does not cross x-axis **37.** no

Practice Exercises, pages 444–445 **1.** −7; −4

3. −3; 1 **7.** $\frac{5}{2}; \frac{3}{2}$ **9.** $\frac{7}{3}; 0$ **11.** −12; −6

13. $x^2 - 10x + 24 = 0$ **15.** $6x^2 - 7x + 2 = 0$ **17.** $x^2 - 2x - 1 = 0$ **19.** $x^2 - 4x + 13 = 0$ **21.** $x^2 - 3x + 1 = 0$ **23.** $16x^2 + 24x - 19 = 0$ **27.** $8x^2 - 12x + 5 = 0$ **29.** $9x^2 + 12x + 8 = 0$ **31.** $25x^2 - 10x + 3 = 0$ **33.** $3x^2 + 6x + 7 = 0$ **35.** $x^2 - 4cx + 4c^2 - 9d^2 = 0$ **37.** $16x^2 + 24ax + 9a^2 - 4c = 0$ **39.** $x^2 + 4x + 49 = 0$ **41.** 10, 14 **43.** $-6 \pm \sqrt{2}$

Practice Exercises, pages 449–450 **1.** $\pm i, \pm i\sqrt{6}$ **3.** $\pm 3, \pm i\sqrt{3}$ **5.** 1 **7.** no solution **9.** $\pm 2, \pm 2i$

11. $\pm 1, \pm\frac{1}{2}\sqrt{2}$ **13.** −5, 1 **15.** $\frac{3 \pm \sqrt{57}}{6}$

17. $\frac{3 \pm \sqrt{21}}{12}$ **21.** $\frac{6 \pm \sqrt{42}}{3}$ **23.** ± 2

25. $\frac{-1 \pm \sqrt{2}}{5}$ **29.** 16 **33.** $\frac{105 - 3\sqrt{1029}}{2}$

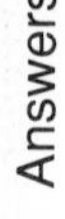

35. $-1, \frac{1}{2}$ **37.** 1 **39.** 2, −1 **41.** 9, −6

Practice Exercises, pages 454–455 **1.** $\{x: x < -2 \text{ or } x > -1\}$ **5.** $\{x: 1 \le x \le 5\}$ **7.** $\left\{x: x < -\frac{7}{3} \text{ or } x > 1\right\}$ **9.** $\{x: x \le -7 \text{ or } x \ge 7\}$
13. $\left\{x: x < 0 \text{ or } x > \frac{3}{2}\right\}$ **15.** $\{x: x < -4 \text{ or } x > 1\}$
19. $\left\{x: -\frac{3}{2} < x < 5\right\}$ **21.** $\left\{x: x \le -1 \text{ or } x \ge \frac{5}{3}\right\}$
25. $\left\{x: -1 < x < \frac{2}{5}\right\}$ **27.** $\left\{x: -\frac{7}{3} \le x \le \frac{5}{2}\right\}$
29. $\{x: -0.1 < x < 8.1\}$ **31.** $\{x: x < -1.6 \text{ or } x > 0.6\}$ **35.** $\{x: x \le -0.6 \text{ or } x \ge 4.1\}$
37. $\sqrt{17}$ **39.** $\{x: x < -2 \text{ or } x > 3\}$

Test Yourself, page 455 **1.** 1 rat. real **2.** 2 complex conj. **3.** 2 irrat. unequal real **4.** crosses x-axis twice **5.** crosses x-axis twice **6.** does not cross x-axis **7.** ±8 **8.** $\frac{9}{2}$ **9.** 3 **10.** −2, 9
11. $3, \frac{5}{2}$ **12.** $\frac{4}{9}, 2$ **13.** $x^2 + 5x - 24 = 0$
14. $x^2 - 8x + 6 = 0$ **15.** $8x^2 - 4x + 3 = 0$
16. $\pm 2, \pm i\sqrt{3}$ **17.** 9 **18.** $-\frac{10}{3}, -\frac{3}{2}$
19. $\frac{7 \pm \sqrt{57}}{2}$ **20.** $\frac{-1 \pm \sqrt{13}}{2}$ **21.** $\frac{2}{3}$
22. $\{x: -5 < x < 2\}$ **23.** $\left\{x: x \le -3 \text{ or } x \ge \frac{1}{4}\right\}$

Summary and Review, pages 458–459 **1.** up; $x = 0$; $v\ (0, 0)$ **3.** up; $x = 0$; $v\ (0, 6)$ **5.** $x = \frac{3}{2}$; $v\left(\frac{3}{2}, 0\right)$; $\frac{3}{2}$, 9 **7.** 10, −2 **9.** $\frac{3 \pm \sqrt{41}}{2}$
11. $\frac{-5 \pm \sqrt{5}}{10}$ **13.** l: 20 ft; w: 5 ft **15.** 2 real rational unequal **17.** 2 complex conjugates
19. $x^2 - 6x + 7 = 0$ **21.** $\pm\sqrt{3}, \pm 2$ **23.** $-\frac{1}{2}$, 2 **25.** $\left\{x: -1 \le x \le \frac{7}{3}\right\}$

Mixed Review, page 462 **1.** 6.538 **3.** 42 **5.** 9.12 **9.** $(x - 3)(x - 2)$ **11.** $(3m - 4)^2$ **13.** $(y + 5)(y - 8)$ **15.** $\pm 2i$ **17.** ±5 **19.** $\pm 2\sqrt{5}$ **21.** 0; 5 **23.** −1; ±3 **25.** 160 ft^3

Chapter 10 Polynomial Functions

Practice Exercises, pages 467–468 **1.** $4x^3 - 5x^2 - 6x$ **3.** $2a^5 - 3a^4b^2 + 6b^4$ **5.** $9d^4 - 6d^3 + 12d^2 - \frac{7}{d}$ **7.** $13x^5 - 9x^3 - 4 + \frac{8}{x^3}$
9. $x - 16 - \frac{4}{x + 2}$ **11.** $a - 10 - \frac{26}{a + 6}$
17. $x + 2 - \frac{2}{x - 2}$ **21.** $x^2 - 4x + 1 + \frac{2x - 5}{x^2 + 3x - 4}$ **23.** $x^2 + 2x - 5$ **25.** $3n^2 - 2n + 3$ **29.** $x^2 + x + 1$ **31.** $16x^2 - 12x + 9$
33. $y^3 - y^2 + y - 1$ **35.** $x^{2a} + x^a + 7 + \frac{23}{x^a - 5}$ **37.** $x^{2a} - 5x^a + 6 - \frac{8}{x}$ **39.** $x^{2a} - x^a - 2$ **41.** $(2x - 1)$ cm

Practice Exercises, pages 472–474 **1.** $x^2 - 10x - 25 - \frac{42}{x - 2}$ **5.** $x^2 + 5x + 2 - \frac{6}{x - 2}$
7. $2x^2 - 7x - \frac{3}{x + 1}$ **11.** $2x^2 + 6x + 4 + \frac{14}{x - 2}$ **13.** $3x^2 + 7x + 11 + \frac{32}{x - 4}$ **17.** $x^3 - 5x^2 + 22x - 115 + \frac{595}{x + 5}$ **19.** $2x^3 + x^2 + 2x + \frac{10}{x - 2}$ **21.** $y^2 + y + 1$ **23.** $x^2-4x+16$
25. $y^2 + y + 1 + \frac{2}{y - 1}$ **27.** $y^2 - 4y + 16 - \frac{128}{y + 4}$ **29.** $3a^2 - 8a - 2 + \frac{6}{2a + 1}$
31. $x^3 + 3x^2 + 2x - 2$ **33.** $4x^2 + 6x + 9 + \frac{54}{2x - 3}$ **35.** 24 **37.** 4 **39.** 7 **46.** $2(3y^2 + 5x^3)(3y^2 - 5x^3)$ **47.** $2y(4xy^2 + 3f)(16x^2y^4 - 12xy^2f + 9f^2)$ **48.** $(2x + 3)(3x - 4)$
49. $6xy^2(2x + y)(3x - 2y)$ **50.** $\frac{y}{2}$ **51.** $xy^2(y + 1)$ **52.** −1 **53.** $4i\sqrt{2}$ **54.** $-7\sqrt{2}$ **55.** $28 - 6i$

Practice Exercises, pages 478–480 **1.** 8; 0 **5.** 0; 0 **7.** yes **9.** yes **13.** no **15.** yes
35. $2x^3 - x^2 - 8x + 4$ **37.** $3x^3 - 8x^2 - 5x + 6$ **39.** $x^3 + 3x^2 - 2x - 6$ **41.** $2x^3 + x^2 + 18x + 9$ **53.** $(x + 4), (2x - 3)$ **55.** $(x + 2), (x - 4)$ **57.** $(ax - 3), (ax - 2)$

Test Yourself, page 480 **1.** $5z^4 - 4z^3 - 12z$
2. $3x^2 - 2x + 3$ **3.** $x^2 - 5x - 27 - \frac{66}{x - 3}$
4. $2x^3 - 3x^2 - 3x + 5 - \frac{4}{x + 2}$ **5.** 2 **6.** 0
7. yes **8.** no **9.** $(x - 3)(x - 1)(x - 5)$
10. $(x - 1)(2x^2 - 8x - 13)$

Practice Exercises, pages 486–487 **1.** −1; 3 **3.** 7; −5 **5.** $2i; -2i$ **7.** 2 pos.; 1 neg. **9.** 2

pos.; 0 neg. **11.** -1; 1; -6 **13.** -2; 2; $-\frac{3}{2}$ **17.** 2; 1; 3 **19.** $-2, -1, 1, 2$ **21.** $2, -4$ **23.** 4; -2 (mult. 2) **25.** $\frac{3}{4}$; -1 (mult. 2) **27.** 1; $2 \pm i$ **31.** -1; $3 \pm 2i$ **33.** -1; 2; $-1 \pm 2i$ **35.** $\frac{3}{2}$; $1 \pm i$ **37.** 1; -2; $2 \pm 3i$ **39.** -1; 2 (mult. 2); $\pm 2i$ **41.** 3, 4, 5 **45.** $2, -2$ **46.** no real solution **47.** -3 **48.** 124° **49.** $x > 11.5$ **50.** $y = x - 9$ **51.** -6; 5; 0; $2 + i$ **52.** $x = \frac{1}{2}$; $y = 2$ **53.** $(-2, \frac{3}{4})$ **54.** $(1, 2, -3)$ **55.** $-\frac{3}{2}, \frac{1}{4}$ **56.** $-6, \frac{1}{6}$

Practice Exercises, page 489 **1.** *E* **3.** *A* **5.** *J* **7.** *D* **9.** *B* **11.** *T* **13.** *T* **15.** *F*

Practice Exercises, pages 494–495 **1.** -2; 1; 3 **3.** -3; -2; 2 **5.** -2.5; 1.5; 3 **7.** -1.7; 1.7; 3 **9.** 1.3 **11.** 1.6 **13.** -2; 0; 3 **15.** -2; -1; 1; 2 **17.** -0.3; 1; 3.3 **19.** 1.79 **21.** 0.35 **23.** -0.75; 0.72; 2.78 **25.** $P(2) < 0$; $P(5) > 0$; thus, there must be a real zero between the points. **27.** 0.6 m

Practice Exercises, pages 499–501 **1.** edge of cube: 3 ft; tank: l: 4 ft; w: 2 ft; h: 5 ft **3.** l: 8 ft; w: 5 ft; h: 2 ft **5.** 1 cm or 4 cm **7.** 6, 3, 1, -1 **9.** 6, 9, 12, 15 **11.** $-3, -5, -6, -2$ **13.** 0.9 cm or 4.8 cm **15.** 6.9 in. **17.** edge of cube: 5.9 ft; box: l: 10.8 ft; w: 3.9 ft; h: 8.9 ft

Practice Exercises, page 505 **1.** no int.; $x = 0$; $y = 0$ **3.** *x*-int.: 0; *y*-int.: 0; $x = 1$ **7.** *x*-int.: 3, -3; *y*-int.: $\frac{9}{4}$; $x = 2$; $x = -2$; $y = 1$ **9.** no int.; $x = 0$; $y = 0$ **13.** $x = 0$; $y = 0$ **15.** *x*-int.: 4; $x = 2$; $x = -2$; $y = 0$ **19.** 4, 5, 7, 13, 14 **21.** *x*-int.: $\frac{2}{3}$; *y*-int.: $\frac{1}{2}$; $x = 2$, $x = -2$; $y = 0$ **23.** *y*-int.: -4; $x = 1$; $x = -1$; $y = 0$

Test Yourself, page 506 **1.** $-\frac{1}{2}$; -2; 3 **2.** $-\frac{1}{3}$; $\frac{1}{3}$; $\frac{2}{3}$ **3.** 6, $\pm i$ **4.** $-3, 1, \pm 2i$ **5.** -2; 3; 5 **6.** $-2, -0.3, 1.2$ **7.** *l*: 10 ft; *w*: 8 ft; *h*: 5 ft **8.** 4, 6, 16 **9.** no int.; $x = 0$; $y = 0$ **10.** both int. 0; $x = \frac{2}{3}$; $y = \frac{1}{3}$

Summary and Review, pages 508–509 **1.** $xy^2 - x^2y^3 + x^2$ **3.** $y^2 - y + 1 - \frac{82}{y + 1}$ **5.** $x^2 - 12x - 18 - \frac{32}{x - 2}$ **7.** -20 **9.** $(x - 1)(x - 2)(x - 3)$ **11.** $1, \frac{3}{2}, 3$ **13.** $-1, \pm 2i$ **15.** $-3, -1, 2$ **17.** 1.2 **19.** edge of cube: 8 ft; box: l: 10 ft; w: 4 ft; h: 7 ft

Cumulative Review (Ch. 1–10), page 512 **1.** 3 **3.** 1, 6 **5.** $x^2 - 2x + 2 - \frac{10}{x - 1}$ **7.** 5; 0 **9.** -162 **11.** $y = 2x + 7$ **13.** $\frac{x + 4}{x - 5}$; $x \neq -2, 1, 3, 5$ **15.** $-1, -2$ **17.** 12 **19.** $(5y - 2)(5y + 2)$ **21.** $(y - 4)(y^2 + 4y + 16)$

Chapter 11 Conic Sections

Practice Exercises, pages 518–519 **1.** $x^2 + y^2 = 4$ **3.** $(x + 2)^2 + (y - 2)^2 = 1$ **5.** $x^2 + y^2 = 49$ **7.** $(x - 5)^2 + (y + 1)^2 = 36$ **9.** (0, 0); 3 **13.** (1, 0); 7 **15.** (0, -7); 6 **19.** (3, 1); $\sqrt{5}$ **21.** (2, -2); $\sqrt{14}$ **23.** $\left(\frac{5}{2}, 3\right)$; $\frac{3\sqrt{5}}{2}$ **27.** $\left(\frac{5}{2}, \frac{3}{2}\right)$; $\frac{\sqrt{2}}{2}$ **29.** (0, 4); $\sqrt{11}$ **31.** $(x + 5)^2 + y^2 = 25$ **33.** (2, -1); $\sqrt{7}$ **35.** $(2a, 3a)$; $\sqrt{13a^2 - 4a}$ **37.** $(x - 3)^2 + (y - 4)^2 = 20$

Practice Exercises, pages 524–526 **1.** $x = \frac{1}{8}y^2$ **5.** $y = \frac{1}{8}x^2$ **11.** $x = -\frac{1}{12}y^2 + 1$ **13.** $v(0, 0)$; $x = 0$; $F(0, 3)$; *d*: $y = -3$ **17.** $v(0, 0)$; $x = 0$; $F(0, -5)$; *d*: $y = 5$ **19.** $v(2, 0)$; $x = 2$; $F\left(2, \frac{1}{16}\right)$; *d*: $y = -\frac{1}{16}$ **23.** $v(-2, -3)$; $x = -2$; $F\left(-2, -\frac{11}{4}\right)$; *d*: $y = -\frac{13}{4}$ **25.** $v(2, -3)$; $y = -3$; $F(3, -3)$; *d*: $x = 1$ **29.** $v(3, 2)$; $x = 3$; $F\left(3, \frac{9}{4}\right)$; *d*: $y = \frac{7}{4}$ **31.** $v(-5, 3)$; $y = 3$; $F\left(-\frac{19}{4}, 3\right)$; *d*: $x = -\frac{21}{4}$ **33.** $y = -\frac{1}{16}x^2$ **35.** $x = \frac{1}{8}y^2$ **39.** $y = \frac{1}{8}(x - 3)^2$ **43.** $v\left(-\frac{1}{4}, \frac{17}{8}\right)$; axis: $x = -\frac{1}{4}$; $F\left(-\frac{1}{4}, 2\right)$; *d*: $y = \frac{9}{4}$; max. **45.** $v\left(\frac{51}{4}, \frac{3}{4}\right)$; axis: $y = \frac{3}{4}$; $F\left(\frac{205}{16}, \frac{3}{4}\right)$; *d*: $y = \frac{203}{16}$; min. **49.** $y = (x - 2)^2 + 1$ **52.** $x = 34$ or $x = -2$ **53.** $(3, 5\sqrt{3} - 2)$ or $(3, -5\sqrt{3} - 2)$ **54.** no **55.** $y = \frac{1}{2}x + 2$ **56.** $x = 3$; $y = -8$ **57.** 0.00000325 **58.** $8 + i$ **59.** $\frac{-4 + 22i}{25}$

60. 8 **61.** $-2 < x < 2$ **62.** $c = 0; x = \pm 2\sqrt{2}$
63. $\frac{5x^2 + 7x}{x^2 + 3x + 2}$ **64.** $x + x + 2 + x + 4$

Practice Exercises, pages 532–533 **1.** $\frac{x^2}{16} + \frac{y^2}{4} = 1$ **3.** $\frac{x^2}{9} + y^2 = 1$ **5.** x-int.: ± 5; y-int.: ± 4; $F(\pm 3, 0)$; major axis: 10 **7.** x-int.: ± 3; y-int.: ± 6; $F(0, \pm 3\sqrt{3})$; major axis: 12 **9.** x-int.: ± 11; y-int.: ± 9; $F(\pm 2\sqrt{10}, 0)$; major axis: 22 **11.** x-int.: ± 5; y-int.: ± 2; $F(\pm\sqrt{21}, 0)$; major axis: 10 **13.** x-int.: ± 1; y-int.: ± 5; $F(0, \pm 2\sqrt{6})$; major axis: 10 **15.** x-int.: ± 3; y-int.: ± 5; $F(0, \pm 4)$; major axis: 10 **17.** x-int.: ± 10; y-int.: ± 5; $F(\pm\sqrt{3}, 0)$; major axis: 20 **19.** x-int.: $\pm\sqrt{6}$; y-int.: ± 2; $F(\pm\sqrt{2}, 0)$; major axis: $2\sqrt{6}$
21. $C(3, 4)$; $F(3, 4 \pm \sqrt{5})$ **23.** $C(-3, -2)$; $F(-3 \pm \sqrt{15}, -2)$ **25.** $C(-1, 2)$; $F(-1 \pm \sqrt{7}, 2)$
27. $C(-3, 5)$; $F(-3, 8)$; $F_2(-3, 2)$ **29.** $\frac{x^2}{16} + \frac{y^2}{12} = 1$ **31.** $\frac{x^2}{16} + \frac{y^2}{25} = 1$ **33.** $\frac{(x + 5)^2}{4} + \frac{(y - 7)^2}{25} = 1$ **35.** $\frac{(x - 3)^2}{18} + \frac{(y - 4)^2}{20} = 1$
37. $\frac{x^2}{4} + \frac{y^2}{3} = 1$ **39.** $\frac{x^2}{9} + \frac{y^2}{5} = 1$
41. $\frac{x^2}{8} + \frac{y^2}{9} = 1$ **43.** $\frac{x^2}{25} + \frac{y^2}{16} = 1$

Practice Exercises, pages 538–540 **1.** $\frac{x^2}{9} - \frac{y^2}{16} = 1$ **3.** $\frac{x^2}{16} - \frac{y^2}{9} = 1$ **5.** x-int.: ± 5; y-int.: none; $F(\sqrt{41}, 0)$; $y = \pm\frac{4}{5}x$ **7.** x-int.: ± 3; y-int.: none; $F(\pm 3\sqrt{5}, 0)$; $y = \pm 2x$ **9.** x-int.: ± 9; y-int.: none; $F(\pm\sqrt{130}, 0)$; $y = \pm\frac{7}{9}x$ **11.** x-int.: none; y = int.; ± 5; $F(0, \pm\sqrt{41})$; $y = \pm\frac{5}{4}x$
13. x-int.: none; y-int.: ± 2; $F(0, \pm\sqrt{13})$; $y = \pm\frac{2}{3}x$ **15.** x-int.: none; y-int.: ± 2; $F(0, \pm\sqrt{29})$; $y = \pm\frac{2}{5}x$

17.

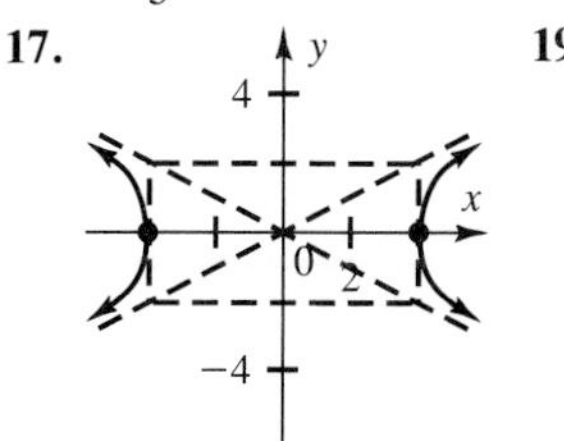

19.

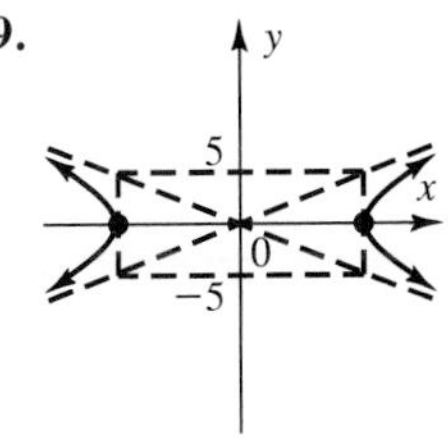

21.

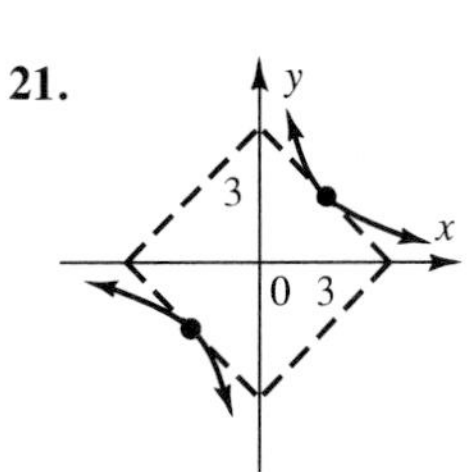

23.

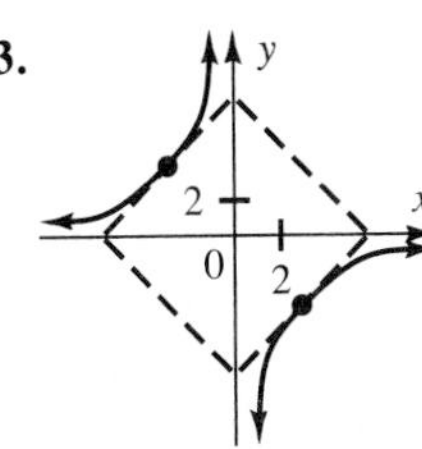

25.

27.

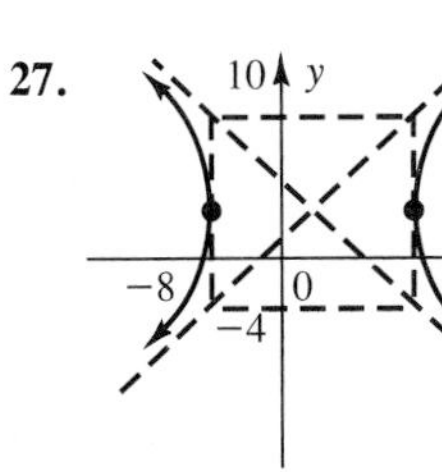

29.

31.

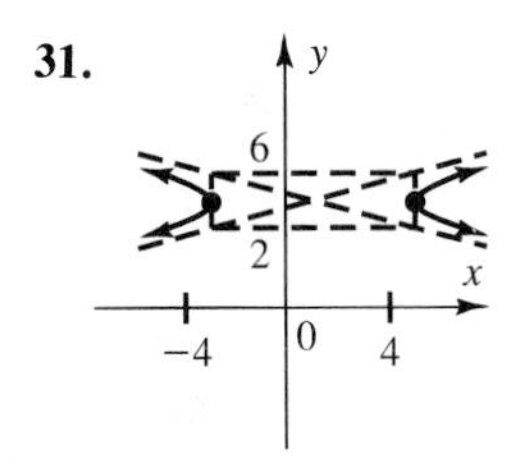

33.

35. ellipse **37.** parabola
39. none
41. $\frac{x^2}{9} - \frac{y^2}{7} = 1$
43. $\frac{y^2}{9} - \frac{x^2}{16} = 1$
45. $\frac{x^2}{16} - \frac{y^2}{4} = 1$

47. $\frac{(y - 1)^2}{25} - \frac{(x + 2)^2}{39} = 1$

Test Yourself, page 540 **1.** $(x - 4)^2 + (y - 3)^2 = 25$ **2.** $x^2 + y^2 = 16$ **3.** $c(-1, 3)$; 6
4. $c(-4, 8)$; $2\sqrt{17}$ **5.** $y = \frac{1}{4}(x - 5)^2 - 1$
6. $v\left(2, \frac{9}{2}\right)$; axis: $x = 2$; $F(2, 5)$; d: $y = 4$ **7.** x-int.: ± 9; y-int.: ± 5; $F(\pm 2\sqrt{14}, 0)$; major axis: 18
8. x-int.: ± 10; y-int.: ± 6; $F(\pm 8, 0)$; major axis: 20

9. **10.**

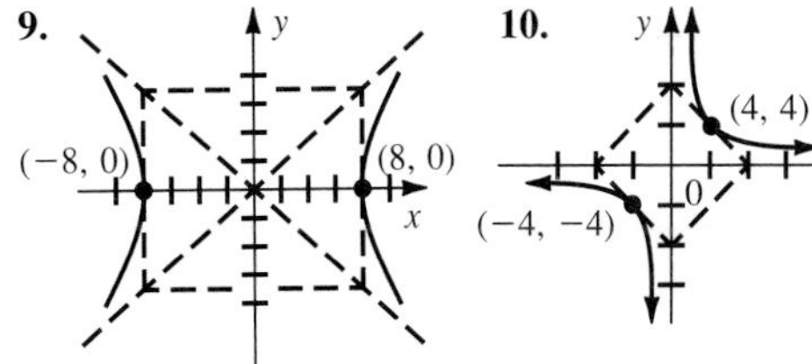

Answers

Practice Exercises, pages 544–546 **1.** circle; line; 2 **3.** hyperbola; line; 2 **5.** hyperbola; line; 0 **7.** hyperbola; line; 2 **9.** circle; line; $(-5, 0)$; $(0, 5)$ **11.** circle; line; $(0, 2)$; $(2, 0)$ **13.** circle; circle; no real solutions

15.

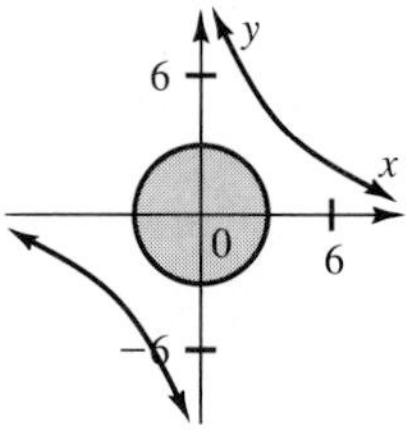

17. circle; circle; $(2, 0)$ **19.** ellipse; hyperbola; $(2, 0)$; $(-2, 0)$ **21.** circle; hyperbola; $(2.8, 2.8)$; $(-2.8, -2.8)$ **23.** circle; line; $(2.9, -0.9)$; $(-0.9, 2.9)$

25. **27.**

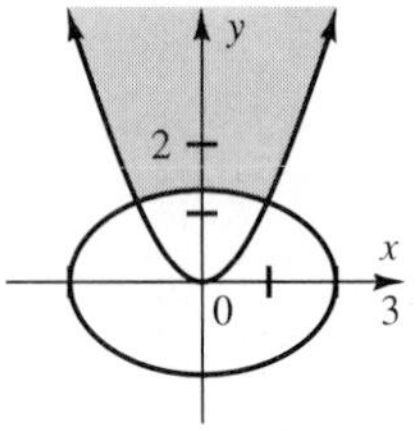

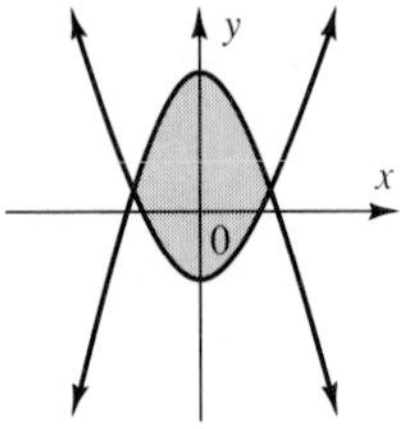

29. $(-1.2, 0.3)$; $(2.8, 2)$

31. **33.** yes; 2

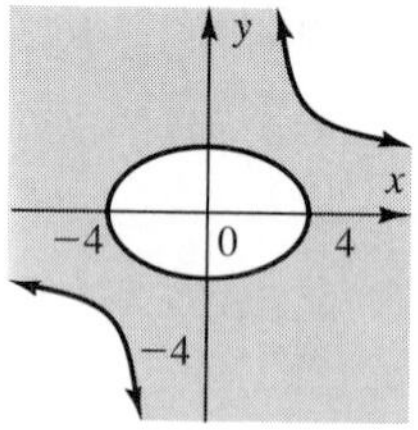

Practice Exercises, pages 550–551 **1.** $(-3, -4)$; $(4, 3)$ **3.** $\left(-\frac{15}{13}, \frac{24}{13}\right)$; $(-3, 0)$ **5.** $(3, 4)$; $(3, -4)$; $(-3, 4)$; $(-3, -4)$ **7.** $(5, 3)$; $(5, -3)$; $(-5, 3)$; $(-5, -3)$ **9.** $(\sqrt{11}, 2\sqrt{3})$; $(\sqrt{11}, -2\sqrt{3})$; $(-\sqrt{11}, 2\sqrt{3})$; $(-\sqrt{11}, -2\sqrt{3})$ **11.** $(8, 0)$; $(-8, 0)$ **13.** $(\sqrt{2}, \sqrt{2})$; $(-\sqrt{2}, -\sqrt{2})$ **15.** $(2, \sqrt{3})$; $(2, -\sqrt{3})$; $(-2, \sqrt{3})$; $(-2, -\sqrt{3})$ **17.** $(0, -1)$; $(-3, 2)$ **19.** $\left(\frac{\sqrt{95}}{5}, \frac{2\sqrt{15}}{5}\right)$; $\left(\frac{\sqrt{95}}{5}, -\frac{2\sqrt{15}}{5}\right)$; $\left(-\frac{\sqrt{95}}{5}, \frac{2\sqrt{15}}{5}\right)$; $\left(-\frac{\sqrt{95}}{5}, -\frac{2\sqrt{15}}{5}\right)$ **21.** no solutions **23.** $(2, -4)$ **25.** $(2, 4)$; $(-4,2)$ **27.** $\left(-2, -\frac{1}{2}\right)$; $(9, 5)$ **29.** $(2,5)$; $\left(\frac{6}{5}, \frac{21}{5}\right)$ **31.** $(4, 3)$ or $\left(-\frac{56}{13}, -\frac{33}{13}\right)$ **33.** $(3, 5)$; $(3, -5)$; $(-3, 5)$; $(-3, -5)$

Practice Exercises, pages 555–557 **1.** 4; $y = \frac{4}{x}$ **3.** neither **5.** inverse; 18 **7.** joint; π **9.** $\frac{4}{3}$ **11.** -2.9 **13.** $\frac{16}{9}$ **15.** 7 **17.** 30 **19.** 576 **21.** 2.25 **23.** $t = \frac{kq}{s}$ **25.** $r = \frac{kt^3}{v^3}$ **27.** $w = \frac{k}{d^2}$ **29.** $r = \frac{kl}{d^2}$ **31.** 105 **33.** 68 **35.** 32 **37.** 1 **39.** $\frac{40}{3}$ **41.** t is doubled **43.** 25 cm **45.** 54 mi/h **46.** Examine the sign of the x-term. **47.** $\begin{bmatrix} 3 & -1 & 1 \\ 12 & 0 & 1 \end{bmatrix}$ **48.** $(2xy^2 + 1)(4x^2y^4 - 2xy^2 + 1)$ **49.** $6x^5 - 12x^4 - 2x + 4$

Practice Exercises, pages 560–562 **1.** 1.1 m **3.** 6.1 kg **5.** 400 cm^3 **7.** 36 in. **9.** 375 J **11.** 1000 rpm **13.** 3 ft **15.** 49.5 cm

Test Yourself, page 562 **1.** 2 **2.** infinite **3.** $\left(\frac{2 + \sqrt{94}}{2}, \frac{-2 + \sqrt{94}}{2}\right)$; $\left(\frac{2 - \sqrt{94}}{2}, \frac{-2 - \sqrt{94}}{2}\right)$ **4.** $(\sqrt{2}, 0)$; $(-\sqrt{2}, 0)$ **5.** 8 **6.** 6.0 m

Summary and Review, pages 564–565
1. $(x - 2)^2 + (y - 3)^2 = 16$ **3.** $y = \frac{1}{12}x^2 - 1$ **5.** x-int.: ± 6; y-int.: ± 10; $F(0, \pm 8)$; major axis: 20 **7.** x-int.: ± 13; y-int.: none; $F(\pm\sqrt{194}, 0)$; $y = \pm\frac{5}{13}x$ **9.** circle **11.** $\left(\frac{4\sqrt{10}}{5}, \frac{3\sqrt{15}}{5}\right)$; $\left(\frac{4\sqrt{10}}{5}, -\frac{3\sqrt{15}}{5}\right)$; $\left(-\frac{4\sqrt{10}}{5}, \frac{3\sqrt{15}}{5}\right)$; $\left(-\frac{4\sqrt{10}}{5}, -\frac{3\sqrt{15}}{5}\right)$ **13.** 24

Mixed Review, page 568 **1.** 37.5% **3.** 9.5% **5.** 60% **7.** 0.125 **9.** 0.105 **11.** 0.555 **13.** 2.89×10^{-5} **15.** 5.632×10^{-5} **17.** 8×10^5 **19.** 27 **21.** 81 **23.** 9 **25.** 32

Chapter 12

Practice Exercises, pages 573–575 **1.** $\sqrt[6]{x}$ **3.** $\sqrt[7]{x^2}$ **5.** $\frac{1}{\sqrt[8]{y^3}}$ **7.** $\sqrt[2]{x^3}$ **9.** $y^{\frac{1}{2}}$ **11.** $x^{\frac{2}{5}}$ **13.** $(3t)^{\frac{1}{6}}$ **15.** $7x^{\frac{1}{3}}$ **17.** 6 **19.** 7 **21.** 16 **23.** 4 **25.** $\frac{1}{3}$ **27.** 8 **29.** $x^{\frac{1}{2}}$ **31.** $x^{\frac{1}{2}}$ **33.** $x^{\frac{1}{6}}y^{\frac{1}{4}}$ **35.** $\frac{4x^7}{9y^9}$ **37.** $\frac{y^3}{2x^5}$ **39.** $-\frac{2y^3}{x^2}$ **41.** 2.178 **43.** 4.287 **45.** 2.508 **47.** 0.027 **49.** -7

51. 64 **53.** 2,097,152 **55.** $\frac{1}{4}$ **57.** 64 **59.** $\frac{1}{36}$

61. $\frac{3}{49}$ **63.** $-\frac{1}{64}$ **65.** $x^{\frac{10}{9}}$ **67.** $x^{\frac{4}{3}}$ **69.** $\frac{1}{3x^{\frac{2}{3}}}$

71. $\frac{5x^{\frac{7}{12}}}{7^{\frac{2}{3}}}$ **73.** $\frac{1}{xy^{\frac{4}{3}}}$ **75.** $\frac{y^9}{x^3}$ **77.** -18 **79.** 8

81. 2 **83.** $x^{\frac{1}{4}}$ **85.** $\frac{1}{x^{\frac{1}{3}}}$ **87.** $x^{\frac{1}{2}}$ **89.** 635.9

91. 71.4 mg

Practice Exercises, pages 579–581 **1.** 9.739
3. 0.019

5.

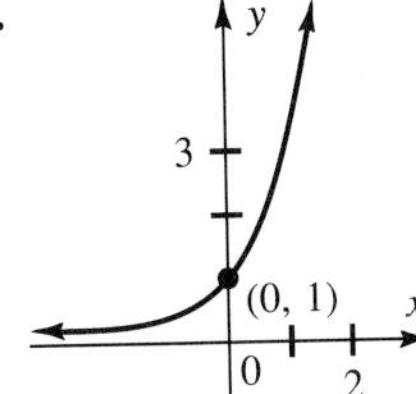

7.

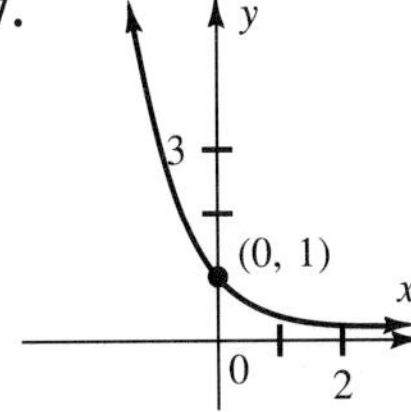

9. 3 **11.** 3 **13.** -2 **15.** -1 **17.** $\frac{5}{2}$ **19.** $\frac{7}{2}$

21. -5 **23.** -6

25.

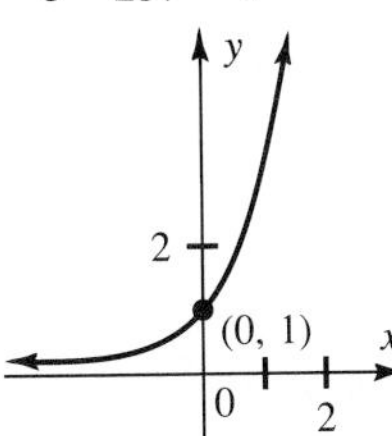

27.

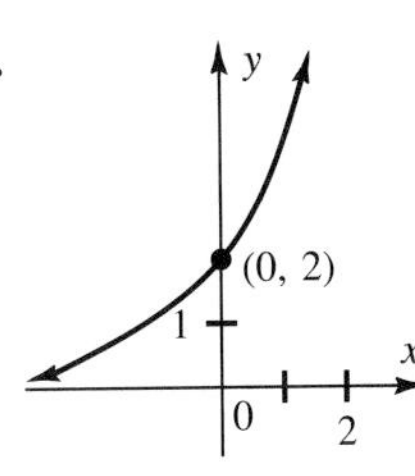

29.

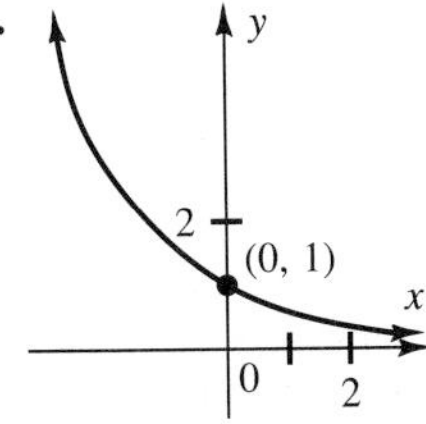

31.

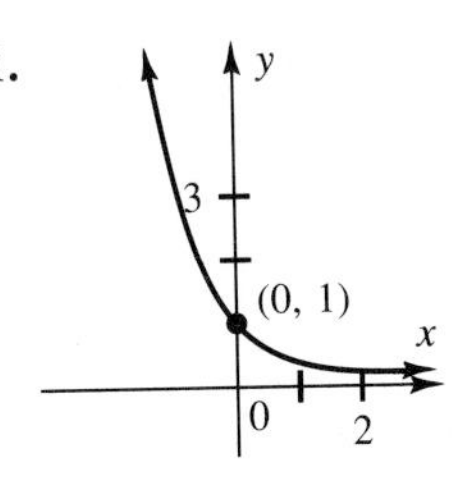

33. -1 **35.** 4 **37.** -7 **39.** $-\frac{3}{2}$

41.

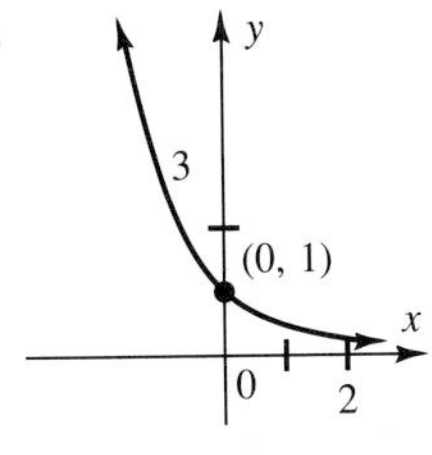

43. 2 **45.** -4 **47.** -4 **49.** 6 **51.** -5
53. 0.032

Practice Exercises, pages 585–586 **1.** $3^3 = 27$

5. $3^{-2} = \frac{1}{9}$ **7.** $b^n = m$ **9.** $\log_3 81 = 4$

13. $\log_{10} 0.1 = -1$ **17.** 3 **21.** -5 **23.** $\frac{3}{2}$

25.

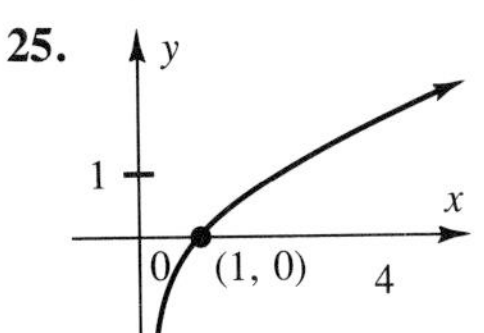

27. -3 **29.** -1
31. 2 **33.** -6 **35.** 0
37. 0.001

39.

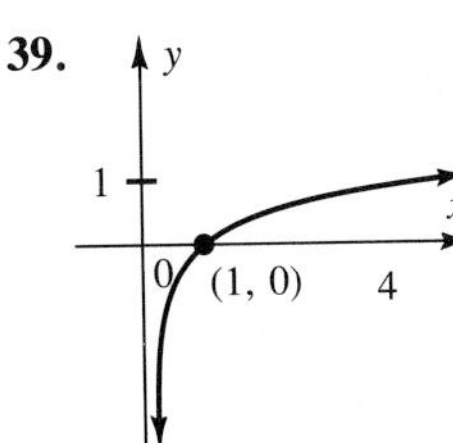

$D = \{x: x \in \text{pos. reals}\}$
$R = \{y: y \in \text{reals}\}$

41.

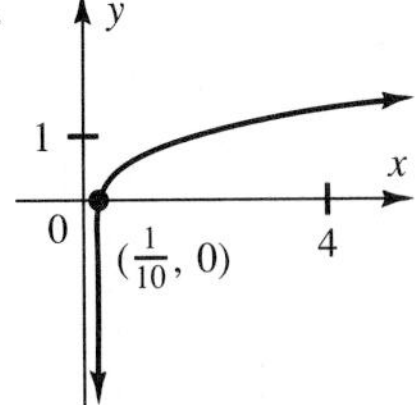

$D = \{x: x \in \text{pos. reals}\}$
$R = \{y: y \in \text{reals}\}$

43.

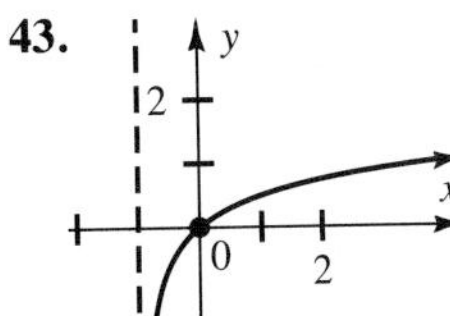

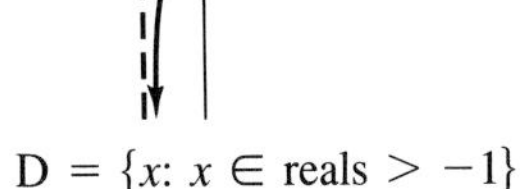

$D = \{x: x \in \text{reals} > -1\}$
$R = \{y: y \in \text{reals}\}$

45. 1 **47.** -6
49. 0 **51.** -1
53. 2 **55.** $\frac{1}{8}$ amp

57. $x = 0$; $(0, 0)$; min.

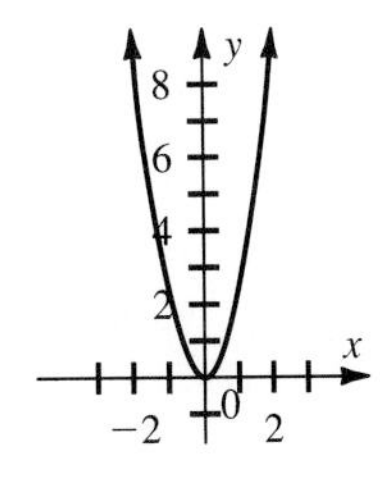

58. $x = 0$; $(0, -2)$; max.

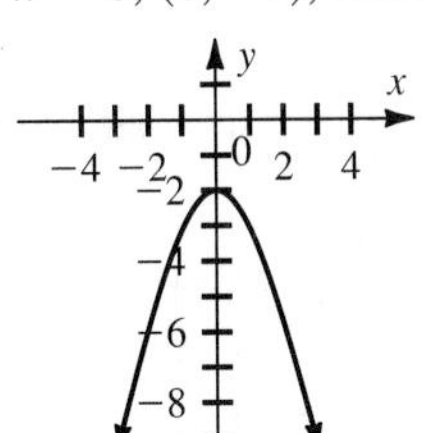

59. $x = 0$; $(0, 1)$; max.

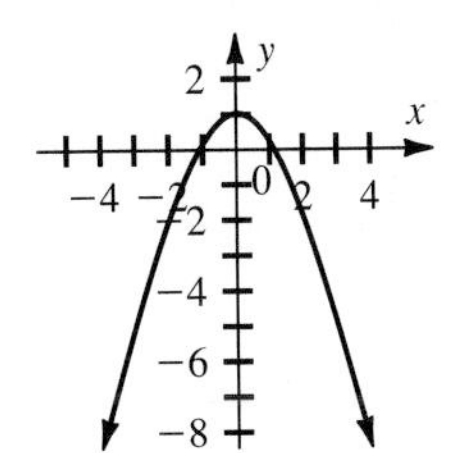

60. $x = 0$; $(0, 2)$; min.

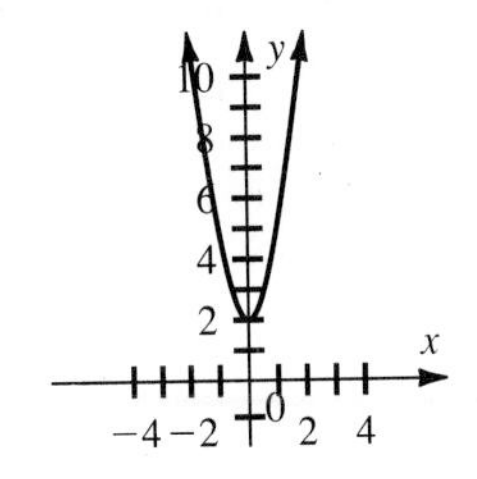

Answers

Practice Exercises, pages 590–592 **1.** $\log_a x + \log_a z$ **5.** $\log_a y - \log_a c$ **9.** $3 \log_6 x$ **13.** $3 \log_2 x$ **17.** $3 \log_3 y + \log_3 z$ **19.** $2 \log_3 2 + 2 \log_3 x$ **21.** $\frac{3}{2} \log_{10} m - \frac{1}{2} \log_{10} n$ **23.** $2 \log_a x + 3 \log_a y + 3 \log_a z$ **25.** $\log_2 yt^4$ **29.** $\log_5 \frac{x}{\sqrt[5]{y}}$ **31.** 1 **33.** 2 **35.** −2 **37.** 8 **39.** $\log_b 2 + 2 \log_b x + 3 \log_b y$ **41.** $4 \log_b 2 + 4 \log_b x + 4 \log_b y$ **45.** $4 \log_b x + 4 \log_b y - \log_b 2$ **47.** $\frac{3}{5} \log_b x$ **49.** $\frac{4}{3} \log_b x + \frac{2}{3} \log_b y$ **53.** $\log_b 7 + 3 \log_b x + 2 \log_b y - \frac{1}{2} \log_b z$ **55.** $\log_2 \frac{xy}{z}$ **57.** $\log_2 \sqrt{x^3}$ **59.** 3 **61.** $\frac{11}{5}$ **63.** ± 7 **65.** $\frac{1}{2} \log_b x + \frac{2}{3} \log_b y - \frac{2}{5} \log_b z$ **67.** $\log_3 x + \log_3 y - \frac{3}{2} \log_3 z$ **69.** $\log_a \frac{\sqrt[3]{x^2}\sqrt[4]{y^3}}{z^5}$ **71.** not possible **73.** 4 **77.** 3 decibels **79.** 7

Test Yourself, page 592 **1.** 8 **2.** $\frac{1}{256}$ **3.** $\frac{1}{x^{\frac{1}{2}}}$ **4.** $\frac{1}{x^{\frac{6}{5}}}$ **5.**

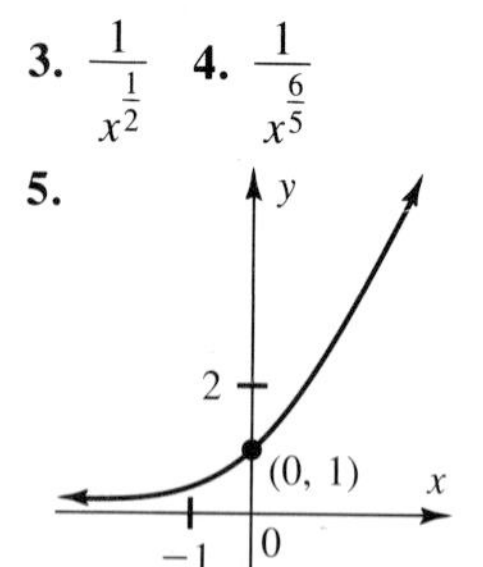

6. 1 **7.** $\log_3 729 = 6$ **8.** 7 **9.** $\log_2 x + 2 \log_2 y$ **10.** −12

Practice Exercises, pages 596–597 **1.** 0.7723 **3.** −0.8697 **5.** 5.5452 **7.** −9.2829 **9.** 8439 **11.** 0.5976 **13.** 8659 **15.** 1.2553 **17.** 0.7737 **19.** 1.7754 **21.** 1.029×10^{13} **23.** 7.558×10^{-15} **25.** 4.106×10^{-15} **27.** 1.3776 **29.** 3.4776 **31.** −31.7525 **33.** enter: 1 INV ln **35.** 1.2465 **37.** 1.9429 **39.** −3.4966 **41.** 13.4058 **43.** 0.00023 **44.** $\frac{-5 \pm \sqrt{29}}{2}$ **45.** $-6 \pm \sqrt{42}$ **46.** $\frac{1}{4}, \frac{1}{2}$ **47.** $\frac{11 \pm \sqrt{201}}{10}$ **48.** Negative discriminate indicates complex conjugate solutions. **49.** $\frac{3}{8}$ **50.** $2, -2, i\sqrt{7}, -i\sqrt{7}$

Practice Exercises, pages 600–601 **1.** 2.32 **3.** 2.12 **5.** 2.73 **7.** 1.02 **9.** 0.975 **11.** 0.201 **13.** 5.68 **15.** 1.98 **17.** 1.50 **19.** 2.56 **21.** 2.67 **23.** 2.57 **25.** 7.42 **27.** 3.16 **29.** 1.05 **31.** 3.38 **33.** 0.799 **35.** 1.71 **37.** 3.90 **39.** 4.05 **41.** −12.2 **43.** −2.46 **45.** −4.31 **47.** 0.827 **49.** 5.82 h **51.** 310 m

Practice Exercises, pages 605–606 **1.** 7.8 yrs **3.** 9.4 yrs **5.** 10,100 yrs **7.** 225 days **9.** 20.2 decibels **11.** 19.5 decibels **13.** 6.8 h **15.** 1396.7 v **17.** 8.4 yr **19.** 5 times as intense

Test yourself, page 607 **1.** 3.7973 **2.** −4.6021 **3.** 0.0770 **4.** −5.4798 **5.** 18.50 **6.** 9.324×10^{-17} **7.** 7.447×10^{11} **8.** 2.41 **9.** 2.80 **10.** −0.793 **11.** 1.77 **12.** 1.74 **13.** 4.22

Summary and Review, pages 610–611 **1.** −2 **3.** $\frac{1}{27}$ **5.** $x^{\frac{4}{5}}$ **7.** $\frac{1}{x^{\frac{5}{14}}}$ **9.**

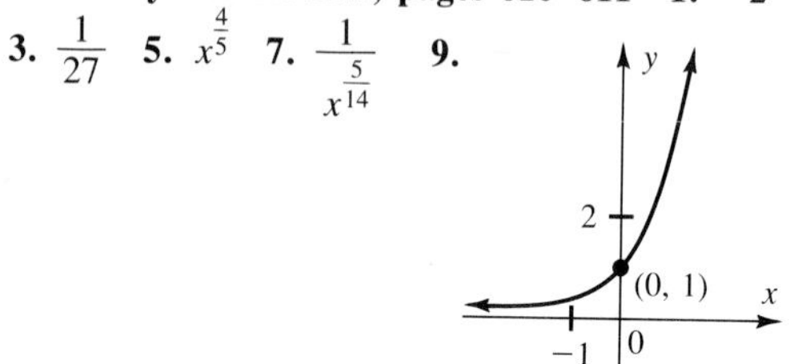

11. 5 **13.** 1 **15.** $2^4 = 16$ **17.** $\log_5 125 = 3$ **19.** 3 **21.** −4 **23.**

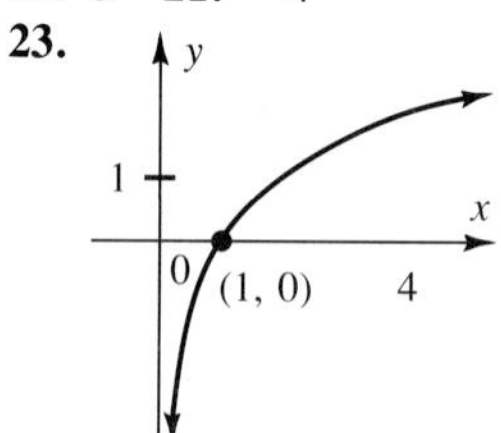

25. $2 \log_b x + \log_b y$ **27.** $\frac{1}{4} \log_b x + \frac{1}{4} \log_b y$ **29.** 5.8932 **31.** 1.674×10^8 **33.** 2.809×10^{14} **35.** 2.35 **37.** 3.17 **39.** 4.11

Cumulative Review (Ch. 1–12), pages 614–616 **1.** no solution **3.** $\{(x, y): 3x + 4y = -4\}$ **5.** 2 **7.** $\frac{1}{4^6}$ **9.** $5r(r - 5s)(r^2 + 5rs + 25s^2)$ **11.** $3a^2(a^2 + 1)$ **13.** $(p - r)(q + t)$ **15.** 1.35×10^{-2} **17.** 2.30×10^{-5} **19.** $\frac{3}{4}$ **21.** $\frac{1}{3}$ **23.** $\begin{bmatrix} -1 & -7 & 8 \\ -8 & 4 & 16 \\ -3 & 14 & -4 \end{bmatrix}$ **25.** $\begin{bmatrix} 35 & 4 & 2 \\ -20 & 13 & -12 \\ -25 & -18 & 14 \end{bmatrix}$

27. undefined **29.** $\begin{bmatrix} 9 & -6 \\ 0 & -12 \\ -15 & 3 \end{bmatrix}$ **31.** $3x^2\sqrt[3]{x}$

33. $\frac{\sqrt{2x}}{2x}$ **35.** $a\sqrt[3]{a}$ **37.** $-4 \le q \le 3$

39. $\{x: x \in \text{real numbers}\}$ **41.** $\frac{x(x + 3)}{2(x - 1)}$

43. $\frac{y + x}{x - y}$ **45.** circle, hyperbola; $(3, -4)$, $(-3, 4)$, $(4, -3)$, $(-4, 3)$ **47.** hyperbola, parabola; $(\sqrt{2}, 1)$, $(-\sqrt{2}, 1)$, $(\sqrt{5}, -2)$, $(-\sqrt{5}, -2)$

49.

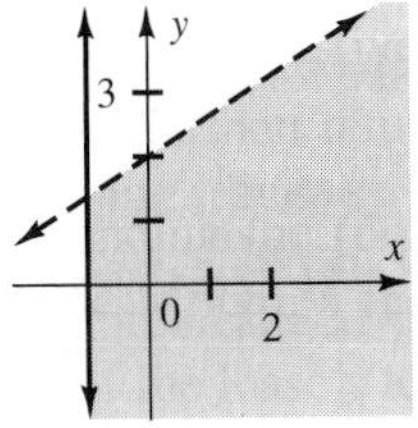

51. 0, 16 **53.** -3 **55.** x-int., 0; y-int., 0; $x = -1$; $y = 1$ **57.** y-int., -1; $x = \frac{1}{2}$; $y = 0$

59. $\frac{4 \pm 2i\sqrt{3}}{3}$ **61.** $2 \pm \sqrt{3}$ **63.** 2.5 mi/h; 0.5 mi/h **65.** 13 **67.** $\frac{3i - \sqrt{5}}{14}$

69. $(-3, 0, -1)$ **71.** 7 **73.** $\frac{15}{4}$ **75.** $V(0, 2)$

77. $V(-4, 6)$ **79.** $3x + y = 0$ **81.** $x^2 - 4x - 44 = 0$ **83.** $y = \frac{x + 4}{3}$; function **85.** $y = \pm\sqrt{\frac{-x - 4}{2}}$; not a function **87.** $y = \frac{A - 2x}{2}$

89. $y = \frac{3}{2a + b}$; $a \neq -\frac{b}{2}$ **91.** $2\sqrt{5}$; $M(-3\sqrt{6}, 2\sqrt{5})$ **93.** $-\frac{1}{2}$ **95.** 12 min

Chapter 13 Sequences and Series

Practice Exercises, pages 620–621 **1.** $-3, -6, -9$ **3.** 10,000, 100,000, 1,000,000 **5.** 31, 50, 81 **7.** 14, 11, 21 **9.** 7, 4, 8 **11.** 301 **13.** 5050 **15.** 2, 6, 12, 20, 30, 42, 56, 72, 90, 110 **17.** n^2

Practice Exercises, pages 625–627 **1.** $a_1 = 2$; $a_{n+1} = a_n + 4$; 18, 22, 26 **3.** $a_1 = 1$; $a_{n+1} = a_n - 1$; $-3, -4, -5$ **5.** $a_1 = -4$; $a_{n+1} = a_n + 12$; 44, 56, 68 **7.** $a_1 = 1$; $a_{n+1} = a_n + \frac{1}{2}$; 3, $3\frac{1}{2}$, 4 **9.** $a_n = 2n$; 20 **11.** $a_n = n + 1$; 12 **13.** $a_n = 2n + 1$; 29 **15.** 4, 5, 6 **17.** 1, 8, 27 **19.** $0, -1, -2$ **21.** $a_1 = 0$; $a_{n+1} = a_n + 0.5$; 2, 2.5, 3 **23.** $a_1 = 1$; $a_{n+1} = \frac{1}{2}a_n$ or $a_n = \frac{1}{2^{n-1}}$; $\frac{1}{16}, \frac{1}{32}, \frac{1}{64}$ **25.** $a_1 = 5$; $a_{n+1} = -a_n$; $5, -5, 5$ **27.** $-1, 2, 5, 26$ **29.** 4, 9, 16, 25 **31.** $\frac{1}{2}, \frac{4}{3}, \frac{9}{4}, \frac{16}{5}$ **33.** $a_1 = 20$; $a_{n+1} = a_n + 4$; 20, 24, 28 **35.** $a_n = 2^{n-1} \cdot 10$ **37.** $a_n = 4(n - 1) + 1$ **39.** $a_{n-2} = a_{n-3} + 2$ **41.** $a_1 = 4$; $a_{n+1} = a_n + 0.5$; \$5.50 **43.** $a_1 = 25$; $a_{n+1} = a_n + 10$

Practice Exercises, pages 632–633

1.

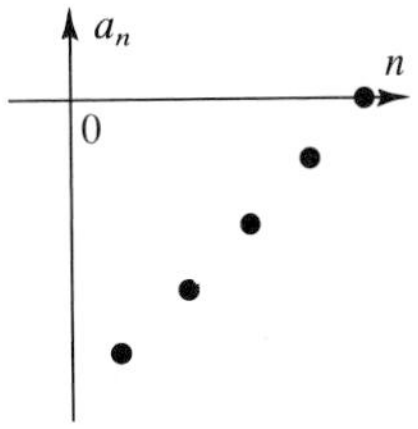

3. 32 **5.** -20 **7.** 26 **9.** $a_1 = 0$; $d = 3$ **11.** $a_1 = -9$; $d = 7$ **13.** $a_1 = -12$; $d = 3$ **15.** 23 **17.** 54, 58 **19.** 6 **21.** $31x$ **23.** 9 **25.** 2 **27.** 22 **29.** 54 **31.** 21st **33.** $9k + 32$ **35.** $a_1 = \frac{7r - 2s}{5}$; $d = \frac{s - r}{5}$ **37.** 4 **39.** (C5 *(C1 + E5))/2 **41.** 89 **42.** $3m^3 - 7m^2 + 26m - 78 + \frac{232}{m + 3}$ **43.** $4x^2 - 5x + 11 - \frac{25}{x + 2}$ **44.** yes **45.** no **46.** $-4, 3$ **47.** -1.3

Practice Exercises, pages 638–639

1.

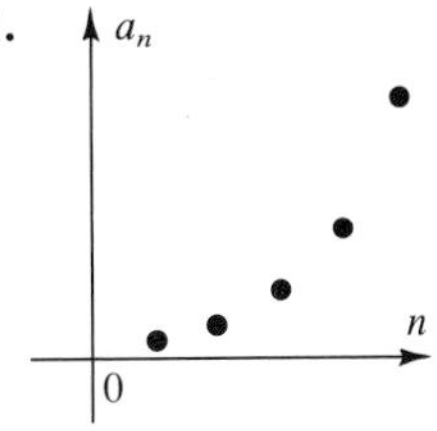

3. 56 **5.** $\frac{1}{9}$ **7.** 15,309 **9.** 131,072 **11.** $-\frac{1}{1024}$ **13.** 4, 16, 64 or -4, 16, -64 **15.** 6, 12, 24, 48 **17.** 36 **19.** 0.00048 **21.** 10^{35} **23.** 10 **25.** $3072x$ **27.** 7 **29.** 3 **31.** 18th **35.** \$12,800

Practice Exercises, pages 644–645 **1.** $4 + 8 + 12 = 24$ **3.** $5 + 8 + 11 + 14 + 17 + 20 = 75$ **5.** 180 **7.** 2300 **9.** -270 **11.** 108 **13.** -392 **15.** -36 **17.** 180 **19.** 15 **21.** $\sum_{n=1}^{6} n$ **23.** $\sum_{n=1}^{8} (n + 7)$ **25.** $\sum_{n=1}^{11} (3n - 2)$ **27.** $a_{20} = -73$ **29.** $a_{40} = 300$ **31.** -15 **33.** 34 **35.** $10x + 45y$ **37.** 1 **41.** 496 **43.** 675 cm

Test Yourself, page 645 **1.** 0, 3, 8 **2.** 0, 3, 6 **3.** 73 **4.** 20, 25 **5.** −0.1875 **6.** 6, 12, 24 or −6, −12, −24 **7.** 474 **8.** 330

Practice Exercises, pages 649–650 **1.** 255 **3.** 1456 **5.** 7,777,777 **7.** −19,682 **9.** $\frac{127}{16}$ **11.** 2186 **13.** $\frac{2047}{1024}$ **15.** 4095 **17.** $\frac{43}{64}$ **19.** arith.; 49 **21.** arith.; 57 **23.** geom.; 1093 **25.** geom.; $x^3 + x^2 + x + 1$ **27.** 4 **29.** 2 **31.** $\frac{x - 256x^9}{1 - 2x}$ **33.** $\frac{y - yx^{15}}{1 - x^3}$ **35.** $\frac{xy^{12} - x^{13}}{y^8 - x^2y^6}$ **37.** 682 sq in. **39.** $94,629.05 **41.** $\frac{x + 2}{x - 2}$ **42.** $\frac{\sqrt{10y}}{2y^3}$ **43.** $2x\sqrt[3]{12xy}$ **44.** $\frac{2x\sqrt{6xy}}{y}$ **45.** $9\sqrt{3} + 9\sqrt[3]{2x} - \sqrt[3]{x}$ **46.** $\frac{6 - 3\sqrt{2}}{2}$ **47.** $-4\sqrt[3]{2x}$ **48.** $2i\sqrt{6y}$ **49.** $-16\sqrt{2}$ **50.** i **51.** $\frac{9}{64}$ **52.** max.; 5 **53.** $\frac{1}{4}$ **54.** $5x^2 + 6x - 8 = 0$ **55.** $\sqrt{5}, -\sqrt{5}, i\sqrt{2}, -i\sqrt{2}$

Practice Exercises, pages 655–656 **1.** $\frac{5}{4}$ **3.** does not exist **5.** $\frac{10}{9}$ **7.** does not exist **9.** $\frac{1}{6}$ **11.** $\frac{9}{4}$ **13.** $\frac{4}{3}$ **15.** $\frac{2}{9}$ **17.** $\frac{19}{33}$ **19.** $\frac{25}{333}$ **21.** $\frac{5}{4}$ **23.** $\frac{3}{4}$ **25.** $\frac{1}{1.2}$ **27.** $\frac{7}{8}$ **29.** 10 **31.** $\frac{83}{99}$ **33.** $\frac{7}{40}$ **35.** $\frac{1}{2}$ **37.** $\frac{(x + 1)^4}{x}$ **39.** 12 **41.** 5000 **43.** 1200 mm

Practice Exercises, pages 660–661 **1.** 11 mi **3.** 45 times **5.** 972 bacteria **7.** $13.25 **9.** 124 quarters = $31.00 **11.** 12.5 mi/gal **13.** 21 rev/min; 320 rev **15.** 175 ft **17.** no; no

Practice Exercises, pages 665–666 **1.** 1; 1, 1; 1, 2, 1; 1, 3, 3, 1; 1, 4, 6, 4, 1; 1, 5, 10, 10, 5, 1; 1, 6, 15, 20, 15, 6, 1; 1, 7, 21, 35, 35, 21, 7, 1; 1, 8, 29, 56, 70, 56, 29, 8, 1; 1, 9, 37, 85, 126, 126, 85, 37, 9, 1; 1, 10, 46, 122, 211, 252, 211, 122, 46, 10, 1 **3.** $x^4 + 4x^3y + 6x^2y^2 + 4xy^3 + y^4$ **7.** $x^6 + 6x^5 + 15x^4 + 20x^3 + 15x^2 + 6x + 1$ **9.** $x^5 + 10x^4 + 40x^3 + 80x^2 + 80x + 32$ **11.** $16x^4 + 96x^3y + 216x^2y^2 + 216xy^3 + 81y^4$ **15.** $32x^5 + 80x^4y + 80x^3y^2 + 40x^2y^3 + 10xy^4 + y^5$ **17.** $x^6 + 18x^5y + 135x^4y^2 + 540x^3y^3 + 1215x^2y^4 + 1458xy^5 + 729y^6$ **19.** $x^{10} - 10x^9y + 45x^8y^2 - 120x^7y^3 + 210x^6y^4 - 252x^5y^5 + 210x^4y^6 - 120x^3y^7 + 45x^2y^8 - 10xy^9 + y^{10}$ **21.** $x^8 + 12x^6 + 54x^4 + 108x^2 + 81$ **25.** $\frac{1}{x^5} + \frac{5}{x^4y} + \frac{10}{x^3y^2} + \frac{10}{x^2y^3} + \frac{5}{xy^4} + \frac{1}{y^5}$ **27.** $\frac{1}{x^3} - \frac{3}{2x^2} + \frac{3}{4x^2} - \frac{1}{8}$ **29.** $x^2 + 4x^{\frac{3}{2}}y^{\frac{1}{2}} + 6xy + 4x^{\frac{1}{2}}y^{\frac{3}{2}} + y^2$ **31.** $-12 + 316i$ **33.** $-7 - 24i$ **35.** $2x^4 + 24x^2 + 8$ **37.** −1194 **39.** 45 **41.** 21

Practice Exercises, pages 670–671 **1.** $x^5 + 5x^4y + 10x^3y^2 + 10x^2y^3 + 5xy^4 + y^5$ **5.** $x^5 - 10x^4y + 40x^3y^2 - 80x^2y^3 + 80xy^4 - 32y^5$ **9.** $x^7 - 7x^6 + 21x^5 - 35x^4 + 35x^3 - 21x^2 + 7x - 1$ **13.** $594x^{10}$ **17.** x^{11} **21.** $262{,}144x^{18}$; $7{,}077{,}888x^{17}$; $90{,}243{,}072x^{16}$ **23.** x^{24}; $12x^{22}y$; $66x^{20}y^2$ **27.** 32,768; $245{,}760xy$; $860{,}160x^2y^2$ **29.** $29{,}568x^{10}y^6$ **33.** y^{28} **35.** 1.2 **37.** 1.4 **39.** 0.7 **41.** x^4, $8x^3$, $24x^2$

Test Yourself, page 671 **1.** −728 **2.** 765 **3.** $\frac{8}{7}$ **4.** does not exist **5.** does not exist **6.** 8 **7.** 71 ft **8.** 144 ft **9.** $x^5 - 5x^4y + 10x^3y^2 - 10x^2y^3 + 5xy^4 - y^5$ **10.** $16x^4 + 96x^3y + 216x^2y^2 + 216xy^3 + 81y^4$ **11.** $243x^5 + 405x^4y + 270x^3y^2 + 90x^2y^3 + 15xy^4 + y^5$ **12.** $-120x^7$

Summary and Review, pages 674–675 **1.** $a_1 = 2$; $a_{n+1} = a_n + 5$; 22, 27, 32 **3.** 22 **5.** −60 **7.** 162 **9.** $\frac{1}{8}$ **11.** −40 **13.** 119 **15.** 44 **17.** 4 **19.** $-\frac{130}{27}$ **21.** 8 **23.** $x^4 + 8x^3 + 24x^2 + 32x + 16$ **25.** $495x^4y^8$

Mixed Review, page 678 **1.** $\overline{AB}$, $\overline{AC}$, $\overline{AD}$, $\overline{AE}$, $\overline{AF}$, $\overline{BC}$, $\overline{BD}$, $\overline{BE}$, $\overline{BF}$, $\overline{CD}$, $\overline{CE}$, $\overline{CF}$, $\overline{DE}$, $\overline{DF}$, $\overline{EF}$; 15 **3.** $\frac{5}{12}$ **5.** $\frac{2}{3}$ **7.** 16 kg **9.** 546.4 lb **11.** 35,000 **13.** April

Chapter 14

Practice Exercises, pages 684–686 **1.** 24 **3.** 4096 **5.** 720 **7.** 362,880 **9.** 40,320 **11.** 125 **13.** 5040 **15.** 6720 **17.** 9,000,000 **19.** 221,184 **21.** 136,080 **23.** 1260 **25.** 40,320 **27.** 120 **29.** 80,640 **33.** 30,240 **35.** 720

Practice Exercises, pages 689–691 **1.** 35 **3.** 220 **5.** 22,100 **7.** 364 **9.** 28 **11.** 36 **13.** 4845 **15.** 28; 45 **17.** 24 **19.** 15 **21.** 0

27. 42 **29.** 462 **31.** 1365 **33.** 120 **35.** 90 **36.** 270,725 **37.** 715 **38.** union; $\{-10, -9, -8, -7, -6, -5, -4, -3, -2, -1, 1, 3, 5, 7, 9\}$ **39.** inter.; $\{0, 2, 4, 6, 8, 10\}$ **40.** inter.; $\{-6, 0, 6\}$ **41.** $y = 2x + 4$; $D = \{1, 2, 3, 4\}$; $R = \{6, 8, 10, 12\}$ **42.** $y = x^2 - 3x + 4$; $D = \{2, 3, 4, 5\}$; $R = \{2, 4, 8, 14\}$ **43.** $y = 7$ **44.** 14 **45.** $\left(1\frac{1}{2}, 3\sqrt{2}\right)$

Practice Exercises, pages 695–697 **1.** $S = \{1, 2, 3, 4, 5, 6, 7, 8, 9, 10\}$ **3.** $\{1, 2, 3, 4\}$ **5.** $\frac{2}{5}$; 2:3 **9.** $\frac{3}{10}$; 3:7 **11.** $\frac{1}{720}$ **15.** $\frac{1}{36}$; $\frac{1}{9}$ **17.** $\frac{99}{54,145}$ **21.** $\frac{105}{506}$ **25.** $\frac{11}{36}$ **27.** $\frac{1}{2}$ **31.** $\frac{1}{2}$

Practice Exercises, pages 703–705 **1.** $\frac{4}{5}$ **3.** $\frac{2}{5}$ **5.** $\frac{1}{12}$ **7.** $\frac{1}{4}$ **11.** $\frac{1}{9}$ **13.** $\frac{4}{9}$ **17.** $\frac{8}{15}$ **19.** $\frac{3}{10}$ **23.** $\frac{3}{4}$ **25.** dep. **27.** $\frac{6}{7}$ **31.** 0.05 **37.** 0.7

Test Yourself, page 705 **1.** 40 **2.** 360 **3.** 3,628,800 **4.** 36 **5.** 204,490 **6.** $\frac{1}{2}$ **7.** 1:1 **8.** $\frac{5}{36}$ **9.** 0.32 **10.** $\frac{3}{5}$ **11.** 0

Practice Exercises, pages 709–711 **3.** 7 **7.** 9 **9.** mode **11.** 5 **13.** Team 1: mean; Team 2: mode or median; Team 3: median **15.** Company *A*: mean or median; Company *B*: mean **17.** 9.4 **21.** 26.3 **23.** 33

Practice Exercises, pages 715–716 **1.** 4 **3.** 26 **5.** 73.6; 71.4; 8.5 **7.** 4.1; 5.0; 2.2 **9.** 14.3; 9.6; 3.1 **11.** 55; 0.8; 0.9 **13.** 7.3; 17.4; 4.2 **15.** 7.5; 26.4; 5.1 **17.** 2655.2 ft; 227,014.9 sq ft; 476.5 ft **19.** 0 **23.** 11.9; 2.9; 1.7 **25.** 151.6; 230.5; 15.2 **26.** $9x^6z^2$ **27.** 5 **28.** $\frac{9x^2}{y^4z}$ **29.** $y = \pm\frac{\sqrt{2x-2}}{2}$; no **30.** $8x^2 + 32x + 32$ **31.** They are symmetric to each other; $y = x$ is the line of symmetry **32.** -500 **33.** $x = 2$; $y = -6$

Practice Exercises, pages 720–721 **1.** 0.4987 **3.** 0.9861 **5.** 0.9615 **7.** 0.6006 **9.** 0.0026 **11.** 0.1587 **13.** 0.0228 **15.** 0.4192 **17.** 59 min **19.** George **21.** 47.72% **23.** 0.0228

Practice Exercises, pages 725–727 **3.** H_0: mean quantity of juice = 64.0 oz; H_a: mean quantity of juice $\neq$ 64.0 oz **5.** $63.57 < x < 64.43$ **7.** H_0: mean number of sheets = 500; H_a: mean number of sheets $\neq$ 500 **9.** $495.35 < x < 504.65$ **11.** no **13.** yes **15.** H_0: mean height = 73.0 cm; H_a: mean height $\neq$ 73.0 cm **17.** $72.20 < x < 73.80$ **19.** 49

Test Yourself, page 727 **1.** 10; 9; 8 **2.** 8.9; 2.98 **3.** 0.4641 **4.** 0.6554 **5.** no

Summary and Review, pages 730–731 **1.** 20; 362,880; 5040; 1320 **3.** 1260 **5.** 56; 480 **7.** 0.20; 0.17 **9.** 4; 1.41; 1.19 **11.** no

Cumulative Review (Ch. 1–14), page 734 **1.** 0.001875 **3.** yes **5.** peanuts: \$1.50; cashews: \$3.50 **7.** 1140 **9.** $\log_6 \frac{1}{36} = -2$ **11.** $\log_b c = e$ **13.** $\frac{255}{256}$ **15.** $C(-4, 0)$; $6\sqrt{2}$ **17.** $2.2y$ **19.** $16x^4 - 32x^3y + 24x^2y^2 - 8xy^3 + y^4$

Chapter 15

Practice Exercises, pages 741–742 **1.** II **3.** IV **5.** II **7.** pos. x-axis **9.** II **11.** III **13.** $\frac{5\pi}{6}$ **15.** $\frac{\pi}{4}$ **17.** 30° **21.** 23°25′12″ **23.** 48°21′0″ **25.** 14.6° **27.** 35.8° **29.** −90° **33.** −540° **35.** $-\frac{9\pi}{4}$ **39.** 7π **41.** $-\frac{10\pi}{3}$ **45.** 171.9° **47.** 45.8° **49.** $-\frac{3\pi}{2}$ **53.** $-\frac{11\pi}{3}$ **57.** 0.75

Practice Exercises, pages 746–747 Functions are given thus: sin; cos; tan; csc; sec; cot **1.** $\frac{3}{5}$; $\frac{4}{5}$; $\frac{3}{4}$; $\frac{5}{3}$; $\frac{5}{4}$; $\frac{4}{3}$ **5.** $\frac{4}{5}$; $\frac{4}{3}$; $\frac{5}{4}$; $\frac{5}{3}$; $\frac{3}{4}$ **9.** pos. **11.** neg. **15.** neg. **21.** $\frac{\sqrt{2}}{2}$; $\frac{\sqrt{2}}{2}$; 1; $\sqrt{2}$; $\sqrt{2}$; 1 **25.** $\frac{2\sqrt{2}}{3}$; $2\sqrt{2}$; $\frac{3\sqrt{2}}{4}$; 3; $\frac{\sqrt{2}}{4}$ **29.** pos. **31.** neg. **35.** 1 **37.** $\frac{16}{25}$ **39.** any quadrant **41.** $\frac{7}{25}$ **43.** $\sqrt{m}$ **44.** $\sqrt[4]{z^3}$ **45.** $\frac{1}{\sqrt[3]{x^2}}$ or $\frac{\sqrt[3]{x}}{x}$ **46.** $3^{\frac{1}{3}}$ **47.** $w^{\frac{1}{3}}$ **48.** $m^{\frac{2}{z}}$ **49.** $\frac{2}{5}$ **50.** 8 **51.** 1 **52.** 4 **53.** −1 **54.** 0 **55.** $\log_3 2 + \log_3 x$ **56.** $\log_m 3 - \log_m c$ **57.** $3 \log_2 m$

Practice Exercises, pages 751–752 The six functions are given in this order: sin; cos; tan; csc; sec; cot **1.** $\frac{\sqrt{2}}{2}$; $-\frac{\sqrt{2}}{2}$; −1; $\sqrt{2}$; $-\sqrt{2}$; −1

Answers

5. $-\frac{\sqrt{2}}{2}$; $\frac{\sqrt{2}}{2}$; -1; $-\sqrt{2}$; $\sqrt{2}$; -1 **9.** 0; 1; 0; undef.; 1; undef. **13.** $\frac{1}{2}$; $\frac{\sqrt{3}}{2}$; $\frac{\sqrt{3}}{3}$; 2; $\frac{2\sqrt{3}}{3}$; $\sqrt{3}$ **17.** 1; 0; undef.; 1; undef.; 0 **19.** $\frac{\sqrt{3}}{2}$; $\frac{1}{2}$; $\sqrt{3}$; $\frac{2\sqrt{3}}{3}$; 2; $\frac{\sqrt{3}}{3}$ **23.** $-\frac{1}{2}$; $-\frac{\sqrt{3}}{2}$; $\frac{\sqrt{3}}{3}$; -2; $-\frac{2\sqrt{3}}{3}$; $\sqrt{3}$ **25.** 210°; 330° **29.** 45°; 315° **31.** 30°; 150°; 390°; 510° **35.** 120°; 480° **39.** $10\sqrt{3}$ ft

Practice Exercises, pages 756–757 **1.** 0.5842 **5.** 0.9537 **9.** 0.5774 **13.** 0.3805 **17.** 1.4020 **21.** 41.2° **25.** 17.8° **27.** −0.1988 **31.** −0.6106 **35.** −0.5544 **39.** 41.5° **45.** 133°0′; 227°0′ **49.** 198°30′; 341°30′ **53.** 93°50′; 266°10′ **55.** 0.6884

Practice Exercises, pages 761–762 **1.** 1; 360° **5.** 3; 2π **9.** $\frac{1}{2}$; 2π **13.** 1; 4π **17.** 2; $\frac{2\pi}{3}$ **21.** $\frac{2}{3}$; π **23.** 2; 720° **27.** 2; $\frac{8\pi}{3}$ **29.** 0.02; 250; 0.004

Test Yourself, page 762 **1.** II **2.** III **3.** IV **4.** III **5.** II **6.** III **7.** $\frac{5\pi}{6}$ **8.** $\frac{7\pi}{6}$ **9.** $\frac{13\pi}{8}$ **10.** 120° **11.** 225° **12.** 240° **13.** $\sin S = -\frac{8}{10}$; $\tan S = -\frac{8}{6}$; $\csc S = -\frac{10}{8}$; $\sec S = \frac{10}{6}$; $\cot S = -\frac{6}{8}$ **14.** $\frac{1}{2}$ **15.** $-\sqrt{3}$ **16.** 0 **17.** $\frac{\sqrt{2}}{2}$ **18.** 0.7112 **19.** −1.4193 **20.** −0.6316 **21.** −1.7730 **22.** 3; 2π **23.** 1; $\frac{2\pi}{3}$

Practice Exercises, pages 765–766 **1.** π **5.** 2π **9.** 2π **13.** odd **17.** $\frac{\pi}{2}$ **21.** 2π **25.** 2π **29.** 2π **33.** 1.771 **35.** 1.699 **36.** $a_{n+1} = a_n + 7$ **37.** $a_n = 3_n - 10$ **38.** 32 **39.** 83 **40.** $\frac{x+2}{2(x-3)}$ **41.** $4x^6 + 16x^4 - 12x^3 + 16x^2 - 24x + 9$ **42.** 3 and 1 **43.** $\frac{2(2x-y)}{3x-4y}$

Practice Exercises, pages 770–771 **1.** $\angle T = 58°$; $s = 23$; $t = 36$ **3.** $\angle S = 30°$; $r = 40$; $t = 35$ **9.** $q = 15$; $\angle P = 53°$; $\angle Q = 53°$; $\angle R = 74°$ **13.** 16 ft **15.** $\angle B = 46°40'$; $a = 17.2$; $b = 18.4$ **19.** $\angle B = 59°30'$; $c = 35.4$; $a = 18.0$ **23.** 35°30′ **25.** $\angle Q = 49°50'$; $PR = 47.5$; $QR = 40.1$; $PQ = 62.2$ **27.** 208 ft **29.** 3.5 mi

Practice Exercises, pages 776–777 **1.** 120 **3.** 45 **5.** $\angle F = 72°$; $e = 20$; $f = 24$ **7.** $\angle D = 80°$; $d = 22$; $e = 21$ **11.** $\angle D = 25°$; $\angle E = 123°$; $e = 32$ **13.** 84.9 **15.** $\angle E = 138°40'$; $d = 15.9$; $f = 10.8$ **19.** no solution **25.** 32 cm

Practice Exercises, pages 780–781 **1.** 15 **3.** 83° **5.** $\angle E = 57°$; $\angle F = 69°$; $d = 17$ **7.** $\angle D = 100°$, $\angle E = 48°$, and $f = 8.6$ **11.** $\angle D = 51°40'$; $\angle F = 63°40'$; $e = 17.4$ **17.** $\angle D = 65°50'$; $\angle F = 65°50'$; $e = 20.3$ **19.** 3.0 **21.** 18 cm

Practice Exercises, pages 784–785 **1.** 60° **3.** 77° **5.** 51° **7.** 6300 mm^2 **9.** 560 m **11.** 2900 cm^2 **13.** 54 **15.** 1st base: 64 ft; 2nd base: 67 ft; 3rd base: 64 ft **17.** 979 ft

Test Yourself, page 786 **1.** 2π **2.** $\angle T = 66°$; $s = 120$; $t = 110$ **3.** 220 **4.** $\angle B = 91°$; $\angle C = 41°$; $b = 22$ **5.** 10 **6.** 250 ft

Summary and Review, pages 788–789 **1.** $\frac{5\pi}{6}$ **3.** $\frac{3\pi}{4}$ **5.** 300° **7.** $\sin K = \frac{4}{5}$; $\cos K = -\frac{3}{5}$; $\csc K = \frac{5}{4}$; $\sec K = -\frac{5}{3}$; $\cot K = -\frac{3}{4}$ **9.** $-\frac{\sqrt{2}}{2}$; $\frac{\sqrt{2}}{2}$; -1 **11.** $-\frac{\sqrt{2}}{2}$; $-\frac{\sqrt{2}}{2}$; 1 **13.** 0.9088 **15.** 1; π **17.** 2; 2π **19.** 2π **21.** $\angle B = 55°$; $a = 14$; $b = 20$ **23.** $\angle B = 84°$; $\angle C = 40°$; $c = 8$ **25.** $c = 9.8$; $\angle A = 53°$; $\angle B = 75°$ **27.** 70 m

Mixed Review, page 792 **1.** $-\frac{1}{9}$ **3.** 6 **5.** $\frac{4}{33}$ **7.** $\frac{23}{35}$ **9.** $4\sqrt{3}$ **11.** $7 + 4\sqrt{3}$ **13.** $\frac{\sqrt{3}}{3}$ **15.** $\frac{4\sqrt{5}}{9}$ **17.** $2\frac{1}{4}$ lb

Chapter 16

Practice Exercises, page 797 **3.** $\pm\sqrt{1 - \cos^2 A}$ **7.** $\pm\sqrt{\csc^2 A - 1}$ **9.** $\sin A$ **11.** $\cot A$ **15.** $\csc A$ **17.** $\sin^2 A$ **19.** $\sec B$ **23.** $1 - \sin A$

Practice Exercises, page 800 **3.** $1 - 2\sin^2 C = 2\cos^2 C - 1$, $1 - 2(1 - \cos^2 C) = 2\cos^2 C - 1$, $1 - 2 + 2\cos^2 C = 2\cos^2 C - 1$, $2\cos^2 C - 1 = 2\cos^2 C - 1$ **5.** $\frac{\sin B + \tan B}{1 + \cos B} = \tan B$,

Answers

$\dfrac{\sin B + \dfrac{\sin B}{\cos B}}{1 + \cos B} = \tan B$, $\dfrac{\sin B \cdot \cos B + \sin B}{\cos B(1 + \cos B)} =$ $\tan B$, $\dfrac{\sin B}{\cos B} = \tan B$, $\tan B = \tan B$

28. $[3(1) + 4] + [3(2) + 4] + [3(3) + 4] + [3(4) + 4]$ **29.** 24,564 **30.** $0.3 + 0.03 + 0.003 + \ldots$

Practice Exercises, pages 804–805 **1.** $\frac{\sqrt{6} + \sqrt{2}}{4}$ **5.** $\frac{\sqrt{6} + \sqrt{2}}{4}$ **9.** $\frac{\sqrt{2} + \sqrt{6}}{4}$ **11.** $\frac{\sqrt{3}}{2}$ **13.** $2 + \sqrt{3}$ **17.** $\frac{33}{65}$ **21.** $\frac{16}{65}$ **23.** $\frac{\tan A + \tan B}{1 - \tan A \tan B}$ **27.** $2 + \sqrt{3}$ **31.** $\sin 3\theta$ **35.** $\tan 11C$ **41.** 75.2 mi

Practice Exercises, pages 809–810 **1.** 1 **3.** undef. **7.** $\frac{120}{169}$ **11.** $-\frac{24}{25}$ **13.** $\frac{\sqrt{2 + \sqrt{3}}}{2}$ **17.** $\frac{\sqrt{10}}{10}$ **21.** $\frac{1}{3}$ **23.** $\sqrt{2} + 1$ **25.** $\frac{240}{289}$ **29.** $\frac{4\sqrt{17}}{17}$ **33.** $\frac{24}{7}$ **43.** $\pm\sqrt{\dfrac{1 - \sqrt{\dfrac{1 + \cos A}{2}}}{2}}$

Practice Exercises, pages 814–815 **1.** $\pm\frac{\pi}{3} + 2\pi k$ **5.** 60° **7.** 0° **13.** −30° **15.** $38° + 360k°$ or $142° + 360k°$ **19.** 21° **23.** $-\frac{\pi}{6}$ **25.** $\frac{3}{5}$ **27.** $\frac{3}{4}$ **31.** $\frac{12}{5}$ **35.** $\frac{5}{4}$ **39.** $-\frac{17}{8}$ **43.** $\frac{\sqrt{2} - \sqrt{6}}{4}$ **45.** 0 **47.** $\frac{63}{65}$ **51.** θ increases as v increases

Test Yourself, page 815 **5.** $\frac{16}{65}$ **6.** $\frac{220}{21}$ **7.** $-\frac{24}{25}$ **8.** $\frac{\sqrt{5}}{5}$ **9.** undefined **10.** 0

Practice Exercises, pages 817–818 **1.** 30°, 150° **3.** 63.43°, 116.57°, 296.57°, 243.43° **5.** 45°, 225° **9.** $0, \frac{\pi}{3}, \pi, \frac{5\pi}{3}$ **13.** $\frac{\pi}{3}, \frac{5\pi}{3}$ **15.** $\frac{3\pi}{2}$ **21.** $0, \pi$ **23.** π **25.** $\frac{\pi}{4}, \frac{5\pi}{4}$ **27.** $0, \pi, \frac{5\pi}{6}, \frac{\pi}{6}$ **29.** $0, \frac{\pi}{3}, \frac{5\pi}{3}$ **31.** $\frac{3\pi}{2}, \frac{\pi}{6}, \frac{5\pi}{6}$ **37.** $\frac{\pi}{2}, 0$ **39.** $\frac{\pi}{4}$ **41.** 604,800 **42.** $\frac{3}{4}$ **43.** $x^{12} + 16x^9 + 96x^6 + 256x^3 + 256$ **44.** $486{,}486x^9$

Practice Exercises, page 822 **9.** (2, 2) **11.** (−12, 14) **19.** speed: 221.2 mi/h; course: 120° **21.** course: 183.5° or 192.5° **23.** 100°

Practice Exercises, pages 825–826 **1.** 104.5° **3.** 310 lb, 45°30′ **5.** 226.8° or 313.2° **7.** 376.2 cm **9.** 274 mi/h; 147° **11.** 2.9 cm

Practice Exercises, pages 830–831 **9.** $P(3\sqrt{2}, 45°)$ **11.** $P(\sqrt{6}, 225°)$ **15.** $P(4, 90°)$ **17.** $P\left(\frac{3\sqrt{3}}{2}, \frac{3}{2}\right)$ **19.** $P(2, -2\sqrt{3})$ **25.** $\sqrt{13}(\cos 123.7° + i \sin 123.7°)$ **27.** $6(\cos 180° + i \sin 180°)$ **29.** $\sqrt{3} + i$ **31.** $-\frac{5\sqrt{2}}{2} - \frac{5\sqrt{2}}{2}i$ **33.** $\sqrt{3}(\cos 54.7° + i \sin 54.7°)$ **37.** $0.995 + 0.105i$ **41.** $2.298 + 1.928i$ **45.** $\frac{\sqrt{6}}{2} - \frac{\sqrt{2}}{2}i$ **47.** $\frac{3}{2} + \frac{3\sqrt{3}}{2}i$ **51.** 3 **53.** $P_2\left(\frac{5}{2}, -\frac{5\sqrt{3}}{2}\right)$ **55.** $P_2(-4.596, -3.857)$, $P_1(-4, 4\sqrt{3})$

Test Yourself, pages 831 **1.** 90°, 270° **2.** $\frac{\pi}{4}, \frac{5\pi}{4}$ **3.** $x = 3$; $y = 2$ **4.** $x = 6$; $y = 7$ **5.** mag. 30 lb; angle: 36°50′ **6.** 0° **7.** $8(\cos 60° + i \sin 60°)$ **8.** $\sqrt{29}(\cos 111.8° + i \sin 111.8°)$ **9.** $(2, 2\sqrt{3})$ **10.** $2\sqrt{3} + 2i$

Summary and Review, pages 834–835 **1.** $\tan^2 B + \cos^2 B$ **5.** $\frac{\sqrt{2} + \sqrt{6}}{4}$ **7.** $\frac{36}{85}$ **9.** $\frac{\sqrt{2 + \sqrt{2}}}{2}$ **11.** $-\frac{120}{169}$ **13.** 5 **15.** $\frac{13}{5}$ **17.** 0°, 120°, 180°, 240° **19.** $x = -6$; $y = 8$ **21.** magnitude: 114.97; 48-lb force: 34.3°; 80-lb force: 19.7° **23.** $1 + i\sqrt{3}$

Cumulative Review (Ch. 1–16), pages 838–840 **1.** $\frac{1}{\frac{1}{y^{12}}}$ **3.** $\frac{y^3}{125x^2}$ **5.** $110 - 28\sqrt{6}$ **7.** $i\sqrt{3}$ **9.** $\frac{3}{13}$ **11.** 4π **13.** π **15.** $\frac{4}{5}$ **17.** $\frac{7}{9}$ **19.** $\frac{32}{9}$ **21.** $18\sqrt{2}$ **23.** $2x + y = 3$ **25.** $\pm\frac{2i\sqrt{6}}{3}$ **27.** $5, \frac{12}{7}$ **29.** $\frac{12y + 16}{-3y + 10}$ **31.** $\sec B$ **33.** 15°, 75°, 135°, 195°, 255°, 315° **35.** 90°, 270° **37.** −4 **39.** no sum **41.** $\frac{1}{16} - \frac{1}{2}b + \frac{3}{2}b^2 - 2b^3 + b^4$ **43.** $4\sqrt{2}$ **45.** 12 **47.** 3003 **49.** 1.3 **51.** rat. **53.** complex **55.** 0 **57.** 16 **59.** 16 × 16 **61.** $F(\pm 2\sqrt{5}, 0)$ **63.** $x = -\frac{1}{4}y^2$ **65.** −1, ±0.7 **67.** $\log_{10} \frac{x}{5}$ **69.** $\log_5 \sqrt{a}$ **71.** $6\sqrt{3}$ sq. units **73.** $2(\cos 60° + i \sin 60°)$ **75.** $2(\cos 135° + i \sin 135°)$ **77.** −5.563

GLOSSARY

The explanations given in this Glossary include definitions and brief descriptions of the key terms used in this book.

abscissa (p. 94) The x-coordinate.

absolute value (p. 8) The absolute value of a real number a, written $|a|$, is its distance from zero on the number line.

absolute value function (p. 110) $f(x) = |x|$

amplitude (p. 759) The amplitude of a periodic function is one-half the difference between the maximum and minimum values of the function.

arithmetic means (p. 630) The terms between any two given terms of an arithmetic sequence.

arithmetic sequence (p. 628) A sequence in which the difference between successive terms is a constant.

arithmetic series (p. 640) The indicated sum of an arithmetic sequence.

asymptote (p. 503) A line that a graph approaches closer and closer but does not intersect.

augmented matrix (p. 222) A matrix formed by writing the constants of the equations in a column and then attaching this column to the matrix of coefficients.

average (p. 82) The average or arithmetic mean of a set of values is the sum of the values divided by the number of values.

base (p. 11) In the expression r^n, r is the base of the exponent n.

bi-modal (p. 708) A bi-modal set of data has two modes.

center of a hyperbola (p. 535) The midpoint of the segment joining the foci.

center of an ellipse (p. 528) The midpoint of the segment joining the foci.

change-of-base formula (p. 599) $\log_b N = \dfrac{\log_a N}{\log_a b}$

circle (p. 515) The set of all points in a plane that are the same distance from a given point, called the center.

closed half-plane (p. 177) A half-plane that includes the boundary.

combination (p. 687) A combination of n elements of a set taken r at a time, denoted ${}_nC_r$, is any r-element subset of the given set.

combined variation (p. 554) A relation that involves both direct and inverse variation.

common difference (p. 628) The constant difference between successive terms in an arithmetic sequence.

common logarithm (p. 593) A base-10 logarithm.

common ratio (p. 634) The constant ratio between the consecutive terms of a geometric sequence.

complement (p. 701) If E is an event in a sample space S, then any outcome in S that is not in E is said to be in E-complement ($\overline{E}$).

complex conjugate theorem (p. 484) If a complex number $a + bi$ is a solution of a polynomial equation with real coefficients, then its conjugate, $a - bi$, is also a solution of the equation.

complex number (p. 392) A number that can be written in the form $a + bi$, where a and b are real numbers and $i = \sqrt{-1}$.

Glossary

complex rational expression (p. 327) An expression with a rational expression in its numerator and/or denominator.

components of a vector (p. 820) Two vectors whose sum is a given vector.

composite function (p. 110) Given two functions, $f(x)$ and $g(x)$, the composite function, $(g \circ f)(x)$ or $g[f(x)]$ is the operation of applying $f(x)$ and then $g(x)$ to the values of x.

compound sentence (p. 57) Two sentences joined by the word *and* or the word *or*.

conic section (p. 514) A curve formed by the intersection of a plane and a cone.

conjugate axis of a hyperbola (p. 535) One of two axes to which the hyperbola is symmetric. It does not contain the foci.

conjugates (p. 266, 368) Two binomials whose first terms are equal and whose last terms are opposites. Example: $a + b$ and $a - b$

conjunction (p. 57) Two sentences joined by the word *and*.

consistent system (p. 155) A system of equations that has at least one solution.

constraint (p. 182) An inequality representing a restriction placed on available resources in a linear programming problem.

converge (p. 652) An infinite geometric series converges if the sequence of partial sums S_n approaches a limit as n increases without bound.

coterminal angles (p. 737) Two angles in standard position whose terminal sides coincide.

Cramer's rule (p. 236) A method of solving systems of equations using determinants.

degree of a monomial (p. 256) The sum of the exponents of the variables.

degree of a polynomial (p. 260) The degree of the term of highest degree.

dependent events (p. 702) Two events are dependent if the outcome of one affects the outcome of the other.

dependent system (p. 155) A system of equations that has an infinite number of solutions.

depressed polynomial (p. 477) A polynomial of degree $n - 1$ obtained by dividing a polynomial of degree n by a binomial of the form $x - a$.

determinant (p. 233) A real number associated with a square matrix.

deviation from the mean (p. 713) The value of $x_i - \overline{x}$, where x_i is a number in a set of data and $\overline{x}$ is the mean of the data.

direct variation (p. 144) A function defined by an equation that can be written in the form $y = kx$, $k \neq 0$.

directrix of a parabola (p. 520) The given line that is the same distance as the focus from each point on the parabola.

discriminant (p. 437) The discriminant of a quadratic equation $ax^2 + bx + c = 0$ is $b^2 - 4ac$.

disjoint sets (p. 699) Sets that have no elements in common.

disjunction (p. 58) Two sentences joined by the word *or*.

distance formula (p. 379) The distance d between $A(x_1, y_1)$ and $B(x_2, y_2)$ is given by $d = \sqrt{(x_2 - x_1)^2 + (y_2 - y_1)^2}$.

diverge (p. 653) A series diverges if the sequence of partial sums increases or decreases without limit as the number of terms gets very large.

domain (p. 98) The set of all the first coordinates of the ordered pairs in a relation.

ellipse (p. 527) The set of all points P in a plane such that the sum of the distances from P to two fixed points is a constant.

empty set ($\emptyset$) (p. 25) A set that contains no elements.

equivalent vectors (p. 819) Vectors with the same length and direction.

even function (p. 765) A function for which $f(-x) = f(x)$ for all x in the domain.

event (p. 692) A subset of a sample space.

explicit formula (p. 624) A rule that expresses the nth term of a sequence as a function of n, where n is a positive integer.

exponent (p. 11) In the expression r^n, n is the exponent.

exponential function (p. 576) The exponential function $f(x) = b^x$, with base b, is defined by the equation $y = b^x$ if x and b are real numbers with $b > 0$ and $b \neq 1$.

extremes (p. 146) The first and fourth terms, y_1 and x_2, of the proportion $y_1 : x_1 = y_2 : x_2$.

factorial notation (p. 667) n factorial, written $n!$, is equal to $n(n - 1)(n - 2) \cdots 3 \cdot 2 \cdot 1$. Also, $0! = 1$.

feasible region (p. 182) The graph of a linear system of constraints.

foci of a hyperbola (p. 534) Two fixed points such that the difference between the distances from those points to any point on the hyperbola is a constant.

foci of an ellipse (p. 527) Two fixed points such that the sum of the distances from those points to any point on the ellipse is constant.

focus of a parabola (p. 520) A fixed point that is the same distance as the directrix from any point on the parabola.

frequency distribution (p. 707) A table showing a set of values and the frequency of each value.

function (p. 99) A relation in which each element in the domain is paired with one and only one element in the range.

fundamental counting principle (p. 681) If one event can occur in m different ways and a second event can occur in n different ways, then together the events can occur in $m \cdot n$ different ways, assuming that the second event is not influenced by the first event.

fundamental theorem of algebra (p. 483) Every polynomial function with complex coefficients has at least one zero in the set of complex numbers.

general term of a sequence (p. 623) The nth term.

geometric mean (p. 637) *See* mean proportional.

geometric means (p. 636) The terms between any two given terms of a geometric sequence.

geometric sequence (p. 634) A sequence in which the ratio of consecutive terms is a constant.

geometric series (p. 646) The indicated sum of the terms of a geometric sequence.

half-plane (p. 177) The part of a plane that lies on one side of a line in the plane.

histogram (p. 707) A vertical bar graph with no spaces between the bars.

hyperbola (p. 534) The set of all points P in a plane such that the difference of the distances from P to two fixed points is a constant.

imaginary number (p. 388, 392) A complex number $a + bi$ is imaginary if $b \neq 0$.

inconsistent system (p. 155) A system of equations that has no solutions.

independent events (p. 702) A and B are independent events if and only if $P(A \cap B) = P(A) \cdot P(B)$.

independent system (p. 155) A system of equations that has exactly one solution.

index (p. 357) In the expression $\sqrt[n]{k}$, the index is n.

infinite sequence (p. 623) A sequence that does not have a last term.

intersection of sets (p. 58) The intersection of two sets A and B, written $A \cap B$, is the set of elements common to both A and B.

inverse function (p. 114) If the inverse of a function f is also a function, it is written f^{-1} and is called the inverse function of f.

Glossary

inverse of a relation (p. 114) The relation obtained by interchanging the first and second coordinates in every pair of the original relation.

inverse variation (p. 552) A function defined by an equation that can be written in the form $xy = k$.

irrational number (p. 3) A number that neither repeats nor terminates.

leading coefficient of a polynomial (p. 481) The coefficient of the term of highest degree.

linear equation (p. 105) A linear, or first-degree, equation is an equation whose graph is a line.

linear programming (p. 182) The minimizing or maximizing of an objective function under given linear constraints.

literal equation (p. 46) An equation that contains more than one letter or variable.

logarithmic function (p. 583) For all positive real numbers x and b, $b \neq 1$, the inverse of the exponential function $y = b^x$ is the logarithmic function $y = \log_b x$.

magnitude (p. 820) The length of a vector.

major axis of an ellipse (p. 528) The longer of two axes to which the ellipse is symmetric. It contains the foci.

matrix (p. 197) A rectangular array of numbers written within brackets.

matrix equation (p. 200) An equation in which the variable represents a matrix.

mean (p. 707) The mean, $\overline{x}$, of a set of data is the arithmetic average of the data.

mean proportional (p. 637) A single geometric mean between two terms of a geometric sequence.

means (p. 146) The second and third terms, x_1 and y_2, of the proportion $y_1 : x_1 = y_2 : x_2$.

median (p. 708) The median of a set of data is the middle value if there is an odd number of values and they are listed in ascending or descending order. It is the mean of the two middle values if there is an even number of values.

midpoint formula (p. 379) The coordinates of the midpoint of the line segment with endpoints (x_1, y_1) and (x_2, y_2) are $\left(\frac{x_1 + x_2}{2}, \frac{y_1 + y_2}{2}\right)$.

minor (p. 234) The minor of an element in an $n \times n$ determinant is the $(n - 1) \times (n - 1)$ determinant found by eliminating the row and column that contain that element.

minor axis of an ellipse (p. 528) The shorter of the two axes to which the ellipse is symmetric.

mode (p. 708) The mode of a set of data is the value with the greatest frequency.

modulus of a complex number (p. 829) The absolute value of the number.

mutually exclusive events (p. 701) Events that have no elements in common.

natural logarithm (p. 595) A logarithm to the base e.

normal distribution (p. 718) A distribution whose graph is a normal, or bell-shaped, curve.

null hypothesis (p. 723) A hypothesis stating that a population mean is equal to some given standard.

objective function (p. 182) The representation of a quantity to be maximized or minimized in a linear program.

odd function (p. 765) A function for which $f(-x) = -f(x)$ for all x in the domain.

odds (p. 694) The odds that an event will occur are expressed as the ratio of the number of successes to the number of failures.

open half-plane (p. 178) A half-plane that does not include the boundary.

opposite vectors (p. 819) Two vectors that have the same length but opposite directions.

ordinate (p. 94) The y-coordinate.

parabola (pp. 412, 520) The graph of a quadratic function, which is the set of all points in a plane that are the same distance from a given line and a fixed point not on the line.

parallelogram rule (p. 820) Using two given vectors as adjacent sides, a parallelogram is completed. The diagonal is the sum, or resultant vector.

partial sum (p. 640) The nth partial sum of a series is the sum of the first n terms of the series.

Pascal's triangle (p. 663) A pattern for finding the coefficients of the terms of a binomial expansion.

periodic function (p. 759) A function that repeats over and over again.

permutation (p. 681) A permutation of some or all of the elements of a set is any arrangement of the elements in a definite order.

point-slope form of a linear equation (p. 127) $y - y_1 = m(x - x_1)$, where m is the slope and (x_1, y_1) are the coordinates of the given point on the line.

polar axis (p. 827) In the polar coordinate system, the polar axis is a ray with the pole as its endpoint.

polar coordinates (p. 827) The polar coordinates of a point P in the polar coordinate system is an ordered pair (r, θ), where r is the distance from the pole to P and θ is the measure of the angle formed by the polar axis and ray OP.

polar form (p. 829) A nonzero complex number is expressed in polar form when it is in the form $r(\cos \theta + i \sin \theta)$.

pole (p. 827) In the polar coordinate system, the pole is the endpoint of the polar axis.

principal values (p. 812) The values of a function that occur within a restricted domain.

probability (p. 693) The probability that an event will occur is the ratio of the number of outcomes in the event to the number of outcomes in the sample space.

pure imaginary number (pp. 389, 392) A complex number $a + bi$ in which $a = 0$ and $b \neq 0$.

Pythagorean identities (p. 795) Fundamental trigonometric identities that are based on the Pythagorean theorem.

quadrantal angle (p. 737) An angle whose terminal side falls on an axis.

quadratic formula (p. 427) The solutions of the quadratic equation $ax^2 + bx + c = 0$, $a \neq 0$, are given by the formula

$$x = \frac{-b \pm \sqrt{b^2 - 4ac}}{2a}.$$

radian (p. 738) The measure of a central angle that intercepts an arc with the same length as a radius of the circle.

radical sign (p. 357) In the expression $\sqrt[n]{k}$, the radical sign is $\sqrt{}$.

radicand (p. 357) In the expression $\sqrt[n]{k}$, the radicand is k.

radius of a circle (p. 515) The distance from the center of the circle to any point on the circle.

random sample (p. 722) A sample in which each member of the set of data has an equal chance of being included and each sample of the same size is equally likely to be selected.

range of a relation (p. 98) The set of all second coordinates of the ordered pairs.

range of a set of data (p. 712) The difference between the highest value and the lowest value.

rational expression (p. 313) An expression that can be written in the form $\frac{P}{Q}$, where P and Q are polynomials and $Q \neq 0$.

rational number (p. 3) A number that can be expressed as the quotient of two integers.

real numbers (p. 3) The set of real numbers is the union of the sets of rational and irrational numbers.

rectangular hyperbola (p. 538) The graph of an equation of the form $xy = k$, where $k \neq 0$.

recursive formula (p. 623) A rule for a sequence that gives the first term, a_1, and a formula for obtaining the $(n + 1)$th term, a_{n+1}, from the preceding term, a_n.

reference triangle (p. 743) A triangle formed by drawing a perpendicular to the x-axis from any point on the terminal side of an angle in standard position.

relation (p. 98) A set of ordered pairs.

remainder theorem (p. 476) If $P(x)$ is a polynomial and a is a number, and if $P(x)$ is divided by $x - a$, then the remainder is $P(a)$.

resultant vector (p. 819) The sum of two vectors.

sample space (p. 692) The set of all possible outcomes of a random experiment.

scalar product (p. 199) The product of a scalar (a real number) and a matrix.

sequence (p. 623) A function defined on the set of consecutive positive integers or a subset of consecutive positive integers.

series (p. 640) The indicated sum of the terms of a sequence.

sigma notation (Σ) (p. 641) A shorthand method of writing a sum.

slope of a line (p. 118) The steepness of the line, which is measured by the ratio of the change in the vertical distance to the change in the horizontal distance.

slope-intercept form of a linear equation (p. 120) $y = mx + b$, where m is the slope and b is the y-intercept.

standard deviation (p. 714) The standard deviation is the positive square root of the variance.

standard form of a linear equation (p. 105) $Ax + By = C$, where A, B, and C are real numbers and A and B are not both zero.

standard position of an angle (p. 737) An angle is in standard position on a Cartesian coordinate plane if its vertex is the origin and its initial side is the positive x-axis.

synthetic division (p. 470) A compact form for dividing a polynomial in one variable by a binomial of the form $x - a$.

transverse axis of a hyperbola (p. 535) One of two axes to which the hyperbola is symmetric. It contains the foci.

union of sets (p. 59) The union of two sets A and B, written $A \cup B$, is the set of elements that are in either A or B or both.

variance (p. 713) A statistic used as a measure of variability in a set of data.

vector (p. 819) A directed line segment.

vector quantity (p. 819) A quantity that has both magnitude and direction.

vertex of a parabola (p. 413) The point where the parabola crosses its axis of symmetry.

vertical line test (p. 100) If no vertical line can be drawn that touches more than one point on a graph, the graph represents a function.

zero matrix (p. 198) The zero matrix, for the set of all $m \times n$ matrices is an $m \times n$ matrix in which every element is 0.

zero of a function (p. 481) The zeros of a function $f(x)$ are the values of x for which $f(x) = 0$.

zero-product property (p. 286) For all real numbers a and b, $ab = 0$ if and only if $a = 0$ or $b = 0$.

INDEX

Photo Credits

Chapter 1 **1:** Jay Freis/The Image Bank. **2:** Bill Gallery/Stock Boston, Inc. **7:** The Granger Collection. **8:** Mike Powell/Allsport USA. **13:** Dan Helms/Duomo. **23:** Adam J. Stoltman/Duomo.
Chapter 2 **45:** Don Smetzer/Tony Stone Worldwide/Chicago Ltd. **46:** Scientific Division Yellow Springs Instrument Co, Inc. **52:** Larry Mulvehill/Science Source/Photo Researchers, Inc. **80:** John Stuart/The Image Bank. **86:** David Alan Harvey/Woodfin Camp & Associates. **90:** John William Banagan/The Image Bank.
Chapter 3 **93:** Wanda Warming/The Image Bank. **103:** The Granger Collection. **104:** Walter Bibikow/The Image Bank. **113:** Obremski/The Image Bank. **123:** Ken Karp. **137:** Mike Maple/Woodfin Camp & Associates.
Chapter 4 **143:** Jon Nicholson/Tony Stone Worldwide/Chicago Ltd. **144:** Breton Littlehales/Folio Inc. **159:** Dan McCoy/Rainbow. **164:** The Granger Collection. **176:** Guido Alberto Rossi/The Image Bank. **187:** Gabe Palmer/The Stock Market.
Chapter 5 **195:** Pete Saloutos/The Stock Market. **196:** Michael Melford/The Image Bank. **226:** The Granger Collection. **232:** Ed Kashi. **246:** ITTC Productions/The Image Bank.
Chapter 6 **253:** Stephanie Stores/The Stock Market. **254:** NASA. **279:** Michael Holford. **285:** NASA. **289:** Courtesy of Constance Reid. **294:** Craig Aurness/Woodfin Camp & Associates. **297:** Ted Streshinsky/The Stock Market. **298:** Richard Megna/Fundamental Photographs.
Chapter 7 **305:** NASA. **306:** Ed Kashi. **311:** Official U.S. Navy Photograph by PH 2 David C. Maclean. **342:** David Stoecklein/The Stock Market. **346:** Stacy Pick/Stock Boston, Inc. **348:** Frank Siteman/Stock Boston, Inc.
Chapter 8 **355:** Wil Blanche/DPI. **361:** Stanford Linear Accelerator Lab. **376:** NASA. **377:** Brownie Harris/The Stock Market. **383:** J.B. Diederich/Contact/Woodfin Camp & Associates. **402:** left: Courtesy of Professor Benoit Mandelbrot; right: Brett Froomer/The Image Bank. **403:** Courtesy of Professor Benoit Mandelbrot.
Chapter 9 **411:** top: The Granger Collection; bottom: Peter Lamberti/Tony Stone Worldwide/Chicago, Ltd. **412:** Robert Frerck/Odyssey Productions. **417:** Stephen Dunn/Allsport USA. **456:** top: Don and Pat Valenti/DRK Photo; bottom; David Barnes/The Stock Market.
Chapter 10 **463:** Don and Pat Valenti/Tony Stone Worldwide/Chicago Ltd. **496:** Ted Horowitz/The Stock Market. **507:** Tony Worldwide/Chicago Ltd.
Chapter 11 **513:** Allen Green/Photo Researchers, Inc. **520:** Roger Ressmeyer/Starlight. **527:** NASA. **563:** NASA.
Chapter 12 **569:** Benn Mitchell/The Image Bank. **575:** Courtesy Dr. Carolyn W. Meyers. **581:** Jonathan Blair/Woodfin Camp & Associates. **582:** Charles West/The Stock Market. **601:** Dan McCoy/Rainbow. **602:** Ellis Herwig/Stock Boston, Inc.
Chapter 13 **617:** David H. Wells/JB Pictures **622:** DPI. **634:** J. Barry O'Rourke/The Stock Market. **651:** Chris Jones/The Stock Market. **657:** Frank Whitney/The Image Bank. **672:** Cesar Paredes/The Stock Market.
Chapter 14 **679:** Rob Nelson/Stock Boston, Inc. **686:** Cliff Feulner/The Image Bank. **688:** Vince Streano/The Stock Market. **707:** G. Contorakes/The Stock Market. **711:** Gabe Palmer/The Stock Market. **715:** Jerry Wachter/Focus on Sports. **721:** UPI/Bettmann Newsphotos. **722:** Ken Karp. **728:** Steve Smith/Wheeler Pictures.
Chapter 15 **735:** Arthur Meyerson/The Image Bank. **752:** The Bettmann Archive. **753:** U.S. Navy official photograph. **758:** Dan McCoy/Rainbow. **787:** James Murray/Photo Researchers, Inc.
Chapter 16 **793:** Bill Frantz/Tony Stone Worldwide/Chicago Ltd. **801:** Seth Thomas/Westclox. **823:** Jean Bichet/The Image Bank. **832:** The Granger Collection.